GREAT SALT LAKE

An Overview of Change

edited by
J. Wallace Gwynn, Ph.D.

Special Publication
of the
Utah Department of Natural Resources

2002

Front cover photograph: NASA #STS047-097-021 Date: September 1992
Back cover photos: compliments of USGS

Book design by Sharon Hamre, Graphic Designer

ISBN 1-55791-667-5

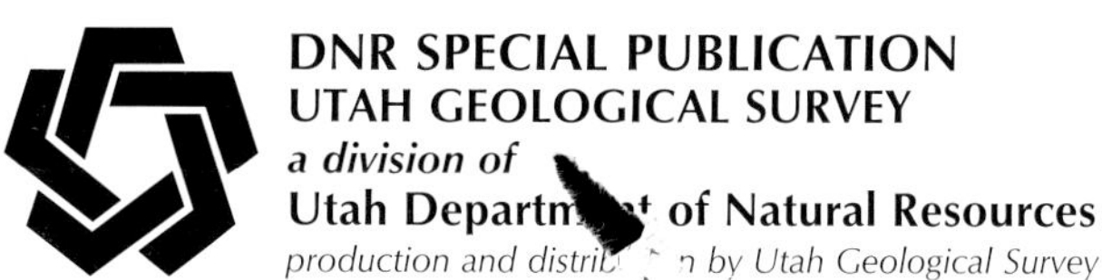

Utah Department of Natural Resources

Mission Statement

Sustain and enhance the quality of life for people today and tomorrow through the coordinated and balanced stewardship of our natural resources.

Values Statement

We serve with integrity, civility, and accountability to make a difference.

STATE OF UTAH

Michael O. Leavitt, Governor

DEPARTMENT OF NATURAL RESOURCES

Robert L. Morgan, Executive Director

Administration
Forestry Fire and State Lands
Oil Gas and Mining
Parks and Recreation
Water Resources
Water Rights
Wildlife Resources
Utah Geological Survey

Department of Natural Resources
1594 West North Temple
P.O. Box 145610
Salt Lake City, Utah 84114-5610
tel. (801) 538-7200
fax (801) 538-7315
http://www.nr.utah.gov

TABLE OF CONTENTS

GEOLOGY/PALEONTOLOGY/ARCHAEOLOGY

CHEMISTRY/HYDROLOGY/LIMNOLOGY

GSL AND ASSOCIATED INDUSTRIES

1980S FLOODING

BIOLOGY

RECREATION

BONNEVILLE SALT FLATS

STATE AND FEDERAL GSL ISSUES

HUMAN HABITATION AND ART

Over the past 30,000 years or so, the Great Salt Lake and its ancestral predecessor, Lake Bonneville, have provided an important nourishing habitat for creatures of all kinds including small organisms, migrating birds, and mammals ranging from mammoths and saber-tooth tigers to prehistorical and modern humans. Early humans settled the riparian lands that feed the lake, and modern man, the Mormon pioneers, chose the lush environs of the majestic Great Salt Lake in which to settle and subsequently flourish. The Utah Department of Natural Resources has the tremendous responsibility and challenge to ensure that the lake continues to sustain a healthy ecosystem, while simultaneously balancing other modern-day, multiple-use activities like brine-shrimp harvesting, mineral/energy development, and recreation. In our efforts to achieve this balance, there is no element as important as having accurate, up-to-date knowledge about the lake's hydrologic, biologic, and geologic systems. Without this knowledge, even the best stewards will make wrong decisions. It is therefore with great interest and excitement that the Utah Department of Natural Resources encourages and supports the study and gathering of scientific data on the Great Salt Lake's natural systems. Only with this infusion of new information can the dynamic and changing nature of this grand system be properly understood and managed for the benefit of all.

Kathleen Clarke,
Executive Director, Utah Department of Natural Resources
April 1998 to December 2001

This volume, *Great Salt Lake - an overview of change* - brings together multi-disciplinary articles on the history, scientific research, artistic aspects, management, development, utilization, and other subjects related to Great Salt Lake and its extended environs including the Bonneville Salt Flats. It is an update to and expansion of its predecessor, *Great Salt Lake - a scientific, historical and economic overview*, published more than 20 years ago by the Utah Geological Survey. This volume is intended to be: (1) a valuable reference for those managing the lake and planning for its future, (2) a credible springboard for future, lake-related research and information volumes, and (3) an up-to-date reference for all who are interested in Great Salt Lake.

Within the past 20 years, Great Salt Lake has changed significantly, both physically and chemically. Most notable of these changes was the lake's spectacular five-year, 12-foot rise from a surface elevation of about 4,200 feet in 1983 to nearly 4,212 feet in 1986 and 1987. During this period of rapid rise, serious and costly flooding took place around Great Salt Lake, especially along its developed south shore. The flooding was particularly damaging because of the continual development on the Great Salt Lake flood plain during the period from about 1890 to 1983. During this time, the high-water elevations did not exceed 4,205 feet, leading planners and developers to underestimate the lake's flooding potential.

Because of the lake's relentless rise and the associated flooding it brought, the State of Utah financed and implemented two flood-control programs: (1) breaching the Southern Pacific Railroad (SPRR) causeway in August 1984 to reduce the water-surface elevation difference (head differential) of about 3.5 feet that had developed across it, and (2) pumping water from the lake into the western desert (the West Desert Pumping Project) to provide an additional 320,000 acres of evaporation area for the lake in the West Pond. Breaching the causeway helped to successfully reduce the head differential to less than one foot, and consequently redistributed the salt content between the north and south arms of the lake. The West Desert Pumping Project (April 1987-June 1989) was successful in reducing the overall surface elevation of the lake more than two feet, but during the process, more than 650 million tons of salt was left in the West Desert. This loss equated to nearly 12 to 14 percent of the lake's total salt load.

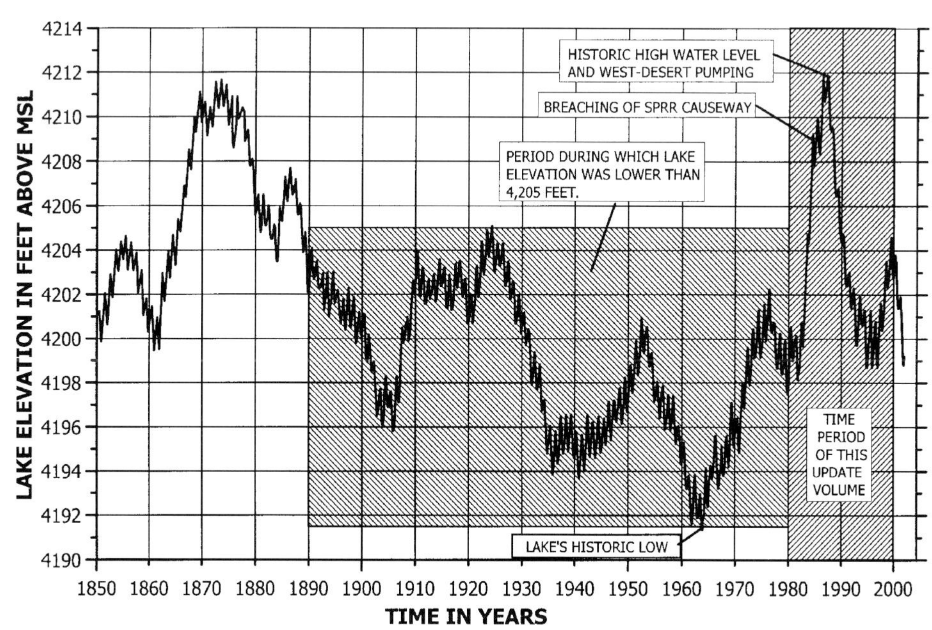

The physical and chemical changes that occurred in Great Salt Lake from 1980 through 2000, particularly during the mid-1980s flooding period, adversely affected the mineral, brine shrimp, transportation, recreation, and other industries. During the flooding, the mineral-extraction companies battled the rising lake levels and declining brine salinities. Solar-evaporation-pond dikes were lost, solar ponds were flooded, and additional evaporation-pond area had to be constructed to compensate for the diluted lake brines. The brine-shrimp industry experienced fluctuating cyst harvests from the mid-1980s through 1997, and had its worst egg-harvest seasons during the late 1990s, presumably due to the south arm's salinity dropping below optimum levels, about 13 to 20 percent salt, that are needed for sustainable shrimp production. Ironically, the in-

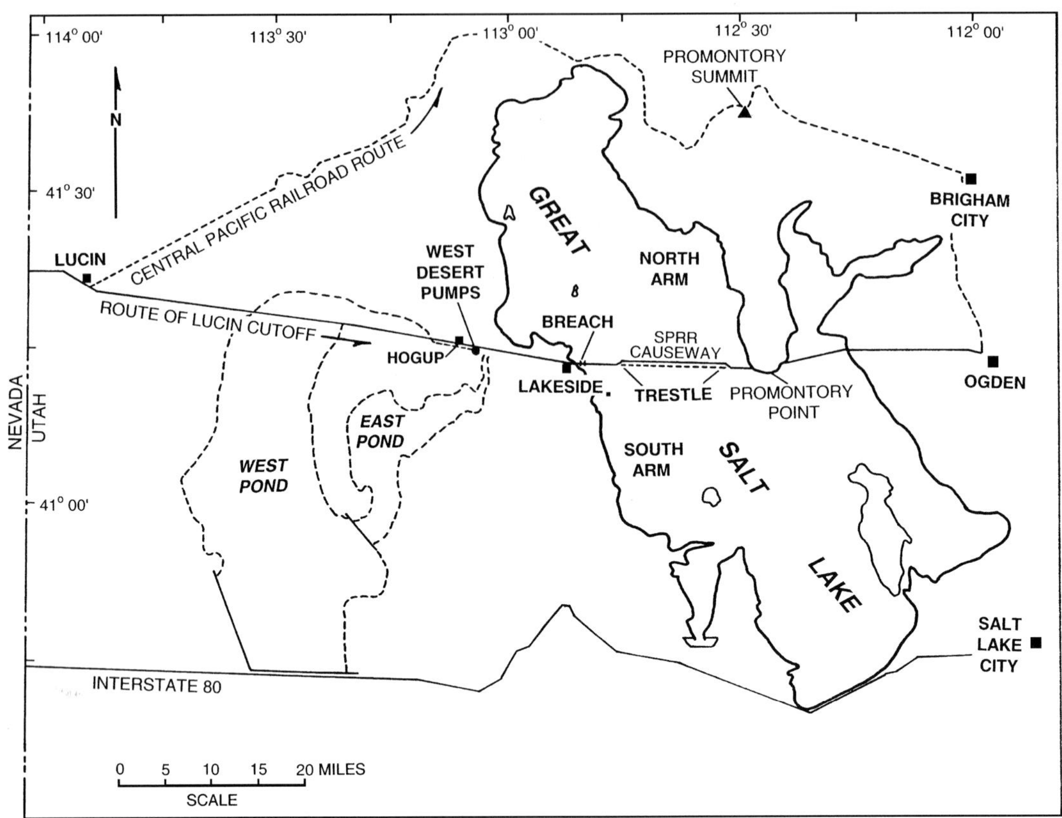

dustry had its two best successive harvest seasons starting with the 2000-2001 season. Transportation facilities, including highways, rural roads, railroads, and even Salt Lake International Airport, were threatened by the 1980s flooding. The highway and railroad at the south end of the lake, and the causeway through the center of the lake, had to be built up to keep above the rising lake and some roads near the lake were simply inundated during this time. The majority of the lake's recreation facilities, including bathing beaches, and lakeside resorts were flooded, and road inundation left Antelope Island State Park isolated from the mainland from 1984 until about 1992 when the roadway was rebuilt at a higher elevation. Nearly all of the state and federal waterfowl-management areas were inundated, and restoration of these facilities has been slow and costly. Thousands of acres of farm lands were flooded and poisoned by the invading salt water. Between 1983 and 1987, it is estimated that lake flooding caused over $240 million in damages (in 1985 dollars).

Since 1980, the various State Divisions within the Department of Natural Resources that have responsibilities for Great Salt Lake have more actively studied and managed the lake. The Utah Division of Forestry, Fire & State Lands, which manages the bed of the lake below the surveyed meander line, as well as the minerals in the water, developed the Great Salt Lake Contingency Plan (1983), a General Management Plan (1987), the Comprehensive Management Plan (1995), and a Mineral Leasing Plan (1996). In 2000, the Department of Natural Resources' multi-divisional Great Salt Lake Planning Team, with input from other state, federal, public, and private interests, developed the Great Salt Lake Comprehensive Management Plan - Resource and Decision Documents. The Utah Division of Wildlife Resources initiated the Great Salt Lake Ecosystem Project, beginning with a study of the lake's brine shrimp (*Artemia franciscana*) population. The Utah Geological Survey, in cooperation with the Utah Divisions of Wildlife Resources and of Water Quality, continued to monitor the lake's chemistry. The U.S. Geological Survey monitored the lake's south- and north-arm surface elevations, and participated in the State's Great Salt Lake Ecosystem Project. Several colleges, universities, and private entities are also actively engaged in lake research.

Interest in basic scientific research and general awareness of Great Salt Lake and its extended environs as an important ecosystem has increased in recent years. Researchers have studied many lake-related topics, including brine-shrimp population dynamics, the geology of Antelope Island, brine chemistry and salt-load dynamics, archaeology and paleontology, lake currents and wind movements, and waterfowl population and habitat. Scientists have also studied the movement and loss of salt from the Bonneville Salt Flats. This volume reflects much of this recent research.

I sincerely thank the many authors, industry representatives, and reviewers who have donated their time, expertise, and other resources to create this volume. The articles in this volume cover a wide variety of topics, range in scope from general interest to very technical, and vary greatly in style and content.

Hopefully, readers of this volume will learn more about Great Salt Lake and gain a greater appreciation of its beauty, mysteries, complexities, and value as a multi-faceted resource. Readers may also gain a greater appreciation of the State's very difficult task of balancing the economic development of the lake with efforts to preserve the lake as a fragile and unique ecosystem, as a scientific laboratory, and as one of Utah's most beautiful treasures.

J. Wallace Gwynn, editor
Utah Geological Survey

MEN OF THE LAKE - IN MEMORIAM

- **JAMES R. PALMER -** *1920-1996*
- **NORMAN B. HELGREN -** *1920-1989*
- **DOYLE STEPHENS -** *1944-2000*

JAMES R. PALMER

(1920-1996)

James R. Palmer, or Jim as many knew him, began his life-long association with Great Salt Lake's salt industry in about 1938. He was out of work at the time and needed money for school, when he heard that a salt plant was being built on the mud flats south of Stansbury Island. He traveled to the job site where he met a Mr. Miller, the project foreman, and asked him for a job. Mr. Miller said that he didn't think they were hiring. Undaunted, Jim immediately traveled to the Crystal White Salt Company office in Salt Lake City (predecessor of the American Salt Company). The job prospect still appeared slim until young Jim told the man behind the desk that he (Jim) was not doing anything at the time, that he had a shovel, and that he would be glad to give two weeks of free work. The man replied that he had never had an offer like that, and that if Jim wanted to work, he would pay him $1.50 a day plus $1.50 worth of stock in the company.

During Jim's long and productive career that followed, he became a world-renowned figure in the salt extraction industry. He owned the Lake Point Salt Company (1977-1982), and traveled internationally as a consultant. He was also the inventor of the Palmer-Richards salt harvester which is used in the salt industry worldwide. Former Utah Governor, Calvin L. Rampton, appointed Jim to the Great Salt Lake Board where he served for more than a decade, helping to plan for the long-term industrial and recreational use of the lake. He was also an active member of the Great Salt Lake Technical Team which met under the direction of the Division of State Lands and Forestry to discuss and investigate lake-related issues, and to develop a management plan for the lake.

Jim Palmer with the Palmer-Richards salt harvester

Jim was always active in community affairs, serving on the Grantsville City Council, two terms as the mayor of Grantsville, and two terms as a Tooele County Commissioner. He served on the Stansbury Sewer and Water Improvement District as a trustee, and volunteered countless hours to the community. In 1995, he was honored as Tooele County's Citizen of the Year. During World War II he served his country as a bombardier in the Army Air Corp. Above all else, though, Jim regarded family, religion, honesty, hard work, and service to others as the most important components of his life.

Written by J.W. Gwynn and Palmer family

NORMAN B. HELGREN

1920-1989

Norman Helgren was born and raised in Omaha, Nebraska, the son of a Swedish immigrant who edited a Swedish newspaper and later became president of the Omaha Loan and Builders Association.

Norm graduated from the University of Michigan with his BS degree in Chemical Engineering in 1942. He spent the next six years working for Nicaro Nickel in Cuba, and for a time in Puerto Rico.

In 1955, Norm joined the Morton Salt Company, first working at Morton's solution-mining operation at Manistee, Michigan as plant engineer. Three months later, he was transferred to Weeks Island, Louisiana at Morton's salt-mining operations, working in the same capacity.

In 1957, Norm was transferred to Morton's Saltair facility on Great Salt Lake as plant engineer, and was named assistant plant manager six months later. He became plant manager in 1960. In 1965, he was transferred to the Chicago office to assist the vice president of Morton International Inc. in coordinating foreign salt production operations, and later became assistant general manager of Morton Salt. He returned to Salt Lake City in 1977 as facility manager, and remained in that position until he retired in 1986 after 31 years of service with Morton Salt.

Norm served as chairman of the Great Salt Lake Section of the American Institute of Chemical Engineers, and was affiliated with the Utah Engineering Council, the American Red Cross, the American Mining Congress, and served on the Board of Utah Manufacturers Association.

Norm is best remembered by his co-workers for his gentle manner, and for the respect he showed all his employees. He considered the employees to be "good friends" and enjoyed their stories and knowledge of Great Salt Lake. During his 19 years working on Great Salt Lake, Norm demonstrated a determination to preserve the natural hydrology and ecology of the lake, and to wisely utilize its natural resources.

Written by J.W. Gwynn with Jeri Helgren

DOYLE STEPHENS

1944 – 2000

SCIENTIST, MENTOR, AND FRIEND

Doyle completed his dissertation on the limnology and invertebrate life of Great Salt Lake in 1974 and thus began a lifelong interest in Great Salt Lake. The lake had to wait, however, as he began his career with the U.S. Geological Survey in 1974 at the Gulf Coast Hydroscience Center in Bay St. Louis, Mississippi. There, Doyle worked with a team of scientists to study the oxygen cycle and reaeration in streams.

After transferring to the USGS - Utah District in 1980, his interests in aquatic biology and chemistry led him into multifaceted studies of many kinds. His innovation, energy, expertise, and ability to communicate gained the respect and trust of people from all parts of Utah's water and wildlife communities. Much of his work was cooperative projects with the U.S. Fish and Wildlife Service and the U.S. Bureau of Reclamation, and he worked at many of the National Wildlife Refuges and State Waterfowl Management Areas. Over a 12-year period, he characterized the hydrology and contamination potential of Utah's major wetland areas, all of Utah's National Wildlife Refuges, and a third of the state-managed Waterfowl Management Areas.

Doyle's major scientific contributions were made in identification and management of wetland contaminants created or augmented by runoff/drainage from irrigation projects, and in the understanding of the limnology/ecology of Great Salt Lake. The focus in much of his wetlands work was the movement of selenium and other contaminants with irrigation drainage into, and within, a number of wetlands in the Green River basin of Utah. Doyle was appointed by the Director of the U.S. Geological Survey in 1998 to represent the USGS on the Salton Sea Science Subcommittee, a group of scientists whose task was to restore the Salton Sea to its original condition. The appointment was made because of his expertise in the understanding of contaminants associated with irrigation projects and research on saline lakes.

On May 8, 2001, Doyle was awarded (posthumous) the Governor's Medal for Science and Technology. This medal, the State's highest scientific honor, was presented to five recipients, who are leading innovators of technology, and educators, who deserve special recognition for their outstanding contributions to knowledge in the physical, biological, mathematical, and engineering sciences and technology.

Doyle's long-time interest in Great Salt Lake and its biology led him into the recent studies on the ecology/population dynamics of the lake's brine shrimp including the related physical and chemical aspects of the lake. As the lake's brine-shrimp industry has grown over the last decade, and the realization that the lake's ecology is vitally important to the western hemisphere's bird migration, the significance of his long-time work became ever more appreciated.

A positive attitude and a desire to challenge all technical obstacles facilitated his success in achieving the difficult goals of his research. He produced 51 technical reports and papers, gave 21 formal presentations at scientific meetings, and was an active participant in many technical committees pertaining to Great Salt Lake, the Salton Sea, and the selenium contamination in the wetlands of the Green River basin of Utah. Internationally renowned for his contributions to the biologic sciences, he was in constant demand for review of papers from the academic community, and he always found time to accommodate these requests in spite of a busy personal workload.

Doyle took pride in assisting students with science projects and proudly displayed their acknowledgments of his assistance. He recently helped several high school students with projects for science fairs and assisted Salt Lake Community College in producing a video on ecology.

Most of all, Doyle was a remarkable human being, and was the kind of person who made it enjoyable to come to work.

Written by Kidd M. Waddell, USGS.

GEOLOGY/PALEONTOLOGY/ARCHAEOLOGY

RECENT DEVELOPMENTS IN THE STUDY OF LAKE BONNEVILLE SINCE 1980

by

Charles G. (Jack) Oviatt
Department of Geology, Kansas State University, Manhattan, KS 66506-3201
and
Robert S. Thompson
U.S. Geological Survey, MS 980, Box 25046, Denver Federal Center, Denver, CO 80225

ABSTRACT

Since 1980, many aspects of Lake Bonneville have been investigated and important discoveries have been made despite over 100 years of careful research by many people since the days of G.K. Gilbert. This includes new work in such fields as stratigraphy and sedimentology, geochronology, isostasy, geomorphology, paleohydrology and paleoclimatology, and other areas.

There is no sign of diminishing efforts, and the Bonneville basin continues to be an excellent natural laboratory for studying the last few million years of Earth history.

INTRODUCTION

Shorelines on the piedmont slopes and mountain fronts (possibly around Utah Lake) were noted by the Spanish explorers Dominguez and Escalante in 1776, and were briefly studied by John C. Fremont and Howard Stansbury in the 1840s and 1850s (Sack, 1989). However, it was not until G.K. Gilbert began his exhaustive research on Lake Bonneville during the 1870s and 1880s that the paleoclimatic significance of the shorelines and deposits of Lake Bonneville was fully appreciated by geologists and residents of the Salt Lake region. Gilbert's (1890) work was monumental, and although some of his interpretations have been modified, and much new information on the lake has accumulated over the century since his work was published, many of his hypotheses have withstood numerous scientific tests. The history of changing ideas on the history of Lake Bonneville since Gilbert has been reviewed by Machette and Scott (1988) and Sack (1989).

The purpose of this paper is to briefly review studies of Lake Bonneville since 1980. When Currey (1980) published an important summary of Lake Bonneville and Great Salt Lake, his paper on the coastal geomorphology of Great Salt Lake included a discussion and detailed map of Lake Bonneville shorelines in the vicinity of Great Salt Lake. Using an approach similar to Gilbert's, Currey recognized that geomorphology and stratigraphy together (morphostratigraphy) can offer excellent and vital clues to lake history that neither the landforms nor the deposits alone can yield. Currey's (1980, figure 8) reconstruction of the history of Lake Bonneville and Great Salt Lake (figure 1A) was simple, yet refined compared to many previously published interpretations of lake-level change in the Bonneville basin. His map of the major shorelines in the vicinity of Great Salt Lake was a major contribution to knowledge about the lake. Many of Currey's students have made contributions to studies of Lake Bonneville and Great Salt Lake, including Burr (Burr and Currey, 1988); Murchison (1989), Sack (1989, 1992, 1994, 1995), and Oviatt (1987).

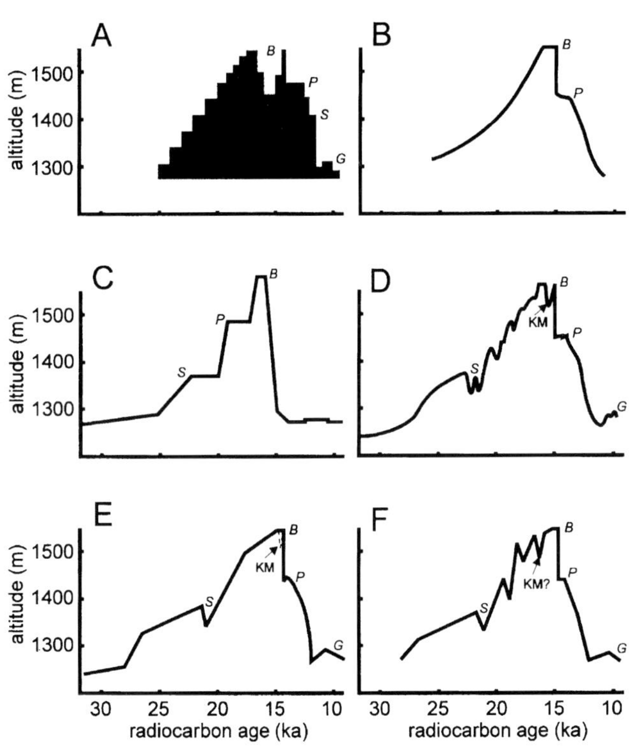

Figure 1. *Changes in the interpretation of the Lake Bonneville chronology from 1980 to 1997, figure redrawn from the references given below. For all diagrams, B=Bonneville shoreline, P=Provo shoreline, S=Stansbury shoreline, G=Gilbert shoreline, KM=Keg Mountain oscillation. Altitude scales are approximate; see original references for precise definitions of shoreline altitudes. A. schematic chronology of Currey (1980); B. from Scott and others (1983); C. from Spencer and others (1984); D. from Currey and Oviatt (1985); E. from Oviatt and others (1992); F. from Oviatt (1997).*

Scientific and public awareness of Great Salt Lake and Lake Bonneville (figure 2) were raised during the early and middle 1980s, when Great Salt Lake rose dramatically to its historic high (about 4,211.85 feet), and flooded many square kilometers of lake plain (Kay and Diaz, 1985; Arnow and Stephens, 1990), initially in response to the strong 1982-83 El Niño. Damage was extensive to highways, railroads, sewage-treatment plants, and many other facilities, and stimulated a need and desire to learn more about the lake's history and the causes of its behavior.

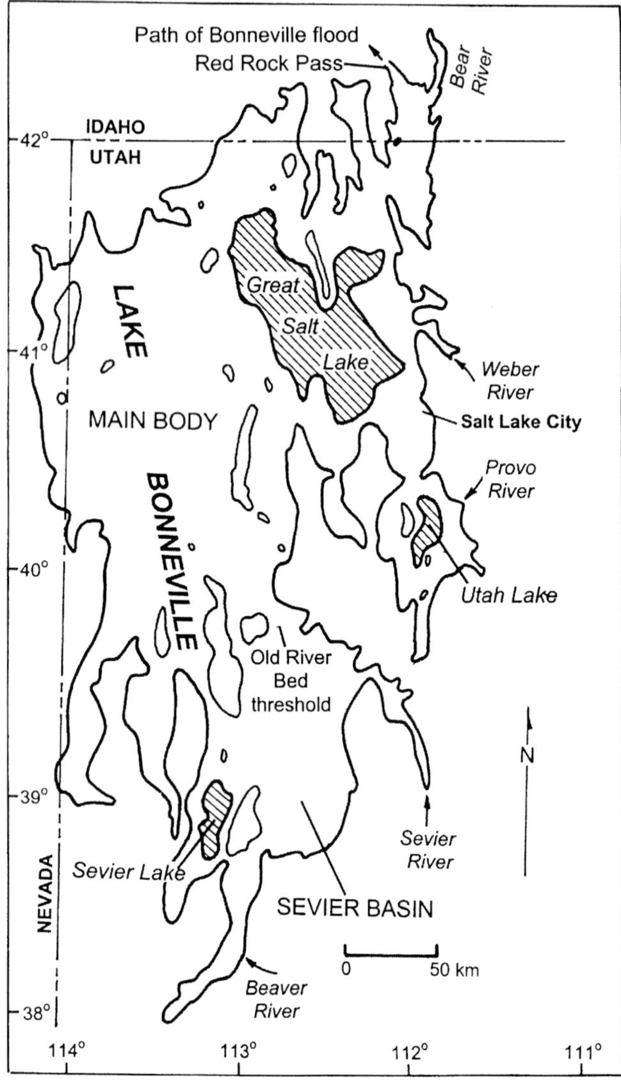

Figure 2. Map of Lake Bonneville (after Currey, 1982). The Old River Bed threshold is a low divide that separates the Sevier basin from the main body of Lake Bonneville.

DEVELOPMENTS SINCE 1980

Major Lake Cycles

In addition to Currey's efforts at revitalization of Lake Bonneville studies, several other people began work on Lake Bonneville stratigraphy in the late 1970s and 1980s. Scott and others (1983) published new interpretations of the exposed stratigraphic record of Lake Bonneville and pre-Bonneville lakes. Based on new observations at numerous exposures in gravel pits along the Wasatch Front, and new radiocarbon ages, they demonstrated that the Bonneville lake cycle essentially consisted of a single major lake transgression and regression roughly coincident with marine oxygen isotope stage 2 (figure 1B). They also showed that deposits, previously referred to as the Alpine Formation, were in some cases actually of Bonneville age, and in other cases pre-Bonneville (deposited during a lake cycle presumably correlative with marine oxygen-isotope stage 6, which was later named the Little Valley lake cycle; McCoy, 1987). The Draper Formation, which had been interpreted as the deposits of a deep-lake cycle during the early Holocene, was reinterpreted by Scott and others (1983) as alluvium in the type area, and as Bonneville deposits (of late Pleistocene age) elsewhere. Although the Draper Formation continued to be mapped in some areas in the 1990s (Morrison, 1991; Varnes and Van Horn, 1991), it is clearly not of lacustrine origin (Oviatt and others, 1994b).

Much of the geochronologic control that allowed the discrimination of these different-aged deposits was provided by the results of amino acid racemization studies of mollusks by McCoy (1987). In addition, Oviatt (1987) revisited some of Gilbert's (1890) stratigraphic sections and showed that the interpretation of lake history derived from those sections required only a single major lake transgression rather that the two or more that Gilbert and later workers had envisioned.

Other developments in the understanding of Lake Bonneville were numerous and came from many different sources during the 1980s and 1990s. During the Cutler Dam lake cycle, which may have been coincident with marine oxygen-isotope stage 4, the lake rose to moderate levels in the basin (Oviatt and others, 1987). Work in the Sevier Desert, south of Great Salt Lake, increased understanding of the pre-Bonneville-Little Valley lake cycle, and demonstrated that multiple stratigraphic tools are sometimes needed to discriminate between units of different ages (Oviatt and others, 1994b).

Currey (1982) and Currey and others (1984) extended mapping of the major shorelines of Lake Bonneville to the entire basin and included precise data on the elevations and map coordinates of hundreds of points on the Bonneville, Provo, Gilbert, and Gunnison shorelines. Currey's mapping allowed precise definition of the isostatic response of the lithosphere to the Lake Bonneville water load and its removal (Bills and May, 1987; Bills and others, 1994). Currey and Burr (1988) published a highly refined model of lake-level changes during the period of intermittently open-basin conditions at the Bonneville and Provo shorelines, and emphasized the isostatic controls on the stratigraphic and geomorphic record.

A major breakthrough in the 1980s was the realization that the Stansbury shoreline, which Gilbert had not studied extensively, but which he suspected had been produced after formation of the Provo shoreline on the regressive phase of Lake Bonneville, actually formed during the transgressive phase (Currey and others, 1983; Green and Currey, 1988; Oviatt and others, 1990).

Spencer and others (1984) reported the results of an extensive study of sediment cores taken from Great Salt Lake (figure 1C), and demonstrated the potential of core studies for deciphering details of lake history that cannot be determined from studies of outcrops. In addition, they (Spencer and others, 1985a, 1985b) made significant contributions to understanding the geochemistry of Lake Bonneville and

Great Salt Lake. Prominent members of the Spencer and others (1984) group who have made other major contributions to studies of Lake Bonneville included R.M. Forester, B.F. Jones, K. Kelts, J. McKenzie, and D.B. Madsen.

Forester's (1987) work on ostracodes in the Bonneville sediments has opened new possibilities for reconstructing the history of Pleistocene environmental changes. Basaltic volcanic eruptions into Lake Bonneville provided ash beds that are useful in intra-basin correlations (Oviatt and Nash, 1989; Oviatt and others, 1992; Miller and others, 1995; Oviatt and Miller, 1997). The deep-water deposits of Lake Bonneville, which Gilbert (1890) referred to as the White Marl, have proven to be a repository of geochemical and paleoclimatic information (Oviatt and others, 1994a; Oviatt, 1997). The hydrodynamics of the catastrophic Bonneville flood, which was caused by the rapid release of approximately 4,750 km^3 of water into the Snake River, were carefully studied and refined by Jarrett and Malde (1987) and O'Connor (1993).

A valuable contribution to Lake Bonneville studies, and to the history of science, was published in the 1980s by one of the pioneers of Lake Bonneville research. Hunt (1982) transcribed G.K. Gilbert's field notes, which Gilbert had taken on a series of trips to Utah between 1875 and 1880. Any serious student of Lake Bonneville can learn a tremendous amount about the lake and Gilbert's field methods by studying Gilbert's notes and retracing his steps. In association with the 1988 Annual Meeting of the Geological Society of America in Salt Lake City, the field trip "In the Footsteps of G.K. Gilbert" (Machette, 1988), focused on Gilbert's contributions to studies of Lake Bonneville and neotectonics in the eastern Basin and Range.

Bonneville Geochronology

The new radiocarbon ages obtained by Scott and others (1983) and numerous other ages, plus extensive morphostratigraphic work, allowed a refined chronology of Lake Bonneville to be produced by the mid 1980s (Currey and Oviatt, 1985) (figure 1D). Dating of dispersed organic carbon by accelerator mass spectrometry (AMS) in one of the cores obtained by Spencer and others (1984) permitted Thompson and others (1990) and Oviatt and others (1992) to begin correlations between offshore, deep-water sediments and shoreline deposits of Lake Bonneville. As a result, the general history of Lake Bonneville, from about 28,000 to 10,000 radiocarbon yr B.P. is known with a high degree of accuracy (figure 1E).

Renewed interest in amino acid geochronology by D. Kaufman and his students (Light, 1993), has produced new data on the ages of stratigraphic units, and on paleotemperatures. The use of amino acids in fossil ostracodes shows particular promise (Kaufman and others, 1997; Oviatt and others, 1999).

Lake Bonneville in a Global Context

A decade after his paper on the coastal geomorphology of Great Salt Lake and vicinity, Currey (1990) published another major contribution to Lake Bonneville research. In this paper, Currey used Lake Bonneville as an example of a typical paleolake in a semidesert basin, and discussed the lessons that can be learned from it in terms of paleoclimate and environmental change (see also Currey, 1994a, and 1994b).

A regional comparison of the chronologies of large Pleistocene lakes in the Great Basin, including those in the Bonneville, Lahontan, Mono, and Searles basins, were similar (Benson and others, 1990), including latest Pleistocene lake-level changes in Lake Bonneville and Lake Lahontan (Benson and others, 1992).

Following Antevs (1948), Benson and Thompson (1987) proposed that the major lake cycles in the Bonneville and Lahontan basins (in western Nevada) were caused by the diversion of storm tracks into the Great Basin region by the huge Northern Hemisphere ice sheets during oxygen isotope stage 2. Oviatt and others (1990) suggested that even smaller-scale lake-level fluctuations, such as the Stansbury oscillation about 21,000 yr B.P. during the transgressive phase, were caused by shifts in storm tracks governed by the ice sheets. Hostetler and others (1994) modeled the hydrology of Lake Bonneville and found that the large size of Lake Bonneville helped maintain and enhance the lake by providing moisture that was precipitated in the mountains east of the lake and fed back to the lake as runoff. Therefore, Lake Bonneville, which was a product of global-scale changes, had a strong influence on the regional-scale climate. Oviatt (1997) attempted to link millennial-scale changes in the hydrologic budget of Lake Bonneville to changes in the size and shape of the Laurentide ice sheet (figure 1F).

Ongoing Work

Since 1995, a series of new sediment cores has been obtained from Great Salt Lake. Studies of these cores are providing information on fine details of the Bonneville lake cycle that previously could not be deciphered from either outcrops or the limited available cores (Thompson and others, 1997; Jolley, 1998). A re-examination of the Burmester core, which was studied by Eardley and others (1973), indicates that there were probably only four deep-lake cycles in the Bonneville basin during the last 800,000 years, rather than the 17 cycles interpreted from the core by Eardley and his colleagues (Oviatt and others, 1999). The reinterpretation was aided by the tephrochronologic work of Williams (1994). Long sediment cores from the Sevier basin, south of the Great Salt Lake, are yielding evidence that there were no deep-lake cycles in the Bonneville basin between about 3,000,000 and 800,000 yr B.P. (Thompson and others, 1995). Studies of cuttings from deep drill holes in Great Salt Lake are revealing a similar conclusion (Cohen and others, 1997; Davis, 1998; Davis and Moutoux, 1998; Kowalewska and Cohen, 1998).

Other ongoing work includes studies of paleovegetation (Madsen and Rhode, 1990; Thompson, 1990; Rhode and Madsen, 1995), fish (Broughton and others, 2000), and small mammals (Grayson, 1993, 1998), which are providing valuable information about the environmental changes in the basin during the regressive phase of Lake Bonneville and the Holocene. Paleomagnetic work on lacustrine sediments shows promise as a means of correlating precisely with stratigraphic sequences outside the Bonneville basin (Thompson and others, 1995; Liddicoat and Coe, 1998). New work on the sedimentology of Lake Bonneville deposits (Chan and Milligan, 1995; Lemons and others, 1996;

Lemons and Chan, 1999) permits viewing the Bonneville record from perspectives traditionally applied to older rocks. Geophysical investigations are beginning to elucidate the internal structures of deltas and other shoreline depositional features that normally cannot be observed except in gravel-pit exposures (Smith and Jol, 1992). Atwood (1994) is actively studying the geomorphic response of late Holocene and historic high stands of Great Salt Lake.

SUMMARY

In summary, the Bonneville basin continues to be studied from a number of different points of view by people attempting to decipher the geologic history of the last few million years, and to view the lake from a global perspective (Thompson and others, 1993). Despite over 100 years of research the number and quality of serious investigations continues to increase and shows no sign of ending or leveling off.

REFERENCES

Antevs, E., 1948, Climatic changes and pre-White man, *in* The Great Basin with emphasis on glacial and postglacial times: University of Utah Bulletin, v. 38, p. 168-191.

Arnow, T., and Stephens, D., 1990, Hydrologic characteristics of the Great Salt Lake, Utah: 1847-1986: U.S. Geological Survey Water-Supply Paper 2332, 32 p., scale 1:125,000.

Atwood, G., 1994, Geomorphology applied to flooding problems of closed-basin lakes . . . specifically Great Salt Lake, Utah: Geomorphology, v. 10, p. 197-219.

Benson, L.V., Currey, D.R., Dorn, R.I., Lajoie, K.R., Oviatt, C.G., Robinson, S.W., and Stine, S., 1990, Chronology of expansion and contraction of four Great Basin lake systems during the past 35,000 years: Palaeogeography, Palaeoclimatology, Palaeoecology, v. 78, p. 241-286.

Benson, L. V., and Thompson, R. S., 1987, The physical record of lakes in the Great Basin, *in* Ruddiman, W. F., and Wright, H. E., Jr., editors, North America and adjacent oceans during the last deglaciation: Boulder, Colorado, Geological Society of America, The geology of North America, v. K-3, p. 241-260.

Benson, L.V., Currey, D.R., Lao, Y., and Hostetler, S., 1992, Lake-size variations in the Lahontan and Bonneville basins between 13,000 and 9000 ^{14}C yr B.P.: Palaeogeography, Palaeoclimatology, Palaeoecology, v. 95, p. 19-32.

Bills, B.G., and May, G.M., 1987, Lake Bonneville - Constraints on lithospheric thickness and upper mantle viscosity from isostatic warping of Bonneville, Provo, and Gilbert stage shorelines: Journal of Geophysical Research, v. 92, p. 11,493-11,508.

Bills, B.G., Currey, D.R., and Marshall, G.A., 1994, Viscosity estimates for the crust and upper mantle from patterns of lacustrine shoreline deformation in the eastern Great Basin: Journal of Geophysical Research, v. 99, no. B11, p. 22,059-22,086.

Broughton, J.M., Madsen, D.B., and Quade, Jay, 2000, Fish remains from Homestead Cave and lake levels of the past 13,000 years in the Bonneville basin: Quaternary Research, v. 53, p. 392-401.

Burr, T.N., and Currey, D.R., 1988, The Stockton Bar, *in* Machette, M.N., editor, In the footsteps of G.K. Gilbert – Lake Bonneville and neotectonics of the eastern Basin and Range Province: Utah Geological and Mineral Survey Miscellaneous Publication 88-1, p. 66-73.

Chan, M.A., and Milligan, M.R., 1995, Gilbert's vanishing deltas: A century of change in Pleistocene deposits of northern Utah, *in* Lund, W.R., editor, Environmental and engineering geology of the Wasatch Front region: Utah Geological Association Publication 24, p. 521-532.

Cohen, A.S., Davis, O.K., Kowalewska, A., and Moutoux, T.E., 1997, Late Cenozoic paleolakes and paleoclimate in the Bonneville basin [abs.]: Geological Society of America Abstracts with Programs, v. 29, p. 254.

Currey, D.R., 1980, Coastal geomorphology of Great Salt Lake and vicinity, *in* Gwynn, J.W., editor, Great Salt Lake: A scientific, historical, and economic overview: Utah Geological and Mineral Survey Bulletin 116, p. 69-82.

—1982, Lake Bonneville-Selected features of relevance to neotectonic analysis: U.S. Geological Survey Open-File Report 82-1070, 31 p.

—1990, Quaternary paleolakes in the evolution of semidesert basins, with special emphasis on Lake Bonneville and the Great Basin, U.S.A.: Palaeogeography, Palaeoclimatology, Palaeoecology, v. 76, p. 189-214.

—1994a, Hemiarid lake basins - Hydrographic patterns, in Abrahams, A.D., and Parsons, A.J., editors, Geomorphology of desert environments: London, Chapman and Hall, p. 405-421.

—1994b, Hemiarid lake basins-Geomorphic patterns, *in* Abrahams, A.D., and Parsons, A.J., editors, Geomorphology of desert environments: London, Chapman and Hall, p. 422-444.

Currey, D.R., Atwood, G., and Mabey, D.R., 1984, Major levels of Great Salt Lake and Lake Bonneville: Utah Geological and Mineral Survey Map 73, scale 1:750,000.

Currey, D.R., and Burr, T.N., 1988, Linear model of threshold-controlled shorelines of Lake Bonneville, *in* Machette, M.N., editor, In the footsteps of G.K. Gilbert – Lake Bonneville and neotectonics of the eastern Basin and Range Province: Utah Geological and Mineral Survey Miscellaneous Publications 88-1, p. 104-110.

Currey, D.R., and Oviatt, C.G., 1985, Durations, average rates, and probable causes of Lake Bonneville expansions, still-stands, and contractions during the last deep-lake cycle, 32,000 to 10,000 years ago, *in* Kay, P.A., and Diaz, H.F., editors, Problems of and prospects for predicting Great Salt Lake levels: Salt Lake City, University of Utah Center for Public Affairs and Administration, p. 9-24.

Currey, D.R., Oviatt, C.G., and Plyler, G.B., 1983, Lake Bonneville stratigraphy, geomorphology, and isostatic deformation in west-central Utah, *in* Gurgel, K.D., editor, Geologic excursions in neotectonics and engineering geology in Utah-Guidebook Part IV: Utah Geological and Mineral Survey Special Studies 62, p. 63-82.

Davis, O.K., 1998, Palynological evidence for vegetation cycles in a 1.5 million year pollen record from the Great Salt Lake, Utah, USA: Palaeogeography, Palaeoclimatology, Palaeoecology, v. 138, p. 175-185.

Davis, O.K., and Moutoux, T.E., 1998, Tertiary and Quaternary vegetation history of the Great Salt Lake: Journal of Paleolimnology, v. 19, p. 417-427.

Eardley, A.J., Shuey, R.T., Gvosdetsky, V., Nash, W.P., Picard, M.D., Grey, D.C, and Kukla, G.J., 1973, Lake cycles in the Bonneville basin, Utah: Geological Society of America Bulletin, v. 84, p. 211-216.

Forester, R. M., 1987, Late Quaternary paleoclimate records

from lacustrine ostracodes, *in* Ruddiman, W. F., and Wright, H.E., Jr., editors, North America and adjacent oceans during the last deglaciation: Geological Society of America, Geology of North America K-3, p. 261-276.

Gilbert, G. K., 1890, Lake Bonneville: U.S. Geological Survey Monograph 1, 438 p.

Grayson, D.K., 1993, The desert's past: A natural history of the Great Basin: Washington, Smithsonian Institution Press, 356 p.

—1998, Moisture history and small mammal community richness during the latest Pleistocene and Holocene, northern Bonneville basin, Utah: Quaternary Research, v. 49, p. 330-334.

Green, S.A., and Currey, D.R., 1988, The Stansbury shoreline and other transgressive deposits of the Bonneville lake Cycle, *in* Machette, M.N., editor, In the footsteps of G.K. Gilbert – Lake Bonneville and neotectonics of the eastern Basin and Range Province: Utah Geological and Mineral Survey Miscellaneous Publication 88-1, p. 55-57

Hostetler, S. W., Giorgi, F., Bates, G. T., and Bartlein, P. J., 1994, Lake-atmosphere feedbacks associated with paleolakes Bonneville and Lahontan: Science, v. 263, p. 665-668.

Hunt, C.B., editor, 1982, Pleistocene Lake Bonneville, Ancestral Great Salt Lake, as described in the notebooks of G.K. Gilbert, 1875-1880: Brigham Young University Geology Studies, v. 29, pt. 1, 225 p.

Jarrett, R.D., and Malde, H.E., 1987, Paleodischarge of the late Pleistocene Bonneville flood, Snake River, Idaho, computed from new evidence: Geological Society of America Bulletin v. 99, p. 27-134.

Jolley, W., 1998, Carbonate precipitation and lake level change in the Bonneville basin, Utah: Manhattan, Kansas State University, M.S. thesis, 90 p.

Kaufman, D.S., Bright, J., and Oviatt, C.G., 1997, Improved amino acid geochronology using rapidly racemizing amino acids in lacustrine ostracodes: Geological Society of America Abstracts with Programs, v. 29, p. 253.

Kay, P.A., and Diaz, H.F., 1985, Problems of and prospects for predicting Great Salt Lake levels - Papers from a conference held in Salt Lake City, March 26-28, 1985: Salt Lake City, Center for Public Affairs and Administration, University of Utah, 309 p.

Kowalewska, A., and Cohen, A.S., 1998, Reconstruction of paleoenvironments of the Great Salt Lake basin during the late Cenozoic: Journal of Paleolimnology, v. 20, p. 381-407.

Lemons, D. R., and Chan, M.A., 1999, Facies architecture and sequence stratigraphy of fine-grained lacustrine deltas along the eastern margin of late Pleistocene Lake Bonneville, northern Utah and southern Idaho: American Association of Petroleum Geologists Bulletin, v. 83, p. 636-665.

Lemons, D. R., Milligan, M. R., and Chan, M. A., 1996, Paleoclimatic implications of late Pleistocene sediment yield rates for the Bonneville basin, northern Utah: Palaeogeography, Palaeoclimatology, Palaeoecology, v. 123, p. 147-159.

Liddicoat, J.C., and Coe, R.S., 1998, Paleomagnetic investigation of the Bonneville Alloformation, Lake Bonneville, Utah: Quaternary Research, v. 50, p. 214-220.

Light, A., 1993, Amino acid paleotemperature reconstruction and radiocarbon shoreline chronology of the Lake Bonneville basin, USA: Boulder, University of Colorado, M.S. thesis, 142 p.

Machette, M.N., editor, 1988, In the footsteps of G.K. Gilbert – Lake Bonneville and neotectonics of the eastern Basin and Range Province: Utah Geological and Mineral Survey Miscellaneous Publication 88-1, 120 p.

Machette, M.N., and Scott, W.E., 1988, Field trip introduction: A brief review of research on lake cycles and neotectonics of the eastern Basin and Range Province, *in* Machette, M.N., editor, In the footsteps of G.K. Gilbert – Lake Bonneville and neotectonics of the eastern Basin and Range Province: Utah Geological and Mineral Survey Miscellaneous Publication 88-1, p. 7-14.

Madsen, D.B., and Rhode, D., 1990, Early Holocene pinyon (*Pinus monophylla*) in the northeastern Great Basin: Quaternary Research, v. 33, p. 94-101.

McCoy, W. D., 1987, Quaternary aminostratigraphy of the Bonneville basin, western United States: Geological Society of America Bulletin, v. 98, p. 99-112.

Miller, D.M., Nakata, J.K., Oviatt, C.G., Nash, W.P., and Fiesinger, D.W., 1995, Pliocene and Quaternary volcanism in the northern Great Salt Lake area and inferred volcanic hazards, *in* Lund, W.R., editor, Environmental and Engineering geology of the Wasatch Front Region: Utah Geological Association Publication 24, p. 469-482.

Morrison, R.B., 1991, Quaternary stratigraphic, hydrologic, and climatic history of the Great Basin, with emphasis on Lakes Lahontan, Bonneville, and Tecopa, *in* Morrison, R.B., editor, Quaternary nonglacial geology - conterminous U.S.: Geological Society of America, The Geology of North America, v. K-2, p. 283-320.

Murchison, S.B., 1989, Fluctuation history of Great Salt Lake, Utah, during the last 13,000 years: Salt Lake City, University of Utah, Ph.D. thesis, 137 p.

O'Connor, J.E., 1993, Hydrology, hydraulics, and geomorphology of the Bonneville flood: Geological Society of America Special Paper 274, 83 p.

Oviatt, C.G., 1987, Lake Bonneville stratigraphy at the Old River Bed: American Journal of Science, v. 287, p. 383-398.

—1997, Lake Bonneville fluctuations and global climate change: Geology, v. 25, p. 155-158.

Oviatt, C.G., Currey, D.R., and Miller, D.M., 1990, Age and paleoclimatic significance of the Stansbury shoreline of Lake Bonneville, northeastern Great Basin: Quaternary Research, v. 33, p. 291-305.

Oviatt, C.G., Currey, D.R., and Sack, Dorothy, 1992, Radiocarbon chronology of Lake Bonneville, eastern Great Basin, USA: Palaeogeography, Palaeoclimatology, Palaeoecology, v. 99, p. 225-241.

Oviatt, C.G., Habiger, G.D., and Hay, J.E., 1994a, Variation in the composition of Lake Bonneville marl - A potential key to lake-level fluctuations and paleoclimate: Journal of Paleolimnology, v. 11, p. 19-30.

Oviatt, C. G., McCoy, W. D., and Nash, W. P., 1994b, Sequence stratigraphy of lacustrine deposits - a Quaternary example from the Bonneville basin, Utah: Geological Society of America Bulletin, v. 106, p. 133-144.

Oviatt, C.G., McCoy, W.D., and Reider, R.G., 1987, Evidence for a shallow early or middle Wisconsin lake in the Bonneville basin, Utah: Quaternary Research, v., 27, p. 248-262.

Oviatt, C.G., and Miller, D.M., 1997, New explorations along the northern shores of Lake Bonneville, *in* Link, P.K., and Kowallis, B.J., editors, Geological Society of America field trip guidebook, 1997 annual meeting, Salt Lake City, Utah, Part 2, Mesozoic to recent geology of Utah: Brigham Young Geology Studies, v. 42, pt. 2, p. 345-371.

Oviatt, C.G., and Nash, W.P., 1989, Late Pleistocene basaltic ash and volcanic eruptions in the Bonneville basin, Utah: Geological Society of America Bulletin, v. 101, p. 292-303.

Oviatt, C.G., Thompson, R.S., Kaufman, D.S., Bright, J., and Forester, R.M., 1999, Reinterpretation of the Burmester core, Bonneville basin, Utah: Quaternary Research, v. 52, p. 180-184.

Rhode, D., and Madsen, D.B., 1995, Late Wisconsin vegetation in the northern Bonneville basin: Quaternary Research, v. 44, p. 246-256.

Sack, Dorothy, 1989, Reconstructing the chronology of Lake Bonneville: An historical overview, *in* Tinkler, K.J, editor, History of geomorphology: London, Unwin Hyman, p. 223-256.

—1992, Obliteration of surficial paleolake evidence in the Tule Valley subbasin of Lake Bonneville, *in* Fletcher, C.H., and Wehmiller, J.F., editors, Quaternary coasts of the United States: Marine and lacustrine systems: Society for Sedimentary Geology (SEPM) Special Publication No. 48, p. 427-433.

—, 1994, Geomorphic evidence of climate change from desert-basin palaeolakes, *in* Abrahams, A.D., and Parsons, A.J., editors, Geomorphology of desert environments: London, Chapman and Hall, p. 616-630.

—1995, The shoreline preservation index as a relative-age dating tool for late Pleistocene Bonneville shorelines - An example from the Bonneville basin, U.S.A.: Earth Surface Processes and Landforms, v. 20, p. 363-377.

Scott, W.E., McCoy, W.D., Shroba, R.R., and Rubin, M., 1983, Reinterpretation of the exposed record of the last two cycles of Lake Bonneville, western United States: Quaternary Research, v. 20, p. 261-285.

Smith, D.G., and Jol, H.M., 1992, Ground penetrating radar investigation of a Lake Bonneville delta, Provo level, Brigham City, Utah: Geology, v. 20, p. 1083-1086.

Spencer, R.J., Baedecker, M.J., Eugster, H.P., Forester, R.M., Goldhaber, M.B., Jones, B.F., Kelts, K., McKenzie, J., Madsen, D.B., Rettig, S.L., Rubin, M., and Bowser, C.J., 1984, Great Salt Lake and precursors, Utah - The last 30,000 years: Contributions to Mineralogy and Petrology, v., 86, p. 321-334.

Spencer, R.J., Eugster, H.P., and Jones, B.F., 1985a, Geochemistry of Great Salt Lake, Utah II: Pleistocene-Holocene transition: Geochemica et Cosmochemica Acta v. 49, p. 739-747.

Spencer, R.J., Eugster, H.P., Jones, B.F., and Rettig, S.L., 1985b, Geochemistry of Great Salt Lake, Utah I - Hydrochemistry since 1850: Geochemica et Cosmochemica Acta v. 49, p. 727-737.

Thompson, R.S., 1990, Late Quaternary vegetation and climate in the Great Basin, *in* Betancourt, J.B., Van Devender, T.R., and Martin, P.S., editors, Packrat middens – The last 40,000 years of biotic change: Tucson, University of Arizona Press, p. 200-239.

Thompson, R.S., Oviatt, C.G., Kelts, K., and Jolley, W., 1997, Late Quaternary history of Great Salt Lake (Lake Bonneville) [Abs.]: Geological Society of America Abstracts with Programs, v. 29, p. 254.

Thompson, R.S., Oviatt, C.G., Roberts, A.P., Buchner, J., Kelsey, R., Bracht, C., Forester, R.M., and Bradbury, J.P., 1995, Stratigraphy, sedimentology, paleontology, and paleomagnetism of Pliocene-early Pleistocene lacustrine deposits in two cores from western Utah: U.S. Geological Survey Open-File Report 95-1, 94 p.

Thompson, R. S., Toolin, L. J., Forester, R. M., and Spencer, R. J., 1990, Accelerator-mass spectrometer (AMS) radiocarbon dating of Pleistocene lake sediments in the Great Basin: Palaeogeography, Palaeoclimatology, Palaeoecology, v. 78, p. 301-313.

Thompson, R.S., Whitlock, C., Bartlein, P.J., Harrison, S.P., and Spaulding, W.G., 1993, Climatic changes in the western United States since 18,000 yr B.P., *in* Wright, H.E., Jr., Kutzbach, J.E., Webb, T., III, Ruddiman, W.F., Street-Perrott, F.A., and Bartlein, P.J., editors, Global climate since the last glacial maximum: Minneapolis, University of Minnesota Press, p. 468-513.

Varnes, D.J., and Van Horn, R., 1991, Surficial geologic map of the Oak City area, Millard County, Utah: U.S. Geological Survey Open-File Report 91-433, scale 1:31,680.

Williams, S.K., 1994, Late Cenozoic tephrochronology of deep sediment cores from the Bonneville basin, northwest Utah: Geological Society of America Bulletin, v. 105, p. 1517-1530.

GEODYNAMICS OF LAKE BONNEVILLE

by

Bruce G. Bills[1, 2], Tamara Jo Wambeam[3], and Donald R. Currey[3]

[1]Geodynamics Branch, Goddard Space Flight Center, Greenbelt, Maryland

[2]Scripps Institution of Oceanography, La Jolla, California

[3]Department of Geography, University of Utah, Salt Lake City, Utah

ABSTRACT

Great Salt Lake is surrounded by shorelines of earlier lakes. Though they appear to be level, the higher shorelines are actually tilted, with elevations that increase toward the center of the Bonneville basin. The water impounded in these earlier lakes represented a load that was sufficient to depress the crust of the Earth by amounts that increase with lake depth, and reach about 70 m for the highest shorelines of Lake Bonneville. These shoreline elevation patterns provide an important probe of the long-term strength of the Earth. Regional viscoelastic structure models, when adjusted to optimally reproduce the observed shoreline deflection pattern, imply an effective lithospheric thickness of 25 km and an effective viscosity of the upper mantle of 10^{19} Pascal seconds (Pa s). The relatively thin lithosphere and low viscosity mantle are both consistent with higher than average temperatures thought to exist in the upper mantle. These models can also be used to interpolate between, and extrapolate beyond, the epochs represented by measured shorelines. An atlas of computed deflection maps for the Bonneville basin is presented, spanning the interval from 2 to 26 thousand years ago.

INTRODUCTION

The shorelines surrounding Great Salt Lake provide a natural laboratory for study of the long-term strength of the Earth. An explanation of how the shorelines provide that information involves excursions into a variety of topics. To help set the stage, we first describe a simple observational experiment and its surprising implications. Following that we describe again, in greater detail, the background, patterns, and implications of the shorelines.

A Simple Experiment

Encountering a large lake in a desert environment is usually a somewhat surprising experience. Great Salt Lake, and its much larger predecessor Lake Bonneville, present many surprises to the careful observer. First, there is evidence that the lake level has changed quite dramatically. Standing at any point along the present shore of Great Salt Lake, if you turn and look away from the lake, you will see ample evidence of higher lake levels. If you go out into the lake and explore the shape of the muddy bottom, you will find evidence of lower lake levels. The historical average level (1847-1998) of the water surface is 1,280.3 m. The lowest of the exposed shorelines and lake sediments (up to 3 m above present level) reflect very recent history. Great Salt Lake reached a level above 1,283.5 m around 1873 and again in 1986-1987, and was as low as 1,278 m during 1962 to 1965. As you go farther from the lake, and higher in elevation, you will continue to encounter lake sediments and shoreline features (gravel bars, spits, tombolos, erosional cliffs, etc.) until you are nearly 300 m above present lake level. In fact, despite the semi-arid climate of the Great Basin, the lower portions of all the valleys from the Wasatch Front west to the Nevada-Utah border are dominated by coastal landforms. However, beyond that point, the landscape changes abruptly, and there is no further evidence of shoreline processes on the landscape. At many points near the top of this wave-formed landscape, the individual shorelines can be traced visually for many tens of kilometers, and appear to be perfectly level lines painted across the hills.

A simple experiment that you and a group of friends could accomplish in a single day would yield a second surprising pattern. First, note that the present shoreline of Great Salt Lake is almost perfectly level. Next, each of you should get an altimeter (or sufficiently accurate topographic maps) and drive away from Great Salt Lake, each in a different direction, until you are above the highest shoreline. Along the way you will encounter many different shoreline features. If you each note the elevations of major shoreline features as you cross them, and then compare these notes with your fellow travelers, you would find an interesting pattern. All of you would note major shorelines at elevations near 1,350 m, 1,450 m and 1,550 m. These are the Stansbury, Provo and Bonneville shorelines, respectively. Many would also note a smaller, yet quite distinctive, shoreline near 1,300 m. It is the Gilbert shoreline. If you then looked more closely at the actual elevations of these features, and the distance from Great Salt Lake at which they were encountered, you would again find an interesting pattern. Despite initial appearances, the heights of these shorelines are not constant. Instead, they all decrease with increasing distance away from the center of the basin. In addition, the amount of variation is roughly proportional to the average height (greatest on Bonneville, least on Gilbert). Lastly, the variation is generally more rapid in an east-west direction than in a north-south direction. Figure 1 illustrates the pattern. Though these higher shorelines certainly appear to be perfectly level, they actually are not. Each of them slopes upward toward the center of the basin. The slopes are far too subtle to discern by just looking, since they never exceed 5 cm/km. However, the actual magnitude of the cumulative departures from horizontal is quite impressive, especially on the Bonneville and Provo shorelines.

This pattern was first observed by G.K. Gilbert (1890), using nothing more than a crude altimeter, and a keen eye for patterns. He also correctly deduced the explanation for this pattern. It is due to a combination of two simple processes. The shorelines all formed as level surfaces, similar to the present shoreline of Great Salt Lake. However, when the lake stood at these higher levels, the load of the water in the lake was sufficient to push the crust of the Earth down by amounts that locally approached 1/3 of the water depth. When the lake level dropped, the crust rebounded to its pre-load level, and the shorelines were bowed up in the middle of the basin.

A proper quantitative understanding of the dynamics of this process has taken a considerable amount of work, but the basic idea remains a simple one - stresses associated with accumulations of water on the surface of the Earth causes hot rock in the interior of the Earth to slowly flow. Careful examination of the spatial and temporal patterns of the crustal deformation recorded in the shorelines provides one of the best means available for determining the long-term strength of rocks within the top few hundred km of the Earth.

On short time scales (seconds to years), the mechanical behavior of rock is essentially elastic. In that case, the strain (deformation per unit length) is proportional to stress (force per unit area). The elastic properties of the Earth are very well known from observations of the speed of transmission of seismic waves associated with earthquakes and explosions. However, on much longer time scales (millions of years), the deep interior of the Earth behaves very much like a viscous fluid. In that case, the rate of strain (deformation per unit length per unit time) is proportional to stress. Comparison of the vertical deformation patterns recorded in the shoreline elevations with the results of mathematical models of the Earth's response to vertical loads, allows viscosity of the rocks in the deep interior to be inferred.

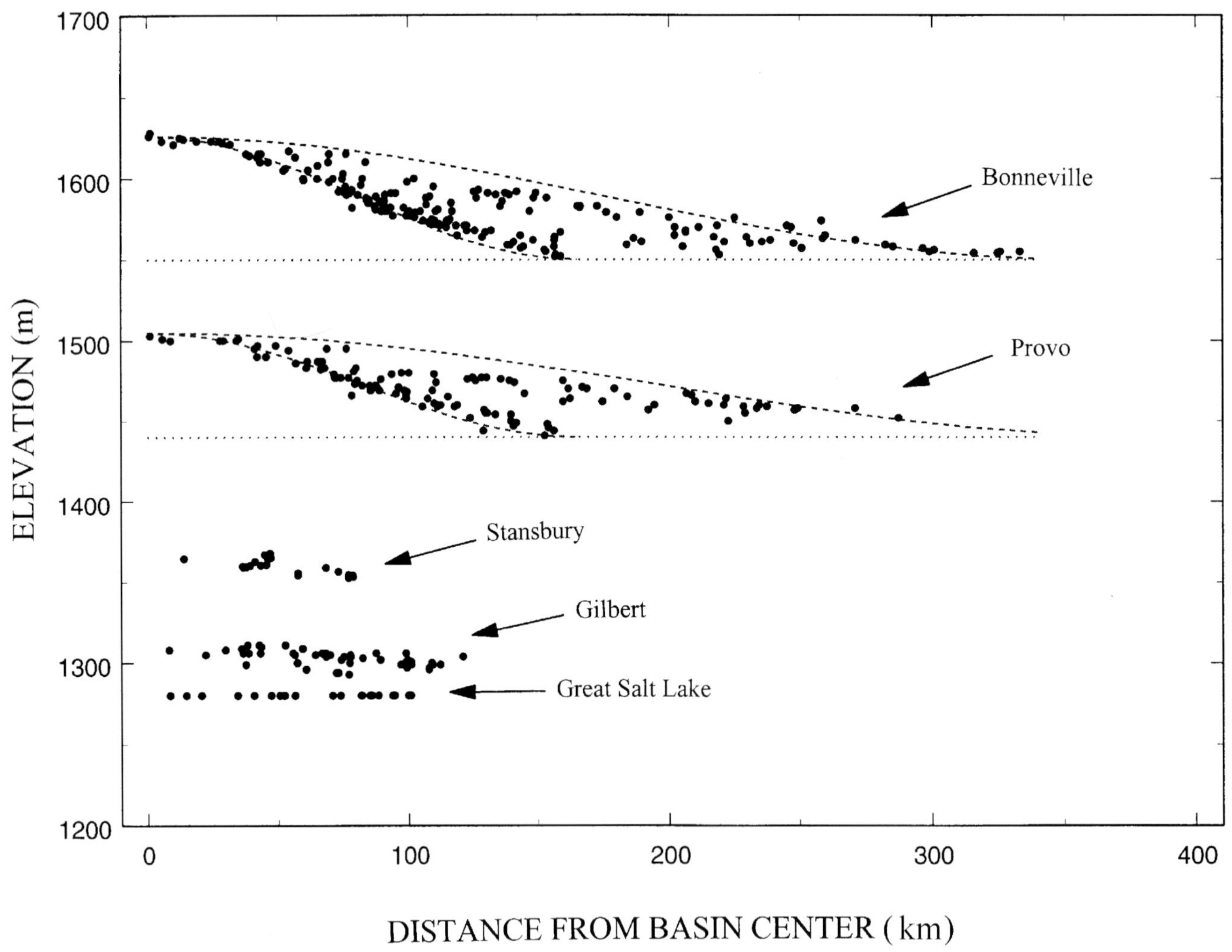

Figure 1. Bonneville basin shoreline elevations. The Bonneville basin contains many separate paleo-shorelines. The higher shorelines occur at variable elevations, which are generally highest near the basin center and lower near the periphery. For each of four prominent shorelines, the pattern of elevation variations, versus distance from the basin center, is illustrated. Data for Bonneville, Provo, and Gilbert shorelines are from Currey (1982). Stansbury data are restricted to eastern half of basin.

General Background

Lake Bonneville was the largest of the late Pleistocene lakes in the Great Basin. With a surface area in excess of 50,000 km^2, and a water volume in excess of 10,000 km^3, Lake Bonneville was large enough that the crust of the Earth was significantly depressed beneath the load. The series of shorelines that formed as the level of Lake Bonneville rose and fell over the last 30,000 years provides a clear indication of this crustal deformation. Each of these shorelines formed as a nearly level surface when the load was present, but subsequent recovery of the load-induced crustal depression has bowed the coastal features such that tracing any single shoreline across the basin yields a pattern that rises in elevation toward the center. The highest shorelines are roughly 300 m above the present level of Great Salt lake, and are about 70 m higher near the center of the basin than on the periphery.

This pattern of shoreline deformation was first noted by Gilbert (1890) and was correctly interpreted by him as a response to the load associated with the lake. A key element in his argument was that the observed deformation pattern resembles the maximum water depth pattern, as would be expected if buoyant support is the primary mechanism at work. Local buoyant support would predict that 1 m of water should displace 1/3 to 1/4 m of rock, depending on rock density. Gilbert also noted that the pattern of deformation is actually smoother than the water depth variations themselves, which he interpreted as reflecting the influence of a near-surface layer of rock with significant elastic strength. Though he did not use the term "lithosphere," he did note that the amplitude and smoothness of the shoreline pattern is consistent with a 50-km-thick layer of strong material in the subsurface, and this is apparently the earliest published estimate of what would today be called lithospheric thickness (Walcott, 1970; Wolf, 1993). He also considered the possibility of viscous or plastic flow at depth, but concluded that too little was known, both in the area of the short term or elastic properties of rocks and the actual chronology of lake level fluctuations, to make quantitative estimates of the longer term rock behavior from the shoreline patterns. Gilbert (1890) concluded his discussion of the shoreline deformation pattern with the comment:

> The application of an analytic theory of these relations could yield the best results only with a better determination than we now have of the elasticities of rocks, and with a better determination of the figure of the deformation of the Bonneville Basin; but even with the imperfect data at hand it might establish a presumption for or against the existence of a liquid substrate beneath the rigid crust, and if the mathematical difficulties were surmounted, there can be little question that the observational data would be supplied, for their procurement is opposed by little beside their expense.

In the years since Gilbert wrote those words, extensive work on the elastic properties of rocks has been done, the mathematical difficulties he referred to have been surmounted, rather extensive collections of shoreline height data have been surveyed, and many dated specimens constrain the chronology of lake level fluctuations. Indeed, the shoreline deformation pattern of Lake Bonneville is probably the best studied case of its type in the world. No attempt will be made here to give a complete history of that study, but reference to earlier work is encouraged. Work on modeling the rebound includes Crittenden (1963a,b, 1970), Passey (1981), Nakiboglu and Lambeck (1982, 1983), Iwasaki and Matsu'ura (1982), Bills and May (1987), May and others (1991), and Bills and others (1994). Work on characterizing the shoreline height pattern is reported in Crittenden, (1963a,b), Passey (1981), and Currey (1982). The latter reference is the primary source of data used in the present study. Characterizing the chronology of lake level fluctuations has proven challenging. Recent efforts are reported by Eardley and others (1973), Scott and others (1983), Spencer and others, (1984), Currey and others (1983, 1984), Currey and Oviatt (1985), Oviatt and McCoy (1988), Currey (1990), Oviatt and others (1990, 1992, 1994), and Oviatt (1997).

BASIN GEOMETRY

The topography of the eastern Great Basin is illustrated in figure 2. The digital elevation model used to generate the figures was obtained from the global archive of Digital Terrain Elevation Data (DTED) maintained by the National Imagery and Mapping Agency (www.nima.mil) and has a horizontal resolution of 30 arc second, which is 0.93 km in latitude, and varies from 0.58 to 0.73 km in longitude. Figure 2a depicts the elevations, with darker shades of gray roughly corresponding to deeper water when Lake Bonneville was at its greatest extent. Figure 2b shows contours of 1,550 and 1,600 m elevation. Those two values roughly correspond to the lower and upper altitudinal limits over which the Bonneville shoreline occurs. Higher values occur near the basin center and lower values occur near the periphery.

LAKE LEVEL CHRONOLOGY

The water surface level of lakes in the Great Basin fluctuates on a variety of time scales, dominantly under climatic control. First, the lake adjusts its surface area so that evaporative losses balance precipitative inputs, both directly on the lake and as runoff from the catchment area (Mifflin and Wheat, 1979; Hastenrath and Kutzbach, 1983, 1985; Mason) and others, 1994). A recent reconstruction of lake surface elevation history for the last 40 kyr (Oviatt, 1997), is illustrated in figure 3. Radiocarbon years have been converted to calendar years using the calibration of Bard and others (1990). Many earlier lake cycles have occurred (Eardley and others, 1973; Williams, 1994), but are not further discussed here.

At very low levels, there were separate lakes in each of several disconnected basins. As the lake level increased, the component basins coalesced. During the Stansbury oscillation (25 to 22 kyr), tufa-cemented gravel and barrier beaches of sand and gravel were deposited (Oviatt and others, 1990). The lake level then generally increased for about 10 kyr, but continued to exhibit significant oscillations. As the level of the Bonneville shoreline was first attained, the lake began exporting water to the Snake River in southern Idaho. The last of the transgressive oscillations may have actually begun after the lake reached threshold control. In that event, it would correspond to the Keg Mountain oscillation of Currey and others (1985).

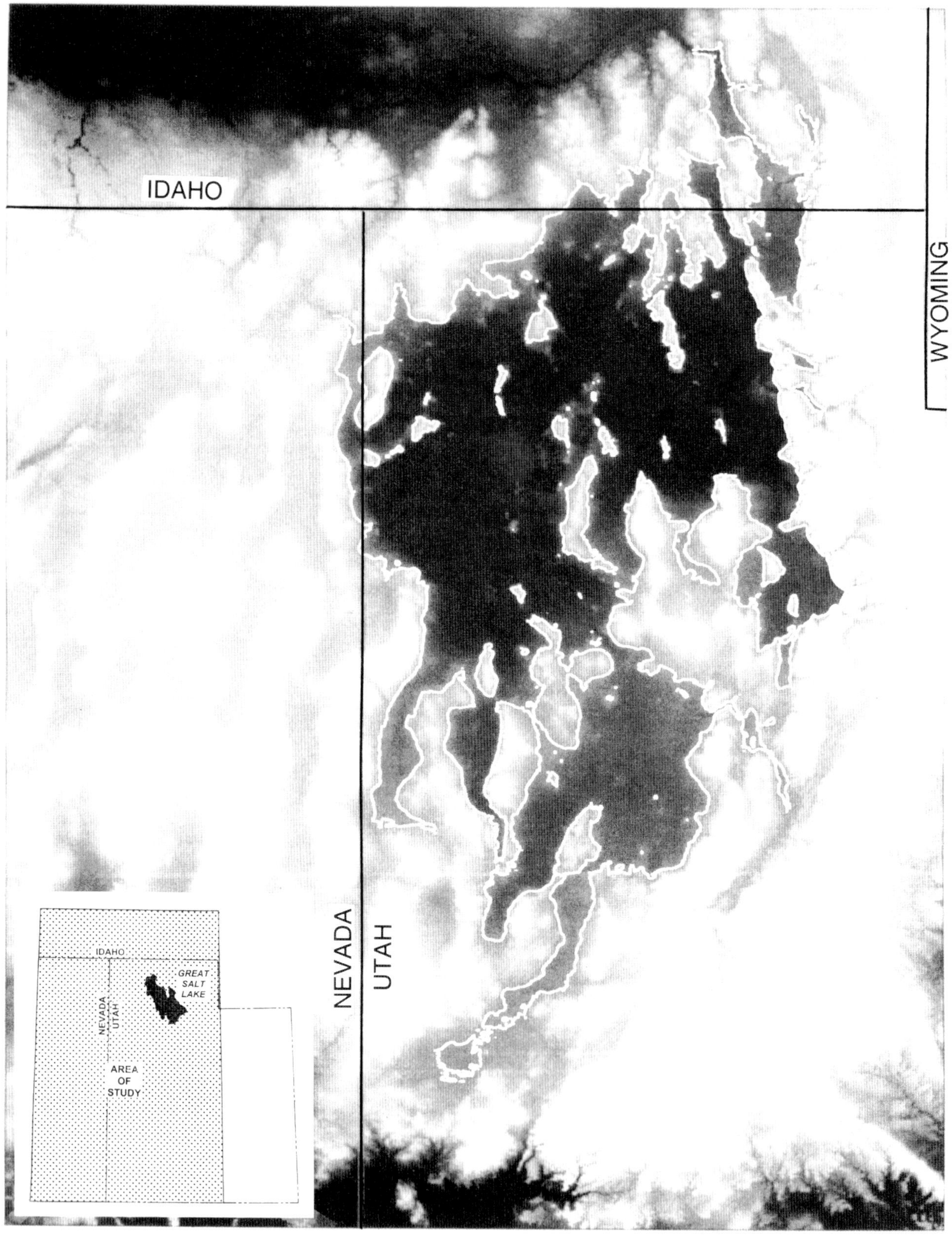

Figure 2a. *Grayscale representation of the topography of the Bonneville basin. Resolution of the digital elevation model is 30 arc second.*

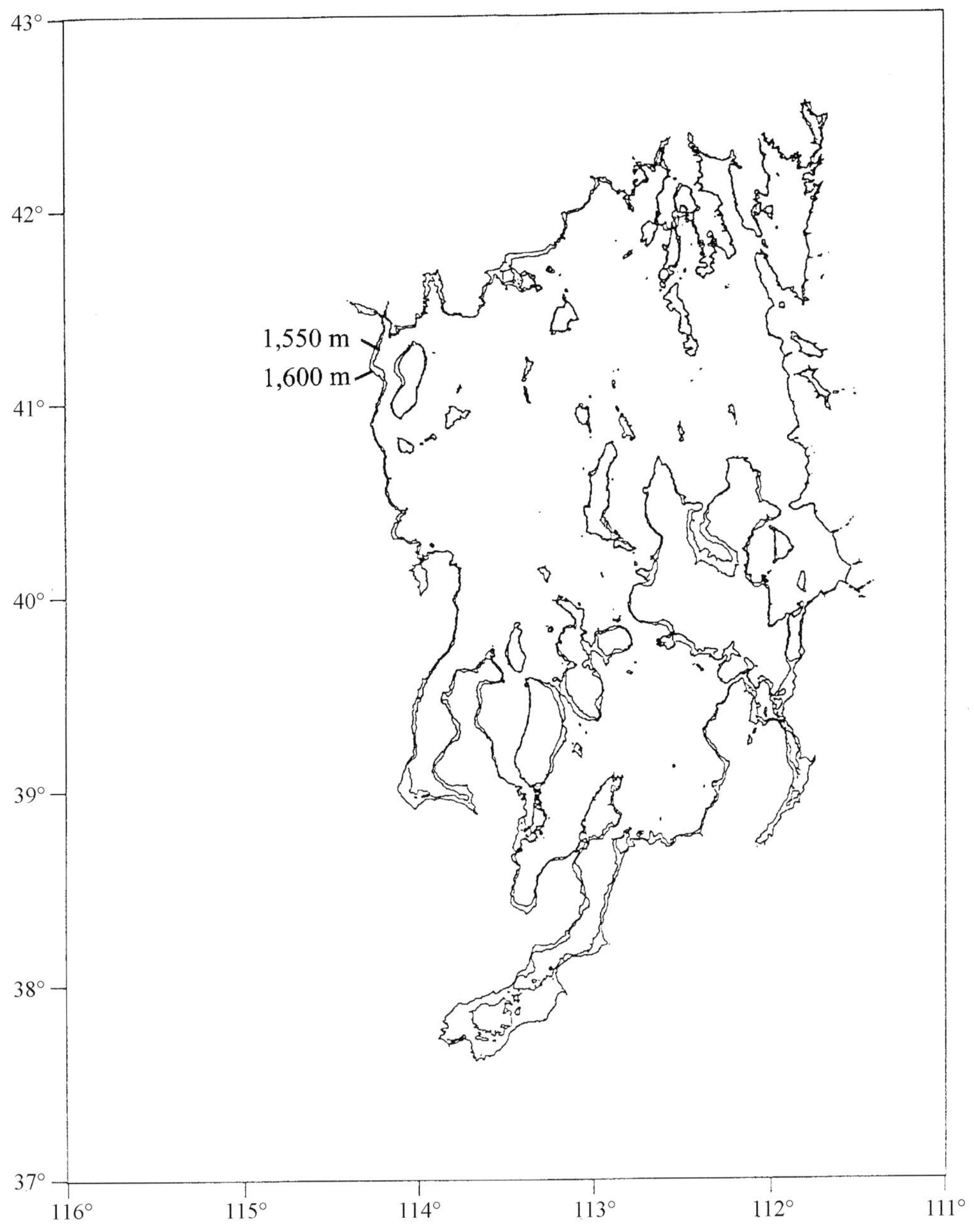

43°
42°
41°
40°
39°
38°
37°
1,550 m
1,600 m
116° 115° 114° 113° 112° 111°

Figure 2b. Approximate outline of Lake Bonneville, at its maximum extent. Contours represent 1,550 m and 1,600 m, which span the range of elevations at which the highest Bonneville shoreline is found.

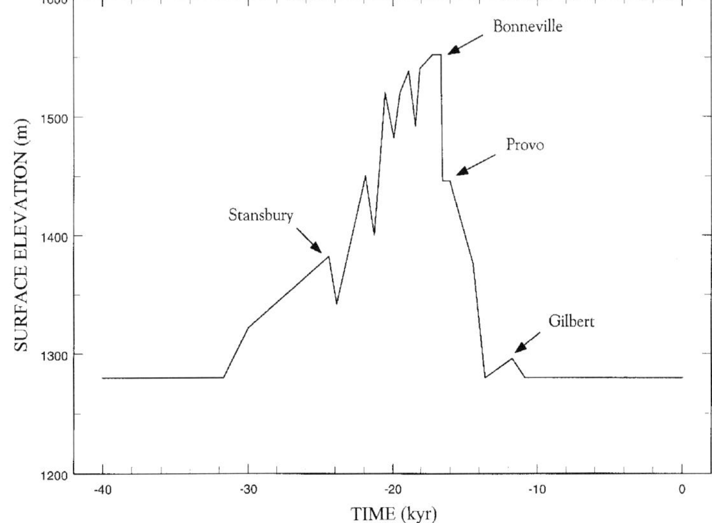

Figure 3. Lake Bonneville surface elevation history. The solid line represents a recent synthesis of several hundred dated samples that constrain the lake surface elevation for the past 40 kyr. Several of the highstands (Stansbury, Bonneville, Provo, Gilbert) are associated with prominent shorelines. From Oviatt (1997), with radiocarbon age calibration from Bard and others (1990).

The Bonneville highstand was abruptly terminated by a catastrophic flood (Gilbert, 1890; Malde, 1968; O'Connor, 1993) when headward erosion at the outlet caused the threshold to fail. The lake level dropped approximately 110 m, most of which probably occurred within a few months, resulting in removal of about 5,000 km^3 of water. Lake level eventually stabilized at the Provo level, and stayed there until the climate dried enough that it fell again. Lake level apparently fell quite quickly to, or even below, historic levels of Great Salt Lake. There was a brief return to pluvial conditions, and the lake rose to nearly 1,300 m elevation and formed the Gilbert shoreline. Throughout the Holocene, levels have varied somewhat, but have remained below the Gilbert level.

SHORELINE ELEVATION DATA

The principal source of shoreline elevation data used in the present study is the compilation by Currey (1982). That study provides shoreline information, including elevations, locations, and geomorphic context, for 181 Bonneville, 112 Provo, and 48 Gilbert sites. Many of the smaller valleys in the Great Basin, to the west and south of Bonneville, also contained lakes about the same time (Mifflin and Wheat, 1979). Though none of these other lakes were large enough to produce significant crustal deflection, they do serve as far-field tilt meters. Bills and others (1994) reported shoreline elevation data for the highest shorelines in the three valleys immediately to the west of Bonneville, including 50 measurements from Lake Waring, 27 from Lake Clover, and 25 from Lake Franklin. These high shorelines are presumed contemporaneous with the Bonneville shoreline. The pattern of deflection seen on these shorelines is depicted in figure 4.

The coordinate system used to display these results deserves some comment. The scale of this pattern is small enough that the spherical geometry of the Earth is not significant. As a result, a local cartesian coordinate system can be used. This dramatically simplifies the calculations involved in simulating the deflection process. The Universal Transverse Mercator (UTM) coordinate system provides a good approximation to a local cartesian system. The locations of the shoreline measurement sites are indicated in figure 4 as offsets east and north from an arbitrary coordinate origin of (33^{0000} E, 445^{0000} N) in UTM zone 12. This corresponds to 40.19° N, 113.00° W.

CRUSTAL DEFORMATION PROCESS

The Earth supports surface loads via a combination of buoyancy and elastic strength, with viscous flow modulating the temporal response. Though the complete mathematical analysis required to accurately simulate these processes may appear rather complex, the fundamental principles are quite simple. Buoyant support is governed by Archimedes' principle: a load placed on the surface of the Earth will displace an amount of matter with a mass equal to that of the load. The

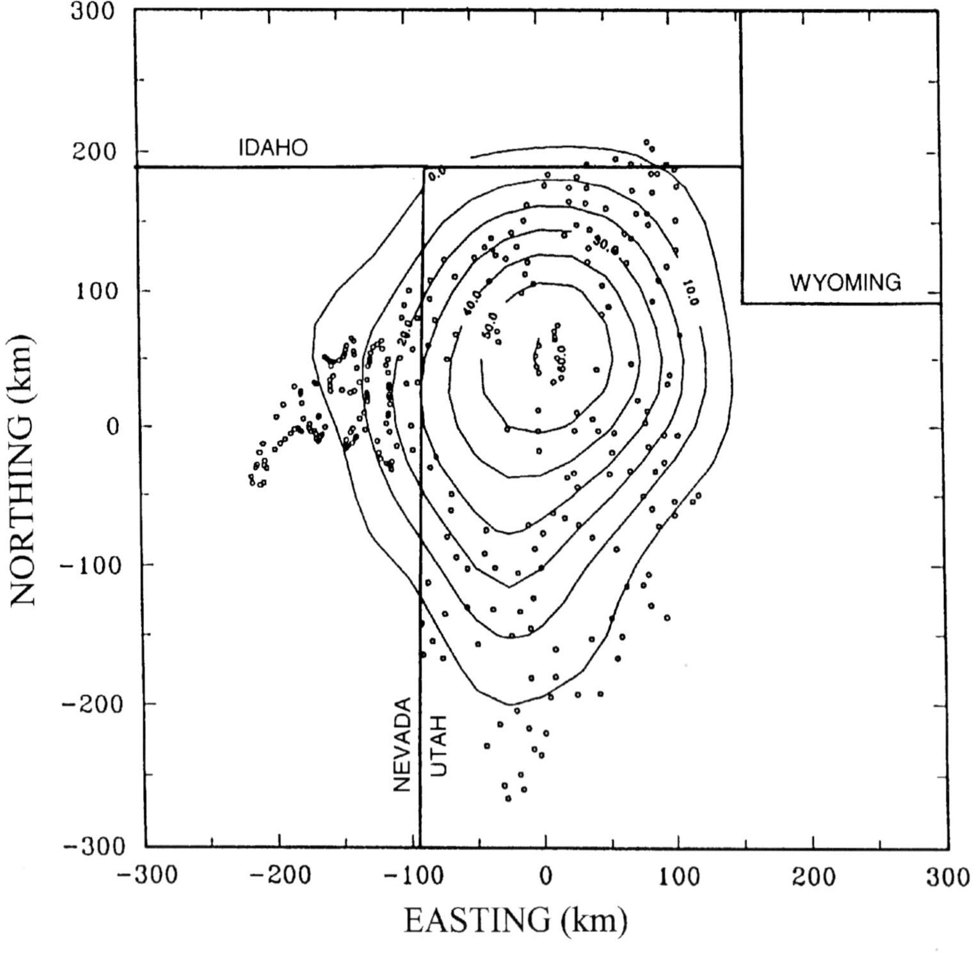

Figure 4a. Observed deflections of the Bonneville shoreline in meters.

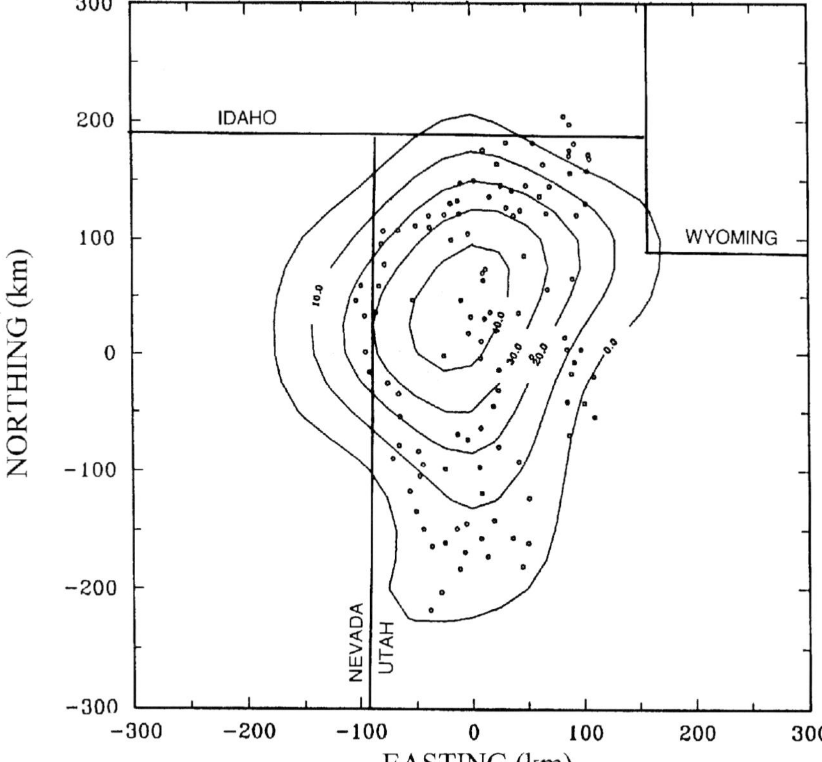

Figure 4b. *Observed deflections of the Provo shoreline in meters.*

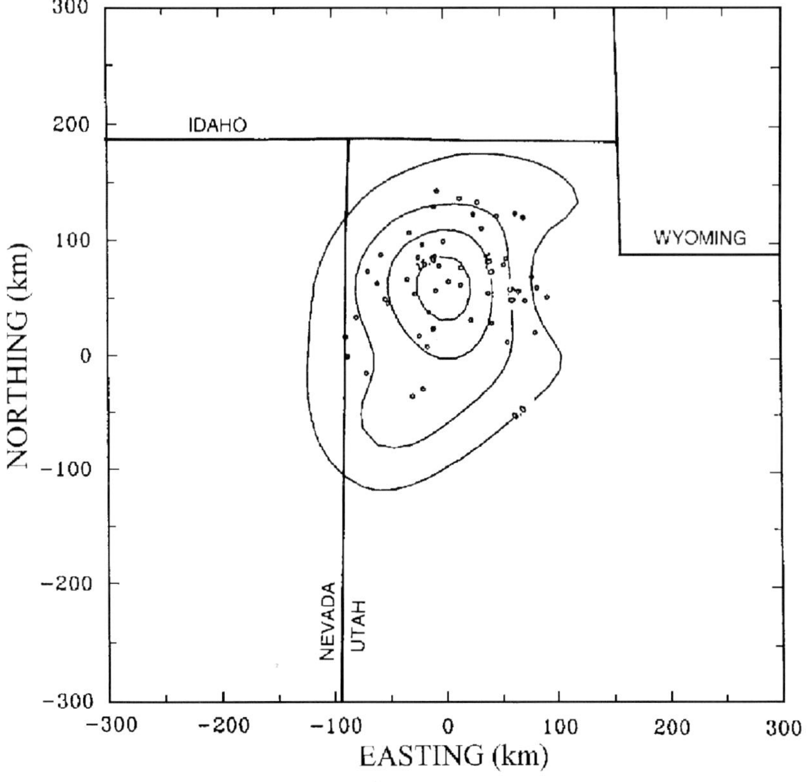

Figure 4c. *Observed deflections of the Gilert shoreline in meters.*

effect of a subsurface layer with significant elastic strength is simply to enlarge the region over which buoyant support of a load occurs. Subsurface viscous flow mediates the transition from prompt regional elastic support to final local buoyant support.

In an elastic solid material, stress (force per unit area) is proportional to strain (displacement per unit length). In a viscous fluid material, stress is proportional to strain rate. The elastic rigidity of crustal or upper mantle rocks can be determined from laboratory experiments or seismic wave propagation velocities, and is typically about 10^{11} Pa (Pascal = Newton/meter^2 = 10^{-5} bar). Estimates of upper mantle rock viscosity values are generally in the range 10^{18} to 10^{21} Pascal seconds (Pa s). For comparison, the viscosities of water and air are about 10^{-3} and about 10^{-5} Pa s, respectively.

On short time scales (seconds to decades), the Earth behaves elastically from the top of the crust to the core-mantle boundary. On long time scales (10^6 year and longer), the entire Earth behaves like a viscous fluid. One advantage of climatically driven lake loads in exploring long-term strength in the crust and upper mantle is that the time scale of the significant load variations (10^3 to10^4 years) is comparable to the time scale in which the transition from elastic to viscous behavior occurs.

Earth materials generally exhibit a combination of elastic and viscous behaviors. The two simplest models of viscoelastic behavior make different suppositions about how to combine the relevant stress and strain contributions. In the Kelvin model, it is supposed that the stresses are additive (total stress is the sum of viscous plus elastic contributions). In the Maxwell model, the strain rates are presumed to be additive (total strain rate is the sum of viscous plus elastic contributions). Using an analogy with electrical circuits, the Maxwell model may be thought of as combining the elastic and viscous elements in series, while the Kelvin model combines them in parallel. The Maxwell model behaves elastically on short time scales and viscously on long time scales. The response of the Kelvin model is just the reverse, with viscous flow dominant on short time scales, and elastic strain dominant on long time scales. In either case, the time required for transition from prompt response to ultimate response is given by the ratio of the viscosity to the rigidity, and is called the Maxwell relaxation time. For typical upper mantle rocks, the relaxation time is in the range from 0.3 to 30 years.

The Maxwell relaxation time characterizes the response of a homogeneous sample. If the Earth structure involves layers with widely differing viscosities, then the transition from elastic to viscous behavior will be more complex. If the spatial pattern of the applied load is represented by the superposition of sinusoidal variations, with a broad spectrum of wavelengths, each wavelength will have a different transition time from elastic to viscous behavior. As a simple, but illustrative example, suppose that the Earth consists of a weak (low viscosity) middle layer, with an elastic plate above and a higher viscosity region below. In such a case, the transition times for all load wavelengths will be longer than the Maxwell time of the weak middle layer. As the thickness of the weak layer is decreased, it will be more difficult to extrude material laterally in the weak channel, and the relaxation times will increase.

INFERRED EARTH STRUCTURE

One application of the Bonneville shoreline deformation pattern to further elucidate the depth dependence of long-term strength of the Earth within the Great Basin. Global studies of glacial isostatic rebound have much the same objective (Peltier, 1974, 1986). The short-term strength, as parameterized by the elastic rigidity, is well known from seismic studies (Priestley and others, 1980). All that remains is to estimate an effective viscosity in each of several layers at depth within the crust and upper mantle.

It is worth noting that though the elastic and viscous models used are strictly linear, the overall process response is slightly non-linear. That is, not only does the deformation depend on the load, but the load also depends somewhat on the deformation. In particular, crustal deflection notably increased the volumetric capacity of Lake Bonneville. If one were to estimate the volume of water contained in Lake Bonneville when its surface elevation was 1,550 m, by naively comparing that value with the present topography of the region, the volume obtained would be an underestimate by roughly 20 percent. This problem can be easily dealt with by an iterative process in which the deflection computed at each step is added to the present day topography before computing the load at the next step.

For a given Earth model (density, rigidity, and viscosity in each of several layers), and a given loading history, the viscoelastic deformation process model yields a calculated deformation pattern corresponding to Bonneville, Provo, and Gilbert epochs. A measure of the misfit between data and model is provided by the sum of the squares of the residuals. By adjusting the loading history and the viscosity model one parameter at a time, and seeing how those adjustments influence the pattern of residuals, it is possible to solve for a linear combination of adjustments that minimizes the misfit. Figure 5 is a cartoon representation of the solution process. Figure 6 illustrates a viscosity model derived from the shoreline data and loading history in this way. The indicated errors only reflect measurement uncertainty in the shoreline elevation pattern.

The upper mantle viscosity values obtained in this way from the Lake Bonneville shorelines are considerably lower than typical global average viscosity estimates obtained from glacial isostatic models. A plausible explanation for that difference is that the upper mantle beneath the Great Basin is hotter than the global average. That interpretation is well supported by observations of high heat flow (Lachenbruch,

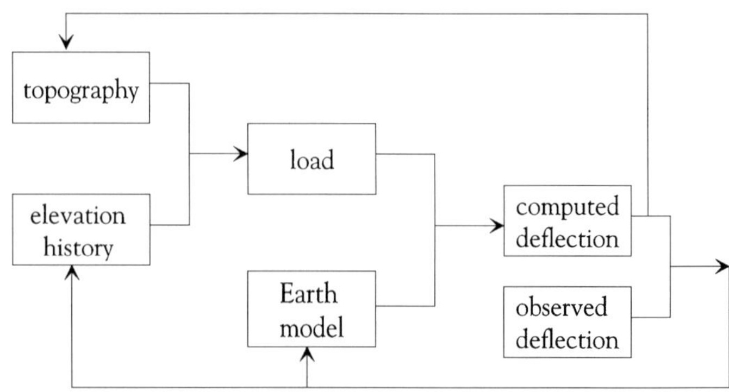

***Figure 5**. Deflection-Simulation Process (see text for further discussion).*

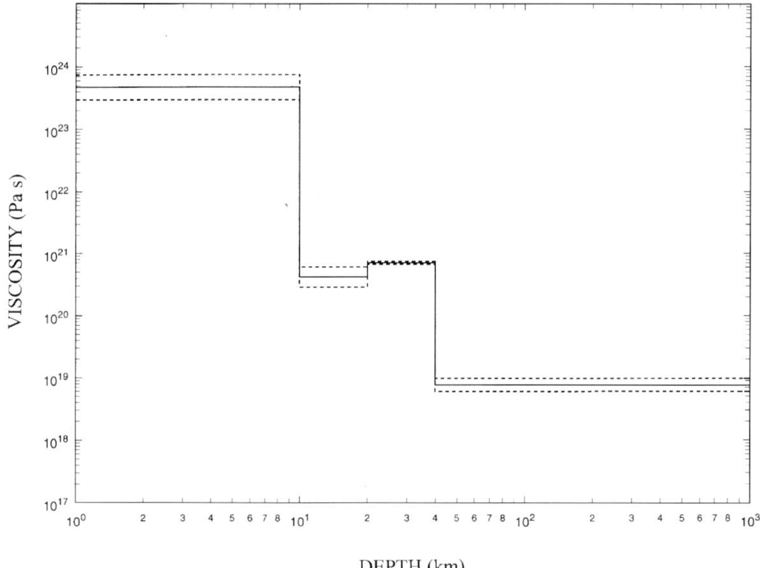

***Figure 6**. Bonneville basin viscosity model. In this model, viscosity is assumed constant within each of four depth ranges. Estimates were obtained by varying each of the values until the computed deflection pattern optimally matched the observed pattern of shoreline heights (see text for further discussion).*

1978), recent volcanism (Condie and Barsky, 1972; Oviatt and Nash, 1989), and extensional tectonics (McKee and Noble, 1986; Minster and Jordan, 1987; Zoback, 1989). For a recent review of the geophysical structure of this region, see Humphreys and Dueker (1994a,b).

DEFLECTION PATTERN ATLAS

The significance of the crustal deformation produced by Lake Bonneville water loads is not limited to geodynamics research into the long-term strength of the Earth. The deformation was large enough to change local and regional topographic gradients, and thus influence hydrologic processes. Several examples will illustrate this concept. The Bonneville salt flats represent a surface accumulation of salt near the western edge of the Bonneville basin. The salt accumulates there, rather than draining into Great Salt Lake, because there is a low topographic divide separating them. However, that divide was not present at the Pleistocene-Holocene transition, since it largely represents isostatic rebound. Hydrologic closure of Utah Valley and formation of Utah Lake have a similar origin. The proximity of the terminal reach of the Bear River to the Wasatch Front is also due to differential uplift of the basin center.

In order to promote additional research on the influence of isostatic rebound on hydrologic systems in the Bonneville basin, figures 7 and 8 present several views of the spatio-temporal pattern of crustal deflection. Figure 7 shows the temporal evolution at selected locations on an east-west transect across the basin. The largest response is obtained at a location near the center of the basin. Progressively more distal sites experience progressively less deflection. Note that the quantity plotted is the absolute deflection, with positive values indicating downward motion of the surface. In the model, there is no deflection prior to the imposition of the first load at -32 kyr. At the present time, there is still about 3.8 m of unrecovered deflection at the center of the basin. Note that some sites have experienced both positive and negative vertical deflections. The occurrence of uplift, rather than subsidence, in response to imposed loads may seem paradoxical. However, the explanation is actually rather simple. If the load geometry and viscosity structure are chosen correctly, it may occur that the subsurface deflection results in material being extruded laterally, rather than moving away from the load on a more nearly vertical trajectory. If that happens, then a peripheral bulge, or region of uplift, will form beyond the edge of the load.

The observed deflection pattern of the shorelines does not represent absolute deflection, but relative deflection. The shorelines were level when they formed, and any departures from horizontality today represent the difference between the deflection existing when the shoreline formed and the deflection remaining today. Figure 8 presents an atlas of computed relative deflection patterns at times from -2 to -26 kyr. The basic plan represents deflections at uniform 2 kyr time steps. The only exception is inclusion of the pattern at -16.5 kyr, which approximates the Bonneville shoreline epoch. The horizontal coordinate system is identical to that used in figure 4. Positive deflections are downward.

CAVEATS

The atlas of computed Bonneville basin deflection patterns presented here should be viewed as a visual representation of a working hypothesis. There are several important simplifications involved in the model that were used to generate these figures.

Other sources of vertical motion

A detailed comparison between the observed and computed patterns of rebound on the Bonneville, Provo, and Gilbert shorelines reveals a number of features that deserve note. In volumetric terms, the largest misfit between the data and model involves a regional flexure pattern that is well approximated by a quadratic trend surface. The observed shorelines are lower in the northeast corner flexure. The cause of this pattern is not well understood. It might represent a long-term response to basinward sediment transport, or the regional tectonic processes that caused the basin to form.

The next largest discrepancy, in terms of volume, between the data and model occurs along the Wasatch Front, in the regions where major tributary streams have deposited large amounts of sediment. Along the Bear, Weber, and Provo River deltas, the observed shoreline elevations are below the model predictions. Sedimentary loads have been emplaced at those localities, which have depressed the crust.

The largest single residual is at Pahvant Butte, where the Bonneville shoreline is about 15 m below the regional trend At that location, a volcanic eruption occurred during the highstand of the lake, between Bonneville and Provo time (Gilbert, 1890; Condie and Barsky, 1972; Oviatt and Nash, 1989). Post-eruptive collapse explains the discrepancy.

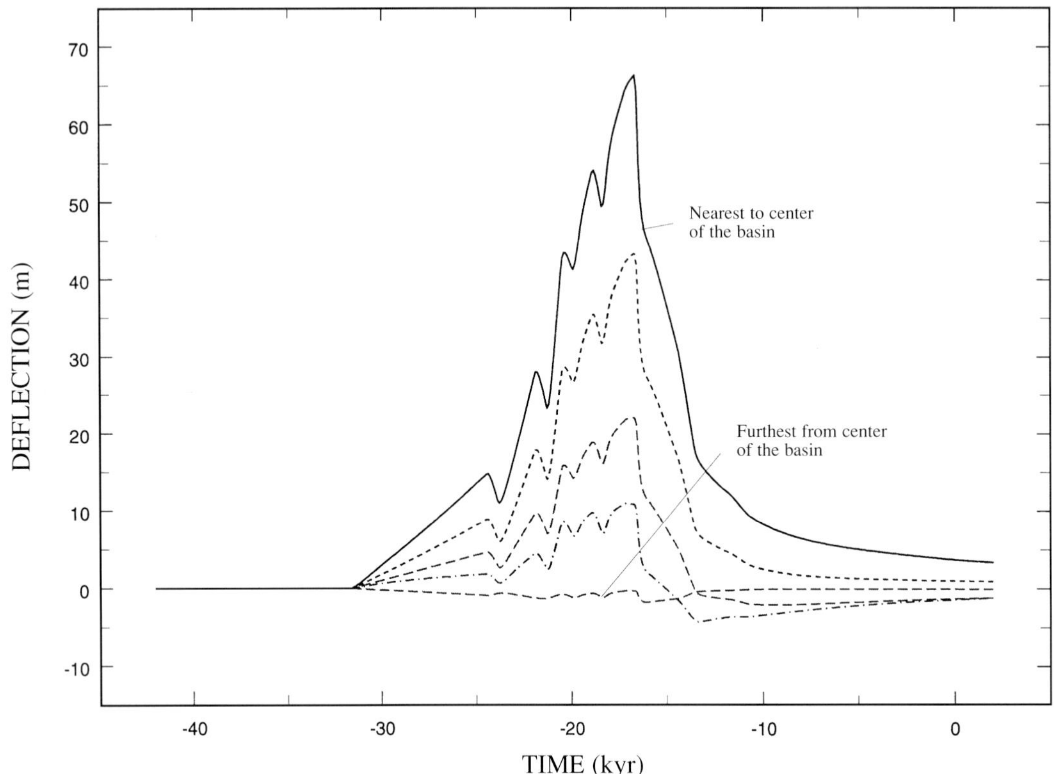

Figure 7. *Bonneville basin deflection histories. Computed vertical motions are shown schematically, as a function of time, for several locations within the basin. Points near the center experience maximum vertical motion. Peripheral points experience minimum motion.*

Shallow intrusive uplift occurred at Cove Creek dome (Condie and Barsky, 1972; Oviatt, 1991).

Alternative lake level histories

The computed deflection history depends quite sensitively on the adopted lake level history. The dependence might be characterized as both direct and indirect. Suppose, for the sake of illustration, that one small interval of the adopted lake level history was changed so that the level decreased by 10 m for 1 kyr. The direct effect would simply be a diminished deflection during, and after, the perturbed interval. In a self-consistent model, an indirect effect would arise from the influence on the estimated viscosity model. The sign and magnitude of that indirect effect would depend on which part of the loading history was changed. As an example of a change that would have large direct and indirect effects, consider the possible inclusion of the Keg Mountain oscillation (Currey and others, 1985). Since it would represent a large decrease in lake load immediately preceding the formation of the Bonneville shoreline, it would require a lower viscosity to produce the same deflection pattern.

SUMMARY

The geometry of the shorelines of ancient Lake Bonneville provide a valuable record of both climatic variability and the long-term strength of the Earth. The fact that there are shorelines much higher than the present level of Great Salt Lake provides evidence of past climates quite different than the present. The fact that these shorelines are not quite level provides evidence of flow and deformation of the earth's crust and upper mantle.

The utility of this record was first recognized by G.K. Gilbert, over a century ago, and it provided one of the first quantitative estimates of long-term crustal strength. Subsequent generations of geologists and geophysicists have continued to refine and interpret the record. As research in this area continues, we can anticipate further insights into climatic and geodynamic processes from this unique record.

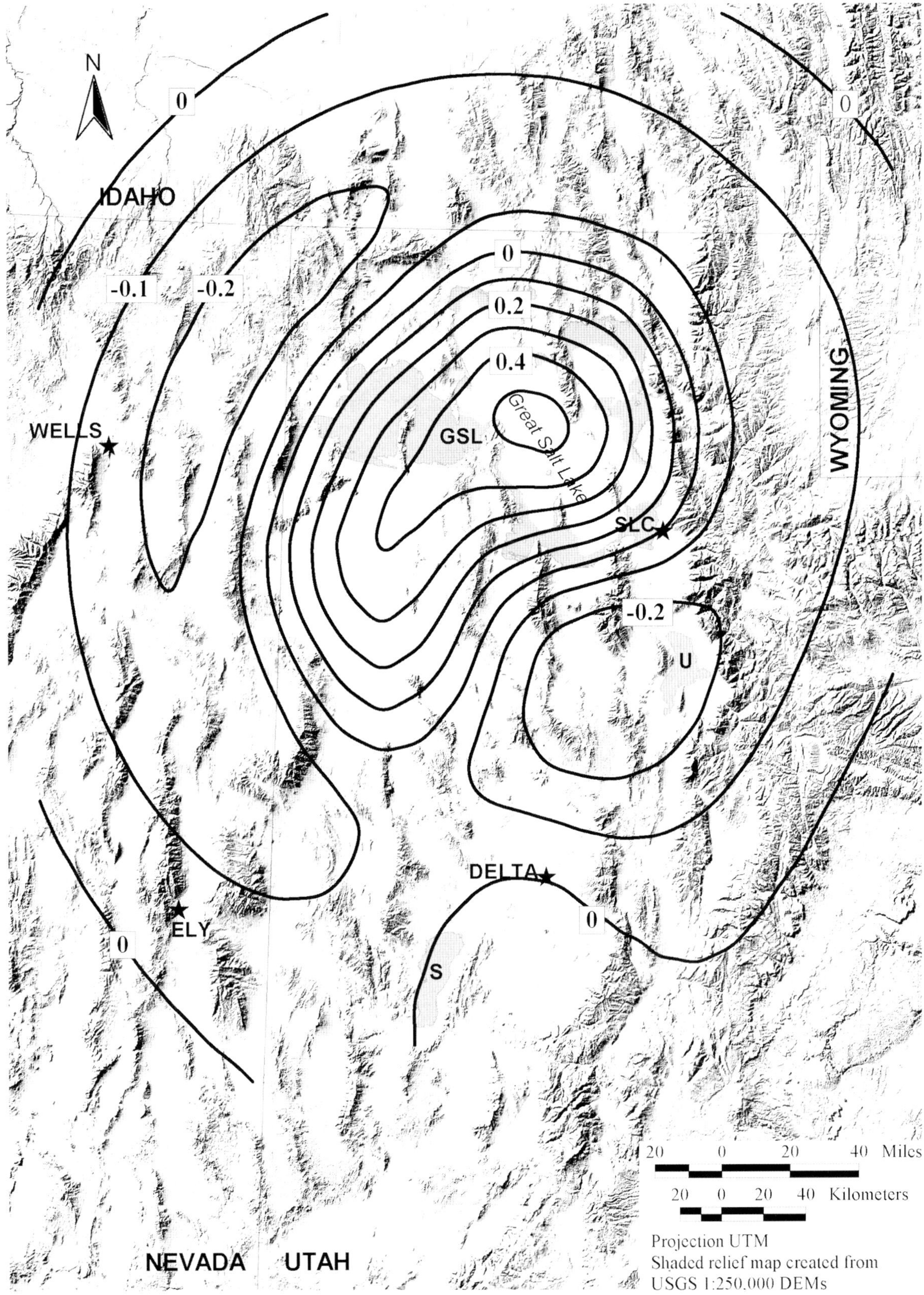

Figure 8a. *Downward deflection contours (0.1 meter interval) at -2 kyr (approximately 2000 radiocarbon years ago). Paleolakes are shown in dark gray. As geodynamically configured at that time, the surface elevation of Great Salt Lake (GSL) was about 1,287 meters above modern sea level (Murchinson, 1989). Utah Lake (U) was little different than in modern time. The surface elevation of Sevier Lake (S) was about 1,383 meters (Oviatt, 1988).*

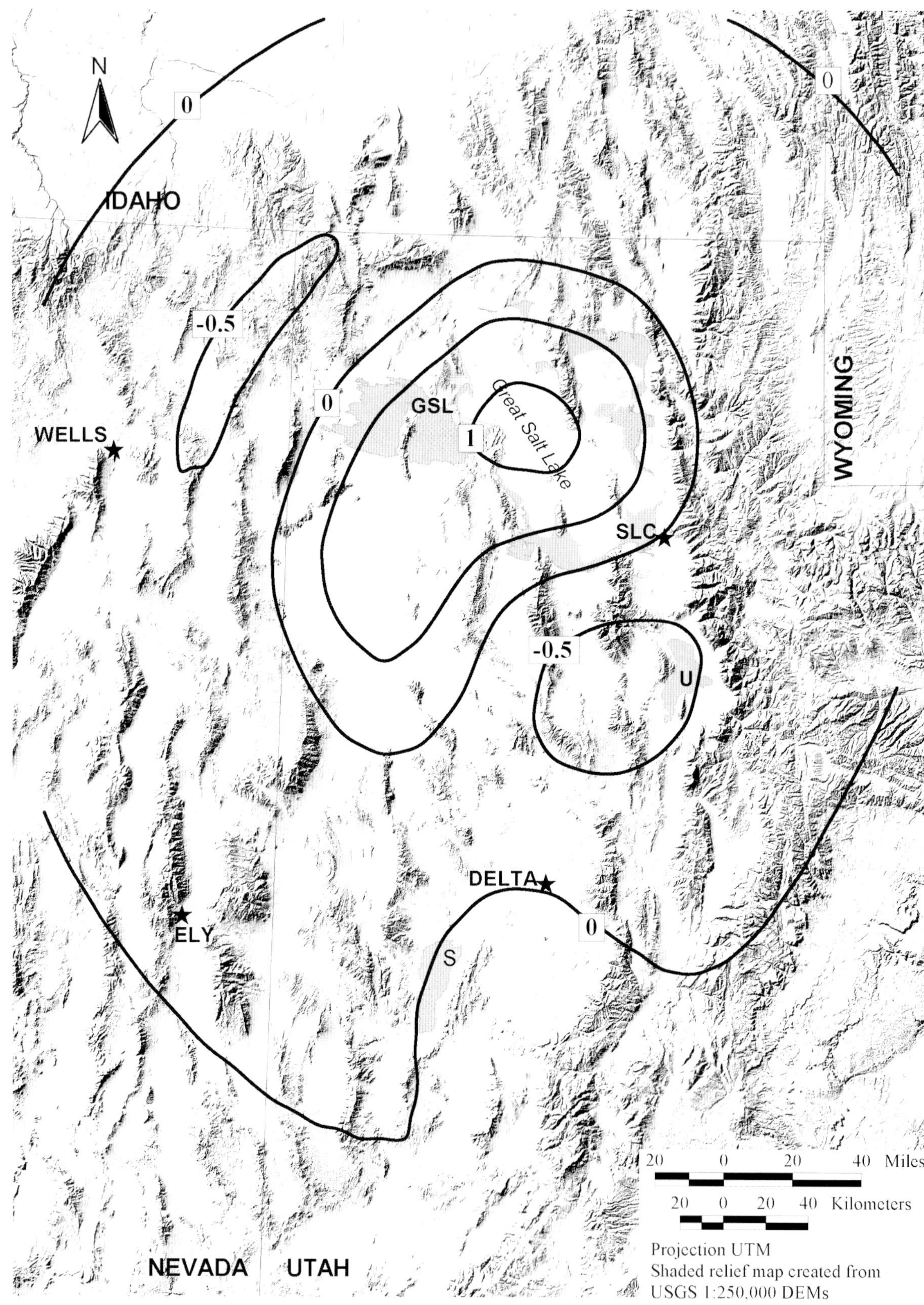

Figure 8b*. Downward deflection contours (0.5 meter interval) at -4 kyr (approximately 4000 radiocarbon years ago). Paleolakes are shown in dark gray. As geodynamically configured at that time, the surface elevation of Great Salt Lake (GSL) was about 1,281 meters above modern sea level (Murchinson, 1989). Utah Lake (U) and Sevier Lake (S) were little different than in modern time.*

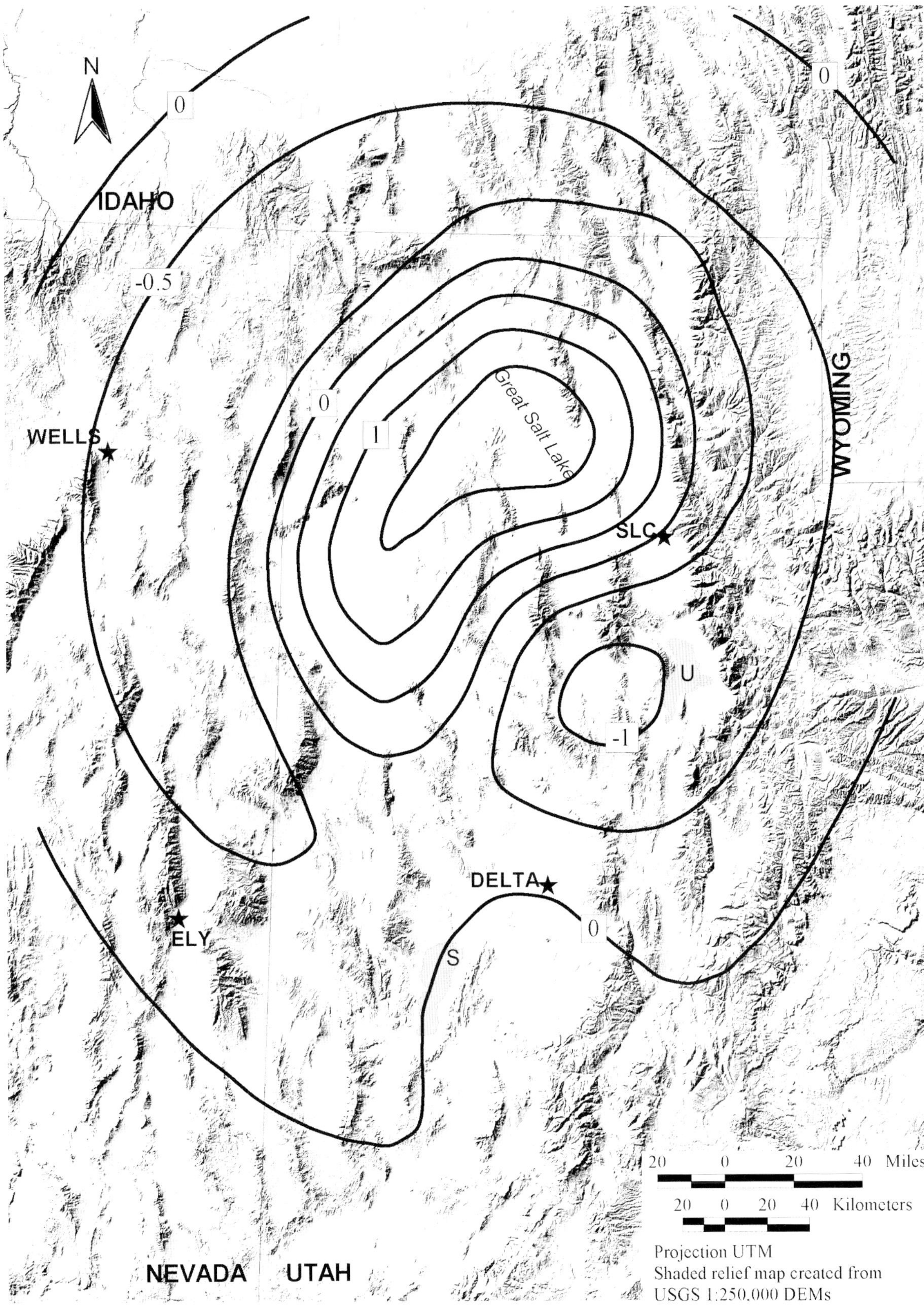

***Figure 8c**. Downward deflection contours (0.5 meter interval) at -6 kyr (approximately 6000 radiocarbon years ago). Paleolakes are shown in dark gray. As geodynamically configured at that time, the surface elevation of Great Salt Lake (GSL) was below the elevation of modern Great Salt Lake (Murchinson, 1989). Utah Lake (U) and Sevier Lake (S) were little different than in modern time.*

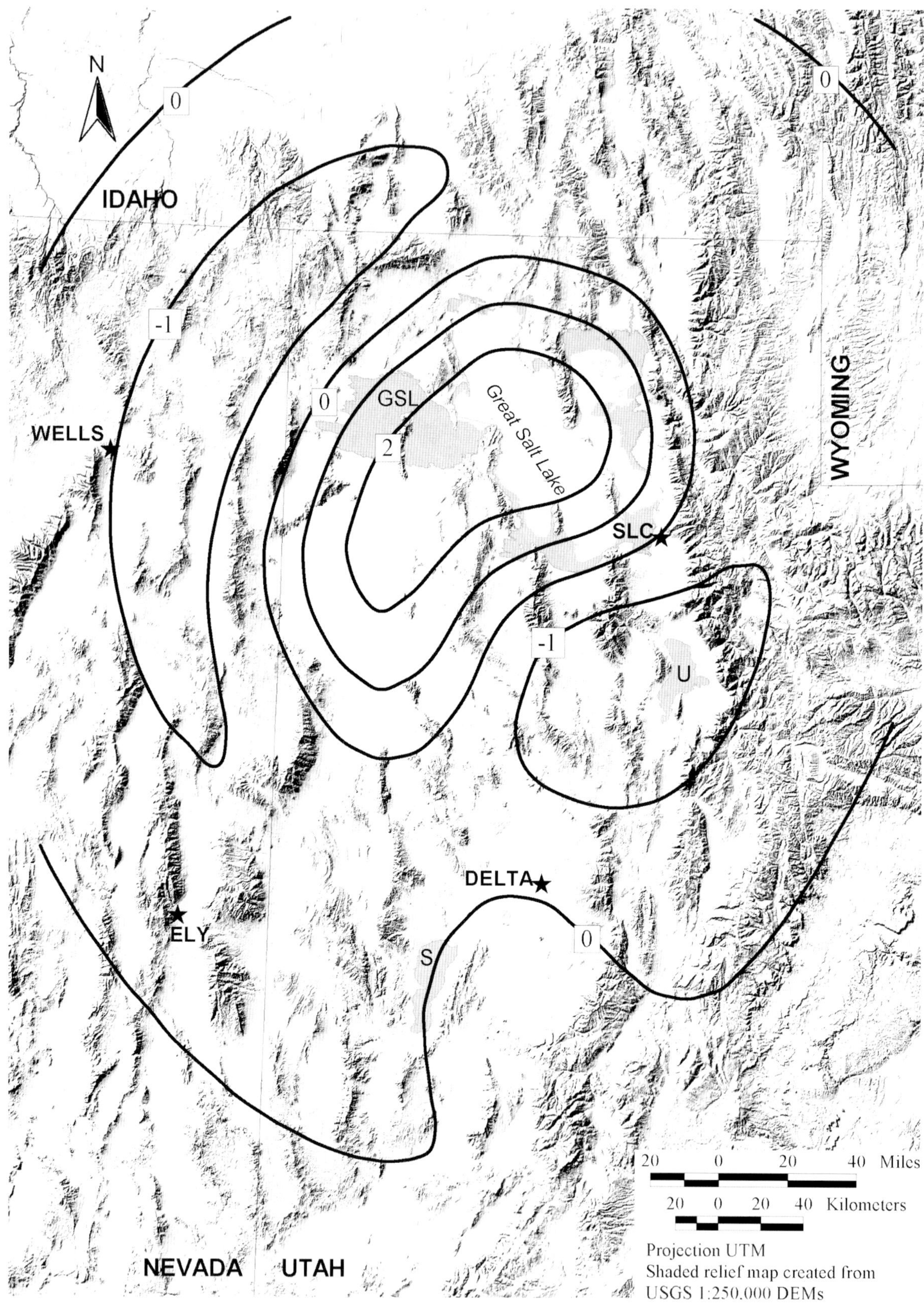

***Figure 8d**. Downward deflection contours (1 meter interval) at -8 kyr (approximately 7,400 radiocarbon years ago). Paleolakes are shown in dark gray. As geodynamically configured at that time, the surface elevation of early Great Salt Lake (GSL) was about 1,283 meters above modern sea level (Murchinson, 1989). Utah Lake (U) and Sevier Lake (S) were little different than in modern time.*

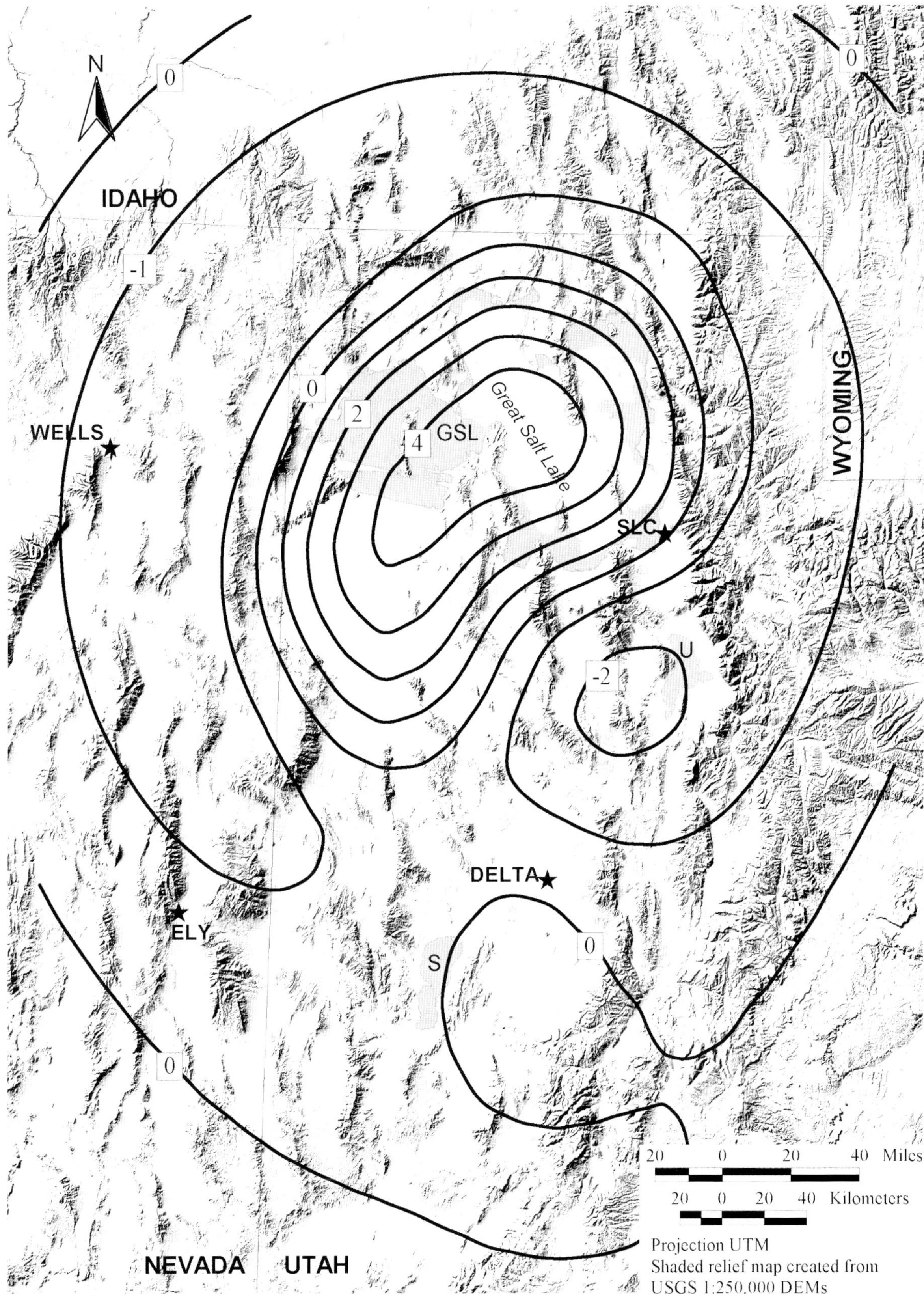

***Figure 8e**. Downward deflection contours (1 meter interval) at -10 kyr (approximately 9,400 radiocarbon years ago). Paleolakes are shown in dark gray. As geodynamically configured at that time, the surface elevation of early Great Salt Lake (GSL) was about 1,288 meters above modern sea level (Murchinson, 1989). Utah Lake (U) and Sevier Lake (S) were little different than in modern time. Structural evolution of Utah Valley permitted containment of Utah Lake starting about this time.*

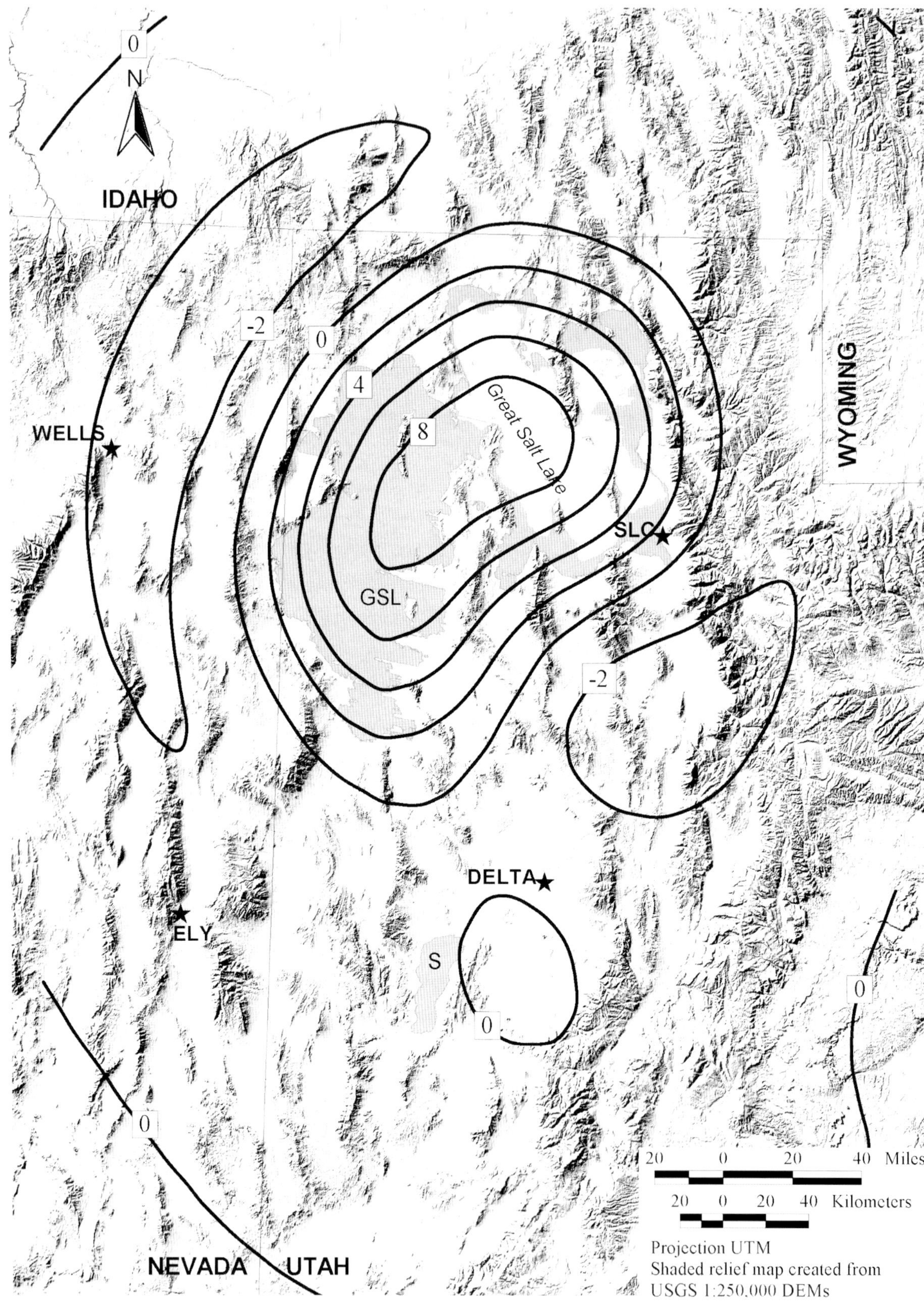

Figure 8f. *Downward deflection contours (1 meter interval) at -12 kyr (approximately 10,500 radiocarbon years ago). Paleolakes are shown in dark gray. As geodynamically configured at that time, the surface elevation of early Great Salt Lake (GSL) was about 1,294 meters above modern sea level (Murchinson, 1989).*

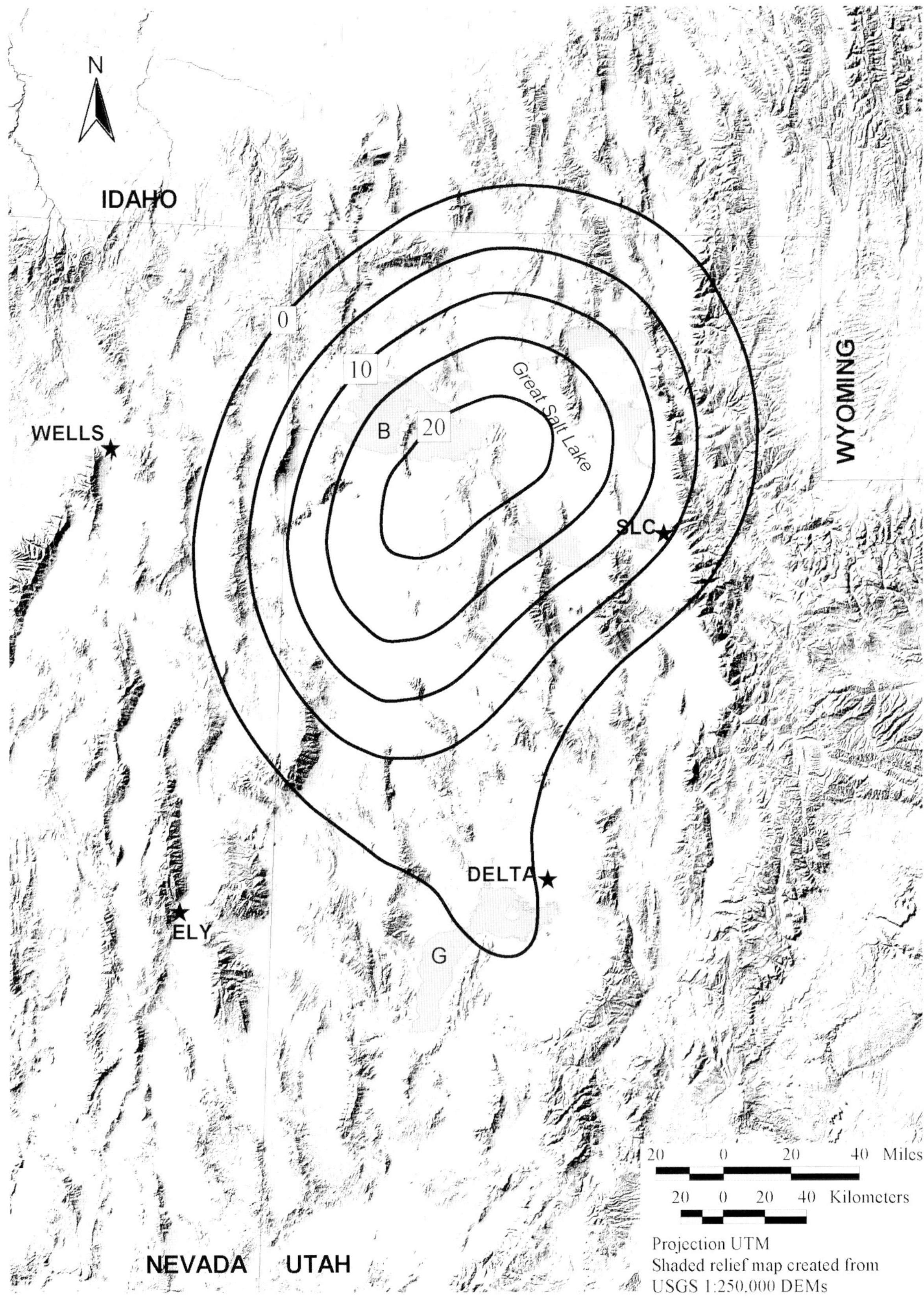

Figure 8g. *Downward deflection contours (5 meter interval) at -14 kyr (approximately 12,150 radiocarbon years ago). Paleolakes are shown in dark gray. As geodynamically configured at that time, the surface elevation of terminal Lake Bonneville (B) was about 1,280 meters above modern sea level (see figure 3). The surface elevation of Lake Gunnison (G) was about 1,390 meters (Oviatt, 1988).*

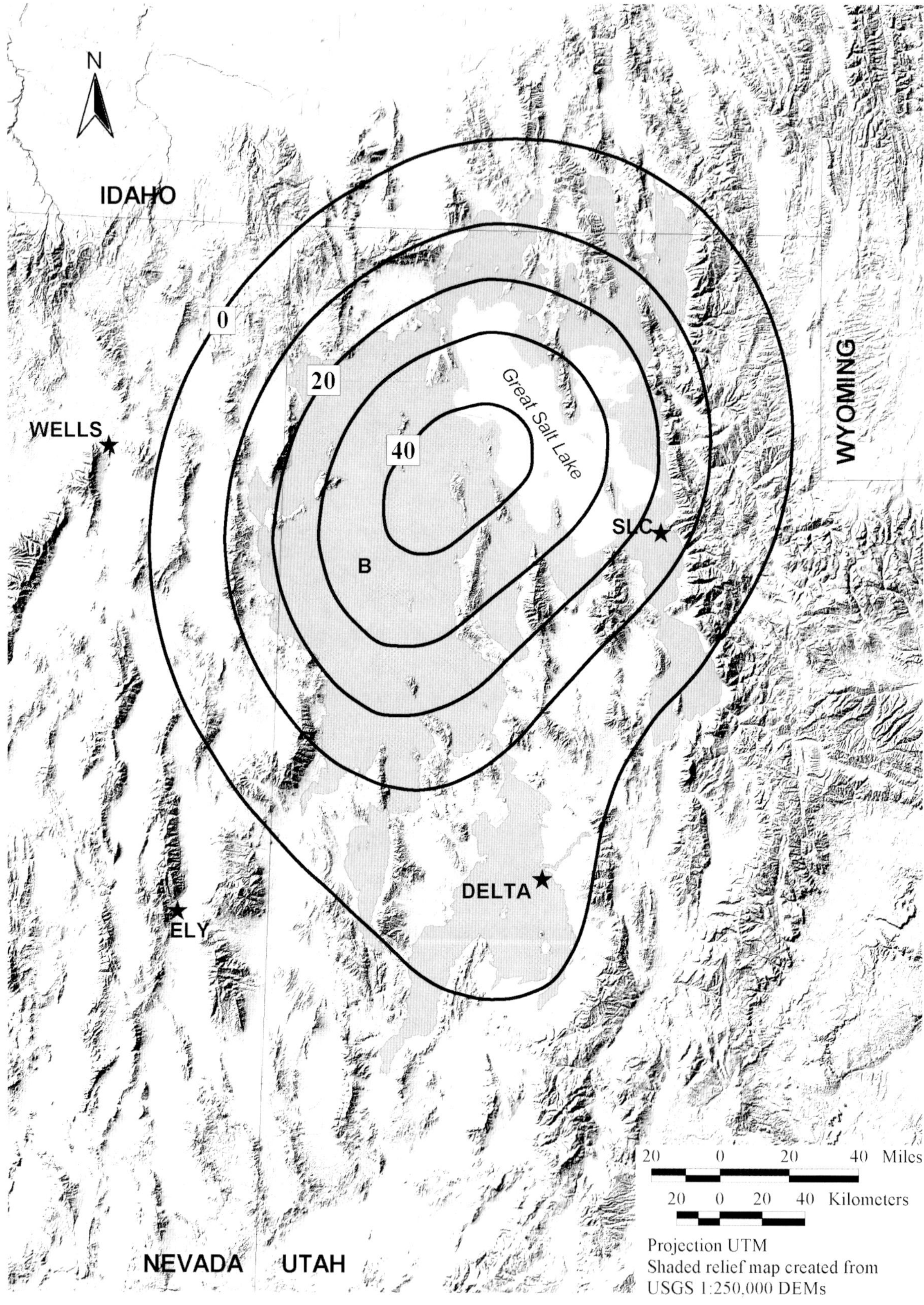

***Figure 8h**. Downward deflection contours (10 meter interval) at -16 kyr (approximately 13,750 radiocarbon years ago) near the lake maximum and before the flood event. Paleolakes are shown in dark gray. As geodynamically configured at that time, the surface elevation of Lake Bonneville (B) was about 1,446 meters above modern sea level (see figure 3).*

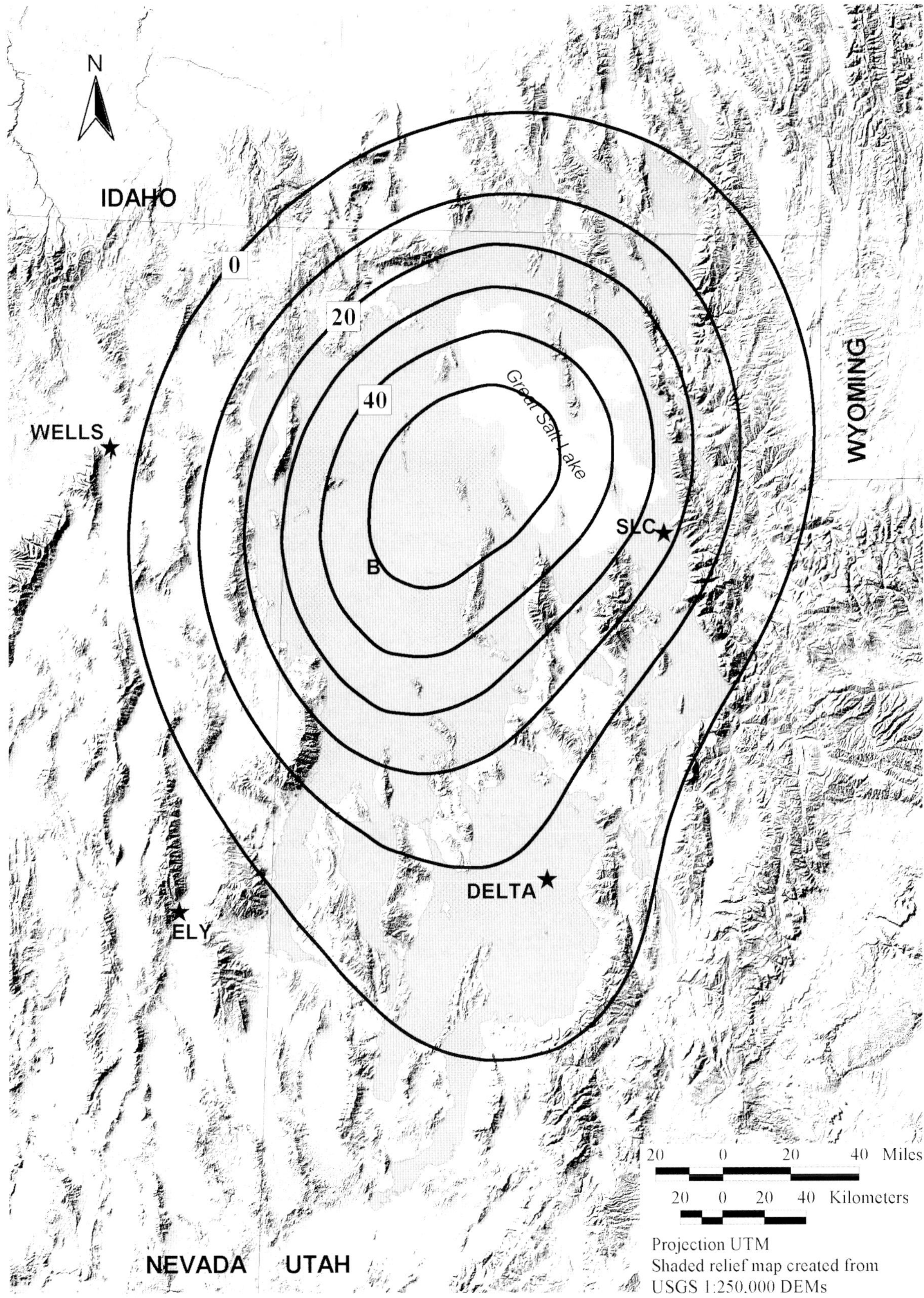

Figure 8i. *Downward deflection contours (10 m interval) at -16.5 kyr (approximately 14,000 radiocarbon years ago). Paleolakes are shown in dark gray. As geodynamically configured at that time, the surface elevation of Lake Bonneville (B) was about 1,552 meters (highstand) above modern sea level (see figure 3).*

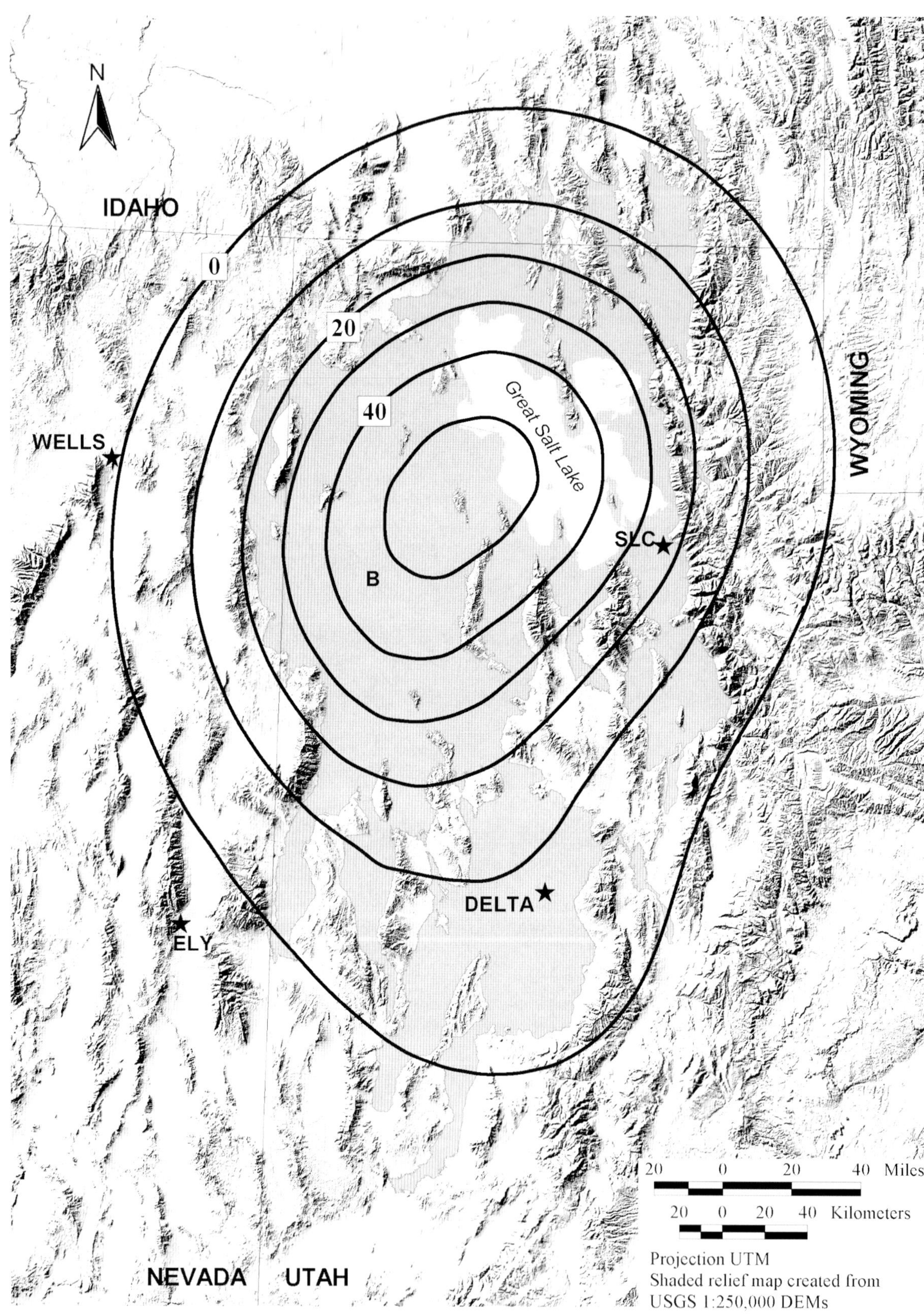

Figure 8j. *Downward deflection contours (10 meter interval) at -18 kyr (approximately 15,000 radiocarbon years ago). Paleolakes are shown in dark gray. As geodynamically configured at that time, the surface elevation of Lake Bonneville (B) was about 1,541 meters above modern sea level (see figure 3).*

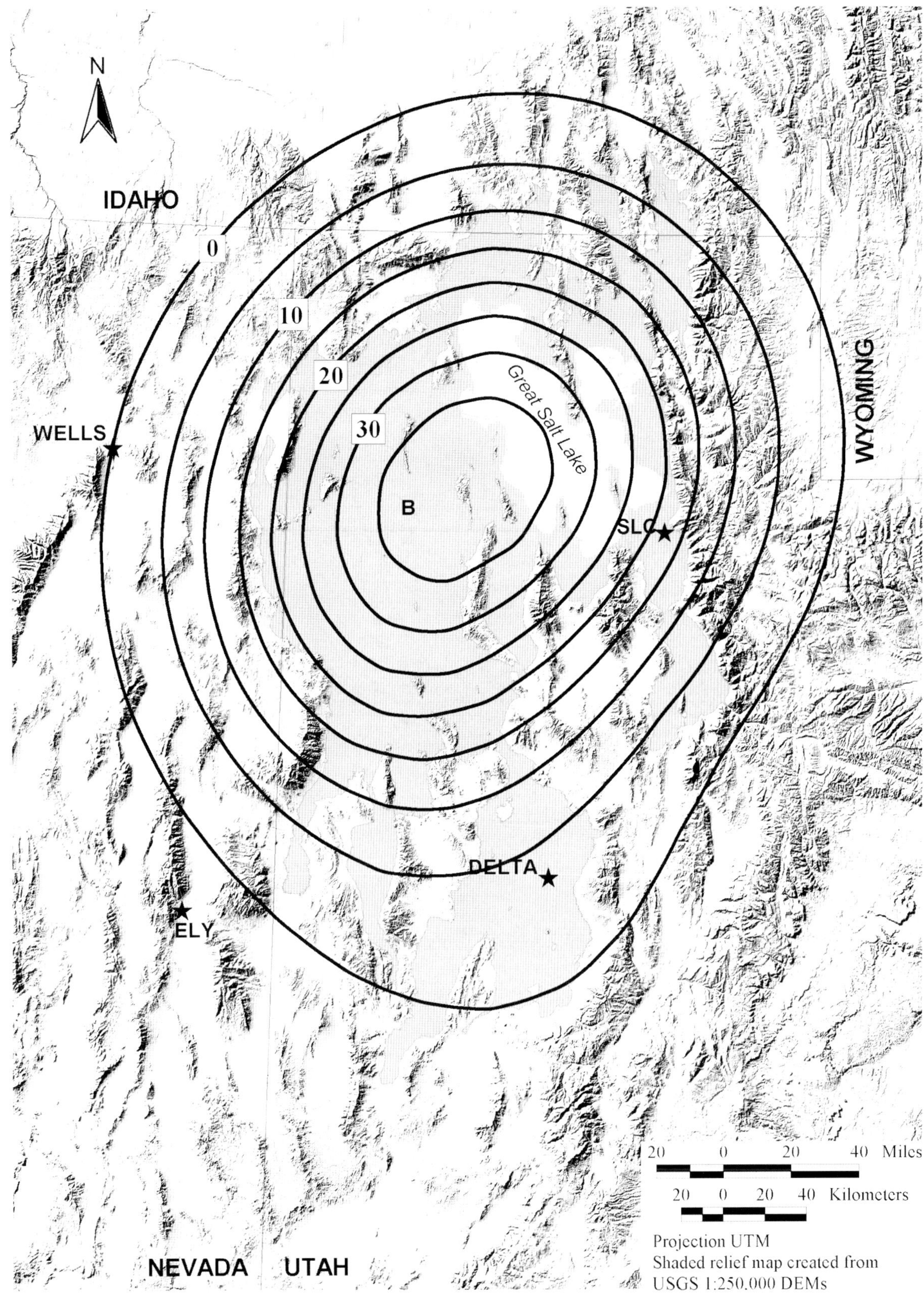

***Figure 8k**. Downward deflection contours (5 meter interval) at -20 kyr (approximately 16,700 radiocarbon years ago). Paleolakes are shown in dark gray. As geodynamically configured at that time, the surface elevation of Lake Bonneville (B) was about 1,484 meters above modern sea level (see figure 3).*

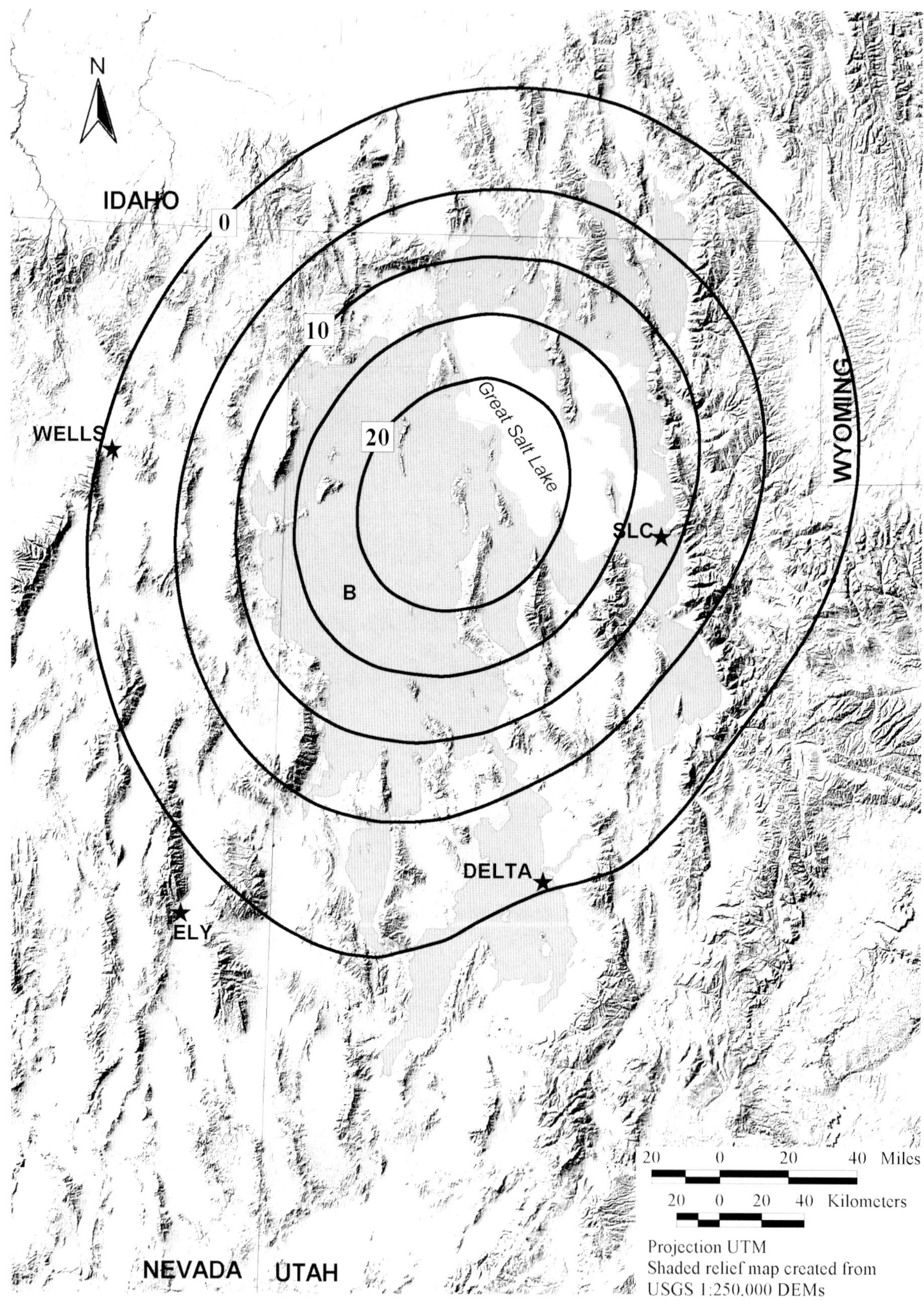

Figure 8l. *Downward deflection contours (5 meter interval) at -22 kyr (approximately 18,800 radiocarbon years ago). Paleolakes are shown in dark gray. As geodynamically configured at that time, the surface elevation of Lake Bonneville (B) was about 1,546 meters above modern sea level (see figure 3).*

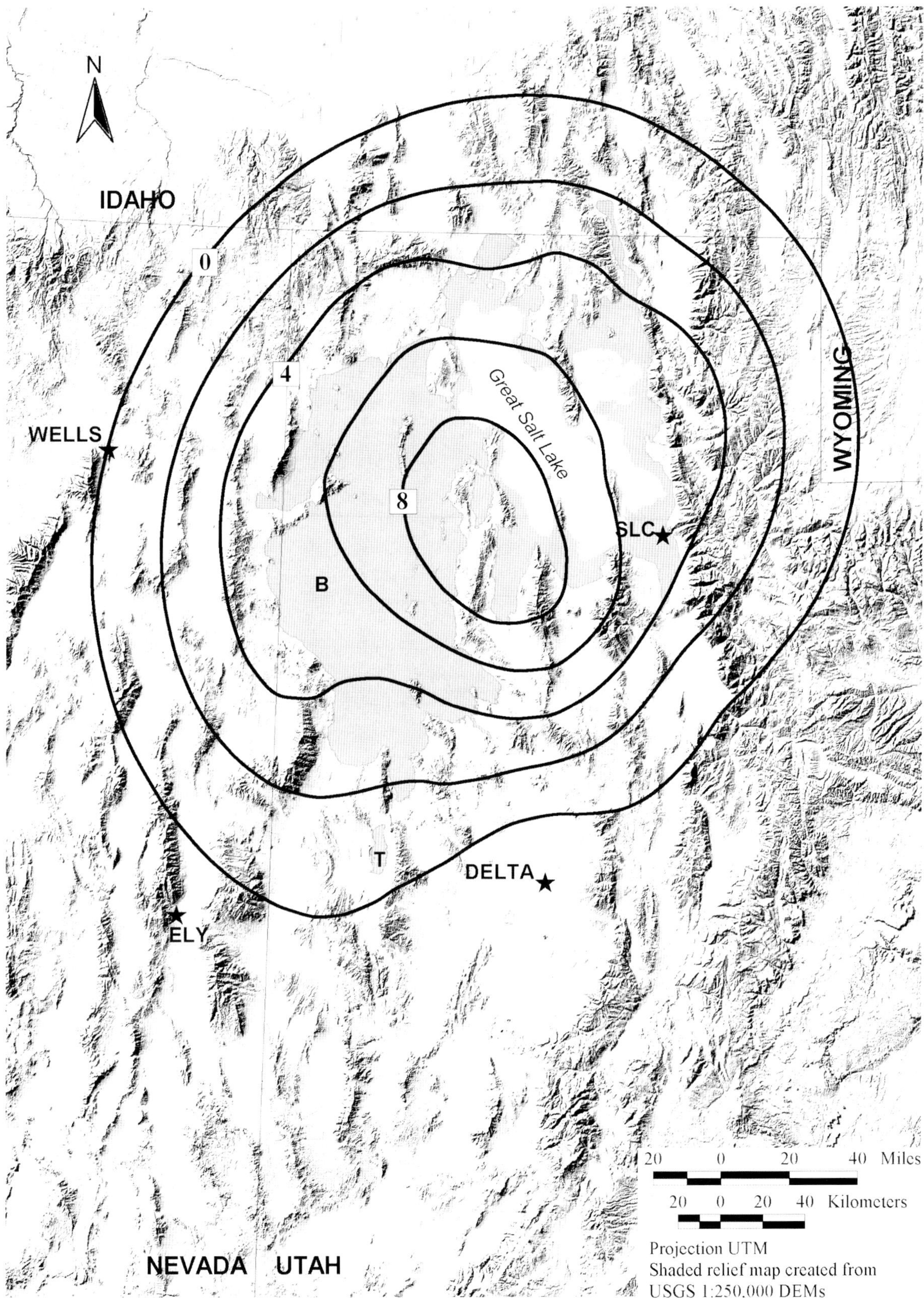

Figure 8m. *Downward deflection contours (2 meter interval) at -24 kyr (approximately 20,950 radiocarbon years ago). Paleolakes are shown in dark gray. As geodynamically configured at that time, the surface elevation of Lake Bonneville (B) was about 1,350 meters above modern sea level (see figure 3). The surface elevation of the lake in Tule Valley (T) was about 1,350 (Sack, 1999).*

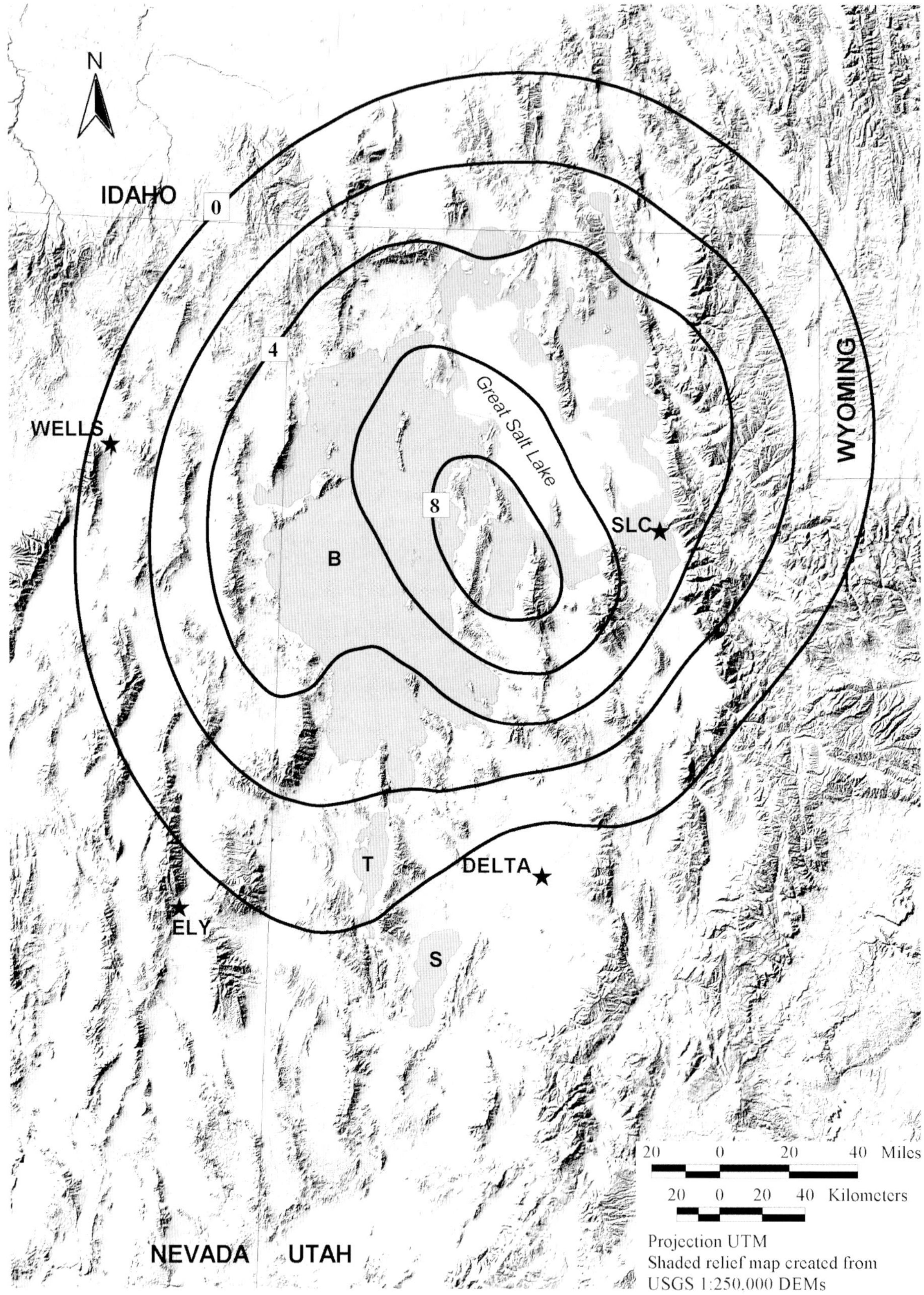

Figure 8n. *Downward deflection contours (2 meter interval) at -26 kyr (approximately 23,100 radiocarbon years ago). Paleolakes are shown in dark gray. As geodynamically configured at that time, the surface elevation of Lake Bonneville (B) was about 1,364 meters above modern sea level (see figure 3). The surface elevation of the lake in the Sevier Desert (S) was about 1,389 meters (Oviatt, 1988). The surface elevation of the lake in Tule Valley (T) was about 1,370 meters (Sack, 1999).*

REFERENCES

Bard, E., Hamelin, B., Fairbanks, R.G., and Zindler, A., 1990, Calibration of the ^{14}C timescale of the past 30,000 years using mass spectrometric U-Th ages from Barbados corals: Nature, 345, 405-410, 1990.

Bills, B.G., and May, G.M., 1987, Lake Bonneville-constraints on lithospheric thickness and upper mantle viscosity from isostatic warping of Bonneville, Provo, and Gilbert stage shorelines: J. Geophys. Res., v. 92, p. 1,493-11,508.

Bills, B.G., Currey, D.R., and Marshall, G.A., 1994, Viscosity estimates for the crust and upper mantle from patterns of lacustrine shoreline deformation in the Eastern Great Basin: J. Geophys. Res., v. 99, p. 22,059-22,086.

Condie, K.C., and Barsky, C.K., 1972, Origin of Quaternary basalts from the Black Rock Desert region, Utah: Geol. Soc. Am. Bull., v. 83 , p. 333-352.

Crittenden, M.D., 1963a, New data on the isostatic deformation of Lake Bonneville: U.S. Geol. Surv. Prof. Pap., 454, p. E1-E31.

Crittenden, M.D., 1963b, Effective viscosity of the Earth derived from isostatic loading of Pleistocene Lake Bonneville: J. Geophys. Res., v. 68, p. 5517-5530.

Crittenden, M.D., 1970, Viscosity and finite strength of the mantle as determined from water and ice loads: Geophys. J. R. Astron. Soc., v. 14, p. 261-279.

Currey, D.R., 1982, Lake Bonneville - selected features of relevance to neotectonic analysis: U.S. Geol. Surv. Open File Rep. 82-1070, 31 p.

Currey, D.R., 1990, Quaternary paleolakes in the evolution of semidesert basins, with special emphasis on Lake Bonneville and the Great Basin: Palaeogeogr. Palaeoclimatol. Palaeoecol., v. 76, P. 189-214.

Currey, D.R., and Oviatt, C.G., 1985, Durations, average rates, and probable causes of Lake Bonneville expansions, stillstands, and contractions during the last deep-lake cycle, 32,000 to 10,000 years ago: Geogr. J. Korea, v. 10, p. 1085-1099.

Currey, D.R., Oviatt, C.G., and Czarnomski, J.E., 1984, Late Quaternary geology of Lake Bonneville and Lake Waring: Utah Geol. Assoc. Publ., v. 13, p. 227-237.

Curey, D.R., Oviatt, C.G., and Plyer, G.B., 1983, Lake Bonneville stratigraphy, geomorphology, and isostatic deformation in west-central Utah: Utah Geol. Miner. Surv. Spec. Stud., no. 62, p. 63-82.

Eardley, A.J., Shuey, R.T., Gvodetsky, V., Nash, W.P., Picard, M.D., Grey, D.C., and G.J. Kukla, 1973, Lake cycles in the Bonneville basin, Utah: Geol. Soc. Am. Bull., 84, p. 211-216.

Gilbert, G.K., 1890, Lake Bonneville: U.G. Geological Survey Monograph 1, 438 p.

Hastenrath, S., and Kutzbach, J., 1983, Paleoclimatic estimates from water and energy budgets of East African lakes; Quater. Res., v. 19, p. 141-153.

—1985, Late Pleistocene climate and water budget of the South American altiplano: Quater. Res., v. 24, p. 249-256.

Humphreys, E.D., and Dueker, K.G., 1994a, Western U.S. upper mantle structure: J. Geophys. Res., v. 99, p. 9615-9634.

—1994b, Physical state of the western U.S. upper mantle, J. Geophys. Res., v. 99, p. 9635-9650.

Iwasaki, T., and Matsu'ura, M., 1982, Quasi-static crustal deformations due to a surface load: J. Phys. Earth, v. 30, p. 469-508.

Lachenbruch, A.H., 1978, Heat flow in the Basin and Range province and thermal effects of tectonic extension: Pure Appl. Geophys., v. 117, p. 34-50.

Malde, H.E., 1968, The catastrophic late Pleistocene Bonneville flood in the Snake River Plain, Idaho: U.S. Geological Survey Professional Paper 596, 52 p.

Mason, I.M., Guzkowska, A.J., Rapley, C.G., and Street-Perrott, F.A., 1994, The response of lake levels and areas to climatic change: Clim. Change, v. 27, p. 161-197.

May, G.M., Bills, B.G., and Hodge, D.S., 1991, Far-field flexural response of Lake Bonneville from paleopluvial lake elevations: Phys. Earth Planet. Int., v. 68, p. 274-284.

McKee, E.H., and Noble, D.C., 1986, Tectonic and magmatic development of the Great Basin of western United States during late Cenozoic time: Mod. Geol., v. 10, p. 39-49.

Mifflin, M.D, and Wheat, M.M., Pluvial lakes and estimated pluvial climates of Nevada: Nev. Bur. Mines Geol. Bull. 94, 57 p.

Minster, J.B., and Jordan, T.H., 1987, Vector constraints on western U.S. deformation from space geodesy, neotectonics and plate motions: J. Geophys. Res., v. 92, p. 4798-4804.

Murchinson, S.B., 1989, Fluctuation history of Great Salt Lake, Utah, during the last 13,000 years, unpublished Univ. of Utah, Ph.D. dissertation.

Nakiboglu, S.M., and Lambeck, K., 1982, A study of the Earth's response to surface loading with applications to Lake Bonneville: Geophys. J. R. Astron. Soc., v. 70, p. 577-620.

Nakiboglu, S.M., and Lambeck, K., 1983, A reevaluation of the isostatic rebound of Lake Bonneville: J. Geophys. Res., v. 88, p. 10,439-10,447.

O'Connor, J.E., 1993, Hydrology, hydraulics, and geomorphology of the Bonneville flood: Geol. Soc. Amer. Spec. Paper 247, 83 p.

Oviatt, C.G., 1988, Late Pleistocene and Holocene lake fluctuations in the Sevier Lake Basin, Utah: USA, J. Paleolimnology, v. 1, p. 9-21.

Oviatt, C.G., 1991, Quaternary geology of the Black Rock Desert, Millard County, Utah: Utah Geol. Miner. Surv. Spec. Stud., 73, 27 p.

Oviatt, C.G., 1997, Lake Bonneville fluctuations and global climate change: Geology, v. 25, p. 155-158, 1997.

Oviatt, C.G., and Nash, W.P., Late Pleistocene basaltic ash and volcanic eruptions in the Bonneville basin: Utah, Geol. Soc. Am. Bull., v. 101, p. 292-303.

Oviatt, C.G., Currey, D.R., and Miller, D.M., 1990, Age and paleoclimatic significance of the Stansbury shoreline of Lake Bonneville, northeastern Great Basin: Quat. Res., v. 33, p. 291-305.

Oviatt, C.G., Currey, D.R., and Sack, D., 1992, Radiocarbon chronology of the Lake Bonneville, eastern Great Basin, USA: Paleogeog. Paleoclim. Paleoecol., v. 99, p. 225-241.

Oviatt, C.G. and McCoy, W.D., 1988, The Old River bed: Utah Geological and Mineral Survey Miscellaneous Publication no. 88-1, p. 60-65.

Oviatt, C.G., McCoy, W.D., and Nash, W.P., 1994, Sequence stratigraphy of lacustrine deposits - a Quaternary example from the Bonneville basin, Utah: Bull. Geol. Soc. Amer., v. 106, p. 133-144.

Passey, Q.R., 1981, Upper mantle viscosity derived from the difference in rebound of the Provo and Bonneville shorelines: Lake Bonneville basin: J. Geophys. Res., v. 86, p. 11,701-11,708.

Peltier, W.R., 1974, The impulse response of a Maxwell Earth: Rev. Geophys., v. 12, p. 649-669.

—1986, Deglaciation induced vertical motion of the North American continent and transient lower mantle rheology: J. Geophys. Res., v. 91, p. 9099-9123.

Priestley, K., Orcutt, J.A., and Brune, J.N., 1980, Higher-mode surface waves and structure of the Great Basin of Nevada and Western Utah: J. Geophys. Res., v. 85, p. 7166-7174.

Sack, D., 1999, Fluvial Linkages in Lake Bonneville Subbasin Integration, unpublished data.

Scott, W.E., McCoy, W.D., Shroba, R.R., and Meyer, R., 1983, Reinterpretation of the exposed record of the last two cycles of Lake Bonneville, western United States: Quat. Res., v. 20, p. 261-285.

Spencer, R.J., Baedecker, M.J., Eugster, H.P., Forester, R.M., Goldhader,M.B., Jones, B.F., Kelts, K., McKenzie, J., Madsen, D.B., Rettig, S.L., Rubin, M., and Bowser, C.J., 1984, Great Salt Lake and predecessors, Utah - the last 30,000 years: Contrib. Minerer. Petrol., v. 86, p. 321-334.

Walcott, R.I., 1970, Flexural rigidity, thickness, and viscosity of the lithosphere: J. Geophys. Res., v. 75, p. 3941-3954.

Williams, S.K., 1994, Late Cenozoic tephrostratigraphy of deep sediment cores from the Bonneville Basin, northwest Utah: Geol. Soc. Amer. Bull., v. 105, p. 1517-1530.

Wolf, D., 1993, The changing role of the lithosphere in models of glacial isostasy - a historical review: Global Planet Change, v. 8, p. 95-106.

Zoback, M.L., 1989, State of stress and modern deformation of the northern Basin and Range province: J. Geophys. Res., v. 94, p. 7105-7128.

PETROLOGY AND GEOCHEMISTRY OF RECENT OOIDS FROM THE GREAT SALT LAKE, UTAH

by

Vicki A. Pedone and Carl H. Norgauer
Department of Geological Sciences, California State University Northridge,
Northridge, CA 91330-8266
vicki.pedone@csun.edu

ABSTRACT

Ooids from shallow subaqueous environments at four localities around the margin of the south arm of the Great Salt Lake show significant differences in cortex microfabric that are related to the overall energy of the environment and not to differences in geochemistry. Low-energy environments are dominated by dark, micrite-rich, finely crystalline, radial-concentric and concentric ooids. High-energy environments are dominated by clear, micrite-poor, coarsely crystalline, radial ooids. Geochemistry of the ooids is similar at all four localities. Abundances of Ca, Na, Sr, K, Mn, and Fe show no significant variations between sample localities. In all localities, Mg abundance is greater in micrite than in coarsely crystalline aragonite and is probably due to the presence of minor amounts of microcrystalline Mg-calcite and/or dolomite. Carbon and oxygen isotope composition of ooids from all localities tightly cluster around an average $\delta^{13}C = +3.8$ o/oo and $\delta^{18}O = -5.0$ o/oo.

INTRODUCTION

An ooid is a carbonate grain that consists of a nucleus of any type of carbonate or siliciclastic grain that is surrounded by concentric growth layers of a carbonate mineral. These growth layers form the cortex of the grain. Ooids on the shoreline of the Great Salt Lake were recognized by the U.S. Geological Exploration of the 40th Parallel in 1877 (Eardley 1938). However, petrologic studies of the ooids were few (Matthews 1930; Eardley 1938; Carozzi, 1962) until the 1970s, when geologists debated the primary versus diagenetic origin of radial cortex fabric in ooids (for example, Shearman and others, 1970; Friedman and others, 1973). This debate was spurred by the observation that the majority of modern marine ooids are aragonite, with a tangential, concentric microfabric, whereas the majority of well-preserved ancient ooids are calcite, with radial and radial-concentric microfabrics. The interpretation of most geologists was that the microfabric of the ancient ooids was originally tangential aragonite, but it was altered from tangential to radial by replacement of aragonite by calcite.

As ooids from different depositional settings were studied, this interpretation was questioned. The ooids of the Great Salt Lake are an exception to the mineralogy-microfabric paradigm because they are aragonite ooids with radial and radial-concentric microfabric. Early studies of the ooids attributed the radial structure to recrystallization of aragonite to calcite (Eardley, 1938), or of aragonite to aragonite (Carozzi, 1962). Using staining, X-ray diffraction, selective dissolution, and scanning-electron-microscopy techniques, Kahle (1974), Sandberg (1975), and Halley (1977) demonstrated that the mineralogy of the Great Salt Lake ooids was aragonite, and that the radial fabric was primary and not the result of diagenesis.

Halley (1977) proposed that the radial fabric was favored by hypersalinity and low-levels of agitation in the lake. The degree of supersaturation in the lake could be raised to a level to overcome inhibitory effects of ion complexing and organic matter without the aid of intense physical agitation, as required in the marine environment to form ooids. Radial growth is favored kinetically, and such growth is maintained because the low to moderate levels of agitation do not flatten crystals into tangential orientations. As a result of studies of the Great Salt Lake ooids, radial microfabric in aragonite ooids became accepted as an indicator of formation in depositional environments where physical agitation of grains is low to moderate.

Once the controversy over the origin of radial fabric in ooids was resolved, interest in additional study of the Great Salt Lake ooids waned. However, the earlier studies did not fully document or explain the origin of the range of cortex fabrics exhibited by the ooids. A question left unanswered was whether the range of cortex fabrics is controlled by differences in agitation intensity or frequency, or by differences in trace-element geochemistry and/or organic content.

This study compares the petrology and geochemistry of ooids from four localities in the Great Salt Lake, which were chosen to contrast varying conditions of wave and wind energy. Variation in grain size, mineralogy, and cortex fabric were documented for ooids at each locality. Geochemical analyses of the medium-grained-sand size fraction, which contains the majority of ooids in each sample, include: $\delta^{13}C$; $\delta^{18}O$; elemental abundances of Na, K, Ca, Mg, Fe, Mn, and Sr; and content of total C (C_{tot}), and organic C (C_{org}). Better definition of the relationship between cortex fabric and physical and chemical conditions in the environment of formation will make ooids more precise indicators of hydrodynamics and/or hydrochemistry.

OOIDS OF THE GREAT SALT LAKE

Ooid beaches rim most of the islands and mainland coast west of a line connecting the Promontory Mountains and Antelope Island. Wave action is more energetic in this part of the lake than in the lake east of this line (figure 1) (Eardley, 1938). The wind in many areas has formed dunes of ooid sand up to several meters high. Lake-bottom samples of ooids have been recovered at elevations as low as 1,276 m, about 1.5 m below the historic lake-level low. Near Badger Island, ooid deposits are about 5 m thick; near the south shore recreation areas, they are greater than 2 m thick (Gwynn and Murphy, 1980).

The age of the ooids around the modern shoreline of the Great Salt Lake is unknown. It is likely that ooids are still actively forming. Evidence supporting recent precipitation of aragonite includes now-exposed crusts of aragonite formed within a four-year period when the shoreline above 1,280 m was submerged during the 1980s highstand of the lake. A grab sample of ooids is likely to contain a mixture of ooids that range in age from 10 years, to as much as a 1,000 years old.

SAMPLE LOCALITIES

Introduction

Four sample localities in the south arm (figure 1) were selected on the basis of differences in shoreface gradient and wind energy, important factors in determining wave energy (table 1). Sites were visited in January and June from 1994 to 1996 to examine seasonal fair-weather conditions. Samples of subaqueous ooids were collected on January 10 and 11, 1995, when the shoreline elevation was 1,280.2 m, near the seasonal low level. Hence, the sample sites are submerged year round and should contain the most actively forming ooids at each site. Samples of subaerial backshore facies were collected at two sites to examine the nature and extent of diagenetic alteration in subaerially exposed ooids.

Northwest End of Stansbury Island and Saltair

The sample site on Stansbury Island is a northeast-southwest trending cove, with a broad, low-gradient sandflat that extends lakeward from the shoreline for more than 1 km. Patches of oolite beachrock were exposed in areas of the subaerial beach face, where loose sediment had been swept off by wind. Steep cliffs of Precambrian quartzite rim the landward side of the cove. Between the shoreline and the cliffs, ooids form a series of low ridges and swales with about 0.5 m relief. The unvegetated ridge closest to the lake is probably an active berm, with a relief of about 1.5 m above the shoreline at time of sampling. The ridges behind the berm are partly vegetated by juniper shrubs. A sample of subaqueous ooids was collected about 30 m from the shoreline at a water depth of 10 cm. Samples also were taken from the subaerial shoreface, about 0.75 m above lake level; from the top of the berm; and from the swale behind the berm crest.

The sample locality at the Saltair public beach consists of a very gently sloping sandflat that extends more than1 km lakeward. Subaqueous ooids were collected about 50 m from the shoreline at a water depth of 5 cm.

Antelope Island Rookery and Bridger Bay

The sample locality at Antelope Island rookery is located on a rocky headland that forms the northern tip of Bridger Bay. The boulder-strewn, 20-m-wide beach terminates landward against quartzite cliffs. The subaqueous shoreface contains numerous quartzite boulders and rounded, gravel-size fragments of oolite, presumably derived from fragmentation of the oolite beachrock that underlies much of the shallow shoreface of Bridger Bay. Samples were collected about 1 m from the shoreline in 15 cm of water and from the subaerial beachface, about 30 cm above the shoreline.

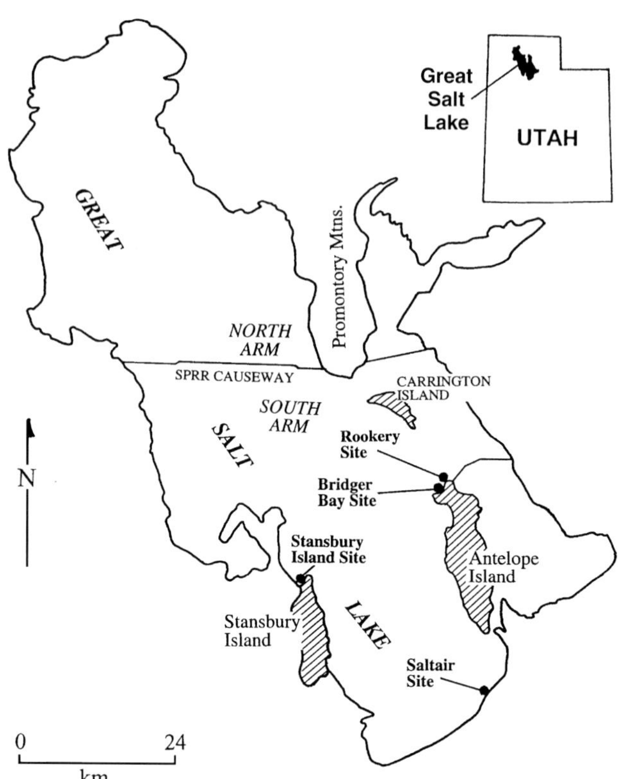

Figure 1. Location of the four sample localities (.) in the south arm of the Great Salt Lake. Shoreline elevation is 1,280.5 m (4,200 ft).

The north-south-trending cove of Bridger Bay on Antelope Island is bounded by rocky headlands and contains a broad, sandy beach. The subaqueous shoreface is covered by slightly sinuous sand waves that extend more than 100 m perpendicular to the shoreline. Crests and troughs are flat and about 2 m wide. Maximum relief of the sand waves is about 40 cm, tapering to 5 cm at the shoreline. At the time of sample collection, when water depth throughout much of the cove was only 20 cm, crests were emergent. Lag deposits of flat cobbles of oolite covered the floors of the troughs. These clasts presumably were derived from fragmentation of beachrock by a major storm. Samples of ooid sand were collected from the crest and trough of one of the sand waves. The subaerially exposed beach rises gently (about 1 m) landward over a distance of about 100 m, where it abruptly rises in an ooid-sand covered ridge that stands about 5 m above the shoreline. Samples were collected from the subaerial beachface, about 0.5 m above the waterline, and from the middle of the ridge.

Fair-Weather Wave Energy at Each Site

Storm energy is similar at all sites, so it is the differences

Table 1. Comparison of physical characteristics of sample localities. Beach gradients (inverses of slope) were determined from topographic maps and wind energy from contour maps of mean kinetic wind energy in the Great Salt Lake Air Basin Wind Study (1980).

Locality	Shoreface gradient	Mean kinetic wind energy (miles²/hr²)				Total wind energy (miles²/hr²)
		summer downslope	summer upslope	winter downslope	winter upslope	
Stansbury Island	480	20	10	< 4	< 4	≅ 38
Saltair	600	20	20	< 4	6	≅ 50
Rookery	80	50	12	4	9	75
Bridger Bay	370	50	12	4	9	75

in intensity of fair-weather wave agitation of ooids that distinguish the study localities. The fair-weather wave energy at each site is dependent on wind conditions and slope of the shoreface. Daily wind data are not available for these specific sites. Instead, the mean kinetic wind energy for summer and winter downslope and upslope winds at each site was obtained from contour maps in the Great Salt Lake Air Basin Study (1980) (table 1). The gradient of the shoreface (the inverse of slope) was determined from 7.5-minute topographic maps. The vertical change included the number of topographic contours less than 4,200 ft at each site.

Based on wind data from table 1, ooids at Bridger Bay and the rookery on Antelope Island should experience more intense wave agitation than those at Saltair and Stansbury Island. Both localities on Antelope Island experience the same wind energy; but because the rookery has a steeper shoreface gradient than Bridger Bay, ooids at the rookery should experience more intense wave energy than those in the adjacent cove. The ooids at Saltair experience slightly higher wind energy than those at Stansbury Island, but Stansbury Island has a slightly greater shoreface gradient than Saltair. However, the shoreface gradients at both sites are very gentle. Wind energy is probably more important than gradient in determining wave agitation at a site. Therefore, ooids at Saltair likely experience greater wave agitation than those on Stansbury Island. Hence, the ranking of relative fair-weather wave energy of the four sites from highest to lowest is: 1) rookery, 2) Bridger Bay, 3) Saltair, and 4) Stansbury Island.

METHODS

Following McAllister's (1958) procedure for the removal of salt, the ooid samples were sieved and divided into standard 1- ϕ (phi) intervals, where grain diameter in phi units equals minus log to the base two of grain diameter in millimeters. In all samples, ooids were the dominant grain in the 2 and 3-ϕ fractions (medium- and fine-grained sand). A Swift Automatic Point Counter was used to count 300 grains in nine thin sections of the medium- to very fine-grained sand fractions. Horizontal and vertical intervals of 3 mm were used so that no grain was counted twice. For each ooid, the nucleus type and cortex fabric were tabulated (Norgauer, 1996).

The 2-ϕ fraction (medium-grained sand) was used for all geochemical analyses. Samples were cleaned repeatedly in deionized water in an ultrasonic bath. Mineralogy was determined by X-ray diffraction (XRD) on compacted powdered samples. In ten samples, Ca, Mg, Na, and K were measured using inductively-coupled-plasma, atomic-emission-spectrometry (ICP-AES), and Fe, Mn, and Sr were measured by inductively-coupled-plasma, mass-spectrometry (ICP-MS) at the University of California Riverside. Ooid samples, along with two internal standards, were pretreated with a buffered cation exchange solution (0.2 M $NH_4C_2H_3O_2$) to minimize leaching of cations from non-soluble minerals (for example, clay minerals), and were prepared for analysis under rigorous clean-lab conditions. The same elements were measured in filtered (0.45 μm) water samples collected at each sample site. External error (1 σ), based on standards, is +/-2 to 3 percent for Ca, Mn, Na, Sr, K, and +/-3 to 5 percent for Mg and Fe.

A Cameca® SX50 electron microprobe at Cambridge University was used to analyze Mg, Sr, and Na in selected areas of peloid nuclei and ooid cortices in samples from Saltair and the rookery. Analyses were made in Ag-coated thin sections, using a 20 kV accelerating voltage, 15 nanoamp (nA) beam current, and 5-μm-diameter spot size. Detection limit is 100 parts per million (ppm) for Mg and 200 ppm for Sr and Na. Data reduction was done by computer using Bence-Albee® Atomic Number/Absorption/Fluorescence (ZAF) matrix-correction techniques, assuming an oxide-type analysis mode. Total C (C_{tot}), and organic C (C_{org}) were measured in ten samples by combustion in a LECO induction furnace at the University of Indiana. Replicate analyses were done on all samples.

$\delta^{13}C$ and $\delta^{18}O$ of 2-ϕ fractions were measured by a Finnegan® MAT 250 Isotope Ratio Mass Spectrometer at the stable isotope laboratory at University of California Los Angeles (UCLA), using off-line digestion and gas-extraction methods. Three randomly chosen aliquots from ten samples were taken to evaluate intrasample variation. Prior to analysis, samples were roasted for 24 hr at 150°C. Replicate analyses were run on six randomly chosen samples. An internal standard was analyzed with each batch of samples. Analyses have a 1-σ uncertainty of ±0.1 ‰ for C and O.

PETROLOGY

Grain Types

Of the 12 samples studied, seven consist only of sand-size grains (particles between 0.0625 mm and 2.0 mm). The other five samples, one from Stansbury Island, two from the rookery, and two from Bridger Bay, consist of sand and gravel-size particles of oolite (beachrock fragments), ranging from 2 up to 30 weight percent of the sample. Nine of the 12 samples studied have a size mode of +2 ϕ (0.25 mm). The gravel-bearing samples from the rookery have modes of -1 ϕ

(1.0 mm) and -2 ϕ (2.0 mm). The one sample from Saltair has a mode of +3 ϕ (0.125 mm).

The composition of the medium-grained sand fraction (0.125 to 0.5 mm) was determined in nine samples by point-counting thin sections of this size fraction. The three samples that contain more than 20 percent gravel were not included. All nine samples are dominated by whole ooids (59 to 82 percent, x = 66 percent), with minor amounts of intraclasts of pelsparite and pelmicrite (2 to 28 percent, x = 15.5 percent), random grains (5 to 23 percent, x =13.5 percent), and peloids (2 to 9 percent, x = 2 percent). Minor amounts (< 5 percent) of broken ooids and extraclasts, mostly quartz grains with minor mica and rock fragments of schist and mudstone, are found in a few samples. Random grains are spherical to subspherical and consist of mixtures of micrite and clear aragonite spar. The spar is equant to bladed, and blades are poorly to randomly oriented (figure 2A). Random grains are common as ooid nuclei and are common as individual grains in the 3 to 4-ϕ size fraction (fine- to very fine-grained sand). They are probably fragments of radial cortices observed in an orientation perpendicular to the long axis of the crystals. Aragonite and quartz were the only minerals detected by XRD analyses in nine samples of the 2-ϕ fraction.

Ooid Nuclei

Dominant ooid nuclei are: 1) peloids, presumably fecal pellets of brine shrimp; 2) quartz grains; and 3) carbonate intraclasts of pelsparite and random grains (table 2). Percentages of quartz nuclei are similar at all localities except Stansbury Island, indicating that this cove receives little windblown detritus from the quartzite cliffs that frame it. The low percentages of ooids with peloid nuclei in subaerial facies at Stansbury Island and Bridger Bay suggest that these ooids are less likely to be transported onshore by wind when sand flats are exposed than ooids with other types of nuclei. They are probably slightly larger and/or heavier than those with quartz or intraclast nuclei.

Ooid Cortices

Cortex crystals in Great Salt Lake ooids range in size and shape from equant micrite (<4 μm) to elongate, narrow rays that range up to 160 μm long and 50 μm wide. Sandberg (1975), and Halley (1977) note that <10-μm crystals are mostly randomly oriented, whereas larger crystals are radially oriented. Our study defines three cortex fabrics based on the length of radial aragonite crystals within growth layers: radial, >80 μm; radial-concentric, 40 to 80 μm; and concentric, <40 μm (figure 2). The boundaries between growth layers are defined by 5 to 20-μm-thick, concentric bands of micrite. Optical continuity is maintained between crystals in successive laminae.

No recrystallization to sparry calcite was observed in any ooid, although some subaerial ooids show minor dissolution (figure 2A). Radially oriented crystals are commonly separated laterally by micrite, termed "inter-ray micrite" in order to distinguish it from micrite in concentric growth bands (figure 2). The percentage of inter-ray micrite was estimated visually and tabulated for all radial and radial-

Table 2. Frequency nuclei types in the 2-ϕ (phi) size fraction.

Locality	% peloid	% intraclast	% quartz
Stansbury Island			
subaqueous	28	63	9
subaerial beachface	5	87	8
swale behind berm	<1	80	19
crest of berm	0	82	18
Saltair			
Saltair subaqueous	35	21	44
Antelope Island rookery subaerial beachface	39	13	48
Bridger Bay			
subaqueous	30	28	42
subaerial beachface	4	40	55
berm	<1	45	55

cocentric ooids. Crystals >5 μm wide are designated as coarsely crystalline, and crystals <5 μm, finely crystalline.

Radial cortex fabric consists of at least one layer of > 80-μm-long crystals, in which concentric growth bands are faint or absent. Most of the crystals are > 5-μm-wide and are separated by inter-ray micrite (figures 2A and B). Some radial ooids have an outer layer or rind of < 80-μm-long crystals. Radial-concentric cortex fabric consists of two or three layers of 40 to 80-μm-long, < 5-μm-wide crystals, usually separated by inter-ray micrite (figure 2C). Layers are separated by well defined concentric bands of micrite. Concentric cortex fabric consists of numerous layers of < 40-μm-long, < 5-μm-wide crystals separated by well-defined concentric bands of micrite (figures 2B and C).

Radial fabric is the most common at all localities except Stansbury Island (figure 3). The least common fabric is concentric fabric, except at Stansbury Island where 46 percent of the ooids are concentric. The amount of inter-ray micrite varies considerably in ooids from the different localities. Ooids at Stansbury Island and Saltair are significantly richer in inter-ray micrite than those at the rookery and Bridger Bay (figure 4). There are no differences in the percentages of different cortex fabrics, or in the amounts of inter-ray micrite, between subaerial and subaqueous ooids at Stansbury Island or at Bridger Bay. This indicates that the subaerial ooids formed in subaqueous settings at these sites where conditions for ooid formation have remained constant.

GEOCHEMISTRY

Elemental Abundances

Ca, Sr, Na, K, Mn, and Fe abundances, and C_{tot} and C_{org} contents show no significant variations related to locality, to subaerial versus subaqueous environments (tables 3 and 4), or to specific fabric within ooids (table 5). This is consistent with the similarity in water chemistry measured at each locality (figure 5). Sr is most likely present as a substitution for Ca in the aragonite crystal structure, as supported by its homogeneous distribution in all fabrics and by its nearly identical abundances, as measured by ICP and microprobe

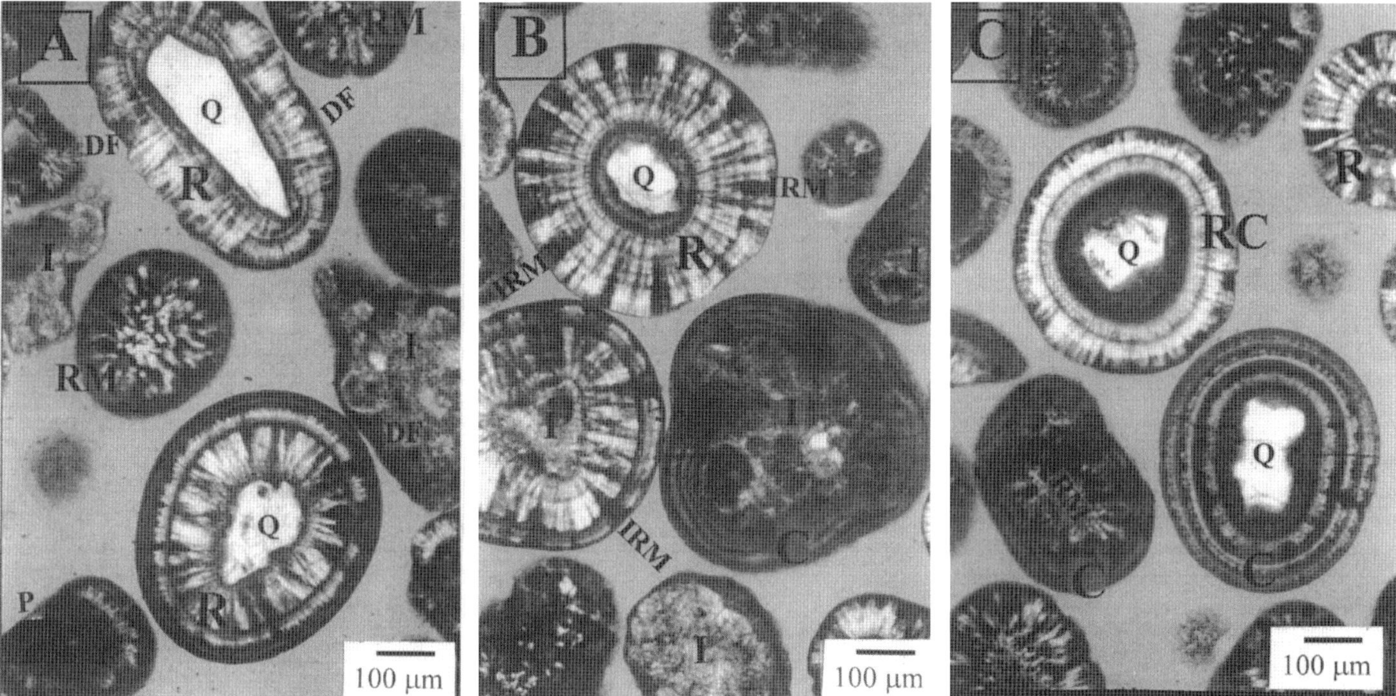

Figure 2. Photomicrographs of Great Salt Lake ooids. R = radial, RC =radial-concentric, C = concentric, RM = random grain, I = intraclast, Q = quartz, P = peloid, IRM = inter-ray micrite, DF = dissolution features. A. Subaerial sample from Stansbury Island, swale behind berm. Note minor dissolution features and the outer layer of small crystals on one of the radial ooids. B. Radial and concentric ooids from Stansbury Island, crest of berm. C. Concentric and radial-concentric ooids from Bridger Bay, subaqueous sample.

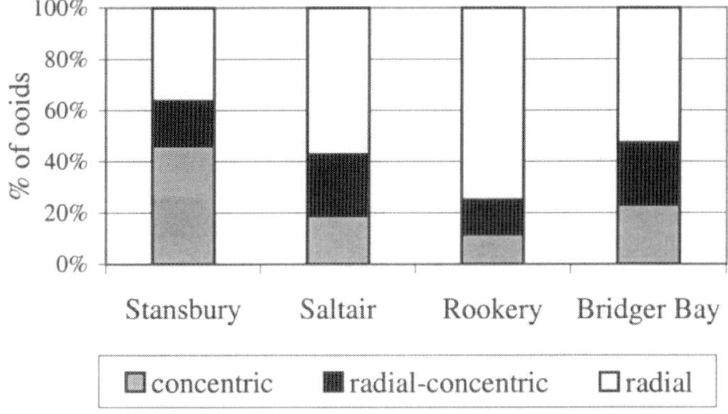

Figure 3. *Cortex fabrics in subaqueous ooids from four localities in the south arm of the Great Salt Lake.*

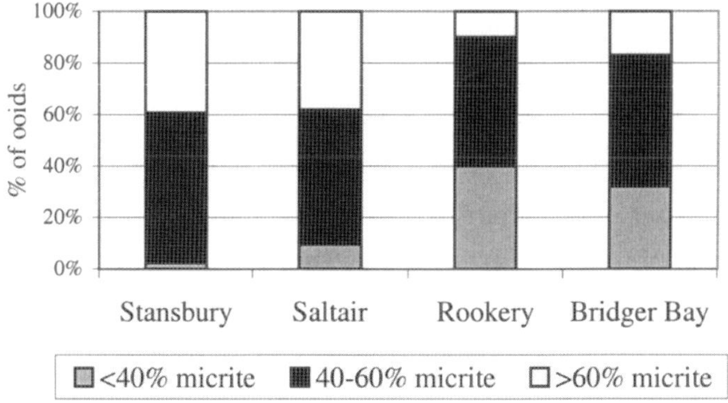

Figure 4. *Percentages of inter-ray micrite in subaqueous ooids from four localities in the south arm of the Great Salt Lake.*

(figure 6). The low abundance of Sr in the Great Salt Lake ooids, about 4,000 ppm compared to 9,800 ppm in modern marine aragonite ooids (Bathurst, 1975), reflects the low abundance of Sr in the Great Salt Lake (3-5 ppm; Pedone, unpublished data) compared to seawater (8 ppm; Brown, 1989).

Because Na does not substitute for any element in the crystal lattice of aragonite, the homogeneous distribution of Na in micrite and coarsely crystalline material suggests that Na is present in fluid inclusions. This mode of occurrence is supported by the lower abundances of Na measured by ICP compared to those measured by microprobe because the pre-wash treatment of ICP samples minimizes contributions from fluid inclusions (I. Montanez, University of California, Riverside, personal communication, 1996). The average Na abundance in Great Salt Lake ooids, 4,000 to 5,000 ppm, is higher than that found in modern marine aragonite, about 2,000 ppm (Tucker and Wright, 1990) because Na abundance in the Great Salt Lake varies from 4 to 10 times higher than that in seawater. The average C_{org} content in the Great Salt Lake ooids is 0.1 wt percent (table 4), which places it at the low end of the C_{org} range in natural ooids, 0.1 to 0.5 wt percent (Morse and Mackenzie, 1990). The low organic carbon content most likely reflects the large volume of coarsely crystalline, clear aragonite in the Great Salt Lake ooids. Micritic ooids, such as modern tangential marine ooids, usually contain more organic matter between the numerous crystals.

Differences in Mg abundances measured by ICP and microprobe analyses are larger than those of other elements (figure 6). Microprobe measurements show that the abundance of Mg is significantly higher in micrite fabrics, particularly peloids, than in clear crystalline aragonite (table 5).

Table 3. Inductively Coupled Plasma-Atomic Emission Spectroscopy (ICP-AES) and Inductively Coupled Plasma-Mass Spectrometry (ICP-MS) data. External error for Ca, Mg, Sr, Na and K is 2-3%; for Mn and Fe, 3-5%.

Sample	Ca ppm (AES)	Mg ppm (AES)	Sr ppm (MS)	Na ppm (AES)	K ppm (AES)	Mn ppm (MS)	Fe ppm (MS)
Stansbury Island							
subaqueous	400204	6968	3851	4233	169	24	242
subaerial beachface	402019	5446	3691	3698	113	15	34
swale behind berm	400372	3542	4065	3547	63	15	30
swale behind berm (replicate)	400313	3434	4060	3471	60	16	60
crest of berm	400839	3633	3670	3656	80	9	35
Saltair							
Saltair subaqueous	401041	3735	3737	3799	233	41	94
Saltair subaqueous (replicate)	400564	5712	3299	3236	196	76	173
Antelope Island rookery							
subaerial beachface	408626	6399	3578	3436	86	40	56
subaerial beachface (replicate)	394607	8008	3607	3719	81	51	66
Antelope Island Bridger Bay							
subaqueous crest of sandwave	402357	6598	3710	3819	146	40	65
subaqueous trough of sandwave	400150	5399	3520	3817	154	23	60
subaerial beachface	402027	4187	3689	3728	105	17	33
berm	400511	3499	3813	4031	138	19	32

Table 4. Carbon content (wt percent) of ooids measured by Leco induction furnace. Abbreviations: C_{tot} = wt percent total carbon, C_{org} = wt percent organic carbon, rep = replicate analysis, avg = average of the two analyses.

Sample	C_{tot}	C_{tot} (replicate)	C_{tot} (avg)	C_{org}	C_{org} (replicate)	C_{org} (avg)
Stansbury Island	11.44	11.52	11.48	0.10	0.11	0.10
subaqueous						
subaerial beachface	11.25	11.51	11.38	0.09	0.09	0.09
swale behind berm	11.20	11.27	11.24	0.07	0.08	0.07
crest of berm	11.49	11.42	11.46	0.09	0.11	0.10
Saltair						
Saltair subaqueous	10.18	10.34	10.26	0.11	0.10	0.11
Antelope Island subaqueous	10.77	10.95	10.86	0.10	0.11	0.11
Bridger Bay						
subaqueous (crest of sandwave)	10.39	10.57	10.48	0.10	0.10	0.10
subaqueous (trough of sandwave)	10.87	10.50	10.69	0.09	0.09	0.09
subaerial beachface	10.74	10.82	10.78	0.08	0.08	0.08
berm	10.73	10.80	10.77	0.05	0.07	0.06

Table 5. Average elemental abundances (ppm) in ooids measured by electron microprobe.

	Bridger Bay			Saltair		
	Mg	**Sr**	**Na**	**Mg**	**Sr**	**Na**
finely crystalline radial	1113	3732	5470	486	3911	5850
coarsely crystalline radial	330	3840	5463	476	3728	4935
inter-ray micrite	1680	3720	5471	23730	4191	4969
concentric-band micrite	4312	4209	5143	1024	4406	4902
peloid nuclei	25740	3043	4400	29450	2950	10185

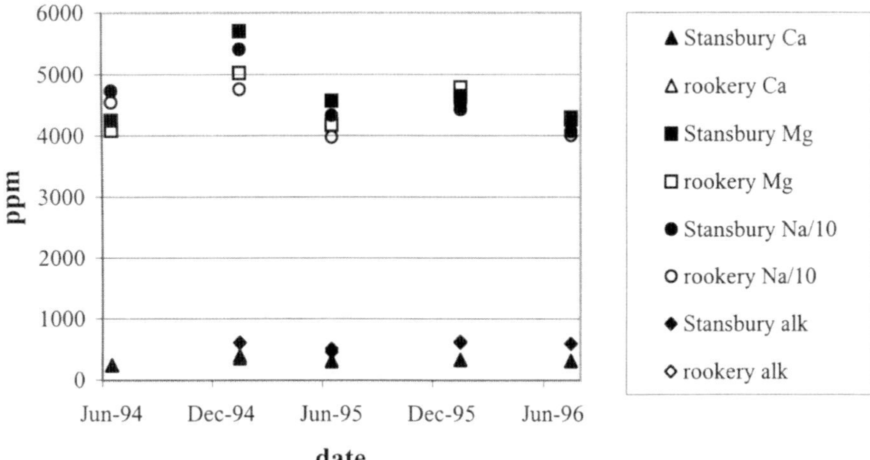

Figure 5. *Alkalinity and abundances of Na, Ca, and Mg in water samples from Stansbury Island and the rookery, 1994 to 1996. Because of the order-of-magnitude greater abundance of Na compared to all other elements, the abundances of Na shown are divided by ten.*

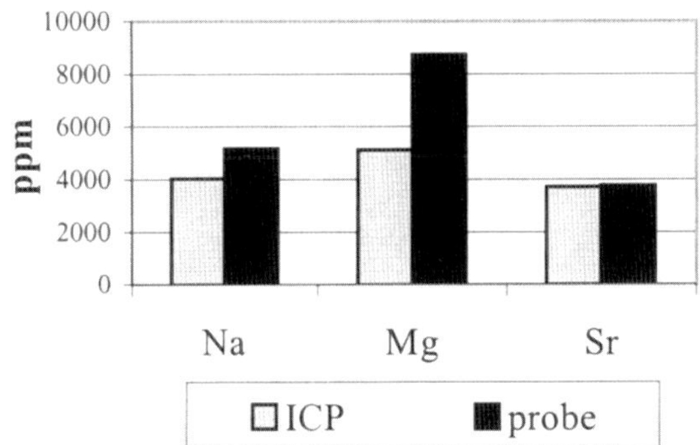

Figure 6. *Comparison of elemental abundances from ICP and electron-microprobe analyses.*

These analyses are similar to the microprobe results of Halley (1977), who found that Mg was undetected in coarsely crystalline aragonite, but varied widely and unsystematically in micrite.

Possible modes of occurrence of Mg in Great Salt Lake ooids include: 1) substitution for Ca in the aragonite crystal lattice, 2) fluid inclusions, 3) Mg calcite, 4) dolomite, and 5) Mg-bearing clay minerals. The 300 to 1,000 ppm Mg abundance in finely and coarsely crystalline aragonite (table 4) most likely results from substitution and fluid inclusions. All aragonite, regardless of crystal size, should have similar baseline values of Mg. Therefore, the high abundances of Mg in some electron microprobe analyses of micrite (2,000-52,500 ppm) must include contributions from other phases, such as Mg calcite, dolomite, and/or Mg-bearing clay minerals. Because Mg abundances measured by ICP analyses are significantly above the 300 to 1,000 ppm level, it is likely that most "excess" Mg is contributed by carbonate minerals because sample preparation for ICP analyses minimizes leaching of cations from clay minerals.

Carbon and Oxygen Isotopes

Three randomly chosen aliquots from ten samples were analyzed to test both intra-sample and inter-sample variation in isotopic composition (same samples as listed in table 3) (figure 7). $\delta^{18}O$ and $\delta^{13}C$ data form a tight cluster around $\delta^{18}O = -5.0$ ‰ (PDB) and $\delta^{13}C = +3.9$ ‰ (PDB) (PDB stands for the Pee Dee Belemnite limestone reference). Inter-sample variation is no greater than the intra-sample variation of < 1 ‰. Similarity in composition suggests that the temperature of formation and the isotopic composition of the water were similar at all localities during aragonite formation in the ooids.

The isotopic composition of the Great Salt Lake shows small seasonal variations and significant decade to century variations (Pedone, this volume). However, the tight cluster of data from the ooids suggests that the average isotopic composition of the Great Salt Lake remained fairly constant during the period of formation represented by the Holocene shoreline ooids. Water samples at each locality were collected in June or July from 1994 to 1996 when the lake elevation was 1,280.6 m to 1,280.7 m, near the historic average of 1,280 m. Samples from Saltair were eliminated from the study because freshwater effluent from nearby drainage canals resulted in anomolously depleted $\delta^{18}O$ and salinity values compared to all other localities. The temperature range of the samples during this three-year period was 12° C to 28°C, with an average temperature of 21°C. $\delta^{18}O$ ranged from -3.2 ‰ Standard Mean Ocean Water (SMOW) to -5.3 ‰ (SMOW), with an average of -4.4 ‰ (SMOW).

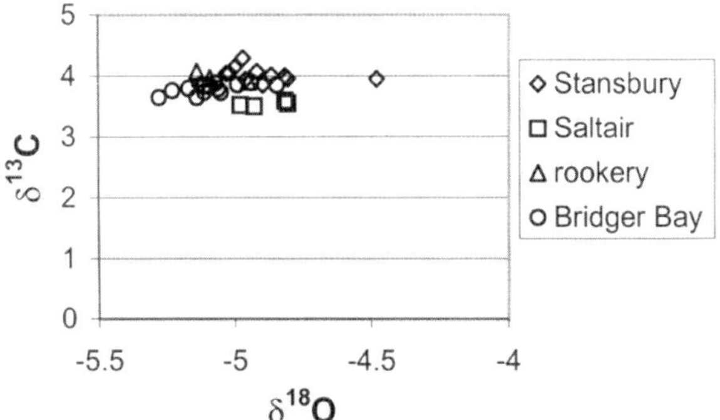

Figure 7. *Carbon and oxygen isotopic composition of ooids from four localities in the Great Salt Lake. The reference standard for both isotopes is PDB.*

The calculated value of aragonite formed in such water, using the equation of Grossman and Ku (1981), is -5.2 ‰ (SMOW), which is similar to the value measured in the ooids. One explanation of the ooid isotopic data is that the average temperature and isotopic composition of the water in which the ooids formed were similar to those experienced in recent years during the late spring/early summer.

DISCUSSION

The cortex fabric of ooids is different for each of the study areas in the south arm of the Great Salt Lake, which suggests that there is little or no mixing of ooids between localities (table 6). Factors that could influence cortex fabric, many of which are inter-related, include: water chemistry, cortex geochemistry, wave energy, abrasion of the growth surface, and mechanism of crystallization (for example, biotic versus abiotic precipitation). The similarities in water chemistry (figure 5) and ooid geochemistry (tables 3, 4, and 5) between localities rule out these factors.

There is a definite relationship between cortex fabric and estimated relative fair-weather wave energy at each site (table 6). The percentage of radial ooids is highest at the rookery, the highest energy site, and lowest at Stansbury Island, the lowest energy site. The opposite is true of concentric ooids, that is, this fabric is most common at Stansbury Island and least common at the rookery. There is little difference in cortex types between the two intermediate-energy sites of Bridger Bay and Saltair. The amount of inter-ray micrite also shows a correlation to the estimated energy at each site. It is lowest at the two highest energy sites and highest at the two lowest energy sites.

Formation of concentric ooids might be the result of a high ratio of biotic to abiotic growth, compared to a low ratio in radial ooids. Folk (1993) reported a high concentration of nannobacteria in dark concentric laminae and a low concentration in clear radial crystals in ooids from the Great Salt Lake, indicating that nannobacteria are important agents in the formation of micrite in the concentric growth bands. Such bands are most common in ooids at Stansbury Island, the lowest energy site, and least common at the rookery, the highest energy site. This association suggests that longer hiatuses between addition of radial cortex laminae and/or less frequent episodes of abrasion in low-energy environments promote formation of microcrystalline aragonite by nannobacterial activity on the ooid surface.

Table 6. Relationship between ooid microfabric and energy of the depositional environment.

Locality	relative energy (1=highest)	% radial fabric	% concentric fabric	% of inter-ray micrite >60%
Rookery	1	74	10	10
Bridger Bay	2	52	22	18
Saltair	3	58	19	38
Stansbury Island	4	37	45	40

CONCLUSIONS

The microfabric of ooid cortices can be used to distinguish differences in relative fair-weather wave energy at different localities in the Great Salt Lake. Ooids formed in high-energy sites have large, clear radial crystals, few concentric growth bands, and little inter-ray micrite. Ooids formed in low-energy sites have small radial crystals, numerous concentric growth bands, and abundant inter-ray micrite. The differences in cortex texture probably reflect differences in the degree of abiotic and biotic precipitation of aragonite. Infrequent agitation in low-energy environments favors retention of surface microbes that induce precipitation of micrite in closely spaced growth bands. Frequent agitation and surface abrasion in high-energy environments favors the inorganic precipitation of large crystals.

ACKNOWLEDGMENTS

We thank John Bleassard for access to the sampling locality on Stansbury Island. Acknowledgment is made to the Petroleum Research fund, administered by the American Chemical Society, and to California State University Northridge for support of this research. Paul Jewell suggested many valuable improvements during review.

REFERENCES

Bathurst, R. G., 1975, Carbonate sediments and their diagenesis (second edition): Elsevier, New York, 620 p.

Brown, J., 1989, Seawater: Its composition, properties and behavior: New York, Pergammon Press, p. 30.

Carozzi, A. V., 1962, Cerebroid oolites: Illinois Academy of Science Transactions, v. 55, p. 239-249.

Eardley, A. L., 1938, Sediments of Great Salt Lake, Utah: American Association of Petroleum Geologists Bulletin, v. 22, p. 1305-1411.

Folk, R. L., 1993, SEM imaging of bacteria and nannobacteria in carbonate sediments and rocks: Journal of Sedimentary Petrology, v. 63, n. 5, p. 990-999.

Friedman, G. M., Amiel, A. J., Braun M., and Miller D. S., 1973. Generation of carbonate particles and laminites in algal mats - example from sea-marginal hypersaline pool, Gulf of Aqaba Sea: American Association of Petroleum Geologists Bulletin, v. 57, p. 541-557.

Great Salt Lake Air Basin Wind Study, 1980: Wasatch Front Regional Council, 420 West 1500 South, Suite 200, Bountiful, Utah 84010.

Grossman, E. T, and Ku, T. L., 1981, Aragonite-water isotopic paleotemperature scale based on the benthic foraminifera Hoeglundia elegans: Geological Society of America Annual Meeting Abstracts with Programs, p. 464.

Gwynn, J. W., and Murphy, P. J., 1980, Recent sediments of the Great Salt Lake basin *in* Gwynn, J. W., editor, Great Salt Lake: a scientific, historical and economic overview: Utah Geological and Mineral Survey, Bulletin 116, p. 84-96.

Halley, R. B.,1977, Ooid fabric and fracture in the Great Salt Lake and the geologic record: Journal of Sedimentary Petrology, v. 47, p. 1099-1120.

Kahle, C. F., 1974, Ooids from Great Salt Lake, Utah, as an analogue for the genesis and diagenesis of ooids in marine limestones: Journal of Sedimentary Petrology, v. 44, p. 30-39.

Matthews, A. A. L., 1930, Origin and growth of the Great Salt Lake oolites: Journal of Geology, v. 38, p. 633-642.

McAllister, F. F., 1958, Rapid removal of marine salts from sediment samples: Journal of Sedimentary Petrology, v. 28, p. 231-232.

Morse, J. W., and Mackenzie, F. T., 1990, Geochemistry of sedimentary carbonates: Developments in Sedimentology 48, Elsevier, Amsterdam, 707 p.

Norgauer, C. H., 1996, Physical and chemical controls on microfabric of Recent ooids from the Great Salt Lake, Utah: California State University Northridge, unpublished MS thesis, 116 p.

Pedone, V. A., 2001, Oxygen-isotope composition of the Great Salt Lake, 1979 to 1996: (this volume).

Sandberg, P. A., 1975, New interpretations of Great Salt Lake ooids and of ancient non-skeletal carbonate mineralogy: Sedimentology, v. 22, p. 497-537.

Shearman, D. J., Twyman, J., and Karmi, M. Z., 1970, The diagenesis of oolites: Proceedings of the Geologists' Association (UK), v. 81, p. 561-575.

Tucker, M. E., and Wright, V. P., 1990, Carbonate sedimentology: Blackwell Scientific Publications, Oxford London, 482 p.

STORM-RELATED FLOODING HAZARDS, COASTAL PROCESSES, AND SHORELINE EVIDENCE OF GREAT SALT LAKE

by
Genevieve Atwood
Earth Science Education

ABSTRACT

Storm-related flooding caused extensive damage above Great Salt Lake's stillwater lake level during the 1986-87 lake highstand. Storm-related flooding of Great Salt Lake results from two coastal processes: lake storm surge and wave runup. Lake coastal processes differ from marine coastal processes. Marine storm surge results from exceptional tides and tidal currents, swell, precipitation, barometric pressure differences, wind drift and other currents, wind set-up, and wind waves. Lake storm surge on Great Salt Lake primarily results from wind set-up, with minor, and poorly understood contributions from barometric pressure differences, precipitation, wind drift and other currents, and lake seiching. Wave runup, dependant on wave energy, is greatly affected by shore conditions. Superelevated shorelines on Antelope Island, the largest island in Great Salt Lake, record the farthest inland high-water line and other expressions of the 1980s highstand shoreline. Shoreline elevation differences from site to site result from complex interactions of coastal processes and local conditions such as fetch, slope, aspect, and beach roughness. Wave runup appears to contribute more to shoreline superelevation magnitude and variability than lake storm surge.

INTRODUCTION

In 1986 and 1987, Great Salt Lake reached high levels almost 4,212 feet (1,283.8 m) above sea level (a.s.l.) after rising 11 feet (3.4 m) in response to a four-year wet cycle of above-average precipitation over the lake's drainage basin. During the 1980s wet cycle, Great Salt Lake inundated hundreds of square miles of lakebed for the first time in over 100 years, destroying or seriously damaging structures built on the lake's historic floodplain. Storm waves broke through dikes constructed higher than 4,212 feet (1,283.8 m) a.s.l. in Stansbury Bay, Gilbert Bay, Farmington Bay, and Ogden Bay (figure 1). Facilities as much as 10 feet (3 m) above the lake's highstand stillwater level were damaged, such as park pavilions at Antelope Island State Park that were undercut by wave action. Storm waves also left geomorphic evidence of storm-related flooding such as debris lines. Shore terms used in this paper are shown in figures 2 and 3 and discussed later.

Defining flood hazard zones of Great Salt Lake has been challenging because two types of flooding cause damage: inundation due to lake level rise, and inundation due to storm-related wind-driven waves. Flood hazards migrate with changes of lake elevation (figure 1). Location-specific variations of storm damage are poorly understood, and therefore poorly predicted, but are of economic consequence as

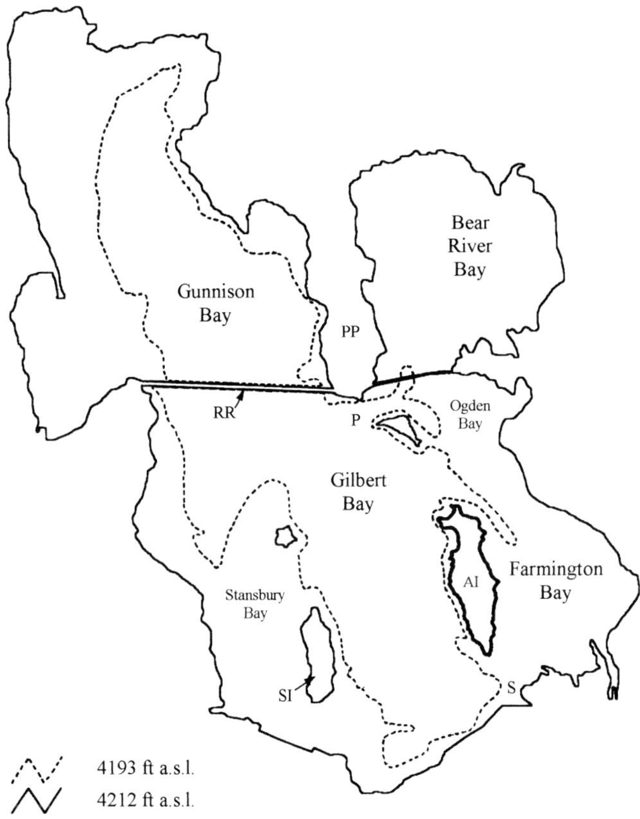

Figure 1. *Historic extent of Great Salt Lake. During the 1960s dry cycle, the lake's lowstand was approximately 4,193 feet (1278.0 m) a.s.l. During the 1860s-70s and 1980s, the lake's highstands were approximately 4,212 feet (1283.8 m) a.s.l. The shape and surface area of Great Salt Lake change dramatically with lake level fluctuations. The area exposed to storm-related flooding hazards migrates as the lake fluctuates. Locations referenced include: AI = Antelope Island, SI = Stansbury Island, RR = railroad causeway, PP = Promontory peninsula, P = USGS Promontory gauge, and S = USGS Saltair marina gauge.*

construction of facilities continues on Great Salt Lake's lakebed. Location-specific information could quantify risks associated with specific projects. It also could provide insight on how to avoid or mitigate damage.

The conventional approach to mapping the flood-hazard zone of the lake is to agree upon an inundation elevation and add a few feet as buffer against storm-related damage. For example, Utah's Division of Forestry, Fire and State Lands (DFFSL) addressed the issue in 2000:

> DFFSL's statutory mandate is to define the lake's flood plain and the legislative policy is to maintain the lake's flood plain as a hazard zone. DNR [Utah Department of Natural Resources] considers the flood plain to extend to 4217. This is based on recent high lake level of roughly 4212, plus three feet for wind tide and two feet for wave action (Great Salt Lake Planning Team, 2000, p. 14).

This policy makes two questionable assumptions: 1) that the historic highstand should be the highstand level for planning purposes and 2) that all coastal areas of Great Salt Lake experience similar storm damage. Holocene history of Great Salt Lake in the past 3,000 years includes at least one lake level rise above 4,217 feet (1,286 m) a.s.l. (Murchison and Mulvey, 2000). As discussed below, storm-related flooding of the 1980s was not uniform.

Storm-related processes of Great Salt Lake are coastal processes. This paper examines coastal processes of Great Salt Lake and compares them to marine coastal processes. Then research on Antelope Island's 1980s shoreline is used to examine how wave energy, lakebed slope, shore aspect, and beach roughness affect storm-related damage. The 1980s shoreline expressions provide useful evidence of flooding extent and storm-related geomorphic changes of materials and landforms. Antelope Island's 1980s highstand shorelines are exceptionally well preserved, relatively continuous, accessible, and distinguished by their anthropogenic debris. By analyzing elevations of these shorelines it is possible to document effects of coastal processes and better define location-specific flooding hazards of Great Salt Lake.

PREVIOUS WORK

Magnitudes of Storm Surge

Marine, storm-related, coastal hazards have been studied extensively because of the extraordinary threats they pose. In A.D. 1421 a marine storm surge killed 10,000 to 100,000 people in Britain. In 1970, a marine storm surge killed 200,000 to 500,000 people in Bangladesh (Pugh, 1987; Carter, 1988; Hansom, 1988). Marine storm surge magnitude varies from 3 feet (1 m) for the Adriatic Sea, 6 feet (2 m) for the Baltic Sea, 10 to 25 feet (3 to 7 m) for the Gulf of Mexico, and 30 feet (> 9 m) for the Bay of Bengal . Lake storm surge has been estimated as 6 feet (2 m) for Lake Erie, and 10 feet (3 m) for Lake Okeechobee (Allen, 1984).

Utah Geological Survey Shoreline Research

From 1987 through 1990, the Utah Geological Survey (UGS), in cooperation with Antelope Island State Park, mapped the island's geology and identified the island's geologic hazards (Doelling and others, 1990; Hecker and Case, 2000). As part of that effort, island shorelines were studied to better define elevation and frequency of high lake levels of Great Salt Lake (Atwood, 1994; Atwood and Mabey, 1995, 2000). Shore profiles were surveyed at 19 locations where series of shorelines recorded multiple flooding events. This work was coordinated with University of Utah researchers identifying and surveying the lake's late Holocene high shoreline (Murchison, 1989; Murchison and Mulvey, 2000). Elevations of shoreline series on the western side of Antelope Island did not match elevations of similar shoreline series on the eastern side of the island. The studies concluded that Antelope Island's Holocene shorelines could not be correlated based on elevation alone, and that storm-related flooding occurred significantly above the lake's stillwater level. Elevations of scattered, prominent expressions of the 1980s shoreline differed by as much as 6 feet (2 m) due to coastal processes, implying that flooding hazards differed by as much as 6 feet (2 m) around the island. The UGS research of the 1980s was followed by research undertaken in the 1990s as part of a doctoral program in the Department of Geography, University of Utah. Aspects of that on-going research are reported in this paper.

Corps of Engineers Estimates of Storm-Related Flooding of Great Salt Lake

The state of Utah and communities of Davis County have called upon the U.S. Army Corps of Engineers for assistance in estimating site-specific flooding hazards and calculating elevations for proposed inter-island dikes (Rollins, Brown, and Gunnell, Inc. and Creamer & Noble Engineers, 1996a, 1996b). Assistance to Davis County included calculating flooding hazards at five shoreline locations on the eastern mainland of Great Salt Lake (U.S. Army Corps of Engineers, 1996a, 1996b). For Davis County locations, the Corps of Engineers assumed a stillwater lake level of 4,212 feet (1,283.8 m) a.s.l. and calculated combined effects of "wind set" (wind set-up) and wave runup. Wind effects were calculated using estimates for wind speed, fetch, average fetch depth, and a friction coefficient. Wave runup was calculated using empirical relationships for significant waves (U.S. Army Corps of Engineers, 1984). Significant wave height was based on assumptions for Great Salt Lake including wind speeds, wind duration, and fetch. Vegetation and beach roughness were factored into a slope-roughness coefficient. Predicted superelevation (combined wave runup and wind set-up) ranged from 1.95 feet (0.6 m) at Syracuse, to 7.64 feet (2.3 m) along a railroad embankment bordering Farmington Bay.

Shorelines as Evidence of Storm-Related Flooding

Shorelines are formed by wave action at the lake-land interface and record storm-related flooding. Shorelines are used to document climate change and tectonic deformation including isostatic deformation related to lake loading often assuming that they originally defined a horizontal datum (Smith and Dawson, 1983). If a paleoshoreline is not horizontal, either it was not horizontal when formed, or it has been deformed since formation, or both.

Gilbert (1890) in his study of Lake Bonneville, a predecessor of Great Salt Lake, recognized that elevation differences among Bonneville shoreline features made estimates of stillwater elevation imprecise. He analyzed spatial distributions of shoreline features:

> At an early stage of the investigation, the writer thought that the coasts facing in certain directions gave evidence of exceptional amounts of wave work, and imagined that he had discovered therein the record of prevalent westerly winds or westerly storms in ancient times. This belief was dissipated by further study; and he discovered, as students of modern shores long ago discovered, that there is a close sympathy between the magnitude of the shore features and the 'fetch' of the efficient waves. The greater the distance through which waves travel to reach a given coast, the greater the work accomplished by them. The highest cliffs, the broadest terraces, and the largest embankments are those wrought by the unobstructed waves of the main body; and opposite coasts appear to have been equally affected (Gilbert, 1890, p. 70).

Several researchers of Great Basin paleolakes have recognized the problem of paleoshoreline superelevation as a source of imprecision when estimating paleolake elevation, but have had limited evidence to quantify its magnitude and pattern . A conventional approach has been to assume original horizontality:

> Because the static surface of a lake is essentially level, landforms created at the lake margins form at basically the same elevation throughout a basin. If coastal landforms are not level throughout a basin, subsequent neotectonic deformation in the form of seismotectonic displacement or isostatic deflection has likely occurred (Lillquist, 1994, p. 143).

DISCUSSION OF COASTAL TERMINOLOGY

Coastal process terms used in marine environments and lakes are not always consistent. Confusion can result. Through time, or in different contexts, words have been used with different meanings, for example, "storm surge." Terminology associated with elevations of features and elevation differences as used in this paper are shown in figure 2. Storm-related coastal processes of Great Salt Lake are shown in figure 3. The lake's undisturbed horizontal surface is its stillwater surface (see figures 2 and 3). As the lake transgressed to, and regressed from, its 1980s highstand stillwater lake level it left littoral materials and features. A typical 1980s shore on Antelope Island is shown in cross-section in figure 2. Wind-driven waves resulted in superelevated shorelines, in other words, features above the lake's stillwater level (figures 2 and 3).

Shoreline expressions: evidence of processes taking place at the lake-land interface. The 1980s highstand shoreline is a set of shoreline expressions. Expressions include erosional and depositional features such as undercut scarps and debris lines. These features are at different elevations at different locations around the lake. The farthest inland high-water line is the depositional or erosional evidence of most extensive flooding associated with the 1980s wet cycle.

Stillwater lake level: elevation of the surface of a lake undisturbed by wind, seiches, or currents. The U.S. Geological Survey (USGS) continuously monitors lake elevations at two (and for some years, three) locations on Great Salt Lake. Lake elevations measured during the 1980s have been adjusted by USGS several times to reflect corrections related to uncertainties in datum ties and instability of gauges. After the datum adjustments of 2001, USGS reports Great Salt Lake 1986 highstand as 4,211.6 feet (1,283.7 m) a.s.l. and 1987 highstand as 4,211.5 feet (1,283.7 m) a.s.l.

Shoreline superelevation: elevation difference between a lake's highstand stillwater elevation and elevation of an expression of its highstand shoreline. This is a conservative estimate. A storm that creates a highstand shoreline does not always occur when the lake is exactly at its highstand elevation. Differential superelevation refers to variations of superelevation along a stretch of shoreline such as along Antelope Island's 40-mile (65 km) shoreline. Absolute superelevation ties differential superelevation to a stillwater lake elevation, such as Great Salt Lake's 1986-1987 highstand elevations monitored by USGS. Lake storm surge together with wave runup cause shoreline superelevation.

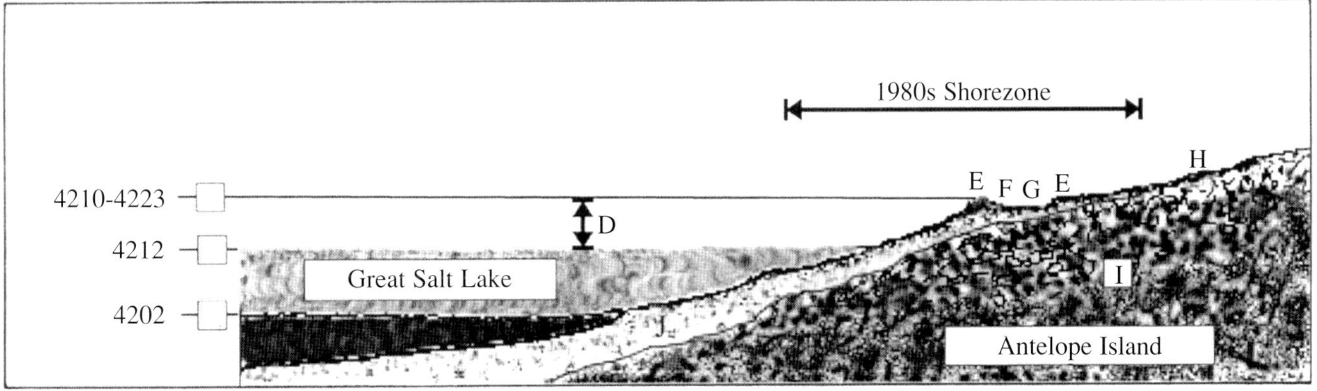

Figure 2. *Shore terminology. The stillwater lake level is the horizontal lake level undisturbed by storm processes or lake seiching. The historic average elevation of Great Salt Lake is 4,202 feet (1,280.8 m) a.s.l. (A). The 1980s highstand stillwater lake level was approximately 4,212 feet (1283.8 m) a.s.l. (B). Elevations of 1980s highstand shoreline expressions range from 4,210.4 to 4,223.4 ft (1283.3 to 1287.3 m) a.s.l. (C). The difference in elevation between the 1980s highstand stillwater level and a 1980s highstand shoreline expression is shoreline superelevation (D). Shoreline expressions include debris lines (E), lagoons (F), and killed vegetation (G). Older shorelines are exposed above the 1980s superelevated shorelines (H). Substrate underlies 1980s shorelines and consists of bedrock, colluvium, and sediments of older lakes (I). 1980s and older lake deposits are exposed in the beach zone by regression of Great Salt Lake from its highstand level and they continue lakeward (J).*

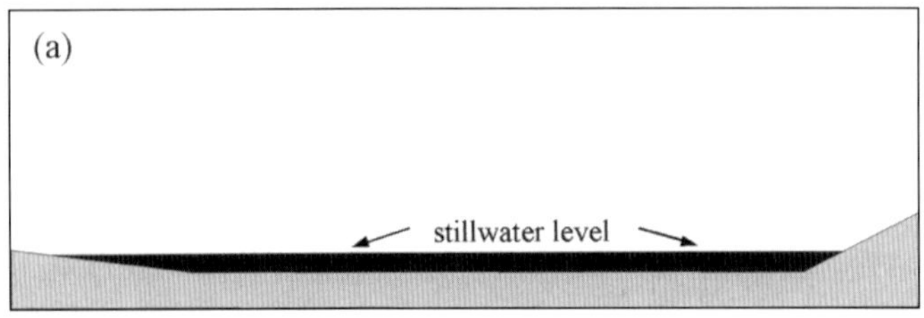

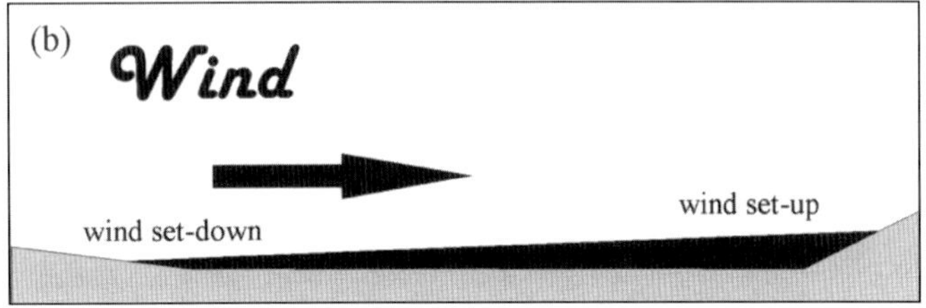

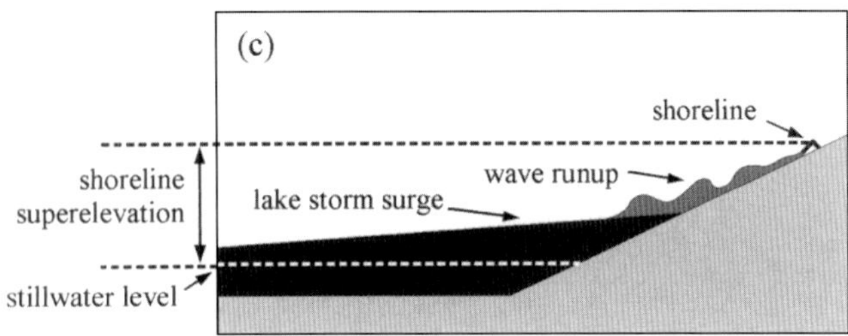

Figure 3. *Storm-related coastal processes. Shorelines of Great Salt Lake are constructed during storms. Upper diagram (a) shows an undisturbed horizontal lake surface, the stillwater lake level. Middle diagram (b) shows effects of wind blowing across the lake, wind set-up. Wind set-up is compensated by wind set-down. Lower diagram (c) is a detail of the downwind shorezone showing the two coastal processes that result in shoreline superelevation: lake storm surge and wave runup. Wind set-up is the major component of lake storm surge. Lake set-up is the general term that refers to lake elevation above stillwater level. The distinction between lake set-up and lake storm surge is that lake set-up does not attribute causation whereas lake storm surge indicates superelevation caused by storm processes. Waves run up the beach face from the set-up or set-down lake level to where they deposit debris. Although both processes are important, wave runup appears to be a greater factor in shoreline superelevation magnitude and variability than lake storm surge.*

Debris line: bands of shoreline deposits. Debris of 1980s shoreline expressions includes natural and anthropogenic materials. Debris lines constructed at the farthest inland high-water line often incorporate floated material, flotsam. Flotsam-rich debris lines have tree limbs, lumber, automobile tires, and locally derived twigs incorporated into deposits of sand and gravel. Debris lines constructed by bedload processes consist primarily of non-floating materials. Materials entrained and transported at wave base include sand, gravel, and anthropogenic materials denser than lake water, such as pottery, asphalt, and concrete. Debris lines formed by high-energy waves are superelevated. Debris lines formed on erosional surfaces or in low-energy wave environments are below, or at, stillwater lake elevation.

Storm surge: elevated surface of a waterbody in response to storm conditions. For marine environments, storm surge is sometimes defined narrowly as the change in elevation above typical fair-weather levels due to meteorological conditions alone , and sometimes more broadly, as the change in elevation during a major storm due to all contributing factors including meteorological conditions, tides, and currents (Heaps, 1985). Lake storm surge is the elevated surface of a lake in response to storm conditions. For Great Salt Lake, the major cause of lake storm surge is wind set-up. Lake set-up is the elevated lake surface caused by any process, whether or not storm-related. Non-storm-related processes include currents, or lake seiching up to several days after a storm. Wind set-up is the component of lake set-up and lake storm surge caused by wind. Lake set-up is compensated by a lowered lake level called lake set-down, or negative lake set-up. Storm surge does not include the amplitude of wind waves above the set-up surface.

Lake seiche: standing-wave oscillation of a lake's surface. For example, figure 4 shows a seiche event of November 1998 when Gilbert Bay's surface alternately rose and fell at its north and south ends. At the standing wave's node, lake level remains constant. Lake set-up, generally lake storm surge, initiates lake seiching. Wind associated with a storm sets up the lake surface, elevating one side of the lake relative to the other. When wind strength diminishes or wind direction changes, oscillation begins. Lake seiching, in turn, results in lake set-up and lake set-down as the lake oscillates. Lake seiching may affect lake storm surge. It is not known whether an already set-up or set-down lake surface affects the progress or magnitude of lake storm surge.

Wave runup: rush of wave-propelled water and entrained sediment from the lake, upward and landward across a beach face. Wave runup begins from a set-up or set-down lake surface.

Wind drift: movement of water or materials at the surface of a waterbody caused by wind. Wind drift currents flow downwind and, in large water bodies such as marine systems, can be affected by Coriolis forces. Geostrophic currents are typically large-volume, slow-moving, marine currents where flow gradients, including curvature of flow paths, balance forces associated with Earth's rotation.

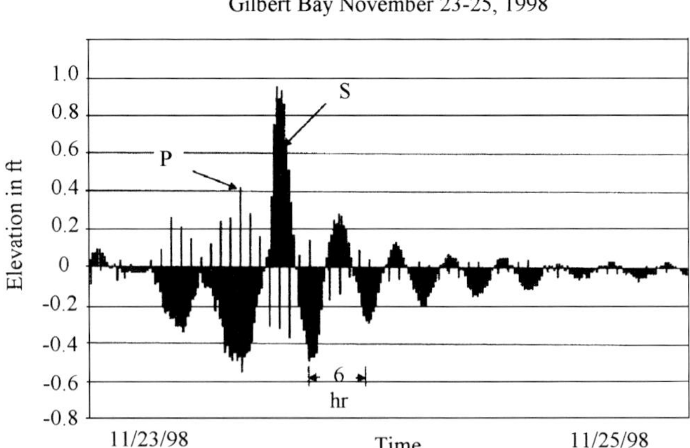

Figure 4. *Seiching of Great Salt Lake, Gilbert Bay, November 23-25, 1998. Data from USGS Promontory gauge (P) and from USGS Saltair marina gauge (S) are shown as elevations above and below a common, adjusted, stillwater lake level. Storm winds from the south caused lake set-down at Saltair during 11/23 and lake set-up at Promontory. Lake set-up and set-down initiated seiching with alternately rising and falling lake levels at Saltair and Promontory gauges. Lake level readings were recorded more frequently at Saltair than at Promontory. Note that basin configuration of Gilbert Bay results in greater magnitude of seiching at Saltair marina than at Promontory.*

Wave train: series of waves of similar energy traveling as a group. Storms impart energy into the surface of a waterbody causing waves of differing energies. Individual wind waves organize into wave trains as they travel from their energy source (Gill, 1982). Wave trains far from their storm of origin are called swell. Waves of an unorganized, confused, sea are called individual wind waves. Individual wind waves of storms across Great Salt Lake construct highstand shoreline expressions. Wave trains modify features of the littoral zone. Swell is not a phenomenon of Great Salt Lake.

STORM SURGE ON GREAT SALT LAKE COMPARED TO MARINE STORM SURGE

Research on lacustrine coastal processes is a subset of coastal science, which is dominated by research on marine processes. Marine systems are vast, more complex, more dangerous, and more difficult to observe than those of Great Salt Lake. Seven coastal processes contribute to marine storm surge: exceptional tides and tidal currents, swell, precipitation, barometric pressure differences, wind drift and other currents, wind set-up, and wind waves (figure 5). Only one process, wind set-up, accounts for most of lake storm surge, with minor, or unknown, contributions from precipitation, barometric pressure differences, wind drift and other currents, and seiching (figure 5). Shoreline superelevation results from the combined effects of storm surge and wave runup.

Tides

Tidal factors, including tidal currents, are major components of marine storm surge. A tidal range of 5 to 10 feet (2 to 4 m) is typical of most marine shorelines, although exceptional tidal ranges of 50 feet (15 m) occur at some locations. A coast's tidal range, along with the timing of marine storm surge with respect to high tide, can greatly affect marine storm surge. Tides have been measured on a few large lakes. Tides have not been measured on Great Salt Lake and tides do not contribute to lake storm surge, shoreline superelevation, or storm-related flooding hazards of Great Salt Lake.

Swell

Swell from distant storms, or swell racing away from a slow-moving storm system, can arrive onshore before or with the main storm and contribute to marine storm surge. Wave trains on Great Salt Lake do not race ahead of a storm system and arrive on shore before the main storm. Great Salt Lake is small compared to an ocean. Large regional storms affect all of the lake. Relatively short storm duration, shallow lake depth, and limited fetch do not provide opportunities for longer wavelength waves to organize and travel as wave trains ahead of a storm on Great Salt Lake. Low-amplitude wave trains associated with regional storms arrive onshore after a storm. Wave trains of low-amplitude, <1 feet (0.3 m), waves also can originate from an area of the lake disturbed by local winds and travel across the lake. Low-amplitude wave trains generally do not constitute a coastal hazard for Great Salt Lake, although waves winnow fine-grained sediments from lower portions of the beach face.

Precipitation

Precipitation associated with tropical storms can increase marine storm surge by 1 to 2 feet (0.5 m). Cloudbursts occur on Great Salt Lake and, theoretically, a deluge over a significant portion of the lake could contribute a few centimeters to lake set-up, or could initiate lake seiching. However, it is unlikely that cloudburst precipitation contributed to 1980s shoreline superelevation on Antelope Island. Cloudbursts tend to be associated with summer storm patterns that occur after the lake has passed its annual peak elevation. Lake level rise of the 1980s was due to above-average precipitation across most of the watershed, not cloudbursts on isolated portions of the lake. Precipitation contributes indirectly to storm damage by saturating unconsolidated materials and making them prone to ground failure.

Barometric Pressure Differences

Barometric pressure differences contribute to storm surges associated with low-pressure systems. Changes in sea surface elevation are inversely related to atmospheric pressure. One-millibar (mb) change in atmospheric pressure causes 0.03 feet (1 cm) elevation change. Atmospheric pressure drops of 100 to 200 mb at the center of a tropical depression cause 3 to 6 feet (1 to 2 m) of sea surface superelevation. For Great Salt Lake, atmospheric pressure differences rarely exceed 10 mb (William Alder, National Weather Service, verbal communication, 1994). Maximum contribution to lake storm surge would be on the order of a few inches. Barometric effects do not appear to contribute significantly to lake storm surge, shoreline superelevation on Antelope Island, or flooding hazards of Great Salt Lake.

		Coastal Process/Agent Contributing to Storm Surge	Potential contribution: Great Salt Lake	Potential contribution: marine coasts
SHORELINE SUPERELEVATION	Storm surge	Exceptional tides and tidal currents	0	major
		Swell	0	major
		Precipitation	minor	significant
		Barometric pressure differences	minor	major
		Wind drift and other currents	not known	significant
		Wind setup/sea surface tilt/surge wave	significant	significant
		Seiching	not known/minor	n/a
		Wind waves	not included in lake storm surge	significant
	Wave runup		major	major

Figure 5. *Comparison of Great Salt Lake and marine storm-related coastal processes that contribute to superelevated shorelines. Two processes cause superelevated shorelines of Antelope Island: lake storm surge and wave runup. Lake storm surge is largely due to wind set-up. Wind set-up contributes up to 3 ft (1 m) to lake storm surge and shoreline superelevation. Wave runup contributes up to 10 ft (3 m). Coastal processes that contribute to marine storm surge are more numerous, more complex, and result in greater storm surges that those that produce lake storm surge on Great Salt Lake. Marine and lake coastal processes not considered include ice, tsunamis, and river discharge.*

Wind Drift and Other Currents

Oceans currents result from wind, density differences, barometric pressure differences, tidal effects, and effects of Earth's rotation. Ocean currents that pile water along marine coasts can contribute significantly to marine storm surge (Allen, 1984). Wind drift currents have been studied on Great Salt Lake (Rich, 1991), but their effect on lake surface elevation has not been quantified. Wind drift currents and other lake currents have not been observed contributing to lake storm surge on Great Salt Lake. Hundreds of automobile tires are incorporated in highstand superelevated shorelines and along shores of Antelope Island. Tires float low in the water. They are transported by wind drift rather than blown across the lake. They may have been carried to Antelope Island by wind drift currents associated with lake storm surge. Geostrophic currents and advective currents associated with river inflow have been hypothesized for Great Salt Lake, but their nature and magnitude have not been quantified (Katzenberger and Whelan, 1975). Wind drift and other currents may be important, poorly defined components of lake storm surge, shoreline superelevation on Antelope Island, and storm-related hazards of Great Salt Lake.

Wind Set-up, Sea Surface Tilt, and Storm Surge Waves

Strong, steady wind blowing across an ocean causes a sea surface tilting, and this tilt is called wind set-up, sea surface tilt, or storm surge wave depending on the scale of the tilt and its location with respect to shore (Gill, 1982; Pugh, 1987). Storm surge waves are associated with large storm systems that have moved from deep water onto ocean shelves, or into shallow water. Storm surge waves and their progress have been documented on Lake Michigan (Johnson, 1919).

Greater wind speed, longer fetch, greater frictional forces, decreased water depth, and decreased water density increase the magnitude of wind set-up, extent of sea surface tilt, and height of storm surge waves (Pugh, 1987). The effect of increased wind speed on wind set-up is exponential. Wind set-up increases as the square of wind velocity. Fetch is calculated as the distance across open water affected by storm winds. Wind set-up across Great Salt Lake is the greatest contributor to lake storm surge. Combined with wave runup, storm surge results in high-elevation storm damage and superelevated shorelines. Calculations using equations from Pugh (1987) estimate wind set-up magnitude for Great Salt Lake as about 3 feet (1 m) for strong, steady winds with significant fetch.

Progression of wind set-up across Great Salt Lake during a storm is poorly understood. Sophisticated remote sensing techniques can monitor small sea-surface elevation differences. Such technological advances may make documentation of the progress of wind set-up across Great Salt Lake possible.

Wind Waves and Storm Surge

Wind waves can be considered separately or as part of storm surge. In marine coastal research, broad definitions of marine storm surge include contributions of wind wave amplitude. Wind waves of geomorphic significance associated with large storm systems attain great height (Davies, 1980). Extreme significant waves can be 25 to 80 feet (8 to 25 m) in marine environments (Komen and others, 1994).

Individual wind waves on Great Salt Lake can have wave height greater than 6 feet (2 m) (Doyle Stephens, USGS, verbal communication, 1998). U.S. Army Corps of Engineers shore protection investigations for Davis County estimated significant wave height as 3 to 4.6 feet (0.9 to 1.4 m) (U.S. Army Corps of Engineers, 1996b). Lake surface elevation differences associated with wave height are the same order of magnitude as lake surface elevation differences from all other contributions to lake storm surge combined. Wave height determines wave runup. Wave runup appears to be the most significant factor determining shoreline superelevation on Antelope Island and storm-related flooding damage of Great Salt Lake. Wind waves rush up the beach face from a set-up or set-down lake surface. As used in this paper, wind waves are considered part of the wave runup coastal process and wave height of wind waves is not included in lake storm surge magnitude.

Wave Runup

Literature concerning wind waves is extensive. It includes wave theory, hydrodynamics, and morphodynamics. Energy from wind is transferred into waves. Shoreline aspect, materials, roughness, and slope affect how energy is reflected, dissipated, or absorbed (Komar, 1998). Wide, gently sloped, sand beaches dissipate wave energy, whereas bedrock cliffs reflect energy. Wetland vegetation and bouldery beaches increase friction and dissipate wave energy.

Wave shape as well as wave energy affects coastal geomorphic processes and coastline evolution. Wind waves on lakes and reservoirs can reach heights of 10 feet (3 m), and may be steeper than marine wind waves (Zenkovich, 1967). Wave steepness, the ratio of wave height to wavelength, and littoral conditions affect wave type Pethick, 1984). Wave type, such as plunging, spilling, and collapsing, affects erosional and depositional processes of marine shores (Komar, 1998).

Wind waves generated by storms are the major agent of shoreline change on Antelope Island. Waves transport energy to Antelope Island's shores where it is reflected, absorbed, or dispersed. Individual wind waves, not swell, associated with a few storms when the lake was at its 1986 and 1987 highstands produced Antelope Island's 1980s highstand shoreline. Wind waves caused much of the storm-related damage of the 1980s including overtopping, eroding, and breaching dikes.

Lake Seiches

Lake seiches are initiated by lake set-up. Once initiated, and if undisturbed, lake seiching can continue for days. Lake seiching, may affect lake storm surge. Seiche amplitude is controlled by initial magnitude of lake set-up, friction forces, and basin configuration. Seiche period is primarily a function of basin configuration. Seiche period increases with increased length of a water body, and decreases with increased water depth (Open University Oceanography course team, 1989.

Seiching of Gilbert Bay is well documented by USGS lake monitoring (figure 4). Lin and Wang studied seiches of Gilbert Bay from 1968 to 1976. They compared seiche period and pattern with hypothetical periods for a geometrically simplified Gilbert Bay (Lin and Wang, 1978; Wang, 1978). Seiching occurred for about 50 percent of their monitored period. They concluded that barometric pressure differences and wind-driven water stacked against the shore initiated lake seiching. They further concluded that wind set-up of major magnitude required winds of 10 knots (10 to 15 mph, 5 m/sec) for 12 hours duration. The rate of movement of barometric pressure differences across the lake, interacting with progression of windwaves, could magnify lake set-up and subsequent seiching. Maximum amplitude of seiching was 2 feet (0.6 m). Longer-than-predicted seiche periods were explained by bottom friction and by Earth's rotation. They hypothesized that seiche currents transported bottom materials. Their research relied on wind observations at Salt Lake International Airport about 30 miles (50 km) from the center of Gilbert Bay.

Seiching of Great Salt Lake was observed and records from USGS gauges at Saltair and Promontory were analyzed as part of 1990s Antelope Island field investigations reported in this paper. Gilbert Bay appeared to oscillate with a 6-hour period when disturbed by wind (figure 4). Wind set-up did not appear to require steady, long-duration, strong winds. Seiche amplitude was higher at the Saltair gauge than at the Promontory gauge apparently due to lake configuration. Wind measurements now available from weather stations installed in and around Great Salt Lake in the late 1990s record complex patterns of wind direction and velocity across Gilbert Bay not evident from wind measurements from Salt Lake International Airport alone. Contribution of seiching to lake storm surge is poorly understood, and is assumed to be minimal for the large storm events that created 1980s highstand shorelines on Antelope Island.

Interaction of Coastal Processes

Coastal processes are further complicated by the interrelatedness of atmospheric, hydrodynamic, and morphodynamic processes. For example, water versus atmospheric temperature differences and locations of islands and promontories add complexities to wind patterns. Characteristics of storms that influence lake storm surge include wind speed, wind direction, wind duration, wind gustiness, the storm path, barometric pressure gradients, and precipitation. Characteristics of the lake that influence lake storm surge include water density, presence of ice, lake stratification, currents, water depth, and wave characteristics such as height, length, energy, speed, sea-surface roughness, and wave type. Characteristics of land that influence lake storm surge include regional and local coastal shape; shoreline aspect with respect to the path of the storm; offshore, nearshore, and on-shore slope; extent of mud flats or abrasion platforms; and roughness of bottom materials. Shoreline superelevation is the net effect of complex processes acting at interfaces of land, water, and atmosphere. The processes transfer energy from wind into water, then transfer energy across the lake, and finally, transfer energy from water to land resulting in changes of shore features.

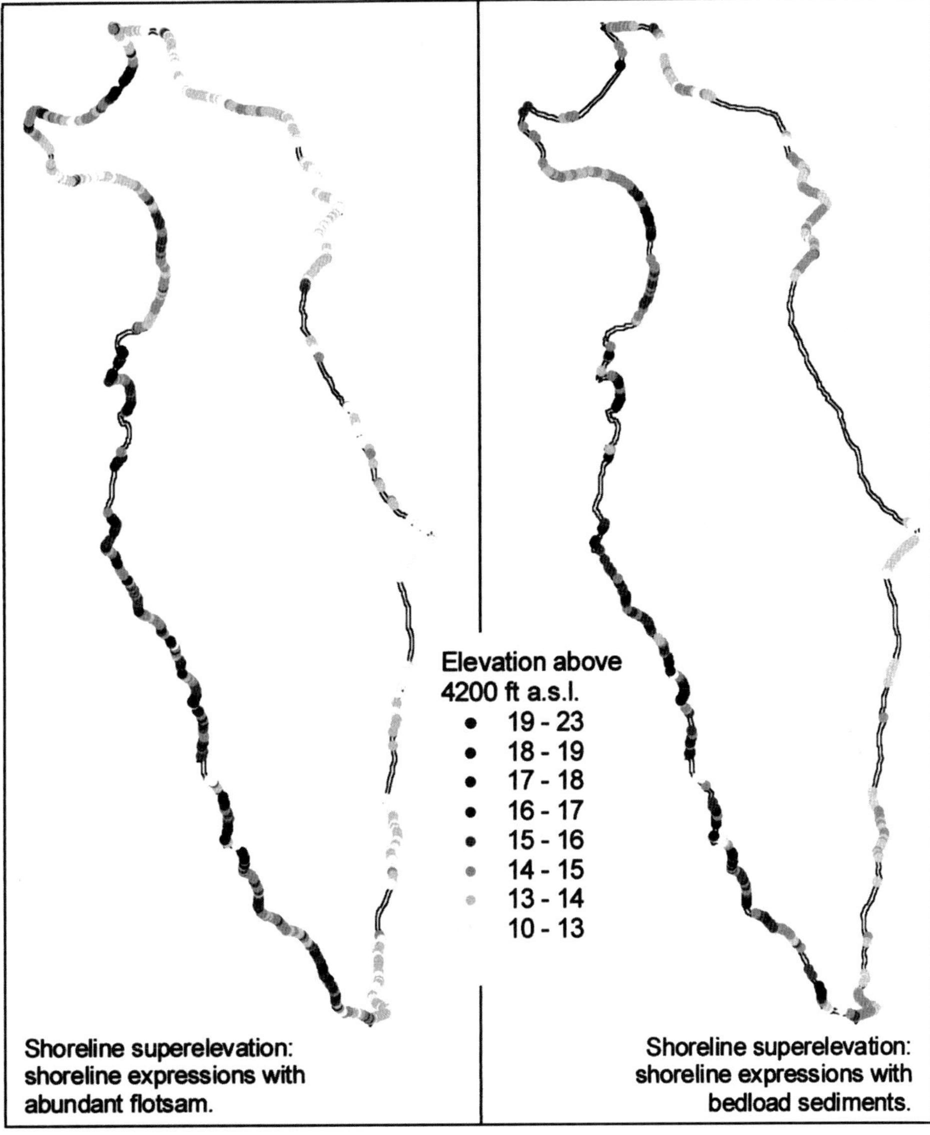

Figure 6. *Shoreline superelevation, Antelope Island's 1980s highstand shoreline. Left map shows surveyed elevations of shoreline expressions with flotsam. Right map shows surveyed elevations of shoreline expressions consisting of bedload sediments. Elevations are shown as feet above 4,200 feet (1280.2) a.s.l. West-side shoreline elevations tend to be higher and more variable than those of the east side of the island.*

EVIDENCE OF SHORELINE SUPER-ELEVATION ON ANTELOPE ISLAND

The 1980s highstand shorelines of Great Salt Lake can be distinguished from older shorelines by their anthropogenic debris (trash) and from younger shorelines by stratigraphic position. The research reported here analyzed characteristics of 1980s highstand shorelines on Antelope Island and documented shoreline expressions as geoantiquities, precious archives of Earth system history endangered by natural and human disturbance (Chan and others, 2001). Investigations included two field components: mapping shoreline materials and surveying shoreline elevations. Natural and anthropogenic shoreline materials and their abundance were mapped. Shoreline expressions were characterized as erosional or depositional. Elevations of all shoreline types, exposures, and expressions with identifiable evidence of 1980s highstand deposition or erosion were surveyed. Additional datasets derived from UGS and USGS geologic and topographic maps included geology, slope, aspect, general shape of the shoreline, and fetch.

Research Results

Elevations of 1980s highstand shoreline expressions are shown in figure 6. Elevations range from extreme low values for bedload shoreline expressions on erosional shorelines, 4,210.4 feet (1,283.3 m) a.s.l., to extreme high values on debris where west-side headlands focused wave energy, 4,223.4 feet (1,287.3 m) a.s.l. Elevations are displayed in eight classes in feet above 4,200 feet (1,280.2 m) a.s.l. using USGS 1998 elevations for Great Salt Lake for 1998 not adjusted for datum corrections of 2001.

Shoreline superelevation data are highly variable. Superelevation results from several coastal processes interacting with local conditions and individual data points should be interpreted in context of superelevation trends. Nonetheless, shoreline superelevation is not a random phenomenon. There are patterns at island-scale and at bay-by-bay scale. Superelevation is higher and more variable on the western side than the eastern side of Antelope Island. The western side has more high extreme values and the eastern side has more low extreme values. Variability of shoreline superelevation follows geographic patterns of superelevation magnitude. Shorelines with higher median elevation have greater superelevation variability than those with low median values.

Bay-by-bay patterns of shoreline superelevation, and patterns of shoreline superelevation on contrasting sides of headlands, indicate that wave runup contributes more to magnitude and variability of shoreline superelevation than lake storm surge. Shorelines with greater fetch have higher superelevation. Greater fetch for shorelines of Antelope Island tends to be to the west and northwest. These also are the directions of some of the strongest winds associated with regional low-pressure storm systems.

Patterns of superelevation were compared with shore characteristics such as aspect, fetch, slope, and location. Shorelines facing west and northwest have higher superelevation. Those facing east and southeast have lower superelevation. Shorelines with steeper beachface slopes have higher superelevation than those with gentle slopes. Lake surface tilt associated with wind set-up is not obvious from shoreline superelevation patterns.

COASTAL PROCESSES AND STORM-RELATED FLOODING HAZARDS OF GREAT SALT LAKE

Shoreline superelevation is direct evidence of location-specific storm-related flooding hazards of Great Salt Lake. Expressions of the 1980s highstand shoreline on Antelope Island provide sufficient evidence to quantify the magnitude of shoreline superelevation for Great Salt Lake at its 4,212 feet (1283.8 m) a.s.l. elevation and to compare storm-related flooding hazards of Great Salt Lake with marine environments.

Scale and Coastal Processes

Great Salt Lake is a shallow, closed-basin lake, small compared to an ocean. Size alone limits storm-related coastal flooding by limiting the surface area available for energy transfer from wind into waves and the time duration of energy transfer. For Great Salt Lake, the effective duration of a storm is the duration of that storm while over the lake. Great Salt Lake is a fetch-limited coastal system. For oceans, fetch is calculated as "the distance the wind blows over the sea in generating the waves" (U.S. Army Corps of Engineers, 1984, p. 1-6). At the scale of Great Salt Lake, effective fetch is the surface of open water that extends from the upwind shore to the downwind shore. It is an area rather than a linear transfer zone.

Great Salt Lake's small size limits the formation of wave trains and swell. Wave type affects coastal hazards (U.S. Army Corps of Engineers, 1984. Marine coastal processes and features associated with swell include edge waves, certain kinds of bars, and certain types of beach cusps (Komar, 1998). In contrast, Antelope Island's highstand shoreline expressions were constructed by individual wind waves, not wave trains.

Time and Coastal Processes

Variability of shoreline superelevation on Antelope Island is due to variations in local conditions, differences in fetch, and differences in storm direction. Also, variability of shoreline superelevation is evidence of the short duration during which coastal processes worked on the features and materials of the highstand shoreline's littoral zone. Marine shores evolve so that coastal configurations reflect coastal morphodynamic equilibrium conditions. Shore configuration, including shoreline superelevation, of Antelope Island's 1980s highstand shoreline is evidence of incomplete development of coastal features, not equilibrium conditions. Superelevation of Antelope Island's highstand shoreline would be less variable if it represented the net effect of several hundred storms rather than the effects of fewer than 10 storms.

Coastal Processes, Wave Energy, and Storm Direction

Fetch and wind strength account for much of the variability of wave energy and storm surge in marine and lake systems. For marine systems, increases in wind velocity result in exponential increases in wind set-up whereas icreases in fetch result in linear increases (Pugh, 1987). Determining effective fetch may not be a simple calculation for Great Salt Lake due to lake bathymetry and lake configuration. However, if the contribution of fetch to shoreline superelevation can be isolated, the remaining differences in superelevation may indicate wind strength. This could be a useful paleoclimate indicator of storm direction. It also is an essential component of a model that predicts location-specific storm-related flooding hazards.

One research approach to isolate the contribution of fetch from wind strength would be to compare shoreline superelevation for shorelines with approximately equal fetch but different storm exposures. For example, Gilbert Bay lies between Stansbury Island and Antelope Island. The eastern side of Stansbury Island is generally the leeward side of that island. The western side of Antelope Island is generally the windward side of that island. Fetch to the respective shores is relatively similar. Significant differences in superelevation of the 1980s highstand shoreline of the eastern side of Stansbury Island and the western side of Antelope Island might clarify relative contributions of wind strength versus fetch to wave energy, wave runup, and lake storm surge.

Changes in the configuration of Great Salt Lake that greatly diminish fetch should diminish shoreline superelevation. In 1959, the Southern Pacific Railroad completed an earth-fill causeway across Great Salt Lake dividing the western portion of the lake into Gunnison and Gilbert Bays and dramatically changing the lake's configuration. Where 1860s-70s highstand shoreline can be distinguished from the 1980s highstand shoreline, a comparison of shoreline elevation differences could quantify effects of diminished fetch.

Shoreline superelevation records general patterns of high-energy versus low-energy waves. Patterns of bedrock exposures also correlate with high-energy and low-energy wave environments on Antelope Island. On the eastern, generally leeward side of the island, only three 1980s shoreline stretches have exposed bedrock. On the western, generally windward, side of the island, dozens of bedrock headlands jut into Gilbert Bay. Antelope Island's distribution of bedrock at shoreline level may be controlled by distribution of Tertiary versus older bedrock units and may not be the result of coastal processes. Further research in the Bonneville basin could test the hypothesis that patterns of bedrock outcrops indicate paleostorm direction.

Coastal Processes, Wave Energy, and Shore Materials

The largest particles moved by wave action on Antelope Island's 1980s farthest inland high-water line did not systematically correlate with wave energy, assuming shoreline superelevation indicates wave energy. Cobble-sized particles correlated with higher superelevation for much, but not the entire island. Sand and fine gravel correlated with lower superelevation for much, but not the entire island. On Antelope Island, cobbles are sourced from bedrock, former shorelines, and from debris-flow deposits. The littoral zone of the 1980s highstand lake is exposed to terrestrial processes when the lake recedes. Coastal processes associated with the 1980s highstand did not have time to work littoral materials into sediment packages indicative of wave energy. Size of beach materials alone is not a good predictor of storm-related flooding hazards of Great Salt Lake.

Great Salt Lake as a Laboratory for Coastal Process Research

Compared to an entire ocean, Great Salt Lake is a convenient waterbody to study. It is a simpler system. During lake highstands, individual storms leave shoreline evidence of coastal processes. Highstand shorelines are exposed after storm events and left undisturbed by coastal processes when the lake recedes. Relative contributions of fetch, wind speed, slope, water depth, and bottom roughness to coastal processes can be studied with little concern for tides, currents, barometric pressure differences, and swell. Antelope Island's coastal processes are dominated by wind waves of storms, in other words; no wind, no waves, no coastal work.

Shoreline superelevation indicates location-specific elevation differences of storm-related flooding of Great Salt Lake. A model that explains location-specific shoreline superelevation will also predict location-specific storm-related flooding hazards of Great Salt Lake. Components of a predictive model are shown in figure 7. Wind provides the

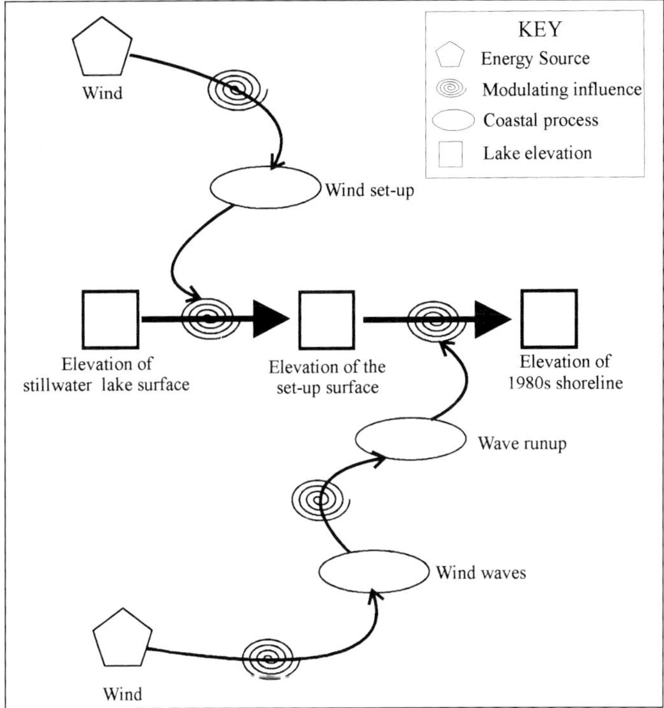

Figure 7. *Simple model of shoreline superelevation. A location-specific predictive model for shoreline superelevation also would predict areas vulnerable to storm-related flooding. Wind imparts energy into the lake's surface: modulating influences include wind strength and lake surface roughness. Wind set-up is the major coastal process contributing to lake storm surge. Lake storm surge results in a set-up lake surface: modulating influences include fetch and water depth. Wind waves cause wave runup. Wave runup begins from the set-up lake surface and results in superelevated shorelines: modulating influences include beach roughness and beachface slope.*

energy for the wind set-up component of lake storm surge and for wave runup. Atmospheric, lake, and shore conditions modulate energy transfers and affect the magnitude of shoreline superelevation. Shoreline evidence of Antelope Island quantifies the magnitude of shoreline superelevation. However, to develop a predictive model of location-specific storm-related flooding hazards, the relative importance of fetch and wind strength to wave energy must be isolated. Evidence of shoreline superelevation of Antelope Island alone is not sufficient to do this.

ACKNOWLEDGMENTS

This paper summarizes dissertation work in progress in the Department of Geography, University of Utah, scheduled for completion in 2002. Marjorie A. Chan, Thomas J. Cova, Paul W. Jewell, Roger M. McCoy, Harvey J. Miller, and C.G. (Jack) Oviatt have shared ideas and constructive criticism on lake research. Donald R. Currey, committee chair and mentor, reviewed this paper. Don R. Mabey, field colleague for 1980s to 1990s mapping and shoreline elevation surveying, offered valuable suggestions. This paper benefited from detailed technical reviews by Roy D. Adams and Alisa Felton, and from UGS reviews by J. Wallace Gwynn, Bryce T. Tripp, David E. Tabet, Kimm M. Harty, and Richard G. Allis.

REFERENCES

Adams, K.D., and Wesnousky, S.G., 1998, Shoreline processes and the age of the Lake Lahontan highstand in the Jessup embayment, Nevada: Geological Society of America Bulletin, v. 110, no. 10, p. 1318-1332.

Allen, J.R.L., 1984, Sedimentary structures: their character and physical basis (second ed.): Developments in sedimentology: Amsterdam, Netherlands, Elsevier, 2 v., 1257 p.

Atwood, Genevieve, 1994, Geomorphology applied to flooding problems of closed-basin lakes... specifically Great Salt Lake, Utah: Geomorphology, v. 10, no. 1-4, p. 197-219.

Atwood, Genevieve, and Mabey, D.R., 1995, Flooding hazards associated with Great Salt Lake, *in* Lund, W.R., ed., Environmental and engineering geology of the Wasatch Front Region: Utah Geological Association, Publication 24, p. 483-493.

Atwood, Genevieve, and Mabey, D.R., 2000, Shorelines of Antelope Island as evidence of fluctuations of the level of Great Salt Lake, *in* King, J.K., and Willis, G.C., eds., The geology of Antelope Island, Davis County, Utah: Utah Geological Survey, Miscellaneous Publication 00-1, p. 85-97.

Carter, R.W.G., 1988, Coastal environments: London, UK, Academic Press, 617 p.

Chan, M.A., Currey, D.R., and Jewell, P.W., 2001, GeoAntiquities & Lake Bonneville: online, <http://www.geog.utah.edu/geohistory/index.html> accessed April 2001, homepage: University of Utah Department of Geography.

Davies, J.L., 1980, Geographical variation in coastal development (second ed.): New York, NY, Longman, 212 p.

Doelling, H.H., Willis, G.C., Jensen, M.E., Hecker, S., Case, W.F., and Hand, J.S., 1990, Geologic map of Antelope Island Davis County, Utah: Utah Geological Survey Map 127, 2 sheets, 1:24,000, 27 p.

Davis, R.A., 1996, Coasts: Upper Saddle River, NJ, Prentice Hall, 274 p.

Gilbert, G.K., 1890, Lake Bonneville: U.S. Geological Survey, Monograph 1, 438 p.

Gill, A.E., 1982, Atmosphere - ocean dynamics: international geophysics series: New York, NY, Academic Press, 428 p.

Great Salt Lake Planning Team, 2000, Great Salt Lake comprehensive management plan resource document: Utah Department of Natural Resources, 217 p.

Hansom, J.D., 1988, Coasts: Cambridge topics in geography: second series: Cambridge, UK, Cambridge University Press, 96 p.

Hecker, Suzanne, and Case, W.F., 2000, Engineering geology considerations for park planning, Antelope Island State Park, Davis County, Utah, *in* King, J.K., and Willis, G.C., eds., The geology of Antelope Island, Davis County, Utah: Utah Geological Survey, Miscellaneous Publication 00-1, p. 150-161.

Heaps, N.S., 1985, Tides, storm surges and coastal circulations, *in* P.P.G. Dyke, A.M., and E.H. Robson, ed., Offshore and coastal modelling: New York, NY, Springer-Verlag: Lecture notes on coastal and estuarine studies, p. 3 - 54.

Hutchinson, G.E., 1957, A treatise on limnology: New York, NY, John Wiley & Sons Inc, v. 1, Geography, physics, and chemistry.

Johnson, D.W., 1919, Shore processes and shoreline development: New York, NY, John Wiley & Sons, 584 p.

Katzenberger, W.M., and Whelan, J.A., 1975, Currents of Great Salt Lake: Utah Geology, v. 2, no. 2, p. 103 - 107.

Komar, P.D., 1998, Beach processes and sedimentation (second ed.): Upper Saddle River, NJ, Prentice Hall, 544 p.

Komen, G., Cavaleri, L., Donelan, M., Hasselmann, K., Hasselmann, S., and Janssen, P.A.E.M., 1994, Dynamics and modelling of ocean waves: Cambridge UK, Cambridge University Press, 532 p.

Lillquist, K.D., 1994, Late Quaternary Lake Franklin: lacustrine chronology, coastal geomorphology, and hydro-isostatic deflection in Ruby Valley and northern Butte Valley, Nevada: Salt Lake City, UT, University of Utah, Ph.D. dissertation, 185 p.

Lin, Anching, and Wang, Po., 1978, Wind tides of the Great Salt Lake: Utah Geology, v. 5, no. 1, p. 17-25.

Murchison, S.B., 1989, Fluctuation history of Great Salt Lake, Utah, during the last 13,000 years: Salt Lake City, UT, University of Utah, Ph.D. dissertation, 137 p.

Murchison, S.B., and Mulvey, W.E., 2000, Late Pleistocene and Holocene shoreline stratigraphy on Antelope Island, *in* King, J.K., and Willis, G.C., eds., The geology of Antelope Island, Davis County, Utah: Utah Geological Survey, Miscellaneous Publication 00-1, p. 76-83.

Open University Oceanography Course Team, 1989, Waves, tides and shallow-water processes (first ed.): Milton Keynes, UK, Pergamon Press, 187 p.

Patton, P.C., and Kent, J.M., 1992, A moveable shore: living with the shore: Durham, NC, Duke University Press, 143 p.

Pethick, J., 1984, An introduction to coastal geomorphology: London, UK, Edward Arnold, 260 p.

Pugh, D.T., 1987, Tides, surges and mean sea-level: Chichester, UK, John Wiley & Sons, 472 p.

Rich, J., 1991, Determination of circulation currents and their velocities in the South Arm of the Great Salt Lake: Salt Lake City, UT, University of Utah, M.S. Thesis, 101 p.

Rollins, Brown, and Gunnell, Inc. and Creamer & Noble Engineers, 1987, Great Salt Lake inter-island diking project - final design report - volumes 1 and 2: unpublished report for the Utah Division of Water Resources, variously paginated.

Smith, D.E., and Dawson, A.G., editors, 1983, Shorelines and isostasy: Institute of British Geomorphological Research Special Publication no. 16, London, UK, Academic Press, 387 p.

Steffen, C.C., 1983, The Great Salt Lake: major economic impacts of high lake levels: Utah Economic and Business Review, v. 43, no. 9-10, p. 1-7.

Stone, G.W., Armbruster, C.K., Grymes, J.M., and Huh, O.K., 1996, Researchers study impact of Hurricane Opal on Florida Coast: EOS, v. 77, no. May 7, 1996, p. 181.

Tackman, G.E., Bills, B.G., James, T.S., and Currey, D.R., 1999, Lake-gauge evidence for regional postglacial tilting in southern Manitoba: Geological Society of America Bulletin, v. 111, no. 11, p. 1684-1699.

U.S. Army Corps of Engineers, 1984, Shore protection manual (4th edition): U.S. Army Corps of Engineers Coastal Engineering Research Center, 2 vol.

—1996a, Davis County, Utah, 100-year stage, Great Salt Lake lakeshore delineation: U.S. Army Corps of Engineers, Sacramento District, memorandum, March 1996, 3 attachments, 5 p.

—1996b, Flood plain management services study for Davis County, Utah: U.S. Army Corps of Engineers, May 1996, Appendix A: hydrology; Appendix B: 62 maps; text 10 p.

U.S. Geological Survey, 2001, Great Salt Lake datum correction: online, <http://wwwdutslc.wr.usgs.about/news/gslcorr.html> accessed April 2001, homepage: USGS Water Resources Division.

Wang, Po, 1978, Seiches in the Great Salt Lake: Salt Lake City, UT, University of Utah, Ph.D. dissertation, 119 p.

Zenkovich, V.P., 1967, Processes of coastal development: New York, NY, John Wiley and Sons, 738 p.

Artist Joe Venus' depiction of a Pleistocene faunal scene in Utah, which is part of a mural he painted for the Paleontological Museum of the College of Eastern Utah in Price, Utah.

QUATERNARY VERTEBRATES OF THE NORTHEASTERN BONNEVILLE BASIN AND VICINITY OF UTAH

by

Wade E. Miller, Ph.D.
Department of Geology,
Brigham Young University, Provo, Utah 84602

ABSTRACT

Quaternary vertebrates continue to be found in the northeastern Bonneville basin, especially along the Wasatch Front. Many new discoveries over the past two decades have made it advisable to provide current information about them. This article serves mainly as a reference paper, updating earlier work. The recorded localities yielding these newly discovered fossils, with their identifications and represented elements, are its major contribution. A number of the sites and their accompanying specimens have not yet been studied in depth and offer significant research opportunities.

INTRODUCTION

A compilation of scientific, historical and economic reports pertaining to the Great Salt Lake and adjacent areas in Utah was published by the Utah Geological and Mineral Survey (Gwynn, 1980). One of the articles contained in that publication (Nelson and Madsen, 1980) listed Pleistocene fossil localities of the northern Bonneville basin (figure 1), and the vertebrates recovered from them. This article serves as an update to that article.

The entire Bonneville basin (area of Lake Bonneville, figure 2) includes an area of approximately 20,000 square miles (Hintze, 1988). In the past two decades, much construction has taken place in Utah, especially along the Wasatch Front (figures 3 - 7), which marked the eastern margin of Lake Bonneville. As a result of this activity, and more active pursuits in fossil collecting, many new fossil localities have been discovered, and in some instances new taxa have been uncovered from previously known localities. These new records are included in the present article. In order to provide continuity between this article and that of Nelson and Madsen (1980), a summary of previously listed Pleistocene localities in northern Utah with their reported fossils is given here in table 1. Also, the formatting used here approximates that used in the earlier article.

In many instances, it has not been possible to determine whether recovered specimens, especially those located in caves, are actual fossils or not. By definition, an organism is not regarded as a fossil unless it is Pleistocene or older in age (that is, older than 10,000 years). Therefore, this report is somewhat broader in scope than the earlier one, as it includes the entire Quaternary epoch of the past 2,000,000 years. It also contains information on cave faunas which are marginal to the northern Bonneville basin proper since the vertebrates recovered from these caves must have inhabited at least parts of the basin. Many of the cave specimens, however, are Holocene in age and offer important data relative to climatic changes. Grayson (1994) mentioned Holocene cave faunas and their importance, based on archaeological sites, in his paper on extinct late Pleistocene mammals from the Great

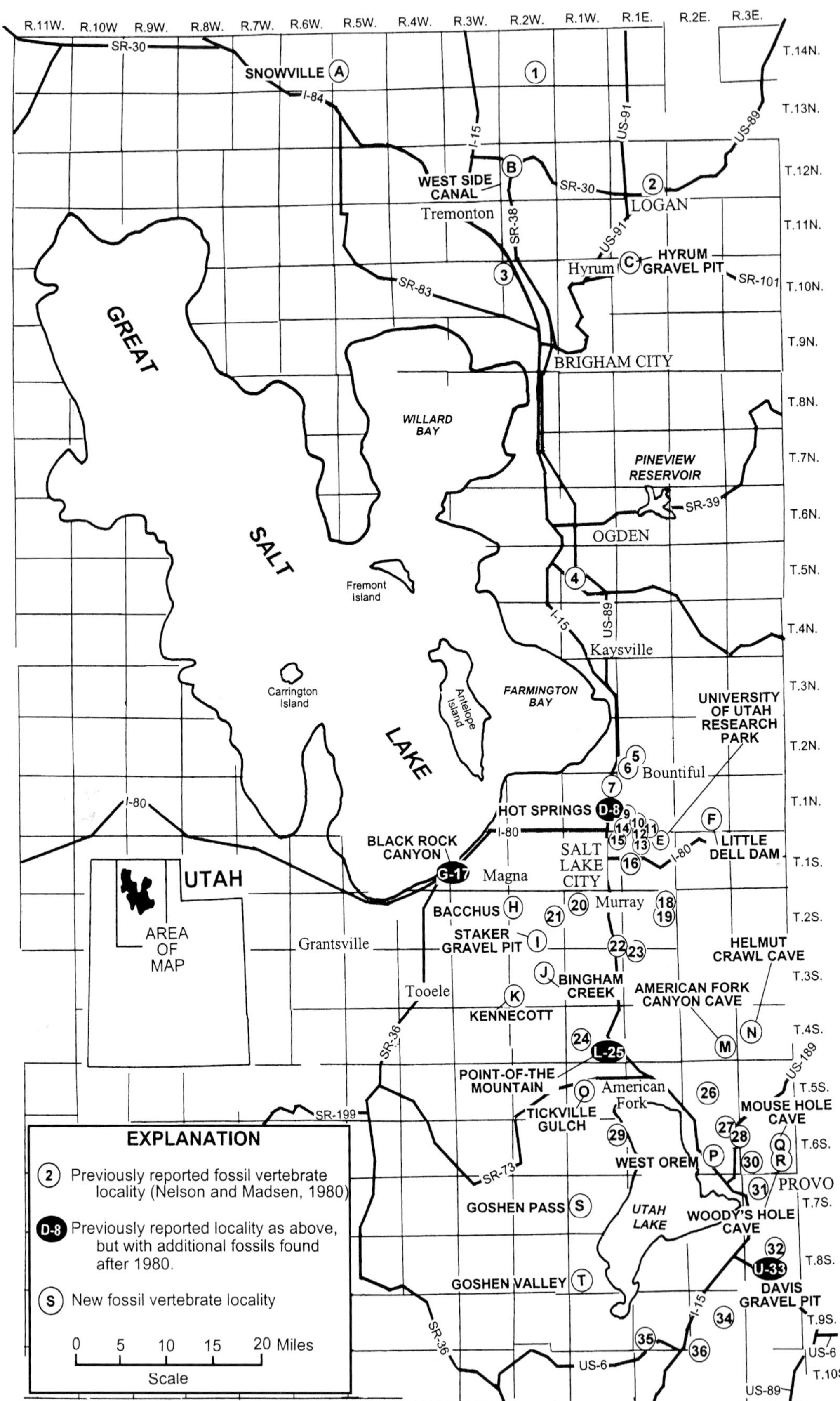

Figure 1. *Outline map showing Quaternary vertebrate localities in the northern Bonneville basin.*

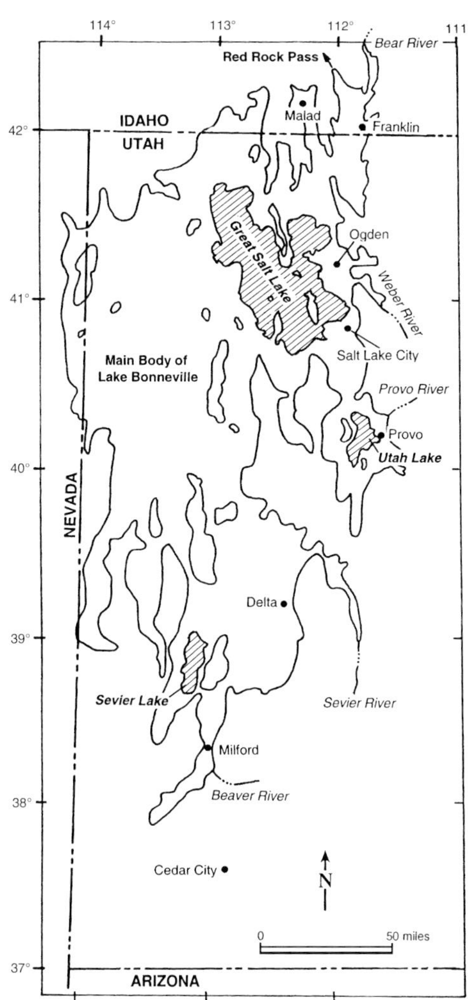

Figure 2. Map of Lake Bonneville, which indicates the present extent of the Bonneville basin, and remnant lakes. Most Quaternary vertebrate sites listed in this paper are located in the northeastern part of this basin (courtesy of James Miller).

Figure 3. View of the Wasatch Mountains east of Mapleton, with Hobble Creek Canyon at far left and Spanish Fork Canyon at far right. Lake Bonneville terrace deposits occur along the mountain base. Pleistocene musk ox and mammoth specimens were found in the deltaic/alluvial deposits shown in the lower right portion of photo (photo courtesy of Lehi Hintze).

Figure 4. *View of a portion of Salt Lake City looking north, with Salt Lake salient extending out from the Wasatch Front. Sand and gravel pits (some of which can be seen at the base of the salient) in this area have yielded various types of Pleistocene vertebrates (photo courtesy of Lehi Hintze).*

Figure 5. *View of the south end of Salt Lake Valley, looking southeast, showing a part of the Wasatch Front. Little Cottonwood Canyon is near the center of the photo. Pleistocene vertebrates have been found in lacustrine and alluvial sediments along the Front as well as in valley deposits (photo courtesy of Lehi Hintze).*

Figure 6. *View looking north at Point-of-the-Mountain which separates Salt Lake Valley (north) from Utah Valley (south). Sand and gravel pits are developed on either side of Interstate 15 (shown in center), with most Pleistocene vertebrates being recovered from pits to the west (left) of the freeway.*

Figure 7. A portion of the Geneva sand and gravel pit at Point-of-the-Mountain, looking west. Among the Pleistocene vertebrates discovered at this locality are: musk ox, mammoth, horse, bison and ground sloth.

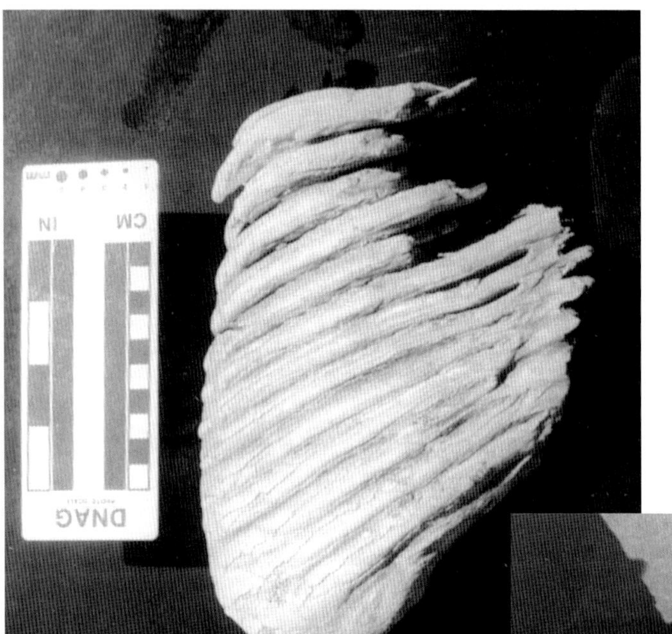

Figure 8. *Lateral view of a* Mammuthus *(mammoth) molar tooth.*

Figure 9. *Dorsal (occlusal) view of* Equus *(horse) lower jaw with nearly complete dentition.*

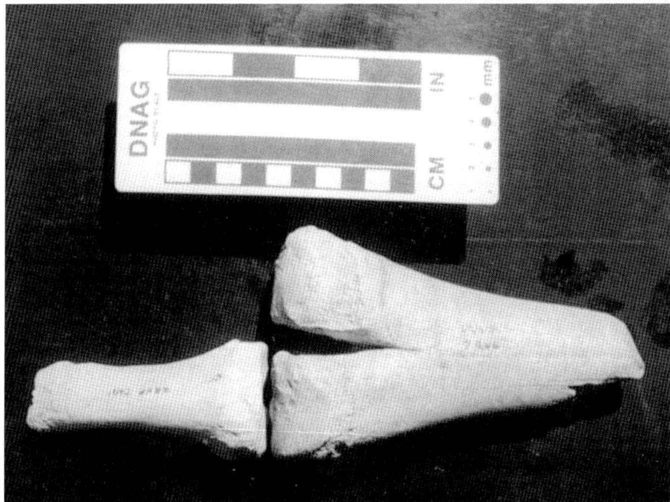

Figure 10. *Distal end of a metapodial and first phalanx (foot bones) of* Camelops *(American camel).*

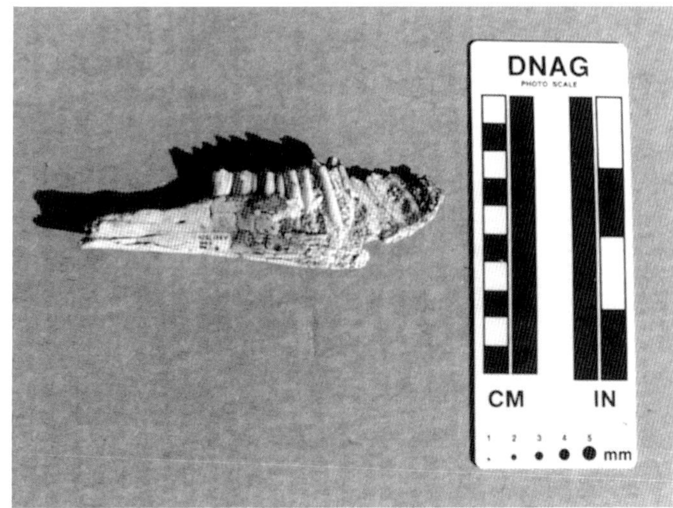

Figure 11. *Lateral view of a lower jaw of cf.* Tetrameryx *(extinct pronghorn).*

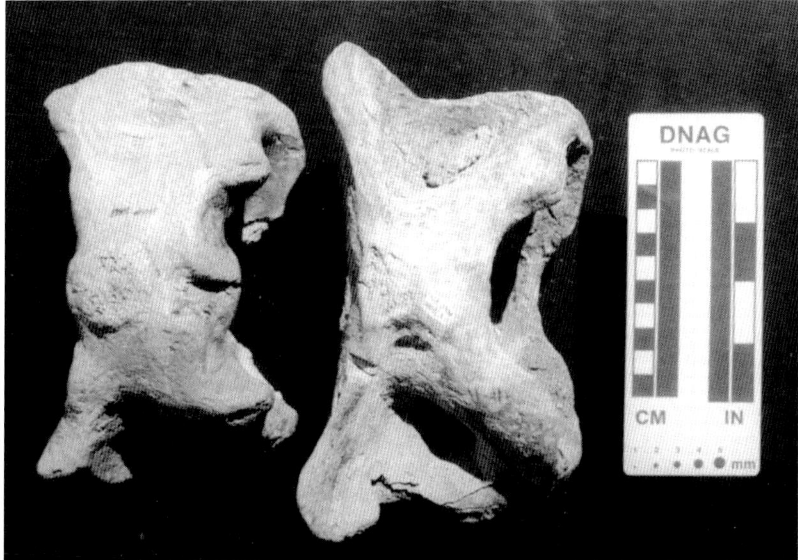

Figure 12. *Ventral view of atlas (first neck) vertebra of* Bootherium *(left) and Bison (right).*

Figure 13. *Dorsal view of male (left) and female (right)* Bootherium *(musk ox) incomplete skulls.*

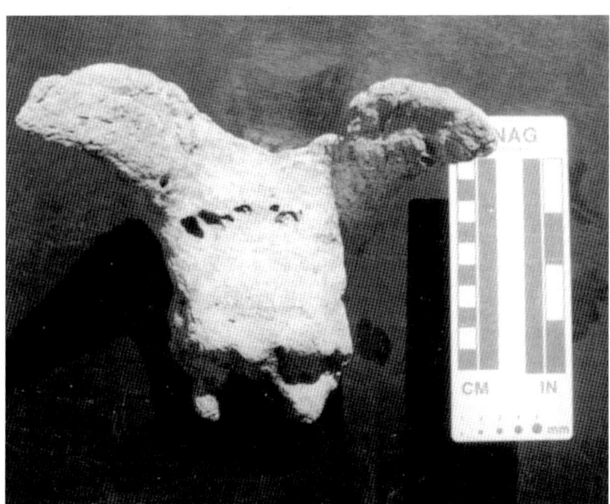

Figure 14. *Dorsal view of partial female* Ovis *(bighorn sheep) skull.*

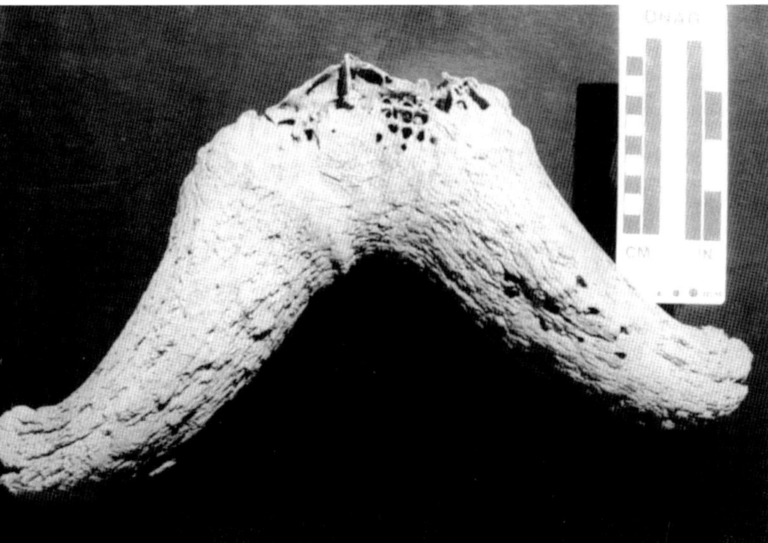

Figure 15. *Dorsal view of male* Ovis *(bighorn sheep) horn cores with attached frontal bones.*

TABLE 1. Updated locality and specimen list of Pleistocene vertebrates (as formatted in Nelson and Madsen, 1980). Site numbers here correspond to those shown as open-numbered circles on figure 1, and are the same as given in the above publication.

Locality	Scientific Name	Common Name	Element	Previous Publication
1. Clarkson, Utah; no exact locality	*Mammuthus* sp.	Mammoth	tusk fragment	Blackwelder, E., 1939
2. Logan City Cemetery	*Bootherium bombifrons* (=*Symbos cavifrons*)	Woodland musk ox	cranial fragment	Nelson, M. E., and Madsen, J. H., 1978
3. Bear River City; no exact locality	*Mammuthus* sp.	Mammoth	molar	Blackwelder, E., 1939
4. Weber River	*Equus caballus*	Horse	none listed	Feth, J. H., 1955
5. Warner Gravel Pit; near Bountiful	*Equus* sp.	Horse	none listed	Pack, F. J., 1939
6. Foss-Lewis Sand and Gravel, Woods Pit; near Bountiful	*Ovis* sp. *Bootherium bombifrons* (=*Symbos cavifrons*)	Bighorn sheep Woodland musk ox	partial skull partial cranium	Nelson, M. E., and Madsen, J. H., 1978
7. Concrete Products, White Hill Pit	*Ovis* sp. *Bootherium bombifrons*	Bighorn sheep Woodland musk ox	skull and horn cores horn core and ? skull fragment	Nelson, M. E., and Madsen, J. H., 1978
8. Monroc Gravel Pit, North Salt Lake City	*Ovis* sp.	Bighorn sheep	partial skull, horn core scapula	Nelson, M. E., and Madsen, J. H., 1978
9. North Bench, North Salt Lake City	*Ovis* sp.	Bighorn sheep	skull and horn cores	none
10. Hardman Gravel Pit, Salt Lake City	*Bison* sp. *Camelops* cf. *hesternus* *Odocoileus hemionus* *Ovis* sp. *Ovis catclawensis* *Ovis canadensis* *Bootherium bombifrons* (=*Symbos cavifrons*)	Buffalo Western camel Mule deer Bighorn sheep Extinct Bighorn sheep Bighorn sheep Woodland musk ox	two skulls jaws and teeth skull (? Recent) skulls at least 10 specimens none listed skull	Pack, F. J., 1939 Stokes, W. L., and Condie, K. C., 1961 Stock, A. D., and Stokes, W. L., 1969 Nelson, M. E., and Madsen, J. H., 1978
11. University of Utah Medical Center	? *Citellus*	Ground squirrel	ulna	none
12. Fort Douglas; east Salt Lake City	*Bootherium bombifrons* (= *Symbos cavifrons*)	Woodland musk ox	tibia	none
13. Fort Douglas, east Salt Lake City	*Ovis*. sp.	Bighorn sheep	horn core fragment	none
14. Mouth of City Creek Canyon	cf. *Mammuthus* *Ovis canadensis*	Mammoth Bighorn sheep	tusk fragments partial skull	none Stokes, W. L., and Condie, K. C., 1961
15. Downtown Salt Lake City	*Bootherium bombifrons* (= *Symbos cavifrons*)	Woodland musk ox	partial skull	Chadbourne, P. A., 1871; Hay, O. P., 1927
16. Ure mammoth locality	*Mammuthus* sp.	Mammoth	partial lower molar	none

Locality	Scientific Name	Common Name	Element	Previous Publication
17. Black Rock Canyon	*Ovis* sp.	Bighorn sheep, bird	partial skull and metapodial	Nelson, M. E., and Madsen, J. H., 1978
18. Southeast Salt Lake City	*Bootherium bombifrons*	Woodland musk ox	dentary	none
19. Harper Gravel Pit	cf. *Mammuthus*	Mammoth	tusk fragments	none
20. Sorensen Gravel Pit	*Bison antiquus*	Extinct buffalo	partial skull with horn cores	Stokes, W. L., et al., 1966
21. Monroc Gravel Pit, Kearns	*Bootherium bombifrons* (= *Symbos cavifrons*)	Woodland musk ox	partial skull	Nelson, M. E., and Madsen, J. H., 1978
22. Sandy mammoth locality	*Mammuthus* cf. *columbi*	Mammoth	partial skeleton	Madsen, D. B., et al., 1976
23. Sandy	*Camelops* cf. *hesternus*	Western camel	axis	none
24. Merrico Gravel Pit, Bluffdale	*Equus* sp. *Bison* sp.	Horse Buffalo	vertebrae vertebrae	none
25. Point of the Mountain	*Bootherium bombifrons* (= *Symbos cavifrons*)	Woodland musk ox	partial skull	none
26. Near Pleasant Grove	*Bootherium bombifrons* (= *Symbos cavifrons*)	Woodland musk ox	cranial fragments	Nelson, M. E., and Madsen, J. H., 1978
27. Orem, no exact locality	*Bootherium bombifrons* (= *Symbos cavifrons*)	Woodland musk ox	vertebrae	Stokes., D. L., and Hansen, G. H., 1937
28. Bonneville shoreline; no exact locality	*Bison latifrons* or *B. antiquus*	Extinct buffalo	skull fragment	King, C., 1878
29. Utah Lake (northwest side)	*Thomomys talpoides*	Gopher	jaw	Hunt, C. B., et al., 1962
30. Provo; no exact locality	*Bootherium bombifrons* (= *Symbos cavifrons*)	Woodland musk ox	partial skull with horn cores	Nelson, M. E., and Madsen, J. H., 1978
31. Slate Canyon	*Bootherium bombifrons* (= *Symbos cavifrons*)	Woodland musk ox	four partial crania	Nelson, M. E., and Madsen, J. H., 1978
32. Springville; no exact locality	*Bootherium bombifrons* (= *Symbos cavifrons*)	Woodland musk ox	atlas and axis	Nelson, M. E., and Madsen, J. H., 1978
33. Spanish Fork; no exact locality	*Mammuthus* sp.	Mammoth	skeleton, but just teeth collected	Pack, F. J., 1939
34a. Payson; no exact locality	*Mammuthus* sp.	Mammoth	limb bones	Pack, F. J., 1939
34b. Payson; no exact locality	*Mammuthus* sp.	Mammoth	tusks and other skeletal elements	Hansen, G. H., 1928
34c. Payson; no exact locality	*Mammuthus* sp.	Mammoth	partial jaw and isolated bones	Bissell, H. J., 1963, 1968
35. Johnson Gravel Pit	*Bootherium bombifrons*	Woodland musk ox	partial cranium with left horn core	none
36. Santaquin Gravel Pit	*Bootherium bombifrons* (=*Symbos cavifrons*)	Woodland musk ox	partial cranium	Bissell, H., J., 1963; Helson, M.E., and Madsen, J.H., 1978

Basin. A catalogue of late Quaternary vertebrates from all Utah, including cave faunas, was recently published (Jefferson, and others, 1994). However, the scope of the Jefferson article precluded some of the details being presented here.

Several Quaternary faunas recovered from caves and rock shelters have been reported from northwestern Utah. These are not listed in the "annotated locality and specimen list" section. While some of the recovered specimens are apparently late Pleistocene in age, most are Holocene. Pole Creek Cave, located in the extreme northwest corner of Utah (Box Elder County), has produced a small fauna consisting of lagomorph (rabbit), rodent, carnivore, bison, and mountain sheep. Hogup Cave, on the west side of the Hogup Mountains in northwestern Box Elder County, has yielded a diverse vertebrate fauna which extends back 8,500 years (Durrant, 1970). Other cave faunas in northwestern Utah come from Tooele County. These are associated caves that occur near Wendover, Utah: Danger Cave (Jennings, 1957; Grayson, 1987), Juke Box Cave (Jennings, 1957), and several smaller caves (Grayson 1987, 1988; Jefferson and others, 1994). Included animals are: Chiropterans, lagomorphs, rodents, carnivores, and ungulates (hoofed mammals) which usually constitute the bulk of the faunas. However, fish, reptiles, and birds are also represented.

A few important Quaternary vertebrate localities exist in northeastern Utah aside from those of the Bonneville basin. The most important of these are the Blonquist Rockshelter (Nelson, 1988), Silver Creek Junction in Summit County (Miller, 1976), and Bear Lake in Rich County (Jefferson and others, 1994). These three localities have produced a moderately diverse vertebrate fauna representing indigenous taxa of northern Utah's highlands. Included among the extinct forms are: a ground sloth, dire wolf, saber-toothed cat, mammoth, horse, camel and giant bison. A sinkhole located on the Wasatch Plateau, in northeast Sanpete County, also yielded numerous elements of two mastodons (Miller, 1987). Smaller animals recovered in the above local faunas, such as fish, frogs, lizards, snakes, birds, and small to medium-sized mammals, still inhabit the region today. Causes for extinctions are currently being debated, but climatic changes during the past 10,000 to 12,000 years must have been a major factor. Man also could have played a significant role according to several paleontologists (Martin and Klein, 1984).

METHODS

Previously published articles pertaining to Quaternary vertebrates of Utah serve as a basis for this article. The articles by Nelson and Madsen (1980, 1987) and Jefferson and others (1994) were especially useful. Additionally, unstudied specimens housed at Brigham Young University in Provo, and at the Utah Geological Survey in Salt Lake City, were examined in order to make the present work more complete. Other state institutions were also contacted to learn whether unreported Quaternary vertebrates from the Bonneville basin existed elsewhere. None were located.

Since the Paleontological Section of the Utah Geological Survey maintains an alpha-numerical fossil-locality numbering system for all sites within the state, regardless of separate institutional numbering systems, this is the numbering system used here. In this system, the number 42 represents Utah (the 42nd state), followed by the county abbreviation, the cumulative fossil-site number for that county, and a terminal letter(s) indicating the broad category of fossils. For example, the letter "V" represents a vertebrate site. Locality data-of-record for the various sites is not always complete.

Available information for each new or revisited site appears under the location headings listed below. Each of the Quaternary vertebrate sites listed in this article is shown alphabetically (by letter in open circle) in figure 1. This figure approximates figure 1 of Nelson and Madsen (1980), but has been updated for use in this article. Figure 1 also shows (by number in open circle) all the previous localities from Nelson and Madsen (1980), and finally, it shows previously reported sites which have produced new taxa since 1980 (solid ovals with white letters).

ANNOTATED LOCALITY AND SPECIMEN LIST

Bacchus (42SL126V) Salt Lake County. H

Nelson, M. E., and Madsen, J. H, 1980; 1986.

Location: NW 1/4, Sec. 8, T2S, R2W

Identified taxa: Mammalia (mammals), *Vulpes vulpes* (fox), jaws; *Bootherium bombifrons* (woodland musk ox), partial skull.

Age: Pleistocene.

Comments: While fossils of the Woodland musk ox have been recovered from several sites along the eastern margins of Lake Bonneville, the presence of red fox at the Bacchus locality makes it the first record of this animal as a fossil in Utah.

Bingham Creek (UT42SL132V) Salt Lake County. J

Jefferson, G. T. and others, 1994. Material undescribed.

Location: NE 1/4, Sec. 14, T3S, R2W.

Identified taxa: Mammalia (mammals), *Bison ?antiquus* (buffalo), partial dentary and phalanx.

Age: Late Pleistocene.

Comments: If upon study it can be determined that the bison material collected here comes from the extinct *Bison antiquus*, then the late Pleistocene age can be confirmed.

Black Rock Canyon (42SL071V) Salt Lake County. G

Nelson, M. E., and Madsen, J. H., 1980; 1987.

Location: SW 1/4, Sec. 20, T1S, R3W.

Identified taxa: Aves (birds), *Phalacrocorax macropus* (cormorant), postcranial bones; Mammalia (mammals), *Navahoceros* cf. *fricki* (deer), metapodial; *Ovis* cf. *canadensis* (sheep), partial skull and partial metapodial.

Age: Late Pleistocene.

Comments: A fossil big-horn sheep was earlier reported from this site. The cormorant found here is a rarity as a fossil from Utah. The extinct large deer recognized at Black Rock Canyon (*Navahocereos*) appears to be the only record of this cervid from the state.

Helmet Crawl Cave (42Ut438V) Utah County. N

Jefferson, G. T. and others, 1994. Fauna unstudied.

Location: Near Tibble Fork Reservoir, American Fork Canyon.

Identified taxa: Aves (birds), wing bones and synsacra; Mammalia (mammals), *Lepus* cf. *americanus* (rabbit), skull, jaws, and postcranial bones; *Sciurus* sp. (squirrel), skull, jaws and postcranial bones; ?*Citellus* (chipmunk) partial skull, jaws and postcranial bones; *Marmota* cf. *flaviventris* (marmot), skulls, jaws and postcranial bones; *Neotoma* cf. *cinerea* (pack rat), skulls, jaws and postcranial bones; ?*Peromyscus* (deer mouse), jaw; *Microtus* sp. (vole), skull fragments, jaws and postcranial bones; *Mephitis* cf. *mephitis* (skunk), skull; *Mustela* sp. (weasel), partial skull; *Vulpes* sp. (fox), partial jaw; *Ursus* cf. *americanus* (bear), metacarpal; *Odocoileus* cf. *hemionus* (deer); ?*Ovis* (sheep), partial maxilla, partial dentary, and postcranial bones; *Bison* sp. (buffalo), phalanges.

Age: Latest Pleistocene and/or Holocene.

Comments: Without radiometric dating the age of the fauna from this locality remains uncertain. The probability of a Holocene age seems likely, however.

Tickville Gulch (42Ut431V) Utah County. O

Jefferson, G. T. and others, 1994. Fauna unstudied.

Location: SW 1/4, Sec. 20, T5S, R1W

Identified taxa: Aves (birds), Ciconiiformes (wading birds), carpometacarpus; Reptilia (reptiles), Lacertilia (lizards), partial jaw; Mammalia (mammals), *Lepus* sp. (rabbit), partial skull, jaws and postcranial bones; *Sylvilagus* sp. (cottontail), premaxilla; *Sciurus* sp. (squirrel), postcranial bones; *Neotoma* sp. (pack rat), tooth and postcranial bones; ?mustelid, deciduous tooth; ?*Odocoileus* sp. (deer), partial humerus.

Age: Latest Pleistocene and/or Holocene.

Comments: The vertebrate material here was mostly recovered from two to three meters below the existing ground surface, but shows no sign of permineralization. However, since many of the known Pleistocene bones also show no permineralization, this cannot be used as a diagnostic condition to determine fossilization.

Woody's Hole Cave (42Ut440V) Utah County. R

Jefferson, G. T. and others, 1994. Fauna unstudied.

Location: Southside of Rock Canyon; SE 1/4, Sec. 28, T6S, R3E

Identified taxa: Aves (birds), postcranial bones; Reptilia (reptiles), Serpentes (snakes), vertebrae; Mammalia (mammals), *Sylvilagus* sp. (cottontail), postcranial bones, *Spermophilus* sp. (ground squirrel), skull fragments, jaw fragments and postcranial bones; *Thomomys umbrinus* (gopher), palate and jaws; *Neotoma* cf. *cinerea* (pack rat), partial skulls, jaws and postcranial bones; *Microtus* cf. *longicaudus* (vole), palates and jaws (?postcranial bones); *Martes* cf. *americana* (marten), jaws: *Odocoileus* cf. *hemionus* (deer), antler fragment and maxilla; *Ovis* cf. *canadensis* (sheep), jaw fragments and postcranial bones.

Age: Latest Pleistocene and/or Holocene.

Comments: The age of this mostly unstudied fauna remains unknown, but since all the identified taxa still live in the region, it appears to be Holocene.

Mouse Hole Cave (42Ut439V) Utah County. Q

Jefferson, G. T. and others, 1994. Fauna unstudied.

Location: North side of Rock Canyon; SE 1/4, Sec. 28, T6S, R3E

Identified taxa: Osteichthyes (bony fish), skull and jaw bones and vertebrae; Aves (birds), partial skulls and poscranial bones; Mammalia (mammals), *Anthozous* cf. *palidus* (bat), partial skull; *Lepus* cf. *americanus* (rabbit), partial skull, jaws and postcranial bones; *Spermophilus* cf. *varigatus* (ground squirrel), skulls, jaws and postranial bones; *Castor* cf. *canadensis* (beaver), partial jaw; *Erethizon* cf. *dorsatum* (porcupine), skull; *Neotoma* cf. *cinerea* (pack rat), skulls, jaws and postcranial bones; *Canis* cf. *lupus* (wolf), maxilla; *Ursus* cf. *americanus* (bear), metapodials and phalanx; *Odocoileus* cf. *hemionus* (deer), maxilla and postcranial bones; *Ovis* cf. *canadensis* (sheep), jaws and postcranial bones.

Age: Latest Pleistocene and/or Holocene.

Comments: A possibility exists that this fauna may extend in chronological range from latest Pleistocene to Holocene. Radiometric dating would be needed to determine exact dates.

West Side Canal/Cutler Dam (42Bo049V) Box Elder County. B

Feduccia, A., and Oviatt, C. G., 1986; Oviatt, C. G. and others, 1987. Fauna partially studied.

Location: One locality 10 km south of town of Fielding, the other two 4.0 to 4.5 km east, and northeast of Fielding

Identified taxa: Osteichthyes (bony fish), *Salmo* sp. (trout), postcranial bones; *Gila atraria* (chub), unlisted bones; Aves (birds), *Cygnus buccinator* (swan), postcranial bones; Mammalia (mammals), *Mammuthus* sp. (mammoth), unlisted bones; unidentified large mammal, bone fragments.

Age: Late Pleistocene.

Comments: Bones listed from this locality were actually recovered at sites several kilometers apart in pre-Bonneville lacustrine deposits. A ^{14}C date run on wood from these beds indicates an age in excess of 36,000 yrs. B. P.

Snowville (42Bo277V) Box Elder County. A

Mead, J. I. and others, 1992; Repenning, C. A., 1987.

Location: Near town of Snowville, no exact locality.

Identified taxon: Mammalia (mammals), *Mictomys borealis* (lemming), two jaws.

Age: Late Pleistocene.

Comments: *Mictomys* is a very rare fossil from the western United States. The one from this locality, the furthest north Pleistocene locality reported from Utah, probably represents a new species. Other unreported small mammal specimens apparently have been discovered at this site (Chris Bell, Personal Communication).

American Fork Canyon Cave (42Ut014V) Utah County. M

Hansen, G. H., and Stokes, W. L., 1941. Fauna not fully studied.

Location: North side of American Fork Canyon, 43 m (140 ft) above river bed, about 5 km (3 miles) up canyon (east) of Timpanogos Cave National Monument offices.

Identified taxa: Mammalia (mammals), unidentified bat; *Cynomys leucurus* (prairie dog); *Marmota flavaventris* (marmot); *Erethizon dorsatum* (porcupine); *Neotoma lepida* or *N. cinerea* (desert or bushytail woodrat); *Lynx rufus* (bob cat); *Martes* sp. (marten); *Mustela frenata* (weasel); *Spilogale* sp., (skunk); *Ursus* sp. (bear); *Cervus elaphus* (elk); *Odocoileus hemionus* (deer); *Ovis canadensis* (sheep).

Age: Late Pleistocene and/or Holocene.

Comments: As with the other cave faunas listed in this article, the age(s) remain uncertain. Unfortunately, the present location of the specimens is unknown. While the cave exists above the highest level of Lake Bonneville, the high water mark would have put the shoreline reasonably close to it.

Davis Gravel Pit (42Ut437V) Utah County. U

Jefferson, G. T. and others, 1994. Fauna unstudied.

Location: Immediately northwest of the mouth of Spanish Fork Canyon

Identified taxa: Mammalia (mammals), ?*Mammuthus* (mammoth), tusk section; *Bootherium bombifrons* (woodland musk ox), partial skull with horn core.

Age: Late Pleistocene.

Comments: The two specimens recovered from this site were not found at the same spot, but apparently came from the same general area (within two hundred meters in the same Bonneville deposit).

Hot Springs (42SL006V) Salt Lake County. D

Smith, G. R and others, 1968; Nelson, M. E., and Madsen, J. H., 1987.

Location: 820 m southwest of the summit of Ensign Peak, and 120 m east of Victory Road; SE 1/4, Sec. 25, T1N, R1W.

Identified taxa: Osteichthyes (bony fish), *Salmo clarkii* (trout), skull and jaw bones and vertebrae; *Prosopium gemmiferum* (whitefish), skull and jaw bones; *Prosopium spilonotus* (whitefish), skull bones and vertebra; *Gila atraria* (chub), skull bones; *Catostomus ardens* (sucker), skull bones; *Cottus bairdii* (sculpin), skull bones; *Cottus extensus* (sculpin), skull and jaw bones.

Age: Late Pleistocene.

Comments: This is one of the few Pleistocene localities in Utah that has yielded identifiable fish. Those recovered here were associated with hot spring activity that coated hard materials such as bone, providing a better chance for fossilization here.

Point-of-the-Mountain (42SL140 & 141V) Utah County. L

Nelson, M. E., and Madsen, J. H., 1980; Jefferson, G. T and others, 1994. Fauna partially studied.

Location: Just west of I-15, southernmost Salt Lake County, in the Geneva Gravel Pit; NE 1/4, Sec. 23, T4S, R1W.

Identified taxa: Mammalia (mammals), *Megalonyx* cf. *jeffersoni* (ground sloth), partial skeleton; *Mammuthus* cf. *columbi* (mammoth), tooth; *Equus* large sp. (horse), postcranial bones; *Bison* sp. (buffalo), partial skeleton; *Bootherium bombifrons* (woodland musk ox), partial skulls and postcranial bones; *Ovis* cf. *canadensis* (sheep), skull fragments and postcranial bones.

Age: Late Pleistocene.

Comments: Fossil material at this Geneva gravel pit has been collected over a multi-year period. Preservation is especially good, with two partial skeletons (ground sloth and bison) being uncovered. These fossils have mostly come from a level about 9 m (30 ft) below the original land surface (before quarrying). Mr. Richard Trotter and Mr. Joseph Miller have been especially helpful in recovering fossils at this gravel pit where they work.

Goshen Pass (42Ut432V) Utah County. S

Jefferson, G. T. and others, 1994. Fauna unstudied.

Location: W 1/2, Sec. 19, T7S, R1W

Identified taxa: Mammalia (mammals), small rodents, teeth and postcranial bones.

Age: Late Pleistocene or Holocene

Comments: Specimens from this locality were collected from anthills.

Goshen Valley (42Ut433V) Utah County. T

Jefferson, G. T and others, 1994. Fauna unstudied.

Location: SW 1/4, Sec. 30, T8S, R1W

Identified taxa: Mammalia (mammals), rodents and lagomorph(s), postcranial bones.

Age: Late Pleistocene or Holocene

Comments: The material collected at this site came from exposures produced by a stream channel cut into possible Bonneville sediments.

Kennecott (42SL135V) Salt Lake County. K

Gillette, and Miller, 1999. Partially studied fauna.

Location: Below Bingham Canyon Open Pit Copper Mine: NE 1/4, Sec. 30, T3S, R2W.

Identified taxa: Reptilia (reptiles), Lacertilia (lizards), partial jaw; Mammalia (mammals), *Neotoma* sp., (pack rat), teeth; *Microtus* sp. (vole), teeth; unidentified rodents, crania and postcranial bones; *Equus* large and small sp. (horse), incomplete jaws with dentition, isolated teeth, and postcranial bones; cf. *Camelops* sp. (camel), postcranial bones; cf. *Tetrameryx* sp. (extinct pronghorn), jaws with dentition, and vertebrae. Additionally, small mammal material is present but unidentified.

Age: Pleistocene.

Comments: The Kennecott local fauna, like that of Little Dell, is above the Bonneville shoreline, but differs from it by having coexisted with the lake. It contains a rare and unusual taxon from Utah, the pronghorn *Tetrameryx*. This site no longer exists due to excavations.

Little Dell Dam (42SL133V) Salt Lake County. F

Gillette, Bell, and Hayden, 1999.

Location: Little Dell Dam in Mountain Dell Canyon, 16 km east of Salt Lake City; NW 1/4, Sec. 36, T1N, R2E.

Identified taxa: Mammalia (mammals), soricid (shrew), jaw fragment; *Spermophilus* sp. (ground squirrel), molar tooth; *Eutamius* sp. (chipmunk), molar teeth; *Thomomys* sp. (pocket gopher), lower dentition; *Neotoma* sp. (pack rat), partial molar; *Peromyscus* sp. (white-footed mouse), molar; *Perognathus* sp. (pocket mouse), molar; *Lemmiscus curtatus* (sagebrush vole), molar; *Microtus paroperarius* (extinct vole), molar; *Allophaiomys pliocaenicus* (steppe vole), upper and lower dentition; *Mictomys* sp. (bog lemming), lower dentition; *Phenacomys* cf. *gryci* (heather vole), molars; *Mimomys* cf. *dakotaensis* (extinct vole), molar; *Zapus* sp. (jumping mouse), molars; many unidentified rodent postcranial bones; proboscidean (mammoth or mastodon), tusk and bone fragments; *Equus* sp. (horse), tooth and postcranial bones.

Age: Pleistocene (possibly Irvingtonian).

Comments: This locality has three closely spaced sites that produce the above fossils. It is unique and important as possibly representing the only reported early Pleistocene (Irvingtonian Mammal Age) fauna in Utah. Plant and invertebrate fossils also were recovered from this site. If it is of

Irvingtonian Age, this fauna would pre-date Bonneville deposits, and at an elevation of 1,768 m, it is well above the highest Bonneville shoreline.

Hyrum Gravel Pit (42Ca210V) Cache County. C

Gillette and Miller, 1999.

Location: Blacksmith Fork Delta of Lake Bonneville, NW 1/4, Sec. 3, T10N, R1E.

Identified taxon: Mammalia (mammals), cf. *Mammuthus* sp. (mammoth), complete tusk.

Age: Late Pleistocene.

Comments: The complete tusk recovered from near Hyrum, Utah is currently on display at the Geology Department of Utah State University in Logan.

Gravel Pit (42Ca211V). Cache County. (No site locality data)

Gillette and Miller, 1999.

Location: No exact locality data available.

Identified taxon: Mammalia (mammals), *Bootherium bombifrons* (woodland musk ox), mandible.

Age: Late Pleistocene.

Comments: A Geology instructor at Utah State University brought this specimen to the Utah Geological Survey office of the paleontologist in Salt Lake City for identification. It now is housed at the former institution.

West Orem (42Ut430V) Utah County. P

Gillette, McDonald, and Hayden, 1999.

Location: South side of University Parkway, 1.1 km east of I-15 exit 272. NE 1/4, Sec. 27, T6S, R2E.

Identified taxon: (Mammalia) mammals, *Megalonyx jeffersoni* (Jefferson's ground sloth), teeth, partial postcranial skeleton).

Age: Late Pleistocene.

Comments: The partial ground sloth skeleton recovered in West Orem is the first record of *Megalonyx* from Utah. It was recovered from the Provo shoreline terrace in high-energy beach deposits of Lake Bonneville.

University of Utah Research Park - Huntsman Site (42SL136V) Salt Lake County. E

Gillette andMiller, 1999.

Location: Below Bonneville Shoreline Trail at mouth of George's Hollow; Sec. 3, T1S, R1E.

Identified taxon: Mammalia (mammals), *Bootherium bombifrons* (woodland musk ox), horn core and partial postcranial skeleton.

Age: Late Pleistocene

Comments: The musk ox material from this location appears to all belong to a single individual. It was discovered while excavating the building site for the Huntsman Chemical Corporation headquarters.

Staker Gravel Pit (42SL138V) Salt Lake County. I

Gillette and Miller, 1999.

Location: East of Utah Highway 111 at 7400 South, Lark, Utah; Center section 27, T2S, R2W.

Identified taxon: Mammalia (mammals), *Bootherium bombifrons* (woodland musk ox), partial skull with horn cores.

Age: Late Pleistocene

Comments: This skull was collected in deltaic sands and gravels on a Bonneville shoreline terrace. Gastropod shells were also collected at this site near the Bingham copper mine.

Kennecott Gravel Pit (42SL137V) Salt Lake County. (No site locality data)

Gillette and Miller, 1999.

Location: No exact locality data available; Kennecott Corporation Property

Identified taxon: Mammalia (mammals), cf. *Bootherium bombifrons* (woodland musk ox), ? Mastoid bone and vertebra

Age: Late Pleistocene

Comments: In the past few years several finds of Pleistocene mammals have been made in the vicinity of Kennecott's Bingham copper mine. Many, like the specimens from this site, were uncovered while excavating sand and gravel.

SUMMARY

In the summary given by Nelson and Madsen (1980), it was stated that, "Although fossil remains of mammals from Lake Bonneville sediments are somewhat common, the fauna is not very diverse." Nearly two decades later, despite the many additional fossils recovered from these sediments, this statement remains valid. This condition relates to the fact that most Quaternary fossil discoveries have come about through chance finds in excavations, especially in sand and gravel pits. Equipment operators rarely would see anything but large vertebrates. Should large-scale, fine-screening operations take place, an array of medium sized to small forms would undoubtedly be found. This prediction is based on reasonably diverse smaller animals being found in cave faunas of northern Utah (Jefferson and others, 1994), and where screening operations of fossil-bearing Pleistocene sediments have been performed (Miller, 1976).

Almost all of the large Pleistocene mammals found in Utah have beeen reported from the area of Lake Bonneville. Represesntative fossils can be seen in figures 8 to 15. The larger Quaternary fossil vertebrates of the northern Bonneville basin include a ground sloth (*Megalonyx*), giant bear (*Arctodus*), black bear (*Ursus*), mammoth (*Mammuthus*), two species of horse (*Equus*), camel (*Camelops*), extinct large deer (*Navahoceros*), mule deer (*Odocoileus*), extinct pronghorn (*Tetrameryx*), two extinct species of buffalo (*Bison*), woodland musk ox (*Bootherium*), and mountain sheep (*Ovis*). Skulls of the latter two taxa are unusually common relative to the numbers of overall Pleistocene fossils known to date in Utah. Both have exceptionally numerous sinus cavities in the skull and horn cores (see figures 13, 14, and 15). Stokes and Condie (1961) commented that mountain sheep might have skulls that floated on Lake Bonneville until they sunk and became buried in sediments; thus preserving them at a higher ratio compared to other bones. As stated above, the medium and small-sized vertebrates are mostly known from cave faunas. These include a variety of fish, amphibians (mostly frogs), lizards, snakes, birds, carnivores, rodents and rabbits. Compared to Pleistocene faunas of other western states (e.g., the Rancho La Brea fauna of southern California), the one from northern Utah must be considered incomplete. Many more vertebrate taxa, including a variety of extinct forms, undoubtedly were present during the Pleistocene Epoch here. By more fully understanding the complete Quaternary vertebrate fauna of northern Utah, information will be gained concerning climatic and ecological conditions that existed during this important chronological interval.

ACKNOWLEDGMENTS

Dr. Michael Nelson and Mr. James Madsen, Jr. are to be commended for compiling data on Pleistocene vertebrates from the northern Bonneville basin, and for encouraging others to further this work in order to gain a more complete understanding of Ice Age life in the region. Special thanks to the dozens of people who, upon finding fossils, brought them to the attention of professional paleontologists, enabling these finds to be reported. Many important fossils have been found by amateurs who are here gratefully thanked.

Dr. David Gillette, and Ms. Martha Hayden of the Utah State Paleontology Office, Utah Geological Survey, provided much valuable assistance in obtaining needed records and materials. Don Burge of the College of Eastern Utah's Paleontological Museum kindly granted permission to use much of a commissioned mural at that museum. The artist, Joe Venus of Sandy, Utah, cooperated in every way making his depictions of restored Pleistocene mammals available for the present paper. Both are greatly thanked for their help. Dr. Lehi Hintze of Brigham Young University kindly provided geological photos of the Wasatch Front used in this article (from his *Geologic History of Utah*, 1988). J. Wallace Gwynn of the Utah Geological Survey updated the index map (figure 1), for which deep thanks are expressed. Dr. Laurel Casjens, and Ms. Janet Gillette of the Utah Museum of Natural History, Salt Lake City, kindly made it possible to use the Pleistocene scene drawn by Ms. Doris Tischler that is shown in the special color section of this publication. The original mural is on display at the Utah Museum of Natural History. Ms. Tischler is greatly thanked for this work. Ms. Leena Rogers typed the manuscript, for which much appreciation is given.

REFERENCES

Bissell, H. J., 1963, Lake Bonneville - Geology of southern Utah Valley, Utah: U. S. Geological Survey Professional Paper, 257-B:101-130.

Bissell, H. J., 1968, Bonnevile, an Ice Age lake: Brigham Young University Geology Studies, 15:1-66.

Blackwelder, E., 1939, Pleistocene mammoths in Utah and vicinity: American Journal of Science, 237:890-894.

Chadbourne, P. A., 1871, The discovery of a musk-ox in Utah: American Naturalist, 5:315-316.

Durrant, S. D., 1970, Faunal remains as indicators of neothermal climates at Hogup Cave, p. 241-245: *in* C. M. Aikens, Hogup Cave: University of Utah Anthropological Papers, 93.

Fedducia, A., and C. G. Oviatt, 1986, A trumpeter swan, (*Cygnus buccinator*) from the Pleistocene of Utah: Great Basin Naturalist, 46:547-548.

Feth, J. H., 1955, Sedimentary features in the Lake Bonneville Group in the east shore area, near Ogden, Utah, *in* Tertiary and Quaternary geology of the eastern Bonneville Basin, Utah: Geological Society Guidebook to the geology of Utah, 10:45-66.

Gillette, D. D., C. J. Bell, and M. C. Hayden, 1999, Preliminary report on the Little Dell Dam fauna, Salt Lake County, Utah (Middle Pleistocene, Irvingtonian Land Mammal Age), *in* Gillette, D.D., editor, Vertebrate Paleontology in Utah,: Utah Geological Survey Miscellaneous Publications, 99-1: 495-500.

Gillette, D.D., and Miller, W.E., 1999, Catalogue of new Pleistocene mammalian sites and recovered fossils from Utah *in* Gillette, D.D., editor, Vertebrate Paleontology in Utah: Utah Geological Survey Miscellaneous Publication 99-1: 523-530.

Gillette, D.D., McDonald, H.G., and Hayden, M.C., 1999, The first record of Jefferson's Ground Sloth, *Megalonyx jeffersoni* in Utah, (Pleistocene, Rancholabrean Land Mammal Age), *in* Gillette, D.D. editor, Vertebrate Paleontology in Utah: Utah Geological Survey, Miscellaneous Publication, 99-1, 509-521.

Grayson, D. K., 1987, The biogeographic history of small mammals in the Great Basin - observations on the last 20,000 years: Journal of Mammalogy, 68:359-375.

Grayson, D.K., 1988, Danger Cave, Last Supper Cave, and Hanging Rock Shelter; the faunas: Anthropological Papers, American Museum of Natural History, 66:1-130.

Grayson, D.K., 1994, The extinct late Pleistocene mammals of the Great Basin *in* Harper, K. T., L. L St. Clair, K.H. Thorne and W. M. Hess (editors), Natural History of the Colorado Plateau and Great Basin: University Press of Colorado: 55-85.

Gwynn, J. W. (Editor), 1980, Great Salt Lake - a scientific, historical and economic overview: Utah Geological and Mineral Survey Bulletin, 116:1-400.

Hansen, G. H., 1928, Hairy mammoth skeleton in Utah: Science, 68:621.

Hansen, G. H., and W. L. Stokes, 1941, An ancient cave in American Fork Canyon: Utah Academy of Science, Arts and Letters, 18:27-38.

Hay, O. P., 1927, The Pleistocene of the western region of North America and its vertebrated animals: Carnegie Institute of Washington Publication, 322:1-346.

Hintze, L. F., 1988, Geologic history of Utah: Brigham Young University Geology Studies Special Publication 7:1-202.

Hunt, C. B., H. D. Varnes, and H. E. Thomas, 1962, Lake Bonneville - Geology of northern Utah Valley, Utah: U. S., Geological Survey Professional Paper, 257-A:1-99.

Jefferson, G. T., W. E. Miller, M. E. Nelson, and J. H. Madsen, Jr., 1994, Catalogue of late Quaternary vertebrates from Utah: Natural History Museum of Los Angeles County, Technical Reports, 9:1-34.

Jennings, D. J., 1957, Danger Cave: University of Utah Anthropological Papers, 27:1-328.

King, C., 1878, Report of the geological exploration of the Fortieth Parallel: U. S. Army Department of Engineers Professional Papers, 1:1-1802.

Madsen, D. B., D. R. Currey, and J. H. Madsen, Jr., 1976, Man, mammoth, and lake fluctuations in Utah: Antiquities Section Selected Papers, Division of State History, 2:45-58.

Martin, P.S., and Klein, R.G. (Editors), 1984, Quaternary extinctions-a prehistoric revolution: University of Arizona Press, Tuscon, 892 p.

Mead, J. I., C. J. Bell, and L. K. Murray, 1992, *Mictomys borealis* (northern bog lemming) and the Wisconsin paleoecology of the east-central Great Basin: Quaternary Research, 37:229-238.

Miller, W. E., 1976, Late Pleistocene vertebrates of the Silver Creek local fauna from north central Utah: Great Basin Naturalist, 36:387-424.

Miller, W. E., 1987, *Mammut americanum*, Utah's first record of the American Mastodon: Journal of Paleontology, 61:171-187.

Nelson, M.E., 1988, A new cavity biota from northeastern Utah: Current Research in the Pleistocene, v. 5, p. 77-78.

Nelson, M. E., and J. H. Madsen, Jr., 1978, Late Pleistocene musk oxen from Utah: Transactions of the Kansas Academy of Science, 81:277-295.

Nelson, M. E., and J. H. Madsen, Jr., 1980, A summary of Pleistocene fossil vertebrate localities in the northern Bonneville Basin of Utah, *in* Gwynn, J. W., editor, Great Salt Lake - a scientific, historical, and economic overview: Utah Geological and Mineral Survey Bulletin, 16:97-113.

Nelson, M. E., and J. H. Madsen, Jr., 1986, Canids from the late Pleistocene of Utah: Great Basin Naturalist, 46:415-420.

Nelson, M. E., and J. H. Madsen, Jr., 1987, A review of Lake Bonneville shoreline faunas (late Pleistocene) of northern Utah, *in* Kopp, R. S., and R. E., Cohenour (editors), Cenozoic Geology of western Utah, Symposium and Field Conference: Utah Geological Association Publication, 16:319-333.

Oviatt, C. G., W., D. McCoy, and R. G. Reider, 1987, Evidence for a shallow early or middle Wisconsian-age lake in the Bonneville Basin, Utah: Quaternary Research, 27:248-262.

Pack, F. J., Jr., 1939, Lake Bonneville, a popular treatise: University of Utah Bulletin, 30:1-112.

Repenning, C. A., 1987, Biochronology of the microtine rodents of the United States, *in* M. O. Woodburne, editor, Cenozoic mammals of North America - Geochronology and Biostratigraphy: University of California Press, Berkeley, CA., p. 236-268.

Smith, G. R., W. L. Stokes, and K. F. Horn, 1968, Some late Pleistocene fishes of Lake Bonneville: Copeia, 4:807-816.

Stock, A. D., and W. L. Stokes, 1969, A re-evaluation of Pleistocene bighorn sheep from the Great Basin and their relationship to living members of the genus *Ovis*: Journal of Mammalogy 50:805-807.

Stokes, W. L., and K. C. Condie, 1961, Pleistocene bighorn sheep from the Great Basin: Journal of Paleontology, 35: 598-609.

Stokes, W. L., and G. H. Hansen, 1937, Two Pleistocene musk-oxen from Utah: Utah Academy of Science, Arts, and Letters Proceedings, 14:63-65.

Stokes, W. L., M. Anderson, and J. H. Madsen, Jr., 1966, Fossil and sub-fossil bison of Utah and southern Idaho: Utah Academy of Science, Arts, and Letters Proceedings, 43:37-39.

ANCIENT AMERICAN INDIAN LIFE IN THE GREAT SALT LAKE WETLANDS: ARCHAEOLOGICAL AND BIOLOGICAL EVIDENCE

by

Steven R. Simms
Department of Sociology, Social Work, and Anthropology
Utah State University, Logan, UT 84322-0730
ssimms@hass.usu.edu

and

Mark E. Stuart
2054 E. 6550 S., Ogden, UT 84405
ms1hummer@aol.com

ABSTRACT

The Wasatch Front was heavily used in prehistoric times, when vast wetlands along the eastern shores of Great Salt Lake were home to Native Americans for over 10,000 years. Flooding in the late 1980s exposed hundreds of villages, campsites, and 85 human skeletons, dating to the Fremont period about 900 years ago when native populations peaked. The Northwestern Band of the Shoshone Nation permitted analyses on the skeletons including radiocarbon dating, carbon isotopes, CAT scans, and DNA. Fremont lifestyles ranged from farming to foraging during the lives of individuals, plus there was movement of people around the landscape and among lifestyles. Activity patterns varied by gender. Farming and population declined after A.D. 1300, but foragers continued to frequent the Wasatch Front until A.D. 1600-1700 when population greatly declined and did not recover, possibly from European disease. Direct lineal descent from the Fremont to the Shoshone cannot be demonstrated at this time, but both ancients and moderns are Native Americans. The cultural resources of Great Salt Lake are rich, nonrenewable, subject to destruction, and hold ethical implications suggesting greater appreciation by managers and Great Salt Lake advocacy groups.

INTRODUCTION

Perhaps most famous for its images juxtaposing water and desert, the Great Salt Lake also supports vast tracts of ecologically varied wetland habitats along its eastern shores (figure 1). These wetlands were a focal point for American Indian life beginning over 10,000 years ago, and continuing to historical times. Flooding of Great Salt Lake in the late 1980s exposed new evidence for ancient human life in the area, including residences, work places, and human remains. The Wasatch Front was likely a populous part of the Utah region in ancient times, just as it is today, but the most intense occupation of the region by ancient people occurred within the last two millennia. It is this period that is best represented in the archaeology of the Great Salt Lake wetlands.

The population in the vicinity of Great Salt Lake markedly increased between A.D. 400 and 1300, peaking between A.D. 900 and 1200. This time is referred to as the Fremont period. "Fremont" is also the name of an archaeological culture named after the Fremont River in central Utah by Noel Morss who first documented the ancient culture during an expedition to that area in 1928-29 (Morss, 1931). This origin of the name is contrary to the popular belief that archaeologists named the culture after the Euro-American explorer, John C. Fremont. The Fremont is best thought of as a farming and foraging lifestyle, with cultural materials such as houses and ceramics being similar to those found across the Colorado Plateau and eastern Great Basin region. The Fremont culture is not presumed to represent a single ethnic group, nor even a single language, and stands in contrast to the millennia of foraging cultures that previously occupied the region. The farming of maize, beans, and squash by the Fremont fundamentally altered the character of life. Ironically, farming was abandoned during the 12th and 13th centuries, marking the onset of the Late Prehistoric period. A sizable foraging population however, remained tethered to the Great Salt Lake wetlands until A.D. 1500 or later. Sometime during the 17th century, the life way may have become more mobile, and the population around Great Salt Lake decreased. While there is evidence for a substantial lake transgression in the 17th century, an event that would have inundated wetlands, it is possible that the decrease in population was caused by some of the dozen or so diseases introduced to the Americas by European colonization. The diseases, first brought to the continent in the 16th century, moved rapidly among Indian populations, and may have spread into northern Utah with the introduction and use of the horse. Regardless of whatever caused the population decline, the American Indian presence along the Wasatch Front never returned to the population levels found during the Fremont and the early phases of the Late Prehistoric period.

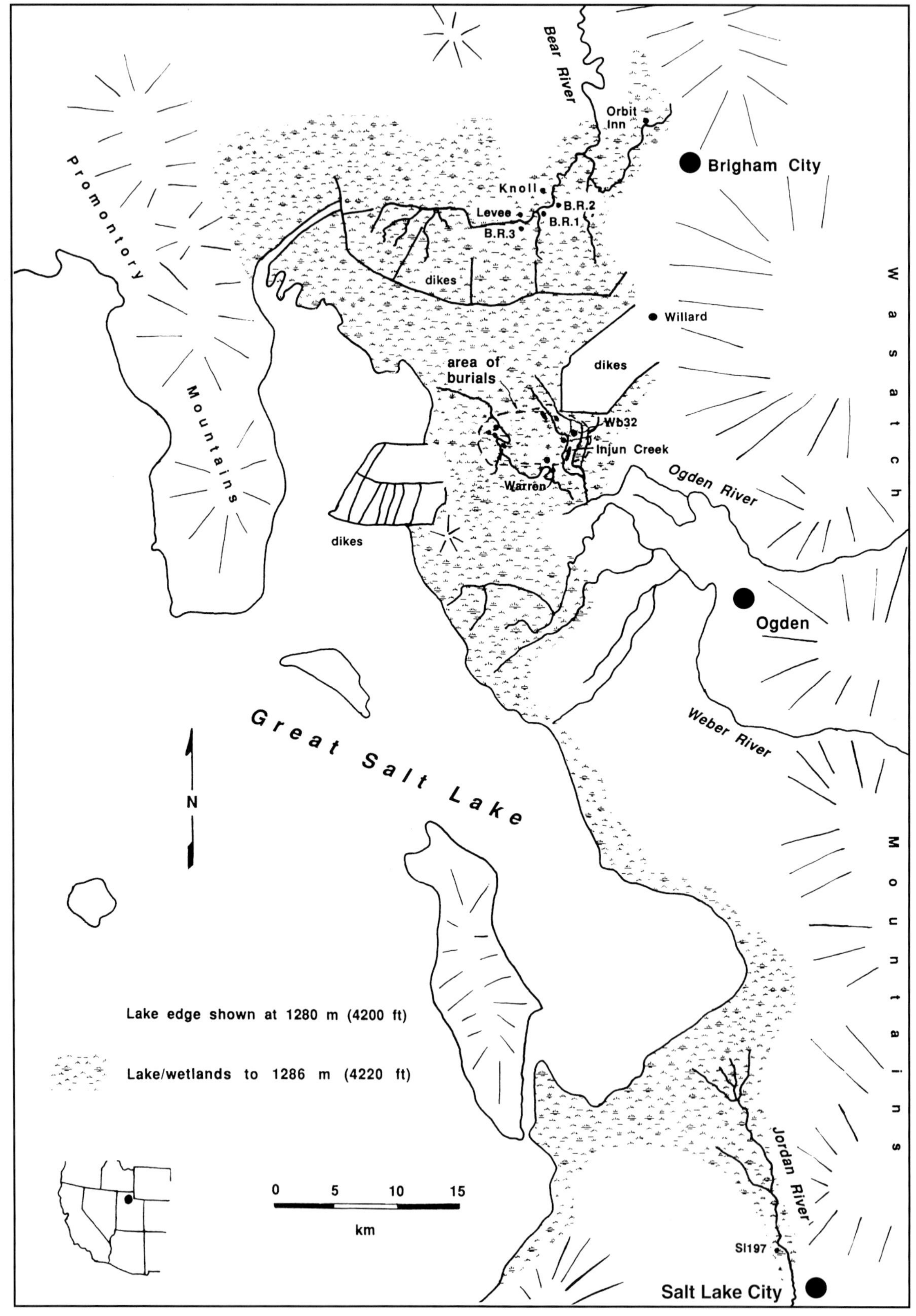

Figure 1. Map of eastern Great Salt Lake wetlands showing surrounding sites (referenced in text) and area of burial recovery and test excavations.

On the eve of European contact with the area in the late 18th and early 19th centuries, the Indian acquisition of horses and firearms had completely altered their lifestyles. Images from this recent period yield a picture of American Indian life on the Wasatch Front that is as romanticized as it is different from the preceding millennia. The relationship between the historical Ute and Shoshone occupants of the Wasatch Front, and the ancient people, is well-established as "Native American." However, the dynamics of interaction between the pre-Columbian and historical Native Americans remains the subject of exciting investigation.

The story of life on the Wasatch Front during the last two millennia is enhanced by new evidence from the Great Salt Lake wetlands. Flooding and subsequent regression of Great Salt Lake during the late 1980s revealed hundreds of archaeological sites (figure 2a-b) and exposed the bones from at least 85 ancient people (figure 3a-d) (Simms and others, 1991; Fawcett and Simms, 1993; Simms, 1999). Through the efforts of avocational and professional archaeologists, several state and federal government agencies, and the wisdom of local American Indian tribes, the skeletons were saved, studied in 1990 – 92 and eventually interred in a dedicated burial vault in 2001 (Simms, 1993; Simms and Raymond, 1999). This combined effort enabled a rare marriage of archaeological evidence with modern, high-tech analyses that gleaned extensive information from human remains, undreamed of only a few decades ago. The archaeology tells how people lived and how they used the landscape. It describes their culture, provides a chronology of events, and reveals the behavior of people living in farming villages, foraging hamlets, short-term campsites, and special task locations. It also yields some information about how people interacted with the supernatural. The studies conducted upon human bone reveal information difficult to obtain by recording or excavating archaeological sites. Studies directly upon the human skeletal remains indicate how much maize (and meat) people ate during their lives (Coltrain and Stafford, 1999), identify health and nutritional status (Bright and Loveland, 1999), characterize lifetime activity patterns (Ruff, 1999), and in a general way, identifies genetic characteristics and relationships (O'Rourke and others, 1999). Over 50 of the 85 Great Salt Lake individuals were dated using Accelerator Mass Spectroscopy radiocarbon dating (AMS C-14), a technique requiring as little as 0.1 gram of bone. By directly dating individual skeletons, and not just the archaeological sites, the age of the skeletons is more precise.

The marriage of archaeological and biological evidence enables our story of the ancient Great Salt Lake wetlands to look beyond the familiar labels that often force our understanding of the past into neatly bound stereotypes of immutable peoples, cultures, or ancient races. We now can describe life where individuals were active agents of dynamism in cultures that exhibited permeability and plasticity as much as they surely exhibited elements of ethnic identity and resistance to change. There were connections among people, there was residential fluidity across space, and there were changes of lifestyle during the lifetimes of individual people. As might be expected, there was genetic exchange across landscapes as well. This creats a complex and woefully incomplete understanding of the connections between ancient and modern American Indians in our region.

Archaeology is the only way to know what the ancient human wilderness of the Wasatch Front was like, and the Great Salt Lake wetlands contain a rich archaeological record. The richness of this resource is not evident to the casual visitor, and is often undervalued in the planning, management, political, and even educational processes. For that reason, we include a statement on the management and polit-

Figure 2a. *Site 42WB40 on natural levee (dark area) and sediment-filled stream channels on either side (light areas).*

Figure 2b. *Site 42WB48 large, square pit house with antechamber exposed at the surface.*

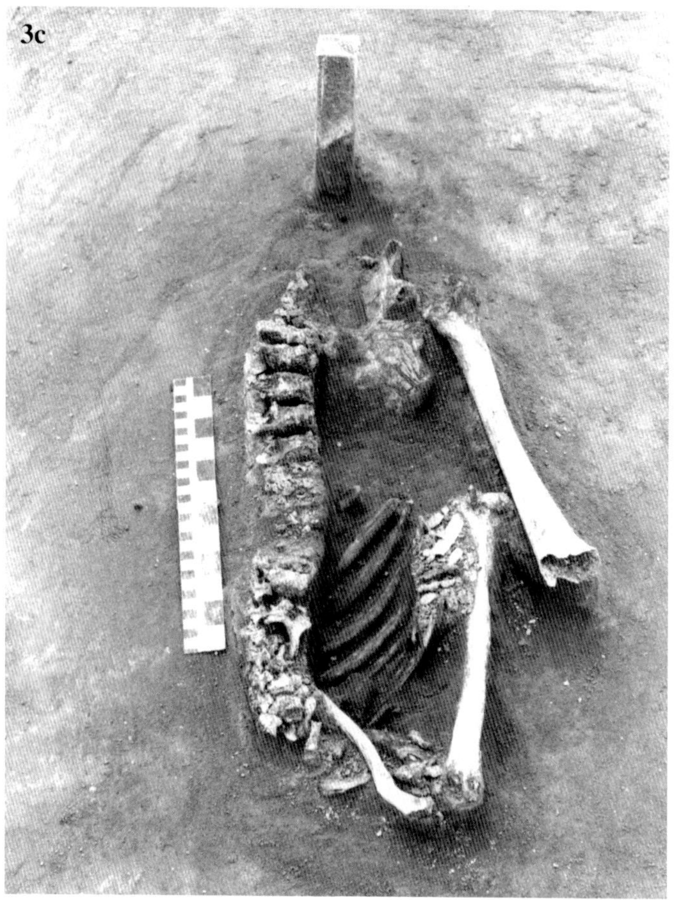

ical implications of this rich and legally protected resource. The archaeological record of Great Salt Lake is a nonrenewable, priceless information resource about our human past.

GEOGRAPHY AND HUMAN USE

The Great Salt Lake wetlands are best developed near the debouchments of the Bear, Ogden, Weber, and Jordan Rivers, and a host of associated streams and springs along the nearby Wasatch Range. The influx of fresh water variously reduces the salinity of the lake depending on seasonal flow levels, wind patterns, lake elevation, and the arrangements among numerous stream channels guiding water into the

Figures 3a-d. *Samples of human burials to illustrate range of skeletal conditions.*

lake. Small fluctuations in water level cause rapid lateral shoreline shifts due to the shallowness of the lake basin. In fact, wind from individual storms can raise lake levels along the eastern shores as much as two meters (Atwood, 1994). These factors combine to form a dynamic mosaic of brackish-to-fresh-water marshes and wetlands, with ponds and slow-moving streams bounded by walls of bulrush and cattails. On slightly higher ground, and along the natural levees flanking every stream channel, salt grass meadows and greasewood plains predominate. These habitats support an abundance of wildlife including fish (suckers and chub), and small mammals (muskrats, rabbits, and beaver). Less evident today are the bison, antelope, mule deer, and bighorn sheep (not restricted to high altitudes in prehistoric times) the ancient landscape supported. Great Salt Lake waterfowl, ducks, geese, and grebes to name only a few, occurred in legendary proportions (Fremont, 1988 [1845]). Hundreds of plants were used as food, including an array of greens, shoots, and especially roots, tubers, and seeds because they are storable. Various plant foods were available from springtime to well into the winter. The plants were eaten fresh, or stored for the hardest and coldest part of winter.

Although plant foods were abundant in the Great Salt Lake wetlands, one should not conjure an image of an ancient "Garden of Eden." For human foragers, or farmer-foragers, the availability of food within the wetlands varied greatly by season, and winter was a time requiring food storage. Wetlands are dynamic, and the subtle topographic relief and high annual variability of effective moisture (temperature, precipitation, and evaporation) require an ever-changing set of foraging strategies.

Foraging strategies shaped the behavior of the Fremont along Great Salt Lake. People had to decide where to locate villages and camps. This required decisions about which resources to exploit, while foregoing other resources. People had to decide which resources to transport and how far. Transport costs shaped behavior then, just as they do today. People had to consider that changing circumstances sometimes made it better to live in central places, deep in the wetlands, or to live on the margins of the wetlands. At other times, it may have been best to live outside the wetlands all together, and use them only as needed. Decisions about these ever-changing situations conditioned the development of social networks that variously brought people together or cast them into relative isolation. These social relationships, in turn, shaped the character of people's ideological beliefs and their perceptions about what was valuable in life.

Wetlands provided ancient foragers with concentrated patches of useable biomass, but the bounty was not uniform across space, nor spans of time as short as a single year. This human habitat was more of a dynamic theater of trade-offs than it was an idyllic and static Garden of Eden.

The advent of farming only added to the behavioral options. While farming could produce a large surplus of food, enabling short-term population increases, the juxtaposition of surplus against inevitable crop shortfalls guaranteed tension between population growth and resource scarcity. Rather than bringing stability, farming brought greater dynamism to the region. People dealt with the tensions of growth by adjusting their group size, and by putting greater labor into farming while also exploiting wild foods.

Decisions about daily life were not limited to food. People had to decide whether to make ceramics, and what kind, depending on their mobility. Ceramics do not transport well, but are useful if one is settled, or has an intention of returning to a place. The debris from manufacturing stone tools also reflects the different approaches taken by the ancient people to managing precious toolstone that had to be transported from outside of the wetlands.

PREVIOUS ARCHAEOLOGICAL INVESTIGATIONS

The Great Salt Lake wetlands have been known to be archaeologically rich since the 19th century (Maguire, 1892), and several generations of professionals and avocationists have produced a wealth of published and unpublished information. In this section, we provide references to basic information sources, but focus on the aspects of the archaeology that indicate what people did. Our goal is not to summarize the archaeology, but to provide some examples of what it tells us about the past, and to explain why systematic study and astute management of archaeological resources is important.

Past excavations provide information on eleven sites, all of Fremont and Late Prehistoric age (A.D. 400 to historic period). These sites are: (1) Willard (Judd, 1926; Kennedy, 1930; Maguire, 1892; Steward, 1933), (2) Warren (Enger and Blair, 1947; Hassell, 1961; Manful, 1938), (3 and 4) Knoll and Levee (Fry and Dalley, 1979), (5, 6 and 7) Bear River numbers 1 (Pendergast, 1961), 2 (Aikens, 1967), and 3 (Shields and Dalley, 1978), (8) Injun Creek (Aikens, 1966), (9) Orbit Inn (Simms and Heath, 1990), (10) site 42SL197 (Schmitt and others, 1994), and (11) site 42WB32 (Fawcett and Simms, 1993). Their general locations are shown on figure 1. Other sites referred here, and by their Smithsonian numbers (such as 42WB48) are reported in the more recent studies associated with the recovery of human remains (Fawcett and Simms, 1993; Simms and others, 1991; Simms, 1999).

Sites that include substantial Fremont residential farming bases, such as Willard, are located on flood plains, near the toes of alluvial fans, in the manner so typical of Fremont agricultural sites along the eastern Great Basin rim. Such sites are located under virtually every modern Utah city and town along Interstate Highway 15 from Brigham City to Cedar City (for overviews of the Fremont Culture, see Jennings, 1978; Madsen, 1989; Marwitt, 1970; Madsen and Simms, 1998). The Willard site, destroyed by ground leveling and dike construction, is largely underneath the picnic and camping areas of Willard Bay State Park. Over 50 house mounds contained superimposed pit structures and/or adobe-surface structures (Judd, 1926; Steward, 1933). Superposition suggests pit houses evolved from circular to square shapes between A.D. 1125 to 1200 (see Talbot, 1997 for a thorough review of Fremont architecture). The move from circular to square housing, in conjunction with other mobility evidence, is a significant transition world-wide because it correlates with increased tethering to places (Flannery, 1972; Gilman, 1987; Hunter-Anderson, 1977; Whiting and Ayres, 1968). Artifacts from the Willard site, in museum and private collections, contain small side-notched and corner-notched arrow points, grinding stones, slate knives, burials, maize, beans, squash, textiles, ornaments, and figurines.

The Warren site near Plain City had 16 mounds (Manful, 1938) containing houses of the Willard type (Enger and Blair, 1947). Assemblage characteristics at the Warren site are similarly broad, with bison bone, maize, bushels of fish bone (Manful, 1938), figurine fragments, large, Utah-type metates, a lignite-bead necklace, and other trappings of residentially stable occupation. The Warren site, along with countless others, has been destroyed by field leveling for modern agriculture, and by urbanization.

The Knoll site, and the late component of the Levee site located on the lower Bear River, yielded round and square pit houses which are shallower than those at Willard. Radiocarbon dates indicate a transition from round to square pit houses after A.D. 1100, with human use possibly extending through A.D. 1200 (Fry and Dalley, 1979:5). Pit houses are also present at Bear River number 3, and at several sites in the area of human skeleton recovery (42WB48, 42WB185a-c, and 42WB324). These pit houses were found at lower elevations than ever before (1,282 m, 4,205-6 feet). In contrast to Willard and Warren, where storage was in surface structures of adobe, storage at these other sites was limited to small, bowl-shaped, subsurface pits indicative of short-term storage and food hiding (Zeanah, 1988). Although the Knoll assemblage of associated artifacts is not as substantial as at the Willard and Warren sites, all of the pit house sites indicate considerable residential stability. This stability may, or may not, have been associated with farming (Madsen, 1982).

The Great Salt Lake Fremont have long been portrayed as a foraging economy, but the presence of farming is supported by the site locations and maize remains at Willard, Warren, Bear River number 3, 42WB32, and Injun Creek. The best places to locate large farming villages was just outside of the wetlands and along streams, precisely where the greatest amount of modern urbanization has occurred. Substantial Fremont sites are known along the Weber River, Ogden River, Layton Creek, the Jordan River, and City Creek, which once flowed through downtown Salt Lake City.

Other sites yield circular mud and pole structures, indicative of less investment than pit houses. These sites include the early component of the Levee site, Bear River numbers 2 and 3, Injun Creek, 42WB185c, and 42WB32 sites. At 42WB32, three such structures are superimposed, and three radiocarbon samples date them from A.D. 1035 to 1155, the heart of the Fremont period (Fawcett and Simms, 1993). Storage at these sites is also in small subsurface pits, again contrasting with deeper storage pits that represent larger scale storage goals (Talbot and Richens, 1996). An exception is Injun Creek near Plain City, where a larger, aboveground, adobe storage structure was found. However, Injun Creek is the only non-pit house site in the area yielding maize remains. The diversity of the artifact assemblage at most of these sites is high, as would be expected at residential bases where the full human technological inventory would be represented.

Higher mobility is indicated at Bear River number 1, a temporary camp with no structures and a less diverse artifact assemblage. It was likely a stopover at the location of a bison kill (Lupo and Schmitt, 1997). The Orbit Inn site, near the Brigham City airport, is a Late Prehistoric residential camp that was intermittently occupied from spring through fall for several weeks each visit over a 50-year span. Here, only light structures were found, either wickiups or brush windbreaks, but there were distinct activity areas: 18 subsurface pits, sizable refuse deposits, and a broad assemblage composition that accumulated through repeated use of the site. Light structures, reflecting brief use, are also found at other sites, (such as 42WB40, 42WB144, 42WB184 and 42BO73) some of which may be Fremont in age. Subsurface pits used for roasting, storage, and ultimately refuse, abound across the study area; one site (42WB317) southwest of Willard Bay contained more than 150 subsurface pits.

Chipped-stone artifacts and ceramics also help reveal the behavior of ancient inhabitants. The remains from manufacturing chipped-stone tools feature a high degree of tool stone conservation, and reflect the absence of naturally occurring tool stone in the wetlands. There was caching of primary and secondary flakes (Cornell and others, 1992), low flake weight, and frequent resharpening of projectile points. The intensive use of tool stone appears to result not only from limited supply, but also from low residential mobility within the food-rich wetlands. The approach to tool manufacture, called bi-polar reduction, also suggests low mobility and tool stone conservation. On the other hand, a technique called biface reduction is evident at a minority of sites, indicating that logistic use of sites by people from larger residential bases did occur (Elston, 1988; Kelly, 1988, 1995; Simms and Whitesides, 1993).

There is high variation among the side-notched arrow points, with many falling at the boundaries of archaeologists' point categories. Projectile points are useful as time markers when sample sizes are adequate, and when other site-dating evidence exists. It remains dangerous, however, to date sites and ascribe them to cultures on the basis of only a single line of evidence, such as a projectile point or two. It is also risky to categorically define ancient peoples' ethnic, linguistic, or tribal affiliations only on the kind of arrow points found (for example, the assumption that if it is a Desert side-notched point, this person must be Shoshone). It is precisely these kinds of unresolved issues that require careful recording of archaeological sites in the Great Salt Lake wetlands because the current sample sizes in publicly available collections are small.

Broken ceramics are ubiquitous at Great Salt Lake wetland sites, and ceramics were adopted not only by farmers during the Fremont period, but by foragers with residential patterns stable enough to make ceramic use worthwhile. Dean (1992) finds significant overlap among northern Utah ceramic types in their attributes of temper, color, and wall thickness, suggesting subtle gradation in forms. Janetski (1994) counters with evidence from Utah Valley that shows there is indeed a chronological sequence of ceramic types from Fremont to Late Prehistoric times. Simms and others (1997) show that the existence of a chronological sequence of ceramic types is not at odds with findings of significant variability in ceramics.

Simms and others (1997) found a relationship between the degree of labor investment in undecorated Great Salt Lake ceramics (by far the most common ceramics found), and independent archaeological measures of residential mobility. Labor investment varies, within culturally defined limits, according to the intended use life of the vessel. Essentially, greater functional diversity and higher quality of vessel manufacture are associated with the large residential farming bases (for example, Willard). Pottery of lesser investment occurs at residential bases where evidence indicates shorter-term stays, reliance on foraged foods, or a mixed farmer-forager diet (for example, the Levee site and 42WB32). High ceramic variability occurs at residential camps (for example, the Orbit Inn) because ceramics at such sites were discarded under various mobility regimes, and thus came from a variety of local sources. We find high-quality pottery at a few small sites, like that found at farming bases, suggesting a logistic connection to such bases. X-ray diffraction was employed to test hypotheses about raw-material sources for ceramic manufacture; analytical results on the types of source materials were consistent with predictions based on the degree of residential mobility. Variability in ceramic quality can be found even at the height of Fremont farming times showing that even then, not all people made the same decisions about how to approach life.

Archaeology yields a picture of the Great Salt Lake Fremont people who farmed and lived in large villages near the toes of the alluvial fans adjacent to the Wasatch Range. These people supplemented their diet with foraged foods and exploited the wetlands, leaving the cultural material they brought from their villages at temporary camps and work stations in the wetlands. People also ventured into the surrounding mountains and, while the archaeology of the Wasatch Range is little known, Fallen Rocks Shelter above Ogden is one example of Fremont occupation along the mountain flanks (Stuart and Pringle, 1988). Some Fremont people were based in substantial residential sites well out onto the lake-bed surface. These sites do not appear as focused on farming as the aforementioned large farming villages. Finally, there are a host of sites with indications of limited occupation, with stays ranging from a few days to several weeks. At these sites, structures were often built, and occupation could be substantial, but the sites were used seasonally and probably repeatedly. Short-term camps and task-specific sites were left either by farmers away from their fields and villages, or by people who were primarily foragers.

It would be easy to leap to the conclusion that there were distinct farming people and foraging people. This certainly occurs elsewhere and may have occurred here, but anthropology is increasingly interested in the connections among people, regardless of ethnic distinction. A Fremont-wide synthesis from this perspective can be found in Madsen and Simms (1998). The archaeology of Great Salt Lake suggests that the contrast between forager and farmer does not represent two peoples. Instead, it appears that people experienced change during the course of their lives and over many generations of people, as they adjusted the mix of farming and foraging depending on socioecological circumstances. Archaeology alone, however, cannot resolve this question, but an answer can be approached with knowledge from the human skeletal remains.

ARCHAEOLOGICAL STUDIES AFTER GSL FLOODING

In 1983, the Great Salt Lake water level began rising and peaked in the spring of 1987, at an elevation of 1,283.8 m (4,211.8 ft), perhaps exceeding the historically recorded maximum reached in the 1870s (the exact year and level is not known, see Mabey, 1986). The lake soon began receding, and by the fall of 1987 large areas of lake bed were freshly exposed. Lake levels continued to decline for the next six years to approximately 1,280.2 m (4,200 ft). During the lake-elevation rise, the submerged land northwest of Ogden was scoured by waves powered by prevailing northwesterly winds along a shoreline unprotected by the dikes that shielded other areas (figure 1). After the lake receded, the scoured landscape was a flat, nearly featureless plain stripped of vegetation. Surface deflation ensued, as wind removed the thin crust of alkali that covered the land surface, revealing hundreds of archaeological sites (figure 2a-b), and thousands of human bones (figure 3a-d). Other stretches of the eastern shores of the lake did not experience the same degree of erosion. The mouth of the Bear River, the lower Jordan River, and the stretch between Farmington Bay and Ogden Bay could also contain archaeological sites and burials under a shallow layer of sediment (Allison, 1997).

Prior to the removal of the human remains by Utah State University, state and federal agencies consulted with several American Indian tribes. In November 1989, representatives of the Northwestern Band of the Shoshone Nation decided that remains jeopardized by erosion and vandalism should be removed. Further consultation regarding disposition and analysis of the human remains continued for several years after 1989. The State of Utah funded construction of a burial chamber at Pioneer State Park in Salt Lake City where the remains were interred by the Shoshone in 2001 (Simms, 1993; Simms and Raymond, 1999).

Characteristics of the Burials

The age of the exposed skeletons ranges from A.D. 400 to 1450, but the overwhelming majority lived during the peak of the Fremont period between A.D. 700 to 1300, when the Indian population peaked. The bones of 85 individuals were recovered. The minimum number of 85 individuals is determined by counting specific bones such as the left clavicle (there can only be one per person). However, of the 85 individuals, 45 percent are represented by fewer than five bone elements, and only 20 percent consist of more than half of the skeleton (Simms and others, 1991).

The bones were affected by natural processes as well as ancient cultural practices. Erosion often scattered the skeletons. In several cases, sediments showed that previous lake-level fluctuations had exposed skeletons, only to rebury them. In some instances, transport of bones by natural lake processes from closely spaced burials often created a single locus of commingled human remains. In other instances, commingling resulted from human practices. In one case, 11 individuals were found clustered in a 4-meter (13-foot) wide area of burned bulrush plants (the human remains were not burned). Several of the 11 individuals appear to have been buried in a single event, but it is also clear that others were buried later. In another case, an infant was secondarily interred upon the lap of a partial, adult female skeleton that exhibited evidence of burning (the infant was unburned). A few cases of commingling appear to result from intentional commingling of one or two elements from several individuals. A burial practice that may produce commingling, but for which we have only sketchy documentation in the Great Basin, is token burial. Token burial is the practice of retrieving a bone(s) of a deceased relative who initially had to be buried away from home (Brooks and Brooks, 1990), and burying the bone(s) at another location.

Cut-marks and drilling are evident on some individuals, indicating postmortem alteration of the bone by humans using sharp tools (Simms and others, 1991: 36, 38, 39, 42, 56, table 7). This practice is highly variable in expression, but occurs widely across cultures around the world over time periods spanning tens of thousands of years. It usually indicates some sort of ritual modification of the bone. While this sort of finding is often uncritically attributed to cannibalism, there is no evidence for this in the sample of Great Salt Lake human remains described here. Further, since ritual postmortem bone modification is far more frequent than cannibalism, it is the most parsimonious interpretation with the evidence at hand.

The clearest case of bone modification is a hole drilled through the ulna (lower arm) of an adult female, whose remains were interred in a ritual burial including red ochre (an iron oxide clay used in the burials of many cultures of the world for thousands of years). An infant was later interred upon the lap of this woman, but whether the wrist bone was drilled at that time or some other time cannot be determined.

Cutmarks were found on two other adult females, around the ankle in one case. However, not all the evidence of cutmarks is conclusive and in one or two cases, the marks appear to be from shovel damage by a known, twentieth century relic hunter who dug up skeletons and sometimes even left telltale broken wine bottles at his "digs."

Gnawing by carnivores and rodents is evident on some bones. Stratigraphic evidence showed that burials had often been exposed to carnivores by erosion associated with post-interment fluctuations of the Great Salt Lake (Simms and others, 1991).

Very few grave goods were found with the Great Salt Lake skeletons, but a low frequency of non-perishable grave goods is consistent with patterns of burial from foraging societies in general. Even in the case of Fremont farmers, grave goods are infrequent (Gunnerson, 1969; Madsen and Lindsay, 1977; Dodd, 1982; Janetski and Talbot, 1997). A few of the Great Salt Lake burials do contain grave goods such as a fragment of deer antler, a tubular bone bead and tiny shell pendant, a large Utah-type metate, a quartzite mortar, and a burial with a mano and fresh-water clam shells. A few burials yield red ochre-staining of human skeletal elements showing that bones had been exhumed by Native Americans after the flesh was gone and rubbed with red ochre. Red ochre is associated with the burial ritual in many societies around the world and throughout ancient and modern history. The above-mentioned interment of 11 individuals yielded numerous grave goods including fan-shaped arrangements of bone awls, an arrangement of 13 bone counters or gaming pieces, a carved bison horn, a Utah-type metate (found almost exclusively at Fremont residential bases), and a carved-bone duck-head effigy (Simms and others, 1991). Most of these grave goods are smeared or stained with red ochre.

Burial position and orientation is variable among the 27 individuals for which this could be determined (Simms and others, 1991). Positions include flexed (reclining, resting on left or right), extended, and unusual postures. Burials face in all directions, but north was the most common (n = 10) followed by east (n = 8), south (n = 3) and west (n = 1). Intermediate orientations included northwest and northeast (n = 2), southeast (n = 1) and southwest (n = 1). The variation suggests there were a variety of interpretations as to the correct placement of the dead during the three centuries of religion represented by the Great Salt Lake sample.

Burial pits were detected in only a few instances. This can be attributed to previous erosion and reburial of remains. However, even the complete or nearly complete skeletons are rarely associated with pits. In some cases, burial was made in aeolian (wind) sediments. Fowler (1992) reports that Northern Paiute from the Stillwater area of western Nevada often buried the deceased in sand hills flanking the wetlands. Burial pits can easily become obscured if the sands are mobile. In other instances, water burial seems plausible and is known to the area, such as the case of the Shoshone chief, Pocatello (Madsen, 1986). If a deceased person were lowered into one of the slow-moving water channels, they would become covered with silts, a situation common among the Great Salt Lake burials; water burial may account for the position and sedimentary context of some skeletal remains (figure 3d).

The Great Salt Lake skeletal sample is extremely valuable in another way. Most human skeletal collections are either found within archaeological sites or are isolated finds, such as those found in rock crevices or caves. The Great Salt Lake sample is not only the second largest skeletal collection ever found in the Great Basin, but it is a sample of all human remains interred over a 30-square-mile (78-square-km) area, and exposed en masse by natural processes. It thus lacks the bias so characteristic of archaeological skeletal samples.

Studies on The Human Remains

With permission from the Northwestern Band of the Shoshone Nation, small amounts of bone, typically fragments or small bones, were analyzed using carbon isotopes, visual examination, CAT scans, and DNA extraction. Stable carbon-isotope analysis identifies the major classes of plants consumed over the life of an individual. Since maize is in one of these groups, an estimate can be made of the amount of maize consumed. This analysis was conducted by Coltrain and Stafford (1999). Bones also reveal information about adequacy of diet, especially in the critical early years of life. They also indicate health factors such as general skeletal fitness, arthritis, and other specific syndromes. Teeth are another indicator of diet and health. Analyses of the health and nutritional aspects of the human remains were conducted by Bright and Loveland (1999). Using CAT scans on femora (hips) and humeri (upper arms), an engineering analysis of the internal bone structure can be made. Bone structure is acquired over an individual's lifetime, and is determined largely by activity patterns. Structural differences in bone can be detected between the weaker lower limbs from decreased lower-body activity that is often associated with farming, and strong lower limbs from a more active life associated with walking steep mountains. These analyses were conducted by Ruff (1999). Finally, one of the most intriguing but most preliminary analyses is the ancient population genetics using DNA extraction and amplification. These analyses were done by Dennis O'Rourke, Ryan Parr, and Shawn Carlyle (O'Rourke and others, 1999; Parr, 1998; Parr and others, 1996). The results of these studies are given in the next section.

LIFESTYLES INFERRED FROM THE ARCHAEOLOGICAL AND BIOLOGICAL EVIDENCE

The Fremont

Native American farming began in Utah about 2,000 years ago and as a consequence population grew. People congregated in larger groups because farming produced a stored supply of food. The production of a surplus provided a hedge against risk, while the reduced mobility of the farming lifestyle and the episodic surpluses from harvests fostered increased population. Crop shortfalls are also a part of farming, and these added a measure of risk, especially in the face of increasing population size. Farming also created new obligations and expectations that accompany a more complex social network. Along the Wasatch Front, full-time farming was a way of life for some, while other people switched back and forth from farming to foraging during their lifetimes. Still others remained foragers in the outlying areas, much like the situation in the 20th century where the last remaining foragers in the world were found only in the most remote and marginal places. Thus, foraging continued in areas not suitable for farming, producing a situation where foragers could either attach themselves to farming centers, or became marginalized in the outback. These circumstances made the Fremont period a time of increased demographic fluidity; movement of individuals, families and bands among lifestyles, and across landscapes. One of the impacts of farming was the intensification of human networks that likely ensnared Native peoples with different cultural backgrounds to a greater extent than in earlier periods.

Carbon isotope data support a demographic mix of farming and foraging. There is increasing dependence on maize from A.D. 400 to A.D. 850 (Coltrain, 1997). From A.D. 850 to A.D. 1250, there is high variation in the stable carbon isotope values indicating that some people obtained perhaps 50 to 60 percent of their lifetime calories from maize (Coltrain, 1997; Coltrain and Stafford, 1999). The same studies show that other people ate very little maize over the course of their lives, regardless of whether there were phases of their lives during which they ate more maize. Variation is even apparent in skeletons from the farming bases at Willard, Warren, and 42SL197 on the Jordan River delta, where people who ate a great deal of maize are found within the same sites as those who ate very little maize. The most common results however, are carbon isotope values in the middle range. These people either spent their lives eating a mixed diet, or experienced fluctuations in the mix of farmed or foraged foods during their lives.

The biomechanical analysis of limb bones (Ruff, 1999) indicates that men moved greater distances, and over difficult terrain, while women moved intermittently. This is consistent with an archaeological record of residential stability in the wetlands, moving among a series of base camps. Men were the primary harvesters of resources outside of the wetlands, while women were more tethered to the wetlands. A similar pattern is reported for the Carson-Stillwater area of western Nevada (Larsen and others, 1995; Larsen and Kelly, 1995). The behavioral inferences from the biomechanical analysis are consistent across the Great Salt Lake skeletal sample. This suggests that the dietary diversity shown by the stable isotope data occurred within a single overall pattern of life, and that the variations in diet are not expressions of two separate cultures of farmers and foragers.

The molecular genetic analysis (O'Rourke and others, 1999; Parr, 1998; Parr and others, 1996) indicates homogeneity across the Great Salt Lake sample. This too, suggests there was a single system with people cycling between farming and foraging lifestyles, producing genetic linkages. Whether people living any particular lifestyle saw themselves as belonging to a distinct culture cannot be known, but regardless of cultural identity, there was a crossing of boundaries in lifestyle and in reproductive relationships. These results are as preliminary as they are seductive, but are buttressed by a strong record of human behavior in the world showing that aboriginal peoples can marry widely and repeatedly, practice multilingualism, modify ethnic and religious identities, move in and out of farming, and move among different lifestyles and places during their lives.

Fremont to Late Prehistoric Transition

By the 12th and 13th centuries, farming was in decline. This was a time of climate change in western North America and for the Fremont, a reduction in summer monsoon rainfall that negatively affected maize farming. While droughts had occurred previously, those of the 12th and 13th centuries occurred in the context of the largest aboriginal population the region had ever supported. Under these circumstances, people were left with fewer options; the landscape could not support more foragers without overexploitation, and most of the locations for rainfall/runoff farming were occupied. Thus, people could not respond to crises by moving and colonizing new land without inviting contention from others. By the end of the Fremont period the Wasatch Front was fully utilized.

Interment of burials in the area essentially terminates after A.D. 1300, about the end of the Fremont period (Simms 1999: 34, 44, Table 3.2). Given that more recent burials would be expected to occur if they were present, the abrupt decline in the size of the sample is consistent with a population decline at the end of or the Fremont period.

A.D. 1300 is the accepted date for the end of the Fremont period (Janetski, 1994), but in terms of human behavior, the transition was likely underway before this time (Coltrain, 1997; Simms, 1990). The stable carbon isotope analysis yields some evidence for a decline in farming in the Great Salt Lake wetlands beginning around A.D. 1150 (Coltrain, 1997; Coltrain and Stafford, 1999). Forager diets predominate after this time, yet molecular genetic analysis indicates genetic continuity between the pre- and post-A.D. 1150 populations (O'Rourke and others, 1999; Parr and others, 1996). This suggests that the decline in farming happened to indigenous people who persisted through a transition lasting over a century. If migrants entered the region to fill the gap, the Great Salt Lake sample indicates that they did so only after the decline in farming after A.D. 1300.

The archaeological record of Great Salt Lake, as well as Utah Valley (Janetski, 1994), also shows clearly that a substantial number of people continued a foraging life way in the wetlands up to A.D. 1500 or later. Thus, there was no mystical disappearance of a farming people, and no vanishing of "the Fremont" (again merely an archaeological term for a period when farming was practiced). A foraging economy however, could not support as many people, and overall population in the region declined.

The Final Decline - The Late Prehistoric to Historic Periods

By the end of the 16th century, the archaeological record in the Great Salt Lake wetlands is sparse. Only three radiocarbon dates later than A.D. 1600 are available. These include dates from the Fire Guard site in Ogden (Stuart, 1993), Injun Creek (Aikens, 1966) near Plain City, and the Fox site on the Jordan River near the north end of Utah Valley (Janetski, 1990). This circumstance invites speculation of depopulation due to European diseases because this has been a successful explanation for similar population anomalies around the continent (for example, Campbell, 1990; Ramenofsky, 1987; Verano and Ubelaker, 1992), and has been proposed for the Great Basin (Beck and Jones, 1992;

Simms, 1990). Diseases such as smallpox, measles, and influenza, to name only a few, were introduced to the Americas via Florida, Mexico, and possibly the west coast during the A.D. 1500s. American Indians had no resistance to these Old World diseases because the two hemispheres had been isolated for so long. Diseases were rapidly transmitted among Indian population centers in the Mississippi Valley (Ramenofsky, 1987), and slightly later to the Southwest (Reff, 1991), California (Preston, 1996), and the Columbia Plateau (Campbell, 1990). It is not known whether these diseases penetrated the Intermountain region with its low population densities. On the other hand, the relatively high concentration of people along the ancient Wasatch Front could have resulted in severe epidemics if diseases were introduced.

The Great Salt Lake wetlands' archaeological record supports a strong demographic anomaly in the A.D. 1600s. This unfortunately occurred when one or more rises of Great Salt Lake level inundated the wetlands between A.D. 1550 and 1700 (Currey, 1990; Currey and James, 1982; Currey and others, 1984; Murchison, 1989) destroying human habitat. Correlations to dendroclimatological evidence suggests the most likely window for one of these transgressions was A.D. 1610 to 1620 (Fritts and Shao, 1992). By the early 1700s, the lake had regressed, once again opening a large wetland for exploitation.

Another significant change is the introduction of the domesticated horse beginning by the middle A.D. 1600s, when the Comanche and Ute raided New Mexico for horses and metal goods. The Pueblo revolt in New Mexico in A.D. 1680 caused even more horses to be obtained by Indian groups. The introduction of the horse, and the increased mobility it brought, possibly contributed to the spread of disease.

After the flooding of the wetlands had receded in the early 1700s, people did not seem to return in any significant numbers. While it is possible that the change in lifestyle, brought about by the horse, lightened the archaeological imprint, foragers of previous centuries certainly left archaeological remains, and horse-mounted tipi dwellers of the plains left obvious archaeological sites. Our impression is that far fewer people lived in the region during A.D. 1600 to 1700 and thereafter. The mystery is heightened by the fact that lake fluctuations had occurred before without depopulating the region. The explanation for these changes remains unclear, but lake levels and climate may be less important factors behind the precipitous population decline in the 17th century than the introduction of European disease to a densely populated strip along the Wasatch Front, and the introduction of the horse that hastened the spread of disease.

By the time Euro-Americans arrived in the late 18th and early 19th centuries, the Ute of Utah Valley possessed horses (Warner, 1976), and early fur trappers and explorers commented on how few Indians occupied the Great Salt Lake wetlands, given the apparent richness of the area (Dewey, 1966). One group of Eastern Ute claimed to have lived near the mouth of the Weber River, but left the area when bison herds declined (Trenholm and Carley, 1964). Other inhabitants included Shoshone groups from Wyoming who annually frequented the Great Salt Lake wetlands for the fall bison hunt (Murphy and Murphy, 1960; Madsen, 1985; Wilson, 1991). The explorer John C. Fremont observed Shoshone fishing at the mouth of the Bear River, and later that summer found Shoshone in the marshes trapping fish and living in groups of two to 10 families (Fremont, 1988).

MANAGING ARCHAEOLOGICAL REMAINS

The Great Salt Lake wetlands, from Brigham City to Salt Lake City, are rich in cultural resources. These resources are protected by federal and state laws dating as early as 1906 that make it illegal to remove ancient objects from the surface or from excavations of any kind. Since the Fremont period population density was so high near Great Salt Lake, relative to much of ancient Utah, human skeletal remains from that period will continue to appear and be susceptible to vandalism and erosion. State law makes it a crime to disturb any suspected human remains prior to contacting local law enforcement. Pressures for development along the eastern edge of Great Salt Lake will surely continue, for example, the proposed Legacy Highway. Anywhere there are ancient stream channels meandering toward the lake, archaeological sites and human remains will be encountered whether they are readily apparent on the surface or not.

Perhaps more important than the law is the respect due the deceased, and the invaluable knowledge contained in the archaeological sites and in the human skeletal remains. In this paper, we have tried to show the value of the knowledge that can be gleaned from archaeological sites, from human skeletal remains, and how technical advances are increasing the amount of knowledge that can be gleaned. The foundation for the preservation is thus not preservation per se, but the respect due to living Native Americans descended from the ancients, and the educational value that comes from preservation, judicious study, and management of archaeological sites and human remains.

Ancient Native American skeletal remains must be protected. Destruction has occurred for decades and is caused by the hunters, ATV enthusiasts, fishermen, hikers, and collectors who use the Great Salt Lake wetlands. Destruction from natural erosion is also persistent, especially in areas more exposed to wind and wave action. Where the impacts of continued urban growth are inevitable, cultural resources should be recorded and in the case of human remains, moved and respectfully re-interred. Utah has made progress toward including Native American views and desires in the process of managing the state's heritage resources. No living person however, can own the dead, and the remains from ancient times are a storehouse of knowledge about America's collective past. Even if we decide not to study human remains now, it may be shortsighted for the living to categorically prevent future generations from making a decision to study. We really cannot speak for future generations of Utahns, Indian or non-Indian. A far-sighted solution was implemented in the case of the Great Salt Lake remains. They are interred in a burial vault with careful record keeping and managed by the tribes under the auspices of the Utah Division of Indian Affairs (Simms and Raymond 1999).

Despite progress in federal and state land managing agencies, the management of cultural resources along Great Salt Lake has primarily been reactive in the face of developments such as highways, parking lots, buildings, dikes and canals, sewers, park development, and vandals. If we could move beyond the view that cultural resources are either a nuisance to development, or are only the object of preservation for preservation's sake, both the educational value of the past and the management task of selective preservation could be approached in a more proactive way. The Great Salt Lake area has a long history of professional and avocational archaeology, and given its proximity to urban areas, land managers should consider the current and potential uses of the archaeological resources by the public. Possibilities include programs of field recording and excavation, laboratory analysis, liaison with local Native American tribes, interpretive displays, and educational programs involving schools and public-service groups. Efforts along these lines are evident at Antelope Island State Park, and in some Wasatch Front towns such as Marriott. The potential for this form of inclusive management is great at the Bear River Migratory Bird Refuge (U.S. Fish and Wildlife Service), the Willard Bay area (Utah State Parks and U.S. Bureau of Reclamation), and in communities such as Layton, Syracuse, and Plain City to name a few.

The cultural resources of Great Salt Lake are a fragile and nonrenewable window to our past that helps us understand the present. The high frequency of ancient human skeletal remains add moral and political ingredients to the challenge of management. Cultural resources deserve a rightful place in the suite of management responsibilities accorded local, state and federal governments, and in the agenda of preservationist groups such as the Friends of Great Salt Lake and the Sierra Club who serve as advocates for the rich natural and heritage resources of the Great Salt Lake wetlands.

ACKNOWLEDGMENTS

Archaeological research along the Great Salt Lake is supported by the National Science Foundation, the State of Utah, the U. S. Bureau of Reclamation, Utah State University, the University of Utah, and the U. S. Fish and Wildlife Service. Knowledge of the ancient Wasatch Front comes from amateur archaeologists, professionals, students, and the public in numbers too great to extend individual thanks despite the magnitude of many personal contributions. We thank members of the Utah Statewide Archaeological Society, the Utah State University Archaeology Field Schools, the Utah Division of State History, a host of scientists who donated time and expertise. We especially thank the people of the Northwestern Band of the Shoshone Nation who continue to seek a voice in the management of their ancestral homelands, and who take an active interest in the ancient history of the region we all call home.

REFERENCES

Aikens, C.M., 1966, Fremont-Promontory-Plains relationships, including a report of excavations at the Injun Creek and Bear River no. 1 sites, northern Utah: Salt Lake City, University of Utah Press, Anthropological Papers 82, 102 p.

—1967, Excavations at Snake Rock Village and the Bear River No. 2 site: Salt Lake City, University of Utah Press, Anthropological Papers 87, 65 p.

Allison, J.R., 1997, Legacy-West Davis Highway Cultural Resource Survey Design: Baseline Data, Inc, Orem, Utah.

Atwood, Genevieve, 1994, Geomorphology applied to flooding problems of closed-basin lakes... specifically Great Salt Lake, Utah: Geomorphology, v. 10, p. 197-219.

Beck, Charlotte, and Jones, G.T., 1992, New directions? - Great Basin archaeology in the 1990s: Journal of California and Great Basin Anthropology, v. 14, p. 22-36.

Bright, J.R., and Loveland, C.J., 1999, A biological perspective on prehistoric human adaptation in the Great Salt Lake wetlands, *in* Hemphill, B.E, and Larsen, C.S., editors, Prehistoric life ways in the Great Basin wetlands - Bioarchaeological Reconstruction and Interpretation: Salt Lake City, University of Utah Press, 394 p.

Brooks, R.H., and Brooks, Sheilagh, 1990, An interpretation of forager burial patterns in the Great Basin, suggested from analysis of human remains in two Nevada cave sites: Paper presented at the 22nd Great Basin Anthropological Conference, Reno, Nevada.

Campbell, S.K., 1990, Post-Columbian culture history in the northern Columbia Plateau A.D. 1500-1900: New York, Garland Publishers, 228 p.

Coltrain, J.B., 1997, Fremont economic diversity - a stable carbon isotope study of formative subsistence practices in the eastern Great Basin: Salt Lake City, University of Utah, Department of Anthropology, Ph.D. dissertation, 212 p.

Coltrain, J.B., and Stafford, T.W., Jr., 1999, Stable carbon isotopes and Salt Lake wetlands diet - towards an understanding of the Great Basin formative, *in* Hemphill, B.E., and Larsen, C.S., editors, Understanding prehistoric life ways in the Great Basin wetlands - bioarchaeological reconstruction and interpretation: Salt Lake City, University of Utah Press, 394 p.

Cornell, A., Stuart, M.E., and. Simms, S.R, 1992, An obsidian cache from the Great Salt Lake wetlands, Weber County, Utah: Utah Archaeology, v. 5, p. 154-159.

Currey, D.R., 1990, Quaternary paleolakes in the evolution of semi-desert basins, with special emphasis on Lake Bonneville and the Great Basin: USA Palaeogeography, Palaeoclimatology, Palaeoecology, v. 76, p. 189-214.

Currey, D.R., Genevieve Atwood, and D.R. Mabey, 1984, Major levels of Great Salt Lake and Lake Bonneville: Utah Geological and Mineralogical Survey, Map 73, scale 1:750,000.

Currey, D.R., and James, S.R, 1982, Paleoenvironments of the northeastern Great Basin and northeastern Basin Rim Region - a review of geological and biological evidence, *in* Madsen, D.B., and O'Connell, J.F., editors, Man and environment in the Great Basin: Washington, D.C., Society for American Archaeology, SAA Papers 2, p. 27-52,

Dean, P.A., 1992, Prehistoric pottery in the northeastern Great Basin - problems in the classification and archaeological interpretation of undecorated Fremont and Shoshoni wares: Eugene, University of Oregon, Department of Anthropology, Ph.D. dissertation, 248 p.

Dewey, J.R., 1966, Evidence of acculturation among the Indians of northern Utah and southeast Idaho - a historical approach: Salt Lake City, Utah Archaeology, v. 12, no. 3. p. 3-10.

Dodd, W.A., 1982, Final year excavations at the Evans mound site: Salt Lake City, University of Utah Press, Anthropological Papers 106.

Elston, R.G., 1988, Flaked stone tools, *in* Preliminary investigations in Stillwater Marsh - human history and geoarchaeology: Portland, Oregon, U.S. Fish and Wildlife Service

Cultural Resources Series No. 1, v. 1, p. 155-183.

Enger, W.D. and Blair, W., 1947, Crania from the Warren Mounds and their possible significance to northern periphery archaeology: American Antiquity, v. 13, p.142-146.

Fawcett, W.B. and Simms, S.R., 1993, Archaeological test excavations in the Great Salt Lake wetlands and associated analyses: Logan, Utah State University Contributions to Anthropology 14, various pagings.

Flannery, K.V., 1972, The origins of the village as a settlement type in Mesoamerica and the Near East - a comparative study, *in* Ucko, P.J., Tringham, R. and Dimbleby, G.W., editors, Man, settlement, and urbanism: London, Duckworth, p. 22-53.

Fowler, C.S., 1992, In the shadow of Fox Peak - An ethnography of the Cattail- Eater Northern Paiute People of Stillwater Marsh: Portland, Oregon, U. S. Fish and Wildlife Service, Cultural Resource Series 5.

Fremont, J.C., 1988, The exploring expedition to the Rocky Mountains: Reprinted by the Smithsonian Institution Press, Washington D.C. (original 1845), 319 p.

Fritts, H.C. and Shao, X.M., 1992, Mapping climate using tree-rings from western North America *in* Bradley, R.S., and Jones, P.D., editors, Climate Since A.D. 1500: New York, Routledge, p. 269-295.

Fry, G.F., and Dalley, G.F., 1979, The Levee site and the Knoll site: Salt Lake City, University of Utah Press, Anthropological Papers 100, 113 p.

Gilman, P.A., 1987, Architecture as artifact - pit structures and pueblos in the American southwest: American Antiquity v. 52, p. 538-564.

Gunnerson, J.H., 1969, The Fremont Culture - a study in culture dynamics on the northern Anasazi frontier: Cambridge, Harvard University, Papers of the Peabody Museum of Archaeology and Ethnology, v. 59, no. 2, 221 p.

Hassell, F.K., 1961, An open site near Plain City, Utah: Salt Lake City, Utah Archaeologist, v. 7, no. 2.

Hunter-Anderson, R.L., 1977, A theoretical approach to the study of house form, *in* Binford, L.R., editor, For theory building in archaeology: New York, Academic Press, p. 287-315.

Janetski, J.C., 1990, Wetlands in Utah Valley prehistory, *in* Janetski, J.C. and Madsen, D.B., editors, Wetland adaptations in the Great Basin: Provo, Brigham Young University Museum of Peoples and Cultures Occasional Paper 1., p. 233-258.

—1994, Recent transitions in the eastern Great Basin - the archaeological record, *in* Madsen, D.B., and Rhode, D., Across the West - human population movement and the expansion of the Numa: Salt Lake City, University of Utah Press, p. 157-178.

Janetski, J.C., and Talbot, R.K., 1997, Social and community organization, *in* Janetski, J.C., Talbot, R.K., Newman, D.E., Richens, L.D., Wilde, J.D., Baker, S.B., and Billat, S.E., editors, Clear Creek Canyon archaeological project, v. 5, - results and synthesis: Provo, Brigham Young University, Museum of Peoples and Cultures Technical Series 95-9. p. 309-328.

Jennings, J.D., 1978, Prehistory of Utah and the eastern Great Basin: Salt Lake City, University of Utah Press, Anthropological Papers 98, 263 p.

Judd, N.M., 1926, Archaeological observations north of the Rio Colorado: Washington D.C., Bureau of American Ethnology Bulletin 82,171 p.

Kelly, R.L., 1988, The three sides of a biface: American Antiquity v. 53, p. 717-734.

—1995, Hunter-gatherer lifeways in the Carson Desert - a context for bioarchaeology, *in* Larsen, C.S., and Kelly, R.L., Bioarchaeology of the Stillwater Marsh - prehistoric human adaptation in the Western Great Basin: New York, American Museum of Natural History, Anthropological papers v. 77, p. 12-32.

Kennedy, O.A., 1930, Old Indian relics of Willard puzzle, Salt Lake Tribune newspaper article. Filed in box 8, file 7, Charles Kelly collection, scrapbook on archaeology: Salt Lake City, Utah State Historical Society.

Larsen, C.S., and Kelly, R.L., editors, 1995, Bioarchaeology of the Stillwater Marsh - prehistoric human adaptation in the western Great Basin: New York, American Museum of Natural History, Anthropological Papers 77, 177 p.

Larsen, C.S., Ruff, C.B., and Kelly, R.L., 1995, Structural analysis of the Stillwater postcranial human remains - behavioral implications of articular joint pathology and long bone diaphyseal morphology, *in* Larsen C.S., and Kelly, R.L., Bioarchaeology of the Stillwater Marsh - prehistoric human adaptation in the western Great Basin: New York, American Museum of Natural History, Anthropological Papers 77, p. 107-132.

Lupo, K.D., and Schmitt, D.N., 1997, On late Holocene variability in bison and human subsistence strategies in the northeastern Great Basin: Journal of California and Great Basin Anthropology, v. 19, p. 50-69.

Mabey, D.R., 1986, Notes on the historic high level of Great Salt Lake: Utah Geological and Mineralogical Survey Notes, v. 20, p. 13-15.

Madsen, B.D., 1985, The Shoshoni Frontier and the Bear River massacre: Salt Lake City, University of Utah Press, 285 p.

—1986, Chief Pocatello, the white plume: Salt Lake City, University of Utah Press, 142 p.

Madsen, D.B., 1982, Get it where the gettin's good - a variable model of Great Basin subsistence and settlement based on data from the eastern Great Basin, *in* Madsen, D.B., and O'Connell, J.F., editors, Man and environment in the Great Basin: Washington, D. C., Society For American Archaeology Papers 2, p. 207-226.

—1989, Exploring the Fremont: Salt Lake City, Utah Museum of Natural History, University of Utah Occasional Publication 8, 70 p.

Madsen, D.B., and Lindsay, L.W., 1977, Backhoe village: Salt Lake City, Antiquities Section Selected Papers 4, 119 p.

Madsen, D.B., and Simms, S.R., 1998, The Fremont complex - a behavioral perspective: Journal of World Prehistory v. 12, p. 255-336.

Maguire, Don, 1892, Report of the Department of Ethnology: Utah World's Fair Commission in Utah at the World's Columbian Exposition, p. 105-110.

Manful, Elvira, 1938, George East - pioneer personal history: Salt Lake City, Utah State Historical Society.

Marwitt, J.P., 1970, Median Village and Fremont Culture regional variation: Salt Lake City, University of Utah Press, Anthropological Papers 95, 193 p.

Morss, Noel, 1931, The ancient culture of the Fremont River in Utah - report on the Claflin-Emerson Fund, 1928-29: Cambridge, Harvard University, Papers of the Peabody Museum of American Archaeology and Ethnology, 81 p.

Murchison, S.B., 1989, Fluctuation history of Great Salt Lake, Utah, during the last 13,000 years: Limneotectonics Laboratory Technical Report 89-2: Salt Lake City, Department of Geography, University of Utah, 137 p.

Murphy, R.F., and Murphy, Yolanda, 1960, Shoshone-Bannock subsistence and society: Berkeley, University of California Anthropological Records v. 16, no. 7, p. 293-338 .

O'Rourke, D.H., Parr, R.L., and Carlyle, S.W., 1999, Molecular genetic variation in prehistoric inhabitants of the eastern Great Basin, *in* Hemphill, B.E., and Larsen C.S., editors, Understanding prehistoric life ways in the Great Basin wetlands - bioarchaeological reconstruction and interpretation: Salt Lake City, University of Utah Press, 394 p..

Parr, R.L., 1998, Molecular genetic analysis of the Great Salt Lake wetlands Fremont: Salt Lake City, University of Utah, Department of Anthropology, Ph.D. dissertation, 115 p.

Parr, R.L., Carlyle, S.W., and D.H. O'Rourke, 1996, Ancient DNA analysis of Fremont Amerindians of the Great Salt Lake wetlands: American Journal of Physical Anthropology v. 99, p. 507-518.

Pendergast, David, 1961, Excavations at the Bear River site, Box Elder County, Utah: Salt Lake City, Utah Archeology, v. 7, no. 2.

Preston, William, 1996, Serpent in Eden - dispersal of foreign diseases into Pre- Mission California: Journal of California and Great Basin Anthropology v. 18, p. 2-37.

Ramenofsky, A.F., 1987, Vectors of death - the archaeology of European contact: Albuquerque, University of New Mexico Press, 300 p.

Reff, D.T., 1991, Disease, depopulation, and culture change in northwestern New Spain A.D. 1518-1764: Salt Lake City, University of Utah Press, 330 p.

Ruff, C.B., 1999, Skeletal structure and behavioral patterns of prehistoric Great Basin populations, *in* B.E. Hemphill, and C.S. Larsen, editors, Understanding prehistoric life ways in the Great Basin wetlands - Bioarchaeological Reconstruction and Interpretation: Salt Lake City, University of Utah Press, 394 p.

Schmitt, D.N., Simms, S.R., and Woodbury,G.P., 1994, Archaeological salvage Investigations at a Fremont site in the Jordan River Delta: Salt Lake City, Utah Archaeology v. 7, p. 69-80.

Shields, W F., and Dalley, G.F., 1978, The Bear River No. 3 site, in Miscellaneous collected papers 19-24: Salt Lake City, University of Utah Press, Anthropological Papers 99, p. 55-104.

Simms, S.R., 1990, Fremont transitions: Utah Archaeology v. 3, p. 1-18.

—1993, The past as commodity - consultation and the Great Salt Lake skeletons: Utah Archaeology v. 6, p. 1-5.

—1999, Farmers, foragers, and adaptive diversity - the Great Salt Lake wetlands project, *in* B.E. Hemphill, and Larsen C.S., editors, Understanding prehistoric life ways in the Great Basin wetlands - bioarchaeological reconstruction and interpretation: Salt Lake City, University of Utah Press, 394 p.

Simms, S.R., and Heath, K., 1990, Site structure of the Orbit Inn - an application of ethnoarchaeology: American Antiquity v. 55, p. 797-812.

Simms, S.R., Loveland ,C.J., and Stuart, M.E., 1991, Prehistoric human skeletal remains and the prehistory of the Great Salt Lake wetlands: Logan, Utah State University, Contributions to Anthropology 6, 92 p.

Simms, S.R., and Raymond, A.W., 1999, No one owns the deceased! - the treatment of human remains from three Great Basin cases, *in* Hemphill B.E., and Larsen C.S., editors, Understanding prehistoric life ways in the Great Basin wetlands - bioarchaeological reconstruction and interpretation: Salt Lake City, University of Utah Press, 394 p.

Simms, S.R., Ugan A., and Bright, J., 1997, Plain-ware ceramics and residential mobility - a case study from the Great Basin: Journal of Archaeological Science, v. 24, p. 779-792.

Simms, S.R., and Whitesides, S.N., 1993, Lithics and behavior, *in* Fawcett W.B., and Simms, S.R., editors, Archaeological test excavations in the Great Salt Lake wetlands and associated sites: Logan, Utah State University Contributions to Anthropology 14., p. 170-184.

Steward, J.H., 1933, Early Inhabitants of western Utah: Salt Lake City, Bulletin of the University of Utah, v. 23, no. 7, p. 1-34.

Stuart, M.E., 1993, Salvage excavations at the Fire Guard site in northern Utah: Utah Archaeology v. 6, p. 71-78.

Stuart, M.E., and Pringle, Lesa, editors, 1988, Fallen rocks shelter: Salt Lake City, Utah Statewide Archaeological Society.

Talbot, R.K., 1997, Fremont architecture, *in* Janetski, J.C., Talbot, R.K., Newman, D.E., Richens, L.D., Wilde, J.D., Baker, S.B., and Billat, S.E., editors, Clear Creek Canyon archaeological project, v. 5 - results and synthesis: Provo, Brigham Young University Museum of Peoples and Cultures Technical Series 95-9, p. 177-238.

Talbot, R.K., and Richins, L.D., 1996, Steinaker Gap - an Early Fremont farmstead: Provo, Brigham Young University Museum of Peoples and Cultures Occasional Papers 2, 237 p.

Trenholm, Virginia, and Carley, Maurine, 1964, The Shoshonis - sentinels of the Rockies: Norman, University of Oklahoma Press, 367 p.

Verano, J.W., and Ubelaker, D.H., editors, 1992, Disease and demography in the Americas: Washington, D.C., Smithsonian Institution Press, 294 p.

Warner, T.J., editor, 1976, The Dominguez-Escalante journal, with Fray Angelico Chavez (translator): Provo, Brigham Young University Press, 203 p.

Whiting, J.W., and Ayres, B.,1968, Inferences from the shape of dwellings, *in* Chang, K.C., editor, Settlement archaeology: Palo Alto, National Press Book, p. 117-133.

Wilson, E.N., 1991, The white Indian boy: Salt Lake City, Paragon Press, 394 p.

Zeanah, D.W., 1988, Pit feature and food storage strategies in the Great Basin. Paper presented at the 21st Great Basin Anthropological Conference, Park City, Utah: Ms. on file with author, Department of Anthropology, Sacramento, California State University.

CHEMISTRY/HYDROLOGY/LIMNOLOGY

GREAT SALT LAKE, UTAH: CHEMICAL AND PHYSICAL VARIATIONS OF THE BRINE AND EFFECTS OF THE SPRR CAUSEWAY, 1966-1996

by
J. Wallace Gwynn

This article is reprinted, with permission, from Utah Geological Association Guidebook 26, 1998.

ABSTRACT

The Utah Geological Survey has monitored total dissolved solids (TDS), chemistry, and temperature of Great Salt Lake brine since 1966. Analyses of these data, along with complementary lake-surface elevation and hypsographic data (lake stage, area, and volume) from the U.S. Geological Survey (USGS), have allowed the recognition of long-term changes of this natural resource.

From 1966 to 1996, the dry-weight percentages of potassium, magnesium, and sulfate have declined in the lake compared to sodium and chloride. This decline is possibly due to re-solution of sodium chloride that has precipitated earlier on the bottom of the lake's north arm (1960s through 1980s) and south arm (1960s through mid-1971).

It is presumed that the lake brines were homogeneous throughout the lake prior to the construction of the SPRR causeway in 1959. Between 1959 and 1966, a stratified-brine condition (light brine overlying a heavy brine) developed in the south arm, but not in the north. In 1984, the railroad causeway was breached as a flood-control measure to alleviate rising flood levels in the south arm. Bi-directional flow through the breach caused a major redistribution of the lake's dissolved-salt load and formed stratified-brine conditions in the north arm. It is assumed that compaction and subsequent addition of fill material to the SPRR causeway, to raise the railroad tracks during the 1980s flooding, resulted in a reduction in the causeway's overall hydraulic conductivity. Declining lake levels after 1987 greatly reduced the bi-directional flow through the breach opening. Combined, these last two changes played a major role in bringing an end to the stratified-brine conditions in both arms of the lake by mid-1991. Lake elevation versus TDS relationships vary and are dependent upon the lake-system's hydrodynamics.

Salt-load calculations suggest that there are about 4.3 billion tons of salt in the entire lake, that the distribution of salt between the two arms is transitory, and that a number of factors make it difficult to accurately determine the individual-arm salt loads, or the combined load for the lake. The West Desert Pumping Project removed over 600 million tons of salt from Great Salt Lake between 1987 and 1989.

INTRODUCTION

The Utah Geological Survey's lake-brine monitoring program supplies chemical, density, and temperature baseline data for the Great Salt Lake over the past 30 years. These data are needed to: (1) measure the effect of changes in these parameters on the lake's ecosystem, (2) resolve conflicts between competing users of the lake, (3) maximize the state's use of, and profit from, the lake's vast resources, and (4) characterize the nature and complexities of the lake's brine system, which might be useful for understanding ancient lacustrine deposits.

Effect of the Causeway on the Lake Brine

Great Salt Lake (GSL) was separated into two distinct brine bodies, the south and north arms, by the Southern Pacific Railroad's (SPRR) rock-fill causeway, constructed from Promontory Point to the west shore of the lake in 1959 (figure 1). The pre-causeway brine density was probably relatively uniform throughout the lake. After construction of the causeway, mixing was restricted to salt movement through the causeway's two 15-foot (4.57 m) wide by 20-foot (6.09 m) deep culverts and the porous structure of the causeway. The salt content of the south arm, which receives fresh water from the Bear, Weber, and Jordan Rivers, varies inversely with lake-surface elevation. The north arm, which receives mostly salty water moving through the causeway from the south arm, stayed near saturation until the 1980s flooding. Brine-density stratification was observed in the Utah Geological Survey's (UGS) brine-sampling program in the south arm from 1966 through mid-1991 and in the north arm from 1983 through mid-1991. Farmington and Bear River Bays, which are separated from the main body of the south arm respectively by the Antelope-Syracuse causeway and the SPRR causeway, both have lower brine densities than the south arm and are brine-density stratified.

In 1984, during a period of severe flooding of facilities around the south arm of the lake, the State of Utah cut a 300-foot (91 m) long breach into the SPRR causeway to allow faster flow of south-arm brines to the north arm, reducing the nearly four-foot head differential that had developed between the two arms (the south arm being the highest). From 1984 until the latter part of 1988 the dissolved-salt load in the south arm of the lake increased while that in the north arm decreased. This movement of dissolved salt was due to density stratified, bi-directional flow with less salty south-arm brine flowing into the north arm, and more salty north-arm brine moving southward, as return flow, through the lower depths of the breach opening and causeway fill. South-to-north and north-to-south exchange of brines decreased after 1987 due to: (1) declining lake elevation primarily from decreased precipitation and (2) an apparent drop in hydraulic

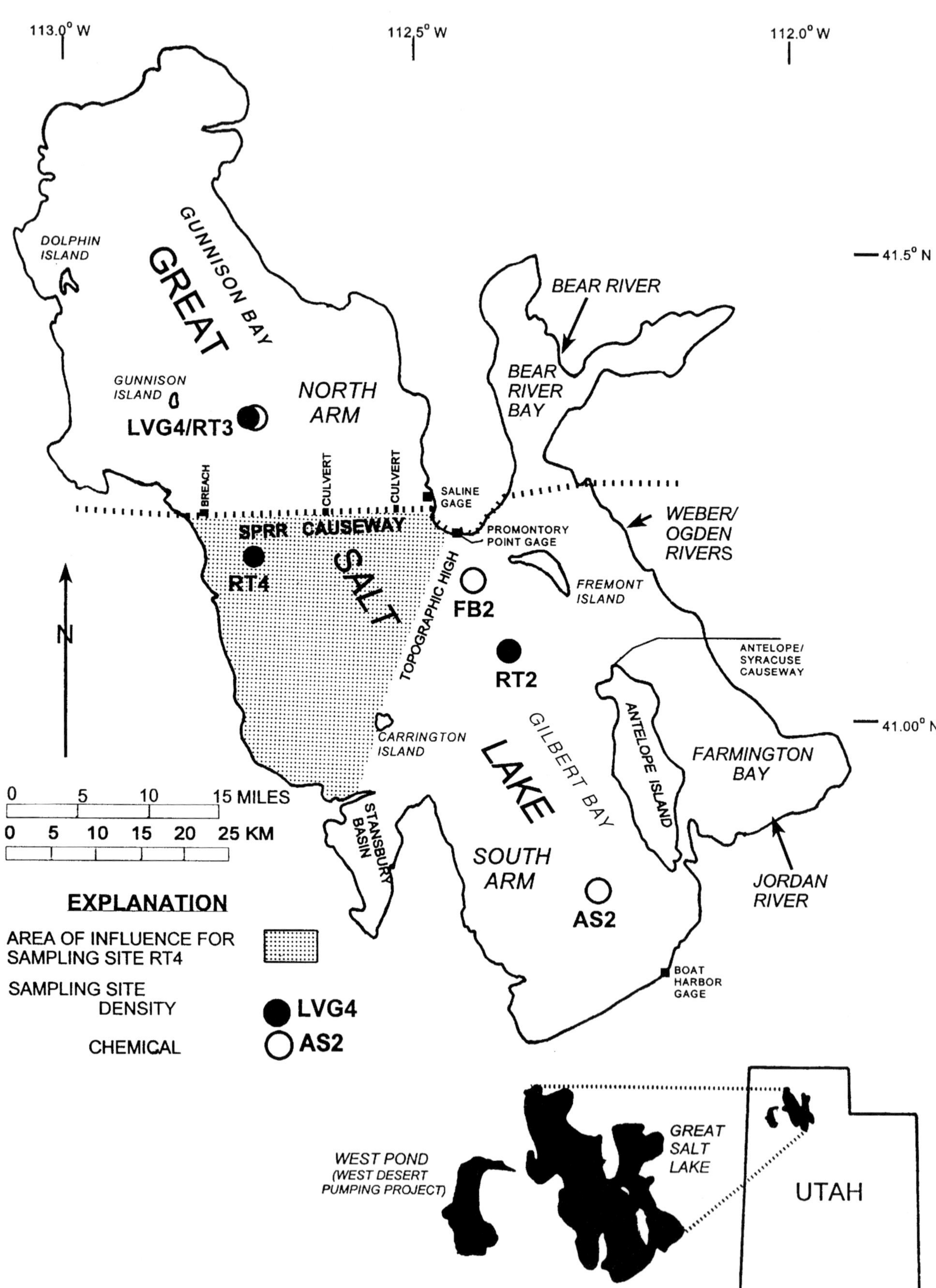

Figure 1. *Great Salt Lake showing locations of brine-chemistry and density sampling-site locations, lake-level monitoring gages, and the area of influence for sampling site RT.*

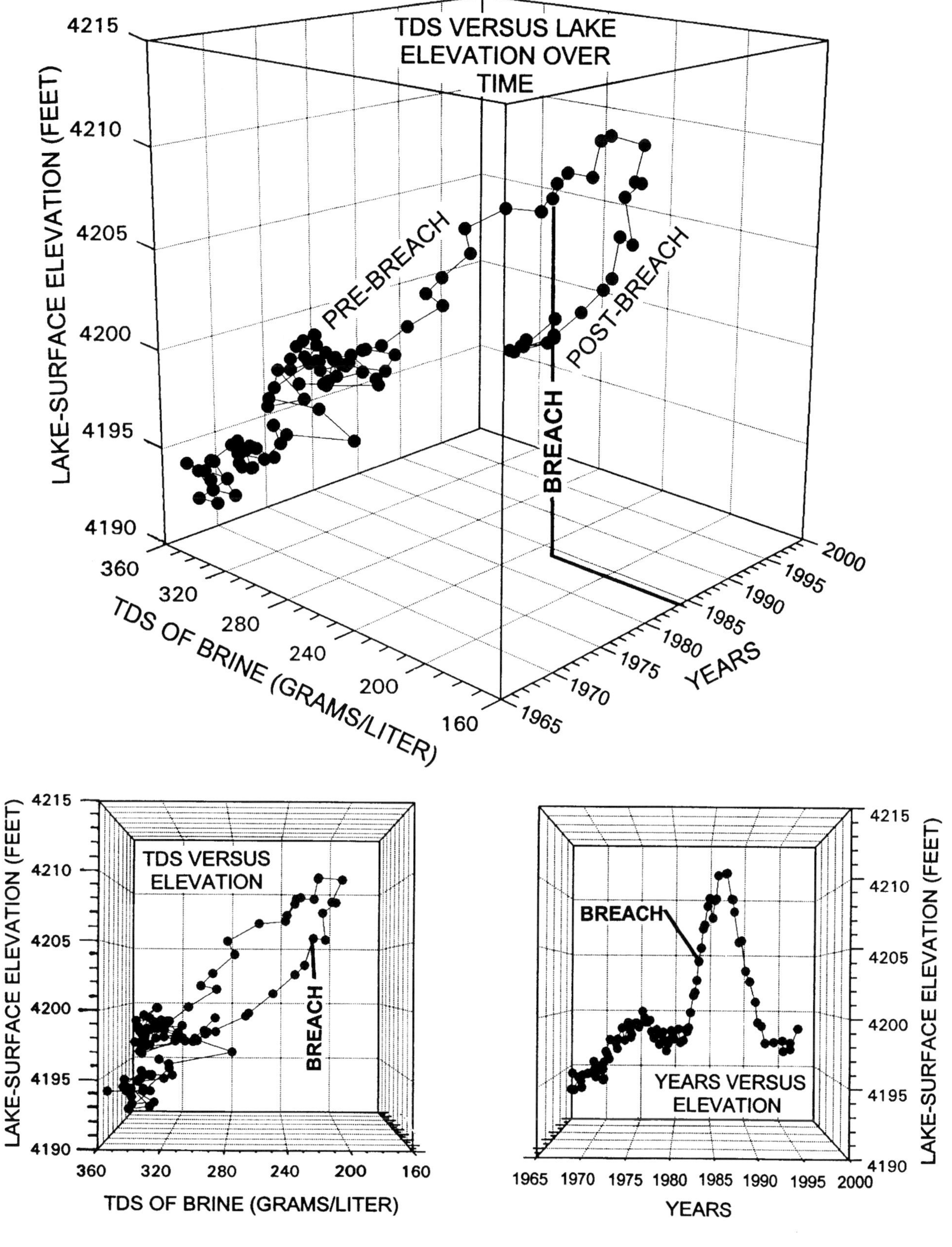

Figure 3. *Three-dimension plot of TDS of north-arm brines versus lake elevation over time (months since December 1965), and XY plots of time and lake elevation versus TDS.*

1984. During the first two months after the causeway was breached, the north arm received about 1-million acre-feet (1,230 hm^3) of dilute south-arm brine as the 3.45-foot (1.05 m) head differential that had developed between the two arms was reduced to 0.95 feet (0.29 m). This influx of dilute brine into the north arm, together with overall increased rainfall, and continued but much lower brine-inflow rates from the south arm, exceeded the concentrating effect of evaporation. For the first time since 1966, a significant inverse relationship existed between brine TDS and north-arm lake level. During the same time, the south arm received large quantities of dense, north-arm return-flow brine which increased the south-arm's average TDS. Thus, at a given lake elevation, south-arm TDS was higher than before the flooding started and the breach was opened. As a result, the TDS versus lake-elevation relationship changed for both arms of the lake during this period of time.

1993 - Present Relationship

The lake-level decreases during 1993 to present caused a large change in the TDS versus lake-level relationship. This change has been due to a number of factors that influence the lake's water and salt balance, to include the following. First, as the level of the lake dropped to the approximate 4,199.5-foot bottom elevation of the breach opening, there was only south-to-north flow into the north arm or no flow at all. Since 1995, however, the south-to-north flow rate has increased each year as the lake level has risen. Second, during the 1980s flooding, the elevation of the SPRR causeway was raised with fines-containing quarry-run fill materials. It is assumed that the addition of these materials, and subsequent compaction have caused a significant decrease in the causeway's hydraulic conductivity. This apparent decrease in hydraulic conductivity has greatly reduced and/or stopped the deep, return flow of north-arm brine into the south arm, as evidenced by the absence of high-density brine within the vertical density profile at sampling site RT4 (figure 1). Combined, all factors are creating a dynamic TDS versus lake-elevation relationship in both arms of the lake. (1) The south-arm brine TDS at a given elevation is steadily declining. This trend will probably continue until the lake elevation increases to the point that bi-directional flow occurs within the breach opening, or additional construction is done to allow more north-south brine-mixing. (2) Since 1992, the north-arm brine has become concentrated to saturation levels during the summer months precipitating salt on the bottom of the lake. Under these conditions, the north-arm TDS does not show an inverse relationship to lake level. This trend will probably also continue into the foreseeable future, regardless of moderate, yearly lake-level fluctuations.

GREAT SALT LAKE BRINE CHEMISTRY

South-Arm Ion Concentrations and Ratios

South-arm concentrations of the six major ions (Na+, K+, Mg++, Ca++, Cl-, and SO_4--), three trace elements Br, Li, and B, and TDS vary inversely with lake elevation (figures 4, 5, and 6). The yearly-average south-arm dry-weight percentages (ratios of the ions in the dissolved salt) of the six major ions are shown in figure 7. These data suggest a slight decline of the sulfate, magnesium, potassium, and calcium ions over time, whereas sodium and chloride show a slight increase (the increase is difficult to see since figure 7 is a logarithmic plot).

North-Arm Ion Concentrations and Ratios

The yearly-average north-arm concentrations of the six major ions (Na+, K+, Mg++, Ca++,Cl-, SO_4--) and TDS, are

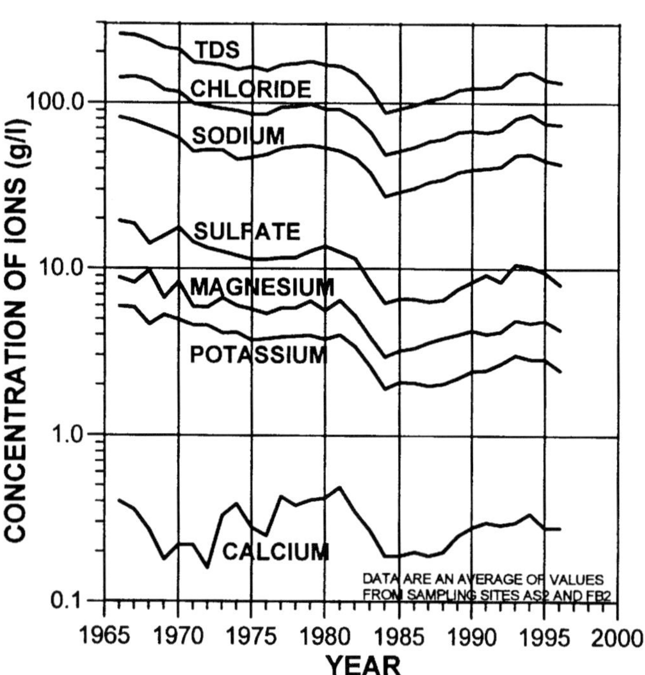

Figure 4. Yearly average of brine TDS and major ion concentrations versus year, for the south arm of Great Salt Lake, Utah.

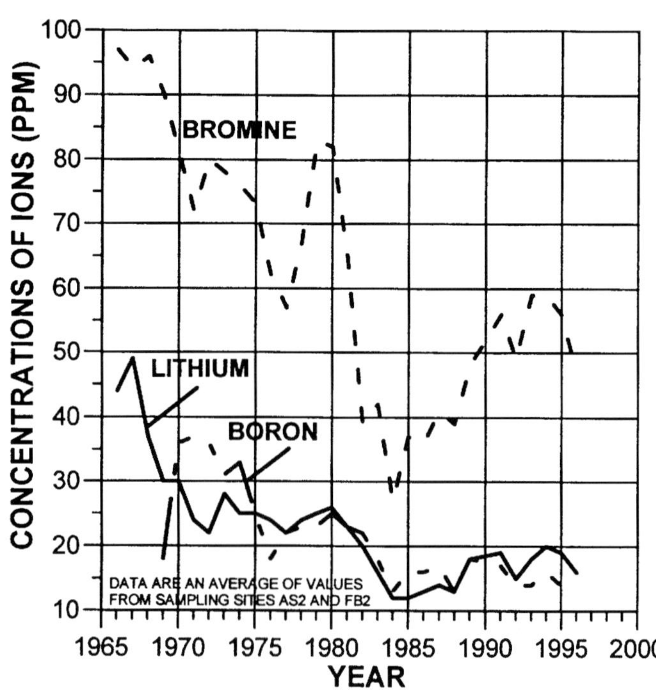

Figure 5. Yearly average bromine, lithium and boron versus year, for the south arm of Great Salt Lake, Utah.

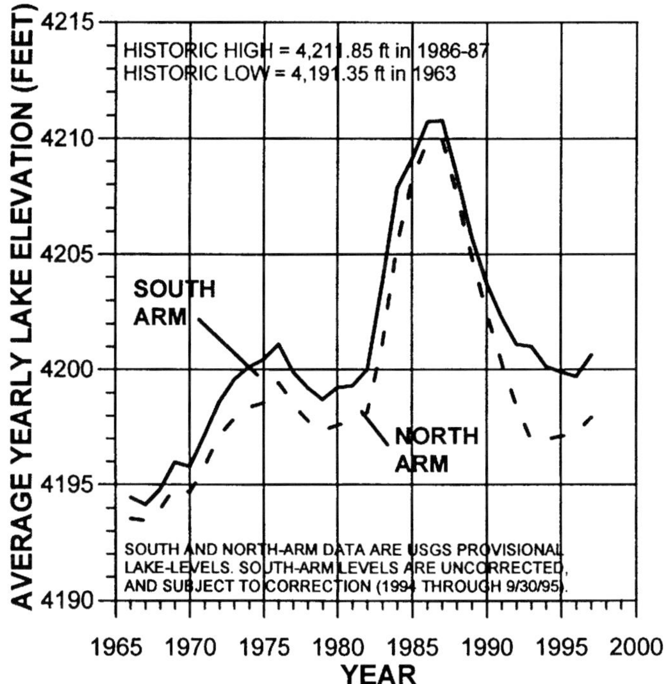

Figure 6. *Average yearly USGS provisional lake elevations, in feet above mean sea level, versus year, for the south and north arms of Great Salt Lake, Utah.*

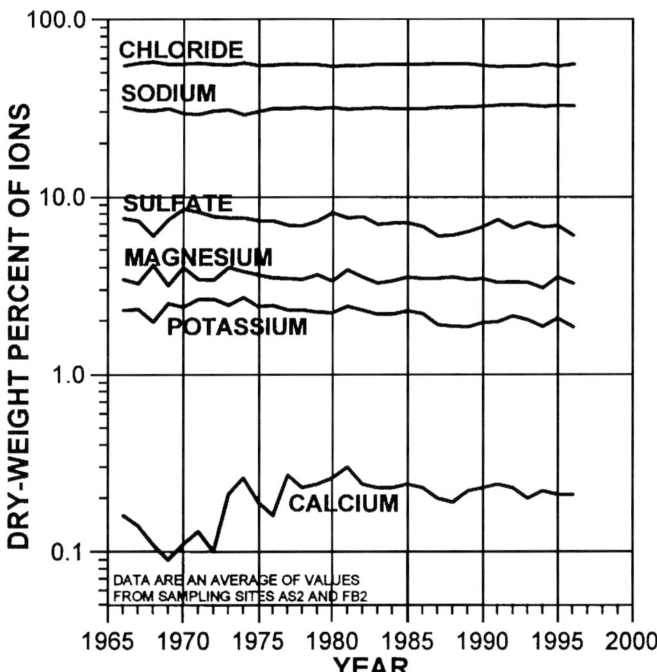

Figure 7. *Yearly average of dry-weight percents for the major ions versus year, for the south arm of Great Salt Lake, Utah.*

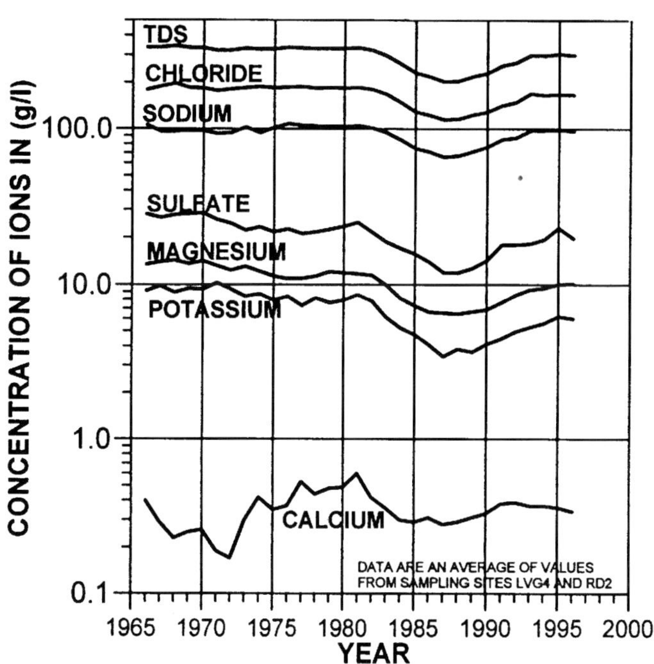

Figure 8. *Yearly average of brine TDS and major ion concentrations versus year, for the north arm of Great Salt Lake, Utah.*

shown in figure 8, and figure 9 shows the changes in bromine, lithium, and boron. Unlike the south arm, these data do not show an inverse relationship of ion and TDS concentrations versus lake levels (figure 6) for most of the period from 1966 through 1981. The north arm of the lake remained at or near the point of sodium chloride saturation because water lost to evaporation equaled or exceeded the small amount of fresh-water inflow. The north-arm TDS and lake-levels were inversely related only when rainfall and dilute inflow from the south arm exceeded the north-arm evaporation rate, starting in 1983.

The yearly-average north-arm dry-weight percent of the six major ions (figure 10) show the same decline of sulfate, magnesium, potassium, and calcium ions over time, as seen in the south-arm brines, whereas sodium and chloride show a very slight increase over time.

Possible Causes of Ion-Ratio Changes

During the low surface-elevation stages of the lake, from 1935 to 1945 and from 1959 into the mid-1960s, sodium chloride precipitated in those portions of the lake now referred to as the north and south arms. Madison (1970) states that salt precipitated at or below an elevation of about 4,195 feet and Whelan (1973) reports that some 1.21 billion metric tons of sodium chloride precipitated throughout the lake. While the precipitated salt in the south arm had dissolved by mid-1972, it took until about 1986 before all the salt had dissolved from the floor of the north arm (Wold and others, 1996). In 1992, salt began to precipitate again in the north arm during the summer months; precipitation continues in 1997. The dry-weight percents of the potassium, magnesium, and sulfate ions in the brine were increased because of the precipitation of great quantities of sodium chloride during the lake's historic-low stages. The present (1997) declining trends seen in these ions is probably due to the pre-1992 dissolution of crystalline sodium chloride on the lake bottom, or the upward diffusion of high-sodium chloride brines from the bottom sediments, thus adding to the increasingly larger percentage of sodium and chloride in the TDS mix. There has been sufficient exchange and mixing of brines between the south and north arms that the same general ion-ratio trends are seen in each arm of the lake.

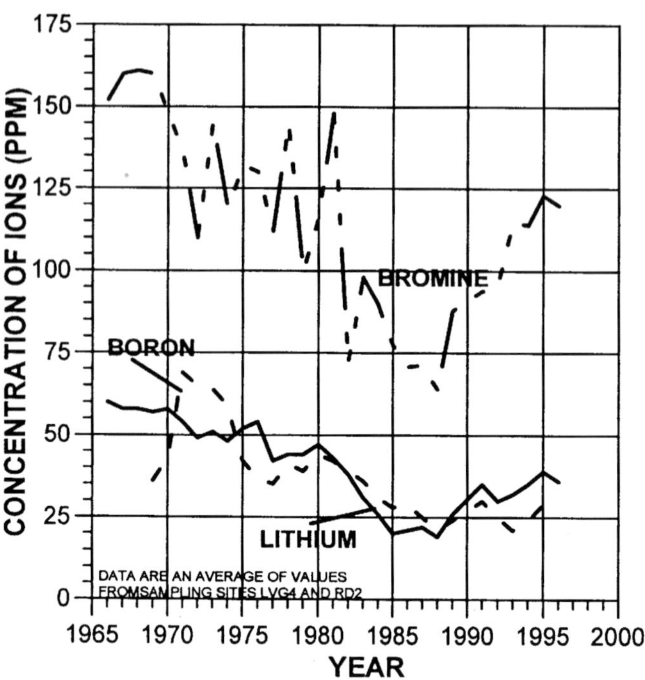

Figure 9. Yearly average bromine, lithium and boron versus year, for the north arm of Great Salt Lake, Utah.

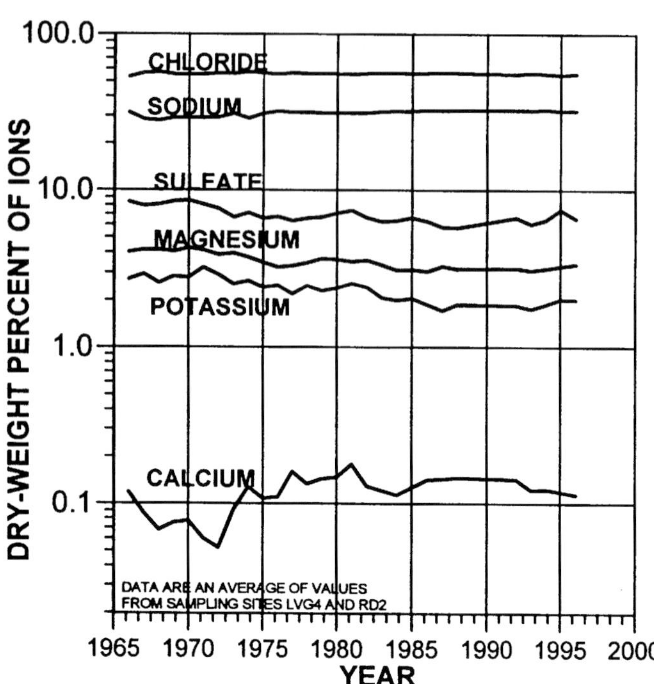

Figure 10. Yearly average of dry-weight percents for the major ions versus year, for the north arm of Great Salt Lake, Utah.

BI-DIRECTIONAL FLOW AND BRINE STRATIFICATION

Bi-Directional Flow

Understanding the parameters responsible for creating and maintaining bi-directional flow is essential to understanding the development and maintenance of brine stratification in the south and north arms of the lake. If the proper head-differential, hydraulic conductivity, and brine-density conditions exist across the causeway, lighter, surficial, south-arm brine moves northward through the culverts, breach, (or the permeable causeway fill) by gravity flow, as the heavier north-arm brine moves southward at depth under hydrostatic pressure (figure 11). The phenomenon of brines of differing densities moving in opposite directions through the same opening is referred to as bi-directional flow. The stratified-brine interface shown in the south arm (figure 11) existed from pre-1966 until mid-1991, but has not been present since that time due to the apparent cessation of return flow through the causeway and vertical mixing.

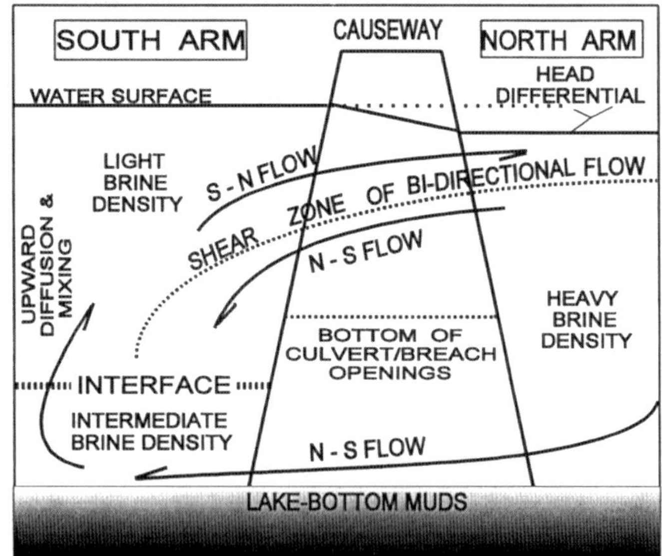

Figure 11. Cross sectional schematic diagram of bi-directional flow of brines through the Southern Pacific Railroad causeway and a typical culvert or breach opening.

Development of Brine Stratification

Pre-causeway Conditions and Early Brine Stratification

Prior to the completion of the SPRR's causeway from Promontory Point to the western shore of the lake in 1959 (figure 1), the water probably circulated freely and mixed throughout the lake. Madison (1970) describes the pre-causeway lake as follows: "Before the causeway was constructed, the dissolved-solids concentration in all parts of the lake probably was generally similar, although differences in inflow, evaporation, currents, wind, and density have always caused some variation in the concentration from place to place." After the completion of the causeway, the north arm began to develop higher densities than the south arm (Hahl and Mitchell, 1963). For example, on April 12 and 19, 1960 and July 19, 1960, three brine densities from samples taken just south of the causeway ranged from 1.192 to 1.214 grams/liter, while three from just north of the causeway ranged from 1.212 to 1.221 grams/liter. It is doubtful that brine stratification existed in the south arm as early as January 22, 1963 (Hahl and Langford, 1964). They determined at that time that all flow through the two causeway culverts was south-to-north, so there was no return flow through the culverts to feed a deep, dense, south-arm brine layer. In October 1965 and May 1966, Hahl and Handy (1969) reported the existence of stratified brine conditions south of the causeway, and in 1966, UGS researchers found that a dense, turbid layer of brine lay on the bottom of the south arm, and was overlain by a clear, less-dense brine. The two brines were separated by a thin transition zone called the interface.

South-Arm Stratification Development

The deep south-arm-brine layer below the interface developed between the time the causeway was completed in 1959 and when stratification was found by the UGS in 1966. Development of the deep, south-arm layer occurred as dense north-arm brine moved southward through the lower portions of the permeable causeway, and its culvert openings, into the depths of the south arm. Madison (1970) reports the top elevation of the deep, south-arm layer to be about 4,175 feet (1,275.5 m). By about 1970, the deep brine layer had grown to about 7 or 8 feet (2.1 to 2.4 m) thick, with its upper surface at about the 4,180-foot (1,274 m) elevation. The area of the south arm occupied by the deep brine layer is shown in figure 12. At about the 4,180-foot elevation, the return-flow rate or influx of north-arm brine to the south was apparently equal to the rates of south-arm wave-induced mixing and upward diffusion, so the deep layer became a stable element at that elevation. Brine stratification in the south arm of the lake began to change in 1984 due to breaching of the causeway, and eventually disappeared in 1991 due to flooding and changes in the SPRR causeway.

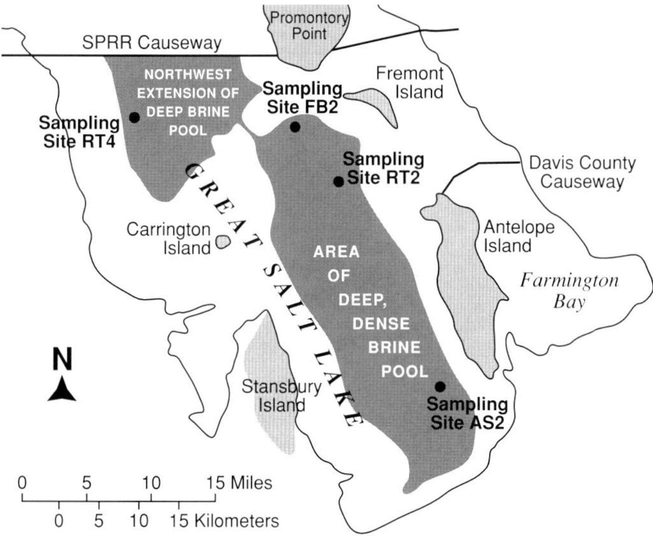

Figure 12. *South arm of Great Salt Lake showing the main and northwestern areas of the pre-1992 deep, dense brine pool below a top elevation of about 4,180 feet.*

North-Arm Stratification Development

UGS geologists noted the development of brine stratification in the north arm in early 1983 when they were doing their incremental sampling of the lake. Initially the north arm's upper waters began to be diluted by increased rainfall and greater inflow from the south arm. Then, with the opening of the breach on August 1, 1984, a flood of nearly a million acre-ft (1,233 hm^3) of light, south-arm brine flowed onto the surface of the north arm. Within a two-month period, the elevation of the south arm dropped 1.2 feet (0.36 m), the north arm rose 1.3 feet (0.39 m), and the head differential dropped to 0.95 feet (USGS provisional lake-level records). As the flood of south-arm brine subsided and the head differential dropped, north-arm brine began to move southward as return flow through the lower 3 feet (0.91 m) of the ten-foot (3.05 m) column of water in the breach opening (USGS breach-flow measurement field notes). Due to the reduced south-to-north inflow rate, the significant differences in brine densities, and a depth-to-the-interface of nearly 24 feet (7.3 m), the upper (lighter) and lower (denser) north-arm brines did not mix quickly and a stratified-brine condition developed which lasted until 1991.

Evolution and Disappearance of Stratification

By mid-1991, south-arm brine stratification had disappeared due to both a reduction of the rate of deep-return flow from the north arm and vertical mixing of upper and lower brines in the south arm. The decrease in return flow is assumed to be due mainly to an overall reduction in the SPRR causeway's hydraulic conductivity. The decrease was probably caused by: (1) the addition of fines-containing quarry-run rock to the causeway to raise it 6 to 10 feet (1.82 to 3.05 m) in height between 1983 and 1987, (2) causeway-fill compaction, and (3) clogging of the two culverts with debris and sediment. It was also due to the lowering of the elevation of the shear zone of bi-directional flow (figure 11) below the bottom of the breach opening as the lake level dropped. As the south arm dropped below a surface elevation of about 4,206 feet (1,282 m), return flow ceased, and only south-to-north flow occurred through the opening; the bottom of the breach opening is at 4,199.5 feet (1,280 m). Another factor in the disappearance of south-arm stratification was that southward movement of brine through the SPRR causeway fill decreased, probably due to compaction, and to the large volumes of fill material that had been added. North-arm brine stratification disappeared in mid-1991 when the inflow rate of lighter south-arm brine into the north arm, through the breach opening and the culverts, was reduced due to declining lake levels. Because of this reduced flow rate, vertical mixing exceeded the inflow rate, and stratification in the north arm disappeared.

Figure 13 illustrates the stratified-brine conditions that existed in the south arm of the lake from 1966 through 1991. The stratified-brine interface elevation was determined by UGS geologists from abrupt density breaks in five-foot-interval sample elevation versus TDS profiles (not included in this report). The south arm of the lake has been completely mixed vertically from mid-1991 to present. Figure 14 represents similar data for the north arm of the lake.

Sample Elevation Versus Brine-Density Plots

Figures 15a-b and 16a-b represent date-specific, sample-elevation-versus-density plots for south-arm sampling sites RT2 and RT4 respectively, and figure 17a-b represents date-specific, sample-elevation-versus-density plots for north-arm sampling site RT3 (figure 1). These data show the progres-

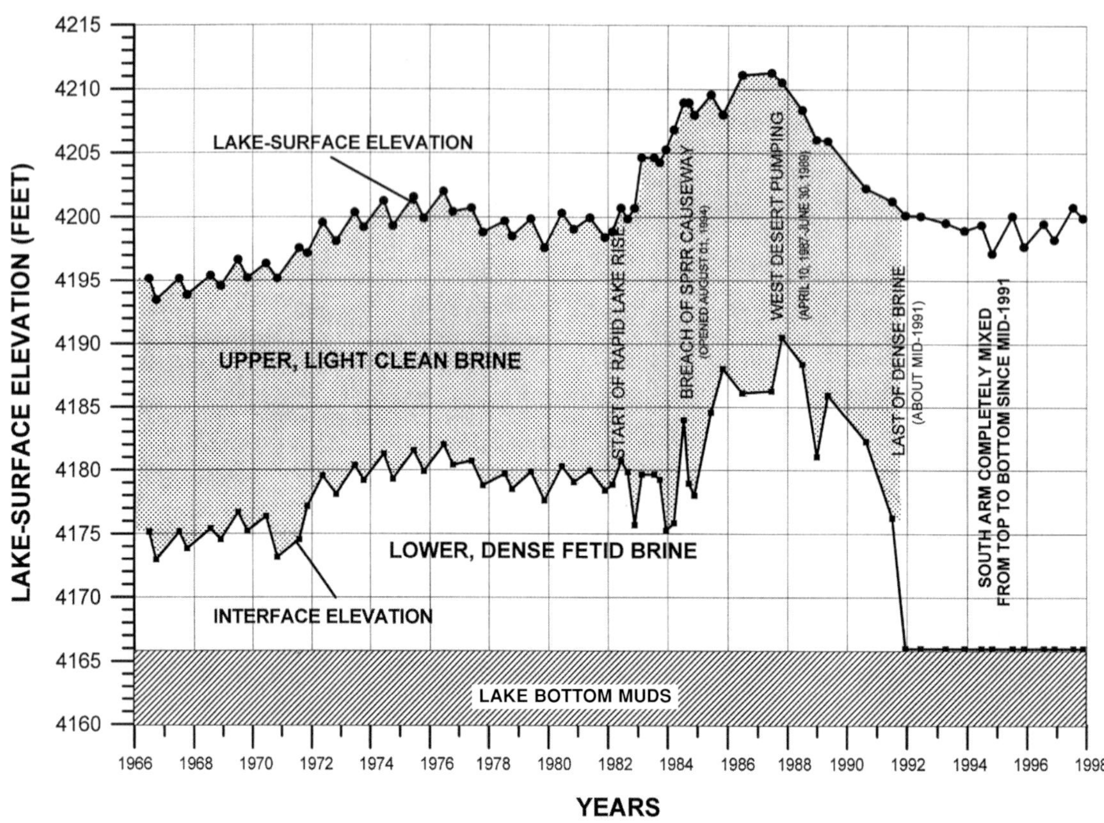

Figure 13. *Lake and interface elevations in the south arm of Great Salt Lake versus years. Specific event dates are shown. Data were collected at sample-site AS2 and FB2 (figure 1).*

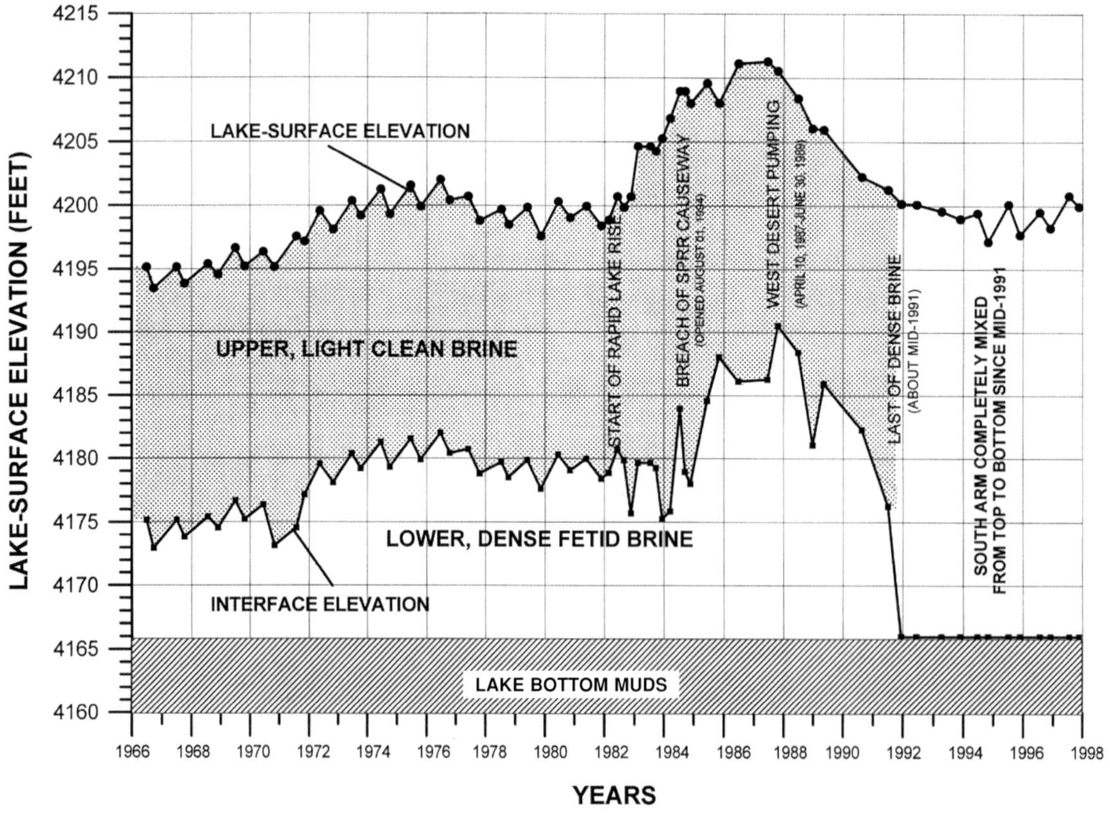

Figure 14. *Lake and interface elevations in the north arm of Great Salt Lake versus years. Specific event dates are also shown. Data were collected at sample-sites LVG4 and RD2 (figure 1).*

sive decline and the final disappearance of both north- and south-arm stratification by mid-1991.

Staged Movement of Dense Return-Flow Brine

The movement of heavy return-flow brine from the north arm into the south is "staged." Heavy brine that has moved southward through the causeway first flows into the area-of-influence for sampling site RT4 (figure 1), or the northwest extension of the deep brine pool (figure 12). From this area, the brine flows over a low, subtle topographic ridge that extends southwestward from Promontory Point to Carrington Island. Once the brine crosses over the ridge, it flows into the main, south-arm deep-brine pool. The denser brine at the bottom of the RT4 profiles (figures 16a-b), compared to the somewhat less-dense brine at the bottom of the RT2 profile (figures 15a-b) supports the staged movement concept. Differences in brine characteristics between the area of influence for sampling site RT4, and the main south-arm brine pool (figure 1) were noted by Hahl and Handy (1969) between 1963 and 1966.

LAKE-LEVEL DATA

In 1984, the head differential between the south and north arms, calculated from the USGS historical lake levels measured at the USGS Boat Harbor and Saline gages (figure 1), became progressively greater than the contemporaneous head differentials surveyed across the causeway during the USGS breach-flow measurements (1984-1994). This was brought to the attention of the USGS by the UGS and lake industry. The USGS recognizes that the north- and south-arm elevations are set to different datums which are many miles apart and have not been tied together, and that part of the problem is the unstable benchmarks available for setting the south arm gage. The cause of the benchmark instability is not known. The USGS resurveyed the gages and in 1995 issued a set of corrections to be subtracted from the south-arm lake-level records from April 15, 1984 to September 30, 1995. The correction factors issued by the USGS with the October, 1995 lake-level reading, are as follows:

April 15, 1984 to September 30, 1989.......... -0.25 feet
October 1, 1989 to September 30, 1990........ -0.35 feet
October 1, 1990 to September 30, 1991........ -0.40 feet
October 1, 1991 to September 30, 1992........ -0.50 feet
October 1, 1992 to September 30, 1993........ -0.55 feet
October 1, 1993 to September 30, 1994........ -0.60 feet
October 1, 1994 to September 30, 1995......... -0.65 feet

The corrections to the south-arm elevations issued by the USGS do not appear to be sufficient to correct for the actual observed head differences, suggesting that the south-arm elevation readings are still too high. A new set of corrections are proposed (figure 18) for past south-arm elevations, based on head differentials reported in USGS culvert and breach-flow measurement field notes. Accurate corrections are crucial to determining salt loads in the lake.

A new south-arm gage was installed by the USGS in December 1996 at the south end of Promontory Point. This gage was tied to the same benchmark (datum) as the north-arm Saline gage. Data from the new gage provide an accurate head differential between the two arms of the lake, but do not necessarily reflect absolute elevation accuracy.

SALT-LOAD CALCULATIONS

Salt-load calculations were made for the south arm, north arm, and total lake, based on the following data: UGS one-foot incremental density determinations made at sites RT2, RT3, and RT4. (see figures 15a-b, 17a-b, and 16a-b, respectively); USGS hypsographic data (Cruff, 1986); and the USGS lake-level records. When determining densities for the incremental elevations that are used in making salt-load calculations, the bottom-density value in each density profile (figure 15-17 series) is applied downward to the 4,170-foot (1,271 m) elevation. The results of the salt-load calculations are shown in figure 19. Wold and others (1996) calculate the 1986 total-salt load for the lake as 4.9 billion tons, compared to the nearly 4.4 billion tons shown on figure 19. The 0.5-billion ton disparity between these two values is probably due to differences in the use and interpretation of available TDS, hypsographic, and lake-level data. It is difficult to attribute increases or decreases in salt loads to specific causes or events because of the complex and dynamic nature of the lake; however, the following factors have all contributed to the changes in the calculated dissolved salt load in GSL:

(1) Dissolution of crystallized salt on the bottom of the north arm during the early 1980s as the lake level rose and brine density declined.

(2) Major redistribution of salt due to the breaching of the SPRR causeway in 1984 (Gwynn and Sturm, 1987).

(3) Precipitation of over 600 million tons of salt on the floor of the West Desert during the State's West Desert Pump-ing Program, 1987-1989 (Waddell and others, 1992; and Wold and Waddell, 1994).

(4) The TDS of the north arm being at or near saturation since 1992, and the precipitation of unknown quantities of salt on the floor of the north arm.

(5) The homogeneous brine conditions in the north and south arms of GSL since 1991.

(6) Changing salinities and salt loads in Farmington and Bear River Bays, depending on lake elevation mixing, and tributary-inflow rates.

(7) Problems with lake level measurements and the development of hypsographic data.

CONCLUSIONS

(1) From 1966 to the present, the dry-weight percentages of the potassium, magnesium, sulfate and calcium ions have declined while sodium and and chloride have increased. This decline is due to dissolution of sodium chloride that had precipitated earlier on the floors of the north arm (1960s through 1980s) and south arm (1960s through mid-1971).

(2) Prior to the construction of the Southern Pacific Railroad causeway in 1959, the brines were probably homogeneous throughout the lake. After 1959, a stratified-brine condition developed in the south arm, but not in the north arm at that time.

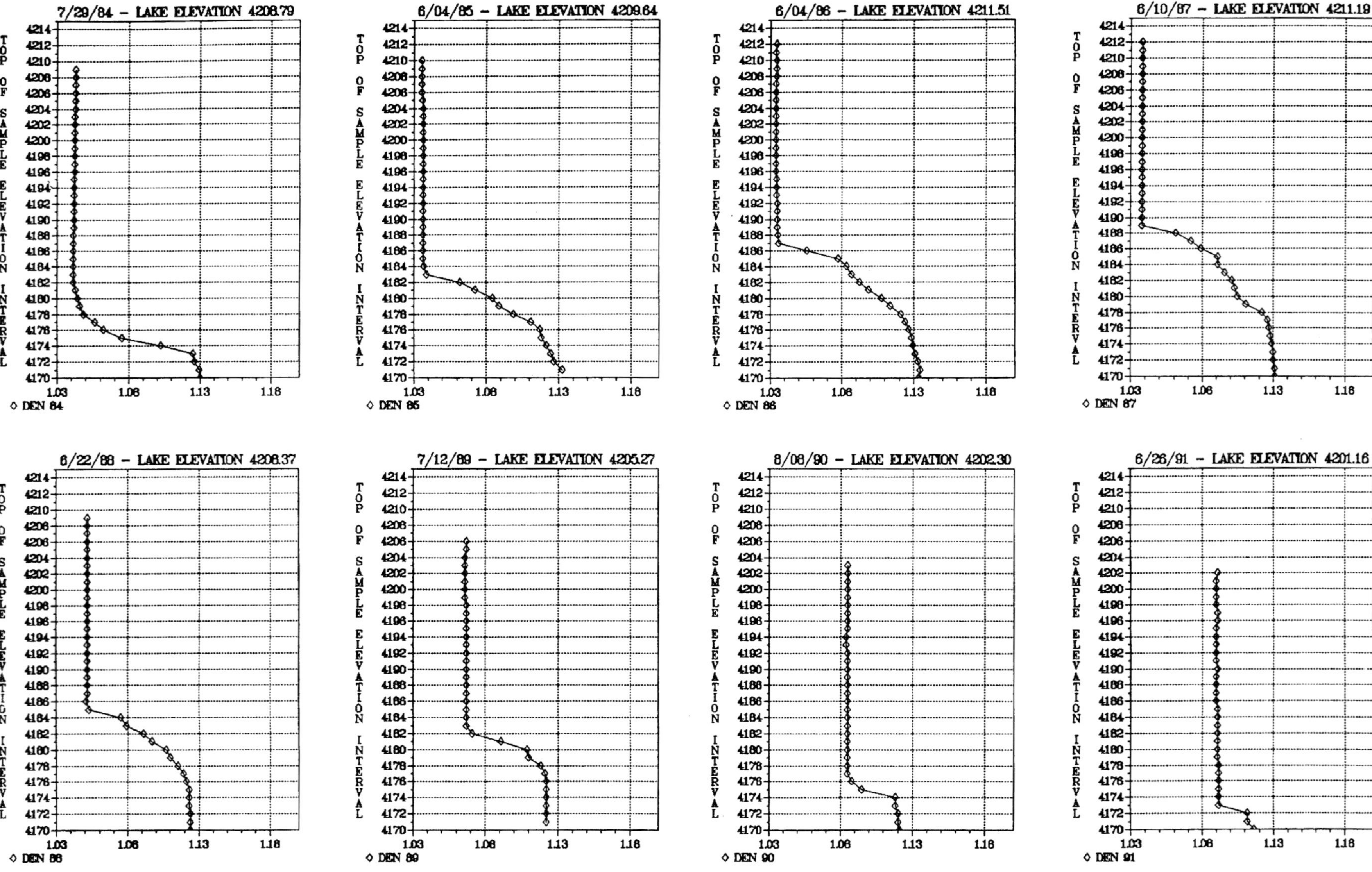

Figure 15a. *Sample-elevation intervals (feet) versus brine density (g/cm³) for incremental sampling site RT2 in the south arm of Great Salt Lake. Plots are date-specific and near the lake's annual high point.*

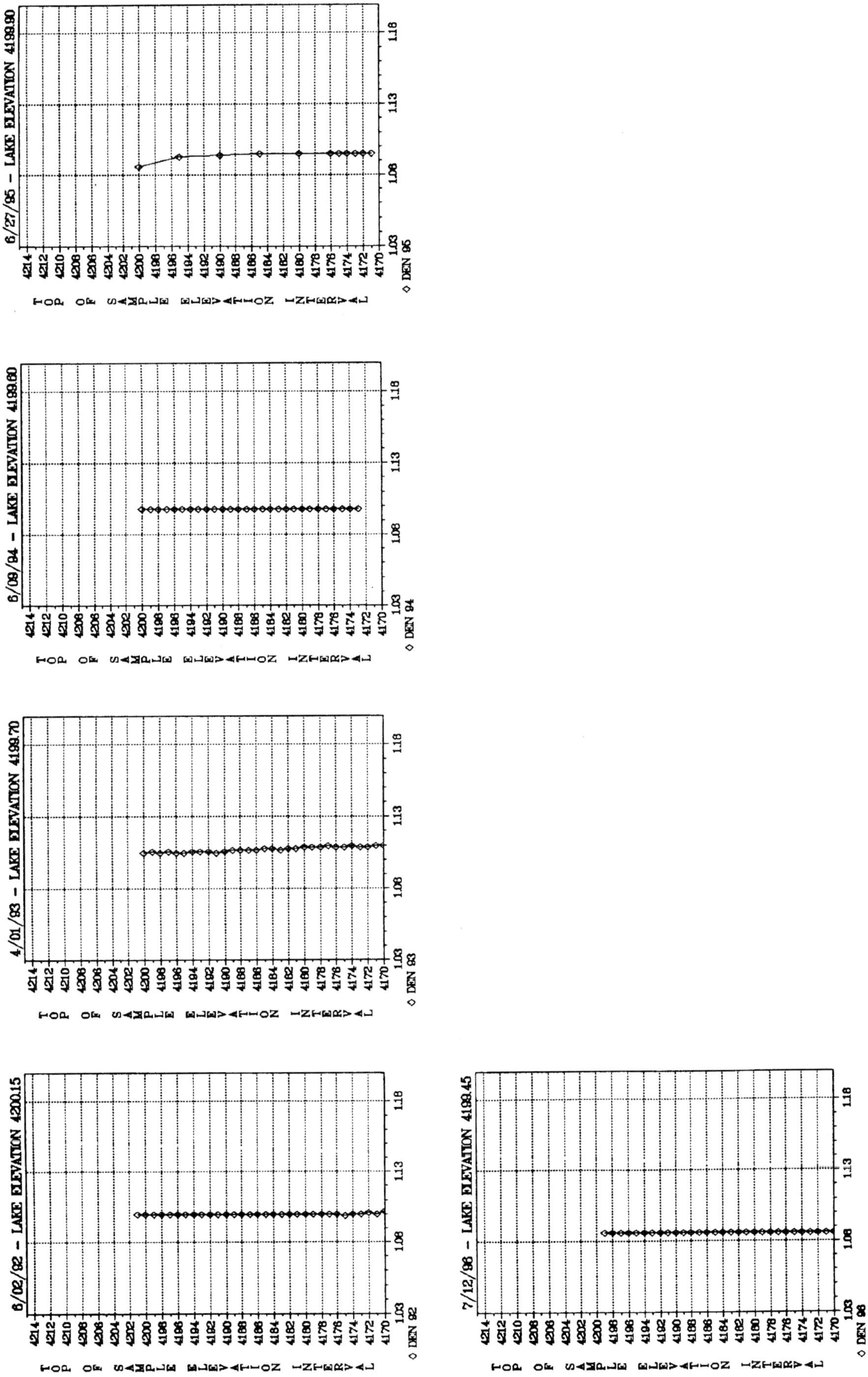

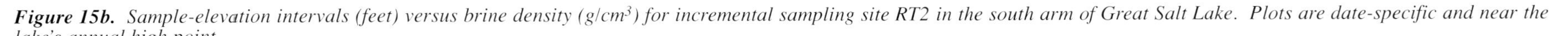

Figure 15b. *Sample-elevation intervals (feet) versus brine density (g/cm³) for incremental sampling site RT2 in the south arm of Great Salt Lake. Plots are date-specific and near the lake's annual high point.*

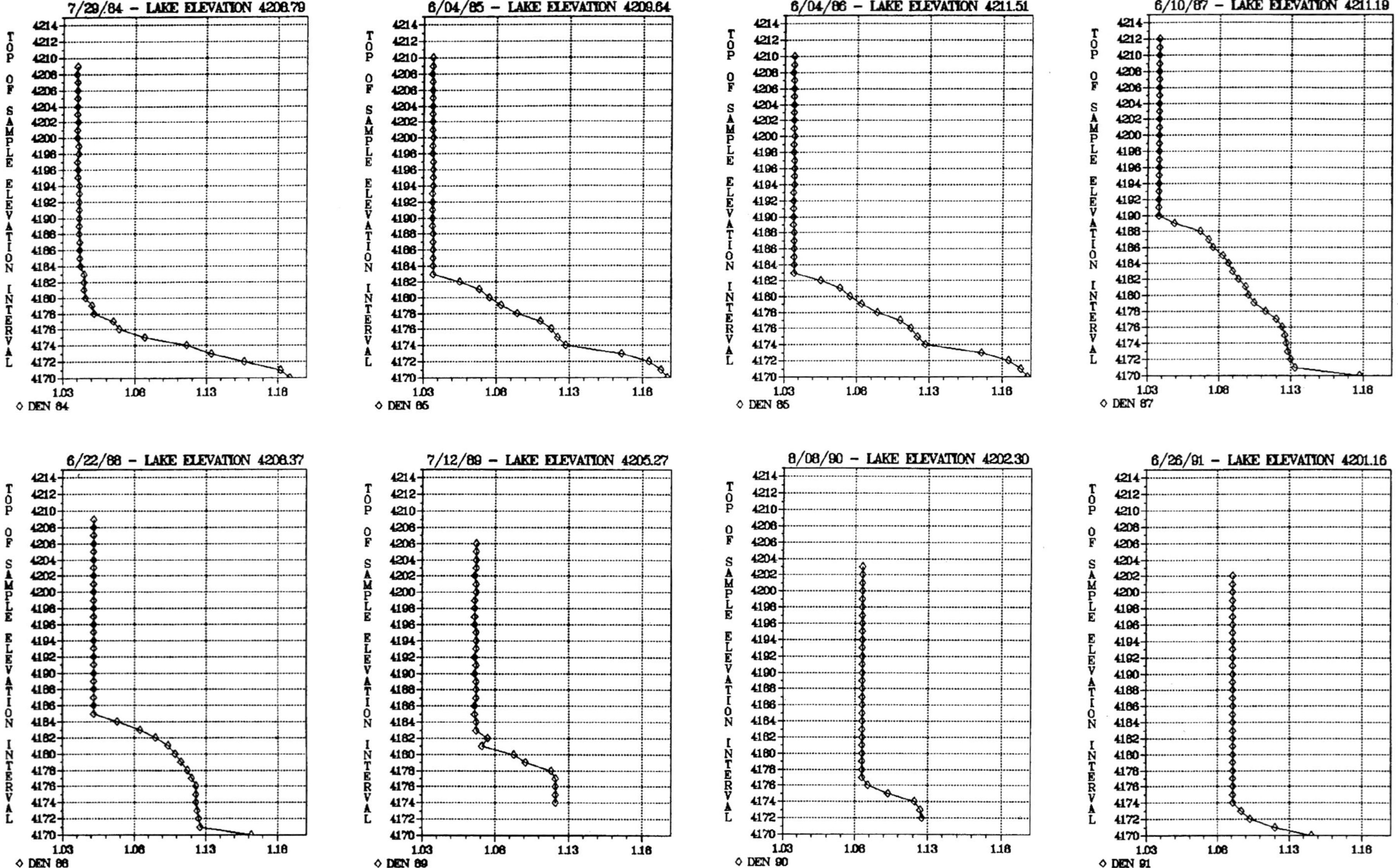

Figure 16a. *Sample-elevation intervals (feet) versus brine density (g/cm³) for incremental sampling site RT4 in the south arm of Great Salt Lake. Plots are date-specific and near the lake's annual high point.*

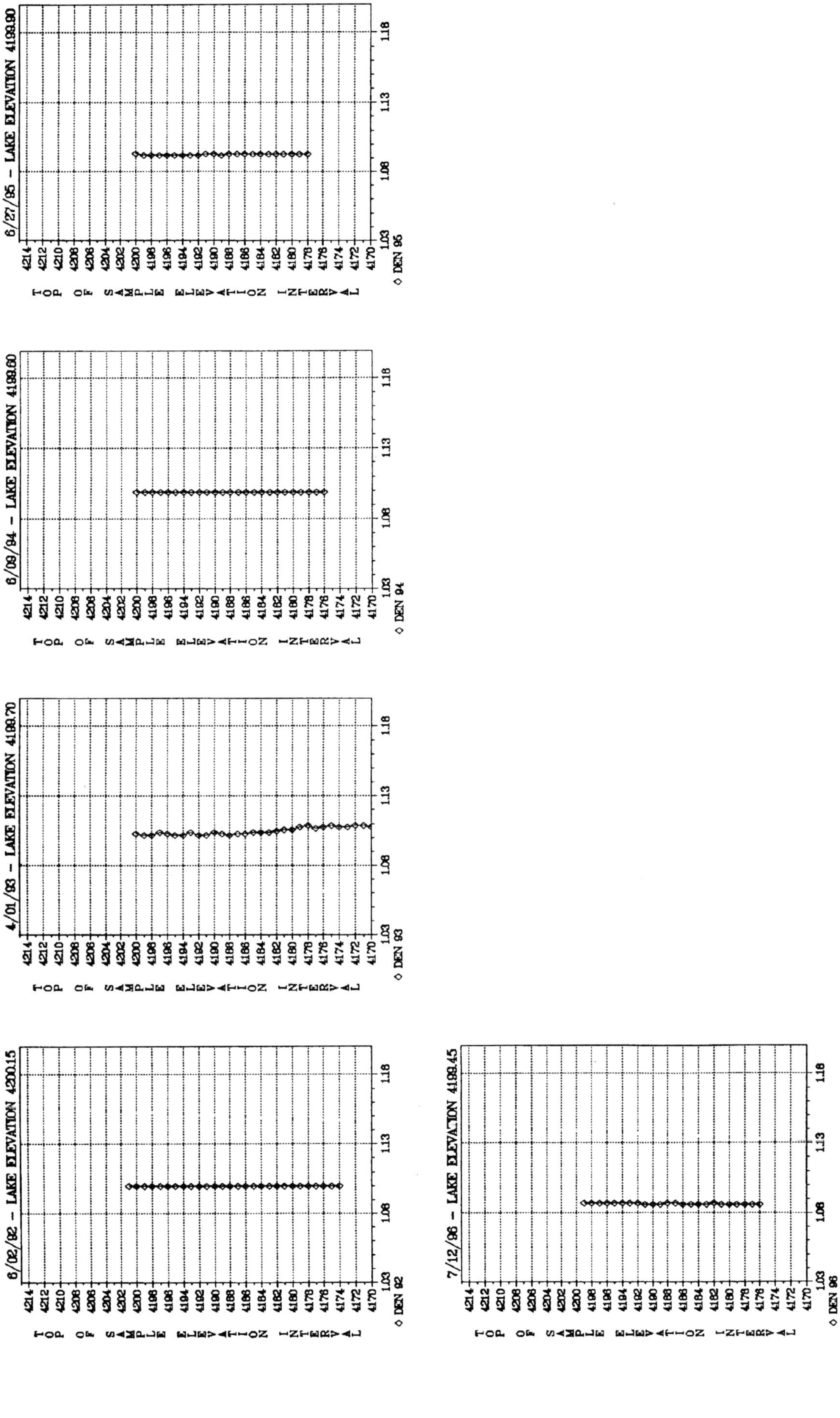
6/27/95 - LAKE ELEVATION 4199.90
6/09/94 - LAKE ELEVATION 4199.60
4/01/93 - LAKE ELEVATION 4199.70
6/02/92 - LAKE ELEVATION 4200.15
7/12/96 - LAKE ELEVATION 4199.45
TOP OF SAMPLE ELEVATION INTERVAL
◇ DEN 95
◇ DEN 94
◇ DEN 93
◇ DEN 92
◇ DEN 96

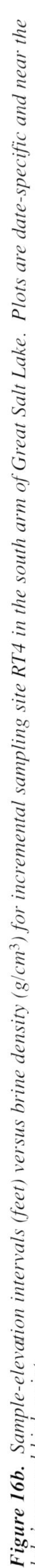

Figure 16b. *Sample-elevation intervals (feet) versus brine density (g/cm³) for incremental sampling site RT4 in the south arm of Great Salt Lake. Plots are date-specific and near the lake's annual high point.*

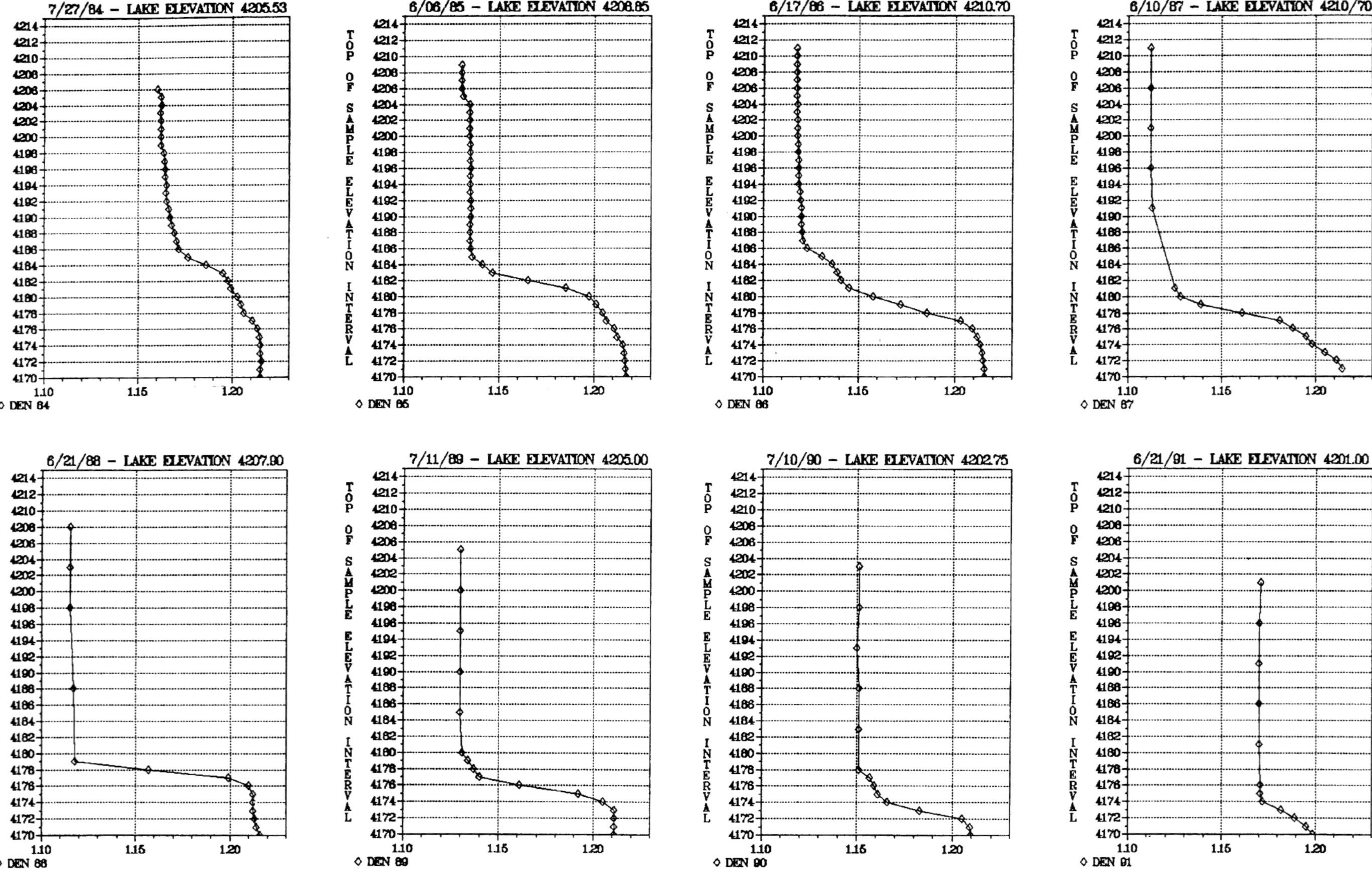

Figure 17a. *Sample-elevation intervals (feet) versus brine density (g/cm³) for incremental sampling site RT3 in the north arm of Great Salt Lake. Plots are date-specific and near the lake's annual high point.*

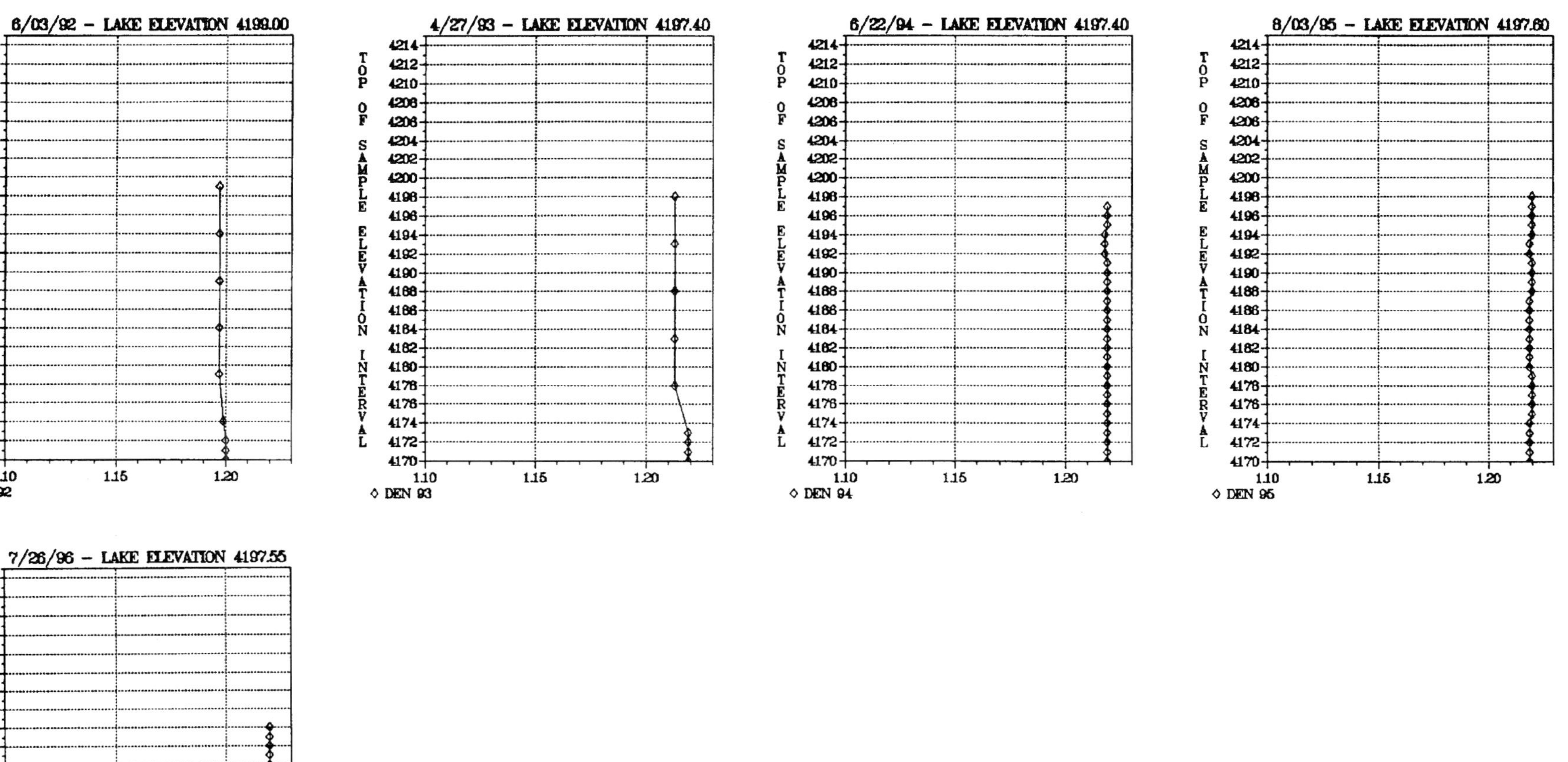

Figure 17b. *Sample-elevation intervals (feet) versus brine density (g/cm^3) for incremental sampling site RT3 in the north arm of Great Salt Lake. Plots are date-specific and near the lake's annual high point.*

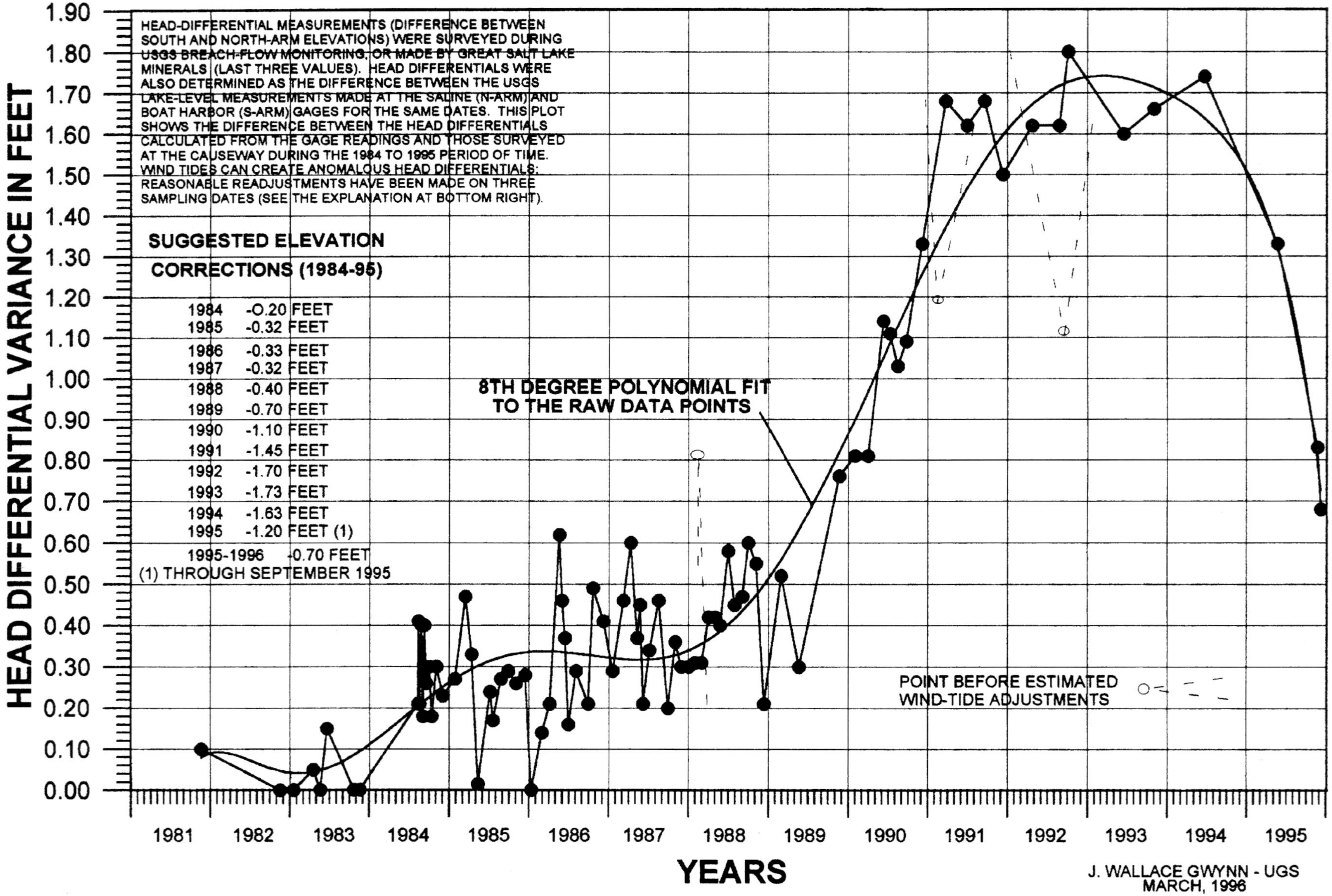

Figure 18. Difference between USGS gage and breach-flow-measurement head differentials: raw data with 8^{Th}-degree polynomial fit, and suggested south-arm elevation corrections to lake-level records.

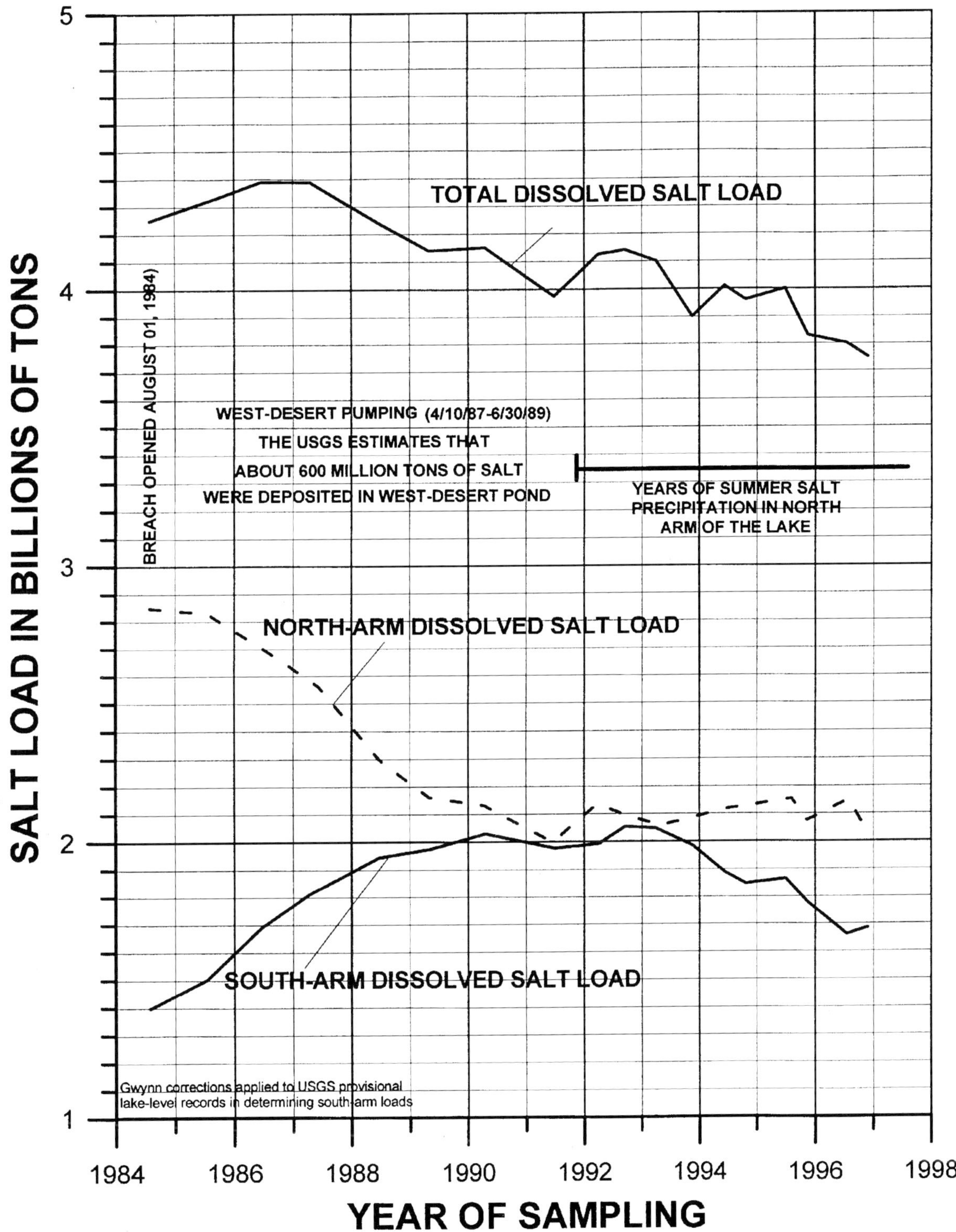

Figure 19. *Salt loads for the south and north arms and total lake from 1984 to 1997.*

(3) The 1984 breaching of the SPRR causeway caused a major redistribution of the lake's dissolved salts and the formation of a stratified-brine condition in the north arm.

(4) It is assumed that the addition of "fines-containing" fill material to the Southern Pacific Railroad causeway during the 1980s to keep it above the rising lake resulted in a great reduction in the causeway's overall hydraulic conductivity. The declining lake level after 1987 greatly reduced the bi-directional flow through the breach opening. Combined, these changes played a major role in bringing an end to the stratified-brine conditions in both arms of the lake by mid-1991.

(5) Salt-load calculations utilizing brine density, and hypsographic and lake-level data, suggest that there is about 4.4 billion tons (3.89 MT) of salt in the lake, that the distribution of salt between the two arms is transitory over time, and that a number of factors make it difficult to accurately determine the individual-arm salt loads or the combined load for the lake.

ACKNOWLEDGMENTS

A sincere thanks is given to Lee Allison, Kimm Harty, David Tabet, Bryce Tripp and Catharine Woodfield, of the Utah Geological Survey, and to Kidd Waddell and Steve Wold of the U.S. Geological Survey for their reviews of this manuscript and their helpful suggestions.

REFERENCES

Arnow, Ted, and Stephens, Doyle, 1990, Hydrologic characteristics of the Great Salt Lake, Utah: 1847-1986: U.S. Geological Survey Water-Supply Paper 2332, 32 p.

Cruff, R.W., 1986, Updated table of hypsographic data for Great Salt Lake: U.S. Geological Survey Memorandum, 8 p.

Gwynn, J.W., and Sturm, P.A., 1987, Effects of breaching the Southern Pacific Railroad causeway, Great Salt Lake, Utah-physical and chemical changes, August 1, 1984-July 1986: Utah Geological and Mineral Survey Water-Resources Bulletin 25, 25 p.

Hahl, D.C., and Handy, A.H., 1969, Great Salt Lake, Utah: chemical and physical variations of the brine, 1963-1966: Utah Geological and Mineralogical Survey Water-Resources Bulletin 12, 33 p.

Hahl, D.C. and Langford, R.H., 1964, Dissolved-mineral inflow to Great Salt Lake and chemical characteristics of the Salt Lake brine, part II: technical report: Utah Geological and Mineralogical Survey Water-Resources Bulletin 3 - part II, 40 p.

Hahl, D.C., and Mitchell, 1963, Dissolved-mineral inflow to Great Salt Lake and chemical characteristics of the Salt Lake brine, part I: selected hydrologic data: Utah Geological and Mineralogical Survey Water-Resources Bulletin 3 - part 1, 40 p.

Madison, R.J., 1970, Effects of a causeway on the chemistry of the brine in Great Salt Lake, Utah: Utah Geological and Mineralogical Survey Water-Resources Bulletin 14, 52 p.

Sturm, P.a., 1985, Great Salt Lake, Utah - Utah Geological and Mineral Survey's Great Salt Lake sampling program, 1966 to 1985 - history, data base, and averaged data: Utah Geological and Mineral Survey Open-File Report 87, 188 pages in various pagings.

Sturm, P.A., McLaughlin, J.C., and Broadhead, Ray, 1980, Analytical procedures for Great Salt Lake brine, *in* Gwynn, J.W., editor, Great Salt Lake, a scientific, historical and economic overview: Utah Geological and Mineral Survey Bulletin 116, p. 175-200.

Waddell, K.M., Gwynn, J.W., Burden, Carol, and Wold, S.R., 1992, Salt budget for west pond, Utah, April 1987 to April 1988: U.S. Geological Survey Water-Resources Investigations Report 91-4117, 29 p.

Whelan, J.A., 1973, Great Salt Lake, Utah: chemical and physical variations of the brine, 1966-1972: Utah Geological and Mineralogical Survey Water-Resources Bulletin 17, 25 p.

Wold, S.R., Blakemore, E.T., and Waddell, K.M., 1996, Water and salt balance of Great Salt Lake, Utah, and simulation of water and salt movement through the causeway: U.S. Geological Survey Open-File Report 95-428, 66 p.

Wold, S.R., and Waddell, K.M., 1994, Salt budget for west pond, Utah, April 1987 to June 1989: U.S. Geological Survey Water-Resources Investigations Report 93-4028, 20 p.

THE WATERS SURROUNDING ANTELOPE ISLAND GREAT SALT LAKE, UTAH

by
J. Wallace Gwynn

Reprinted with permission from Utah Geological Survey Miscellaneous Publication 00-1, 2000

ABSTRACT

Antelope Island and its northern and southern causeways to the mainland separate two distinctive and dynamic bodies of water, both part of the Great Salt Lake system. The south causeway, from the island to the mainland was constructed in about 1952. The Davis County causeway was constructed in 1969, and rebuilt in 1992 after the 1980s flooding.

The salinity of the south arm fluctuates inversely with lake elevation. At lake elevations below the top elevations of the two causeways, Farmington Bay is isolated and its salinity is less than half that of the south arm. At lake elevations above the tops of the two causeways, Farmington Bay becomes one with the south arm and its salinity is about equal. Density stratification was observed in the south arm in 1966 and existed there until mid-1991 when it disappeared due to diminished heavy, deep return flow brine from the north arm through the Southern Pacific Railroad causeway. The ion ratios of the major ions and cations (sodium, potassium, magnesium, calcium, chloride and sulfate) are about the same in both bodies of water. Farmington Bay acts as a biological treatment lagoon for the plant nutrients nitrogen and phosphorus in the Jordan River. The biota in the two bodies of water are significantly different because of the different salinities. Raw sewage was emptied into Farmington Bay from Salt Lake City for many years prior to the construction of treatment plants during the 1950s to 1960s.

Six mineral-extraction industries presently operate on Great Salt Lake, three each on the south and north arms. Products generated include sodium chloride, magnesium chloride, potassium sulfate, magnesium metal, chlorine gas, and dietary supplements.

From the 1930s to the present, there have been numerous proposals to convert Farmington Bay or the entire east embayment into freshwater bodies through diking between the islands and the mainland. During the 1980s flooding period, diking proposals were aimed at protecting individual entities, such as sewage-treatment plants, or large areas of shoreline from flooding. To date, none of the major proposals has come to fruition, due mainly to high project costs, and environmental and health-issue concerns.

INTRODUCTION

Antelope Island is the largest of the four major islands in the south arm of Great Salt Lake, covering about 23,850 acres (figure 1). It is 15.5 miles in length, trending in a north-south direction, with a maximum width of 4.5 miles. The island's highest peak reaches an elevation of 6,596 feet above mean sea level, or about 2,396 feet above a 4,200-foot nominal lake level (USGS, 1974). To the east of Antelope Island lies Farmington Bay, a shallow body of water having a maximum depth of seven to eight feet at a lake elevation of 4,200 feet. Earth-fill causeways extend from the north and south ends of the island to the mainland; these causeways isolate Farmington Bay from the south arm of Great Salt Lake (GSL). The surface area enclosed by these two dikes, the eastern shore of Antelope Island and the mainland, covers about 76,700 acres. The northern and western shores of Antelope Island border the main south-body arm of GSL, which covers about 550,505 acres, and has a maximum depth of about 32 feet at the 4,200-foot lake elevation (Waddell and Fields, 1977). Numerous Davis County communities lie along the narrow, densely populated strip of land between the east shore of Farmington Bay and the Wasatch Range to the east.

The main body of the south arm, including Farmington and Bear River Bays, receives nearly all of the freshwater inflow to the lake (figure 1). The Bear and Weber Rivers contribute about 79 percent of the total inflow (Arnow and Stephens, 1990) and enter the Ogden Bay portion of the lake, located to the northeast of Antelope Island, and north of Farmington Bay. The Jordan River contributes about 13 percent of the total inflow and empties into the southern end of Farmington Bay. Numerous streams from the Wasatch Range enter the bay along its eastern shore (figure 1) and contribute an additional five percent of the freshwater inflow. Montgomery (1984) summarizes the average annual surface inflows to the lake from 1931 to 1976 and for the wet year of 1983.

The purposes of this paper are to: (1) present the basic chemistry and yearly changes in salinity of the main body of the south arm of the lake, (2) provide a brief history of the two causeways that isolate Farmington Bay and their influence on the enclosed water, (3) recount the history of Farmington Bay as a receptacle for both untreated and treated sewage, (4) summarize the basic chemistry and changes in salinity of Farmington Bay, (5) discuss the south-arm extractive mineral industries, and (6) review the proposals to convert Farmington Bay into a body of fresh water. Discussion of the north arm of Great Salt Lake is minimal.

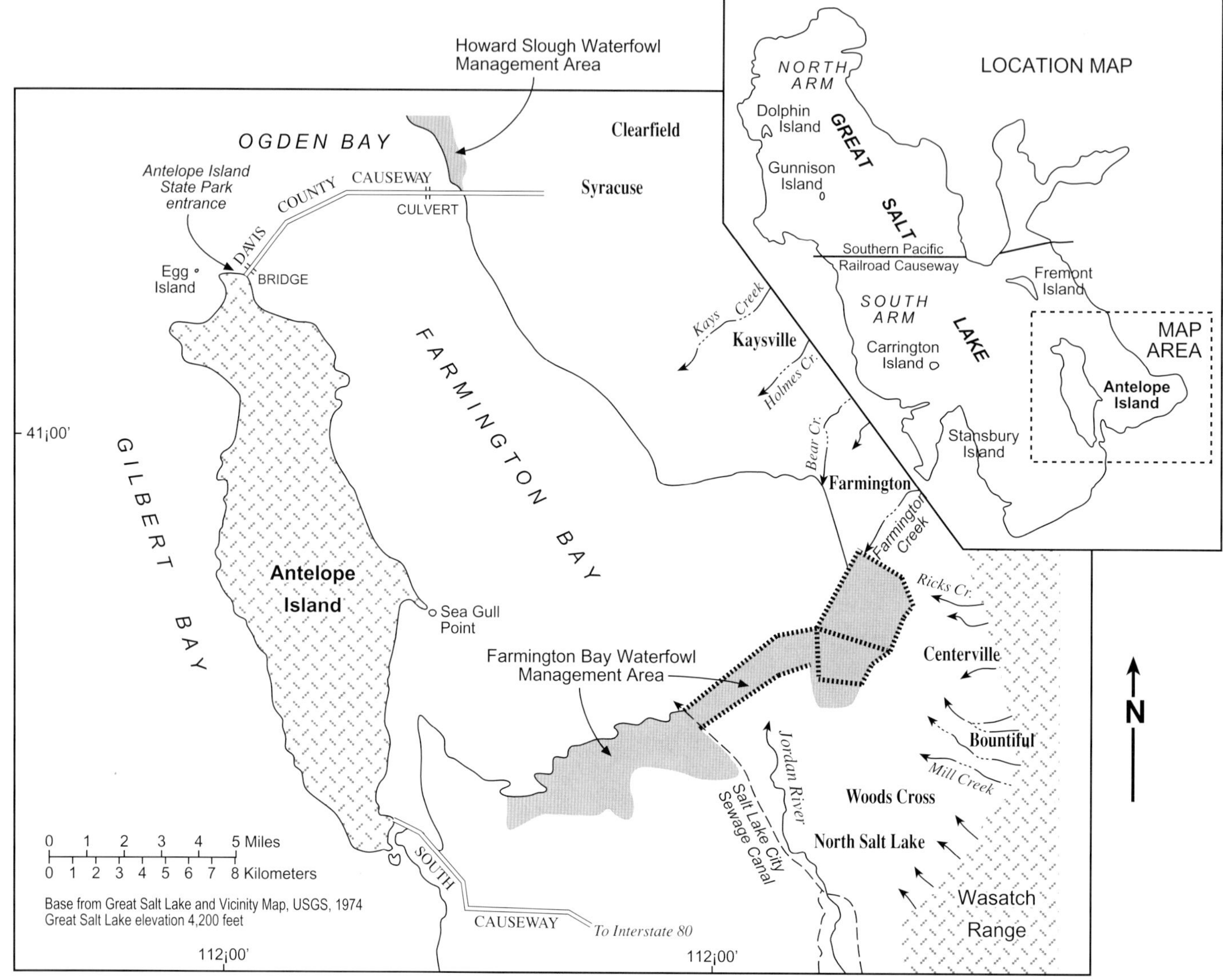

Figure 1. *General location map of Antelope Island, Farmington Bay, the south arm of Great Salt Lake, South and Davis County causeways, and waterfowl management areas.*

SALINITY, STRATIFICATION, AND CHEMISTRY OF THE SOUTH ARM

South-Arm Salinity

Lake Elevation Versus Salinity

Variation of the salinity of the south arm has an inverse relationship with the fluctuation of lake-surface elevation; salinity is lower when the lake elevation is higher. Figure 2 shows how both lake elevation and total dissolve solids (expressed in grams-per-liter [g/L]) vary annually between 1966 and 1997. During the early 1960s, when the lake was at its historic low (4,191.35 ft), the south-arm waters were nearly eight times saltier than typical ocean waters. During 1987, when the lake was at its historic high (4,211.85 ft), the south-arm waters were only 1.4 times saltier than ocean waters. At a nominal south-arm elevation of 4,200 feet, south-arm waters contain about 130 grams-per-liter (11 to 12 percent salt) and are about 3.3 times as salty as the ocean.

Stratified Brines

In 1966, Utah Geological Survey (UGS) personnel observed, through vertical-density profiling of the water column, the existence of a stratified brine condition in the south arm. A dense, dirty, fetid brine was observed at the bottom of the lake, initially below an elevation of about 4,175 feet, but below an elevation of about 4,180 feet after about 1972. Above this dense brine lay a clean, lighter brine. Figure 3 shows the area underlain by the dense brine at the 4,180-foot top elevation. This deep, dense brine originated from the return flow of heavy north-arm brine that migrated from north to south through the lower portion of the Southern Pacific Railroad (SPRR) causeway which separates the south and north arms of the lake. South-arm stratification continued through the 1980s flooding period until mid-1991 when it disappeared. The disappearance was caused in large measure by the greatly reduced volume of heavy north-arm brine migrating through the SPRR causeway into the depths of the south arm. This was due to a reduction in the causeway's

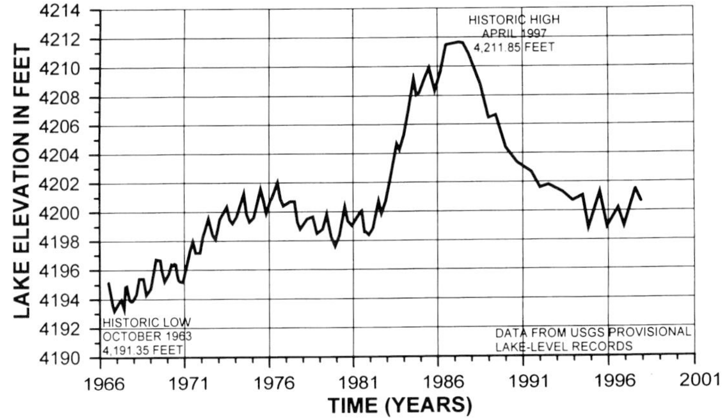

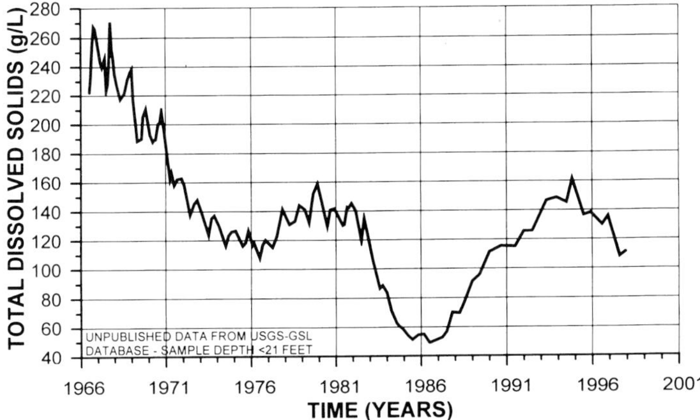

Figure 2. *(Top) Annual variation of the surface elevation of south arm of Great Salt lake, Utah (1966-1997). (Bottom) Annual variation of the total-dissolved-solids of south arm of Great Salt Lake, Utah (1966-1997).*

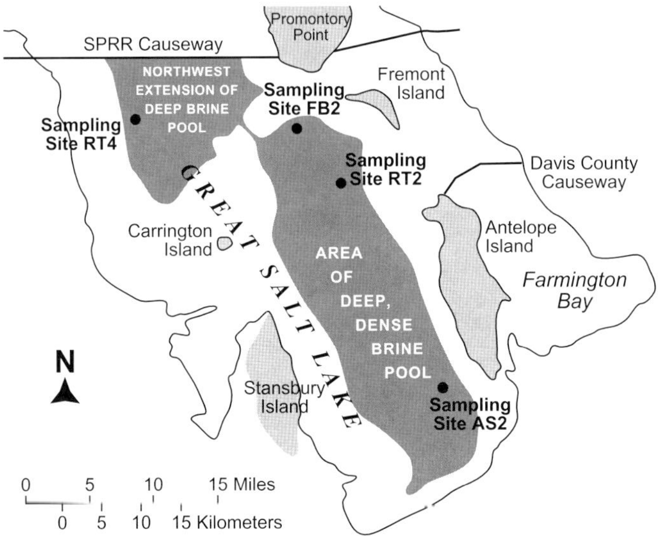

Figure 3. *South arm of Great Salt lake showing the main and northwestern areas of the pre-1992 deep, dense brine pool.*

hydraulic conductivity, caused by the addition and compaction of fill material used to raise the tracks six to eight feet during the flooding period (1983 to 1987). Without a sufficient flow of north-arm brine southward, the volume of dense south-arm brine could not be maintained, and stratification disappeared through vertical mixing.

South-Arm Chemistry

While the salinity of the south arm changes with lake-surface elevation, its chemical composition (ion ratios) remains relatively constant. Table 1 shows the composition of the dissolved salts in Great Salt Lake. The composition of the dissolved salts in typical ocean water, and water from the Dead Sea are given for comparison.

Two earth-fill causeways provide access to Antelope Island but also isolate Farmington Bay from the main south arm of Great Salt Lake, especially when the lake's elevation is below the top (4,206-ft) elevations of the two dikes (figure 4). Because of this isolation, the bay's physical, chemical, and biological properties have become different than those of the south arm.

Table 1. Chemical composition of dissolved salts from Great Salt Lake, typical ocean, and Dead Sea waters on a dry-weight-percent basis (Gwynn, 1997).

ION	GREAT SALT LAKE	TYPICAL OCEAN	DEAD SEA
Sodium	32.1	30.8	12.3
Potassium	2.3	1.1	2.3
Magnesium	3.7	3.7	12.8
Calcium	0.3	1.2	5.2
Chloride	54.0	55.5	67.1
Sulfate	7.6	7.7	0.1
Totals	100.0	100.0	99.8

South Causeway

Between 1951 and 1952, the South causeway (figure 5) was built over an existing trail on the mud flats from the south end of Antelope Island eastward to the mainland (Harward, 1996). The work was done as a cooperative project between Davis County and the Island Improvement (Ranching) Company to prevent raw sewage, then being discharged into Farmington Bay, from polluting the southernmost beaches of the south arm (R. Harward, island resident, verbal communication, September 1997). The causeway also provided southern access to the island. Starting in 1978, a conveyor-belt route was constructed by the S.J. Groves Construction Company to move fill material from the island to the mainland for freeway construction. The conveyor belt was built atop of the existing causeway southeastward from the island for a distance of about 3 miles. The present top elevation of the causeway is estimated to range from 4,205 (Clinton Baty, Great Salt Lake State Park harbor master, verbal communication, October 1997) to 4,206 feet (Tim Smith, Antelope Island State Park manager, verbal communication, August 1997). An elevation marker (USGS, 1972) on the causeway shows an elevation of 4,205 feet. Thus, when the lake-surface elevation is below about 4,205 to 4,206 feet, Farmington Bay is separated from the main body of the south arm at the southern end; when the lake elevation is above this elevation, there is some mixing of south arm and Farmington Bay waters.

Davis County Causeway

Prior to 1969, there was no northern causeway separat-

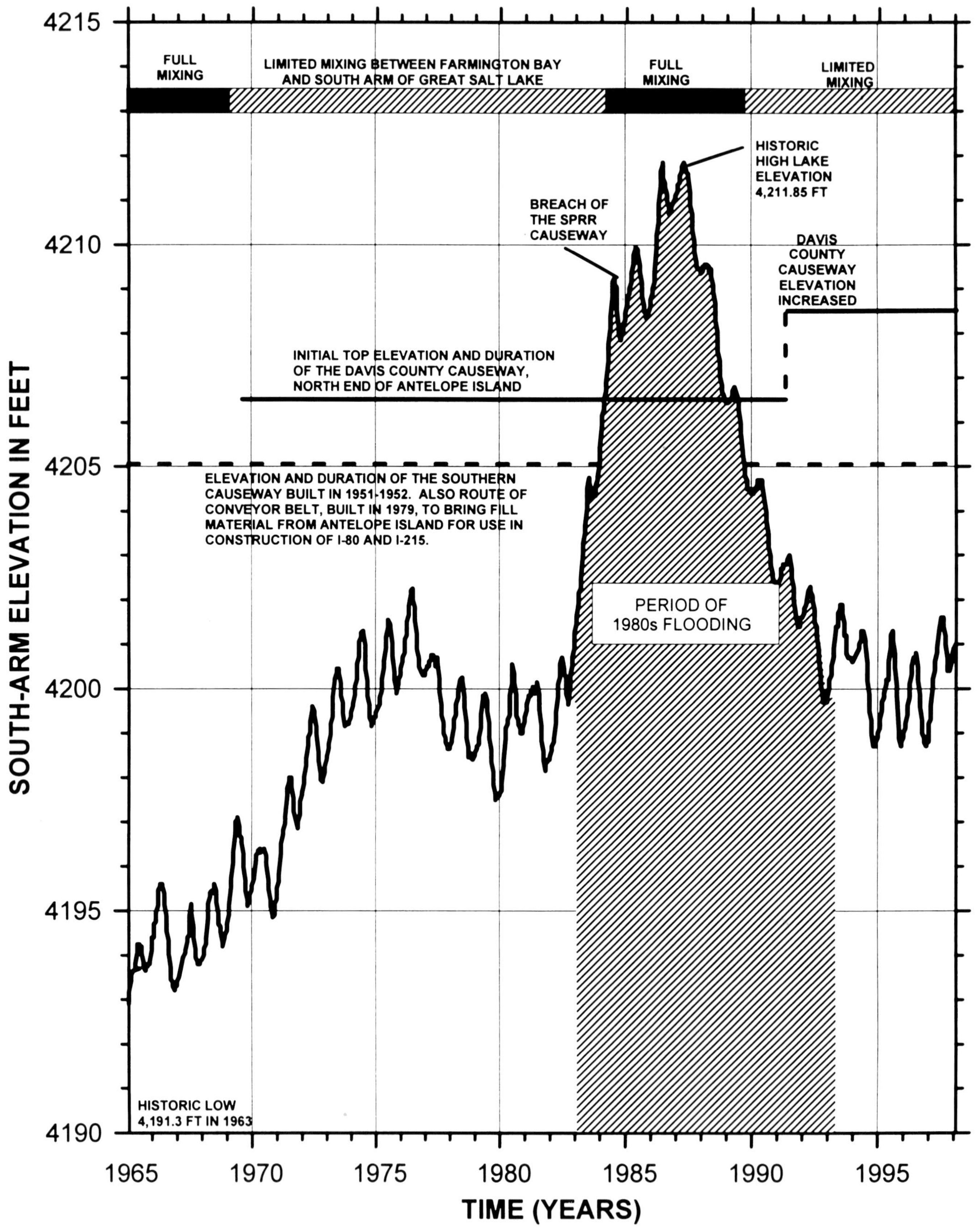

Figure 4. *Top elevations of the Davis County and South causeways, and the periods of full and limited mixing between Farmington Bay and south arm waters of Great Salt lake, Utah.*

Figure 5. Aerial photograph of south causeway between the south end of Antelope Island eastward to the mainland.

ing Farmington Bay and the main body of the south arm, thus allowing the free movement and mixing of water between the two areas. During 1969, an earth-fill causeway was constructed by Davis County and the Utah Department of Transportation from the east shore of the lake, near Syracuse, westward to the northern end of Antelope Island (figure 6). The causeway separated Farmington Bay from the south arm proper and restricted mixing between the two areas except for flow through one narrow bridged opening, consisting of 10 precast box-culvert-type sections bolted together. The original top elevation of the northern causeway was about 4,206.5 feet (Smith, 1997).

Figure 6. Aerial photograph showing the Davis County causeway that extends between the north end of Antelope Island and the mainland.

In early 1984, the Davis County causeway was flooded when the lake surface rose from about 4,200 feet in elevation (1983) to its historic high of 4,211.85 feet (1986-1987); it remained submerged until mid-1989 when the lake receded below 4,206.5 feet. During that time, it is assumed that the waters of the south arm and Farmington Bay were free to mix; and that mixing was probably the greatest when the lake level was at its higher elevations. After 1989, when the lake level had receded, mixing again became restricted. In 1992, the causeway was rebuilt, through the combined efforts of the Utah Department of Transportation and Davis County, to an elevation of 4,208.5 feet along the eastern portion of the causeway (Saunders, 1994, and Saunders and Smith, 1994). During the same year, a new bridge was built near Antelope Island at the west end of the causeway. The bridge has a concrete deck supported by five 100-foot-long steel girders with a deck elevation of 4,213.5 feet. The road from the bridge west to the island has a top elevation of about 4209.85 feet. For additional control of the waters between Farmington Bay and the main body of the south arm, a box culvert was constructed about one mile from the east end of the causeway. This structure consists of two 10- by 15-foot box culvert sections laid side by side.

FLOW OF RAW AND TREATED SEWAGE INTO FARMINGTON BAY

From the time that the first sewers were installed in Salt Lake City (time uncertain) until the construction of sewage treatment plants in the 1950s and 1960s (figure 7), Farmington Bay was the recipient of Salt Lake City's raw sewage. Today, with rare exception, only treated effluent from sewage treatment plants servicing Salt Lake County and Davis County currently enters the bay. Under emergency conditions, individual sewage treatment plants are authorized by the Utah Division of Water Quality (UDWQ) to temporarily discharge raw sewage to the Jordan river or into Farmington Bay (Arne Hultquist, UDWQ, verbal communication, 1997).

Numerous investigators have studied the water quality and pollution of Farmington Bay caused by the introduction of both untreated and treated sewage. These include studies by: the Utah Department of Health and Davis County Health Department (1965), Bott and Shipman (1971), Carter and Hoagland (1971), Colburn and Eckhoff (1972), Searle (1976), Ford Chemical Laboratory (1977), Maxell and Thatcher (1980), Sturm (1983), Israelsen and others (1985), Chadwick and others (1986), and Sorensen and Riley (1988). Studies by these investigators include: types and distribution of pollutants in both water and sediments; types, distribution and survival of bacteria; water chemistry and water quality including the presence and distribution of heavy metals, nutrients and organics; odor-causing mechanisms; health and safety issues; and pollution-control measures.

SALINITY, STRATIFICATION, AND WATER CHEMISTRY OF FARMINGTON BAY

Farmington Bay Salinity

Salinity Changes Related to Lake Elevation

The overall salinity of Farmington Bay is affected by lake level, much like the south arm of the lake; as the lake rises, salinity drops. Salinity is also influenced by the lake's elevation relative to the 4,205- to 4,206-foot top-elevation range of the South and Davis County causeways. Below this elevation range, Farmington Bay is isolated from the main

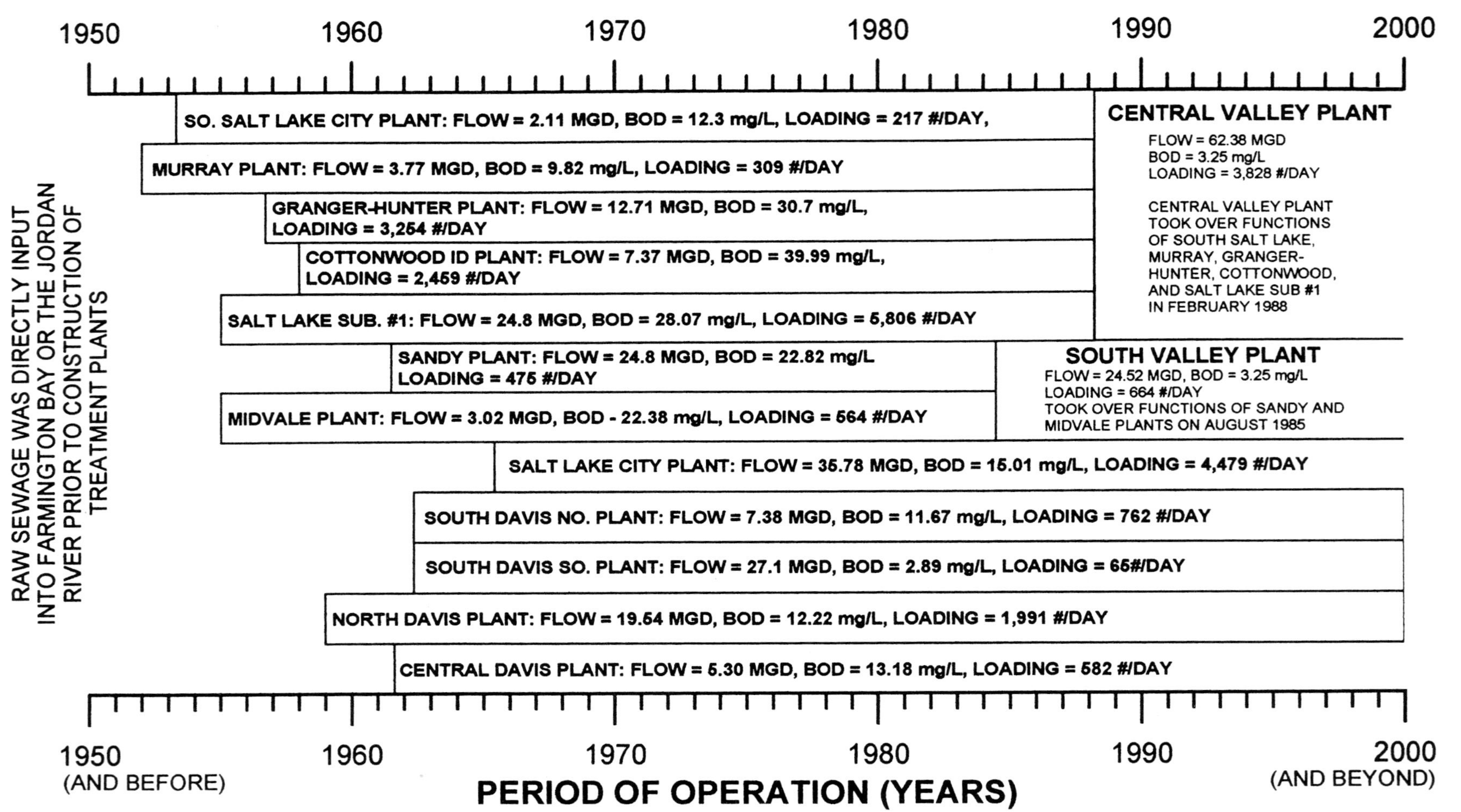

Figure 7. *Periods of operation, flow, BOD and loading, for individual sewage treatment plants in Salt Lake and Davis Counties which empty treated-sewage effluent into the Jordan River or directly into Farmington Bay.*

south arm of the lake by the two causeways, with the exception of low-volume, bi-directional flow through the opening(s) in the Davis County causeway. Bi-directional flow is the phenomena of two brines of differing densities flowing in opposite directions through the same opening. Averaged water-quality data from Bott and Shipman (1971), Maxell and Thatcher (1980), Hansen (1981), and the Utah Division of Water Quality (1994-1996) suggest that at lake elevations below 4,205 feet, the salinity of Farmington Bay is about half that of the south arm. Above the 4,205 to 4,206-foot elevation range, there is direct communication between Farmington Bay and the south arm which allows increased mixing as the lake rises. Under these sustained conditions, data from Sorensen and Riley (1988) suggest that the salinities of the south arm and Farmington Bay become the same.

South-to-North Salinity Gradient

When the water of Farmington Bay is isolated from the south arm, a salinity gradient exists which increases from south to north within the bay (Utah Department of Health and Davis County Health Department, 1965; Bott and Shipman, 1971; Maxell and Thatcher, 1980). This gradient is thought to be attributable to two factors: (1) fresh water enters the bay at the south end, mixing with the resident salty water. As the mixed water moves towards the openings in the Davis County causeway, water is removed at the surface by evaporation causing it to be come saltier; and (2) the upper portion of Farmington Bay water probably increases in salinity as it mixes with the underlying tongue of higher-density brine (discussed below) that extends into the bay from the bridged opening in the Davis County causeway.

Farmington Bay Stratification and Bi-Directional Flow

Stratification

Bott and Shipman (1971), and Sturm (1983) documented the existence of brine stratification in Farmington Bay, consisting of a layer of dense brine at the lake's bottom overlain by a layer of less-dense brine. The source of the dense brine was the north-to-south underflow of heavier south-arm water through the bridged opening in the Davis County causeway. Bott and Shipman's (1971) data suggest that the interface between the two layers of brine within Farmington Bay was at about 5 feet deep during June 1971; Sturm's (1983) data suggest the interface depth was between 5 and 6 feet deep during July 1981.

Bi-Directional Flow

Bi-directional flow exists at the Davis County causeway, and is discussed by Carter and Hoagland (1971), Colburn and Eckhoff (1972), Hansen (1981), and Sturm (1983). The north-to-south flow of denser south-arm brine through the Davis County causeway bridge is driven by hydrostatic pressure, while the south-to-north flows of overlying lighter water through this opening are driven by the existing hydrostatic head. Sturm (1983) shows that the depth to the bi-directional-flow interface within the bridged opening, during 1982 measurements, fluctuated between 4.75 feet and 6.25 feet under calm conditions. Under south-wind conditions, the depth to the interface became much deeper (8 to 10 feet), and the upper, south-to-north flows greatly increased. Under north-wind conditions, the depth to the interface diminished, sometimes to nothing, and the deep, north-to-south flows increased.

Farmington Bay Chemistry and Nutrients

(Nutrient section contributed by Doyle Stephens, USGS, 1997)

A comparison of a limited number of chemical analyses from Farmington Bay, mainly from the north end near the Davis County bridge, to analyses from the main south arm of the lake (figure 8), shows that there is no significant difference between the major-ion ratios within the two bodies of water. This is because the main source of salt for Farmington Bay comes from the south arm as return flow through the bridged opening.

Farmington Bay acts as a biological treatment lagoon for the plant nutrients nitrogen and phosphorus in the Jordan River. Typically, the Jordan River caries 3-4 mg/L of dissolved nitrogen (as nitrate and ammonium) and 1-2 mg/L of dissolved phosphorus into Farmington Bay. Most of these dissolved nutrients are removed from the water by algae, aquatic macrophytes, and bacteria within Famington Bay. Between January and August 1997, sampling in Farmington Bay by the USGS at a site 5 miles south of the Davis County causeway found dissolved nitrate nitrogen always less than the reporting level (detection limit) of 0.05 mg/L, dissolved ammonia less than 0.02 mg/L, and total dissolved phosphorus less than 0.3 mg/L. Throughout the summer of 1997, there were massive blooms of green algae and cyanobacteria (blue-green "algae"). Salinity, at 1-meter depth, ranged from 3.8 percent in January to 6 percent in August.

South Arm and Farmington Bay Biota

(Section contributed by Don Paul, Utah Division of Wildlife Resources, 1997)

Great Salt Lake (GSL) biology varies between locations that have different salinities. Antelope Island and its causeways divide two locations of differing salinities, namely, Farmington Bay and the south arm of the lake (figure 1). The balance between inflow to and evaporation from the lake, coupled with the terminal (no outlet) nature of the GSL cause frequent shifts in lake elevation and salinity, and consequently, the biology of various isolated lake locations.

Living organisms have difficulty coping with the osmotic stress of high concentrations of salt ions; only a few are successful. However, even with high salt concentrations, different ecological systems can exist side by side where these salinities are locally variable. An excellent example of such local variation is the south arm of GSL and Farmington Bay which are separated by the Davis County causeway, as described in the following examples. During May 1996, GSL elevation was 4,200.7 feet and the salinity of the south arm was 12.2 percent, whereas the salinity of Farmington Bay was and 5.6 percent (Doyle Stephens, USGS, personal communication, November, 1997).

The south arm fosters a relatively simple system of some

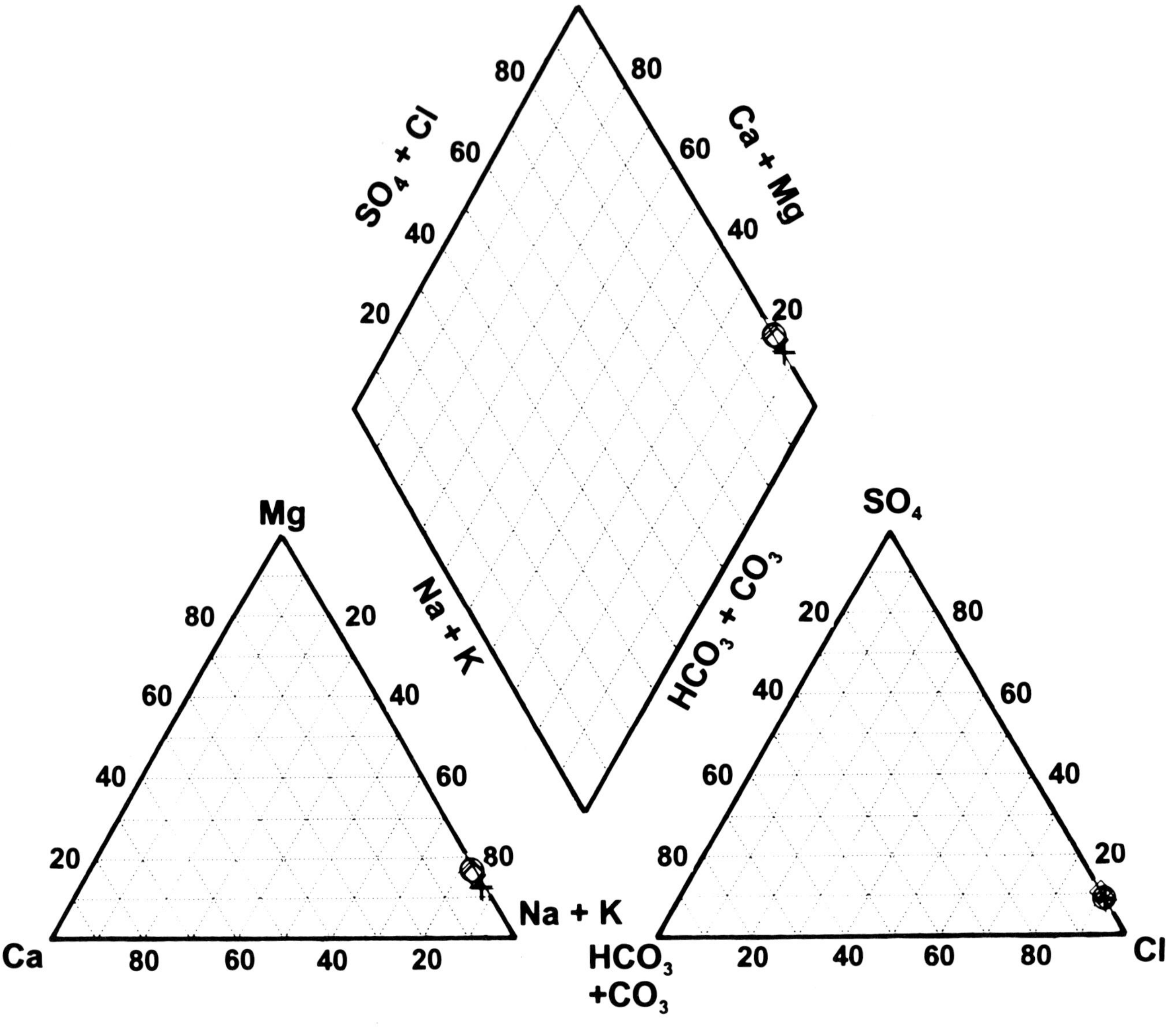

EXPLANATION

DIAMONDS (Sample analyses from Utah Department of Health, unpublished

Average of 8 samples from Farmington Bay, 100 feet south of the Antelope-Syracuse causeway, 1977-1978.

Average of 59 samples from Farmington Bay, at the Antelope-Syracuse causeway bridge (south side), 1980-1982.

Average of 6 samples from Farmington Bay, in the main deep channel, 1977-1978.

Average of 8 samples from Farmington Bay, NE of Seagull Point, 1977-1978

CROSSES (Samples analyses from UGS-GSL database, unpublished)

Average of 2 samples from Farmington Bay, south of Antelope-Syracuse bridge, 0.5 foot depth, 1976-1977.

Average of 2 samples from Farmington Bay, south of Antelope-Syracuse bridge, 15 foot depth, 1976-1977.

Average of 2 samples from Farmington Bay, south of Antelope-Syracuse bridge, 25 foot depth, 1976-1977.

CIRCLES (Sample analyses from UGS-GSL database, unpublished)

Average of 48 samples from south arm, sites AS2/FB2, 0.5 foot depth, 1977-1983.

Average of 48 samples from south arm, sites AS2/FB2, 15 foot depth, 1977-1983.

Average of 48 samples from south arm, sites AS2/FB2, 25 foot depth, 1977-1983.

Figure 8. *Trilinear plot showing chemical compositions of waters from Farmington Bay and the south arm of Great Salt Lake.*

20 species of algae and 10 species of protozoa when the salinities are at 115 g/L (Rushforth and Felix, 1982). Four macro invertebrates coexist with these microscopic species. These macro invertebrates include two species of brine flies which typically inhabit the area along the shoreline, small numbers of a corixid, and a species of brine shrimp that occupies the entire reach of the south arm. The constant shifting of the GSL elevation and shoreline fosters macro invertebrates that are reproductively flexible and seasonally abundant, that is, brine flies and brine shrimp. At salinities above 115 g/L the south arm ecosystem is relatively predator free and the brine shrimp and fly populations go unchecked, increasing within the limits of the algae production, which seems to be nitrogen limited. The brine shrimp species in GSL, *Artemia fransiscana*, is an effective filter-feeder capable of reducing the algae in the south arm water column by mid-summer to an extent that the shrimp become food-stressed.

In contrast to 1997 salinity conditions, the high-lake-level conditions during the 1980s significantly reduced the salinity to 50 g/L in the south arm. At that time, the south-arm biology was similar to that of the current (1997) Farmington Bay condition. These diluted conditions effectively reduced brine shrimp production and densities enough that both the avian species and the brine-shrimp-harvest industry, reliant on the *Artemia* resource, were nonplussed in their activities in the south arm. Insect predation (corixids), on what few brine shrimp were produced, was significant in nearshore areas, especially around underwater structures. The ecology became more complex with dilution, shifting away from a brine shrimp- and brine fly-dominated system to a more competitive ecosystem in the south arm. North of the Southern Pacific Railroad causeway, conditions improved for some brine shrimp production as salinities fell below 20% and consequently there was a limited brine shrimp harvest and some Eared Grebe and Wilson's phalarope use of the macro invertebrates in that region of the lake during high lake years. Bird-use estimates were considerably lower during these years for species keying on summer and fall populations of brine shrimp and flies at Great Salt Lake.

Farmington Bay, with its lower salinities, supports a larger number of species and a more complex interaction between trophic (nutrient) levels with many more checks and balances. The predacious corixid *Trichocoixa verticalis* is a flighted insect that is known to prey on brine shrimp especially the smaller nauplii and sub-adults. This predator flourishes in lower salinities of 50 g/L. At these salinities, *T. verticalis* mean densities were 52 individuals/m^3 (Wurtsbaugh, 1992). In Farmington Bay during April 1997, *T. verticalis* individuals numbered 100/m3 (Doyle Stephens, USGS, personal communication, 1997). Subsequently, the population of *Artemia* in Farmington Bay was suppressed by predation. In addition, predation by other macro invertebrates and zooplankton in the Farmington Bay complex ties up much of the available invertebrate food resources in the water column.

When the south arm is at 115 g/L salinity and the system is brine-fly and brine-shrimp dominated, these invertebrates are released to the next trophic level and predatory opportunists, the birds. Significant numbers of some species and suites of species of birds depend upon these macro-halofiles (invertebrates). The Great Salt Lake was designated a Hemispheric Shorebird Reserve by the Western Hemispheric Shorebird Reserve Network in 1990 because it supports millions of shorebirds. These shorebird numbers are supported by significant wetland resources located on the east side of the lake, including Farmington Bay Wildlife Management area (figure 1), and by the copious halophiles of the Great Salt Lake. Besides the recognition for shorebird populations, the Great Salt Lake ecosystem supports notedly high numbers of other species as compared to continental, hemispheric, and world populations. The waters surrounding Antelope Island are often the producers of the food that fuels these birds.

MINERAL EXTRACTION INDUSTRIES

Common salt was extracted from Great Salt Lake by trappers and explorers as early as 1825, and became the principal source of salt for the early Mormon settlers beginning in 1847. Since that time, there have been numerous salt industries on the lake (Clark and Helgren, 1980), mainly along the lake's southern shore. Small operations were also located in Spring Bay on the extreme north end of the lake, at Promontory Point, and on the mud flats west of Syracuse (within the present bounds of Farmington Bay). Today, six active mineral-extraction industries operate on the lake which contribute significantly to the local economies. These companies, the source of their brine (arm of the lake), and the products they produce, are shown in table 2.

Table 2. Active mineral-extraction industries on Great Salt Lake, Utah (after Gwynn, 1997).

COMPANY NAME	ARM OF LAKE	PRODUCTS
Magnesium Corporation of America (MAGCORP)	South	Primary magnesium metal, alloys, and chlorine gas
Cargill Salt Company	South	Sodium chloride products
Morton Salt Company	South	Sodium chloride products
IMC Kalium Ogden Corporation	North	Potassium sulfate (fertilizer), magnesium chloride products
IMC Salt	North	Sodium chloride products
North Shore Ltd. Partnership - Mineral Resources International	North	Concentrated brines that are processed into dietary supplement

CHRONOLOGY OF DIKING STUDIES AND PROPOSALS RELATED TO THE SOUTH ARM OF GREAT SALT LAKE

Studies, and Early Proposals for Freshwater Embayments (1930-1965)

Numerous proposals have been put forth over the years to create freshwater bodies out of the entire east embayment (the general area east and north of Antelope and Fremont Islands), or out of Farmington Bay, through the construction of dikes between the islands and the mainland. Richards (1965, 1972) reviews the past studies of the east embayment as a freshwater reservoir, none of which have been implemented. The chronology and a short description of these studies up to 1972 are summarized in appendix A and illustrated in figure 9.

Flooding Years (1983-1987)

Lake-Level Control Alternatives

In water year 1983, Great Salt Lake began its historic rise from a low elevation of about 4,200 feet to its historic high of 4,211.85 feet in 1986-1987. As the lake rose, extensive flooding occurred around the lake, causing many millions of dollars in damage, mainly around the south arm. During this time, several lake-level control measures were investigated, including upstream development, breaching the SPRR causeway, West Desert pumping, and diking (Allen and others, 1983; and Utah Division of Water Resources, 1984). Of these alternatives, the SPRR causeway breach was constructed in 1984, and West Desert pumping was implemented in 1987.

Diking Alternatives

Diking alternatives were proposed during the flooding years (1983-1987) mainly to protect large areas around the perimeter of the lake or to protect individual entities such as sewage-treatment plants. A number of these proposals included the construction of dikes between Antelope Island and the mainland, or between Antelope and Fremont Islands and the mainland. The major diking alternative proposals presented during the flooding years include: Creamer and others (1985), Holland and Spiers (1985), Chadwick and others (1986), Rollins and others (1987), and Sorensen and Riley (1988). The lake began to drop after 1987, and none of the above-listed diking schemes was implemented.

Some dikes were constructed within or in proximity to Farmington Bay (Anderson, 1985). For example, a dike around the South Davis North Wastewater Treatment Plant was built during 1985 and 1986. Two sections of dike were constructed through the eastern part of Farmington Bay west of Farmington and Centerville by Utah Power and Light Company during late 1984 and early 1985. These dikes were constructed to protect power-transmission towers from ice loading. Phillips Petroleum also raised the dikes around its wastewater holding ponds west of Bountiful.

Post-Flooding Years (1987-present)

After the flooding years, attention was again returned to the creation of freshwater impoundments, within Farmington Bay or the entire east bay, through various diking schemes. The first of these was proposed in 1989 by the Lake Wasatch Coalition (Johnson, 1989). The proposed freshwater Lake Wasatch would have been formed through diking from the south shore of the lake to the south tip of Antelope Island, from the north end of Antelope Island to the south tip of Fremont Island, and from the north end of Fremont Island to Promontory Point (figure 9). This proposal failed because of the high construction costs and concern about disrupting the GSL ecosystem (Woolf, 1993).

Also in 1989, The Utah Legislature, through the "Great Salt Lake Development Authority Act" (Smedley and others, 1989), formed the Great Salt Lake Development Authority (GSLDA) to investigate the technical, economic, and environmental feasibility of an inter-island diking project that would create the freshwater Lake Wasatch in the eastern half of Great Salt Lake. The recommendations of the GSLDA (Great Salt Lake Development , 1990) were as follows:

(1) Development of "Lake Wasatch" through a Great Salt Lake interisland diking project, as defined in Utah Code Ann. §17A-2-1603 (9), does not appear to be economically or environmentally feasible.

(2) Proposals to create small freshwater lake or water storage reservoir in Farmington Bay deserve further study.

(3) Public access to Antelope Island should be restored as quickly as possible.

In mid-1990, a preliminary design report was issued by the Utah Division of Water Resources (Palmer and others, 1990) for a diked water impoundment, called the Davis County Pond, as part of the Bear River Development Project. The purpose of the work was to study, test, and recommend an area or areas for final pond design. Two recommended areas were: (1) the eastern end of Farmington Bay from the Farmington Bay Waterfowl Management area on the south to Shepards Lane on the north, and (2) a triangular area along the northeastern shore of Farmington Bay, bounded on the south by a western extension of Shepards Lane and on the west by a southerly extension of 1,000 West street in Syracuse (figure 9).

In January 1993, Davis Lake (figure 9) was proposed by the Davis County Commission (Rosebrook, 1993). This proposed project, a scaled-down version of Lake Wasatch, would have converted Farmington Bay into a freshwater body by using the existing dikes at both the north and south ends of Antelope Island.

In early 1995 (pre-February), the Bonneville Bay proposal emerged, which called for the creation of a freshwater reservoir similar to Davis Lake (figure 9). The unique feature put forth in this proposal was the use of new technology, the "Wasatch Wall Pile Dike System," developed by Don D. Johnson in 1985, (promotional literature) in the construction of both the north and south dikes.

To date, none of the proposals to change either Farmington Bay or the entire east embayment into freshwater reservoirs has come to fruition. In spite of the many advantages cited by the proponents of these projects, questions related to sediment pollution, health hazards, eutrophication potential, wastewater-treatment plant costs, water salinity, mosquito production potential, and flooding issues (Chadwick and others, 1986; Sorensen and Riley, 1988; Stauffer and Austin, 1995) remain understudied and unanswered.

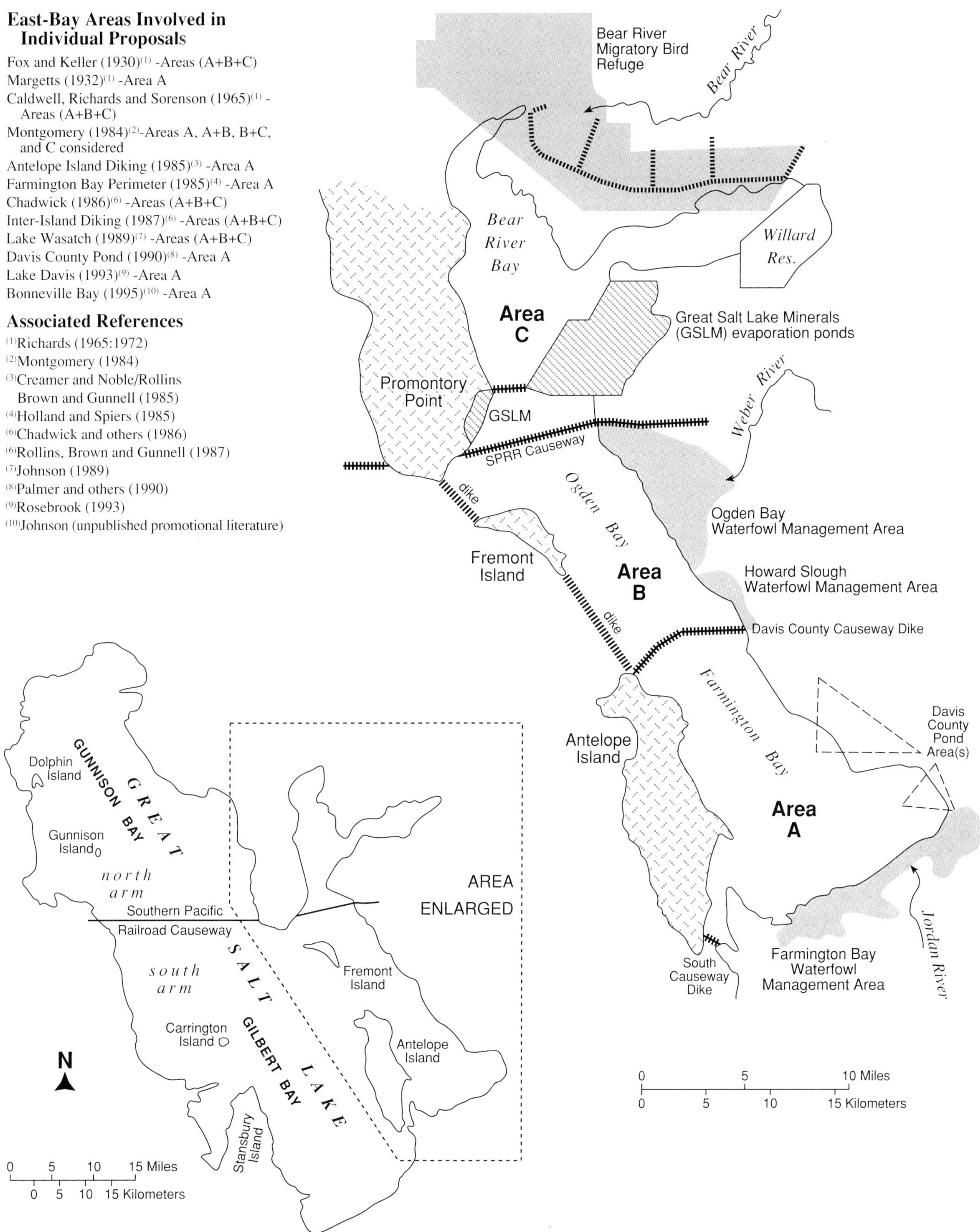

Figure 9. *Areas of east embayment involved in the various proposed diking schemes to create freshwater bodies, and the location of proposed dikes, solar-evaporation-pond areas, and waterfowl-management areas.*

REFERENCES

Allen, M.E., Christensen, R.K., and Riley, J.P., 1983, Some lake level control alternatives for the Great Salt Lake: Utah Water Research Laboratory, Utah State University, Water Resources Planning Series UWRL/P-83/01, 101 p.

Anderson, D.L., 1985, Executive summaries - Great Salt Lake studies: Utah Department of Natural Resources, Division of Water Resources, 47 p.

Arnow, Ted, and Stephens, Doyle, 1990, Hydrologic characteristics of the Great Salt Lake, Utah- 1847-1986: U.S. Geological Survey Water-Supply Paper 2332, 32 p.

Bott, Cordell, and Shipman, S.T., 1971, Water chemistry and water quality of Farmington Bay, *in* Carter, C.K., project director, Some ecological considerations of Farmington Bay and adjacent Great Salt Lake State Park: University of Utah, p. B1-B27.

Carter, James, and Hoagland, Timothy, 1971, Geological and hydrological characteristics of Farmington Bay, *in* Carter, C.K., project director, Some ecological considerations of Farmington Bay and adjacent Great Salt Lake State Park: Salt Lake City, University of Utah, p. A1-A-27.

Chadwick, D.G., Jr., Riley, J.P., Seierstad, A.J., Sorensen, D.L., and Stauffer, N.E., Jr., 1986, Expected effects of in-lake dikes on water levels and quality in the Farmington Bay and the east shore areas of the Great Salt Lake, Utah: Utah Water Research Laboratory, Utah State University, 91 p.

Clark, J.L., and Helgren, Norman, 1980, History and technology of salt production from Great Salt Lake, *in* Gwynn, J.W., editor, Great Salt Lake - a scientific, historical and economic overview: Utah Geological and Mineral Survey Bulletin 116, p. 203-217.

Colburn, A.A., and Eckhoff, D.W., 1972, Pollution input from the lower Jordan Basin to Antelope Island Estuary, *in* Falkenborg, Donna, editor, The Great Salt Lake and Utah's water resources: Proceedings of the first annual conference of the Utah Section of the American Water Resources Association, Salt Lake City, Utah, November 30, 1972, p. 104-120.

Creamer & Noble Engineers, and Rollins, Brown & Gunnel, Inc., 1985, Final design report for Great Salt Lake-Antelope Island diking, executive summary: Unpublished consultants report, prepared for the Utah Department of Transportation, pp. I1-I24.

Ford Chemical Laboratory, 1977, An investigation of bacteria concentrations in specific areas of the Great Salt Lake: Ford Chemical Laboratory, Unpublished report prepared for Utah State Division of Health, Department of Social Services, 30 p.

Great Salt Lake Development Authority, 1990, Report of the Great Salt Lake Development Authority to Governor Norman Mangerter, Senator Arnold Christensen, and Representative Craig Moody: Great Salt Lake Development Authority Board of Directors (December 10), 5 p.

Gwynn, J.W., 1997, Brine properties, mineral extraction industries, and salt load of Great Salt Lake, Utah: Utah Geological Survey Public Information Series 51, 2 p.

Hansen, D.E., 1981, A hydrologic model to determine water quality management of the Farmington Bay and Farmington Bay bird refuge: Logan, Utah State University, M.S. thesis, 67 p.

Harward, M.R., 1996, Where the buffalo roam - life on Antelope Island: M.R. Harward, 382 W. 3300 S., Bountiful, Utah 84010, 87 p.

Holland, F.L., and Spiers, K.L.,1985, Farmington Bay area perimeter diking alternative - final design report, executive summary: James M. Montgomery Consulting Engineers, Inc. published report prepared for the Utah Division of Water Resources, Utah Department of Natural Resources, 15 p.

Israelsen, C.E., Sorensen, D.L., Seierstad, A.J., and Brennard, Charlotte, 1985, Preliminary identification, analysis, and classification of odor-causing mechanisms influence by decreasing salinity of the Great Salt Lake: Utah Water Research Laboratory, Utah State University, prepared for the Utah Department of Natural Resources, Utah Division of Water Resources, 63 p.

Johnson, D.D., 1989, Lake Wasatch - the great freshwater lake: 1071 E. 250 N., Bountiful, Utah, 97 p. (unpublished collection of promotional information packets on Lake Wasatch).

Maxell, M.H., and Thatcher, L.M., 1980, Coliform bacteria concentrations in Great Salt Lake waters, *in* Gwynn, J.W., editor, Great Salt Lake - a scientific, historical and economic overview: Utah Geological and Mineral Survey Bulletin 116, p. 323-335.

Montgomery, J.M., 1984, Great Salt Lake diking feasibility study - volume I: James M. Montgomery, Consulting Engineers, Inc. unpublished report prepared for Utah Department of Natural Resources, Project No. 1609.003, Utah Division of Water Resources, 293 p.

Palmer, Jim, Leeflang, Bill, Everitt, Ben, and Aubrey, Dan, 1990, Preliminary design report for proposed reservoir in Farmington Bay of the Great Salt Lake: Utah Department of Natural Resources, Utah Division of Water Resources, 125 p.

Richards, A.Z., Jr., 1965, A preliminary master plan for the development of the Great Salt Lake--over a period of the next 75 years: Caldwell, Richards & Sorensen, Inc., unpublished report prepared for the Great Salt Lake Authority, 32 p.

—1972, Review of past studies of the east embayment as a freshwater reservoir, *in* Falkenborg, Donna, editor, The Great Salt Lake and Utah's water sources: Proceedings of the first annual conference of the Utah Section of the American Water Resources Association, Salt Lake City, Utah, November 30, 1972, p. 180-189.

Rollins, Brown and Gunnell, Inc., and Creamer & Noble Engineers, 1987, Great Salt Lake inter-island diking project: unpublished consultants report, v. 1, section 1, 9 p.

Rosebrook, Don, 1993, Davis has fresh plans for Antelope Island: Deseret News, January 6-7 1993 edition, p (?).

Rushforth, S.R. and Felix, E.A., 1982, Biotic adjustments to changing salinities in the Great Salt Lake, Utah: Microbial Ecology (US), v. 8, p. 157-161.

Saunders, G.E., 1994, Massive spread footing key to causeway reconstruction: Roads & Bridges, August edition, p. 26-27.

Saunders, G.E,. and Smith, S.W., 1994, Causeway reopening benefits people and wildlife: Public Works, February edition, p. 34-36.

Searle, R.T., 1976, The Great Salt Lake: Office of Legislative Research, State of Utah, Research Report No. 7, 274 p.

Smedley, S.M., and Bain, Walt, Burningham, K.R., Woddoups, M.G., and 24 other legislators, 1989, Great Salt Lake Development Authority: 1989 General Session [of the Utah Legislature], H.B. No. 208, 57 p.

Smith, S.W., 1997, Memo from S.W. Smith, Director of Davis County Public Works, to D.R. McConkie, Davis County

Commissioner, 1 p.

Sorensen, D.L. and Riley, J.P., 1988, Great Salt Lake inter island diking - water quality considerations, final report: Utah Water Research Laboratory, Utah State University, prepared for Utah Department of Natural Resources, Division of Water Resources, 261 p.

Stauffer, Norm and Austin, Lloyd, 1995, Proposal to convert Farmington Bay into a freshwater bay: Memorandum to D. Larry Anderson, Director of Utah Division of Water Resources, dated July 17, 1995, 4 p.

Sturm, P.A., 1983, Farmington Bay project report: Utah Geological and Mineral Survey Open-File Report 86, 56 p.

U.S. Geological Survey, 1972, Antelope Island South Quadrangle, Utah: U.S. Geological Survey, 7.5 minute series orthophotomap (topographic), scale 1:24,000.

U.S. Geological Survey, 1974, Great Salt Lake and vicinity, Utah: U.S. Geological Survey topographic map, scale 1:125,000.

Utah Division of Water Quality, 1994-1996, unpublished laboratory-analysis reports: Utah Division of Water Quality, Utah Department of Environmental Quality (20 analyses).

Utah State Department of Health and Davis County Health Department, 1965, Preliminary investigation of pollution of Great Salt Lake east of Antelope Island: Utah State Department of Health and Davis County Health Department, unpublished report 13 p.

Waddell, K.M., and Fields, F.K., 1977, Model for evaluating the effects of dikes on the water and salt balance of Great Salt Lake, Utah: Utah Geological and Mineral Survey Water-Resources Bulletin 21, 54 p.

Woolf, Jim, 1993, Davis wants to convert bay into freshwater reservoir: Salt Lake Tribune, January 7, 1993.

Wurtsbaugh, W.A., 1992, Food-web modifications by an invertebrate predator in the Great Salt Lake (USA): Oecologia., v. 89, p. 168-175.

APPENDIX A

Chronology and description of past studies of the east embayment as a freshwater reservoir (modified from Richards 1965, 1972)

May, 1930: L.E. Fox and T.L. Keller, in a work titled "Putting Great Salt Lake to work," propose a three-part dike extending westerly from the mainland to the south end of Antelope Island, and northerly from the north end of Antelope Island to the south end of Fremont Island, and from the north end of Fremont Island to the Promontory Point (figure 9).

March, 1932: S.G. Margetts, in a work titled "Report on proposed freshwater lake," recommended a smaller freshwater lake be developed in Farmington Bay by damming the 146 square-mile area between the mainland and the south end of Antelope Island and between the north end of Antelope Island and the mainland at Syracuse (figure 9).

June, 1933: R.A. Hart and N.E. McLaughlan, published a work titled "Abstract of report on laboratory tests of feasibility of freshening the proposed diked-off portion of Great Salt Lake."

July, 1933: D.A. Lyon, published a work titled "Report of sub-committee appointed to investigate technical problems that present themselves in connection with the proposed diking of Great Salt Lake for the purpose of creating a body of fresh water."

December, 1933: J.L. Crane, provided a summary report on Great Salt Lake diking project, and concluded that the "smaller reservoir" known as the Farmington Bay Project would be feasible from the standpoint of engineering construction. He recommended that the diking project be undertaken by the State of Utah.

July, 1935: T.C. Adams, reported to the Utah State Planning Board on additional engineering studies that he had been commissioned to make by said board to substantiate certain recommendations of the 1933 Crane report.

1955: The Utah Legislature asked the Utah State Road Commission to initiate a study of the advisability of constructing the inter-island dikes, and particularly to address the availability of water.

1956: Dames & Moore and Caldwell, Richards & Sorensen, Inc, provided a joint report to the Utah State Road Commission on the availability of water.

1957: Caldwell, Richards & Sorensen, Inc. prepared a report for an advisory committee to the State Road Commission titled "An engineering report and opinion on the quantity and quality of water which may become available for use by industry, agriculture and for improved recreation from the east embayment reservoir of Great Salt Lake if the inter-island dikes were to be built."

1963: The Utah State Legislature enacted the "Great Salt Lake Authority," with the responsibility to plan, formulate and execute a program for the development of the mainland, the islands, the minerals, and the water within Great Salt Lake meander line as established by the United States Surveyor General for industrial, recreational, agricultural, industrial, and chemical purposes.

January, 1963: D.R. Mabey published an article in the Utah Engineering and Science Magazine titled "Tailings--a new resource?" which addressed the possibility of using tailings from Kennecott Copper Corporation to construct the inter-island dikes.

January, 1965: Caldwell, Richards & Sorensen, under the direction of the Great Salt Lake Authority, published a report titled "Preliminary master plan for the development of Great Salt Lake over the next 75 years. This plan envisioned constructing a massive landfill in the southern half of Farmington Bay using Kennecott tailings, and through the construction of dikes, creating three freshwater bays: (1) a south bay east of Antelope Island, (2) a central bay east of Fremont Island, and (3) a northern bay within the present Bear River Bay (figure 9).

OXYGEN-ISOTOPE COMPOSITION OF GREAT SALT LAKE, 1979 TO 1996

by
Vicki A. Pedone
Department of Geological Sciences,
California State University Northridge, Northridge, CA 91330-8266,
vicki.pedone@csun.edu

ABSTRACT

The oxygen-isotope composition of the hydrologically closed Great Salt Lake is a function of balance between inflow and evaporation. Contemporaneous geographic and vertical differences in south arm of the Great Salt Lake were typically <1‰ between 1993 and 1996 and probably result from local variations in evaporation rate and water circulation. Annual differences of <1‰ result from variation in the isotopic composition of inflow water. Shifts in $\delta^{18}O$ of 1‰ to 1.5‰ occur seasonally in the lake, decreasing with the addition of spring snowmelt and increasing during evaporative concentration in late summer and fall.

The addition of large volumes of snowmelt and rainfall between 1983 to 1986 caused lake volume to increase 120 percent and $\delta^{18}O$ values to decrease in all water masses of the lake. The $\delta^{18}O$ of surface and bottom water decreased 4‰ and 1.6‰, relative to Standard Mean Ocean Water, (SMOW), respectively, in the north arm; and 7‰ and 3‰ (SMOW), respectively, in the south arm. The $\delta^{18}O$ of the water showed a strong negative covariance with lake volume during lake expansion. Water samples were not available for the period of lake contraction between 1987 and 1993, but by 1994, lake volume and $\delta^{18}O$ of water masses had returned their pre-rise values.

An important question is if, and how, carbonate minerals, formed in ancient lakes similar to the Great Salt Lake record these variations in oxygen-isotope composition. This study suggests that time-equivalent layers could exhibit lateral variations up to ~1‰ owing to geographic differences in evaporation rates and water circulation and/or where the carbonate mineral precipitated in the water column. However, the seasonal and lake-expansion shifts in $\delta^{18}O$ are greater than the contemporaneous isotopic heterogeneity in the lake and have potential to be recognized in the geologic record of carbonate minerals formed in a lake similar to the Great Salt Lake.

INTRODUCTION

Lacustrine carbonate minerals are potentially sensitive recorders of high-frequency changes in the isotopic composition of lake waters because they form at geologically rapid rates. Oxygen-isotope compositions of tufa and lacustrine stromatolites have been particularly useful in reconstructing paleoclimatic and paleohydrologic variations in lakes (Stuiver, 1970; Abell and others, 1982; Casanova and Hillaire-Marcel, 1993; Benson and others, 1996; Cohen and others, 1997). Factors that control the oxygen isotopic ratio of lake water and its record in lacustrine carbonates are complex and dependent on the specific geologic, geographic, hydrologic, and climatic setting of the basin (e.g., McKenzie, 1985; Gasse and others, 1987; Kelts and Talbot, 1989; Talbot, 1990; Benson, 1994; Benson and others, 1996; Li and Ku, 1997).

The balance between inflow and evaporation controls the $\delta^{18}O$ isotopic composition of the water on a seasonal and year-to-year basis in a hydrologically closed, shallow lake. However, neither the magnitude of these changes nor the geographic and vertical isotopic heterogeneity in the Great Salt Lake has been investigated. This study examines changes in oxygen isotopic composition of the Great Salt Lake during its largest expansion of the 20th century and investigates isotopic heterogeneity in the lake during a period of average lake volume. Analyses include water samples from the north and south arms of the Great Salt Lake collected between 1979 and 1986, and from shoreline and lake-center localities collected between 1994 and 1996. Information on isotopic heterogeneity, seasonal variations, and annual variations in the Great Salt Lake will improve the understanding of variation in $\delta^{18}O$ of carbonate minerals formed in shallow, hydrologically closed saline lakes.

HYDROLOGY OF THE GREAT SALT LAKE

Because the Great Salt Lake has no outflow, the lake volume changes according to the amount of water removed by evaporation and the amount added by direct precipitation, surface flow, and subsurface flow. Lake-elevation, lake-area, and lake-volume data have been tabulated by Cruff (1986), and are available from the Water Resources Division of the U.S. Geological Survey in the computer file GPYPER> GSL.HYPSO. Historical lake-elevation data are also available from the U.S. Geological Survey, which began measurements of the elevation of the Great Salt Lake in 1875. The USGS currently operates one lake-elevation gage in the south arm (boat harbor gage), located at the Great Salt Lake State Park, and one gage in the north arm (Saline gage), located at Little Valley Harbor. In addition, observational records of lake level provide estimates of lake elevation back to the late 1840s (Arnow, 1980).

The average historical volume of Great Salt Lake is 13,440,000 acre-feet. The historical lake-volume maximum of 29,720,000 acre-feet occurred in 1986, and historical lake-volume minimum of 6,969,000 acre-feet occurred in 1963. Seasonal fluctuations in lake volume range from 3.5 percent to 7 percent, with peak water volume in late spring and the lowest water volume in late autumn.

Urbanization along the Wasatch Front and construction projects in the Great Salt Lake have severely impacted the hydrology of the lake. One of the most striking alterations resulted from construction of the Southern Pacific Railroad causeway in 1959, which divided the Great Salt Lake hydrologically into the north and south arms (figure 1). Hahl and Hardy (1969), Madison (1970), Waddell and Bolke (1973), Whelan (1973), and Gwynn and Sturm (1987) have described the effects of this construction. Prior to construction of the Southern Pacific Railroad causeway, most of the Great Salt Lake was well mixed from top to bottom. After construction, water in the north and south arms became chemically distinct and salinity stratified. Construction of a breach in the causeway in 1984 caused the deep brine in the south arm to gradually disperse. By the early 1990s, salinity stratification was no longer detectable in either arm (Wally Gwynn, Utah Geological Survey, personal communication, 1999).

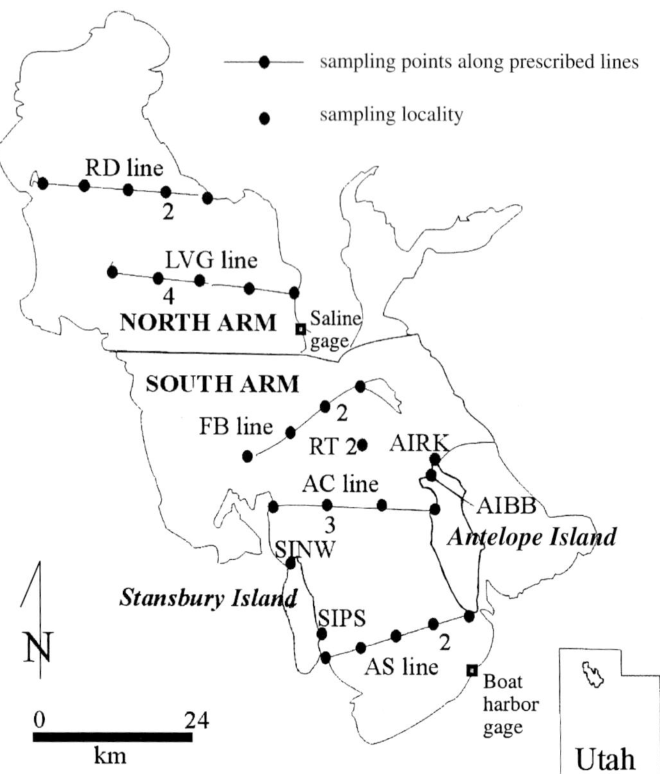

Figure 1. Location of shoreline and lake-center sampling sites in the north and south arms of the Great Salt Lake. Abbreviations (AIRK, AIBB, SINW, AND SIPS) are found in the text. The shoreline elevation is 1,280.5 m (4,200 ft).

SAMPLING SITES

Shoreline sampling localities on Stansbury Island were located in a small cove on the northwest end of the island (SINW), and near the pumping station of the Morton Salt Company on the southeast end of the island (SIPS) (figure 1). Both localities are very shallow, low-energy sandflats. The two localities on Antelope Island were in Bridger Bay on the northwest end of the island. One site was on the rocky headland that forms the northern limit of the bay (AIRK). The other was in the cove in which the public beach is located (AIBB). These sites have steeper shoreface gradients than those on Stansbury Island and experience higher average wave energy and more active water circulation.

The study also analyzed samples collected by the Utah Geological Survey and the Utah Division of Water Quality. These two agencies have been monitoring the composition of the water column at many localities, shown in figure 1. Samples for this study came from LVG 4 and RD2 in the north arm, and FB2, RT2, AC3, and AS2 in the south arm. Because sampling trips to the lake center were scheduled on short notice when weather permitted, and winter sampling was done during the fall academic semester, the author was not able to sample the shoreline localities on the same days that the boat crew sampled the lake center.

METHODS

Water samples from four shoreline sites in the south arm of the Great Salt Lake were collected between January 1994 and July 1996 (figure 1). Winter sample collection is near the time of lake-elevation minimum, and summer sample collection near the time of lake-elevation maximum. Shoreline samples were drawn into a 60-cc syringe and expelled through a 0.45-mm filter into acid-washed, 30-ml, screw-top high-density polyethylene (HDPE) bottles. Bottles were filled completely and kept refrigerated until sent for analysis.

The Utah Division of Water Quality collected surface and bottom water samples for this study between 1994 and 1996 in screw-top, volatile-organic-chemical (VOC) bottles. This study also took aliquots of surface and bottom water samples, collected between 1979 and 1986 from the north and south arms, and archived by the Utah Geological Survey. The archived samples were stored in completely filled, tightly capped, 500-ml plastic bottles.

The oxygen isotope composition of the water samples was analyzed by Coastal Science Laboratories in Austin, Texas. Oxygen isotopic composition was obtained by equilibrating about 10 ml of the water sample with CO_2 in a 20-cc plastic syringe for at least two hours at 25°C (Yoshiro and Mizutani, 1986). The gas extracted from the sample was purified through a series of cold traps, and the oxygen isotopic composition of the pure CO_2 collected was measured on a VG Micromass SIRA Series II. Accuracy of the method, as determined by analyses of reference materials, is better than 0.3‰. All data in this paper are given relative to the Standard Mean Ocean Water (SMOW) standard for $\delta^{18}O$.

RESULTS

Lake Center of the South Arm, 1979-1995

The lake was near its average historical volume between 1979 and 1982 (figure 2). During this period, the annual values of the oxygen isotopic composition of the south arm

record these short-term variations in oxygen-isotope composition. This study suggests that time-equivalent layers could exhibit lateral variations up to ~1‰ owing to geographic differences in evaporation rates and water circulation and/or where the carbonate mineral precipitated in the water column. Seasonal changes in lake volume produce $\delta^{18}O$ variations of 1 to 2‰. Such changes could be detected in samples taken in closely spaced time-sequential layers, assuming that carbonate minerals were formed in the lake during the extremes of seasonal variations.

Events like the significant, though short-lived, lake expansion of the 1980s have high potentials of being preserved and detected in the geologic record. This expansion caused a decrease in $\delta^{18}O$ of >2‰ for at least four years (figure 2), and possibly as long as 10 or 11 years, if $\delta^{18}O$ negatively co-varied with lake volume during the contraction phase. A similar expansion occurred in the 1870s. High-spatial-resolution sampling of carbonate deposits would likely capture the record of these ~100-year frequency events.

CONCLUSIONS

Contemporaneous geographic and vertical differences in south arm of the Great Salt Lake were typically <1‰ between 1993 and 1996, and probably result from local variations in evaporation rate and water circulation. Annual differences of <1‰ result from variation in the isotopic composition of inflow water. Shifts in $\delta^{18}O$ of 1‰ to 1.5‰ occur seasonally in the lake, decreasing with the addition of spring snowmelt and increasing during evaporative concentration in late summer and fall. Increase in lake volume of 120 percent between 1983 and 1986 caused decreases of >2‰ in both the surface and bottom layers of the then-salinity-stratified north and south arms. The seasonal and lake-expansion shifts in $\delta^{18}O$ are greater than contemporaneous isotopic heterogeneity in the lake, and have potential to be recognized in the geologic record of carbonate minerals that were formed in lakes similar to the Great Salt Lake.

ACKNOWLEDGMENTS

I thank the Morton Salt Company and Mr. John Bleassard for access to the sampling localities on Stansbury Island. Acknowledgment is made to the donors of the Petroleum Research fund, administered by the American Chemical Society, and to California State University Northridge for support of this research. The final draft of the manuscript was improved by suggestions made by reviewers Paul Jewell and Robert Gloyn.

REFERENCES

Abell, P.I., Awramik, S.M., Osbourne, R.H., and Tomellini, S., 1982, Plio-Pleistocene lacustrine stromatolites from Lake Turkana, Kenya - morphology, stratigraphy and stable isotopes: Sedimentary Geology, v. 32, p. 1-26.

Arnow, T., 1980, Water budget and water surface fluctuations of Great Salt Lake, *in* Gwynn, J. W., editor, Great Salt Lake - a scientific, historical and economic overview: Utah Geological and Mineral Survey, Bulletin 116, p. 255-263.

Benson, L., 1994, Stable isotopes of oxygen and hydrogen in the Truckee River-Pyramid Lake surface-water system. 1. Data analysis and extraction of paleoclimatic information: Limnology and Oceanography, v. 39, p. 344-355.

Benson, L.V, White, L.D., and Rye, R., 1996, Carbonate deposition, Pyramid Lake Subbasin, Nevada, 4. Comparison of the stable isotope values of carbonate deposits (tufas) and the Lahontan lake-level record: Palaeogeography, Palaeoclimatology, Palaeoecology, v. 122, p. 45-76.

Casanova, J., and Hillaire-Marcel, C., 1993, Carbon and oxygen isotopes in African lacustrine stromatolites - paleohydrological interpretation: Climate Change in Continental Isotopic Records, Geophysical Monograph 78, p. 123-133.

Cohen, A.S., Talbot, M.R., Awramik, S.M., Dettman D.L., and Abell, P., 1997, Lake level and paleoenvironmental history of Lake Tanganyika, Africa, as inferred from late Holocene and modern stromatolites: Geological Society of America Bulletin, v. 109, p. 444-460.

Cruff, R.W., 1986, Updated table of hypsographic data for Great Salt Lake: Memorandum, U.S. Geological Survey, Water Resources Division, 8 p.

Gasse, F., Fontes, J.C., Plaziat, J.C., Carbonel, P., Kacmarska, I., De Deckker, P., Soulie-Marsche, I., Callot, Y., Dupeuble, P., 1987, Biological remains, geochemistry and stable isotopes for the reconstruction of environmental and hydrological changes in the Holocene lakes from North Sahara: Palaeogeography, Palaeoclimatology, Palaeoecology, v. 60, p. 1-46.

Gwynn, J. W., and Sturm, P. A., 1987, Effects of breaching the Southern Pacific Railroad causeway, Great Salt Lake, Utah - physical and chemical changes August 1, 1984 - July, 1986: Utah Geological and Mineralogical Survey, Water-Resource Bulletin 25, 25 p.

Hahl, D.C., and Hardy, A. H., 1969, Great Salt Lake, Utah - chemical and physical variations of the brine, 1963-1966: Utah Geological and Mineralogical Survey, Water-Resource Bulletin 12, 33 p.

Kelts, K.R., and Talbot, M.R., 1989, Lacustrine carbonates as geochemical archives of environmental change and biotic-abiotic interactions, *in* Tilzer, M.M. and Serruya, C., editors, Ecological Structure and Function in large Lakes: Science and Technology Publ., Madison, WI, p. 290-317.

Li, H.-C., and Ku, T.-L., 1997, $\delta^{13}C$-$\delta^{18}O$ covariance as a paleohydrological indicator for closed-basin lakes: Palaeogeography, Palaeoclimatology, Palaeoecology, v. 133, p. 69-80.

Madison, R.J., 1970, Effects of a causeway on the chemistry of the brine in Great Salt Lake, Utah: Utah Geological and Mineralogical Survey, Water-Resource Bulletin 14, 52 p.

McKenzie, J.A., 1985, Carbon isotopes and productivity in the lacustrine and marine environment, *in* Stumm, W., editor, Chemical Processes in Lakes: Wiley, New York, p. 99-118.

Stuiver, M., 1970, Oxygen and carbon isotope ratios of fresh water carbonates as climatic indicators: Journal of Geophysical Research, v. 75, p. 5247-5257.

Talbot, M.R., 1990, A review of the palaeohydrological interpretation of carbon and oxygen isotopic ratios in primary lacustrine carbonates: Chemical Geology (Isotope Geosciences Section), v. 80, p. 261-279.

Waddell, K.M., and Bolke, E. L., 1973, The effects of restricted circulation on the salt balance of Great Salt Lake, Utah: Utah Geological and Mineralogical Survey, Water-Resources Bulletin 18, 54 p.

Whelan, J.A., 1973, Great Salt Lake, Utah: Chemical and physical variations of the brine, 1966-1972: Utah Geological and Mineralogical Survey, Water-Resources Bulletin 17, 24 p.

Yoshiro, N., and Mizutani, Y., 1986, Preparation of carbon dioxide for oxygen-18 determination of water by use of a plastic syringe: Analytical Chemistry, v. 58, p. 1273-1275.

APPENDIX

Appendix - Oxygen isotopic composition of water from the south arm, 1993 to 1996. Locations of the sample sites are shown in figure 1.

Sample locality	Date of collection	Depth (m)	$\delta^{18}O$ (SMOW)
RT2	1 Apr-93	0	-2.7
RT2	1 Apr-93	9	-2.7
AS2	16 Nov-93	0.15	-2.5
AS2	16 Nov-93	9	-2.9
FB2	16 Nov-93	0.5	-2.9
FB2	16 Nov-93	9	-2.8
SIPS	15 Jan-94	0.1	-2.9
AS2	9 Jun-94	0.15	-3.7
AS2	9 Jun-94	9	-3.6
RT2	9 Jun-94	0	-3.7
RT2	9 Jun-94	8.2	-3.6
AIRK	26 Jun-94	0.4	-3.2
SINW	25 Jun-94	0.25	-3.3
RT2	15 Nov-94	0.15	-2.0
RT2	15 Nov-94	8.8	-2.2
AIBB	11 Jan-95	0.15	-3.4
AIRK	11 Jan-95	0.2	-3.1
SINW	10 Jan-95	0.2	-2.8
SIPS	10 Jan-95	0.2	-2.9
AC3	27 Jun-95	0	-4.7
AC3	27 Jun-95	8.8	-4.3
AS2	27 Jun-95	0	-4.7
AS2	27 Jun-95	8.8	-5.5
AIBB	9 Jun-95	0.3	-5.3
AIRK	9 Jun-95	0.4	-5.2
SINW	8 Jun-95	0.1	-4.5
SIPS	8 Jun-95	0.25	-4.4
AIBB	12 Jan-96	0.2	-3.8
AIRK	12 Jan-96	0.15	-3.5
SINW	11 Jan-96	0.1	-3.3
SIPS	11 Jan-96	0.1	-3.4
AIBB	10 Jul-96	0.25	-5.0
AIRK	10 Jul-96	0.2	-4.8
SINW	9 Jul-96	0.1	-4.4
SIPS	9 Jul-96	0.15	-4.3

LIPID COMPOSITION OF RECENT SEDIMENTS FROM THE GREAT SALT LAKE

by

James W. Collister
Department of Civil and Environmental Engineering
University of Utah

and

Steven Schamel
Petroleum Research Center
Department of Chemical and Fuels Engineering
University of Utah

ABSTRACT

Lipids extracted from recent sediments in the Great Salt Lake were studied to assess organic matter inputs into this hypersaline depositional environment. Because many organic-rich source rocks are deposited under hypersaline lacustrine conditions, this study has important implications for petroleum genesis and systems analysis. These environments commonly have density-stratified water columns with anoxic bottom waters that restrict metazoan grazers. Consequently, excellent preservation of organic material is common in these settings. Distributions of lipids in extracts clearly indicate variable inputs from indigenous algae and bacterial communities along with wind-borne terrestrial lipids and plant debris transported into the watershed. In addition, water surface temperatures during the algal bloom can be estimated from alkenone distributions. This provides a new method for reconstructing paleoclimatic conditions for the Great Salt Lake.

In the most extreme environments (evaporation ponds at Stansbury Island), lipids from halophilic photosynthetic bacteria dominate relative to those derived from algae. A high abundance of lipids from wind-borne terrestrial higher plants is observed at these sites presumably due to low primary production by algae. Sediments from open-lake environments are dominated by lipids from algae with lesser contributions from bacteria and terrestrial plant material. Distributions of acyclic isoprenoids and functionalized lipids such as sterols are highly variable, reflecting both differences in organic matter input as well as variations in redox conditions (Eh and pH) at the various sampling sites. Extremely oxidizing conditions and elevated salinities occur at the Stansbury Island evaporation ponds and at Antelope Island, north of the causeway. Moderately oxic conditions appear to prevail at Saltair Beach and at Antelope Island, south of the causeway where the water is less saline. Extremely reducing conditions are indicated for a bacterial layer in the evaporation pond as well as bacterial-dominated organic material from a saline marsh at White Rock Bay. These samples contain abundant pink-colored organic material produced by halophilic photosynthetic bacteria.

INTRODUCTION

The Great Salt Lake is a remnant of an extensive Pleistocene lake that once covered 51,200 km^2 of the northeast Basin and Range Province (Morrison, 1991). The modern lake (figure 1- showing southern half only) is less than one-tenth that size, covering only 4,403 km^2 at the historical average water surface elevation of 1,280m. At this elevation, the maximum depth of the lake is just 10.4 m. Nevertheless, this is the fourth largest terminal lake in the world. About two-thirds of the water input to the lake is from three rivers (Bear, Weber and Jordan) entering from the east, one-third is from precipitation falling directly on the lake, and a very small contribution is from groundwater flow. Evaporation from the lake surface is the only means of water removal; at present there are no outflows from the lake. As a consequence, the lake has been saline to hypersaline since at least the late Pleistocene-early Holocene (Kowalewska and Cohen, 1996).

Hypersaline lakes are highly productive, but have low species diversity. Compared to other modern saline lakes (table 1), the Great Salt Lake ranks in the middle range of biomass production rates. Annual primary production in the Great Salt Lake during 1973 was 145 g carbon (C)/m^2 (Stevens and Gillespie, 1976), and the maximum recorded primary production is 6,147 mg C /m^2/day (table1).

The construction of a rock-fill railroad causeway (figure 1) in 1959 resulted in two ecologically distinct parts of the lake, the Gunnison Bay north of the causeway and Gilbert Bay to the south. Inasmuch as 90 percent of the fresh water enters the south arm, the north arm has a relatively uniform salinity of 25 percent or higher. Salinity in Gilbert Bay is highly variable ranging between 6 percent and 27 percent in recent decades. Salinity, as well as the areal extent of the lake, changes with lake level, the result of relative rates of fresh water inflow (infall) and surface evaporation.

Saline and hypersaline lakes are well known as ideal settings for the deposition and preservation of organic-rich sediments that, through burial and heating, become oil-prone source rocks (Katz, 1990). Indeed, marine and lacustrine hypersaline depositional environments are responsible for

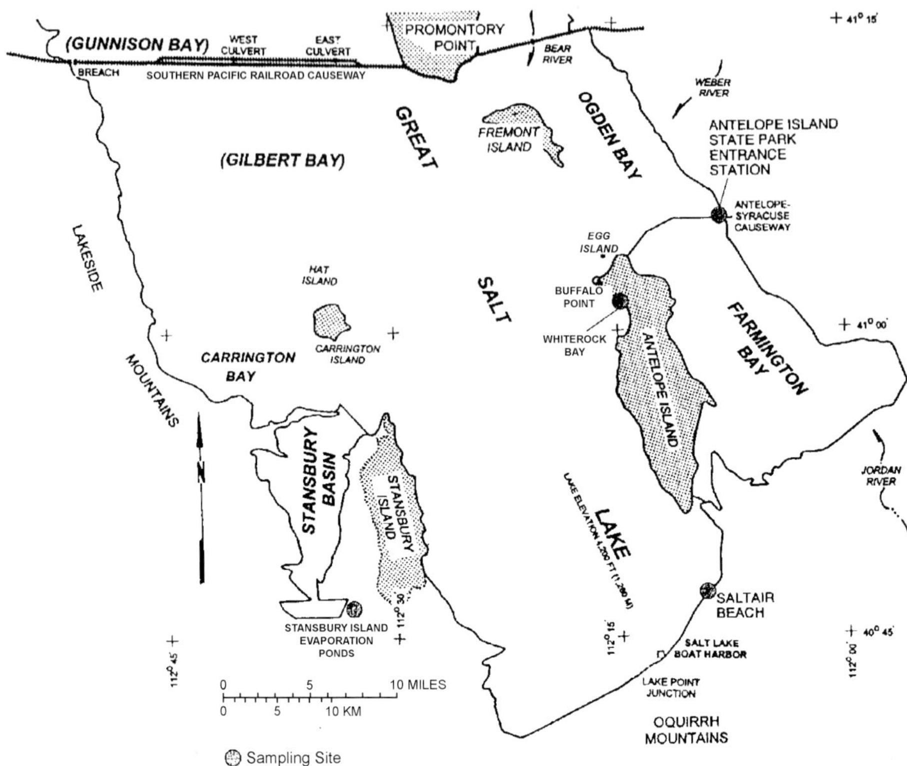

Figure 1. Map showing sampling sites within Gilbert Bay, Great Salt Lake, Utah.

Table 1. Primary productivity and environmental parameters for modern hypersaline settings (Data from Hammer, 1981; Warren 1986).

Location	pH mg / L	PO_4 mg / L	NO_3 Productivity Mg C m^{-2} / day	Maximum	Algal Characteristics
Africa:					
Solar Lake (Sinai)	-	-	-	5,000-12,000	Cyanobacterial mat
Egypt, Mariut Lake	7.8-9.35	0-0.003	0-0.56	9,611	Various algae
Ethiopia, Agranguadi Lake	10.3	3.2	<0.005	19,000	*Spirulina platensis*
Kilotes Lake	9.6	6.5	<0.005	4,133	*Spirulina chroococcus*
Kenya, Nakuru Lake	10.5	4.4	-	2,867	Cyanobacteria
Simbi Lake	10.55	2.0	-	-	
Tanzania, Rcshitani Lake	10.1	-	-	6,733	
Australia:					
Corongamite Lake	8.7-9.2	0-0.05	0.001-0.019	4,425	*Nodulari spumigena*
Pink Lake	7.6-8.6	0-0.007	<0.001	184	*Dunaliella salina*
Red Rock Lake	9.4-9.7	17.6-27.4	<0.001	7,531	*Anabena spiroides*
Werowrap Lake	9.8	15.9-27.2	0.4-4.6	7,683	*Anabena spiroides*, G. *aeruginosum*
Spencer Gulf	-	-	-	3,111	Cyanobacteria
Shark Bay	-	-	-	480	Algal mat
Canada:					
Humbolt Lake	8.7	0.37-1.27	0.5-1.5	7,968	*Aphanizomenon flasaquae*
Little Manitou Lake	8.8	0.07-0.83	0-1.5	1,188	*Chaetoceros elmorei*, *Rhizoclomium* sp.
Manitou Lake	9.4-9.6	0.7-1.13	0-0.28	2,620	*N. spumigena*
Waldesa Lake	8.5	0.06-0.09	0-0.03	345	Various algae
United States of America:					
Borax Lake	9.55-9.8	-	-	525	Various algae
Devils Lake	8.3-9.1	0.04-0.67	0-1.045	140	Various algae
Great Salt Lake	7.8-8.2	0.3-1.2	0.01-0.05	6,147	*Dunaliella* sp.
Soap Lake	-	>0.144	>4.0	2,919-7,000	*Chlorella* sp., cyanobacterial mats

many of the world's most productive petroleum basins (Aizenshtat and others, 1999). Although the geochemistry of lacustrine source rocks has been intensively investigated, little is known about the early stages of organic diagenesis that starts the decaying remains of a hypersaline lake's biota on the path towards oil-generating source rocks. Of particular interest is the manner in which the molecular components of the biota are altered, or preserved, during the early stages of decay and diagenesis. How much of the chemical substance of the organisms is preserved as, or converted to, lipids from which oil is generated? This pilot study investigates the feasibility of extracting and meaningfully interpreting lipids from the modern sediments of the Great Salt Lake, which is a typical hypersaline depositional environment.

AQUATIC ECOSYSTEMS OF THE GREAT SALT LAKE

Due to the stress of saline water, the hypersaline Great Salt Lake supports only a few aquatic organisms, which nevertheless form a highly productive and complex organic community. The number of species reported in the lake decreases from 32 at 6-7 percent salinity to just 6 at 28 percent salinity (Stephens, 1998). Annual variation of water temperature and salinity control the time of appearance and relative abundance of the various species. A winter bloom of phytoplankton is followed in the early spring by a hatch of brine shrimp (*Artemia franciscana*), which graze the phytoplankton. As the phytoplankton are eaten and the water clears, a small population of benthic macrophyte green algae and cyanobacteria begins to bloom.

In Gilbert Bay, Felix and Rushforth (1980) describe seven species of green algae (*Chlorophyta*), four species of "blue-green" algae or cyanobacteria (*Cyanophyta*), one species of dinoflagellate (*Pyrrhophyta*) and 17 species of diatoms. Among these species, the most abundant were the green alga *Dunaliella veridus* and the diatoms *Amorpha coffeiformis, Navicula graciloides, Navicula tripunctata*, and *Ropalodia musculus*. Evans (1960) reported 15 distinct bacterial taxa and, more recently, Stephens (1998) describes eleven species of halophylic bacteria.

In the more saline Gunnison Bay, bacteria constitute the major organic fraction by mass (Post, 1975, 1981). A huge density of halophylic bacteria (7 x 10^{13} bacteria/m^3; Post, 1975) gives the water of the north arm a reddish-purple color due to carotenoid pigments in the bacteria that serve as antioxidants and provide protection from UV radiation. Most of the bacterial consortia are obligate extreme halophiles with growth occurring only at salinity levels above 15 percent NaCl, and maximum rates occurring for some at 22 to 27 percent NaCl (Crane, 1974; Post, 1980). Commonly, these organisms aerobically decompose organic material to produce CO_2. However, under anaerobic conditions, bacterial photosynthesis occurs where the pigment bacteriorhodopsin replaces chlorophyll (Stoechenius, 1976) that is used in aerobic photosynthesis. Production of methane, ethane and propane observed by Stube (1976) indicates the presence of methanogenic bacteria and an active methane cycle in the sediments of the North Arm. The presence of hydrogen sulfide in the deeper, anaerobic sediments of the lake indicates the presence of sulfate-reducing bacteria. The interplay between these organisms and the methanogens is expected to vary with variations in SO_4 supply and salinity conditions (Lovely and Klug, 1983).

Algae constitute the second largest group of organisms in Gunnison Bay in terms of biomass and are comprised of two principle planktonic species, the red-pigmented *Dunaliella salina* and the closely related green-pigmented species *D. viridus* (Post, 1980). *Dunaliella salina* occurs in the highest numbers and is the larger in size. Its reddish orange color derives from a carotenoid (Post, 1980). Algal blooms are occasionally observed in this part of the lake where red-orange *D. salina* swirls contrast against a purple-red bacterial background (Post 1977, 1980). During these blooms, cell densities may approach 10,000 cells/ml but typically average of 170 cells/ml (Post, 1981). In contrast to the red algae, the green algae (*D. viridus*) are much less abundant; they average 48 cells/ml but are sometimes absent from water samples (Post, 1981).

The organic matter in the lake sediment is not restricted to decayed remnants of the aquatic biota. Sedge (*Cyperaceae*) and cattails (*Typha-Sparganium*) grow in the fresh and brackish wetlands bordering the lake and are especially abundant on the eastern shores. Epicuticular waxes and other components from these plants may enter into the lake floor sediments as the lake level rises. Sagebrush (*Artemisia*), salt sage (*Chenopodiaceae-Amaranthus*) and pine (*Pinus*) are abundant in the arid flatlands and mountains surrounding the lake. Lake-sediment cores show significant quantities of pollen from all of these genera increasing in abundance in the Pleistocene and Holocene (Davis and Moutoux, 1998). Terrestrial plant material, mainly wind-borne pollen and other plant debris, contribute significantly to the organic matter preserved in much of the lake sediment.

METHODS OF INVESTIGATION

Description of Sampling Sites

Nine samples of organic-rich sediments were collected at four sites in the southern half of the Great Salt Lake (figure 1; table 2). The sites were selected to sample diverse shoreline settings and for ease of access.

Stansbury Island evaporation ponds: Four samples were collected from evaporation ponds operated by the Magnesium Corporation of America in the broad mud flats west and southwest of the southern end of Stansbury Island. The site is at the end of the public access portion of a road and causeway leading westward from the southwest tip of Stansbury Island. The ponds that were sampled are 2.0 km west of the main Stansbury Island access road. At the time the samples were gathered, the ponds were flooded with hypersaline water and the bottom sediment consisted of interbedded halite and black sapropel. Three of the samples are the organic-rich mud or sapropel (SL002, SL006 and SL007). The fourth sample is a thin film of pink halophylic bacteria that was on halite at the sediment-water interface (SL009).

Saltair Beach: A single sample (SL001) was taken from the outer portion of the broad beach immediately east of the Saltair State Park on the south shore of the lake near the mouth of Lee Creek. The sample is organic-rich, fine-

Table 2. Location of sampling sites and recent sediment information for the southern Great Salt Lake.

Sample ID	Sample Site/Comments	Latitude	Longitude	Date	Collector
SL001M	Saltair Beach	40.7690ºN	112.1588°W	April, 1998	S. Schamel
SL002M	Stansbury Island evaporation ponds	40.8038°N	112.5728°W	April, 1998	S. Schamel
SL003M	Antelope Island State Park Entrance (south of causeway)	41.0889°N	112.1199°W	June, 1998	J. Collister & S. Schamel
SL004M	Antelope Island State Park Entrance (south of causeway)	41.0889°N	112.1199°W	June, 1998	J. Collister & S. Schamel
SL005M	Antelope Island State Park Entrance (north of causeway)	41.0889°N	112.1199°W	June, 1998	J. Collister & S. Schamel
SL006M	Stansbury Island evaporation ponds (pond S)	40.8038°N	112.5728°W	August, 1998	S. Schamel
SL007M	Stansbury Island evaporation ponds	40.8038°N	112.5728°W	August, 1998	S. Schamel
SL008M	White Rock Bay; tributary stream; red photosynthetic bacteria	41.0243°N	112.2449°W	June, 1998	J. Collister & S. Schamel
SL009M	Stansbury Island evaporation ponds; bacterial layer	40.8038°N	112.5728°W	August, 1998	S. Schamel

grained, oolitic sand with a minor content of decayed terrestrial plant fragments. Here, as elsewhere on the lake shore, the shoreface oolitic sands locally contain large quantities of brine shrimp (*Artemia franciscana*) fecal pellets that are washed ashore. The brine shrimp graze on phytoplankton following the spring hatch, and on benthic green algae and cyanobacteria during the summer and fall.

White Rock Bay: A single sample (SL008) of sapropelic mud was gathered from a small, isolated hypersaline pond at the edge of the lake, adjacent to a spring-fed marsh in White Rock Bay, Antelope Island State Park. The site is 2.0 km southeast of Buffalo Point on the northwest shore of the island. The sample consisted of the top centimeter of gelatinous sapropel together with a well-developed coating of pink, photosynthetic halophylic bacteria.

Antelope Island State Park entrance station: Three samples came from near the entrance station to the park, at the east end of the causeway leading out to Antelope Island. This site is on the southeast shore of Ogden Bay, 7.0 km west of Layton and 10.5 km ENE of the northern tip of the island. Two of the samples (SL003 and SL004) are organic-rich silt and fine quartz sand from the exposed lake floor about 200 m southwest of the entrance station (figure 2). The third sample (SL005) is sapropelic mud from the lake floor sediments about 400 m west of the entrance station (figure 3). All three samples contain minor amounts of decayed terrestrial plant remains. Each sample is from a depth of at least 10 cm within the undisturbed sediment.

Figure 2. *Organic-rich silt and fine quartz sand of sample SL003 from the lake floor sediments 200 m southeast of Antelope Island State Park entrance station.*

Analytical Methods

Sediment samples were extracted ultrasonically using chloroform-methanol (2:1, v:v) for a minimum of three times and the supernatants were combined to yield a total extract.

Figure 3. *Sapropelic mud of sample SL005 on the lake floor about 400 m west of the Antelope Island State Park entrance station. The sediment surface is covered with shorebird guano.*

The extract was subsequently added to a separatory funnel (250 ml) and methanol and distilled water (5 percent NaCl solution) was added to bring the chloroform-methanol-water ratio to 1:1:0.9 (v:v:v). After separation of the water from the lipid phase, the chloroform phase was recovered and excess solvent was removed by rotary evaporation at 40°C; extraction yields are listed in table 3. Aliquots of the total extracts were chromatographically fractionated using a stationary phase of silica gel (10g of 100 to 200 mesh activated at 110°C overnight, then deactivated with 5 percent distilled water). Ten fractions (designated F1 through F10) were collected using mobile phases of increasing polarity: (F1) 40 ml hexane, (F2) 25 ml of 25 percent toluene in hexane, (F3) 25 ml of 50 percent toluene in hexane, (F4) 25 ml of 5 percent ethyl acetate in hexane + 25 ml of 10 percent ethyl acetate in hexane, (F5) 25 ml of 15 percent ethyl acetate in hexane, (F6) 25 ml of 20 percent ethyl acetate in hexane, (F7) 25 ml of 25 percent ethyl acetate in hexane, (F8) 25 ml of 75 percent ethyl acetate in hexane, (F9) 25 ml of ethyl acetate and (F10) 40 ml of methanol.

Table 3. Extraction yields for Great Salt Lake sediments.

Sample ID	Moist Sample Weight (mg)	Extract (mg)	Extract Yield (percent)
SL001M	41,820	26.0	0.06
SL002M	40,510	34.1	0.08
SL003M	11,987	21.5	0.18
SL004M	14,316	28.4	0.20
SL005M	9,325	35.2	0.38
SL006M	5,548	11.8	0.21
SL007M	9,400	17.2	0.18
SL008M	56,150	32.8	0.06
SL009M	48,123	15.6	0.03

The chromatographically separated lipid fractions were analyzed using a Hewlett Packard 6890 gas chromatograph (GC) equipped with a flame ionization detector (FID). Split injection (50:1) was employed at 300°C (detector temperature = 350°C) onto a non-polar Restek column (30 x 0.25 µm x 0.25 µm; Rtx-1). The GC column temperature was programmed from 35°C (2 minutes isothermal) to 310°C at 4°C/minute. Autosamplers were used to increase efficiency and achieve reproducible injections. Polar fractions (F7-F9) were run as trimethyl-silate (TMS) derivatives. Fractions were placed in two dram vials; 200 µl pyridine and BSTFA [bis(trimethylsilyl)trifluoroacetamide] were added to the sample. Samples were purged under a stream of nitrogen gas and were placed in a heating block at 60°C for one hour. Excess solvent was removed under a stream of nitrogen gas. Samples were solvated in 100 ml dichloromethane and injected into the GC-FID or the gas chromatograph-mass spectrometer (GC-MS).

Quantification of the lipids was performed using a series of external standard solutions (5β-cholane and O-terphenyl) ranging from 4 to 16,500 parts per million (ppm). A linear calibration ($y = 8.811 \times 10^{-4} + 1.496 \times 10^{-3}$; $r^2 = 0.999$) was determined for the GC-FID detector response (y=peak area in mV) versus ppm of analyte (x). Known quantities of fractions for each sample (1/100 µl) were injected, and the FID response was recorded. Concentrations of unknown compounds (ppm) were measured from the linear correlation between ppm and peak area for the external standard. Analyte concentrations (ppm) were subsequently converted into ng/mg extract according to the amount injected and the fractional mass of each chromatographically isolated specimen.

Full scan GC-MS analyses were done using a Hewlett Packard 6890 gas chromatograph equipped with a Hewlett Packard 6890 Mass Selective Detector (MSD). A non-polar J&W column (60 m x 0.25 mm x 0.25 mm; Rtx-1) was used and the GC column temperature was increased from 35°C (2 minutes isothermal) at 2°C per minute to 310°C (70.5 isothermal), for a total run time of 210 minutes. Systematic dilution factors were used to obtain consistent syringe injection and an autosampler was used to increase the efficiency of the procedure. Data were collected and processed with Hewlett Packard HP ChemStation® software.

RESULTS AND DISCUSSION

Extraction yields for recent sediments from the Great Salt Lake are low, ranging from 0.03 to 0.38 by weight percent (table 3). These low yields are typical of immature recent sediment samples where the majority of the organic material is present as insoluble macro-molecular tissues from terrestrial plants, algae, and bacteria. During early diagenesis, functionalized lipids polymerize to form polymeric, solid, humic acids (Peters and Moldowan, 1993). With increasing maturation, this material is converted into kerogen and increasing amounts of bitumen, which is the source of crude oils pooled in sedimentary basins (Hunt, 1996). Hence, extractable lipids (free lipids) are usually present in relatively low concentrations in recent sediments. Despite low extraction yields, the compound classes that are present can be used to infer the biologic sources of organic material in the sediment. This is possible because different organisms (e.g., algae, bacteria, higher plants) are known to produce diagnostic suites of compounds, called biological markers, which are related to the organism's metabolism and environmental conditions (Peters and Moldowan, 1993; and references therein). Mackenzie (1984) defines biological markers as: "*organic compounds detected in the geosphere whose basic carbon skeleton suggests an unambiguous relationship to a known contemporary natural product.*" For example, *sterols* are produced by eukaryotic organisms exclusively, whereas *hopanoids* are only produced by bacteria.

Saturated Hydrocarbons

Both *n*-alkanes and acyclic isoprenoids dominate the GC-FID traces of the saturated hydrocarbon fraction (F1) in the majority of the sediments collected from the Great Salt Lake (figure 4a and 4b). An exception to this is the sample collected from the evaporation ponds near Magnesium Corporation (SL009M; figure 4c). The *n*-alkane profiles are generally bimodal with a primary maximum at n-C_{27} or n-C_{29}, and a secondary maximum at n-C_{17}. Strong odd-carbon preferences are observed at n-C_{17}, and in the long-chain region of the chromatograms (figures 4a, 4b). The predominance of n-C_{17} is consistent with organic matter from algal and bacterial sources whose lipids contain odd-chain homologues in this part of the chromatogram (Clark and Blumer, 1967; Blumer and others, 1971). Even-chain alcohols and acids (particularly n-C_{16} and n-C_{18} homologues) also are produced by these organisms, but they are diagenetically converted to odd-chain alkanes in oxidizing depositional environments (Didyk and others, 1978) producing the C_{15} and C_{17} normal alkanes.

Long-chain *n*-alkanes displaying strong odd-carbon preferences are commonly used to infer significant inputs of terrestrial organic matter in sedimentary environments (e.g., Collister and others, 1992; Collister and others, 1994a, b). This interpretation is based upon observations that alkanes in the epicuticular waxes of higher plants have elevated odd-carbon preferences (Eglinton and others, 1962; Eglinton and Hamilton, 1963; Rieley and others, 1991; Collister and others, 1994b). Accordingly, the abundant long-chain *n*-alkanes in sediments from the Great Salt Lake might infer elevated inputs of wind-borne plant waxes from terrestrial vegetation growing near the lake. However, Nichols and others (1988) noted high abundances of long-chain *n*-alkanes in extracts of sea-ice diatoms from Antarctica where terrestrial input should be minimal. They conclude that algae could be a source of long-chain *n*-alkanes, although their data do not display the strong odd carbon preferences we observe for long-chain *n*-alkanes in samples from the Great Salt Lake.

A strong correlation between n-C_{27} and n-C_{29} (figure 5) suggests a single source for these compounds, such as terrestrial plant waxes. Abundant long-chain *n*-alkanes observed in sapropels from the evaporation ponds west of Stansbury Island (SL002, SL006 and SL007) suggest that these samples contain the largest amounts of terrestrial organic matter (table 1, figure 5). Extremely high salinity of water in these ponds probably inhibits the growth of algae, resulting in proportionally greater contributions of organic matter from wind blown pollen to the sediments. Thus, any wind-blown pollen input would be less diluted. Alternatively, the strong correlation between the quantitative abundance of n-C_{17} and n-C_{29} shown in figure 5 may indicate a source of long-chain *n*-alkanes from algae. Typically, these alkanes are thought to derive from mixed bacteria/algal and terrestrial higher plants, respectively (Collister and others, 1992, 1994a; and references therein). Nichols and others (1988) demonstrated that algae are a potential source of *n*-alkanes and may contribute to what traditionally is interpreted as a terrestrial signal in GC-FID traces for sediments and crude oils. Be-cause terrestrial higher plants and aquatic algae have different CO_2 metabolisms, compound-specific isotope analysis of individual *n*-alkanes may resolve the uncertain origins of these long-chain *n*-alkanes. Different isotope fractionation factors associated with contrasting metabolic pathways will produce distinctive $\delta^{13}C$ values for the resultant lipids, which are clearly observed in geologic samples (Collister and others, 1992, Collister and others, 1994a, b).

Unlike the other samples we examined, the sample of the halophylic bacterial layer (SL009) from the evaporation ponds at Stansbury Island is dominated by hexadecene and octadecene (positions of double bonds not determined) and heptadecane (figure 4c and 5b). These compounds are of likely bacterial and algal origin since they (and their precursors) are abundant in microbial lipids (Goossens and others, 1989a, b; Ben-Amotz and others, 1985; Blumer and others, 1971; Clark and Blumer, 1967). Long-chain *n*-alkanes with a strong odd-carbon preference are present in this sample, albeit in trace amounts.

Most recent sediment samples contain a secondary series of homologous compounds that elute before the *n*-alkanes; these compounds are identified as *n*-alkenes, which range from C_{17} to C_{27} (figure 4). The *n*-alkenes are moderately abundant in sediments from the Stansbury Island evaporation ponds, but are absent or present in trace amounts, in the remainder of the samples. Unlike the *n*-alkanes, these compounds lack an odd-to-even predominance. Recent studies (Gelin and others, 1999) have shown that pyrolysis of *alge-*

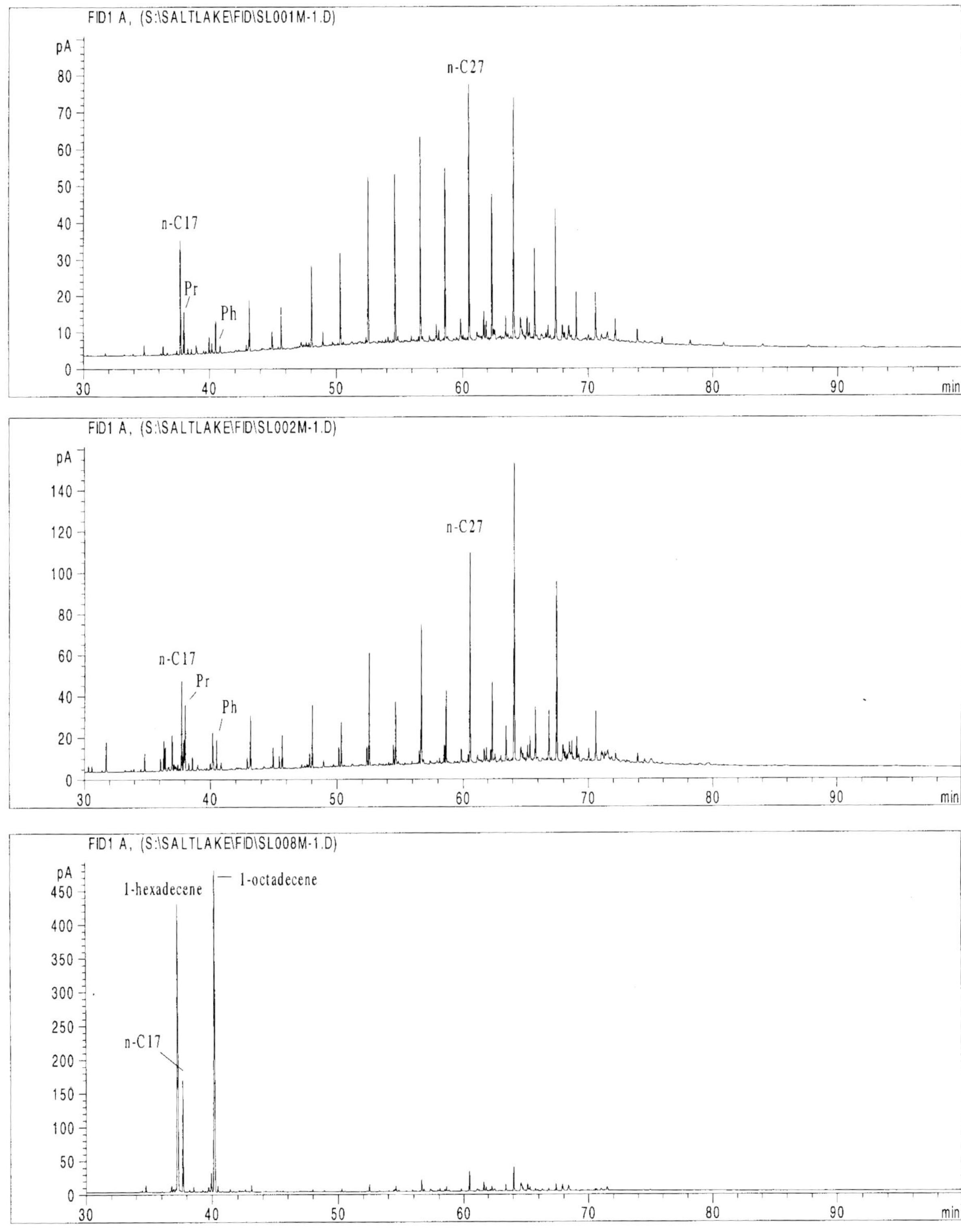

Figure 4. *Typical GC-FID traces for saturated hydrocarbon fractions of recent sediments from the Great Salt Lake: (a) SL001M; (b) SL002M; (c) SL008M. Pr=pristine, Ph=phytane, numbers refer to the carbon number of normal (straight-chain) alkanes.*

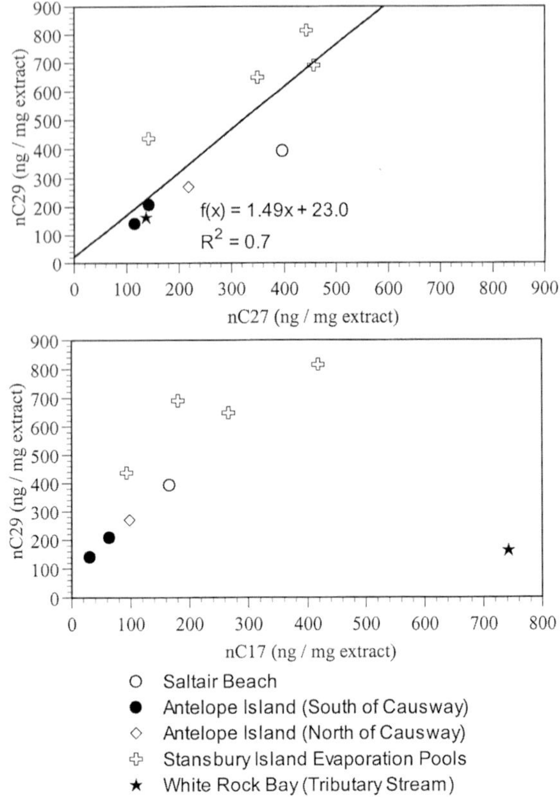

Figure 5. *Cross plots of the quantitative abundances of saturated hydrocarbons in the Great Salt Lake extracts: (a)* n-C_{27} *(ng/mg extract) versus* n-C_{29} *(ng/mg extract); (b)* n-C_{17} *(ng/mg extract) versus* n-C_{29} *(ng/mg extract).*

nans (resistant biopolymers in algal membranes) can produce large amounts of mid- to long-chain hydrocarbons with no odd-even predominance. Thus, these algal materials may represent a major source of oil-prone sedimentary organic matter.

Pristane (i-C_{19}) and phytane (i-C_{20}) are the dominant acyclic isoprenoids in the sediment extracts. Pristane to phytane ratios for the Great Salt Lake sediment samples are variable, ranging from 0.54 to 9.48 (table 4). Pristane and phytane are commonly derived from the phytol side chain of the chlorophyll molecule produced by algae and higher plants. Variations in diagenetic conditions (Didyk and others, 1978) result either in oxidation of phytol to pristane (i-C_{19}) by removal of a carboxyl group, or in reduction phytol to phytane (i-C_{20}). Accordingly, the ratio of these compounds commonly is used as an indication of the redox potential of the depositional environment. A pristane/phytane (Pr/Ph) less than one indicates reducing conditions, whereas a Pr/Ph greater than one indicates oxidizing conditions (Didyk and others, 1978). Hence, the wide range of Pr/Ph ratios observed in the Great Salt Lake recent sediments might correspond to redox conditions in the various depositional settings that our samples came from (table 4). The highest Pr/Ph values are observed for samples from the hypersaline evaporation ponds at Stansbury Island and those collected north of the Antelope Island causeway; these values suggest oxidizing conditions. Moderate Pr/Ph values are observed at Saltair Beach and south of the Antelope Island causeway where salinities are moderate. The lowest Pr/Ph values (<1.0) are

Table 4. Lipid distributions for the saturated hydrocarbon fractions of Great Salt Lake sediment extracts.

Sample	Pr/Ph	Pr/n-C_{17}	Ph/n-C_{18}	$2*n$-C_{17}/(n-C_{16}+n-C_{18})	$2*n$-C_{27}/(n-C_{26}+n-C_{28})	$2*n$-C_{29}/(n-C_{28}+n-C_{30})	Concentration (ng C/mg extract) n-C_{17}	Pr	n-C_{18}	Ph	n-C_{25}	n-C_{27}	n-C_{29}
SL001M	3.89	0.39	0.35	5.29	1.58	2.08	0.165	0.066	0.049	0.019	0.339	0.395	0.394
SL002M	7.86	0.71	0.28	3.85	2.96	4.84	0.181	0.129	0.060	0.018	0.298	0.457	0.691
SL003M	0.54	0.21	0.71	2.18	1.81	3.76	0.030	0.009	0.018	0.014	0.123	0.114	0.142
SL004M	5.25	0.16	0.16	6.53	2.88	5.68	0.063	0.012	0.014	0.004	0.114	0.141	0.209
SL005M	9.11	1.21	0.23	2.66	3.80	6.20	0.097	0.117	0.056	0.015	0.127	0.217	0.271
SL006M	9.48	1.12	0.21	3.17	4.18	9.5	0.267	0.297	0.153	0.037	0.208	0.349	0.648
SL007M	7.71	0.85	0.28	2.81	5.23	10.71	0.418	0.355	0.165	0.050	0.257	0.443	0.815
SL008M	0.22	0.00	0.18	17.58	4.24	6.71	0.742	0.004	0.041	0.009	0.082	0.137	0.163
SL009M	1.95	1.00	1.02	3.14	1.61	7.12	0.093	0.093	0.049	0.050	0.112	0.142	0.437

Pr = pristane (2,6,10,14 - tetramethylpentadecane)
Ph = Phytane (2,6,10,14 - tetramethylhexadecane)

associated with the halophylic bacteria-rich samples at Stansbury Island and at White Rock Bay. These halophilic bacteria are photosynthetic under anoxic, reducing conditions. Although systematic variation of Pr/Ph ratios are observed in the samples, these ratios should be interpreted with caution since besides phytol from chlorophyll, other sources of these compounds exist. These include tocopherols (vitamin E) and membrane lipids of methanogenic bacteria. Finally, diagenetic alteration of phytols to isoprenyl thiophenes in sulfate-rich environments (discussed below) may complicate the use of this ratio as an environmental indicator (Goossens and others, 1984; ten Haven and others, 1986, 1987).

Alkenones

A series of compounds known as alkenones are present in all samples. These straight chain ketones contain 37 to 39 carbons with two to four double bonds (figure 6). Prymnesiophyte alga biosynthesize these compounds with variable numbers of double bonds in order to maintain membrane fluidity in response to temperature variation in the environment. Indeed, Marlowe and others (1990) and Volkman and others (1980) found that planktonic alga of the class *Prymnesiophyceae* produce alkenones having increasing unsaturation with decreasing growth temperature. Subsequently, Brassell and others (1986a, b) observed that the degree of unsaturation of long-chain ketones in surface sediments varies with sea surface temperature (SST). Brassell and others (1986a) calculated a parameter called U^k_{37} to indicate the degree of unsaturation using the equation:

$$U^k_{37} = (C_{37:2} - C_{37:4}) / (C_{37:2} + C_{37:3} + C_{37:4})$$

where, for example, $C_{37:2}$ designates the relative abundance of an alkenone with 37 carbon atoms and 2 double bonds. More recently Prahl and Wakeham (1987) simplified the expression to define the $U^{k'}_{37}$ ratio as:

$$U^{k'}_{37} = C_{37:2} / (C_{37:2} + C_{37:3})$$

Numerous calibrations from batch cultures indicate large variation of the U^k_{37} versus temperature relation between algal strains (e.g., Brassell and others, 1986a) and species (Volkman and others 1995). Nonetheless, field calibrations using this approach have demonstrated that, despite these variations, (Conte and Eglinton, 1993; Brassel, 1993) the relationship between $U^{k'}_{37}$ and temperature is remarkably consistent with sea surface temperature between 5-25ºC (Sikes and Volkman, 1993; Rosel-Melé and others, 1995). Prahl and others (1988) proposed the relationship between temperature and $U^{k'}_{37}$ to be: $U^{k'}_{37} = 0.034°C + 0.039$. In a more recent study Sikes and Volkman (1993) suggested that a more realistic calibration for cold waters in polar regions is: $U^{k'}_{37} = 0.0414°C - 0.156$.

The abundance of alkenones in the Great Salt Lake samples varies between 3.7 and 1750 ng carbon / mg extract with the lowest values observed for the bacterial-rich samples (SL008 and SL009; table 5). Alkenone distributions are dominated by the C_{37} homologues with progressively lesser amounts of the C_{38} and C_{39} homologues respectively. The C_{37} alkenone distributions in the Great Salt Lake sediments are dominated by the $C_{37:3}$ compound with lesser, near-equal abundances of the $C_{37:4}$ and $C_{37:2}$ components. Values for the $U^{k'}_{37}$ parameter in the Great Salt Lake sediments (table 5) range between 0.16 and 0.257 which correspond to temperatures of 7.6° to 10.0° C based on the calibration of Sikes and Volkman (1993). Stephens (1990) describes the brine shrimp hatch as occurring between 0 and 5°C. The resulting brine shrimp consume the winter algal bloom. The preliminary temperature estimates from the alkenone distributions may represent the temperature during which the highest algal productivity occurred at each respective site, or records the temperature during the winter algal bloom. However, extensive studies of $U^{k'}_{37}$ for lake-systems have not been published. Calibration of this parameter for the Great Salt Lake remains a goal for future research. Reliable calibration for this hypersaline setting should aid paleoclimatic studies of cores from the Great Salt Lake.

Sterols

Sterols originate from a variety of eukaryotic organisms, including phytoplankton, zooplankton and vascular plants (Huang and Meinschein, 1979; Volkman, 1986). Sterols with 27 carbons (e.g., cholesterol) and 28 carbons (24-methylcholesta-5,22E-dien-3β-ol and 24-methylcholesta-5,24(28)-dien-3β-ol) are dominant in plankton and invertebrates. In contrast, C_{29} sterols are generally dominant in higher plants.

Table 5. Alkenone distributions for the F5 fractions of Great Salt Lake sediment extracts

Sample	Σ C_{37} Alkenones ng/mg extract	U^k_{37}	$U^{k'}_{37}$	Temp°C[1]	Temp°C[2]
SL001M	207.6	0.069	0.205	4.9	8.7
SL002M	47.5	-0.011	0.209	5.0	8.8
SL003M	451.3	0.003	0.178	4.1	8.1
SL004M	1750.0	-0.010	0.186	4.3	8.3
SL005M	18.3	0.038	0.209	5.0	8.8
SL006M	115.0	0.145	0.257	6.4	10.0
SL007M	102.5	0.106	0.236	5.8	9.5
SL008M	3.7	-0.058	0.160	3.6	7.6
SL009M	9.6	0.097	0.250	6.2	9.8

[1] Prahl and Wakeham 1987
[2] Sikes and Volkman 1993

(a)

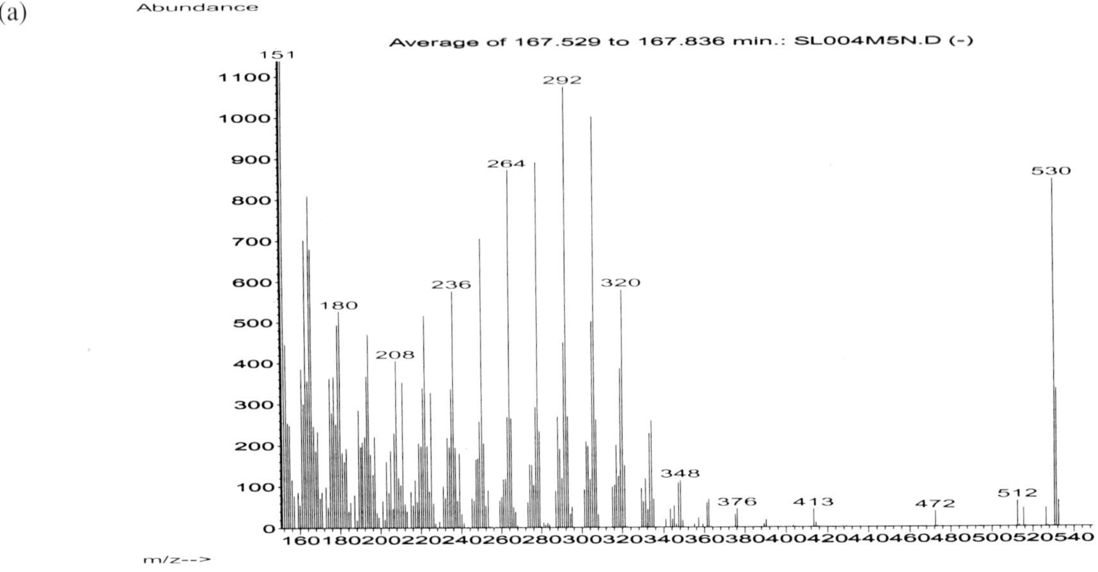

(b)

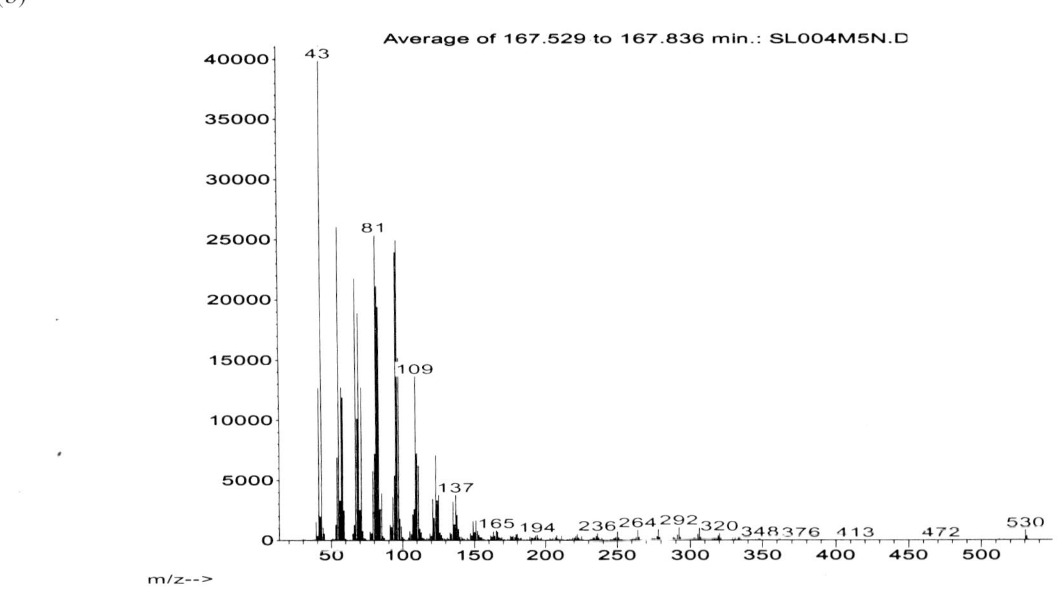

(c)

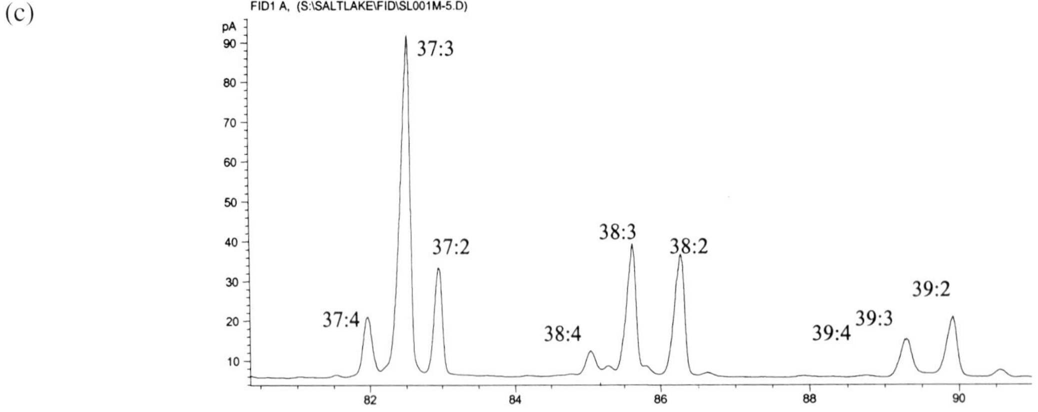

Figure 6. *Mass spectra (a) and (b) of the 37:3 alkenone and (c) the distribution of C37, C38, and C39 alkenones in Great Salt Lake sediment extracts.*

However, Volkman (1986) concludes that much of 24-ethylcholesta-5-en-3β-ol (a C_{29} sterol) in particulate matter in oligotrophic waters derives from green algae, *Prymnesiophyte algae*, and cyanobacteria. More recently Volkman and others (1994) identified simple distributions of C_{28} and C_{29} Δ^5-sterols in the green algal class *Parinophyceae* and suggested that these organisms may contribute significant C_{29} sterols to marine sediments.

The sterols identified in the current study are listed in table 6. Distributions of sterols in the extracts are highly variable and are likely to reflect differences in source inputs (e.g., algal versus terrestrial organic matter) and diagenetic alteration in response to differing environmental conditions at the various localities. The C_{29} sterols are most abundant in the sediment samples collected from south of the Antelope Island causeway (SL003 and SL004). There are near equal abundances of C_{29} and C_{27} sterols in samples collected at Saltair beach, from north of the Antelope Island causeway, and from the Stansbury Island evaporation ponds (figure 7). The elevated abundance of C_{29} homologues in samples from south of the Antelope Island causeway clearly reflects elevated input of terrestrial organic material (Volkman, 1986). Indeed, sedges and grasses are growing along the shoreline at this locality and degraded plant remains are observed in the samples SL003 and SL004.

The sample of sapropel from White Rock Bay contains sterols that are dominated by the C_{28} compounds (figure 7). Elevated C_{28} steranes in sediments and petroleums are often interpreted to derive from algae. This sample was composed of decayed algal material on which a film of pink photosynthetic bacteria was growing. Because eubacteria are not known to produce sterols (Volkman, 1986), the most likely precursor organisms are cyanobacteria ("blue-green algae"). This interpretation is consistent with the low abundance of alkenones in the sample (figure 8) suggesting that prymnesophyte algae are not an important source of organic material at this site (figure 8).

Quantitative abundances of sterols are highly variable ranging from 39 to 443 ng/mg extract (table 7; figure 8). The lowest abundances occur in samples from the Stansbury Island evaporation ponds. This is consistent with the reduction of algal input associated with extreme salinities and the higher concentration of lipids from halophilic bacteria. Samples collected from more open, less saline depositional environments contain higher total sterol concentrations suggesting the greater influence of algae and terrestrial plant debris in the lipid profile.

Other Classes of Compounds

Two sulfur-containing compounds positively identified as 3-methyl-2-(3,7,11-trimethyldodecyl) thiophene and 3-(4,8,12-trimethyldodecyl) thiophene (figure 9) are observed in varying abundances. These and other sulfur-containing hydrocarbons commonly form in reducing depositional environments where H_2S produced by sulfate-reducing bacteria exceeds available reactive iron (Sinninghe-Damsté and others, 1989; Brassell and others, 1986b). Such compounds are commonly found in extreme environments, such as the Dead Sea (Brassell and others, 1986a,b). They have also have been identified in peculiar high-sulfur crude oils generated from source rocks thought to been deposited in hypersaline depositional settings, such as the oils seeping to the surface at Rozel Point on the eastern shore of the Gunnison Bay (Sinninghe-Damsté and others, 1989). In these environments, sufficient organic matter is produced such that bacterial degradation is incomplete and anaerobic degradation by sulfate reduction continues below the oxic zone of the sediments. Subsequent incorporation of H_2S into double bonds of algal and bacterial lipids leads to the combining of sulfur with organic matter. For the compounds identified in samples from the Great Salt Lake, incorporation of sulfur into phytol (the side chain of chlorophyll in both plants and algae) and subsequent abiotic cyclization probably accounts

Table 6. Sterols identified in Great Salt Lake extracts.

Compound Name (as TMS derivatives)	Parent Mass	Carbon number
27-nor-24-methylcholesta-5,22E-dien-3β-ol	M+ = 456	C_{27}
27-nor-24-methylcholesta-22E-en-3β-ol	M+ = 458	C_{27}
Cholest-5-en-3β-ol	M+ = 458	C_{27}
5α-Cholestan-3β-ol	M+ = 460	C_{27}
24-methylcholesta-5,22E-dien-3β-ol	M+ = 460	C_{28}
24-methyl-5α-cholesta-22E-dien-3β-ol	M+ = 470	C_{28}
24-methylcholesta-5,24(28)-dien-3β-ol	M+ = 470	C_{28}
24-methyl-cholest-5-en-3β-ol	M+ = 472	C_{28}
24-methyl-5α-cholestan-3β-ol	M+ = 474	C_{28}
23,24-dimethylcholesta-5,22-dien-3β-ol	M+ = 484	C_{29}
24-ethylcholesta-5,22E-dien-3β-ol	M+ = 484	C_{29}
23,24-dimethyl-5α-cholest-22E-en-3β-ol	M+ = 486	C_{29}
24-ethyl-5α-cholest-22E-en-3β-ol	M+ = 486	C_{29}
23,24-dimethylcholes-5en-3β-ol	M+ = 486	C_{29}
24-ethylcholesta-5α-cholestan-3β-ol	M+ = 488	C_{29}

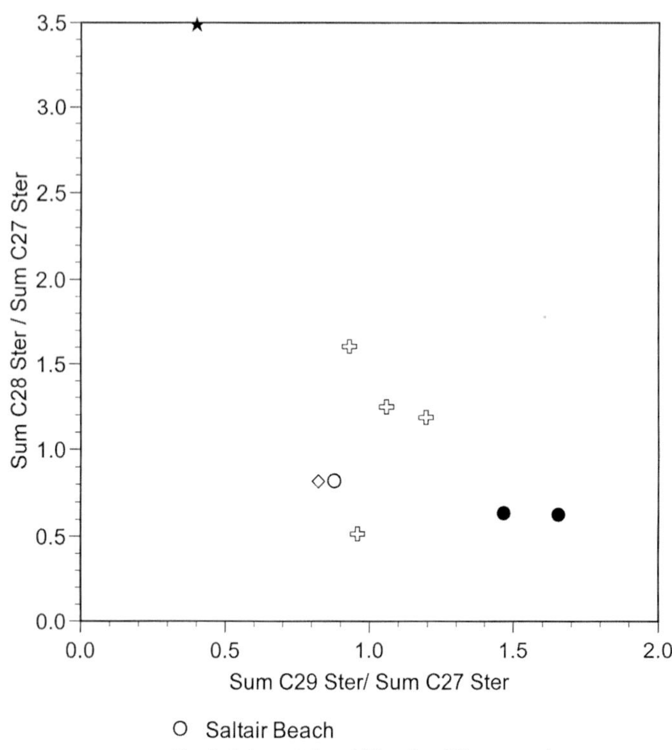

Figure 7. *Cross-plot showing the variation of sterols in the Great Salt Lake sediment extracts. The* ΣC_{29} *sterols and* ΣC_{27} *sterols versus the* ΣC_{28} *sterols* ΣC_{27} *sterols.*

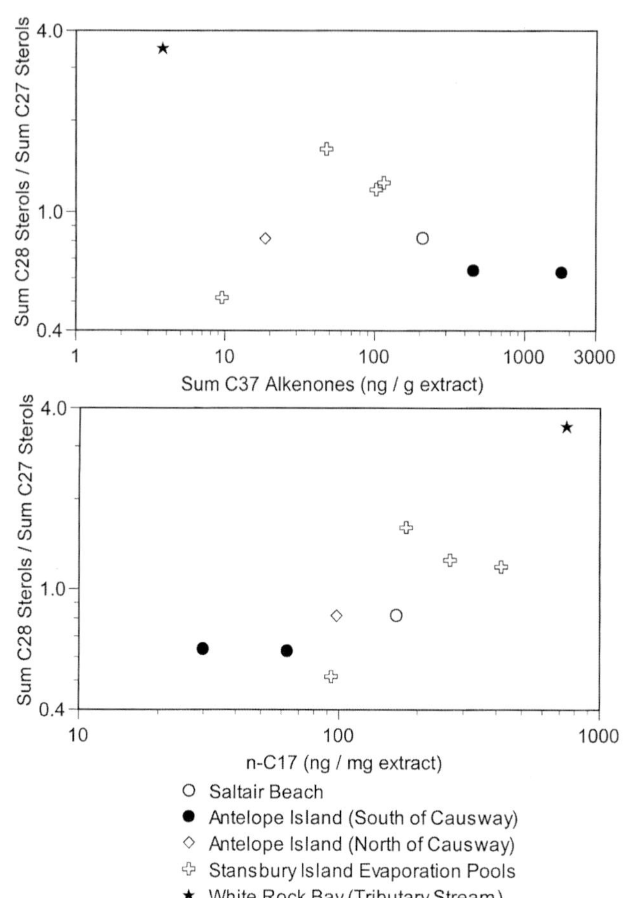

Figure 8. *Cross plots showing the distributions of algal biomarkers in the Great Salt Lake extracts: (a) Sum* C_{37} *Alkenones (ng/g extract) versus the Sum* C_{28} *Sterols/Sum*C_{27} *Sterols ratio; (b)* n-C_{17} *(ng/mg extract) versus the Sum* C_{28} *Sterols/Sum* C_{27} *sterols ratio.*

Table 7. Sterol distributions for the F7 fractions of Great Salt Lake sediment extracts.

Sample	Cholesterol / Cholestanol	Sum Sterols ng/mg extract	Σ C29 Sterols / Σ C27 Sterols	Cholestanol / Cholesterol
SL001M	0.78	160.0	0.877	1.29
SL002M	1.42	67.5	0.931	0.70
SL003M	0.46	138.7	1.466	2.19
SL004M	0.92	443.4	1.653	1.09
SL005M	2.75	212.2	0.820	0.36
SL006M	0.42	39.0	1.059	2.39
SL007M	0.57	39.9	1.195	1.76
SL008M	2.88	188.2	0.401	0.35
SL009M	0.97	34.6	0.959	1.04

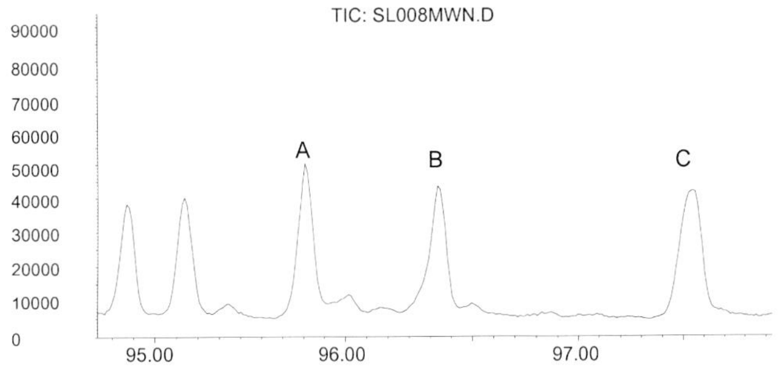

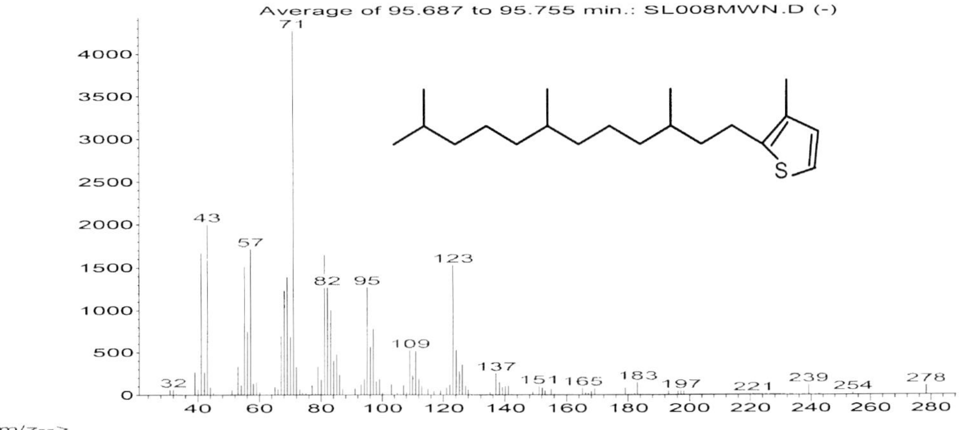

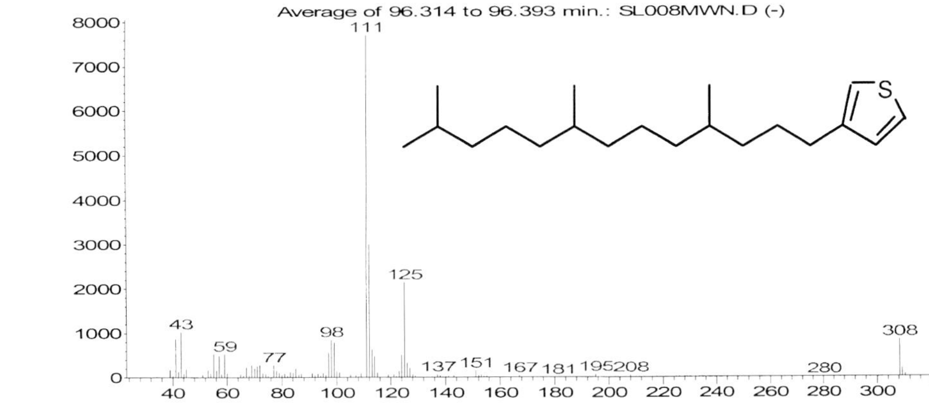

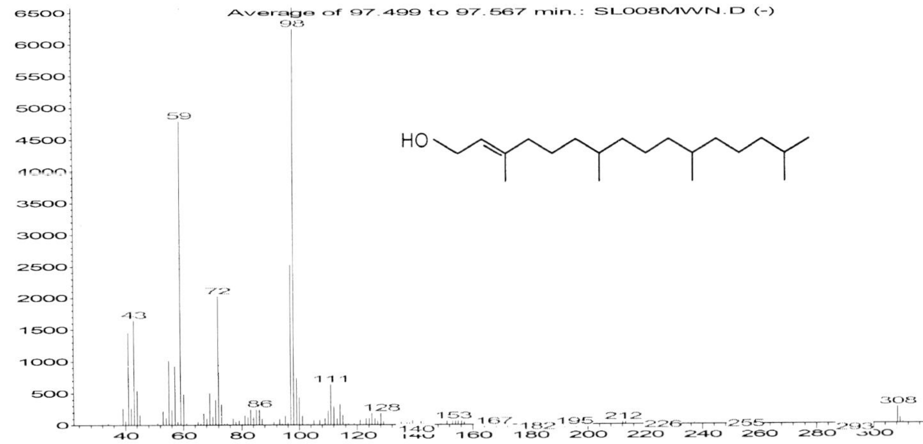

Figure 9. *GC-MS full scan trace for SL008M. The letters designate the peaks for which the mass spectra are shown in (b)-(d). Compounds are (b) 3-methyl-2-(3,7,11-trimethyldodecyl) thiophene; (c) 3-(4,8,12-trimethyldodecyl) thiophene; and (d) phytol.*

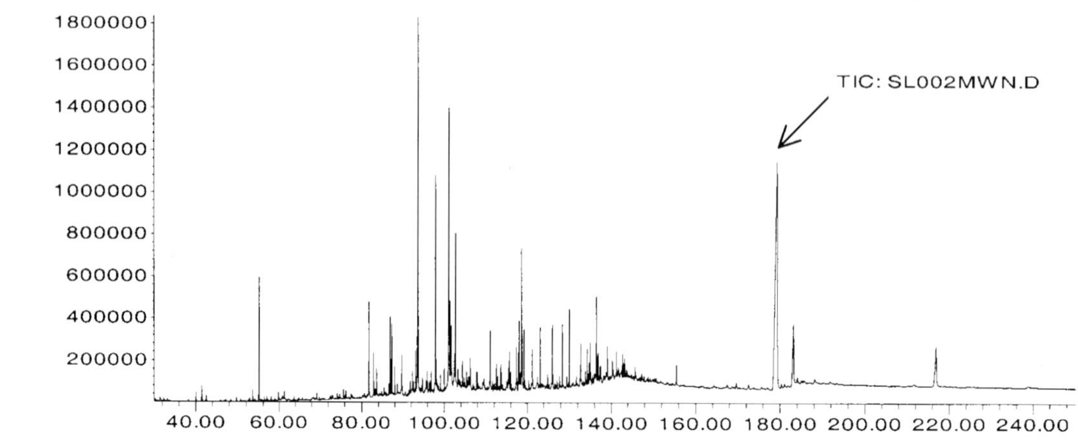

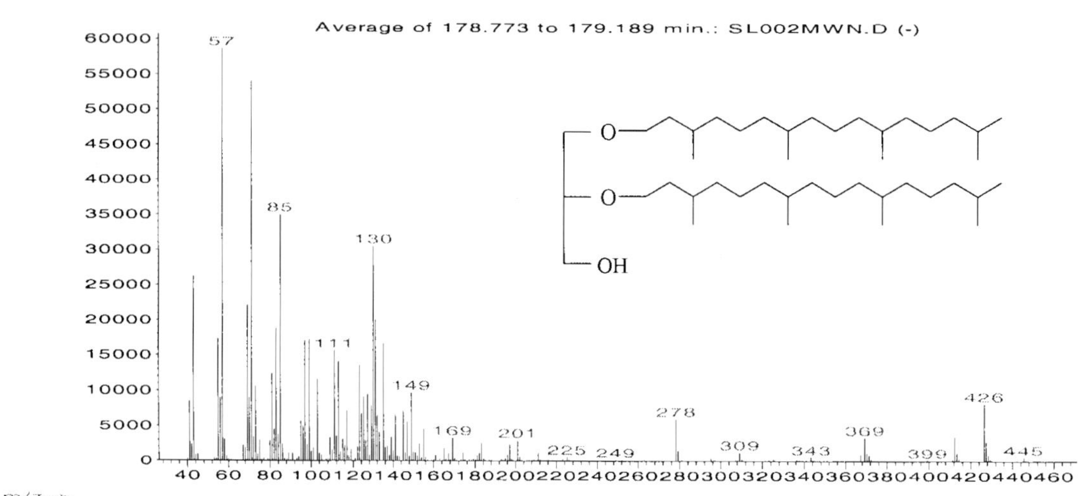

Figure 10. *Full scan GC-MS data for SL002M extract. The top panel is the total ion trace. The peak labelled with the arrow is the peak for which the spectra is shown in panel (b). The structure of bis-O-phytanylglycerol as a TMS derivative is shown. The following two peaks are longer chain homologues. These compounds are biomarkers for halophilic photosynthetic bacteria.*

for the observed isoprenoid thiophenes (Brassell and others, 1986; Sinninghe-Damsté and others, 1989).

Samples from the Stansbury Island evaporation ponds and from Saltair Beach contain biological markers specific for halophilic bacteria (figure 10). Two compounds in these samples were positively identified as isopranylglycerol diethers, based on comparison with published GC-MS spectra for these compounds (Teixidor and others, 1993).

With the exception of the bacterial layer (SL009) in the Stansbury Island evaporation ponds, tocopherol (Vitamin E) was identified in low abundance in all samples. The concentrations of tocopherol tend to increase with increased abundances algal lipids, such as alkenones and total sterol concentration. This relationship may suggest that these compounds derive from the indigenous algal population, although cyanobacteria and higher plants also are known to synthesize these compounds (Goosens and others, 1984).

CONCLUSIONS

Highly variable lipid distributions are observed for samples collected from several sites in the Great Salt Lake. These differences reflect variations in the dominant type of organic material, variations in redox (Eh, pH) conditions in the sediment or water column, as well as temperature variations. Extremely oxidizing conditions appear to prevail north of the Antelope Island causeway and in the Stansbury Island evaporation ponds where salinities are high. Moderately oxidizing water columns prevail at Saltair Beach and south of the Antelope Island causeway in conjunction with normal saline waters. Reducing conditions are indicated for the samples from the bacterial films in the evaporation pond at Stansbury Island and at White Rock Bay where abundant pinkish organic material from anaerobic, photosynthetic bacteria is observed.

In the most extreme environments (evaporation ponds at Stansbury Island), lipids from halophilic photosynthetic bacteria dominate relative to those derived from algae. A high abundance of lipids from wind-borne terrestrial higher plants is also observed at these sites, presumably due to low primary production by algae. Open-lake environments are dominated lipids from algae and lesser contributions wind-borne plant waxes and bacterial lipids. Distributions of acyclic isoprenoids and functionalized lipids, such as sterols, are highly variable reflecting both differences in precursor organic matter, as well as variations in redox (Eh and pH) conditions at the various sampling sites.

REFERENCES

Aizenshtat, Z., Miloslavski, I., Ashengrau, D., and Oren, Aharon, 1999, Hypersaline depositional environments and their relation to oil generation, *in* Oren Aharon, editor, Microbiology and Biogeochemistry of Hypersaline Environments - The Microbiology of Extreme and Unusual Environment , CRC Press, p. 89-108.

Ben-Amotz, A., Tournabene, T.G., and Thomas, W.H., 1985, Chemical profile of selected species of microalgae with emphasis on lipids, Journal of Phycology, v 21, p. 72-81.

Blumer, M., Guillard, R.R.L., and Chase, T., 1971, Hydrocarbons of marine phytoplankton, Marine Biology, v. 8, p. 183-189.

Brassell, S.C., 1993, Application of biomarkers for delineating marine paleoclimate fluctuations during the Pleistocene, *in* Engel M.H. and Macko S.A., editors, Organic Geochemistry-Principles and Applications: New York, Plenum Press, p. 699-738.

Brassell, S.C., Eglinton, G., Marlowe, I.T., Pflaumann, U., and Sarnthein, M., 1986a, Molecular stratigraphy - a new tool for climate assessment: Nature, v. 320, p. 129-133.

Brassell, S.C., Lewis, C.A., de Leeuw, J.W., de Lange, F., and Sinninghe-Damsté, J.S., 1986b, Isoprenoid thiophenes - novel products of sediment diagenesis: Nature, v. 320, p. 160-162.

Clark, R.C., and Blumer, M., 1967, Distributions of *n*-paraffins in marine organisms and sediment: Limnology and Oceanography, v. 12, p. 79-87.

Collister, J.W., Summons, R.E., Lichtfouse, E. .and Hayes, J..M., 1992, An isotopic biogeochemical study of the Green River oil shale: Organic Geochemistry, v. 19, p. 265-276.

Collister, J.W., Lichtfouse, E., Hieshima, G., Hayes, J.M., 1994a, Partial resolution of sources of n-alkanes in the saline portion of the Parachute Creek Member, Green River Formation (Piceance Creek Basin, Colorado): Organic Geochemistry, v. 21, p. 645-659.

Collister, J.W., Rieley, G., Stern, B., Eglinton, G. ,and Fry, B., 1994b, Compound-specific $\delta\ ^{13}O$ analyses of leaf lipids from plants with differing carbon dioxide metabolisms: Organic Geochemistry, v. 21, p. 619-627.

Conte, M.H., and Eglinton, G., 1993, Alkenone and alkenoate distributions within the euphotic zone of the eastern North Atlantic - correlation and production temperature: Deep-Sea Research, v. 40, p. 1931-1965.

Crane, J.L., Jr., 1974, Characterization of selected bacteria from the north arm of the Great Salt Lake: Master of Science Thesis, Utah State University, Logan, Utah, 40 p.

Davis, O.K., and Moutoux, T.E., 1998, Tertiary and Quaternary vegetation history of the Great Salt Lake, Utah, USA: Journal of Paleolimnology, v. 19, p. 417-427.

Didyk, B.M., Simoneit, B.R.T., Brassell, S.C., and Eglinton, G.E., 1978, Organic geochemical indicators of palaeoenvironmental conditions of sedimentation: Nature, v. 272, p. 216-222.

Eglinton, G., and Hamilton, R.J., 1963, The distribution of *n*-alkanes, Swain, T., editor, Chemical Plant Taxonomy: Academic Press, London-New York, p. 187-217.

Eglinton, G., Gonzales, A.G., Hamilton, R.J.,and Raphael, R.A., 1962, Hydrocarbon constitution of the wax coatings of leaves - A taxonomic survey: Phytochemistry, v. 1, p. 89-102.

Evans, F.R., 1960, Studies on growth of protozoa from Great Salt Lake with special reference to Cristigera: Journal of Protozoology, v. 7, p. 14-15 (Supplement).

Felix, E.A., and Rushforth, S.R., 1980, Biology of the South arm of the Great Salt Lake, Utah, *in* Gwynn, J.W., editor,Great Salt Lake - A scientific, historical and Economic Overview: Utah Geological and Mineral Survey Bulletin 116, 400 p.

Gelin, F., Volkman, J.K., Largeau, C., Derenne, S., Sinninghe-Damsté, J.S., and De Leeuw, J.W., 1999, Distribution of aliphatic, nonhydrolyzable biopolymers in marine microalgea: Organic Geochemistry, v. 30, p. 147-159.

Goossens, H., de Leeuw, J.W., Schenck, P.A., and Brassell, S.C., 1984, Tocopherols as likely precursors of pristane in ancient sediments and crude oils: Nature, v. 312, p. 440-442.

Goossens, H., Düren, R.R., de Leeuw, J.W., and Schenck, P.A., 1989a, Lipids and their mode of occurrence in bacteria and sediments-II. Lipids in the sediment of stratified, freshwater lake: Organic Geochemistry, v. 14, p. 27-41.

Goossens, H., de Leeuw, J.W., Ripjstra, W.I.C., Meyburg, G.J., and Schenck, P.A., 1989b, Lipids and their mode of occurrence in bacteria and sediments - I. A methodological study of the lipid composition of Acinetobacter calcoaceticus: LMD, p. 79-41.

Hammer, U.T., 1981, Primary production in saline lakes: Hydrobiologia, v. 81, p. 47-57.

Huang, W.Y., and Meinschein, W.G., 1979, Sterols as ecological indicators: Geochimica et Cosmochimica Acta, v. 43, p. 739-745.

Hunt, J.M., 1996, Petroleum Geochemistry and Geology, 2nd edition: New York, Freeman, 743 p.

Katz, B.J., 1990, Controls on distribution of lacustrine source rocks through time and space, *in* Katz, B.J., editor, Lacustrine Basin Exploration - Case Studies and Modern Analogs: American Association of Petroleum Geologists Memoir 50, p. 61-76.

Kowalewska, A., and Cohen, A.S., 1996, Reconstruction of paleoenvironments of the Great Salt Lake Basin during the late Cenozoic: Journal of Paleolimnology, v. 13, p. 1-26.

Lovely, D.R., and Klug, M.J., 1983, Sulfate reducers can outcompete methanogens at freshwater sulfate concentrations: Applied and Environmental Microbiology, v. 45, p. 187-192.

Mackenzie, A.S., 1984, Applications of biological markers in petroleum geochemistry, *in* Brooks, J. and Welte D., editors, Advances in Petroleum Geochemisty, Volume 1: London, Academic Press, p. 115-214.

Marlowe, I.T., Brassell, S.C., Eglinton, G., and Green, J.C., 1990, Long-chain alkenones and alkyl alkenoates and the fossil coccolith record of marine sediments: Chemical Geology, v. 88, p. 349-375.

Morrison, R.B., 1991, Quaternary stratigraphic, hydrologic, and climatic history of the Great Basin, with emphasis on Lake Lahontan, Bonneville and Tecopa, *in* Morrison, R.B. , editor, Quaternary nonglacial geology - conterminous U.S., The Geology of North America K-2 (DNAG): Boulder, CO, The Geological Society of America, p. 283-320.

Nichols, P.D., Volkman, J.K., Palmisano, A.C., Smith, G.A., and White, D.C., 1988, Occurrence of and isoprenoid C_{25} diunsatruated alkene and high neutral lipid content in Antarctic sea-ice diatom communities: Journal of Phycology, v. 24, p. 90-96.

Peters, K.E., and Moldowan, J.M., 1993, The Biomarker Guide - Interpreting Molecular Fossils in Petroleum and Ancient Sediments: Englewood Cliffs, Prentice Hall, 363 p.

Post F.J., 1975, Life in the Great Salt Lake, Utah: Science, v. 36, p. 43-47.

—1977, The microbial ecology of the Great Salt Lake: Microbial Ecology, v. 3, p. 143-165.

—1980, Biology of the North Arm, *in* Gwynn, J.W., editor, Great Salt Lake - a scientific, historical and economic overview: Utah Geological and Mineral Survey Bulletin 116, 400 p.

—1981, Microbiology of the Great Salt Lake north arm: Hydrobiologia, v. 81, p. 59-69.

Prahl, F.G., and Wakeham, S.G., 1987, Calibration of unsaturation patterns in long-chain ketone compositions for paleotemperature assessment: Nature, v. 330, p. 367-369.

Prahl, F.G., Muelhausen, L.A., and Zahnle, D.A., 1988, Further evaluation of long-chain alkenones as indicators of paleoceanographic conditions: Geochimica et Cosmochimica Acta, v. 52, p. 2303-2310.

Rieley, G., Collier, R.J., Jones, D.M., Eglinton, G., Eakin, P.A., and Fallick, A.K., 1991, Sources of sedimentary lipids deduced from stable carbon isotope analyses of individual *n*-alkanes: Nature, v. 352, p. 425-427.

Rosel-Melé, A., Eglinton, G, Pflaumann, U., and Sarnthein, M., 1995, Atlantic core-top calibration of the Uk37 index as a sea-surface palaeotemperature indicator: Geochimica et Cosmochimica Acta, v. 59, p. 3099-3107.

Sikes, E.L., and Volkman, J.K., 1993, Calibration of alkenone unsaturation ratios ($U^{k'}_{37}$) for paleotemperature estimation in cold polar waters: Geochimica et. Cosmochimica Acta, v. 57, p. 1883-1889.

Sinninghe Damsté, J.S., Eglinton, T.I., de Leeuw, J.W., and Schenck, P.A., 1989, Organic sulfur in macromolecular sedimentary organic matter - I. Structure and origin of sulphur-containing moieties in kerogen, asphaltenes and coals as revealed by flash pyrolysis: Geochimica et Cosmochimica Acta, v. 53, p. 873-889.

Stephens, D.W., 1974, A summary of biological investigations concerning the Great Salt Lake, Utah (1861-1973): Great Basin Naturalist, v. 34, p. 221-229.

—1990, Changes in lake level, salinity and the biological community of Great Salt Lake (Utah, USA), 1847-1987: Hydrobiologia, v. 197, p. 139-146.

—1998, Salinity-induced changes in the aquatic ecosystem of the Great Salt Lake, Utah, *in* Pitman J.K., and Carroll A.R., editors, Modern and Ancient Lake Systems - New Problems and Perspectives: Utah Geological Association Guidebook 26, p. 1-8.

Stephens, D.W., and Gillespie, D.M., 1976, Phyloplankton production in the Great Salt Lake, Utah, and a laboratory study of algal response to enrichment: Limnology and Oceanography, v. 212, p. 74-87.

Stoeckenius, W., 1976, The purple membrane of salt-loving bacteria: Science, v. 234, p. 38-46.

Stube J., 1976, Microcosm simulation of nitrogen cycling in the Great Salt Lake: Unpublished Master of Science Thesis. Utah State University.

Teixidor, P., Grimalt, J.O., Pueyo, J.J., and Rodriguez-Valera, F., 1993, Isopranylglycerol diethers in non-alkaline evaporitic environments: Geochimica et Cosmochimica Acta, v. 57, p. 4479-4489.

ten Haven, H.L., de Leeuw, J.W., and Schenck, P.A., 1986, Organic geochemical studies of a Messinian evaporitic basin, northern Appenines (Italy) I - Hydrocarbon markers for a hypersaline environment: Geochimica et Cosmochimica Acta, v. 49, p. 2181-2191.

ten Haven, H.L., de Leeuw, J.W., Rullkötter, J., and Sinninghe-Damsté, J.S., 1987, Restricted utility of the pristane/phytane ratio as a palaeoenvironmental indicator. Nature, v. 330, p. 641-643.

Volkman, J.K., 1986, A review of sterol markers for marine and terrigenous organic matter: Organic Geochemistry, v. 9, p. 83-99.

Volkman, J.K., Barrett, S.M., Dunstan, G.A., and Jeffrey, S.W., 1994, Sterol biomarkers for microalgae from the green algal class Prasinophyceae: Organic Geochemistry, v. 21, p. 1211-1218.

Volkman, J.K., Eglinton, G., Corner, E.D.S., and Sargent, J.R., 1980, Novel unsaturated straight-chain C^{37}-C^{39} methyl and ethyl ketones in marine sediments and a coccolithophore *Emiliania huxleyi*, *in* Douglas, A.G., and Maxwell, J.R., editors, Advances in Organic Geochemistry 1979, p., New York, Pergamon Press, 219-227.

Volkman, J.K., Barrett, S.M., Blackburn, S.I., and Sikes, E.L., 1995, Alkenones in *Gephyrocapsa oceanica*; implications for studies of paleoclimate: Geochimica et Cosmochimica Acta, v. 59, p. 513-520.

Warren, J.K., 1986, Shallow water evaporitic environments and their source rock potential: Journal of Sedimentary Petrology, v. 56, p. 442-454.

White III, W. W. 1998. Depth of ponded brine at shear-strength test hole MB-1, Bonneville Salt Flats: Unpublished field notes. p. 37. Available upon request from BLM Salt Lake Field Office.

WATER AND SALT BALANCE OF GREAT SALT LAKE, UTAH, AND SIMULATION OF WATER AND SALT MOVEMENT THROUGH THE CAUSEWAY, 1963-98

by

Brian L. Loving[1], Kidd M. Waddell[1], and Craig W. Miller[2]

[1]Hydrologist, U.S. Geological Survey, Salt Lake City, Utah
[2]PE, Engineer Specialist, Utah Department of Natural Resources, Water Resources Division, Salt Lake City, Utah

ABSTRACT

The Southern Pacific Transportation Company completed a rock-fill causeway across Great Salt Lake in 1959. The causeway changed the water and salt balance of Great Salt Lake by creating two separate but interconnected parts of the lake, with more than 95 percent of freshwater surface inflow entering the lake south of the causeway.

The water and salt balance of Great Salt Lake primarily depends on the amount of inflow from tributary streams and the conveyance properties of the causeway that divides the lake into south and north parts. The conveyance properties of the causeway consist of two 15-foot-wide culverts, a 290-foot-wide breach, and permeable rock-fill material.

The dissolved-solids concentrations of the north and south parts were approximately equal at the time the causeway was completed in 1959, but by 1972, the concentration was about 200 grams per liter greater in the north part than in the south, and by December 1998, the concentration was 250 grams per liter greater in the north than in the south. The theoretical concentration that would have occurred in an undivided lake in December 1998 was 190 grams per liter. In 1998 the concentration in the south part was about 90 grams per liter, or 100 grams per liter less than the theoretical concentration for an undivided lake. In 1998 the concentration in the north part was about 340 grams per liter, or 150 grams per liter more than the theoretical concentration for an undivided lake.

The balance of salt between the north and south parts can be changed either by redistribution of some of the freshwater surface inflow to the south part into the north part, or by changing the conveyance properties of the causeway. For example, diversion of some of the flow of the Bear River into the north part would change the balance of salt between the north and south parts. However, in this study modification of causeway conveyance properties was considered as the only practical means of changing the salt balance for a specified net freshwater inflow.

A water and salt balance model of Great Salt Lake, developed by the U.S. Geological Survey in 1973, was modified in 2000 to incorporate the effects of changes in the conveyance properties of the causeway and withdrawals from the lake as part of the West Desert Pumping Project from 1987 through 1989. Modifications to the 1973 model included added capability to simulate: (1) stratified flow through submerged culverts, and (2) loads and concentrations of chloride, magnesium, potassium, and sodium in the lake.

The calibrated model was used to simulate the effects of several combinations of breach depths and widths on the dissolved-solids concentration in each part of the lake. The simulations indicated that deepening the breach is more effective in reducing the difference in concentration between the two parts of the lake than widening the breach without deepening. In December 1998, the dissolved-solids concentration of the south part of the lake was 28 percent of the concentration in the north part. If the breach had been deepened from 4,200 feet to 4,190 feet in January 1987, the dissolved-solids concentration of the south part would have been 55 percent of the north part by December 1998. By comparison, widening the breach from 290 feet to 600 feet, without deepening it, would have increased the dissolved-solids concentration of the south part to 33 percent of that of the north part.

From 1987 through 1992, about 500 million tons of salt, including 17.5 million tons of magnesium and 16 million tons of potassium, were removed from the lake as part of the West Desert Pumping Project. The pumps operated only during April 1987 to June 1989, but there was some return flow from the West Pond to the lake from 1990 through 1992. Model simulation indicated that without the West Desert Pumping Project, the dissolved-solids concentration in the north part would have been 3 grams per liter higher, and the concentration in the south part 9 grams per liter higher by December 1998.

INTRODUCTION

According to Wold, Thomas, and Waddell (1997, p.1):

> Prior to construction of a railroad causeway from 1957 through 1959, the hydrologic characteristics of Great Salt Lake, Utah, were typical of a closed lake having no outlet to the sea. After completion of the causeway (figures 1 and 2) in 1959, the water and salt balance of the lake changed. The causeway divides the lake into a south and a north part (figure 1). Slightly more than one-third of the surface area of the lake is north of the causeway. Because more than 95 percent of freshwater surface inflow enters the lake south of the causeway, the

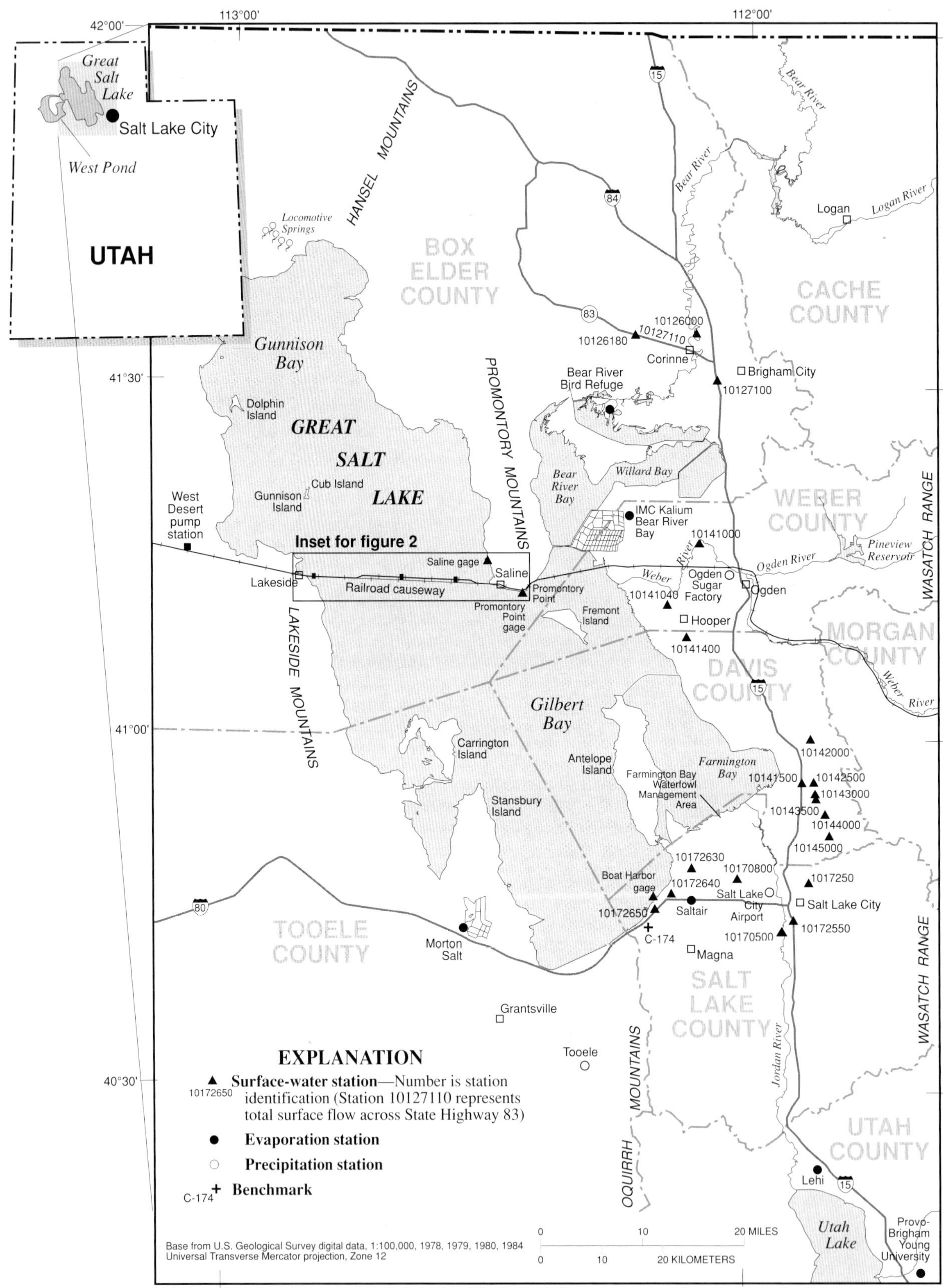

Figure 1. Location of study area and data-collection sites used to estimate inflow, water-surface altitude, and evaporation for Great Salt Lake, Utah, 1987–98 (Loving, Waddell, and Miller, 2000).

causeway has interrupted the circulation and caused substantial changes to the hydrology and chemistry of the lake.

Previous modeling of the water and salt balance incorporated the causeway conveyance properties and hydrologic conditions that existed from 1959 through 1986. Waddell and Bolke (1973) described the effects of the causeway on the water and salt balance of Great Salt Lake and developed a model to simulate the effects of the causeway on the salt balance for variable culvert widths and tributary inflows to the lake. This 1973 model of Waddell and Bolke was calibrated for causeway conveyance and hydrologic conditions existing from 1969-1972. This model was valid until about 1981. During the 1980s, fill material was frequently added to the causeway (figures 2 and 3) to maintain the top of the causeway above the water surface. During 1983, the two 15-ft-wide culverts became submerged beneath the rising water surface and eventually filled with debris; in August 1984, a 290-ft-wide breach was opened near the western end of the causeway. Because these new conditions warranted revision of the model, Wold, Thomas, and Waddell (1997) modified the original model to include the breach and changes in the dimensions and hydraulic characteristics of the causeway fill and recalibrated the model using the water balance from 1980 through 1986.

From 1987 through 1998, additional changes in the causeway conveyance properties and withdrawals for the West Desert Pumping Project made it necessary to revise the model again. Loving, Waddell, and Miller (2000) modified the model and recalibrated it to incorporate the changes that occurred from 1987 through 1998. The purpose of the later study was to update earlier mathematical models of the lake in order to: (1) understand the causes of historical changes in the water and salt balance, and (2) predict the effects of modifying the conveyance properties of the causeway on the water and salt balance of the lake. The study of Loving, Waddell, and Miller (2000) was done because continued freshening of the south part of the lake during the 1990s caused concern about the ecology of the lake.

Description of the Study Area

Great Salt Lake is a closed lake located in semiarid northwestern Utah in the Basin and Range Province (Fenneman, 1931). The lake is bordered on the west by desert and on the east by the Wasatch Range. Great Salt Lake is a remnant of freshwater Lake Bonneville, which existed about 30 to 10 thousand years ago. Lake Bonneville covered much of western Utah and small parts of Idaho and Nevada, and was about 1,000 ft deep at the deepest part. In 1963, when Great Salt Lake was at its lowest water-surface altitude in recorded history, at about 4,191 ft, it covered about 950 mi^2 and had a maximum depth of about 25 ft. In 1986, when Great Salt Lake was at its highest water-surface altitude in recorded history, at about 4,212 ft, it covered about 2,400 mi^2 and had a maximum depth of about 45 ft.

Purpose and Scope of Present Study and Past Modeling

The purpose of modeling the water and salt balance of Great Salt Lake is to provide a tool to predict the effects of changes in the hydraulic conveyance properties of the causeway on this balance. This report summarizes the work of Loving, Waddell, and Miller (2000), who updated the original model developed by Waddell and Bolke (1973). Wold, Thomas, and Waddell (1997) made updates of the original model based on data from 1980 through 1986, and Loving, Waddell, and Miller (2000) made further revisions to include data from 1987 through 1998. Included are a description of the model and results of simulations for different breach dimensions, and the effect of the West Desert Pumping Project on the water and salt balance of the lake.

During the study of Waddell and Bolke (1973), the altitude of the culvert bottoms were about 4,180 ft for the east culvert and about 4,183 ft for the west culvert. The altitudes of the ceilings of the culverts were not documented, but are believed to have been near 4,203 ft for the east culvert, and 4,206 ft for the west culvert. As a result of settling, by 1998 the altitude of the culvert ceilings had subsided to about 4,195 ft for the east culvert and about 4,198 ft for the west culvert. Since about 1983, the settling in combination with higher lake levels (figure 4) has submerged the culverts.

The model was modified to incorporate changes to the hydraulic conveyance properties of the causeway that have occurred since 1986, and for the effects of the withdrawals from the lake as part of the West Desert Pumping Project during 1987-89. The major changes to the hydraulic conveyance properties include provision for flow through submerged culverts and reduced flow through the causeway fill. To check the calibration, the capability to compute the concentration and load of four major ions was added to the model.

The equations of Holley and Waddell (1976) were modified to account for flow through submerged culverts (E.R. Holley, University of Texas, written communication, 1998; see Loving, Waddell, and Miller, 2000, appendix D). Submerged culvert flows were measured during 1996-98, but because the culvert openings were partially or fully plugged with debris, the cross-sectional areas of the culverts could not be defined, and measurements could not be used to verify the theoretical equations for flow through the submerged culverts.

The causeway-fill flow was computed with the procedures presented by Wold, Thomas, and Waddell (1997, p. 38), but the flow through the deeper, older fill was reduced by a constant factor (see section "Water and Salt Balance Model and causeway-fill flow"). Wold, Thomas, and Waddell (1997, p. 4) used the two-constituent solute-transport model of Sanford and Konikow (1985) to simulate flows through the fill (referred to in this report as the fill-flow model). No additional field data were collected during the study of Loving, Waddell, and Miller (2000, p. 6) to directly evaluate the hydraulic properties of the fill. The seepage computed by the fill-flow model was indirectly evaluated and revised by comparing the water and salt balance computed by the model with independent computations of the water and salt balance from measured data.

Monitoring data, to include freshwater surface inflow, precipitation on the lake surface, and evaporation were compiled for Great Salt Lake for monthly intervals from 1987 through 1998 and used as input to the model (Loving, Waddell, and Miller, 2000, appendix A). The water and salt balances were computed from these data. Dissolved salt load

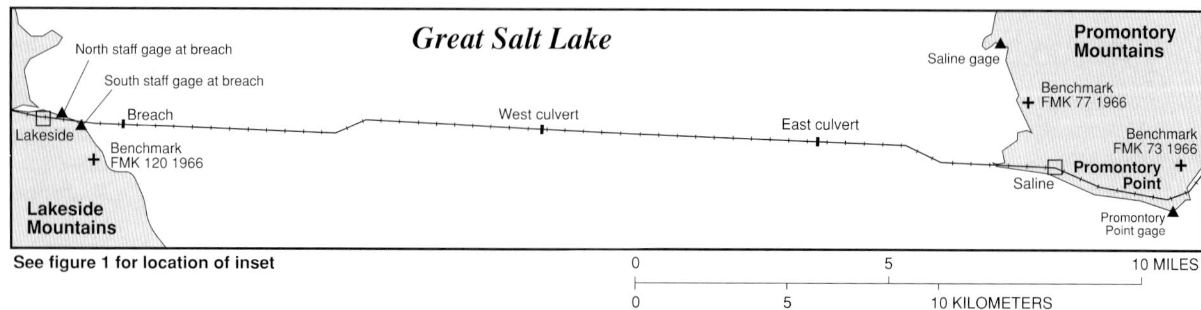

Figure 2. *Location of breach and culverts in the causeway across Great Salt Lake, Utah (Loving, Waddell, and Miller 2000).*

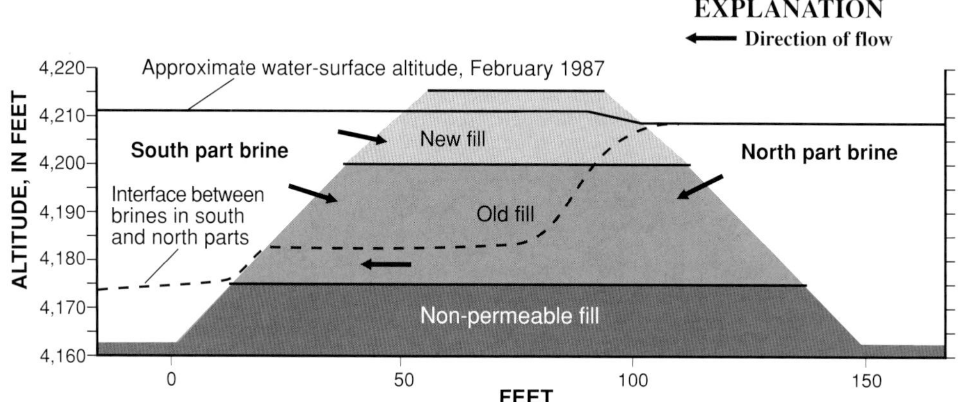

Figure 3. *Diagrammatic cross section of the causeway across Great Salt Lake, Utah (Loving, Waddell, and Miller, 2000).*

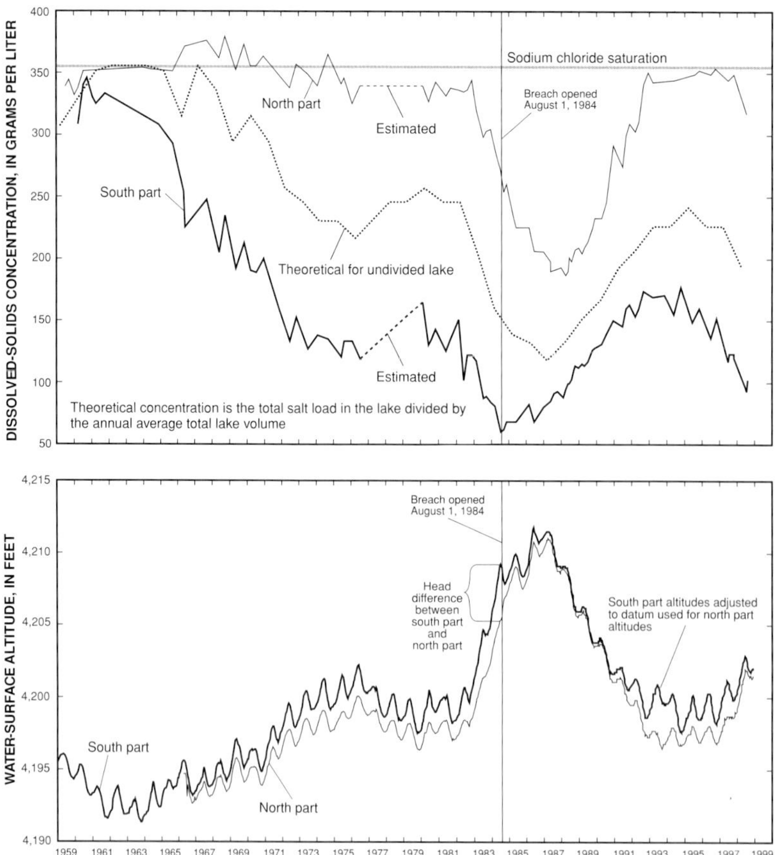

Figure 4. *Dissolved-solids concentration of the south part, north part, and theoretically undivided lake, and water-surface altitude of the south and north parts of Great Salt Lake, Utah, 1959–98 (Loving, Waddell, and Miller, 2000).*

was computed for the south and north parts of Great Salt Lake and used to evaluate the model for the calibration period, 1987 through 1998.

WATER AND SALT BALANCE OF GREAT SALT LAKE

Pre-causeway Conditions

Prior to completion of the causeway in 1959, the hydrologic characteristics of Great Salt Lake were typical of a closed lake having no outlet to the sea. The water-surface altitude rose and fell in response to the balance between surface evaporation, and inflow to the lake from runoff, ground-water, and precipitation. During periods when the water-surface altitude fell, surface area and volume decreased, and dissolved-solids concentration increased. During periods when the water-surface altitude rose, surface area and volume increased, and dissolved-solids concentration decreased. The lake was thought to be well mixed and have no density stratification, and salt precipitation was thought to occur throughout the entire lake when the lake volume was small enough for the salt concentration to exceed the saturation level for sodium chloride.

Effects of Causeway Construction

Construction of the causeway in 1959 changed the water and salt balance of Great Salt Lake by creating two separate but interconnected parts of the lake with different water-surface altitudes and dissolved-solids concentrations. During the period from 1959 through 1984, the railroad causeway's conveyance properties consisted of two 15-ft-wide by 23-ft-high culverts (Loving, Waddell, and Miller, 2000, appendix D) and the hydraulic properties of the causeway rock-fill material. A 290-ft-wide breach was added in 1984, to provide for additional circulation when the surface altitude of the south part was above 4,200 ft.

Because almost all surface-water inflow is to the south part (figure 1), the water in the south part is less saline than the water in the north part (figure 4), and the water-surface altitude of the south part (ES, figure 5) is higher than the north part (EN). The differences between the water-surface altitudes and densities of the south and north parts provide the potential for brine to flow in both directions through the causeway conveyances (QS and QN, figures 5 and 6). In general, the less dense brine from the south part flows northward through the upper part of the causeway conveyances (culverts, breach, and causeway fill) and the more dense brine from the north part flows southward through the lower part of the causeway conveyances.

Restricted lake circulation is indicated by a theoretical water and salt balance for an undivided lake, as well as historical trends in the dissolved salt loads for the north and south parts. The theoretical dissolved-solids concentration for an undivided Great Salt Lake was computed to compare with concentrations that actually occurred in the north and south parts of the lake from 1959 through 1998 (figure 4). The dissolved-solids concentrations of the north and south parts were approximately equal at the time the causeway was completed in 1959, but began diverging by 1961 (figure 4). In 1972, the concentration was about 200 g/L greater in the north than in the south part, and by 1998 the concentration was 250 g/L greater in the north than in the south. The theoretical concentration that would have occurred in an undivided lake in December 1998 was 190 g/L. In 1998, the concentration in the south part was about 90 g/L, or 100 g/L less than the theoretical concentration for an undivided lake. In 1998, the concentration in the north was about 340 g/L, or 150 g/L greater than the theoretical for an undivided lake.

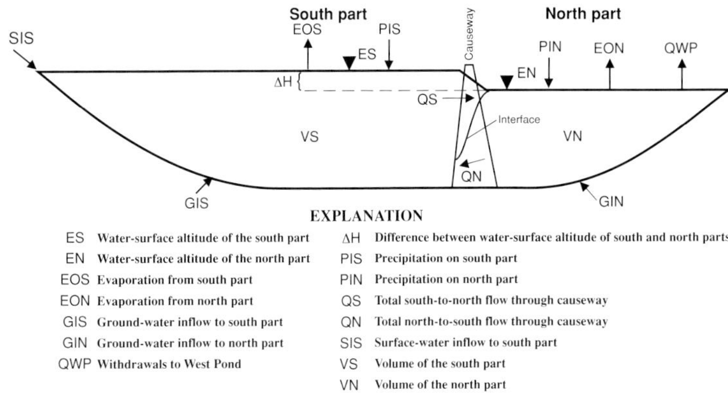

Figure 5. *Schematic diagram of water balance for Great Salt Lake, Utah (Loving, Waddell, and Miller, 2000).*

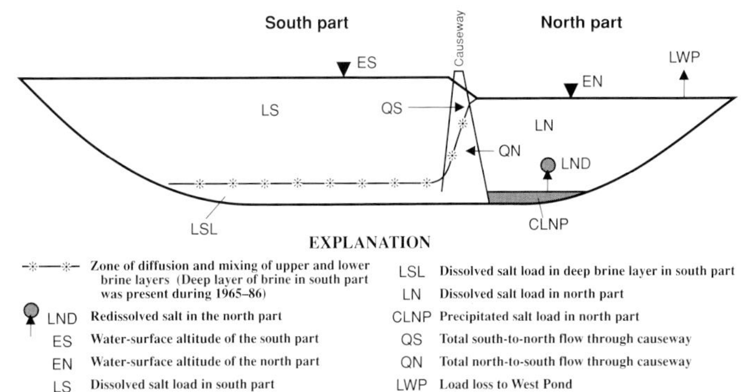

Figure 6. *Schematic diagram of salt balance for Great Salt Lake, Utah (Loving, Waddell, and Miller, 2000).*

Dissolved Salt Loads

The dissolved solids in the freshwater inflow to Great Salt Lake contributes a relatively insignificant part of the total salt load contained in the lake; thus, the total salt load is assumed constant for a large number of years. Hahl (1968, p. 20) determined the salt load contributed from freshwater inflow to be less than 0.0035 billion ton per year. During a 100-year period, surface-water inflow would add about 0.35 billion ton of salt to the lake, which is 8 percent of the current total load of 4.5 billion tons. Also, the annual addition of dissolved salt from inflow to Great Salt Lake is roughly equal to the amount of salt mined annually from the lake (J. W. Gwynn, Utah Geological Survey, oral communication, 2000). The chemical composition of the inflowing dissolved salt, however, is probably different than that extracted by mining.

Total salt loads were determined during periods when salinity was well below saturation with respect to sodium chloride. During those periods, it can be assumed that essentially all of the salt is dissolved and that the load of salt can be computed as the product of the lake volume and salt concentration. During 1985-86, when the water-surface altitude was near historic highs and the salinity was below saturation with respect to sodium chloride in both parts of the lake, the total salt load was estimated by Loving, Waddell, and Miller (2000, p. 8) to be 5.0 billion tons. A prior study by Wold, Thomas, and Waddell (1997, p. 7) estimated the total salt load to be 4.9 billion tons on the basis of salinities measured in 1976. The figure of 5.0 billion tons of total salt load was used in this study for purposes of water and salt balance. About 0.5 billion ton of dissolved salt was pumped to the West Desert from 1987 to 1989, thereby reducing the total load of salt in Great Salt Lake to about 4.5 billion tons (see section on "Effects of the West Desert Pumping Project on Loads of Ions").

Theoretical Equations of Water and Salt Balance for the Divided Lake

Loving, Waddell, and Miller (2000, p. 8) derived a general relation between the water and salt balance of the north and south parts in terms of the total flows north-to-south (QN) and south-to-north (QS) through the causeway. The change in the dissolved-solids load for the north part (figure 6) was expressed as:

$$QS * CS - QN * CN = \Delta LN + \Delta CLNP + LWP \quad (1)$$

where:
CS = dissolved-solids concentration of the south part;
CN = dissolved-solids concentration of the north part;
QS = fill flow + culvert flow + breach flow (all from the south to north part);
QN = fill flow + culvert flow + breach flow (all from the north to south part);
ΔLN = change in the dissolved-solids load of the north part;
ΔCLNP = change in the precipitated salt load of the north part due to precipitation (+) or re-solution (-); and
LWP = load of salt pumped from (+), or returned to (-) the north part of Great Salt Lake as part of the West Desert Pumping Project.

Solving equation 1 for QS:

$$QS = \frac{1}{CS}[\Delta LN + \Delta CLNP + LWP] + QN * \frac{CN}{CS} \quad (2)$$

From figure 5, the water balance for the north part is:

$$\Delta VN = PIN + GIN - EON + QS - QN - QWP \quad (3)$$

where:
PIN = precipitation on the north part;
GIN = ground-water inflow to the north part;
ΔVN = change in volume in north part;
EON = evaporation from the north part;
QWP = volume of water pumped from (+) or returned to (-) north part of Great Salt Lake as part of the West Desert Pumping Project.
Solving equation 3 for QS:

$$QS = \Delta VN - PIN - GIN + EON + QN + QWP \quad (4)$$

Equating equations 2 and 4 and solving for QN:

$$QN = \frac{CS}{CN\text{-}CS}[\Delta VN - PIN - GIN + EON + QWP - \frac{\Delta LN}{CS} - \frac{\Delta CLNP}{CS} - \frac{LWP}{CS}] \quad (5)$$

Most of the variables in equation 5 are dynamically related; when the value of one variable changes, the values of the other variables change. The exceptions to this are ground-water inflow to the north part (GIN), outflow of water (QWP) and salt load (LWP) to West Pond as part of the West Desert Pumping Project (see section titled "Effects of the West Desert Pumping Project on Loads of Ions"), and the conveyance properties of the causeway (dimensions of culverts and breach, and hydraulic properties of the fill). South-to-north flow, (QS), and north-to-south flow, (QN), are largely dependent on the specific hydraulic characteristics of the fill, culverts, and breach (see Loving, Waddell, and Miller, 2000, appendices C, D, and E, respectively).

The balance of salt between the north and south parts of the lake can be changed either by redistribution of some of the freshwater surface inflow to the south part (SIS, figure 5) into the north part of the lake, or by changing the conveyance properties of the causeway. For example, diversion of some of the flow of the Bear River into the north part would change the balance of salt between the north and south parts. However, in this study modification of causeway conveyance properties was considered as the only practical means of changing the salt balance for a specified net freshwater inflow. A water and salt balance computer model was used to facilitate the calculations needed to evaluate the effects of changing the conveyance properties on the water and salt balance of the lake.

Trend of Dissolved and Precipitated Salt Loads in North and South Parts, 1963-98

The trend of total dissolved salt load (figure 7) can be used as an indication of salt precipitation (CLNP) or re-solution (LND) (see figure 6). Because the total amount of salt available in the lake is assumed to be constant for a large number of years, an increase in total dissolved salt load represents re-solution of salt precipated on the lake bed, and a decrease in total dissolved salt load represents precipitation of salt. The only exception to this occurred during the West Desert Pumping Project from 1987 through 1989, when 0.5-billion net tons of dissolved salt was pumped from the lake.

In 1963, shortly after completion of the causeway, the water surface declined to its lowest recorded altitude (4,191.35 ft) and volume. At this low volume, the south and north parts of the lake were saturated with respect to sodium chloride, and a salt crust formed on the lake bed south and north of the causeway (Madison, 1970, p. 12). The estimated total dissolved salt load during 1963 was about 4.1 billion tons, and the precipitated salt load was about 0.9 billion tons.

During 1964-71, the water-surface altitude of the lake generally rose (figure 4) and the dissolved-solids concentration in the south part decreased because of increased surface-water inflow (and associated increase of lake volume) and the net movement of the dissolved salt load to the north part (figure 7). The load loss from the south part ceased during 1972 and was nil from 1972 through 1980, which indicates that the balance of dissolved salt loads between the south and north parts was near equilibrium for the inflow conditions

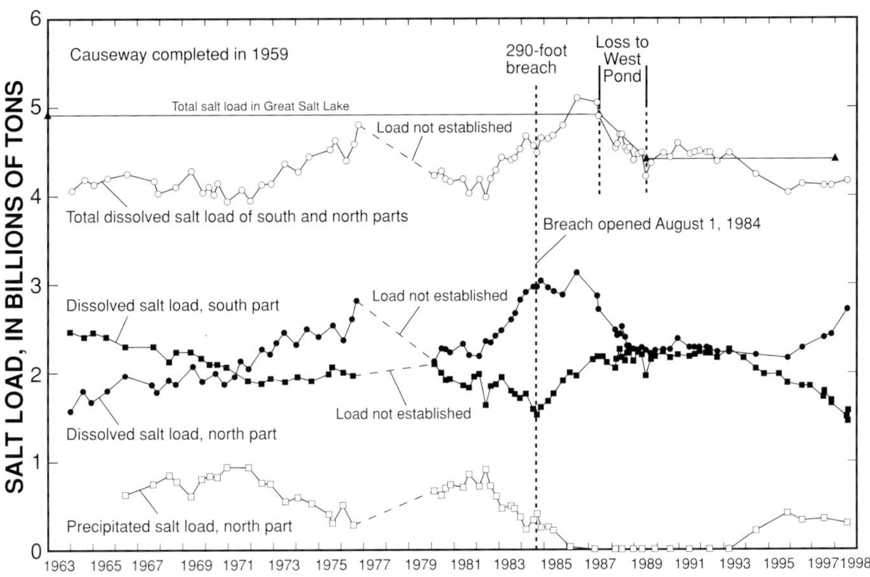

Figure 7. *Dissolved and precipitated salt load in Great Salt Lake, Utah, 1963–98 (Loving, Waddell, and Miller, 2000).*

and causeway conveyance properties existing during that period. The dissolved salt load in the north part increased from1972 through 1976 because of re-solution of the salt crust on the bottom of the north part as the water-surface altitude rose and the volume of the lake increased. Most of the salt that precipitated throughout the entire lake from 1959 through 1963 probably dissolved from the south part by 1972, but the precipitated salt in the north part did not dissolve because the brine there was at, or near, saturation.

From 1980 through 1986, average annual inflow was about 240 percent greater than the long-term average inflow for the period 1931 through 1976 (Wold, Thomas, and Waddell, 1997, p. 8). This record inflow caused the water-surface altitude of the south part to rise about 14 ft, reaching a historic high of about 4,212 ft on June 3, 1986 (figure 4). To keep up with the rising water surface, the causeway fill was raised and widened during this period, changing the hydraulic characteristics of the fill and affecting the salt balance of the lake (Loving, Waddell, and Miller, 2000, appendix C). The record inflows from January 1980 to July 1984 caused about 0.5 billion tons of dissolved salt to move from the south to the north part (figure 7). The dissolved salt load of the north part was also increased as a result of the re-solution of precipitated salt. The precipitated salt was dissolved because the increase in the lake volume diluted salt concentration of the north-part brines below saturation.

In August 1984, to combat the rising level of Great Salt Lake, the State of Utah completed construction of a 290-ft opening (breach) on the western edge of the causeway (figure 2). The breach, which increased causeway conveyance for water-surface altitudes above 4,200 ft, was designed to reduce the head differential that had developed between the north and south parts (figure 4). The effect of the breach was to increase south-to-north flow by about 50 percent, and more than double north-to-south flow when compared with average flow through the causeway from January 1980 to July 1984 (Wold, Thomas, and Waddell, 1997, p. 8).

Because of the increased flow through the causeway, the head difference between the south and north parts decreased from an average of about 3.5 ft prior to the breach opening, to a range of about 0.5 to 1.0 ft within 3 months after the breach was opened (figure 4). Associated with this rapid decrease in head difference was an increase in the altitude at which the pressure gradient was conducive to north-to-south flow through both the fill (QNF) and the breach (QNB). The increased bi-directional flow through the causeway generally caused the dissolved salt load to increase in the south and decrease in the north (figure 7). From August 1, 1984, when the breach was opened, to May 1989, a net dissolved salt load of about 0.5 billion ton moved from the north to south part.

From 1987 through 1994, the water-surface altitude of the lake generally decreased because annual evaporation was greater than inflow (figure 4). As water-surface altitude decreased, flow through the causeway generally decreased because of less depth of flow above the breach bottom and within the permeable part of the causeway fill (above 4,175 ft) (figure 8). By 1992, the water-surface altitude of both parts was near the altitude of the bottom of the breach (4,200 ft); consequently, there was almost no breach or culvert flow, and north-to-south fill flow was less than 25 percent of what it was in early 1987 (figure 8).

Effect of Causeway Conveyance Properties, 1987-98

For a specified inflow to the lake, increases in the dimensions of the culverts and breach, and increases in the permeability of the fill material result in increases in flow through the causeway (QS and QN) and decreases in the difference between the dissolved-solids concentration of north and south parts (CS and CN). To illustrate the effects of the conveyance properties on net load gain or loss, equation 5 was simplified for the conditions of no West Pond pumping (QWP and LWP = 0) and a constant inflow of freshwater:

$$QS*CS = QN*CN$$

or

$$QS/QN = CN/CS \qquad (6)$$

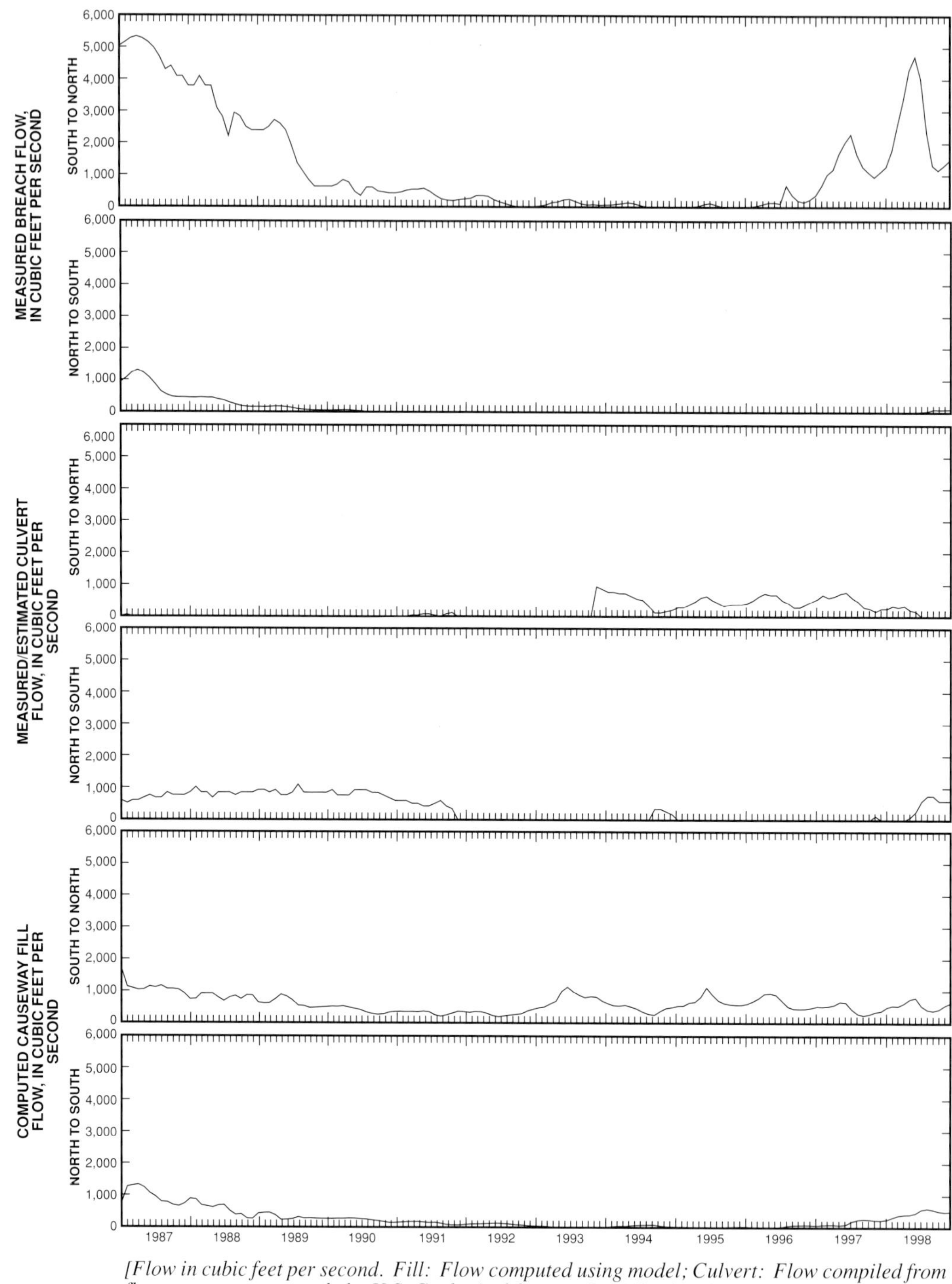

[Flow in cubic feet per second. Fill: Flow computed using model; Culvert: Flow compiled from flow measurements made by U.S. Geological Survey (ReMillard and others, 1986; Herbert and others, 1997, 1998, 1999), No culvert flow is inferred from January 1, 1992, through November 30, 1993, as a result of debris plugging the culverts; Breach: Flow compiled from measurements made by U.S. Geological Survey (ReMillard and others, 1987, 1988, 1989, 1990, 1991, 1992, 1993; Herbert and others, 1997, 1998, 1999)].

Figure 8. *Flow through breach, culverts, and fill of causeway across Great Salt Lake, Utah, 1987-98 (Loving, Waddell, and Miller, 2000).*

If CS = CN, then QS = QN. This condition would exist with no causeway, assuming complete mixing between the north and south parts.

If QS>QN, which generally has occurred since the causeway was constructed, then CS<CN. For example, if the ratio of QS to QN were 2.0 for a constant freshwater inflow, then the ratio of CN to CS would eventually approach 2.0. Therefore, if the concentration of the north part (CN) were constant at 350 g/L, the concentration of the south part (CS) would approach about 175 g/L.

The general trends of salt loads and how they are related to the conveyance properties of the causeway and the water-surface altitude of the lake (inflow) can be explained with equation 6. The water-surface altitude of the south part ranged from about 4,212 ft in 1987, to about 4,202 ft in 1990. During much of this period, the causeway conveyance properties were increased as a result of the water-surface altitude being several feet above the bottom (4,200 ft) of the 290-ft-wide breach. The causeway fill was also raised and widened, which changed the hydraulic properties for flow through the fill (Loving, Waddell, and Miller, 2000, appendix C). The flow through the breach allowed the south-to-north head difference to decrease from an average of about 3.5 ft in 1984, to less than 1.0 ft within 3 months of the opening of the breach.

Two-way, or bi-directional flow through the openings in the causeway occurs below that point at which the hydrostatic head on both sides of the causeway is equal. When bi-directional flow exists, there is an interface between the two flows. The interface between these flows is not a sharp boundary but a diffuse zone that can vary in thickness depending on the densities, velocities, and medium, or opening through which it is flowing (figure 3). The zone of diffusion, or interface, between the north and south brines near the middle of the causeway fill ranges from about 2 to 5 ft in thickness, and in the breach ranges from about 1 to 3 ft in thickness.

For the purpose of explaining the interface depth and its relation to bi-directional flow, assume that the causeway is replaced by an impermeable wall separating the south and north parts. Under this assumption, there would be no flow and pressures on either side of the wall would be hydrostatic. Because of the higher elevation of the south part, the hydrostatic pressure gradient on the upper part of the wall is from south to north. Because the brine in the north part is denser than that of the south part, there is a certain depth at which the pressure gradient on both sides of the wall would be equal. Above this depth of equal pressure, conditions would favor south-to-north flow, and below this depth conditions would favor north-to-south flow.

In the causeway, however, flow is occurring and static conditions do not exist. The interfacial depth slopes from near the water surface on the north edge of the causeway, to a greater depth where the north-to-south flow exits the causeway into the south part (figure 3). With equation 7, the altitude of the interface can be approximated by the Ghyben-Hertzberg principle (Badon-Ghyben, 1888; Herzberg, 1901) (Loving, Waddell, and Miller, 2000, appendix C). The altitude computed with equation 7 approximates the upper surface of the north-to-south flow near where it exits on the south edge of the fill or breach.

$$\text{Altitude of Interface} = ES - \Delta H * \rho n / (\rho n - \rho s) \quad (7)$$

where:

ES = water-surface altitude of the south part, in ft;
ΔH = difference between the water-surface altitudes of the south and north parts, in ft;
ρs = density of the south part in g/mL; and
ρn = density of the north part in g/mL.

The head difference (ΔH) remained at about 1.0 ft or less until 1991, and as a result, the altitude below which the pressure gradient was conducive to north-to-south flow (QN) through the culverts, breach, and fill was above 4,200 ft during most of the period from 1987 through 1990 (figure 9).

The interface altitude ranged from about 4,205 ft in 1987 to about 4,176 ft in 1993. The altitude of the top of the no-flow zone for the causeway fill was determined to be about

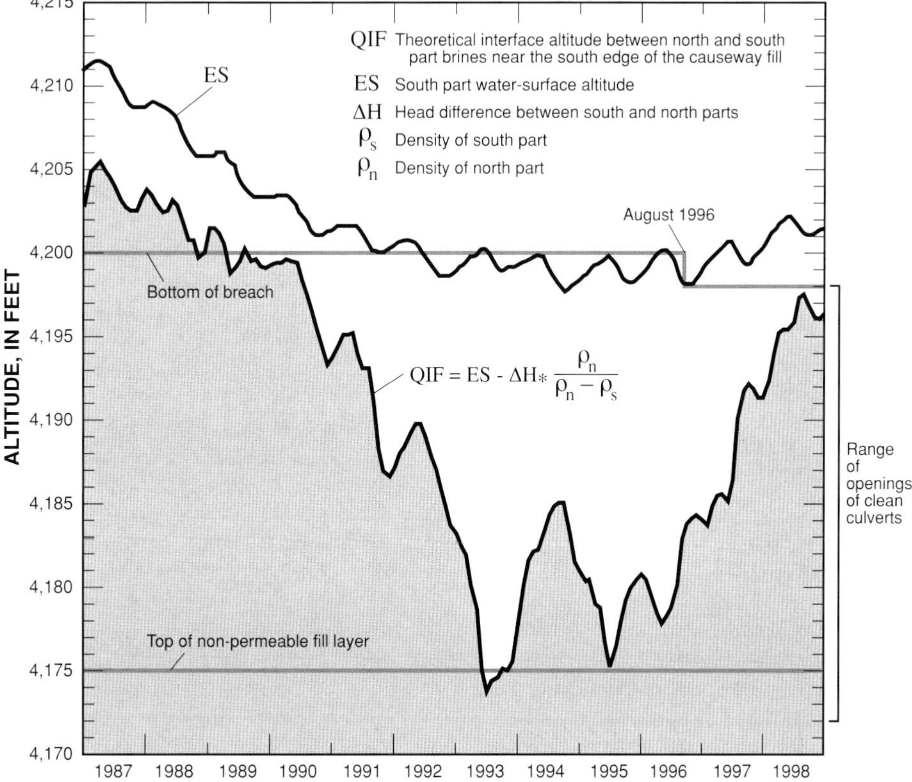

Figure 9. *Altitude of water surface in the south part and theoretical interface between north and south part brines in the causeway across Great Salt Lake, Utah (Loving, Waddell, and Miller, 2000).*

4,175 ft. Thus, for an interface altitude near 4,205 ft, there is about 30 ft of cross-sectional depth through which north-to-south flow can occur through the fill, and for an interface altitude of 4,176 ft, there is only 1 ft of cross-sectional depth through which north-to-south flow can occur. Also, the higher the interface, the more the potential exists for north-to-south flow to occur through the culverts, if open.

Although ΔH, ρs, and ρn are interdependent, ΔH responds must faster to change in the conveyance properties than do ρs, and ρn. The south-to-north flow through the breach occurs in an open channel with a free surface and therefore maintains a low head difference (ΔH) when the altitude of the water surface in the south part is above the bottom of the breach. Eventually, with mixing in the north and south parts, the salinities and densities, ρs and ρn, reach a new equilibrium with ΔH, depending on the conveyance properties and amount of inflow. Generally, the ratio of QS to QN decreases with decreasing ΔH, and increases with increasing ΔH. A decrease in the ratio of QS to QN favors net movement of salts to the south part, and an increase in the ratio favors net movement to the north part.

During 1987-90, the ratio of QS to QN ranged from about 1.97 to 1.45, compared to the ratio of CN to CS of about 2.1 to 1.8, which caused considerable net transfer of dissolved salts back to the south part. In 1992, the water-surface altitude of the south part dropped below 4,200 ft and remained so during most of 1992-96. During this time, there was little breach flow (figure 8) and the head difference began increasing, peaking at about 3.5 ft in 1993. Associated with the increase of head difference was a rapid decline in the altitude at which the pressure gradient was conducive to north-to-south flows (figure 9), and by 1993 the depth had plunged to about 4,176 feet, at which time the ratio of QS to QN peaked at 65.

The water-surface altitude of the south part began increasing in about 1997 and the ratio of QS to QN had decreased to about 4 by 1998. The ratios of QS to QN were much greater than the ratio of CS to CN during much of 1991-98 and resulted in considerable net movement of dissolved solids to the north part through 1998. The net loss of dissolved load from the south part from 1991 through 1998 occurred primarily because of the decreased conveyance created by the water-surface altitude being near or below the bottom of the breach, and the interface being so deep into the fill for much of the period.

By 1998, the water-surface altitude of the south part increased to about 4.5 ft above the bottom of the breach. South-to-north breach flows increased, the head difference dropped below 1.0 ft, and the altitude of the pressure gradient conducive to north-to-south flow within the causeway rose to about 4,197 ft.

Stratification

Before the causeway was built, adequate data are not available to determine whether the salinity of the lake was stratified. Since the construction of the causeway, the lake has, at times, been vertically stratified (figures 10 and 11). A lower layer of denser brine, with relatively constant volume, was first observed in 1965 in the south part of the lake (LSL, figure 6). This layer had chemical characteristics similar to the brine in the north part. Madison (1970, p. 12) observed that the lower layer of brine in the south part occurred everywhere the lake bottom was below 4,175 ft. Stratification in the south part occurred because the causeway reduced circulation in the lake. The causeway created the conditions necessary for the north part to achieve and maintain a higher density than the south part. Because of the higher density in the north part, brine could flow slowly from the north to south part through the culverts and permeable fill, thereby providing a constant supply of high-density brine to maintain the deep, south part brine layer. Data collected by the U.S. Geological Survey (USGS) and the Utah Geological Survey (UGS) during 1973 and in 1974 indicate that the volume of the lower layer of brine in the south part and the altitude of the interface with the overlying low-density brine were essentially unchanged, even though the water-surface altitude had increased by several feet. From 1972 through July 1984, the volume of the deep layer of brine in the south part and the altitude of the interface with the overlying brine increased, and the interface in the south part became more diffuse.

During 1984, when the breach was opened, increased causeway conveyance, combined with record surface-water inflow, produced a large amount of bi-directional flow through the causeway. There was little or no stratification in the north part prior to 1984, but during 1984 both parts of the lake became stratified. This condition persisted through 1987. Stratification decreased from 1988 through 1991, but persisted in both parts through 1991. As surface-water inflow and causeway conveyance decreased from 1988 through 1991, flow through the causeway (QS and QN) also decreased and the salinity stratification gradually disappeared and did not recur between 1992 and 1998. In October 1998, samples were collected only to a depth of 4,172.5 ft and stratification below this depth could not be detected. We assume stratification in both the north and south parts occurred because the increases in rates of flow from south-to-north (QS) and north-to-south (QN) were greater than the rates of mixing in either part.

Effects of the West Desert Pumping Project on Loads of Ions

From April 10, 1987, to June 30, 1989, water was pumped from the north part of the lake into West Pond in an effort to lower the lake-surface altitude during a period of substantially higher than normal inflow. On the basis of flow measurements at the West Desert pump station (figure 1), the West Desert Pumping Project (WDPP) removed more than 2.5 million acre-ft of water and 695 million tons of salt from the lake during a 27-month period. From January 1990 to June 1992, after the pumping had ceased, 200,000 acre-ft of water and 94 million tons of salt returned to the lake from West Pond through the Newfoundland Weir (Wold and Waddell, 1994). Estimates based on direct measurements of salt flowing to and from West Pond indicate that about 0.6 billion ton, or 12 percent of the dissolved-solids load for the lake, were left in West Pond as a result of the WDPP. As an independent check of the amount of salt lost to West Pond, the dissolved-solids load of the lake just before April 1987 was compared to the load just after the return flow had ceased in 1992. The loss of total salt to West Pond was estimated to be about 0.5 billion ton or 10 percent of the total salt in the lake

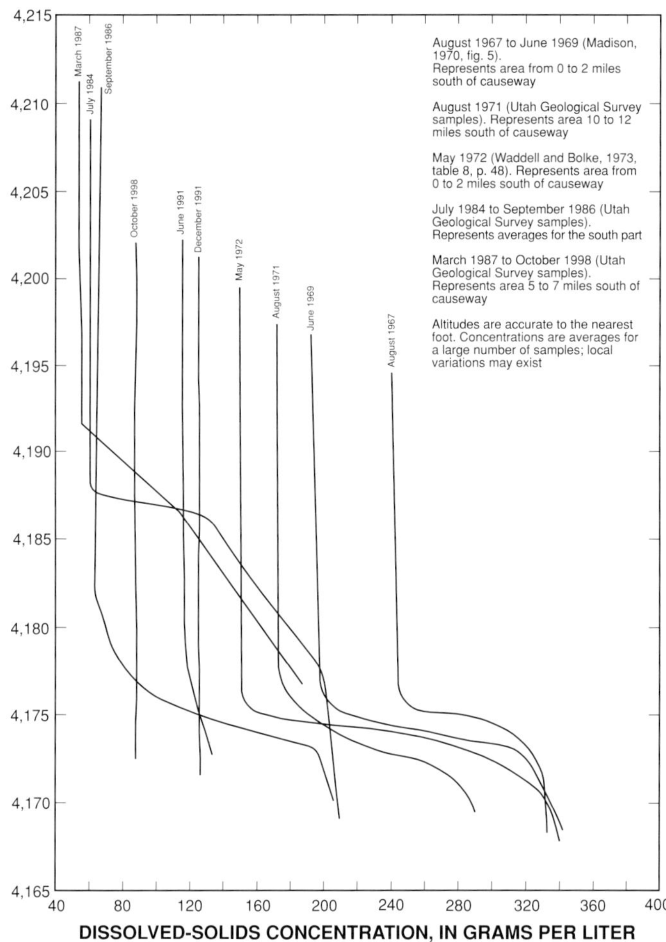

Figure 10. Approximate dissolved-solids concentration gradients for south part of Great Salt Lake, Utah on selected dates, 1967-98 (Loving, Waddell, and Miller, 2000).

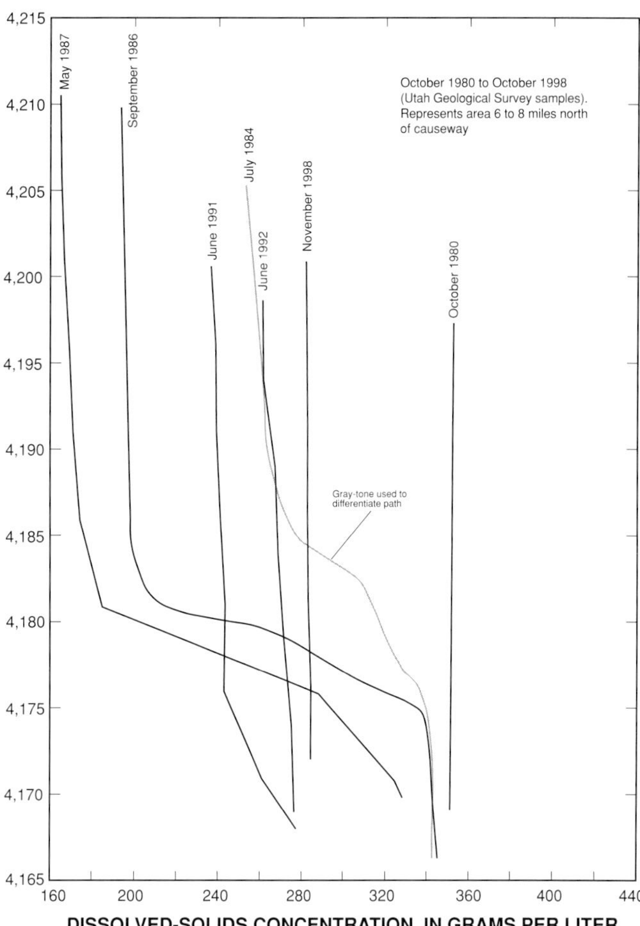

Figure 11. Approximate dissolved-solids concentration gradients for north part of Great Salt Lake, Utah, on selected dates, 1980-98 (Loving, Waddell, and Miller, 2000).

before the WDPP (figure 7), and this amount was used in this study in the salt balance equation calculations.

Because of the importance to the mineral industry of the loss of magnesium and potassium to West Pond, this loss was also estimated from historical chemical data for Great Salt Lake. The most complete, continuous record of spatially distributed ion-concentration data available for Great Salt Lake are the unpublished chemical data collected by the UGS. The loads of magnesium and potassium in Great Salt Lake were computed for the period from 1966 through 1999 using the UGS data. The methods used for computing dissolved-solid loads are described by Loving, Waddell, and Miller (2000, appendix B).

	Prior to West Desert Pumping Project (1966-86)	**During West Desert Pumping Project (1987-89)**	**After West Desert Pumping Project (1991-98)**
Magnesium load, loss in millions of tons	11	17.5	7
Potassium load, loss in millions of tons	12.5	16	3.5

Prior to the WDPP (1966-86), as well as after (1991-99), both magnesium and potassium loads had a declining trend (preceeding summary and figure 12). The trend lines during the period from 1966 to1986 indicate a loss of 11 million tons of magnesium and 12.5 million tons of potassium. The trend lines during the period from 1991 to 1999 indicate a loss of 7 million tons of magnesium and 3.5 million tons of potassium. The only known factor that could contribute to this decline is withdrawal by the mining industry.

Comparison of magnesium and potassium loads just before and after the WDPP indicate that about 17.5 million tons of magnesium (12 percent of the pre-WDPP total magnesium) and 16 million tons of potassium (18 percent of the pre-WDPP total potassium) were lost from the lake to West Pond as a result of the WDPP. Because all of the lake's salt was in solution due to undersaturated brine conditions during the WDPP, the percentage of the lake's total loads of magnesium, potassium, and total dissolved solids (sum of all ions) lost to West Pond were assumed to be the same. The 12-percent loss of magnesium agrees with the 10- to 12-percent loss computed for dissolved solids, but the computed potassium loss was 6 percent greater than the computed dissolved-solids loss.

The computed loads of magnesium and potassium are highly variable from measurement to measurement (figure 12). Computed loads of magnesium and potassium each had

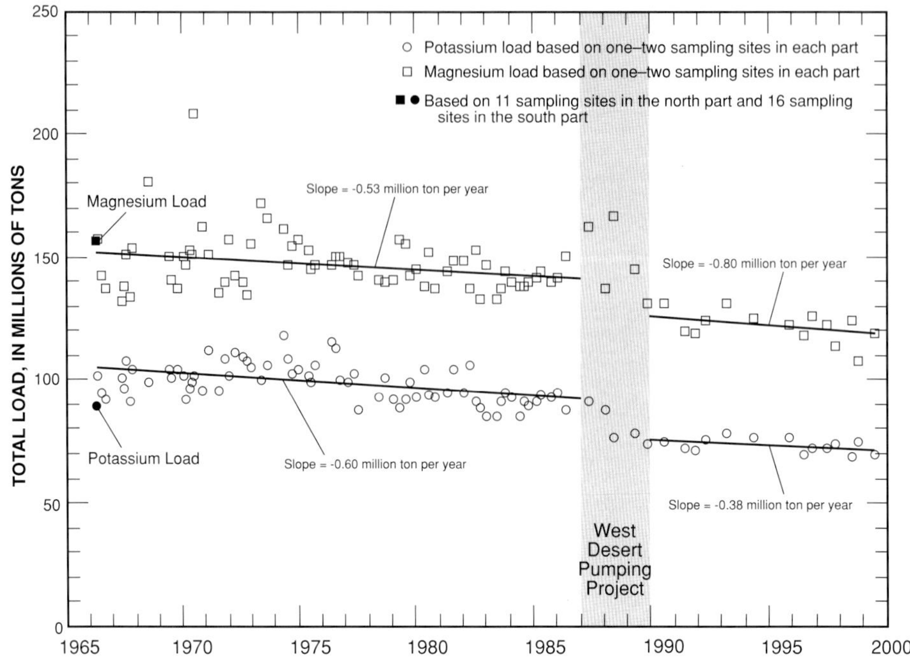

Figure 12. *Total magnesium and potassium loads in Great Salt Lake,Utah, 1966-99 (Loving, Waddell, and Miller, 2000).*

standard errors from the trend lines of 8 percent of the mean loads during the period from 1966 to 1986. Because of the magnitude of the variability in loads is comparable to the variability of the computed losses to West Pond, it cannot be inferred that the computed losses of total magnesium and potassium during the WDPP are substantially different from those computed for total dissolved solids.

Sources of Error in Computing Loads of Ions

Spatial sampling error associated with the use of a small number of samples to represent large parts of the lake, and analytical error associated with analysis of brines, are sources of error in computing loads or masses of salts in Great Salt Lake (Loving, Waddell, and Miller, 2000, p. 15). If the brine samples used for a load calculation do not represent an average vertical profile of the entire lake, then the load calculation can be in error because of the large volume of the lake (load = concentration * volume). In general, the greater the number of spatially distributed sample sites, the smaller the spatial error.

Spatial sampling error can result from both areal differences and stratification. The bay areas, where most of the freshwater inflow occurs, typically have lower dissolved-solids concentration than the main body of the lake and are the primary cause of areal differences in the south part of the lake. Historically, stratification has been minor in the north part of the lake, but at times substantial in the south part. The major exception was during the historic rise in lake levels during 1983-87, and particularly after opening of the causeway breach during 1984, when both parts were stratified.

Loads of dissolved solids computed by Loving, Waddell, and Miller (2000, p. 16) were determined from measurements taken from three to nine sites in each part of the lake, whereas individual ion loads were determined from measurements at only one to two sites in each part of the lake. Specific gravity measurements, corrected to 20°C, are a simple and reliable way to estimate dissolved-solids concentration; therefore, a large number of specific-gravity measurements were made to estimate dissolved-solids loads, rather than using summation of individual ion loads.

To provide an estimate of how much spatial sampling error affects the computation of loads in the lake, a comparison for two different sampling periods was made between loads based on vertical profiles at one sampling site in each part of the lake and the loads at three to nine sites in each part. One comparison period was from 1987 when the lake was stratified, and the other period was from 1992, when the lake was well mixed and had very little areal or vertical stratification.

During 1987, the load of dissolved solids computed from a vertical profile of samples at a single site in each part was 4.94 billion tons, in comparison with a load estimate from four to nine sampling sites in each part at 4.83 billion tons. Thus, the load computed from vertical profiles at two sites was about 2 percent higher than the load computed from 13 sites. In comparison, during 1992 when the lake was well mixed both vertically and areally, the load based on measurements from two sites was only 1 percent higher than the load based on measurements from eight sites.

Only in the south part of the lake has a significant areal difference in salinity occurred. This difference is present in the bay areas, where freshwater from the Bear, Weber, and Jordan Rivers enters the lake. Although the bay areas generally are fresher than the main south part of the lake, their small volume compared to that of the main south part results in only small error in computing loads of dissolved solids in the south part of the lake (Loving, Waddell, and Miller, 2000, appendix B).

Inherent in the computation of dissolved solids from the sum of the ions is the cumulative error associated with the determination of each ion. Laboratories and methods for analyzing brines changed from 1966 through 1998, and may have contributed to the differences between the dissolved solids computed from specific gravity and the sum of the ions (figure 13). Methods used for measuring specific gravity from 1966 through 1998 were basically the same and are considered to have only a small analytical error due to the simplicity of the measurement and the excellent relation between specific gravity and dissolved solids (Waddell and Bolke, 1973, p. 35).

The differences between dissolved solids as computed from the sum of the ions and specific gravity were evaluated by Loving, Waddell, and Miller (2000, p. 16) in an effort to determine the possible causes and effects these differences might have on the calculation of losses of specific ions, such as potassium and magnesium resulting from the WDPP, or the exchange of dissolved salts between the north and south parts of Great Salt Lake. Possible causes of these differences in computed dissolved solids are ion-balance errors and (or) analytical bias. Ion-balance errors were evaluated as the difference between the sum of the cations (sodium, potassium, and magnesium) and anions (chloride and sulfate) in terms of equivalents per liter. During 1976-98, the error was about plus or minus 6 percent and was not large enough to cause the differences seen in the dissolved solids; thus, these differences may be a result of analytical bias.

Irrespective of the analytical techniques used by the laboratories, the brines generally are diluted substantially prior to analysis. Small errors in dilution can lead to large errors in the final computation of ion concentrations even if the analysis of the diluted brine is quite accurate. Because of the viscous nature of the concentrated brines, small aliquots pipetted to volumetric flasks for large dilution can result in underestimating the actual dilution. Standard pipettes are calibrated for use with fluids of viscosity less than that of Great Salt Lake brine, and as a result, may deliver less than the amount indicated for the pipette.

Potassium and magnesium concentrations are too low relative to the differences in dissolved-solids concentrations as computed from the sum of the ions and from specific gravity (figure 13) to account for these differences. Sodium and chloride compose about 85 to 90 percent of the dissolved-solids concentration in the lake. For the cations and anions to balance and still have a discrepancy in the dissolved-solids concentration, sodium and chloride would have to be biased low in approximately the same ratio.

The higher differences between dissolved-solids concentrations as computed from the sum of the ions and from specific gravity (figure 13) in the north part as compared to that of the south part may indicate dilution errors. From 1988 through 1998, the concentrations of all ions in the north part were about 2 to 3.5 times greater than those in the south part and would require more dilution and possibly result in greater sources of error than for brines from the south part.

Prior to about 1991, the dissolved-solids concentration computed from the sum of the ions in the north part ranged from about 0 to 25 g/L less than that computed from specific gravity (figure 13). After about 1991, the difference increased to about 50 g/L. Prior to 1991, in the south part the difference ranged from about 0 to 20 g/L, increased to about 30 g/L after 1991, and then slowly decreased to almost 0 by 1998.

In consideration of the possible analytical bias, perhaps from dilution errors, the losses of potassium and magnesium from the lake after 1988 may be overestimated. Because there is indication that the analytical bias increases as the dissolved-ion(s) concentration increases (figures 4 and 13), the samples collected from the north part of the lake would have a greater analytical error or bias than those from the south. The error biasing concentrations low would also appear as load losses, especially in the north part, and may account for the difference between the expected and computed losses of potassium during the WDPP.

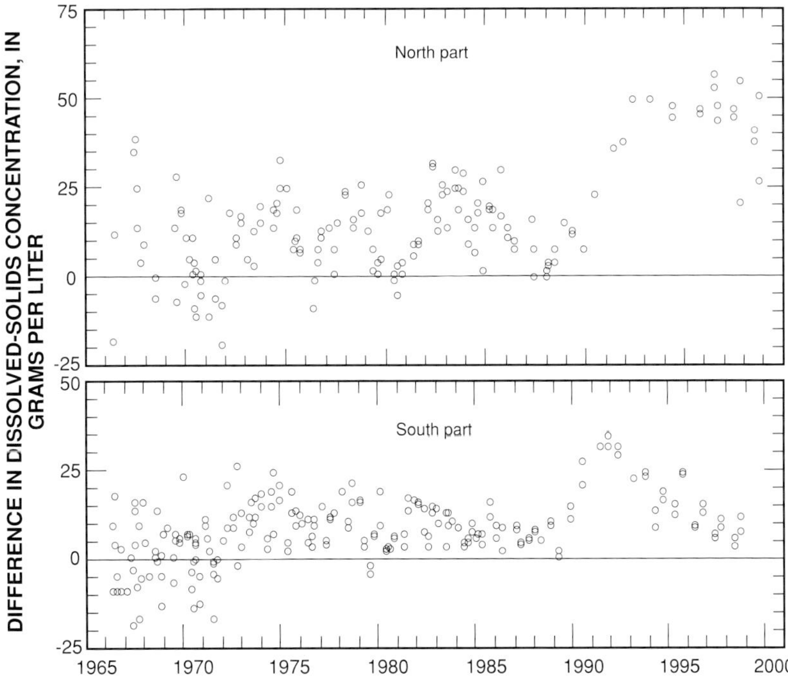

Figure 13. *Difference between dissolved-solids concentration computed from the sum of individual ion concentrations and as computed from specific-gravity measurements, 1966-99, for the north and south parts of Great Salt Lake, Utah (Loving, Waddell, and Miller, 2000).*

SIMULATION OF WATER AND SALT MOVEMENT THROUGH THE CAUSEWAY

The original model developed by Waddell and Bolke (1973) was used during the 1970s by the Utah Department of Natural Resources to simulate water and salt movement through the causeway and to determine the effects of causeway modifications on the salt balance between the two parts of the lake. The original model also was used to simulate the effects of different widths of culvert or breach openings on the head difference between the south and north parts of the lake. The simulations made by the original model were used to help design the breach that was constructed in the causeway in 1984 to help alleviate flooding along the shores of the south part of the lake.

The original model simulates the effects of the causeway on the water and salt balance of the lake for variable rates of inflow (Waddell and Bolke, 1973). The main components of, and modifications to, the original model are shown in table 1. The current model (Loving, Waddell, and Miller, 2000) has all the simulation capabilities of the original model but also can simulate effects of the breach, flow through submerged culverts, and loads of four major ions.

Calibration of the Water and Salt Balance Model, 1987-98

Model calibration involved two primary computer programs. First, a water-balance program was used to calibrate the hydrologic and climatic variables that are input to the water and salt balance model. The water and salt balance model was then used to calibrate the causeway-fill flows (figure 14).

Monthly values of measured and estimated surface inflow, ground-water inflow, precipitation on the lake surface, and evaporation were compiled for 1987-98 by Loving, Waddell, and Miller (2000, appendix A). The input data were evaluated with a water-balance program (Wold, Thomas, and Waddell, 1997, appendix A). The purpose of the water-balance program was to test, and if necessary, modify the independent estimates of monthly inflow, precipitation, and evaporation. The water-balance program computes water-surface altitudes from the monthly input and compares them to measured altitudes from 1987 through 1998.

Computations of the water-surface altitude made with the independent estimates of inflow and evaporation indicate that there was either too much evaporation or too little inflow (figure 15). Most of the surface-water inflow (SIS) and pumpage to the West Pond (QWP) were measured, and ground-water inflow (GIS and GIN) is very small when compared to other inflow sources. Thus, most of the error in the water balance was probably from a combination of error in precipitation and evaporation. Although there is probably some error in the amount of precipitation, only evaporation was adjusted as part of the water-balance data evaluation. The maximum adjustment to evaporation during 1 year was 8 percent, and adjustment averaged 4 percent from 1987 through 1998.

Water and Salt Balance Model and Causeway-Fill Flow

Loving, Waddell, and Miller (2000, p. 19), used the water and salt balance model to evaluate the causeway-fill flows for 1987-98. Evaporation, as adjusted in the water balance program, and the same inflow, were used as input to the water and salt balance model (figure 15).

Table 1. Major modifications to the original model of Waddell and Bolke (1973) (Loving, Waddell, and Miller, 2000, table 1).

Component	First modification (Wold, Thomas, and Waddell, 1997)	Second modification (Loving, Waddell, and Miller, 2000)
Initial Conditions	• Revised water-surface altitude, area, and volume relations (George Pyper, U.S. Geological Survey, written commun., 1986)	• Revised water-surface altitude, area, and volume relations to account for MagCorp dike that was breached in June 1986 and repaired in January 1994 • Added dissolved loads of individual major ions
Input Data	• Compiled inflow and evaporation for 1980-86	• Compiled inflow and evaporation for 1987-98 • Compiled withdrawals and return flows of water and dissolved solids to and from West Pond
Open-channel Flow	• Replaced energy equations for open-channel flow with equations of Holley and Waddell (1976) • Added the breach as a channel for which flow could be computed	• Added equations to account for flow through submerged culverts • Revised the equations that convert flow computed for a rectangular-shaped breach to flow through a trapezoidal-shaped breach
Flow Through Causeway Fill	• Computed fill flow with solute-transport model of Sanford and Konikow (1985) • Modified the fill-flow computations to account for changes in the hydraulic conductivity of the fill	• Modified the fill-flow computations to account for changes in the hydraulic conductivity of the original fill

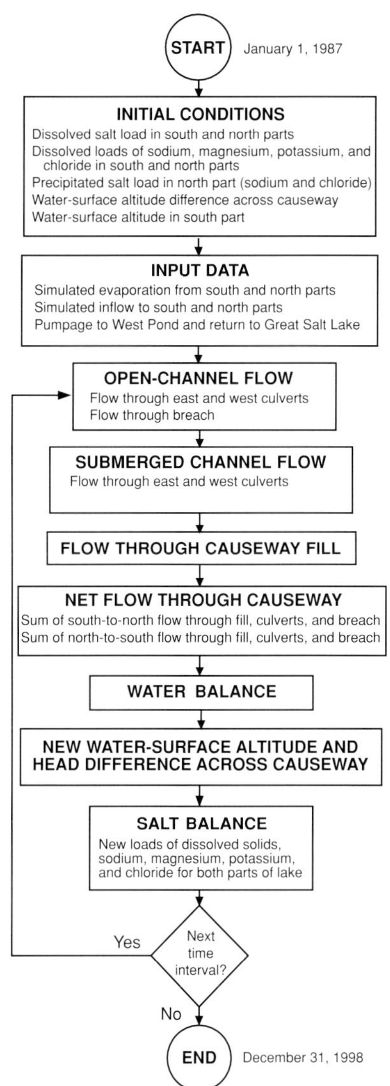

***Figure 14.** Flow chart of Great Salt Lake water and salt balance model (Loving, Waddell, and Miller, 2000).*

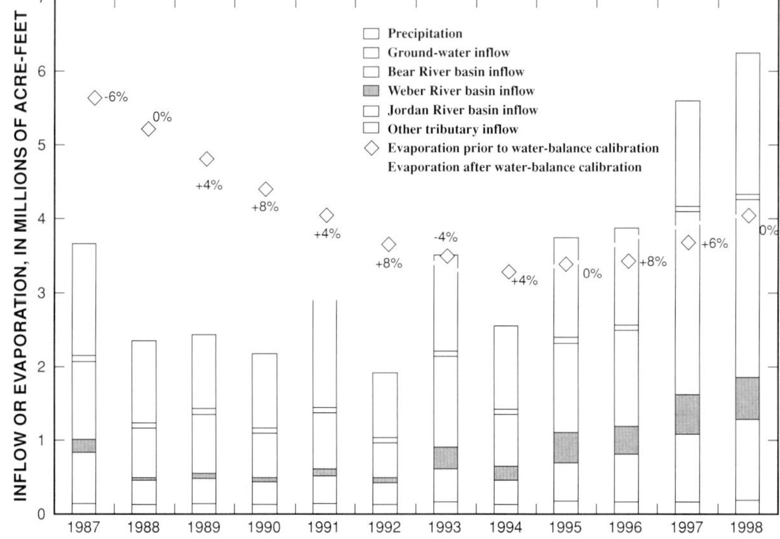

***Figure 15.** Inflow and evaporation for Great Salt Lake, Utah, 1987-98 (Loving, Waddell, and Miller, 2000, figure A2).*

South-to-north and north-to-south causeway-fill flows (QSF and QNF) were considered to be the least accurate components of the causeway flows (QS and QN). During the calibration of the water and salt balance model, it was assumed that the hydraulic-conductivity values used in the fill-flow model (Loving, Waddell, and Miller, 2000, appendix C) were the cause of the errors between the simulated and measured water-surface altitudes (ES and EN), salt loads (LS, LN, and CLNP), and densities (ρ_s, and ρ_n). The model was calibrated by assuming that a change in the hydraulic conductivity was directly proportional to a change in the fill flow. With this assumption, constant factors were used to reduce the fill flows and were evaluated by comparing simulated water-surface altitudes, salt loads, and densities to measured values during 1987-98.

To calibrate the model for 1980-86, Wold, Thomas, and Waddell (1997, p. 12) reduced the hydraulic conductivity of the fill to 40 percent of what Waddell and Bolke (1973) determined it to be. Comparison of simulated and measured water-surface altitudes (ES and EN), salt loads (LS, LN, and CLNP), and densities (ρs, and ρn) indicated that the hydraulic conductivity of the fill continued to decrease during the 1980s and 1990s (figure 16). By reducing the causeway-fill flows by a constant factor, the simulated and measured water-surface altitude, salt loads, and densities were in closer agreement.

After several trials of assigning a single hydraulic-conductivity value to the entire vertical profile of causeway fill, Loving, Waddell, and Miller (2000), decided to assign the new fill added during the high lake levels of the 1980s a different hydraulic conductivity than the original fill below it (figure 3). The best match of simulated and measured water-surface altitudes, salt loads, and densities for the 1987-98 calibration occurred when the new fill was assigned the same hydraulic conductivity used by Waddell and Bolke (1973) and the original fill below it was reduced to 10 percent of the 1973 conductivity. No other components of the water and salt balance were changed as part of the calibration. The final calibration values of breach, culvert, and fill flow, and the water balance parameters for 1987- 98, are shown in figure 17.

After the calibration of causeway fill flows, the sensitivity and accuracy of the simulated water and salt balance components were evaluated by comparison with measured data (figure 18). The components most sensitive to adjustment of the hydraulic conductivity of the fill are head difference, density difference, and precipitated salt load in the north part. The maximum differences between simulated and measured head differences were + 0.5 ft and -0.9 ft, with the largest differences occurring during 1992-96. Simulated density differences closely matched measured values, with the largest difference being about 0.008 g/mL.

Although all error in the water and salt balance was attributed to the causeway-fill flow, the differences between simulated and measured values result from the combined error from the causeway fill, culvert, and breach flow. Most of these differences can be attributed to errors in the estimated culvert flow during 1987-96, when culvert flow was not monitored.

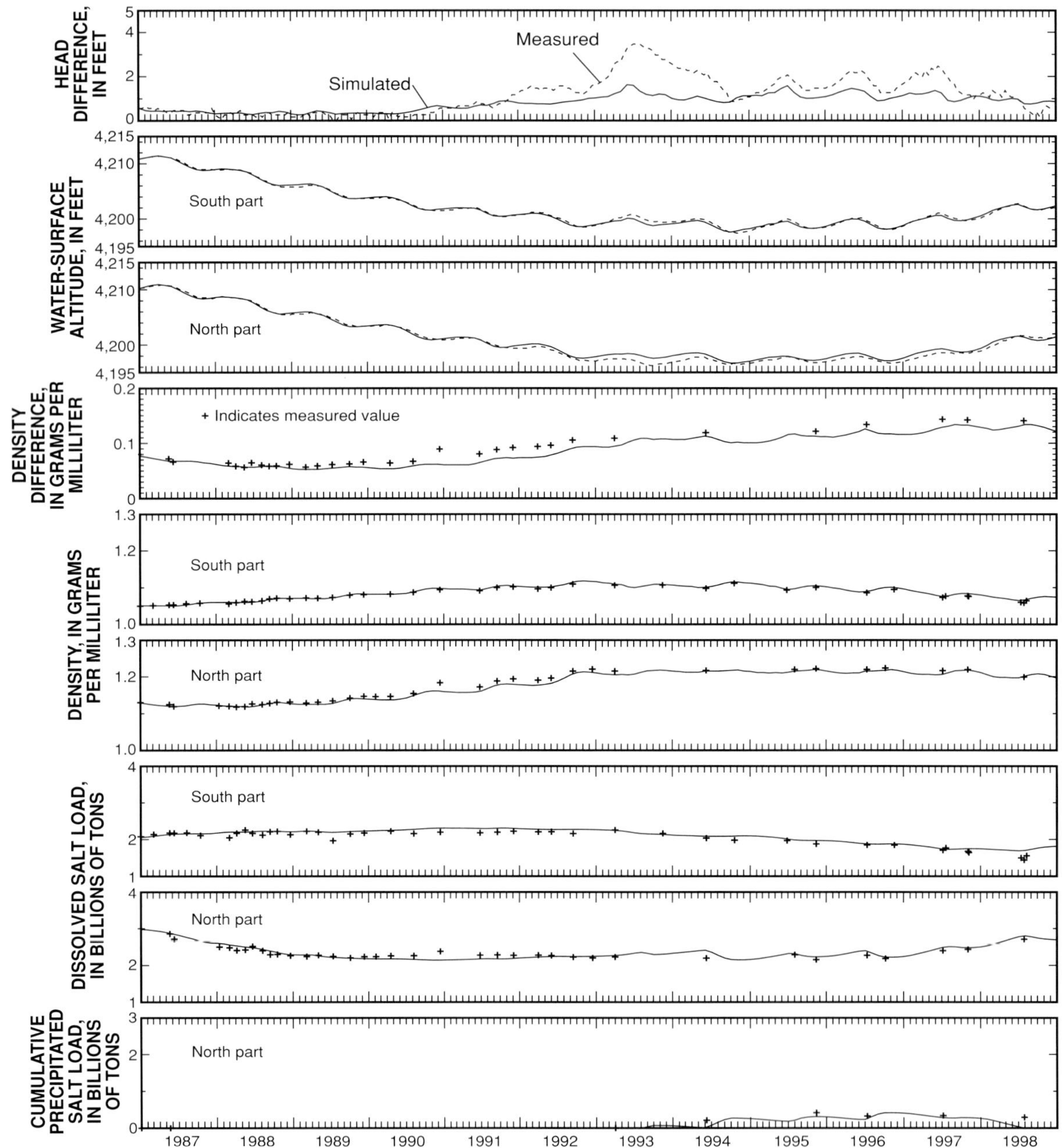

__Figure 16.__ Simulated and measured head difference, water-surface altitude, density difference, density, and dissolved and cumulative precipitated salt load in the south and north parts of Great Salt Lake, Utah before model calibration, 1987-98 (Loving, Waddell, and Miller, 2000).

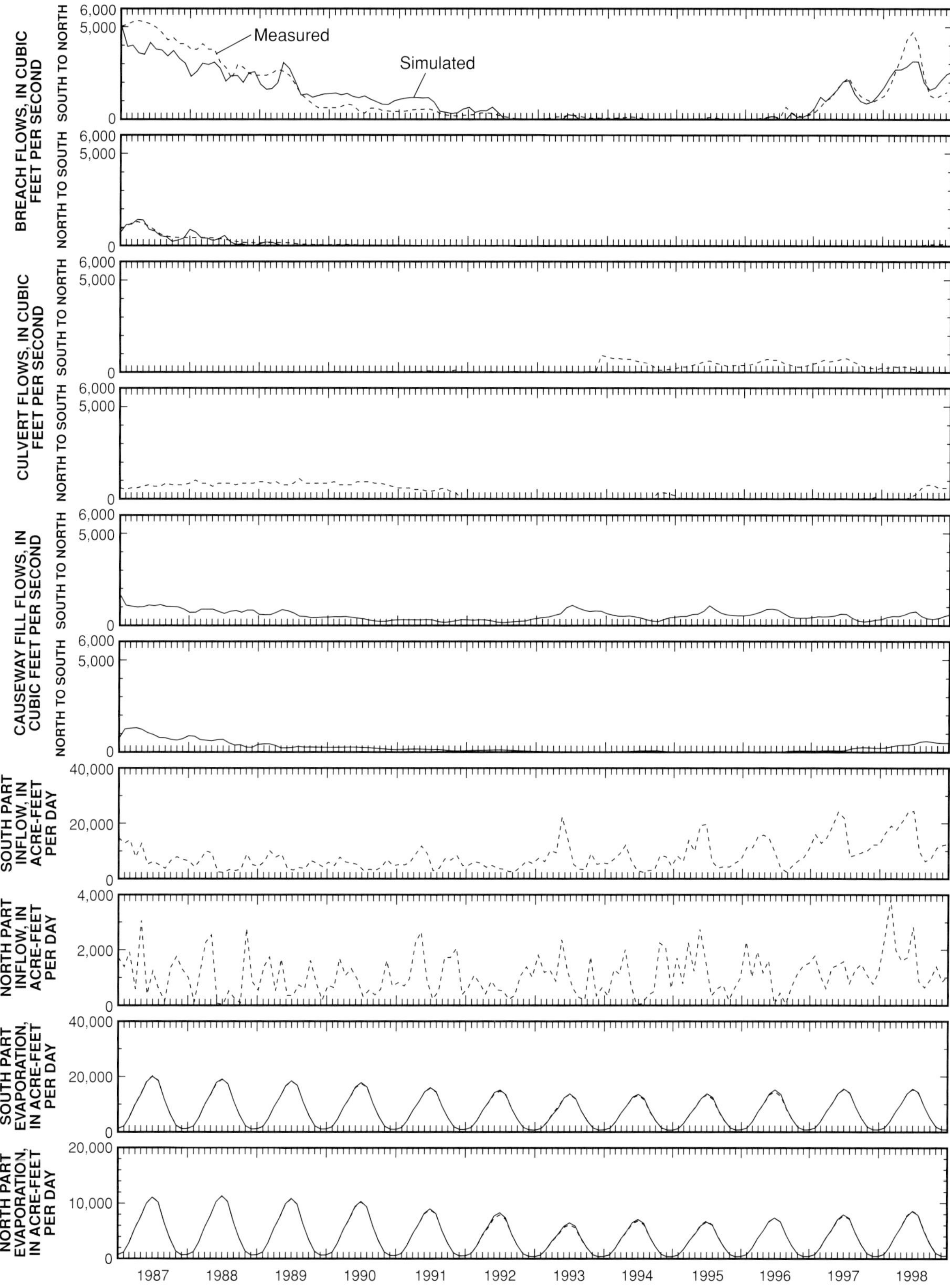

Figure 17. Simulated and measured breach flow, measured and estimated culvert flow, computed causeway-fill flow, total measured and estimated inflow, and estimated and computed evaporation in the south and north parts of Great Salt Lake, Utah, after model calibration, 1987-98 (Loving, Waddell, and Miller, 2000).

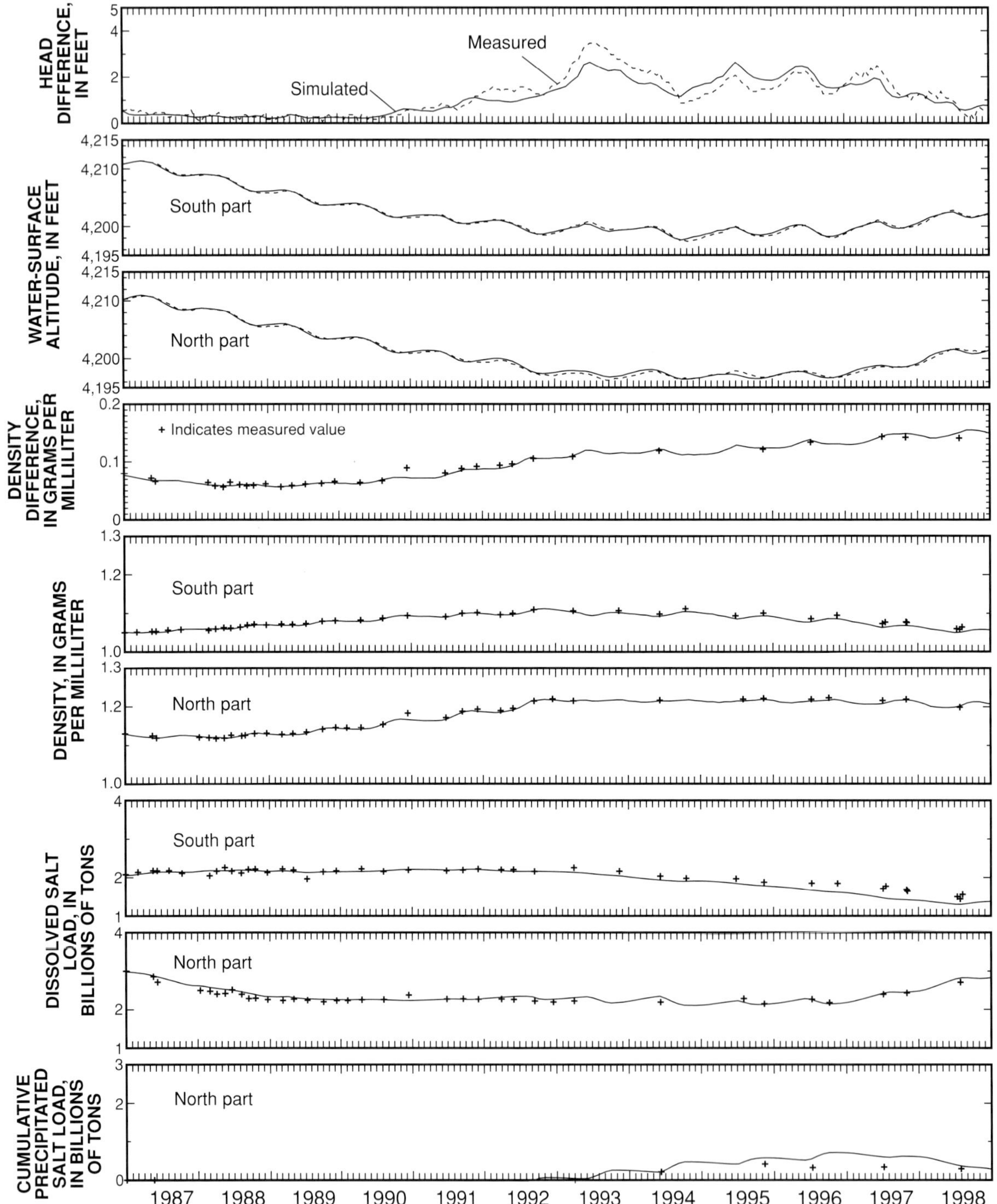

Figure 18. *Simulated and measured head difference, water-surface altitude, density difference, density, and dissolved and cumulative precipitated salt load in the south and north parts of Great Salt Lake, Utah, after model calibration, 1987-98 (Loving, Waddell, and Miller, 2000).*

Simulation of Major-Ion Loads as an Independent Check of Model Calibration

The dissolved loads of four major ions (sodium, magnesium, potassium, and chloride) were simulated by Loving, Waddell, and Miller (2000, p. 20) to independently check the model accuracy after the model had been calibrated. As discussed earlier in the section "Sources of error in computing loads of ions," ion loads vary widely. Because of this variation, the model did not match the measured loads of individual ions as well as the total dissolved-solids loads computed from specific-gravity measurements. Simulated south part loads matched measured loads reasonably well for all four ions (figure 19). Simulated north part loads matched measured loads well for sodium and chloride, but not as well for magnesium or potassium.

Simulated Effects of Breach Dimension on the Salinity Balance Between the North and South Parts

The calibrated model was used by Loving, Waddell, and Miller (2000, p. 25) to simulate the effect of several combinations of breach depth and width on the salinity in each part of the lake. The 1987-98 inflows and evaporation were used as a baseline simulation (along with estimated culvert flows and West Pond withdrawals for the period). During 1987-98, the lake experienced a wide range of surface-water inflows and water-surface altitudes, providing hydrologic conditions

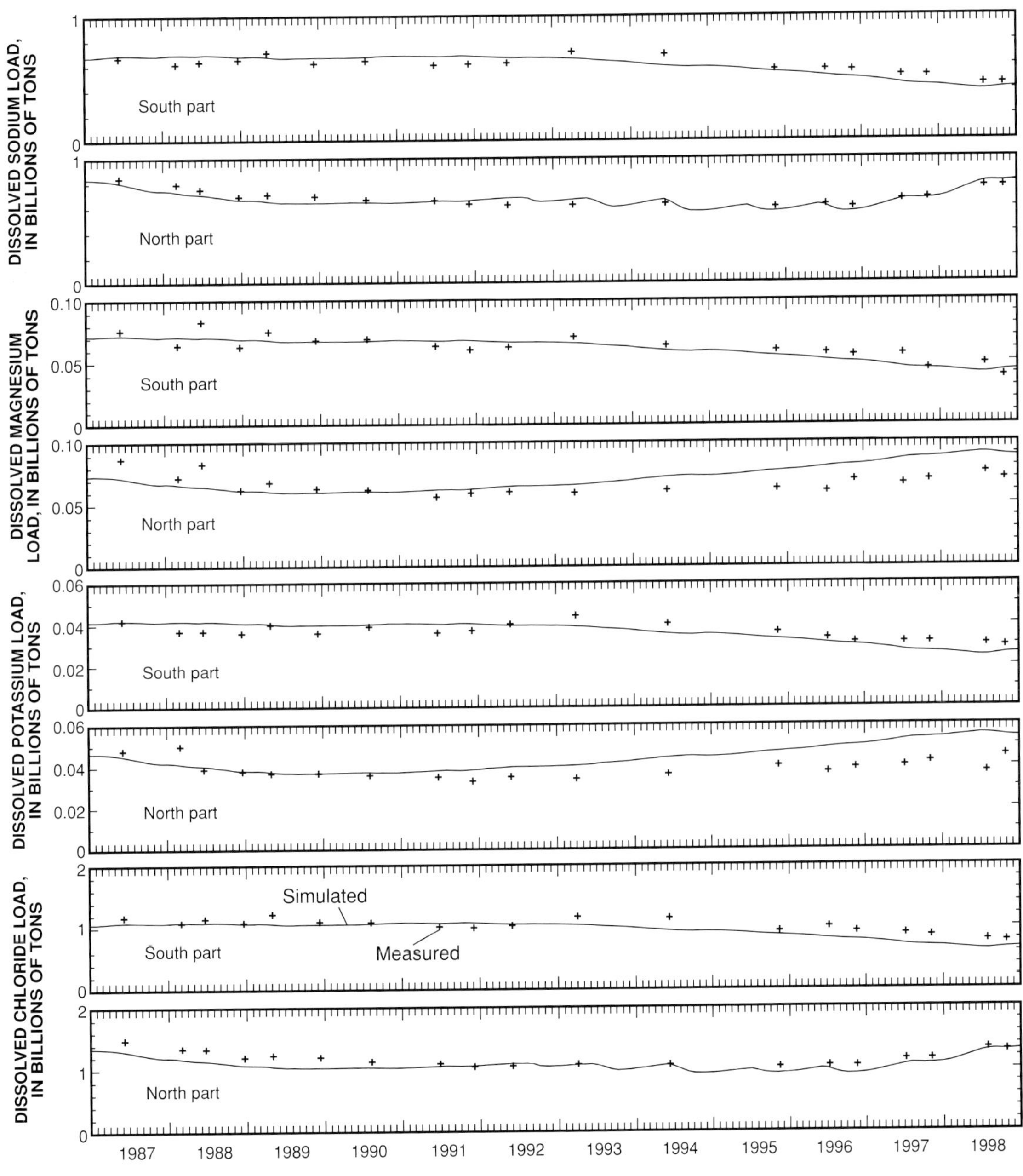

Figure 19. *Simulated and measured loads of sodium, magnesium, potassium, and chloride in the south and north parts of Great Salt Lake, Utah, after model calibration, 1987–98 (Loving, Waddell, and Miller, 2000).*

suitable for evaluating the effect of variable breach dimensions.

During 1987-98, the general shape of the breach was trapezoidal, with the widest opening (about 290 ft) at the top of the breach and the narrowest opening (about 200 ft) at the bottom (figure 20a). The altitude of the bottom of the breach was about 4,200 ft until August of 1996, when it was deepened to an effective altitude of 4,198 ft. For the simulations of different breach dimensions, the breach shape was assumed to be rectangular with the same width at the bottom as at the top (figure 20b).

Model simulations (tables 2 and 3) indicate that deepening the breach is more effective in reducing the difference in dissolved-solids concentration between the two parts of the lake than widening without deepening the breach. In December 1998, the dissolved-solids concentration of the south part of the lake was 94 g/L, or 28 percent of the concentration in the north part. Deepening the breach bottom to an altitude of 4,190 ft in January 1987 would have increased the dissolved-solids concentration of the south part to 146 g/L, or 55 percent of the north part concentration by December 1998. In comparison, widening the breach from 290 ft to 600 ft without deepening it would only have increased the dissolved-solids concentration of the south part to 110 g/L, or 33 percent of the north part concentration.

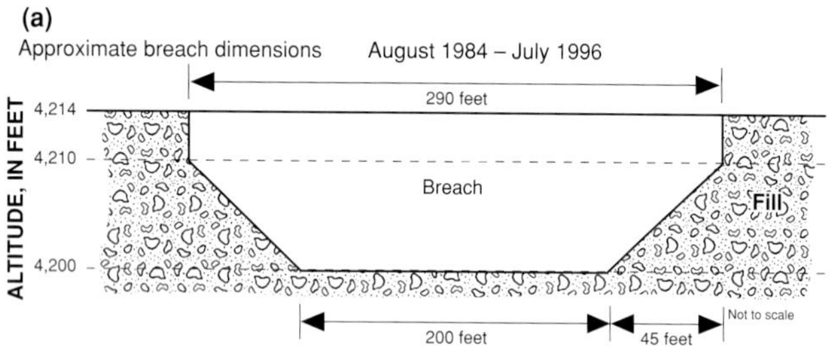

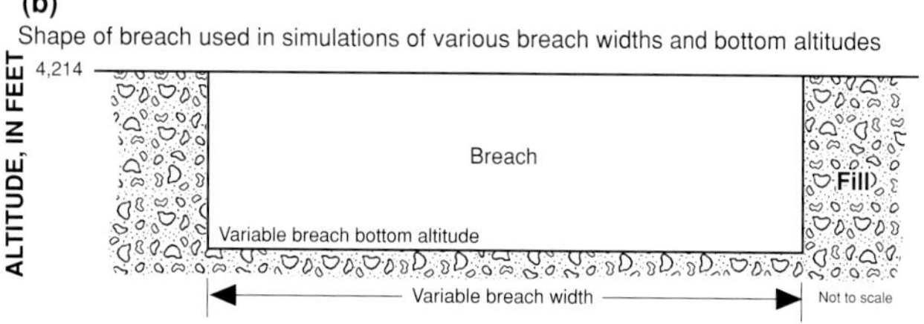

Figure 20. Approximate shape and dimensions of the breach (a) during August 1984 to July 1996, and (b) used in simulations of selected breach widths and depths, Great Salt Lake, Utah (Loving, Waddell, and Miller, 2000).

Table 2. Simulated dissolved-solids concentration of the south part as a percentage of that of the north part for selected breach dimensions, Great Salt Lake, Utah, December 31, 1998 (Loving, Waddell, and Miller, 2000).

[Assumes breach dimensions were modified from existing dimensions on January 1, 1987; breach dimensions produced head differences too small for the model to simulate breach or fill flows].

Breach width, in feet	Altitude of the bottom of breach, in feet						
	[1]4,198	4,195	4,193	4,190	4,185	4,180	4,175
100	27	30	31	38	48	55	60
150	28	32	35	44	54	-	-
200	29	33	40	49	59	-	-
250	30	35	44	53	-	-	-
[2]290	30	38	46	55	-	-	-
400	31	43	-	-	-	-	-
500	33	-	-	-	-	-	-
600	33	-	-	-	-	-	-

[1]Approximate altitude of the bottom of the breach during August 1996 through 1998; the altitude was approximately 4,200 feet August 1984 through July 1996.

[2]Approximate width of the top of the breach August 1984 through 1998.

A dash indicates breach.

Table 3. Results of model simulations showing comparison of salinities in Great Salt Lake, Utah, with a 290-ft-wide rectangular breach of selected bottom altitudes (Loving, Waddell, and Miller, 2000).

[g/mL, grams per milliliter; g/L, grams per liter]

	Initial Conditions January 1987	Breach bottom altitude of 4,198 feet Dec. 1998	Breach bottom altitude of 4,195 feet Dec. 1998	Breach bottom altitude of 4,193 feet Dec. 1998	Breach bottom altitude of 4,190 feet Dec. 1998
South part					
Percent Salinity	7.7	9.4	11.1	12.3	13.4
Density (g/mL)	1.051	1.063	1.075	1.084	1.092
Concentration (g/L)	81	100	119	133	146
North part					
Percent Salinity	18.4	27.5	26.3	24.4	22.8
Density (g/mL)	1.131	1.209	1.198	1.181	1.167
Concentration (g/L)	208	332	314	287	265

Simulated Effects of the West Desert Pumping Project on the Salinity of the North and South Parts

The calibrated model was used to simulate effects of the WDPP on salinities of the north and south parts of the lake. Model simulations indicate that had there been no WDPP, salinity in December 1998 would have been 3 g/L or 1 percent higher in the south part, and 9 g/L or 0.3 percent higher in the north part (table 4). In addition, model simulations indicate that if there had been no WDPP, the net movement of salt from the south part to the north part from 1987 through 1998 would have been about 0.5 billion ton instead of 0.7 billion ton.

Sensitivity and Error Analyses

A sensitivity and error analysis was performed by Loving, Waddell, and Miller (2000, p. 27) to determine the influence of the model components on the accuracy of the water and salt balance model results. Their results include:

Water Balance

Most sensitive to surface-water inflow (60 percent of total inflow during 1987-98), and evaporation (100 percent of total outflow, except during the WDPP).

After evaporation was calibrated in the water balance program, simulated south part water-surface altitude matched measured within an average of 0.2 ft, and with a maximum difference of 0.7 ft.

Fill Flow

For a specified set of boundary conditions (ES, EN, ρs, ρn), fill flow is most sensitive to hydraulic conductivity of the fill material.

During 1987-98, the fill flow computed by the water and salt balance model averaged 611 acre-ft/d, and the fill flow computed using a theoretical equation of the water and salt balance averaged 503 acre-ft/d, or about 21 percent less than the model. The greatest differences between flows computed by the water and salt balance model and the theoretical equation of the water and salt balance occurred during 1987-91 when the boundary conditions were rapidly changing from the decline in inflow and lake altitude.

Breach and Culvert Flows

Model-computed breach flow is most sensitive to head difference (ΔH), density difference (Δρ), and the physical dimensions of the breach.

The standard error of estimate for computed and measured breach flow was 722 ft^3/s, or 30 percent of the mean for south-to-north flow, and 294 ft^3/s, or 116 percent of the mean for north-to-south flows.

The standard error of estimate for computed and measured flow through unsubmerged culverts during 1980-83 was 12 percent of the mean for south-to-north flow, and 62 percent of the mean for north-to-south flow (Wold, Thomas, and Waddell, 1997, p. 54). Culvert flow was not computed as part of the model calibration for the period 1987 through 1998 because the culverts were frequently plugged, preventing the use of theoretical equations to compute flow.

Water and Salt Balance Model

The model was calibrated by reducing the causeway-fill flows by a constant factor. The model components most sensitive to this adjustment are head difference (ΔH), density difference (Δρ), and precipitated load of salt in the north part (CLNP).

After the model was calibrated, the maximum difference between simulated and measured head differences was 0.9 ft, density differences, 0.008 g/mL, and precipitated loads of salt in the north part, 0.220 billion ton.

The calibrated model was used to simulate the effect of several breach bottom altitudes on the dissolved-solids concentration in each part of the lake. Model simulations indicate that had the breach been deepened from an altitude of 4,200 to 4,195 ft in January 1987, the December 1998 dissolved-solids concentration of the south part would have been 38 percent of the north part dissolved-solids concentration. To determine the degree of accuracy involved when a deeper breach is considered in the model, simulations were made deepening the breach to 4,195 ft, with breach flows increased and decreased by the standard error found between measured and simulated breach flows. The standard error for south-to-north breach flow was determined to be 30 percent, and for north-to-south flow, 116 percent (Loving, Waddell, and Miller, 2000, appendix E). The results of the simulations are:

If the south-to-north breach flows computed by the model were 30 percent too low, a breach bottom altitude of 4,196 ft, instead of 4,195 ft, would have been required to achieve a dissolved-solids concentration of the south part of 38 percent of the north part concentration in December 1998. Similarly, if the south-to-north breach flows computed by the model were 30 percent too high, a breach bottom altitude of 4,193.5 ft, instead of 4,195 ft, would have been required.

If the north-to-south breach flows computed by the model were 116 percent too low, a breach bottom altitude of 4,196 ft, instead of 4,195 ft, would have been required to achieve a dis-

Table 4. Results of a model simulation showing comparison of salinities in Great Salt Lake, Utah, with and without the West Desert Pumping Project (Loving, Waddell, and Miller, 2000).

[WDPP, West Desert Pumping Project; g/mL, grams per milliliter; g/L, grams per liter]

	Initial Conditions January 1987	With Effects of WDPP December 1998	Without Effects of WDPP December 1998
South part			
Percent Salinity	7.7	8.8	9.8
Density (g/mL)	1.051	1.059	1.066
Concentration (g/L)	81	94	105
North part			
Percent Salinity	18.4	27.4	27.7
Density (g/mL)	1.131	1.209	1.211
Concentration (g/L)	208	332	335

solved-solids concentration of the south part of 38 percent of the north part concentration in December 1998. Similarly, if the north-to-south breach flows computed by the model were 116 percent too high, a breach bottom altitude of 4,192.5 ft, instead of 4,195 ft, would have been required.

SUMMARY

A rock-fill causeway across Great Salt Lake was completed in 1959. The effect of the causeway was to change the water and salt balance of Great Salt Lake by creating two separate but interconnected parts of the lake, with more than 95 percent of freshwater surface inflow entering the lake south of the causeway. The water and salt balance of Great Salt Lake depends primarily on the amount of inflow from tributary streams and the conveyance properties of the causeway that divides the lake into south and north parts. The causeway restricts circulation between the south and north parts. The conveyance properties of the causeway originally consisted of two 15-foot wide culverts and permeable rock-fill material, but the causeway has since been modified by the addition of a breach and modifications to the causeway fill. During the 1980s, fill material was added to the causeway as lake levels rapidly rose, changing the overall dimensions of the causeway fill. In August 1984, a 290-ft-wide breach with a bottom altitude of about 4,200 ft was opened near the western end of the causeway to reduce the head difference between the two parts of the lake.

The dissolved-solids concentrations of the north and south parts were approximately equal at the time the causeway was completed in 1959, but by 1972, the concentration was about 200 g/L greater in the north part than in the south, and by December 1998, the concentration was about 250 g/L greater in the north than in the south. The theoretical concentration that would have occurred in an undivided lake in December 1998 was about 190 g/L. In 1998, the concentration in the south part was about 90 g/L, or 100 g/L less than the theoretical concentration for an undivided lake. In 1998, the concentration in the north part was about 340 g/L, or 150 g/L more than the theoretical concentration for an undivided lake.

Previous modeling of the water and salt balance incorporated the causeway conveyance properties and hydrologic conditions that existed during 1959-86. Waddell and Bolke (1973) described the effects of the causeway on the water and salt balance of Great Salt Lake and developed a model to simulate the effects of the causeway on the salt balance for variable culvert widths and tributary inflows to the lake. This original model of Waddell and Bolke was calibrated for causeway conveyance and hydrologic conditions existing during 1969-72. This model was valid until about 1981. During the 1980s, fill material was frequently added to the causeway to maintain the top of the causeway above the water surface. During 1983, the two 15-ft-wide culverts became submerged beneath the rising water surface and eventually filled with debris; in August 1984, a 290-ft-wide breach was opened near the western end of the causeway. Because these new conditions warranted revision of the model, Wold, Thomas, and Waddell (1997) modified the original model and recalibrated it for 1980-86.

Since 1986, additional changes in the causeway conveyance properties and withdrawals for the West Desert Pumping Project have made it necessary to revise the model of Wold, Thomas, and Waddell (1997). In Loving, Waddell, and Miller (2000), the U.S. Geological Survey, in cooperation with the Utah Department of Natural Resources, Division of Forestry, Fire and State Lands, and Tooele County, Utah, modified and recalibrated the 1997 model to incorporate the changes that occurred during 1987-98.

For simulations of several hypothetical breach dimensions with 1987-98 boundary conditons, deepening the breach was more effective in reducing the difference in dissolved-solids concentration between the two parts of the lake than widening without deepening the breach. In December 1998, the dissolved-solids concentration of the south part of the lake was 94 g/L, or 28 percent of the concentration of the north part. Simulations indicate that had the altitude of the bottom of the breach been deepened from 4,200 ft to 4,195 ft in January 1987, the December 1998 dissolved-solids concentaration of the south part would have been 119 g/L or 38 percent of the north part concentration. Deepening the breach bottom to 4,190 ft in January 1987 would have increased the dissolved-solids concentration of the south part to 146 g/L or 55 percent of the north part concentration by December 1998. Widening the breach from 290 ft to 600 ft without deepening it would only have increased the dissolved-solids concentration of the south part to 110 g/L, or 33 percent of the north part concentration.

During 1987-92, about 500 million tons of salt, including 17.5 million tons of magnesium and 16 million tons of potassium, were removed from the lake as part of the West Desert Pumping Project. The pumps only operated during April 1987 through June 1989, but there was some return flow from West Pond to the lake during 1990-92. Model simulation indicated that had there been no West Desert Pumping Project, the dissolved solids concentration in the north part would have been 3 g/L higher, and the concentration in the south part 9 g/L higher by December, 1998.

REFERENCES

Badon-Ghyben, W., 1888, Nota in Verband met de Voorgenomen Putboring Nabij Amsterdam (Notes on the probable results of well drilling near Amsterdam), Tijdsdar, Kon. Inst., Ing., The Hague, 1888/9, p. 8–22.

Fenneman, N.M., 1931, Physiography of the Western United States: New York, McGraw- Hill, 534 p.

Hahl, D.C., 1968, Dissolved-mineral inflow to Great Salt Lake and chemical characteristics of the Salt Lake brine: Summary for water years 1960, 1961, and 1964: Utah Geological Survey Water-Resources Bulletin 10, 35 p.

Herbert, L.R., Tibbetts, J.R., Wilbert, D.E., and Allen, D.V., 1997, Water resources data, Utah: U.S. Geological Survey Water-Data Report UT– 97–1, 328 p.

—1998, Water resources data, Utah: U.S. Geological Survey Water-Data Report UT– 98–1, 300 p.

—1999, Water resources data, Utah: U.S. Geological Survey Water-Data Report UT– 99–1, 340 p.

Herzberg, A., 1901, Die Wasserversorgung einiger Nordseebaden (The water supply on parts of the North Sea coast in Germany), Z. Gasbeleucht, Wasserversorg, 44, p. 815– 819, 824–844.

Holley, E.R., and Waddell, K.M., 1976, Stratified flow in Great Salt Lake culvert: Journal of the Hydraulics Division, American Society of Civil Engineers, v. 102, no. HY7, Proceedings Paper 12250, July 1976, p. 969–985.

Loving, B.L., Waddell, K.M., and Miller, C.W., 2000, Water and salt balance of Great Salt Lake, Utah, and simulation of water and salt movement through the causeway, 1987-98: U.S. Geological Survey WRIR 00-4221, 32 p.

Madison, R.J., 1970, Effects of a causeway on the chemistry of the brine in Great Salt Lake, Utah: Utah Geological Survey Water-Resource Bulletin 14, 52 p.

ReMillard, M.D., Herbert, L.R., Sandbert, G.W., and Birdwell, G.A., 1986, Water resources data, Utah: U.S. Geological Survey Water-Data Report UT–86–1, 404 p.

—1987, Water resources data, Utah: U.S. Geological Survey Water-Data Report UT– 87–1, 368 p.

—1988, Water resources data, Utah: U.S. Geological Survey Water-Data Report UT– 88–1, 364 p.

—1989, Water resources data, Utah: U.S. Geological Survey Water-Data Report UT– 89–1, 383 p.

—1990, Water resources data, Utah: U.S. Geological Survey Water-Data Report UT– 90–1, 383 p.

—1991, Water resources data, Utah: U.S. Geological Survey Water-Data Report UT– 91–1, 375 p.

—1992, Water resources data, Utah: U.S. Geological Survey Water-Data Report UT– 92–1, 343 p.

—1993, Water resources data, Utah: U.S. Geological Survey Water-Data Report UT– 93–1, 333 p.

Sanford, W.E., and Konikow, L.F., 1985, A two-constituent solute-transport model for ground water having variable density: U.S. Geological Survey Water-Resources Investigations Report 85–4279, 88 p.

Waddell, K.M., and Bolke, E.L., 1973, The effects of restricted circulation on the salt balance of Great Salt Lake, Utah: Utah Geological Survey Water-Resource Bulletin 18, 54 p.

Wold, S.R., and Waddell, K.M., 1994, Salt budget for West Pond, Utah, April 1987 to June 1989: U.S. Geological Survey Water-Resources Investigations Report 93-4028, 20 p.

Wold, S.R., Thomas, B.E., and Waddell, K.M., 1997, Water and salt balance of Great Salt Lake, Utah, and simulation of water and salt movement through the causeway: U.S. Geological Survey Water-Supply Paper 2450, 64 p.

ADJUSTMENTS TO 1966-2001 GREAT SALT LAKE WATER-SURFACE ELEVATION RECORDS, DUE TO CORRECTED BENCHMARK ELEVATIONS

by
Brian Loving
U.S. Geological Survey

INTRODUCTION

Great Salt Lake is divided into north and south parts by a rock-fill causeway (figure 1) that was completed in 1959. The U.S. Geological Survey (USGS) operates gages that collect water-surface elevation data on the north part of the lake "Saline gage" (figure 2, USGS station 10010100), and on the south part of the lake at " Boat Harbor gage" (figure 3, USGS station 10010000). Fom the mid-1980s to 2001, it was known that the difference in water-surface elevation between the two parts of the lake, as determined from the Boat Harbor and Saline gage readings, was greater than the difference measured directly at the causeway. Since the lake surface is considered to be relatively "flat" on calm days, and since the gages were periodically checked against benchmarks with surveying levels, the difference was assumed to be an error in the given elevations of the benchmarks to which the gages are referenced. During the periods 1969 through 1982, and from 1997 through 1999, a second gage was operated on the south part of the lake at Promontory Point (figure 1, USGS station 10010050). This gage was referenced to the same line of benchmarks as the Saline gage. The difference in water-surface elevations between the two parts of the lake, as measured at the Promontory Point and Saline gages, generally agreed with the difference measured directly at the causeway. The purpose of this paper is to summarize the discrepancies between the gage readings north and south of the causeway, present the results of a Global Positioning System (GPS) survey which revealed the source of the discrepancies, and present the changes made to the Great Salt Lake elevation records in 2001 for the period from 1966 through 2001.

Until the late 1990s, there was no economically feasible way to verify the given elevations of the reference benchmarks of the Great Salt Lake elevation gages. In 1999, the National Geodetic Survey (NGS) conducted a high-resolution (GPS) survey in Utah. The USGS, and the Utah Water Resources, participated in this survey to find the elevations of five benchmarks around Great Salt Lake that are used to determine the water-surface elevations of the lake. The final calculations from this survey were presented by the NGS in March 2001. This survey provided the first direct check and comparison of the elevations of all five benchmarks.

When the Boat Harbor and Saline gages were adjusted to the new benchmark elevations, the difference in water-surface elevation between the two parts of the lake measured by the gages generally agreed with the difference measured directly at the causeway. The records of water-surface elevation have been adjusted for the Boat Harbor, Saline, and Promontory Point gages based on the 1999 NGS benchmark elevations.

FINDINGS

Water-surface elevations reported at the USGS Great Salt Lake gages are considered to be accurate to within +/- 0.10 ft of the datum in use. Of the five benchmarks surveyed as part of the 1999 GPS survey, only three were considered by the NGS to be accurate to within 0.10 feet: "WES DES PUMPS" "FMK 77 1966" and "SALTAIR" (figures 1, 2, and 3).

The elevation of the 77 FMK 1966 benchmark, located near the Saline gage (figure 2), was found by the GPS survey to be 4,231.155 feet, National Geodetic Vertical Datum of 1929 (NAVD 29). Data from the establishment of the Saline gage in 1966 to 2001 had been adjusted to the 77 FMK 1966 benchmark with an elevation of 4,230.888 feet (0.267 feet lower).

The Promontory Point gage was referenced to the FMK 73 1966 benchmark, which is on the same line as the FMK 77 1966 benchmark. Since the GPS survey adjusted the FMK 77 benchmark 0.27 feet higher, and since the datum of the FMK 73 and FMK 77 benchmarks have historically agreed, it is assumed that the given elevation of FMK 73 should also be raised 0.27 feet.

The Boat Harbor gage has been tied to two different benchmarks since the 1960s. The first, "BM H-39 1922," was used from sometime before the 1960s until 1985. Sometime between 1985 and 1989 it was destroyed during the construction of Interstate Highway 80. After 1985, the primary reference benchmark for the Boat Harbor gage was "C-174 (1970)." Using the new GPS survey elevation for the Saltair benchmark (located at the Boat Harbor gage), and the surveyed height differences between Saltair, C-174, and BM H-39 from previous levels, corrected elevations for C-174 and BM H-39 were computed. From the new survey, it was found that the previously given elevation for the BM H-39 was 0.14 feet too high, and the elevation for the C-174 benchmark was 0.42 feet too high.

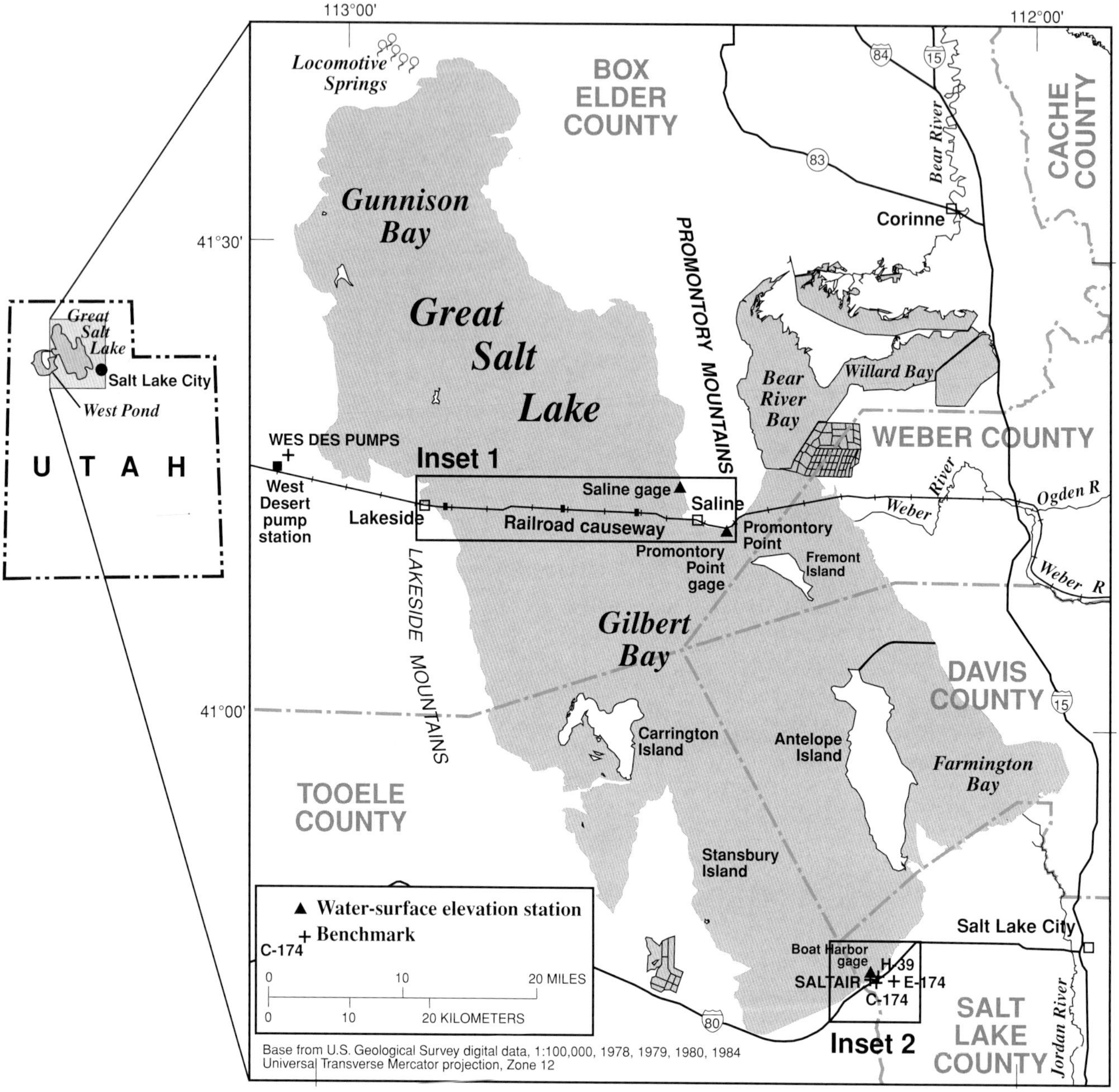

Figure 1. Location of water-surface elevation stations and reference benchmarks near Great Salt Lake, Utah.

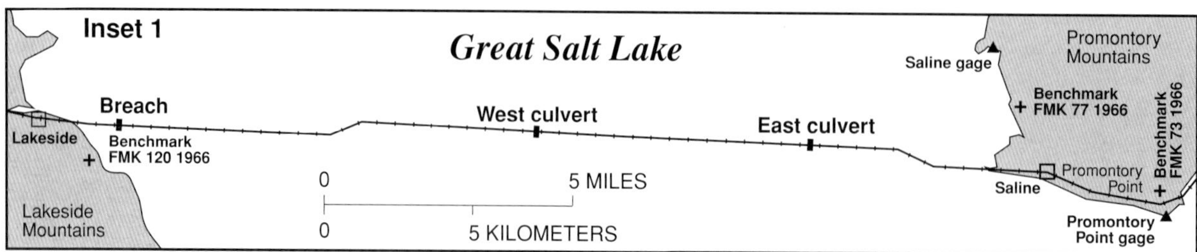

Figure 2. Location of water-surface elevation stations and reference benchmarks near the causeway across Great Salt Lake, Utah.

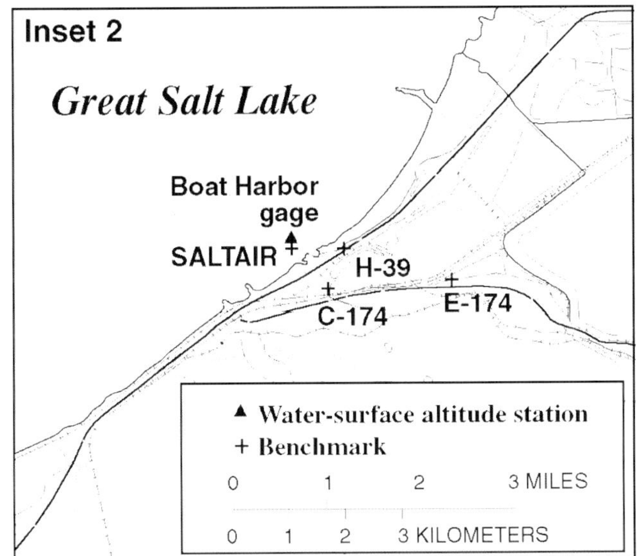

Figure 3. *Location of water-surface elevation station and reference benchmarks near Saltair, Utah.*

In addition to the changes in given elevations for the reference benchmarks, all three of the gages used at the Boat Harbor during 1980-2001 settled. The following is a synopsis of the findings on the settling of Boat Harbor gages from 1980 to 2001. During the period from 1981 through 1983, the Boat Harbor gage settled 0.25 feet. This was discovered in 1983 and the affected data were corrected then. In 1985, the Boat Harbor gage became inundated by the rising lake and was moved to a temporary location, where it was attached to a large concrete sign. The sign and the gage settled about 0.44 feet during the period it was operated, from 1985 to 1989. This problem was not discovered until 2001, because the gage datum was established with BM H-39 1922, which had a given elevation 0.14 feet too high and when the gage was discontinued, its datum was checked against benchmark C-174 (1970) which had an original given elevation that was 0.42 feet too high. So, even though the gage settled about 0.44 feet, it appeared to the surveyors at the time to be off by only about 0.12 feet, and no changes were made. A prorated correction for this settling had to be applied to the Boat Harbor water-surface elevation data from July 1985 to August 1989. Because the current gage was established off of the settled temporary gage, a constant -0.44-foot correction had to be applied to the data from August 1989 to September 1994, when the 0.44-foot error was removed.

The present gage, installed in August of 1989 (figure 4), also settled during the first six years it was used. Levels indicate that the gage settled about 0.55 feet from September 1989 to July 1993, and about 0.11 feet from July 1993 to June 1995. The record was adjusted for part of this settling in 1995.

In 1995, the 0.25-foot settling correction, applied in 1983, was mistakenly applied to the records from 1984 through 1995. No evidence was found in 2001 that this 0.25 correction was needed during the period 1984 through 1995.

There is no indication that the gage has moved since 1995.

Changes to the Great Salt Lake Elevation Records made in 2001

Saline Gage: To adjust for the change in the given elevation for benchmark FMK 77 1966, 0.25 feet was added to all Saline gage water-surface elevation data from the establishment of the gage in 1966, to August 15, 1989, when data were rounded to the nearest 0.05 feet. From August 16, 1989 to 2001, when data were rounded to the nearest 0.10 feet, 0.30 feet was added.

Promontory Point Gage: To adjust for the change in the given elevation for benchmark FMK 73 1966, 0.25 feet was added to the Promontory Point gage water-surface elevation data from 1969 to 1982, when data were rounded to the nearest 0.05 feet. From 1997 to 1999, when data were rounded to the nearest 0.10 feet, 0.30 feet was added.

Boat Harbor Gage: While GPS and other levels indicate that benchmark BM H-39 had a given elevation that was 0.14 feet too high, the record was not adjusted for this apparent error for the years prior to 1984. This 0.14-foot error likely entered the record in the 1950s (or earlier), and not enough information is available to justify an adjustment of such a small amount back to the 1950s. Table 1 is a summary of the corrections that were applied to the Boat Harbor gage water-surface-elevation data in 2001. These corrections are actually the sum of a combination of corrections to the problems described above in the "Findings" section.

The corrected water-surface elevation data is published in the report "Water Resources Data, Utah, Water Year 2001", U.S. Geological survey Water-Data Report UT-01-1, Herbert, L.R. and others, 2002. For more information on these changes, please contact the U.S. Geological Survey, Hydrologic Data Section, in Salt Lake City at 801-908-5000.

Table 1. Summary of corrections to Boat Harbor gage water-surface elevation data, to be applied on May 1, 2001.

Period of time	Correction applied to Boat Harbor gage record May 1, 2001 (in feet)
4/16/1984-6/30/1985	+0.10
7/1/1985-6/30/1986	0.00
7/1/1986-6/30/1987	-0.10
7/1/1987-6/30/1988	-0.20
7/1/1988-8/21/1989	-0.35
8/22/1989-9/30/1990	-0.40
10/1/1990-9/30/1991	-0.40
10/1/1991-9/30/1992	-0.50
10/1/1992-9/30/1993	-0.50
10/1/1993-9/30/1994	-0.50
10/1/1994-9/30/1995	-0.40
10/1/1995-4/30/2001	-0.40

CONCLUSIONS

Since the mid-1980s, it has been known that the difference in water-surface elevation between the two parts of the lake, as determined from the Boat Harbor and Saline gage readings, was greater than the difference measured directly at the causeway. These differences were due to several factors including settling of the south-arm gages, benchmarks, and the mistaken application of a settling correction. The results of a GPS survey conducted on 1999 provided correct elevations for the reference benchmarks and made it possible to correct the lake elevation data. A single elevation correction was applied to the Saline and Promontory Point gage data from 1966 through 2001. Multiple elevation corrections were applied to the Boat Harbor gage data, from 1984 to 2001. USGS water-surface elevation data for Great Salt Lake is now adjusted for datum corrections discovered by the 1999 GPS survey.

Figure 4. *Photo of present Boat Harbor gage near Saltair, Utah.*

GREAT SALT LAKE SOUTH ARM CIRCULATION: CURRENTS, VELOCITIES AND INFLUENCING FACTORS

by

Julie Rich
Department of Geography, Weber State University, Ogden, Utah

ABSTRACT

Water circulation within Great Salt Lake has been the subject of several scientific reports, all of which present differing ideas as to the location, cause, and direction-of-flow of the brine currents. The research discussed in this chapter involves the use of satellite-tracked drifter buoys that were used to map the currents in the south arm of the lake as well as to determine their direction and velocity. A digitized map was compiled from telemetrically transmitted geographic positions relayed from two drifter buoys through TIROS (Television and Infrared Observation Satellite) and NOAA (National Oceanic and Atmospheric Administration) satellites. In addition, remote-sensed imagery was obtained from Landsat and the Space Shuttle Large Format Camera, taken during previous summer seasons. These two systems were found to complement and support one another, providing information about the spatial location, size, direction, and rate-of-flow of the circulation patterns in Great Salt Lake's southern arm. The results obtained indicate that the south arm of the lake has unique circulation patterns and velocities, which this report attempts to explain.

The most important finding was the relative motion of the largest brine currents in the south arm, for the summer periods investigated. The largest gyres in the lake exhibited an unusual and pronounced counterclockwise rotation. This movement is contrary to that found in the average, large water body in the northern hemisphere, which has an induced rotation in a clockwise direction due to Coriolis effect. The present research suggests that the explanation for the lake brine's counter clockwise motion results from cyclonic wind stress created near the right-hand shore (looking downwind). This phenomenon, known as Eckman Drift, results from the lake's thermal energy being radiated to the air flowing across its surface, inducing increased wind stress as the temperature differential is narrowed between air and water. Water-surface temperatures within the south arm, as monitored by the drifter buoys, and used in conjunction with contemporary and historical wind data, support this explanation as the motive force that produces the counterclockwise circulation pattern found in Great Salt Lake. In addition, the speed of the currents was determined by the time and distance between drifter buoy locations, and was found to vary with the relative size of the gyres created by the movement of the brine. The fastest flowing currents were those in the two large mid-lake gyres, moderate speeds occurred in the medium-sized circulating brine systems located in sheltered bays, and the slowest speeds in small eddies, induced by the larger gyres, along the lake's shorelines. The propagation, direction, location, and speed of the currents in Great Salt Lake's south arm, discussed in this study, are generally at odds with the information and the conclusions about the lake's currents found in previously published studies.

INTRODUCTION

The principle objectives of this research were: (1) to determine brine circulation patterns and velocities, and (2) the reason for the counterclockwise motion of the brine in the south arm of Great Salt Lake. Previous attempts, undertaken to measure these movements, lacked the long-term empirical data necessary to analyze brine velocity and direction.

Research for this chapter involved a 48-day period of data collection undertaken during the summer of 1991. Statistical information on the circulation patterns of Utah's inland sea was gathered using equipment supplied by NOAA and the Department of Defense. NOAA satellite-tracked drifter buoys were employed to collect empirical data on the direction-of-movement, velocity, and surface temperature of the brine in the south arm. Supporting information on weather was collected using a Department of Defense portable automatic weather station located at Buffalo Point, on the north end of Antelope Island (figure 1). Data gathered by this station assisted in understanding the relationship between wind sheer and brine movement. South-arm circulation patterns deduced from the drifter buoys were compared to circulation patterns visable on satellite imagery taken during a previous summer season. Brine-current patterns are visable by satellite because suspended material in the brine, and carried by the lake's movements, exhibits a spectral signature that can produce a visual image of the currents. The orientation and locations of the gyres (circulation patterns), as seen on the satellite image, support the pattern obtained in the present study done with satellite-tracked drifters. What the remote-sensed image could not show was the temperature and velocity of the moving brine, nor the reason for its predominant counterclockwise direction. This research establishes the temperature, direction, speed, and reason for the long-term, counterclockwise motion in the south arm of Great Salt Lake. The principal reason for the movement of the lake's brine, opposite to that produced by

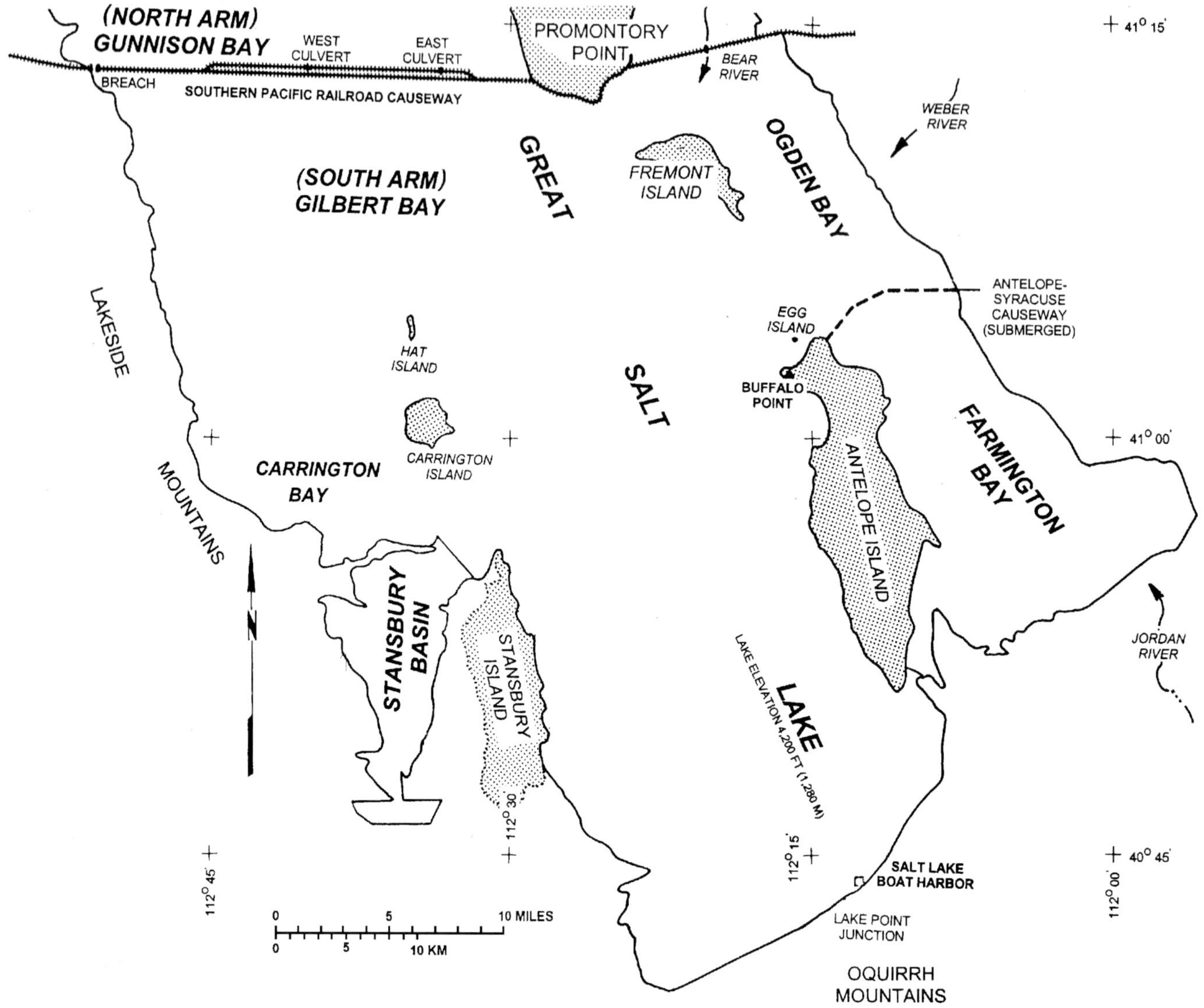

Figure 1. *Great Salt Lake south arm region.*

the Coriolis effect in the northern hemisphere, is the principle known as Ekman Drift, which is created by wind stress (drag). This will be discussed later in the chapter.

REVIEW OF LITERATURE

A pioneering lake-current study, completed in the late-1960s by Hahl and Handy (1969), was the first scientific attempt to explain the brine movement in Great Salt Lake. Qualitative lake-current information was assembled while the authors collected brine-chemistry data between October 1965 and May 1966. Hahl and Handy observed that the current flow on and parallel to the south side of the Southern Pacific Railroad causeway indicated a westward movement of the brine. From this they inferred that a general counterclockwise circulation was occurring in the two arms of the lake (figure 2). This conclusion was supported by the location and orientation of spits and sand features around the shoreline (Hahl and Handy, 1969).

A map of Great Salt Lake completed by Greer (1971) illustrates a large cyclonic cell and a smaller cell west of Hat Island within Gilbert Bay (south arm). It also shows a large cell in Gunnison Bay (north arm). All three of these circulating cells are shown moving in a counterclockwise direction (figure 2).

Lin (1976) carried out a drogue study to examine the water circulation within Great Salt Lake. On October 17, 1973, Lin placed a pair of surface drogues and a pair of deeper drogues in the lake 6 mi. (9.65 km) north of the Salt Lake Boat Harbor. These drogues were observed as they drifted for approximately four hours. The information obtained from Lin's field experiment was included in a memorandum to the Great Salt Lake Advisory Board (Lin, 1974) that described two disparate, bi-level currents in the lake. First, Lin suggested that the deep brine layer in Gilbert Bay had a southern cell and a northern cell. The southern cell rotated in a clockwise direction and the northern cell moved counterclockwise. Second, Lin identified a set of surface cells located directly above the deep brine cells that rotated in the exact opposite direction of the underlying cells. Lin also reported that Gunnison Bay had two near-surface cells that were wind driven. The northern cell rotated in a clockwise direction while the southern cell moved in a counterclockwise direction (figure 3). According to Lin, these lake-current patterns in Gilbert and Gunnison Bays are created by northwesterly or southwesterly winds, and the direction of rotation changes with wind direction. It should be noted that Lin's field measurements were taken over the course of only a few hours. They could have been influenced by temporary water movements as well as the surface brines being moved by wind at a faster rate than those deeper in the lake.

Katzenburger and Whelan (1975) conducted a more detailed study of the currents of Great Salt Lake. They stated that the major propagating factor for the lake's currents was river inflow. Their research inferred that the configuration of the lake's topography channeled the inflow of the rivers and thus establishes the counterclockwise rotation found in the south arm (figure 3). Katzenburger and Whelan concluded that inflow from the Bear and Jordan Rivers creates two major counterclockwise currents in the south arm. These currents, they said, have a velocity of approximately 0.115 mph (0.184 kph) during the late summer and early fall with a 20-fold increase to about two knots in the spring during peak runoff. They further indicated that flow through the causeway openings from Gilbert Bay into Gunnison Bay moves in a north-northwesterly direction then diminishes, and that no pattern of rotation has been noted in the north arm (figure 3). Their report also stated that wind-generated currents are rare and diminish rather quickly. The findings in their study were based on data from the researchers' long-time experience on the lake, along with measurements taken with current meters at selected sites.

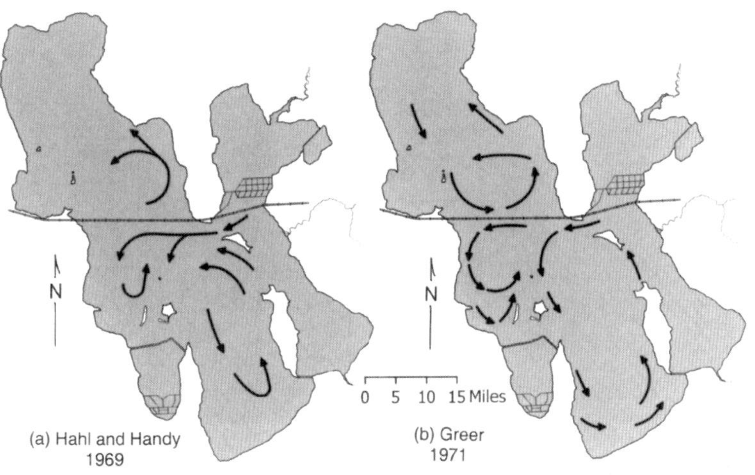

Figure 2. Previous lake circulation studies by a) Hahl and Handy (1969), and b) Greer (1971).

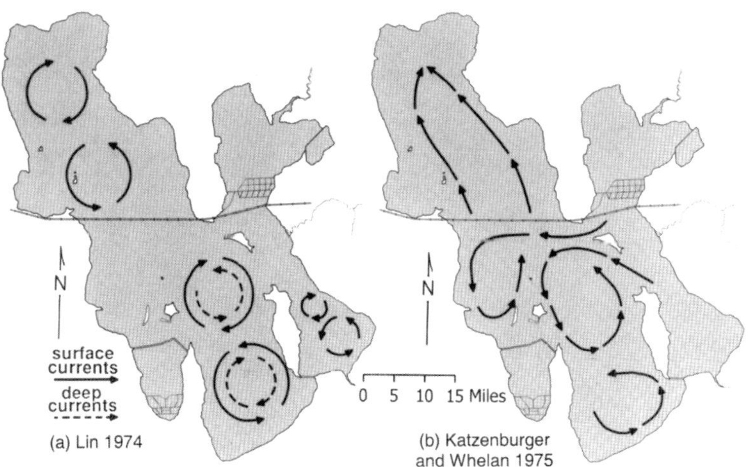

Figure 3. Previous studies concerning lake circulation by a) Lin (1974), and b) Katzenburger and Whelan (1975).

J. Wallace Gwynn of the Utah Geological Survey used a rather unique approach to determine the circulation in the north arm of Great Salt Lake. An experiment conducted on February 22 and 23, 1978, involved tracking dark blue, biodegradable dye with Landsat imagery. The dye was introduced into the north arm of the lake at the Southern Pacific Railroad causeway's west culvert (figure 1). Following a lapsed time of 2 hours the satellite passed overhead recording the position of the dye mark, which had traveled about 1 mile (1.61 km) north and 1.4 mi. (2.26 km) west of the input site. The following day Landsat made a second pass, recording the new position of the dye mark at a location 8 mi. (12.88 km) directly north of the initial entry location. No indication of current movement in either a clockwise or counterclockwise direction could be discerned since there was simply a northerly advancement of the dye from its origin. The dye mark then dissipated and was undetected on subsequent satellite imagery. The report concluded that surface currents and not wind were the major cause for movement of the dye (Utah Geological and Mineral Survey, 1978).

Watters and Kincaid (1979) developed a mathematical, hydrodynamic circulation model of Great Salt Lake, which calculated the flow rate and direction of the lake's currents. The authors applied the circulation model to the south arm of Great Salt Lake, configuring the parameters in relation to water depth, surface elevation, inflow, outflow, barometric pressure, eddy viscosity, fluid density, Coriolis coefficient, and bottom-boundary shear. Due to numerical instability problems, a wind factor was not included as part of the hydrodynamic model. Figure 4 illustrates the current-movement diagram adapted from their report, which shows a distinct gyre in the southern portion of the south arm rotating in a counterclockwise direction. Immediately north of this cell is a U-shaped flow. It begins at the northern tip of Antelope Island and flows southwest. It then curves to the right and flows in a northwesterly direction before turning to the left and traveling in a westerly direction north of Carrington Island. Finally, it bends northward flowing to the railroad causeway where it is calculated to enter the north arm through seepage and flow. The authors referred to their work as premature due to a lack of field data to verify the model. Their recommendations for further study called for intensive gathering of lake velocity and directional measurements in conjunction with simultaneous collection of wind velocity.

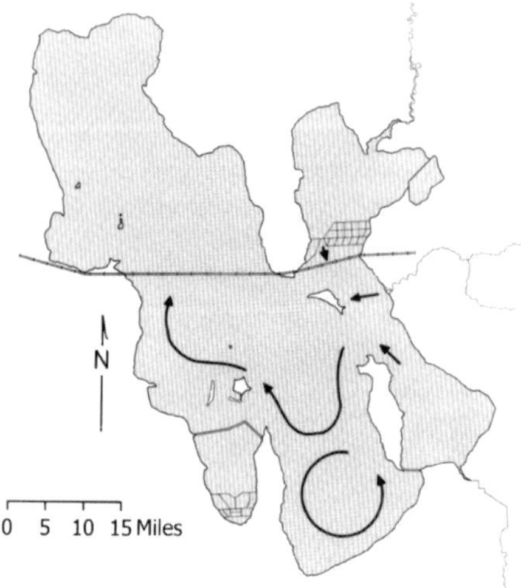

Figure 4. Mathematical hydrodynamic circulation model by Watters and Kincaid (1979).

TECHNIQUES AND DATA ANALYSIS

Satellite-Tracked Drifter Buoys

A 48-day field study was carried out to gather data in-situ on the spatial variations and velocities of lake circulation. The National Oceanic and Atmospheric Administration (NOAA) furnished two satellite-tracked drifter buoys (numbered 2493 and 2494) constructed of steel and containing electronics, and a battery capable of powering the system for up to three months. A transmitting antenna was attached to the top of the drifters with the main housing positioned below (figure 5). A temperature probe was located inside the buoys to record water surface temperature. The drifters were equipped with "window-shade" type drogues which measured 6 by 8 feet (1.83 by 2.44 meters), and acted like underwater sails which caught the currents about 10 feet (3.05 meters) below the water surface. Design of the buoys was such that only a small percentage of the body of the instrument was above the surface of the lake, which dramatically reduced the effect of wind upon it. The surface area of the drogue measured 48 ft^2 ($4.6m^2$) whereas the instrument capsule above water is rounded and < 2 ft^2 (0.18 m^2), resulting in the surface area of the drogue and instrument below the water 25.5 times greater than the exposed portion above the lake surface. Add to this the fact that brine is over 1,000 times heavier than the air above creates a force 25,000 times greater upon the instrument by the brine currents than that of the wind. This ensures that the drifters measure brine direction and speed rather than wind. Each drifter transmitted its position to a satellite six to eight times each day. The transmitted data included the drifter's identification number, total number of messages transmitted from that location (drifter repeats its position), latitude and longitude, data collection time (based on Greenwich Mean Time), location valid time, and water temperature up to a maximum limit of 77°C (24°C). The position of the drifters was determined from the Television and Infrared Observation Satellite (TIROS) and other NOAA satellites. Satellites tracked the buoys using Doppler-shifted radio signals that were transmitted from the drifter system.

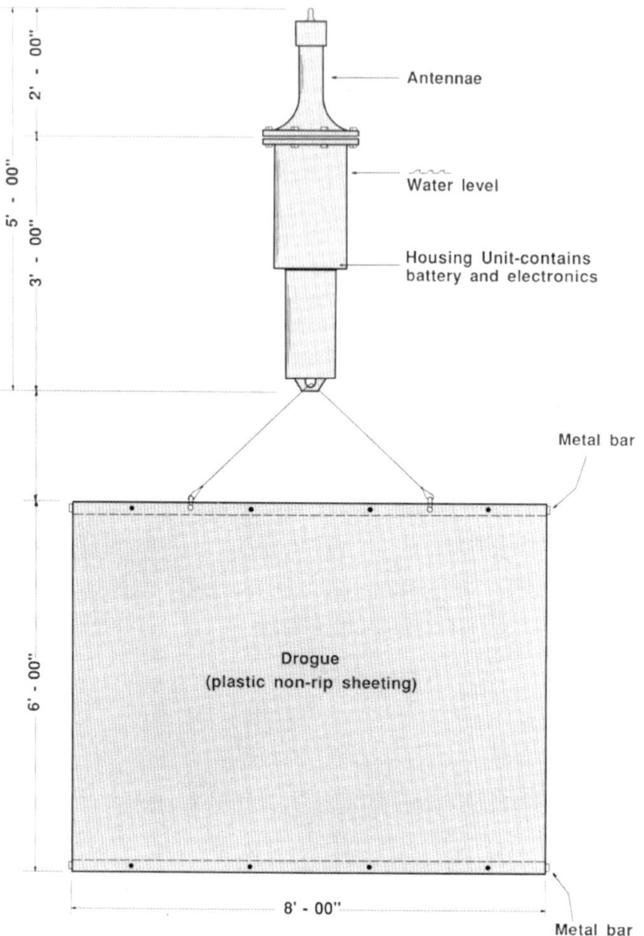

Figure 5. National Oceanic and Atmospheric Administration drifter buoy.

The six-to-eight different locations per day per drifter, transmitted over a period of 48 days, provided a considerable amount of data concerning the two drifter's movements in the south arm. The drifter-position data were plotted on U.S. Geological Survey 7-1/2 minute quadrangle maps (1:24,000 scale) covering Great Salt Lake. The data plots created a series of vectors that delineated the paths of the drifters and thus the various circulation currents and cells. In addition, a 1:125,000 map of Great Salt Lake was utilized to plot daily drifter vectors and provided a more generalized pattern of the drifters' progression. Plotting at 1:125,000 scale required that only one location per day (at approximately the same time) be marked on the map; the time chosen was between the hours of noon and 5:00 p.m. Mountain Daylight Time (MDT). During this time period, the lake normally experienced minimal on shore or off shore breezes that could possibly affect the drifter's position. At times, however, it was necessary to plot more than one position per day on the 1:125,000-scale map to reveal a particular circulation pattern that was seen on the 1:24,000 scale map plots. Once the drifter locations had been plotted at the two different map scales, the spatial-point data distribution could then be analyzed and compared with remotely sensed images. Images were chosen that revealed a distinct circulation pattern made visible by suspended matter in the lake brines. Some generalization of sub-diurnal data points was done to identify basic circulation patterns. Figure 6 is a reduction of the drifter vectors on the 1:125,000 scale map and offers a visual representation of the tracks for both drifters.

Circulation Characteristics

Velocities were calculated for each drifter vector and were subsequently divided into three distinct categories: (1) macro-velocities, those with a calculated rate greater than 0.150 mph (0.214 kph); (2) meso-velocities, those with a calculated rate between 0.075-0.150 mph (0.121 - 0.241 kph); and (3) micro-velocities, those with a calculated rate below 0.075 mph (0.121 kph). The categories were based upon the tabulated velocities and the plotted drifter tracks, revealing varying speeds at different locations in the lake.

The following is a synopsis of the NOAA drifters' movements that occurred over the 48 days from June 11 through July 29, 1991 (calendar day 162 through calendar day 210). The 48-day time period began when the lake was still rising, due to spring run-off, and continued into mid-summer when the lake had peaked and was approaching its lowest level of the year. Figure 6 shows the movement of each drifter divided into three distinctive tracks, each covering a specific period of time.

On June 11, 1991 (day 162), the two NOAA drifters and a weather-monitoring instrument were taken by boat from the shoreline near the city of Syracuse to the north end of Antelope Island. The weather station was set up on top of Buffalo Point, a position near the center of the research area, where wind speed and direction could be recorded simultaneously with the drifter movements.

The drifters were taken out by boat and positioned in the lake at specific latitude and longitude positions using a LORAN locating device. The first track for drifter 2493 began on June 11 (day 162) when it was placed in the lake at 41° 00' 50" N and 112° 19' 00" W. Immediately upon entering the lake the device moved off in a northwesterly direction, dragging its retrieval rope behind it and leaving a series of small eddies in its wake. After its positioning, drifter 2493 followed a north-northwesterly course traveling between Carrington and Fremont Islands. This area of the lake had some of the highest current velocities recorded for drifter 2493 during track 1. These macro-velocities were monitored from June 13 to June 15 (day 164 to 166), and had an average speed of approximately 0.221 mph (0.354 kph). Drifter 2493 then followed a west-northwesterly course to a point just south of the railroad causeway where it turned left 110° and traveled in a south-erly direction into Carrington Bay. Here, a counterclockwise micro-velocity current was found with a measured velocity of 0.067 mph (0.108 kph). The first track of drifter 2493 was finished on July 1 (day 182). A total of 20 days had transpired with 51.55 mi. (82.99 km) being traversed at an average calculated velocity of 0.109 mph (0.175 kph) (figure 6).

Drifter 2494 was placed in the lake 1 mi. (1.61 km) northwest of drifter 2493's starting point, at 41° 02' 30" N and 112° 18' 30" W. It responded to the current in the same manner as drifter 2493. Drifter 2494 moved from its launch site near Buffalo Point into a clockwise eddy northwest of Antelope Island. Within this meso-velocity current, the drifter moved at a rate of approximately 0.118 mph (0.190 kph). The drifter then slid off the edge of this eddy and was carried westward across Gilbert Bay to Carrington Island. Macro-velocities were recorded June 15 to 16 (days 166 to 167) in the mid-section of the south arm with 7.4 mi. (11.9 km) being traveled in 21.7 hours, a rate of 0.341 mph (0.549 kph). The drifter was then swept into a clockwise current off the north end of Carrington Island where it circulated at a micro-velocity rate of 0.042 mph (0.068 kph). Drifter 2494 was subsequently beached on the north shore of Carrington Island and ended its first track on June 22 (day 173). It covered a total of 29.55 mi. (47.58 km) over a period of 11 days recording an average calculated velocity of 0.112 mph (0.180 kph).

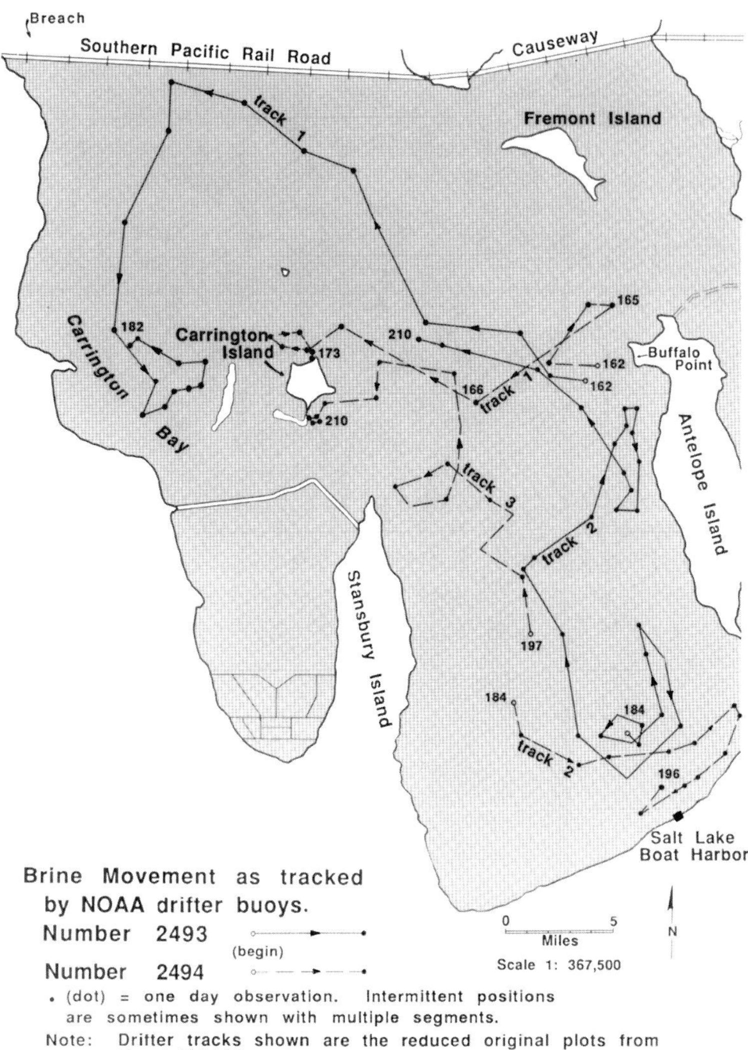

Figure 6. Plotted tracks of NOAA drifter buoys for June 11 - July 29, 1991 (calendar days 162 - 210).

An aerial reconnaissance was undertaken on June 26 (day 177) to locate the drifters. The wind was gusting to 30 mph from the northwest creating rough water-surface conditions. After flying for 1.5 hours over the lake, west of Carrington Island, the reconnaissance finally located drifter

2494 beached on the island's north shore. Drifter 2493 could not be located, even though its position was recorded by satellite as somewhere in Carrington Bay at approximately 40° 59' 42" N and 112° 34' 26" W. Numerous flights were made over the general area but the light-colored device could not be found in the rough waves. A second aerial reconnaissance over the west side of the lake was undertaken on June 28 (day 179). The water surface was then quite calm and it looked hopeful that the lost drifter would be located in Carrington Bay. The area was scanned for 45 minutes without result.

On July 1, the author and field crew ventured across a rough lake in a small boat hoping to retrieve the drifters. Drifter coordinates were entered into the portable LORAN unit, which aided in locating the drifters' positions. After a difficult two-hour mid-lake crossing to Carrington Island, drifter 2494 was retrieved, beached in a cove on the north shore of the island. To locate drifter 2493, lost in Carrington Bay, NOAA representatives were immediately contacted by cellular phone to obtain the latest updated position. With new coordinates programmed into the LORAN unit, the author and crew were guided to the drifter location 7 mi. west of Carrington Island.

Since the drifters had monitored the current patterns and velocities in the region of the lake between Antelope Island and Carrington Bay, they were retrieved and taken to the extreme southern portion of Gilbert Bay near the Salt Lake Boat Harbor. Here both drifters began their second track on July 3 (day 184). Drifter 2493 was placed in the lake 4.5 mi. (7.2 km) northwest of the harbor (40° 46' 44" N and 112° 45' 54" W), and drifter 2494 was taken to a point 5.5 mi. (8.8 km) to the northwest of the starting point of drifter 2493 at coordinates 40° 47' 42" N and 112° 22' 16" W.

Over the next three days, drifter 2493 tracked a small counterclockwise, meso-velocity current in the south portion of Gilbert Bay (figure 6). This meso-velocity current had a rate of 0.084 mph (0.135 kph). Drifter 2493 was then carried north for approximately 5 mi. (8 km) where it made a sharp right-hand turn traveling southward, then southwestward, passing by the Salt Lake Boat Harbor. This drifter was then pulled northward through the center of Gilbert Bay where it reached macro-velocities as high as .173 mph (.279 kph). It then slowed down in a right-hand turn and accelerated again northeastwardly toward Antelope Island. Upon reaching the western shoreline of Antelope Island, drifter 2493 was caught in an elongated micro-velocity clockwise eddy. This circulating current had a calculated rate of 0.049 mph (0.079 kph). Drifter 2493 moved out of this shoreline cell where it was then caught up in a northwesterly macro-velocity current moving through the mid-lake region towards Carrington Island. During the second track of drifter 2493, completed on July 29 (day 210), it traveled a total of 62.15 mi. (100.06 km) in 26 days at an average rate of 0.101 mph (0.163 kph) (figure 6).

Drifter 2494 began its second track (day 184) moving southward from its input location (figure 6). As it approached the Salt Lake Boat Harbor it veered into an elongated clockwise current. Within this micro-velocity clockwise eddy, the drifter traveled at an average speed of 0.070 mph (0.113 kph). Track 2 for this drifter had to be terminated earlier than anticipated (July 15, day 196) because it was circulating too close to the boat harbor. During its second track, a total of 21.95 mi. (35.2 km) was covered over the course of 12 days at an average calculated velocity of 0.076 mph (0.122 kph). On July 16 (day 197) a return trip was made to the area to retrieve and reposition drifter 2494 east of Stansbury Island at approximately 40° 50' 00" N and 112 ° 22' 00" W.

Track 3 for drifter 2494 began in the center of the south arm at coordinates 40° 52' 42" N and 112° 21' 29" W and appeared to follow a large-scale current in the middle of Gilbert Bay that trended in a counterclockwise direction. After traveling nearly 5 mi. (8 km) in this current, the drifter made a sharp left-hand turn and seemed to track a different counterclockwise cell, which was located northeast of Stansbury Island. The drifter then moved out of this macro-velocity flow and pursued a northerly and then westerly track, eventually beaching itself on Carrington Island. On July 29 (day 210), an aerial reconnaissance was made over the island, and after a prolonged search the device was discovered in a shallow area off the southern shore. On the return flight back to the airport, drifter 2493 was spotted between Carrington and Antelope Island moving in a westerly direction. During the course of track 3, drifter 2494 covered 32.65 mi. (52.57 km) in 13 days, finishing its course on July 29 (day 210). The average calculated velocity for this track was 0.108 mph (0.174 kph).

On July 31, 1991 (day 212), the research boat was launched for the final time to collect the drifters. The boat departed from the causeway near Syracuse and cruised 1.5 hours to arrive in the vicinity of Carrington Island. Approximately 1 mi. (1.61 km) before reaching the island, drifter 2493 was sighted and retrieved. About 2 mi. (3.22 km) southwestward from Carrington Island, in a very shallow area of the lake, drifter 2494 was located. When this device was brought into the boat, it was discovered that the complete drogue was missing having probably been torn off crossing the island's shoals.

An immediate departure took place due to an approaching storm front. Arriving at Antelope Island some fifty minutes later, a quick hike was made to the top of Buffalo Point to collect the portable weather station. Once the drifters and weather equipment were secured in the boat, the return trip was made back to the Syracuse launch site. Even though a storm system was moving into the area, the water surface of Great Salt Lake was very calm, making conditions for this final research expedition favorable.

In reviewing the various tracks it was evident that when the drifters became caught in eddies, principally near the shoreline of the lake, they moved at a much slower rate than they did in the currents found in the mid-lake region. The reduced velocity resulted from shoreline friction with the lake brines. It may be argued that the drifter and its underwater sail moving near the shoreline could have dragged lake bottom causing reduced velocities to be measured. Fortunately, the drogues are specially designed to tear away when dragging occurs, causing the drifter to loose stability and be blown onshore. This was demonstrated when drifter 2494 came too close to Carrington Island shoals, lost its drogue, and was beached.

The locations where the speed of the drifters registered the most rapid lateral movement occurred in the mid-lake region of Gilbert Bay, in the area between Carrington and Fremont Islands, and in the region of the lake between the railroad causeway southward to Carrington Bay. These relatively swift movements could be the result of reduced shoreline friction, and also the greater momentum of a large, open body of water. It appears that a major confluence for the lake currents exists around Carrington Island. Even though the drifters were placed in the lake at five different locations they were retrieved near or on Carrington Island four times.

There was no evidence for any strong attraction of the drifters to the 300-foot (90-meter) breach at the western end of the Southern Pacific Railroad causeway, where water flows from the south arm to the north arm. The first track for drifter 2493 followed a macro-velocity current towards the breach, but when it came within 5 mi. (16.4 km) of that area the drifter turned in a counterclockwise direction to the left and followed the current into Carrington Bay. The researcher believes that the relative size of the breach is too small to affect or alter the course of the broader circulation pattern found in the south arm.

Spatial Pattern of Drifter Buoys Compared to Satellite Imagery

Satellite images from Landsat and Large Format Camera (LFC) pictures taken from the space shuttle (figure 7) were obtained for this study from various sources. Some of the images show quite distinct spatial patterns of horizontal diffusion in the Great Salt Lake's surface movement. Different types of suspended matter carried by the lake's currents have different spectral responses than the surrounding water and are characterized by different tones on the image. The contrast created by these spectral differences produces a distinguishable pattern that can be outlined on the image. The suspended matter could be brine fly casings, sediment, brine shrimp, duneliella, bacteria, or possibly variations in salinity.

The pattern of surface movement within Great Salt Lake can be seen quite clearly on some remotely sensed imagery. Figure 8 is a reproduction of a Landsat Thematic Mapper

Figure 7. *Large Format Camera image of Great Salt Lake (May, 1985).*

(TM) image that reveals some very distinct circulation patterns in the south arm. This image was obtained by the Landsat 5 satellite on July 27, 1984, and recorded spectral energy from Great Salt Lake using TM Band 2. Band 2 senses wavelengths emitted in the 0.52 - 0.60 micrometer range of the electromagnetic spectrum. These shorter wavelengths, rather than being absorbed, are more readily scattered by suspended particles in the water, which allows the sensor to detect differences in the spectral response of the foreign matter versus the water (Campbell, 1987).

The plots of the NOAA drifter movements recorded from June 11 to July 31, 1991 were overlaid onto the Landsat Thematic Mapper image to compare the location of the circulation cells seen on the photo to the position of drifter tracks. Similarities exist between the two individual figures, with cells that were detected by the drifters matching to a fairly high degree the circulation patterns in the Landsat image. Figure 9 offers a visual representation of the relationship between the drifter tracks and the Landsat circulation patterns.

Analysis of the relationship between the Landsat TM image and the drifter tracks seems to reveal two large counterclockwise circulation cells in Gilbert Bay, and another counterclockwise circulation cell in Carrington Bay. Two elongated clockwise cells are evident, with one near the southern shore of the south arm and the other along the western shore of Antelope Island.

Remotely sensed images taken on other dates show only one large counterclockwise cell in the south arm. The reasons for this difference could be due to seasonal variations or many other factors, but since the drifter movements for this June 11 through July 31 study period corresponded with the two-cell image, it could possibly be a phenomena that occurs during the summer months. One of the aims of this study was investigating the general long-term counterclockwise motion and not the factors that influence one or two-cell circulation in the south arm.

Figure 8. *Landsat Thematic Mapper (TM) band 2 taken 27 July 1984.*

CYCLONIC WIND STRESS THEORY

The long-term motion of the south arm of Great Salt Lake, as evidenced by the NOAA drifters, is one of circulation by macro-velocity currents in a counterclockwise direction. This overall counterclockwise movement can be explained, in part, by research conducted by Emery and Csanady (1973). Their research involved reviewing the surface circulation of 40 lakes located in the northern hemisphere, with all but one found to have a long-term counterclockwise movement. The authors explain that inflow and lake bottom topography seem to have very little affect on the lake circulation patterns. They state instead that water movements in some lakes are generated mainly by the drag of wind blowing across the water surface. The authors mention, for example, that on a windy day one can observe a smooth area of water adjacent to an area studded with capillary waves on the surface. This phenomenon occurs where a stable layer of air resting on top of the water surface, which is colder by several degrees than the wind above, suppresses turbulence and reduces wind drag to zero. As the wind warms the overriding layer of air and thus the upper portion of the water surface, the temperature differential between

Figure 9. Landsat (TM) image with overlay of NOAA drifter buoy tracks.

with colder temperatures found near the left-hand shore, then a progressive warming as one moves towards the right-hand shore. The temperature of the water has a direct relationship to the temperature of the layer of air residing directly above the surface. Under these conditions a cyclonic wind-stress curl will normally be created. The left-hand shore will have wind drag suppressed by the cool layer of air directly above the cold water surface. Warmer water surface and air temperatures near the right-hand shore allow surface drag to occur, propelling the water downwind. A vacancy is created near the right-hand shore as the surface water moves downwind. Subsequently, this void will be filled by surface water moving in from the left, thus establishing a cyclonic (counterclockwise) wind stress curl (figure 10).

South Arm Water Surface Temperatures

To help substantiate the idea of a cyclonic wind-stress curl, which could be the genesis of a long-term counterclockwise rotation, Great Salt Lake water surface temperatures recorded by the NOAA drifters were analyzed. The drifters contained thermal sensors that recorded surface water temperature data simultaneously with position data. Since the sensor inside the drifters had an upper limit of 77°F (24.88°C) any temperature exceeding that limit was not recorded. Temperatures were, therefore, recorded at the beginning of the research project on June 11, but as summer progressed and Great Salt Lake warmed, the recorded temperature data became sporadic. Temperature data ceased completely on July 11 as upper water surface temperatures exceeded the drifters' limits.

Figure 11 shows the isotherm map that was constructed from the temperature data for the northern portion of Gilbert Bay between Antelope Island and Carrington Bay. It was evident from the data that a consistent temperature difference existed during the period of research between the east (right-hand) and west (left-hand) shore of the lake. Higher temperatures were recorded on the east shore of the study site near Antelope Island, which then began decreasing westward, with the lower temperatures being found on the west shore near the Lakeside Mountains. This temperature difference, according to Emery and Csanady (1973), is the result of air movement across the lake that warms the water surface, decreasing temperature differential allowing wind stress to move the water downwind. Ekman Drift deflects this warmer water, which is being propelled forward, to the right.

these layers decreases and greater drag is produced. With the wind dragging the warmed water Ekman Drift will then force this water to the right of the airflow (looking downwind).

Ekman Drift is the effect of Coriolis on water that is being propelled forward by the wind (Pickard and Emery, 1982). Coriolis results from the rotation of the earth and has an apparent tendency to alter the course of any moving object that is on or near the earth's surface. In the northern hemisphere, the alteration or deflection will be to the right of the path of travel. With the shifting of the upper water layers to the right, due to Ekman Drift, an up-welling of colder water near the left shore will occur. The up-welling will create an increased surficial temperature difference across the lake

Since higher water-surface temperatures were found near Antelope Island, this would indicate that the principal wind direction over the lake was from a general southerly direction during this summer period. Because this study deals with the lake's circulation patterns, it must also, of necessity, explore the wind patterns in and around the southern portion of the lake in order to better explain how they interact with the circulating motion of Great Salt Lake.

Lake Circulation and its Relationship with Wind

To determine the principal wind direction over the south arm of Great Salt Lake, both contemporary and historical wind data were reviewed. Contemporary wind data were collected by a portable weather station set up on Buffalo Point near the research site. This weather data indicated that the area near the northern tip of Antelope Island has two basic wind patterns: (1) a high-velocity wind flowing from the southwest, and (2) a lighter flow from the northeast. The wind rose for the Salt Lake International Airport by contrast indicates a general wind direction from the southeast that is due principally to air flowing out of the Jordan River drainage basin. Farmington Bay on the other hand has a general wind direction from the south-southeast resulting from both frontal activity and wind drainage out of the Jordan River Basin to the south of the lake (Peck and Richardson, 1966). Wind patterns for the extreme southern portion of the lake at Lake Point Junction (Silver Sands), near the northern end of the Oquirrh Mountains, reveal a summer wind condition that is principally from the west-southwest (Wasatch Front Regional Council, 1980).

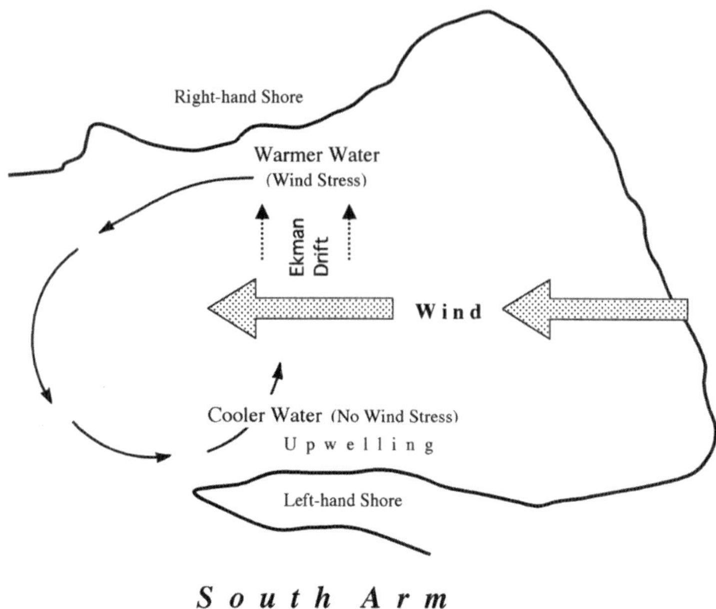

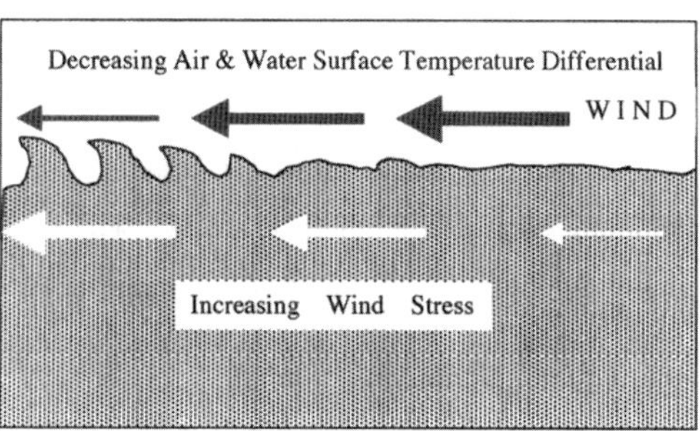

Figure 10. Schematic illustrates decreasing temperature differential and increasing wind stress as wind flows across a water surface. Map shows the effect of Eckman Drift, wind stress localities, and the induced cyclonic wind stress curl on the south arm brines.

The composite wind conditions for the south arm of Great Salt Lake, as interpreted from both the Buffalo Point station and historical wind data is southerly in direction. As this wind moves northward across the south arm of the lake, the temperature differential between wind and water decreases, creating surface drag that begins to move the brine downwind. The warmed brine is then forced to the right by Ekman Drift causing the east shore (Antelope Island) to be warmer and upwellings of cooler water to occur near the west shore (Lakeside Mountains). This temperature differential causes increased wind drag along the right-hand shore (looking downwind), as opposed to the water near the left-hand shore with cooler temperatures that suppress wind stress. Hence as air flows from south to north, wind drag is at its greatest near Antelope Island setting up conditions for an induced cyclonic wind-stress curl that forces the water into a counterclockwise direction (figure 10).

The second highest wind component on Great Salt Lake is from the north as measured by the Buffalo Point weather station. In accordance with the cyclonic wind stress theory (Emery and Casandy, 1973), the same counterclockwise circulation will be induced whether the wind flow is from the north or south. When a northerly wind moves across the lake the differential in air temperature versus water temperature decreases, creating surface drag that begins to move the brine downwind. Ekman Drift then forces the warmed water to the right (Lakeside Mountains) increasing drag along the western margin of the lake. The cooler water on the left-hand shore (Antelope Island) suppresses wind drag and will circulate toward the right-hand shore to replace the brine that was curled to the right on the western shore setting in motion a counterclockwise circulation. Thus, wind from the two dominant compass directions flowing across the lake each produce a similar cyclonic wind stress curl that induces a counterclockwise movement of the brine in the south arm of the lake.

COMPARISON OF RESULTS TO PREVIOUS RESEARCH

Analysis of the plotted tracks from the NOAA drifter buoys, used in conjunction with satellite imagery, revealed not only spatially distinct circulation patterns throughout the south arm of Great Salt Lake, but currents and cells that were distinct due to velocity. Three different rates of movement within the south arm were established: (1) macro-velocities; (2) meso-velocities; and (3) micro-velocities. The circulation currents and cells that were tracked by the drifters were

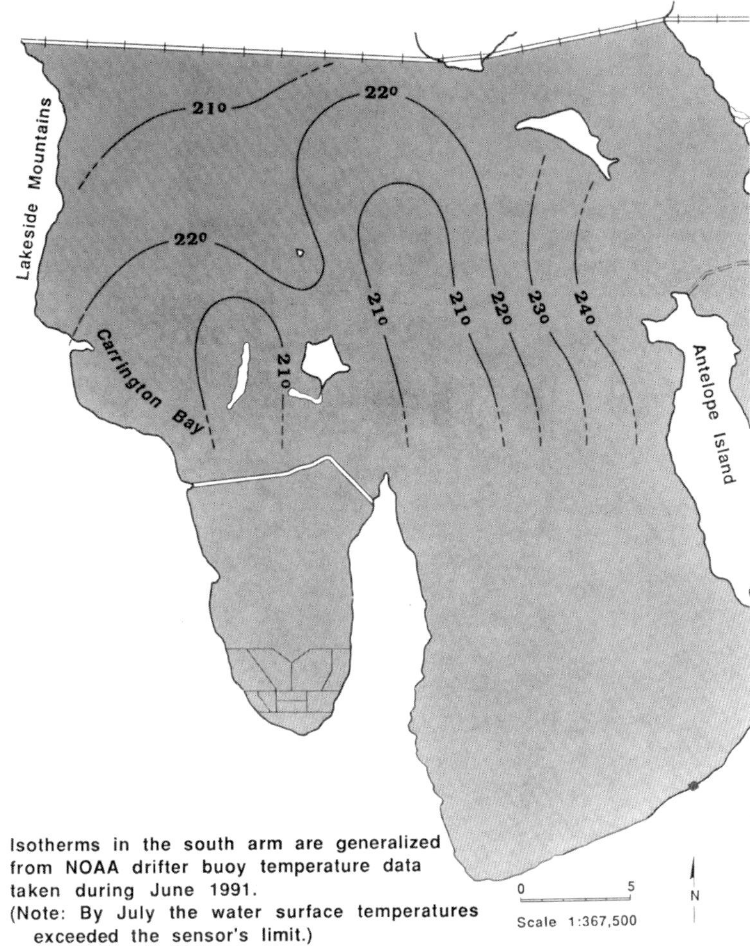

Figure 11. *South arm surface temperatures (°C) between northern Antelope Island and Carrington Bay.*

assigned a particular rate based upon individual speed of movement. Four micro-velocity circulation cells were found along the periphery of the south arm. One was located along the western shore of Antelope Island, a second was just 2 mi. (3.2 km) north of Carrington Island, a third along the southern shore near the Salt Lake Boat Harbor, and a fourth in the southern portion of Carrington Bay. In addition, two meso-velocity circulation cells were observed in Gilbert Bay. One was positioned approximately 4 mi. (6.4 km) north of the Salt Lake Boat Harbor, and the second near the northwest point of Antelope Island. Three macro-velocity circulation patterns were evident from the data. Two of these were situated between Antelope and Stansbury Islands, with the northern circulating cell defined in both the imagery and the drifter tracks. The second circulating cell located directly south of the northern gyre is evident on the Landsat imagery, but was only partially shown by the drifters' tracks. It can be seen in figure 9 that drifter 2494 began a partial track of this cell, but was probably located on the outer edge and was subsequently deflected to the right and propelled out of the fast-moving current by centrifugal effect. The third macro-velocity flow was tracked moving towards the railroad causeway in a northwest flow between Fremont and Carrington Islands. The drifter then followed a counterclockwise flow toward the west shore of the lake and ended in a micro-velocity cell. This current has the potential of being one of the largest circulating cells in the lake, but due to the location of Carrington Island and the dike connecting it to Stansbury Island, a barrier is created which curtails the flow of brine and allows only a moderately sized circulation cell to be formed in Carrington Bay. Analysis of the imagery and drifter tracks shows this flow beginning as a westward movement from an area between Fremont and Antelope Islands, consistent with the inferred currents suggested by the literature. The macro-scale counterclockwise currents found in Gilbert Bay and in the northern region of the south arm make up the principal long-term motion of the lake, as shown by the fact that the drifters traveled the greatest distances at the highest velocities in these areas. These macro-scale currents flow at an average rate of 0.215 mph (0.346 kph), nearly three times the velocity of the micro-scale currents that exist mainly near the shorelines.

The micro-velocity cells that exist near the periphery of the macro-velocity currents seem to be by-products of the long-term counterclockwise motion found in Great Salt Lake. These micro-velocity cells probably begin as water that, due to centrifugal effect, slides off the main cyclonic cell and is deflected to the right, establishing a clockwise cell that flows in the opposite direction to the main circulation cells. All but one of the four observed micro-scale circulation cells have a clockwise direction and are located near the shoreline. A counterclockwise micro-velocity cell can be found in Carrington Bay, and is part of a fast-flowing current that moves along the western shore before entering this confined area. The friction from the surrounding topography reduces the velocity of the counterclockwise cell. It is probable that part of the macro-velocity current moves around the outer edge of this cell and into the area northwest of Carrington Island. The micro-velocity cell positioned north of Carrington Island could possibly be generated from brines deflected from this current. Figure 12 graphically displays the results of this research.

A majority of the previously mentioned published reports on circulation patterns in the south arm of Great Salt Lake agree that the long-term motion here has a counterclockwise direction. Some of the circulating cells described in the south arm have different locations compared to this study, but nonetheless, they concur on the general direction of flow.

Two works deserving special note, however, are the reports by Lin (1974) and Watters and Kincaid (1979). Both reached findings that are contradicted by the data in this study. Lin states that a clockwise cell exists in the south arm, located between Antelope and Carrington Islands (figure 3). During the course of this NOAA drifter field study, however, a counterclockwise cell was detected in that particular region. The mathematical, hydrodynamic circulation model adapted to Great Salt Lake by Watters and Kincaid calculated a current that flows clockwise from between Fremont and Antelope Islands to Carrington Island. The flow continues northerly toward the causeway (figure 4). This direction is contradictory to the movement tracked by the NOAA drifters in this portion of the lake. The tracks of the drifters indicate

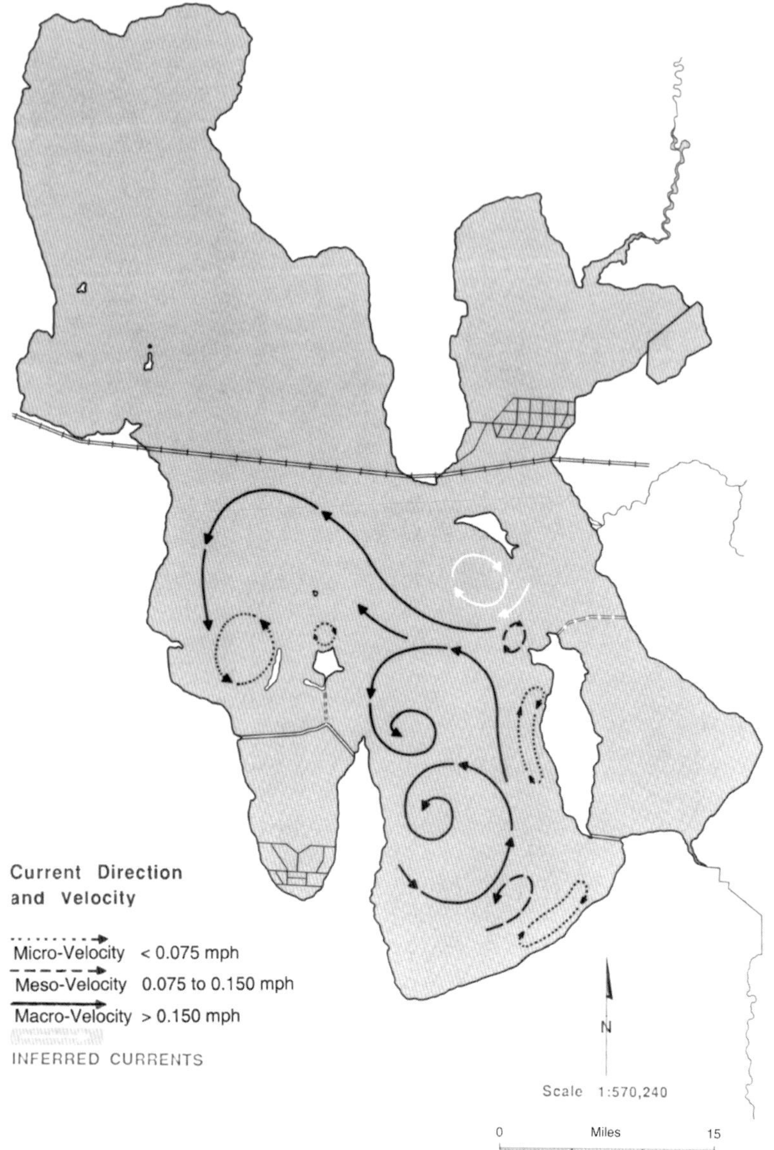

Figure 12. Great Salt Lake late spring and mid-summer circulation as interpreted by author.

a counterclockwise current and a southerly flow away from the causeway.

Two of the five previously published reports on Great Salt Lake circulation include velocity measurements for the south arm currents. Lin (1976) monitored the mid-lake current with a velocity of 0.22 mph (0.356 kph). This rate is greater than the velocities indicated by the NOAA drifter data for the same vicinity. Analysis of the drifters showed a counterclockwise cell in this region of the south arm with an average velocity of 0.102 mph (0.164 kph). Lin's research covered a time span of 4 hours whereas the research using NOAA drifter buoys was conducted over a period of 48 days.

Katzenburger and Whelan's (1975) study stated that the circulation of the lake had a velocity of 0.115 mph (0.184 kph) during late summer, a decrease from the 2.3 mph (3.7 kph) during spring run-off. The satellite-tracked drifters determined that the overall late summer velocity was close to 0.100 mph (0.161 kph), which is relatively close to Katzenburger and Whelan's calculations. However, their calculated spring run-off velocity of 2.3 mph (3.7 kph) seems rather high. Nowhere in the south arm did the NOAA drifters detect velocities at this speed. The highest velocity rating was 0.341 mph (0.549 kph) for the period of June 11 through July 31. It is believed by this author that Katzenburger and Whelan's high-velocity measurements, taken in the south arm near the causeway, were in fact measuring fast-flowing surface water coming from the Bear River, which was lensing westward across the higher density lake brines.

The views of previous authors differed as to why the lake circulates in a counterclockwise direction. Hahl and Handy (1969) stated that an interrelationship exists between inflow, wind, and currents. Lin (1976) said that northwesterly or southwesterly winds tend to set up the rotation of the lake, and Katzenburger and Whelan (1975) concluded that tributary inflow drives the currents in a counterclockwise direction.

CONCLUSIONS

The circulation patterns and velocities described in this investigation were determined principally from data compiled during late spring and early summer 1991. The data, which were relayed from two NOAA satellite-tracked drifter buoys, were used to compile digitized maps. A portable weather station was placed on Antelope Island to determine mid-lake air temperature, direction, and velocity during the study period. Information gained from these instruments was supplemented by satellite imagery from Landsat and the space shuttle. The data obtained over a 48-day collection period were used to determine the method of propagation, temperature, speed, and direction of the near surface brines in the south arm of Great Salt Lake. Analysis of the NOAA drifter plots was used in conjunction with satellite imagery to create a composite map that revealed spatially distinct circulation patterns and velocities of brine movement throughout the south arm of Great Salt Lake. This information contradicted most of the previous research on rate of flow, location of gyres and the motive force propelling the lake brine. The strengths of the study were that it was based upon a long-term (48 day) analysis of the lake circulation patterns using geo-positioning buoys supported by satellite imagery and weather data.

The most important finding of the study was the relative motion of the largest gyres in the south arm of the lake. Given the northern hemisphere location of the lake it is expected that the movement of brine would be in a clockwise direction due to the Coriolis effect. This research found a counterclockwise motion in the central portion of the south arm of Great Salt Lake driven by Ekman Drift, which induced wind stress curl. This counterclockwise circulation results from the lake's thermal energy being radiated to the air above it, equalizing the temperature between the air and the brine. Wind stress is increased as the temperature difference narrows forcing the water to be pulled along by the air's movement. As the brine begins to move downwind, Eckman Drift deflects it towards the right-hand shore (looking downwind) inducing upwelling near the left-hand shore, which results in a temperature differential between the two. As air moves along the lake surface wind stress readily carries the right hand brines down wind. Alternatively, the cooler brines on the left hand suppress drag and move in to replace the vacancy created by the shifting right-hand brines establishing a counterclockwise rotation. Water surface temperatures within the south arm as monitored by the drifter buoys, in conjunction with recorded wind velocity and direction, support this explanation for the motive force producing the counterclockwise circulation pattern found in the lake. This explanation for the direction of the major south arm currents was a conclusion not reached by any previous study on the lake currents. In addition, the speed of the currents was determined from drifter buoy locations and was found to vary with the relative size of the gyres created by the brine movement. The fastest currents were the mid-lake gyres, moderate speeds were typical of medium-sized gyres located in sheltered bays, and the slowest speeds were in eddies generated by the larger gyres near the lake's shorelines. This study is the first to provide long-term synoptic data on the location, velocity and direction of brine currents in the south arm of Great Salt Lake. Because the circulation patterns and velocities determined in this investigation were based upon data collected during late spring through mid-summer no claim is made for other seasons or times. Further research into the circulation patterns of the lake is necessary in order to broaden our understanding of the dynamic processes of this unique body of water under long-term, year round conditions.

ACKNOWLEDGMENTS

The author wishes to acknowledge and thank the following individuals and institutions who made this research possible: R. Barbaro, D.C. Greer, G. Miller, T. Miller, R. Muzzi, R. McCoy, M. Ridd, J.W. Gwynn, Department of Defense, Great Lakes Environmental Research Laboratory, National Oceanic and Atmospheric Administration (NOAA), Utah Geological Survey, and Weber State University.

REFERENCES

Campbell, J.B., 1987, Introduction to remote sensing: New York, Guilford Press, 551 p.

Emery, K.O. and Csanady, G.T., 1973, Surface circulation of lakes and nearly land-locked seas: Proceedings National Academy of Science, v. 70, no. 1, January 1973, p. 93-97.

Greer, D.C., 1971, Great Salt Lake, Utah: Annals of the Association of American Geographers, Map Supplement 14, v. 61, no. 1.

Hahl, D.C. and Handy, A.H., 1969, Great Salt Lake, Utah; chemical and physical variations of the brine, 1963-1966: Utah Geological and Mineral Survey Water-Resource Bulletin 12, 33 p.

Katzenburger, W.M. and Whelan, J.A., 1975, Currents of the Great Salt Lake, Utah: Utah Geology, v. 2, no. 2, p. 103-107.

Lin, Anching, 1974, Circulation patterns of the lake, a state of knowledge: Memorandum to State Interagency Technical Team of the Great Salt Lake Policy Advisory Board, dated July 26.

—1976, A survey of physical limnology of Great Salt Lake: Utah Division of Water Resources, Department of Natural Resources, 82 p.

Peck, E.L. and Richardson, E.A., 1966, Hydrology and Climatology of Great Salt Lake, *in* Stokes, W.L., editor, Guidebook to the Geology of Utah, The Great Salt Lake: Utah Geological Society no. 20, 173 p.

Pickard, G.L. and Emery, W.J., 1982, Descriptive Physical Oceanography - an introduction, 4th edition: New York, Pergamon Press, 249 p.

Utah Geological and Mineral Survey, 1978, Great Salt Lake study of surface currents: Utah Geological and Mineral Survey, Survey Notes, v. 12, no. 3, p. 1.

Wasatch Front Regional Council, 1980, Great Salt Lake Air Basin Wind Study, February 25, 1980: Wasatch Front Regional Council, 79 p.

Watters, G.Z., and Kincaid, C.T., 1979, A mathematical hydrodynamic circulation model of Great Salt Lake for resource management: Utah Water Research Laboratory WR-196, Logan, Utah, 74 p.

GSL AND ASSOCIATED INDUSTRIES

THE ECONOMICS OF GREAT SALT LAKE

by

Alan E. Isaacson
Research Analyst

Frank C. Hachman
Associate Director - Retired

R. Thayne Robson
Director

Bureau of Economic and Business Research
University of Utah
1645 East Campus Center Drive, Room 401
Salt Lake City, UT 84112

INTRODUCTION

Great Salt Lake (figure 1) is one of the most distinguishing and prominent features of the Utah landscape, and one of the features for which Utah is most widely known. This inland lake receives drainage from surrounding mountain ranges and has no outlet other than through evaporation. Water in the lake has a high saline content which fluctuates as does the lake level with cycles of wet and dry periods. The lake ecology has fascinated explorers, residents, scientists, and Utah citizens as long as people have lived near Great Salt Lake.

This essay discusses in relatively broad terms the economics of Great Salt Lake. The term "economics" is used to mean the volume and value of transactions involving uses of the lake for mineral production, recreation, tourism, wildlife habitat, as well as the expenditure of public and private resources to sustain related ecosystems. Some may suggest that there are economic values for which no transactions have occurred in the past, and for which may not occur in the future either. Economists are likely to argue that alleged values for which no market tests exist in the form of actual transactions are uncertain and difficult to quantify. However, changes are occurring in the willingness to spend money and resources on the preservation of ecological assets, and in the methods of assessing the economic value of species, habitats, and ecosystems that will, in all probability, change the willingness of both public and private parties to spend money and resources on environments such as Great Salt Lake.

This essay reviews the economic transactions related to mineral production, recreation, wildlife management, and ecosystem management as reflected by actual economic transactions. These transactions take many forms, including the production and sale of minerals, the harvesting and sale of artemia (brine shrimp), the construction and operation of duck clubs, wildlife preservation and enhancement operations, and the use of the lake for boating and other forms of recreation. The existence of the lake and the islands within the lake have provided beaches and a broad range of recreation and wildlife activities. For some of these activities, both state and federal agencies have budgets and expenditures to promote recreation, wildlife, and scenic experience that the general public appears willing to support.

The reader should be aware that the nature of the transactions vary considerably from governmental budgetary expenditures to actual production, prices, and sale of minerals. Some economic activities, such as the operation of private duck clubs, are interesting not only for the size and extent of investments and the costs of maintaining and operating, but also for the apparently private use of a publically owned resource. But the same principles apply to mineral production and other forms of recreation. Given the apparent private use of public resources, some important economic questions arise regarding the use of these public resources. There are many important economic questions that could be evaluated. How much should the public charge for the various economic uses made of the lake in terms of royalty payments, and user fees, etc.? Do those who manage the lake have reasonable market tests that are applied in setting levels of use and the prices charged? How could or should the management of the resources associated with Great Salt Lake be enhanced in sustainable economic value terms? Have traditional economic interests been able to preserve a favored status that may be inconsistent with the best protection of the resource over the foreseeable future? However, many of these questions go beyond the scope of this essay. The objective here is to identify and describe economic transactions related to Great Salt Lake resources and to raise some questions that result from these economic activities.

GREAT SALT LAKE MINERALS PRODUCTION

Mineral production is perhaps the area where Great Salt Lake has the greatest direct impact on Utah's economy. Several mineral operations depend on the lake for feed material. Salt (sodium chloride) quickly comes to mind when minerals from Great Salt Lake are mentioned. However, several other minerals and by-products are also produced from the lake (table 1). These include sulfate-of-potash (SOP) or potassium sulfate, magnesium chloride brine and solid flake, magnesium metal, and chlorine gas. Five companies currently produce minerals directly from Great Salt Lake (table 1). Great Salt Lake provides feed material for 50 percent of the U.S.'s primary magnesium production, and is one of two locations in the U.S. where agricultural sulfate-of-potash is produced, in addition to the more commonly known salt recovery operations. The total value of minerals produced from the lake is approximately $300 million annually, or about 17 percent of the value of all minerals produced annually in the state of Utah (U.S. Geological Survey, 1996).

IMC Global purchased the former Great Salt Lake Minerals Company, part of the Harris Chemical Group, Inc., from D. George Harris and Associates in April, 1998. IMC Global operates in Utah through its IMC Kalium and IMC Salt subsidiaries. Cargill Salt is an agricultural products firm based in Minneapolis, MN. Cargill purchased the North American Salt operations of Akzo Nobel in 1996. Morton International is a commodity and speciality chemical company based in Chicago, IL with a long history in the salt business. Magnesium Corporation of America (Magcorp) is a wholly owned subsidiary of New York-based Renco Group, Inc. North Shore Limited Partnership produces magnesium chloride brines which it markets as nutritional supplements.

Soluble Salts

Sodium chloride, common salt, was the first mineral commodity produced from Great Salt Lake. Currently, there are three salt producers operating around the lake. These companies are IMC Global in Weber County, and Morton International and Cargill Salt in Tooele County. Several other salt operations have gone out of business over the past several decades. In 1978, there were seven salt companies operating on Great Salt Lake. These companies were: Great Salt Lake Minerals in Weber County; American Salt Company, Hardy Salt Company, Lakepoint Salt Company, and Utah Salt Company, all in Tooele County; Lake Crystal Salt Company in Box Elder County; and Morton Salt Company in Salt Lake County. Several of these operations have closed, while others have either been purchased or moved.

Lakepoint Salt was sold to AMAX, Inc., in 1986, to market salt produced in AMAX Magnesium's evaporation ponds. AMAX renamed the company Sol-Aire Salt and Chemical Company. AMAX sold Sol-Aire Salt on January 1, 1987 to Diamond Crystal Salt of St. Clair, MI, for $800,000 and future royalties. In July of 1987, severe market competition and other factors prompted Diamond Crystal to sell its salt division, including the newly purchased Utah operation, to International Salt Company, a unit of Akzo NV of the Netherlands. Akzo continued to operate the facility until July, 1997, at which time Akzo sold its Akzo Nobel Salt Inc., subsidiary to Cargill Inc.'s salt division. Cargill is a large, privately held agricultural products company based in Minneapolis, MN. Cargill is supplied brine for salt recovery through an agreement with Magcorp, and does not remove brine or water directly from the main body of the lake. The facility is located at the southwest corner of the lake, just west of the Timpie Springs Waterfowl Management Area (figure 1).

Morton Salt operated in Salt Lake County until 1991. During 1991, Morton sold its Saltair plant in Salt Lake County to allow for the Kennecott Copper tailings pond expansion and purchased the North American Salt Plant farther west in Tooele county. North American Salt invested the proceeds of the sale in the Great Salt Lake Minerals Company facility (a sister company to North American Salt) and, in return, received the right to market salt by-products produced by Great Salt Lake Minerals. Morton Salt subsequently moved its production facility farther west in Tooele County. The current facility is located on the southern shore of the lake, south of Stansbury Island (figure 1).

In addition to sodium chloride (common salt) several other soluble salts are being produced, or have been produced in the past, from Great Salt Lake brines. These salts

Table 1. Currently operating mineral producers on Great Salt Lake.

Operator	Product					Location
	Chlorine Gas	Magnesium Chloride	Magnesium Metal	Salt	SOP	County
IMC Global		X		X	X	Weber
Cargill Salt				X		Tooele
Morton International				X		Tooele
Magnesium Corp. of America	X		X			Tooele
North Shore Limited Partnership		X				Box Elder

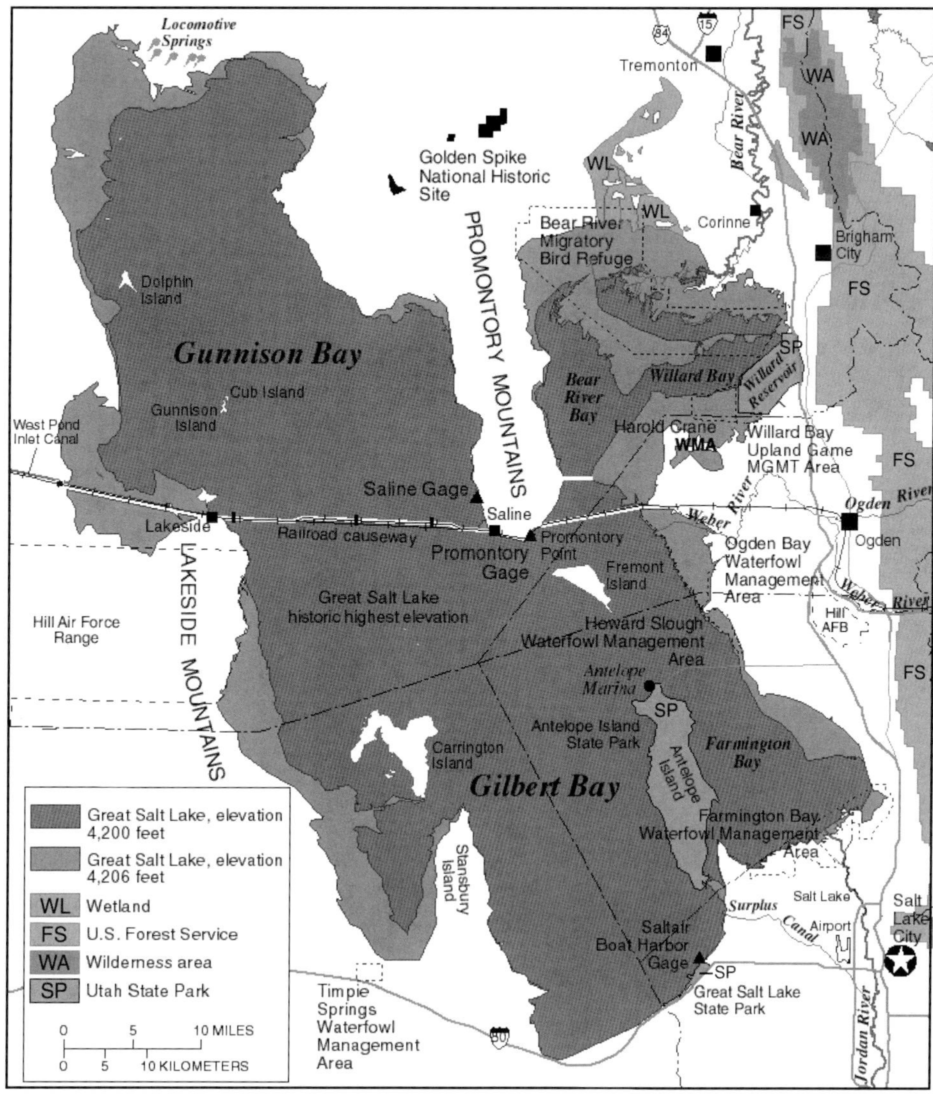

Figure 1. *Great Salt Lake and surrounding area.*

include: SOP (potassium sulfate), and magnesium chloride brine. In the past, sodium sulfate or salt cake was also produced. Salt Lake Sodium Products operated a plant on the south shore during the 1940s with a capacity of 100 short tons per day of sodium sulfate (Morgan, 1973). Great Salt Lake Minerals also produced sodium sulfate but discontinued production in 1994.

IMC Global is currently the only company to produce soluble salts other than sodium chloride from Great Salt Lake. IMC Global's facility is located in Weber County, just north of the Union Pacific railroad causeway where it crosses the lake (figure 1). The project was originally conceived by the Lithium Corporation of America (Lithcoa) with the aim of recovering dissolved lithium from the lake water. Since lithium exists in the lake water at parts-per-million levels, lithium recovery was determined to be uneconomic, even from evaporation pond bitterns, and the project was redirected to recover the more concentrated potassium salts. Laboratory studies were conducted during 1963 and 1964 and were followed by three years of pilot plant testing and pilot pond construction. Full-scale production was started in 1967 by Great Salt Lake Minerals and Chemicals, which was a joint venture between Lithcoa and the German firm Salzdetfurth. Lithcoa was subsequently purchased by Gulf Resources and Chemicals Corp., which became the sole owner of Great Salt Lake Minerals in 1973.

On May 5, 1984, rising lake water breached the outer dike of GSL's evaporation pond system. In addition to diluting the brine and rendering it unsuitable for mineral production, significant physical damage occurred to dikes, pond floors, pumping stations, and other structures. This tragedy occurred despite the company spending $8.1 million in 1983, and $8.3 million in early 1984 to raise the dikes. The company collected $10 million from a business interruption insurance policy and proceeded to repair the dikes. Repairs were completed in 1987 and full-scale production subsequently resumed.

Gulf Resources & Chemicals owned the facility until February, 1989, when the facility was sold for $34.5 million to the privately held Harris Chemical Group, Inc. During 1991, Great Salt Lake Minerals made plans to construct additional evaporation ponds on the west side of the lake and dredged a twenty-mile-long underwater channel, the Behrens Trench, to carry concentrated brine eastward across the lake to the processing facility. Work on the additional evaporation ponds and underwater channel commenced in 1992. Current production capacity for SOP is approximately 210,000 short tons per year and is expected to increase to about 315,000 short tons per year by the year 2003.

In April, 1998, IMC Global, Inc., headquartered in Northbrook, IL, purchased Harris Chemical Group, Inc., including Great Salt Lake Minerals. While the price for Great Salt Lake Minerals was not reported, the purchase price for the entire Harris Chemical Group was $1.4 billion. This included a $450 million cash payment for equity and assumption of $950 million in debt. Along with Great Salt Lake Minerals, IMC Global received salt, soda ash, boron chemicals, and other inorganic chemical operations as part of the Harris Chemical Group purchase. Therefore, it would be inappropriate to infer a value for Great Salt Lake Minerals from the price for the complete Harris Chemical Group. IMC Global renamed Great Salt Lake Minerals, "IMC Kalium Ogden Corp.," and also arranged for IMC Salt to market all salt products produced at the IMC Kalium Ogden facility. Current products are SOP, salt, magnesium chloride brine, and solid hydrated magnesium chloride flake.

An estimated $154 million worth of soluble salts were produced from the lake in 1997 (figure 2). This is an increase from just under $70 million in 1991. These estimates were developed using published capacity figures, industry capacity utilization data, and after conversations with industry personnel. Flooding during the early 1980s heavily damaged the lake's various producers' facilities and dramatically reduced production until the early 1990s. Production at Great Salt Lake Minerals (now IMC) was greatly reduced from 1983 until 1988 with only salt and brine products being produced. Potash was not produced from early 1984 until 1989, primarily due to damage to the evaporation pond's floors. Similar problems plagued the other salt producers on the lake.

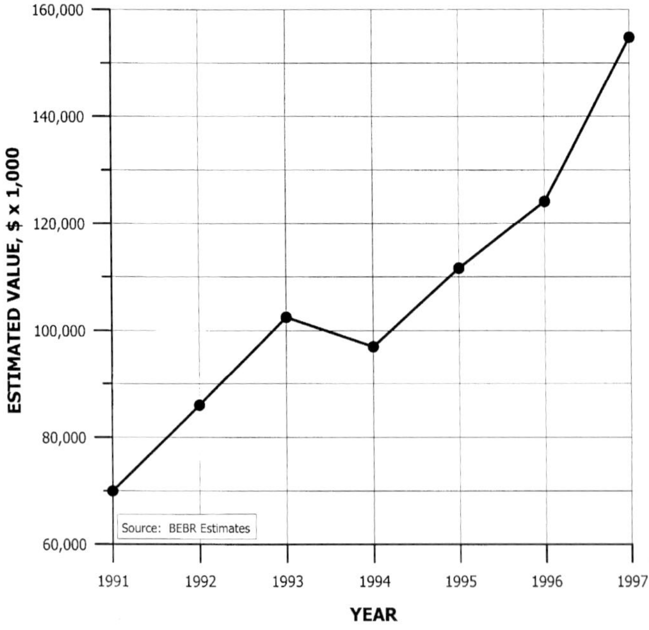

Figure 2. *Estimated value of soluble salts produced from Great Salt Lake from 1991 through 1997.*

The increased value of soluble salts produced from the lake is due primarily to increased production while prices for the commodities produced remained fairly stable. Between 1991 and 2001, estimated salt prices have varied between $30 to $50 per short ton and published prices for agricultural SOP varied between $175 and $200 per short ton (Chemical Market Reporter, 1991-2001).

Employment at the companies producing soluble salts from Great Salt Lake varies greatly. IMC Kalium is the largest employer, with employment ranging from 300 to 399. Morton is the second largest employer with employment ranging from 100 to 199, while Cargill employs from 50 to 99 persons (Division of Workforce Service, FirmFind, 2000). Wages at the salt producers tend to be higher than average for Utah. The average monthly wage for Standard Industrial Classification Code 281 - Manufacturing Industrial Inorganic Chemicals, which includes IMC Kalium, was $3,739, while the average monthly wage for SIC Code 2899 - Manufacturing Other Chemical Preparations, which includes Morton and Cargill, was $3,328 during 2000. For comparison, the average monthly wage in Utah for 2000 was $2,401, while the average monthly wage for Manufacturing Nondurable Goods was $2,585 (Utah Division of Workforce Services, 2001).

Magnesium

The variety and economic value of minerals produced from Great Salt Lake changed dramatically during the early 1970s. NL Industries, Inc., opened a magnesium-recovery plant during 1972 at Rowley, UT, on the western shore of the lake. The first metallic magnesium was produced in 1973. Production continued at a low level, caused by ongoing technical problems in the plant through April 1976, when production was suspended for a plant overhaul. Production resumed in 1977, after $55 million had been spent on modifications. This included a vastly redesigned electrolysis section which used technology developed by Norsk Hydro A/S of Oslo, Norway. The modification decreased plant capacity from 40,000 tons of magnesium per year to 25,000 tons per year. However, maintenance costs and chemical reagent consumption were also drastically reduced. The plant operated at nearly design capacity after the overhaul.

NL Industries sold the plant to AMAX, Inc., in 1980 for $58 million. AMAX continued operating the plant at near capacity levels while undertaking a project to reinforce the dikes around the evaporation ponds during 1983 and 1984. On June 7th, 1986, the main dike separating the plant's evaporation ponds from Great Salt Lake ruptured during a storm; lake water flooded the ponds, diluting the brine, and rendering it unuseable. AMAX responded by purchasing magnesium chloride brine from the Kaiser Aluminum & Chemical Corp. potash plant (now owned by Reilly Industries) near Wendover, UT and from Leslie Salt Company, located near San Francisco Bay. AMAX also began construction of evaporation ponds west of Great Salt Lake, and north of Knolls, UT. These ponds were operated from 1988 to 1995. The Stansbury Basin ponds were brought back on-line in 1993. Since there is a three-year production cycle in the ponds, Magcorp started harvesting from Stansbury Basin ponds again in 1996.

The facility was sold to Renco Group, Inc., during September, 1989 for an undisclosed price. Since that time, Renco has steadily increased the plant capacity from 32,000 short tons per year in 1989 to the current capacity of 44,000 short tons per year. The operation is currently in the midst of upgrading its electrolytic cells. These modifications are intended to (1) reduce operating costs, and (2) comply with both Title III regulations of the Clean Air Act as well as magnesium refinery-specific regulations expected from the U.S. Environmental Protection Agency during November, 2000. The projected capital expenditures for the years 1999 through 2001 are $20 million, $35 million, and $17 million, respectively (Platt's Metals Week, 1999).

Up to the end of 1998, Magcorp was one of three primary magnesium producers in the United States; the other two producers were Northwest Alloys at Addy, WA, and Dow Chemical at Freeport, TX. Dow Chemical closed the Texas facility at the end of 1998, after winter storms caused extensive damage to the plant. The future of the Texas facility had been in question, with Dow offering the facility for sale before the damage occurred. The Dow Chemical facility represented 45 percent of the U.S. production capacity of 160,000 short tons per year. With the closure of the Dow Chemical facility, domestic U.S. primary magnesium capacity declined to 88,000 short tons per year, with Magcorp representing half of this capacity.

In recent years the U.S. has become a net importer of magnesium metal. Prior to 1996, the U.S. exported more magnesium than it imported; however, since 1996 more magnesium has been imported into the U.S. than was exported. Closing the Dow Chemical facility will undoubtedly increase U.S. dependency upon imported magnesium. Northwest Alloys is owned by Alcoa, Inc., and essentially all of its production is used internally by Alcoa. Therefore, Magcorp is the only merchant primary magnesium producer in the U.S.

The estimated value of Magcorp's magnesium production (figure 3) was erratic throughout the late 1980s and early 1990s. The value increased during the mid-1990s, coincident with an expansion program at Magcorp. A peak value of about $151 million was reached in 1996, with production

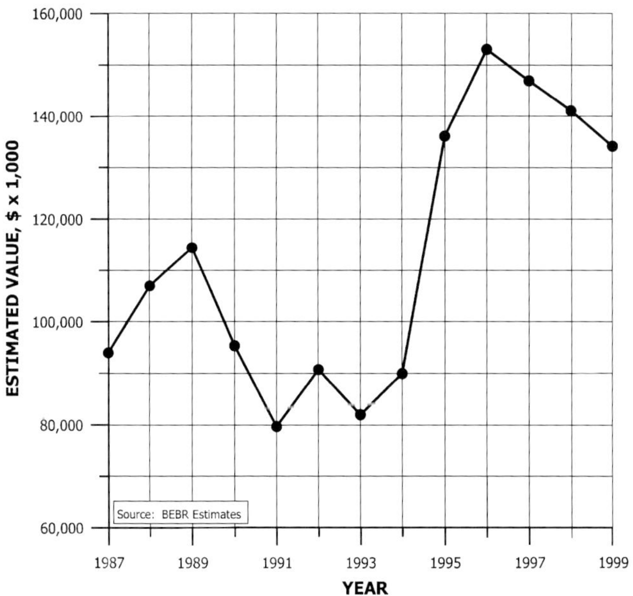

Figure 3. *Estimated value of MagCorp's magnesium production from 1987 through 1999.*

values dropping since then. This drop coincided with decreasing magnesium prices and the U.S. becoming a net magnesium importer.

As with the soluble salts, increased production capacity was the primary reason for the increased value of magnesium and chlorine. Estimated prices for magnesium varied between $2,240 per short ton 1993 and $3,060 per short ton in 1996 (U.S. Geological Survey, 1987-1996). However, the average estimated price during 1997 was $2,520 per short ton, which is only $20 more than the average 1987 price of $2,500 per short ton. Given the large fluctuations and overall sideways trend in the price of magnesium, the size of the magnesium industry's contribution to Utah's economy, over the long term, is most dependent upon production volume. In the future, these fluctuations in price can be expected to continue, with the industry being profitable when prices are high and profit margins contracting when the prices are low. As of June, 2000, worldwide magnesium prices are low and expected to decline in the short term, with mixed predictions for longer time periods (Platt's Metals Week, 2000). Both MagCorp and the parent company Renco Metals filed for bankruptcy during August 2001 citing low magnesium prices, although there was no immediate effect on employment at the company as a result of the filing and the plant continued to operate (Twyman, 2001). In 1998, the U.S. exported 38,940 short tons of magnesium and imported 90,750 short tons of magnesium. Both of these figures are of the same order of magnitude as the Magcorp production capacity, which indicates that Magcorp's profitability, and its impact on Utah's economy, are closely related to the world economy and worldwide demand and supply of magnesium.

Besides metallic magnesium and chlorine gas, calcium chloride and ferric chloride are produced in lesser quantities. Magcorp produces ferric chloride that is marketed by a third party. Hill Brothers Chemical Company operates a small plant adjacent to the Magcorp facility and recovers solid calcium chloride from an aqueous stream from the Magcorp plant.

As of October 31, 1999, Magcorp employed 503 persons (Renco Metals Inc., 2000). The majority of these were employed at the plant at Rowley, UT, although several dozen were employed at the company's headquarters in Salt Lake City, UT. Employment at Magcorp tends to pay higher than the average wage in Utah. The average monthly wage for Standard Industrial Classification (SIC) Code 333 - Primary Nonferrous Metals, the SIC Code that includes Magcorp, during 2000 was $4,130. This compares to an average monthly wage for the state of Utah of $2,401, and for Manufacturing Durable Goods in Utah of $3,123 for 2000 (Utah Department of Workforce Services, 1999). The company is also responsible for a significant portion of the Utah employment in SIC Code 333. Average annual employment for this SIC Code for 2000 was 1,183. Therefore, Magcorp is responsible for about 40 percent of the employment in SIC Code 333 in Utah.

ARTEMIA (BRINE SHRIMP)

Great Salt Lake is a major source of the world's artemia cysts, or brine shrimp eggs. Artemia were first commercially harvested from Great Salt Lake in 1950. Historically, the major use of artemia was to feed tropical fish in home aquariums. However, the development of large-scale prawn farming operations in southeast Asia and Latin America during the 1980s has created a large demand for artemia cysts. Currently, approximately 80 to 85 percent of Utah-produced artemia cysts are sold for prawn feed. The remainder is used for tropical fish food and feeding fish raised for meat (Woolf, 1998). As a result of this changing market, the Utah artemia industry is more dependent upon global events than in the past. A major market for Great Salt Lake artemia was Ecuadoran aqua-culture. Unfortunately, El Nino-related storms destroyed a majority of the Ecuadoran aqua-culture during 1996 (Thomas Troy, Brine Shrimp Direct, oral communication, 1998). A major competitor is Russian artemia, mostly from the Caspian sea, but also from other portions of European Russia and Siberia. Other artemia-producing countries include Argentina, Brazil, Mexico, and Vietnam (Howard Newman, Inve Aquaculture, written communication,1998; Don Leonard, Utah Artemia Association, oral communication, 1998). Domestically, artemia are also harvested from San Francisco Bay and Mono Lake in California, and from Lake Abert, Oregon (McCrae, 1996; Mono Lake Committee, 2000). However, these other sources are small compared to Great Salt Lake. For example, Lake Abert is currently producing about 42,000 pounds of harvested artemia a year, as compared to an annual harvest from Great Salt Lake in the millions of pounds.

There is a four-month-long artemia harvesting season on Great Salt Lake, starting on October 1. However, the Utah Division of Wildlife Resources can close the season early to prevent populations from reaching critically low levels. In 1997, the season closed after 27 days and in 1998 it closed after only 22 days. In 1999, the season opening was delayed until October 25, and harvesting continued through the end of the season on January 31, 2000. However, harvesting was allowed only in the northern half of the lake, north of the railroad causeway. The harvested cysts originated in the southern part of the lake, and were transported into the northern part via wind and wave action through the breach in the causeway. In past years, harvesting has occurred for as long as seven months.

During 1999, 32 companies had licenses to harvest artemia (table 2), up from 21 in 1995 and only 4 in 1985 (Clay Perschon, Utah Division of Wildlife Resources, oral communication, 2000). Approximately half of the companies both harvest and process artemia; the remaining companies sell what they harvest to both these dual producing and processing companies. To legally harvest artemia, companies must purchase an annual Certificate of Registration from the Utah Division of Wildlife Resources. The current cost of a Certificate of Registration is $10,000, and most companies hold more than one Certificate. Each Certificate of Registration gives the holder the exclusive right to harvest around a point marked by a large float. The placement of the floats is temporary and determined by the company, which places the float in the water when a productive area is located. Therefore, holding multiple Certificates of Registration gives companies the ability to harvest from a larger area. During the 1999-2000 harvest season, 79 Certificates of Registration were sold, representing an outlay of $790,000. The number of Certificates of Registration is currently capped at 79.

Changing lake salinity is a major factor in the decreasing harvest. Artemia are ideally suited for a salinity of about 13 to18 percent, because algae, which is their main food supply, grows best within this salinity range. However, the south arm of the lake dropped from 12 to about 7 percent salinity from 1996 to 1999 (J. Wallace Gwynn, Utah Geological Survey, oral communication, 2000), with associated detrimental effects on the algae and the artemia. The north arm, north of the Union Pacific Causeway, is saturated, or nearly saturated, with salt. Such high salt concentrations are harmful to artemia.

Table 2. Reported Utah artemia harvest, 1985 through 2000.

Year	Licensed Companies	Certificates of Registration	Harvested, Pounds**
1985	4		298,035
1986	4		1,887,300
1987	4		7,012,775
1988	9		6,806,415
1989	12		10,268,232
1990	11		8,927,818
1991	11	26	13,532,797
1992	12	20	10,172,399
1993	12	18	8,864,092
1994	14	29	6,485,954
1995	21	63	14,749,596
1996	32	79	14,679,498
1997	32	79	6,113,695
1998	32	79	4,606,352
1999	32	79	2,631,853
2000	32	79	19,963,087

**Unprocessed, wet biomass.
Source: Utah Division of Wildlife Resources.

The actual amount of harvested material that is processed and sold varies greatly. The harvested weight varies by as much as 100 percent annually. The maximum harvested weight in recent years was 14.7 million pounds in 1995. By 1999, the amount harvested had declined to 2.6 million pounds, before rebounding to 19 million pounds in 2000. Don Leonard (Utah Brine Shrimp Association, oral communication, 1998) estimated that the weight of artemia cysts sold averages 25 percent of the harvested weight reported to the State. The remaining 75 percent of the harvested weight is non-cyst debris that are removed during processing. Process recovery varies greatly by operator and depends highly upon operator skill. For some operators, the marketable weight is as little as 10 percent of the harvested weight.

The price received for artemia cysts varies greatly (figure 4), with prices during recent years varying as much as 300 percent in one year (Howard Newman, Inve Aquaculture, written communication, 1998). Furthermore, there are several different grades of cysts, which also vary in price. The grading of cysts is determined by the percentage of cysts that can be expected to hatch under standardized conditions. The three highest grades are, in descending order, Premium, Grade A, and Grade B, with hatch rates of 90, 80, and 70 percent, respectively. There are also lower grades, which are more difficult to market, with hatch rates of 60 percent and 50 percent. The supplies of Premium and Grade A cysts are almost always completely sold, while the sale of the lower

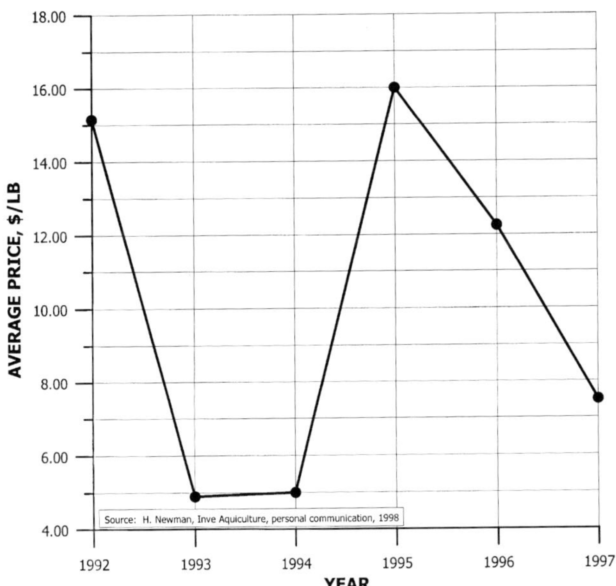

Figure 4. Average worldwide price of artemia from 1992 through 1997.

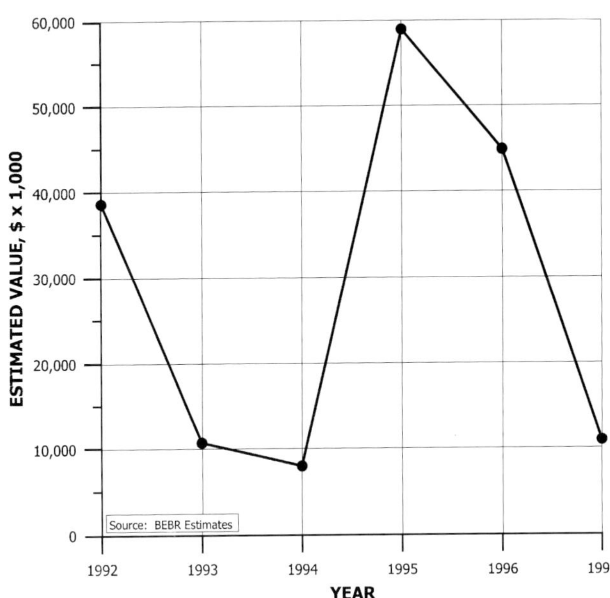

Figure 5. Estimated market value of artemia produced from Great Salt Lake from 1992 through 1997.

grades (B and lower) is more variable, depending upon market conditions (Don Leonard, Utah Brine Shrimp Assoc., oral communication, 1998). During September, 1998, Premium, Grade A, and Grade B cysts were quoted at $30, $25, and $20 per 425-gram can respectively (M & M Distributing, 1998). These quotes were from a distributor, and not a harvester that operates on the lake; therefore, they may not be indicative of values received by local harvesters. They do, however, indicate the differential values for the various grades of artemia cysts.

The amount of cysts available for marketing as well as cyst grade and availability of alternative feed materials for aquaculture influence the market price. The lower harvests in 1998 and 1999 resulted in increased prices. In July, 2000, the quoted price for cysts with a hatch rate of 80-85 percent was $64.95 for a 15 oz can (Brine Shrimp Direct, 2000). Suppliers also adopted such measures as rationing purchases and refusing new customers (Brine Shrimp Direct, 2000; Artemia International, 2000).

The Utah State Tax Commission initiated a royalty on artemia production in 1997. The royalty is imposed on raw, wet, unprocessed cysts. Total wet weight of the harvested biomass is reported by the companies. The Tax Commission assumes that 40 percent of the reported weight is debris such as algae, bird feathers, floating wood, etc, and sets a value for the remaining 60 percent of the material from which it collects a 3.5 percent royalty. Royalties collected to date are $60,791 in 1997, $61,942 in 1998, $77,121 in 1999, and $247,792 in 2000. All revenue generated by this royalty is deposited in the Department of Natural Resources Species Protection Account (Utah State Tax Commission, 2001).

Combining the reported harvested weight with the 25 percent processing efficiency estimate, and price data, allows for estimating the market value of artemia produced from Great Salt Lake (figure 5). Market value of Utah-produced artemia peaked at an estimated $58 million in 1995, and declined to $11 million in 1997. During the same time, the price dropped from $16 to $7.50 per pound and the amount harvested fell by 58%. While quoted prices increased from 1998 to 2000, the decreased harvest was sufficient to prevent a rise in total market value. Currently, biological factors, rather than market forces, are limiting production from Great Salt Lake.

The drastically fluctuating artemia harvest and prices make forecasting the future of the Utah artemia industry tenuous at best. The estimated value of Utah-produced artemia increased by over 600 percent from 1994 to 1995. It then decreased by 80 percent from 1995 to 1997. Obviously, the future of Utah artemia production depends on many factors, including demand for shrimp prawns, lake water salinity, competition from other artemia sources, and alternative foods for aqua-culture. At least one Utah producer also harvests artemia in several other countries. Without changes in the current salinity of the lake, biological factors will continue to limit the artemia harvest by forcing shortened seasons. Under this scenario, market values in the $8-$11 million range is the most likely forecast. Future events, such as a drought and related increasing salinity of the south arm, could drastically change the size and economic value of the artemia harvest.

RECREATION

Many recreational opportunities exist on Great Salt Lake. It is a center for waterfowl hunting in the western U.S. There are numerous state wildlife refuges, one federal wildlife refuge, and two state parks on the edges of, or within, the lake.

State Parks

The Utah Division of Parks and Recreation operates two parks, Antelope Island State Park, and Great Salt Lake State Park, on Great Salt Lake. Antelope Island State Park is located on Antelope Island in the southeast portion of the lake (figure 1). Most of the visitor facilities are located on the north end of the island. Great Salt Lake State Park is located on the south shore of the lake.

Antelope Island State Park is accessed via a causeway that crosses Farmington Bay. The causeway was built by the State of Utah in 1969. The causeway was repeatedly washed out by heavy wave action and was completely flooded in 1985 by rising lake waters. It remained unusable for the next eight years. The state legislature transferred ownership of the causeway to Davis County in 1991 and contributed $4 million to rebuild it. In 1992, the State contributed an additional $500,000 for causeway reconstruction. Davis County rebuilt the causeway, which reopened in 1993, at a cost of $5 million. A $7 toll is assessed for every vehicle that uses the causeway. Of this $7, $5 is considered Day Use Fees for using the state park and transferred to the State Division of Parks and Recreation. Davis County uses the remaining $2 to maintain the causeway (Utah Division of Parks and Recreation, 1999).

State Park Expenditures

Expenditures for Antelope Island have steadily increased from 1993 through 1999 (table 3), while Great Salt Lake State Park expenses have remained relatively constant. Antelope Island expenses increased from $205,488 in 1993 to $884,525 in 1999. The increase in expenditures for Antelope Island coincided with more active management of the bison herd, the introduction of pronghorn antelope and bighorn sheep, and increased visitors' access to the southern reaches of the island, especially the Fielding Garr ranch house. Expenses for Great Salt Lake State Park were $292,530 in 1993 and $272,215 in 1999 (Lana Hadlock, Utah Division of Parks and Recreation, written communication, 2000). The relatively stable expenditures for Great Salt Lake State Park (marina) indicate little change in activities in recent years, although the marina usually has all three hundred boat slips rented. This indicates a possible market for additional marina services on the lake.

State Park Revenues

Revenues are collected from visitors to state parks in various ways. In addition to fees common to state parks such as entrance and camping fees, additional fees are collected for boat docking at both parks, and for bison hunting on Antelope Island.

Antelope Island: Revenue collected at Antelope Island increased from $46,977 in 1993 to $635,307 in 1998, or 1,252 percent, before dropping to $434,468 in 1999 (table 4). The most significant revenue category was Day Use Fees from 1993 through 1996; in 1997 the Special Fees category was the largest, accounting for $319,494 or 50.3 percent of the annual revenue.

The dramatic increase in special fees collected at Antelope Island is due to increased value of bison sold from the island's herd. The Utah Division of Parks and Recreation assumed management of the bison herd in 1987. Prior to 1987, the herd was not managed and suffered from inbreeding and malnutrition. State park personnel round up the herd annually and sell individual bison at auction to breeders. During the mid-1980s, the average price per animal was about $700. Managing the herd increased the average price per animal to $3,800 in 1998. The 1998 bison auction netted $225,000 for the state park system (Loomis, 1998). Table 4 also includes fees for bison hunting on the island; for 1998 these included five resident permits at $1,100 each and one nonresident permit at $2,600.

In addition to revenue collected by the Division of Parks and Recreation, two concessionaires operate at Antelope Island State Park. These are Buffalo Point, Inc., which operates a restaurant, souvenir shop, and R & G Horse and Wagon Outfitters, LLC, which offers horse rentals and guided horseback tours of the island. Revenue from these concessionaires is included in the Lease category (tables 4).

Great Salt Lake Park: Revenue collected by Great Salt Lake State Park decreased from $365,872 in 1993 to $313,705 in 1997, or by 14.2 percent (table 5).

The primary revenue source at Great Salt Lake State Park is boat mooring, which is included under Special Fees in table 5. From the mid-1980s until 1997, Great Salt Lake State Park included several beaches east of the marina. Receding lake water resulted in the visitors facilities at these beach areas being a noticeable distance from the shoreline, and visitor use subsequently declined. This is reflected in the declining Day Use revenue. In 1997, these beach areas were transferred from the Division of Parks and Recreation to the Division of Forestry, Fire, and State Lands.

In addition to revenue collected by the Division of Parks and Recreation, Salt Island Adventures operates the Island Serenade cruise boat from the Great Salt Lake State Park marina. Revenue from Salt Island Adventures is included in the Special Fees category.

State Wildlife Management Areas

The Utah Division of Wildlife Resources maintains seven Wildlife Management Areas on the edge of Great Salt Lake. These are, in a clockwise direction from the northwest corner of the lake, Locomotive Springs, Public Shooting Grounds, Harold S. Crane, Ogden Bay, Howard Slough, Farmington Bay, and Timpie Springs. The State budgets for these areas are in table 6.

Budgeted amounts for the state Wildlife Management Areas increased by 38 percent from 1996 to 1999. The largest increase was for Farmington Bay and Timpie Springs, which had a combined budget increase of nearly 100 percent. The remaining areas had budget increases of about 20 percent from 1996 to 1999.

The Utah Division of Wildlife Resources has proposed a $3 million visitor center at Farmington Bay (Associated Press, 1999). A feasibility study was completed in December 2000, at a cost of $19,000 (Bill Douglas, Utah Division of Wildlife Resources, oral communication, 2001), to determine the type of audience the visitor center would attract, what facilities are needed, and where the center should be located. Current visitation to Farmington Bay is estimated to be 50,000 visitors annually. This estimate includes both bird watchers and hunters.

Bear River Bay Migratory Bird Refuge

The U.S. Fish and Wildlife Service, an agency in the U.S. Department of Interior, operates the Bear River Bay

Table 3. State expenditures ($) for state parks bordering Great Salt Lake from 1993 through 1999.

State Park	1993	1994	1995	1996	1997	1998	1999
Antelope Island	205,488	406,799	478,991	498,797	606,795	758,474	884,525
G S L State Park	292,530	269,848	259,131	224,309	238,754	287,361	272,215
Total	498,018	676,647	738,122	723,106	845,549	1,045,835	1,156,740

Source: Lana Hadlock, Utah Division of Parks and Recreation

Table 4. Revenue ($) collected by Antelope Island State Park from 1993 through 1999.

Revenue Category	1993	1994	1995	1996	1997	1998	1999
Day Use	30,438	180,022	183,682	218,195	251,572	249,904	276,398
Camping		15,031		22,287	23,019	21,686	24,029
Resale	6,894	9,111	4,788	704	22,297	31,171	38,587
Lease	241	3,767	8,511	14,432	18,925	15,594	14,178
Special Fees	9,404	93,935	138,303	101,642	319,494	113,413	81,276
Total	46,977	301,866	335,284	357,240	635,307	431,768	434,468

Source: Utah Division of Parks and Recreation

Table 5. Revenue ($) collected by Great Salt Lake State Park.

Fee Category	1993	1994	1995	1996	1997	1998	1999
Day Use	15,814	36	6,037	934	63	19	11
Camping	1,038	996		594	734	212	
Resale		197	92	284	305	46	
Lease	15,733	21,624	32,025	25,206	24,297	6,624	8,918
Special Fees	333,287	333,203	307,165	291,237	288,306	284,340	315,706
Total	365,872	356,056	345,319	318,255	313,705	291,241	324,635

Source: Utah Division of Parks and Recreation

Table 6. Budgeted amounts ($) for state Wildlife Management Areas on the Great Salt Lake from 1996 through 1999.

Management Area	1996	1997	1998	1999
Locomotive Springs/Public Shooting Grounds	118,882	112,268	140,771	144,454
Harold S. Crane	5,150	5,150	5,150	6,150
Ogden Bay/Howard Slough	112,946	100,371	131,073	134,628
Farmington Bay/Timpie Springs	70,613	66,011	141,570	141,056
Total	307,591	283,800	418,564	426,288

Source: Utah Division of Wildlife Resources

Migratory Bird Refuge on the shore of Great Salt Lake west of Brigham City (figure 1). The budget for the refuge for the years 1987 through 1999 is shown in table 7. The low budgets during the 1980s were a result of flooding which occurred during 1983 and subsequent damage to wildlife areas. For several years, the refuge was closed and management operated out of the Ouray wildlife refuge near Vernal, Utah. Since 1990, the budget has steadily increased from $375,000 to $1,517,010 in 1999.

The refuge has plans to build a new headquarters and visitors center. This center will include a visitors center, structured classrooms for visiting school classes, a research laboratory, and the refuge headquarters office. The facility will be located adjacent to Interstate-15, at exit 366 near Brigham City, UT. Projected cost for the facility is approximately $7 - 9 million, with $1.5 million of this amount being raised by the Friends of the Bear River Bird Refuge, and the remainder coming from federal appropriations. The projected date for opening the new visitors center is 2003. The refuge is also obtaining approximately $1.3 million in additional federal funding to rebuild the access road from Brigham City to the refuge (Steve Hicks, Bear River Bird Refuge, oral communication, 2001).

Total Government Expenditures

Total estimated expenditures on government facilities on Great Salt Lake increased from $1,866,000 in 1996 to

Table 7. Bear River Refuge Budgets ($) from 1987 through 1999.

Year	Budget
1987	40,000
1988	10,000
1989	55,000
1990	375,000
1991	501,700
1992	530,993
1993	632,260
1994	630,800
1995	840,520
1996	834,890
1997	1,170,280
1998	1,336,100
1999	1,517,010

Source: U.S. Fish and Wildlife Service

$2,800,000 in 1998. During these years, about half of the government expenditures was for the Bear River Migratory Bird Refuge. This represents federal tax dollars that probably would not have been spent in Utah in the absence of the Great Salt Lake. The remaining half was State of Utah expenditures, which represented tax money collected from Utah residents. In the absence of Great Salt Lake, these expenditures may have been spent in other parts of the state, or may not have been collected as taxes from Utah residents. Therefore, the money expended in Utah in connection with the Bear River Bird Refuge represents a positive economic impact on Utah's economy. In contrast, expenditures related to the state facilities do not represent additional funds being added to the Utah economy, but is a redistribution of money within Utah.

WATERFOWL HUNTING

In addition to the government's expenditure for maintaining waterfowl hunting areas along the lake, sizable expenditures are made by hunters. During 1998, 57 percent of all waterfowl hunting in Utah was associated with Great Salt Lake (Table 8). Essentially all of this hunting was done by Utah residents. During 1998, 138,000 person-days were spent hunting, with an estimated $3.3 million spent on hunting related travel and about $4.5 million spent on hunting equipment. The amount spent on waterfowl hunting has increase dramatically in the past fifteen years. During the early 1980s, much of the prime hunting areas were flooded, drastically reducing the available hunting opportunities.

Hunting Expenditures

The number of person-days spent waterfowl hunting on Great Salt Lake for the years 1987 through 1998 is presented in table 8 (Tom Aldrich, Utah Division of Wildlife Resources, written communication, 1998, 2000). The number of hunters was relatively low (approximately 40,000 to 50,000) throughout the 1980s due to flooding. Since 1990, the number of water fowl hunters along Great Salt Lake has increased, to a 1996 high of 184,229. Furthermore, the percentage of waterfowl hunting in Utah that occurs around Great Salt Lake increased from 41 percent in 1987, to 61 percent in 1996.

The estimated average daily travel expenditures for waterfowl hunting in the United States increased from $9.46 in 1980, to $15.73 in 1991, to $21.33 in 1996. Simultaneously, estimated daily equipment expenditures increased from $5.41 in 1980, to $15.45 in 1991, to $26.66 in 1996 (USFWS, 1980; USFWS, 1991; USFWS, 1996). This daily expenditure rate was then multiplied by the hunter-days to obtain total expenditures for water fowl hunting along Great Salt Lake (tables 8-11).

Annual trip expenditures from waterfowl hunting along Great Salt Lake were estimated to be in the range of $500,000 to $600,000 during the 1980s, when hunting diminished due to high water levels. Estimated trip expenditures have steadily increased during the 1990s, from $725,936 in 1990 to $3,313,344 in 1998. Estimated annual equipment expenditures were between $400,000 and $500,000 during the 1980s, and increased from $671,561 in 1990 to $4,451,316 in 1998. Total expenditures were just over $1 million during the late 1980s, and increased to just under $8 million in 1998.

Hunting occurs at numerous locations around Great Salt Lake's shoreline. In addition to the state and federal waterfowl refuges, private, state, and U.S. Bureau of Land Management lands around the lake are used for hunting. There are also approximately three dozen private duck clubs located along the south, east, and north shores. The size and economics of these clubs varies tremendously. Most are organized as corporations, with members purchasing a share which gives them membership in the club. The clubs basically provide access to private land for hunting, and are subject to all wildlife laws. There are no rules or regulations specifically governing duck clubs other than the laws pertaining to all businesses, such as business licenses and zoning regulations. Most clubs have a fixed number of shares, therefore, new members must purchase a membership from an existing member. Membership prices range from a few hundred dollars to approximately $50,000, depending upon the individual club. The clubs also have annual dues that range up to several thousand dollars. Some clubs, predominantly the smaller ones, lease ground from private land owners, usually agricultural interests. Other clubs have extensive land holdings and make significant expenditures managing wetlands and maintaining clubhouses and similar facilities. For example, one of the larger clubs recently spent $260,000 to build a boat house, and received a $175,000 grant from the federal government for reconstructing dikes (Ed Richards, Chesapeake Duck Club, oral communication, 1998).

Economic Impact of Waterfowl Hunting

Essentially all waterfowl hunting in the state of Utah is done by state residents (USFWS, 1996; USFWS, 1991; USFWS, 1980). The low number of non-resident hunters indicates that most of the expenditures are disposable income of Utah residents. Therefore, one can argue that hunting does not have an economic impact on the Utah economy. In the

Table 8. Days spent waterfowl hunting along the Great Salt Lake.

Hunter Days	1987	1988	1989	1990	1991	1992	1993	1994	1995	1996	1997	1998
Locomotive Springs	281	200	394	257	252	907	1,284	162	769	477	454	505
Public Shooting Grounds	1,721	3,373	1,819	4,282	3,020	3,762	3,948	3,539	4,065	4,879	2,175	3,676
Harold Crane	190	300	547	2,350	2,198	2,374	5,273	5,524	7,932	8,422	3,418	3,702
Ogden Bay	1,551	2,077	3,155	3,263	7,238	10,529	9,461	10,484	13,792	18,561	10,289	7,068
Howard Slough	344	33	153	758	1,302	1,894	1,705	2,480	3,017	4,895	1,702	2,735
Farmington Bay	233	880	1,311	1,618	2,238	7,013	5,115	5,934	8,864	12,951	10,213	11,107
Timpie Springs	328	356	191	219	221	211	724	382	396	604	167	376
Bear River Migratory Bird Refuge	690	356	114	963	1,249	1,537	3,177	3,415	5,405	6,992	5,485	8,481
Salt Lake County	10,625	9,978	7,468	10,356	13,579	17,020	19,294	15,321	16,576	20,325	10,809	13,259
Davis County	8,082	5,564	5,982	5,027	7,762	13,703	16,367	19,853	25,591	34,039	23,934	22,841
Weber County	5,054	8,515	6,366	6,484	8,229	13,268	17,260	19,185	21,153	26,793	14,160	18,316
Box Elder County	19,296	11,247	14,708	13,419	13,293	16,731	25,038	27,771	34,432	45,291	33,348	45,703
Total Great Salt Lake Days Hunting	48,395	42,879	42,208	48,996	60,581	88,949	108,648	114,050	141,992	184,229	116,154	137,769
State Wide Days Hunting	118,427	91,087	82,411	86,857	105,692	155,946	180,100	189,436	222,925	300,316	203,295	243,537
GSL Days Hunting as % of State Total	41	47	51	56	57	57	60	60	64	61	57	57

Source: Utah Division of Wildlife Resources

Table 9. Estimated travel expenditures ($) for waterfowl hunters along the Great Salt Lake by year and hunting area.

Hunter Travel Expenditures	1987	1988	1989	1990	1991	1992	1993	1994	1995	1996	1997	1998
Travel Expenditure per day, $*	12.49	13.19	13.97	14.82	15.73	16.71	17.76	18.88	20.07	21.33	22.66	24.05
Locomotive Springs	3,509	2,639	5,505	3,808	3,964	15,158	22,813	3,059	15,436	10,174	10,286	12,145
Public Shooting Grounds	21,491	44,507	25,414	63,443	47,505	62,872	70,132	66,830	81,595	104,069	49,277	88,408
Harold Crane	2,373	3,958	7,642	34,818	34,575	39,675	93,671	104,314	159,215	179,641	77,439	89,033
Ogden Bay	19,368	27,406	44,079	48,345	113,854	175,966	168,056	197,977	276,840	395,906	233,110	169,985
Howard Slough	4,296	435	2,138	11,231	20,480	31,653	30,290	46,832	60,559	104,410	38,561	65,777
Farmington Bay	2,910	11,612	18,316	23,973	35,204	117,205	90,870	112,056	177,923	276,245	231,388	267,123
Timpie Springs	4,096	4,697	2,669	3,245	3,476	3,526	12,858	7,214	7,949	12,883	3,784	9,043
Bear River Migratory Bird Refuge	8,616	4,697	1,593	14,268	19,647	25,687	56,428	64,488	108,492	149,139	124,270	203,968
Salt Lake County	132,680	131,660	104,337	153,437	213,597	284,447	342,742	289,318	332,722	433,532	244,892	318,879
Davis County	100,924	73,417	83,576	74,481	122,096	229,011	290,747	374,899	513,675	726,052	542,255	549,326
Weber County	63,112	112,355	88,941	96,068	129,442	221,741	306,607	362,285	424,594	571,495	320,813	440,500
Box Elder County	240,958	148,404	205,489	198,819	209,099	279,617	444,776	524,421	691,136	966,057	755,541	1,099,157
Total Trip Expenditures, $	604,332	565,788	589,698	725,936	952,938	1,486,559	1,929,989	2,153,691	2,850,135	3,929,606	2,631,616	3,313,344

*Source: 1980, 1991, 1996 from the National Survey of Fishing, Hunting, and Wildlife-Associated Recreation. Other years estimated using linear regression. Estimated expenditures are left unrounded for convenience rather than to denote accuracy.

Table 10. Estimated expenditures ($) for equipment made by waterfowl hunters along the Great Salt Lake by year and hunting area.

Equipment Expenditures	1987	1988	1989	1990	1991	1992	1993	1994	1995	1996	1997	1998
Equipment Expenditure per day, $*	9.47	10.72	12.13	13.71	15.45	17.36	19.44	21.68	24.09	26.66	29.40	32.31
Locomotive Springs	2,662	2,144	4,779	3,523	3,893	15,745	24,960	3,512	18,522	12,717	13,348	16,317
Public Shooting Grounds	16,303	36,151	22,063	58,691	46,659	65,307	76,733	76,717	97,908	130,074	63,945	118,772
Harold Crane	1,800	3,215	6,635	32,210	33,959	41,212	102,486	119,747	191,048	224,531	100,489	119,612
Ogden Bay	14,692	22,261	38,267	44,724	111,827	182,780	183,871	227,267	332,190	494,837	302,497	228,367
Howard Slough	3,259	354	1,856	10,389	20,116	32,879	33,141	53,760	72,667	130,501	50,039	88,368
Farmington Bay	2,207	9,432	15,901	22,177	34,577	121,743	99,422	128,634	213,496	345,274	300,262	358,867
Timpie Springs	3,107	3,816	2,317	3,002	3,414	3,663	14,068	8,281	9,538	16,103	4,910	12,149
Bear River Migratory Bird Refuge	6,536	3,816	1,383	13,199	19,297	26,682	61,739	74,029	130,183	186,407	161,259	274,021
Salt Lake County	100,649	106,943	90,580	141,944	209,795	295,461	374,997	332,121	399,244	541,865	317,785	428,398
Davis County	76,559	59,634	72,556	68,902	119,923	237,879	318,109	430,364	616,377	907,480	703,660	737,993
Weber County	47,876	91,263	77,214	88,873	127,138	230,328	335,462	415,883	509,485	714,302	416,304	591,790
Box Elder County	182,788	120,544	178,394	183,927	205,376	290,444	486,634	602,006	829,318	1,207,459	980,431	1,476,664
Total Equipment Expenditure, $	458,437	459,571	511,944	671,561	935,975	1,544,124	2,111,622	2,472,321	3,419,975	4,911,549	3,414,928	4,451,316

*Source: 1980, 1991, 1996 from the National Survey of Fishing, Hunting, and Wildlife-Associated Recreation. Other years estimated using linear regression.
Estimated expenditures are left unrounded for convenience rather than to denote accuracy.

Table 11. Total Great Salt Lake related waterfowl hunter expenditures by year.

Hunter Expenditures	1987	1988	1989	1990	1991	1992	1993	1994	1995	1996	1997	1998
Total Trip & Equipment	1,062,769	1,025,359	1,101,642	1,397,497	1,888,913	3,030,683	4,041,612	4,626,012	6,270,110	8,841,155	6,046,543	7,764,661

Source: BEBR Calculations.
Estimated expenditures are left unrounded for convenience rather than to denote accuracy.

absence of hunting opportunities, this disposable income would undoubtedly be spent on other recreational activities. However, it does not follow that spending this disposable income on other (recreational) activities would not alter the Utah economy. In the absence of hunting opportunities on Great Salt Lake, these expenditures may be made on hunting opportunities in other parts of Utah, or in surrounding states. The distribution of waterfowl hunting in Utah demonstrates this. In 1987, when the lake level was high and much of the hunting areas flooded, 41 percent of the days spent waterfowl hunting in Utah were related to the lake. By 1993, when the lake level had dropped and hunting areas were restored, 60 percent of Utah waterfowl hunting was related to the lake. Obviously, hunters' travel expenditures are spent in the areas with hunting opportunities. During the high water years from 1983 until 1989, hunting travel expenditures were lower than they would have been had the lake not flooded. Assuming that 60 percent of Utah waterfowl hunting is related to Great Salt Lake under normal hunting conditions, then a reduced hunting level caused by high water from the mid-1980s to 1989 resulted annually in a nearly $540,000 decrease in hunting travel expenditures made on Great Salt Lake.

PRECIPITATION EFFECTS

It is often stated that Great Salt Lake may have an effect on the weather patterns in the surrounding vicinity (Miller and Millis, 1989). Quantifying the extent of these changed weather patterns and their effects on the local economy is difficult. The major effect on the weather occurs from the "lake effect," when storms form due to the lake water being warmer than the nearby air. The lake effect results in locally heavy precipitation, usually in the form of snow. Substantially all of the "lake effect" snowfall resulting from Great Salt Lake falls in the Salt Lake Valley, Tooele, and on the Bountiful bench. An estimated 10 percent of the annual snowfall at the Salt Lake City Airport is a result of the lake effect. Since the majority of the water supply for the area surrounding the lake is supplied by rivers and creeks originating in the Uintah and Wasatch mountain ranges, which receive little lake effect precipitation, it can be concluded that lake effect has little influence on the water supply and related economic development of northern Utah (Daryl Onton, University of Utah Meteorology Dept., oral communication, 1998).

Perhaps the greatest economic effect from "lake effect" storms is damage in the Salt Lake Valley and lost business from businesses closing during storms. One of the largest snow storms to hit the Salt Lake Valley, which was attributed to lake effect, occurred on October 18, 1984, and caused an estimated $1 million in damage. Unfortunately, distinguishing between "lake effect"-induced precipitation and that due to general weather patterns is difficult. Quite often a storm which is part of the general weather pattern will pass over the lake and be strengthened by moisture it acquires from the lake water. Determining the portion of storm precipitation that is due to lake effect and the portion due to the original storm is difficult (Daryl Onton, University of Utah Meteorology Dept., oral communication, 1998).

CONCLUSIONS

The economics of Great Salt Lake is manifest in several different forms. Perhaps the most obvious, with the largest impact on Utah's economy, is mineral production from the lake waters. Several mineral commodities including salt, potash, and magnesium are produced, with total estimated annual sales approaching $300 million. The lake is the only domestic source of commercially marketed primary magnesium, and one of two locations in the United States that produces agricultural sulfate-of-potash.

Artemia (brine shrimp) harvesting has been associated with Great Salt Lake for several decades. During this time, the industry has transformed from a domestic industry marketing to home aquarium hobbyists to an industry marketing to aqua-culture operations worldwide. Current estimated sales is $10 million annually, with artemia supply and not market demand being the limiting factor. The future of the brine shrimp industry depends on several factors including lake ecology, competing artemia sources, and development of alternative feed supplies for aquaculture.

The willingness of the public to make expenditures for recreational facilities around the lake appears to be increasing. Current government expenditures for managing facilities are approaching $3 million annually. About half of this is federal government expenditures, with most of the remaining being state or local funding. Two visitors centers at wildlife refuges are in the construction or planning phases, indicating the public's heightened ecological awareness and support for facilities.

REFERENCES

Artemia International, 2000, Artemia International - purveyors of quality artemia cysts: Accessed at Internet site http://www.artemia-international.com.

Associated Press, 1999, County urged to support plan for Farmington Bay: Deseret News, (Feb. 28).

Brine Shrimp Direct, 2000, Directory of Products: Accessed at Internet site http://www.brineshrimpdirect.com

Chemical Market Reporter, 1991-2001 issues, Schnell Publishing Co., New York, NY.

Loomis, Brandon, 1998, Island's Bison are Fat and Sassy for 12th Roundup: Salt Lake Tribune, (Nov. 12).

McCrae, Jean, 1996, Oregon Developmental Species Brine Shrimp Artemia sp: Oregon Department of Fish and Wildlife: Accessed at Internet site http://www.hmsc.orst.edu.odfw/devfish/sp/brine.html

M & M Distributing, October 1998: Accessed at Internet site http://www.mnmonline.com

Miller, Woodruff., and Millis, E.L.,1989, Estimating Evaporation from Utah's Great Salt Lake Using Thermal Infrared Satellite Imagery: American Water Resources Association, Paper no. 88083, Water Resources Bulletin, v. 25, no. 3, 10 p..

Mono Lake Committee, 2000: Accessed at Internet site http://www.monolake.org.

Morgan, D. L., 1973, The Great Salt Lake: Albuquerque, University of New Mexico Press, 432 p.

Platt's Metals Week, 1999, Maccorp Magnesium Cell Upgrade Slight Delay So Far, Platt's Metals Week, v. 70. no. 16, (April 19), p. 5.

—2000. Magnesium Market's Better Days Still to Come, Platt's Metals Week, v. 71. no. 26, (June 26), p. 2.

Twyman, Gib, 2001, MagCorp Filing Raises a Red Flag: Deseret News, (Aug. 5).

Renco Metals Inc., 2000, U.S. Securities and Exchange Commission (SEC) Form 10-K405. (January 31, 2000).

U.S. Fish and Wildlife Service, 1980, National Survey of Fishing, Hunting, and Wildlife-Associated Recreation, p. 22

—1991, National Survey of Fishing, Hunting, and Wildlife-Associated Recreation, p. 29.

—1996, National Survey of Fishing, Hunting, and Wildlife-Associated Recreation, p. 25.

U.S. Geological Survey Minerals Yearbooks, 1987-1996, Volume I, Metals and Minerals, Magnesium-chapter editions.

Utah Department of Workforce Services, 1999, Annual Report of Labor Market Information, 1998.

Utah Department of Workforce Services, 2000, FirmFind: Accessed at http://firmfind.dws.state.ut.us/pgMain.asp.

Utah Division of Parks and Recreation, 1999: Accessed at Internet site http://www.nr.state.ut.us/parks/www1/ante.htm

Utah State Tax Commission, 2000, Annual Report Fiscal Year 1998-99, p 86.

—2001, Annual Report Fiscal Year 1999-2000, p 107.

Woolf, J., "Utah Brine Shrimp Feed Asia Industry" Salt Lake Tribune, May 24, 1998.

THE EXTRACTION OF MINERAL RESOURCES FROM GREAT SALT LAKE, UTAH: HISTORY, DEVELOPMENTAL MILESTONES, AND FACTORS INFLUENCING SALT EXTRACTION

by

J. Wallace Gwynn,
Utah Geological Survey, P.O. Box 146100,
Salt Lake City, Utah, 84114-6100, USA

This paper is reprinted, with permission from Utah Geological Survey Miscellaneous Publication 01-2, 2001

ABSTRACT

Salt from Great Salt Lake was probably used by the Native Americans prior to the Jim Bridger Expedition to the lake in 1824-25, and was used by the region's early explorers and trappers after that time. From the arrival of the Mormon pioneers in 1847, salt has been produced from Great Salt Lake, starting first with the collection of crude salt from along its shores, and later by boiling down the lake's brine to obtain crystalline salt.

Since these early beginnings, salt production from the lake has become a substantial commercial enterprise. Since the early 1960s, research and development has led to the economic production of potassium sulfate, magnesium metal, chlorine gas, magnesium chloride products, and nutritional supplements. Sodium sulfate has been produced, but is no longer marketed. Many of the salt-producing techniques and equipment that were developed prior to 1964 are still being used today.

Morton Salt, Cargill Salt, and IMC Salt currently extract sodium chloride, which is precipitated during the early stages of evaporation when the brine concentration reaches saturation, or about 26 to 27 percent total dissolved solids. Magnesium Corporation of America produces magnesium metal and chlorine gas, and the IMC Kalium Ogden Corporation produces potassium sulfate, as well as magnesium chloride brine and flake products. These salt, gaseous, and metallic products are processed from the sodium-chloride-saturated, high-magnesium-chloride brines, and from the potassium and magnesium salts that are precipitated from the lake brine after it has been further concentrated.

One obstacle to salt production is the fluctuating level of Great Salt Lake. In 1983, the lake level began a dramatic rise from an elevation of 4,200 feet (1,280.16 m) to its historic high of 4,211.85 feet (1,283.77 m) in 1986-87 (uncorrected USGS provisional lake-level records). The State of Utah employed flood-control measures, including breaching the Southern Pacific Railroad causeway in 1984, and pumping lake water to a shallow basin west of the lake in 1987. During the high-water years, the lake's mineral-extraction industries faced many challenges including the need to raise dikes to prevent flooding of facilities, broken dikes which resulted in the inundation of solar ponds, and dealing with the lake's greatly reduced brine salinities which resulted in lower annual salt and concentrated-brine production.

The years following the record-high lake levels have been a time of adaptation, innovation, and change for Great Salt Lake industries. After its solar ponds were flooded in 1986, AMAX, Inc. built a new solar-pond complex near Knolls, and used the concentrated brines from the West Pond, generated by the State's West Desert Pumping Project,. Great Salt Lake Minerals built a 21-mile-long (33.8 km) open, underwater canal, called the Behrens Trench, to convey concentrated brines from a new, remote solar pond on the west side of the lake to the east side of the lake, to help increase its production of sulfate-of-potash. There were also numerous changes in corporate ownership and production-facility locations.

EARLY RECOVERY OF SALTS FROM THE LAKE

Pre-1847 Salt Procurement and Use

Before the Mormon pioneers entered the Salt Lake Valley in 1847, Great Salt Lake was probably a source of salt for the Native Americans living around its shores. However, no evidence of any extensive Native American exploitation of the salt resource remains (Clark and Helgren, 1980).

The first Euro-American settlers known to use salt from the lake were mountain men from Ashley's Rocky Mountain Fur Company, including Jim Bridger. During the late fall of 1825, a rendezvous site was established near the present location of Ogden, Utah. While camped in the area, the mountain men boiled away some of the lake brine in a kettle to obtain salt (Clark and Helgren, 1980).

John C. Fremont's memoirs (Fremont, 1887) of his second expedition west (1843-1844) describe a trip in a specially prepared rubber raft from a point near the outlet of the

Weber River to what is now called Fremont Island. While returning to the mainland the next morning, Fremont filled a 5-gallon (18.9 L) bucket with brine from which he intended to make salt. Fremont described the salt-making process:

> Today we remained at this camp, in order to obtain some further observations and to boil down the water which had been brought from the lake for a supply of salt. Roughly evaporated over the fire, the five gallons of water yielded fourteen pints of very fine-grained and very white salt, of which the whole lake may be regarded as a saturated solution.

1847-1870 Mormon-Pioneer Period

On July 28, 1847, the Mormon leader, Brigham Young, and some of his associates made a trip to Great Salt Lake to satisfy their curiosity as to the nature of this well-known landmark, and to bathe in its buoyant water. While there, they gathered some salt off the rocks, which was pure, white, and fine. The salt found on the shore of the lake proved initially to be as important to the pioneers as that found in the water. Later that year, a committee assigned to extract salt from the lake water and to gather salt from along the lake shore left Salt Lake City on August 9th and returned August 13th. During this time they prepared 125 bushels (4,405 L) of coarse white salt, probably from a large bed of salt 6 inches (15.2 cm) deep lying between two sand bars (Clark and Helgren, 1980). The committee reported enough pure salt in this bed to provide at least 10 wagon loads without further refining. These shore deposits, however, yielded a poor-quality, bitter-tasting salt due to the other minerals found in them. During this trip, they also boiled down four barrels of salt water to one barrel of fine white table salt (Clark and Helgren, 1980).

The earliest known, permanent salt-boiling operation was established in the spring of 1850 by Charley White, according to reports by Lieutenant J.W. Gunnison and Captain Howard Stansbury, who were conducting government surveys on the lake's south shore. Gunnison reported that White could produce 300 pounds (136.2 kg) of salt per day by boiling brine in six 60-gallon (227 L) kettles. Charley White operated this salt company until 1861. In 1870, the Ninth Census reports only one facility producing salt in Utah. This operation may have been owned by the Joseph Griffith and William F. Moss families of "E.T. City" or Lake Point. The Moss and Griffith salt works were a small home industry, most likely run as a sideline to a farm or ranch (Clark and Helgren, 1980).

EARLY DEVELOPMENT OF SALT-EXTRACTION METHODS AND EQUIPMENT

Dikes, Solar-Evaporation Ponds, and Use of Pumps

By 1873, the level of Great Salt Lake had risen enough to inundate many of the natural salt beds. A new brine-concentration method was tried; dikes were constructed across the entrances of coves and along the south shore of the lake so that the periodic rise and fall of the lake could fill the diked areas with brine. Early salt makers depended on wind tides resulting from northwest winds, which raised the water level on the southern shore of the lake from 1 to 1.5 feet (.3 to .48 m), thereby filling the diked areas with fresh brine. However, the wind tides were not dependable and some of the stronger storms washed away the dikes and dissolved the salt that had been deposited (Clark and Helgren, 1980).

With experience, the early salt makers learned that earth alone was unsuitable for constructing dikes, and planks would not bear the weight of the heavy brine waves. Jeremy and Company, organized in 1870, successfully constructed its dikes by driving two rows of cottonwood stakes into the ground every 2 feet (0.6 m), placing the two rows of stakes 7 feet (2.1 m) apart. A latticework of willows was then woven on each row of stakes and the willows were backed by several inches of bulrushes. The area between the two rows of stakes was filled with earth, making a substantial dike that proved effective for constructing ponds from 5 to 100 acres (2 to 40.5 hm^2) in extent (Clark and Helgren, 1980). The remains of a later version of this type of dike are shown in figure 1 wherein wooden planks were used instead of willows and bulrushes.

Natural wind-tide fluctuations of the lake were too unreliable for filling the ponds, and by 1880, some of the salt companies began using steam- or horse-powered pumps to fill their ponds with brine. In 1888, the Inland Salt Company had established a central (steam?) power source to run a 10-inch centrifugal pump and the machinery in its mill (Clark and Helgren, 1980).

Figure 1. *Abandoned remains of soil-filled, "stake and plank" solar pond dike from Morton Salt's original ponds at Burmester.*

Fractional Crystallization

The original method of producing solar salt by flooding an area and allowing the brine to completely evaporate produced very bitter-flavored salt. Salt producers learned that by not evaporating all of the water they could precipitate a high-quality salt and discard the remaining brine or "bittern" (Clark and Helgren, 1980). This process of controlled evaporation and precipitation, called fractional crystallization, was developed and first implemented by the Inland Salt Company in 1888. Fractional crystallization consists of moving the brine through a series of interconnected evaporation ponds and precipitating only the desired salts in specific ponds as the brine concentration increases (figure 2). First, lake brines are pumped into settling ponds where most of the sediment and debris settle out, and the small amounts of calcium and magnesium carbonates and sulfates are precipitated. Second, the brines are moved from settling ponds into "evaporation" ponds where they are concentrated until the "salt point" is reached. Third, the brines from evaporation ponds are finally moved to the "garden" or "crystallizer" ponds where large quantities of halite (sodium chloride) are precipitated. By draining the remaining brine off the "garden ponds" at the optimum concentration, or before the amount of potassium, magnesium, and sulfate becomes too high, a non-bitter salt is produced. If the ponds are drained below the optimum concentration, good salt is lost (Clark and Helgren, 1980).

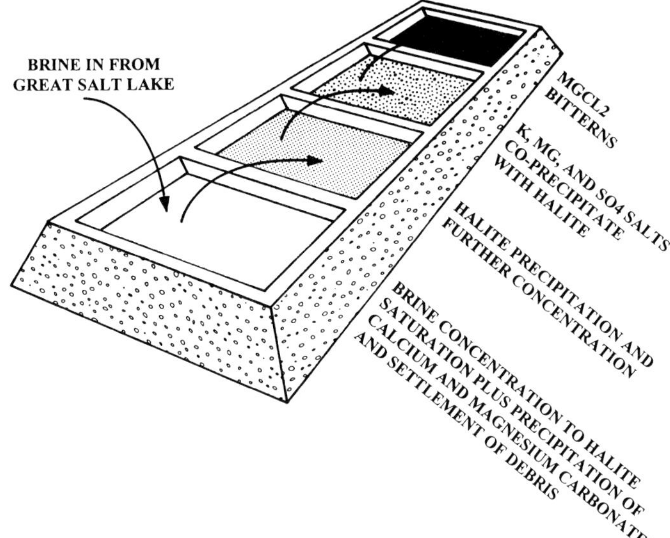

Figure 2. *Schematic diagram illustrating the concept of fractional crystallization. As brine is concentrated, it is moved from pond to pond where the desired minerals are precipitated.*

The Split

The Inland Salt Company also developed a procedure referred to as creating the "split" or forming a cleavage plane between the pond's salt floor and the new crop of salt. This was done to improve the ease of harvesting new crops of salt. Without a split, the crystals from a new crop would interlock with the large salt crystals on the pond floor, making a hard, continuous layer of salt with no way of breaking the new salt loose. Two types of splits were developed: the mechanical split and the sun split. A "mechanical split" is formed by dragging a heavy object, such as a piece of railroad rail, over the pond floor (figure 3). This knocks the edges off the large existing salt crystals and forms a layer of fine salt crystals termed the "split" or "cleavage plane" (Clark and Helgren, 1980). A "sun split" is made by draining or filling the ponds until just a small amount of highly concentrated brine covers the floor. The split is then created by allowing a layer of very fine crystals to precipitate to a depth of 1/8 inch (.32 cm) over the large, jagged salt crystals on the pond floor. After the fine crystals are deposited, fresh, highly concentrated brine is brought into the ponds. The large crystals of the new crop of salt then precipitate upon the layer of fine crystals or the "cleavage plane" (Clark and Helgren, 1980).

Figure 3. *Tractor pulling "drags" to form a "mechanical split" in a solar pond (photo courtesy of Morton Salt Company).*

Salt-Harvesting Equipment

The use of tractors was introduced at Morton Salt Company's harvesting plant in 1923, when Ed Cassidy brought his farm tractor to the Burmester ponds to replace the horses used to pull the plows. Machinery had not previously been used for fear that its heavy weight would break through the thin salt floors. Following Cassidy's successful venture, the company purchased some Fordson tractors to plow the salt, though salt was still stockpiled by hand (Clark and Helgren, 1980).

Local inventors contributed various other ideas that helped mechanize salt production. When Morton Salt consolidated its production facilities at Saltair in 1933, salt was still harvested by hand. To implement mechanical harvesting, the salt-floor thickness had to be increased to 18 inches (.46 m), and the dikes around the ponds had to be raised. After the pond floors had been thickened, local inventors modified a small farm tractor to scrape salt into a bin that was pulled across the ponds. In 1938, the modified tractor was replaced by a local invention called a "Hootin Nanny." In 1949, another local invention, called a "Jackrabbit," was used until it was replaced by a commercially manufactured machine called the "Scoop-Mobile" (figure 4). The "Scoop-Mobile" was replaced in 1964 by a revolutionary new machine called the "Palmer-Richards Salt Harvesting Combine" (figure 5). This machine was developed locally by James Palmer and A.Z. Richards, Jr. of the Solar Salt Company (Clark and Helgren, 1980). The Palmer-Richards Salt Harvesting Combine, or versions of it, are still in common use today.

Figure 4. "Scoop-Mobile" loading salt (photo courtesy of Morton Salt Company).

Figure 5. Palmer-Richards salt harvester (center) being moved by tractor (right), loading salt into truck (left) (photo courtesy of Morton Salt Company).

Cutoff Trench

Brine leakage and loss into porous sediments underlying the ponds at Lake Point had presented problems to salt producers since 1901. The Weir Company corrected this problem by digging trenches around the ponds to an impervious clay stratum. The material that was originally excavated from the trenches was then thoroughly mixed and replaced into the trenches, thus forming a "cutoff" trench (figure 6). This trench-fill material interrupted the lateral continuity of permeable strata by lowering its permeability relative to the undisturbed sediments. This reduced or prevented brine movement outward from under the pond. If the material excavated from the trench was too sandy, clay from borrow areas was added. Pond dikes were built on top of the cutoff trench (Clark and Helgren, 1980; Gwynn and Sturm, 1980). To prevent leakage through typically coarse dike-construction material, the material used to fill the trench, or other low-permeability material, was extended upward into the dike as well.

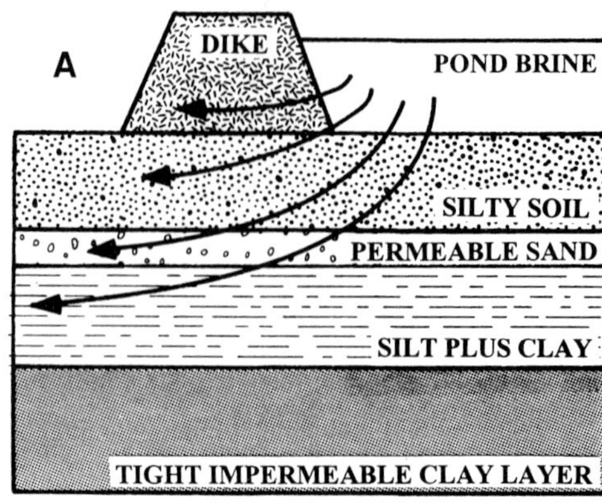

Brines escaping down through pond floor or laterally out of pond area through permeable layers of silty soil, sand and silty clay, or through permeable dike material.

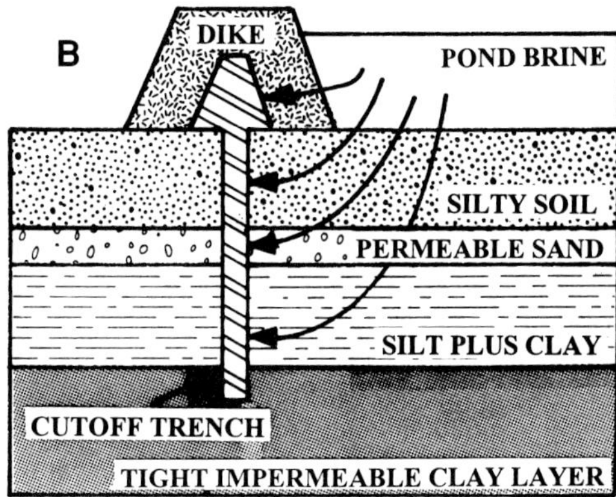

Brine seeping down through pond floor or laterally through the permeable dike material, comes up against cutoff trench or dike core filled with impermeable back fill material and cannot move laterally from pond area.

Figure 6. Schematic cross section showing leakage of brine from solar pond (A) without cutoff trench, and containment of brine in solar pond when a cutoff trench is present (B).

THE MINERAL-EXTRACTION INDUSTRY TODAY

Sodium Chloride Production and Companies

From the late-1800s until the present, numerous salt-producing companies have come and gone around Great Salt Lake (see Clarke and Helgren, 1980). Today, three salt producers operate on Great Salt Lake. Morton Salt and Cargill Salt operate at the south end of the lake, while IMC Salt operates at the north end (figure 7). These producers use the basic principles of solar evaporation, developed and perfected over many years, to produce a high-purity product.

In modern practice, lake brine is concentrated in large evaporation ponds to the point of sodium chloride saturation, which is about 26 to 27 percent total dissolved salts, equivalent to a brine density of about 1.224 g/cc, or approximately

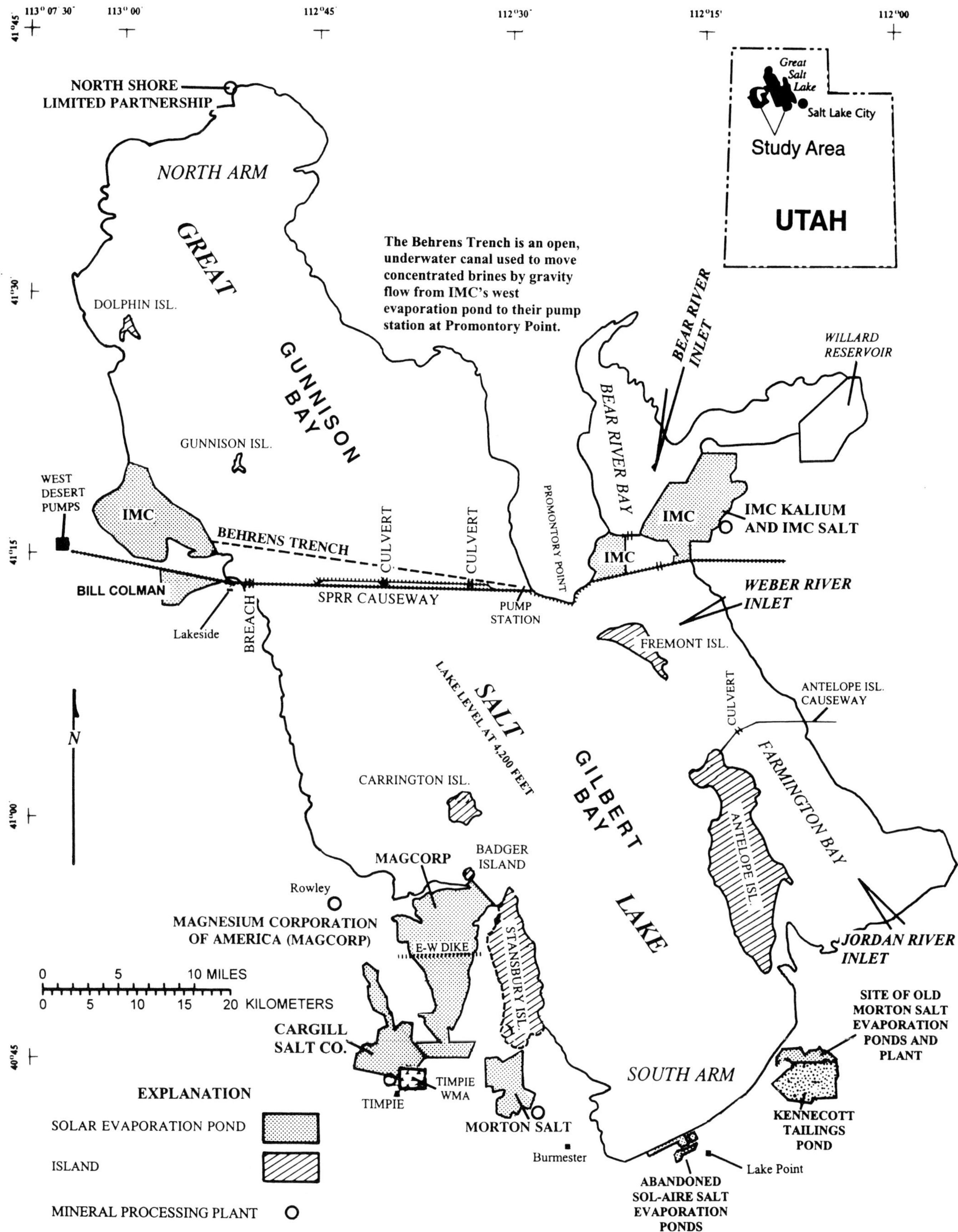

Figure 7. *Locations of Great Salt Lake mineral-extraction industries, the Southern Pacific Railroad (SPRR) causeway, and the Behrens Trench.*

26.53 degrees Baumé (degrees Baumé are often reported by the salt industry instead of density units [g/cc]). The conversion from specific gravity (g/cc) to degrees Baumé is made as follows (Mannar and Bradley, 1983): degrees Baumé = 145 - (145 ÷ specific gravity of the brine at 15.6°C). Salt producers using the less concentrated south-arm brine need much larger ponds for pre-concentration than those using the more concentrated north-arm brine, to obtain equivalent yields.

When the brine reaches the saturation point, it is moved to the "crystallizers" where sodium chloride precipitates and accumulates on the bottom of the ponds. In general, halite is allowed to precipitate during the major part of the summer until the brine reaches a density of 1.245 g/cc or about 28.53 degrees Baumé (Nate Tuttle, Morton Salt Co., verbal communication, April 1999). At this concentration, the remaining brine is drained from the ponds and returned to the lake. The cutoff point used by the different salt producers may vary slightly from this value, however. If the brines are drained much before reaching a density of 1.245 g/cc, good sodium chloride is discarded, and if drained much after this point, elevated amounts of magnesium, potassium, and sulfate in the precipitated salt lower its quality.

The salt industry currently uses more than 80,000 acres (32,376 ha) of solar-evaporation ponds around Great Salt Lake for the annual production of more than two million tons (1,814,000 MT) of sodium chloride and other products.

Morton Salt Company

Clark and Helgren (1980) relate the history of the Morton Salt Company and its predecessors from the early 1900s until about 1980. Morton's history, since 1980, includes two episodes of flooding, and the relocation of its operation. Morton Salt first experienced flooding in the early 1980s, not from the lake, but from a breach in Kennecott Utah Copper Company's tailings-pond dike south of Morton's (old) operations. Then, starting in 1983, Morton Salt experienced minor flooding as the lake level rose, forcing them to raise some of their dikes. As the lake level continued to rise, dilution of the lake brine reduced the amount of salt they were able to produce. In September 1991, the Morton Salt plant and its solar ponds adjacent to the Kennecott tailings pond were abandoned, and its operations were transferred to the North American Salt facilities near the south end of Stansbury Island (figure 7). This move was brokered by Kennecott as part of its tailings pond expansion program. Morton's old evaporation ponds and salt-processing facilities have since been covered by Kennecott's tailings-pond expansion.

IMC Salt Company

The history of the IMC Salt Company's predecessor, North American Salt Company (preceded by American Salt), from the early 1900s up until about 1980 is reported by Clark and Helgren (1980). During the 1980s, when lake levels rose, American Salt's solar ponds and processing facilities at the south end of the lake were situated high enough above the lake that its dikes were not breached. American Salt was forced to expand its evaporation ponds to the west, however, in order to recover lost production capacity due to diluted lake brines. For a period of time during the high-water years, when American Salt could not produce salt, it contracted with AMAX Magnesium Company (a magnesium-from-brine company located northwest of American Salt) to obtain salt from its Stansbury Basin solar evaporation ponds.

In 1991, American Salt was purchased by the Harris Group of New York, and renamed the North American Salt Company. When Morton Salt sold its facilities to Kennecott and took over the south-shore operations of North American Salt, North American Salt's management moved northward and consolidated operations with a sister company's salt production and processing facilities located west of Ogden, Utah. There, North American Salt harvested salt from the early-stage solar ponds of its sister company, Great Salt Lake Minerals (figure 7). In April 1998, North American Salt was sold to IMC Global, and renamed IMC Salt, Inc. Today, IMC Salt is the largest producer of sodium chloride products from Great Salt Lake.

Cargill Salt Company

Cargill Salt Company's Lake Point facility was operated by many predecessors (Clark and Helgren, 1980), including AKZO Salt, and the Lakepoint Salt Company, who had purchased the facilities from Hardy Salt Company in 1977. The Lake Point operation (figure 7) was then sold to Domtar Salt Company in 1982, who installed new salt-washing and stockpiling facilities. In about 1982-1983, Domtar's entire solar pond system was inundated when the Union Pacific Railroad installed a culvert through its causeway to equalize the water elevations on the north and south sides. In 1984, Domtar sold its salt-processing facilities to AMAX and the name of the operation was changed to Sol-Aire Salt and Chemical Company. Sol-Aire Salt immediately prospered as it used the salt that had accumulated in AMAX's Stansbury Basin solar ponds over the years. This prosperity did not last long, however, because on June 7, 1986, AMAX's Stansbury Basin ponds flooded when its northern main dike, west of Badger Island, broke during a storm. This left Sol-Aire without a source of salt (Bauer, this publication). Salt was then hauled from Kaiser Chemicals (now Reilly Chemicals) near Wendover, Utah, until AMAX operations resumed on the lake in the fall of 1988.

Diamond Crystal Salt purchased the Sol-Aire Salt plant facilities at Lake Point from AMAX in 1987, and also purchased land west of AMAX's solar ponds that would be suitable for a plant site adjacent to the Timpie Waterfowl Management Area (figure 7). A washing facility was soon constructed at the Timpie site. Diamond Crystal also constructed an east-west dike within the main body of AMAX's Stansbury Basin solar pond. After the new dike was completed, water was pumped from the southern part of the divided pond to expose the residual salt floor. This residual salt was harvested, washed at the Timpie washing plant, and shipped by truck to the Lake Point facility for further processing (Bauer, this publication).

In 1989, AKZO Salt, a large European company, purchased International Salt Company, who had purchased Diamond Crystal Salt Company, and changed the name of the Timpie operation to AKZO Salt of Utah. From 1991 to 1994, the present-day facility at Timpie was constructed, and the Lake Point facility was shut down and partially reclaimed. In 1995, AKZO merged with Nobel, a Swedish Corporation, and the resulting company was named AKZO-Nobel. In 1997, Cargill Salt purchased all of the AKZO-Nobel's salt producing facilities in the U.S., and is the current owner and operator of the Timpie facilities (Bauer, in preparation).

Non-Sodium Chloride Salt Production and Companies

Lake industries that produce non-sodium chloride salts, or more specifically magnesium and potassium salts and highly concentrated magnesium chloride brines, begin their processes the same way as the sodium-chloride salt industries, that is, through the solar evaporation of lake brine. The major differences among the evaporation processes are the greater amounts of water evaporated, the degree to which fractional crystallization is used, and the small percentage of the original total-dissolved-salts and/or lake brine that is produced as final product. To produce high-magnesium-chloride brines, more than 95 percent of the original water must be evaporated from the lake brine to yield a 7 to 8 percent magnesium chloride brine. These highly concentrated brines are: (1) used by Magnesium Corporation of America (MagCorp) as raw materials for a process that ends in the electrowinning of magnesium metal and chlorine gas, (2) sold by IMC Kalium as a dust suppressant, and (3) marketed by North Shore Limited Partnership for making nutritional supplements. By contrast, just over 50 percent of the water in a brine of equivalent concentration must be evaporated to precipitate sodium chloride. In the production of potassium and magnesium salts, such as those used by IMC Kalium for the manufacture of potassium sulfate, (kainite, schoenite, carnallite, etc.) less than 5 percent of the original dissolved salts remain in the brines that are fed to the final stages of the ponding operations. Here they precipitate as salts with the proper chemistry and sufficient purity to be harvested and used.

The evaporation process required to produce the potassium and magnesium salts used by IMC Kalium is basically the same as that required to produce the high-magnesium-chloride brine produced by MagCorp. MagCorp uses this brine to produce magnesium metal and chlorine gas, and IMC Kalium uses the potassium and magnesium salts to produce sulfate-of-potash.

To obtain highly concentrated potassium- and magnesium-bearing brines economically, the solar ponds must be very large relative to the pond size required by the sodium chloride industries. Companies also carefully monitor the brine chemistry and the chemistry of salts that precipitate as the brine is moved sequentially from pond to pond.

Magnesium Corporation of America

During World War II, National Lead Industries Inc. began to develop technology for producing magnesium using the ferro-silicon process while operating a magnesium plant for the government at Lucky, Ohio. In 1951, National Lead continued to gain expertise in the production of magnesium metal through the formation of a jointly owned company with Allegheny-Ludlum to form the Titanium Metals Corporation of America (TIMET) at Henderson, Nevada (Toomey, 1980).

In the 1960s, National Lead Industries joined with Hogle-Kearns Inc., a Salt Lake City investment firm, and Hooker Chemical, to investigate the possibilities of producing and selling commercial quantities of magnesium metal. In searching for additional sources of magnesium, they became aware of the potential at Great Salt Lake. A review of the various sources of magnesium and economical sources of electrical power led to the selection of the Rowley site on the southwest side of Great Salt Lake as the preferred location to locate a plant.

During 1965 and 1966, National Lead conducted pilot-plant operations to select the best process for use with Great Salt Lake brines. Solar ponds were constructed at Burmester and a pilot plant for producing magnesium chloride brine was built at Lake Point (figure 7). Magnesium chloride from this pilot plant was fed to a prototype electrolytic cell at TIMET in Henderson, Nevada, and in 1966, the first 200-pound (90.7 kg) ingot of magnesium metal made from the brine of Great Salt Lake was produced. By 1967, Hooker Chemical unilaterally left the consortium, and in September 1969, National Lead purchased Hogle-Kearns' interest in the facility, leaving National Lead alone in the project (Haws, 1993). In 1969, after this successful pilot program, National Lead decided to build a full-scale magnesium plant at Rowley to utilize brine from Great Salt Lake. Plant construction began in 1970, and the actual start up of the magnesium operation took place in the summer of 1972. Operations were shut down in 1975 to make process modifications with the assistance of Norsk Hydro, a Norwegian magnesium producer. In the mid-1970s, National Lead changed its name to NL Industries, and in November 1980, NL Industries was sold to AMAX, Inc., a diversified mining and natural resources company, for approximately $60 million (Toomey, 1980; Tripp, this volume).

In 1983, three years after the change in ownership, the lake began a rapid rise, and millions of dollars were expended by AMAX to fortify the main dike that separated its Stansbury Basin ponds from Great Salt Lake (figure 7). On June 7, 1986, however, the main northern dike of AMAX's solar-ponding complex was breached during a severe storm, allowing the dilute lake waters to enter the Stansbury Basin and flood the entire solar-ponding complex. With its solar ponds flooded, AMAX was unable to produce the highly concentrated magnesium chloride brine essential to its magnesium-production process. For the next two years, in order to maintain its operations, AMAX purchased magnesium chloride brine from Reilly Chemicals, 90 miles to the west near Wendover, Utah, and from Leslie Salt (now owned by Cargill Salt) near San Francisco, California (Tripp, this volume).

After its ponds were flooded in 1986, AMAX conducted an extensive investigation to find a new source of magnesium chloride brine. Engineers determined that a solar-evaporation complex could be constructed in a timely manner near Knolls, adjacent to the southern end of the state's West Pond, which contained nearly saturated lake brines (figure 8). By December 1987, the Knolls project was completed and concentrated brine was being pumped from the West Pond into the new solar ponds. After the brine was concentrated to the proper density, it was transported by a 41-mile (66 km) pipeline to the magnesium production facility at Rowley. This new source of brine enabled it to continue production of magnesium metal and chlorine gas (Tripp,this volume).

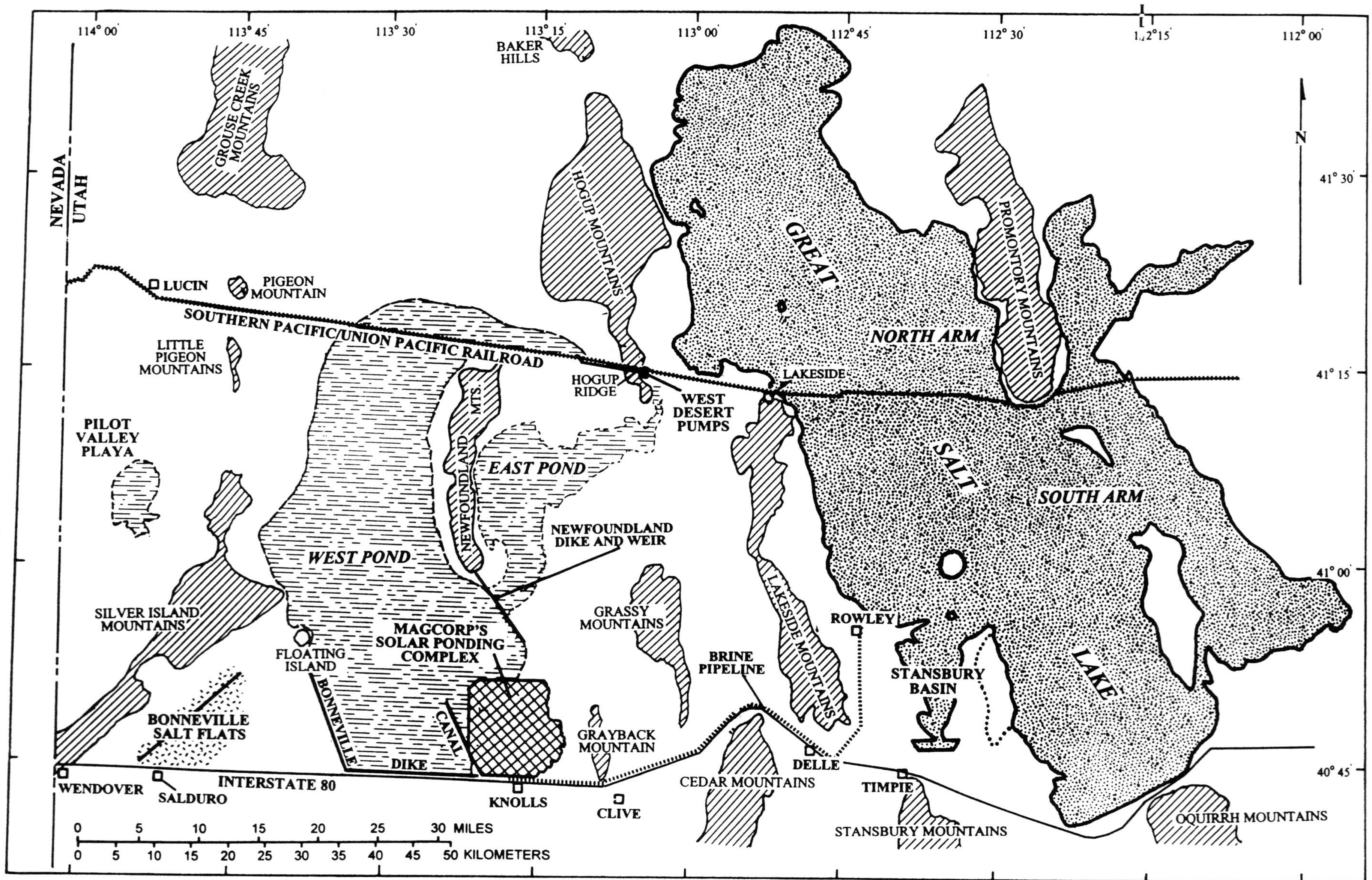

Figure 8. *Elements of the State's West Desert Pumping Project, and MagCorp's Knolls solar-ponding complex; Bonneville Salt Flats, Great Salt Lake, and transportation routes are also shown.*

On September 1, 1989, AMAX sold the Rowley magnesium facility to Renco Inc., a privately held company from New York, and the operation was renamed Magnesium Corporation of America (MagCorp) (Tripp, this volume). In 1995, MagCorp resumed brine-concentration activities in the Stansbury Basin, as well as periodically utilizing the brines from the Knolls facility.

IMC Kalium Ogden Corp.

During the early 1960s, many of the foremost chemical companies, including Dow Chemical Company, Monsanto Chemical Company, Stauffer Chemical Company, Lithium Corporation of America (Lithcoa), and Salzdetfurth A.G., scrambled to reserve acreage for lakeside developments on Great Salt Lake (Kerr, 1965). Of these, Lithcoa and Salzdetfurth A.G. were the first to develop commercial brine/salt operations. The potash facility now operated by IMC Kalium Ogden Corp. (formerly Great Salt Lake Minerals & Chemicals Corp.) was constructed after an exploration project and feasibility study was carried out by Lithcoa. Laboratory studies were conducted in 1963 and 1964, and these were followed by three years of pilot plant testing and construction of pilot evaporation ponds (Industrial Minerals, 1984). During 1964, Lithcoa representatives appeared before the Utah State Land Board (the State agency that regulated lake development, now the Division of Forestry, Fire, and State Lands) in order to acquire permission to extract minerals from Great Salt Lake (Lewis, 1965; Woody, 1982). Within the next year or so, permission was granted.

In 1965, studies continued on methods for extracting minerals from Great Salt Lake. During that same year, Lithcoa entered into a partnership with Salzdetfurth, A.G., of Hannover, West Germany, an important producer of potash and salt, (Lithcoa 51 percent and Salzdetfurth 49 percent ownership) to develop the land and mineral rights on the lake held by Salzdetfurth (Lewis, 1966; Engineering and Mining Journal, 1970).

In 1967, Lithcoa and Chemsalt, Inc., a wholly owned subsidiary of Salzdetfurth, A.G., proceeded with plans to build facilities on the north arm of Great Salt Lake to produce potash, sodium sulfate, magnesium chloride, and salt from the lake brine (Lewis, 1968). Lithcoa was acquired that same year by Gulf Resources and Minerals Co. (Houston, Texas), and at that point, Gulf and Salzdetfurth began developing a $38 million solar evaporation and processing plant west of Ogden (Knudsen,1980). In October 1970, the new facility began operating. The plant was designed to produce 240,000 short tons (217,680 MT) of potassium sulfate, 150,000 short tons (136,050 MT) of sodium sulfate, and up to 500,000 short tons (453,500 MT) of magnesium chloride annually (Gulf Resources & Chemical Corporation, 1970; Eilertsen, 1971).

In May 1973, Gulf Resources bought its German partner's share of the Great Salt Lake project, and changed the name of the operations to Great Salt Lake Minerals and Chemicals Corp. (GSLM). At that time, the German partner had also undergone some changes and was known as Kali und Salz A.G. (Gulf Resources & Chemical Corporation, 1973; Behrens, 1980; Industrial Minerals, 1984).

As Great Salt Lake rose to its historic high of 4211.85 feet (1,283.77 m) in the 1980s, GSLM spent $8.1 million in 1983, $8.1 million in early 1984, $3.0 million in 1985, and $4.8 million in 1986 to protect its evaporation pond system against the rising lake level. On May 5, 1984, a northern outer dike of the system breached, resulting in severe flooding and damage to about 85 percent of the pond complex. The breach resulted in physical damage to dikes, pond floors, bridges, pump stations, and other structures. In addition, brine inventories were diluted, making them unusable for production of sulfate-of-potash (Gulf Resources & Chemical Corporation, 1986). During the next five years, GSLM pumped the water from its solar ponds, reconstructed peripheral and interior dikes and roads, replaced pump stations, and laid down new salt floors in order to restart its operation.

In 1994, as part of GSLM's most recent plans to increase the production of potassium sulfate, a large evaporation pond complex was constructed on the west side of the lake (figure 7). The question of how to move concentrated brines from the new western ponds to the main evaporation-pond complex on the east side of the lake (figure 7), produced a unique answer. A 21-mile (33.8 km) long, open, underwater canal, called the Behrens Trench, was dredged in the lake's north-arm floor, from the western pond's outlet near Strongs Knob to a pump station located just west of the southern tip of Promontory Point. The heavy brine from the west pond is fed into the low-gradient canal, where it flows slowly by gravity eastward, beneath the less-dense Great Salt Lake brine, to the primary pump station. From there, the heavy brine travels around the south end of Promontory Point, then northward, where it begins its journey through the final series of solar-evaporation ponds.

On April 1, 1998, GSLM was purchased by IMC Global and renamed IMC Kalium Ogden Corp. IMC Kalium is presently the largest producer of sulfate-of-potash in North America.

North Shore Limited Partnership

In 1996, North Shore Limited Partnership refurbished a small, 20-acre (8.1 ha) solar pond complex that had been used for salt harvesting in the early to mid-1900s, with the intent of producing a highly concentrated brine (figure 7). Brine was pumped into the ponds in 1997, and the first concentrated bitterns were produced by the end of that year. These brines are processed and refined by a sister corporation, Mineral Resources International, who manufactures an entire line of human dietary supplements. Another sister company, Trace Minerals Research, markets these products.

FACTORS INFLUENCING BRINE CONCENTRATION AND SALT EXTRACTION

The initial concentration of the Great Salt Lake brine that is evaporated to produce salts and concentrated brines is of prime importance to mineral producers. This is because: (1) the higher the feed-brine concentration, the greater the amount of dissolved salt there is per unit volume of brine that is pumped, and (2) at higher concentrations, less water must be evaporated to reach saturation so saturation can be achieved faster. The concentration of Great Salt Lake brines is influenced by two main factors: first, the natural long- and short-term balance between total inflow to the lake and evaporation, and second, the influence of the Southern Pacific Railroad's (SPRR) rock-fill causeway, which divides the main body of the lake into two parts, a north and a south arm (figure 7). Two State-financed flood-control projects of the mid-1980s have influenced the salinity of the lake's two arms and their salt-load distributions. These two projects were: (1) breaching the SPRR causeway in 1984, and (2) pumping lake water to a shallow desert basin west of the lake, commonly known as the West Desert. This latter project operated from April 1987 through June 1989. Additional information about the Great Salt Lake West Desert Pumping Project is found in Utah Division of Water Resources (1999).

Natural Rise and Fall of the Lake

The level of Great Salt Lake fluctuates on both an annual and "long-term" basis (five to 20 years). Annually, the average fluctuation of the lake has been approximately 1.61 feet (.49 m) for the south arm and about 1.33 feet (.41 m) for the north arm, according to U.S. Geological Survey (USGS) provisional lake-level records. The lake is normally at its highest level in about May or June, and at its lowest in October or November. Records kept since the arrival of the Mormon pioneers in the Salt Lake Valley in 1847 show Great Salt Lake has exhibited long-term lake-level fluctuations that have a range of about 20 feet (6.1 m), from a low of 4,191.35 feet (1,277.52 m) above mean sea level (msl) in 1963, to a high of 4,211.85 feet (1,283.77 m) above msl in 1987.

The salinity of the lake brines changes with both long- and short-term lake-elevation fluctuations. From 1966 to 1999, the south arm's salinity fluctuated between 4.6 and 26 percent total dissolved solids (TDS), and the north arm's salinity fluctuated between about 16 and 28 percent TDS (Utah Geological Survey, unpublished data).

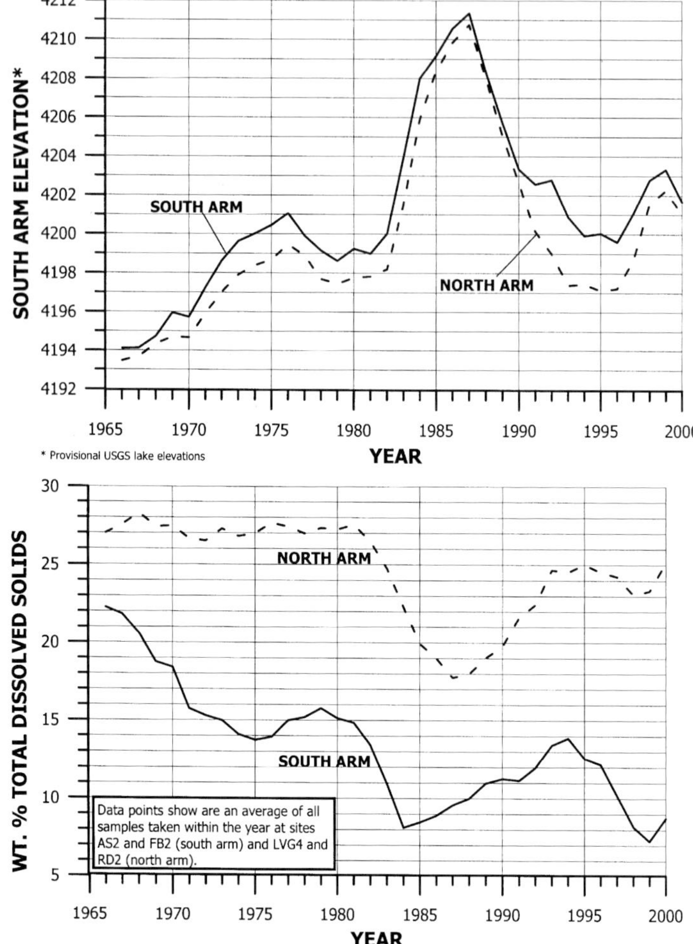

Figure 9. *Water levels of the south and north arms of Great Salt Lake from 1966 to 2000 (top) and average weight percent of dissolved salts in the brine (bottom).*

Influence of the SPRR Causeway

In 1960, the SPRR completed construction of a rock-fill railroad causeway across the lake from Promontory Point on the east to Lakeside on the western shore (figure 7). Once constructed, the causeway prevented the free circulation of water within the main body of the lake, and by 1966 had created a significant difference in the brine salinities of the south and north arms of the lake. The main reason for the difference in salinities is that the south arm of the lake receives the majority of the freshwater inflow to the lake (see figure 7), while the north arm of the lake receives mainly salty water from the south arm. There was limited exchange of water between the two arms of the lake through the somewhat porous causeway fill and its two 15 x 20-foot (4.6 x 6.1 m) culverts, but not enough to prevent a salinity difference from developing. A hydraulic head differential also formed between the two arms of the lake across the causeway, as the south arm developed a higher surface elevation than the north. The south arm also became much less saline than the north arm. Figure 9 shows the average lake elevations (uncorrected USGS provisional lake-level records) and lake salinities (UGS brine-chemistry database) for both the south and north arms of Great Salt Lake, respectively, since mid-1966.

The salinity of the south arm of the lake rises and falls inversely with lake elevation, that is, as the lake level increases, the salinity decreases. It is interesting to note, however, that from 1966 to 1983, the salinity of the north arm did not follow that relationship as closely as the south arm, and remained relatively close to saturation during this time. This was because evaporation was equal to or greater than the diluting effects of the northward flow of lower salinity, south-arm water through the causeway into the north arm. During the lake's high-water years (1983-1987), increased precipitation and the lowered salinity of the south-arm water flowing into the north arm overcame evaporation, and the north arm's salinity dropped as the lake level increased. From 1987 to the present time, the north arm elevation has declined, and its salinity has increased to a level somewhat less than pre-1983 levels. Because the higher inflow of dilute south-arm brine through the breach opening equals or exceeds evaporation, north-arm brine salinity is still exhibiting an inverse relationship with lake elevation.

Breaching the SPRR Causeway

As the lake rose during the 1980s, it flooded, or threatened to flood, many public, private and industrial facilities around the lake. This prompted the State of Utah to investigate a number of flood-control measures, which included: (1) breaching the SPRR causeway, (2) upstream storage, (3) diversions of the Bear River, and (4) pumping water from the lake to the West Desert.

In 1984, the State of Utah first opted to breach the SPRR causeway by constructing a 300-foot (91.4 m) opening (figure 7). This action was designed to allow a greater rate of brine flow from the south arm to the north, and thereby reduce the 3.5-foot (1.06 m) head differential that had developed between the two arms of the lake by the end of 1983. The large head differential between the two arms was caused by the abnormally high inflows of fresh water into the south arm, and by the rather impermeable nature of the SPRR causeway which restricted south-to-north flow. Within about two months after the causeway was breached on August 1, 1984, the head differential between the south and north arms had been reduced to less than 1 foot (0.3 m).

Shortly after the causeway was breached, bi-directional flow began to take place through the breach opening. Large amounts of low-salinity, south-arm water flowed northward through the upper part of the breach opening into the north arm, while at the same time, high-salinity, north-arm brine moved southward through the bottom part of the breach opening into the depths of the south arm, adding considerable volume to the existing deep, dense south-arm brine (Gwynn and Sturm, 1987). As a result of this bi-directional exchange of brines, the overall salt concentration of the south arm increased while that of the north arm decreased.

From the time the breach was constructed in 1984 until about December 2000, south-to-north flow through the breach opening occurred when the level of the south arm was above the 4,199.5-foot to 4,196-foot (1,280.00 m to 1,278.94 m) bottom elevation of the breach. Bi-directional flow occurred when there was more than six feet of water in the breach opening, and the head differential was low (usually less than 1 foot (0.3 m). In December 2000, the breach was deepened to a bottom elevation of about 4,193.0 feet (1,278.02 m) by the State. This will allow for greater north-to-south return flow.

Pumping to the West Desert

After the causeway was breached in 1984, the lake continued to rise, and the State of Utah again reviewed its flood-control options. In 1986, it opted to pump lake water to a 320,000-acre (129,504 ha) impoundment area in the West Desert, commonly known as the West Pond. The West Pond is located about 6 miles (9.65 km) east of the Bonneville Salt Flats, or 25 miles (40.23 km) east of the Utah-Nevada state line, and just north of Interstate Highway 80 (Figure 8). The main purpose of this flood-control measure was to provide additional evaporative area for the lake, in addition to removing up to 690,000 acre-feet (850.8 hm^3) of water from the lake. Pumping to the West Pond started on April 10, 1987, ended on June 30, 1989, and probably lowered the level of the lake by 1.5 to 2 feet (.46 to .61 m) of the nearly 5.5-foot (1.67 m) drop that took place during that time. During the 26-month pumping period, 695 million tons (630.4 million MT) of dissolved salt was removed from Great Salt Lake amounting to about 14 percent of the lake's original salt load (Wold and Waddell, 1994). This removal of salt reduced the lake's salt load and its overall salinity at a given lake elevation. About 386 million tons (350 million MT) of the salt that was pumped to the West Pond remained there as brine or precipitated as crystalline salt when pumping ceased in June 1989. This unusually large amount of salt was precipitated in the West Pond because the north-arm brine was relatively close to saturation when it was pumped from Great Salt Lake. Had south-arm water been pumped, as was specified in the initial pumping-project design, less salt would have precipitated and more of the salt would have returned to the lake after the pumping stopped. North arm water was pumped rather than south-arm water as a project cost-saving measure. In addition to the estimated 386 million tons (350.1 million MT) of dissolved salt retained in the West Pond, about 88 million tons (80 million MT) of salt was diverted from the West Pond by AMAX, 10 million tons (9.07 million MT) seeped into the ground, about 123 million tons (111.56 million MT) flowed out of the West Pond onto the Air Force Test and Training Range, and about 88 million tons (79.81 million MT) could not be accounted for. Only 94 million tons (85.26 million MT) flowed back from the West Pond into Great Salt Lake (Wold and Waddell, 1994).

SUMMARY

Salt from Great Salt Lake was probably used by the Native Americans prior to Jim Bridger's visit to the lake in 1824-25, and was used by the region's early explorers and trappers after that time. The arrival of the Mormon pioneers in 1847 signaled the beginning of the earliest, rudimentary commercial efforts to collect crude salt from along Great Salt Lake's shores and to boil down the lake's brine to obtain crystalline salt (halite).

From these early beginnings, the production of salt from the lake has grown to a large commercial industry, with development prior to 1964 of many of the salt-production techniques and equipment used today. Beginning in the early 1960s, research led to the development of techniques to produce other commodities such as potassium sulfate, magnesium metal, chlorine gas, magnesium chloride products, and nutritional supplements. Sodium sulfate was produced at one time, but is no longer marketed.

The Morton, Cargill, and IMC Salt Companies currently extract sodium chloride, which is precipitated during the early stages of evaporation when the brine concentration reaches saturation or about 26 percent salt. MagCorp produces magnesium metal and chlorine gas, and IMC Kalium produces potassium sulfate and magnesium chloride brine and flake products. These salt, gaseous, and metallic products are processed from the high-magnesium-chloride brines, and from the potassium and magnesium salts that are precipitated from the lake brine after it has been concentrated to the point where more than 95 percent of the original water has been removed.

In 1983, the lake's surface began its dramatic rise from a level of 4,200 feet (1,280.16 m) above msl to its historical high of 4,211.85 feet (1,283.77 m) above msl in 1986-87. The State of Utah employed two flood-control measures to reduce damage caused by the rising lake waters: (1) breaching the SPRR causeway in 1984, and (2) pumping lake water to the West Desert in 1987. During the lake's high-water years, the mineral-extraction industries faced many challenges such as: (1) the need to raise dike heights to prevent flooding of facilities, (2) broken dikes that resulted in the inundation and loss of solar ponds, and (3) reduced brine salinities that resulted in lower annual salt production. Adaptations and innovations developed during and since the flooding years include the following: AMAX Corporation's utilization of the concentrated brines from the West Pond created through the state's West Desert Pumping Project, and Great Salt Lake Mineral's underwater Behrens Trench which conveys concentrated brines from a remote solar pond on the west side of the lake to the east side of the lake. There have also been numerous changes in corporate ownership and production-facility locations during the past 20 years. It must be recognized that Great Salt Lake is a dynamic, hydrologic system that is capable of supplying not only a long-term supply of minerals to the extractive industries, but its share of challenges for them as well.

REFERENCES

Behrens, Peter, 1980, Industrial processing of Great Salt Lake brines by Great Salt Lake Minerals and Chemicals Corp., *in* Gwynn, J.W., editor, Great Salt Lake - a scientific, historical and economic overview: Utah Geological and Mineral Survey Bulletin 116, p. 223-228.

Clark, J.L., and Helgren, Norman, 1980, History and technology of salt production from Great Salt Lake, *in* Gwynn, J.W., editor, Great Salt Lake - a scientific, historical and economic overview: Utah Geological and Mineral Survey Bulletin 116, p. 203-217.

Engineering and Mining Journal, 1970, Utah company gets set to tap mineral wealth of Great Salt Lake: Engineering and Mining Journal, v. 171, no. 4, p. 67-70.

Eilertsen, D.E., 1971, Potash: U.S. Bureau of Mines Minerals Yearbook, 1970 - Metals, Minerals, and Fuels, v. 1, p. 951-964.

Fremont, J.C., 1887, Memoirs of my life: Belford, Clarke and Company, Chicago, p. 230-231.

Gulf Resources & Chemical Corporation, 1970, Annual Report, p. 8.

—1973, Annual Report, p. 12.

—1986, Annual Report, p. 6.

Gwynn, J.W., and Sturm, P.A., 1980, Solar ponding adjacent to the Great Salt Lake, Utah, *in* Gwynn, J.W., editor, Great Salt Lake - a scientific, historical and economic overview: Utah Geological and Mineral Survey Bulletin 116, p. 365-376.

—1987, Effects of breaching the Southern Pacific Railroad causeway, Great Salt Lake, Utah - physical and chemical changes, August 1, 1984-July, 1986: Utah Geological and Mineral Survey Water Resource Bulletin 25, 25 p.

Haws, Bill, 1993, At 34,500 tpy, Magcorp is third largest producer: Rocky Mountain Pay Dirt, p. 4A-8A.

Industrial Minerals, 1984, Great Salt Lake Minerals & Chemicals: Industrial Minerals, London, February issue, no. 197, p. 47-49

Kerr, H.S., 1965, An analysis of Utah's potash industry: Salt Lake City, University of Utah, M.S. thesis, 122 p.

Knudsen, M.B., 1980, Lake's brines were his passport to Utah: Deseret News, January 12, 1980, section S, p. 5.

Lewis, R.W., 1965, Potash: U.S. Bureau of Mines Minerals Yearbook, 1964 - Metals, Minerals, and Fuels, v. 1, p. 865-871.

—1966, Potash: U.S. Bureau of Mines Minerals Yearbook, 1965 - Metals, Minerals, and Fuels, v. 1, p. 747-749.

—1968, Potash: U.S. Bureau of Mines Minerals Yearbook, 1967 - Metals, Minerals, and Fuels., v. 1, p. 975-986.

Mannar, M.L.V., and Bradley, H.L., 1983, Guidelines for the establishment of solar salt facilities from seawater, underground brines and salted lakes: Industrial and Technological Information Bank (INTIB), Industrial Information Section, United Nations Industrial Development Organization (UNIDO) Technology Programme, 173 p.

Toomey, Robert, 1980, Production of magnesium from the Great Salt Lake, in Gwynn, J.W., editor, Great Salt Lake - a scientific, historical and economic overview: Utah Geological and Mineral Survey Bulletin 116, p. 219-222.

Utah Division of Water Resources, 1999, The Great Salt Lake West Desert Pumping Project - its design, development, and operation: Utah Department of Natural Resources, Utah Division of Water Resources, various pagings.

Wold, S.R., and Waddell, K.M., 1994, Salt budget for West Pond, Utah, April 1987 to June 1989: U.S. Geological Survey Water-Resources Investigations Report 93-4028, 20 p.

Woody, R.H., 1982, New faces won't alter Gulf Resources' operations much locally: Salt Lake Tribune, June 20, 1982, section B, p. 10.

A BRIEF HISTORY OF THE MORTON SALT COMPANY

by

Nathan Tuttle and James Huizingh
Morton Salt Company
P.O. Box 506
Grantsville, UT 84029-0506

INTRODUCTION

The Morton Salt Company is the oldest salt company, operating under its original name, now producing salt from Great Salt Lake. Morton Salt Company's roots trace back to 1848 with the founding of Richmond & Company, agents for Onondaga Salt, Syracuse, New York. In 1886, Joy Morton, having acquired a major interest in Richmond & Company, renamed it Joy Morton & Company. In 1910, the company was renamed the Morton Salt Company, and in 1914, the famous "Morton Umbrella Girl" logo, and "When It Rains It Pours" slogan first appeared on the famous blue packages of table salt (Morton Salt, 2001). Beginning in 1918, the Morton Salt Company established itself on Great Salt Lake, and has become one of the lake's stalwart salt-extraction industries.

MORTON'S EARLY HISTORY ON GREAT SALT LAKE

The following, pre-1980 history of the Morton Salt Company is taken from Clark and Helgren, (1980), and sets the stage for the events that have taken place from 1980 to the present.

Inland Crystal Salt Company did not share its dominant position in Utah's salt industry with any serious competitor until Morton Salt Company leased a potash plant at Burmester, Utah, in 1918, and established a competitive foot hold. In 1923, Morton Salt Company purchased controlling interest in the Inland Company from the Mormon Church. By 1927, the remaining stock was acquired, and Inland Crystal Salt Company was reincorporated as a wholly-owned subsidiary under the name of Royal Crystal Salt Company. Morton Salt Company produced salt from its plant at Burmester (figure 1), and also from its subsidiary plant at Saltair until 1933, at which time production and refining facilities were combined at the Saltair location. Although both companies operated from the same plant, the separate identity of Royal Crystal Salt (figure 2) was maintained until that company was dissolved in 1958.

Growth of the Saltair facility enabled the Morton Company to retain a dominant position in the intermountain salt market. Its monopoly faced a temporary threat from several new developments around the lake during the late 1930s and the early 1940s; however, none of the new companies endured more than three or four years. In the 1950s and 1960s, Lake Crystal Salt Company from the north shore, Deseret Livestock Salt Company, and Stansbury Salt Company from the south and southwest shores gained a foothold in the salt business and retained it. Fortunately, these new companies organized at a time when the market was expanding. In the decade following 1950, the market increased 50 percent. It doubled again in the next 10 years. Morton Salt Company increased its production in spite of the competitive pressure.

Over the years, Morton employees were responsible for numerous technological innovations, including the early use of motorized salt-harvesting equipment. Morton's Saltair plant site was also the home of one of the state's long-running, federal cooperative weather stations around Great Salt Lake, which operated continuously from 1956 through 1991. Weather data from this station has been valuable in designing solar-pond facilities on the lake.

MORTON SALT COMPANY 1980 - 2000

The 1980s were a tumultuous time for Morton Salt, starting in 1982, when heavy rainfall just south of the solar ponds caused a flood in the C-7 canal that ran between the ponds and the Kennecott tailings pile. It is estimated that 5.5 inches of rain fell within a couple of days. The flood washed over the crystallizing ponds, wiping out dikes, and dissolving much of the salt that had been laid down in previous years. The flood also flowed through the salt stockpile area, dissolving much of that salt too. All of the drains in and around the plant were overflowing and the runoff could not get to the lake. In response to this disaster, the south periphery dike was raised to stop water from entering the pond area. The Ritter drain was rerouted around the east side of the pond area instead of through the middle of the pond area.

In addition to the flooding just described, the lake was rising at the same time. To combat the rising lake, Morton's north periphery dike was raised. The rising lake was also responsible for a significant dilution of the feed brine to the solar ponds. The dilution of the brine, in turn, prompted the construction of an additional 670 acres of concentrator ponds to evaporate excess water, although that was far less acreage than was actually needed to keep salt production up to pre-flood levels. The brine intake pumps were raised three times during this period. All of this work cost in excess of two million dollars. Because of these problems, Morton was unable to produce enough salt to meet its market demands, and a supplemental supply of salt was obtained from the AMAX ponds in the Stansbury basin west of Stansbury Island (figure 3). This supply of salt carried Morton Salt through the

Figure 1. *W.F. Schuessler's rendition of Morton Salt's plant at Burmester, Utah (date unknown).*

Figure 2. *Old Royal Crystal Salt Company facilities, view looking southeast (about 1936).*

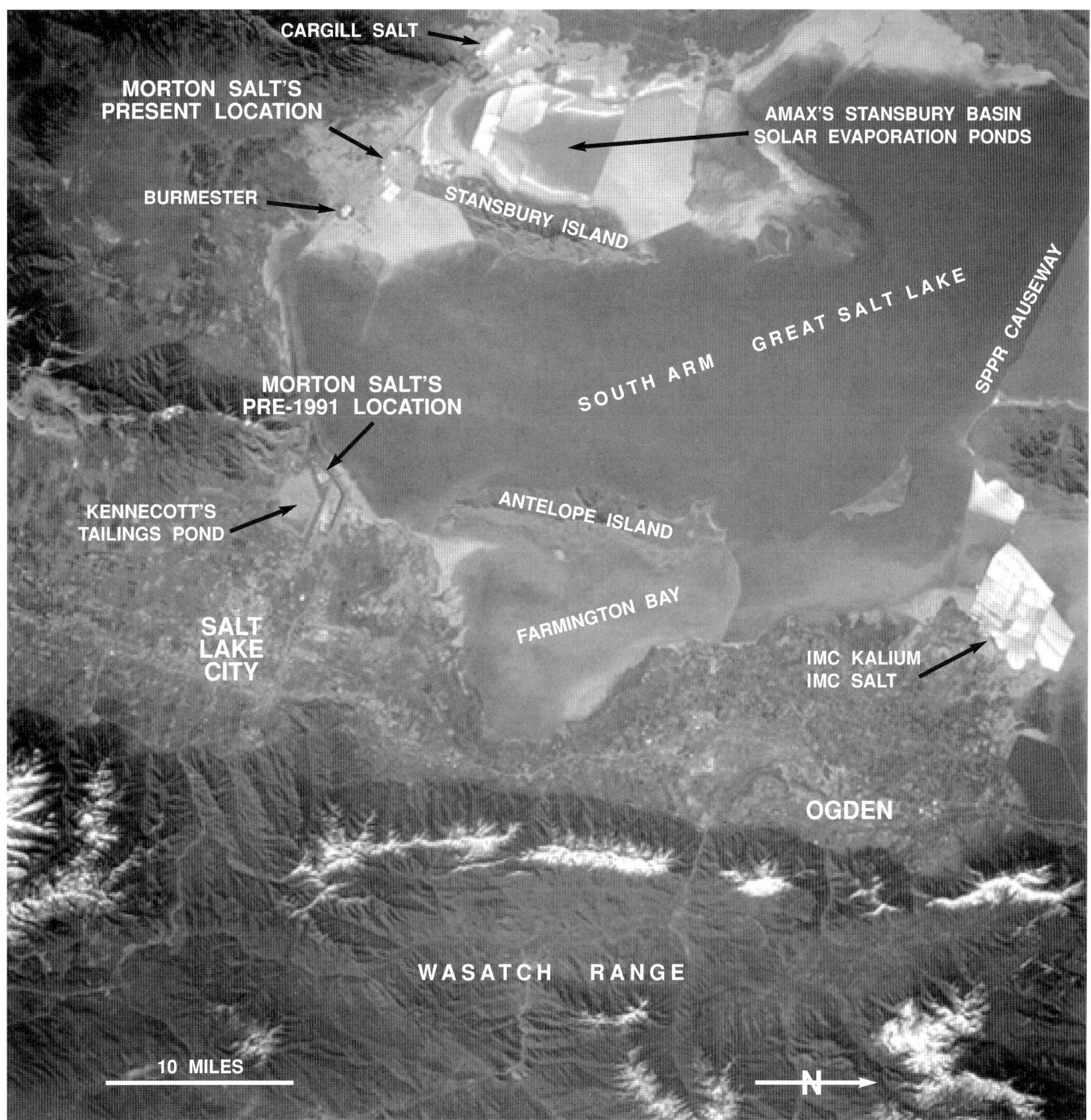

Figure 3. *Location of pre- and post-1991 Morton Salt Company facilities, and other lake features and industry sites. NASA photo STS084-706-14, May 1997.*

lean times. There was a time during the flooding that Morton's corporate office in Chicago was wondering if it was worth the effort of staying in business on the lake. Finally, by 1989, salt production had recovered to a reasonable level, but now, unforseen changes were afoot on the lake.

In 1991, Kennecott was in need of land to expand its tailings pond area. Since Morton Salt's large solar-pond area was just north of the Kennecott tailings area (figures 3 and 4), it was a logical choice for acquisition. Morton Salt, however, was not about to pack up and leave the lake, having just gone through the roughest of times and survived at Saltair. So, Kennecott struck a deal with the North American Salt Company, who at that time owned the salt plant near Grantsville (the old American Salt Plant). Kennecott bought the North American plant and traded it to Morton Salt for its Saltair plant in September 1991. Through this trade, Morton Salt was able to continue producing salt from Great Salt Lake.

Some of the equipment at Morton's Saltair plant was moved to the "new" Morton Salt - Grantsville facility (figure 5). Equipment in the old North American mill was upgraded

Figure 4. *Aerial view of pre-1991 Morton Salt facilities, view looking southwest (date unknown).*

Figure 5. *Recent aerial view of present-day Morton Salt facilities, view looking northeast.*

and automated, and new warehouse and maintenance facilities were built. Fire protection equipment was added to the facilities, and a new septic tank and drain field were built. Many of the older solar-pond dikes were repaired and new dikes were added. The railroad spur off the Union Pacific tracks was improved, and a middle track was added to provide easier movement of rail cars. A much-improved system for producing treated road salt was built behind the mill-stockpile stacking system, which was brought over from Saltair, and a new salt-washing and drying system was installed to provide salt of high purity. Finally, a new office complex was built (figure 5), and the entire area around the facility was paved. All of this work cost in excess of five million dollars. At the end of 2000, Morton Salt and its 140 employees are well positioned to continue providing quality salt products to a large area of the United States.

REFERENCES

Clark, J.L., and Helgren, Norman, 1980, History and Technology of salt production from Great Salt Lake, *in* Gwynn, J.W., editor, Great Salt Lake - a scientific, historical and economic overview: Utah Geological and Mineral Survey Bulletin 116, p. 203-217.

Morton Salt, 2001, Growing from our roots: Online,http//www.mortonsalt.com/mile/prflmile.htm

HISTORY OF THE CARGILL (FORMERLY AKZO) SALT COMPANY FROM EARLY PIONEER TIMES TO THE PRESENT

by
Danny Bauer
Cargill Salt, now Utah State Science Advisor

INTRODUCTION

The Cargill Salt Company, formerly AKZO Nobel, is located near the Timpie Waterfowl Management Area (WMA), at the southern end of Great Salt Lake (figure 1). The Cargill Salt Company is preceded by a long history of changes in corporate ownerships and, in recent years, a change in its physical location from Lake Point, Utah, where it began 150 years ago during the early pioneer period, to its present location.

EARLY HISTORY (PIONEER TIMES TO 1977)

The early history of Cargill's predecessors, up through the creation of the Lakepoint Salt Company in 1977, was recounted by Clarke and Helgren (1980) as follows:

> Salt has been produced from the waters of Great Salt Lake near Lake Point, Utah, since the pioneer period. Salt boilers were set up as a small home industry to supplement the meager income of local farmers. Subse-

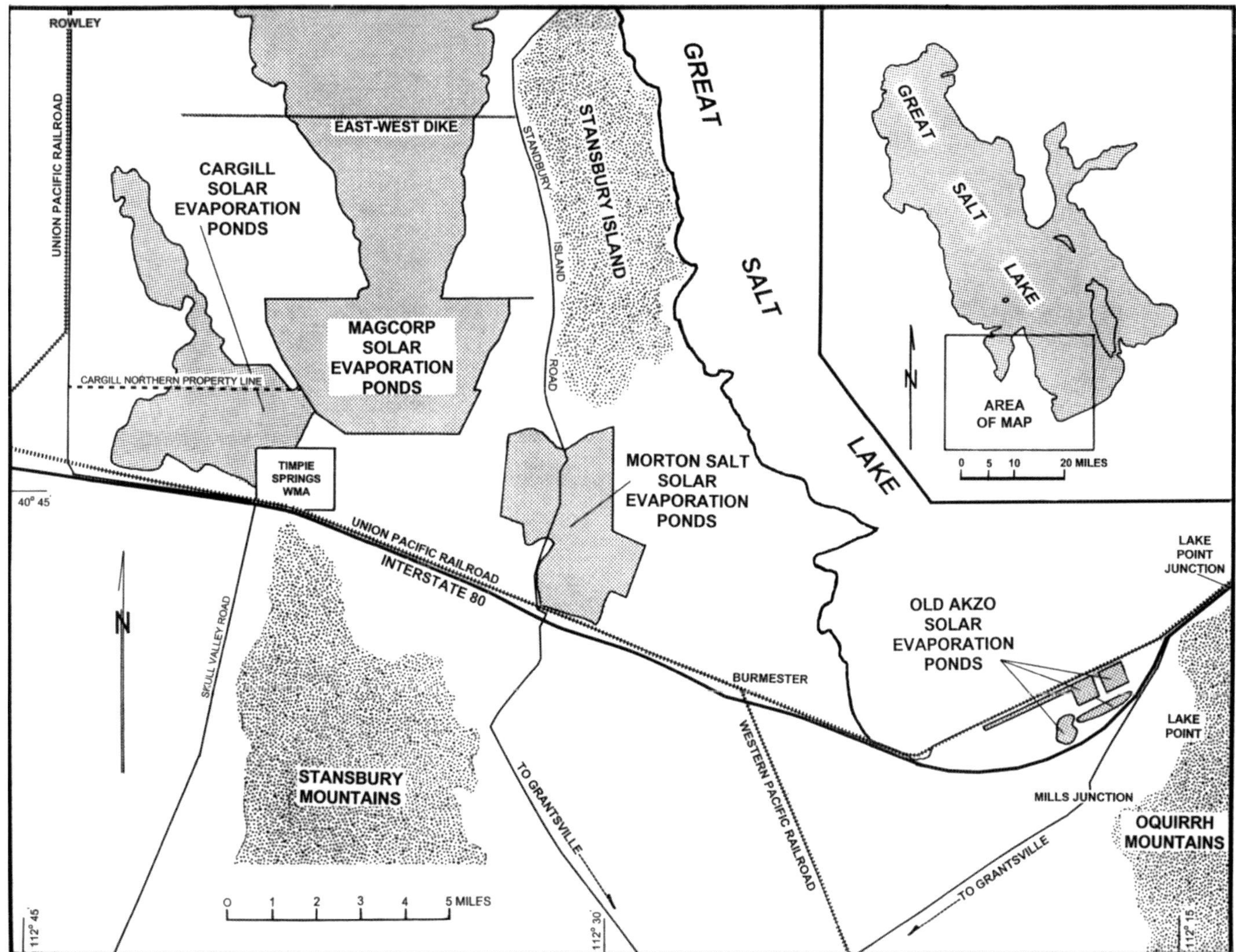

Figure 1. Locations of the Cargill, MagCorp, Morton, and old AKZO solar-ponding sites, along the southern end of Great Salt Lake, Utah.

quently, half a dozen companies have used the location in an attempt to make a profitable enterprise out of salt extraction.

Salt production using the solar-evaporation process at the Lake Point site was first introduced by Weir Salt Company in 1901. Nearly fifty years of dormancy followed Weir's abortive attempt to enter the industry before Deseret Livestock Company reactivated the site in the spring of 1949.

The porous soil underlying the floor of the ponds has presented problems to the salt producers at the Lake Point site since 1901, when Weir Salt Company encountered this condition, and the resulting seepage of brine from the ponds was a contributing factor in the demise of that company. This condition was corrected by digging trenches to an impervious clay under-strata and creating a bond between the dike material and the clay; a seal was achieved that prevented seepage.

As with other salt makers, men at Lake Point experimented with different methods of production. One of the first salt-harvesting machines used at Lake Point was designed by this salt company. Deseret Livestock Salt Company began using a central stockpile, and each of its successors has followed suit.

In late 1952 or early 1953, ownership of Deseret Livestock Company, including the salt works, was sold to David Freed and David Robinson. Knowing little about salt production, they offered that part of their holdings for sale. Council McDaniel purchased the company and reincorporated it under the name of Deseret Salt Company. McDaniel operated the salt works until the latter part of 1958, when he sold to Leslie Salt Company.

In 1961, Leslie Salt Company, largest salt producer on the west coast, was charged by the Federal Trade Commission with creating a monopoly. The complaint alleged that Leslie's acquisition of Deseret Salt Company tended to create a monopoly in the production and sale of salt in the west. The proceeding was settled through a divestiture order requiring Leslie to sell its Utah holdings. On November 2, 1965, Hardy Salt Company, of St. Louis, Missouri, purchased Leslie's Lake Point plant.

In 1977 the Lakepoint Salt Company was formed by purchasing the Hardy Salt operation at Lake Point. A group of local investors teamed up with the former local management of the American Salt Company plant to start this operation. The existing Hardy plant was closed down for several months while extensive revisions were accomplished to return this plant to a profitable operation.

RECENT HISTORY (1977-PRESENT)

In 1982, a Canadian company by the name of Domtar negotiated purchase of Lakepoint Salt Company, retaining the existing management. Domtar installed a new salt-washing and stockpiling facility, and planned on increasing the plant's output, but the sudden increase of the lake level, caused by unseasonably heavy precipitation, put pressure on their dikes. While they were attempting to strengthen their dikes, they received notice from the Union Pacific Railroad, which bordered their pond system on the north, that they (U.P. Railroad) were going to install a culvert through the railroad's base in order to equalize the water on both sides of the railroad. This action inundated the entire Lake Point pond system. Domtar decided not to contest this action.

In 1984, Domtar entered into an agreement to purchase wet solar salt from the AMAX Corporations' pond system at Timpie Point, Tooele County, at exit 77, I-80 West. The plant operated for one year under this arrangement. Domtar then decided to sell the Lake Point operations to AMAX. AMAX changed the name of the operation to Sol-Aire Salt and Chemical Company.

The AMAX solar pond system, built principally for magnesium chloride brine production, also produced millions of tons of by-product sodium chloride. This salt was harvested at the AMAX pond system and shipped by truck to Sol-Aire's Lake Point facility to be processed. This provided Sol-Aire with an extremely attractive situation and under the new ownership, Sol-Aire immediately prospered.

In 1986, Diamond Crystal Salt Company from Michigan entered the picture and proposed a partnership with AMAX to build a large salt complex at Timpie Point. Negotiations were being finalized when the north dike of AMAX's large Stansbury Basin pond system failed, resulting in the complete flooding of their salt and magnesium chloride ponds. Negotiations to form the partnership were put on hold.

When AMAX decided to abandon their flooded Stansbury Basin ponds in 1987, and relocate their pond operations to the West Desert near Knolls, Utah, Diamond Crystal entered into an agreement with AMAX to purchase the Sol-Aire salt plant facilities at Lake Point, and also land that would be suitable for a plant site at Timpie. They also agreed to assume a portion of the solar-pond land leases with the government, in the Stansbury Basin. During the period from 1987 to 1988, Sol-Aire did not have any productive solar ponds and purchased salt from Kaiser Chemicals at Wendover near the Utah-Nevada border. The salt was shipped by a truck and processed at the Lake Point facility.

Starting in February 1987, Diamond Crystal constructed an east-west-trending dike within AMAX's large Stansbury Basin evaporation pond (figure 1) to provide access to large quantities of salt left from the AMAX's pre-flood operations. This work was completed in early 1988, and the lower, flooded, southern third of the pond system was pumped down to expose the residual salt. In the fall of 1988, salt was harvested from the reclaimed ponds, washed at the newly constructed Timpie washing plant, and shipped by truck to the Lake Point facility for further processing.

During this time, preliminary planning for the new salt refinery at Timpie began. In 1989, events far from Utah suddenly changed the direction of the program. AKZO, a large European company with salt operation in the USA (known as International Salt) and in Europe, purchased the Diamond Crystal Salt Company. As a result, the construction activity at Timpie was drastically curtailed. The Sol-Aire name was changed to AKZO Salt of Utah.

In 1991, increased demand for solar salt prompted AKZO to install a small drying operation and bulk-rail loading facility at Timpie. Further investment was made in 1993, which resulted in the completion of the current salt-producing facility at Timpie in 1994. The Lake Point facility (figure 2) was shut down and partially reclaimed. A slight

Figure 2. *Old Lake Point facilities before being shut down and partially reclaimed.*

change in management was made in 1995 when Nobel, a Swedish Corporation merged with AKZO, with the resulting company being called AKZO Nobel.

In 1995 again, events far from Utah would cause major changes in the ownership of the Timpie facility. An underground mine owned by AKZO Nobel, located in Foster, New York, started to flood. The mine, which had been worked for over fifty years, would take over two years to completely flood. This mine was a large supplier to the very profitable highway-salt business in the east. After production had stopped at the mine for two years, AKZO Nobel, which by now was the largest producer of salt in the USA, based on volume, decided to sell all its production facilities in the USA, including the Timpie solar facility.

In 1997, Cargill Salt, which for a time in the early 1960s operated the Lake Point facility under the name of Leslie Salt, purchased all the AKZO Nobel salt-producing facilities in the USA. Cargill, who currently has solar salt operations in Utah, California, Oklahoma, the Island of Bonaire in the Dutch Antilles, Venezuela, and Australia, also mines rock salt in Louisiana and Ohio. Cargill also produce evaporated food-grade salt from facilities in Kansas, Michigan, and New York. At this time, Cargill is investing heavily in the Timpie facility near the Timpie Waterfowl Management Area (WMA) (figures 1, 3 and 4) to bring its salt-drying and processing capacity up to the capacity of the salt-producing solar ponds.

Figure 3. *New Cargill salt-processing facilities near the Timpie Springs WMA.*

Figure 4. *Salt stockpiles at the new Cargill salt-porcessing facility near Timpie WMA.*

REFERENCE

Clark, J.L., and Helgren, Norman, 1980, History and technology of salt production from Great Salt Lake *in* Gwynn, J.W., editor, Great Salt Lake - a scientific, historical and economic overview: Utah Geological and Mineral Survey Bulletin 116, pp 203-217.

PRODUCTION OF MAGNESIUM FROM THE GREAT SALT LAKE

by

G.T. Tripp
Magnesium Corporation of America

ABSTRACT

The relatively high concentration of magnesium chloride in the Great Salt Lake provides for commercial extraction of magnesium metal. Using natural geographical features and drawing on previous industrial experience, a consortium of commercial and manufacturing entities put together a system to extract and purify the magnesium minerals from the lake and generate magnesium metal. The process for producing metal is described. Products, commercial by-products of the process, and their uses are briefly described.

INTRODUCTION AND HISTORY

Through historical time, the Great Salt Lake has been recognized as a valuable source of minerals. In addition to the sodium chloride "salt" incorporated in its name, the lake's waters also serve as the raw material for the magnesium facility operated by Magnesium Corporation of America (MagCorp) at Rowley, Utah. That facility accounts for all of the United States' magnesium-metal production, and about 14 percent of the free world's magnesium-metal production.

Production of magnesium metal from Great Salt Lake has its roots from other places. During World War II, National Lead Industries began to develop technology for the production of magnesium by operating a government-owned magnesium plant at Lucky, Ohio, using the ferro-silicon process. The company gained additional expertise in the production of magnesium metal, in 1951, with the formation of a jointly owned company, Titanium Metals Corporation of America (TIMET), at Henderson, Nevada. Magnesium metal was used as the chemical-reducing agent in the production of titanium. After reduction, the magnesium forms magnesium chloride ($MgCl_2$), which is electrolyzed to recover both magnesium and chlorine.

In the early 1960s, National Lead began investigating the possibility of producing and selling commercial quantities of magnesium metal. In searching for sources of magnesium, National Lead Industries joined with Hogle-Kearns, a Utah investment firm, and Kerr McGee, a diversified chemical company, in a venture to assess the Great Salt Lake's potential for producing magnesium metal. The venture partners sought rights from the State of Utah to Stansbury Basin because they recognized its potential to be developed into an economical, natural solar pond system. The Stansbury basin is a shallow depression in the lake bed which is bounded on the east by Stansbury Island, on the west by the Lakeside Mountains, and on the south by the naturally increasing topography. To the northwest, the planned solar ponds would be bounded by a mud flat that is somewhat higher in elevation than the majority of the basin. To the northeast, the area between Badger Island and Stansbury Island consisted of relatively shallow water that could be diked, and still allow access to deep water for a constant supply of feed brine for solar evaporation.

During 1965 and 1966, National Lead Industries conducted pilot operations to select the best magnesium-producing process using Great Salt Lake brines. Scale-model solar ponds were constructed at Burmester, Utah, and a manufacturing pilot plant, designed for producing electrolytic-cell feed, was built near Lakepoint, Utah. Magnesium chloride product from this pilot plant was trucked 450 miles and fed into a prototype cell at TIMET in Henderson, Nevada. Based on the successes and data obtained from this early work, the decision was made in 1969 to build a magnesium plant at Rowley, Utah, utilizing brine from Great Salt Lake. By the time construction plans were finalized, National Lead had acquired sole ownership of the proposed magnesium operation.

Construction of the integrated facility began in 1970, with Ralph M. Parsons as the general contractor. The new magnesium manufacturing plant site was located ten miles north of Interstate 80 on the west side of the Stansbury basin. The exact location was fixed to straddle two power-district boundaries. This provided alternative sourcing of electrical power, a substantial component of the manufacturing cost. The "Rowley" plant site was named for Jeff Rowley, who was the CEO of National Lead during the period of construction. In addition to normal construction costs, it was necessary to construct 15 miles of paved highway, a railroad spur, a natural gas line, and a dedicated 138,000-volt power line to service the new plant.

Actual start up of the magnesium operations occurred in the summer of 1972. Initially, operations experienced substantial difficulties as viable operating systems were developed. It was necessary to completely shut down operations in 1975 to re-engineer some parts of the process. Norsk Hydro, a magnesium producer in Norway, was contracted to assist in this process, which included developing a system for the removal of boron. Boron, which occurs naturally in Great Salt Lake, had unexpected and adverse affects on the quality of the final magnesium product.

After the 1975 shut down, routine operations suffered economically from the higher than expected capitalization cost, and lower than expected plant productivity. In 1980, NL Industries (National Lead changed its name to NL Industries in the mid-70s) sold the magnesium operation to Amax Inc., a diversified mining and natural resource company.

Shortly after the transfer of ownership to Amax, a change in the weather pattern caused an unprecedented rise in the level of the Great Salt Lake. Amax spent millions of dollars to continue operation; much was spent to raise and fortify the dikes that separated the Stansbury basin ponds from Great Salt Lake. Expenditures were also required to expand the solar evaporation pond area to compensate for the mineral content dilution of the lake brine. Modifications were also made to the Rowley magnesium manufacturing process in an attempt to compensate for weaker than normal feed brines.

In spite of Amax's efforts to raise and fortify its dikes, a storm on June 7, 1986, coupled with record lake elevations, breached the main dike separating the solar ponds from the lake. Over the course of a week, the Stansbury basin ponds filled with 7 feet of lake water while the elevation of the main south arm of the lake dropped by an estimated 5.5 inches.

The extensive time and expense required to bring the flooded Stansbury basin ponds back into production, coupled with the uncertain future lake levels, caused Amax to examine alternative brine sources and solar-ponding sites. During the flooding, magnesium production continued at a reduced rate using concentrated magnesium chloride brines purchased and trucked in from Reilly Chemical near Wendover, Utah, and Leslie Salt located near San Francisco Bay, California.

After substantial analysis, it was determined, that an area near Knolls, Utah, comprised of mud flats interspersed with sand dunes, could be quickly converted into solar ponds at a reasonable cost. Brine from the New Foundland Bay part of the state's West Desert Pumping Project was available to fill the proposed ponds. The permits to use the Knolls site were obtained, and construction with an expedited completion schedule commenced in May of 1987. The engineering firm of Morrison-Knudsen acted as the general contractor. The construction included a six-mile feed canal, six pump stations, a maintenance shop and office facility, over 60 miles of containment dikes, and a 41-mile pipeline from the new ponds to the plant at Rowley. Sufficient construction was completed by December of 1987 to allow the Knolls Solar Evaporation Ponds to be filled.

The initial construction of the West Desert Pumping Project presumed that the high elevation of the lake would continue for an extended period. The project was somewhat abbreviated in order to conserve resources. Changes in the weather at the completion of construction caused the lake elevation to drop faster than was anticipated. With the lower lake level, the original pump station inlet canal would not supply a sufficient volume to the pumps to circulate lake water through the West Desert and fully utilize the newly constructed Knolls solar pond facility. In 1988, Amax assisted the State of Utah in extending the inlet canal to the Hogup Pump Station to provide suction-head protection and capacity to the West Desert Pumping Project. This was accomplished using a large cutter-suction dredge which was then owned and operated by Amax.

The Knolls ponds were designed with the anticipation that the West Desert Pumping project would operate for only a limited period of time. As such, they were built to produce and store enough concentrated brine to supply feedstock to the Rowley plant for 10 to12 years after the State's pumping to the West Desert ceased.

The West Desert Pumping Project operated from April 1987 through June 1989, and then was shut down. Brine was generally accessible to Amax from the Newfoundland Bay until mid-1990 when the brine supply dried up. Since that time, brine has occasionally been available from the Newfoundland Bay during wet periods, or rain events, because of the re-dissolution of the salts left on the surface by the West Desert Pumping Project.

In 1989, Amax sold the magnesium facility to Renco Inc., a privately held company in New York. The magnesium operation was renamed Magnesium Corporation of America (MagCorp).

In the late 1980s and early 1990s, the level of the Great Salt Lake receded as quickly as it had risen in the early 1980s. By 1992, the lake level had retreated to a level that allowed MagCorp to begin the process of re-commissioning the ponds in the Stansbury basin. In 1995, the first brine harvest from the Stansbury basin ponds was brought into the plant.

ROWLEY MAGNESIUM PROCESS

Costs and reliability are both critical factors in any manufacturing process. The same is obviously true for magnesium operations at the Rowley plant. Operating decisions are made with the intent of reducing energy costs and maximizing equipment efficiency. The present Rowley process is described below.

Solar Evaporation

The raw material used in the manufacture of magnesium metal is a concentrated magnesium chloride brine. Magnesium, however, occurs naturally in the Great Salt Lake in very small quantities. The natural concentration of magnesium in the south arm of the lake ranged from a low 0.18 percent in 1986, when the lake level was at its zenith, to approximately one percent at the lake's historic low level in 1963. The nominal concentration of magnesium is about 0.4 percent by weight. To be an economically acceptable feed to the Rowley magnesium manufacturing process, a concentrated feed brine of greater than 8.0 percent (by weight) magnesium is required. To achieve this substantial increase in concentration from 0.4 percent to 8.0 percent magnesium, MagCorp employs the world's most extensive industrial use of solar energy in its Stansbury basin evaporation ponds. The ponds, constructed on lands leased from the State, occupy approximately 65,000 acres, with the actual area covered by water amounting to about 48,000 acres. Between 20- and 35-billion gallons of lake water are annually brought into the Stansbury basin ponds, depending on evaporative performance, and brine requirements. The basin is divided into smaller individual ponds, where efficient operation and maximum recovery are achieved by operating the ponds in a continuous or sequential mode, rather than letting individual ponds evaporate to the desired concentration. When operated this way, the brine advances through the ponds like a slow moving river that becomes shallower as concentration increases. The progressive concentration of magnesium is illustrated in table 1, which starts with the relative ion concentrations of Great Salt Lake brine, and then shows the effluent from three of the sequential ponds. The magnitude of this evaporation step is illustrated by the fact that less than one percent of the original volume of Great Salt Lake brine finally reaches the plant to be used in the manufacture of magnesium metal. During the concentration of the brine, about 5 million tons of mixed salts are deposited in the ponds each year.

Table 1. Percent of each ionic constituent in GSL brine and pond effluents (weight percent of ion in brine).

Ion	Great Salt Lake Brine	Effluent Pond No. 1S	Effluent Pond Pond No. 2WE	Effluent Pond No. 3C to Holding Pond
Mg	0.4	2.0	4.8	8.5
K	0.3	1.5	3.6	0.15
Na	4.0	7.0	2.6	0.2
Li	0.002	0.01	0.024	0.07
B	0.0018	0.009	0.021	0.06
Cl	7.0	14.0	16.0	22.6
SO_4	1.0	5.0	5.3	4.2

Because of Utah's seasonal variations in weather and temperature, and because the rate of evaporation is inversely related to the concentration of the brine, it is only possible to achieve the desired final brine concentration in the two or three hottest and driest months of the year, typically starting in the month of June. When the proper brine concentration is achieved, the brine is pumped to "deep-storage" holding ponds near the plant, that in aggregate can store up to three years supply of brine. Deep storage is required to avoid dilution by precipitation. In addition to the magnesium-rich brine product, which serves as the feed stock to the magnesium operation, sodium chloride, and potassium and magnesium salts are also recovered and sold, as will be described later.

Feed Preparation

The preparation of the concentrated magnesium chloride ($MgCl_2$) brine for use as feed to the electrolytic cells entails the removal of unwanted impurities, and further concentration. The process steps are outlined in figure 1.

The concentrated $MgCl_2$ brine is pumped from the deep-storage holding ponds into the production plant where it enters a series of reaction tanks where calcium chloride ($CaCl_2$) is added. Here, gypsum is precipitated and collected in a thickener. This reaction may be simplified as follows:

Equation 1. $MgSO_4 + CaCl_2 \rightarrow CaSO_{4(ppt)} + MgCl_2$

This step removes most of the sulfate, which may be unstable in molten systems.

Next, the brine next passes through a solvent extraction (liquid-liquid) step where a long-chain alcohol in a kerosene carrier is used to remove the naturally occurring boron from the solution. Boron is removed because it adversely affects the purity of the final product.

In a third step, the brine passes through a pre-heater vessel which utilizes waste process-heat to further concentrate the brine prior to its being fed into large spray dryers. The spray dryers convert the concentrated brine to a dry, gray $MgCl_2$ powder. The intent of this step is to remove all of the water, but some residual amount remains. During the drying process, some of the $MgCl_2$ is oxidized to magnesium oxide (MgO) through the hydrolysis of magnesium chloride, as shown in a simplified way as:

Equation 2. $MgCl_2 + H_2O \rightarrow MgO + 2HCl$

The spray-drying process is energized by the exhaust gases from natural-gas fired turbine/generators in a true utilization of co-generation technology. The electricity produced by the generators is used in the electrolysis of $MgCl_2$ described later. Gas burners are available to operate the spray dryer when the turbines are not available for use.

In the cell-feed preparation step, the spray-dried $MgCl_2$ powder is pneumatically conveyed to, and stored in, large bins until it is fed to the reactor, which is locally referred to as the "reactor" process. In the reactor, the $MgCl_2$ powder from the spray dryers is melted and further purified by introducing chlorine and carbon to remove MgO, and any remaining water. This is a continuous process where the spray-dried $MgCl_2$ powder is first fed into a brick-lined furnace known as a "melt cell" where it is melted by electrical-resistance heating. Then the molten salt overflows the melt cell, moves through a launder, and into a "reactor cell" where additional retention time brings the intended chemical reactions to completion. The melt temperature in the reactor cell is kept at 1,500°F by providing sufficient alternating current.

Removal of impurities is complex, but can be summarized by the following simplified chemical equations:

Equation 3. $2MgO + C + 2Cl_2 \rightarrow 2MgCl_2 + CO_2$

Equation 4. $2H_2O + C + 2Cl_2 \rightarrow 4HCl\uparrow + CO_2$

The finished, molten $MgCl_2$ product continuously overflows from the reactor cells and is transported to the electrolytic cells.

Hydrogen chloride (HCl), produced from the chlorination of water (H_2O) as shown in equation 4, is recovered as concentrated hydrochloric acid. This acid is subsequently reacted with limestone ($CaCO_3$) to produce the $CaCl_2$ (see equation 5) that is needed for desulfation process, and in the production of ferrous and ferric chlorides, which are sold as by-products.

Equation 5. $2HCl + CaCO_3 \rightarrow CaCl_2 + CO_2 + H_2O$

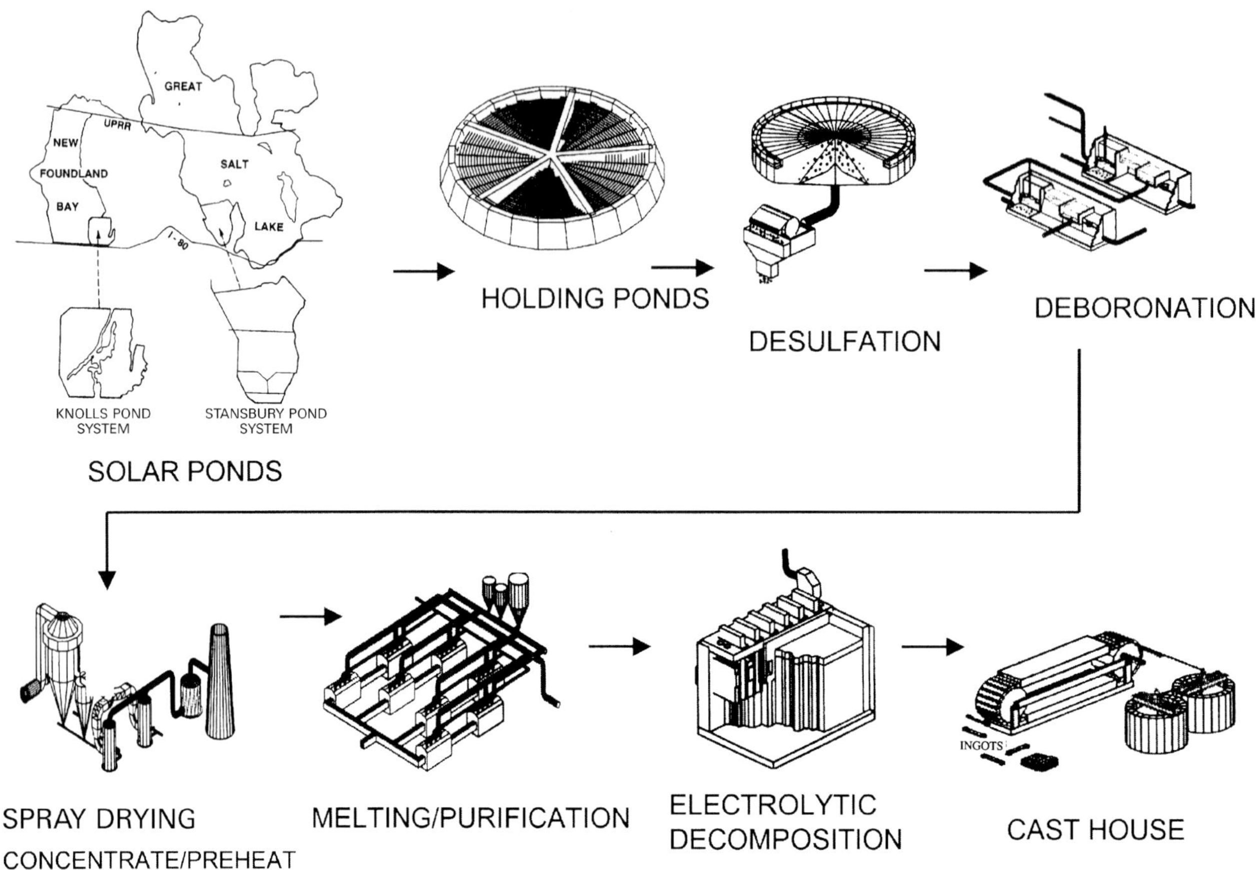

Figure 1. *Magnesium process flow diagram.*

Production and Handling of Magnesium Metal

Molten salt containing about 94 percent $MgCl_2$ is transferred to the electrolytic cells on a rigid schedule. In the electrolytic cells the molten magnesium chloride is electrochemically decomposed to molten magnesium and gaseous chlorine. The electrolytic cells are constructed as brick-lined steel furnaces equipped with negatively charged steel electrodes (cathodes), and postively charged graphite electrodes (anodes). Magnesium forms on the cathode and migrates to the top of the molten salt due to buoyancy. Chlorine gas forms on the anodes and is removed from the electrolytic cells by induced draft. The chlorine is then drawn to the chlorine recovery plant where it is cleaned, compressed and liquified. The magnesium metal is removed from the cells and it sent to the foundry. The design of electrolytic cells separates the molten magnesium metal from the gaseous chlorine to prevent chemical recombination. The majority of the electrolytic cells at Rowley (2002) are a modern design that was developed at Rowley over a period of years, and are commonly referred to as M Cells. The electrolytic cells are dependent on careful construction and electrode alignment for efficient operation, magnesium and chlorine recovery, and extended operation. External cooling is provided for temperature control. In addition to the "M Cells," Magcorp also utilizes the "Amax Sealed Cell" design which was developed under Amax's term of ownership in the 1980s.

The molten metal is transferred from the electrolytic cells to the foundry via mobile transports referred to as vacuum wagons. These wagons utilize a stationary vacuum system to draw molten metal from the cell's molten-salt surfaces into a steel vessel. Air pressure is subsequently applied to force the molten magnesium from the steel vessel and transfer it to holding/refining furnaces. Here, the magnesium is refined and/or alloyed as required. The magnesium is then cast into ingots of varying shapes appropriate for the customer's needs. Ingots may vary in weight from 15 to 1000 pounds. Most magnesium shipments to MagCorp's customers are done by truck, with a lesser portion being shipped by rail.

USES OF MAGNESIUM

In 1997, the total world consumption of magnesium was estimated at 368,170 short tons. The plant at Rowley is among the largest in the free world. Its output represents about 14 percent of the world production, and over 50 percent of the domestic production.

The largest single use of magnesium is for alloying with aluminum to provide strength, malleability, and corrosion resistance. The aluminum beverage can is the largest single user, but significant quantities of magnesium are present in almost every structural use of aluminum ranging from window frames to aircraft components.

The second most important market for magnesium is in structural applications via die casting. The largest, and ever-increasing, volume of magnesium die castings is used in the automotive industry, where magnesium's strength and light weight are being used to improve fuel economy. Magnesium die castings are also used in housings for many computers, cell phones, and other electronic components because of its electrical dampening properties. Other important uses of magnesium die-cast parts are in the manufacture of manual and power hand tools, where magnesium's light weight improves safety and operator performance.

Magnesium's electro-chemical potential is ideal for use in corrosion protection applications. A substantial amount of magnesium is consumed each year as high, electrical-potential sacrificial anodes which are installed for corrosion protection on oil, gas, and other pipelines, hot water heaters, and in other applications where cathodic protection is required.

When magnesium is added to iron, it causes brittle gray iron to be transformed into higher strength-ductile iron. Automotive crankcases, which used to be forgings, are now manufactured with ductile iron. The steel industry's preferred method of removing embrittling sulfhur compounds involves injecting powdered magnesium into the molten steel, after which the sulfur can be removed as a slag.

Magnesium is also used as a reducing agent in the production of titanium, zirconium, hafnium, and beryllium. Other chemical uses of magnesium include production of grignard reagent catalysts, motor oil additives, pyrotechnic materials, and as a pharmaceutical material.

CO-PRODUCTS AND BY-PRODUCTS

Chlorine

In addition to magnesium metal, MagCorp also produces and sells a number of co-products and by-products. Chlorine is a co-product of electrolytic magnesium production. For each pound of magnesium produced, approximately three pounds of chlorine are also produced. Much of the chlorine produced at MagCorp's Rowley operation is recycled in the process (see equations 3 and 4). Excess chlorine is transported by rail and sold in the Salt Lake area and throughout the Western United States. MagCorp accounts for about 0.3 percent of the United States' chlorine production.

The chlorine produced at Rowley has been used in the production of ferric chloride, plastics, leaching of gold-bearing minerals, and other general commercial uses. Other uses include water purification, medicines, and many chemicals, just to name a few.

Calcium Chloride

MagCorp produces calcium chloride by reacting limestone (previously oolitic sand was used) with hydrochloric acid in environmentally controlled equipment (see equation 5). One-third of the calcium chloride produced is used in MagCorp's brine-conditioning process for the removal of sulphate (see equation 1). The balance is available for outside sales. Elsewhere, calcium chloride is used as an additive to specialized concrete, as a dust suppressant, ice melting, in chemical manufacturing, in the paper industry, and for heavy-media purposes in oil field service.

Ferrous and Ferric Chloride

Iron chlorides (ferrous and ferric) are also produced at MagCorp's Rowley facility for outside sales. The by-product hydrochloric acid is reacted with a variety of economially acceptable metallic iron sources to produce ferrous chloride. Ferrous chloride can then be treated with chlorine to convert it to ferric chloride. Iron chlorides are sold mainly for sewage treatment. Other uses include metal treatment, producing iron-based chemicals, and as a chlorinating and oxidizing agents.

Brines and Sodium Chloride

By contractual arrangement, about 25 percent of MagCorp's total brine volume in its solar pond system is diverted to an adjacent sodium chloride operation where high-quality, commercial grades of salt (NaCl) are deposited. The bitterns (concentrated brines exiting the sodium chloride operation) are then returned to MagCorp's ponds for inclusion with the brine being concentrated for use in the production of magnesium metal. In MagCorp's solar ponding operation, a significant amount of NaCl is precipitated. Much of this is not commercially useful, however, due to impurities and remoteness of location.

Potassium and Magnesium Salts

As the brine in MagCorp's solar-evaporation ponds approaches the final desired magnesium concentration, mixtures of potassium and magnesium salts such as kainite, schoenite, and carnallite are being precipitated. These salts, deposited in the final ponds, are acceptable as feed stock for the commercial production of sulfate-of-potash fertilizer, and are, on occasion, harvested and sold for that purpose.

Potential By-Products

Other by-products are being considered to accompany the production of magnesium at Rowley, Utah. These products include commercial grades of hydrochloric acid, lithium metal or lithium compounds, and bromine compounds.

CONCLUSIONS

Magnesium is a versatile metal for which demand has grown in recent years due to its properties of good strength and low density. The minerals in the Great Salt Lake serve as a source of raw material for one of the world's significant and successful magnesium producers. The unique nature of the Great Salt Lake provides substantial advantages to the recovery process as well as creating special challenges. The abundant volume of magnesium (even at dilute concentrations) in the Great Salt Lake, in combination with the use of solar energy, provides the basis for an attractive raw-material source. The chemistry of the lake, and the extraction process used, result in production of magnesium metal with few impurities. The challenge of developing a commercially viable extraction process has taken years to refine due to the unique and distinctive nature of the manufacturing process.

IMC KALIUM OGDEN CORPORATION - EXTRACTION OF NON-METALS FROM GREAT SALT LAKE

by

David Butts
DSB International, Inc., formerly with
IMC Kalium Ogden Corporation

INTRODUCTION

Each spring and summer, IMC Kalium Odgen Corporation (IMC) pumps millions of gallons of water from Great Salt Lake to begin a multi-year journey through 90 solar-evaporation ponds. These ponds span over 40,000 acres of mud flats and other areas that were previously considered non-productive wasteland. This land is now supplying work for hundreds of employees, increasing the tax base of local counties and the state, and furnishing naturally occurring mineral products to countries on nearly every continent.

IMC's plant is located 15 miles west of Ogden at the north end of Little Mountain (figure 1). Canals and solar ponds stretch over 50 miles from the east side of Great Salt Lake at Little Mountain, all the way to Clyman Bay on the west side of the lake. A one-of-a-kind canal constructed along the lake bottom connects the two widely separated solar-pond areas. The canal, known as the Behrens Trench, is named after Peter Behrens (figure 2), president of Great Salt Lake Minerals (now IMC Kalium) from 1971 to 1989, who promoted the project.

IMC produces three economically profitable minerals from the numerous salts that are extracted from the waters of Great Salt Lake. These minerals are halite, arcanite, and bischofite. Halite is the mineral name for common salt or sodium chloride; arcanite is the mineral name for potassium sulfate, also called potash or sulfate-of-potash (SOP); and bischofite is the mineral name for hydrated magnesium chloride. The salts used to produce all three of these minerals are precipitated (deposited) from the lake brine in IMC's solar-pond complex by utilizing the sun's energy to evaporate water. Sodium sulfate can also be produced from the lake water.

SODIUM CHLORIDE (SALT)

Common salt is known for its use in foodstuff, but it has other hidden uses. Tens-of-thousands of manufactured products are associated with this important mineral. Vinyl, plastics, synthetic fibers, and bleach, to name a few, use chlorine extracted from sodium chloride. Caustic soda (lye) is also produced from sodium chloride and has many uses in the

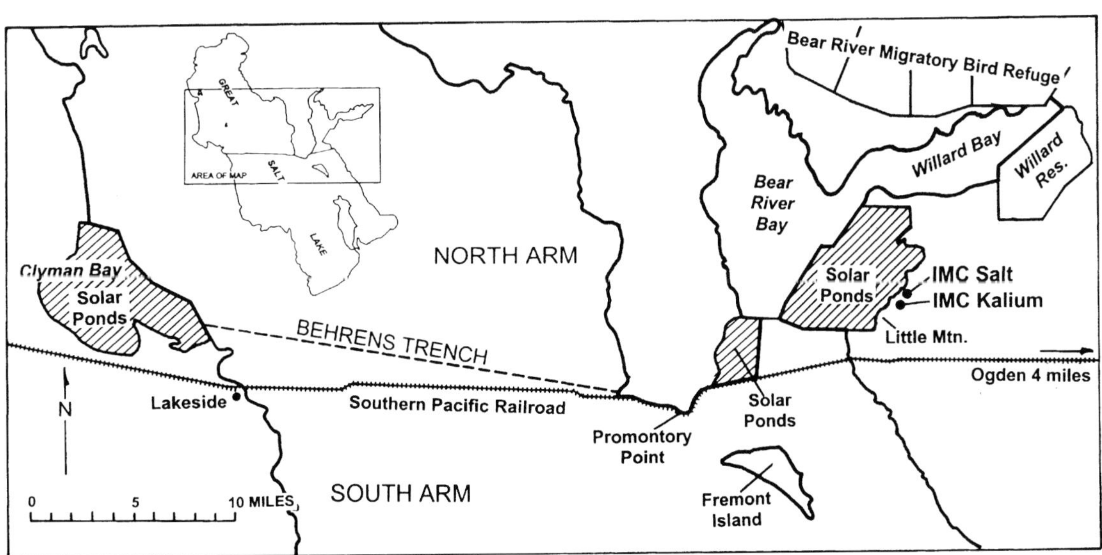

Figure 1. *Area of Great Salt Lake covered by IMC's mineral-extraction operations.*

Figure 2. Peter Behrens, President of Great Salt Lake Minerals and Chemicals from1971 to 1989, and Vice Chairman until 1992.

processing of soap, paper, and the manufacture of other chemicals. Salt is also used in oil field drilling operations, leather tanning, and making preservatives. Hundreds of thousands of tons of salt are used each winter to de-ice roads.

Three salt-production facilities currently operate around Great Salt Lake. IMC is the largest; its products are marketed by IMC Salt Inc., a separate IMC business unit. One and a half million tons of salt are processed through the plant annually, making it the largest producer of solar salt in the United States. The salt is harvested from the solar ponds then washed, dried, screened, and packaged in a variety of sizes and formulations. It is also compressed into blocks for livestock use, and into cubes and pellets for use in water softening. Salt from the lake is produced under many labels (brand names).

It is estimated that there are about 4.5 billion tons of salt in the Great Salt Lake, and over one million tons are added each year through inflow to the lake. At the present rate of extraction, it would take 2,000 years to remove all salt from the lake.

POTASSIUM SULFATE

Salt (halite) is the highest-tonnage product of IMC's mineral and chemical complex, but it is not the main revenue-producing product. Salt is considered a by-product that is created as Great Salt Lake water is concentrated to produce potassium and magnesium-bearing salts such as epsomite, schoenite, kainite, and carnallite. These salts are used in manufacturing IMC's main product, sulfate of potash (SOP) or arcanite, which is used mainly as a commercial fertilizer. Producing these salts in economic quantities requires large expanses of solar ponds, miles of canals, numerous pumping stations, and a carefully controlled operation. Production is often compared to a large farm where minerals rather than crops are grown, and growth is dependent on hot, dry weather. Excessive rain can significantly reduce production.

SOP plays an important role in the economics of Utah. IMC's Ogden facility is the largest producer of solar-based SOP in the world. During the latter part of the evaporative process, the potassium salts that crystallize in the solar ponds have been co-mingled with large quantities of impurities (salts containing sodium, magnesium, and chloride) that must be removed. For every ton of potassium salts harvested from the solar ponds, multiple tons of impurities are also mined. These impurities are removed in the refinery through a process called leaching, using water as the leaching media. Salts are moved through a series of thickeners and tanks, where the impure salts are dissolved. Near the end of the process, the remaining potassium salts are converted into SOP under carefully controlled conditions of temperature and chemistry. The process, which uses only water as the leach agent, makes the plant effluent environmentally safe. The unwanted or impure salts which came from the lake initially are returned to Great Salt Lake.

SOP is an excellent fertilizer. Besides containing potassium, an essential nutrient for all plants and animals, it also contains sulfur that provides added nutritional value to the salt. In addition to being shipped domestically, SOP is also exported to Pacific Rim countries where Japan and China receive a large portion of the volume.

Most large terminal lakes of the world, such as the Dead Sea, have a chloride-based brine and only potassium chloride can be extracted from their waters. The Great Salt Lake is different in that it contains appreciable amounts of sulfate which makes it possible to produce SOP as a final product. Not only does the sulfate in Great Salt Lake brine make it possible to produce this double fertilizer, but there is enough sulfate left over after SOP is produced, to convert potassium chloride (KCl) to SOP, as follows:

$$2KCl + SO_4^{=} = K_2SO_4 + 2Cl^{=}$$

This makes it possible to convert lower priced potassium chloride from IMC's facilities in Canada into a more valuable product.

MAGNESIUM CHLORIDE

Magnesium chloride ($MgCl_2$) constitutes a large percentage (30 to 36 percent) of the bittern (the remaining brine) that is formed as Great Salt Lake water spends two years moving through and concentrating in IMC's solar ponds. From this brine, IMC produces two magnesium chloride products: a solid salt, bischofite ($MgCl_2.6H_2O$), and a liquid containing 30 to 35 percent magnesium chloride.

To make bischofite, the brine is first pumped from a holding pond into a refinery where sulfate is removed as the mineral gypsum, and other impurities such as dirt and coloration are removed with filters and activated charcoal. Second, the brine is heated and more water is removed by evaporation. Third, the hot magnesium chloride liquid is then collected and cooled on the surface of a cooled, rotating drum where it solidifies. The solid bischofite is removed from the drum as a solid-flake product, which is used as an industrial chemical, de-icer, and in making specialty magnesia cements.

The magnesium chloride liquid product is used for a variety of purposes. It makes an excellent dust suppressant when sprayed on dusty roads. Due to its ability to absorb and retain moisture, magnesium chloride will stop dirt roads from dusting for months during dry weather. It can also be sprayed directly on icy roads, bridges, or other structures to quickly and effectively remove ice. Magnesium chloride is also used in numerous other applications. It is a component in wall board, is used in specialty products such as oxychloride cements, and is used as a fertilizer where magnesium is deficient in the soil.

SODIUM SULFATE

Another product that can be produced from the lake brine is anhydrous sodium sulfate (Na_2SO_4), also known as "salt cake." This material is used in the Kraft paper industry and as a filler in powdered detergents. Salt cake was made by Great Salt Lake Minerals Corporation (GSL), but was discontinued after the purchase of GSL by IMC. IMC had signed a noncompetitive agreement with North American Chemicals Corporation (NACC), a subsidiary of IMC, who also make anhydrous sodium sulfate at Trona, California.

In the winter, mirabilite ($Na_2SO_4 \cdot 10H_2O$), also known as Glauber's Salt, crystallizes in the solar ponds from the cold brine, along with some sodium chloride. When salt cake is produced, the dual-salt mixture is harvested from the ponds and hauled into the plant by trucks where it is purified and converted to pure, anhydrous sodium sulfate, a form required by the buyers. Even though it is no longer produced by IMC, it remains a potential product.

IMC KALIUM HISTORY AND OPERATIONS

History and Pond Expansions

IMC's operation began in 1965 based on lithium extraction. At that time, research and process development was made by Lithium Corporation of America (Lithcoa), based in North Carolina. When it became apparent it was not economical to extract lithium alone, the corporation joined with a German potash company, Salzdetfurth, who were experts in producing potassium fertilizers (potash). From 1968 to1998, the potash operation was known as Great Salt Lake Minerals and Chemicals Corporation or simply GSL. In April of 1998, GSL was acquired by IMC Global and was renamed IMC Kalium Ogden Corporation. Prior to 1998, the several IMC salt (sodium chloride) production facilities were know as North American Salt Company, owned by D. George Harris and Associates. On April 1, 1998, the North American Salt Company was also purchased by IMC Global and renamed IMC Salt Inc.

The objective of Lithcoa and Salzdetfurth was to extract lithium and potash jointly. By 1968, construction of a 12,000-acre solar-pond complex was begun, the production of lithium was placed on hold, and the production of potassium sulfate became the primary goal. To increase potash production, the solar-pond complex was first expanded during 1970 and 1971 from 12,000 acres to 19,000 acres; a second large expansion program in 1991 brought the total ponding area to 35,000 acres. Still a third expansion program in 1998 and 1999 added several thousand more acres bringing the total to over 40,000 acres. The main body of IMC's solar ponds and the processing plant are located between Little Mountain, some 15 miles west of Ogden, and the Promontory Mountains. Figure 3 shows an aerial photo of IMC's plant and main solar ponds, looking west. The Bear River Migratory Bird Refuge lies north of the solar ponds, with the south arm of Great Salt Lake to the south.

Clyman Bay Pond and Behrens Trench

With ever-increasing world market demands for potassium sulfate, IMC sought to increase the area of its solar ponds in order to increase production. With no additional expansion area available near the main ponds (east of the Promontory Mountains), the problem was solved by constructing solar ponds on the west side of the lake, covering a mud-flat area some 40 miles west of the Little Mountain plant site (figure 1). This west-side pond expansion presented a challenge, however. How could concentrated brine, produced on the west side of the lake, be transported over a 21-mile distance covered by lake water? This problem was solved by dredging a gently eastward-sloping canal, later named the Behrens Trench, under the lake to convey the brines (figure 1). It has long been known that as water from the lake evaporates, it becomes progressively heavier. Brine concentrated in the west solar ponds is heavier than the lake brine, and, if carefully injected into the high end of the underwater canal, it will sink and flow downhill eastward within the canal under the lighter lake water.

Flowing heavy brine in an open canal covered overhead with lighter brine presented unique problems and challenges because no one in the world was doing it, and no hydraulic information was available. Solutions to this engineering problem had to be researched by GSL. Research began in 1987, and by 1990 enough data had been gathered that the president of GSL, Peter Behrens, made the decision to have the 21-mile-long canal dredged under the lake (figures 4 and 4a). Mr. Behrens was a key engineer in the development and operation of the GSL processing facilities since their beginnings in the late 1960s and early 1970s.

Pond Coloration and Harvesting

The east-side ponds are a prominent landmark easily seen by airline passengers on approach to the Salt Lake International Airport. The red, green, and brown-coloration of the ponds is the result of algae and bacteria whose colors change with changing brine concentration and with the seasons. The red color in highly concentrated salt ponds is caused by betacarotene formed in the algae *Dunanella*. Research has

Figure 3. *Aerial view looking west towards IMC's plant, stock pile, and solar ponds with the Promontory Mountains in the background.*

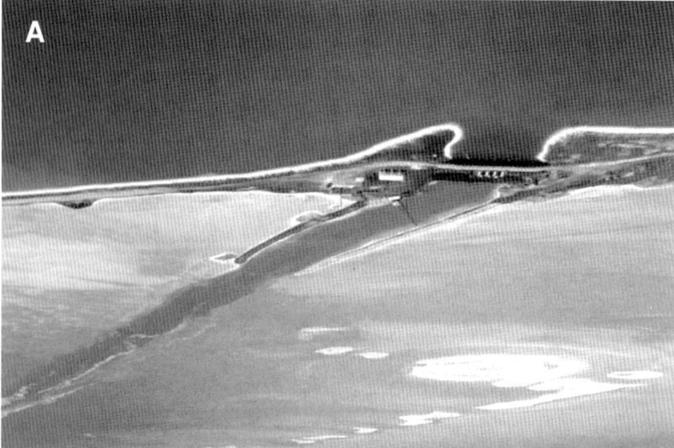

Figure 4a and 4b. *Behrens Trench leaving the east-dike control structure of IMC's west solar evaporation pond.*

been done by Eastman Kodak to extract betacarotene from the algae, but so far it is not commercially viable. Green and brown coloration is usually found in the potash ponds, and white comes from the highly reflective salts in shallow or empty ponds.

All of the minerals precipitated in the solar ponds are recovered by an operation called harvesting. Each year, minerals grown in the ponds are removed from the pond floors and brought by truck into the plant area where they are stockpiled (see figure 3) to await processing. Figure 5 shows a truck being loaded by a specially designed harvester.

PROCESSING THE CHEMICALS

The brines of Great Salt Lake contain five main ionic components: sodium, magnesium, potassium, chloride, and sulfate. These ions, when concentrated in solar-evaporation ponds, combine to form many mineral salts from which economic products are produced. Table 1 shows the order in which the minerals are precipitated as water is evaporated, their chemical formulas, and common names, and table 2 shows the products made from these minerals by IMC, and other common uses.

Figure 5. *Potash salts being loaded into truck by special harvester.*

Table 1. Order of mineral crystallization from Great Salt Lake brine, common names, and uses.

Mineral name	Chemical formula	Common names
Halite	$NaCl$	Salt
Mirabilite	$Na_2SO_4 \cdot 10H_2O$	Glauber's Salt
Epsomite	$MgSO_4 \cdot 7H_2O$	Epsom Salts
Schoenite	$MgSO_4 \cdot KCL \cdot 6H_2O$	
Kainite	$MgCl_2 \cdot KCL \cdot 3H_2O$	
Carnallite	$MgCl_2 \cdot KCL \cdot 6H_2O$	
Bischofite	$MgCl_2 \cdot 6H_2O$	

Table 2. Products made by IMC and other uses.

Mineral	Products made by IMC	Other common uses
Halite	Purified salt	Foodstuffs, de-icers, chlorine gas, caustic soda, water softening - over 14,000 uses
Mirabilite	Anhydrous sodium sulfate	Detergent filler, paper manufacture
Epsomite Schoenite, Kainite, and Carnallite	SOP (potassium sulfate or potash)	Epsomite is used in water conditioning, as a laxative, and in bath salts
Bischofite	Purified mineral	De-icer, dust suppressant and a specialty chemical

Figure 6 is a flow sheet showing the numerous products produced by IMC's Ogden operation; however, anhydrous sodium sulfate is not currently produced. Possible mineral products that might be produced in the future are sodium sulfate, magnesium sulfate (Epsom salts), magnesium hydroxide, and magnesium oxide. Great Salt Lake waters also contain many other dissolved ions and metals. Gold is found dissolved as a gold chloride, but the amount is only eight parts-per-billion or less. Other ions, present in small concentrations, but large enough that extraction has been considered, include lithium, boron, bromine, and bicarbonate.

WEATHER

Weather has a profound effect on Great Salt Lake's mineral-extraction industries. The average rainfall in the vicinity of IMC's solar ponds is around 13 inches per year. Even a few inches of rainfall more than average cause significant decreases in mineral production. With normal rainfall and spring runoff, the lake rises two to three feet each spring. The hot, dry summers then cause the lake water to evaporate and the lake level to drop again. Since the lake is terminal (having no outlet), an increase in rainfall causes the lake level to rise. This decreases the lake's salinity because the same amount of salt is dissolved in more water.

Heavy precipitation in 1983 and 1984 caused the lake to rise an unprecedented four to five feet each of the two years. This rapid rise was faster than IMC's predecessor, GSL, could raise its 20-plus miles of perimeter dikes. In 1984, GSL's south perimeter solar-pond dike breached, and the entire pond complex was flooded. This breach forced the company to stop SOP production from 1984 to 1989.

FLOOD-CONTROL PROJECTS

Beginning in 1983, the lake level began to rise from an elevation of about 4,200 feet (msl), and by 1984 it had risen to an elevation of 4,209. State officials were concerned about the flooding that was occurring, mainly around the lake's southern shore, and they worried that the lake would continue to rise. Numerous flood-control options were evaluated (see article by Austin, this volume), including breaching the Southern Pacific Railroad causeway (figure 1), upstream developments, and pumping water to the western desert. It was determined breaching the causeway would bring that the most immediate relief. This would reduce the nearly 3.5-foot head differential (difference in lake level) that had developed across the causeway (the south arm was higher). The breach was completed in August 1984, and within about two months, the head differential across the causeway was reduced to less than one foot.

In spite of the effectiveness of the breach, the lake continued to rise, and the State opted to implement the West Desert Pumping Project. This project was designed to pump water to a large depression west of the lake, known as the West Pond. The West Pond, located on the west side of the Newfoundland Mountains, was nearly one third the area of Great Salt Lake. Pumping would initially remove nearly a million acre-feet of water, just to fill the pond, and the pond's 320,000-acre surface area would greatly increase the effective evaporative area of Great Salt Lake. In 1987, a large pumping station was completed by the State of Utah on the west side of the lake near Hogup (about 12 miles west of Lakeside). At full capacity, the station's three pumps could move nearly 3,000 cubic feet per second (cfs) from the lake to the West Pond. In 1987, 1988, and 1989, nearly 700-million acre-feet of lake water was pumped from the north arm of the lake to the West Pond. The West Desert Pumping Project provided a safeguard to help prevent an uncontrolled rise in the lake level, and to lower the lake level after it peaked in 1987. With proceeds from an insurance policy, GSL's dikes and other facilities were repaired, and the production of fertilizer was started again in 1989.

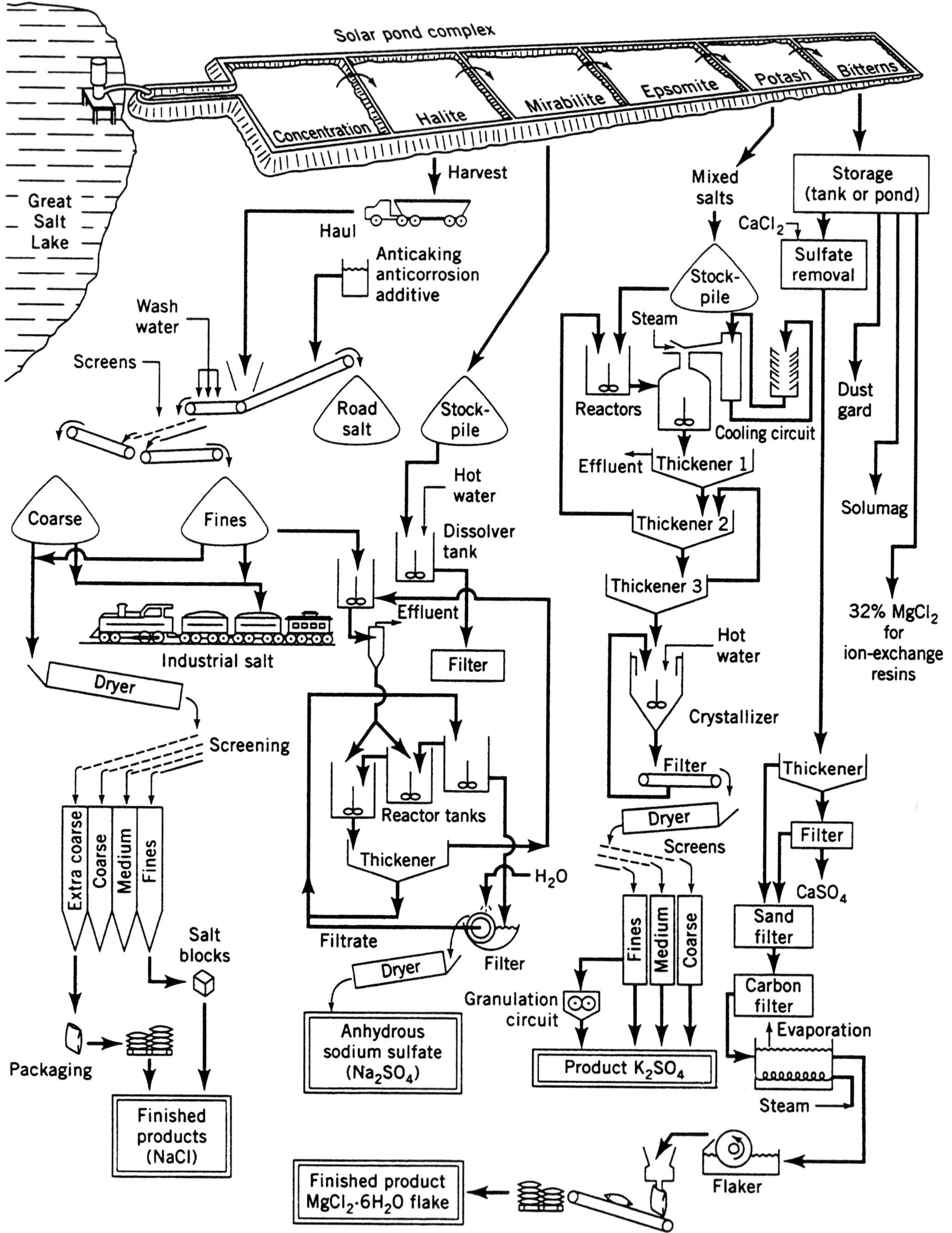

Figure 6. *Abbreviated flow sheet of the multiproducts of IMC (formerly Great Salt Lake Minerals and Chemicals Corporation). Reprinted by permission of John Wiley & Sons, Inc. (Chemicals from Brine, David Butts, Kirk-Othmer Encyclopedia of Chemical Technology, Fourth Ed., v. no. 5, copyright 1993 by John Wiley & Sons, Inc.)*

SUMMARY

IMC produces sodium chloride, potassium sulfate and magnesium chloride products from Great Salt Lake. From 1965 to 1968, Lithcoa was investigating the extraction of lithium from Great Salt Lake. When it became apparent that the sole extraction of lithium was not economical, Lithcoa was joined by a German potash company, Salzdetfurth, who were experts in producing potash. From 1968 to1998, the potash operation was known as GSL. In April of 1998, GSL was acquired by IMC Global, and was renamed IMC Kalium Ogden Corp.

The solar evaporation of brines from Great Salt Lake produces a variety of salts. Halite, the first salt to precipitate, is produced and marketed by three companies around the lake, IMC being the largest producer. Through further evaporation of the brine beyond the saturation point of halite a number of potassium and magnesium salts are formed including epsomite, schoenite, kainite, and carnallite. From these salts, SOP, or the mineral arcanite, is produced using a leaching and crystallization process. Other IMC products include magnesium chloride as both a solid and as a brine. Sodium sulfate or salt cake was produced in the past by GSL.

High-water conditions during the early 1980s caused the breach of dike separating the solar ponds from the lake, which resulted in the flooding of GSL's solar ponds. This forced the company to stop SOP production from 1984 to 1989, though some production of salt continued. In 1987, the State of Utah implemented the West Desert Pumping Project to transfer water from the lake to the western desert. The pumping project provided a safeguard to help prevent the lake level from rising uncontrolled. By 1989, dikes and other facilities were repaired, and GSL resumed production of fertilizer. In the early 1990s, additional evaporation ponds were constructed on the west side of the lake. Concentrated brines are conveyed from these ponds to the main solar-pond complex east of the Promontory Mountains by a 21-mile, underwater canal called the Behrens Trench.

Editor's Note: On November 28, 2001, IMC Global Inc. sold its business unit (IMC Salt) and with it, IMC Kalium Ogden Corp. to Apollo Management, LP. This new business entity is now known as Compass Minerals Group with headquarters in Overland Park, Kansas. As part of this transaction, IMC Kalium Ogden Corp. will change its name to Great Salt Lake Minerals Corporation or GSL. There has been no change in personnel, telephone numbers, addresses or other aspects of the business.

NUTRITIONAL ENTERPRISES ON GREAT SALT LAKE - NORTH SHORE LIMITED PARTNERSHIP AND MINERAL RESOURCES INTERNATIONAL

by

Corey D. Anderson and Val Anderson
Mineral Resources International

ABSTRACT

An ample supply of natural minerals and trace elements in the human diet is necessary for maintaining good health. Sufficient amounts of these constituents are often lacking in food but can be supplied by mineral supplements. More than 30 years ago, Hartley Anderson realized that Great Salt Lake water could provide many of the constituents lacking in food, and started selling lake water as a nutritional supplement. The business he started has continued and evolved, so that today, Mineral Resources International (MRI), and its sister company, North Shore Limited Partnership (North Shore), produce both liquid and powdered-mineral supplements produced from Great Salt Lake brine. North Shore produces concentrated brines in its 20-acre solar-evaporation complex, located at the north end of the lake. The concentrated north-arm brine that is used is taken from a remote setting, located miles from any sources of pollution. MRI processes this brine into final products at its West Haven facilities, just west of Ogden. MRI reports that its testing reveals the presence of no contaminants in its brines, only naturally occurring trace elements and the lake's natural flora and fauna, all of which it considers to be beneficial. These products are distributed nationally and internationally through a number of marketing channels, which include several sister companies. This paper discusses the need for minerals and mineral supplementation, the major minerals present in the lake, North Shore's brine production and harvesting activities, its products, their nutritional values, and its product marketing.

HISTORY OF ANDERSON'S NUTRITIONAL ENTERPRISE

More than 30 years ago, Hartley Anderson, living near Ogden, read a series of newspaper articles written by Dr. George W. Crane with headlines such as "The Ocean's 44 Antidotes for Deficiency Ailments" and "Trace Chemicals Essential to the Body" (Crane, 1973, 1979, and pre-1968). These articles piqued Hartley's interest with information about the amazing benefits people were receiving from drinking a little bit of sea water each day. This led him to research Great Salt Lake, sometimes referred to as an inland sea. He found that Great Salt Lake brine not only had the same minerals, and in nearly the same proportions, as discussed by Dr. Crane, but that Great Salt Lake brine was approximately six times more concentrated than regular sea water. Hartley knew there was a need for these minerals and thus, the idea for a company was born. He now jokes that he sells the Great Salt Lake by the pint for a living. Some appreciate the good-natured irony in that statement, but when he first launched his unusual business in 1969, there were others who thought he had lost his mind. What started as simply an idea has since become a thriving enterprise.

Hartley's next step was to get some Great Salt Lake brine, share it with some friends who suffered from some of the ailments listed by Dr. Crane, and see what happened. While his observations did not constitute a truly scientific study, the results were still very educational, and very positive. Science has been slow to vindicate why the minerals from Great Salt Lake caused such dramatic and varied results in people (for example, the relationship between the high levels of magnesium and boron in the brine and the nutritional significance of these elements to bone and joint health is discussed later in this paper, but it was not well understood at the time Hartley started his venture). Nevertheless, he knew from his growing stack of testimonials that the company's first product, Inland Sea Water™, appeared effective. In 1969, Hartley and his wife, Gaye, founded their company and started selling pure Great Salt Lake water to the public. In the intervening time, as their seven children have entered the business, additional companies have been formed by separating manufacturing from marketing, entering new markets, and creating an additional enterprise to focus on the brine-harvesting operation. The Anderson family owns controlling interest in the harvesting and manufacturing operations, which now have third-generation family members in the businesses.

MRI and its sister companies now produce, harvest, process, and sell several distinct, liquid- and powdered-mineral products from Great Salt Lake. They also produce a host of additional liquid, powdered, and tableted products using the same minerals as core ingredients. North Shore does the initial concentrating and harvesting of the brine in their ponds at the north end of the lake, and is presently the only producer licensed to produce a food-grade product from Great Salt Lake. MRI processes the brine, and manufactures the entire line of dietary supplements.

ELEMENTS NEEDED IN HUMAN NUTRITION

Recommended Dietary Allowances (National Research Council, 1989) lists Recommended Dietary Allowances (RDA), and other definite dietary requirements, for 15 elements besides the foundation elements carbon, hydrogen, oxygen, nitrogen, and sulfur. It also lists probable or possible dietary functions for another 10 elements. Additional, less-well-proven research exists showing possible nutritional benefits for up to 72 of the elements at one level or another (Bergner, 1997; Heinerman, 1998; Shauss, 1999; various published and unpublished works on file at MRI) Certainly, additional research is needed to verify these benefits.

The American Medical Association also recognizes the importance of minerals in our diet. "Variations in the distribution of certain minerals in the environment are known to have an effect on health" (American Medical Association, 1989). The lack of certain minerals in our soil is evidenced by the need for constant fertilization, as plants need nitrogen, hydrogen, oxygen, chlorine, carbon, boron, sulfur, potassium, magnesium, phosphorous, iron, zinc, copper, manganese, and molybdenum. Some of these elements are commonly replaced through fertilizers to provide maximum crop yields through minimum investment. Humans, however, are known to additionally need calcium, sodium, fluorine, bromine, chromium, iodine, silicon, selenium, beryllium, lithium, cobalt, vanadium, and nickel, which are not necessarily replaced through plant fertilization (Schauss, 1982 and 1983).

MRI MINERAL SUPPLEMENTS AND REPORTED HEALTH BENEFITS

MRI and its sister companies harvest, process, and market minerals and trace elements from the Great Salt Lake for dietary supplementation. The minerals are harvested from the lake as ions in liquid form, and thus maintain the complete spectrum and natural equilibrium of the elements found in the lake. These minerals are the basis for each of MRI's unique products, and help provide a strong foundation for balanced supplementation.

MRI and its sister companies have also developed additional products by adding other nutritional ingredients, such as vitamins, herbs, enzymes, and certain additional minerals, to the lake's minerals. Product development has been influenced by feedback from people using the lake's minerals. Over the years, the single, most-reported benefit has been in the relief of arthritis-related symptoms. Other user-reported benefits include bone, joint, and calcium-related issues, to include the elimination of unwanted calcium deposits, and improved healing of damaged bone structures. These benefits can be explained in part by looking at the nutritional interrelationships among magnesium, boron, calcium, and other trace elements. For example, over half the magnesium in the human body resides in the skeleton (National Research Council, 1989). In addition to being involved in the functioning of more than 300 enzymes, magnesium is also important to calcium and potassium homeostasis (National Research Council, 1989). Recently, it has been shown that boron is important for optimal calcium absorption and utilization, and thus, bone metabolism (Nielson, 2000). Boron has emerged as being nutritionally important throughout the life cycle. Clinical evidence is emerging showing that boron may be an important nutrient in preventing and treating osteoarthritis (Schauss, 1999). Copper, zinc, manganese, fluorine, and silicon have also been shown to be complementary to the proper utilization of calcium (National Research Council, 1989). All of the above-mentioned elements are present in Great Salt Lake brine, and in nutritional concentrates made from the brine. Nevertheless, MRI's products are sold only as dietary supplements, and their benefits are considered to be entirely nutritional. To date, none of the potential disease-related effects have been sufficiently proven to allow for any drug claims to be made for any of the products, nor have any of the companies applied to government to be allowed to make any such claims.

BRINE PRODUCTION AND HARVESTING

Location of Solar-Evaporation Facilities

North Shore owns a narrow strip of land one mile long, just above the shoreline between Locomotive Springs and Monument Point (figure 1), in the northern half of section 9, T. 11 N., R. 9 W., SLBM. Within this area, North Shore operates 20 acres of solar-evaporation ponds (figures 1, 2) that were used in the early to mid-1900s for salt harvesting. The ponds are located just over a hundred yards north of the lake's 1986-87 north-arm, high-water line (4,210.9 feet). At the lake's 1963 historic low-water level of 4,191.35 feet, a large mud flat stretched south for a distance of two to three miles. The lake's actual shoreline has fluctuated back and forth within these two extremes. Great Salt Lake's established meander line, a line surveyed in 1887 below which is State sovereign land, is only about 150 yards to the south of North Shore's ponds. North Shore is responsible for the physical operation of the ponds that are technically classified as a mine by state regulators. North Shore operates its ponds within environmental and safety parameters and regulations, and pays royalties and lease fees to the State of Utah.

Brine Production

During high-water years, brine is pumped directly from the lake, up a small rise, and into the ponds. During low-water years, the brine is pumped from a canal that reaches southward into the lake for more than two miles (figure 2) to bring brine into the pump station. After the brine is pumped into the solar ponds, it is concentrated through controlled solar evaporation as the brine is moved from pond to pond. During the brine's journey through the ponds, its chemistry is carefully and regularly monitored, and the ponds are managed to maximize the brine's content of magnesium and trace elements. By doing this, North Shore produces a brine that contains the maximum nutritional value. Concentrated brines are North Shore's primary products rather than the salts that are precipitated. Further testing and processing of the concentrated brines take place at MRI's facility to ensure the quality of the finished consumer product.

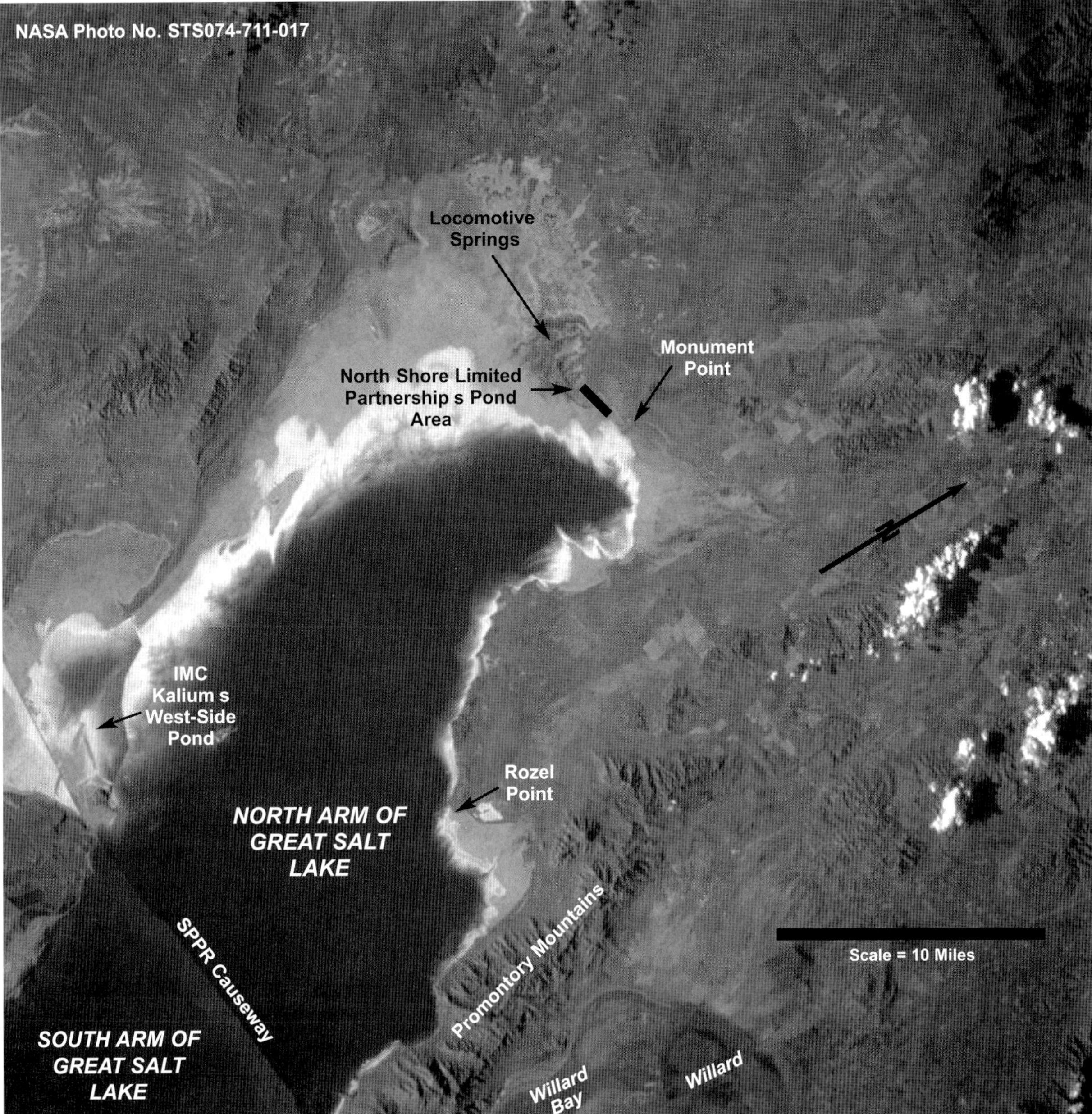

Figure 1. *Location of North Shore Limited Partnership's pond area, and other features around the north arm of Great Salt Lake, Utah.*

BRINE QUALITY

Benefits of Solar Pond Location

North Shore's location on the northern tip of Great Salt Lake (figure 1) presents several advantages in concentrating the lake brines and keeping them free of contaminants. First, the high concentration of north-arm brines provides an obvious production advantage since less water has to be removed in the solar ponds. Second, but more important, the remote and pristine nature of the north arm is an ideal setting for producing a high-purity brine for nutritional purposes.

The north arm of the lake is many miles from any significant sources of pollution. It is separated from human habitation by the Promontory and North Promontory Mountains to the east, the relatively uninhabited areas to the north and west, and it has limited mixing with the south arm of the lake because of the Union Pacific Railroad fill. Additionally, no major tributaries flow directly into the north arm. The only significant source of water flowing into the north arm comes from the south arm, through the breach, the two small culverts in the railroad causeway, and through the causeway's rock fill.

Since the concentrated brines produced and harvested by North Shore are used for nutritional purposes, North Shore's emphasis is focused more on brine quality than on brine quantity. Consequently, the advantages of remoteness and freedom from industrial and population pollution far outweigh the disadvantages, which include additional costs associated with hauling the heavy brine across miles of dirt roads, and the lack of utilities and other services that are normally considered essential.

Figure 2. *North Shore Limited Partnership's solar-evaporation ponds located on the extreme northern end of the north arm of Great Salt Lake, view is looking southward. The pump station is located near the southeast corner of pond one. During low lake-level periods, water is brought from the lake to the pump through the diversion canal (photo courtesy of Matthew Anderson).*

Avoidance of Process Contamination

North Shore's grounds, ponds, and equipment are well maintained, and are operated to avoid the use of chemicals and other sources of contamination. The presence of any potential contaminants is monitored to ensure an end product that is free of undesirable elements. North Shore's production of brine for human consumption could be compared to an organic farm, where food is produced without the use of pesticides, chemicals, or fertilizers.

Great Salt Lake Organic and Metallic Contaminants

Tayler and others (1980) documented the lake's ability to clean itself. Much of the organic matter entering the south arm of the lake through its tributaries is removed by an extensive system of wetlands. Any remaining organic matter is cleaned by the lake's effective biosystem, which includes algae and brine shrimp.

The lake also has had the ability to rid itself of heavy metals. Heavy metals have been shown, in the past, to precipitate out of the lake brine as metallic sulfides. Tayler and others (1980) state that "... the total soluble concentrations of heavy metals in the water are extremely low. The heavy metals in the lake, along with clays, organic materials and carbonates, are precipitating to the sediments... The lake thus avoids accumulation of heavy metals in the lake waters, and is non-toxic and self-cleansing. The unique saline condition of [the] Great Salt Lake determines the precipitation and immobilization of heavy metals in the lake." They further state that the levels of certain metals are actually lower in the lake water than in the average levels of the streams flowing into the lake.

MRI has performed regular contaminant testing on the lake water through independent laboratories for a number of years, and similar tests on its concentrated brine during every major brine-harvest period. Tests are run for the following contaminants: organics and petroleum chemicals, agricultural chemicals and pesticides, non-halophyllic bacteria (the type that could survive in a non-salt-saturated environment and hence be of any consequence to human health), and heavy metals. With the exception of heavy metals, which are naturally occurring at very low levels, and extremely low-non-halophyllic-bacteria levels, MRI's tests have shown a complete absence of the above contaminants.

Heavy-metal concentrations, determined by Environmental Protection Agency (EPA) testing method SW-846/3010A (EPA, 2001b), which is an EPA-approved test for metal contaminants in drinking water, and other standard testing methods, have been run on the high concentration brine products produced by MRI. These tests have shown the following four metals of primary concern to be present, but below the following levels: aluminum <1 ppm, lead <0.5 ppm, cadmium <0.5 ppm, and mercury <0.1 ppm.

The National Academy of Sciences acknowledges that certain metals, normally considered toxic, are probably essential in trace amounts, although nutritional requirements have not been set. Other metals may be essential, and have been shown to be important to animal nutrition at very low, naturally occurring levels (National Research Council, 1989). The list of these metals includes nickel, arsenic, cadmium, lead, and others. When Great Salt Lake brine and MRI's concentrated brines are tested using EPA testing method SW-846/3010A (EPA, 2001b), these elements are usually below detection levels. The numbers listed in the previous paragraph are based largely on tests using these methods. Other more sensitive tests had, at least in some instances, revealed their presence in very small amounts. The fact that these elements are present in such small amounts would suggest that there has been no buildup of these elements in the lake brine, at least in the area where North Shore's harvesting takes place. This is consistent with earlier findings (Tayler and others, 1980).

PRODUCTS AND NUTRITIONAL VALUES

During North Shore's production of concentrated brines, sodium, chloride, sulfate, and potassium, are partially removed as crystalline salts. As this occurs in the solar ponds, the levels of magnesium chloride and the trace elements become enriched. From these concentrated brines, North Shore produces a variety of basic concentrated-brine products. North Shore's sister company, MRI, trucks these brine products in a dedicated tanker to MRI's production facility in West Haven, Weber County. Here, the concentrated brine is tested, polished using a proprietary multi-step procedure, and processed into final products.

ConcenTrace®

The first product is a concentrated brine sold under the name ConcenTrace® by both MRI and sister company Trace Minerals Research. MRI claims that the nutritional values of this product, are as follows: One-half teaspoon of this brine contains about 250 mg of magnesium, which is 62 percent of the U.S. Daily Value or "DV." DV is defined as the official-for use on food labels - U.S. Recommended Dietary Intake or Daily Reference Value for adults and children over 12, as set by the U.S. Food and Drug Administration (FDA) 21CFR.101.9(c)(8)(iv) (FDA, 2001a). Sodium and potassium are each less than 10 mg, or 0 percent DV. Lithium ranges between 1.4 to 2.5 mg, and boron between 1 to 2 mg. Based on more recent research (Schauss, 1999; Nielson, 2000) these quantities of lithium and boron are nutritionally significant, but no DVs have been set by FDA. Sulfate ranges between 40 to 100 mg, and chloride between 640 to 800 mg; 680 mg of chloride is 20 percent of the DV. Although sulfur is vital to nutrition (National Research Council, 1989), no DV has been set for either sulfur or sulfate.

Sulfate and chloride contents are standardized later on in the processing to specifications which vary, depending on the desired use, thus creating several different finished products. An additional variation of ConcenTrace®, which is sold specifically as a magnesium supplement, is a version with a somewhat higher magnesium content and a sodium content low enough to qualify it as "sodium free" under FDA guidelines 21CFR 101.61(b)(1) (FDA, 2001b).

Omni-Min™ and Other Brine Products

Omni-Min™, a slightly less-concentrated product than ConcenTrace®, has a claimed magnesium level of 200 mg per 1/2 tsp., and sodium and potassium levels between 20-35 mg per 1/2 tsp. This product is currently being sold under the name Omni-Min™ as a raw material for dietary supplement and food manufacturers, and for use by other companies not affiliated with MRI.

Using a combination of both harvesting and polishing techniques, a variety of brines similar to those mentioned above are produced that optimize the levels of different combinations of magnesium, sodium, potassium and sulfate. Brines are currently being produced for several different uses, including agricultural, culinary, and external uses for the human body, such as bath salts and cosmetic.

Inland Sea Water™

The company still produces and sells its original product, Inland Sea Water™, which is pure Great Salt Lake brine. To minimize the lake's concentration variations between high-water and low-water years, harvesting is being done directly from North Shore's pre-concentration ponds. In this way, the brine is harvested at a concentration that is equal to the lake's natural maximum concentration.

Inland Sea Water™ is still used as a dietary supplement, as a salt seasoning, and as a source of nutritional electrolytes, where higher sodium levels are desirable. It contains significant levels of a number of major ions in addition to its trace-mineral content (Supportive, non-MRI data on trace element analysis of Great Salt Lake brine generally can be found in Arnow and Stephens, 1990; EPA, 2001a; Gwynn, 1996). Two teaspoons of average Great Salt Lake brine in a low-lake level, high-concentration year, contains approximately 40 percent of the DV of magnesium in a form highly usable by the human body. Two teaspoons also contain about 45 percent of the DV for sodium, four percent of the DV of potassium, and 32 percent of the DV of chloride, all usable by the body. Although sulfur is also recognized as essential to the body, no specific requirements have been set, but sulfur levels in the brine, in the form of sulfate, also seem to be quite nutritionally significant. Care should be taken, however, not to consume excessive amounts of sodium or sulfate, since too much sodium has been shown to have a negative effect on blood pressure and too much sulfate can have a laxative effect.

In its pure form, Great Salt Lake brine can be used in place of processed salt at a ratio of about 4:1(liquid : dry) salt requirement. It can be sprayed as a fine mist directly on foods for a more immediate salt taste with a lower level of sodium intake. Used like this, with regular salt intake as a guide, both sodium and sulfate consumption can be kept to appropriate levels.

Major and Trace Elements Present

The following list is a compilation of those elements identified by MRI as being present in ConcenTrace® and Omni-Min™, and by extension, Inland Sea Water™. Results were obtained from a number of different independent

Table 1. Listing of elements/ions found in *ConcenTrace®*.

Symbol	Element	Symbol	Element
Cl	Chloride	Te	Tellurium
Mg	Magnesium	Ba	Barium
$SO4^{-2}$	Sulfur as total sulfate	Ag	Silver
Na	Sodium	Cd	Cadmium
K	Potassium	U	Uranium
Br	Bromide	Ga	Gallium
$HCO3^{-1}$	Carbon as total carbonate	Zr	Zirconium
Li	Lithium	Be	Beryllium
B	Boron	Bi	Bismuth
Ca	Calcium	Hf	Hafnium
F	Fluoride	Sm	Samarium
Si	Silicon	Ce	Cerium
N	Nitrogen	Cs	Cesium
Se	Selenium	Au	Gold
P	Phosphorus	Tb	Terbium
I	Iodide	Eu	Europium
Cr	Chromium	Gd	Gadolinium
Fe	Iron	Hg	Mercury
Mn	Manganese	Dy	Dysprosium
Ti	Titanium	Th	Thorium
Rb	Rubidium	Ho	Holmium
Co	Cobalt	Lu	Lutetium
Cu	Copper	Tm	Thulium
Sb	Antimony	Er	Erbium
As	Arsenic	Yb	Ytterbium
Mo	Molybdenum	Nd	Neodymium
Sr	Strontium	Pr	Praseodymium
Zn	Zinc	Nb	Niobium
Ni	Nickel	Ta	Tantalum
W	Tungsten	Tl	Thallium
Ge	Germanium	Re	Rhenium
Al	Aluminum	In	Indium
Sc	Scandium	Pd	Palladium
V	Vanadium	Pt	Platinum

laboratories using a number of standard testing methodologies. Actual test results are on file at MRI.

Hydrogen and oxygen, as dissolved gasses, and the noble gases (helium, neon, argon, krypton, Xenon, and radon) have never been tested for, but are believed to be present due to the fact that they are present in sea water (Lide, D.R., 1993).

PRODUCT MARKETING

North Shore sells the brines primarily to MRI who further tests, standardizes and polishes them. MRI then packages the polished brines for sale as dietary supplements in their liquid form. MRI also uses the concentrated brines as an important ingredient in the manufacture of many other dietary supplements in liquid, powder, tablet, and capsule form.

MRI sells these finished, packaged products to sister companies Trace Minerals Research, Marine Minerals, Marine Biotherapies, and Bio-Nativus Wholistic Resources, that were established to sell MRI's products to different markets in the United States. Trace Minerals Research (TMR) sells its products primarily through health food stores. Marine Minerals sells by way of network marketing and mail orders. Marine Biotherapies markets to the beauty industry, mostly through salons and aestheticians. Bio-Nativus Wholistic Resources serves natural health care professionals, focusing mainly on chiropractors and naturopaths.

MRI sells bulk and private labeled products to a number of other companies who have incorporated the minerals into part of their existing product lines. MRI also manufactures products for distributors and trading partners all over the world.

NUTRITION - THE BEST USE OF GREAT SALT LAKE?

When the pioneers entered the Salt Lake Valley in 1847, they found a large body of salt water, which seemed too salty to support any life. After their long and arduous journey, this must have felt like a crowning insult. Since that time, this wonderfully unique body of water has, in many ways, been slow to reveal her treasures. In the future, however, it may yet be shown that the highest and best use of the lake's brines may be in providing nutritional elements to the human body. The nourishing, balancing, and stimulating effect of these minerals to both body and mind may prove to be of immense value to the human family. This provides yet another incentive to protect, and even jealously guard this very special treasure that nature has given us.

EDITOR'S NOTE

Neither the Utah Geological Survey nor the State of Utah endorse or recommend any commercial products, nor do they make any judgement or guarantees about the efficacy or safety of any products. For further information, MRI can be contacted at: 1990 W. 3300 S. Ogden, Utah, 84401, Phone 801-731-7040, fax 801-731-7985, or online: http://www.mineralresourcesint.com

REFERENCES

American Medical Association, 1989, The American Medical Association Encyclopedia of Medicine, Charles B. Clayman editor: Random House (see Electrolyte, p 396; Environmental Science - Minerals, p. 409; Ion, p. 605; Minerals and Main Food Sources, p. 690.)

Arnow, Ted, and Stephens, Doyle, 1990, Hydrologic Characteristics of the Great Salt Lake, Utah, 1947-1986: U.S. Geological Survey Water-Supply Paper 2332, 32 p.

Bergner, P., 1997, The Healing Power of Minerals, Special Nutrients, and Trace Elements: Rocklin, Ca, Prima Publishing, 312 p.

Crane, G.W., 1973, The Worry Clinic - sea water proves to have 44 beneficial elements: Ogden, Utah, The Ogden Standard Examiner, May 8, p. 9A,

—1979, The Worry Clinic - ocean's vital trace chemicals essential to body health: Ogden, Utah, The Ogden Standard Examiner, March 22.

—Pre-1968, The Worry Clinic. Author's note: Dr. Crane wrote his syndicated column for many years, beginning sometime prior to 1968. MRI has many of his articles on file. It is unclear which article on which date originally started Hartley Anderson's interest.

EPA, 2001a, Modernized Storet database: U.S. Environmental Protection Agency: Online, http://www.epa.gov/storet.

—2001b, SW-846 On-Line, Method 3010A, Acid digestion of aqueous samples and extracts for total metals for analysis by FLAA or ICP spectroscopy: Online, http://www.epa.gov/epaoswer/hazwaste/test/under.htm

FDA, 2001a, Nutritional labeling of food, 21CFR101.9(c)(8)(iv): Online,http://frwebgate.access.gpo.gov/cge-bin/multidb.cgi

—2001b, Part 1 - Food Labeling, Nutrient content claims for sodium content of foods, 21CFR 101.61(b)(1): Online, http://www.access.gpo.gov/nara/cfr/waisidx_99/21cfr101_99.html

Gwynn, J.W., 1996, Commonly asked questions about Utah's Great Salt Lake and ancient Lake Bonneville: Utah Geological Survey Public Information Series 39, 22 p.

Heinerman, J., 1998, Heinerman's Encyclopedia of Nature's Vitamins and Minerals: Paramus, NJ, Prentice Hall, 398 p.

Lide, D.R. (editor), 1993, CRC Handbook of Chemistry and Physics: Boca Raton, FL, CRC Press, 74th Edition p. 14-9.

National Research Council, 1989, Recommended dietary allowances, 10th edition, by National Research Council, Commission on Life Sciences, Subcommittee on the Tenth Edition of the RDAs, Food and Nutrition Board: Washington, D.C., National Academy Press, p. 174-259, 267.

Nielson, F.H., 2000, Emergence of Boron as Nutritionally Important Throughout the Life Cycle: Nutrition, v. 16 (7/8), p. 512-14.

Schauss, A.G., 1982, Keynote Lecture, Texas Conference on Nutrition and Behavior, University of Texas at Austin, October 28.

—1983, Nutrition and Behavior, Journal of Applied Nutrition, v. 35, p. 30-43.

—1999, Minerals, Trace Elements and Human Health: Biosocial Publications, Tacoma, WA, p. 1, 5, 33-34.

Tayler, P.L., Hutchinson, L.A., and Muir, J.K., 1980, Heavy metals in the Great Salt Lake, Utah, in Gwynn, J.W., editor, Great Salt Lake, a scientific, historical and economic overview: Utah Geological and Mineral Survey Bulletin 116, p. 194-200.

U.S. Government Code of Federal Regulations [21CFR.101.9 (c)(8)(iv)] - Section 21, Part 101, Food Labeling, §101.9 Nutrition labeling of food., subsection (c)(8)(iv); 21 CFR Ch. 1 (4-1-01 Edition) p. 30. Online, fttp://frwebgate.access.gpo.gov/cgi-bin/multidb.cgi

HEAVY-OIL DEPOSIT, GREAT SALT LAKE, UTAH

by

L.C. Bortz
AMOCO Production Company
Denver, Colorado

L.C. Bortz, - AAPG Studies in Geology Series No. 25, AAPG 1987

ABSTRACT

The western portion of the Great Salt Lake contains two large Neogene basins, informally called the "North" and "South" basins. These basins are separated by an arch that trends northeast between Carrington Island and Fremont Island. Both basins are filled with Miocene, Pliocene, and Quaternary sediments and volcanic rocks. Each basin has an estimated maximum thickness of over 4,300 m (14,000 ft) of Tertiary rocks. Palynology indicates the oldest Tertiary sedimentary rocks present in both basins are Miocene, but a radiometric date indicates the presence of Oligocene rocks.

Structurally, the basins are slightly asymmetric, deeper on the east with an obvious boundary fault zone on the east flank of each basin. Faulting is present on the western flanks but of a lesser magnitude. The most common structural traps found in these basins are anticlinal closures, faulted noses, and fault closures. These structures are probably the result of continued differential subsidence of pre-Miocene blocks throughout Neogene time.

A total of 15 exploratory wells were drilled by Amoco in the Great Salt Lake, from June 1978 to December 1980, resulting in an oil discovery at West Rozel and oil and/or gas shows in eight other wildcat wells.

The West Rozel oil field produces from fractured Pliocene basalts at a depth of 640-730 m (2,100-2,400 ft). The trap is a faulted, closed anticline covering approximately 2,300 acres. The discovery well, Amoco No. 1 West Rozel Unit (NW NW Sec. 23, T8N, R8W, Box Elder County), has an oil column of 88 m (290 ft) but produced at rates of only 2-5 BOPH with a gas-lift system. The oil is 4° API gravity, 12.5% sulfur, and has a pour point of 75°F. Two development wells that have smaller oil columns (No. 2, 26 m [85 ft]; No. 3, 60 m [194 ft]) were equipped with a downhole hydraulic pump and produced oil at rates up to 90 BOPH. Additional development of the field was not initiated because of the high water cut and the high cost of operating an "offshore" field.

INTRODUCTION

This paper is a case history of a recent exploration program in the Great Salt Lake area, a Neogene fluvial and lacustrine basin in the Basin-and-Range Province of the western United States. This province is characterized by numerous horsts and grabens separated by tensional faults.

The first wildcat well drilled in offshore Great Salt Lake was spudded in June 1978. During the next 2.5 years a total of 15 wells (13 wildcats and two development wells) were drilled. In addition to the heavy-oil discovery at West Rozel, valuable geologic data were obtained in this program. This paper will: (1) summarize the geology of the Great Salt Lake area, (2) discuss the results of Amoco's exploration program, and (3) describe the heavy-oil deposit at the West Rozel field.

GEOLOGIC SUMMARY

The Great Salt Lake is near the eastern edge of the Basin-and-Range Province, defined as the western flank of the Wasatch Range in this area (figure 1). Salt Lake City is near the southeastern corner of the lake.

Stratigraphy

Outcrops are generalized in figure 1 and range in age from Precambrian to Recent. Palynology was used to subdivide and correlate Tertiary rocks within the Great Salt Lake basins (figure 2). On this section, the Tertiary sedimentary rocks are all Neogene. However, a volcanic tuff near the bottom of well No. 6 was determined to be 29.9 ± 1.3 m.y. by the zircon fission-track method. This indicates that locally some Oligocene rocks may underlie the Neogene in the Salt Lake area. The deepest portion of the north basin is near wells 5 and 6. Here the total Tertiary section is between 4,300 and 4,600 m (14,000-15,000 ft) thick. In the south

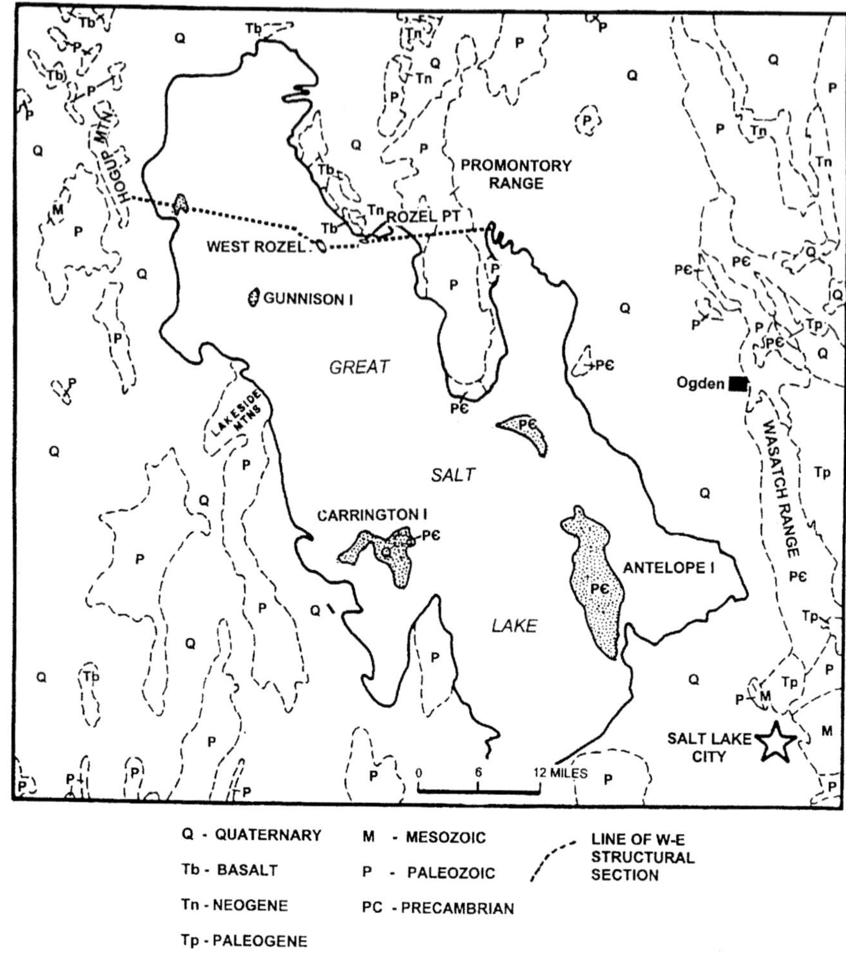

Figure 1. *Generalized geologic map, Great Salt Lake area. Map modified after Hintze (1980).*

basin (well 8) the thickest Tertiary section is about 4,300 m (14,000 ft).

Miocene and Pliocene sediments were deposited in fluvial, marginal lacustrine, and lacustrine environments. Deposition during middle Pliocene to Pleistocene was more lacustrine, as evidenced by the abundance of aquatic pollen in these units (figure 2). The Pliocene "West Rozel" basalt (wells 1, 2, and 3) is composed of numerous flows that probably came from volcanic vents in the northwestern part of the Rozel Hills (figure 3). The areal extent of the "West Rozel" basalt is also shown on this figure.

The main reservoir in the Salt Lake area is the "West Rozel" basalt, which contains heavy oil at Rozel Point and West Rozel (figure 3). Many of the fluvial and lacustrine Neogene sandstones have good reservoir quality, but none were saturated with heavy oil.

Source Rocks

Neogene sediments are the most likely source rocks for the heavy oil found in the Great Salt Lake. Numerous analyses of surface and subsurface Paleozoic and Mesozoic sedimentary rocks in the area indicate absence of source potential or expended oil source rocks. In the vicinity of West Rozel, Miocene sediments are lake-margin shales and carbonates, a thick tuff, and some red beds. Pliocene rocks below the "West Rozel" basalt are anhydritic dark shales and carbonates and organic, dark, calcareous shales and silts. These organic-rich, fine-grained clastic sediments are the probable source rock for the heavy oil at West Rozel and Rozel Point. Vitrinite reflectance in these rocks show that this section is in the peak oil generation stage.

Rozel Point Oil Seep

One of the reasons that the petroleum industry has been attracted to the Great Salt Lake area is the Rozel Point oil seep, at the south end of Rozel Point (figure 3), described by Eardley (1963, 1966). Heavy, viscous oil flows from a probable fault zone, which has uplifted the Rozel Hills relative to the lake to the southwest. Since the early 1900s there have been many attempts to produce oil from shallow wells drilled near the seeps. Oil and water have been recovered from Pliocene basalt (probably the "West Rozel" basalt) at depths of 45-90 m (150-300 ft). The area is currently leased and has been intermittently produced. Oil from the seeps and shallow wells has been intermittently produced, and has characteristics similar to those of the oil found at West Rozel field.

Oil shows in the northern Basin-and-Range Province, including the Great Salt Lake area, were reported by Bortz (1983).

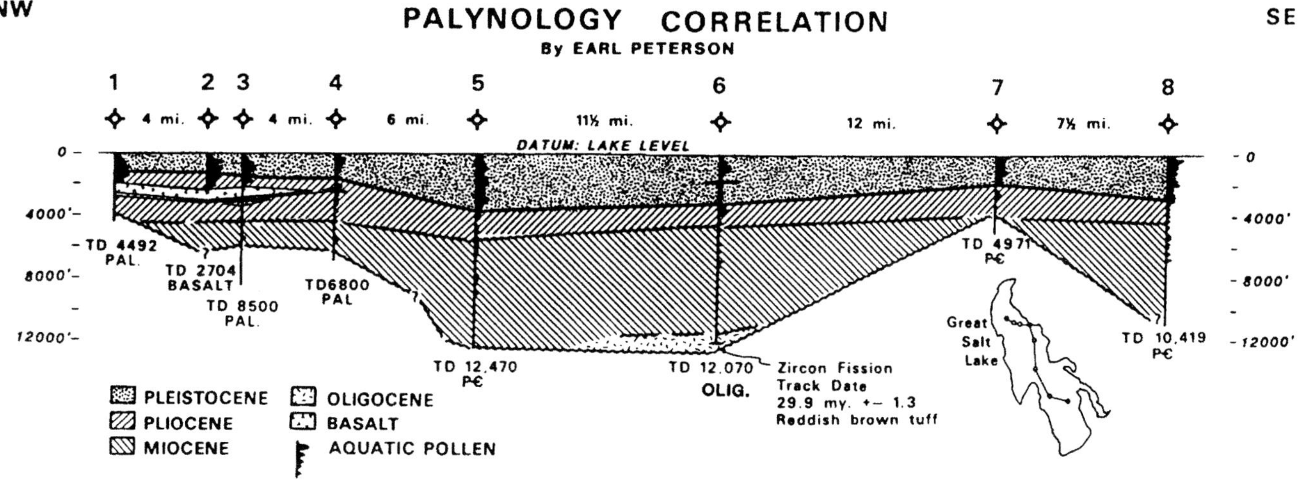

Figure 2. Tertiary correlation section. Line of section also shown on figure 8.

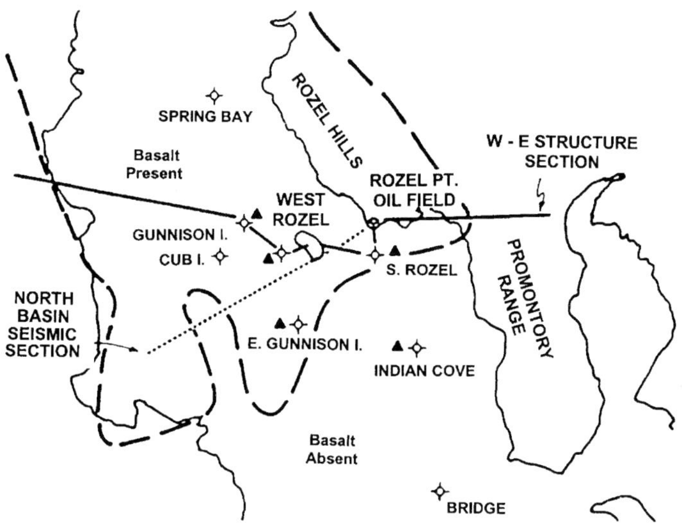

Figure 3. Location map for the West Rozel field area.

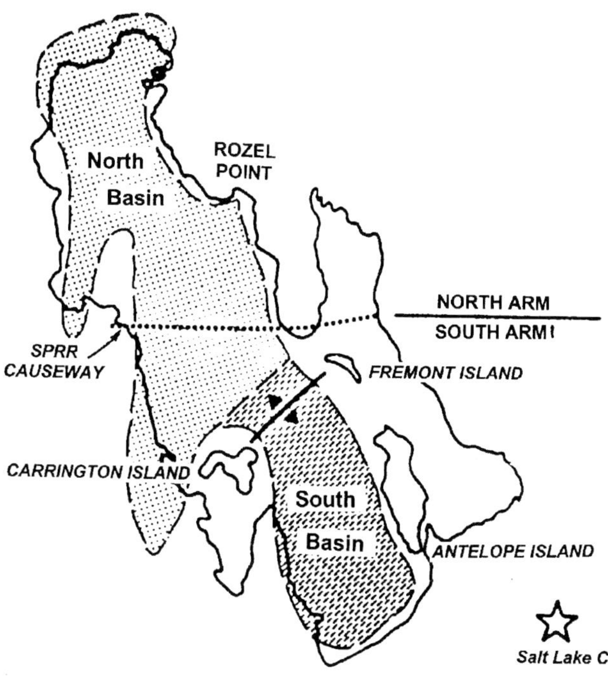

Figure 4. Map showing major basins in the western part of the Great Salt Lake.

Structure

The western portion of the lake contains two large basins that are separated by an arch (horst) that extends northeast from Carrington Island to Fremont Island (figure 4). There are several small subsidiary basins within these two basins. Other major basins are present east of Antelope Island.

The generalized structure of the north basin is shown on figure 5. At the west end, the Hogup Mountains are composed of folded and faulted upper Paleozoic rocks. Immediately to the east there is an area of what may be Paleozoic rocks covered by a thin veneer of Quaternary rocks. Next is a subsidiary basin that has an estimated 2,400 m (8,000 ft) of Tertiary sediments and volcanics (no wells have been drilled in this basin). The main north basin is between the north extension of Gunnison Island and Lakeside horst and Rozel Point. In this portion of the basin the maximum Tertiary thickness is about 2,100 m (7,000 ft). Numerous normal faults extend from the Paleozoic "basement" into the Tertiary and some are mappable at the surface; for example at Rozel Point is an intrabasin horst that separates the main north basin from the Rozel Graben, which contains 2,100-2,400 m (7,000-8,000 ft) of Tertiary rocks. East of the Rozel Graben is the Promontory Range, composed of lower and middle Paleozoic rocks.

The seismic sections in this paper show the structure of these Neogene basins. Figure 6 is an east-west line in the south basin west of Antelope Island. At the east end of the line, the basin-margin fault separating Antelope Island from the basin has a steep west dip. Other nearby seismic sections show this fault to be almost vertical. Neogene sediments (and possibly Oligocene volcanic rocks) overlie Precambrian rocks and show clearly on this line, which has good Tertiary reflections. The Antelope Island well (figure 8) was drilled on the eastern structure. Figure 7 is a seismic section in the north basin that goes through the West Rozel field. The structural interpretation at the top of the "West Rozel" basalt is reliable, but reflections from below the basalt are general-

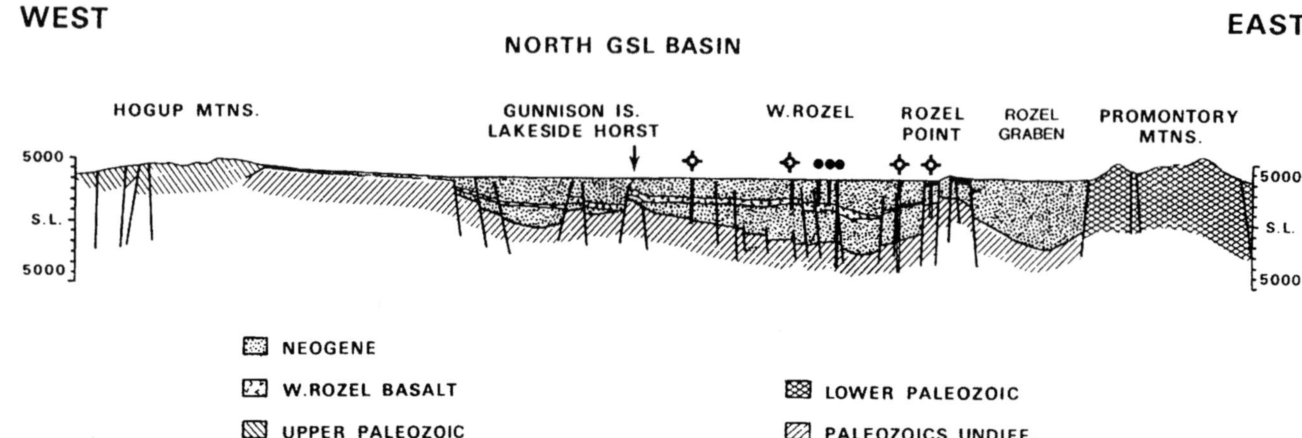

Figure 5. *Generalized structure section of the North Basin. See figures 1 and 3 for line of section.*

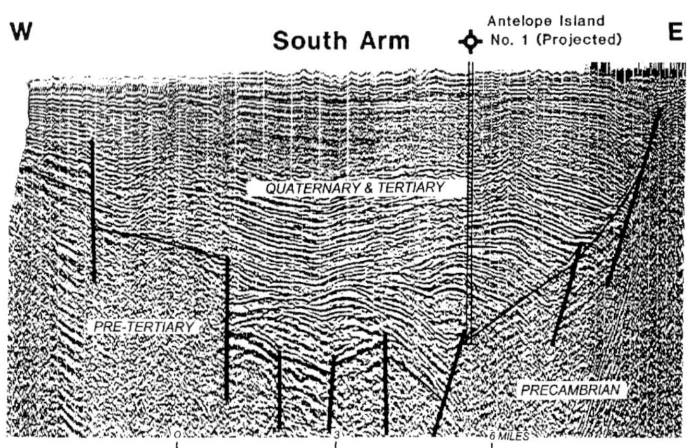

Figure 6. *Seismic section in the basin west of Antelope Island. See figure 8 for line of section.*

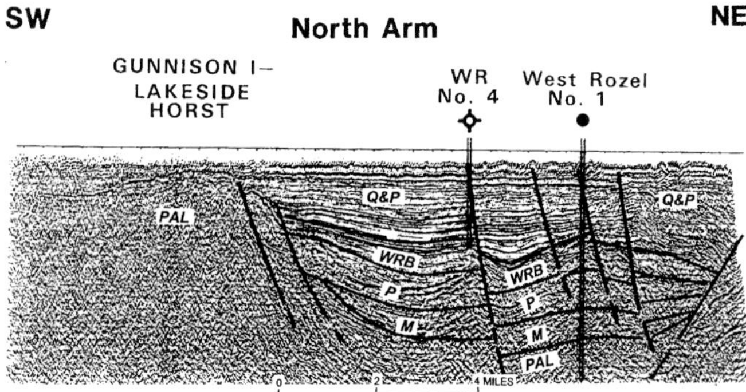

Figure 7. *Seismic section in the North Basin through the West Rozel field. See figures 3 and 8 for line of section. Q & P, Quaternary and Pliocene; WRB, West Rozel basalt; P, Pliocene; M, Miocene; PAL, Paleozoic.*

ly not reliable, so the interpretation below the basalt relies heavily on well control. The faulted, anticlinal structure at West Rozel field is obvious. The basalt is over 335 m (1,100 ft) thick at West Rozel and thins rapidly to the east (at Rozel Point it is 46 m [150 ft] thick); it also thins to the west, where it is in fault contact with Paleozoic rocks on the east flank of the Lakeside-Gunnison Island horst. Miocene sediments are present, but wedge out or thin toward the western basin range.

AMOCO'S EXPLORATION PROGRAM

To evaluate a 620,000-acre lease block in the western part of the Great Salt Lake, Amoco conducted extensive offshore reconnaissance and detailed seismic surveys and drilled a total of 15 wells. The first well was spudded during June 1978, and operations were terminated in January 1981. The 15 wells, including the West Rozel heavy-oil discovery and two development wells, are summarized on table 1, and their locations are shown on figure 8. Nine of the 13 wildcats had shows of oil and or gas. All of the oil shows were heavy oil, similar to the oil found at West Rozel field and Rozel Point. The best shows of oil were found in the North basin where the "West Rozel" basalt is present in the subsur-

Table 1. Summary of Amoco wells drilled in the Great Salt Lake, listed in chronological order.

Well	Total Depth	Fm. At total depth	Remarks
Indian Cove	12,470	PC	S/O S/G
W. Rozel - 1	8,500	Pal.	Oil disc.
S. Rozel	6,802	Pal.	S/O
Gunnison Island	4,492	Pal	S/O
W. Rozel - 2	2,704	Plio.	Oil well
Antelope Island	10,419	PC	S/O
Carington Island	4,971	Plio.	No shows
Bridge	12,070	Olig. (?)	S/O S/G
Midlake	2,450	Plio.	No shows
Sandbar	7,864	Pal.	S/O S/G
W. Rozel - 3	2,790	Plio.	Oil well
E. Gunnison	7,843	Pal.	Free oil on 2 DSTs
W. Rozel - 4	2,210	Plio.	S/O
Spring Bay	4,883	Pal.	No Shows
Cub Island	1,716	Pal.	No shows

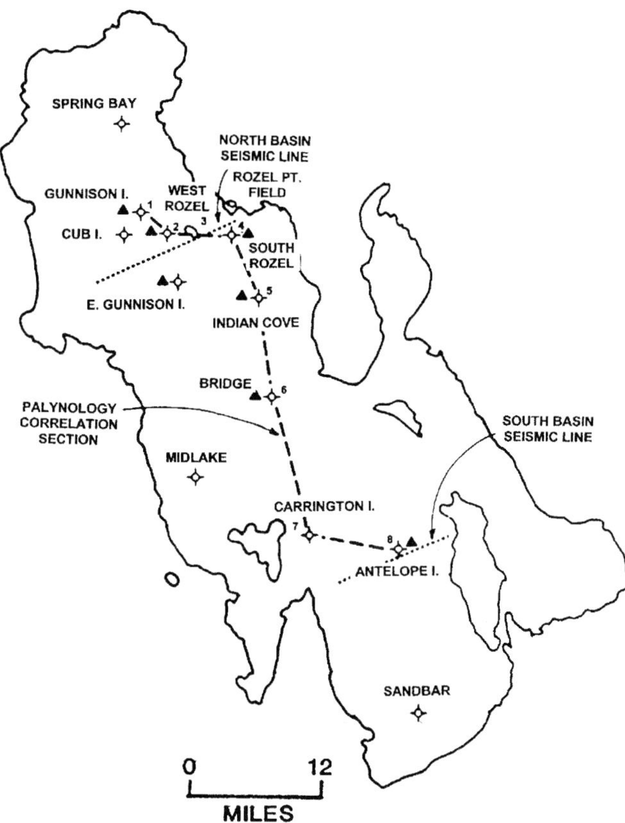

Figure 8. Location map for Amoco exploratory wells and seismic and palynology lines of section.

face (figure 3). Nine wells were drilled to the pre-Tertiary with seven penetrating Paleozoic rocks and two penetrating Precambrian rocks. The Indian Cove well was the deepest and reached total depth in Precambrian at 3,800 m (12,470 ft).

Amoco drilled the 15 wells from a floating barge (figure 9) that was disassembled and trucked from one side of the lake to the other because there is no access over or through the Southern Pacific Railroad causeway (figure 4). Woodhall (1980) has described some of the engineering problems of this marine drilling operation.

WEST ROZEL HEAVY-OIL FIELD

The second well drilled in the lake discovered heavy oil at West Rozel field (NW NW Sec. 23, T8N, R8W). Shows of oil were found in the mud as the bit approached the basalt that is now informally called the "West Rozel" basalt. At times as Amoco drilled into the basalt, the shale shaker was overloaded with heavy oil similar to that found at Rozel Point oil field. Open-hole drillstem tests (DSTs) in the top of the basalt recovered full strings of mud with only slight shows. An evaluation of the formation pressures indicated the reservoir has darcys of permeability. A DST of the basalt through perforations in production pipe from 695 to 735 m (2,280-2,410 ft) recovered a full string of oil. At that point Amoco had a discovery, but the next question remained unanswered: Was it a commercial discovery?

Figure 9. Three views of the drilling barge used by Amoco in the Great Salt Lake.

Structure

The oil field as mapped on the top of the "West Rozel" basalt is a faulted, closed anticline covering approximately 2,300 acres, with a vertical closure of about 90 m (300 ft) (figure 10). Closure is defined by the 2,400-ft contour (below the lake level). The highest part of the structure is very near the West Rozel No. 1. The discovery well and two development wells were drilled in the central closure (stippled area). Other closures are outlined by a stippled band.

Effective closure is provided by dip to the west and north. On the south and east, fault contact with down-thrown Pliocene clays and silts forms the seal. There is also regional dip in these directions. Faults placing basalt against basalt do not form effective seals at West Rozel field. This was part of the reason that the structure west of the field (figure 10) was not productive when tested by the West Rozel No. 4. Another problem was that a tight sandstone at the top of the basalt filled most of the closure.

"West Rozel" Basalt Reservoir and Production

The three productive wells at West Rozel field are shown on a structure section (figure 11). The datum, which is the approximate oil-water contact, is 735 m (2,410 ft) below the lake surface. The oil column for each well is shown and beside each log:

W.R. No. 2	26 m	(85 ft)
W.R. No. 3	59 m	(194 ft)
W.R. No. 1	88 m	(290 ft)

Formation water containing 92,200 mg/l chloride has been recovered on DSTs and from perforations below the oil column. Pliocene clays and silts seal the basalts.

Core, pressure, and production data indicate the basalt is primarily a fractured reservoir. Fractures vary in abundance and orientation (vertical to horizontal) within the reservoir. The zones with diagonal lines shown on figure 11 are interpreted to have only fracture porosity. Effective porosity may be enhanced where the fractures are in contact with vugs and vesicles. A 15% porosity line drawn on each log shows that much of the column has greater than 15% porosity. This is not effective porosity because the density log is measuring vugs and vesicles that are saturated with oil only where they are in contact with a fracture. The stippled zones on figure 11 are interpreted to have significant effective fracture-matrix porosity. Effective porosity for these zones estimated from core and log data are:

Fractured only	1-3%
Matrix and fractures	7-9%

Three core photographs from West Rozel No. 3 are typical of the olivine basalt within the oil column. Figure 12 shows two types of basalt: (1) the core to the right is a dense basalt with fractures filled with calcite and zeolite minerals; (2) the core to the left is a vesicular basalt. These samples are near the top of the oil reservoir but have little or no oil saturation. Figure 13 is basalt with some oil saturation in matrix and fractures near the middle of the oil reservoir. Figure 14 shows basalt "rubble" that was recovered by use of a rubber sleeve. The "rubble" apparently is a conglomerate with good matrix porosity and some matrix permeability. This core sample near the bottom of the oil reservoir was completely saturated with heavy oil.

At West Rozel field, oil-in-place reserves, using an average effective porosity of 8% for those zones with fracture-matrix porosity, are estimated to be in excess of 100 million barrels of oil. The recovery factor is not known, but for a fractured reservoir with some matrix porosity it is expected to be somewhere between 1 and 10% for primary production.

The West Rozel field and the Rozel Point heavy oils are very similar. Both are dark brown and have low gravities (4-9° API). The API gravity of these oils is difficult to determine because of entrained water; measured gravities at West Rozel field have ranged from 4° to 9° API, with 4° API considered the most accurate. The two oil characteristics that are perhaps the least encouraging are the high pour point and high sulfur content (12-13%). The reservoir temperature at West Rozel is 140°F, which improves the mobility of the oil when compared to the 115°F reservoir temperature at Rozel Point. The specific characteristic of the West Rozel field heavy oil are:

Gravity	4° API
Color	Dark brown
Sulfur	12.5%
Pour point	75°F
Viscosity	3000-4000 cp @ 140°F

The whole oil chromatogram (figure 15) shows that the West Rozel oil is biodegraded.

Oil Production Summary

Three wells capable of producing heavy oil were completed in the West Rozel field (figs. 10, 11). The discovery well (W.R. No. 1) was not equipped to be pump tested and

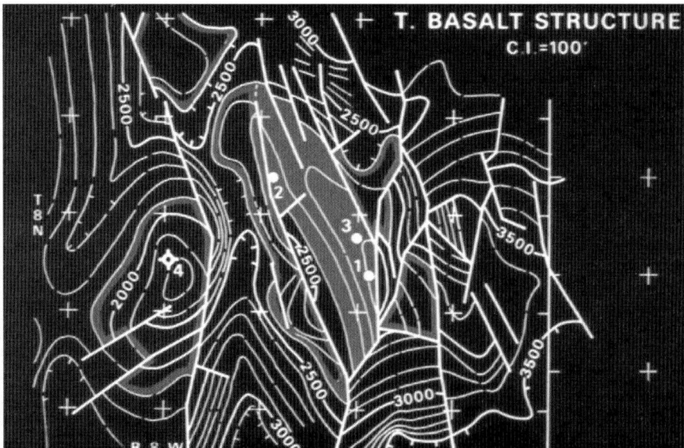

Figure 10. Top-of-basalt structure map in the West Rozel field area. Datum is the surface of the lake.

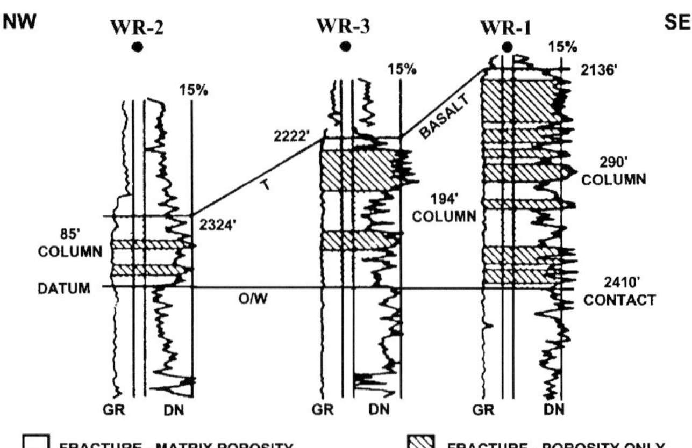

Figure 11. Structure section at West Rozel showing basalt reservoir. See figure 10 for well locations.

Figure 12. *West Rozel No. 3. Photograph of core near top of oil column.*

Figure 13. *West Rozel No. 3. Photograph of core near middle of oil column.*

Figure 14. *West Rozel No. 3. Photograph of "rubble" recovery in core near base of oil column.*

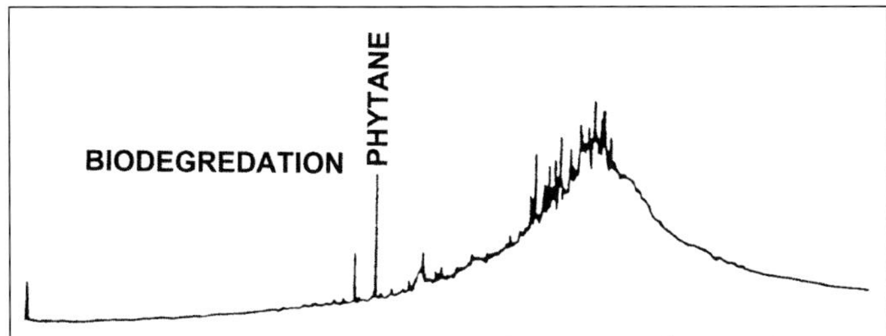

Figure 15. *Whole-oil chromatogram of West Rozel heavy oil.*

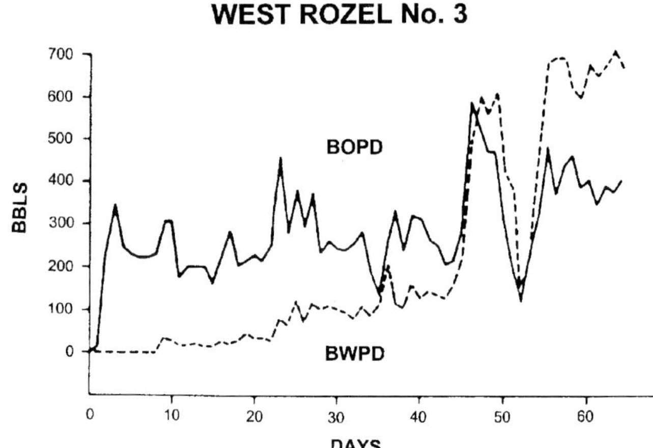

Figure 16. *Graph showing oil and water production for West Rozel No. 3.*

only a limited amount of testing was done on this well using a nitrogen gas-lift system. Maximum production was 2-5 BOPH during short-term tests, and this well produced a total of about 148 BO. The other two wells were pump tested, using a Kobe downhole hydraulic pump. The oil production from these wells is summarized:

Well	Days produced	Bbl water	Bbl oil
W. R. No. 2	46	16,400	14,100
W.R. No. 3	64	14,191	18,780
		Total oil	32,880

The No. 3 was produced continuously for 64 days, and a graph of the water and oil production is shown on figure 16. Note that at the end of the test period, this well was producing over 1000 barrels of fluid per day: 300-400 BO and 600-700 BW. On a short-term test the W.R. No. 2 was pumped at rates up to 90 BOPH.

Figure 17 is a photograph of the floating production platform used at West Rozel. The transport barge used to take the heavy oil to storage tanks at Little Valley Harbor is shown on figure 18.

Figure 17. *Floating production platform used at West Rozel.*

Figure 18. *Transport barge.*

CONCLUSION

Additional development of the field was not initiated because of the high water cut and the high cost of operating an "offshore" field. All of the West Rozel oil wells were plugged and abandoned when operations were terminated in January 1981. It is hoped that improved primary and enhanced recovery techniques and systems will soon allow the economic development of the heavy-oil deposit at West Rozel field.

ACKNOWLEDGMENTS

My thanks to Amoco Production Company for permission to publish this paper. Special thanks to Ron Calvert for critically reading the paper, the many helpful suggestions, and the use of some of his concepts on Tertiary stratigraphy. Also thanks to Don Nixon, Earl Peterson, Bill Krebs, Tom Hemler, Jerry Daly, Keith Farmer, Marilyn Cowhick, and Deb Skelton.

REFERENCES CITED

Bortz, L.C., 1983, Hydrocarbons in the northern Basin and Range, Nevada and Utah: Geothermal Resources Council Special Report 13, p. 179-197.

Eardley, A.J., 1963, Oil seeps at Rozel Point: Utah Geological and Mineralogical Survey, Special Study 5, 28 p.

—1966, Sediments of Great Salt lake, *in* Guidebook to the Geology of Utah No. 20, The Great Salt Lake: Utah Geological Society, p. 105-120.

Hintze, L.F., 1980, Geologic map of Utah: Utah Geological and Mineralogical Survey, 1:500,000.

Woodhall, R.J., 1980, Engineering problems of Great Salt Lake, Utah, marine oil drilling: Utah Geological and Mineralogical Survey Bulletin 116, p. 372-392.

ROZEL POINT OIL FIELD, BOX ELDER COUNTY, UTAH: GEOLOGY, DEVELOPMENT HISTORY, AND CLEANUP

by

Gilbert L. Hunt
Utah Division of Oil, Gas and Mining, Salt Lake City, Utah

and

Thomas C. Chidsey, Jr.
Utah Geological Survey, Salt Lake City, Utah

ABSTRACT

The Rozel Point oil field has been of interest to prospectors, entrepreneurs, and others for more than 100 years. Natural oil seeps found at Rozel Point were indications of what existed below the surface waiting to be exploited. The oil from Rozel Point field accumulated in a shallow Pliocene (?) basalt in a combination fault and unconformity trap. This low-gravity oil contains high amounts of sulfur and was probably generated from Tertiary lacustrine strata to the west, migrating upward along faults and fractures to the reservoir rocks or forming seeps. At least 10,000 barrels of oil have been produced from the field, but its unique properties make production difficult and refining costly.

Its location near the shore of Great Salt Lake provided many challenges for those attempting to develop this hydrocarbon resource as the lake level fluctuated. High lake level, disrepair of wells and production equipment, and new environmental laws resulted in concerns about possible pollution and adverse effects on wildlife in the mid-1990s. During 1995, reconnaissance studies by the Utah Division of Oil, Gas and Mining and the U.S. Environmental Protection Agency (EPA) looked at options for remediating leaking wells at Rozel Point field. During January and February 1996, the EPA funded and supervised a well-capping program at Rozel Point.

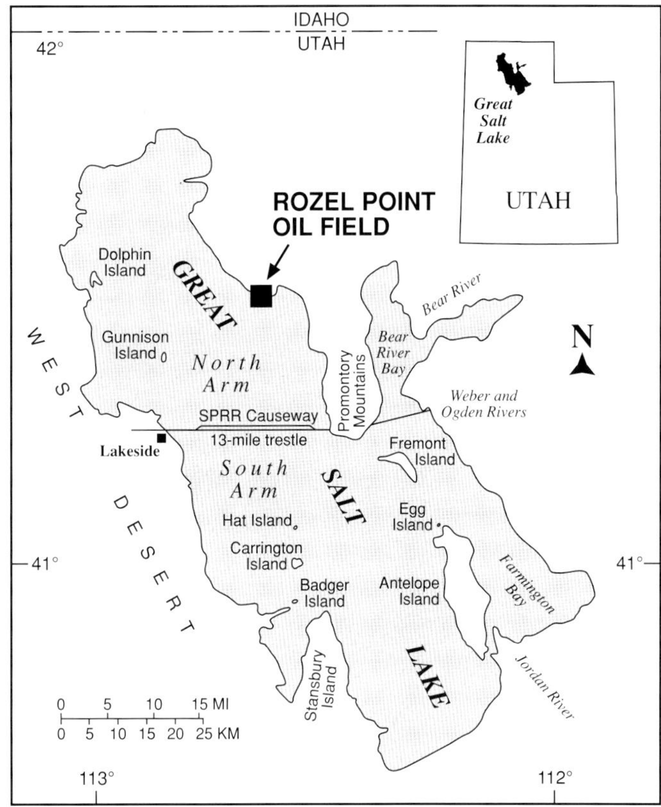

Figure 1. *Location of Rozel Point oil field on north arm of Great Salt Lake, Box Elder County, Utah.*

INTRODUCTION

Rozel Point oil field, located along the north arm of Great Salt Lake to the west of the Promontory Mountains (figure 1), is the oldest oil field in Utah with production activity dating back to around 1904. Naturally occurring oil seeps observed at low lake levels led to its discovery, possibly by Howard Stansbury during his expedition to the valley of Great Salt Lake in 1852 (Stansbury, 1852). Rozel Point field is unusual in several ways: (1) it is one of only two oil fields in the entire Basin and Range physiographic province of Utah (the other being West Rozel, also in Great Salt Lake to the west-southwest [see Bortz, this volume]); (2) wells are now offshore; (3) the oil reservoir is a basalt; (4) drilling depths are less than 300 feet (91 m); and (5) the oil contains a very high percentage of sulfur and other constituents yielding unique properties.

Due to the rising level of Great Salt Lake during the 1980s, concerns about alleged pollution from abandoned oil wells, and the adverse effects leaking wells may have had on wildlife, a reconnaissance and documentation program was undertaken in 1995 by the Utah Division of Oil, Gas and Mining (DOGM). The objectives of the program were to collect available well information, locate wells and related oil-field equipment on the ground and in the lake, and determine water depths. The information gathered was used by contractors and others to evaluate remediation alternatives. A well-capping project was implemented in 1996, funded and supervised by the U.S. Environmental Protection Agency (EPA).

GEOLOGY OF THE ROZEL POINT FIELD AREA

Successful oil fields possess three important geological features: (1) a well-developed trapping mechanism, (2) a well-developed reservoir rock unit, and (3) a hydrocarbon source. The Rozel Point field lacks a well-developed reservoir rock and well-developed trapping mechanism, as is evident from naturally occurring oil seeps in the area. Without a trap, the hydrocarbons at Rozel migrate to the surface.

Geologic Setting

Rozel Point and Great Salt Lake are located in the Basin and Range physiographic province. The nearby mountain ranges and valleys formed along generally north-south-trending normal faults resulting from regional extension. Many of these faults are present beneath valley-fill deposits under the lake and have created potential traps for hydrocarbons. Potential reservoir rock in the area consists of fractured vesicular basalt. Organic-rich sediments of the Salt Lake Group have been buried deep enough and long enough to generate hydrocarbons as evidenced by the natural oil seeps at Rozel Point.

The Rozel Hills just to the north of the oil field are composed of the Miocene-Pliocene Salt Lake Group and Tertiary basalt flows (figure 2). The Salt Lake Group consists of light-colored fanglomerate, conglomerate, and tuffaceous sandstone

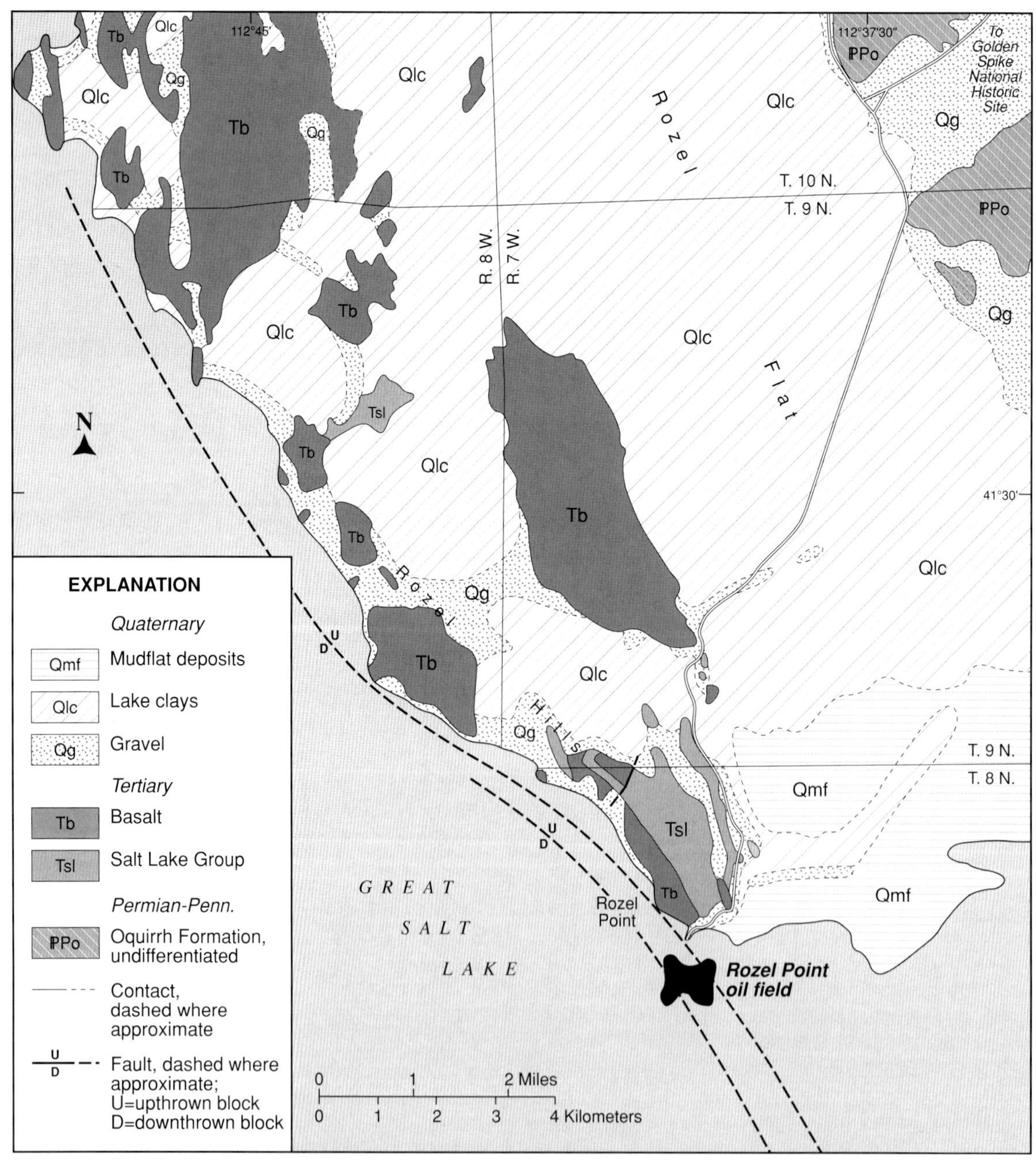

Figure 2. Geologic map of the Rozel Hills and Rozel Flat area, Box Elder County, Utah (modified from Doelling, 1980).

and limestone (Doelling, 1980). The basalt is up to 250 feet thick (76 m), dark gray to brown, crystalline, and vesicular (Doelling, 1980). The bedrock formations and basalt flows throughout the Rozel Hills and Rozel Flats area are typically covered by Quaternary gravels, lake clays, and mudflat deposits (figure 2). The Rozel Hills are formed by a small horst adjacent to the Rozel graben in the lake to the west.

Rozel Point Reservoir, Trap, and Source

Rozel Point field might best be described as a leaky combination fault and angular unconformity trap. The main oil reservoir is a 2- to 3-foot-thick (0.6-1 m), vesicular and fractured olivine basalt of probable Pliocene age that lies 125 to 300 feet (38.1-91.4 m) below the bottom of the lake (Heylmun, 1961). The Tertiary rocks exposed in the Rozel Point area dip east-northeast 15° (Jensen and Doelling, 1987). They have been uplifted and tilted as part of the south end of the Rozel Hills horst block which is bounded on the west by an inferred northwest-trending normal (down-to-the-west) fault zone (figure 3). Rozel Point field lies within this fault zone. The reservoir has likely been sliced up by small faults and displaced upward against less permeable rocks (limestone and claystone), but probably lacks true closure along the length of the fault zone. In addition, the top of the reservoir is poorly sealed in an angular relationship by unconsolidated lake muds and silts.

The oil at Rozel Point was generated from immature source beds, probably lacustrine strata of the Miocene-Pliocene Salt Lake Group (Doelling, 1980), in the Rozel graben area to the west. The oil migrated upward along faults

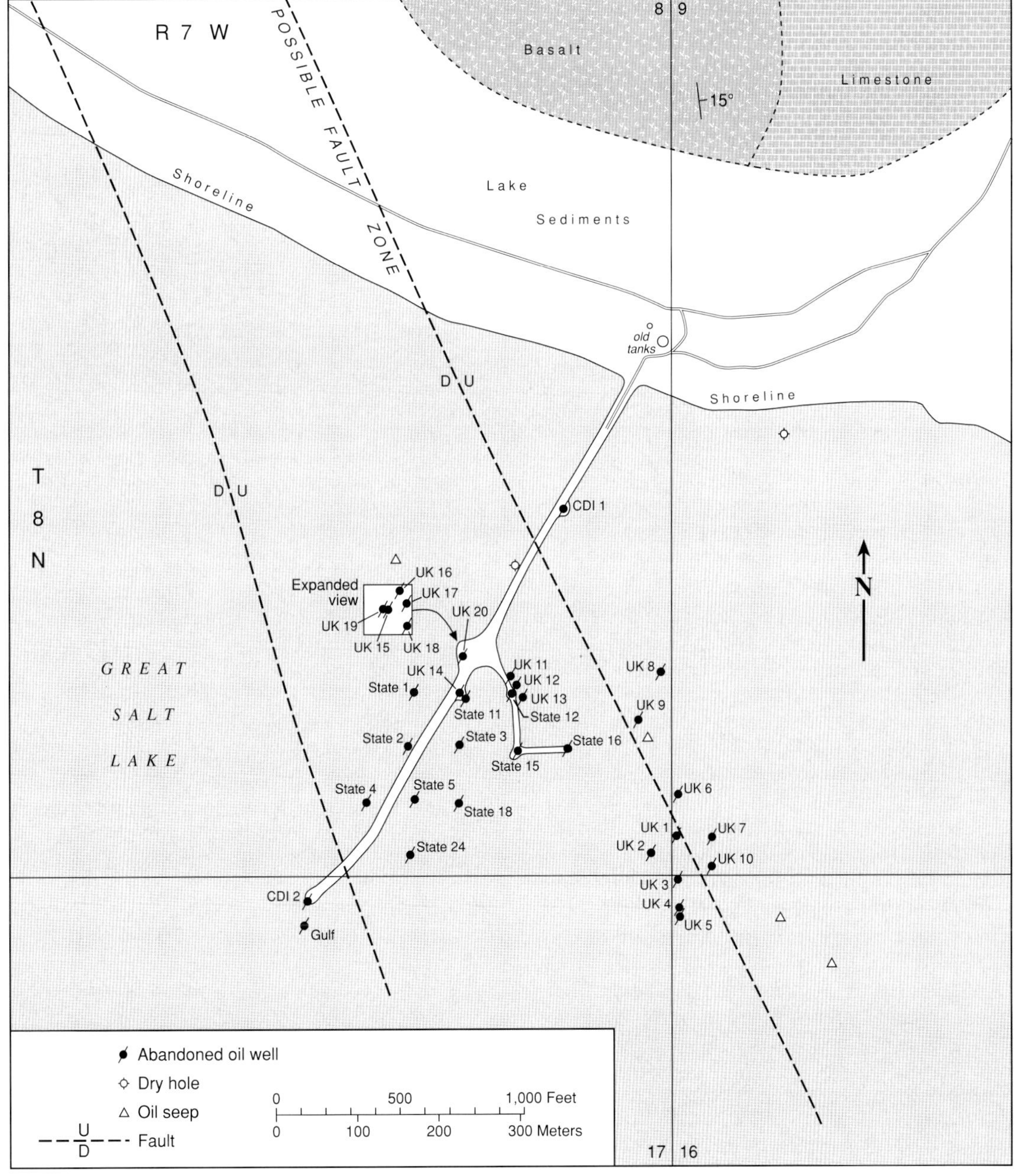

Figure 3. *General geology of Rozel Point area and oil well locations within T. 8 N., R. 7 W., SLBL&M, Box Elder County, Utah (modified from Eardley, 1963; Kendell, 1993).*

and fractures to the reservoir rocks: (1) from the source beds directly, or (2) from a larger, deep oil accumulation elsewhere beneath the lake. Because of the high viscosity of the Rozel Point oil and the low lithostatic pressure due to shallow depths, the migration process is relatively slow.

Asphalt Oil Seeps

Naturally occurring oil seeps are found throughout the world. For example, the La Brea tar pits of Los Angeles are natural seeps that have existed for over 12,000 years. Oil (asphalt) seeps in the Rozel Point area, on the north shore of Great Salt Lake, have been known since the late 1800s. Thick, black, tar-like asphaltic oil has flowed from these seeps for hundreds, if not thousands of years. Several natural seeps have been documented in the immediate vicinity of the Rozel Point oil field. Oil issues from craterlets having cones that are 12 to 18 inches (30-46 cm) high (figure 4A) (Slentz and Eardley, 1956). The seeps cover areas 20 to 40 feet (6.1-12 m) in diameter at the bottom of the lake. These were observed during the low-lake-level years of 1901, 1934, and 1963 (Jensen and Doelling, 1987). The usually submerged seeps are irregularly distributed from the shore in an east-southeasterly direction for about a half mile (figure 3).

Oil migrates upward along faults and fractures in the bedrock beneath the lake before emerging in soft lake clay (figure 4B). There are no estimates of seepage rates at which the oil flows. When the area is covered by water, the oil issues from the sediments as small, worm-like stringers (Eardley, 1963). Oil also appears as numerous globules 1/4 to 1/2 inch (0.6-1.3 cm) in diameter over areas several square feet in size (Slentz and Eardley, 1956). The oil is very low in volatile components, resulting in a dense, semi-solid material that commonly sinks to the bottom of the lake. The oil is carried away from the seeps by waves and deposited along the shoreline. The oil may also have been deposited in lake beds over time and then reworked by wave action. In the heat of the summer, the shoreline deposits of oil warm up and flow, covering parts of the rocks, beach sands, and abandoned oil-field equipment. Most of the rocks in the Rozel Point area are boulders of black basalt, which can give viewers the false impression they are completely covered with oil.

Figure 4. *Oil seeps at Rozel Point. (A) Craterlet of oil exposed during low lake level, photographed circa 1937 (contributed by Jack N. Conley, Petroleum Geologist [from Doelling, 1980]). (B) Oil seep under about a foot of water, photographed in 1995.*

Oil Characteristics

The oil produced and seeped at Rozel Point is brownish-black, viscous, gives off a strong asphaltic odor, and has some unusual characteristics. It contains up to 13 percent sulfur, has anomalously high nitrogen and carbon residue concentrations, and is highly asphaltic (Doelling, 1980). The high percentages of these constituents result in an oil having a very low gravity (5 to 9° API), which makes it denser than water (Kendell, 1993). Bacterial alteration has likely affected the composition and character of the oil (Patton and Lent, 1980).

The oil from Rozel Point is extremely difficult to produce and costly to refine. According to Heylmun (1961), this oil is chemically similar to ichtyol, a rare substance used for medicinal purposes, and thus, has the potential to be an extremely valuable commodity. Higher fractions of the asphalt, when added to motor oil, are known to increase the lubricity of the oil thus making it more useful (Eardley, 1963). The oil from Rozel Point has been used in the past to resurface roads, for waterproofing, and for impregnating the cords of automobile tire fabric prior to the introduction of latex (Doelling, 1980).

The unique chemical components of the oil at Rozel Point offer several possible commercial products. From time to time, various petroleum companies and entrepreneurs pursue new methods to economically remove the sulfur from the oil, which would result in a lighter, more easily refined oil and therefore a more marketable product. Others look for ways to use the oil as it is, for example, in making shingles.

Reserves, Production, and Potential

Reserve estimates for the field have not been determined because the trap size and reservoir parameters are so poorly defined. However, at least 10,000 barrels (1,590 m^3) of oil have been produced from the field (Doelling, 1980). Initial well production rates were between five and 10 barrels (0.8-1.6 m^3) of oil per day (Heylmun, 1961; Eardley, 1963). These facts suggest that significant oil accumulations may exist in the area representing potential exploration targets. Using improved well completion techniques, this potential could be prudently exploited if an economical method for removing sulfur from the oil can be developed, or if as-is uses are found.

OIL DEVELOPMENT SUMMARY

Development of the hydrocarbon resources at Rozel Point has been documented since 1904. Several periods of exploration and development have resulted in the drilling of 30 to 50 wells (figures 3 and 5), and the accumulation of associated production equipment (the actual number of drilled wells is difficult to accurately determine because early drilling records are poor and incomplete). The field has been inactive since the mid-1980s, partly due to high lake level at that time. As a result of years of neglect, the field has fallen into disrepair and some wells were leaking an undetermined volume of tar-like oil into Great Salt Lake (Doug Howard Consulting, Inc., 1995). Reports in the late 1800s of asphaltum seeps on the northeast shore of the lake stimulated interest in the hydrocarbon potential of the area. Several articles were written between 1904 and 1905 discussing these occurrences of asphalt (Maguire, 1904, 1905; Gibbs 1905). The U.S. Geological Survey published more comprehensive studies by Richardson (1905) and Boutwell (1905). Later publications by Schneider (1921), Eardley and Haas (1936), Hansen and Bell (1949), and Heylmun (1963), discussed development and stimulated interest in the hydrocarbon potential of the area.

Well Construction

The objective of drilling at Rozel Point was to facilitate the collection of oil. Conventional wells are designed to handle high formation pressure, and to prevent interformation transfer of fluids which protects ground water. None of the conventional design features are evident at Rozel Point. It appears that the primary objectives in constructing the Rozel Point wells were to provide a conduit for oil migration, prevent sloughing of unconsolidated materials into the hole, and provide a mechanical attachment for pumps and other production equipment to accelerate oil flow.

Figure 5. Oil production operations at Rozel, 1965 (photo from Utah Division of Oil, Gas and Mining files).

Drilling History

The seeps at Rozel prompted drilling as early as 1897, according to Joseph Lippman (Boutwell, 1905), when a borehole was drilled to a depth of 130 feet (39.6 m) at a point in the lake about a quarter mile from shore. Also, according to J.J. Driver (Boutwell, 1905), president of the Utah Oil and Liquid Asphalt Company, which controlled the property at that time, "five wells have been sunk" on one of the islets, "two of which are about 140 feet deep, and the others from 80 to 100 feet." He also states that an experiment to facilitate the flow of the viscous asphalt was made by running a steam coil down one of the wells for about 100 feet (30 m) and forcing the steam into the asphalt bed. The reported result was that the asphalt was partially liquefied and flowed up more readily for a few hours, until it was cut off by apparent caving of the walls around the bottom orifice of the well casing. Today we can only guess where the location of these wells might be. One possibility is at a site where standing pilings and several well casings are found. Maguire (1905) made reference to a 1,400-foot-deep (426 m) well drilled by either Great Salt Lake & Promontory Oil & Asphalt Company or Utah Asphaltum & Oil Company.

Additional drilling is documented from 1921 through 1922 (Leonora-Saboda well), 1937 (Eller & Snyder wells), and 1956 (Stoddard well). Slentz and Eardley (1956) reported that 15 to 20 wells had been drilled around Rozel Point.

Further development in the Rozel Point area was done by Gulf Oil Corporation and Charles E. King (current Pixley wells) in 1963 through 1967. King drilled and completed 11 wells on state lands between July 1964 and April 1967.

During July 1964, Harry Reginald (lessee) and Charles E. King drilled and completed the Rozel State No.1 well. Due to the extremely high density and viscosity of the oil (7° to 9° API gravity), a special chain pump was used to produce the well. Records for this well indicate that 13-3/8-inch (34 cm) surface casing was set at 153 feet (46.6 m) and cemented to surface. A 12-inch (30.5 cm) hole was drilled in the vesicular basalt section from 153 to 178 feet (46.6-54.3 m) (open hole interval).

In April 1965, Harry Reginald and Charles E. King (applicants), accompanied by David H. James, Consulting Petroleum Engineer (expert witness), appeared before the Utah Oil and Gas Conservation Commission (referred to as the Commission) seeking an exception to the well spacing requirements to allow the applicants to drill four wells adjacent to the Rozel State No.1 well. Following the hearing, the Commission approved the applicants' request, granted the exception, and gave permission to drill additional wells to a depth not to exceed 500 feet (152 m) within a 5-acre (2 ha) area to include the previously drilled Rozel State No. 1 well.

Mr. Charles E. King, as operator, proceeded to drill four wells as follows:

Rozel State No. 4	Spudded May 20, 1965
Rozel State No. 3	Spudded June 1, 1965
Rozel State No. 2	Spudded June 12, 1965
Rozel State No. 5	Spudded June 30, 1965

The four wells were each drilled and completed for production in essentially the same manner. An 8-3/4-inch (22 cm) hole was drilled to depths of 150 to 160 feet (45.7-48.8 m) and 7-inch (18 cm) casing was set and cemented to surface. A (±) 6-inch (15 cm) hole was drilled from below the surface casing to about 250 feet (76.2 m) total depth through the vesicular basalt section (approximately 100 feet [30.5 m] of open hole). The wells were equipped with 2-3/8-inch (6 cm) tubing (2-inch [5 cm] inside diameter) to a depth of about 200 to 245 feet (61.0-74.7 m), rods and insert bottom-hole pumps, and bottom-hole heaters. On the surface, small pumping units were used as well as a central heating plant to move the oil (figure 5). Initial production rates from the wells varied from 5 to 10 barrels (0.8-1.6 m^3) of oil per day. Approximately 650 barrels (103 m^3) of oil were reported to have been produced through early 1966.

In February 1966, Mr. Reginald and Mr. King received approval from the Commission for 1-acre (0.4 ha) drilling and spacing units for further development and oil production from the vesicular basalt section in the area of the existing wells. Subsequently, Mr. Charles E. King, as operator, drilled and completed the following wells:

Rozel State No. 11	Spudded July 19, 1966
Rozel State No. 12	Spudded October 14, 1966
Rozel State No. 16	Spudded November 14, 1966
Rozel State No. 15	Spudded December 17, 1966
Rozel State No. 18	Spudded March 30, 1967
Rozel State No. 24	Spudded April 20, 1967

Again, these six wells were each drilled and completed for production in essentially the same manner. A 12-1/4-inch (31 cm) hole was drilled to depths of 150 to 160 feet (45.7-48.8 m) and 8-5/8-inch (22 cm) casing was set and cemented to surface. A (±)7-7/8-inch (20 cm) hole was drilled from below the surface casing to about 250 to 260 feet (76.2-79.2 m) total depth through the vesicular basalt section (approximately 100 feet [30.5 m] of open hole). The wells were equipped with 2-7/8-inch (7.3 cm) tubing (2-1/2 inch [6.4 cm] inside diameter) and rods, and 2-1/4-inch (5.7 cm) insert down-hole pumps. Small pumping units were apparently used at the surface.

Table 1. Wells in the Rozel Point project area (see figure 3).

Well ID*	Approx. Location†	Comments	Leaking Status‡	Operator
State 1	726' FSL, 983' FEL, sec. 8	-	I	Pixley
State 2	497' FSL, 1028' FEL, sec. 8	-	I	Pixley
State 3	506' FSL, 818' FEL, sec. 8	-	I	Pixley
State 4	291' FSL, 1205' FEL, sec. 8	-	I	Pixley
State 5	298' FSL, 1005' FEL, sec. 8	-	L	Pixley
State 11	706' FSL, 818' FEL, sec. 8	-	I	Pixley
State 12	706' FSL, 610' FEL, sec. 8	-	I	Pixley
State 15	497' FSL, 610' FEL, sec. 8	-	L	Pixley
State 16	497' FSL, 401' FEL, sec. 8	-	L	Pixley
State 18	289' FSL, 818' FEL, sec. 8	-	I	Pixley
State 24	80' FSL, 1028' FEL, sec. 8	-	I	Pixley
CDI 1	1424' FSL, 416' FEL, sec. 8	-	N	Pixley
CDI 2	104' FNL, 1424 FEL, sec. 17	-	N	Pixley
St. Rozel No.1	unknown	-	N	Gulf Oil
UK 1	160' N of SW corner of sec. 9	Leonora Saboda Well	L	Lenora Mining & Milling
UK 2	110' SW of UK No. 1, sec. 8	-	L	unknown
UK 3	25' E of NW corner of sec. 16	Eller Well	L	H.M. Eller
UK 4	125' S SE of NW corner of sec. 16	Stoddard No. 2	N	Ratner Oil
UK 5	25' S of UK No. 4, sec. 16	-	N	unknown
UK 6	150' N of UK No. 1, sec. 9	-	L	unknown
UK 7	150' E of UK No. 1, sec. 9	-	N	unknown
UK 8	200' NE of UK No. 9, sec. 8	Flows some water	N	unknown
UK 9	300' E NE of State No. 16, sec. 8	-	N	unknown
UK 10	150' E NE of UK No. 3, sec. 9	-	N	unknown
UK 11	65' N of State No. 12, sec. 8	-	L	unknown
UK 12	40' NE of State No. 12, sec. 8	-	L	unknown
UK 13	50' E SE of State No. 12, sec. 8	-	I	unknown
UK 14	25' S SE of State No. 11	-	L	unknown
UK 15	250' NE of State No. 1	Center of wooden pillars	L	unknown
UK 16	16' S of UK No. 15	-	I	unknown
UK 17	15' S of UK No. 15	-	I	unknown
UK 18	21 1/2' W of UK No. 17	-	I	unknown
UK 19	4' W of UK No. 15	-	N	unknown
UK 20	15' SE of UK No. 18	-	N	unknown

* ***Bold italics*** typing indicate wells leaking oil that were capped.
† FSL = from south line, FEL = from east line, FNL = from north line of the section.
‡ Leaking Status: L = Leaks; I = Indeterminate leak; N = Does not leak
Data from Utah Division of Oil, Gas and Mining (1995), and Doug Howard Consulting, Inc. (1995)

During June 1982, Mr. Kenneth Pixley, DBA K.P. Enterprises, drilled the C.D.I. No.1 well located in NW1/4NE1/4, section 8, T. 8 N., R. 7 W., Salt Lake Base Line and Meridian, to a depth of 263 feet (80.2 m), and the C.D.I. No. 2 well located in NW1/4NE1/4, section 17, T. 8 N., R. 7 W., to a depth of 30 (?) feet (9.1 m).

In June 1984, Utah Petrochemical Corporation submitted drilling applications to DOGM for 12 proposed wells located in the NW1/4NW1/4 section 16, T. 8 N., R. 7 W, SLBLM. The wells were located to achieve a 1-acre (0.4 ha) drainage pattern. However, the wells were not drilled for economic reasons and the drilling-permit approvals lapsed.

CLEANUP EFFORTS

Due to concerns over wells leaking oil into Great Salt Lake, and the possibility that birds using the area would be adversely affected, investigations by the DOGM were initiated in the spring of 1995 to look into remediation of the problem. Since the situation involved oil and navigable waters, the EPA became involved under authority of the Federal Oil Pollution Act. Studies by DOGM were completed during the spring and summer of 1995, evaluating the existing conditions at Rozel Point, and the options available for responding to oil releases into the lake (Utah Division of Oil, Gas and Mining, 1995). Remedial alternatives ranged from extensive subsurface well work, to well capping, to temporary alternatives (Doug Howard Consulting, Inc., 1995). Capping the wells with a steel plate covered with several inches of concrete was the alternative chosen by the EPA. The EPA funded and supervised a well-capping project during January and February of 1996. Casing heads from 34 of the wells drilled at Rozel Point were found. There are 17 documented wells in the Rozel Point project area, and 17 wells for which no prior documentation could be found. Data for the 34 wells, including well identification (ID), approximate location, leaking status, and operator are given in table 1.

During the remediation work, 18 of the 34 wells listed in table 1 that were documented as leaking oil were capped, and are indicated by ***bold italics*** typing of the Well ID. Figure 6A shows one of the typical well-capping operations in the lake. The procedure involved: (1) excavating around the well to a depth of several feet below the lake bottom (figure 6B), (2) cutting off the casing and any tools or equipment stuck in it (figure 6C), (3) welding a metal cap over the casing (figure 6D), and then (4) covering the casing and cap with several inches of concrete. There was no attempt to cap or seal the natural seeps beneath the lake's surface, and tar-like oil will

Figure 6. *Typical well-capping operations in Great Salt Lake: (A) operations from a distance, (B) excavating around an abandoned well prior to capping, (C) cutting off wellhead, and (D) welding cap on casing.*

still appear on the beach sands and shoreline rocks of Rozel Point.

The work was not without its challenges. During the project a track hoe became mired in the mud and water entered its engine and other vital components. The result was a large added expense to the project.

CONCLUSIONS

Rozel Point field, the oldest oil field in Utah, was discovered because of naturally occurring oil seeps in the area. Rozel Point field is one of only two oil fields in the entire Basin and Range physiographic province of Utah. The oil was produced from a shallow basalt, a very unusual reservoir. The oil has unique properties that make it difficult to produce and costly to refine. Several periods of exploration and development resulted in 30 to 50 wells, and the accumulation of associated production equipment. The field has been inactive since the mid-1980s, partly due to high lake level at that time. As a result of years of neglect, the field has fallen into disrepair and some wells were leaking tar-like oil into Great Salt Lake.

Capping the leaking wells at Rozel Point field has successfully prevented the continued flow of oil from those wells into Great Salt Lake. This effort will have no effect on the flow of oil from the nearby natural oil seeps beneath the lake. Capping the wells should reduce the overall amount of oil that gets into the lake, but deposits of tar-like oil will still appear on the beach sands and shoreline rocks of Rozel Point. Oil seeps are naturally occurring phenomena found worldwide and that have existed through most of geologic history. The oil seeps at Rozel Point will likely continue to flow for hundreds, perhaps thousands of years to come.

REFERENCES

Boutwell, J.M., 1905, Oil and asphalt prospects in Salt Lake Basin, Utah: U.S. Geological Survey Bulletin 260, p. 468-479.

Doelling, H.H., 1980, Geology and mineral resources of Box Elder County, Utah: Utah Geological and Mineral Survey Bulletin 115, p. 209-214.

Doug Howard Consulting, Inc., 1995, Rozel Point oil field - remedial alternatives report, Box Elder County, Utah: Unpublished consultant's report for Ecology and Environment, Inc., 50 p.

Eardley, A.J., 1963, Oil seeps at Rozel Point: Utah Geological and Mineralogical Survey Special Study 5, 26 p.

Eardley, A.J., and Haas, Merrill, 1936, Oil and gas possibilities in the Great Salt Lake Basin: Utah Academy of Science Proceedings 13, p. 61-80.

Gibbs, J.F., 1905, The question of petroleum near Salt Lake: Salt Lake Mining Review 6, January 15, p. 19-20.

Hansen, G.H., and Bell, M.B., 1949, The oil and gas possibilities of Utah: Utah Geological and Mineralogical Survey, 341 p.

Heylmun, E.B.,1961, Rozel Point, Box Elder County, Utah, *in* Preston, Don, editor, Oil and gas fields of Utah: Intermountain Association of Petroleum Geologists, non-paginated.

—1963, Oil and gas possibilities of Utah, west of the Wasatch and north of Washington County, *in* Crawford, A.L., editor, Oil and gas possibilities of Utah re-evaluated: Utah Geological and Mineralogical Survey Bulletin 54, p. 287-301.

Jensen, M.E., and Doelling, H.H., 1987, Road log - an excursion around the Great Salt Lake, Utah, *in* Kopp, R.S., and Cohenour, R.E., editors, Cenozoic geology of western Utah: Utah Geological Association Publication 16, p. 678-679.

Kendell, C.F., 1993, Rozel Point, *in* Hill, B.G., and Bereskin, S.R., editors, Oil and gas fields of Utah: Utah Geological Association Publication 22, non-paginated.

Maguire, Don, 1904, Oil and asphaltum in the Great Salt Lake Basin: Salt Lake Mining Review 5, February 15, p. 13-15.

—1905, Oil and asphaltum on the shores of Great Salt Lake, Utah: Mining and Scientific Press 90, p. 302.

Patton, T.L., and Lent, R.L., 1980, Current hydrocarbon exploration activity in the Great Salt Lake, *in* Gwynn, J.W., editor, Great Salt Lake - a scientific, historical and economic overview: Utah Geological and Mineral Survey Bulletin 116, p. 115-124.

Richardson, G.B., 1905, Natural gas near Salt Lake City, Utah: U.S. Geological Survey Bulletin 260, p. 480-483.

Schneider, Hyrum, 1921, Possibilities of oil and gas in Salt Lake and Cache Valleys: Utah Academy of Science Transactions 2, p. 39.

Slentz, L.W., and Eardley, A.J., 1956, Geology of Rozel Hills, *in* Eardley, A.J., and Hardy, C.T., editors, Geology of parts of northwestern Utah: Utah Geological Society Guidebook 11, p. 32-40.

Stansbury, Howard, 1852, Expedition to the Valley of the Great Salt Lake in Utah: Philadelphia, Lippincott, Grambo and Company, 487 p.

Utah Division of Oil, Gas and Mining, 1995, Rozel Point oil field - site reconnaissance and documentation, Box Elder County, Utah: Utah Department of Natural Resources, Division of Oil, Gas and Mining unpublished report, 74 p.

THE BRINE SHRIMP INDUSTRY IN UTAH

by

David Kuehn

Sanders Brine Shrimp Company, L.C.
3850 South 540 West
Ogden, UT 84405
(801) 393-5027

INTRODUCTION

The brine shrimp industry on Great Salt Lake (GSL) got off to an inauspicious start in 1949, when an aquarium enthusiast from Ogden, Utah, Mr. Cleon C. Sanders, heard of a nutritious food, brine shrimp, that he could feed to his tropical fish. Initially, Mr. Sanders purchased frozen brine shrimp from San Francisco. Then he heard from a friend that GSL was teeming with the tiny shrimp (*Artemia franciscana*, the species of brine shrimp in GSL), and upon further investigation found they were highly nutritious for tropical fish.

In 1950, Sanders sent an article to "The Aquarium" magazine describing this new fish food. Shortly after this article appeared, requests for frozen adult brine shrimp arrived from various parts of the United States (Sturm, and others, 1980). Soon thereafter, he formed the Sanders Brine Shrimp Company to harvest, process, and market, adult brine shrimp; in later years, the harvest of brine shrimp eggs (cysts) was added.

The early market for frozen brine shrimp was small, and was limited to tropical fish enthusiasts and a few aquaculturists. Sanders had few competitors on the lake for the first 30 years. Then in the 1980s, the farming of various Penaeid species of shrimp, and fish began to expand globally. Newly hatched *Artemia* nauplii were found to be an ideal food for the larval shrimp and fish. The focus of the GSL brine shrimp industry switched from harvesting adult brine shrimp to harvesting their eggs or cysts. Cysts do not require freezing, but are dried and canned for long-term storage. The brine shrimp industry in Utah has evolved from a small part-time hobby into a lucrative business with 32 companies currently harvesting cysts on GSL.

HARVESTING

Adult Brine Shrimp

Adult *Artemia* can usually be found throughout the south arm of GSL for approximately a nine-month period (April-December). *Artemia* populations vary from year to year due to two main factors, water salinity, and the concentration of the algae, on which the *Artemia* feed. Scientific studies have shown that *Artemia* populations fluctuate throughout the summer months. Populations peak and over-graze their food supply, resulting in a subsequent population crash to levels that the lake can support. The Utah Division of Wildlife Resources (UDWR) has prohibited the direct harvest of adult *Artemia* since 1983. The harvest of adult *Artemia* is now only allowed as a result of by-catch during the cyst-harvest season.

Brine Shrimp Eggs

Shore Harvest

Brine shrimp eggs or cysts have been harvested since 1952. Cysts have been collected from both the north and south arms of GSL, however, the majority of the cysts have been collected from the south arm in recent years. The majority of cysts are annually produced by adult brine shrimp during the period from August to November.

Cysts typically float on the water surface, and congregate into patches or steaks of varying thickness. Due to wind and wave action, these masses eventually wash up on shore. Harvesting from the shore was the first method utilized to collect the cysts (figure 1), and is still used today. The cysts are collected by raking, shoveling, and scraping them into piles. From these piles, the cysts are put into small (60 lb) sacks made of burlap, or polypropylene (figure 2). Figure 3 shows an aerial view of a beach-harvest operation.

In the early 1980s, brine shrimp companies began har vesting cysts offshore. This was done by intercepting the cysts before they reached the shore. Harvesters would wade out as far as possible, pulling a length of oil-containment boom behind them, and then attempt to completely encircle or "corral," a streak. Once encircled, the cysts would be pumped out of the boom, and into the small bags. At that time, harvest locations were often miles from the nearest road, and a new type of vehicle, the all terrain vehicle (ATV), became the preferred way to transport the cysts from the harvest site to a road. The cysts were then loaded onto railcars,

Figure 1. *Harvesting* Artemia *cysts from a GSL beach. Photo courtesy of Prime Artemia, Inc.*

Figure 2. *A beach harvest crew with 60 lb. bags of* Artemia *cysts. Photo courtesy of Prime Artemia, Inc.*

Figure 3. *An aerial view of a beach-harvest operation.* Artemia *cysts are captured by an oil-containment boom in shallow water.*

or tractor trailers for transport back to local processing facilities. Harvesters found that currents and prevailing winds tended to concentrate more cysts at certain shoreline locations around the lake than others. Companies attempted to obtain exclusive land leases from public and private landowners for these special locations, and were often successful.

In the mid-1980s, boat harvesting was developed. This new cyst-harvesting method is extremely effective. However, shore harvesting continued at a smaller scale through the late 1980s. During the 1990s, shore harvests grew due to the explosion of new companies entering the industry. Swarms of ATVs can again be seen along the shorelines of GSL during the harvest season. In fact, UDWR has estimated that shore-based harvesting accounted for 44.6 percent of the total harvest during the 1997-98 season.

Aquatic Harvest

In 1986, Sanders Brine Shrimp Company became the first company to use boats to harvest cysts offshore. This aquatic harvest first started on the north arm of GSL. This harvest method was so successful that other companies soon followed suit, and a new land rush developed. Instead of procuring leases and ownership of areas around the lake where the cysts accumulated on the shorelines, sites suitable for boat harbors were sought (figure 4). Access points for the launch, retrieval, and moorage of boats are extremely limited due to the area's topography. These access points are closely guarded company assets.

Harvesting off the water's surface presented new challenges. Aquatic harvesting devices and methods have evolved from open-ended suction hoses to elaborate patented devices like skimmers, bucket-like conveyor belts, specially designed filter belts, and cupped paddlewheel-like devices (figures 5 to 8). The characteristic that all of these devices has in common is the need for the cysts to be concentrated in some way before entering the device. Oil-containment booms are frequently used to gather and concentrate the cysts.

During the 1990s, the brine shrimp industry underwent some drastic changes. Harvesting moved from the north arm back to the south arm due to salinity changes, and the number of companies, and more importantly, the number of Certificates of Registration (COR), greatly increased (table 1). Each COR, costing $10,000, allows a company to harvest the cysts that fall within a 300-yard radius of one location. Until 1995, anyone could purchase a GSL harvesting COR, and any company could obtain more than one COR. However, in 1995 UDWR placed a moratorium on the number of COR (then numbered at 79) available for the *Artemia*-cyst fishery. The purpose of the moratorium was to check industry growth, collect data on the GSL ecosystem, and develop management strategies to help maintain a sustainable resource. There has been almost industry-wide consensus that the institution of the moratorium was a very wise move.

The cyst harvest season mandated by UDWR can run from October 1 through January 31. However, during the last three harvest seasons (1996-1998), the season length has ranged from 22 to 54 days. The shortened seasons of 1996-98 (average season length = 34.3 days) are the result of UDWR setting a harvest quota, and stopping the harvest when the quota is reached. The brine shrimp operators have started each season without knowing what the quota will be.

During the 1997-98 and 1998-99 seasons, harvesting was done on both the north and south arms of the lake. The 1999-2000 harvest was limited to the lake's north arm, except for harvesting from MagCorp's south arm solar evaporation ponds. During the summer of 1999, the UDWR's Wildlife Board opened the rules and regulations of the fishery at the request of UDWR and the operators. One major change switched the harvest from daylight hours only to a 24-hour harvest. Other rule changes allow the UDWR to temporarily delay the start of the harvest season for up to ten days, and also to suspend the harvest twice during the course of the season for up to seven days. These suspensions and closures were used at the start of the 1999 harvest season when low cyst densities were found in the south arm of GSL. Facing a possible complete closure of the 1999 season, the operators approached the UDWR with an agreement, signed by all 32 companies, to only harvest on the north arm. The UDWR found this acceptable because

Figure 4. *Brine shrimp company boat harbors built on the Magnesium Corporation of America's (MagCorp) Badger-Stansbury dike.*

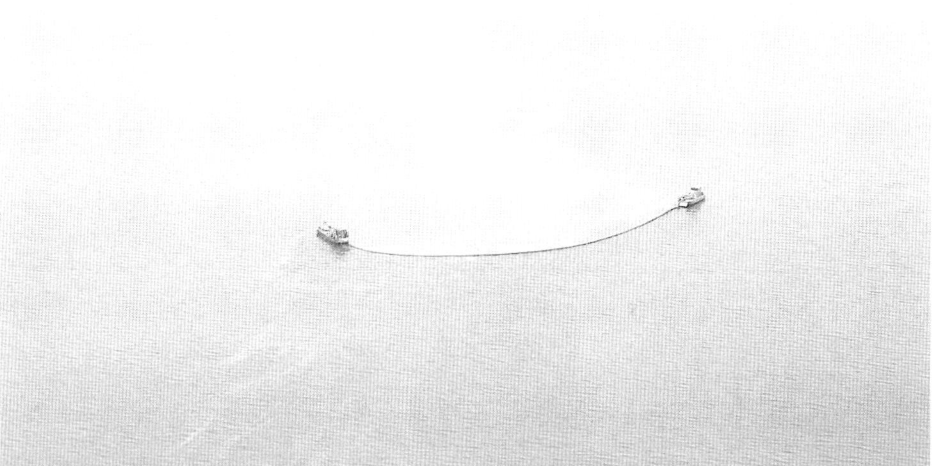

Figure 5. *Two boats attempting to "corral" a patch of* Artemia *cysts.*

Figure 6. *A boat operated by Sanders Brine Shrimp Company harvesting a streak of* Artemia *cysts with a skimming device.*

Figure 7. *A boat operated by JK Brine Shrimpers Company harvesting a streak of* Artemia *cysts with a conveyor-type device.*

Figure 8. *Boats operated by Golden West Artemia Inc. that have "corralled" some* Artemia *cysts, and are preparing to pump the cysts into "super sacks."*

Table 1. Year, number of companies, number of Certificates of Registration (COR's), total pounds harvested (includes brine shrimp cysts, adult biomass, and debris), and harvest location. Data courtesy of Utah Division of Wildlife Resources.

YEAR	NO. OF COMPANIES	NO. OF COR	TOTAL POUNDS HARVESTED	ARM OF LAKE
1985-86	4	-	298,035	N
1986-87	4	-	1,887,300	N
1987-88	4	-	7,012,775	N
1988-89	9	-	6,806,415	N
1989-90	12	-	10,268,232	N
1990-91	11	-	8,927,818	N
1991-92	11	26	13,532,797	N & S
1992-93	12	20	10,172,399	S
1993-94	12	18	8,864,092	S
1994-95	14	29	6,485,954	S
1995-96	21	63	14,749,596	S
1996-97	32	79	14,679,498	S
1997-98	32	79	6,113,695	N & S
1998-99	32	79	4,606,352	N & S
1999-00	32	79	2,542,958	N=

- Data not available
= Harvest also from MagCorp evaporation ponds

cysts found on the north arm are believed "lost to the system," and deemed a harvestable surplus. In other words, cysts and the resulting nauplii cannot survive the hypersaline waters of the north arm, and thus are not thought to contribute to the *Artemia* population of the south arm.

Due to the increase in the number of participants of this fishery, the level of competition has intensified. It has become essential to locate streaks quickly. Airplanes used for spotting cyst concentrations, once considered a luxury in the early 1990s, are now a necessity. Spotter planes traverse the lake 24 hours a day looking for cyst concentrations to which they direct their harvest boats (figure 9). Once a location is found, the pilot determines the latitude and longitude coordinates, and radios the coordinates to a waiting boat. The boat operator enters the coordinates into the boat's Global Positioning System, or Loran-C unit, and races to the site. The first boat to reach the desired harvest location deploys an orange marker, thus staking a claim. All other companies are excluded from harvesting within a 300-yard radius around the boat presenting the orange marker. There are often many spotter planes in the air at the same time, and the number of quality harvest locations is limited. The speed of the boat in reaching a harvest location is often the deciding factor as to which company harvests a particular streak.

Boat speed, carrying capacity, and durability are all requirements of the brine shrimp industry that have presented design challenges for naval architects around the world. Boats must travel over 65-plus miles per hour, carry over 30 tons of cysts, and be built from specialized aluminum alloys to be used in cyst harvesting operations.

Super sacks, which typically have a capacity of 2,000 pounds, are used on the boats to hold the cysts (figure 10). A streak being pumped into the super sack typically contains a slurry of cysts, adult *Artemia*, brine-fly casings, algae, and other debris, which is called "raw product." Most companies separate the cysts from the remaining biomass with a pre-wash. Some companies pre-wash on board their boats, others complete the pre-wash after the sacks are unloaded from the boats, either on shore or at their processing facilities, and still others sell their

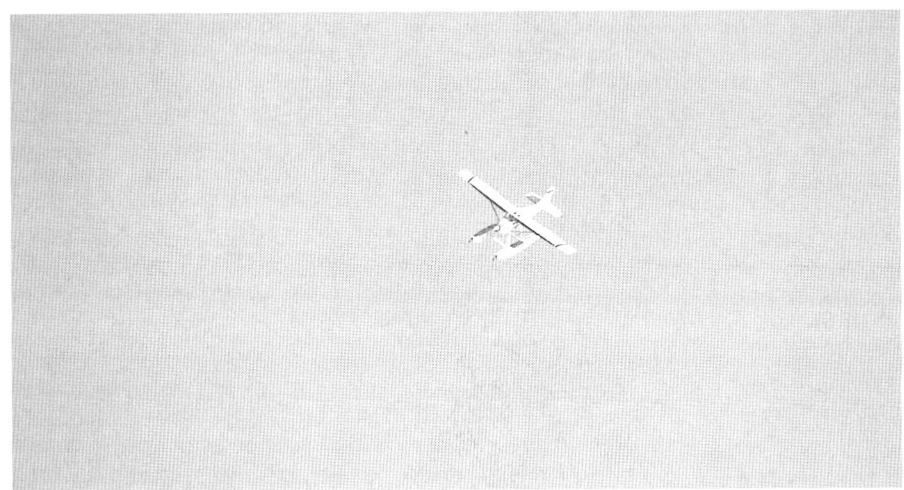

Figure 9. *A spotter pilot flying over the North Arm of GSL in search of* Artemia-*cyst streaks.*

Figure 10. *An* Artemia-*cyst harvest operated by JAW Brineshrimping Inc. "Super sacks" of harvested* Artemia *cysts are shown near the pilothouse.*

sacks of "raw product" to other companies. The product from this pre-wash cleaning yields 99 percent pure cysts.

PROCESSING

The processing techniques in the brine-shrimp industry are highly proprietary and in some cases patented. As listed below, cysts generally go through a four-step process from the raw, pre-washed form to the finished marketable product:

(1) cysts must be aged for a period of time to allow the eggs to mature;
(2) cysts then go through a fresh-water wash to remove any excess debris, and to rinse salt off the cysts;
(3) the cysts are then dried to a low total moisture content; and
(4) the dried cysts are then usually vacuum packed into cans or buckets.

The actual dry yield of marketable cysts compared to the weight that is reported to the UDWR is often less than 50 percent. In other words, of the 14,679,498 total pounds harvested in the 1996-97 season, approximately 7,339,749 pounds or less would be available for the world market from GSL. During the late 1990s, the yield was much worse, ranging from 10 to 35 percent. This weight reduction is due to the removal of excess debris during the pre-wash and processing stages, and the removal of moisture during drying.

MARKETING

Worldwide, the aquaculture industry is experiencing double-digit growth annually. Subsequently, the demand for *Artemia* cysts has followed suit. *Artemia* from GSL are widely regarded as some of the best in the world due to their small size and consistent nutritional quality. Extensive quality control testing is performed on the cysts to determine their hatch rate, and nutritional content. Based on the hatch rate, cysts are then graded (table 2). However, the industry does not have a standard that all brine shrimp companies must follow in terms of grading.

Companies cannot depend on a fixed percentage of their harvest to fall into a particular grade category from year to year. The quality of cysts and grades of cysts change annually. During the 1997-98 harvest season, it was estimated that over 50 percent of the harvest was not of good enough quality to fall into any of the grades listed in table 2. *Artemia* cysts are a variable commodity, and their price can have wild swings annually.

As previously stated, the aquaculture industry has generated a great demand for *Artemia* cysts. *Artemia* are a highly desired food source for several reasons. When re-hydrated, the dried cysts usually hatch within a 24 hour period, resulting in live food for predatorial shrimp, and fish. Cyst exports from Utah go primarily to east and southeast Asia, Latin America, and Europe. The primary usage for cysts in the aquaculture industry is as a larval diet for Penaeid species of shrimp, and increasingly for larval marine finfish and flatfish. There are still limited sales in the aquarium fish market.

Table 2. Grade, counted hatch rate, historical wholesale price range (1985-2000).

Grade	Counted Hatch Rate (%)	Historical Wholesale Price Range ($ USD)/lb.
Premium	85-100	$5 to $35
A	70-85	$3 to $30
B	60-70	$2 to $20

ISSUES FACING THE BRINE SHRIMP INDUSTRY

The brine shrimp industry faces a multitude of different challenges including lake access, government regulation, and cooperation between companies with specific regard as to what is best for the resource. However, the single largest issue facing the industry is the widening salinity gradient between the north and south arms of the lake. Two factors causing this disparity in salinity were consecutive years of above normal precipitation during the late 1990s, and the decreased permeability of the Southern Pacific causeway separating the two arms. Unfortunately, the amount of fresh water that enters GSL cannot totally be controlled. However, the permeability of the causeway, which at one time allowed for a bi-directional exchange of brines between the north and south arms of the lake, could be altered to re-establish the historical level of brine exchange. Since the early 1990s, the north arm salinity has remained near saturation, while the south arm salinity has generally declined annually. The difference in salinity between the north and south arms is at its greatest level since the solid fill causeway was built in 1959.

During July 1999, the salinity of the north arm was approximately 270 parts per thousand (ppt), and the south arm was approximately 80 ppt. The optimal range for harvesting high-quality *Artemia* cysts is between 180-230 ppt. This range is important because cysts exposed to salinities of less than 14 percent become hydrated prior to processing. Hydrated cysts, which contain a high concentration of internal water, are more likely to attempt to hatch prior to processing than if they have a lower internal water content. If the cysts expend hatching energy at this time, the energy available to them to hatch after processing is reduced. In some cases, the available energy may be reduced to a level that is insufficient for hatching to occur. In this case, the lower hatching percentage is equated with a lower quality product.

The change in the south arm's salinity has also changed the balance of the south arm's aquatic fauna on which *Artemia* feed. It has been speculated that this shift in diet has adversely impacted the *Artemia* population, resulting in lower cyst densities, and lower overall cyst quality.

At one time, GSL satisfied over 90 percent of the world's requirement for marketable cysts. However, due to increased demand in the world market, high retail prices, and below-average quality and quantity of cysts from GSL in the 1994 - 1995, and 1997 to 2000 seasons, alternative cyst sources have been developed in China, Russia, and Turkmenistan, among other countries, to accommodate the world's demand. Lower harvesting and production costs in these countries have produced a less expensive, but usually lower-quality product. These cysts, however, are gaining market acceptance. Consequently, GSL cysts may never regain the market share that they once enjoyed.

CONCLUSION

The future of the Utah brine shrimp industry on GSL will always be tied to the lake's salinity due to its prominence as an elementary factor in the GSL ecosystem. With the growth of the global aquaculture industry, however, the potential for Utah's brine shrimp industry looks bright.

REFERENCE

Sturm, P. A., Sanders, G. A., Allen, K. A., 1980, The Brine Shrimp Industry on the Great Salt Lake, *in* Gwynn, J.W., editor, Great Salt Lake, A Scientific, Historical and Economic Overview, Utah Geological and Mineral Survey, p. 243 to 248.

Post Script - February 12, 2002

After the above conribution had gone to press, two subsequent harvest seasons have occurred predominantly on the south arm of the Great Salt Lake. In 2000, 19,963,087 lbs. of cysts, biomass, etc. were reported harvested to the UDWR. And in 2001, 18,111,584 lbs. (to date) of cysts, biomass, etc., were reported harvested to the UDWR. Both of these harvests eclipse the previous record harvest of 1995-1996 by approximately 30 percent. This is particularly impressive due to the fact that the south arm salinity was 87-118 ppt during these two seasons. This is well below the 140 ppt previously considered necessary for high quality cysts. In fact, cycst from these last two harvests have provided some of the consistently highest hatch rates the industry has seen in years. These record harvests have also depressed the wholesale price of cysts to near the lower limits noted in table 2. The more we think we know about *Artemia*, the more we are surprised by its ability to adapt, survive and flourish.

KENNECOTT UTAH COPPER - THE LEGEND CONTINUES

by

Louie Cononelos and Paula Doughty

Kennecott Utah Copper Corporation
8315 West 3595 South
Magna, Utah 84040-6001

INTRODUCTION

Kennecott's beginning dates back to 1863. After learning of ore discoveries in the Oquirrh Mountains, Colonel Patrick Connor sent soldiers of his Third California Infantry, stationed at Fort Douglas in Salt Lake City, to explore for minerals. Their positive findings in and around a ranching area called Bingham Canyon prompted Connor to form Utah's first mining district, earning him the title "Father of Utah Mining."

For the next 40 years, initially placer and then underground mining for lead, silver and gold minerals reigned in Bingham Canyon (figure 1). Large low-grade copper deposits existed, but could not be mined at a profit. As most of the underground mines played out, a report by two young engineers, Daniel Jackling, a metallurgical engineer, and Robert Gemmell, a mining engineer, revolutionized mining and changed Bingham Canyon forever. Their report indicated that surface mining and mass production of low-grade copper porphyry ore could be profitable. Jackling's plan called for the mining of ores from the surface with steam shovels - a process called open-pit mining - and transporting the ores in steam-powered trains to industrial-sized mills or concentrators. Jackling raised a substantial amount of money, and in 1903, formed the Utah Copper Company. A 300 ton-per-day pilot mill was built in Bingham Canyon and named the Copperton Mill. In 1906, steam shovels from the Boston Consolidated Mining Company and Utah Copper made the first cuts in the mountain, "The Hill," which contained low-grade copper ore. In 1910, Utah Copper purchased Boston Consolidated, a move that consolidated most of Bingham Canyon's mining claims and all of the copper concentrating capacity under Utah Copper's direction. By the 1930s, "The Hill" started to disappear and had become the "Pit" as larger and more efficient mining and processing equipment boosted daily production rates. Utah Copper Company was purchased by Kennecott Copper Corporation in 1936 and later became the Utah Copper Division.

During World War II, the Bingham Canyon Mine met the nation's demand for the important metal by producing one-third of the copper used by the Allies (figure 2). The 1950s saw even more growth as the company constructed a refinery and purchased the smelter, near the Great Salt Lake, from ASARCO. With the addition of the refinery and smelter, Kennecott's copper production became completely integrated - from mining through concentrating, smelting and refining.

Figure 1. *Ore cars in Bingham Canyon, 1892.*

Figure 2. *Rail Lines in Bingham Canyon Mine "The Pit," 1961.*

During the 1970s, Kennecott spent $300 million modernizing the smelter complex. The project included new smelting furnaces and new sulfur dioxide emission control technology, along with associated equipment to produce sulfuric acid. This expansion also brought about the erection of a 1,215 foot stack (just 35 feet shorter than the Empire State Building).

The 1980s began with a world-wide copper recession. Standard Oil of Ohio (SOHIO) bought Kennecott in 1981, but low productivity, high production costs, and a decrease in metal prices caused the Bingham Canyon Mine to close in 1985. In 1986, operations resumed and Kennecott started a further $400 million modernization project that included in-pit crushing and conveying facilities at the Bingham Canyon Mine, and the construction of the ultra-modern Copperton Concentrator with three grinding lines and flotation circuits. In 1989, Rio Tinto, the world's largest mining company purchased Kennecott. In 1992, Kennecott Utah Copper (KCU) added a fourth grinding line and flotation circuit at the Copperton Concentrator at a cost of $220 million. Also in 1992, Kennecott began construction of the most technically advanced and cleanest smelter in the world as part of an $880-million modernization of the smelter and refinery facilities. This project was the largest privately financed construction project in the history of Utah. The new smelter became operational in the spring of 1995 (figure 3).

As part of Kennecott's ongoing modernization activities, the 5,700-acre tailings impoundment is undergoing an expansion and seismic upgrade, ensuring continued tailings storage capacity for an additional 25 to 30 years. To offset or mitigate impacts to wetlands and other wildlife habitat relat-

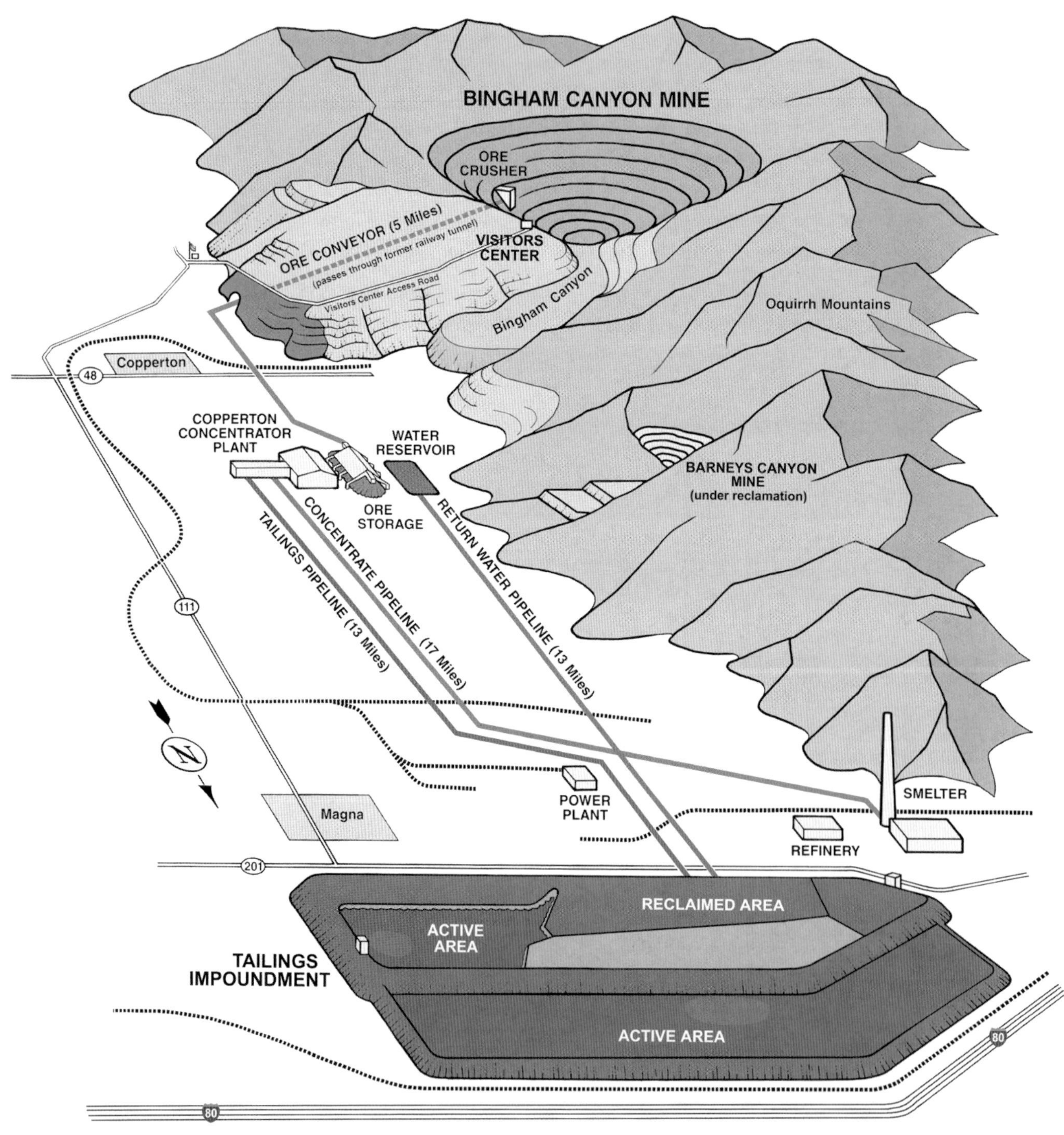

Figure 3. *Schematic of the Kennecott facilities and plants.*

ed to construction of the expanded tailings impoundment, Kennecott purchased approximately 2,500 acres of land north of Interstate 80, and west of 8800 West, immediately north of the expansion property, and adjacent to the southern shore of Great Salt Lake. Construction of the wetland mitigation plan started in May of 1996 and was completed in January 1997. Water was introduced to the site to start operation of the wetlands in February 1997 and the wetland mitigation site became officially referred to as the Inland Sea Shorebird Reserve. The reserve is a breeding, nesting, and feeding site for Great Salt Lake shorebirds - including the American Avocet, Wilson's Phalarope, Cinnamon Teal, and Snowy Plover.

MINING

It has taken over 90 years... and a long-term team effort... for KCU's Bingham Canyon Mine to become the largest man-made excavation on earth, with six billion tons of material removed since 1906 (figure 4). The open pit mining methods invented at the turn of the century, such as loading and hauling, are still used today. The equipment, however, has grown in size and complexity with advances in technology. Today, the monstrous haulage trucks can carry from 255 to 360 tons per load (figure 5). The mine's largest electric shovels have 56-cubic-yard dippers that can scoop up to 98 tons of material in a single bite (figure 6). Computer models help with mine planning and sophisticated communications systems monitor all truck, shovel and rail movements (figure 7).

Based on data received from geologists, mining engineers develop a complex mining plan on a daily, weekly, monthly, yearly, and multi-year basis. The plan divides the mine into ore and waste zones. Ore is material containing minerals that can be mined and processed at a profit. Waste, or overburden, is material that is uneconomic for processing,

Figure 4. *The Bingham Canyon Mine is currently over two and one-half miles across at the top and three-quarters miles deep - the largest man-made excavation on the earth.*

Figure 5. *The enormous haul trucks can carry 255 to 360 tons per load.*

Figure 6. *Large electric shovels scoop up 98 tons of material in a single bite to load waiting haulage trucks.*

Figure 7. *The open-pit mining methods employed are some of the most technologically advanced in the world. Computer models help with mine planning, and sophisticated communications systems monitor all truck, shovel and rail movements.*

but must be removed to expose the ore. Economics, therefore, determine what is ore and what is waste. The mining operation at Bingham Canyon includes drilling and blasting, loading and crushing, and conveying (figure 8). The crushed ore is transported by a five-mile conveyor system: three miles through a tunnel in the mountain and two miles above ground to the Copperton Concentrator. Approximately 173,000 tons of copper ore and 330,000 tons of waste rock are removed on a daily basis (figure 9). Bingham Canyon Mine is currently over two and one-half miles across at the top and three-quarters miles deep. If you took one of the world's tallest buildings (Chicago's Sears Tower) and placed it in the bottom of the pit, it would only reach half way to the top of the mine. By the year 2015, the mine will be an additional 650 feet deeper.

Figure 8. *Haul trucks dump metal-bearing ore into an in-pit crusher.*

Figure 9. *Crushed ore is transported by a five-mile conveyor system to the Copperton Concentrator coarse-ore stockpile.*

CONCENTRATING

Concentrating, the process of separating and gathering the valuable minerals from the ore, is the second metals production step after mining. KCU operates two concentrators - the Copperton and the North Plants.

The state-of-the-art Copperton concentrator includes four Semi-Autogenous Grinding (SAG) mills, eight ball mills, and corresponding flotation cells (figure 10). Copperton, which opened in 1988 and expanded in 1992, contains some of the largest SAG and ball mills in the world. In a SAG mill, much of the grinding is done by 5 1/4" steel balls, each weighing 22 pounds. As the mill revolves, the balls and the basketball-size pieces of ore pound against one another in an aqueous slurry. The ore fragments leave each SAG mill smaller than 5/8" in diameter, and are then transported as a slurry to two adjacent ball mills where 3" steel balls, each weighing 4 pounds, further grind the slurried ore to the consistency of face powder. Slurry from the ball mills flows to the flotation cells where it is mixed with reagents and agitated to produce a bubbly froth. Ore particles containing most of the valuable minerals adhere to the bubbles, which float over the sides of the flotation cells. This material, called "concentrate," is then collected and filtered. The remaining waste particles sink to the bottom to become "tailings."

The concentrate, which contains about 28 percent copper and by-products such as gold and silver, is pumped through a pipeline to the KCU smelter, located about 17 miles to the north, for further processing. The gold and silver follow the copper through the smelter production process and are recovered at the refinery.

Figure 10. *The state-of-the-art Copperton Concentrator includes four Semi-Autogenous Grinding (SAG) mills, eight ball mills, and corresponding flotation cells.*

SMELTING

The third step in KCU's production cycle is smelting. Copper, like silver and gold, can be produced directly from naturally occurring minerals by simple heating and oxidation.

Smelting was first practiced over 6,000 years ago in the Middle East. Smelting has taken place on the south shore of the Great Salt Lake since 1906, when American Smelting and Refining Company, now ASARCO, built a smelter to process ore from Utah Copper's Bingham Canyon Mine. In 1959, Kennecott bought the smelter from ASARCO. The original smelter has been rebuilt and modernized several times to meet the demands of improved emission control and higher productivity.

In 1992, Kennecott began construction of the most technically advanced and cleanest smelter in the world as part of an $880-million modernization of the smelter and refinery facilities (figure 11). This project was the largest privately financed construction project in the history of Utah. The new smelter became operational in the spring of 1995.

Figure 11. *Kennecott's smelter with flash smelting and flash converting is the most technically advanced and cleanest smelter in the world.*

The smelting process treats copper concentrates from the Copperton and North Concentrators. The process starts with drying the copper concentrate and injecting it and oxygen-enriched air into a modern flash-smelting furnace. The copper concentrate - containing approximately 28% copper and similar quantities of sulfur and iron - burns, providing most of the heat to sustain the furnace temperature. Most of the iron and about 67 percent of the sulfur are oxidized. The iron forms a slag which is skimmed from the furnace, cooled, and treated in a slag concentrator at the smelter plant to recover additional copper. The sulfur dioxide is cooled in a boiler to produce steam and then sent to a double contact sulfuric acid plant. In the acid plant, the sulfur dioxide is converted to high-purity sulfuric acid, which is sold and transported in tank cars to agricultural and industrial customers.

Copper produced in the smelting step is a molten copper sulfide called matte, which contains 70 percent copper. The molten matte is tapped from the furnace and quenched in a water spray to form a sand-like solid. This granulated matte is then dried and ground to a fine powder so it can be injected into a second flash furnace. This step was patented by Kennecott. The combination of matte granulation and flash converting, developed by Kennecott and a Finnish company, Outokumpu, represents a major improvement in smelting technology and environmental efficiency because the molten metal transfer, and the resulting gas emissions, are eliminated. The powdered matte burns to provide the heat for the process, liberating sulfur dioxide, and producing molten copper metal, which is 98.6 percent pure. The sulfur dioxide gas is cooled and sent to the double contact sulfuric acid plant.

The copper is transferred to one of two refining furnaces where it is upgraded to 99.5 percent copper (figure 12). Casting wheels then cast the copper into 730 pound plates called anodes, which are transported by rail to the refinery. There, production of the final copper product, called cathodes, is completed. Gold and silver are also recovered during this sequence.

Figure 12. *At the smelter, molten copper is transferred to anode refining furnaces where it is upgraded to 99.5 percent copper.*

REFINING

Refining is the fourth and final step in the copper production cycle. It is also the stage in which precious metals are recovered. Refining takes place at KCU's refinery located about two miles east of the smelter.

The refinery, which began operating in 1950, completes the production cycle of mining, concentrating, smelting, and refining necessary to produce copper ready for sale to fabricators. In 1954 and 1955, the refinery was expanded to better meet customer demands in a highly competitive copper market. In 1994, it was again expanded and modernized. However, expansion and modernization have not changed the basic methods by which copper is refined.

The modernized, fully automated refining process begins when 730-pound anodes arrive from the smelter and are weighed, pressed flat and milled. Stainless steel cathode blanks are interleaved with anodes and transported to the tank house by robotic Automated Laser Guided Vehicles.

In the tank house, the cathode blanks are placed in cells containing an acidic solution called electrolyte (figure 13). Electric current is applied to the cells for 11 days. The anode copper dissolves in the electrolyte and migrates to the cathode, where the copper ions are deposited as 99.99 percent pure copper. Impurities and precious metals settle to the bottom of the cells and are collected and later refined on site at the precious metals plant.

Figure 13. In the refinery tank house, copper dissolves and migrates to cathode blanks where it is deposited as 99.99 percent pure copper.

One 730-pound anode will produce two cathodes, each weighing about 280 pounds. The remaining copper in the anode is resmelted into additional anodes. KCU's modernization has resulted in annual copper production of about 310,000 tons of copper metal (figure 14).

The new precious metals plant technology allows for production of gold in only five days in contrast with the 45-day cycle required in the old refinery. Everyday, approximately 129,000 gallons of anode slime from the tankhouse are pumped to the precious metals plant as a slurry mixture of solids, electrolyte, cell washwater, floor washwater, and anode/cathode washwater. This slurry is pumped into receiving tanks where the precious metals (anode) slime is settled, and the electrolyte is filtered and returned to the tankhouse. The settled solids are processed through a series of autoclaves, tanks, and filters resulting in the extraction and purification of precious metals.

Chlorination leaching is the heart of the precious metals recovery process. This process oxidizes and solubilizes the precious metals in the slime using a hydrochloric acid leach. This leach solution removes essentially all of the gold in the electrolytic slime. The dissolved gold is further refined using solvent extraction to produce a gold filter cake that is melted in an induction furnace and cast into 400-troy ounce bars with a purity of 99.99 percent. In a similar process, silver is produced and cast into 1,000-troy ounce bars.

The precious metals plant annually produces approximately 500,000 ounces of gold and 4,000,000 ounces of silver. The plant also produces small amounts of platinum, palladium and selenium and, because of new technology, lead and tellurium.

Figure 14. Approximately 310,000 tons of refined copper products are produced annually.

TAILINGS

KCU's tailings are the uneconomic by-product of the ore crushing, grinding, and flotation concentrating process (figure 15). After processing, this barren material (essentially devoid of metals values), is transported as slurry in a 60-inch concrete pipe from the Copperton and North Concentrators to the company's tailings impoundment, adjacent to the southeast arm of the Great Salt Lake near Magna, Utah. Since its construction in 1907, more than 1.5 billion tons of tailings have been stored in the 5,700-acre impoundment. To accommodate the approxi-

Figure 15. Kennecott's tailings impoundment, located adjacent to the southeast arm of the Great Salt Lake, stores the uneconomic portion of the ore crushing, grinding, and flotation concentrating process.

mately 56-million tons of tailings deposited annually, the impoundment dike is raised seven to eight feet per year. A peripheral discharge system wets the entire surface area to minimize dust. Process and stormwater are collected on the surface of the impoundment, decanted, and recycled to the plants for reuse. Exterior slopes of the impoundment are contoured and revegetated as the dike is raised.

As part of Kennecott's ongoing modernization activities, the 5,700-acre impoundment is undergoing an expansion and seismic upgrade, ensuring continued tailings storage capacity for an additional 25 to 30 years. Over that time, the company's operations are expected to generate an estimated additional 1.6 billion tons of tailings. In early 1996, the expansion of the impoundment in a northerly and westerly direction was initiated. By about 2002, tailings storage activities will transfer from the existing impoundment to the new impoundment area. The expansion site is an area previously used primarily for salt evaporation, salt processing, and fertilizer production. The expansion consists of a seven-mile-long embankment that will encompass approximately 3,200 acres.

As part of the expansion project planning, KUC identified, extensively investigated, and addressed many environmental issues - air quality, surface water management and water quality, ground water quality, wetlands and wildlife habitat. An Environmental Impact Statement (EIS) was required by the U.S. Army Corps of Engineers (Corps) under Section 404 of the Clean Water Act of 1977. To offset or mitigate impacts to wetlands and other wildlife habitat related to construction of the expanded tailings impoundment, Kennecott purchased approximately 2,500 acres of private land north of Interstate 80, and west of 8800 West, immediately north of the expansion property, and adjacent to Great Salt Lake. A mitigation plan was developed with input and expertise from the Nature Conservancy, National Audubon Society, U.S. Fish and Wildlife Services, Utah Division of Wildlife Resources, Environmental Protection Agency, and the U.S. Army Corps of Engineers.

To compensate for the loss of wildlife habitat resulting from the North Expansion, Kennecott started developing and managing a shorebird reserve in 1996, to increase and improve the overall wildlife habitat in the area. The reserve contains about 4,000 acres of wetlands, saline playas (sandy areas), open water, uplands, and marshes that are home to numerous birds, small mammals, and other animals. In particular, the reserve is a breeding, nesting, and feeding site for shorebirds - including the American Avocet, Wilson's Phalarope, Cinnamon Teal, and Snowy Plover.

An important food for the birds is the bloodworm, the larval form of the midge (chironomid), one of several aquatic insects (macroinvertebrates) found in the area. The life cycles of both midges and the birds are closely interrelated. The seasonal fluctuation of the wetland water level is a key factor, affecting both the quantity of the bloodworms and the ability of the birds to reach them. To enhance food production, the reserve manager "farms" the midges by manipulating the water levels in the reserve's five ponds to encourage breeding and ensure a steady food supply available to the birds.

Construction of the wetland mitigation plan started in May of 1996 and was completed in January 1997. Water was introduced to the site to start operation of the wetlands in February 1997 and the wetland mitigation site became officially referred to as the Inland Sea Shorebird Reserve (ISSR) (figure 16).

Figure 16. *Birds take flight at Kennecott's Inland Sea Shorebird Reserve (ISSR).*

THE ENVIRONMENT

Mining, by its very nature disturbs the environment. However, KUC is dedicated to minimizing these disturbances wherever possible, and to restoring the environment to the maximum extent feasible where disturbances occur. KUC's philosophy is to balance society's need for metals with an environmentally responsible approach to mining. Every day, the company's environmental engineers and scientists work on programs involving air quality, water quality, waste management, land reclamation and revegetation.

The air quality in and around the company's facilities is constantly monitored to ensure compliance with KUC's own standards, as well as those of the government. Steps are continually being taken to enhance air quality. KUC's programs range from using low-sulfur diesel fuels in haulage trucks to construction of the world's cleanest smelter, which captures 99.9 percent of the sulfur contained in copper concentrates.

Water quality programs abound. Rainwater and snowmelt, for example, are diverted around the company's facilities to avoid contact with mining or processing operations. Water used in the operations is treated and recycled, and water quality is the subject of around-the-clock monitoring.

KUC's commitment to land restoration programs is demonstrated by the fact that, during any given year, KUC's engineers reclaim more acreage (from such sites as obsolete facilities, old tailings areas and overburden dumps) than typical operations disturb. Revegetation goes hand in hand with reclamation, and more than 100,000 trees have been planted on company property, as well as thousands of acres of bushes, flowers and grasses. The tailings impoundment is reclaimed and revegetated on an on-going basis.

KUC is committed to the preservation and protection of the environment and believes that mineral mining and processing can co-exist with an attractive, safe clean environment. One of KUC's core values is to conduct its business in an environmentally sound and responsible manner.

About $300 million and innumerable man-hours have been spent on the cleanup of historic mining sites, many of which were created prior to Kennecott's involvement in the Oquirrh Mountains. To date KUC has cleaned up over 20,000,000 cubic yards of mining-related material. KUC aims to remain proactive in minimizing environmental disturbances and preventing environmental problems for future generations.

THE RAILROADS PROXIMATE TO GREAT SALT LAKE, UTAH

by
J. Wallace Gwynn
Utah Geological Survey

ABSTRACT

Since the signing of the Pacific Railway Act by President Abraham Lincoln in 1862, and the driving of the golden spike at Promontory Summit on May 10, 1869, railroads have played a significant role in Utah's economic development and progress. The Central Pacific tracks, taken over by the Southern Pacific in 1884, and the Union Pacific tracks, were initially built north of Great Salt Lake in 1868-69. The Southern Pacific tracks were rerouted across Great Salt Lake in 1902. This route was called the Lucin Cutoff. All of the original trackage north of the lake, from Lucin to Corinne, was removed in 1942. The Western Pacific was built south of Great Salt Lake from Salt Lake City westward to San Francisco, California, and completed in 1909. Both the Southern Pacific and Western Pacific Railroads are major links between the east and west coasts. The Utah Central Railroad from Ogden to Salt Lake City, completed in 1870, was sold to the Union Pacific's Oregon Short Line in 1889. The Denver & Rio Grande Western, completed from Denver, Colorado to Salt Lake City, and on to Ogden in 1883, has served the communities along the Wasatch Front and towns south of Salt Lake City. The Salt Lake, Garfield &Western (SLG&W) Railroad, completed in 1893, and the Bamberger Railroad, completed to Ogden in 1908, provided passenger service to two of Utah's most famous resorts, Saltair and Lagoon, respectively. Both of these railroads were converted to electrical power between 1910 and 1916, and provided both passenger and freight service. The Bamberger route was abandoned in 1958, but the SLG&W continues to operate. The Southern Pacific was purchased by Rio Grande Industries in 1988. The Union Pacific ultimately purchased all of these railroads except the SLG&W, and has become one of the largest and fastest growing transportation companies in the United States. Railroads have played an important role in building the state's saline-mineral industries, and transporting a multitude of products to market destinations. As Great Salt Lake made its dramatic 12-foot rise during the 1980s, the railroads around the lake were forced to spend millions of dollars to raise their tracks in order to maintain service.

INTRODUCTION

Since the completion of the first transcontinental railroad in 1869, railroads have played an important part in our nation's as well as Utah's growth and economic development. Within Utah, railroads have been a necessary element in developing its multitude of mineral resources, including the saline resources of Great Salt Lake and the Bonneville Salt Flats. This paper presents a brief history of the railroads that have been built and operated adjacent to, and across, Great Salt Lake. It also describes some of the lake-related problems that the railroads have encountered over the years, and some of the changes in the lake that the railroads have caused. An understanding of these problems and changes is necessary for planning for the lake's future.

THE FIRST TRANSCONTINENTAL RAILROAD (1832-1869)

By 1832, the nation realized it needed a transcontinental railroad to tie California to the rest of the states. In 1849 and 1850, Howard Stansbury surveyed a route for a transcontinental railroad through the Black Hills and south of Salt Lake City, and in March of 1953, Congress approved measures for the War Department to survey various transcontinental railroad routes. However, it wasn't until 1862, after years of hostile debate over the most favorable route, that Congress approved a route over Donner Summit, California, and funded the project with appropriations and land. Within the year, President Abraham Lincoln signed the Pacific Railroad Act (U.S. Statutes at Large, Vol. XII, p. 489 ff.), which named and directed two companies, the Union Pacific (building west from Omaha, Nebraska) and the Central Pacific (building east from Sacramento, California) to construct the long-awaited Transcontinental Railroad (Linecamp, 1999).

The Pacific Railway Act insured the transcontinental railroad's loans of $16,000 on the plains, $32,000 in the Great Basin, and $48,000 through the mountainous terrain for each mile of track laid. In addition, the Act gave the Transcontinental Railroad 10 alternate sections of public land on each side of the track for each mile of track built. In 1864, the Pacific Railway Act was changed to 20 alternate sections, and provided for land-grant bonds utilizing the land grants as backing. In the end, the Transcontinental Railroad Companies acquired 33 million free acres of land (Linecamp, 1999; Box-Elder, 1999; Union Pacific Railroad, 1999a).

On May 10, 1869, a locomotive from each railroad ("Jupiter" represented the Central Pacific, and "No. 119" represented the Union Pacific) pulled up to a one-rail gap left between the two railways. The Central Pacific had laid 690 miles of track from the west while the Union Pacific had laid 1,086 miles from the east. After a symbolic golden spike was tapped, a final plain-iron spike was used to actually connect the railroads at Promontory Summit. This site was later designated as the Golden Spike National Historic Site (figure 1). These words were engraved upon the symbolic golden spike: "May God continue the unity of our Country as this Railroad unites the two great oceans of the world" (Box-Elder, 1999; Neticus, 1999).

Central Pacific Railroad

A young railroad engineer in California, Theodore Judah, had his own ideas about transcontinental railroad travel. By 1862, he had not only surveyed a feasible route across the Sierra Nevada, but had persuaded a group of wealthy Sacramento businessmen to invest in a new company, the Central Pacific Railroad (CPRR). That same year, Congress authorized the new CPRR to build eastward from Sacramento. Construction of the railroad began in January 1863, but little headway was made as money, raw material, and manpower were diverted to the cause of the Civil War (Box-Elder, 1999).

The Pacific Railroad Act of 1862 set aside a land grant that aided the CPRR in defraying the costs, but during the high-profit epoch of the Civil War, it was just not enough. The solution lay in the Pacific Railroad Act of 1864, which liberalized the funding available to construction by doubling the land grant and providing for land grant bonds, utilizing the land grant as backing. Most of the iron and anything made by heavy industry, including the locomotives, rail, cars and the wheels, had to be made in the east and shipped 15,000 miles around Cape Horn to California, a five- to six-month voyage. The CPRR's labor problems were partially solved by importing approximately 10,000 Chinese laborers. Crews were drilled, pushed, and prodded until they could lay 2 to 5 miles of track per day (Box-Elder, 1999).

By the middle of 1868, the CPRR work crews had laid close to their final 690 miles of track. As they drew closer to meeting up with the Union Pacific in Utah, work crews labored to grade as many miles as possible and thus claim the greater share of land grants. Unfortunately, in the process, they pushed right past the UPRR. For more than 200 miles, the two railroads worked parallel to each other, in opposite directions. Congress finally ordered the two to meet at Promontory Summit, Utah (Box-Elder, 1999). The original route of the CPRR (now abandoned) is shown on figure 1.

In 1869, when the railroad was being rushed to completion, exploration survey parties of both the CPRR and Union Pacific Railroads (UPRR) considered crossing Great Salt Lake. The engineers for both companies came to the same decision. Construction directly across the lake was not feasible because of cost, and because information was lacking

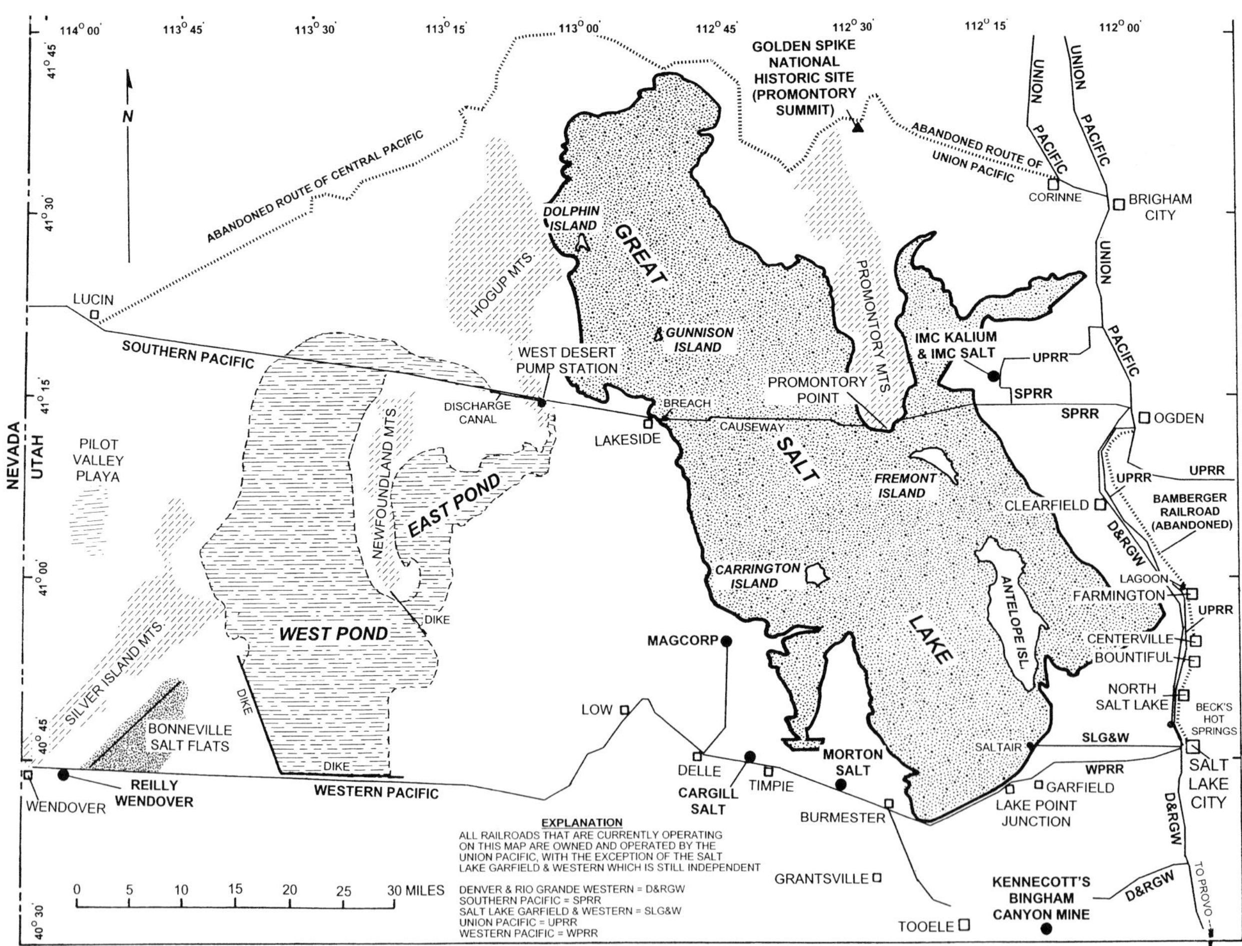

Figure 1. Active and abandoned railroad routes, mineral companies, elements of the West Desert Pumping Project, and other natural and cultural features proximate to Great Salt Lake. Lake elevation is at 4,200 feet above mean sea level.

regarding the lake bottom, depth of the lake, weight of water, and long-term changes in lake levels. The lake's water level at that time was relatively high (about 4,210 feet). As speed of construction was essential, a route around the north end of the lake was adopted, although the terrain on this route caused problems for the railroad, and resulted in grades up to 2.2 percent and many sharp curves (Newby, 1980).

In late 1869, the CPRR purchased the UPRR's section of the railroad from Promontory to Ogden, because Promontory was not suitable as a terminus due to its remote location. In 1884, the CPRR was taken over by the Southern Pacific Railroad (SPRR), although the actual incorporation procedures were not completed until 1958. The original section of the transcontinental railroad around the north end of Great Salt Lake, from Lucin to Corinne, was removed in 1942 during World War II, and replaced by the Lucin Crossing or Cutoff. Most of the rails were taken to California and relaid on numerous wharves and piers in the San Francisco/Oakland area to serve the many shipping needs there during the war (Stephen Carr, M.D., written communication, July 19, 1999).

Union Pacific Railroad

The UPRR, working under the same Pacific Railroad Act terms as the CPRR, began laying track westward in December of 1863, and was plagued by the same financial, labor, and other problems due to the Civil War (Union Pacific Railroad, 1999d). Construction was first led by Thomas C. Durant, president and general manager of the UPRR, and then by Grenville Dodge, a young Union general and civil engineer (Union Pacific Railroad,1999e).

Constructing a railroad line across the plains was impeded by many problems, including raids by American Indians, and the fact that all supplies had to be brought across the plains from the Missouri River, which added greatly to transportation distances as the work progressed westward. Pushing west through Wyoming was exhausting and treacherous. A 650-foot-long bridge constructed across Dale Creek, rising 150 feet up from the bottom of the canyon, swayed in the wind and was terrifying to cross. Weather was also a constant opponent, especially during the winter months (Union Pacific Railroad, 1999e).

Once the Civil War ended, the manpower shortage ended in the east. The UPRR hired veterans from both sides of the war, freed slaves, and Irish, German, and Italian immigrants. The work force combination was racially and politically volatile, and bloodshed was common (Box-Elder, 1999). Near the completion of the work, there were ill feelings between the UPRR's Irish workers and the CPRR's Chinese workers which resulted in a number of deaths (Union Pacific Railroad, 1999e).

INTRASTATE RAILROAD DEVELOPMENTS (1869-1999)

Southern Pacific Railroad

Lucin Cutoff

After 1873, the level of Great Salt Lake began to decline, and by 1900 it had dropped nearly 12 feet to about 4,200 feet (above mean sea level). Also, during the period from 1869 to 1903, the "as constructed" CPRR between Reno and Lucin had been greatly improved. However, the section of railroad between Lucin and Ogden (now abandoned, figure 1) was a bottleneck with many curves and steep grades. The decision was made to investigate alternate routes to Salt Lake City, including several routes across, and others to the south of the lake. Since the desired terminus was at Ogden, the directors of the CPRR required that the lake be crossed on the most direct route. The new route, known as the "Lucin Cutoff," crossed the lake from Lucin to Ogden, and greatly reduced the distance, grade and curvature that existed in the original northern route (Southern Pacific Lines, 1992).

Construction of the Lucin Cutoff began on August 21, 1902, and the line was turned over to the CPRR Operating Department on December 8, 1904. The lake crossing required approximately 28 miles of embankments and a 12-mile trestle. The original 12-mile trestle, that part of the project built over the main deep part of the lake, was the longest bridge across open water in the world. The trestle's top-of-the-tie elevation was built at 4,214.75 feet, and the embankments were constructed at 4,209.75 feet. Over 7.8 million cubic yards of fill material was placed in the embankments from a quarry about four miles northwest of Promontory Point. Nineteen pile drivers were used to build the 12-mile timber trestle. Some 38,000 piles, varying in length from 80 to 130 feet, were driven into the lake bottom. Five piles were initially driven per bent (group of piles), with bents placed every 15 feet. More than 4,100 bents were required to construct the 12-mile trestle. Pile caps, decking, stringers, and then the rails were laid on top of the piles. Brace pilings and two additional piles per bent were driven in the 1930s and 1940s. Over 3,000 men were employed during the construction of the crossing. It was estimated that with traffic of 10 trains every day, the cutoff would save $214,000 per year (1902 dollars) (Stephen Carr, M.D., written communication, July 19, 1999; Southern Pacific Lines, 1992). For additional information on the construction of the Lucin Cutoff, see Cannon (this publication), Newby (1980), and Southern Pacific Lines (1992).

Replacement of Wooden Trestle with Rock-Fill Causeway

By 1950, the timber trestle was nearly 50 years old, and maintenance costs on the structure were increasing steadily under heavier loads and the wear and deterioration of the timber. Train speeds had to be restricted to 20 miles per hour to reduce impact and sway. In 1953, a detailed engineering study of the trestle indicated that the superstructure above the pile caps would need replacement within seven years, and that the piles would require replacement within 25 to 30 years. Since traffic had to be maintained during the repair period, the option of rehabilitating the trestle was not acceptable. A feasibility and economic study was made to determine if the trestle should be replaced with a new concrete trestle or an embankment fill. A detailed engineering study concluded that a sand, gravel, and rock-fill embankment was feasible, and would be the most economical alternative to replace the trestle. In May 1955, a contract was awarded to International Engineering to design the embankment under the direction of a board of consultants and review by SPRR (Southern Pacific Lines, 1992).

Construction of the embankment was awarded to Morrison-Knudsen Company of Boise, Idaho in 1956. The total cost of the project, including the embankments, culverts, exploration, engineering, and railroad track and signals, was about $53 million. The work began in March 1956, and the roadbed was completed July 1, 1959 (Southern Pacific Lines, 1992).

The Rising Lake

A rapid increase in lake elevation starting in September 1982 impacted the railroad routes across and immediately south of the lake. Over nine months, from September 1982 to May 1983, the lake level rose a little more than 5 feet. The rise in lake level continued from 1983 through 1984, as the lake rose another 4.5 feet, and from 1985 through 1986, as the lake rose another 3.5 feet. The lake rose a total of more than 12 feet over three and one-half years, from an elevation of 4,199.5 feet to over 4,211.5 feet. Much of this increase was due to two unusually wet years, with heavy mountain snow packs followed by rainy, cooler-than-normal summers (Union Pacific Railroad, 1986) (for additional climatological information see Alder, this publication). During this time, the SPRR was faced with the monumental problem of raising the track and fills, constructing slope protection, constructing bridges, repairing storm damage, and maintaining freight service. Between Hogup (Milepost [MP] 724) and Little Mountain (MP 767), nearly 59 miles of main line and sidings had to be raised between 6 and 10 feet to keep a bare minimum freeboard of 2 feet (height between the static lake level and the top of the railroad fill). The material used to raise the track elevations and to provide slope protection came from the SPRR's large limestone quarry located on the west side of the lake at Lakeside (figure 1) (Southern Pacific Lines, 1992).

The Box Car Seawall

In the spring and summer of 1983, the lake was rising at such a rapid rate that an interim "quick fix" solution was required to protect the extremely vulnerable north slope of the causeway. The decision was made to utilize scrap railroad box cars to create a "box car seawall" on the north side of the Great Salt Lake causeway. From mid-August to September 1983, 1,430 partially dismantled scrap cars were placed end to end on the causeway and then filled with quarry-run rock. The box cars provided significant protection for a two-year period, during which the tracks were raised (Southern Pacific Lines, 1992).

The Causeway Breach

By the spring of 1984, the very large inflows to the south arm of the lake (the area south of the railroad causeway) had created a difference of water levels across the causeway of nearly 3.5 feet, the south arm being higher than the north arm. As a flood control measure, the State of Utah requested that a 300-foot bridge opening or breach be constructed in the causeway fill (figure 1), just offshore east of Lakeside (see Austin, this publication, and Southern Pacific Lines, 1992 for additional information). The bridge, as constructed, is a double-track structure with an adjacent highway bridge. The bridge is about 300 feet long and is supported by alternating 10- and 14-pile bents that are supported on a 10-foot-thick cemented oolite layer located about 45 feet below the bottom of the bridge cap (Southern Pacific Lines, 1992). The breach was completed on August 1, 1984. The first south-arm water gushed into the north arm on October 1, 1984, and reduced the head differential to less than a foot within about two months. In addition to reducing the difference in water levels across the causeway, the breach allowed for a major redistribution of the differing salt loads of the north and south arms of the lake (Gwynn and Sturm, 1987).

West Desert Pumping Project

After the breaching of the causeway in 1984, the lake continued to rise, reaching an elevation of nearly 4,210 feet in 1985 and continuing to climb to nearly 4,212 feet in 1986. In response to the threat of additional flooding, the state looked at other alternatives to slow or stop the rise of the lake. These alternatives included diking low areas around the lake, diverting and developing water upstream before it entered the lake, and pumping water into a natural basin or depression in the Great Salt Lake Desert west of the lake (the "West Desert"). The concept of pumping water into the West Desert was determined to be the only timely, viable alternative (Southern Pacific Lines, 1992).

The main operational features of the pumping project consist of a pumping plant, a discharge canal, and the West Pond (figure 1). The project also includes a 10-mile-long access road (jointly used by the state and SPRR), four railroad bridges to allow circulation of the brine, a 24-mile-long raised railroad embankment through the West Pond, and three dikes to contain and control the 500-square-mile West Pond (Southern Pacific Lines, 1992). For additional information on the West Desert Pumping Project, see Austin (this publication), Waddell and others (1992), and Wold and Waddell (1994).

Changes in Ownership

Rio Grande Industries purchased SPRR in 1988, but because the SPRR name was more historic and had better public name recognition, the Rio Grande Railroad part of Rio Grande Industries was merged into the Southern Pacific Transportation Company. Then, in the latter part of 1996, the UPRR took over the SPRR, which included the section from Ogden to San Francisco across the Lucin Cutoff, and the ex-Denver & Rio Grande Western line from Salt Lake City to Ogden. As part of this merger, Burlington Northern/Santa Fe (BNSF) was granted trackage rights over UPRR. BNSF trains use the UPRR line from Salt Lake City to Oakland. They also have switching rights between Salt Lake City and Ogden, but mostly utilize Utah Railway's locomotives to do those chores.

The SPRR started essential rail service to one of Great Salt Lake's largest mineral extraction companies, IMC Kalium (formerly Great Salt Lake Minerals and Chemicals Corporation) in about 1969 to 1970.

Utah Central Railroad

Brigham Young, leader of the Mormon Church, foresaw the impact that the coming of the railroad would have and wanted the transcontinental rail line built through Salt Lake City. Representatives of both the CPRR and the UPRR met with him and explained the difficulty and expense of a route through Salt Lake City. In 1869, prior to the completion of the transcontinental railroad, Mormon Church leaders began work on the organization of a connecting railroad between Ogden and Salt Lake City. In January 1870, that line, called the Utah Central Railroad, was completed, connecting Salt Lake City to the national rail system (Stephen Carr, M.D., written communication, July 19, 1999; Media.Utah, 1999). The Utah Central Railroad was purchased by the UPRR's affiliate Oregon Short Line in 1889, and has been a part of the UPRR ever since. A second track was laid from Salt Lake to Ogden between 1909 and 1912 (Stephen Carr, M.D., written communication, July 19, 1999).

Denver and Rio Grande Western Railway

The Denver & Rio Grande Western Railway was incorporated in 1881, and completed its narrow-gauge line between Colorado and Salt Lake City in March 1883, running through the Utah towns of Green River, Price, and Provo along the way. It was extended to Ogden two months later (figure 1). The company was reorganized in 1889 as the Rio Grande Western Railway to enable it to finance the conversion of its line from narrow gauge to standard gauge in 1890. In 1901, the Rio Grand & Western came under the control of the Denver and Rio Grande Railroad in Colorado, and the two companies were merged under the name Denver and Rio Grande Western (D&RGW) in 1921. Before this merger, the two independent companies succeeded in completing a network of branch lines that put them in direct competition with the UPRR in Utah. In addition, the Rio Grande also had a virtual monopoly on the movement of coal out of the state (Media.Utah, 1999).

In 1985, the UPRR and D&RGW combined trackage rights with the D&RGW using the UPRR's main line between North Salt Lake and Ogden, and the UPRR using D&RGW's main line between Salt Lake City and Provo. The D&RGW line between North Salt Lake and Ogden is now owned by the UPRR, but has lain dormant since 1985 (Stephen Carr, M.D., written communication, July 19, 1999). In 1988, the D&RGW and the SPRR merged, retaining the SPRR name (Union Pacific Railroad, 1999a).

Western Pacific Railroad

On Tuesday, March 3, 1903, eleven men sat down around a table in the Safe Deposit Building in San Francisco and organized a new transcontinental railroad to be named the Western Pacific Railroad (WPRR); the new company was incorporated on March 6, 1903. It was to run from San Francisco, California, eastward through the canyons of the Feather River and Beckwourth Pass, and on to Salt Lake City. The route through Feather River Canyon had been mentioned as a railroad route for some time, particularly by one man, Arthur W. Keddie, who had earlier surveyed the route (Kneiss, 1953).

Construction conditions were difficult, particularly in the Feather River and Bonneville Salt Flats sections of the route. Construction camps were established by the WPRR's contractors at points all along the line. Some of the camps were accessible by rail or by wagon road, but much of the rugged Feather River Canyon wasn't accessible even by a foot path. At the Utah end of the route, crossing the Bonneville Salt Flats (figure 1) in 1906 to 1907 was a nightmare due to temperature extremes and the glare from the white salt, which often blinded men after a few hours of work (Kneiss, 1953). A beneficial side effect of construction across the Bonneville Salt Flats (which was called the Salduro Salt Marsh at that time) was that it brought national attention to the salt beds and eventually to the subsurface-brine potash resources. Potash production was begun during World War I by Bonneville Limited from the flat's subsurface brines and continues today by Reilly Wendover. The finished products were transported by the WPRR (Gwynn, 1996).

During the early construction period, it was difficult to hold men under such adverse working conditions while more pleasant work was plentiful elsewhere, and turnover was terrific. After the depression of 1907 set in, however, there were plenty of men available, and at lower wages. Had it not been for this unexpected break, all of the contractors would probably have gone bankrupt, since the work proved considerably more costly than originally estimated (Kneiss, 1953).

On November 1, 1909, the track gangs working from the east and west met, and the final railroad spike was driven. In contrast to the gold spike ceremonies on the first overland railway just forty years before, no decorated engines met head to head before a cheering crowd, and no magnums of champagne were broached. The only spectators were two local women and their little girls. The WPRR was now operative, but far from finished (Kneiss, 1953).

Through-freight service on the WPRR was inaugurated on December 1, 1909. It was disappointingly slow, and the winter of 1910 was very hard. To make matters worse, the waters of Great Salt Lake began to rise, eroding the earth-fill roadbed, and seriously threatening eight miles of line on fill and trestles. The WPRR considered abandoning this track and obtaining trackage rights over the San Pedro, Los Angeles & Salt Lake route farther south, and building a 10-mile connection west of the lake. It was not until the later part of May 1911 that operations returned to normal (Kneiss, 1953).

The WPRR has provided transportation to Utah's saline-mineral industries since 1938, when Bonneville Limited shipped the first potash from its new plant on the Bonneville Salt Flats just east of Wendover (Gwynn, 1996). The WPRR has provided essential rail service to mineral-extraction companies on the south end of Great Salt Lake including MagCorp (formerly AMAX), Cargill Salt (formerly AKZO Salt), IMC Salt (formerly North American Salt prior to their move to the north end of the lake), and Morton Salt. It has also provided rail service to a number of industrial mineral operations (lime and limestone) along its westward route.

On January 23, 1980, an agreement by directors of both companies was announced that gave control of the WPRR to the UPRR, and an application for authority to consummate that transaction was shortly filed with the Interstate Commerce Commission. Two years later in mid-October 1982, the Commission rendered its decision granting the requested authority. Since its purchase, the WPRR has been known as the Feather River Division of the UPRR (Bridges, 1983; Stephen Carr, M.D., written communication, July 19, 1999).

Union Pacific Railroad

Bankruptcy, Control, and Purchases

In 1889, the UPRR consolidated the control of its interests in Utah and Idaho through the organization of the Oregon Short Line & Utah Northern Railway. The UPRR was forced into bankruptcy in 1893, along with its subsidiary railroad companies. The Oregon Short Line emerged from bankruptcy in 1897 as an independent company, and the reorganized UPRR emerged from bankruptcy in 1898. The former Oregon Short Line had controlled much of the traffic that the UPRR depended on, and the new situation was no different. Within two years, however, the new Oregon Short Line was again under the full control of the reorganized UPRR (Media.Utah, 1999).

The UPRR has gained much of its trackage in Utah, and particularly around Great Salt Lake, through purchase, or merger with, other railroad lines. In summary, the Utah Central Railroad was purchased by the UPRR's affiliate Oregon Short Line in 1889, and has been a part of the UPRR ever since. In 1980, the WPRR was bought out by the UPRR; this transaction was finalized in 1982. In 1988, the D&RGW and the SPRR merged, retaining the SPRR name, but subsequently in 1996, the UPRR took over the SPRR.

Today, the UPRR is one of the largest and fastest growing transportation companies in the United States. It is also the oldest railroad company in continuous operation under its original name west of the Mississippi River (Union Pacific Railroad, 1999f).

Problems with Great Salt Lake

As the level of Great Salt Lake began its dramatic rise in 1982 and 1983, an 11-mile portion of the former WPRR main line lying along the south shore of the lake from MP 899, just east of Burmester, to MP 910, just west of Garfield, was threatened by the rising lake level. WPRR records indicate that this stretch of track was originally constructed on a level grade in 1902, but over the years subsidence and settlement occurred with the result that at MP 901.4, the elevation of the top of the rail was 4,207.9 feet (original elevation not given). As the lake level rose, it became apparent that the height of the track must quickly be raised (Union Pacific Railroad, 1986).

Early in 1983, UPRR studied the threat to the trackage along the south shore of Great Salt Lake. A profile was obtained between MP 895 and MP 912, which was the area of concern. The lake elevation was 4,203.6 by May 1983, and was projected to reach an elevation of 4,205. At elevation 4,203, even with low wind velocity, the waves were coming over the track, saturating the ballast, and causing traffic signals to malfunction. The waves penetrated voids in the riprap and washed fill material out from under the track (Union Pacific Railroad, 1986).

In June 1983, the UPRR started the first project to raise the track's elevation and construct a protective berm. The project called for a top-of-rail elevation of 4,210, and a berm (beach) elevation of 4,207. The berm would extend 35 feet northward into the lake. By the time the project was nearing completion in December 1983, it was decided to raise the top-of-rail elevation an additional foot because of continuing projections of higher lake levels. This additional one foot put the top-of-rail elevation at 4,211, and the berm at 4,208 (Union Pacific Railroad, 1986).

Early in 1984, with the lake still rising, UPRR started the next phase of the project which called for raising the track to an elevation of 4,213 and the berm to 4,210.8. This project was completed in April 1984. Even before this phase was completed, it was evident the berm would have to be raised to an elevation of 4,214, and the top-of-rail to 4,216. This project was approved and work started in March 1984. During 1985, the railroad raised the berm one additional foot to an elevation of 4,215 (Union Pacific Railroad, 1986).

With continuation of the wet cycle, the railroad started yet another berm-raising project in May 1986, to take the berm to an elevation of 4,217 and the track to 4,219 feet above sea level. The lake reached its historical high elevation of 4,211.85 in early June 1986 (Union Pacific Railroad, 1986).

On June 7, 1986, the AMAX company's solar-pond dike broke, allowing Great Salt Lake water to flow up against the UPRR's main line as far west as Timpie (MP 886.7) and causing the railroad to extend the berm-raising project an additional 10 miles to the west. This west beach project raised the top-of-rail between Burmester, MP 897.1, and Delle, MP 871.6, to an elevation of 4,218.5 and widened the existing shoulder (Union Pacific Railroad, 1986).

With the completion of the 1986 beach projects, UPRR had spent more than $29 million since 1983 in its efforts to maintain this section of trackage above the rising water of Great Salt Lake (Union Pacific Railroad, 1986).

INTERURBAN RAILROAD DEVELOPMENT (1891-1999)

Bamberger Electric Railroad

The Bamberger Railroad was started in January 1891, when a brawny track gang commenced spiking down the rails northward from Salt Lake City toward Ogden, about 36 miles away. This was the first tangible manifestation of the long-promoted "local" railroad that Mr. Simon Bamberger had been advocating to serve the rich farm communities between the two cities. The major steam railroads (the Oregon Short Line, and the Denver & Rio Grande) aimed only at through traffic, not local business. The name selected for Bamberger's local railroad was "The Great Salt Lake & Hot Springs Railway," and it had as its first goal a popular resort four miles north of Salt Lake City known as "Beck's Hot Springs." As soon as rail was laid down to Beck's Hot Springs, the company started service to that point. The little steam engines that were used provided rapid transportation, and soon the Great Salt Lake & Hot Springs Railway was carrying a sizeable number of people to the resort (Swett, 1974).

In 1892, Simon Bamberger and his associates announced plans to extend the railroad north to a point near Ogden, and from there, proceed east through Weber Canyon to Coalville to provide rail service to the local coal mines. The total length of this line was to be 68 miles, with a 10-mile branch to Ogden. Construction continued, reaching Bountiful in 1892, Centerville in 1894, and Farmington in 1895. Financial difficulties forced a complete reorganization, and on

October 29, 1896, a new company emerged called the "Salt Lake & Ogden Railway," which continued construction toward Ogden (Swett, 1974).

One very important stop on the Salt Lake & Ogden Railroad was the "Lagoon" amusement park. Lagoon, built in 1896, is located in Farmington (figure 1) in what had been a large swamp. The swamp was drained, an artificial lake was created, and the area was made into one of the finest amusement parks in the west (Swett, 1974). Some of the buildings from Bamberger's Lake Park resort, built in 1866 and located on the east shore of Great Salt Lake, west of Farmington, were later moved to Lagoon (Miller, 1980). All Lagoon patrons had to ride the train, and this income became one of the road's most important sources of revenue (Swett, 1974). Lagoon still operates today as Utah's largest amusement park.

The line was aggressively pushed northward from 1902 to 1908, reaching Kaysville in 1903, Layton in 1904, Sunset in 1905, and Ogden in 1908. Steam-operated passenger trains began service from Salt Lake City to Ogden on August 5, 1908, terminating at 31st Street in Ogden. Economic conditions forced the Salt Lake & Ogden Railway to scrap plans to extend the line east to Coalville. By 1910, it was evident to the railroad's directors that unquestioned economy and superior service could be given the public by converting the Salt Lake & Ogden to electric operation. The first day of electric operation was May 28, 1910. Quickly the new interurban became popular, and the steam railroads (D&RGW and UPRR) found themselves faced with a formidable competitor (Swett, 1974).

In 1917, Simon Bamberger gave up the management of his railroad when he was elected as the state's fourth governor (1917-21); his son, Julian Bamberger, succeeded him as the railroad's head. During the same year, the operation was renamed "The Bamberger Electric Railroad," but not everything went smoothly. On May 7, 1918, flames consumed the entire Ogden car house and the adjoining substation. More than half the company's cars were destroyed (Swett, 1974). In 1923, the company built a modern station in Salt Lake City (Hilton and Due, 1960).

When much of the passenger interurban traffic shifted to automobiles, the Bamberger Railroad decided to follow the trend and formed a bus line paralleling the rail route. Bus operation was started on May 15, 1927, directed by a subsidiary company, "The Bamberger Transportation Company" (Swett, 1974).

The Bamberger Railroad was also designed to accommodate freight operations as well as passenger service, and successfully entered the freight business during the late 1920s. In 1933, the company was forced to enter receivership, which continued until 1939. In July of 1939, reorganization took place, and "The Bamberger Electric Railroad" became "The Bamberger Railroad." In World War II, during the first half of the 1940s, the railroad's passenger business tripled, and freight traffic increased eightfold (Swett, 1974).

After a disastrous fire on March 11, 1952, in North Salt Lake, and again in June 1952 at the Ogden substation, the company applied to the Utah Public Service Commission on July 10 to end all rail passenger service, and the Commission agreed. On September 6, 1952, the final interurban trip was made between Salt Lake City and Ogden under electric power, after which Bamberger substituted diesel locomotives for all freight service. At the same time, buses owned by The Bamberger Transportation Company provided all of the company's passenger service. On August 1, 1953, The Bamberger Transportation Company sold its bus operations to the Lake Shore Motor Coach Company (Swett, 1974). In March 1958, the Bamberger Railroad tracks were abandoned, and the corporation was dissolved on September 28, 1959 (Robertson, 1986). The Salt Lake terminal properties were purchased by the D&RGW, and the line north of Hill Field in Ogden was purchased by the UPRR (Hilton and Due, 1960). Since that time, the remaining tracks have been removed and the right-of-way has been assimilated into the region's rapidly growing urban infrastructure.

Salt Lake, Garfield & Western Railroad

Public demand for construction of a railroad to Great Salt Lake was widespread among early Utah residents, and on September 25, 1891, the Saltair Beach Railway was incorporated. On May 31, 1892, it was renamed the "The Salt Lake & Los Angeles Railway" (see figure 2) and construction of the 16.31-mile, "straight-as-an-arrow" route to the west, following the Salt Lake Base Line, was completed to the lake in 1893. The track was eventually extended over the shallow waters of the lake another half mile to the piling-supported Saltair Pavilion, a locally famous amusement park. The railroad began service with three, small, American-type steam locomotives plus an assortment of second-hand coaches. A station was established about a half-mile west of downtown Salt Lake City (just south of the Utah State Fairpark) (Hilton and Due, 1960; Media.Utah, 1999; Swett, 1974).

On October 8, 1916, the Salt Lake & Los Angeles Railway was renamed "The Salt Lake, Garfield & Western Railway" (SLG&W), and new plans were made. The ambitious objectives of the new company were to electrify the entire line (a move discussed as early as 1913), buy new, modern electric trains, purchase the Saltair resort property, and build a branch line to Garfield, Utah. The electric trains commenced operations on August 4, 1919, when the interurban started running on a 15-minute schedule. This popular service attracted large crowds which required the continued use of the steam coaches, but they were hauled solely by the electric locomotives (Swett, 1974).

Freight continued to be hauled by steam engines until 1926, when the SLG&W was completely converted to electric locomotion. The Saltair line hauled freight from several sources. The chief item hauled was, of course, salt, which came from the large Royal Crystal Salt Company plant (predecessor to the Morton Salt Company), whose huge drying vats (solar ponds) lined the railroad right-of-way for several miles. A cement plant, a power plant, and the Saltair resort itself were the other SLG&W patrons. A vital service of the railroad was carrying fresh water to the Saltair Pavilion. While traffic to Saltair provided about two-thirds of the railroad's revenue, it fluctuated inversely with the rise and fall of the lake. The railroad's freight business, although smaller, was much more consistent. In 1930, the railroad abandoned the short Garfield branch, which had never carried much traffic (Hilton and Due 1960; Swett, 1974).

With long-term wear and a depression economy, the railroad's physical plant, cars, and track deteriorated. Thus, Saltair's officials were attracted to conversion to diesel

INCORPORATED UNDER THE LAWS OF UTAH

NUMBER 385

SHARES

CAPITAL STOCK $300,000. 3,000 SHARES $100.00 EACH.

SALT LAKE AND LOS ANGELES RAILWAY COMPANY.

This is to Certify that ______ is the owner of ______ Shares of the Capital Stock of the SALT LAKE AND LOS ANGELES RAILWAY COMPANY Transferable only on the Books of the Company in Person or by Attorney upon the surrender of this Certificate properly endorsed.

WITNESS, the Seal of the said Company and the signatures of the President & Secretary at Salt Lake City Utah this ______ day of ______ A.D. 189_

______ Secretary. ______ President.

Figure 2. *Copy of Salt Lake and Los Angeles Railway Company capital stock certificate (courtesy of Donald M. Hogle, President, Salt Lake, Garfield & Western Railway Co.)*

power. With the purchase of two diesel locomotives and a new passenger car in 1951, the way was clear for abandonment of electric operation which came about on August 16, 1951 (Swett, 1974).

Today, the SLG&W operates freight service between their yard at 1200 West in Salt Lake City, and the rail terminus at 7200 West. Since 1991, the company has experienced substantial growth, and plans carefully controlled growth and addition of industries to their system. SLG&W does not plan for a resumption of passenger service (Donald M. Hogle, SLG&W Railway Co., written communication, September 1999).

SELECTED BIBLIOGRAPHY

Box Elder, 1999, The Central Pacific: www.box-elder.com/gold-spike/abtcprr.html.

Bridges, R.W., 1983, Eighty candles on the final cake: Western Pacific Mileposts, March issue, p. 42-51.

Carr, S.L., 1989, Utah ghost rails: Western Epics, Salt Lake City, Utah, 208 p.

Gwynn, J.W., 1996, History of potash production from the Salduro Salt Marsh (Bonneville Salt Flats), Tooele County: Utah Geological Survey Survey Notes, v. 28, no. 2, 18 p.

Gwynn, J.W., and Sturm, P.A., 1987, Effects of breaching the Southern Pacific Railroad Causeway, Great Salt Lake, Utah - physical and chemical changes, August 1, 1984 - July, 1986: Utah Geological and Mineral Survey Water-Resource Bulletin 25, 25 p.

Hilton, G.W., and Due, J.F., 1960, The electric interurban railways in America: Stanford University Press, Stanford, California, 463 p.

Kneiss, G.H., 1953, Fifty candles for Western Pacific: Western Pacific Mileposts, March issue, p. 4-41.

Linecamp,1999,TranscontinentalRailroad:www.linecamp.couseums/americanwest/western_clubs/transcontinental_railroad/transcontinental_railroad.html.

Media.Utah, 1999, RailroadsinUtah: www.media.utah.edu/medsol/ucme/r/railroad.html.

Miller, D.E., 1980, Great Salt Lake - a historical sketch, *in* Gwynn, J.W., editor, Great Salt Lake - a scientific, historical and economic overview: Utah Geological and Mineral Survey Bulletin 116, p. 1-14.

Neticus, 1999, Golden Spike National Historic Site: www.neticus.com/users/lsessions/Pages/Nps/Golden_Spike.html.

Newby, J.E., 1980, Great Salt Lake railroad crossing, *in* Gwynn, J.W., editor, Great Salt Lake - a scientific, historical and economic overview: Utah Geological and Mineral Survey Bulletin 116, p. 393-400.

Robertson, D.B., 1986, Encyclopedia of western railroad history - the desert states (Arizona, Nevada, New Mexico and Utah): Caldwell, Idaho, The Caxton Printers, Ltd., 318 p.

Southern Pacific Lines, 1992, Southern Pacific Lines welcomes American Railway Engineering Association Committee to the Great Salt Lake: Southern Pacific Lines, 25 p.

Swett, I.L., 1974, Interurbans of Utah: Interurbans, 17309 Alexandra Ave, Cerritos, California 90701, Special Publication 55, 140 p.

Union Pacific Railroad, 1986, Great Salt Lake Beach Project: Union Pacific Railroad unpublished internal report, p. ?

—1999a, Union Pacific Railroad - chronological history: www.uprr.com/uprr/ffh/history/uprr-chr.shtml.

—1999b, Union Pacific Railroad - historical overview: www.uprr.com/uprr/ffh/ history/hist-ovr.shmtl.

—1999c, Union Pacific Railroad - historical overview, building a road: www.Uprr.com/Uprr/ffh/history/ hist-ov2.shtml.

—1999d, Union Pacific Railroad - historical overview, financing: www.Uprr.com/Uprr/ffh/history/hist-ov3.shtml.

—1999e, Union Pacific Railroad - historical overview, construction: www.Uprr.com/Uprr/ffh/history/hist ov4.shtml.

—1999f, Union Pacific Railroad - historical overview, post construction:www.Uprr.com/Uprr/ffh/history/hist-ov5.shtml.

Waddell, K.M., Gwynn, J.W., Burden, Carole, and Wold, S.R., 1992, Salt budget for West Pond, Utah - April 1987 to April 1988: U.S. Geological Survey Water-Resources Investigations Report 91-4177, 29 p.

Wold, S.R., and Waddell, K.M., 1994, Salt budget for West Pond, Utah - April 1987 to June 1989: U.S. Geological Survey Water-Resources Investigations Report 93-4028, 20 p.

THE SOUTHERN PACIFIC RAILROAD TRESTLE - PAST AND PRESENT

by

John S. Cannon and Mary Alice Cannon
Trestlewood Division of Cannon Structures, Inc.

ABSTRACT

Construction of the Lucin Cutoff, a railroad route across Great Salt Lake, began in 1902, and by 1904, trains were crossing the newly constructed structure. The 12-mile-long, open wooden trestle, which connected rock-fill approaches from the eastern and western shorelines, needed major repair or replacement by the early 1950s. Construction of an adjacent rock-fill causeway, 1,500 feet north of the aging trestle, began in 1955. After completion of the causeway in 1959, the old trestle remained idle and largely intact until 1993 when Cannon Structures, Inc. obtained salvage rights to the trestle. Since that time, wood recovered from the old trestle has been brought to shore where it is typically processed or remanufactured for sale. Cannon has designed its remanufacturing process around the physical characteristics of the salvaged wood. Marketing the salvaged wood is challenging, and often requires educating potential customers, and finding niche markets that fit each type of wood from the trestle. Uses for Trestlewood include construction, other industrial applications, architectural and other high-end applications such as: timbers, flooring, decking, poles, millwork, doors, siding, furniture, and cabinets. Cannon's salvage efforts give the wood a new life, and continue the legend of the trestle.

INTRODUCTION

Add wood to the list of products coming out of the Great Salt Lake. No, the lake does not grow wood. It is not home to a great underwater forest, at least not a living one. It is, though, the locale of the historic Lucin Cutoff railroad trestle and its tens of millions of board feet of Douglas Fir timbers, piles, and Redwood decking. Decades after the trestle was replaced by a solid-fill causeway built parallel to it, the wood of the trestle is being reclaimed and reused.

From spike holes in re-sawn timbers to the unique coloring of flooring produced from "pickled" piles, the wood reclaimed from the trestle bears the stamp of the Great Salt Lake. This wood is now as much a product of the Great Salt Lake as it is of the forests from which it was originally cut.

Wood reclaimed from the trestle is being used in a wide variety of ways. Its uses include construction, other industrial applications, architectural and other high-end applications such as timbers, flooring, decking, poles, millwork, doors, siding, furniture, and cabinets.

This article examines the fascinating history of the Great Salt Lake's Lucin Cutoff railroad trestle, from its construction at the beginning of the 1900s, to some of the applications that its wood is being used in today. First, though, an effort should be made to put this article into the proper context by briefly describing the situation that preceded the construction of the trestle.

BEFORE THE TRESTLE

> The simple message was dispatched at 12:47 P.M.: "Done." That unusually brief telegraphic notice on May 10, 1869, set off what may have been the most widespread celebration the United States had witnessed to that time. Fifty tugboats whistled salutes as they paraded along the lakefront in Chicago; New Yorkers shouted with glee at the conclusion of a 100-gun salute; the national capital staged banquets, parades, and a spectacular fireworks display; and prayers and toasts were intermingled throughout the thirty-seven states and territories. . . The event. . . was the completion of the country's first transcontinental railroad. The rails joined at Promontory, a desolate, windswept, and heretofore unremarked spot in what became the state of Utah (Hofsommer, 1986).

The golden spike would not have been driven at Promontory had the Union Pacific Railroad not encountered an obstacle as it worked its way to a junction with the Central Pacific Railroad. That obstacle was the Great Salt Lake. Union Pacific toyed with the possibility of crossing the lake, but ultimately chose a less challenging alternative:

> They [Union Pacific engineers] discussed a little, though perhaps more jocularly than seriously, the feasibility of driving straight across the lake, or at least across its eastern arm. Of course they gave it up. The idea then was almost chimerical. There was neither the genius in finance bold enough to undertake such a stupendous work, nor the traffic to warrant such an expenditure. It may be doubted, too, if there was engineering faith equal to the task. So the line was built up through the hills around the north end of the lake (Davis, 1906).

One might say that Promontory owed its moment of glory on the nation's stage to the Great Salt Lake. Promontory's place in the limelight came to an end when Southern Pacific successfully crossed the Great Salt Lake via the Lucin Cutoff in 1904.

CONSTRUCTION

Shortly after the turn of the century, conditions were ripe for an attempted conquering of the Great Salt Lake. Rail traffic had increased. The Promontory Line "had developed into the chief bottleneck of the whole transcontinental line" (Miller, 1994). Southern Pacific was led by visionaries who believed the railroad could cross the lake. William Hood, Southern Pacific's chief engineer, "had always dreamed of routing the lines straight across the lake" (Dant, 1985). He found an ally in the head of Southern Pacific, Edward H. Harriman, "a man whose financial ability and boldness matched the engineering skill and pluck of Mr. Hood" (Davis, 1906). The abilities and vision of these men were able to take root in a business environment that had become more conducive to bold, long-term investment:

> The times had changed. The day of great and bold enterprises had come. The old era of pinching and often false economy, that let road-bed and rolling-stock run down in order to squeeze out an unjustified dividend, was ended. The condition had been reached where it was only necessary for the engineer to show how the interest on the investment could be made to be told to go ahead (Davis, 1906).

Apparently Mr. Hood was able to show Mr. Harriman that the interest on the investment could be made, because the work on the Lucin Cutoff began. This was an ambitious project. It involved 103 miles of new track between Ogden and Lucin, including an almost 12-mile-long permanent wooden trestle and several more miles of rock and gravel fill through shallow lake brine. Temporary trestles were used to help construct much of the fill (Hofsommer, 1986; Miller, 1994).

Perhaps even more impressive than the magnitude of the Lucin Cutoff was the speed with which it was constructed. Construction of the approaches to the east and west sides of the lake started in February, 1902. The first piles were driven into the bed of the Great Salt Lake in August, 1902. By October, 1903, the permanent trestle was completed. The Lucin Cutoff was put into service on March 8, 1904, just over two years after the start of construction.

Southern Pacific made this aggressive schedule possible by throwing massive amounts of resources at the project. The project was not to be delayed by a shortage of materials, equipment or manpower.

A "perfect forest of piles," (figure 1) not to mention millions and millions of board feet of timbers, were diverted from more typical destinations to the Great Salt Lake for the construction of the permanent and temporary trestles (Davis 1906). Almost unfathomable amounts of rock and gravel (fortunately, available locally) were used to construct the fills.

Large amounts of equipment were needed to handle these materials. Southern Pacific gathered as much equipment as it could. It commissioned the fabrication of 25 huge pile drivers in San Francisco (figure 2). It bought, borrowed or begged over 800 dump rail cars and lined up 80 locomotives with which to pull them. It purchased eight, five-cubic-yard steam shovels.

Of course, large numbers of workers were also needed. At times, over 3,000 men were working on the cutoff, about

Figure 1. *Piles on a flat car. Over 25,000 piles with lengths up to 125 feet were used in the construction of the trestle (courtesy of KUED).*

Figure 2. *Pile drivers working from both sides of the lake met at "Mid-Lake" in 1903 (courtesy of Anne M. Eckman, original source: Southern Pacific Railroad).*

1,000 of them on the trestle (figure 3). Workers were paid between $2.00 (unskilled laborers) and $4.00 or $4.50 (skilled mechanics, carpenters, bridge-workers and engineers) per day (Davis, 1906). Men working on the trestle quickly settled into a very predictable routine:

> A station was erected at each mile-end of the projected road. There, two pile drivers went to work back to back, driving away from each other. Five bents of five piles each, or seventy-five feet in all, was a good day's work. At each station a boardinghouse was built on a platform raised on piles well out of the way of storm-waves. There the men lived until their work was finished. The company furnished supplies and cooks, and the men paid four dollars a week for their board. They worked in10-hour shifts, day and night, Sundays and holidays (Davis, 1906).

Bad weather provided about the only break in this routine:

> There was never a hindrance on the permanent trestle, save when now and then a heavy storm smashed a log-boom and sent the scattered timbers and piles cruising about the lake on their own account, to be slowly and painfully collected again by the launches and towed back to new booms, while the men in the boarding-houses played cards, read, smoked and talked, and drew their pay in idleness (Davis, 1906).

Figure 3. *A few of the thousands of men working on the cutoff at any given time (courtesy of Anne M. Eckman, original source: Southern Pacific Railroad).*

Of course, bad weather was not the only source of challenges. The sheer magnitude of the project, combined with its remote location, created mind-boggling logistical issues:

A constant problem resulted from the need to supply water–in the amount of 500,000 gallons daily–for the locomotives, pile drivers, steam shovels, and boats employed on the project. Much of it was hauled from Deeth, Nevada, to Lakeside, 145 miles (Hofsommer, 1986).

Driving the thousands of piles of the permanent and temporary trestles was not an easy task:

> Water in the permanent trestle section varied from 30 to 34 feet in depth and piles had to be driven many feet into the lake bottom in order to insure a stable structure. When "soft spots" were struck a 100-foot pile could often be driven out of sight without striking solid footing. In such places it was necessary to lash two piles together and drive them into the lake in order to make a solid trestle (Miller, 1994).

In other areas of the lake, just the opposite problem was encountered:

> The progress would be much slower either at this side of the western arm of the lake or at the other side when the 3,200-pound hammers could drive a pile only a few inches. Sometimes, when the pile was already thirty to forty feet deep, it would rebound two or three feet after being struck. Then a hole had to be steam-blasted (Dant, 1985).

All things considered, the construction of the permanent and temporary trestles went remarkably smoothly. The project's greatest challenge, contending with the settling of the fill, was not even trestle-related:

> Here the real work of building the Lucin Cut-Off came in. For a year and nine months that thing [settlement] kept up. . . That first sink began a fight the like of which has not been seen in railroad engineering. It became, apparently, the stupendous task of filling up the bottomless pit. Twenty-five hundred men were at it day and night without cessation. Every hour saw at least one great material-train thrust out on the crazy track to pour its tons of rock and gravel into the greedy, yawning hole (Davis, 1906).

Eventually the 2,500 men prevailed. The result was the completed Lucin Cutoff, "one of the most remarkable and courageous engineering accomplishments of the time" (Hofsommer, 1986, p. 17). Thomas A. Edison affixed his stamp of approval: "The Salt Lake cut-off is certainly a bold piece of engineering and well worth seeing" (Hofsommer, 1986).

SERVICE

The Lucin Cutoff was a resounding success (figures 4-7). It rendered the Promontory Line around the north end of the Great Salt Lake obsolete, and proved worthy of its name:

It is a "cut-off" indeed. . . Forty-three miles in distance are lopped off, heartbreaking grades avoided, curves eliminated, hours of time in transit saved, and untold worry and vexation prevented, at the same time that expenses of operation are reduced more than enough to pay interest on the whole cost twice over (Davis, 1906).

The cutoff was hailed as both an engineering and a financial victory:

> The Lucin Cut-Off is complete, and Mr. Hood, the engineer, is justified for his faith. So, too, is Mr. Harriman, the financier; for in January, 1905, the operating expenses of the new road were sixty-one thousand dollars less than the operating expenses of the old road in January, 1904, although the traffic was greater (Davis, 1906).

Figure 4. *Lucin Cutoff railroad trestle track, view looking east with promontory Point on the left and Fremont Island on the right (courtesy of Utah Railroad Museum).*

Despite dire predictions by pre-construction skeptics, the trestle portion of the cutoff performed admirably. In fact, the six-day period that the trestle was out of commission following a May 4, 1956, fire was the first time in its life that it had been out of service.

Which is not to say that the trestle was maintenance-free. Southern Pacific beefed up the trestle with several thousand additional piles throughout its service life. By the 1950s, the trestle contained over 38,000 piles. Many of the original Douglas Fir cap and deck timbers and Redwood deck planks had been replaced by this time, also.

Figure 5. *Locomotive crossing a single-track portion of the Lucin Cutoff railroad trestle (courtesy of Utah Railroad Museum).*

Figure 6. Double-track section of the Lucin Cutoff railroad trestle (courtesy of Anne M. Eckman, original source: Southern Pacific Railroad).

Figure 7. *Mid-Lake, about 1908. Partially because of the fire danger, the Mid-Lake station was removed long before the trestle was replaced by the solid-fill causeway (courtesy of Anne M. Eckman, original source: Southern Pacific Railroad).*

The beginning of the end of the trestle's life came in the 1950s, when Southern Pacific decided to replace it with a solid fill causeway. Construction of this causeway, which ran parallel to, and 1,500 feet north of the trestle, began in 1955. The new causeway handled its first traffic in July, 1959. Southern Pacific continued to maintain the trestle as a back-up for a few years, but the trestle had seen its last significant traffic by the beginning of the 1960s.

SALVAGE

The trestle may have been near the end of its life as the 1960s started, but its new story had just begun. The 1960s through the early 1990s brought over thirty years of well-deserved rest from the heavy train traffic of its almost 60 years of service. Nature was not so kind. The wind and the waves accompanying the intermittent fierce storms on the Great Salt Lake began to take their toll on the trestle, which was no longer being maintained as it had been when it was the railroad's only means of crossing the lake. Piece by piece, handrail and deck materials broke free and blew or washed into the lake. In not too many years, the trestle was no longer fit as a back-up means of crossing the lake.

The trestle was given a new purpose in the early 1990s. In March of 1993, Cannon Structures, Inc. obtained salvage rights to the trestle from T.C. Taylor Co., Ltd., which had previously acquired these rights from Southern Pacific. Cannon soon thereafter established its Trestlewood Division, through which it has been salvaging, remanufacturing, and marketing the wood from the trestle ever since.

A 48 x 165-foot work barge (figure 8), freight barges, cranes, excavators, a pile driver/extractor attachment, tugboats, other equipment (figures 9-11), workers, and a subcontractor have all played key roles in Cannon's trestle salvage efforts. The work barge was salvaged and restored to serviceable condition after being abandoned in the early 1960s following its use in the building of the causeway.

The basic salvage process is a relatively simple one: workers and equipment on the work barge load salvaged wood onto the barge and/or freight barges. When these barges are full, they are pushed to shore by a tugboat, where they are unloaded and pushed back to the trestle for another load.

Cannon has conducted the salvage of the trestle in two stages. It first salvaged the above-water Douglas Fir timbers and Redwood decking. It then turned to salvaging the Douglas Fir piles (figure 12). A subcontractor, Hein Timber Products, has salvaged most of these piles.

The pile extractor attachment plays an important role in salvaging the piles. This attachment, suspended from the crane by cables, is placed over the butt (large diameter) end of the pole to be pulled. It vibrates the pole until the suction between the tip (small diameter) end of the pole and the sediment on the bottom of the lake is broken. The pole can then be pulled out of the lake by the crane and placed on the work barge for eventual transportation to shore.

Cannon has faced many of the same issues in its salvaging of the trestle as Southern Pacific faced in building it. The most difficult issues always seem to come back to the Great Salt Lake.

Weather is a big issue. The Great Salt Lake features some fierce storms, the fiercest component of which is typically the wind. The only real option in dealing with storms of any magnitude is to get off the lake and wait them out. Even this option does not eliminate the risk of damage–Cannon has had 5 or 6 docked boats sunk by these storms. Spring's south winds are especially tough on the trestle salvage operation. Winters have been surprisingly mild: (1) the Great Salt Lake and Promontory Point have received less snow than the Wasatch Front in general (the dreaded "lake effect" does not seem to impact the lake as much as it does surrounding areas) and (2) the salvage operation is largely protected from winter's north winds by the causeway.

Figure 8. *The 48-foot by 165-foot work barge and equipment working on the north side of the trestle. Standing piles are seen to the left of the barge.*

Figure 9. *View, looking south from the rock-fill causeway, of a tugboat with the Lucin Cutoff trestle beyond, mid-1990s.*

Figure 10. *A track loader helps with the salvage of above-water decking and timbers.*

Figure 11. *Crane, pile extractor attached (mostly hidden at arrow point), and excavators with grapple attachments pulling piles.*

Figure 12. *Partially salvaged trestle, view looking east. Primary elements visible in this photo are piles, and 14-inch by 14-inch pile caps. Fremont Island is seen at right of center, and the Wasatch Range is in the background.*

The Great Salt Lake also exacts its tolls in the form of high maintenance costs. The fluctuating levels of the lake result in significant ongoing dock maintenance costs. The lake's salt is very corrosive on metal equipment, barges and boats. Boat electrical systems have been especially hard hit.

One of the largest challenges posed by the Great Salt

Lake is transportation-related. The lake does not make it easy to get employees to work, or salvaged wood to shore. Getting workers to and from the trestle in the mornings and evenings typically takes thirty minutes to an hour. Pushing barges full of salvaged wood to shore can take several hours. In addition, transporting the poles to shore (and then unloading them on the shore) can be one of the most time-consuming aspects of the pile-salvage operation.

If it were not for the work barge, pile transportation would be very difficult and time consuming. Salvaged poles are typically 50 to 80 feet long; a few poles are longer than 100 feet. These poles are water and salt saturated as they come off the lake. A "typical" 65-foot brine-soaked pole weighs about 4,000 pounds. It would take only 12 such poles to reach the 48,000 pound maximum payload of an over-the-road tractor-trailer. Fortunately, the work barge can typically handle loads of 200 to 250 poles. The barge's million-pound payload has been a real boon to Cannon's pile-salvage efforts.

By mid-2000, over seven years into its salvage efforts, Cannon has brought essentially all of the Douglas Fir timbers and Redwood decking to shore, and about three-quarters of the Douglas Fir piles. In other words, it has taken Cannon over seven years to take down less material than Southern Pacific put up in one year. Of course, Cannon's two-to six-person salvage crew (less than 1 percent of the number of men Southern Pacific typically had working on trestle construction) has something to do with this disparity. The trestle salvage schedule has also been

impacted by: (1) a desire to handle the salvaged wood as carefully as possible in order to preserve its value and (2) remanufacturing and marketing issues.

REMANUFACTURING

Once the wood salvaged from the trestle reaches the shore, it typically requires some processing, or remanufacturing, before it can be sold. When Cannon Structures started salvaging trestle materials, it did not picture itself doing much remanufacturing. After all, it had over 30 million board feet (BF) of wood to deal with (approximately 10 million BF of Douglas Fir timbers; 2 million BF of Redwood decking; and 20 million BF of Douglas Fir piles).

Why not focus its efforts on salvaging the wood and loading it onto trucks and railcars for immediate shipment to companies who were better equipped to manufacture it into products? As it turns out, the manufacturers who were large enough to handle the volume of trestle materials being salvaged were only used to dealing with new logs and lumber. They were looking for standard stock that they could run through standard processes and turn into standard products. Trestlewood is not standard stock. Cannon discovered that it needed to take an active role in identifying and creating markets and producing products that were ideally suited to the unique features of Trestlewood.

The wood from the trestle has presented several challenges to Cannon's remanufacturing efforts. Probably the greatest challenge is removing imbedded metal (figure 13). Virtually all of the salvaged timbers contain nails, bolts and/or spikes. Cannon uses a variety of approaches in dealing with this metal. In some cases, it can produce metal-free

Figure 13. *As pile caps come from the trestle, they look nearly impossible to demetal.*

timbers by cutting off ends or ripping timbers into one metal-free piece and one metal-infested piece. Where markets do not require timbers to be free of metal, Cannon often just cuts any protruding metal flush (or leaves it protruding for more rustic applications). In many cases, Cannon entirely removes metal from timbers. The most noteworthy example of this involves the trestle's 14- by 14-inch pile caps. As the square-shaped caps come off the trestle, they have numerous 3/4- to 1-inch diameter bolts/spikes protruding from all four faces at a variety of angles. They seemed impossible to demetal until Emil Hein of Hein Timber Products designed and built a special hydraulic extractor (figure 14) that successfully removes all of the metal from a high percentage of the pile caps. The result is a metal-free, albeit holey, timber that has become very popular in "distressed" timber applications.

Figure 14. *The Hein-designed hydraulic mill removes the pile cap metal.*

In general, metal and other woodworking complications (checking, imbedded grit, etc.) make reclaimed wood more difficult to process than new wood. The Great Salt Lake environment introduces additional characteristics that are unique even for reclaimed wood. This is especially true of the Douglas Fir piles that have been submerged in the Great Salt Lake. These poles are water and salt saturated as they come off the lake. Tests show that the poles typically consist of more than 20 percent salt by weight, all the way to the core.

The salt and other minerals in the piles result in unique, colorful flooring and timber products (see "Marketing" and "Applications" sections). They also create remanufacturing hurdles. As might be expected, the salt corrodes saws and other wood-processing equipment. As previously mentioned, the weight of the piles creates handling and transit challenges. Water and salt-saturated lumber cut from the piles is more difficult to dry than "normal" Douglas Fir lumber (Cannon's experience has been that it takes about twice as long to kiln dry 4/4 (1 in.) lumber cut from the piles as it does to kiln dry typical green 4/4 Douglas Fir lumber.) Cutting into a brine-soaked pile unleashes some very strong odors that tend to dissipate as the wood dries, but that make processing wet poles a bit unpleasant.

One of the most important advantages that Cannon has had in its Trestlewood remanufacturing efforts is its complete lack of experience in processing new lumber. Cannon has not faced the temptation of trying to apply new-lumber manufacturing techniques to its Trestlewood remanufacturing processes. It decided not to put reclaimed wood through conventional, expensive, high-production machinery. Instead, Cannon designed its Trestlewood remanufacturing processes and machinery from the ground up, based upon the physical characteristics of the wood from the trestle.

Cannon has increased its involvement in the Trestlewood remanufacturing process incrementally. It started by bringing on line the circular and cut-off saws necessary to produce 3-foot crib blocks for mines. It has since upgraded its sawmill capabilities and invested in such improvements as the pile cap metal-removal mill discussed previously; a rough planer; a 30,000 BF dehumidification kiln; and storage facilities. Since early in the project, Cannon has done most of its rough, industrial-grade processing at Promontory Point and most of its finish, higher-end processing at its headquarters in Blackfoot, Idaho. Cannon has also supplemented its own remanufacturing efforts with the processing capabilities of a handful of small, but talented, manufacturing partners.

Cannon's incremental improvements to its own remanufacturing capabilities, and its selective use of remanufacturing subcontractors, has significantly increased the breadth and depth of its product lines. In the early years of the trestle project, a very high percentage of Cannon's revenues came from crib blocks and rough timber stock. By mid-2000, revenues were spread across many more product lines, with Trestlewood's focus moving increasingly in the direction of higher-end, further-processed products like flooring and architectural timbers.

MARKETING

Once the wood from the trestle is brought to shore and is remanufactured into a useable product, one key task, perhaps the most important one of all, remains: marketing. Actually, this statement is oversimplified. It implies that marketing follows neatly on the heels of salvage and remanufacturing. In reality, the marketing of the wood from the trestle has driven the salvage and remanufacturing of the wood, at least as much as it has been driven by them. Cannon's salvage and remanufacturing efforts have been heavily influenced by the company's marketing successes and failures.

Marketing the wood from the trestle can be both simple and difficult at the same time. On one hand, the wood from the trestle is beautiful, unique, full of history and character and 100 percent reclaimed. Much of it is very tight-grained. Trestlewood's uniqueness and other characteristics, while tremendously advantageous, make it different from standard products on the market. As a result, Cannon takes a very active role in identifying, and sometimes even creating, niche markets that fit each element coming off the trestle and then in educating the customers in each market sector about Trestlewood. New Trestlewood products benefit to a certain extent from previous marketing efforts associated with other Trestlewood products, but each one requires their own specific market development efforts.

Much of the product-development process boils down to getting customers comfortable with a product line that evolves over time with the salvage effort. It means identifying markets which are likely to value the characteristics of a particular element, and then getting the element in front of the right people in the market.

Consider the history of Trestlewood's custom-cut timber product lines. Cannon's first significant custom timber sales in 1994 were of timbers cut from the trestle's 8- by 16-inch by 30-foot Douglas Fir stringers. Over time, these timbers became very popular.

When the stringers were depleted, the primary custom timber source became the de-metalled 14- by 14-inch by 18-foot Douglas Fir pile caps. The pile caps were from the same project and of the same species as the stringers, but they were a very different product. Most importantly, they were full of holes (and large holes, at that). Initially, there was a large drop-off from stringer timber sales to pile cap timber sales. Given time and a handful of impressive projects which featured the pile caps, however, the situation reversed itself. Eventually, the pile cap timbers became very popular in their own right. The holes in the pile cap timbers became a selling point, especially among customers who wanted a rustic, "distressed timber" look.

By the end of 1998, pile cap quantities were also dwindling, and Cannon's new primary custom timber source became the Douglas Fir piles. Timbers cut from the piles are as different from the pile cap timbers as the pile cap timbers were from the timbers cut from the stringers. The salt and other minerals in the piles create their own unique challenges and opportunities. Sales of timbers cut from the piles (which Cannon calls its "Trestlewood II" timbers) started slowly while the pile cap timbers were still available. By mid-2000, Trestlewood II timber sales were picking up steam and are well on their way to becoming Trestlewood's best custom timber source yet.

There are no better examples of the Great Salt Lake's role in creating unique wood products than the Trestlewood II products produced from the Douglas Fir piles that have been submerged in the lake for close to a century. When these poles were first driven into the lake bed, they were long and straight, but otherwise rather ordinary. By the end of their stay in the lake, however, the poles were not ordinary. Flooring, timbers and other products cut from the "pickled" piles boast unique, stunning coloring that sets them apart from other wood products.

Finally, it should be noted that Trestlewood distributors play an important role in Cannon's marketing efforts. The quantity of materials coming off the trestle make it imperative that Cannon involve others in selling the trestle's wood. Over time, Cannon has assembled a solid, core group of distribution partners who help get Trestlewood into the hands of its ultimate users. Intermountain Wood Products and its Antiquus Wood Products division have been an especially valuable distribution partner.

APPLICATIONS

Nearly a hundred years after the wood for the Lucin Cutoff railroad trestle was driven into the bed of the Great Salt Lake, some of it is popping up in very prominent places. Wood reclaimed from the trestle is being used in a wide variety of ways. Its uses include construction, other industrial applications, architectural and other high-end applications such as: timbers, flooring, decking, poles, millwork, doors, siding, furniture, and cabinets. The list of Trestlewood applications is long, and growing longer.

Wood from the trestle has been shipped as far away as the east coast of the United States, the Philippines and Japan. Such shipments have been the exception rather than the rule, however. The western United States accounts for a very high percentage of Trestlewood sales. Utah has consistently anchored Trestlewood's western United States market, which seems appropriate given the important role the trestle has played in Utah's past. Various public buildings in Utah incorporate Trestlewood products and allow people to view its unique characteristics.

As of mid-2000, a list of prominent public Utah projects using Trestlewood includes:

1. Historic Cove Fort (Cove Fort)
2. Ogden Nature Center (Ogden) (figure 15)
3. This Is the Place Heritage Park (Salt Lake City) (figure 16)
4. Desert Pearl Inn (Springdale–mouth of Zions National Park)
5. Antelope Island Visitors Center (west of Syracuse)
6. Museum of Church History and Art (Salt Lake City)
7. Union Station (Ogden)
8. Marriott Aquacade (Park City) (figure 17)
9. Department of Natural Resources building (Salt Lake City)
10. Matheson Wetlands Preserve (near Moab) (figure 18)
11. American Stores headquarters (Salt Lake City) (figure 19)
12. The Canyons Ski Resort (Park City)
13. Park City Visitors Center (Park City)
14. Rivers Restaurant (Salt Lake City)
15. Soldier Hollow Day Lodge (Midway) (figure 20)
16. Maddox Ranch House pole building (Brigham City, under construction)

Figure 15. *Redwood siding from Lucin Cutoff trestle deck - Ogden nature Center, Ogden, Utah.*

Figure 16. *Timbers resawn from 14-inch by 14-inch Lucin Cutoff trestle pile caps - This is the Place State Park, Salt lake City, Utah.*

Figure 17. *Marriott Aquacade constructed from 14-inch by 14-inch Lucin Cutoff trestle pile caps - Park City, Utah.*

CONCLUSION

The wood for the Lucin Cutoff trestle saw its last significant use by the railroad shortly after the trestle was replaced by a solid fill causeway in 1959. That could have been the end of the story for the trestle wood; fortunately, it was not. The salvage, remanufacturing and marketing of the trestle have given the wood new life. Now, wood from the trestle is showing up in a lot of prominent places, where it exposes an increasing number of people to the trestle's story. In a way, this wood makes the trestle more accessible to the general public than it has ever been.

Figure 18. *Rough-planed redwood from Lucin Cutoff trestle deck - Matheson Wetlands Preserve, Moab, Utah.*

Figure 19. *Colorful flooring from Lucin Cutoff trestle piles - American Stores headquarters, Salt Lake City, Utah.*

Figure 20. *Soldier Hollow Day Lodge, Midway, Utah (courtesy of HFS Architects and Hogan Construction).*

REFERENCES

Dant, D.R., 1985, Bridge: A Railroading Community on the Great Salt Lake: Utah Historical Quarterly 53 (No. 1), Winter Issue, p. 54-73.

Davis, O.K., 1906, The Lucin Cut-off, a remarkable feat of engineering across the Great Salt Lake on embankment and trestle, Century Magazine, January Issue, p. 459-468.

Hofsommer, D.L., 1986, The Southern Pacific, 1901-1985: College Station - Texas A&M University, 373 p. (Contains foreword by Overton, R.C.)

Miller, D.E., 1994, Great Salt Lake past and present: Publishers Press, 5th edition, 48 p. (updated by Eckmann, A.M.).

We wish to express sincere thanks to John George for his beautiful photos of the Great Salt Lake. He is truly a professional. You will, without a doubt, agree when you view his photos in the first 9 pages of this color section. John started photography as a hobby 30 years ago. His daytime job is program manager at Alliant Techsystems in Magna, Utah where he works with space and strategic rocket motors but, whenever possible, he is out doing what he loves best - hiking to remote areas of the country, patiently waiting for that precise time to click his camera for unique and breathtaking shots. He realizes Utah is a photographer's paradise, but he has travelled throughout the West and mid-western states photographing nature's scenes, wildlife, and wild flowers. His collection - over 125,000 photos. The Great Salt Lake has drawn John to its shores and islands approximately 50 times. These trips have resulted in over 2000 photos of the lake.

Accomplishments:
John's photos have been published and displayed by the the Audubon Society, National Wildlife, Nature Conservancy, National Geographic, Utah Travel Council, Sierra Club, and the Salt Lake International Airport.

Magnificent sunset and reflections on the Great Salt Lake. Stansbury Mountains in distance.
Photo by John George.

Lichens and limestone on Gunnison Island.
Photo by John George.

Ice-flow breakup and lake fog. Wasatch Mountain Range in distance.
Photo by John George.

Brilliant pickleweed along the Great Salt Lake shoreline.
Photo by John George.

Indian rice grass, near Lakeside at west end of causeway.
Photo by John George.

Rabbitbrush in bloom along the Great Salt Lake shoreline.
Photo by John George.

Reeds framing Antelope Island at sunset.
Photo by John George.

Slough and inlet reflections on the Great Salt Lake. Snowy peaks of Antelope Island in distance.
Photo by John George.

Wind-battered foam along shoreline of the Great Salt Lake. Fremont Island and Antelope Island in distance.
Photo by John George.

Summer grasses on Antelope Island. Whiterock Bay in distance.
Photo by John George.

Great Salt Lake shoreline from Stansbury Island ridge.
Photo by John George.

Salt patterns in the Great Salt Lake desert. Hogup Mountains in the distance.
Photo by John George.

Storm over the Wasatch. Fremont Island and saline mudflats in foreground.
Photo by John George.

Evening light on Antelope Island.
Photo by John George.

Dried and buckling geometrical salt patterns, northwestern shores of the Great Salt Lake desert.
Photo by John George.

Mud flats and boulders along the shore of Great Salt Lake.
Photo by John George.

Rock outcroppings on Fremont Island.
Photo by John George.

Artist Doris Tischler's rendition of a Pleistocene faunal scene along the northeastern shoreline of Lake Bonneville, with a portion of the Wasatch Mountains in the background. The original mural is in the Utah Museum of Natural History. (see Miller's paper)

From a small rock shelter above Parrish Creek, painted Fremont anthropomorphs holding hands, shields, or heads, and two mountain sheep filling out the lines of figures. (see Sasson's paper)

NASA Photo STS062-100-063
Panoramic east-northeast-looking, high-oblique photograph of Colorado, Nevada, Utah, and Wyoming
March 1994

NASA Photo STS51B-35-0096
Northeast-looking, low-oblique photograph of Great Salt Lake, Great Salt Lake Desert, Utah
May 1985

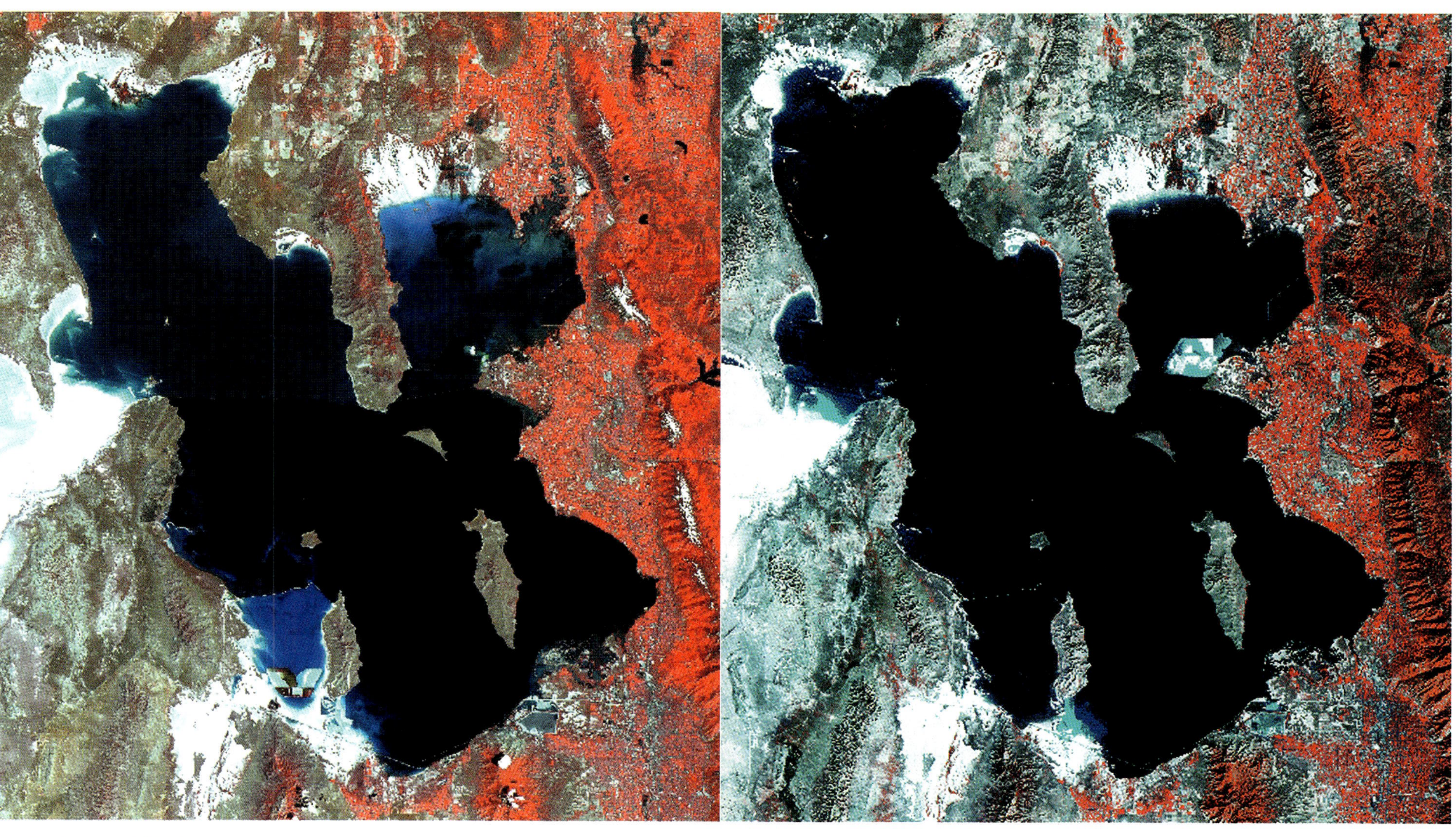

Zastrow and Ridd

Figure 1. *A 1984, false-color composite of Great Salt Lake with Landsat 5 TM bands, 4, 3, and 2 assigned to red, green, and blue color guns, respectively. The east half of the image was collected on June 25, 1984, and the west half on July 2, 1984.*

Zastrow and Ridd

Figure 5. *Near-historic high-water image of the lake, collected on September 22, 1987. On this date the lake water stood at an elevation of about 4,209 feet, near its historic high.*

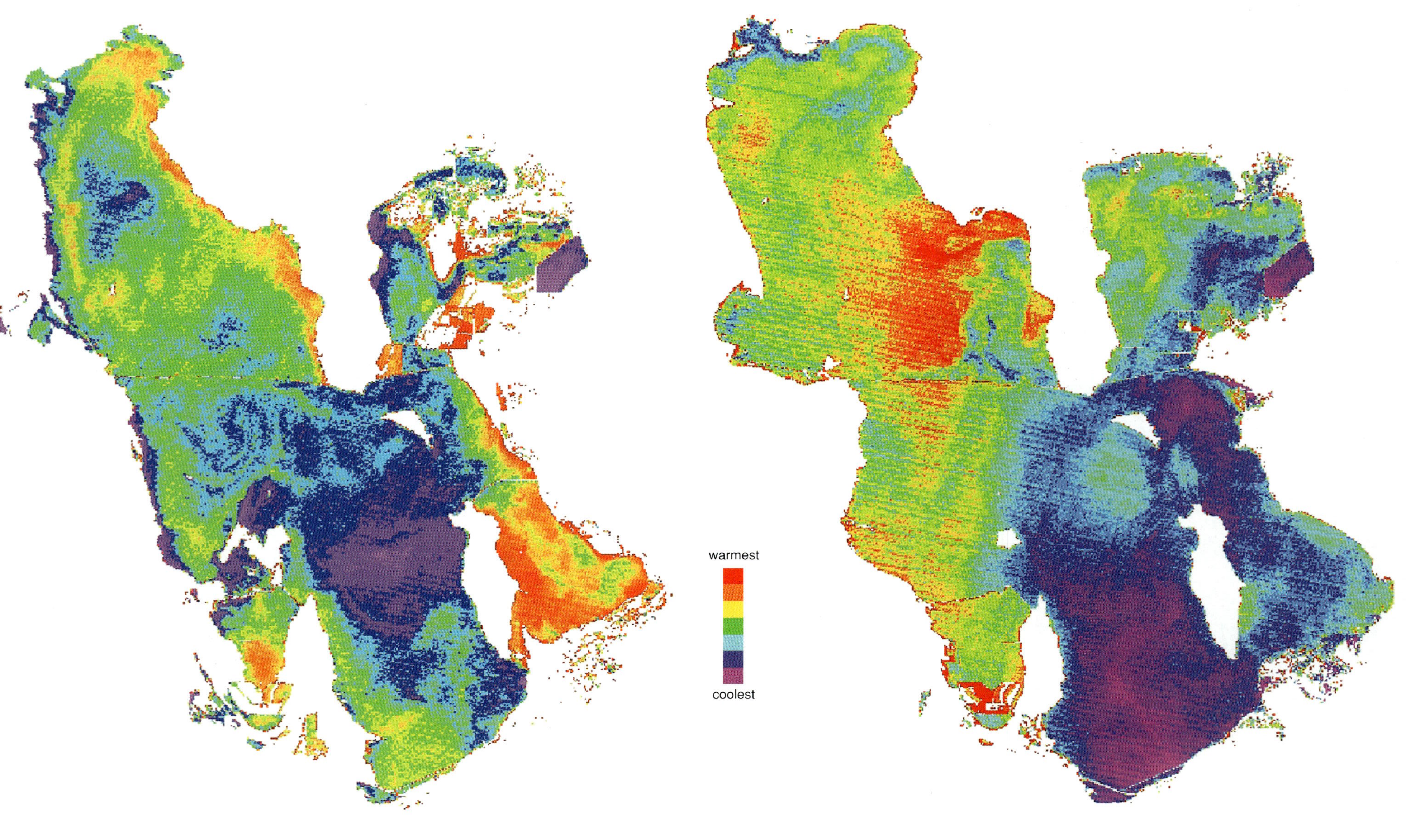

Zastrow and Ridd
Figure 7A. *Thermal image of Great Salt Lake, 1996, at a south-arm elevation of about 4,200 feet. The color range is from red (warmest) to violet (coolest).*

Zastrow and Ridd
Figure 7B. *Thermal image of Great Salt Lake, 1984, at a south-arm elevation of about 4,209 feet. The color range is from red (warmest) to violet (coolest).*

Zastrow and Ridd

Figure 8A. *Bear River Bay, 1996, at a south-arm elevation of about 4,201 feet. Note the exposure of dikes and some pond areas in the Bear River Migratory Bird Refuge system (center of image). Note also the large bar extending southward from the refuge dikes, and in the areas surrounding the eastern portion of IMC Kalium's evaporation ponds (bottom-center of image).*

Zastrow and Ridd

Figure 8B. *Bear River Bay, 1984, at a south-arm elevation of about 4,209 feet. Note the complete inundation of the Bear River Migratory Bird Refuge dike system (center of image), water backing into the mouth of Bear River (upper-right quadrant), and covering much of IMC Kalium's evaporation pond system (bottom left of center of image). The dike around Willard Bay Reservoir remains essentially intact (center of bottom-right quadrant).*

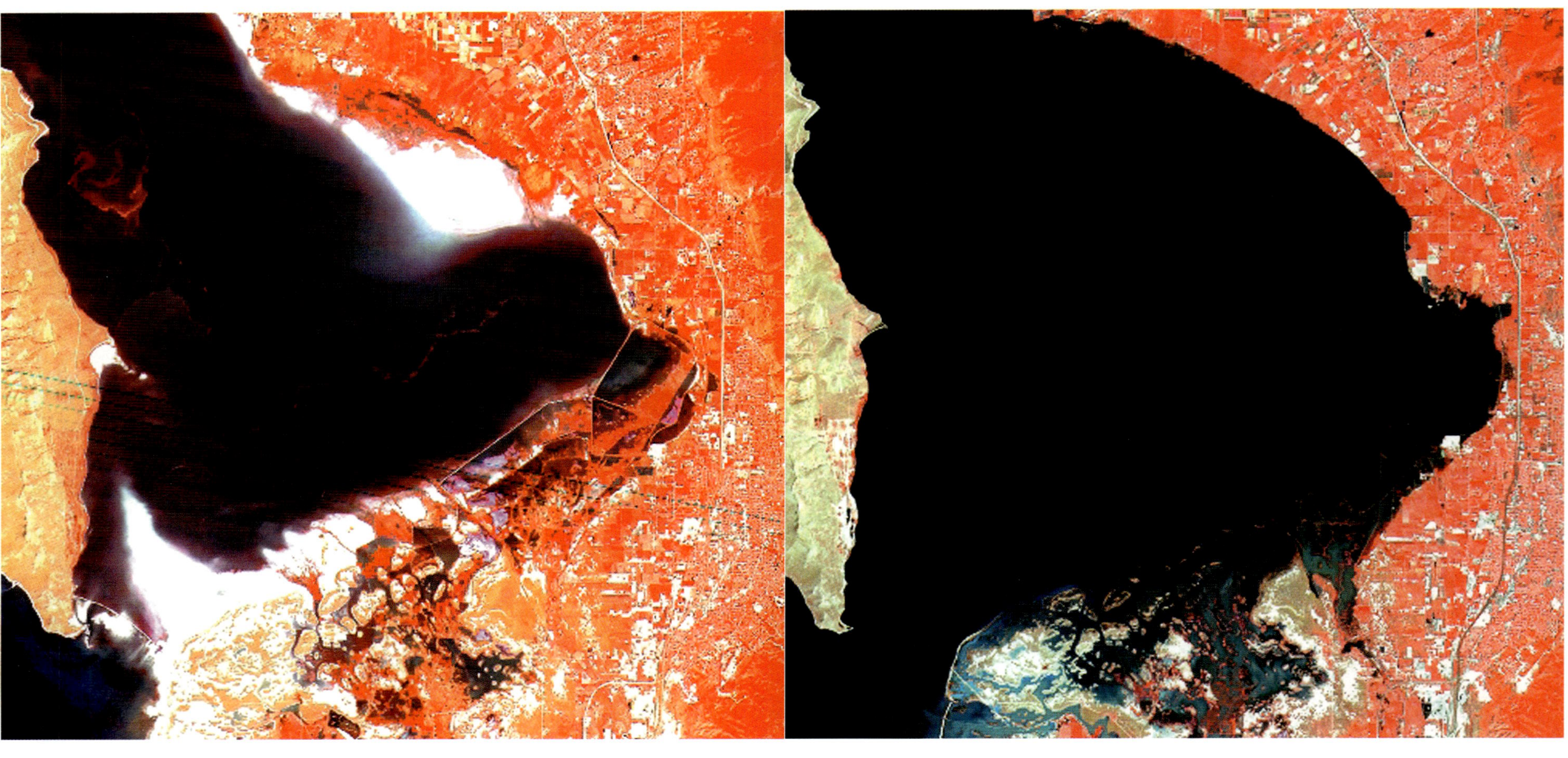

Zastrow and Ridd

Figure 10A. *Farmington Bay, 1996, with a south-arm elevation of about 4,200 feet, the historic norm. Note the Farmington Bay Bird Refuge dike system (top of bottom-right quadrant), a large bar extending into the bay from the northeast (top-center of image), and in the southwest toward Antelope Island (below center of bottom-left quadrant).*

Zastrow and Ridd

Figure 10B. *Farmington Bay, 1984, with a south-arm elevation of about 4,209, more than two feet below the historic high-water line. Note the inundation of the two bars identified in figure 10A, as well as the complete inundation of the bird refuge dikes (top of bottom-right quadrant). Note also the water backing up across the wetlands toward the Salt Lake International Airport (bottom-center of image).*

1980S FLOODING

- **THE NATIONAL WEATHER SERVICE, WEATHER ACROSS UTAH IN THE 1980S, AND ITS EFFECT ON GREAT SALT LAKE** - *by William Alder*

- **PROBLEMS AND MANAGEMENT ALTERNATIVES RELATED TO THE SELECTION AND CONSTRUCTION OF THE WEST DESERT PUMPING PROJECT** - *by Lloyd. H. Austin*

- **SATELLITE IMAGING AND ANALYSIS OF GREAT SALT LAKE** - *by Lori Zastrow and Merrill Ridd*

THE NATIONAL WEATHER SERVICE, WEATHER ACROSS UTAH IN THE 1980S, AND ITS EFFECT ON GREAT SALT LAKE

by
William Alder
National Weather Service, Salt Lake City
1/03/2000

ABSTRACT

The mission of the National Weather Service consists of providing weather and hydrological forecasts, warnings of hazardous and/or severe weather, and gathering climatic information for the United States, its territories, and adjacent coastal waters. Protection of life and property and the enhancement of the national economy are the goals behind the mission statement. The weather patterns of the 1980s severely tested the National Weather Service's ability to meet its stated goals. From excess precipitation (with associated Great Salt Lake flooding) to extreme drought, the weather in the 1980s had a significant impact on life, property, and commerce in Utah. Changes in the 1990s brought on by the National Weather Service modernization efforts have greatly improved forecasters' ability to serve the public, both in terms of improved routine forecasts and more accurate and timely warnings of hazardous weather events.

NATIONAL WEATHER SERVICE (NWS)

NWS Mission

The National Weather Service (NWS) provides weather, hydrologic, and climate information for the United States, its territories, adjacent waters and ocean areas. The goal of NWS is to protect life and property and to enhance the national economy. NWS data and products form a national information database and infrastructure which can be used by the general public, other governmental agencies, the private sector, and the global community. NWS offices are staffed with professionals with degrees from a variety of disciplines with most in meteorology or hydrology.

Salt Lake City Forecast Office Mission

The Salt Lake City Forecast Office of the NWS issues warnings for severe local storms, floods, flash floods, and winter storms along with daily public, aviation, and fire weather forecasts. These services cover 27 zones in Utah, two zones in extreme southeast Idaho, and one zone in southwest Wyoming. The Colorado Basin River Forecast Center issues water-supply forecasts, peak-flow forecasts, flash-flood guidance, and daily-stage forecasts for the eastern Great Basin and upper/lower Colorado Basin. These forecasts cover Utah, Arizona, western Colorado, southern Nevada, extreme western New Mexico, and southwest Wyoming.

NWS Modernization Programs

Doppler Weather Radar

During the 1990s, the NWS launched a major program to modernize its technology. The backbone of the modernization was the Doppler weather radar that was installed across the country. Here in Utah, the Salt Lake City Doppler radar on Promontory Point (west of Brigham City) was installed in the summer of 1994 (figure 1). The Cedar City Doppler radar on Blowhard Mountain (east of Cedar City) came on-line late in the summer of 1996.

Advanced Weather Interactive Processing System (AWIPS)

Another major part of the modernization is the new high-speed workstation and communication network called the Advanced Weather Interactive Processing System (AWIPS). This technology is the center of forecast operations. The AWIPS system receives, processes, and displays huge amounts of weather data from: the network of Doppler radars, various forms of satellite imagery, data from the network of Automated Surface Observing Systems (ASOS), and output from numerical models.

Utah Mesonet

Another recent advancement is the Utah Mesonet. This is a cooperative project between researchers at the University of Utah and forecasters at the Salt Lake City Forecast Office. The goal of this project is to provide access to real-time weather observations in Utah and nearby states. The Utah Mesonet relies upon weather observing networks that are managed by federal, state, and local agencies as well as private firms. Additional stations have been installed at key locations such as near and over the Great Salt Lake and various mountain ski areas. Real-time weather observations of temperature, relative humidity, wind speed and direction, precipitation, and other parameters are available at more than 300 stations.

The Utah Mesonet is used by the NWS to monitor weather conditions around the region. The Mesonet is also

Figure 1. *Salt Lake City Doppler radar on Promontory Point (west of Brigham City), installed in the summer of 1994 (photo courtesy of William Alder, National Weather Service).*

used extensively by researchers to understand severe weather events such as winter snow storms and damaging winds.

UTAH'S VARIABLE WEATHER

Utah has variable weather from flash floods in the summer monsoon season to major snow storms during winter months. Snowmelt flooding can be a problem in the spring and early summer due to a heavy mountain snowpack and unfavorable weather conditions. Tornadoes are rare; when they do occur, they are short-lived and only do minimal damage. Waterspouts have been observed over Utah Lake, the Great Salt Lake, and Bear Lake. Hurricane force canyon (downslope) winds occasionally occur in local areas along the west slopes of the Wasatch Mountains of northern Utah.

The Dreaded Lake Effect (DLE)

The "Dreaded Lake Effect" or DLE, is a phenomenon that enhances precipitation along the southern and eastern shores of the Great Salt Lake. DLE is most often observed during the fall and spring. The high salt content of the lake water (which averages 15% on the south arm and 25% on the north arm of the lake, compared to about 3% for the world's oceans) prevents most of the lake's surface from freezing during the winter months. The open water acts as a relative "hot spot" and lowers the pressure over the lake compared to the surrounding cold ground. This causes air to converge over the lake and produces heavy snow bands. What typically triggers these "lake effect" storms is when a cold airmass flows over the warmer waters of the lake and a 20 degree F temperature difference exists (i.e., the water is 50°F and the air temperature 30°F). Salt Lake, Tooele, and south Davis Counties are most prone to snow squalls off the lake.

On October 18, 1984 a classic "lake effect" snowstorm dropped a record 18-24 inches of snow on the east benches of the Salt Lake Valley, causing over a million dollars of damage to utility lines, homes, businesses, and cars.

1980s Weather Events

El Niño and La Niña Weather Patterns

The weather events in the 1980s were influenced to a degree by both El Niño and La Niña weather patterns. El Niño and El Niño are extreme phases of naturally occurring climate cycles referred to as El Niño/Southern Oscillation. Both terms refer to large-scale changes in sea-surface temperature across the eastern tropical Pacific. Usually, sea-surface readings off the west coast of South America range from the 60s to 70s°F, while they exceed 80°F in the "warm pool" located in the central and western Pacific. This warm pool expands eastward to cover the tropics during El Niño. During La Niña, the easterly trade winds strengthen and cold upwelling along the equator and the west coast of South America intensifies. Sea-surface temperatures off the coast of Peru fall as much as 7°F below normal.

El Niño and La Niña result from interaction between the surface of the ocean and the atmosphere in the tropical Pacific. Changes in sea-surface temperature impact the atmosphere and climate patterns around the globe. In turn, changes in the atmosphere impact the ocean temperature and currents. The system oscillates between warm (El Niño) to neutral and cold (La Niña) conditions on an average roughly every 3-4 years.

For the southwestern United States, El Niño conditions mean wetter than normal conditions from late summer through winter. Opposite conditions occur during La Niña events, as weather conditions are generally warmer and drier than normal. Southern Utah often follows this pattern, but in northern Utah there is no definite signature.

In the 1980s there were two occurrences of El Niño (1982-83 and 1986-87) and two of La Niña (1984-85 and 1988-89). The El Niño in 1982-83 was one of the strongest ever on record with the warmest-ever water temperatures along the equator.

The 1980s Volatile Weather Period

Weather in the 1980s over Utah was a volatile period with unprecedented wet conditions especially in the northern portions. Residents of Utah wondered if we were experiencing a long-term climate change from desert to a wetter regime.

The wet cycle of the 1980s commenced in the fall of 1982. A major fall storm (September 24th-28th) produced phenomenal amounts of rainfall in northern Utah with the heaviest precipitation in the Salt Lake Valley. Widespread severe flooding occurred along both Big and Little Cottonwood Creeks and the Jordan River with damage estimates in excess of $15 million. The rains were attributed to considerable tropical moisture that moved into Utah from the remnants of Hurricane Olivia and the dynamics supplied from a vigorous cold front. It was estimated that flows in the Cottonwood Creeks ranged from 1300-1600 cfs in stream channels with capacities of about 500 cfs. Rainfall amounts for this period in the southeast portion of the Salt Lake Valley and adjacent mountains were 5-6 inches. The Salt Lake Airport recorded a record 7.04 inches of rain for the month which made September 1982 the wettest month ever.

1983 Snowpack and Flooding

The most extensive snowmelt flooding in the history of Utah occurred during the spring months of 1983. The resulting widespread flood and mudslide damage along the Wasatch Front impacted major population areas of the state. Catastrophic floods and mudslides occurred from April through June in many areas of the state. Damage estimates from the resultant disaster were approximately $300 million.

In 1983, the westerlies remained strong with frequent Pacific storms moving into the Great Basin the latter portion of the winter through spring. This produced above normal amounts of precipitation with temperatures remaining below normal. The maximum snowpack in Utah's mountains normally occurs the first week in April, but in the spring of 1983 the snowpack reached its peak in late May. The snowpack in some areas of the state was near record to above record levels with 40-60 inches of water in the snow around 8,000 feet and above.

The worst-case scenario was realized when the heavy snowpack started rapidly melting when temperatures dramatically warmed into the upper 80's and 90's. Moderate to sometimes heavy rains fell during this period, aggravating the runoff from the snowpack.

At the end of May, flooding erupted from the record flows of City Creek (360 cfs) in Memory Grove when the culvert that transports water under South Temple to the Jordan River was clogged by debris. The water had to be diverted through downtown Salt Lake City along State Street (figure 2) south to 13th South. From there the stream was routed westward down 13th South to the Jordan River.

Record flows were measured in five of the six creeks in the Salt Lake Valley. City Creek carried over twice the peak snowmelt flow ever recorded. In addition, Chalk Creek near Coalville, the Sevier River at Hatch, and both Ashley and Dry Creeks near Vernal, registered record flows. Numerous other creeks and rivers in the state were near record or well above record levels. These record high flows allowed the rivers to scour and accumulate debris from the banks. This exacerbated the already severe flood situation.

Landslides and Dam Failures

Saturated soils and a very high water table resulted in additional problems compounding the critical widespread flood situation. In mid-April, the first major land failure

Figure 2. *View looking south of flood waters from City Creek being diverted down State Street through downtown Salt Lake City to 13th South. The water was then directed westward down 13th South to the Jordan River (photo by Steve Thiriot).*

Figure 3. *View looking southwest of Thistle landslide (light-colored mass near center right of photo) blocking the Spanish Fork River in mid-April, 1983, and creating Lake Thistle (photo courtesy of R.L. Schuster, U.S. Geological Survey). This landslide began moving in the spring of 1983 in response to groundwater buildup from heavy rains the previous September and the melting of deep snowpack for the winter of 1982-83. Within a few weeks the landslide dammed the Spanish Fork River, obliterating U.S. Highway 6 and the main line of the Denver and Rio Grande Western Railroad. The town of Thistle was inundated under the flood waters rising behind the landslide dam. Total costs (direct and indirect) exceeded $400 million, the most costly single landslide event in U.S. history.*

Figure 4. *Home being inundated by waters of Lake Thistle (photo by B.N. Kaliser).*

Figure 5. View looking east of homes being destroyed by a major mudflow in Rudd Canyon that occurred on Memorial Day 1983 in the Farmington area of Davis County (photo by Earl Brabb, U.S. Geological Survey).

occurred in Spanish Fork Canyon near Thistle. Lake Thistle was created when this massive landslide dammed the creek (figures 3 and 4). This landslide also closed a major Utah highway and the east-west line of the Denver and Rio Grande Railroad. Many more slides occurred in May and June with the next major mudflow in Rudd Canyon (figure 5). This occurred on Memorial Day in the Farmington area of Davis County.

The persistent rains and high water levels combined with a heavy runoff to create concern about many of the earthen dams in the state. On June 23rd, the Delta-Melville-Abraham-Deseret (DMAD) Dam in Delta breached (figure 6 and 7) and by evening water flooded the town of Deseret and thousands of acres of valuable farmland.

By the end of June, 21 Utah counties had been declared federal disaster areas.

Figures 6 and 7. View looking northeast of the Delta-Melvill-Abraham-Deseret (DMAD) Dam failure on June 23, 1983 (center of photo 6). By evening water had flooded the town of Deseret and thousands of acres of valuable farmland (photos courtesy of Robert C. Rasely, USDA Soil Conservation Service).

Rise in Lake Levels

The extreme precipitation of 1982-1983 caused a record 5-foot rise (fall of 1982-July 1st 1983) in the Great Salt Lake to a level of 4204.70 feet. The previous greatest seasonal rise of the lake was 3.40 feet in 1907. By 1983, the lake rose to the highest level since 1924 when it was 4205.10 feet. Utah Lake reached its highest level ever recorded, 4.93 feet above compromise (4489.34 feet) toward the end of June (figure 8).

Lake Powell nearly overflowed the top of the reservoir in early summer of 1983. Lake Powell receives all of the inflow from the Colorado River and its three major tributaries: (1) the Green River (drains southwest Wyoming and eastern Utah), (2) the Gunnison River (drains the west/central part of Colorado), and (3) the San Juan River (drains southwest Colorado and northwest New Mexico). There are four major reservoirs (Fontenelle and Flaming Gorge on the Green River; Blue Mesa on the Gunnison River; and Navajo on the San Juan River) upstream of Lake Powell which the U.S. Bureau of Reclamation operates in conjunction with Lake Powell. During the spring of 1983 all of these reservoirs filled in June, causing very large unregulated releases into Lake Powell.

During the period of October 1982 to April 1983 Lake Powell and the four upstream reservoirs were all releasing

Figure 8. Utah Lake reached its highest level ever in recorded history, 4.93 feet above compromise elevation (4,489.34 feet) toward the end of June 1983 (photo, from east side of lake, by B.N. Kaliser).

more than their inflow, creating space for the April-July forecast runoff. The weather stayed cool and extremely wet through April into the first half of May. Thereafter, temperatures went from below normal to way above normal causing snow to melt from the valley floors to the mountain peaks.

Most of the snowpack in 1983 melted off in about three weeks from mid-May to early June which allowed for a very efficient runoff of 14% (average efficiency is about 8%).

Lake Powell's water surface elevation rose from 3,685 feet the end of April, to 3,695 feet the end of May and to 3,707 feet by the end of June. The top of Lake Powell's spillway gates is at 3,700 feet. As the water poured into the spillway tunnels the operators noticed red coloring in the water which meant was that the tunnel lining was falling apart and the water was eating away at the sandstone canyon walls. Some very quick design studies were done to see if extensions could be put on the spillway gates in an effort to store more water and to spill less. A decision was made that 10-foot extensions could be installed on the gates, allowing for minimum releases through the spillway tunnels.

The runoff of 1983 was very unusual and exceeded a 200-year event in terms of probability of occurrence. The years of 1983 and 1984 are the two highest consecutive years of runoff since the early 1900s when records were first regularly kept of river flow. The year 1984 had a bigger volume than 1983 but was managed in a more controlled fashion by power plant releases without using spillways.

1986 Flooding

The next major flood episode was February 12th-20th in 1986. As with many significant flooding events in Utah, antecedent conditions played an important role in this case. A period of abnormally warm temperatures, moderate to locally heavy rainfall, and strong wind affected the mountain snowpack. At first, heavy wet snow and strong winds triggered literally hundreds of avalanches in the northern mountains. Then snow turned to rain at higher elevations, and the combination of rain and melting snow produced record runoff and mudslides.

A combination of factors lead to the extreme amount of moisture. The "Pineapple Connection" or a tropical feed of moisture extended from the Hawaiian Islands through California across Nevada and into Utah. The jet-stream was positioned over northern Utah, and strong orographics (mountain affects) consisted of precipitation along the west slopes of the Wasatch. Another problem was that the storm began with high density snowfall. Then, as warmer air moved into northern Utah, the snow level rose to between 7,500 and 8,500 feet. Also, in many areas the ground was frozen, which made it impossible for the water to penetrate the soil.

Rainfall ranged from 2-5 inches along the Wasatch Front, and from 6-12 inches in the Cache Valley, Morgan, and Wasatch Counties. From 60-100 inches of the "wettest snow on earth" fell in the Uintah and Wasatch Mountains at about 8,000 feet. Approximately 60-80 percent of the low elevation snowpack and 15-30 percent of the mid-elevation snow melted.

In the evening of February 17th, the Weber River peaked at 6,140 cfs just shy of the record peak of 7,980 cfs. On February 20th, the Bear River near Collinston peaked at a record 12,600 cfs or 1,000 cfs higher than the previous peak back in 1907.

Rise of Great Salt Lake

A good indication of the volume of the runoff produced during this wet period, is the rise in elevation of the water in the Great Salt Lake. From February 15th - March 1st, 1986, the lake rose a record 0.55 feet or 6.6 inches (USGS provisional lake-level records). This tied the previous record rise of 0.55 feet from March 15th-April 1st in 1983. However, in 1983 the lake was at 4204.7 feet and now was about 6 feet higher. So, due to its greater size and elevation, this is the greatest volumetric increase the Great Salt Lake has ever recorded in a two- week period.

The Great Salt Lake in the 1980s experienced a modern-day record rise and reached a historic high of 4211.60 feet in 1987. The lake started at a relatively low level in 1980 of 4200.50 feet and by 1989 was at 4206.50. The lake rose a record 5.0 feet from the low point in 1982 to the high point in 1983 and another 4.80 feet from the low point in 1983 to the high point in 1984. If that wasn't enough, the lake rose another 3.40 feet from the low point in 1985 to the peak in 1986 (figure 9).

As the Great Salt Lake was inundating the lower portions of Box Elder, Davis, Salt Lake, Tooele, and Weber Counties (figures 10 through 13), the most practical short-term solution was to install gigantic pumps (figures 14 and 15) and pump water from the lake to the west desert to lower the lake level. Work began on the pump project in 1986 with the pumps becoming operational in April of 1987. Water was

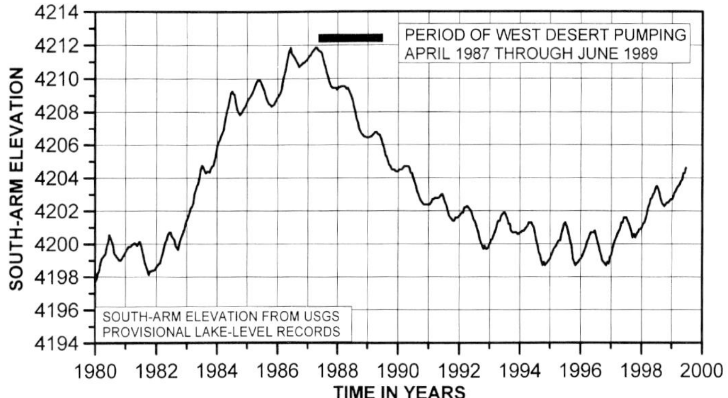

Figure 9. Hydrograph of south arm of Great Salt Lake (1980 - 1999), Utah Geological Survey-Great Salt Lake brine-chemistry database.

Figure 10. Section of the Southern Pacific Railroad tracks destroyed by storm-driven waves (photo by P.A. Sturm on June 17, 1986).

Figure 11. *View looking west of flooding of pasture land, building, and power-line right-of-way along the east shore of Farmington Bay. The dike seen beyond the power-transmission tower was built by Utah Power and Light Company to prevent wind-blown ice on Farmington Bay from damaging the towers (photo by J.W. Gwynn).*

Figure 14. *Aerial view looking southwest of the West Desert Pumping Project's inlet canal (forefront), pumping plant (center), and the 4.1 mile canal (upper) leading west to the West Pond (photo by J.W. Gwynn).*

Figure 12. *View looking west of lake water inundating the Syracuse/Antelope Island causeway, and the Antelope Island State Park's information and toll booth, located on the east end of the causeway, just west of Syracuse (photo by J.W. Gwynn).*

Figure 15. *View looking west at the inlet side of the nearly completed West Desert pumping plant (photo by J.W. Gwynn).*

Figure 13. View looking north of the Saltair pavilion and waterslide facilities being flooded by the rising lake (photo by J.W. Gwynn).

pumped from the south arm of the Great Salt Lake to an area just west of the Newfoundland Mountains called Newfoundland Bay.

Statewide Changes in Climate

In 1980, southern Utah stream flows were very high, averaging over 200% of normal flows in some places, while in the north they were near to slightly above average. The following year was generally a dry year statewide with stream flows much below normal. The lowest values were in the Green River basin of eastern Utah.

As mentioned previously, the very wet cycle began in the fall of 1982. Streamflow values in 1983 were very high statewide with many record April-July volumes established at that time. Many sites had 170-200% of normal runoff. In 1984, it was very wet in the north with stream flows over 150% of normal. In southern Utah, flows were closer to normal. It was wet again in the north in 1985 with streamflow volumes 100-125% of normal with below normal readings in the southwest. In 1986, runoff values exceeded 200% in the north to well below normal in the south; the Virgin River being 50% of normal.

As the decade of the 1980s came to a close the weather turned very dry statewide. The prolonged period of dry was approaching drought conditions in various parts of the state. Runoff volumes were near all time lows in the southern part of the state. In 1989 volumes below 35% of normal were common in the south. Northeast Utah was drier than north/central Utah, but volumes in both areas seldom exceeded 75% of normal.

This much drier regime reduced water stored in the major bodies of water in the state. Reduced inflow allowed the Great Salt Lake to recede to 4204.10 feet in elevation by late 1989 after experiencing a 3.10 foot decline during 1988. The water level of Lake Powell, by the fall of 1989 (October 1st), had dropped to 3,665 feet or 42 feet lower than the high point in 1983.

CONCLUSIONS

The El Niño and La Niña events of the 1980s greatly influenced weather patterns over Utah. Precipitation that was much above normal through the middle of the decade gave way to extreme drought towards the end of the decade. Heavy precipitation brought record streamflows and even flooded streets in downtown Salt Lake. A historic high elevation was recorded on the Great Salt Lake, and Lake Powell saw pool elevations seven feet above the spillway at Glen Canyon Dam. By the end of the decade, streams and lakes had receded with drought conditions wreaking economic havoc to many locations in Utah.

Improvements in technology in the 1990s along with improved modeling techniques have given National Weather Service forecasters the tools to make more accurate and timely forecasts. These improvements have allowed the forecasters to better fulfill their mission of protecting life and property and enhancing the Utah and national economies.

PROBLEMS AND MANAGEMENT ALTERNATIVES RELATED TO THE SELECTION AND CONSTRUCTION OF THE WEST DESERT PUMPING PROJECT

by

Lloyd. H. Austin
Utah Department of Natural Resources
Division of Water Resources
1594 West North Temple
P. O. Box 146201
Salt Lake City, UT 84114-6201

ABSTRACT

Great Salt Lake is a large terminal lake located in northwestern Utah. The lake has historically experienced wide cyclic fluctuation of its surface elevation which has continually plagued those who have utilized its shores. In September 1982, the Great Salt Lake elevation was 4,200 feet above sea level. In June 1986, the lake elevation measured 4,211.85 feet. This rapid rise to a new historic record caused $240 million in flood damages and dramatically impacted the economy of the State of Utah. The State, through this period, struggled to find an acceptable alternative for dealing with the rising Great Salt Lake. In August 1984, the State breached the Southern Pacific Railroad causeway to lower the southern part of the lake. With the continued rise of the lake, they finally selected and funded a $60 million alternative to built a project to pump large volumes of brine into a 325,000-acre pond west of Great Salt Lake where approximately 825,000 acre-feet of net evaporation per year could occur to help lower the lake.

INTRODUCTION

From September 1982 to June 1986, the level of the Great Salt Lake rose from 4,199.8 to 4,211.85 feet. The 4,211.85-foot level exceeded the previous historic high of 4,211.50 feet set in 1873. These and other data used in this document are the lake level data that were available in the 1980s. The U.S. Geological Survey has since made adjustments to some of these data).

During this short period of time the lake rose approximately 12 feet, doubling its volume and flooding some 500,000 acres around its shore. This rise of approximately 12 feet caused $240 million in damages to public and private resources and facilities, dramatically impacting the economy of the State of Utah. Even under the most optimistic lake level reduction scenario, the impact was projected to be felt for at least the next ten years. It is difficult to document or describe the multi-layer, interwoven process between the government, public, and the technical people that reacted to the problem, struggled to find an acceptable alternative for dealing with it, and finally committed $60 million to build it.

THE GREAT SALT LAKE

Great Salt Lake, a remnant of the prehistoric Lake Bonneville, is a large terminal lake located in northwestern Utah. At the elevation of 4,200 feet above sea level, the average level about which the lake has fluctuated during historic times, the lake covers approximately 1,700 mi^2, contains about 15,370,000 acre-feet of brine, which contains about 4.9 x 10^9 tons of salts (Arnow 1983). The average annual inflow (surface, groundwater and precipitation) is about 3.0 million acre-feet. The present Great Salt Lake Drainage Basin is approximately 22,000 mi^2 and includes drainage from Wyoming, Idaho and Nevada as well as much of the northwestern part of Utah (see figure 1). Since 1847, when Great Salt Lake Valley was first settled, the lake has been continually studied. Over the years, hundreds of reports have been published about the lake. These studies have shown Great Salt Lake and its environment to be very interesting, unique and complex.

East of the lake lies the Wasatch mountain range which reaches elevations of more than 11,000 feet. These mountains contribute most of the water inflow to Great Salt Lake. Some 65 percent to 70 percent of Utah's 1.4 million population live along the base of these mountains near Great Salt Lake. West of the lake is the Great Salt Lake Desert. It is a vast, flat desert area nearly barren of vegetation that was once flooded by Lake Bonneville. The Bonneville Speedway and the Hill Air Force Range are located in this area.

The lake is characterized by islands (some with rookeries, some with horses, buffalo, deer and/or antelope), salt water many times saltier than the ocean, beaches, and vast mud flat areas. It also has large marsh areas with thousands of game and non-game birds around the inlets of the three major rivers, brine shrimp, which are the largest form of life living in the lake, a railroad causeway which crosses the lake and significantly affects the salinity and natural circulation of the lake, water related recreation, and striking vistas and sunsets. The vastness and uniqueness of the lake are largely responsible for its importance to Utah and the surrounding

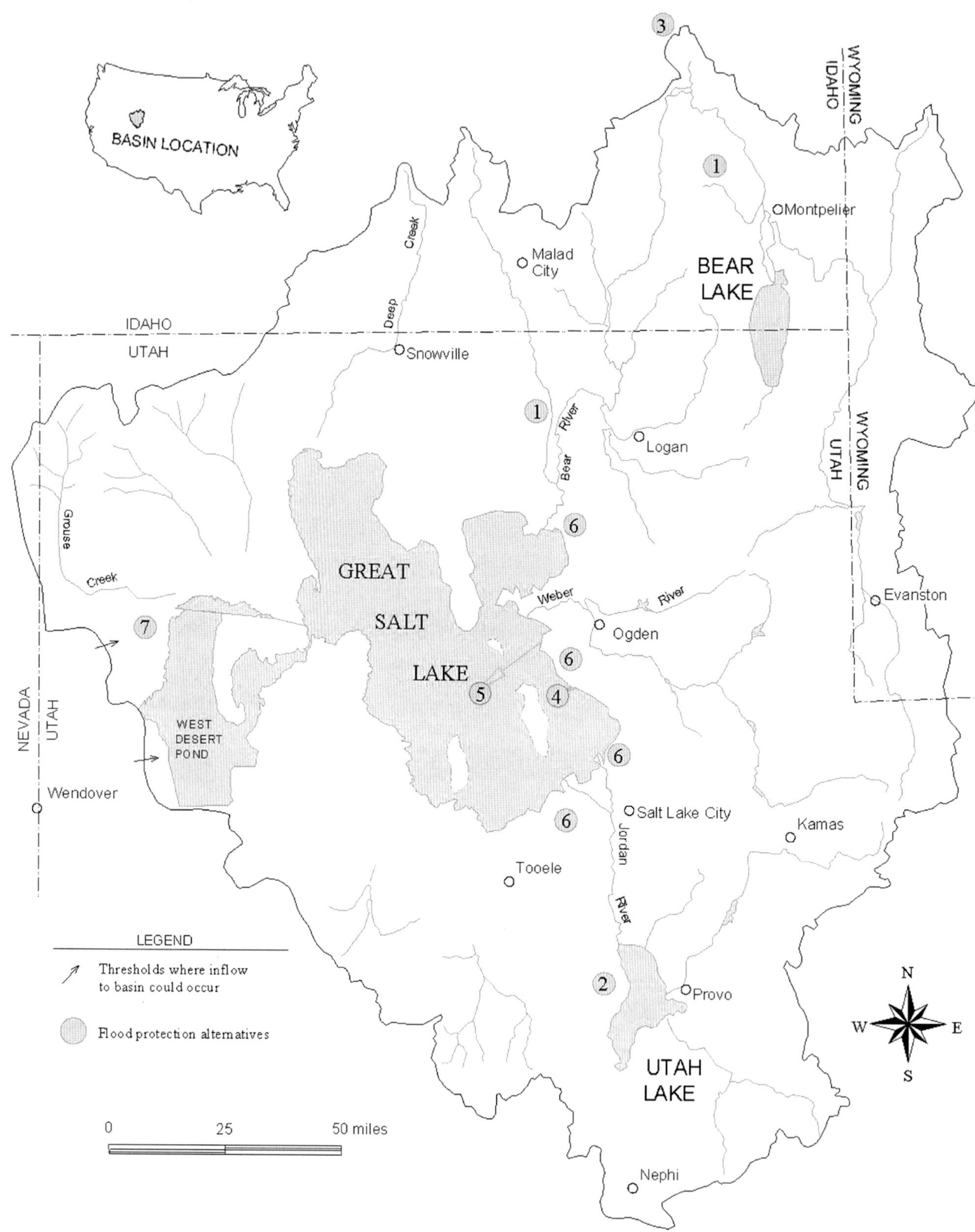

Figure 1. *Great Salt lake drainage basin.*

area for wildlife, hunting, recreation, mineral extraction and tourism (see figure 2). In general, the lake contributes significantly to the economy of the area.

FLOODING PROBLEMS

Great Salt Lake is similar in characteristics to other lakes that are the terminal point of large drainage areas. These lakes are generally shallow bodies of water that experience cyclic fluctuation in their surface elevation.

Great Salt Lake has historically experienced wide cyclic fluctuation of its surface elevation which has continually plagued those who have utilized its shores. Since 1851, the total annual inflow (surface, groundwater and precipitation directly on the lake surface) to the lake has ranged from approximately 1.1 to 9.0 million acre-feet. This wide range of inflow to the lake has caused its surface elevation to vary as much as 20 feet. Historically, the surface elevation of the lake reached a high of 4,211.5 feet in 1873 and a low of 4,191.35 feet in 1963 (see figure 3). A new historic high elevation of 4,211.85 feet was set in 1986 and matched again in 1987.

During the 1940s, 50s and 60s when the lake was relatively low, it was thought by many that the lake would remain low or even dry up. During this period, much of the present development occurred in the lower areas around the lake. This development included large wildlife management areas at the mouths of the rivers, large evaporation ponds in low areas for the salt extraction industries, major roads and railroads across and along the shores, recreation facilities, and a causeway connecting the east shore to Antelope Island to the mainland. Perhaps the real opportunity to reduce the present flooding was missed in this period when development took place so near the lake.

A peak elevation of 4,202.3 feet, reached in 1976 after a steady rise since the 1963 low, prompted a renewal of public awareness of the lake and the problems associated with high levels of the lake. This provided new legislative support to state agencies and universities to address problems related to flooding problems around the lake. In the three years following 1976 (1977-78 being one of the lowest precipitation years on record), the lake level receded more than two feet. In September 1982, however, the lake began rising rapidly again due to abnormally high rainfall and a shortened evaporation season. The continued high precipitation throughout 1983 and 1984 caused inflows of 7.5 and nine million acre-feet respectively to the lake. This caused the lake to peak at elevation 4,204.70 feet in June 1983, and at elevation 4,209.25 feet in June 1984. The two successive, approximate five-foot rises of the lake were the two largest in the historical record. Approximately $200 million of the $240 million of damage occurred during this two-year period. The following year (1985), the lake peaked at 4,209.95 feet.

The 1982-83 rapid rise of the lake triggered hurried renewal of activities by the state, counties, and universities to investigate alternatives for dealing with high levels of Great Salt Lake. A Contingency Plan, printed in January 1983 (Utah Department of Natural Resources, 1983), assembled information about alternatives that had been previously studied, such as pumping water from the lake to the western desert area, breaching the Southern Pacific Railroad causeway, diking low areas around the lake and/or developing/storing water upstream before it enters the lake. The plan, however, stressed that additional information, along with engineering feasibility studies, were needed before any recommendation for action could be made. One of the initial alternatives considered was breaching the Southern Pacific Railroad causeway that divides Great Salt Lake into two sec-

Figure 2. *South shore of Great Salt Lake.*

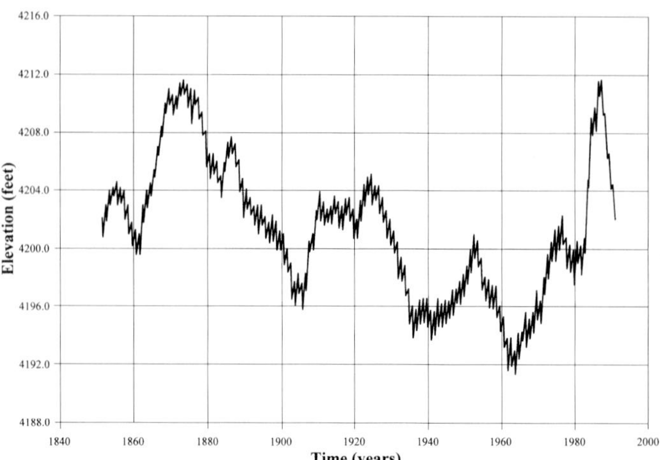

Figure 3. *Historical hydrograph of Great Salt lake (1851-1990).*

tions or arms. The Southern Pacific Railroad causeway is now part of the Union Pacific Railroad system and is generally called the "railroad causeway." It was determined that the elevation of the south arm of the lake, where most of the damage was occurring, could be lowered about one foot by installing a 300-foot bridge in the causeway. The causeway was breached on August 1, 1984 at a cost of about $3 million (see figure 4). By this time, the lake was so high that the breach was viewed as an interim measure until a more permanent solution could be found. The breach, although constructed to reduce flooding, caused some significant changes in the salinity balance between the north and south arm of the lake. These changes, however, are not addressed in this document.

After completing more studies funded by the 1984 Legislature on upstream diversion and storage, West Desert pumping, West Desert EIS, and numerous other projects, it was even more evident to the technical people that the West Desert Pumping Project was the best means by which the lake could be lowered. Philosophically, however, the idea of pumping water to a desert to simply waste it was distasteful to most people. Without a demonstrated ability to forecast future lake levels, there was little support to build the pumping project. This lack of good forecasting was in fact the major flaw in trying to prepare management strategies to deal with the problems caused by the rising level of Great Salt Lake. The status of forecasting future levels of the lake in 1985 was no different from the following conclusion reached in 1980 and published in Bulletin 116 of the Utah Geological and Mineral Survey: "The general consensus of researchers

Figure 4. *August 1, 1984 Southern Pacific Railroad breach activities.*

and climatologists was that future lake-level predictions cannot yet be made with any degree of assurance (Austin 1980)."

The general public, although realizing something should be done, viewed the pumping concept as unacceptable. Politically, this made support of the project a risk. All of the political decision makers found themselves in a classic dilemma of having the most technically feasible but the least politically acceptable solution in the same project. By the time the Utah Legislature met in January-February 1985, the reality of the problem had set in. With much debate and concern, the 1985 Legislature authorized and set aside in a special account of $96 million to design and construct flood-control facilities around the lake. By the end of 1985, the state had completed the design for the West Desert Pumping Project (see figure 5) and completed studies for diking of some critical facilities on the east side of the lake. By this time, millions of dollars had also been spent by private industry and local governments to protect resources and facilities near the lake.

Heavy rains during the last two weeks of February1986 caused the lake to rise from 4,209.35 feet to 4,209.90 feet. This rise represented the largest volumetric increase in a two-week period in the historic record of the lake. The lake was then predicted to peak at 4,211.0 feet, 1.05 feet higher than the year before. The status of flooding around the lake at that time was that resources and facilities near or below elevation 4,210.0 feet had either been protected or had been destroyed. The protection in most cases had been extended through raising dikes, roadways, and causeways etc., up to 4,212 feet or above. More flooding of unprotected farmland and facilities was expected during the spring of 1986. A request to the U.S. Corps of Engineers had been made to again assist in the protection of several small sewage lagoons and a sewage treatment plant that were not protected up to 4,212 feet.

The lake actually peaked at 4,211.85 feet in June 1986. This established a new historic peak surpassing the 4,211.5 foot elevation in 1873. Concern over the lake's continued rise was widespread, but many people still harbored hope the lake would reverse its upward trend. The West Desert Pumping Project eventually won approval from the State Legislature by a substantial margin as the most cost-effective and technically sound solution with the greatest public benefit. In May, the Second Special Session of the 1986 Utah State Legislature authorized $60 million from the $96 million for the Utah Division of Water Resources to construct the West Desert Pumping Project.

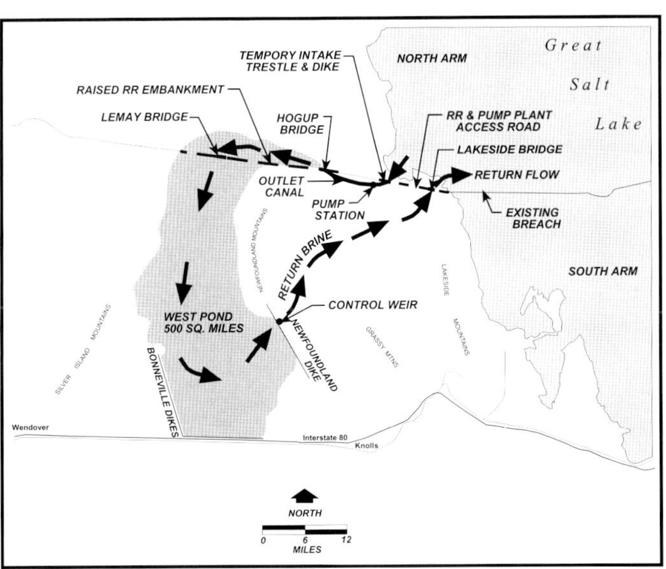

Figure 5. *West Desert Pumping Project - first phase, bare-bones alternative (source: Eckhoff, Watson and Preator Engineering).*

ALTERNATIVES

The search for some relief from high levels of Great Salt Lake did not begin in September 1982 with the recent flooding, but in historic times dating back to Brigham Young, the leader of the Mormon Pioneers who settled near the lake in 1847. During the 1860s when Great Salt Lake was rising steadily, Brigham Young is said to have sent out an expedition to determine whether the lake might have a natural outlet which would halt its rise.

The present search began modestly in the early 1970s as the lake rose steadily about one foot per year from its historic low in 1963. During the early 70s, work was done by several researchers along with state and federal agencies to define the hydrology of the lake, develop computer models of it, and to investigate alternatives for dealing with high lake levels. As this information was refined and used, the uniqueness of the large, shallow, saline terminal lake became clear. Great Salt Lake with its variable surface area and evaporation rate was much different from a fresh water reservoir or lake. In 1973 and 1974, the Utah Division of Water Resources published two reports that summarized much of the work done in the early 70s. They were titled "Great Salt Lake Climate and Hydrologic System" (Utah Division of Water Resources, 1974) and "Hydrologic System Management Alternatives Report" (Utah Division of Water Resources, 1977). The latter report concluded that the types of management programs that had the best potential to deal with rising lake levels were: (1) those that can be used year after year. This generally eliminates storage upstream that would fill quickly and then not be a factor during the rest of the flood problems; and (2) those that would be able to dispose of large volumes of water in short periods of time.

As mentioned earlier, the drought years of 1977 and 1978 slowed the interest in preparing for problems with high lake levels. The flooding that occurred after 1982, however, quickly refocused concern on what could be done, if anything, to deal with the rising lake level. Through this period of time, hundreds of additional ideas for dealing with the lake were sent to the state. Although some ideas were not well thought out, most were variations of alternatives already under investigation. Each new idea, however, was broken down into its basic principals and compared with alternatives already under investigation to evaluate what affect they would have on lowering the lake. The basic categories for alternatives were: (1) Reduce inflow to the lake by (a) exporting water from the basin upstream from the lake; (b) storing water upstream of the lake and/or consumptively using the water by developing large new irrigation projects; and (2) Deal with large volumes of water and flooding at the lake by (a) diking to protect critical resources and facilities in the lake or along the shore; and (b) creating storage areas in valleys or low-lying areas near the lake where water or lake brine could be stored and evaporated. The key to the last alternative, as to whether it was a storage or an evaporation system, was the volume and surface area of the new body of water. With these basic concepts, the hundreds of ideas presented during the 1982-1986 period were reduced to a number that could be evaluated. Engineering reconnaissance, feasibility and, in some cases, preliminary design was conducted on these alternatives. Table 1 that follows, briefly summarizes the alternatives selected for detailed investigation from the many that were suggested. The general location of the alternatives is shown on figure 1 and identified by the same number as used in table 1.

Politically, projects such as building upstream reservoirs and putting water to beneficial use, or diverting water from the basin during periods of flooding were strongly favored. But the West Desert Pumping Project, with each analysis, continued to show the most promise from both a cost and an effectiveness point of view. The West Desert Pumping Project met the basic conditions identified earlier. They were: (1) The project could be used year after year as long as flooding persisted; and (2) The project could remove relatively large volumes of water in a short period of time. This is particularly true during the first year of operation when the pond (storage area) is filled. Filling the pond quickly did not require over-designing the pumping capacity. The design capacity of the pumps was based on the flow necessary to prevent a build-up of salts in the large pond by returning concentrated brine back to the lake.

WEST DESERT PUMPING PROJECT

Concern immediately turned to the construction of the project. Although early design reports estimated an 18-month construction period, pressure was applied to every element of the project to cut the construction time to less than one year. This would allow the project to begin removing brine from the lake prior to the next year's peak of the lake which normally occurs in May. With a tremendous effort from consultants, contractors, suppliers and state agencies, the goal was met and the first of three giant pumps started up on April 11, 1987, the second pump on May 8 and third pump on June 4 (see figures 6 and 7).

Although the name, West Desert Pumping Project, implies a pumping project, it is a flood control action to help reduce the damaging high water level of Great Salt Lake. The project actually expands the evaporation from the terminal lake system by 20 to 30 percent.

The West Desert Pumping Project consists of a 10-mile-long access road along the railroad causeway, a pumping station, a canal, trestles, dikes, a 37-mile-long natural gas pipeline, and a 325,000-acre (500 square miles) evaporation pond in the desert area west of Great Salt Lake (see figure 8).

Three large pumps lift up to 3,150 cfs of brine from the north arm of the lake and discharge it into a 4.1-mile-long outlet canal before it enters the West Desert Pond. The project is designed to pump approximately two million acre-feet of brine a year from Great Salt Lake into the West Desert Pond with a surface area of 325,000 acres. The pond can evaporate up to 825,000 acre-feet (net) of water each year (see figure 9).

The 24.4-mile Bonneville Dike (maximum height of six feet) retains the southwest portion of the evaporation pond and prevents brine from the project from flooding Interstate Highway 80 or the famous land speed area called Bonneville Speedway. A second dike (Newfoundland Dike) some 8.1 miles long (maximum height of seven feet) extends southwest from the southern tip of the Newfoundland Mountains and is used to restrict the surface flooding of a military range used by the U.S. Air Force. A weir in this dike is used to regulate the pond's surface level between 4,216 and 4,217 feet, and to regulate the return of concentrated brines to Great Salt Lake.

Monitoring around the project was established to collect real-time climatic data and transmit it through satellite relay to a central location where operating decisions were made and where emergencies related to high winds or other problems could be addressed.

The heart of the project is three pumping units, each consisting of a natural gas fired engine, gear drive and clutch, right angle drive and a vertical axis mixed flow pump. Each

Table 1. Great Salt Lake Flood Protection Alternatives.

Option	Capitol Cost (million)	Effect (inches)	Implementation (years)
1. Bear River Storage	$140	8	10-20
2. Cedar Valley Irr.	$47-$128	3-6/year	3
3. Portneuf River Div.	$148	2-7/year	5
4. Antelope Island Diking	$38	0	1
5. Inter-Island Diking	$230	0	3
6. Selected shore Diking			
--To elevation 4,213	$5	0	.25
--To elevation 4,216	$30	0	.75
7. West Desert Pumping Project	$60	13 first yr, 7/yr after	1

Figure 6. *First 3,500 horse power engine being moved into place.*

Figure 7. *West Desert pumping plant nearing completion.*

***Figure 8.** Landsat image of the West Desert Pumping Project with return flow to Great Salt Lake.*

Figure 9. *West Desert Pumping Project inlet channel, pumping plant, and 4.1-mile outlet channel.*

engine weighs 162,800 lbs., is about 28 feet long, 12 feet wide, 18 feet high, has 16 cylinders and is rated at 3,500 hp. The engines operate at approximately 330 rpm. Each gear drive weighs 36,000 lbs. and reduces the speed delivered to the pump down to 139 rpm. Each pump consists of a 119-inch diameter impeller with three blades. The pumps have a total weight of 152,000 lbs. and are made of an aluminum-bronze alloy to resist corrosion from the heavy, corrosive brines of Great Salt Lake. The units are designed to operate through a lift of eight to 22 feet and pump from 900 to 1,050 cfs each.

The West Desert Pumping Project was shut down June 30, 1989, after more than two years of successful operation. The pumping project received praise from industry, recreation and transportation interests, wildlife agencies, and awards from professional engineering societies for helping lower the level of the lake more than five feet. The lake's level on June 15, 1989 was 4,206.45 feet. The project pumped about 2.2 million acre-feet of water and was responsible for about 26 inches of the five feet.

The pumping project was funded with a $60 million appropriation from the Utah Legislature. Utah's Division of Water Resources, the agency charged with designing, implementing and operating the project, spent about $58 million to construct the project. In addition, the division spent about $1.6 million per year to operate the pumps.

The long-term shutdown of the pumping project took about eight weeks and cost about $200,000. The shutdown process included (1) securing the pumping station, (2) dismantling, preserving and storing tools, systems and control devices, (3) planning for periodic inspection and maintenance of the project site, and (4) releasing water from the evaporation pond back to the Great Salt Lake. Pumps and engines have remained in place as insurance to reduce the impact of flooding, should the Great Salt Lake again rise to elevations similar to those of the mid-1980s.

Due to the termination of pumping well into the evaporation season (June 30) and other factors, some 600 million tons of salts (this is approximately 12 percent of the 4.9 billion tons of salt in the lake) remained in the pond and West Desert area. Output from a computer model of the system estimated that less than 300 million tons of salt would have remained in the pond and West Desert area had the pumping ended at the end of the evaporation season rather than in the middle. Some of the 600 million tons of salt that remained in the West Desert area have since returned to the lake, but the removal of these minerals from the lake has affected the salinity of the lake. This effect to the salinity of the lake, however, is not addressed in this document.

REFERENCES

Allen, M.E., Christensen, R.K., and Riley, J.P., 1983, Some Lake Level Control Alternatives: UWRL/P-83/01, Utah Water Research Laboratory, Logan, Utah,101 p.

Arnow, Ted., 1983, Water Level and Water Quality Changes in the Great Salt Lake, 1847-1983: U.S. Geological Survey Circular 913, Salt Lake City, Utah, 22 p.

Austin, L.H., 1980, Lake Level Predictions of the Great Salt Lake *in* Gwynn, J.W., editor, Great Salt Lake - a scientific, historical and economic overview: Utah Department of Natural Resources, Utah Geological & Mineral Survey, Salt Lake City, Utah, p. 273-277.

James, L.D., Bowles, D.S., James, W.R., and Canfield, R.V., 1979, Estimation of Water Surface Elevation Probabilities and Associated Damages for the Great Salt Lake: UWRL/P-79/03, Utah Water Research Laboratory, Logan, Utah, 182 p.

Utah Department of Natural Resources, 1983, Recommendations for a Great Salt Lake Contingency Plan for Influencing High and Low Levels of Great Salt Lake: Utah Department of Natural Resources, Utah Division of State Lands and Forestry, Salt Lake City, Utah, 47 p.

Utah Division of Water Resources, 1986, Final Design Report, West Desert Pumping Project (Northern Part) Preferred Alternative, Great Salt Lake, EWP Project No. EU19-04-01 and MKE Project No. 5027, Salt Lake City, Utah, 419 p.

—1974, Great Salt Lake Climate and Hydrologic System: Utah Department of Natural Resources, Utah Division of Water Resources, Salt Lake City, Utah, 84 p.

—1984, Great Salt Lake Diking Feasibility Study: James M. Montgomery, Consulting Engineers, Inc., Salt Lake City, Utah, 308 p.

—1977, Great Salt Lake Hydrologic System Management Alternatives Report: Utah Department of Natural Resources, Utah Division of Water Resources, Salt Lake City, Utah, 111 p.

—1984, Great Salt Lake Summary of Technical Investigations, Water Level Control Alternatives: Utah Department of Natural Resources, Utah Division of Water Resources, Salt Lake City, Utah, 114 p.j313

SATELLITE IMAGING AND ANALYSIS OF GREAT SALT LAKE

by

Lori Zastrow and Merrill Ridd
University of Utah
Department of Geography

ABSTRACT

From the earliest days of imaging the earth from space, Great Salt Lake has been an object of observation. Numerous images, both photographic and radiometric, have been taken over the past three decades. Yet, there has not been a comprehensive study of the lake through remote sensing.

This brief exploration into the possibilities for further investigation is based on Landsat data alone. Launched in 1972, Landsat provides the longest-running set of satellite data for continuous observation and mapping. Another paper in this volume (Rich) deals with detection of circulation patterns of the lake through remote sensing.

This paper deals with six other biophysical variables. They are: identifying and mapping shoreline position, estimating water depth, investigating evidences of salinity, determining thermal patterns, relating brine shrimp production to salinity, and examining the effects of water-level fluctuations on waterfowl habitat.

Three of the six mini-investigations yielded a degree of success and promise for further study: the first two, and the final one. Two others require more sophisticated investigation, namely thermal and brine shrimp/algal detection, and salinity will likely remain relegated to indirect observation, pending new innovations in remote sensing technology. With emerging sensor systems, a coordinated and concurrent ground truth effort may yield significant new information and monitoring opportunities.

INTRODUCTION

This paper explores the use of satellite-based, multispectral, digital imagery to identify patterns of selected biophysical variables of Great Salt Lake, Utah. Specifically, the variables under consideration are: shoreline position, water depth, salinity, thermal properties, and brine shrimp and waterfowl habitation. These variables are intrinsically intertwined in processes such that their concentration and distribution present a complex pattern that varies through space and time. The body of literature defining the response of specific spectral bands to specific conditions of water quality is limited. Because of this, and the fact that empirical data regarding these variables in the lake are limited, our analysis will be less definitive than desired. However, some key relationships are established, and significant research questions raised to support further investigation.

Several kinds of images, both digital and photographic, have been obtained from orbital satellite platforms since the first launch of Landsat on July 23, 1972. Since that time, digital imaging experiments have included the Return Beam Vidicon (RBV) panchromatic sensor mounted on Landsat 1 and the Skylab experiment on a separate spacecraft. Photographic experiments have included the Large Format Camera and hand-held cameras (both analog and digital) by astronauts on orbiting shuttles. Many such photographs have been taken of Great Salt Lake.

This investigation uses only Landsat imaging as it is the longest existing operational (as distinct from experimental) unmanned satellite system orbiting earth designed explicitly for earth observation (as distinct from weather or atmospheric observations). Two instruments have provided the ongoing operational data from Landsat - the Multi Spectral Scanner (MSS) and Thematic Mapper (TM). The MSS sensor, introduced on Landsat 1, has four spectral bands: green (0.5-0.6 µm), red (0.6 0.7 µm), and near infrared (0.7-0.8 µm) and (0.8-1.1 µm). The original Landsat band numbers were 4, 5, 6, and 7, later renumbered to bands 1-4, respectively, with a nominal 80-meter spatial resolution or pixel size. An MSS has been placed on all five Landsat platforms and is still operating (year 2000) on Landsats 4 and 5. The TM sensor has seven spectral bands: blue (0.45-0.52 µm), green (0.52-0.6 µm), red (0.63-0.69 µm), near-infrared (0.76-0.90 µm), mid-infrared (1.55-1.75 µm) and (2.08 2.35 µm), and thermal infrared (10.4-12.5 µm), numbered 1, 2, 3, 4, 5, 7, and 6, respectively. A TM was introduced on Landsat 4 in 1982, but ceased acquiring data in 1993. A TM was incorporated into Landsat 5 in 1984, and is still operating on a limited basis (year 2001). Landsat 6 failed in orbit, and Landsat 7 was successfully launched on April 15, 1999. The TM sensor has a 30 meter nominal spatial resolution, or pixel size for the reflective bands 1, 2, 3, 4, 5, 7, and a 120 meter nominal spatial resolution for the thermal band, band 6. Landsats 4 and 5 are placed opposite each other in orbit, each on a 14-day cycle, thus providing MSS or TM data on a 7-day interval per site.

Given the near-polar orbital path of Landsat and the segmentation of imagery into frames (individual images or scenes), no single scene covers the entire Great Salt Lake. Although a single scene is large enough to cover the lake, this is one of those cases where four scenes join in the middle of the area of interest. Figure 1 is a mosaic of four images (also see colored photo section). The east half of the lake was imaged from Landsat 5 on June 25, 1984 (path 32,

Figure 1. A 1984 grayscale Landsat 5 image of the Great Salt Lake. The east half of the image was collected on June 25, 1984, and the west half on July 2, 1984. (see figure 1 in the color section of this book).

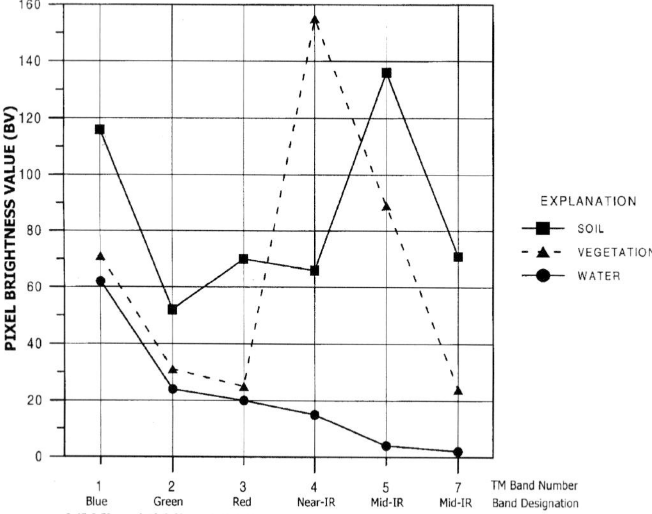

Figure 2. Reflectance responses of typical soil, vegetation, and water features at wavelengths from 0.45 to 2.35 μm.

rows 38 and 39), and the west half on July 2, 1984 (path 33, rows 38 and 39).

The separate spectral bands of differing wavelengths are the essential qualities that make multi-spectral imaging a diagnostic tool. Each reflective band (TM 1,2,3,4,5, and 7) has its own reflectance or brightness value represented as a brightness measure of solar energy reflected from objects on the earth, passing upward through the atmosphere to the sensor on the satellite. The thermal band (TM 6) represents energy emitted directly from the earth. Different objects on the ground absorb and/or reflect solar energy of different wavelengths in different amounts (figure 2). The amount of energy received at the sensor is recorded on-board and/or transmitted to earth by telemetry. Thus, for MSS data, there are four brightness values (BVs) per pixel. For TM data, there are six BV_S for the reflective bands (1, 2, 3, 4, 5, and 7) and one BV for the thermal band.

The mosaic in figure 1 is a false color composite created by assigning the TM spectral bands 4, 3, and 2 to the red, green, and blue color guns, respectively. In this combination, the near-infrared band (TM 4), which is highly reflected from green vegetation (figure 2), shows vegetation in various shades of red. For water, in wavelengths greater than visible, (>0.7 μm) most solar energy is absorbed. This allows for a sharp distinction of water versus land in the near- and mid-infrared regions, wherein water (at least clear water) appears dark on the image. Within the visible range, the elevated blue reading is a consequence of atmospheric scattering of the shortest of the visible wavebands, such that, although dark, water does have a blue cast. Because of its short wavelength, blue light may be transmitted through water, and may reflect from the bottom or from suspended material, and be transmitted back through the water and upward to the satellite sensor, altering the shade of blue. To some degree, green light and even red light may penetrate water and return a signal to the sensor. These variable properties in light behavior are particularly valuable in remote sensing of water.

Can bottom reflection be distinguished from suspended sediment reflection? Can near infrared energy (TM 4), which is highly reflective from green vegetation, become a tool in detecting algae in water, emergent vegetation in shallow water, or even near-surface submergent vegetation? The literature yields few definitive answers to these and other questions regarding the use of remote sensing for determination of particular conditions of water character and quality. As spectral and spatial resolution improves with forthcoming sensors, new studies may provide a better understanding of water character and quality.

This paper presents an examination of six biophysical characteristics of Great Salt Lake utilizing Landsat MSS and TM data. The six are: determination of the shoreline position as the lake rises and falls, estimation of water depth, indications of water salinity, comparative thermal patterns, indications of algae and brine shrimp, and conditions of waterfowl habitat.

SHORELINE POSITION

Since the first Landsat images were obtained in July 1972, the surface elevation of the south arm of Great Salt Lake has varied from a low of about 4,198 feet (September 30, 1972) to a high of 4,211.7 feet (June 3, 1986) above mean sea level (the latter elevation was the historic high). Typically the surface of the north arm of Great Salt Lake has been two to three feet lower than the surface of the south arm due to differences in inflow and evaporation. Elevation differences between the south and north parts of the lake have lessened in recent years after construction of a breach along the western edge of the railroad causeway that separates the two arms. In this section, three applications of shoreline-position determination are demonstrated.

In an unpublished study for the Utah Division of Wildlife Resources by the University of Utah's Center for Remote Sensing and Cartography, shoreline fluctuations in Farmington Bay were examined to define changes in waterfowl habitat. Using MSS data collected from 1972 to 1979 (TM data were not yet available), shoreline positions were determined at one-foot intervals from 4,198 to 4,202 feet inclusive, by selecting the satellite date matching each of five measured lake levels (figure 3). Using the near infrared band (MSS 4) to define the water/land interface, computer-print character maps were produced to show the position of the shoreline for each lake level. Figure 4 displays the positions of the five shorelines in Farmington Bay.

Figure 5 presents a mosaic of the entire lake from TM data collected in September 1987, as the water level stood at about 4,209 feet elevation, near the historic high. Where the slope along the edge of the lake is high, the horizontal position of the shoreline was observed to change very little with lake level fluctuations. Conversely, the shoreline position shifts substantially where the slope is gentle. Traces of Interstate 80 can be seen in figures 1 and 5 where the southern tip of the lake extends beyond the highway. In subsequent figures, other features such as fresh water impoundment dikes for waterfowl management can be seen following regression of the lake.

As the surface elevation of Great Salt Lake rose through the 1980s, inundation of highways and railroads near the south shore of the lake continued, as did the flooding of sewage treatment plants, farmlands, and extensive solar salt plants dependent on solar evaporation. With the loss of those facilities, and anticipated additional losses, a decision was made to install a massive pumping system to move lake water to the western desert. Figure 6 , modified from a portion of a satellite image map (U.S. Geological Survey [USGS], 1995) shows the extent of the water moved by the pumping project. The complete satellite image map is a mosaic of 23 Landsat TM images recorded in 1988-89.

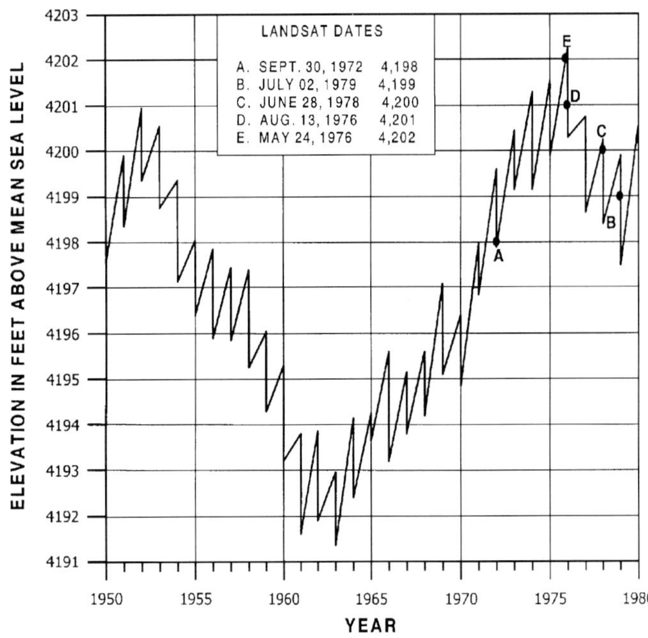

Figure 3. Generalized hydrograph indicating dates of MSS data collection (hydrograph from U.S. Geological Survey).

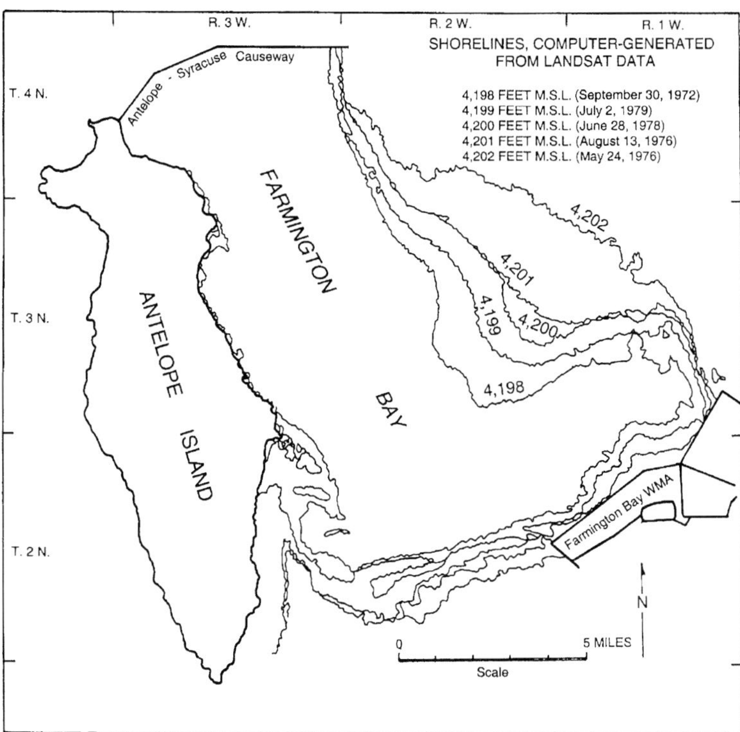

Figure 4. Shoreline position of Farmington Bay at one-foot intervals from 4,198 to 4,202 feet elevation as determined from MSS data (Center for Remote Sensing and Cartography, University of Utah Research Institute).

DEPTH ESTIMATION

To determine if water depth could be determined using TM data, we created a simple depth estimation model for the

Figure 5. *Near-historic high-water image of the lake collected on September 22, 1987. On this date the lake water stood at an elevation of about 4,209 feet, near its historic high (see figure 5 in the color section of this book).*

north arm of the lake based on spectral reflectance in TM bands 1, 2, and 3 (blue, green, and red). Longer wavelengths are almost totally absorbed by water, as shown in figure 2. For purposes of this experiment, we assumed that turbidity, salinity, and biota were constant at the time. This assumption is based on the fact that there is no permanent fresh-water stream inflow to the north arm, only a small contribution at Locomotive Springs at the extreme northern end of the lake, and the fact that there are few interruptions in the shallow basin, presumably resulting in a relatively homogeneous water body north of the earth-fill railroad causeway. At several places, a transect orthogonal to the shoreline was run into the lake, recording the blue, green, and red BVs at locations of known depths on increments of one foot. The source for bathymetry was the Great Salt Lake and Vicinity, Utah map (USGS, 1974) with underwater contours below 4,200 feet. The TM image for this task was taken from data collected June 25/July 2, 1984, when the surface of the south arm was at 4,209.05 feet. The surface of the north arm was assumed to lie at 4,207 feet elevation. As the bathymetric map shows no contour lines above 4,200 feet, the north arm depth model could apply only to water at depths greater than seven feet. To reduce atmospheric effects on BVs at the sensor, a dark object subtraction technique was applied to the visible bands. The darkest pixel for each band was identified and reduced to zero, and the recorded difference for this pixel was subtracted from all other pixels band by band.

The depth model is indicated in table 1. As one would expect, there is a reduction of reflectance (BV) with depth, however, the increment of BV reduction diminishes with increasing depth. Also expected was a greater reflectance in the shorter wavelengths (blue) and lesser reflectance in the longer wavelengths (red). An inconsistency in the reduction of BV with depth in the blue band suggests a turbidity factor may have been affecting blue light more than green or red light. There is also a slight inconsistency in the green band BV sequence with depth, perhaps for the same reason. The red band shows a smooth BV reduction down to 13 feet where it appears to level off.

The model was inverted and applied at other transects in the north arm. That is, using the BV for the three bands as a determinant, the depth was estimated and recorded and then checked against the reference map. The error was recorded at each point of estimated depth and flagged as being underestimated (-) or overestimated (+). Table 2 displays the results of the model application.

We next decided to test which of the visible bands is "most sensitive" for depth determination. Table 3 summarizes those results. The green band shows the least numerical error in five depth classes (bold type), the red band in two, and the blue band in one, although in the 10-foot category the blue band is lowest in arithmetic error (-0.09 feet). A possible explanation for the poor performance in the blue band in general is that the shorter light waves penetrate (are transmitted) more readily through water than longer wavelengths and may be reflecting more brightly from bottom sediments or turbidity. This appears to be supported by the fact that five of the depth categories are underestimated by the blue band (exhibiting greater brightness values per depth category) and the error increases with depth.

An effort to create a depth estimate model for the south arm was unsuccessful. It is clear that the other factors affecting reflectance in the visible wave bands are too variable spatially to assume constancy as seemed to be possible in the north arm. Fresh water inflow from three perennial rivers, in temporally and spatially varying quantities, create a variable distribution of salinity, turbidity, and biota. These factors are

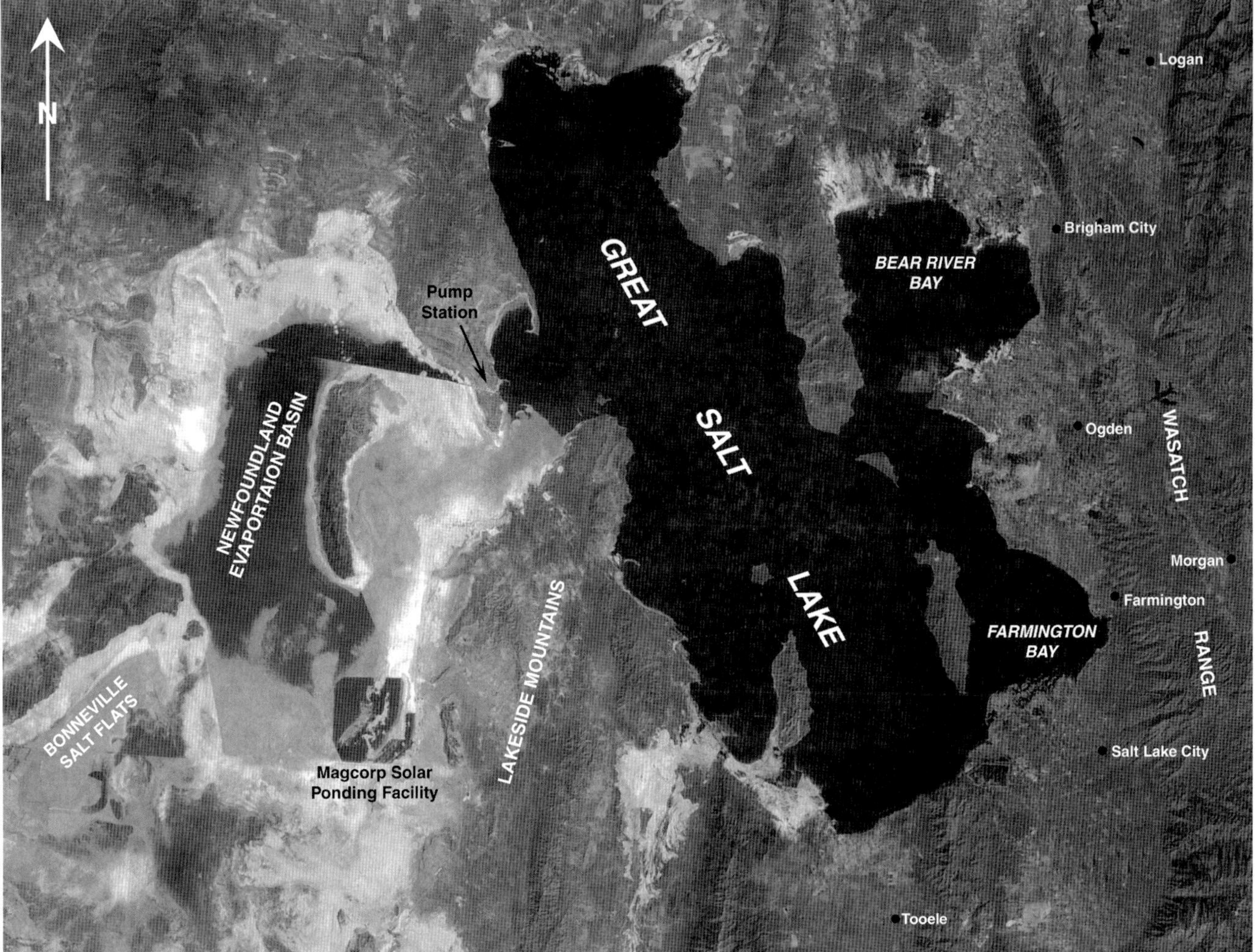

Figure 6. Expanded lake due to pumping into west desert to alleviate flooding of roads, railroads, and other facilities. Modified from satellite image map (of Utah), U.S. Geological Survey, 1995.

further fragmented by island barriers and other physical features.

Note: Subsequent analyses presented in this paper are based on a multi-spectral classification of the lake water itself using all six TM reflective bands. The first step was to accurately determine the water versus land position with the 30-meter resolution imagery. All the features except the lake itself were masked out of the imagery to limit analysis to the water alone.

SALINITY

The most conspicuous feature of space images of Great Salt Lake is the contrast between the north arm and the south arm regardless of the spectral waveband exhibited, the elevation of the water surface, or the season of the year. The completion of the earth-fill causeway across the lake in 1959 changed the lake dynamics, most particularly the distribution of salinity. Until 1984 there were only two culverts connecting the two arms of the lake. To help control of the rising water level associated with an extended wet cycle, a 300-foot breach was cut into the causeway toward the west side of the lake, permitting greater interchange of water between the two arms. Even so, a sharp contrast in salinity remains between the two arms of the lake. In December 2000, the breach opening was deepened to allow a greater return flow of north-arm brine into the south arm. Hopefully, this will reduce the contrast in salinity between the two arms.

Pre-1995 salinity data for the lake are available from the Utah Geological Survey for about nine locations throughout the lake (unpublished electronic database). Since 1995, however, investigations of brine shrimp ecology, and associated water-quality sampling by the USGS, in cooperation with the Utah Division of Wildlife Resources (unpublished data), have greatly increased the frequency and aerial distribution of salinity measurements in the lake. Unfortunately, much of that salinity data postdates the imagery used in this investigation.

For the June 1984 image representing a high-water period, three sample sites in the north arm show a mean density value of 1.160 g/L (21.12 weight-percent salt), in contrast with a mean density of 1.038 g/L (5.2 weight-percent salt) for six samples sites in the south arm. For the June-July 1996 image representing low water period, two salinity samples from the north arm average 1.22 g/L (27.0 weight-percent salt) as compared to the south arm four-site sample of 1.086 g/L (11.9 weight-percent salt).

Table 1. Depth estimation model, north arm Great Salt Lake, based on TM images of June 25 and July 2, 1984.

Water Depth (feet)	BV Blue Band (0.4 -0.5μm)	BV Green Band (0.52 - 0.6μm)	BV Red Band (0.63 - 0.69μm)
7	36	34	34
8	32	28	25
9	29	24	21
10	26	22	19
11	25	21	17
12	27	21	16
13	27	20	15
14	25	19	15

Table 2. Errors in depth estimates.

Actual Depth (feet)	Number of Samples per depth	Numerical Mean Error (a) feet	Arithmetic Mean Error (b) feet
7.0 - 7.9	5	0.60	0.60
8.0 - 8.9	19	0.44	0.23
9.0 - 9.9	18	0.83	0.37
10.0 - 10.9	19	0.99	-0.78
11.0 - 11.9	25	1.2	0.59
12.0 - 12.9	20	1.23	0.36
13.0 - 13.9	25	1.01	0.74
14.0 - 14.9	14	2.14	-2.14

(a) Without regard to underestimate or overestimate, an overall statistical statement of accuracy in depth estimation at the stated depth.
(b) Mean error for all samples in the depth category as to whether understated (-) or overstated (+).

Table 3. Errors in Depth Estimates by Spectral Band.

	Blue Band		Green Band		Red Band	
Actual Depth	Numerical Error (ft)	Arithmetic Error (ft)	Numerical Error (ft)	Arithmetic Error (ft)	Numerical Error (ft)	Arithmetic Error (ft)
7.0 - 7.9	-	-	**0.50**	**0.50**	0.85	0.85
8.0 - 8.9	**0.33**	**-0.16**	0.72	0.72	0.38	0.38
9.0 - 9.9	1.38	0.13	0.89	0.83	0.49	**0.06**
10.0 - 10.9	1.58	**-0.09**	**0.42**	**-0.42**	0.99	-0.99
11.0 - 11.9	1.33	0.67	1.39	1.09	0.99	**0.19**
12.0 - 12.9	1.63	-1.63	**0.85**	**-0.17**	1.03	-1.03
13.0 - 13.9	1.50	-1.50	**0.85**	**-0.17**	0.98	-0.98
14.0 - 14.9	2.72	-2.72	**0.00**	**0.00**	1.83	-1.83

Because salinity differences have major, direct effects on biota (salt-tolerant bacteria, algae, and brine shrimp) it is not clear to what extent the spectral signatures are influenced by salinity. Other studies have been inconclusive as to spectral response to variations in salinity (Ikeda and Dobson; 1985 Bukata, 1995). What does seem clear from the Great Salt Lake study is that the contrast in biota across the causeway, both in biomass density and type of biota, yields a greater influence on spectral signature differences than does salinity itself. That is, the spectral contrast is caused principally by differences in the type and population of biota, and only indirectly by salinity.

THERMAL PROPERTIES

Obtaining meaningful readings of temperature of the lake by remote sensing is not a simple matter because of many factors affecting temperature of the water which are not generally recorded at the time of overpass of the sensor. These factors include, but are not limited to, wind speed, wind direction, relative humidity, air temperature, variations in suspended sediment, variation in dissolved solid concentration, floating debris such as brine shrimp cysts, or other materials at or near the surface.

Figures 7A and 7B represent thermal images taken the same time of year to eliminate seasonal temperature differences for both low- and high-water periods, respectively (also see colored photo section). The color range varies from red (warmest) to violet (coolest), and red, orange, and yellow are warm colors as compared to green, blue, and violet as cooler colors. In general the low-water image is warmer, with a mean BV of 130 versus 128 for the high-water image. The minimum BV is 122 for the low-water image, and for the high-water image is 103. Maximum values are similar at 147 and 146, respectively. Actual water temperatures were not recorded when these thermal images were taken so the BV values represent only relative temperatures.

In examining various areas of the lake, relative temperature differences are much as one would expect. In Farmington Bay water temperatures were clearly warmer in 1996 (low water) when the maximum depth of the bay was about 8 feet. It is warmest near the shoreline on both dates due to shallow water depths and greater solar heating.

Willard Bay Reservoir is the coolest part of the Bear River Bay region, probably because it is fed by a canal whose distance from the source in the Uinta Mountains is much less than that of either the Bear River or the Provo/Jordan River system, both of which also derive from the same area of the western Uinta Mountains. In Bear River Bay it is not clear at this time why the high-water image (7B) appears to be warmer than low-water image (7A). A partial explanation may be that the image from 1984 (high water) was collected two weeks later in the season than the low-water image of 1996, June 25 versus June 10, respectively, allowing two more weeks of solar heating.

North arm temperatures are generally warmer than those of the south arm for both dates. Two reasons for this are that thermal capacity increases with salinity, and that the north arm water is more distant from sources of fresh, cool water. The reason for the broad area of warm water to the north and south of Rozel Point, along the east shore of north arm, is the extensive area of shallow water (figure 9), seen as a mud flat on the USGS, 1:125,000-scale topographic map of Great Salt Lake (1974), and the two additional weeks of warming. The cooling influence of Locomotive Springs at the extreme north-northwest end of the lake is evident, especially in the high-water image (7B). The cooling influence of Timpie Spring, located at the south end of the southwestern embayment of the lake, is evident in the low-water image (7A).

The cool cell north of center in the north arm is an unexplained anomaly in the low-water image (7A). Also the cool strip of water against the west shore, extending nearly the length of the lake, is unexplained at this point in time, as are the cool areas around Carrington Island, which is shown as the white area, west-of-center in the south arm.

The principle problem with thermal imaging of water bodies is the need for better calibration of sensor data from either concurrent field-data collection or through calibration of the sensor with known reference standards. This investigation did not attempt to provide field or reference calibration for the sensor and did not attempt to correct for atmospheric conditions.

Some pioneering remote sensing work has been done on the thermal characteristics of Great Salt Lake by Baskin (1990) using the Thermal Infrared Multispectral Scanner (TIMS) flown to specification over the East Shore area of Great Salt Lake, in-cluding part of Farmington Bay. Baskin used TIMS internal reference data and an atmospheric model to calibrate the sensor data and provide calibrated surface temperatures at 5 and 30 meter spatial resolutions. His work identified and verified locations of submarine groundwater inflow in the areas where the digital thermal data were obtained.

As improved sensor data and calibration methods become available for water investigations, perhaps a new initiative may be undertaken to investigate and monitor changing thermal characteristics of the entire Great Salt Lake.

BRINE SHRIMP CYSTS AND ALGAE

The distinct relations between season, salinity, algae, and brine shrimp cysts are variables requiring further research. Optimum brine shrimp growth in the south arm occurs when water temperature is about 20-25°C (68-77°F) with salinity between 12 and 17 weight percent, and with an abundant food supply (Doyle Stephens, USGS, personal communication, 2000). Fluctuations in water level and salinity are highly affected by spring snowmelt runoff and ambient air temperature. In normal water years, the south arm can remain within the 12-17 weight-percent salt range. Algal availability as brine shrimp food is also dependent upon temperature and salinity. With all elements being optimal, algae grows throughout the winter, is typically at its peak in February, and is visible within TM images. Young brine shrimp generally appear from over-wintering eggs between late February and early March when water temperatures are between 0 and 5.5°C (32 and 41°F). Their appearance follows the winter bloom of phytoplankton that provides food for the young brine shrimp for about one month. When all of the variables come together and the brine cysts begin to hatch and feed, algae is quickly depleted and is almost completely gone by the end of May. This observation is corroborated by

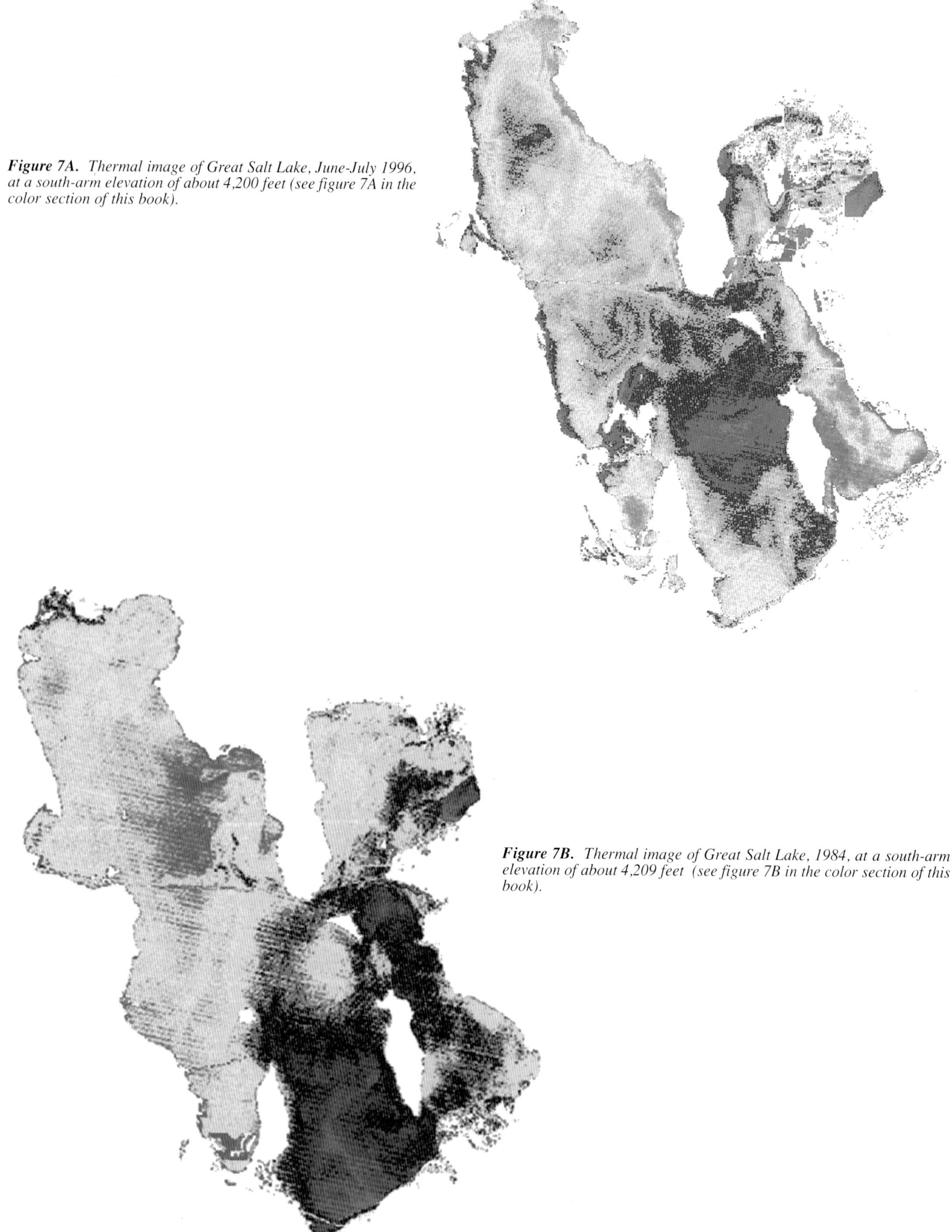

Figure 7A. *Thermal image of Great Salt Lake, June-July 1996, at a south-arm elevation of about 4,200 feet (see figure 7A in the color section of this book).*

Figure 7B. *Thermal image of Great Salt Lake, 1984, at a south-arm elevation of about 4,209 feet (see figure 7B in the color section of this book).*

the lack of algal signatures in TM imagery for June and July.

WATERFOWL HABITAT

Major investments in waterfowl management have been made in Bear River Bay and Farmington Bay over many decades. Figure 8A illustrates the Bear River Bay area during low water at 4,201 feet (4,200 is often referred to as the long-term mean). Figure 8B shows the bay at 4,209 feet, approaching the historic high (also see colored photo section). This eight-foot difference is less than half of the historic range (4,191 in 1963 to 4,211.7 in 1986 and 1987). The reader should refer to the USGS Great Salt Lake and Vicinity map (figure 9) for the location of the various features named in this section. In the images of figure 8 in the color section, the red color indicates healthy green vegetation, as seen in the many farm fields to the east, northeast, and southeast. Red within the bay likewise indicates healthy green vegetation. It is TM band 4 that provides this information, as vegetation is highly reflective in the near infrared band.

In figure 8A, the dikes and ponds intended to hold fresh water in the waterfowl management area are clearly seen, while they are inundated in the high-water image, figure 8B. The dike encompassing fresh-water Willard Bay Reservoir (figure 9), however, remained basically intact when the lake stood at 4,209 feet. At that level, lake water backed up well into the mouth of the Bear River. Maximum water depth at the exit in the southwest corner of the image is about 13 feet at this date (figure 8B), as compared to about five feet in figure 8A. The lighter shades near the mouth of the river in figures 8B are due principally to suspended sediment as the river continues to build its delta. The lighter shade along the northwestern shore is due to shallow water over a bright bottom surface. The outer dike encompassing the solar evaporation ponds along the central-lower portion of figure 8 (IMC Kalium) is visible at this date, but many interior dikes (figure 8) are inundated. The state-owned Harold S. Crane Waterfowl Management Area (WMA) is also totally inundated on this date.

At a water level of 4,201 feet, considerable land area is exposed in Bear River Bay, as seen in figure 8A. Water entering from the river is almost totally utilized in various routing patterns across the management ponds. The major flow line is westward between North Bay and South Bay (figure 9). The Bear River Migratory Bird Refuge, managed by the U.S. Fish and Wildlife Service, encompasses the major impoundment units in both bays. It is evident that during this time of low water, the North Bay units and the western-most South Bay unit

Figure 8A. *Bear River Bay, 1996, at a south-arm elevation of about 4,201 feet. Note the exposure of dikes and some pond areas in the Bear River Migratory Bird Refuge system (center of image). Note also the large bar extending southward from the refuge dikes, and in the areas surrounding the eastern portion of IMC Kalium's evaporation ponds (bottom-center of image) (see figure 8A in the color section of this book).*

Figure 8B. *Bear River Bay, 1984, at a south-arm elevation of about 4,209 feet. Note the complete inundation of the Bear River Migratory Bird Refuge dike system (center of image), water backing into the mouth of Bear River (upper right quadrant), and covering much of IMC Kalium's evaporation pond system (bottom left-of-center of image). The dike around Willard Bay Reservoir remains essentially intact (center of bottom-right quadrant) (see figure 8B in the color section of this book).*

are favored, while units in the center area are essentially bypassed. At the east end of South Bay, a reduced flow from Bear River serves the impoundment units and proceeds into the open bay to join the main flow westward and southward into the main body of the lake.

The distribution of vegetation in figure 8A (red color) is a significant factor in waterfowl habitat. The species distribution and abundance of vegetation are closely tied to depths and distribution of water. The southward-sloping Bear River delta, which underlies the entire waterfowl management area, is expressed in each impoundment unit as water depth increases to the south side in each case. During this low water stage, emergent vegetation is abundant. Plant species include hardstem bulrush (*Scirpus acutus*) Olneys bulrush (*Scirpus olneyi*), and cattail (*Typha domingensis*), and others. At the lower end of units with abundant water, submergent vegetation occurs, perhaps sago pondweed (*Potamogeton pectinatus*). This pattern is also evident in the Harold S. Crane WMA to the south of, and bordering, Willard Reservoir.

Figure 9. *USGS area-location map of Bear River Bay area. Taken from U.S. Geological Survey, 1974.*

Beyond the outer dike of the Bear River Migratory Bird Refuge, emergent vegetation is very extensive as this lower delta area is broadly exposed at low water. This affords extensive habitat for shore birds such as avocets and stilts. In the deeper water, between Willard Bay Reservoir and the Bear River Migratory Bird Refuge, an extensive area of submergent vegetation is displayed where the secondary supply of Bear River water reaches beyond the outer dike and proceeds westward. Note the large patch of deep red in this arm, marked as "Willard Bay" on the USGS Great Salt Lake map (1974). In this area, water depth is about 8 to10 feet. Under these conditions, the submergent vegetation, anchored on the bottom, rises to the surface and displays green leaf surfaces to be readily detected by the near infrared band of TM imagery.

Focusing now on Farmington Bay, figures 10 A and B (also see colored-photo section) display low water at 4,200 and high water at 4,209, respectively. Here again the Bay is considerably enlarged at high water. In the low-water image many features are evident. To begin, several "arcs" of algae bloom are seen in the open water as red patches. This is evidence of relatively fresh water. In essence Farmington Bay at low water is a "fresh water trap." At this and lower stages, Antelope Island is connected to the mainland by an old causeway (pioneer access to the island). This blocks any flow of saline water west of the island, as does the newer causeway at the north end of the island. During this low-water stage, large mud flats and sand bars reach into the bay on the south shore, and along the east shore. The Farmington Bay WMA impoundment units are clearly seen lying on the Jordan River delta with its gentle downward slope to the north. The same pattern of emergent vegetation can be seen on exposed land downslope from each dike.

West and southwest from the Farmington Bay WMA is a great stretch of hummocky near-shore features with intervening playas, containing seasonal water. In each of these small basins emergent vegetation is abundantly displayed in the image. As the lake rises to the level of figure 10B (4,209 feet), just as in Bear River Bay, the waterfowl management structures are inundated. Concurrently, the bay advances southward on land, inundating the many small basins, and generally reduces the vegetation habitat of shore birds, such as avocets, stilts, and snow plovers. Upland vegetation such as greasewood on the hummocks still remains, but foraging species are diminished. It is very likely that this advancement of high water decreases the carrying capacity of shore birds, both in quantity and diversity. Ironically, salinity in the bay increases when the water rises above the southern causeway allowing lake brines to enter Farmington Bay.

On the east shore of the bay another transition takes place. Shore birds which are ground nesters (avocets, stilts, and others) must migrate with the changing water level. Much of the foraging habitat is reduced as the lake reaches higher levels, and the shore bird populations generally decrease. More mobile birds such as egrets and great blue herons which nest in the bulrushes and Phragmites are less affected, but may diminish in numbers as fish supplies may be reduced by changes in water level.

Figure 10A. *Farmington Bay, 1996, with a south-arm elevation of about 4,200 feet, the historic norm. Note the Farmington Bay Bird Refuge dike system (top of bottom-right quadrant), a large bar extending into the bay from the northeast (top center of image), and in the southwest toward Antelope Island (below center of bottom-left quadradant) (see figure 10A in the color section of this book).*

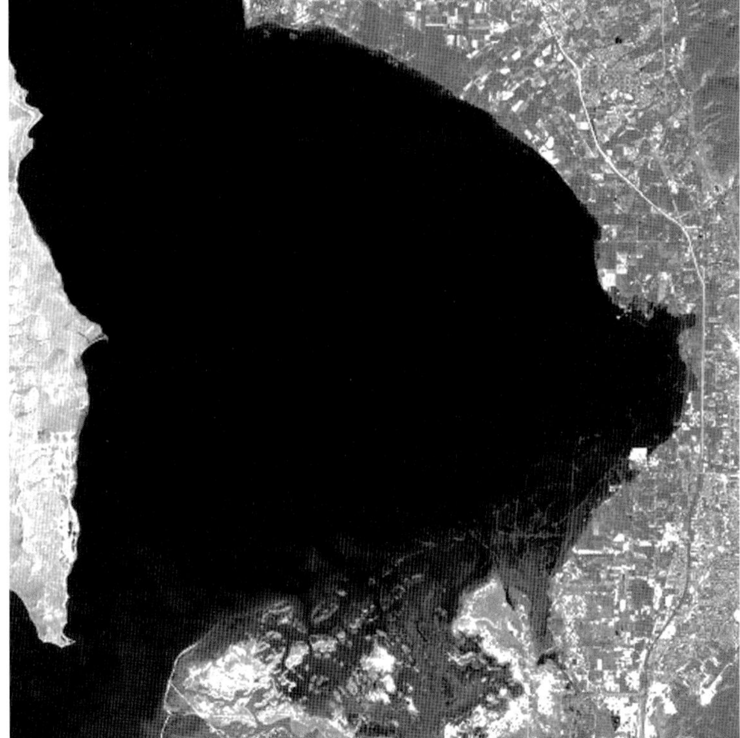

Figure 10B. *Farmington Bay, 1984, with a south-arm elevation of about 4,209 feet, more than two feet below the historic high-water line. Note the inundation of the two bars identified in figure 10A, as well as the complete inundation of the bird refuge dikes (top of bottom-right quadrant). Note also the water backing up across the wetlands toward the Salt Lake International Airport (bottom center of image) (see figure 10B in the color section of this book).*

CONCLUSIONS

This exploration into the potential application of satellite remote sensing to terminal lakes such as Great Salt Lake has been partially successful. Of the six biophysical parameters examined, some were quite successful and potentially useful, some show sufficient promise to warrant further investigation, while others show little promise of future use of the technology. The following provides a brief summary of the usefulness of the biophysical parameters.

1. Shoreline position. First, the delineation of water bodies, within the limits of the spatial resolution or pixel size, is useful in most environments. A single waveband in the near infrared range (for example, 0.7-1.1 μm) is sufficient because of the almost total absorption of infrared energy by water. Determining the shoreline position is a simple procedure providing accurate delineation. Applications at meso-scales include accurate perimeter and area determination. Given multiple applications at various levels of the water body, the technique could serve the computation of area-volume ratios, particularly in the case of terminal lakes and reservoirs.

2. Depth estimation. When the water is homogeneous this procedure seems to be successful, provided some bathymetric data are available from contour maps, or from empirical sources. Variations in suspended material, surface debris, or bottom brightness can decrease the accuracy.

3. Salinity determination. Research suggests little success for direct, accurate measures of salinity under general conditions. Indirect evidence of variations in salinity may be evident from various phenomena that are influenced by the dissolved solids content of the water. Investigation of these "secondary" phenomena may prove to be useful as indicators of dissolved-solids concentration.

4. Temperature determination. This study evaluated only TM data with one broad thermal band at 120-meter spatial resolution. Results are useful to display general patterns and comparisons between extreme water levels, and, presumably, different seasons. More definitive results would require controlled or monitored conditions, including concurrent temperature measurements from the field for calibration. Also, airborne sensors designed specifically for thermal applications can be employed on call according to ambient conditions. TIMS (Thermal Infrared Multispectral Sensor) and ATLAS (Advanced Thermal Land Applications Sensor), both developed at NASA's Stennis Space Flight Center for aerial surveillance, offer the additional advantage of finer spatial resolution to about 5m, as determined by flight altitude.

5. Brine shrimp cysts and algae investigation. TM data may be useful for detecting changing patterns of algal populations with changing seasons. The extent to which this may be helpful in estimating or predicting brine shrimp cyst production is unknown.

6. Waterfowl habitat evaluation. This evaluation showed that remote sensing may be a useful tool in characterizing changing conditions of waterfowl habitat in time and space. This brief exploration dealt only with broad-band TM data in six channels. It seems likely that hyperspectral imaging from orbiting spacecraft, or from aircraft, should be a valuable management tool. This possibility is worthy of further investigation to determine whether indicator species of vegetation for habitat assessment can be detected from hyperspectral, narrow-band instruments.

Remote sensing technology is advancing rapidly. While spatial resolution is becoming more refined (to 1 meter or less from aircraft and spacecraft), it is improved spectral resolution (hyperspectral imaging) that will be more important to water quality investigations. AVIRIS (Airborne Visible Infrared Imaging Spectrometer) is such an instrument with 224 narrow-band channels. AVIRIS data is very useful for vegetation and soil discrimination, and perhaps water as well. Those involved in managing the resources of Great Salt Lake and vicinity should keep abreast of advancements in remote sensing to better understand the many demands on the lake, its resources, and surrounding environment.

REFERENCES

Baskin, Robert L., 1990, Determination of ground-water inflow locations in Great Salt Lake, Utah using the Thermal Infrared Multispectral Scanner: University of Utah Master's Thesis, Salt Lake City, Utah, 67 p.

Bukata, R.P., 1995, Optical Properties and Remote Sensing of Inland and Coastal Waters: CRC Press, Inc., Florida, 362 p.

Cracknell, A.P. (editor), 1983, Remote Sensing Applications in Marine Science and Technology: D. Reidel Publishing Company, Holland, 466 p.

Great Salt Lake Yacht Club, 1992, Great Salt Lake - south arm, soundings in feet (map): Great Salt Lake Yacht Club, scale 1:90,000.

Gwynn, J.W., 1986, An approximation of the physical and chemical characteristics of Farmington Bay and Bear River Bay, Great Salt Lake, Utah: Utah Geological and Mineral Survey Open-File Report 211, 22 p.

—1988, Great Salt Lake brine sampling program 1985-1987. Utah Geological and Mineral Survey Open-File Report 117, various pagings.

Ikeda, Motoyoshi, and Dobson, F.W. (editors), 1985, Oceanographic Applications of Remote Sensing: CRC Press, Inc., Florida, 492 p.

Jensen, J.R., 1996, Introductory Digital Image Processing-a Remote Sensing Perspective: Prentice Hall, New Jersey, 316 p.

Lin, Anching, 1976, A survey of the physical limnology of Great Salt Lake: Utah Division of Water Resources, 82 p.

Sturm, P.A., 1986, Utah Geological and Mineral Survey's Great Salt Lake brine sampling program 1966-1985-history, database, and averaged data, Great Salt Lake, Utah: Utah Geological and Mineral Survey Open-File Report 87, various pagings.

U.S. Geological Survey, 1974, Great Salt Lake and vicinity: U.S. Geological Survey, scale 1:125,000.

—1984, Water Resources Data, Utah - Water Year 1984: U.S. Geological Survey Water-Data Report UT-84-1, 463 p.

—1995, Satellite Image Map: Prepared and published by the U.S. Geological Survey, compiled by the U.S. Fish and Wildlife Service and National Biological Survice, National Gap Analysis Program, and the Department of Geography and Earth Resources, College of Natural Resources, Utah State University, scale 1:750,000.

BIOLOGY

POPULATION DYNAMICS OF THE BRINE SHRIMP, *ARTEMIA FRANCISCANA*, IN GREAT SALT LAKE AND REGULATION OF COMMERCIAL SHRIMP HARVEST

by

Doyle Stephens
U.S. Geological Survey

and

Paul W. Birdsey, Jr.
Utah Division of Wildlife Resources

ABSTRACT

Brine shrimp (*Artemia franciscana*) from Great Salt Lake have been commercially harvested as adults since 1950, and as cysts since 1952. Reported cyst harvest (total unprocessed harvest) has varied widely from 265 pounds (120 kg) in 1968, to about 7,400 tons (6,600 metric tons) in 1995 and 1996. The harvests are directly affected not only by market demand for the product, but by changes in environmental conditions within the lake.

A comprehensive monitoring and analytical program, begun in 1994, employs frequent surveys of *Artemia* populations, measurement of environmental data, and mathematical modeling to understand the dynamics of the *Artemia* population in Great Salt Lake. This information is then used to establish the timing and harvest limits for the commercial fishery in the lake. Since 1995, there have been two years with large populations of *Artemia* and record cyst harvests, and two years with relatively poor cyst harvests.

Record *Artemia* harvests of 7,400 tons (6,600 metric tons) raw weight of all life forms occurred in 1995 and again in 1996. Both years were characterized by salinities of 12 to 15 percent, rapidly warming water temperature in springtime, abundant but small-sized phytoplankton, and presence of small numbers of *Artemia* in March. Small *Artemia* populations and poor harvests occurred in 1997 and 1998. Conditions during these years included salinities of 9.3 to 13 percent, slowly warming water temperature in 1998, phytoplankton dominance by large diatoms in 1997, and presence of large numbers of *Artemia* in March of 1997.

The 1996 moratorium on new shrimping permits has been extended indefinitely pending review of the data. Important relations are being investigated to better understand and protect the resource: the quantity of brine shrimp required by migrating birds, the contribution of beached cysts to subsequent populations, the role of changing water levels and salinity in determining algal and brine shrimp population dynamics, and the loss of cysts from the lake during winter.

INTRODUCTION

The zooplankton community of Great Salt Lake exhibits large annual variations in timing and abundance of life stages, and is dominated by the brine shrimp *Artemia franciscana* Kellogg. These variations are due in part to changes in salinity, water temperature, nutrients, and algal populations. Nauplii typically appear from over-wintering eggs (hard-walled cysts containing an embyro in diapause) between late February and early March when water temperatures are between 0° and 5° Celsius (°C) (32° and 41°F). Their appearance follows by about one month the winter bloom of phytoplankton that provides food for the developing nauplii. Peak numbers of nauplii may occur as early as mid-April, or not until mid-May. The availability of sufficient food is critical because nauplii undergo eight molts before reaching the juvenile stage, and another three molts before reaching adult stage (Lenz, 1984). Mortality rates for larval shrimp can be very high, and few become reproducing adults. Studies with *A. monica* from Mono Lake, California, indicate that nauplii have a 95 percent mortality rate when water temperatures are 5°C (41°F) or lower (Dana and others, 1990).

If food availability and environmental conditions are suitable, the first generation of adults produced from cysts reproduces sexually and will produce the second generation ovoviviparously (live birth). If conditions are less favorable, oviparous reproduction produces cysts. Since 1995, there have been two or three generations of *Artemia* produced each year. Typically, the *Artemia* graze the phytoplankton to near extinction by May and become food limited. The adult females begin oviparous reproduction at this time. The number of adults reaches a maximum about mid-May and gradually declines until December when water temperatures drop below about 3°C (37°F) and they die.

Brine shrimp from Great Salt Lake have been commercially harvested as adults since 1950, and as cysts since 1952. In 1999, the harvest was primarily for cysts. Reported cyst harvest from 1964 to 1978 varied widely from 265 pounds (120 kg) in 1968, to 83 tons (75 mt tons) in 1966 (Sturm and others, 1980), to about 7,400 tons (6,600 mt tons) during 1995 and 1996. (Reported amounts include total unprocessed biomass of which about 3/4 is cysts.) Historically, the salinity of the lake has been the most important parameter associated with the abundance and distribution of *Artemia* and the success of the harvest. Between 1982 and 1989, when the salinity of Gilbert Bay, the southern part of the lake, varied from 6 percent to 10 percent, cyst production was poor, and most harvesters moved to Gunnison Bay, the northern part of the lake, where the salinity was 15 to 17 percent. Since about 1990, reproducing populations of *Artemia* have been limited to Gilbert Bay, although during the 1998 to 99 season, limited harvest efforts resumed in Gunnison Bay (figure 1).

Although much is known about the many species of *Artemia*, little is known about the population in Great Salt Lake, and the biotic and abiotic variables that interact to determine the health and fecundity of the population. A comprehensive monitoring and analytical program was started by the Utah Division of Wildlife Resources (UDWR) in 1994 that included participation by researchers at Utah State University, Logan, and the U.S. Geological Survey. This report discusses how changes in environmental conditions resulted in large variations in the *Artemia* population during 1995 to1998 and how information about the *Artemia* standing crop is used to establish harvest limits for the commercial fishery in the lake.

Location and General Hydrology

Great Salt Lake is a terminal lake, and Holocene successor to Pleistocene Lake Bonneville. Great Salt Lake is the fourth largest terminal lake in the world, covering 1,700 mi^2 (4,403 km^2) at the historic average surface elevation of 4,200 ft (1,280 m) with a maximum depth of 34 ft (10.4 m). The area of the lake is dependent upon the balance between inflow and evaporation, and has varied during historic times from 950 mi^2 (2,460 km^2) in 1963, to 2,300 mi^2 (5,950 km^2) at its peak elevation of 4,211.85 ft (1,283.77 m) in 1986 and 1987 (Arnow and Stephens, 1990).

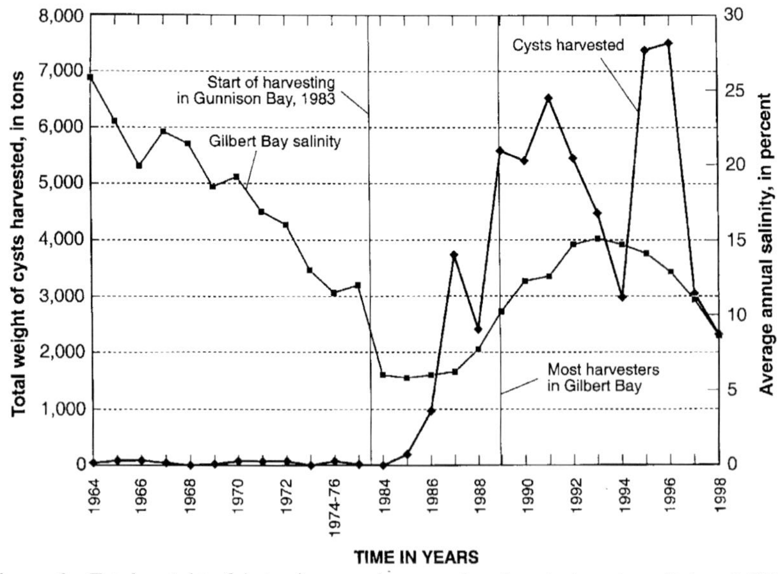

Figure 1. *Total weight of* Artemia *cysts harvested and variations in salinity of Gilbert Bay of Great Salt Lake, 1964-98.*

About 92 percent of the average inflow of 1.9 million acre feet (2.3 billion m^3) to Great Salt Lake is from the Bear, Weber, and Jordan Rivers, with small amounts contributed by precipitation and springs. Water-budget studies indicate that surface flow from rivers contributes about 82 percent of the annual dissolved-solids load of 2.1 million tons (1.9 million metric tons), and springs contribute 18 percent (Waddell and Barton, 1980). The lake is saline due to evaporation that concentrates the dissolved solids. There are about 4.5 billion tons (4.1 billion metric tons) of salt in the lake, and commercial harvest of salt removes about 2.3 million tons (2.1 million metric tons) annually (Gwynn, 1997).

Much is known about the major ions (salts) in Great Salt Lake (Hahl and Langford, 1964; Gwynn, 1998). However, due to analytical difficulties with brines, there are few references to trace elements in the lake (Taylor and others, 1980; Domagalski and Eugster, 1990). Large amounts of nitrogen and phosphorus are discharged by the Jordan River, and to a lesser extent by the Bear and Weber Rivers. Most of the nutrients are removed by wetlands surrounding the east shore of the lake. Nitrogen has been reported to be the nutrient that limits algal production and subsequent *Artemia* nutrition in Great Salt Lake (Porcella and Holman, 1972; Stephens and Gillespie, 1976; and Wurtsbaugh, 1988). Ratios of total nitrogen to total phosphorus in the lake have been less than 15 during the 1995 to 1998 study period, and are indicative of nitrogen limitation for algal growth (Stumm and Morgan, 1970).

Methods

Sampling sites were selected using a stratified, randomized design. Digital map data for Gilbert Bay of Great Salt Lake at a scale of 1:100,000 (U.S. Geological Survey Digital Line Graph series) were assembled to create a Geographic Information System (GIS) coverage and overlaid with a 1-square-mile (2.59 km^2) grid pattern. Each cell of the grid was then sequentially numbered and a computer algorithm was used to generate a table of randomly selected sampling sites. The first seven sites (randomly selected by computer) that were accessible and had depths less than about 13 feet (4 m) were designated as shallow. The first seven sites with depths greater than 16.5 feet (5 m) were designated as deep (figure 2). Since September 1995, three additional nonrandom sites have been added.

Beginning in 1995, *Artemia* were sampled biweekly or monthly at each site using a 1.64-ft diameter (0.5-m) plankton net with a length of 2 m (6.6 ft) and a 153-µm mesh. Each net haul consisted of lowering the net to the bottom of the lake, recording the depth, and quickly raising the net to sample a vertical cylinder of water that had a diameter of 1.64 ft (0.5 m) and length equal to the water depth. Each meter of water depth corresponded to 52 gal (196 L) of water passing through the net. Three replicate hauls were made at each site to provide an estimate of the within-site variability.

Brine shrimp, like all zooplankton, are notoriously patchy in their distribution within a lake. This is partially due to the presence of life

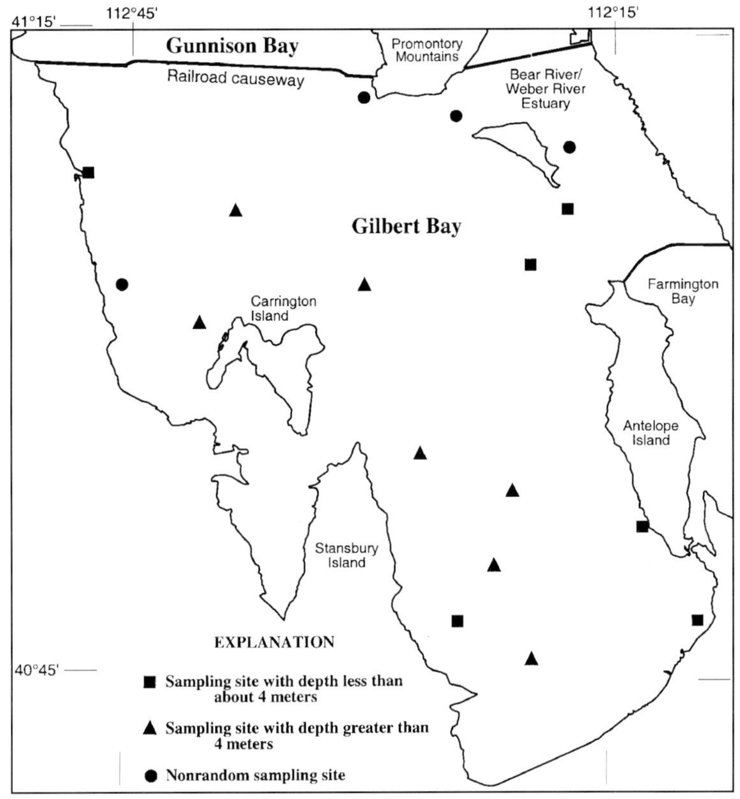

Figure 2. *Location of brine shrimp sampling sites in Gilbert Bay of Great Salt Lake.*

stages such as cysts that float and are totally dependent on currents to move them throughout the water. Other life stages, such as adults, are actively motile and able to move toward food sources and sexual partners with a limited dependency on currents. The patchy distribution makes it difficult to estimate accurately the numbers of various life stages present at any one time in the lake. For example, counts of shrimp cysts collected at each of 17 sites during November 2 through 4, 1998 had an average within-site coefficient of variation (standard deviation expressed as a percentage of the mean) of only 24 percent. However, the between-sites coefficient of variation was 123 percent. Reduction of the between-site variability is being investigated using stratification of sites by depth or area.

Artemia collected in the field and preserved with buffered formalin were counted in the laboratory by using a dissecting microscope. *Artemia* were separated by development stage and size into nauplii (length less than 1.5 mm [.059 in.]), juveniles (length >1.5 mm [.059 in.] [and filtering appendages present), and adults (presence of ovisac on females and claspers on males). Generally, numbers of *Artemia* adults and juveniles were determined from subsamples of a minimum of 50 individuals (if available). Nauplii and free cysts were counted from 5-ml (.31 in.^{3}) subsamples. Because *Artemia* are not distributed randomly, variability of population counts among sites is often high. Therefore, the median is used throughout this discussion to best represent the central tendency of the population.

Environmental measurements conducted at each site consisted of a vertical profile of water temperature and dissolved oxygen, photosynthetically active radiation, and salinity. Water samples at a subset of sites were collected using a pump and filtered (0.45-µm) for nutrient determinations (dissolved nitrate and nitrite, dissolved silica, total phosphorus, total ammonia nitrogen). Chlorophyll was extracted on 0.8 - µm filters and analyzed by fluorometry using procedures in Wetzel and Likens (1979). Concentrations of chlorophyll a were corrected for phaeophytin a. Phytoplankton samples were collected seasonally at several sites, counted, and analyzed for taxonomic identification by Dr. Samuel Rushforth at Brigham Young University.

Environmental Variables and Artemia Populations

Salinity and temperature significantly affect survival and development of all stages of *Artemia*, but effects on reproduction are of greatest significance. Laboratory studies by Wear and others (1986) showed that maturation and reproduction for *Artemia franciscana* (New Zealand population) occurred at temperatures between 20°C and 28°C and salinities between 120 and 200 parts per thousand (ppt). Similarly, Browne and others (1988) reported that the longest reproductive period for *Artemia franciscana* (San Francisco Bay population) occurred when reared at 24°C. Studies of the Great Salt Lake strain of *Artemia franciscana* by Von Hentig (1971; cited in Wear and others, 1986) showed maturation was greater at 30°C than at 20°C, but temperatures between 10°C and 30°C were not investigated. Gliwicz and others (1995) found that under laboratory conditions, some measures of reproduction in *Artemia franciscana* (Great Salt Lake population) required up to five times longer at water temperatures of 10°C than at 25°C. Warmer temperatures early in the year allow nauplii to take advantage of the abundant winter phytoplankton bloom. First generation females that develop when temperatures are warm and food is abundant will reproduce ovoviviparously (eggs that hatch within the ovisac rather than dormant cysts), thereby saving about 22 percent of the additional energy required for cyst production (Clegg, 1962).

The optimum salinity for growth and reproduction of Artemia in Great Salt Lake is not known because it is a function of an interaction with temperature, food supply, and competitors. The maximum salinity for adult *Artemia* determined by laboratory studies of cultures from the lake is about 300 g/L (Croghan, 1958); nauplii cannot tolerate salinities in excess of 146 to 175 g/L (Conte and others, 1972, 1973). As lake salinity declines, the number of species expands, increasing the potential for predation, competition, and disruption in the food supply, as occurred lakewide during 1986 to 1989 (Wurtsbaugh and Berry, 1990) and within Gilbert Bay during 1997 (Stephens, 1998).

Since 1995, declining salinity and abundant dissolved silica (typically near 6 mg/L) allowed many species of diatoms to increase their abundance in the lake. During much of 1997, the phytoplankton community in Gilbert Bay was dominated by pennate diatoms, many of which were larger than 30 µm. The nauplii of *Artemia franciscana* is a non-discriminant filter feeder that will ingest anything with a diameter that does not exceed about 30 µm (Dobbeleir and others, 1980). Stephens (1998) hypothesized that the large numbers of nauplii present in the lake during the spring of

1997 consumed the smaller phytoplankton leaving only the large diatoms. Consequently, the nauplii and early instars of juvenile *Artemia* were physically unable to ingest sufficient numbers of phytoplankton to sustain them, and populations of *Artemia* declined.

Additional support for this hypothesis is based on specific nutritional requirements of *Artemia*. The superspecies of *Artemia franciscana* found in Great Salt Lake is unique within the genus *Artemia* in that it is unable to synthesize the components of some proteins (purine ring) unless the amino acid tryptophan is present in its diet (Hernandorena, 1987). The lack of amino acids needed to synthesize purine ring results in poor formation of appendages in the nauplius stage, and may result in the appearance of black spots on various stages of *Artemia*. Black spot is, therefore, a potentially fatal disease of nutritional deficiency and was observed infrequently in the lake prior to 1997 (Wurtsbaugh, 1995). The replacement of small chlorophyte phytoplankton by large pennate diatoms in late 1996 and 1997 was associated with the appearance of black spot disease in 11 samples collected from the lake between July and November 1997. Additionally, there was a complete loss of laboratory cultures of *Artemia* nauplii due to black spot at Utah State University during 1997, when lake water was used for rearing (Belovsky and Mellison, 1998). The dominance of large diatoms during 1997 probably was responsible for the poor nutritional state of the *Artemia* population and the appearance of black spot disease.

POPULATION DYNAMICS OF ARTEMIA IN GREAT SALT LAKE

Life-history information about *Artemia* in Great Salt Lake has been available since 1918 (Jensen, 1918), but study of the population dynamics of *Artemia* in the lake did not begin until the 1970s (Wirick, 1972; Stephens and Gillespie, 1972; Montague and others, 1982). Subsequently, most of the available information on *Artemia* dynamics in Great Salt Lake is from work done since 1990 (Wurtsbaugh and Berry, 1990; Wurtsbaugh, 1992 and 1995; Gliwicz and others, 1995; Stephens, 1998). Since 1995, there have been two years with large populations and record cyst production and two years with relatively poor cyst production.

A record *Artemia* harvest of 7,400 tons (6,600 metric tons) in 1995 followed the relatively small harvest of 3,000 tons (2,700 metric tons) in 1994. Although the number of cysts present in February 1995 at the start of the season is not known, the adult *Artemia* population in mid-April was only 1,000 per cubic meter (m^3), which indicates that a generally small standing crop of cysts was present. Concentrations of chlorophyll in April and early May were near 30 µg/L and the phytoplankton population throughout the spring was dominated by *Dunaliella viridis* with only small numbers of diatoms present (Wurtzbaugh, 1995). Water temperature increased rapidly from a winter low of 1°C to 8°C by mid-March and the salinity was between 12.7 and 15 percent during 1995. The total phosphorus concentration was near 200 µg/L during 1995, but measurements of nitrogen compounds were not made (Wurtzbaugh, 1995). Female Artemia produced large numbers of nauplii from ovoviparous eggs in April and May and began cyst production in June. The phytoplankton were heavily grazed through the summer, but by September there were 26,600 *Artemia* cysts/m^3 in the water prior to the start of the shrimp harvest. The year can be described as one with relatively high salinity, rapid warming of the water in March, moderately high standing crop of chlorophyte phytoplankton, and low *Artemia* density at the start of the season. These conditions allowed early hatching of nauplii while the winter phytoplankton bloom was still underway, providing sufficient food particles of suitable size for the small number of nauplii, and generally favorable conditions for ovoviparous reproduction of the second generation. The cumulative effect of these conditions was record production of *Artemia* cysts during the summer and fall (due to limited amount of data available prior to August 1995, a population profile for 1995 is not presented).

The population structure in 1996 was similar to that of 1995. There were about 6,300 cysts/m^3 in the lake in mid-February. About 38 percent of these hatched and survived to produce the first generation nauplii population (figure 3). Phytoplankton, as indicated by chlorophyll concentrations, was abundant through early April, as shrimp populations were small. Although the phytoplankton was dominated by diatoms, cell sizes were sufficiently small to be ingested by nauplii, and generally there was high recruitment of these nauplii to juveniles by early May. As the shrimp population increased, the phytoplankton decreased. There was high mortality of the juveniles and by mid-May, the adult population reached about 3,000 individuals/m^3 as the phytoplankton population continued to decline.

A very large number of nauplii were released in mid-May, primarily from cysts produced in response to rapidly declining food supply. The adult population declined throughout the summer and phytoplankton chlorophyll increased slightly through July. A third generation of nauplii were produced primarily from cysts in late July. During August and September, large numbers of nauplii grazed the

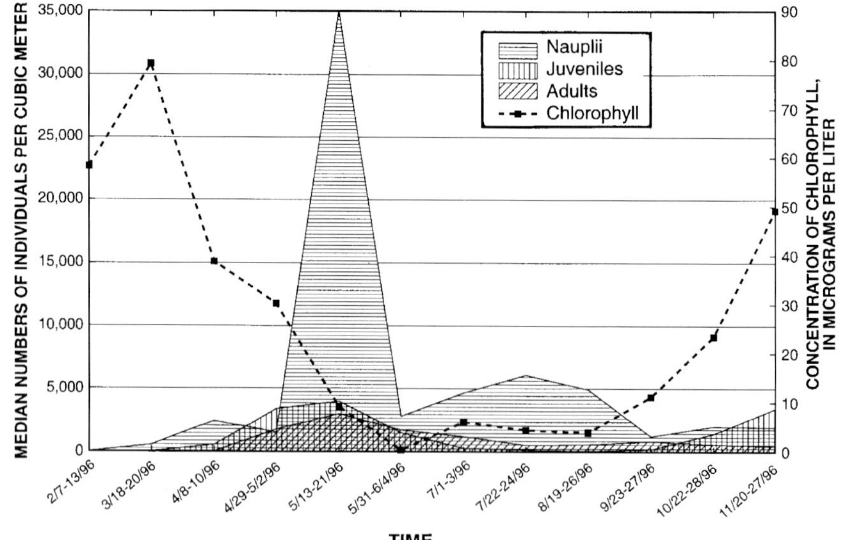

***Figure 3.** Median number of* Artemia *and concentration of chlorophyll in Great Salt Lake, 1996.*

smaller phytoplankton cells that were competing with the large diatoms for nutrients and light, which eliminated the competing smaller phytoplankton and allowed the larger pennate diatoms to flourish and resulted in increased chlorophyll in the lake in October and November. However, the dominance of large pennate diatoms in the phytoplankton in late fall provided little suitable food for the third generation of nauplii, and although some of them reached juvenile stage, it is unlikely that many became reproducing adults. Large numbers of cysts were produced from July through September and the preharvest standing crop of cysts reached nearly 68,000/m^3 (figure 4). Salinity ranged from 12.8 percent to about 14 percent during 1996. The year was characterized by rapidly warming water temperatures in March, a relatively small number of Artemia cysts at the start of the season, spring phytoplankton populations consisting of small diatoms or chlorophytes, oviparous reproduction producing the second generation, mid- and late-summer blooms of phytoplankton, and production of a large quantity of cysts in late summer. The 1996 commercial harvest was 7,400 tons (6,600 metric tons).

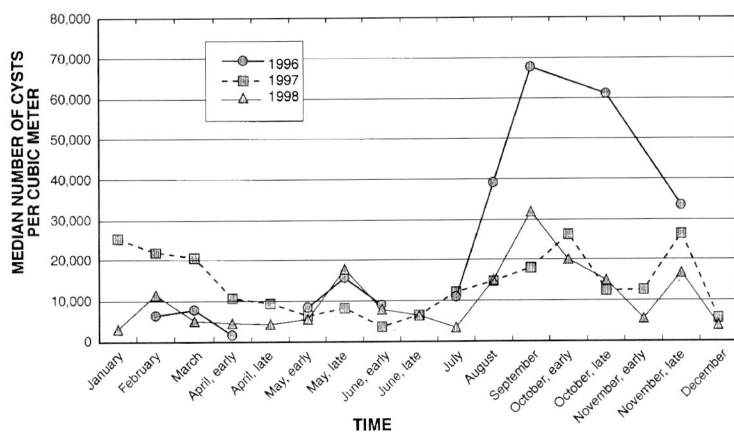

Figure 4. *Median number of* Artemia *cysts in Great Salt Lake, 1996-98.*

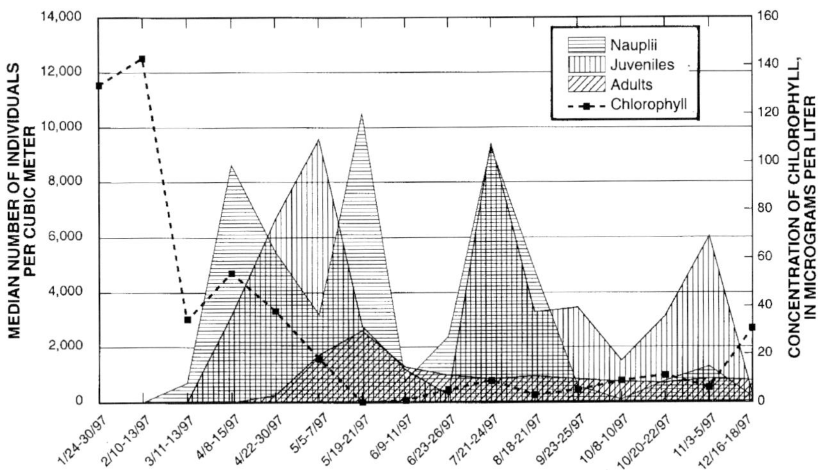

Figure 5. *Median number of* Artemia *and concentration of chlorophyll in Great Salt Lake, 1997.*

The January and February 1997 bloom of pennate diatoms resulted in very large concentrations of chlorophyll in the lake (figure 5). A large number of cysts (20,470/m^3) were present in March 1997 (figure 4), and the first generation of nauplii hatching from the cysts reached nearly 8,600/m^3 by mid-April (figure 5). A large number of these nauplii successfully recruited to juvenile stage by early May. Although large pennate diatoms dominated the spring phytoplankton population through much of March, small centric diatoms also were available. Recruitment of the juveniles to adult stage was poor, and by late May the adult population was only 2,300 individuals/m^3. A second generation of nauplii was produced in May by ovoviparous and oviparous reproduction, but few of these recruited to juvenile stage, likely because of poor food supply. By July, only 35 percent of females were ovigerous, with 33 percent of these females carrying eggs and 67 percent carrying cysts. The lake salinity declined from 13 percent in January to 11 percent in July. A third generation of nauplii produced in July successfully recruited to juvenile stage, but numbers of adults remained near 800/m^3 through September. The phytoplankton remained dominated by large pennate diatoms throughout the summer. The percentages of females carrying cysts were quite high through September and October, but there were fewer cysts per ovisac than there were in 1996. By the end of September, the free cyst density in the water was only 18,000/m^3, much lower than in 1995 or 1996 (figure 4). The harvest began October 1, but was halted by UDWR on October 27 after 6.1 million pounds (2.7 million kg) had been harvested.

The *Artemia* population did poorly in 1997 because a series of environmental and biological conditions occurred that resulted in a food base with phytoplankton cells too large to ingest. The salinity of the lake declined to 11 percent in the summer of 1996, allowing a greater number of phytoplankton species to occur. The large number of *Artemia* that were present in late 1996 and early 1997 preferentially grazed phytoplankton cells that were smaller than 30 µm in diameter, leaving large pennate diatoms to become the dominant food crop. This resulted in poor production of cysts and a poor harvest.

The median number of cysts available to start the population cycle in February 1998 was about 11,400/m^3. The first appearance of nauplii was early February, when the water temperature was 4°C and nauplii increased to a median concentration of about 13,000/m^3 by the third week in May, a month later than in previous years (figure 6).

Water temperatures during the winter of 1998 were not as cold as in the previous two years and never dropped below 2°C at a depth of 1 m at a deepwater mid-lake site.

However, water temperatures were slower to warm in 1998 than in 1996 and 1997. Water temperatures from early May to early July 1998 were as much as 5°C colder than during the same period in 1997. The cold temperatures slowed the maturation of the *Artemia* by about 4 to 6 weeks relative to previous years.

The initial hatch of nauplii in March 1998 did not reach adult stage until May (figure 6). A second large hatch of nauplii, from ovoviparous eggs, occurred in late May and moderate numbers of nauplii were also present in late June. The ratio of number of juveniles to number of nauplii was large

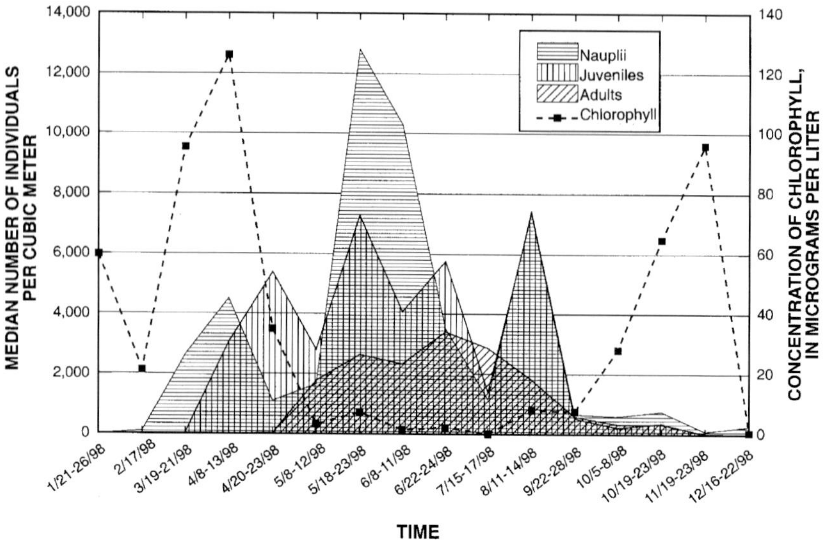

Figure 6. *Median number of* Artemia *and concentration of chlorophyll in Great Salt Lake, 1998.*

in April and May 1998. Large ratios of juveniles to nauplii indicate high survival of the nauplius stage and recruitment into juveniles. By June, the numbers of adults (especially females) were larger than in the previous two years, indicating high survival and recruitment likely due to sufficient food supply during springtime.

However, the percentage of females carrying eggs or cysts was considerably lower during 1997 and 1998 than in previous years. In early May, less than 1 percent of females carried ovoviparous eggs as compared to 30 percent in 1997. By the end of June, only 1 percent of females carried cysts and the average number of cysts per female had dropped to 14. The median number of cysts present in the water at this time was 6,263/m^3. Lake salinity declined from 11 percent in January to 9.3 percent in July.

Cyst production began increasing in late July, and by the end of September more than 30 percent of the females carried cysts with an average of 20 cysts per ovisac. A third group of nauplii, produced primarily from cysts in August, successfully recruited to juvenile stage, but adult populations continued to decline through the late summer. When the commercial shrimp harvest started on October 1, more than 30,000 cysts/m^3 were present in the water. At the start of the harvest in 1997 there were only 19,000 cysts/m^3, but in 1996 there were more than 65,000/m^3.

The year was characterized by generally abundant phytoplankton, but cold water temperatures from May through June retarded development of the *Artemia* population. This resulted in production of most nauplii by oviparous (cyst) mode early in the year and large numbers of female *Artemia*, although few carried ovoviparous eggs throughout the year. Due to low density of cysts in the water, the harvest was suspended after only 22 days and the estimated harvest was between 4.6 and 5.2 million pounds (2,090 to 2,350 metric tons).

REGULATION OF COMMERCIAL HARVEST

The commercial harvest of brine shrimp began in 1950 when Sanders Brine Shrimp Company was formed to supply adult brine shrimp for use in the tropical fish market (Sturm and Sanders, 1980). Brine shrimp cysts were first harvested in 1952 and are the only life stage currently commercially harvested because of their long shelf life and ease of collection and processing. Since its inception, the commercial harvest of brine shrimp cysts has blossomed into an industry reportedly worth between 10 and 60 million dollars annually, depending on the quantity and quality of harvest. Cysts from Great Salt Lake are packaged and shipped around the world for use primarily in the rearing of prawns and finfish by the aquaculture industry. Industry representatives have stated that Great Salt Lake is the world's largest supplier of *Artemia* cysts to the aquaculture industry because the high-quality cysts are of small size and the resulting small nauplii make them readily usable as food by the early life stages of cultured animals.

Permits and Season Length

The UDWR has the legislative authority to regulate the taking of all protected wildlife species within the state. Harvest of brine shrimp and brine shrimp cysts is regulated through Title 23 of the Utah State Code (Annotated) and Administrative Rule R-657-14.

Harvest companies are issued permits called Certificates of Registration (COR) that entitle the company to harvest in only one location, on the lake or beach, at a given time. Harvest locations are defined as a 300-yard (274 m) radius around the COR-bearer's location that cannot be encroached upon by any other company while harvest activities are underway. Unlike in Mono Lake in California where shrimp cysts sink, brine shrimp cysts float in Great Salt Lake, often forming large streaks on the lake surface. Because of movement of the cyst streaks caused by wind and wave action, harvest locations are not restricted to a specific geographic location and can be moved as the cysts move. Originally, fees for the harvest permits consisted of a small annual fee and a royalty payment of $0.04/lb (0.45 kg) of cysts harvested as reported by the companies. No monitoring program existed to verify the company-reported harvest, and it was suspected that not all companies were reporting accurately. The presumed inaccurate reporting led UDWR to change the fee structure to a flat fee of $3,000 per COR in 1985. In 1991, the fee was increased to $10,000 per COR. A single company may have more than one COR, and the total number of CORs available was not limited prior to 1996. Currently, CORs are not transferable between companies.

Changes in the world aquaculture market in the early 1990s led many new companies to participate in the brine shrimp harvest and older companies to expand their operations. As a result, the number of companies participating in the harvest and the number of CORs increased dramatically in a short period of time (figure 7). The harvest companies approached UDWR in early 1996 with concerns about the possibility of over-harvesting if the number of CORs were not limited. As a result of extensive work by industry and UDWR representatives in 1996, a ceiling of 79 CORs distributed among 32 companies was recommended to the Wild-

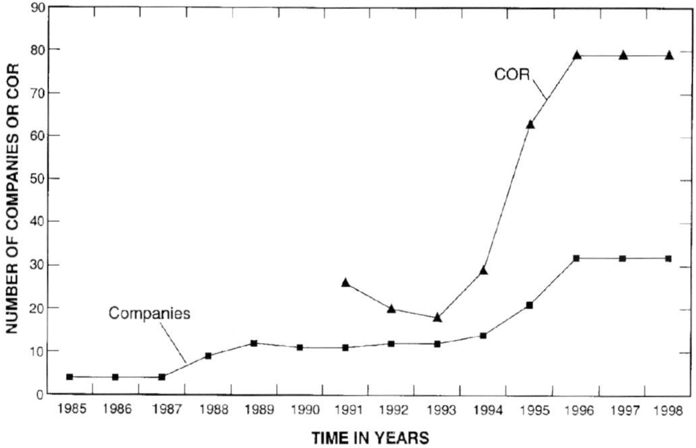

Figure 7. Increase in number of companies harvesting Artemia *on Great Salt Lake, and numbers of Certificates of Registration (COR) for harvesting, 1985-98.*

life Board. Furthermore, the Wildlife Board established a 3-year moratorium on additional CORs in order to allow UDWR to acquire the data necessary to determine at what level a harvest could be sustained. As of February 1999, the COR moratorium was extended indefinitely.

Currently, the harvest season extends from October 1 through January 31. Fishing hours extend from one hour before sunrise to sunset. Originally, the season was open all year long. Concerns about possible overharvesting caused the Wildlife Board to shorten the season to seven months (September through March) in 1991. The number of harvest locations per COR also was reduced from two to one at that time. In 1994, the season length was further reduced to five months (October through February). The current four-month season was established in 1996 as part of the moratorium.

Harvest Limits and Brine Shrimp Population Modeling

Kuehn (this volume) details the evolution of the brine shrimp harvest technology. The use of technology such as airplanes, fast boats, radar, global positioning system receivers, radios, and night vision equipment has become common in the industry. Pettengill (1996) reported that some companies stated that harvesting efficiencies had increased fourfold during the preceding few years as a result of changing technology. The application of new technology coupled with the increased number of companies participating in the harvest have led to difficult management problems during the past several years.

Until recently, many resource managers believed that the brine shrimp fishery in Great Salt Lake would be self-regulating. That is, the amount of work required to gather reduced supplies of cysts during years of poor cyst production would make it economically prohibitive for harvesters to deplete the resource below a certain level. Therefore, brine shrimp would not be over-harvested. Although this may have been true in the past, increased competition and worldwide demand for the cysts led managers to fear that a "tragedy of the commons" (Hardin, 1968) scenario found in many shared resources might occur. Under this scenario, companies will continue to harvest despite declining profitability for fear that their competitors are harvesting a product that rightfully belongs to them. This was demonstrated in 1997 when the companies informed UDWR that the cysts were of poor quality, yet only one of the companies was willing to suspend harvest operations.

The tragedy of the commons mentality combined with changing technologies have created many of the same overcapitalization problems exhibited by other commercial fisheries (Gordon, 1954; McCay and others, 1998) i.e., apparently declining catch per unit effort (Belovsky and Mellison, 1997), subsequent longer working hours, and unsafe working conditions. It is not uncommon for boat crews to spend 20 hours per day on the water, and many crews remain on the lake overnight to be closer to anticipated streaks of cysts in the morning. Spotter planes frequently take off shortly after midnight to begin the search for cysts and remain in the air for many hours at a time. Air traffic controllers have reported less than 200 feet (61 m) of separation between spotter planes on occasion. Ironically, UDWR first recommended the moratorium on additional CORs to avoid many of these situations.

In 1996, UDWR imposed the first restriction on the quantity of cysts harvested. At that time, preliminary modeling information from researchers at Utah State University indicated that it was possible to over-harvest the fishery, and that an over-harvested fishery would result in a precipitous decline in the population to a level that would no longer support a commercial fishery. Unfortunately, the modeling effort was too rudimentary at that time to accurately predict what quantity of cysts was necessary to fully reseed the population. Because no better information was available, the UDWR decided to terminate the season when the harvest total had reached the 1995 level of approximately 7,500 tons (6,800 metric tons) of raw biomass. It was believed that this level was safe because the fishery had sustained that level of harvest previously and recovered. However, the large number of cysts remaining after the harvest in 1996 resulted in an abundant shrimp population that overgrazed the small forms of algae and may have contributed to the problems experienced in 1997.

The poor production and survival experienced during 1997 made it apparent that a harvest quota would need to be established in order to protect the fishery. A simple model to predict the amount of allowable harvest was developed by UDWR that used data from the 1995 and 1996 harvest seasons combined with cyst-density data collected by the U.S. Geological Survey in September prior to the harvest. Further research by investigators at Utah State University indicated that a minimum of approximately 500,000 pounds (225,000 kg) of viable cysts needed to be present in the lake following harvest in March to fully reseed the population.

Several factors complicated the modeling process. First, the period of maximum cyst production overlaps the beginning of the harvest period. Therefore, allowances needed to be made for any additional production that occured during the harvest. Additionally, no data were available on the amount of cysts that could be expected to be lost through the winter. Finally, there were concerns that the value of 500,000 pounds (225,000 kg) represented an estimate based on 100-percent viability of the cysts present. Because brine shrimp cysts must undergo a period of diapause after release from the females, but before hatching (Lavens and Sorgeloos, 1987), the viability of the cysts in the lake is unknown in September. However, the large number of cracked and hydrated cysts in the population in 1997 indicated that viability was probably low.

Data from the previous two harvest seasons indicate that total cyst production was equivalent to 115-130 percent of the quantity of cysts present in the lake in September. The greater of these two values was used to calculate the amount of intrinsic production remaining in the population in September 1997. Future modeling efforts will likely refine this relationship, perhaps as a function of number of females in the population and water temperature.

Because no reliable estimates of cyst loss over a winter could be obtained from an extensive literature search, a loss factor of 50-percent was used to calculate the amount of cysts necessary in December to generate the desired 1.5 million pounds (700,000 kg) the following spring. December was used as the critical period because cyst production typically should have ceased by then.

To compensate for the suspected reduced cyst viability in 1997, resource managers at UDWR decided to attempt to match the quantity of cysts present in the lake in March 1996, about 1.5 million pounds (700,000 kg), to ensure full reseeding of the population. This value represented an almost threefold increase compared to the theoretical minimum and was believed sufficient to accommodate the suspected reduced viability. It should also be noted, that under the conditions existing in 1996, a record harvest was obtained that year with the starting value of 1.5 million pounds (700,000 kg) of cysts present in March.

Using the methods described above, the UDWR set a harvest quota of 4.5 million pounds (2.05 million kg) of raw biomass in 1997. In 1998, a quota of 7.3 million pounds (3.32 million kg) was established initially. Because of the uncertainty in predicting total cyst production during the fall reproductive season, and the differences in the brine shrimp population dynamics in 1998, the standing crop of cysts was monitored weekly. The season was terminated when the standing crop declined below the minimum acceptable value of 2.3 million pounds (1.01 million kg) rather than when the established quota was reached.

An access-point survey (Hayne, 1991) has been used by UDWR to track total harvest since 1996. Values estimated by the survey were compared with the harvest values reported weekly by the companies. The survey was originally implemented because there was the perception in the industry that some companies were not reporting accurately and an unbiased estimate of the harvest was needed. However, biased reporting by the companies did not appear to be occurring. In 1996 and 1997, the two values agreed within 15 percent, slightly more than one standard error, for the harvest from the lake surface. Since the 1997 access-point survey did not adequately sample the beach harvest, it could not be used to estimate the total harvest. The total 1997 harvest was estimated by combining the open-lake harvest generated by the access-point survey with the beach harvest reported by the companies. In 1998, the estimated harvest from the access-point survey was approximately 25 percent higher than the value reported by the companies. It is believed that an incorrect stratification of the survey data was used and resulted in biased estimates. Further modifications of the technique are underway to resolve this problem.

While unbiased, the access-point survey has the drawback of requiring several weeks of input before a relatively precise estimate can be generated. In 1997, the harvest progressed at a faster rate than originally anticipated and had reached approximately 6.1 million pounds (2.8 million kg) before the season was terminated on October 27, 1997. Based on data from the 1995 and 1996 harvest seasons, the harvest quota was expected to be approximately 4 million pounds (1.8 million kg) from the lake and the remainder collected from the beach. However, as previously mentioned, the beach harvest was a much larger percentage of the total than in the previous years. The increased beach harvest accounted for nearly all of the additional harvest in 1997. Approximately 4.2 million pounds (1.9 million kg) of biomass was collected from the lake and the remainder of the total from the beach. Because the harvest quota was originally calculated using data from the lake, not the beach, and the contribution of beached cysts to subsequent populations is unknown, UDWR considered its management goals fulfilled even with the larger total harvest.

Future Harvest Management

The future of the management of the fishery is uncertain at this time. The moratorium on additional CORs ended after the 1998 harvest season and the rule governing the harvest process needs to be rewritten. A number of management options are being considered including extending the moratorium. Studies are continuing to quantify important variables such as the quantity of brine shrimp required by migrating birds, the contribution of beached cysts to subsequent populations, the role of changing water levels and salinity in determining algal and brine shrimp population dynamics, and the winter loss of cysts from the lake. It may take several more years under the current harvest regime to fully understand these factors.

REFERENCES

Arnow, Ted, and Stephens, Doyle, 1990, Hydrologic characteristics of the Great Salt Lake, Utah: 1847-1986: U.S. Geological Survey Water-Supply Paper 2332, 1 plate, 32 p.

Belovsky, G.E., and Mellison, Chad, 1997, Brine shrimp population dynamics and sustainable harvesting in the Great Salt Lake, Utah: Progress report to Utah Division of Wildlife Resources, Salt Lake City, Utah, 20 p.

—1998, Brine shrimp population dynamics and sustainable harvesting in the Great Salt Lake, Utah: Progress report to Utah Division of Wildlife Resources, Salt Lake City, Utah, 19 p.

Browne, R.A., Davis, L.E., and Sallee, S.E., 1988, Effects of temperature and relative fitness of sexual and asexual brine shrimp *Artemia*: Journal of Experimental Biology and Ecology, v. 124, p. 1-20.

Clegg, J.S., 1962, Free glycerol in dormant cysts of the brine shrimp *Artemia salina*, and its disappearance during development: Biological Bulletin, v. 123, p. 295-301.

Conte, F.P., Hootman, S.R., Harris, P.J., 1972, Neck organ of *Artemia salina* nauplii. A larval salt gland: Journal of Comparative Physiology, v. 80, p. 239-246.

Conte, F.P., Peterson, G.L., Ewing, R.D., 1973, Larval salt gland of *Artemia salina* nauplii. Regulation of protein synthesis by environmental salinity: Journal of Comparative Physiology, v. 82, p. 277-289.

Croghan, P.C., 1958, The survival of *Artemia salina* (L.) in various media: Journal of Experimental Biology, v. 35, p. 213-218.

Dana, G.L., Jellison, R., and Melack, J.M., 1990, *Artemia monica* cyst production and recruitment in Mono Lake, California, USA: Hydrobiologia, v. 197, p. 233-243.

Dobbeleir, J., Adam, N., Bossuyi, E., Brudggeman, E., and Sorgeloos, P., 1980, New aspects of the use of inert diets for high density culturing of brine shrimp, *in* Persoone, G., Sorgeloos, P., Roels, O., and Jaspers, E., editors, The Brine Shrimp Artemia, v. 3, Ecology, Culturing, Use in Aquaculture: Universa Press, Belgium, p. 165-174.

Domagalski, J.L., and Eugster, H.P., 1990, Trace metal geochemistry of Walker, Mono, and Great Salt Lake, *in* Spencer, R.J., and Chou, I-Ming, editors, Fluid-mineral interactions: A tribute to H.P. Eugster: The Geochemical Society, Special Publication no. 2, p. 315-353.

Gliwicz, A.M., Wurtsbaugh, W.A., and Ward, A., 1995, Brine shrimp ecology in the Great Salt Lake, Utah: Performance Report to Utah Division of Wildlife Resources for June 1994 to May 1995, Salt Lake City, Utah, 83 p.

Gordon, H. S., 1954, The economic theory of a common property resource: the fishery: Journal of Political Economy, v. 62, p.124-142.

Gwynn, J.W., 1997, Brine properties, mineral extraction industries, and salt load of Great Salt Lake, Utah: Utah Geological Survey Public Information Series no. 51, 2 p.

Gwynn, J.W., 1998, Great Salt Lake, Utah: Chemical and physical variations of the brine and effects of the SPRR causeway, 1966-1996, *in* Pitman J. and Carroll A., editors., Modern and Ancient Lake Systems: New Problems and Perspectives: Utah Geological Association Guidebook 26, p. 71-90.

Hahl, D.C., and Langford, R.H., 1964, Dissolved-mineral inflow to Great Salt Lake and chemical characteristics of the Salt Lake brine: Utah Geological and Mineralogical Survey Water-Resource Bulletin 3, pt. 2, 40 p.

Hardin, Garrett, 1968, The tragedy of the commons: Science, v. 162, p. 1243-1248.

Hayne, D.W., 1991, The access point creel survey: procedures and comparison with the roving-clerk creel survey: American Fisheries Society Symposium 12, p. 123-138.

Hernandorena, Arantxa, 1987, An increased dietary tryptophan requirement induced by interference with purine interconversion in *Artemia*, *in* Decleir, W., Moens, L., Sleggers, H., Jaspers, E., and Sorgeloos, P., editors, *Artemia* research and its applications, v. 2, Physiology, Biochemistry, Molecular Biology: Universa Press, Belgium, p. 233-242.

Jensen, A.C., 1918, Some observations on *Artemia gracilis*, the brine shrimp of Great Salt Lake: Biological Bulletin, v. 34, p. 18-25.

Lavens, P., and Sorgeloos, P., 1987, The cryptobiotic state of Artemia cysts, its diapause deactivation and hatching- a review, *in* Sorgeloos P., Bengtson D., Decleir W., and Jaspers E., editors., *Artemia* Research and its applications, v. 3, Ecology, Culturing, Use in Aquaculture: Universa Press, Belgium, p. 27-63.

Lenz, P.H., 1984, Life-history analysis of an Artemia population in a changing environment: Journal of Plankton Research, v. 6, p. 967-983.

McCay, B.J., Apostle, R., and Creed, C.F., 1998, Individual transferable quotas, co-management, and community: Fisheries v. 23, p. 20-23.

Montague, C.L., Fey, W.R., and Gillespie, D.M., 1982, A causal hypothesis explaining predator-prey dynamics in Great Salt Lake, Utah: Ecological Modeling, v. 17, p. 243-270.

Pettengill, T., 1996, Commercial brine shrimp harvesting: 41 years of DWR involvement: Utah Division of Wildlife Resources unpublished report, Salt Lake City.

Porcella, D.B., and Holman, Anne, 1972, Nutrients, algal growth, and culture of brine shrimp in the southern Great Salt Lake, *in* Riley J.P., editor, The Great Salt Lake and Utah's Water Resources, Proc. First Ann. Con. Utah Section American Water Resources Assoc.: Utah Water Research Lab., Logan, p. 142-155.

Stephens, D. W., 1998, Salinity-induced changes in the aquatic ecosystem of Great Salt Lake, Utah, *in* Pitman J. and Carroll A., editors, Modern and Ancient Lake Systems: New Problems and Perspectives: Utah Geological Association Guidebook 26, p. 1-7.

Stephens, D.W., and Gillespie, D.M., 1972, Community structure and ecosystem analysis of the Great Salt Lake, *in* Riley J.P., editor, The Great Salt Lake and Utah's Water Resources, Proc. First Ann. Con. Utah Section American Water Resources Assoc.: Utah Water Research Lab., Logan, p. 66-72.

Stephens, D.W., and Gillespie, D.M., 1976, Phytoplankton production in the Great Salt Lake, Utah, and a laboratory study of algal response to enrichment: Limnology and Oceanography v. 21, p. 74-87.

Stumm, Werner, and Morgan, J.J., 1970, Aquatic chemistry: New York, John Wiley and Sons, 583 p.

Sturm, P.A., and Sanders, G.C., 1980, The brine shrimp industry on the Great Salt Lake, *in* Gwynn J.W., editor, Great Salt Lake, a scientific, historical and economic overview: Utah Geological and Mineral Survey Bulletin 116, p. 243-248.

Taylor, P.L., Hutchinson, L.A., and Muir, M.K., 1980, Heavy metals in the Great Salt Lake, Utah, *in* Gwynn J.W., editor, Great Salt Lake, a scientific, historical and economic overview: Utah Geological and Mineral Survey Bulletin 116, p. 195-200.

Waddell, K.M., and Barton, J.D., 1980, Estimated inflow and evaporation for Great Salt Lake, Utah, 1931-76, with revised model for evaluating the effects of dikes on the water and salt balance of the lake: Utah Division of Water Resources Cooperative Investigations Report 20, 57 p.

Wear, R.G., Haslett, S.J., and Alexander, N.L., 1986, Effects of temperature and salinity on the biology of *Artemia fransiscana* Kellogg from Lake Grassmere, New Zealand. 2. Maturation, fecundity, and generation times: Journal of Experimental Marine Biology and Ecology, v. 98, p. 167-183.

Wetzel, R.G., and Likens, G.E., 1979, Limnological analyses: Philadelphia, W.B. Saunders Co., 357 p.

Wirick, C.D., 1972, *Dunaliella-Artemia* plankton community of the Great Salt Lake, Utah: University of Utah, Salt Lake City, Utah, unpub. masters thesis, 44 p.

Wurtsbaugh, W.A., 1988, Iron, molybdenum and phosphorus limitation of N^2 fixation maintains nitrogen deficiency of plankton in the Great Salt Lake drainage (Utah, USA): Verh. Internat. Verein. Limnol., v. 23, p. 121-130.

—1992, Food-web modification by an invertebrate predator in the Great Salt Lake (USA): Oecologia, v. 89, p. 168-175.

—1995, Brine shrimp ecology in the Great Salt Lake, Utah: Performance Report to Utah Division of Wildlife Resources, 35 p.

Wurtsbaugh, W.A., and Berry, T.S., 1990, Cascading effects of decreased salinity on the plankton, chemistry, and physics of the Great Salt Lake (Utah): Canadian Journal of Fisheries and Aquatic Sciences, v. 47, p. 100-109.

BRINE SHRIMP IN ANTIQUITY

by

James S. Clegg and Susan A. Jackson
University of California, Davis, Section of Molecular and Cellular Biology, and
Bodega Marine Laboratory, Bodega Bay, CA 94923 USA
Phone: 707 875–2211; Fax 707 875-2009; jsclegg@ucdavis.edu

ABSTRACT

Geologic cores document the presence of the brine shrimp *Artemia* in the Great Salt Lake for at least 600,000 years. Their presence coincides with periods of hypersalinity. We have examined fragments of encysted brine shrimp embryos (cysts) recovered from a 27,000 year-old core section. Their ultrastructure is remarkably good considering the history of these particular cysts. Such structural preservation may reflect the extraordinary stability of *Artemia* cysts, in general, under extreme environmental conditions. We believe this unique system provides novel opportunities for the study of "ancient DNA" and the longevity of dormant organism in nature. However, study of those interesting questions will require the collection of cysts from freshly taken cores.

INTRODUCTION

Pioneering geological studies by Eardley and others (1957) and Eardley and Gvosdetsky (1960) and subsequent work by Spencer and others (1984) have shown that brine shrimp have been present in the Great Salt Lake (GSL) for at least 600,000 years, and possibly much longer. However, their presence has been restricted to periods when salinities were considerably higher than seawater, presumably because they cannot exist in the presence of substantial fish populations (Browne and others, 1991). Although no quantitative estimates can be made from these cores concerning brine shrimp population densities, it may be assumed that their numbers have fluctuated over geologic time.

Because *Artemia* cysts can be recovered from a defined time sequence, depending on the depth (and age) of the core in which they occur, one can study (1) their longevity under natural conditions and (2) the stability of their biological macromolecules, including DNA. The latter study could be of substantial value in view of the current interest and controversy about how long DNA can exist in a variety of fossils.

The first studies on DNA fragments extracted from tissues of long-dead animals were performed by Pääbo (1989), Hagelberg and others, (1989) and Pääbo and others, (1989) using the now well-known polymerase chain reaction (PCR). This pioneering work on "ancient DNA" has been followed by studies on a wide variety of specimens (reviewed in Eglinton and Curry, 1991; Herrmann and Hummel, 1994; Brown and Brown, 1984), including 17-20 million year old magnolia leaves (Golenberg and others, 1990; Pääbo and Wilson, 1991), insects entombed in amber whose ages range from 25-30 and 120-135 million years (DeSalle and others, 1992; Cano and others, 1993), and 80-million-year-old dinosaur bones (Woodward and others, 1994). These claims have been challenged, particularly by Lindahl (1993 a, b) who argued that DNA is not sufficiently stable to survive over such long periods. Responses to Lindahl have appeared (Poinar, 1993; Golenberg, 1994). The question of DNA stability continues to be of interest (Höss and others, 1996; Poinar and others, 1996). Of several uncertainties surrounding the study of ancient DNA, the extent of autolysis and microbial attack (and contamination) following death loom large. We pointed out previously (Clegg and Jackson, 1998) that brine shrimp cysts provide a unique system for investigation where: (1) their cells are protected from autolysis and microbial attack during and following death, (2) they can be sampled repeatedly at the same location of known burial age, (3) they have contemporary specimens living at the same geologically well-known site, and (4) their DNA is currently being studied. In this paper, based chiefly on previous work (Clegg and Jackson, 1998), we present electron microscopic description of brine shrimp cysts taken from a core section whose age has been determined to be 27,000 years old (Spencer and others, 1984). Although these observations are not of unique interest since much older specimens have been studied this way (see Herrmann and Hummel, 1994), our Discussion section presents the case that brine shrimp and their geological setting provide a unique opportunity to test, among other things, claims for the longevity of DNA over geological time.

BRINE-SHRIMP CYST PRODUCTION

A brief review of the life history of brine shrimp seems appropriate. This organism reproduces by development and release of swimming larvae or through the production of 4,000-cell embryos which are encased within a tough chitinous shell ("cysts") and released from the maternal female (Clegg and Conte, 1980). Cyst production is truly massive in the GSL, amounting to hundreds or thousands of metric tons per year (Clegg and Jackson, 1998). Although many of these cysts float on the surface, some sink and are buried in the sediments.

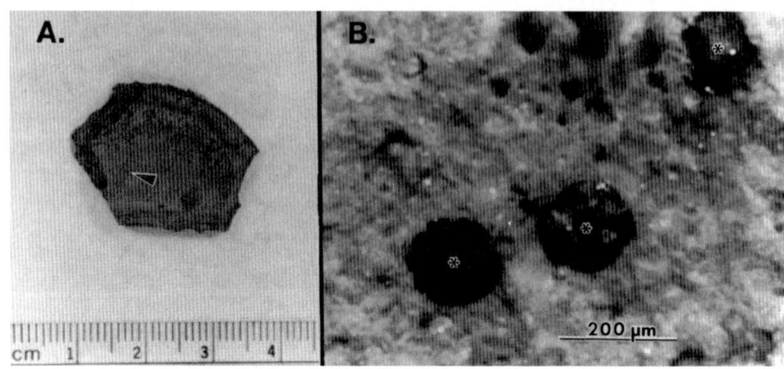

Figure 1. *Section from a core taken by Spencer and others, (1984) estimated to be 27,000 years old (part A). The arrowhead points to embedded fragments of* Artemia *sp. cysts which are shown in enlargement in part B (asterisks). Reprinted with permission from the International Journal of Salt Lake Research.*

MATERIALS AND METHODS

Spencer and others (1984) reported the presence of *Artemia* cysts in a number of cores taken in 1979 from various regions of the Lake. Dr. Spencer kindly sent us a section of one of these cores dated at close to 27,000 years old (Thompson and others, 1990), from which we isolated cyst fragments as follows (Clegg and Jackson, 1998). First, the core-section surface was scraped under a microscope with a razor blade to remove cyst fragments. Second, the remainder of the cysts were freed by soaking the core material in 0.4N acetic acid at 4°C for 48 hours. Next, the resulting suspension was saturated with NaCl, allowing cyst fragments to float free from the dense mineralized particles. Approximately 40 cyst fragments were thus obtained from the core section, and prepared for transmission electron microscopy using methods previously described in detail by Clegg and Jackson (1998). Finally, blocks of resin containing the cysts were sectioned with a diamond knife (Diatome, Switzerland) on a Reichert-Jung Ultracut E ultramicrotome and the sections viewed and photographed using a Zeiss 902B transmission electron microscope. The same procedures as those given above were used on contemporary cysts, except that the hydrated cysts were first punctured with a 26G needle to allow fixative to enter.

RESULTS

Figure 1A shows part of the core section provided by Dr. Spencer and several cyst fragments embedded in it (figure 1B). The core was taken in the south arm of the Great Salt Lake at Spencer's site C (see Spencer and others, 1984) and its age was determined by accelerator-mass spectrometer radiocarbon dating (Thompson and others, 1990).

The cysts in this core section are broken, probably due to the compaction and distortion that took place when the cores were air-dried for storage. Nev-

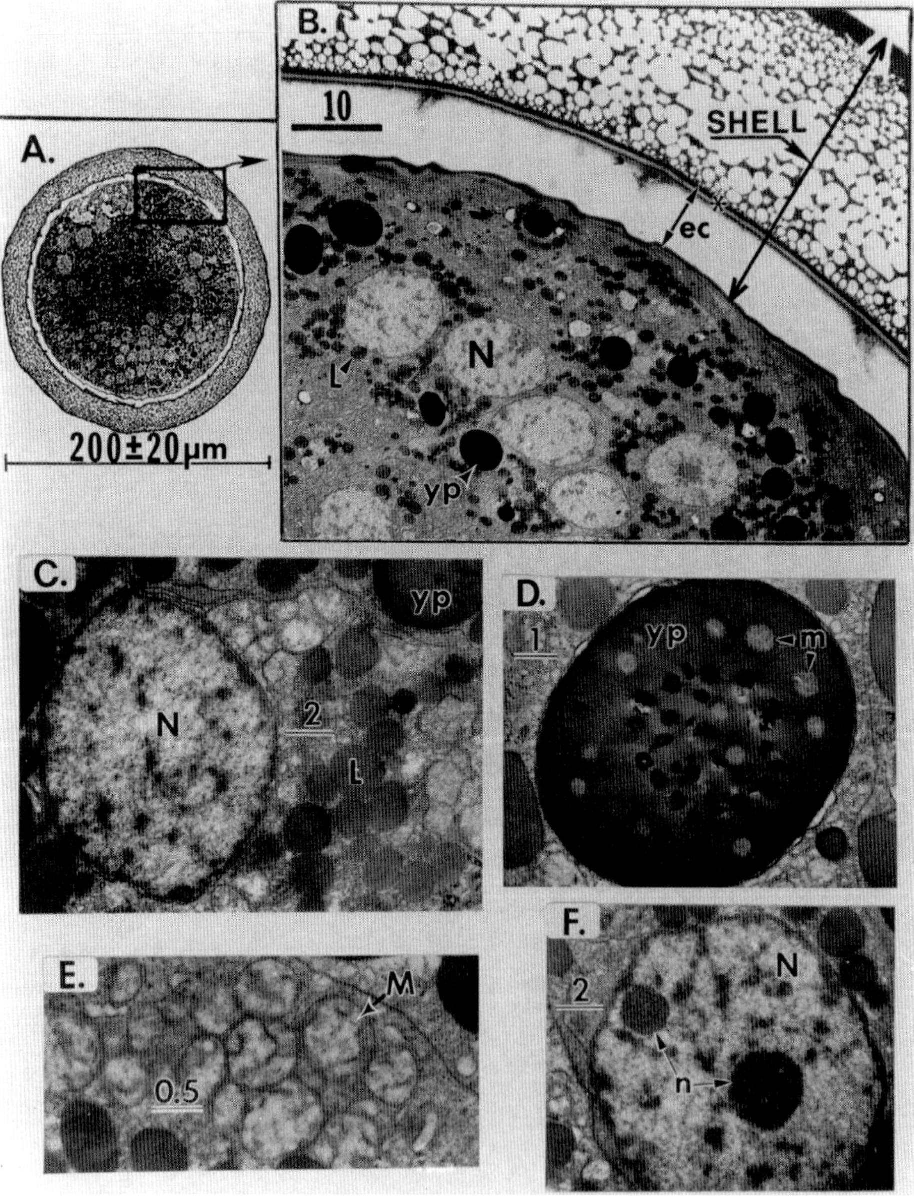

Figure 2. *Transmission electron microscopy of the structure of a contemporary cyst of* Artemia franciscana. *Part A shows a cross section of an entire cyst, the insert of which is enlarged in part B which, among other things, illustrates the relationship of the surrounding thick shell to the inner cell mass. Parts C-F describe the ultrastructure of intracellular components. Embryonic cuticle (ec), nuclei (N), nucleoli (n), lipid droplets (L), yolk platelets (yp), mitochondria (M), mitochondrial inclusions (m). The numbers above the bars are in units of micrometers. Reprinted with permission from the International Journal of Salt Lake Research.*

ertheless, approximately 40 cyst fragments were obtained from this core section (figure 1A) and prepared for transmission electron microscopy. Before showing the photographic results, we first describe, for comparison, the structure of a contemporary cyst (figure 2). The inner cell mass (figure 2A) is surrounded by a thick tough shell (figure 2A, B). Figure 2C-F describes the major organelles present in these cells.

Figure 3 illustrates the ultrastructure typical of the cyst fragments recovered from the 27,000-year-old core section (figure 1). As expected, extensive damage is evident at the level of the embryonic cell mass (figure 3A, B). However, sub-cellular (organelle) integrity is impressive considering the harshness of core dispersion and the fragmented condition of the cysts which had been stored dry at room temperature for over 15 years (see Spencer and others, 1984). Comparisons of the organelles in these cyst fragments with those in contemporary intact cysts (figure 2) indicate the favorable extent of preservation. We believe it is likely that freshly taken moist or wet cores will contain intact cysts. Thus, the observations described in figure 3 appear to represent the worst-preservation case.

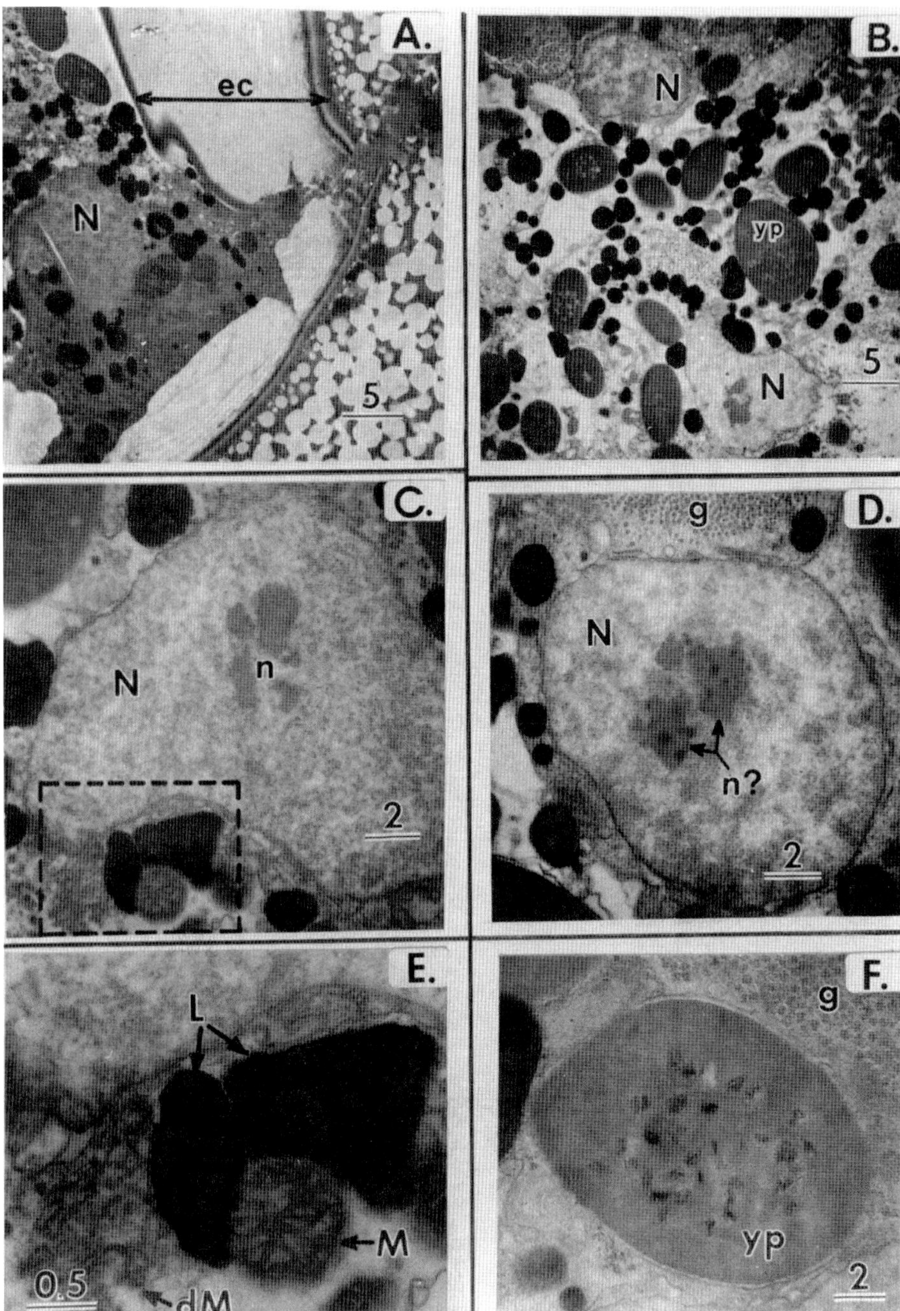

Figure 3. *Transmission electron microscopy of the ultrastructure of cyst fragments of* Artemia franciscana *estimated to be 27,000 years old. Part A shows a region of fragmented cell mass adjacent to the shell and part B describes another area containing disintegrated cells. Parts C-F illustrate selected organelles at increased magnification. The dashed rectangle in part C is enlarged in part E. Abbreviations are the same as in figure 2, plus glycogen (g) and damaged mitochondrion (dM). Numbers above the bars are micrometers. Reprinted with permission from the International Journal of Salt Lake Research*

DISCUSSION

Our current observations are not of particular interest in the context of the ultrastructure of ancient remains (see Herrmann and Hummel, 1994). What is of importance is the well-documented stability of brine shrimp cysts when exposed to a wide variety of harsh environmental insults (Clegg, 1974; Clegg and others, 1978; Clegg and Conte, 1980; Sorgeloos and others, 1987; Browne and others, 1991; Clegg, 1994). For example, when dehydrated, a normal circumstance for these cysts, they easily survive the conditions of outer space (Gaubin and others, 1983). Those abilities reside in a variety of biochemical and biophysical adaptations that stabilize the cells and their components (Clegg and Conte, 1980; Sorgeloos and others, 1987; Browne and others, 1991). Included in this repertoire is the reversible cessation of DNA synthesis and cell division throughout the encysted period (Nakanishi and others, 1962; Finamore and Clegg, 1969). Those observations lead to the conclusion that the DNA of these cysts is stabilized and protected from breakdown during initial burial in the lake sediments. Even though the cysts eventually loose viability it seems possible that many of their stabilizing adaptations remain largely intact. We have argued (Clegg and Jackson, 1998) that the inner-cell mass of intact cysts is protected from microbial attack following burial by the tough shell (figures 2 and 3). Thus, microbial invasion and hydrolytic degradation are of importance in the study of DNA extracted from long-dead biological systems, and that is part of the controversy over the stability of ancient DNA: the unknown microenvironmental conditions that prevail during and after death. Therefore, it has been proposed that the DNA of these cysts represents the upper boundary condition for maximum DNA stability in situ (Clegg and Jackson, 1998).

We note also that the DNA of contemporary cysts is well known (Kim and others, 1992; Landsberger and others, 1992; Escalente and others, 1994; Franco and others, 1994; Jellie and others, 1996; Ortega and others, 1996) and the complete sequence of their mitochondrial DNA has recently been published (Valverde and others, 1994). If sufficiently old cysts can be obtained in the future it may be possible to evaluate the rate of nucleotide substitutions in selected genes over "real time." Because of damage that is evident in cysts from dried cores (figure 1) it will be necessary in future work to obtain fresh cores. Likewise, obtaining cysts from newly taken cores will enable an evaluation of their longevity after burial. That study will be of considerable interest.

An interesting feature of the cores taken by Eardley and others (1957), and especially Eardley and Gvosdovsky (1960), is that *Artemia* undergoes periods of presence and absence in their cores. Where are the *Artemia* during the latter? We propose two explanations: (1) they continue in outlying ponds that experience high salinities seasonally, even desiccation, and maintain the populations via cysts; (2) the region is re-populated by cysts, attached to birds for example, from more remote populations. These are not exclusive possibilities.

CONCLUSIONS

We propose that the combination of biological and geological ingredients described in this paper for *Artemia* in the GSL provides an extraordinary opportunity for the study of ancient DNA. But there are additional questions that can be examined because of this happy combination of biology and geology. For instance, a determination of the longevity of these cysts following burial in the sediments is certainly worth pursuing. Fresh cores will be required to examine that interesting aspect.

ACKNOWLEDGMENTS

We are grateful to Professor R.J. Spencer for providing the core section used in this study, and to Professor D.R. Curry for participating in exchanges of information about the geology of Great Salt Lake. Professor P. Sorgeloos and Dr. F. Mees are thanked for numerous discussions about the topic of this paper. In particular, we appreciate the help of Dr. J. Wallace Gwynn during a visit to Salt Lake City, and for extended discussions about plans to obtain fresh cores from Great Salt Lake. Diane Cosgrove is thanked for manuscript preparation. Finally, we appreciate the assistance of Professor W.D. Williams in obtaining permission to reproduce figures 1-3.

REFERENCES

Benson, L.V., Curry, D.R., Dorn, R.I., Lajoie, K.R., Oviatt, C.G., Robinson, S.W., Smith, G.I. and Stine, S., 1990, Chronology of expansion and contraction of four Great Basin lake systems during the past 35,000 years: Palaeogeography Palaeoclimatology Paeleoecology 78: 241-252.

Brown, T.A. and Brown K.A., 1994, Ancient DNA: using molecular biology to explore the past: BioEssays 16: 719-726.

Browne, R.A., Sorgeloos, P. and Trotman, C.N.A (Eds), 1991, Artemia Biology: CRC Press, Boca Raton, Florida. 374 pp.

Cano, R.J., Poinar, H.N., Pieniazek, N.J., Acra, A. and Poinar, Jr., G.O. 1993, Amplification and sequencing of DNA from a 120-135 million year-old weevil: Nature 363: 536-538.

Clegg, J.S., 1974, Interrelationships between water and metabolism in *Artemia* cysts: hydration-dehydration from the liquid and vapour phases: Journal of Experimental Biology 61: 291-308.

—1994, Unusual response of *Artemia franciscana* embryos to anoxia: Journal of Experimental Zoology 270: 332-334.

Clegg, J.S. and Conte, F.P., 1980, A review of the cellular and developmental biology of *Artemia* G. Persoone, P. Sorgeloos, O. Roels, and E. Jaspers, (Eds)., In: The Brine Shrimp, Artemia, vol. 2. pp. 11-54, Universa Press, Wetteren, Belgium.

Clegg, J.S. and Jackson, S.A., 1998, Significance of cyst fragments of *Artemia* sp. recovered from a 27,000 year old core taken under the Great Salt Lake, Utah, USA: International Journal of Salt Lake Research 6: 207-216.

Clegg, J.S., Zettlemoyer, A.C. and Hsing, H.H., 1978, On the residual water content of dried but viable cells: Experientia 34: 734-737.

DeSalle, R., Gatesy, J., Wheeler, W. and Grimaldi, D., 1992, DNA sequences from a fossil termite in Oligo-Miocene amber and their phylogenetic implications: Science 257: 1933-1936.

Eardley, A.J. and Gvosdetsky, V., 1960, Analysis of Pleistocene cores from the Great Salt Lake, Utah: Bulletin of the Geological Society of America 71: 1323-1344.

Eardley, A.J., Gvosdetsky, V. and Marsell, R.E., 1957, Hydrology of Lake Bonneville and sediments and soils of its basin: Bulletin of the Geological Society of America 68: 1141-1201.

Eglinton, G. and Curry, G.B., 1991, Molecules through time: fossil molecules and biochemical systematics: Philosophical Transactions of the Royal Society, B. 333: 307-433.

Escalente, R., Garcia-Sáez, A., Asunción —A. and Sastre, L. 1994, Gene expression after resumption of development of *Artemia franciscana* embryos. Biochemical Cell Biology 72: 78-83.

Finamore, F.J. and Clegg, J.S., 1969, Biochemical aspects of morphogenesis in *Artemia salina*, *in* G.M. Padilla, G.L. Whitson and I.L.Cameron (Eds) The Cell Cycle: Gene-Enzyme Interactions, pp. 249-278. Academic Press, New York.

Franco, E., Behrens, M.M., Diaz-Guerra, M. and Renart, J. 1994, Structure and expansion of a polyubiquitin gene from the crustacean *Artemia*. Gene Expression 4: 19-28.

Gaubin, Y., Planel, H., Gasset, G., Pianezzi, B., Clegg, J., Kovalev, E.E., Nevzgodina, L.V., Maximova, E.N., Miller, A.T. and Delpoux, M., 1983, Results on Artemia cysts, lettuce and tobacco seeds in the Biobloc 4 experiment aboard the Soviet satellite Cosmos 1129: Advances in Space

Research 3: 135-140.

Golenberg, E.M., 1994, Antediluvian DNA research: Nature 367: 692.

Golenberg, E.M., Giannasi, D.E., Clegg, M.T., Smiley, C.J., Durbin, M., Henderson, D. and Zurawski, G., 1990, Chloroplast DNA sequence from a Miocene Magnolia species: Nature 344: 656-658.

Hagelberg, E., Sykes, B.C. and Hedges, R. 1989. Ancient bone DNA amplified. Nature 342: 485.

Herrmann, B. and Hummel, S. (Eds), 1994, Ancient DNA: Springer-Verlag, New York. 263 pp.

Höss, M., Jaruga, P., Zastawny, T.H., Dizdaroglu, M. and Pääbo, S., 1996, DNA damage and DNA sequence retrieval from ancient tissues: Nucleic Acids Research 24: 1304-1307.

Jellie, A.M., Tate, W.P. and Trotman, C.N.A., 1996, Evolutionary history of introns in a multidomain globin gene: Journal of Molecular Evolution 42: 641-647.

Kim, W., Min, G.S. and Kim, S.H., 1992, The 18S ribosomal RNA gene of a decapod crustacean *Oedignathus inermis*: comparison with *Artemia salina* gene: Nucleic Acids Research 20: 4658.

Landsberger, N., Cancelli, S., Carettoni, D., Barigozzi, C. and Badaracco, G., 1992, Nucleotide variation and molecular structure of the heterochromatic repetitive AluI DNA in *Artemia franciscana*: Journal of Molecular Evolution 35: 486-491.

Lindahl, T., 1993a, Recovery of antediluvian DNA: Nature 365: 700.

Lindahl, T., 1993b, Instability and decay of the primary structure of DNA: Nature 362: 709-715.

Nakanishi, Y.H., Iwasaki, T., Okigaki, T. and Kato, H., 1962, Cytological studies of *Artemia salina*. I. Embryonic development without cell multiplication after the blastula stage: Annotationes Zoologia Japonensés 35: 223-228.

Ortega, M-A, Diaz-Guerra and Sastre, L., 1996, Actin gene structure in two *Artemia* species, *A. franciscana* and *A. parthenogenetica*: Journal of Molecular Evolution 43: 224-235.

Pääbo, S., 1989, Ancient DNA: extraction, characterization, molecular cloning and enzymatic amplification: Proceedings of the National Academy of Sciences USA 86: 1939-1943.

Pääbo, S. and Wilson, A.C., 1991, Miocene DNA sequence _ a dream come true?: Currrent Biology 1: 45-46.

Pääbo, S., Higuchi, R.G. and Wilson, A.C., 1989, Ancient DNA and the polymerase chain reaction. Journal of Biological Chemistry 264: 9709-9712.

Poinar, G.O., 1993, Recovery of antediluvian DNA: reply: Nature 365: 700.

Poinar, H.N., Höss, M., Bada, J.L. and Pääbo, S., 1996, Amino acid racemization and the preservation of ancient DNA: Science 272: 864-866.

Sorgeloos, P., Bengston, D.A., Decleir, W. and Jaspers, E. (Eds), 1987, *Artemia* Research and its Applications. Vols 1-3. Universal Press, Wetteren, Belgium. 1498 pp.

Spencer, R.J., Baedecker, M.J., Eugster, H.P., Forester, R.M., Goldhaber, M.B., Jones, B.F., Kelts, K., McKenzie, J., Madsen, D.B., Rettíg, S.L., Rubin, M. and Bowser, C.J., 1984, Great Salt Lake, and precursors, Utah: the last 30,000 years: Contributions to Mineralogy and Petrology 86: 321-334.

Thompson, R.S., Toolin, L.J., Forester, R.M. and Spencer, R.J., 1990, Acceleration _ mass spectrometer (AMS) radiocarbon dating of Pleistocene lake sediments in the Great Basin. Palaeogeography, Palaeoclimatology, Palaeoecology 78: 301-313.

Valverde, J.R., Batuecas, B., Moratilla, C., Marco, R. and Garesse, R., 1994, The complete mitochondrial DNA sequence of the crustacean *Artemia franciscana*: Journal of Molecular Evolution 39: 400-408.

Woodward, S.R., Weyand, N.J. and Brunnell, M., 1994, DNA sequence from Cretaceous period bone fragments: Science 266: 1229-1232.

AVIAN ECOLOGY OF GREAT SALT LAKE

by

Thomas W. Aldrich
Utah Division of Wildlife Resources
1594 West North Temple
Salt Lake City, Utah 84116-3154
nrdwr.taldrich@state.ut.us

and

Don S. Paul
Utah Division of Wildlife Resources
515 East 5300 South
Ogden, Utah 84405
avocet@utah.uswest.net

ABSTRACT

Great Salt Lake is recognized regionally, nationally, and hemispherically for its extensive wetlands, and its tremendous and often unparalleled values to migratory birds. These values are derived from the lake's unique physical features, including its immense size, dynamic water levels, diversity in aquatic environments, extensive wetlands, and geographic position in avian migration corridors. These features create a mosaic of habitat types that are attractive to literally millions of migratory birds that use the lake extensively for breeding, staging, and in some cases, a wintering destination. Great Salt Lake also has a rich history of wildlife management activities that were initiated in the late 1890s by private hunting clubs, but were followed by substantial state, federal, and private investments in conservation programs. More recently, agency activities have focused on improving our understanding of migratory bird abundance and ecology, and development of cooperative, comprehensive, and integrated management strategies for lake resources. In spite of historical and contemporary conservation activities, the migratory bird resources of Great Salt Lake remain vulnerable to threats from a myriad of natural and anthropogenic factors. The management challenges to sustain migratory bird resources during the 21st century will only increase as competition for lake resources increases.

INTRODUCTION

Great Salt Lake (GSL) is rich in a number of natural resources, most notably its abundant minerals, a unique brine shrimp population, and a diverse and abundant migratory bird population. The lake has also been heavily utilized for the recreation, open space, and tourism opportunities it offers. As demands on the lake's natural resources have increased, so has the need for information necessary for making wise management decisions that will sustain those natural resources into the new millennium. This paper compiles recent data on migratory bird populations using the lake, which can serve as a general reference for interested parties and decision makers. Our intent is to not only enumerate migratory bird resources, but to convey the significant ecological role GSL plays in North American and hemispheric bird conservation. Much of the information provided is derived from recent unpublished Utah Division of Wildlife Resources surveys, but we also include general summaries from other authors where we felt a historical perspective is important to the reader.

GREAT SALT LAKE ECOLOGICAL SETTING

Physical Setting and Its Relationship to Wildlife Habitat

The physical dimensions, geographic setting, geology, and climate of GSL are described in great detail elsewhere (Gwynn, 1980). This article briefly discusses GSL area as it influences bird life associated with the lake. The scale of the ecological setting is immense. The entire system occupies roughly 3,011 mi^2 (7,800 km^2), consisting of lake, uplands, wetlands and drainage stems. Birds use all of these areas year-round.

Climate

GSL sits in a cold desert environment with summer and winter temperatures often reaching 100°F (38°C) and 0°F (-18°C), respectively. The lake area receives an average of 15 inchs (38 cm) of moisture in the vicinity of the Wasatch Front and less than 10 inchs (25 cm) on the west side of the lake.

Hemispheric Setting

GSL is one of several terminal lake systems in the Great Basin and is the fourth largest terminal lake in the world (Stephens, 1997). With its impressive size and extensive associated wetlands, the presence of GSL, in an otherwise xeric environment, underscores the "oasis" effect of the system. Consider the lake's position in the western hemisphere. There is a broad sweep of land that lies between the Cascade Mountains and the Mississippi River valley extending from the arctic rim to nearly Central America that receives less than 20 inches (50 cm) of precipitation annually. Except mountain environments, this is a vast arid region of the Americas. In this setting, productive sites for feeding, fattening, and molting of large numbers of migratory birds are isolated and often vast distances apart. These sites are always the exception and not the dominant habitats. From breeding grounds in the arctic, many species of birds travel more than 1,860 miles (3,000 km) by the time they arrive at GSL. Some will need energy to travel farther south to reach their wintering grounds. An American avocet (*Recurvirostra americana*) travels approximately 1,300 miles (2,100 km) to winter on the west coast of Mexico at Marismas Nacionales. A Wilson's phalarope (*Phalaropus tricolor*) doubles its weight on GSL brine flies (*Ephydra cinerea*) and brine shrimp (*Artemia fransiscana*) before flying nonstop, nearly 5,400 miles (8,800 km), to Laguna mar Chiquita in central Argentina. Other species migrate as near as the Salton Sea or as far as the Straits of Magellan from GSL.

GSL Ecological Setting

There are some notable physical features of the GSL region that are particularly important to birds and other wildlife; many are natural, while others are constructed (figure 1). The close proximity of GSL to the Wasatch Range and the riverine environments of the Bear, Weber, Ogden, and Jordan Rivers, and other smaller drainages plays a principal role in the local ecology. The terminal nature of the lake, with its various saline systems and associated halophiles, contributes greatly to the uniqueness of its natural wonders. GSL is a playa lake with an extremely low-gradient bottom. When its surface elevation is 4,202 feet (1,280.8 m) above sea level, the average depth of the lake is 13 feet (4 m). With the seasonal recharge of water from rivers and other drainages, and the subsequent evaporation, the effect of this shallow flat bottom is most apparent in the highly transitory shoreline. This results in ephemeral pools, expansive mud flats, and sand bars that warm quickly in spring, and easily reach temperatures around 84°F (29°C) in summer. Some parts of the lake shoreline migrate more than 880 yards (805 m) from spring to fall depending on the water and evaporation year. These water depths and shoreline fluctuations are fundamental ingredients in the creation of highly productive habitats for wading waterbirds.

All associated systems (bays, wetlands, etc.) are dominated by alkaline soils. The major ions in GSL's chemistry are sodium and chlorine. The associated plant community consists primarily of halophytes, many of which must stand the rigors of inundation and desiccation. The wetlands associated with these plants are usually lacustrine, sometimes modified by riparian systems, wet meadows, mud bars, sand beaches, and playas. There are brackish- and fresh-water marshes located within the systems dominated by fresh water hydrophytes. GSL drainage structure is sediment driven (Gwynn and Murphy, 1980). Clastic sediments comprised of fine-grained material, are deposited in bay areas of the lake (for example, Bear River Bay, Ogden Bay and Farmington Bay). Some of these form into baymouths, bars, alluvial fans, and crowfoot deltas. These landforms provide an ideal substrate for interspersed wetland vegetation, thus generating habitat edge, and sites for colonial nesting birds.

Within the GSL basin there are essentially four water regions with differing ecologies. These regions are Bear River Bay, Farmington Bay, the Gilbert Bay (southern arm) and Gunnison Bay (north arm) (figure 1). The major difference between these regions is salinity, although there are other features that play roles in defining their ecological peculiarities.

Bear River Bay is the freshest region and receives the largest volume of riverine inflow. Its near-surface salinity is similar to that of the Bear River. This system is bounded on the north and east by state, federal, and private wetlands; on the south by industry; and to the west by the Promontory Mountains. This bay is fresh enough to support a community of submergent hydrophytes including sago pondweed (*Potamogeton pectinatus*) and widgeon grass (*Ruppia maritima*). There are also significant islands of emergent wetlands here, especially in the east part of the bay in the Willard Spur (figure 1). An ecological element of vital importance to pisciverus birds in this area is the fishery that persists when the lake elevation is higher than 4,200 feet (1,280.2 m) above sea level. The avian community at Willard Spur is exceptionally complex. With its species richness, diversity and overall abundance, this area continually provides one of the most magnificent displays of bird life on the lake. Although the smallest region on the lake, it makes an exceptional contribution to the lake's avian population.

Farmington Bay is the next freshest region, with a salinity of approximately 60 parts per thousand (ppt) when GSL is at 4,200 feet (1,280.2 m) above sea level. It does not provide a submergent vegetation community due to the elevated salinity, but has a relatively complex plankton community compared to the south and north arms of the lake. It supports some important halophiles (for example, brine flies) in conjunction with some macroinvertebrates that tolerate diluted brine environments, but it is too saline to support a fishery. There is a significant wetlands influence in this region, often modified by the associated transitory shoreline, especially in the Layton/Kaysville Marsh area (figure 1). This effect is also visible at the south end of the bay, west of Farmington Bay Waterfowl Management Area (WMA). The open-water reaches of this region are an important waterfowl habitat in late fall and winter. The shorelines are important to breeding and migrating shorebirds.

Farmington Bay is bounded on the west by Antelope Island, to the north by the Antelope Island causeway, and to the east by extensive wetlands, including Nature Conservancy property and those of Farmington Bay WMA. The south boundary is formed by an extensive shoreline with portions controlled by the State, duck clubs, private ranching, and the National Audubon Society. A part of this shoreline has been developed into the southern causeway to Antelope Island.

Gilbert Bay (south arm) of GSL is the largest reach of water on the lake, with a salinity of 100 to 140 ppt at 4,200

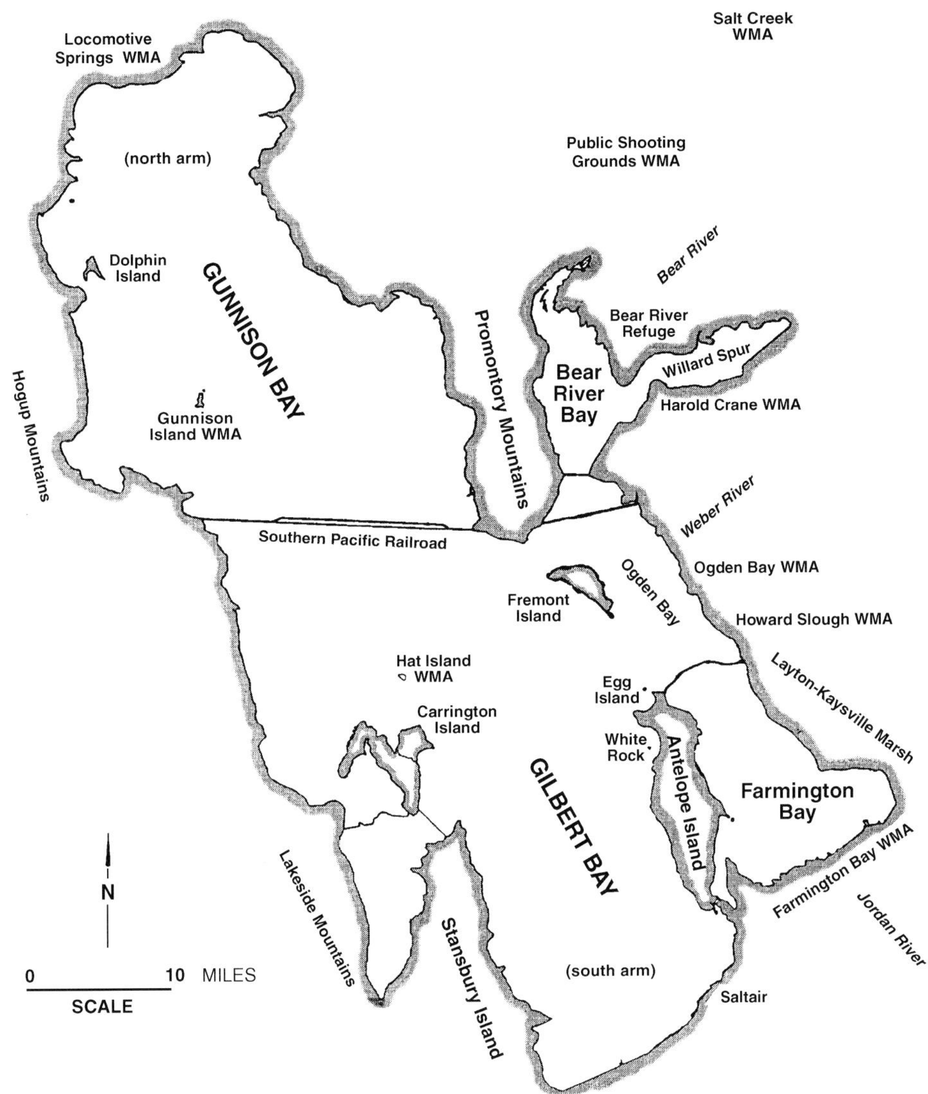

Figure 1. *Map of Great Salt Lake and notable features associated with the lake important to waterbird life.*

feet (1,280.16 m) above sea level. At this salinity, the lake's chemical and biological properties eliminate most in-lake predatory influences (predatory macroinvertebrates, fish, etc.) on the obligate halophiles (for example, brine shrimp and brine flies). These macroinvertebrates flourish under ideal conditions (especially at salinities in excess of 120 ppt), producing millions of pounds of potential protein for birds that have special behaviors or adaptations to exploit this food source. The lake can be unusually productive; during one study, the average production of brine fly biomass was 7.9 g/m^2 (Collins, 1980). During some harvest seasons, the brine shrimp industry removes more than 13.9 million pounds (6.3 million kg) of brine shrimp cysts and detritus from Gilbert Bay (south arm).

The south arm contains several islands crucial to colonial nesting bird species. Among the four water regions of the lake, the south arm has the most variable levels of salinity, especially near points of fresh water discharge. The south arm is framed by the Southern Pacific Railroad causeway and Promontory Point to the north; the Lakeside Mountains to the west; and the Weber River wetlands, Antelope Island. and the Jordan River delta to the east. The south shore of GSL and Stansbury Island are the main geographic features forming the south arm's southern boundary. Here natural areas are interspersed with mining and mineral industries, and the railroad causeway.

Gunnison Bay is the second largest expanse of open water. The north arm is bounded by salt playas and the Locomotive Springs WMA on the north; by the Promontory Mountains on the east; by the Hogup Mountains on the west; and by the Southern Pacific Railroad causeway on the south. It is also the most remote and xeric region of the lake and, perhaps more importantly, the most saline. Salinity ranges between 240 and 280 ppt. Gilbert Bay is presently a hypersaline system dominated by halophytic bacteria; it is too salty for brine shrimp and brine flies to persist. Currently, it appears to be a salt trap with possibly 3 to 5 vertical feet (1 to 1.5 meters) of salt precipitate on the lake floor. It possesses several minor wetlands that function to support small populations of unique birds. Gunnison Island has provided a nesting habitat for an impressive number of American white pelicans (*Pelecanus erythrorhynchos*) and California gulls (*Larus californicus*) before, and since settlement of the area by Anglo-Americans. For birds choosing to nest on this remote island, the tradeoff is a predator-free nesting environment for marathon flights to forage and gather food for young.

Life forms that persist in these GSL environments as residents or tran-

American white pelican adults and young find refuge on Gunnison Island - the only current pelican colony in Utah and one of the largest and most protected in North America.

Incubating pelicans at Gunnison Island. During this period, each adult spends an average of 72 hours at the nest before being relieved by the other parent. When the young hatch, each adult takes a turn airlifting fish from fresh-water sites to feed the young. The closest fishery, Bear River Bay, is 30 miles distant.

sients must do so under harsh and extreme conditions, with seasonal temperature variances exceeding 100°F (38°C), whimsical moisture cycles, and adjacent to brine-laden waters. Many organisms which survive such conditions, especially plankton, have evolved to be seasonally productive and reproductively flexible. Many avian species pattern their arrival, breeding, nesting, and departure times around these seasons of high macroinvertebrate productivity to take advantage of the unique food chain provided by GSL.

Numerous birds utilize the nearby wetlands and uplands associated with GSL. The proportional use is determined by the species, time of year, and current environmental conditions. Some species use only certain elements of the ecosystem. For instance, tundra swans (*Cygnus columbianus*) are dependant on submergent vegetation and rarely frequent GSL; northern shovelers (*Anas clypeata*) use both the lake and wetlands; and common goldeneyes (*Bucephala clangula*) rely heavily on the open lake environment. Understanding habitats and habitat relationships of avian species in the ecosystem is vital in this discussion.

Habitats

In addition to the lake regions listed in the Ecological Setting section of this article, there are many critical and important habitats. Several of these are discussed here.

Wetlands

The wetlands of GSL have long been recognized for their importance to migratory birds. Journal accounts of explorers, as well as market hunters, at the turn of the 19th century reflect some of the earliest written descriptions of these wetland resources and their associated wildlife. More recently, the importance of GSL wetlands has been formally recognized in numerous local, international, and hemispheric planning and management efforts.

In spite of the long-standing acknowledgment of GSL wetland importance, quantification and classification of these wetland resources are imperfect. Jensen (1974) summarized Utah wetland resources based on criteria reflecting their importance to waterfowl. He accounted for approximately 558,000 acres (225,822 ha) of wetlands statewide, with about 85 percent (474,139 acres; 191,884 ha) occurring around GSL (table 1). Photographic coverage of GSL, as part of the U.S. Fish and Wildlife Service (USFWS) National Wetland Inventory (NWI), was conducted in 1981, but coverage of GSL is incomplete. Several concurrent efforts by various agencies are underway to classify and quantify wetland resources within their jurisdictions using NWI data, but as of this writing, no published summary of lake-wide wetland resources by wetland class is available.

The nature, location, and extent of GSL wetlands are driven largely by two factors; the availability of fresh water and the surface elevation of GSL. Because GSL is a terminal basin of ancient Lake Bonneville, salts have accumulated in it since prehistoric times. Because of this, the sediments and water of the current lake bed are highly saline and cannot support vascular plants. Wetland plant communities are therefore distributed where precipitation, surface inflow, and ground-water discharge are sufficient to freshen sediments and allow for plant germination, growth, and reproduction.

Stauffer (1980) modeled the water budget for GSL and estimated an annual fresh water supply to the lake of 3,250,000 acre-feet (4,007 hm^3). From this, surface water flow, precipitation, and groundwater discharge account for 65 percent, 28 percent and 8 percent, respectively, of total fresh water inflow. Of the surface flow, the Bear River supplies the greatest amount of water (55 percent), followed by the Weber River (23 percent) and the Jordan River (14 percent). Ungaged flows from sewage plants and other small tributaries account for the remainder of surface inflow (8 percent). Wetlands of GSL are associated mainly along the deltas of these major rivers, with smaller and more isolated wetland complexes located near drains and small tributaries and ground-water discharge areas at Locomotive and Timpie

Table 1. Wetland acreages and estimated water needs of the Great Salt Lake (Jensen, 1974).

County	1st Magnitude Marsh	2nd Magnitude Marsh	3rd Magnitude Marsh	Total Marsh	Annual Water Needs (Acre-feet)
Box Elder	188,465	53,188	17,000	258,653	800,641
Davis	96,886	—	—	96,886	502,471
Salt Lake	25,130	2,560	7,780	35,470	103,163
Tooele	1,440	—	7,640	9,080	48,657
Weber	28,122	8,903	37,025	74,050	88,928
Total	340,043	64,651	69,445	474,139	1,543,860

springs. Although relatively small in proportion to river flows, precipitation plays an important role in freshening sediments on small playas and mud flats that are removed from other sources of fresh water.

Given that fresh water is essential for wetland persistence, the factor that potentially limits the distribution and extent of wetlands is the surface elevation of GSL. GSL elevations are extremely dynamic and change in response to long-term precipitation cycles, seasonal changes in evaporation and inflow, and daily influences from wind-driven seiches. Since 1851, the surface elevation of GSL has varied more than 20 feet (6 m) from a low of 4191.4 feet (1,277.5 m) above sea level in 1963 to a high of 4211.8 feet (1,283.8 m) in 1986 and 1987. These extremes produce a change in the surface area of GSL of approximately 872,000 acres (352,898 hm^2) (table 2) which more than doubles its size (Cruff, 1986). The relative change is even more pronounced along the eastern shore of GSL, where differences in surface area change more than 21-fold between lake elevations of 4191 and 4205 (1,277.4 and 1,281.7 m) (Waddell and Fields, 1977). The most dramatic example of impacts from long-term lake level fluctuations was the flooding of the 1980s when an 11-foot (3.4 m) rise of GSL occurred over a four year period. The lake inundated approximately 80 percent of GSL wetlands with brines, killing essentially all vegetation and many tuber and seed banks. Flooding resulted in more than $30 million of damage to State-managed facilities. Lake-wide waterfowl use and production also declined 80 percent to 90 percent.

Seasonally, GSL levels change in response to evaporation and inflow producing an annual high during May-July and a low during October-November. On average, the historical seasonal difference in GSL elevations is 18 inches (0.46 m) but has changed as much as five feet (1.5 m) in one year. On average, each foot of change in GSL elevation changes the surface area of GSL by nearly 44,000 acres (17,807 hm^2) lake-wide, and approximately 17,500 acres (7,083 hm^2) along the eastern shore (table 2). However, changes in the surface area in response to lake levels are not uniform. The greatest change in surface area for any given change in lake level is between lake elevations of 4,195 and 4,200 feet (1,278.6 and 1,280.2 m). Winter and spring increases in GSL elevations therefore result in inundation of tens of thousands of acres with brine in most years, and hundreds of thousands of acres in exceptionally wet years.

Wind-driven seiches also have dramatic impacts on GSL elevations. "The magnitude of the setup on the speed, direction, fetch, depth of lake at that point, and direction of the

Table 2. Surface area of the Great Salt Lake relative to lake elevation (Cruff, 1986).

GSL	Surface Area (Acres)				Change in Surface Area/Foot Elevation			
Elevation	North Arm	South Arm	East Shore	GSL	North Arm	South Arm	East Shore	GSL
4191	221,200	378,900	12,000	600,100	8,200	12,100	2,861	20,300
4192	229,400	391,000	14,861	620,400	8,300	12,200	5,285	20,500
4193	237,700	403,200	20,146	640,900	19,600	22,200	9,092	41,800
4194	257,300	425,400	29,238	682,700	19,700	22,400	19,336	42,100
4195	277,000	447,800	48,574	724,800	21,600	40,200	23,530	61,800
4196	298,600	488,000	72,104	786,600	21,700	40,200	24,025	61,900
4197	320,300	528,200	96,129	848,500	21,600	40,300	24,738	61,900
4198	341,900	568,500	120,867	910,400	21,600	40,200	24,572	61,800
4199	363,500	608,700	145,439	972,200	21,700	40,200	29,150	61,900
4200	385,200	648,900	174,589	1,034,100	15,100	21,300	25,411	36,400
4201	400,300	670,200	200,000	1,070,500	15,200	21,200	18,000	36,400
4202	415,500	691,400	218,000	1,106,900	15,200	21,300	14,000	36,500
4203	430,700	712,700	232,000	1,143,400	15,200	21,300	10,000	36,500
4204	445,900	734,000	242,000	1,179,900	15,200	21,300	14,968	36,500
4205	461,100	755,300	256,968	1,216,400	12,200	18,200		30,400
4206	473,300	773,500		1,246,800	12,300	18,300		30,600
4207	485,600	791,800		1,277,400	12,200	18,500		30,700
4208	497,800	810,300		1,308,100	12,200	69,400		81,600
4209	510,000	879,700		1,389,700	7,300	28,900		36,200
4210	517,300	908,600		1,425,900	7,200	15,700		22,900
4211	524,500	924,300		1,448,800	7,400	15,900		23,300
4212	531,900	940,200		1,472,100	40,800	53,300		94,100
Average					15,977	27,936	17,498	43,914

wind. Wind setup exceeding two feet is not uncommon.... The combined effects of wind setup and high waves (wave runup) can produce adverse impacts to elevations five to seven feet above the static lake elevation" (Utah Department of Natural Resources, 1999). These events can push brines onto tens of thousands of acres of mud flats and into emergent marsh areas, and de-water similar acreage on the upwind side of the lake. These events tend to be short lived, however, and normally do not impact established vegetation. Unlike seasonal and long-term changes in GSL levels, wind seiches are accompanied by heavy wave action that can erode shorelines and greatly disrupt sediments. As a consequence, seed germination and plant establishment along lake boundaries can be hindered and, in some cases, prevented by frequent wind events. In extreme cases, established vegetation can be uprooted by heavy, sustained wave action.

Breeding birds that construct nests on mud flats, floating vegetation or emergent cover near water surfaces can also be impacted by wind seiches. Shorebirds, ibis, grebes, geese, and diving ducks are examples of species groups whose nests are commonly destroyed by wind-generated flooding. These impacts are not restricted to the open body of GSL, since many of the larger impoundments on management areas also experience wind tides.

Whereas the boundaries of wetland plant communities around GSL are limited by sediment and water salinities, species composition within wetland environments is influenced by plant tolerances to salinity, hydroperiod and water depth. Christensen and Low (1970) demonstrated that most common wetland plants found within GSL basin germinate and grow best in fresh water, but differences in salt tolerance among plants exist. As a consequence, fresh water areas around GSL tend to be dominated by the most aggressive and competitive plant species while more salt-tolerant plants occupy the brackish areas. Among submersed aquatic species, widgeon grass and muskgrass (*Chara* spp.) are most salt tolerant, while sago pondweed and horned pondweed (*Zannichellia palustris*) are slightly less tolerant. Among emergent species, alkali bulrush (*Scirpus paludosus*) and Olney's bulrush (*Scirpus olneyi*) are most salt tolerant, while cattail (*Typha* spp.) and hardstem bulrush (*Scirpus acutus*) tend to be restricted to fresher environments.

These differences in salt tolerance, in part, explain the general patterns in wetland plant distribution around GSL. Wetlands associated with the freshest riverine inflows (Jordan, Weber and Bear Rivers) tend to be dominated by cattail, sago pondweed, and hardstem bulrush. Wetlands fed by higher salinity inflows (Salt Creek, Locomotive and Timpie Springs) or are located below impoundment dikes or along the lake shore interface, tend to be occupied by widgeon grass, chara, alkali and Olney's bulrush.

Uplands

Similar to GSL wetland habitats, upland communities are extensive and geographically vary throughout the GSL basin. Rawley (1980) described and mapped these communities so details will not be provided here. In general, uplands of GSL on the south, west and north shores tend to be dominated by shadscale (*Atriplex confertifolia*)-greasewood (*Sarcobatus* spp) associations adjacent to barren or sparsely vegetated salt flats and beaches. Combined with little to no fresh water, these upland areas provide relatively less benefit to migratory birds. Uplands along the east shore, however, are closely associated with expansive wetlands, due to greater annual precipitation and runoff, deeper soils, and extensive irrigation systems. Although some greasewood and sagebrush (*Artemisia* spp) persist along benches, upland communities are made up of more numerous grasses, forbs and shrubs with higher wildlife values. Extensive agricultural areas are also present, with cereal grains and alfalfa being the predominant crops.

Upland areas along the east shore serve several important migratory bird functions. They provide cover for upland nesting birds, including several species of dabbling ducks, Canada geese (*Branta canadensis moffitti*), long-billed curlews (*Numenius americanus*), and willets (*Catoptrophorus semipalmatus inornatus*). Rank and dense upland shrub cover is preferred by late-nesting species such as gadwall (*Anas strepera*), while salt grass and other short-stature plants are used by cinnamon teal (*Anas cyanoptera*). Upland areas do not need to be close to wetlands to provide valuable cover; as some species, most notably pintails (*Anas acuta*), nest in areas up to a mile away from water.

Natural upland areas provide substantial food resources for a host of species. Canada geese graze extensively on high-protein young shoots from a variety of wild grasses. Seeds and invertebrates from upland plants that are backflooded during late summer and fall in managed impoundments are also important food resources, particularly to fall migrants.

Agricultural areas on state WMAs, as well as private lands, are used extensively by migratory birds. Canada geese and greater sandhill cranes (*Grus canadensis tabida*) graze heavily on fall and spring-seeded cereal grains and alfalfa. Waste grain in harvested fields is used by several species. Flood irrigated pastures provide enhanced access to both invertebrates and small rodents during summer for a variety of birds. White-faced ibis (*Plegadis chili*) and Franklin's gulls (*Larus pipixcan*) are commonly seen in flocks of hundreds exploiting flooded agricultural fields.

Upland habitats along the east shore are also tremendously important in buffering wetland areas and dependent wildlife from disturbances and activities on adjacent lands. High residential populations, much of the commercial and industrial use, a web of transportation facilities including rail, interstate highway and airport facilities, all exist within the narrow corridor between the Wasatch Range and GSL. Most future municipal- and county-planned expansions are also confined to this limited area. Noise, free-roaming pets and a host of other human activities can reduce the wildlife values in wetlands, and upland buffers can reduce these impacts and provide protection from future encroachment.

Islands

Nine islands on GSL have significant bird habitat values. These islands are Antelope, Hat, Carrington, Egg, Fremont, Gunnison, Landing Rocks, Stansbury, and White Rock (figure 1). These habitat values change with lake volume and elevation. Most islands are accessible by land at low lake elevations, and conversely, are covered or compromised by water at high lake elevations. Of these islands, five are currently colonial nest sites and were this way in the past (Behle, 1958; Rawley, 1976). These islands range in size from 2,821 acres (1,142 ha) at Antelope Island to less than 1.2 acres (0.5

ha) at White Rock. Some of these islands persist as colonial nest sites through time regardless of historical fluctuations in lake elevation. Others are used intermittently, due mostly to topographic characteristics or disturbance from predators or man. Some islands are comprised simply of Precambrian rock with some soil substrate and sparse vegetation, while others are heavily vegetated with shrubs, forbs and grasses. Consistent colonial nesting species of the island environments are California gulls, great blue herons (*Ardea haeredes*), double-crested cormorants (*Phalacrocorax auritus*), American white pelicans, and Caspian terns (*Sterna caspia*). Even though they are not considered colonial nesting species, the Canada goose and some duck species nest within, or near, colonial nesting sites on some of the islands.

Insulation from terrestrial predators is, in part, why colonial nesting species choose islands for nesting and rearing young (Connell, 1975). However, some colonial species have nested in large numbers on the GSL islands that have well-established populations of coyotes (*Canis latrans*), bobcats (*Felis rufus*), and badgers (*Taxidea taxus*). This is an exception for most sites when the lake is 4,202 feet (1,280.8 m) above sea level or higher as water levels prevent access of most land predators.

The islands of GSL provide other habitat values to birds in addition to nesting. For example, Gunnison Island was found to be used for foraging and resting by 110 different species between 1972 and 1974 during a study of American white pelicans (Knopf, 1975). Fremont and Antelope Islands provide thermal and uplift draft for pelicans, assisting them in energy conservation as they travel to and from foraging locations. The peninsula of the Promontory Mountains, along with these two islands, are used by migrating raptors. Many islands provide extensive shoreline for waterbird foraging and nesting.

Bars, mud flats, and hummocks are important substrates for colonial species for nesting, loafing, and foraging, although they are not islands. During years of low lake elevation (<4,200 feet (<1,280.2 m) above sea level), an impressive number of gulls and shorebirds are raised on these natural structures. These structures are a phenomena resulting from the low-gradient lake bottom. Dikes, levees and other manmade structures have become important habitat components to colonial nesting species on the lake. Many have been partly responsible for the expanding population of gulls. These structures could be considered artificial islands. Dikes are crucial nesting substrates for gulls when lake elevations are higher.

GSL WATERBIRD ECOLOGY AND BIOLOGY

Waterfowl

Waterfowl, for the purpose of this discussion, include ducks, geese, and swans, which taxonomically form the family Anatidae. Worldwide, there are believed to be more than 160 species of waterfowl, with 50 species occurring in North America.

As a group, waterfowl are incredibly diverse in structure, size, physiology, and behavior. These characteristics are responsible for a strong presence of waterfowl in Utah and their ability to exploit the diversity of habitats available in and around GSL. Thirty-five species of waterfowl are known to use the lake and its periphery, although their timing of use, distribution, and abundance varies considerably among species (table 3), seasons, years, and varying GSL elevations. Although GSL is recognized mostly as a waterfowl migration corridor, many species are found in abundance during the summer breeding and molting periods, as well as in winter.

Waterfowl Migration

Due to their classic chevron flight patterns, large flock sizes, and frequent vocal arrival, migrating waterfowl are often signals for changing seasons. Waterfowl use of GSL occurs throughout the year, but populations are highest during migrations in late summer, and again in spring. Recent population surveys on state WMAs and the Bear River Migratory Bird Refuge (BRMBR) indicate that waterfowl

Table 3. Waterfowl species and their abundance during different times on the Great Salt Lake (unpublished data, DWR files).

Species	Breeding	Migration	Wintering
Tundra Swan	N	C	C
Trumpeter Swan	N	R	R
Greater White-fronted Goose	N	R	O
Snow Goose	N	C	O
Ross' Goose	N	C	O
Brant	N	O	O
Canada Goose	C	C	C
Wood Duck	R	R	R
Green-winged Teal	N	C	C
American Black Duck	N	O	O
Mallard	C	C	C
Northern Pintail	C	C	C
Blue-winged Teal	U	U	U
Cinnamon Teal	C	C	U
Northern Shoveler	C	C	C
Gadwall	C	C	C
Eurasian Wigeon	N	O	O
American Wigeon	O	C	C
Canvasback	U	C	U
Redhead	C	C	C
Ring-necked Duck	O	C	U
Greater Scaup	N	R	R
Lesser Scaup	O	C	C
Harlequin Duck	N	O	O
Oldsquaw	N	O	O
Black Scooter	N	O	O
Surf Scooter	N	O	O
White-winged Scooter	N	O	O
Common Goldeneye	N	C	C
Barrow's Goldeneye	N	U	U
Bufflehead	N	C	C
Hooded Merganser	N	O	O
Common Merganser	N	C	C
Red-breasted Merganser	N	O	O
Ruddy Duck	C	C	C

N=Not known to be present
C=Common (Found consistently in fair numbers)
U=Uncommon (Found occasionally in small numbers in appro priate habitat)
R=Rare (Found infrequently but regularly in very small numbers in proper habitat and season)
O=Occasional (Seldom found in state and not reported annually)

begin arriving from northern breeding areas in June and peak in September (figure 2). Banding studies have shown that Utah receives most of its migrating ducks from the northwestern and mid-continent breeding areas in Alaska and Canada, but some exchange of birds with other breeding populations does occur. Ducks migrating through GSL predominately winter in California and the west coast of mainland Mexico, but some species, such as redheads (*Aythya americana*) and cinnamon teal, can reach the coast of Texas, and Central and South America (Bellrose, 1976). Green-winged teal (*Anas crecca*) and pintail are the first migrants to arrive, followed by shovelers, wigeon (*Anas americana*), gadwall, mallard (*Anas platyrhynchos*), and canvasback (*Aythya valisineria*). Mergansers and sea ducks, including bufflehead (*Bucephala albeola*), goldeneye, and scaup (*Aythya affinis*), are the last to arrive in November. Cinnamon teal are among the first ducks to leave Utah, with the majority gone by the end of September.

Populations begin returning from southern wintering areas in February and peak again in March. The spring migration period through GSL is markedly shorter and peak populations substantially lower than in the fall.

Although current surveys suggest nearly 700,000 waterfowl are present in the GSL area in September (figure 2), this number does not include populations of birds on many unsurveyed public and private lands. It also does not account for population turnover during the migration period and movement of waterfowl off surveyed areas in response to hunting seasons. Population counts on management areas are therefore a gross underestimation of true populations of waterfowl migrating through GSL. Because GSL is positioned in a major migration corridor between the mid-continent prairie breeding areas and major wintering areas in California, impressive numbers of waterfowl pass through Utah annually. Bellrose (1976), using several sources of information, estimates 3 to 5 million ducks pass through GSL each year.

Migrating waterfowl prefer large, open marshes that are free from disturbance. Key areas that offer these features include impoundments on state and federal management areas, private hunting clubs, and the fresh water areas in Willard Spur, Bear River Bay and the Layton/Kaysville marsh. Once hunting seasons begin in October, waterfowl make extensive use of shallow brackish areas along the GSL shoreline and salt water areas offshore. In November, the salty open lake body also hosts tens of thousands of common goldeneye, green-winged teal, and northern shovelers which appear to be foraging on brine shrimp and their eggs. Research is being initiated by the Utah Division of Wildlife Resources (DWR) to validate the role of cysts in winter waterfowl diets. Waterfowl can make use of the salt water areas due to the presence of salt glands which facilitate nasal excretion of excess salts ingested in water and food.

Numerically, ducks represent the lion's share of migrating waterfowl, but GSL also serves as a major staging area for tundra swans and Canada geese (figure 3). Approximately 75 percent of the western population of tundra swans migrates along an interior route from their Alaska breeding areas through GSL, eventually reaching their wintering areas in the central valley of California. Swans stage for a substantial time in marshes of GSL before moving on. Peak fall populations of 40,000 to 60,000 swans regularly occur. Tundra swans loaf in shallow mud flats in the sanctuary of the BRMBR and mainly feed on sago pond weed tubers, which are found primarily in the deep, persistent fresh-water impoundments of GSL. Prior to the GSL flooding in the 1980s, swan use was distributed over much of the east shore, but currently is focused on the BRMBR. Little swan use of salt water areas occurs on GSL. Although wintering tundra swans feed heavily in agricultural environments in California, their use of fields around GSL during migration has only rarely been documented (Nagel, 1965). Trumpeter swans (*Cygnus buccinator*) also use GSL during migration, but information on timing and abundance is scarce and inconsistent. Trumpeter swans appear to prefer small, more vegetated wetland complexes.

Most Canada geese migrating through Utah are part of the Rocky Mountain Population (RMP) which breeds throughout the intermountain west and largely winters in Idaho, Utah, and the lower Colorado River region of Arizona and southern California. In recent years, smaller races of

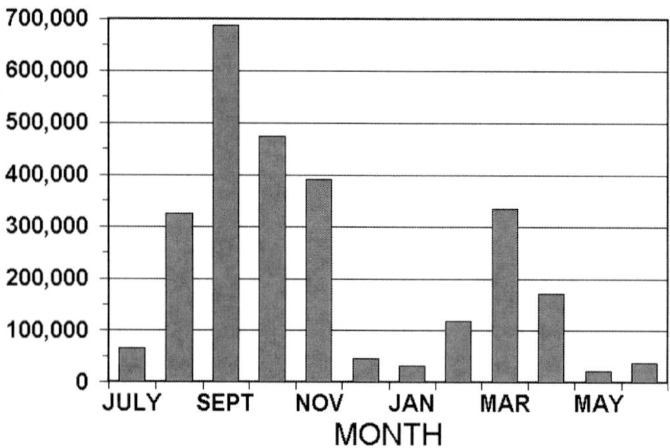

Figure 2. *Monthly duck populations on the Great Salt Lake as measured by ground surveys on state and federal management areas (averaged for years 1993-1998).*

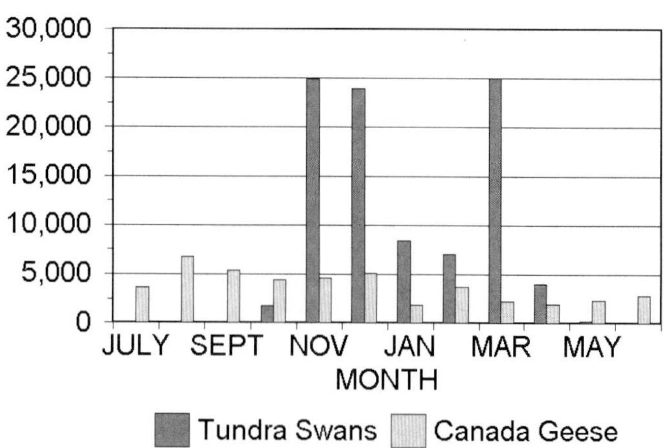

Figure 3. *Monthly Tundra swan and Canada goose populations on the Great Salt Lake (averaged for years 1993-1998).*

Canada geese, likely from interior Alaska and Canada, have become more prevalent during the fall migration, but remain a small proportion of all geese passing through. Based on midwinter surveys, it is likely that about 35,000 (30 percent of the population) RMP geese migrate through GSL annually. Lesser snow geese (*Anser c. caerulescens*), and Ross' geese (*Anser rossii*) historically frequented GSL in spring and fall, but recently their use has been restricted mostly to the spring period. Much conjecture exists regarding the loss of fall use by lesser snow geese, but reasons remain unclear. Theories include disturbance from increased aircraft use (particularly from helicopter use in the 1960s at Hill Air Force Base), habitat changes, and short-stopping of geese on large reservoirs in southern Canada.

The value of high-quality habitat in key migration areas, such as GSL, is important for several reasons. Flight is energetically demanding and many waterfowl travel thousands of miles between breeding and wintering areas. Consequently, waterfowl need periodic stops to replenish energy and water reserves used in migration. Also, most waterfowl need to carry reserves necessary for egg development to their nesting areas. Good-quality habitat along migration routes is therefore important for enhancing successful migration and reproduction. GSL marshes are expansive and diverse, have large areas of open water, and are rich in a variety of invertebrate and plant food resources. These features are the keystone of GSL's value to migrating waterfowl. The evolution of waterfowl migrations through GSL is likely related to its historic importance in providing high-quality habitat during fall and spring migration periods.

Waterfowl Breeding and Production

Although 13 species of ducks potentially breed in GSL, seven are common nesters. Ducks generally select emergent or upland cover for nest construction and remain secretive during the nesting period. For these reasons, duck populations can be difficult to inventory and estimate. Standardized, aerial, breeding bird transects conducted by DWR from 1956-98 within the GSL region, indicated an average of 75 ducks per square mile of habitat, but range from 24 to 276 ducks per square mile (2.59 km^2) among these years (figure 4). Using 1995-98 data (corrected for visibility bias by species), and a habitat base of 404,694 acres (163,779 ha^2) (Jensen, 1974, 1st and 2nd magnitude marsh), the estimated total breeding duck population of the lake exceeds 230,000 in an average year (table 4). These densities of breeding ducks compare favorably with the highest known duck densities on broad landscapes in North America. For cinnamon teal, redheads, and gadwall, GSL is nationally recognized as a key breeding area.

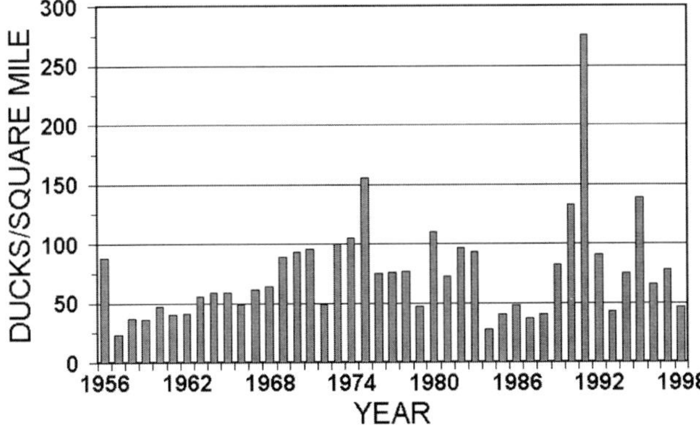

Figure 4. Densities of ducks counted during May aerial surveys on the Great Salt lake by the Utah Division of Wildlife Resources.

Production of ducks is also difficult to estimate for GSL due to the secrecy of hens with young broods. No surveys are currently conducted to try to estimate lake-wide production. However, crude estimates of production can be made. Using an average clutch size of eight eggs, and assuming there are 115,000 breeding pairs of ducks (table 4), the potential annual duck production from GSL could equal 920,000 ducklings. The realized production is undoubtedly lower due to nest and brood predation, and other factors that limit reproductive success. Recent studies of duck nesting success at the Salt Creek, Public Shooting Grounds, and Harold Crane WMAs from 1988-92, indicate nest success rates vary from 2 percent to 59 percent. Avian and mammalian predation is believed to be the major causes of nest failure (Huener and Manes, 1992a, b). No studies of brood survival have been conducted recently in GSL marshes, but the substantial predation of duck nests would suggest broods are also heavily impacted. Using a 20 percent nest success rate for each hen, and a 50 percent brood loss rate reduces the estimated GSL production to about 92,000 fledged ducks. A second method to estimate lake-wide duck production would be to use nest densities from these same studies (rather than using aerial breeding bird densities), and apply the same hen and brood success rates. Using an average nest density of 0.20 nests/acre (DWR studies indicate a range from 0.04 to 1.29 nests/acre), a habitat base of 404,694 acres (163,779 ha^2), the same clutch size, and hen and brood success rates as above, yields a production estimate of about 81,000 fledged ducklings. Sanderson (1980) reports that production from the Bear River and Weber River marshes alone exceeds 100,000 fledglings annually, but does not say how this estimate was generated. Although these estimates are overly simplistic and apply parameter estimates from small study areas to the entire GSL system, they nevertheless are useful in demonstrating the potential and realized production from GSL.

Table 4. Estimated breeding ducks on the Great Salt Lake based on aerial transects conducted by the utah Division of Wildlife Resources, 1995-98. (unpublished data, DWR files).

Birds	Indicated Breeding Birds	Visibility Correction	Expansion Factor	GSL Estimated Breeding Birds
Mallard	131	11	32	48,099
Pintail	33	9	32	9,436
Shoveler	128	6	32	26,510
Gadwall	343	5	32	59,944
C Teal	97	13	32	40,702
Redhead	214	4	32	29,642
Canvasback	3	4	32	347
Ruddy	302	2	32	16,389
Total				231,068

Note:
Indicated breeding birds determined from aerial transects
Visibility correction based on DWR studies in the 1980s
Expansion factor = Acres of 1st and 2nd magnitude marsh /aerial transect acreage

The timing of nest initiation and hatching periods for ducks in GSL is influenced by weather, but some patterns persist year to year. Mallards and pintail are the first ducks to initiate nesting, with mean initiation dates during the first two weeks in May. Shovelers, cinnamon teal, and redheads follow in late May, with gadwall the last to nest. Nest initiation, however, has been observed as early as the end of March, and as late as the beginning of July in some years. The incubation period for ducks ranges from 20-30 days, so hatching dates follow initiation dates by about a month.

GSL is also a major breeding area for Canada geese (table 5). Lake-wide surveys conducted annually by the DWR from 1980 to 1998 indicate that an average of 438 pairs of geese produced about 1,663 young each year. These figures represent 60 percent of the breeding geese in Utah and 53 percent of the statewide production of young. Because we likely do not see all pairs or young during survey flights, these estimates represent minimum numbers. Geese are the earliest nesting waterfowl species in Utah, with most nests initiated in March and April.

The role of endogenous nutrient reserves and food resources on waterfowl reproduction is well documented (Ankney and MacInnes, 1978; Krapu, 1979; Owen and Reinecke, 1979; Raveling, 1979; Ankney and Afton, 1988). Ducks and geese rely heavily on lipid reserves acquired in winter and spring, which are carried to breeding areas for egg production. Much of the protein required for eggs, however, comes from food resources in the breeding areas. Food acquired during incubation recesses is also believed to be important for successful reproduction, particularly in ducks, although most energy for incubation comes from fat reserves. After hatch, growth of young birds, as well as recovery of body condition in females, is dependent on high-quality foods. Abundance and availability of invertebrates are particularly important during these reproductive periods. Wetlands of GSL offer a diversity of food-producing plants and associated invertebrate communities to fuel the energetic needs of reproduction for thousands of locally breeding waterfowl, and millions of migrating waterfowl that breed farther north.

Waterfowl Molting

All birds molt to replace damaged and worn feathers. Among waterfowl, there are numerous differences in the timing and extent of the molting process, and these differences in molt patterns are used, in part, to taxonomically group waterfowl species. Generally, geese and swans molt once annually, whereas ducks go through at least two partial molts each year. A unique characteristic of waterfowl, not shared with many other bird groups, is a complete, simultaneous wing molt that leaves individuals flightless for three to five weeks during summer. In geese, breeding-age adults molt while raising their young, often forming aggregations of mixed flightless families called "creches." Non-breeding geese (one and most two year olds) and failed breeders commonly depart the breeding areas all together and form large "molt migrations" to areas farther north where they are relatively free from predators. Leaving the breeding areas to molt in higher latitudes is also believed to reduce competition for food with successful breeders and their young, as well as to provide molting geese more available time to forage due to increased day lengths. Male ducks show similar patterns by molting wing feathers soon after the breeding period. Some species form large flocks and migrate to traditional areas before molting. Female ducks generally molt after their broods fledge in local breeding areas.

Replacing feathers requires nutrients above the normal basic metabolic needs. The complete wing molt is particularly taxing on protein resources due to the need for replac-

Table 5. Breeding pairs and production of Rocky Mountain Population of Canada geese in Utah (from unpublished UDWR data files).

	GSL		UTAH		GSL as % of Utah	
Year	Breeding Pairs	Production	Breeding Pairs	Production	Breeding Pairs	Production
1980	446	2,025	654	2,926	68.2%	69.2%
1981	382	1,754	674	3,122	56.7%	56.2%
1982	597	2,802	789	3,740	75.7%	74.9%
1983	412	1,974	581	2,733	70.9%	72.2%
1984	139	629	289	1,343	48.1%	46.8%
1985	153	715	369	1,564	41.5%	45.7%
1986	238	1,033	458	2,172	52.0%	47.6%
1987	294	1,024	590	2,428	49.8%	42.2%
1988	283	920	601	2,279	47.1%	40.4%
1989	263	791	651	2,827	40.4%	28.0%
1990	644	1,479	912	2,866	70.6%	51.6%
1991	378	1,317	681	2,633	55.5%	50.0%
1992	738	1,844	1,030	3,430	71.7%	53.8%
1993	927	1,552	1,225	3,014	75.7%	51.5%
1994	552	2,992	889	5,137	62.1%	58.2%
1995	354	1,931	741	4,355	47.8%	44.3%
1996	524	2,442	991	5,106	52.9%	47.8%
1997	414	1,890	743	3,656	55.7%	51.7%
1998	582	2,487	947	4,677	61.5%	53.2%
Average	438	1,663	727	3,158	60.2%	52.7%

ing all wing feathers in as short of a period as possible. Most waterfowl species rely, in part, on stored lipid reserves to fuel the energetic costs during molt, but rely heavily on protein from food resources for nutrient requirements of feather development.

Geese generally prefer large, open, water bodies to molt, whereas ducks take advantage of both open water and dense emergent marshes. Both strategies are believed to be mechanisms to avoid predators during the flightless period. Molting areas are also generally rich in food resources because the high nutrient demands of feather development coincide with a time when mobility and access to food is limited by the inability to fly. Availability of invertebrates as a protein source is a particularly important feature of high-quality molting environments.

GSL is host to substantial numbers of molting waterfowl, including both local breeding birds, and large numbers of molt migrants. Tundra swans do not molt in Utah because they are not present during the summer molting period.

The RMP Canada geese, including some from Utah, migrate north to other intermountain states and to Canada to molt during summer (Krohn and Bizeau, 1979; Krohn and Bizeau, 1980). Some are believed to travel as far north as the Northwest Territories in Canada. Utah has three, recognized, major molting areas for Canada geese that are important for local geese, as well as migrants from other states. Two of these areas are located on GSL (table 6). Bear River Bay, along the east side of Promontory, is the largest molting area in Utah, with over 7,000 geese recorded during the June survey in recent years. This area represents one of the largest concentrations of RMP Canada geese molting in the Rocky Mountain region and its attractiveness is likely related to the lack of disturbance and sheer size of this open body of water. Bear River Refuge also receives substantial use from molting geese, although use in recent years has declined and possibly shifted to the Promontory region.

Pintail and green-winged teal are the most common ducks molting within GSL. Molt migrants begin arriving in mid-June, with most completing the wing molt in July and early August. Unlike geese, most of the molting teal and pintail in GSL are males arriving from northern breeding areas. Although large concentrations of molting pintail and green-winged teal occur on the marshes of the Bear River delta, these ducks can be found molting throughout the fresh-water marshes of GSL. Local breeding ducks begin the wing molt in July, while late-nesting species, such as redheads and ruddy ducks (*Oxyura jamaicensis rubida*) can still be found flightless in September.

Wintering Waterfowl

Waterfowl have relatively large and compact body forms compared to many bird groups. They also feature thick feathers and an abundant down layer. These characteristics afford them the thermal capacity to winter in higher latitudes and colder climates than many other waterbirds. Because migration is energetically expensive and often risky, waterfowl tend to winter as far north as they can to find adequate food and open water. Availability of open water and food are consequently the limiting factors controlling the distribution and abundance of waterfowl in winter.

The fresh-water marshes of GSL normally freeze by the first week in December, eliminating access to most aquatic food sources important to waterfowl. Major movements of waterfowl to other wintering areas accelerates during this ice-up period and populations on GSL reach their annual low by January. Although GSL holds the majority of waterfowl in the state during winter, populations represent only 2 to 8 percent of ducks, geese, and swans wintering in the Pacific flyway (table 7). The Pacific flyway is a waterfowl migra-

Table 6. Use of major molting area by the Rocky Mountain Population of Canada geese in the Intemountain West. (from unpublished UDWR data files).

	Wyoming			Montana	Utah		
Year	**Wheatland Res.**	**Yellowstone Lk.**	**Jackson Lk.**	**Lima Res.**	**Neponset Res.**	**Promontory**	**Bear River Refuge**
1980	8,500	3,500		10,000	1,716	4,156	1,954
1981	7,500	3,000		3,000	2,293	3,823	2,429
1982	5,000	7,275		4,800	2,275	3,929	2,903
1983	5,000	7,470		5,500	1,312	5,934	636
1984	4,500	7,685	780	9,000	1,750	7,212	3,394
1985	6,500	7,298	300		528	1,642	1,712
1986	7,000	2,810	900	10,958	935	3,885	1,723
1987	6,815	6,860	750	10,300	628	2,738	2,549
1988	8,965	6,900	1,500		565	3,101	1,202
1989	9,250	5,035	1,380		771	3,017	3,134
1990	7,563	3,955	225				
1991	7,420	1,990	220		626	2,911	1,312
1992	6,210	1,539			866	3,373	1,215
1993	9,430	1,907	653		991	4,155	78
1994	10,600	1,055	474		1,455	3,587	16
1995					878	7,136	1,418
1996	6,574	2,929	435		739	7,016	109
1997					982	7,252	267
1998					994	11,893	110
Average	7,302	4,451	692	7,651	1,128	4,820	1,453

Table 7. Great Salt Lake mid-winter waterfowl populations relative to Utah and Pacific Flyway populations, 1994-98 (from unpublished UDWR data files).

Species Group	1994-98 MID-WINTER AVERAGE			GSL as % of	
	GSL	UTAH	PACIFIC FLYWAY	Utah	Flyway
Ducks	104,821	151,778	4,688,383	70.2%	2.3%
RMP Canada Geese	3,876	13,378	98,007	31.0%	4.2%
Swans	7,028	7,140	91,003	98.4%	7.7%

tion route located within the Intermountain West and Pacific coast states.

Some species of waterfowl can, however, switch to alternative habitats and food resources, and remain in the GSL area in significant numbers during winter. Canada geese and mallards exploit waste grains in adjacent agricultural areas as long as they remain snow-free. Mallards and pintail utilize riverine habitats and impoundment discharge areas where flowing fresh water remains open and food persists. An apparently unique food resource on GSL, utilized by large numbers of green-winged teal, shovelers, and goldeneye, are brine shrimp cysts found in the salt-water areas of the south arm which remain open long after the fresh-water areas freeze. Flocks of ducks are seen regularly feeding within wind-generated egg masses concentrated on the lake surface, along ice edges, and against shorelines. Some debate exists regarding the food value of cysts to water birds, but MacDonald (1980) reported assimilation efficiencies of 20 to 35 percent for shelducks and flamingos experimentally fed cysts. Research by DWR to determine the value of cysts in the diets of waterfowl on GSL is underway. Shovelers and green-winged teal appear to still require fresh water based on their heavy use of salt-water habitats that are in close proximity to fresh-water sources. Two particularly important areas where these resources occur is the area from Saltair to the south shore of Antelope Island, and the off-shore area between the Ogden Bay WMA and Fremont Island.

Shorebirds

There are 47 species of shorebirds which breed, or winter, in the Pacific flyway, of which 28 are relatively rare, or utilize non-wetland habitats (Page and others,1992). Twenty-three of the 47 species occur with reasonable regularity within the GSL ecosystem. Of these 23, 13 species breed in the arctic or sub-arctic. The other 10 species breed on the prairies, or in the intermountain west. In addition to these 23 species which occur regularly, 11 other species have been recorded on rare occasions on GSL (Paton and others, 1992) (table 8).

By examining records from the U.S Fish and Wildlife Service (USFWS), and various issues of *American Birds*, Skagen and Knopf (1990) suggest that the Intermountain West is frequented more often by species of large shorebirds during migration, for example, the American avocet, black-necked stilt (*Himantopus mexicanus*), and marbled godwit (*Limosa fedoa*), than by small species of neotropical migrant shorebirds. A preponderance of peeps (small sandpipers) move across the Great Plains. These authors also discovered a general pattern among shorebird migration pathways through the continental United States. The distance that individual species migrated also provided a key to shorebird size. In spring, the intermountain area hosts primarily short-distance migrants whose breeding range extends south of 60 degrees north latitude. These are usually large shorebird species. By contrast, long-distance migrants and species that breed north of 80 degrees north latitude pass through the Central Plains. These are most often smaller species. In fall, migratory species that winter in the contiguous 48 states have greater representation in the intermountain region. Species that winter south of the equator have greater representation as migrants through the plains. These findings were substantiated for the GSL ecosystem through several survey efforts and observations in recent years (Halpin and Paul, 1989; Fellows and Edwards, 1991; Page and others, 1992; Paton and others, 1992, Shuford and others, 1995; Paul, unpublished data).

Shorebird Surveys

In 1988, a Pacific flyway project was initiated by Point Reyes Bird Observatory, cooperating agencies, and non-profit organizations. The scope of the project included all coastal and interior wetlands within the Pacific flyway west of the Rocky Mountains, from Alaska through Baja, California and along the west coast of northern Mexico. The project methods were designed to survey the wetlands suspected to possess high species richness or high numbers of individuals. Surveys were coordinated to canvass these areas in a one- or two-day intensive effort, using a large, organized group of experienced birders. In Utah, the GSL ecosystem was the primary focus area. From 1988 to 1993, the surveys were mostly accomplished by ground crews. During the 1994 and 1995 seasons, more comprehensive aerial and ground surveys of this area were achieved. The DWR and USFWS were principle organizers in these two years. On August 10 and 11, 1994, 436,000 shorebirds of 21 species were recorded within the GSL ecosystem. Noteworthy, there were 252,000 American avocets, 65,000 black-necked stilts, 32,000 long-billed dowitchers (*Limnodromus scolopaceus*), and 16,000 marbled godwits. The numbers of avocets and stilts on GSL proved to be many times larger than those reported from any other wetland in the Pacific flyway. The comparative data also bore out that GSL is the only major staging site for marbled godwits in the interior of North America. The 1995 survey included a spring survey on April 25 that accounted for 68,000 shorebirds consisting of 17

Table 8. Shorebirds of the Great Salt Lake (Paton and others, 1992).

Arctic/Subarctic Breeders	Prairie/Intermountain Breeders*	Rare Visitors/Number of GSL Records
Black-bellied plover	Snowy plover	Lesser golden plover/28
Semipalmated plover	Killdeer	Mountain plover/1
Greater yellowlegs	Black-necked stilt	Solitary sandpiper/19
Lesser yellow legs	American avocet	Hudsonian godwit/1
Whimbrel	Willet	Semipalmated sandpiper/15
Red Knot	Long-billed curlew	Pectoral sandpiper/26
Sanderling	Marbled godwit*	Dunlin/59
Western sandpiper	Spotted sandpiper	Curlew sandpiper/1
Least sandpiper	Common snipe	Buff-breasted sandpiper/1
Baird's sandpiper	Wilson's phalarope	Short-billed dowitcher/1
Stilt sandpiper		Red phalarope/6
Long-billed dowitcher		
Red-necked phalarope		
*All breed on GSL unless indicated by *		

The Great Salt Lake's low-gradient bottom provides extensive shallow water covered flats for Western Sandpipers, and other shorebirds, to forage for brine flies and other invertebrates.

Wetlands provide habitat for these Long-billed Dowitchers and many other water birds. These wetlands, when associated with Great Salt Lake, provide a unique habitat relationship that supports millions of water birds.

species. It is well known that spring shorebirds returning north to their breeding grounds move through, and into, the United States in a more progressive pattern than in the fall. In the fall, shorebirds will stage in one place for several days, or even weeks, in order to add body fat for fuel for a marathon flight south. (staging is when time is spent in molt and foraging to prepare for a continued migration along a migratory path). The fall 1995 survey was conducted on August 18, a week later than in 1994. The survey accounted for 340,000 shorebirds comprised of 18 species. The most numerous species were American avocets, Wilson's phalaropes, black-necked stilts, long-billed dowitchers, and western sandpipers *(Calidris mauri)* (table 9), but counts were slightly lower than in 1994. The smaller number of shorebirds counted in the fall of 1995 could be due to missing the peak migration of early August.

Table 9. Numbers of shorebirds recorded during surveys of wetlands at the Great Salt Lake, Utah. Surveys occurred on August 10 and 11 in 1994 and on August 18 in 1985. (Point Reyes Bird Observatory, 1995)

Species	1994	1995
Black-bellied plover	15	40
Snowy plover	384	58
Semipalmated plover	2	2
Killdeer	198	133
Black-necked stilt	65,446	41,470
American avocet	251,880	188,186
Stilt/Avocet	600	0
Greater yellowlegs	91	3+
Lesser yellowlegs	151	5+
Yellowlegs spp.	177	215
Solitary sandpiper	3	0
Willet	152	35+
Spotted sandpiper	37	7
Long-billed curlew	2	8+
Marbled godwit	16,315	10,979
Western sandpiper	330	714+
Least sandpiper	3	10
Semipalmated sandpiper	2	5
Baird's sandpiper	21	27
Peep sandpiper spp.	1,930	20,723
Dowitcher spp.	31,940	28,125
Wilson's phalarope	748	13,692+
Red-necked phalarope	119	0
Phalarope spp.	65,195	35,725
Shorebird spp.	69	0
Total	435,810	340,162

The BRMBR possesses the largest maintained database that includes shorebird numbers for a portion of the ecosystem. They have records spanning several decades; compiled data is complete from 1948 to 1965 and from 1990 to present (V. Roy, personal communication).

There is a (DWR) database for Wilson's phalaropes and red-necked phalaropes (*Phalaropus lobatus*) from 1982 to present with the exception of three years. Studies were conducted in cooperation with Hubbs Seaworld Research Institute in San Diego, a principal investigator of Mono Lake, California and its avian resources. Mono Lake, a significant salt lake on the west side of the Great Basin, and GSL are important locations for phalaropes and other birds associated with salt-driven environments of the Great Basin. The studies of phalaropes at GSL demonstrate that the lake is the most important stopover site in the world for these two species. In a single day, there may be in excess of one-half million Wilson's phalaropes and more than one-quarter million red-necked phalaropes. Wilson's phalarope numbers typically peak in late July. This is relatively early compared with other shorebirds during their southerly migration. The Pacific flyway project did not detect large phalarope numbers.

Several studies of shorebirds, many of them thesis and dissertation work from graduate school research projects, have provided information regarding population biology, habitat selection, diets, seasonal occurrence, and ecological relationships. In addition, significant findings have been recorded in field notes of researchers as an ancillary addendum to their specific research projects. Paton and Edwards (unpublished data), have compiled extensive field notes on shorebirds from 1990 through 1993. Their study of the ecology of snowy plovers (*Charadrius alexandrinus*) on GSL estimated a population of 10,000 individuals for 1992, the largest concentration of snowy plovers in North America and possibly the world (Paton and Edwards, 1992; Paton, 1997). In 1997, Paton returned to GSL with a field crew and conducted an intensive population census. He concluded that there was no significant difference in snowy plover abundance between the early 1990s and in 1997 (Paton,1997). Coastal populations of snowy plovers are in decline and listed as threatened under the Endangered Species Act, making the protection of the large interior population of GSL acutely important.

During 1990 and 1991, Suzanne Fellows carried out graduate research on the temporal and spatial distribution of GSL shorebirds along maintained travel routes, mostly within associated wetlands found on the northeast and northern parts of the lake. Certain methods used in Fellows' study proved to be quite useful for censuses of wildlife management areas with established roads and dikes. This work provided the basis for a more comprehensive waterbird survey in 1997 (Fellows and Edwards, unpublished report). Until 1997, no such appraisal of shorebirds and other waterbirds had been conducted at GSL.

In 1996, plans were developed to establish an ecosystem-based waterbird survey by biologists associated with avian management and research within the Great Salt Lake area. Several different survey protocol types were used in order to conduct a census of all habitats, that is shoreline, wetland, and open water. This community-based effort sought to plot the estimated numbers of individual species populations during breeding season and migration, their periods of use, location and habitat characteristics of use areas against GSL elevations over a five-year period. Survey methods included total counts, area counts, point samples, and aerial counts. This waterbird survey was conducted by a cadre of 85 surveyors representing multiple agencies, nonprofit organizations, and interested publics. The survey area covered 44 sites encompassing the wetlands, shores and open-water areas of the south, east and north portions of GSL. As close as possible, surveys were conducted every tenth day, with a designated or "preferred" day on which the survey should be completed. Seventeen surveys were conducted each year except for 1997, producing a five-year mean for the peak survey period (August 24 to September 2) of 1,104,946 water birds. From these counts bird-use days were calculated at 86,000,000 for 1997-1999. This number reflects the estimated number of birds present and their duration of stay at GSL. Over the five-year study period GSL elevation fluctuated in excess of 4.5 feet. The relatively large difference between lake elevations brought about a significant difference in overall numbers of birds and species between years. At some individual survey sites, there were also significant differences observed in species composition, as well as in species numbers. The changes noted in species composition and numbers is considered an artifact of the profound change in habitat types driven, by the highly transi-

Ribbons of Wilson's phalaropes rise from the Layton Wetlands Preserve and nearby state lands. These concentrations of hundreds of thousands of individuals were a common sight during the years immediately following the 1980s high-lake years. They foraged on copious numbers of brine flies that in turn fed on the detritus left behind as the lake receded from decaying stands of wetland emergent vegetation.

Closeup of flocking Wilson's phalaropes.

tory shoreline during lake fluctuations. The waterbird survey data continue to confirm the importance of GSL for breeding and migratory shorebirds, colonial nesting birds, and other waterbirds.

Some Noteworthy Shorebirds of the Great Salt Lake

The ecosystem diversity, magnitude, and setting of GSL's ecosystem help support an astounding variety and abundance of shorebird species (table 8). Five species stand out as important examples of the ecosystem resources, especially when regarding the relationship between salt and freshwater environments. The red-necked phalarope, Wilson's phalarope, American avocet, black-necked stilt, and the snowy plover are these species.

Wilson's and Red-necked phalaropes

Of the three species of phalaropes in the world, two occur on GSL in considerable numbers and represent a substantial proportion of the world's population. The Wilson's phalarope population on GSL in July frequently represents more than a third of the world's population (figure 6). The magnitude of the Wilson's phalarope staging population was the main factor in the designation of GSL as a hemispheric site within the Western Hemisphere's Shorebird Reserve Network, currently one of six sites in the United States. To qualify as a site of hemispheric importance, the area in question must host at least 500,000 shorebirds or 30 percent of a flyway population. GSL met both criteria for just the Wilson's phalarope.

The Wilson's phalarope breeds from the prairie potholes to the Intermountain West. Outside of its breeding season, it is a salt lake specialist where it exploits halophiles as a food resource during the respective warm seasons of North and South American summers. This species is often polyandrous with reversed sex and parenting roles. The female defends a territory, courts males, breeds, lays eggs for the male to incubate and brood, and, under ideal conditions, does the same with other males in the same nesting season. At GSL, females are the first to arrive in June after breeding and leaving the males to care for the nest and young. On July 3, 1982, a flock of 118,000 Wilson's phalaropes were 95 to 98 percent females. During the early weeks of staging, Wilson's phalaropes are often associated in same sex flocks. Males filter in during July, and by the end of the month this species reaches peak numbers at GSL (table 10). The timing, forage conditions, and behavior of the Wilson's phalarope at GSL is similiar to Wilson's phalarope activity at Mono Lake, California (Jehl, 1988).

During their stay at GSL, Wilson's phalaropes spend their time eating, and molting into post-breeding plumage. Eating takes the form of hyperphagia. It was estimated that these birds put on two grams of body weight per day during their stage at Mono Lake (Colwell and Jehl, 1994). Small samples suggest that these weight gains occur at GSL as well (DWR, unpublished data). By the end of their stage at GSL and Mono Lake, the birds have nearly doubled their body mass, with the additional weight stored as fat. They need the fat reserves gained at staging sites for the substantial flight to the South American wintering grounds. By mid-August, most Wilson's phalaropes have departed for the saline lakes of Bolivia, central Argentina, and other suitable saline environments. In North America in the spring, they return to GSL around late-April and May on their way to their breeding grounds. At GSL and saline lakes of South America, they forage on the same halophile groups (brine shrimp and brine fly), but different species. GSL brine flies and brine shrimp achieve their greatest abundance during the nesting and migration periods of the Wilson's phalarope and many other shorebirds (Collins, 1980; Stephens, 1997).

Table 10. Results of annual surveys of Wilson's phalaropes conducted during peak staging period on the Great Salt Lake, Utah (1982-1998), some years not surveyed.

Year	Date	Lake Elevation (ft.)	Number	Survey Method
1982	Jul 27	4200.25	403,770	Aerial
1984	Jul 12	4208.80	54,000	Aerial
1986	Jul 25	4210.90	387,157	Aerial
1987	Jul 27	4210.45	193,770	Aerial
1988	Jul 29	4207.80	56,215	Aerial
1989	Jul 27	4205.15	85,850	Aerial
1990	Jul 26	4203.15	579,132	Aerial
1991	Jul 26	4201.90	603,333	Aerial
1992	Jul 31	4200.40	115,600	Aerial
	Jul 31		200,000	Boat
1993	Aug 5	4200.95	229,000	Aerial
1994	Jul 29	4199.30	110-125,000	Aerial
1995	Jul 29	4200.15	267,000	Aerial
1996	Jul 31	4199.90	250-300,000	Aerial, Boat & Land
1997	Jul 30	4200.70	191,733	Aerial
1998	Jul 29	4203.10	262,636	Aerial

Wilson's phalaropes are especially in need of a readily accessible, easily taken, and abundant protein source from June through mid-August. At saline lakes, brine flies and brine shrimp are foraged at different intensities by the different sexes of the Wilson's phalarope (Jehl, 1988). The vast south arm of GSL, with its extensive open water and shallow shorelines, provides the primary foraging habitats of Wilson's phalaropes during the critical period prior to their long migration. This part of the lake supports a robust population of brine shrimp and brine flies. On these shores, the brine fly population is estimated to be at 370 million adult flies per linear beach mile (Oldroyd, 1964). Farmington Bay, Fremont Bay, and Bear River Bay are also used by staging phalaropes. During the flood years of the 1980s, the north arm of the lake was the key habitat for both Wilson's and red-necked phalaropes, however, the numbers of individuals were fewer at GSL than previous years, and for some years that followed in the 1990s (Paul, unpublished data).

Red-necked phalaropes arrive at GSL from their sub-arctic breeding grounds later than Wilson's phalaropes, and they linger on into September. Red-necked phalarope biology is similar to that of the Wilson's phalarope with regard to reversed sex roles. Red-necked phalaropes often use different parts of GSL than the Wilson's. The bay west of Stansbury Island (before the development of the salt extraction industry ponds), Carrington Bay, the lake area around Hat Island, and the open water west of Antelope Island (figure 1) have been preferred habitats of the red-necked phalarope in the past.

In general, red-necked phalaropes seem to be more pelagic than Wilson's phalarope while at GSL. The distribution of Wilson's and red-necked phalaropes within the GSL ecosystem is a fitting example of how lake dynamics affect the distribution and abundance of avian food resources, thus in turn, affecting the distribution of birds, like phalaropes, whose principle mission is to locate plentiful and accessible nourishment. The red-necked phalarope diet seems to consist largely of adult brine flies (table 11) (Paul and Jehl, unpublished data). The two phalarope species are the only two shorebirds to utilize open, deeper lake water and its resources. After leaving GSL, red-necked phalaropes spend the winter offshore in open seas of the Pacific Ocean.

American Avocets and Black-necked Stilts

Of the seven species of avocets and stilts in the world, all are associated with saline or alkaline lakes at some point in their life history. American avocet and black-necked stilt populations are intrinsically tied to western North America, and are found in the greatest numbers in salt-rich terminal lake systems and saltwater estuaries. The largest populations are located in the Great Basin, with the highest number of individuals occurring at GSL. Here, a single-day count of American avocets and black-necked stilts have exceeded 250,000 and 65,000, respectively. These numbers are larger than those reported from any other wetland in the Pacific flyway (Shuford and others, 1995).

GSL provides both breeding and migratory habitats for avocets and stilts. Starting in March, American avocets begin their arrival on the lake. They have just come from their wintering grounds, mostly from coastal areas in California, Texas, and western Mexico - for example, Marismas Nacionales in Nayarit (Morrison and others, 1994; Shuford and others, 1998; M. Cervantes, Wetlands International coordinator for Mexico, personal communication, 1998). For some avocets, nesting begins in early April, but may occur as late as July. They are ground nesters, often forming loose colonies. Nests are positioned in relatively open terrain and typically are comprised of vegetation and other debris, or simply occur in a shallow depression. These nest sites are often near or surrounded by water. The mean clutch size is four eggs. Hatching rates are not well understood on GSL, but can range from 43 to 90 percent nest success in neighboring states (Gibson, 1971; Sidle and Arnold, 1982; Dole 1986). Avocet nesting on GSL is often concurrent with that of black-necked stilts. An estimate of 52,000 to 53,000

Table 11. Percentages of adult, pupae and larvae brine flies in stomach contents of red-necked phalaropes from the Great Salt Lake, Utah, 1992. Diets of all red-necked phalaropes sampled comprised completely of the brine fly species (Paul and Jehl, unpublished data).

Date	Weight, grams (sex/age)	Percentages
July 31	37.1 (F/Ad)	100% pupae
August 10	38.1 (M/Ad)	100% adult
	35.3 (F/Ad)	95% adult, 5% pupae
	34.5 (M/Ad)	95% adult, 5% pupae
August 27	30.0 (M/Ad)	larvae, pupae-well digested
	37.2 (F/Ad)	larvae, pupae-well digested
	37.9 (M/Juv)	larvae, pupae-well digested
	34.5 (?/Juv)	larvae, pupae-well digested
	43.8 (F/Ad)	95% larvae, 5% pupae
	41.3 (F/Ad)	100% larvae
	44.2 (F/Juv)	95% pupae, 5% larvae
	38.9 (?/Juv)	60% larvae, 40% pupae
	33.1 (M/Ad)	100% larvae
October 5	42.8 (M/Ad)	70% pupae, 30% larvae
	36.6 (M/Ad)	90% pupae, 10% larvae
	49.2 (F/Ad)	100% pupae

American avocets at Ogden Bay WMA with Antelope Island in the background. Single-day counts of American Avocets occurring within the Great Salt Lake ecosystem exceed 250,000 - more than in all other surveyed wetlands within the Pacific flyway together.

breeding American avocets was ascertained from the 1997-98 GSL Ecosystem Waterbird Survey Project (Paul and others, 1998b).

GSL breeding avocets utilize various species of Chironomidae, including brine flies in all life stages, and Corixid nymphs and adults. However, Chironomidae larvae often provide 63 to 79 percent of their diet (Osmundson,1990). During late summer and early fall, staging avocets assemble in massive flocks along the shores of GSL. These flocks often number in the tens of thousands, and often occur in association with large numbers of Franklin's gulls, California gulls, and black-necked stilts (Paul, unpublished data). These aggregations of avocets and other waterbirds are most frequently found feeding upon dense populations of brine flies. These foraging areas are situated in open water with depths around 3-9 inches (7.5-23 cm) deep. The avocet is a consummate example of the rich bird life that utilizes the shallow, warm, food-rich, and low-gradient shorelines of GSL. Avocets, like many shorebird species, follow the shoreline as it fluctuates through, and between, the seasons. During high lake periods, as during 1984 through 1987, shoreline foraging and nest habitats are significantly impacted.

In 1997 and 1998, the American avocet was chosen as the avian ambassador for the Linking Communities, Wetlands, and Migratory Birds Project. This project was initiated by Wetlands for the Americas of Ottawa, Canada, and by the Western Hemispheric Shorebird Reserve Network of the Manomet Center for Conservation Science in Manomet, Massachusetts, USA. The project is an effort to strengthen local and international community involvement surrounding wetlands' biological, social, cultural, and economic values. GSL was chosen as the site to represent the project for the United States. Chaplin and Quill Lakes, located in Saskatchewan, Canada, and Marismas Nacionales, Mexico, were the sites selected to represent the other two countries involved. The American avocet was chosen because of its notable relationship with all these areas, its extensive breeding population in the prairies of southwestern Canada, its breeding and significant staging population at GSL, and its large wintering population at Marismas Nacionales. The avocet connects these sites through its flight and life history, just as the project hopes to connect the sites through wetlands appreciation and conservation for all their intrinsic values.

Black-necked stilts are closely allied with American avocets ecologically, and associate with them at breeding sites and some winter sites. In comparison to the avocets on GSL, black-necked stilts arrive a week or two later and leave several weeks before (Paton and others, 1992; Paul, unpublished data). Black-necked stilts winter along the west coast of Mexico and along the coast of Baja, California. There are large numbers that winter in the central valley of California (Shuford and others, 1998). Their breeding ecology, clutch size, and nesting habitats are similar to those of American avocets at GSL. Also like avocets, they rely heavily on brine flies and brine shrimp for food (Hamilton, 1975). Staging flocks tend to be monotypic even though they are sometimes folded into a conglomerate of other waterbirds as they forage and roost. There is often some mixing of species on the margins of the individual flocks. With varying GSL elevations, some areas important to stilts and avocets can shift because of inundated or exposed foraging habitat (figure 5).

Snowy Plover

Studies in the early 1990s, and later in 1997, suggest that more snowy plovers nested at GSL than anywhere in North America (Paton, 1997; Paton, unpublished data). The snowy plover and six other subspecies are distributed throughout the

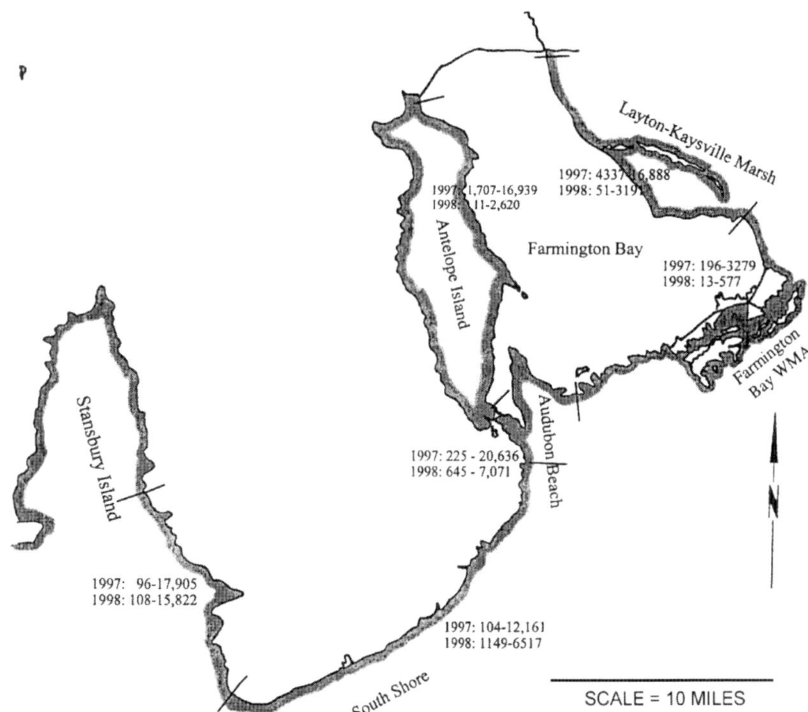

Figure 5. *Combined counts of American avocets and black-necked stilts observed during waterbird surveys on the south arm of the Great Salt lake in 1997 and 1998. For each year, numbers given represent the range of smallest to largest counts of total individuals recorded for the survey season during a single visit. Lines on the map demarcate the boundaries of the survey sites. All surveys were conducted along the lake shoreline.*

world. The western snowy plover (*C. a. tenuirostris*) has two continental populations: one resident population on the Pacific coast, and one migratory population in the Great Basin, inhabiting the salt lakes during the breeding season. The migratory population is the snowy plover of GSL (Page and others, 1986). Populations of Great Basin snowy plovers, including those from GSL, winter on the Pacific coast along the beaches of the Gulf of California. Snowy plovers banded in Utah were observed during the non-breeding season at three locations, Puerto Penasco near La Paz, the Gulf of California, and Scammon's Lagoon on the west coast of Baja, California (P. Paton, Utah State University Ph.D. candidate, personal communication, 1993). In the spring of 1988, the DWR organized surveys to gather baseline information on plover distribution for the GSL ecosystem (Halpin and Paul, 1989). These surveys produced counts for 1988 and 1989 of 487 adults, 26 juveniles and 845 adults, 53 juveniles, respectively.

Plovers utilize the open shorelines and mud flats of GSL where they forage considerably on brine flies and other insects associated with saline soils. They breed and feed in the same habitat type, even though they may not do both in the same location (Paton, 1994). GSL is a dynamic system and it is within the most dynamic component of the lake that the snowy plover subsists, the shoreline and near-shore playas.

Plovers begin to arrive at GSL in late March. Males tend to arrive in the period from late March through early May, females in late March to mid-June (Paton, 1994). Eggs are laid mid-April through mid-July. Much detail has been written about the breeding ecology of snowy plovers (Paton, 1994; Paton and Edwards, 1992). Factors of particular importance to the conservation of snowy plovers on GSL are: 1) all plovers that breed at GSL are migratory and dependent on habitats outside the scope of GSL management; 2) besides the breeding population, GSL hosts significant numbers of transients; and 3) nesting sites and foraging sites for breeding adults are separate and distinct. Some plovers have been observed greater than 1.9 miles (3 km) from the nest. It appears that the snowy plover is a pioneering species that readily adopts suitable nesting habitat when it becomes available. Paton's study sites were all underwater two years before his research began, yet he found significant breeding and foraging activity in these ephemeral habitats.

Impacts of red fox (*Vulpes vulpes*) on snowy plovers and other ground-nesting species of GSL are significant. Predation seems to be a major contributor to plover mortality. The mean survival rate for an adult plover at GSL is 2.7 years (Paton, 1994). Paton suggests that if there is a "good" reproductive year every three to five years, the population probably will persist at GSL.

The most important conservation action for this species, which at 10,000 individuals represents the single largest population in the Americas, is the maintenance of undisturbed shoreline that is allowed to fluctuate through seasons and years. The plover is as flexible as the shoreline is transitory. Paton (1997) pointed out two conditions that have impacted snowy plover populations along shorelines of GSL since his original study in the early 1990s. First is the increased activity of vehicles for recreation or other use. Two important areas for snowy plovers, Lee Creek and Salt Wells Flat, have had increased incidences of disturbance by vehicles. The second condition is the de-watering of some habitats. De-watering, which has occurred at places like Deep Creek at Locomotive Springs, has consequences for plovers as they seek out sites with available fresh water.

Shorebirds Less Understood

There is a plethora of shorebird species for which we have sketchy information, but for which coarse-scale observations lead one to believe that the GSL ecosystem plays an important role in their breeding and migratory life history. Some of these species are the long-billed dowitcher, willet, black-bellied plover (*Pluvialis squatarola*), killdeer (*Charadrius vociferus*), long-billed curlew, greater yellowlegs (*Tringa melanoleuca*), lesser yellowlegs (*Tringa flavipes*), marbled godwit, sanderling (*Calidris alba*), least sandpiper (*Calidris minutilla*), and western sandpiper. On occasion some of these species are present in the tens of thousands, others in the hundreds (Hayward and others, 1976; Paton and others, 1992; Paul, unpublished data; Paton, unpublished data; Fellows and Edwards, unpublished data). As the quilt work of migration is revealed through future studies, there will most likely be many new revelations concerning GSL's function and value to the peregrinations of shorebirds.

Colonial Waterbirds

For the purposes of this paper, colonial waterbirds birds are those birds for which water resources are an integral part of their life history, that nest together, and that belong to the families Podicipedidae, Phalacrocoracidae, Pelecanidae, Laridae, Ardeidae and Threskiornithidae (Recurvirostridae, Charadriidae and Scolopacidae are discussed in the shorebird section). The colonial waterbirds that exist within the GSL ecosystem are varied and utilize several unique habitats (table 12). There are 16 species of colonial nesting birds at GSL, some of which represent the largest single population known to occur in the world, while others represent a significant proportion of the Pacific flyway or Great Basin populations (table 13).

Colonial waterbirds were among the first birds to be noted by early Anglo settlers and explorers (Behle, 1958). A cadre of naturalists, egg collectors, and ornithologists have explored the GSL islands, starting with John C. Fremont and Captain Howard Stansbury in the 1840s and continuing with modern-day biologists, have carried out numerous studies pertaining to the biology and ecology of GSL colonial waterbirds. The foremost ornithologist studying colonial waterbirds on GSL was William H. Behle. His popular book, *The Bird Life of Great Salt Lake* (1958), and other works, have been foundations upon which many productive studies have been built.

Colonial waterbirds demonstrate the full range of adaptability to environmental conditions spanning the stenotopic to the catholic. Some have narrow nesting requirements and others broad. Some have highly specialized diets and others are less limited. Their scale of habitat use varies in a single season from local lake environments to the entire landscape of the Great Basin and beyond. It is for these varied reasons that GSL colonial waterbirds are such compelling subjects.

The GSL eared grebe (*Podiceps nigricollis*) staging population represents nearly half of the North American continental population. American white pelican, white-faced ibis,

Table 12. Breeding colonial waterbird species and nesting habitats/substrates used at the Great Salt Lake (colonial shorebirds not included) (from unpublished UWDR data files).

Nesting Habitats/Substrates						
Islands	**Mud Flats/ Sand Bars**	**Matted Submergant Vegetation**	**Emergant Vegetation**	**Shrubby: in or near water**	**Trees: near water**	**Human-made**
American white pelican	Ring-billed gull	Western grebe	Snowy egret	Great blue heron	Double crested cormorant	Double-crested cormorant
Double-crested cormorant	California gull	Clark's grebe	Cattle egret	Snowy egret	Great blue heron	Great blue heron
Great blue heron	Caspian tern	Forster's tern	Black-crowned night heron	Cattle egret	Snowy egret	Ring-billed gull
California gull		Black tern	White-faced ibis	Black-crowned night heron	Cattle egret	California gull
Caspian tern			Franklin's gull		Black-crowned night heron	Caspian tern
			Forster's tern			Forster's tern
			Black tern			Black tern

Table 13. Population size of notable adult breeding colonial water birds for the Great Salt Lake, Utah (unpublished UDWR data files).

Species	Peak Numbers	Ten-Year Range/Numbers	Data Source
California gull	156,000	1982-1991/76,000-156,000	DWR, (Paul 1990)
American white pelican	20,270	1988-1998/8,326-20,270	DWR unpubl. reports
White-faced ibis	60,000	1988-1998/7,400-60,000	(data missing) USFWS/DWR unpubl. data

and California gull populations at GSL each represent the largest, or nearly largest, in the world (table 13). There are several elements that contribute to this. The magnitude of the ecosystem, the richness of food resources, and the security from terrestrial predators are all paramount reasons for the success of these populations and the motivation for birds to nest in colonies. Other species' populations are also large, but comparative data to assess their value is not available. Some of these include the western grebe (*Aechmophorus occidentalis*), double-crested cormorant, snowy egret (*Egretta thula*), Franklin's gull, and Forster's tern (*Sterna forsteri*). For some of these species, baseline data from GSL is missing. Colonial waterbird species that are often found in significant numbers during the post-breeding season include Franklin's gull, Bonaparte's gull (*Larus philadelphia*), ring-billed gull (*Larus delawarensis*), and black tern (*Chlidonias niger*) (Paton and Edwards, unpublished data; Paul, unpublished data).

The 1980s Flood Years' Effect on Colonial Waterbirds

All 16 species of GSL colonial waterbirds that regularly breed here are subject to the vagrancies of lake dynamics. These species colonial nesting habitats were profoundly affected by the flood years of the 1980s. In 1987, the lake reached its historic high of 4,211.85 feet (1,283.78 m) above sea level, compared to the average long-term elevation of approximately 4,200 feet (1,280.16 m) above sea level. Because of the lake's low-gradient bottom and the amount of shore, the upland and wetland area covered with water increased by 46 percent. This caused extensive flooding of all the state WMAs near the lake and the BRMBR.

The rapid lake-level rise in the 1980s, brought on by unseasonably wet winters and wet cool summers, was unprecedented. Historically, the lake was dynamic, but had never reached elevations higher than 4,207 feet (1,282.29 m) above sea level during the 20th century. Salt water inundated many low-relief nesting islands, flats, bars, and dikes that hosted emergent, wetland nest sites (the south arm's salinity was 57 ppt). Riparian forests were flooded, and shoreline bird colony sites were covered. The complexion of the lake shoreline changed from a varied ecosystem matrix of saltwater habitats and diverse associated wetlands, to a lake with a well-defined shoreline.

The effects of flooding on most colonies were negative , but for some, it was beneficial. In 1982, 15 colonies of California gulls contained 49,862 breeding adults. During the 1983 breeding season, 14 colonies were occupied by 80,487 breeding adults (Paul and others, 1990). Between the 1982 and 1983 nesting seasons, GSL had risen approximately two feet and flooded several of the 1982 colonies, but in 1983, several new colonies were created and some historic nest sites were reoccupied after years of dormancy. There was an amazing 61 percent total increase in breeding adults at GSL. While the rising lake negatively affected some nest colonies by covering them with water, it appears that nesting substrate was not a limiting factor and the wetter conditions may have enhanced food resources for this highly adaptable species.

Another suite of bird species that seemed to benefit, or experienced little impact, from the flooding was herons and egrets (Family Ardeidae). The flooding of riparian forests and tamarisk-lined dikes provided a more secure nesting environment, especially in the Ogden Bay and Howard Slough areas of the Weber River Delta, the flooded Jordan

River Delta, and other lake-wide scattered sites (figure 1). Many of these sites had dikes and roads which, when available to human traffic and disturbance, precluded nesting before the high lake years. This disturbance was eliminated as travel ways were flooded. Since 1987, as the lake receded, roads were made useable again and herons abandoned nesting in these areas. There is no baseline nesting data prior to 1983 to determine if great blue herons, in particular, may have simply been displaced from secure island sites to flooded trees. Unfortunately, alternative nest sites were not available for most ground and emergent vegetation nesting colonial species such as the white-faced ibis, Franklin's gull, Forster's tern, American avocet, and black-necked stilt.

Food resources for colonial birds are affected in significant ways when the lake waxes and wanes with climatic conditions. Piscivorous colonial waterbirds are influenced geographically with regard to flooding on GSL (Flannery, 1988). At the lake's peak level in 1987, state WMAs in the south arm that normally held healthy populations of freshwater fish, carp (*Cyprinse carpio*) in particular, were inundated with brines of 60-70 ppt which killed fish. In the northeast, Bear River Bay, Willard Bay, Harold Crane WMA, BRMBR, and associated private duck hunting clubs all became one enormous fishery. Consequently, this was the only remaining extensive pelican foraging habitat on the lake (Flannery, 1988). With the exception of rivers and small streams, this was true for most other fish-eating birds. Some species, such as the western grebe and Clark's grebe (*Aechmophorus clarkii*), moved to nearby Mantua Reservoir and other freshwater fisheries. These alternative sites, and the anthropogenically influenced GSL regions like the Gunnison Bay, provided some diversity for birds at this time of ecosystem turbulence. The north arm played a role as a waterbird haven during the high lake years. In 1986 and 1987, lowered salinities in the north arm hovered around 160-170 ppt, becoming the only part of the lake which produced a robust brine shrimp population. Consequently, the Gunnison Bay was the primary staging site for Wilson's phalaropes and eared grebes during those years (DWR, unpublished data).

Some GSL breeding species had more trouble contending with changing or adverse conditions because they couldn't find alternative sites and were unable to breed, or forced to simply disperse. They are the white-faced ibis, Franklin's gull, American avocet, black-necked stilt, snowy plover, and Forster's tern. Even though some of these species managed to nest within the system during high lake years, their numbers were dramatically diminished. All of these negatively affected waterbirds were easily impacted by flooding or high-lake conditions because they either nest on or near the ground, or in emergent vegetation near GSL.

A long-term prospectus is necessary to appropriately assess the effect of fluctuating lake conditions for these breeding species. The saltwater intrusion which periodically damages emergent vegetation during high-water cycles is the same disturbance that maintains open beaches and mud flats along other lake margins. For those species nesting in, or on the margins, of hardstem bulrush, alkali bulrush, cattail, and common reed (*Phragmites australis*), such as the white-faced ibis, Franklin's gull and Forster's tern, flooding can be devastating during a nesting season. Snowy plovers, American avocets and black-necked stilts, however, benefit in the long term from disturbances that create or maintain open shorelines. Saltwater inundation also can have some long-range positive effects on wetlands. Flooded and salt-burnt vegetation most often dies after prolonged salt water exposure. A productive emergent marsh is one in flux, or one in an early stage of vegetative succession. These pioneering wetlands provide more edge and structural diversity than do rank old stands.

An excellent example of the evolution of a saltwater inundated wetland is the West Layton/Kaysville Marsh (figure 1) that occupies a near-lake location in western Davis County. For several decades this wetland had an immense emergent component of hardstem and alkali bulrush providing nesting habitat for white-faced ibis, Franklin's gulls, and many other marsh birds. In 1978, several thousand nests of ibis and Franklin gulls were observed by D. Paul and S. Manes (DWR, personal communication). In 1983, there were no nesting ibis at the Layton/Kaysville Marsh. By 1986, this emergent nesting habitat was essentially lost to lake water inundation. The only emergent vegetation remaining rimmed the area to the north and east. This was a typical scene for many of GSL's wetlands. (See photographs in the color photograph section for a visual account of the effects of lake dynamics on shoreline and wetland habitats in the west Layton area). By 1987, the lake had swelled to 4,211.85 feet

Photo 6. *White-faced ibis family at Layton Wetlands during the 1998 nesting season. Ibis colonize flooded wetland stands during years when the lake elevation is above the long-term-elevation average. During average low-lake years (<4,200 feet), this Layton area has an extensive bar that extends into the lake away from the wetlands which, under those conditions, supports hundreds of avocets and other shorebirds.*

Photo 7. *White-faced Ibis provides shade for young nestlings.*

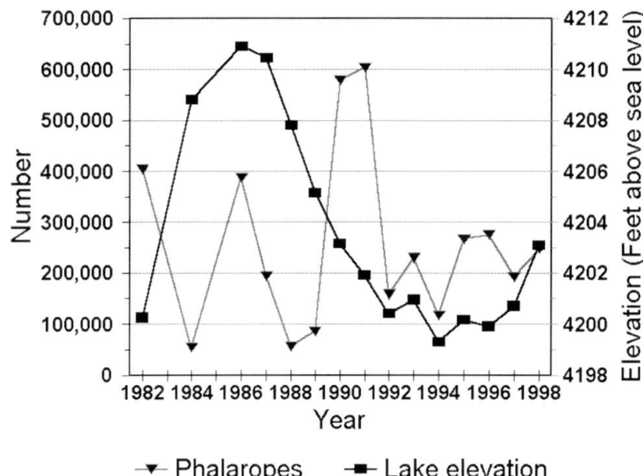

Figure 6. *Total counts from annual surveys of Wilson's phalaropes during their peak staging period in relation to lake elevation on the Great Salt Lake, Utah, 1982-1998 (some years not surveyed).*

(1,283.77 m) above sea level. By February 1989, it had dropped to 4,206.45 feet (1,282.13 m) above sea level and it continued to drop near 4,200.00 feet (1,280.16 m) in 1992.

In 1990, the Layton/Kaysville Marsh was a barren plain of dead bulrush stubble transected by drainages from irrigation run off. Soon, the moist salt-saturated hydric soils became alive with brine flies. These flies used acres of bulrush stems as seven-inch vertical habitat upon which they foraged, rested, and in places where the stems were still underwater, pupated. In less than a decade, this once dead emergent wetland metamorphosed back into one of the most spectacular shorebird habitats on the lake, and perhaps in North America. Hundreds of thousands of Wilson's phalaropes, American avocets, black-necked stilts, snowy plovers, and other species glutted themselves on brine flies. It was during these post-flood years that the greatest numbers of Wilson's phalaropes were recorded for GSL (figure 6), some 200,000 of which foraged on this vast salt plain of brine flies in the Layton/Kaysville Marsh.

In time for the arrival of northern shovelers and green-winged teal at the end of the 1992 growing season, green ribbons of emergent vegetation defined the drainage courses as they wound through the open flats of the Layton/Kaysville Marsh. By the end of 1993, the landscape had significantly changed as riparian strings of rushes broadened out into stands that began to fill open spaces. Paton's record of the rapid and profound changes in snowy plover habitat in the Layton/Kaysville Marsh area during the period from 1990 through 1993 is a stellar example of the lake's wetland dynamics. The snowy plover is the consummate mud flat, open-shoreline shorebird of GSL. From 1990 to 1993, potential plover nesting habitat had declined by 74.7 percent. The reason nesting habitat declined was primarily due to the re-establishment of vegetation (Paton, 1994). By 1995, the Layton/Kaysville marsh looked very much like the emergent marshes of 1978-1982. In 1998, a thorough count of white-faced ibis nests in two colonies at the marsh site came to over 6,000 individual nests (DWR, unpublished data). Also found nesting in the two ibis colonies were Franklin's gulls and Forster's tern. Snowy egrets and cattle egrets (*Bubulcus ibis*) nested nearby in a common reed bed. Currently, fall numbers of waterfowl associated with the West Layton wetland, and near-lake environments, are in the tens of thousands. Within two recent nesting seasons and a lake elevation difference of two feet, there have been changes in this area. In 1997, there was a defined shoreline with a sand/mud bar on the edge of the Layton/Kaysville Marsh which was used by thousands of avocets and other shorebirds. In 1998, the lake had risen, obliterating the shoreline and establishing excellent nesting conditions for ibis (figure 7) (Paul, unpublished data). The most notable difference between the 1978 West Layton emergent vegetation (hardstem and alkali bullrush dominant) and 1997-98 vegetation was the increase in common reed.

The circumstances surrounding the event of migration provide the greatest problems in estimating populations of migrating colonial nesting birds. What happens to displaced, emergent vegetation, nesting species during high-lake elevation years is a conundrum. It is certain that ibis continued to nest in the eastern and northern portions of Bear River Bay, and in other small wetlands, when the lake was above 4,211 feet (1,283.51 m) above sea level, but not in the same numbers as in years before the flood years of the 1980s. In 1985, 150 pairs of white-faced ibis nested in the Cutler Marsh of Cache Valley, Utah. In 1989, at this same location, nesting pairs were estimated at greater than 2,000 (Bridgerland Audubon Society survey coordinator, personal communication, 1989). Between 1984 and 1988, increases in white-faced ibis numbers were noted at wetland sites in Nevada and Oregon, and to a lesser degree in Idaho (B. Sharp, USFWS, personal communication, 1999). At these western Great

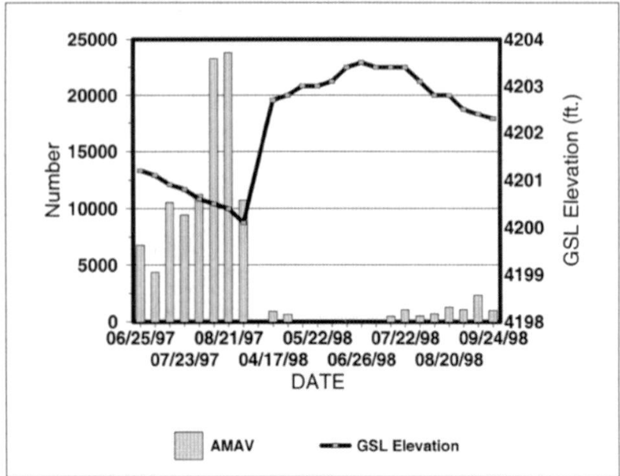

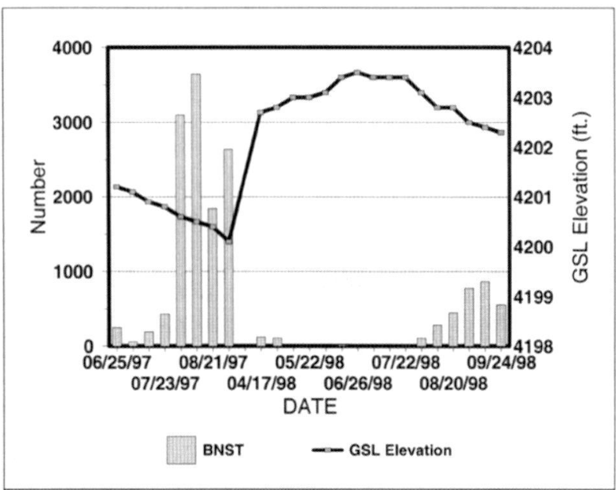

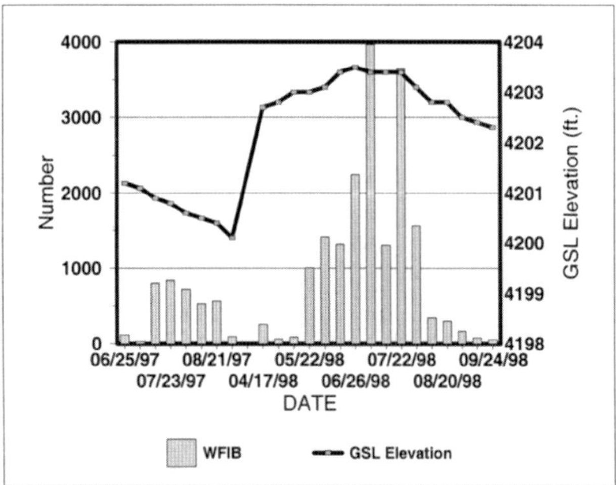

Figure 7. *Numbers of black-necked stilts (BNST), American avocets (AMAV), and white-faced ibis (WFIB) in relation to lake elevation at the Layton-Kaysville Marsh, Great Salt Lake, Utah, 1997 and 1998.*

Basin wetland sites, an increase in moisture occurred from the same wet cycle affecting GSL. Was the increase in ibis numbers due to improved local wetland conditions at these areas of the Great Basin, displacement of birds from the large breeding colonies of GSL when wetlands were so flooded, or a combination of these two situations? Recent cooperation and coordination among states of the Great Basin will assist in better understanding how colonial waterbirds play out their life strategies within the limits of their movement patterns during and between seasons (Haig, 1998).

Important Colonial Waterbirds of Great Salt Lake

Eared grebe: Eared grebes are circumpolar in their distribution with the North American population, breeding from the Great Plains of Canada and the United States into the Great Basin. They winter from the Salton Sea, south into the Gulf of California. Between the breeding grounds and wintering sites, most eared grebes spend time at large saline lakes during the molt migration which takes place when birds move away from the nesting grounds, at which time they molt flight feathers and replace them with fresh ones. Recent population studies demonstrate that most of the eared grebes on the continent molt and stage at two salt lakes in the Great Basin - Mono Lake, California and the Great Salt Lake, Utah. Slightly larger numbers occur at Mono Lake at this time; in 1996 and 1997, the Mono Lake grebe population was estimated at 1.5 million and 1.6 million grebes, respectively, with a 5 percent error (Boyd and Jehl, 1998). During 1997 on GSL, using similar survey methods, 1.5 million grebes were estimated within a 4 percent error (Paul and others, 1998a, b). Eared grebes breed and nest within the GSL system in areas with submergent and emergent vegetation (where nest material and substrate can be found) and where optimal conditions exist for wave and wind abatement to prevent disturbance to their nests. Subsequently, there are eared grebes associated with the lake in all seasons except midwinter. The known breeding population ranges between 1,000 and 3,000 adults (DWR and USFWS, unpublished data). However, GSL's most important role for eared grebe conservation occurs during the fall migration. Because of the vast numbers involved, eared grebes at GSL must accumulate from many localities in North America. Evidence for this includes radio-marked, eared grebe adults from British Columbia being relocated at GSL in fall. A two-year old bird that was banded by J. R. Jehl, Jr. at Mono Lake in 1996 was recovered on GSL in October 1998, suggesting that eared grebes move between salt lakes within the Great Basin.

Eared grebe molt migration begins at the end of August. GSL peak numbers are believed to occur between mid-October and the first of November (Paul and others, 1998b). From eared grebe downing data and field observations, we know that eared grebes depart GSL, in most years, between early December and early January (Jehl and others, 1998). Eared grebe downings are well-documented reoccurring events that most often are associated with winter storms at night. Eared grebes are night migrants and appear to become confused in inclement weather. They often land under these conditions on lighted roads and other hard surfaces resulting in death in large numbers. These lighted wet surfaces appear to give the appearance of water to landing grebes.

Data from 1992 and 1997 describe body composition of eared grebes at GSL (Jehl, 1998). This data indicates a mean body mass of 375 ± 50g for males, and 325 ± 50g for females upon arrival in August. By December 1, males and females are 575 ± 25g and 500 ± 50g, respectively. During this time period, the data demonstrated an increase in leg and heart mass through the fall, as well as a concomitant increase in the size of digestive organs. At the same time, all remige and most covert feathers were dropped simultaneously, rendering the grebe flightless for a period of 35 to 40 days. The synchronous molt and body mass increase, aided by increased size of viscera, demands an abundant food source to sustain these processes. Before grebes depart, they may lose 25 percent of peak body mass as they condition for migration (Jehl, 1998). Migrating weights were obtained from eared grebes that came down in a January 1998 storm in the vicinity of Fillmore, Utah. The mean for males was 420.8 grams (46.82 standard deviation) and 377.1 grams for females (35.60 standard deviation) (Jehl, 1998). It appears that since eared grebes do not use these weight gains for energy during migration, they are amassed to hedge against nutritional needs while flightless for a month or more as they generate new flight feathers.

Stomach contents of eared grebes collected from GSL in 1992 and 1997 during September through December revealed, with rare exception, that grebes fed exclusively on brine shrimp. Some birds examined in late August and early September had fed on brine fly larvae as well as brine shrimp (Jehl, 1998). Brine shrimp cysts were found in gizzards of grebes examined in December, but all were associated with bits of tissue from brine shrimp adults, and were most likely a result of foraging on gravid adult female shrimp and not selected as such. Intestinal contents of ten grebes were determined. Of these grebes, there were thousands of cysts in the gizzard, but there were no more than two to three cysts in any intestine of a single bird. Cysts are most likely retained and digested along with adult shrimp. The food value of cysts remains to be determined through bomb calorimetry.

Years of fall GSL grebe surveys have demonstrated that this species keys in on areas of the lake with high densities of brine shrimp. At present, the lake's grebe population continues to be large and represents a significant portion of the western hemisphere's population.

California gull: There are three species of gulls (Family Larinae) that breed at GSL, Franklin's, ring-billed, and California gulls. The California gull, Utah's state bird, breeds within the GSL ecosystem in larger numbers than anywhere else in the world. Recent breeding season estimates are greater than 150,000 breeding adults (DWR unpublished data). This is a significant increase over the 94,000 to 99,000 breeding adults of the late 1980s (Paul and others, 1990). Behle (1958) noted an increase in California gull nesting activity along the eastern edge of GSL, especially at the state WMAs and other man-made environments. This increase was noted in the late 1940s and 1950s in comparison to his original surveys in 1938. Behle's 1931 estimate of 80,000 adult California gulls was the first attempt to enumerate the breeding population on the lake. Behle's impression was that the lake's California gull population had not increased over the time period from 1930 to 1950, but he noted increased gulls nesting in new east-lake habitats. Fifty years after Behle's first all-lake breeding adult survey, another all-lake survey was completed by Paul and other DWR personnel wherein nearly 50,000 adults were observed nesting on the lake in 1982. Then in 1983, 80,000 adults were counted (Paul, 1984), and by 1991 there were 156,000 gulls (DWR unpublished data).

There are several reasons that have contributed to the recent increase in California gull habitat and breeding adults. California gulls, like many gull species, seem to thrive in the presence of man. They follow the plow, exploit parks and parking lots for food scraps, and forage at landfills. In addition to exploiting food resources in urban and rural environments, they readily accept appropriate man-made structures as nesting habitat. Some examples of these man-made structures include dikes and habitat islands at state WMAs, levees along solar evaporation ponds and dikes and protective breakwater structures associated with sewage lagoon ponds. The largest California gull colony in the world, comprised of 57,265 breeding adults, occurs within the dike and island complex of a salt company on the south end of GSL. Therefore, the combination of readily available food and close proximity to secure anthropogenically enhanced nest sites is a major reason for increased breeding activity at GSL. A study in the early 1980s (Winkler, 1983) of the two California gull populations at GSL and at Mono Lake, California found that the Mono Lake population consistently produced smaller clutch sizes. The mean clutch size at Mono Lake was two, while at GSL it was three. The study suggested that the reduced clutch size at Mono Lake was due to the lack of nearby early spring food resources necessary for the production of additional eggs. Mono Lake sits at a higher elevation near the Sierra Nevada range, and has a paucity of human-provided food resources available to gulls.

Aside from the man-induced values, GSL resources are still paramount in supporting the ecosystem's robust California gull population. California gulls are omnivorous and opportunistic foragers. They exploit lucrative resources in large flocks and with considerable tenacity. They rely on GSL's macroinvertebrates, particularly as they occur in periods of seasonal abundance. Remote colonies of GSL, like Bird and Gunnison Islands, forage for brine shrimp and brine flies in the same, or nearby, locale to reduce energy expenditure in flight travel associated with food gathering. During summer and fall, when and where there are dense populations of brine flies, large numbers of California gulls, often in excess of 20,000 individuals, will gather to forage with masses of other waterbirds. At times, gulls can be seen running down the shoreline, mouths agape, through swarming adult brine flies, gathering up flies much like a vacuum cleaner.

Migration patterns of California gulls from GSL were determined by an intense banding effort in the 1930s and 1940s. Approximately 13,700 banded gulls produced 236 returns. Most birds were banded at Farmington Bay WMA and at Egg Island. The results of this banding demonstrated that birds winter on the west coast from Vancouver down to the Baja Peninsula, with the majority found in the states of California, Oregon, and Washington (Behle, 1958).

It is important to keep in mind that a large population of a single bird species seen at GSL is not necessarily an indicator that all is well. What one sees at GSL may be the largest concentration in the world of that species at that time, as is the case for the California gull. However, observers are not seeing the same species in the same large numbers on the Great Lakes, or the Caspian Sea, or any other place in the world where said species may occur.

American white pelican: Currently there is only one colony of breeding American white pelicans in Utah. It occurs on Gunnison Island and represents one of the largest breeding populations in North America. In surveys conducted in the 1970s of colonies, it ranked second (Rawley, 1976). During the late 1980s and early 1990s, record numbers of breeding adults occurred on Gunnison Island (figure 8). The GSL colony is important not only because of its population size, but also because of its security. Gunnison Island is sequestered in a remote location in the north arm of GSL, safe from boat traffic and other potential human disturbances. The island is protected by law as a wildlife management area. Under statute, it is closed to trespass during the nesting season except for purposes of wildlife conservation as administered by DWR.

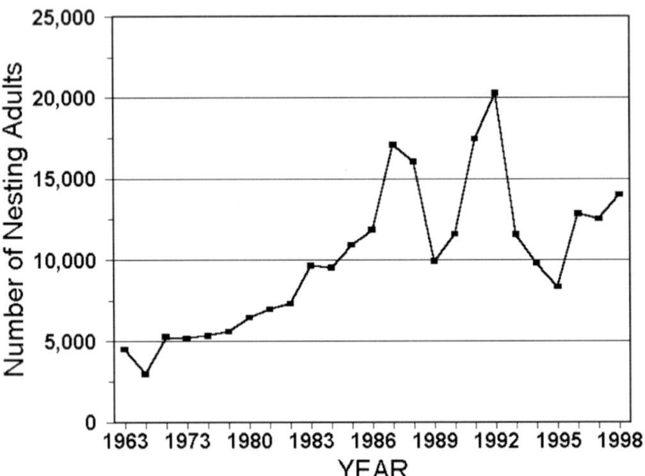

Figure 8. *Estimates of nesting adult American white pelicans at Gunnison Island, Great Salt Lake, Utah, 1963-1998 (some years missing).*

Human disturbance is the main cause of colony desertion by pelicans in North America and other parts of the world. Another major cause of desertion is the presence of terrestrial predators at times when colonies become accessible via land bridges at low-water levels. These two reasons are the cause of desertion by other American white pelican colonies in Utah (Behle, 1944; Knopf, 1975). Other Utah colonies that have since been deserted are: 1) Hat Island (a.k.a., Bird Island), which was last known to be active in 1943; 2) Egg Island, which was observed as a colony by Howard Stansbury in May of 1850; and 3) Badger Island, which was reported to have nesting pelicans by Dr. H. A. Whytock in the 1880s (Behle, 1944). Rock Island on Utah Lake was reported in 1904 as possessing 200 pelican nests with young (Goodwin, 1904). Each of these potential colonies have been impacted by proximal human or predator presence.

Pelicans are piscivorous, thus presenting a logistic challenge to Gunnison Island inhabitants. The nearest fishery is roughly 30 miles (48 km) away, across the Promontory Mountains into Bear River Bay. This means a minimum 60 miles (96 km) round-trip flight for foraging adults (which must negotiate the 2,300-foot (700 meter) vertical lift of the Promontory Mountains with a stomach full of fish) to feed their young on Gunnison Island. Pelicans at Gunnison Island are known to make fishing forays as far north as American Falls Reservoir, and as far south as Utah Lake (Flannery, 1988). These are trips in excess of 186 miles (300 km).

Breeding population information is sketchy through time, but recent Gunnison Island surveys showed increases in numbers since 1980 (figure 8). These increases may reflect the importance to pelicans of introduced carp, gizzard shad (*Dorosoma cepedianum*), and other non-indigenous fishes into the Wasatch Front fisheries. Studies at foraging areas of white pelicans further confirm the current reliance on carp as an important component of the diet of GSL ecosystem pelicans (Flannery, 1988; Paul and others, unpublished data). Carp thrive in nutrient-rich shallow-water systems of the waterfowl management areas and freshwater bays, for example, the Willard Spur and Bear River Bay. Pelicans are communal foragers and excel in the exploitation of shallow fisheries. Before carp were present, there were small indigenous chubs and shiners found in the ecosystem. This helped to support the local pelican population, but probably not to the extent that the large freshwater impoundments of the waterfowl management areas did with their fecund populations of non-native fishes. This enriched environment and the secure nest site of Gunnison Island are the most likely factors lending to the increased numbers of breeding American white pelicans at GSL.

Since the 1930s and 1940s, when a significant effort was made to band pelicans at Gunnison Island by Wm. Behle, Jess Low, and others, there has been some additional banding in the region by others. Pelicans nesting at GSL winter along the west coast and interior of Mexico. Of 1,502 young banded during the 1930s and 1940s, more than 80 band-return observations were made (Behle, 1958). In 1996 and 1997, notable efforts to learn migration patterns of pelicans included a project involving tracking by satellite telemetry from Stillwater NWR in Nevada (Fuller and others, 1998). After one month, a radio-marked pelican flew from Stillwater NWR, Nevada, to Malheur NWR, Oregon, and then to Payette, Idaho. In mid-September this same bird flew to GSL, to the Salton Sea, California, and finally reached Colima State, Mexico by the following December. Seven weeks after capture, another radio-marked pelican flew from Stillwater NWR, Nevada to Cache Valley, Utah. By October, it was at the Salton Sea. Of 17 birds marked with satellite radio transmitters, five wintered at the Salton Sea and six made it to Mexico. This telemetry study has strengthened an observation made from Behle's banding project 50 years ago - the curious behavior of a northward movement from Great Basin banding sites before a southward winter migration. Several birds banded on GSL traveled north to southeastern Idaho (Behle, 1958), just as several birds banded at Stillwater NWR traveled north before finally wintering south.

White-faced ibis: GSL harbors the largest breeding population of white-faced ibis in the Great Basin, and perhaps in the continental United States. White-faced ibis are emergent marsh nesters, often keying in on hardstem bulrush as a nest substrate. Alongside Franklin's gulls, they nest in colonies which range from a few pairs to as large as several thousand pairs (Sharp, 1985). Nesting habitats, comprised of stands of emergent vegetation, occur in standing water. Water serves at least two functions in the colony. It reduces access by mammalian predators and supports growth of emergent vegetation for nest material and substrate. The nature of this nesting habitat type fosters nomadic and opportunistic nesting behavior due to the ephemeral nature of the lacustrine emergent wetlands within the GSL ecosystem, and at other wetland sites characteristic to the Great Basin. As shorelines flood and recede, and as weather patterns shift between wet and dry, emergent stands gain or lose their attractiveness to ibis as colonial nest sites. For these reasons, basin nesting ibis seem to shift colony sites often within local marshes as well as between Great Basin wetlands (Sharp, 1985; Kelchin,

1998). An example of intermittent white-faced ibis breeding at the west Layton/Kaysville wetland complex is described earlier in this article.

Even under the worst nesting conditions, as in the high-water years of the mid-1980s, GSL's breeding population of ibis persisted, but often in reduced numbers. During the mid-1980s, the lake population was reduced to hundreds of pairs, but the same wet water cycle improved emergent vegetation nesting conditions around the region. Sites such as Carson Lake, the Ruby Marshes and Stillwater NWR, Nevada; Malheur Lake, Oregon; Mud Lake, Idaho; and the Cutler Marsh, Utah all witnessed an increase in available nesting habitat and subsequent breeding white-faced ibis. The evidence is only anecdotal, but it does suggest that as conditions change, ibis move within the region to colonize and nest. Since the early 1990s, white-faced ibis have nested in large numbers at several wetlands around GSL. Thousands of nests have been identified in the Bear River Bay complex, the west Layton/Kaysville complex, and Farmington Bay wetlands. An estimated 25,000 to 30,000 nesting pairs occurred at GSL annually during 1996 through 1998 (DWR and USFWS, unpublished data).

White-faced ibis often forage away from their nest colonies in flood-irrigated pastures and alfalfa fields, which are nearly all located on private land. Stomachs of white-faced ibis in Utah revealed a preference for insects, especially the larvae and adults of Diptera and Coleoptera and earthworms (Capen and Low, 1974; Steele, 1980). Discovering this selective foraging behavior, conservation efforts need to address possible impacts to upland agrarian habitats such as urbanization of farmlands, and switching from flood irrigation to pressurized water systems, and how these changes may affect white-faced ibis.

White-faced ibis winter around inland lakes of north-central Mexico, on the coastal plains of Mexico, and in the Sacramento, San Joaquin, Coachella, Imperial and Colorado River valleys of California (Shuford and others, 1996). It is known that significant private lands are being used by wintering ibis in these areas.

In summary, it is apparent when considering conservation strategies for white-faced ibis that several factors are especially important to keep in mind. Historical colony data demonstrates that colonies reoccur at the same location through time even when the site is flooded or desiccated periodically. Judgements of a nest's site value must always be made with regard to its potential and not its present condition. For this reason, it is imperative to document colony locations from year to year. Most ibis colony sites are used by several other species as well, such as Franklin's gulls, Forster's terns, snowy and cattle egrets and black-crowned night herons (*Nycticorax nycticorax hoactli*). So, protecting a site for ibis protects the site for multiple species. A second important consideration is the preservation of farmland and farming activities that utilize flood irrigation of pastures and fields. These unique flooded areas provide an important and unique foraging habitat for white-faced ibis and Franklin's gulls. Flood irrigation provides an abundance of earth worms and other invertebrates that rise to the soil surface and are easily made available to foraging birds. Outside of the California gull, the white-faced ibis may be the most readily recognized and admired marsh bird in Utah because it is so often seen flying through our towns to and from its breeding colonies as it forages near rural towns and on farms.

GSL ECOSYSTEM AND AVIAN CONSERVATION

History Of Habitat Management On GSL

GSL has a rich history of management activities related to waterbirds and their habitats. Although harvest regulation is a major and important management activity for waterfowl species, we will not describe regulation processes or their history in any detail in this discussion.

Early actions to preserve wetland habitats around GSL are detailed by Nelson (1966). Sportsmen's groups began organizing hunting clubs in the 1890s and were among the first to initiate land acquisitions with the intent to protect waterfowl habitat. The Jordan Fur and Reclamation Company was likely the first hunting club organized upon statehood in 1896, hence its commonly used name "Newstate." Other prominent clubs were organized around 1900 and still operate today including the Bear River and Chesapeake clubs on the Bear River delta, and the Burnham Club on the Jordan River delta. Numerous other clubs were established over the next 50 years, many utilizing water sources from smaller streams and agricultural/municipal return flows and surpluses. Because the state does not require clubs to be licensed, no lake-wide accounting of duck club acreage is available today. Information from Jensen (1974) is the most recent estimate and indicates about 50,000 acres (20,235 hm^2) of land is incorporated in private hunting clubs on GSL.

The state initiated its first acquisition and management of important wetlands in 1923, with the purchase and development of Public Shooting Grounds WMA. As its name implies, preserving public hunting opportunities during a time when private clubs were incorporating the best areas was the motivating reason for this development. In the 1930s, the state expanded its formal presence with establishment of Farmington Bay, Ogden Bay, and Locomotive Springs WMAs. New state areas were created, and existing areas expanded during the period from 1950 through the 1990s. The DWR currently manages eight WMAs encompassing over 80,000 acres (32,376 hm^2) on the margins of GSL (figure 9).

The DWR also has state management authority to use any, and all, unsurveyed state-owned lands below the 1855 surveyed meander line of GSL, within certain townships, for the creation, operation, maintenance, and management of wildlife management areas, fishing waters and other recreational activities (Utah Code Ann. Sec. 23-21-5). This management authority now covers approximately 487,000 acres (197,089 hm^2) of lake bed (figure 9). This statutory dedication of sovereign state lands for wildlife management extends as far back as 1943, even though ownership of the bed of GSL was unresolved until 1976, when a U.S. Supreme Court decision gave clear title to the state for lands below the meander line.

In the late 1970s, action was taken by the DWR and state legislature to purchase two important bird colony islands on GSL, Gunnison, and Hat (Bird) islands. These islands were purchased and are managed as WMAs by DWR. Utah rules protect them from human disturbance during critical nesting seasons.

Federal management on GSL centers around the BRMBR, which was established in 1929, in part, to help con-

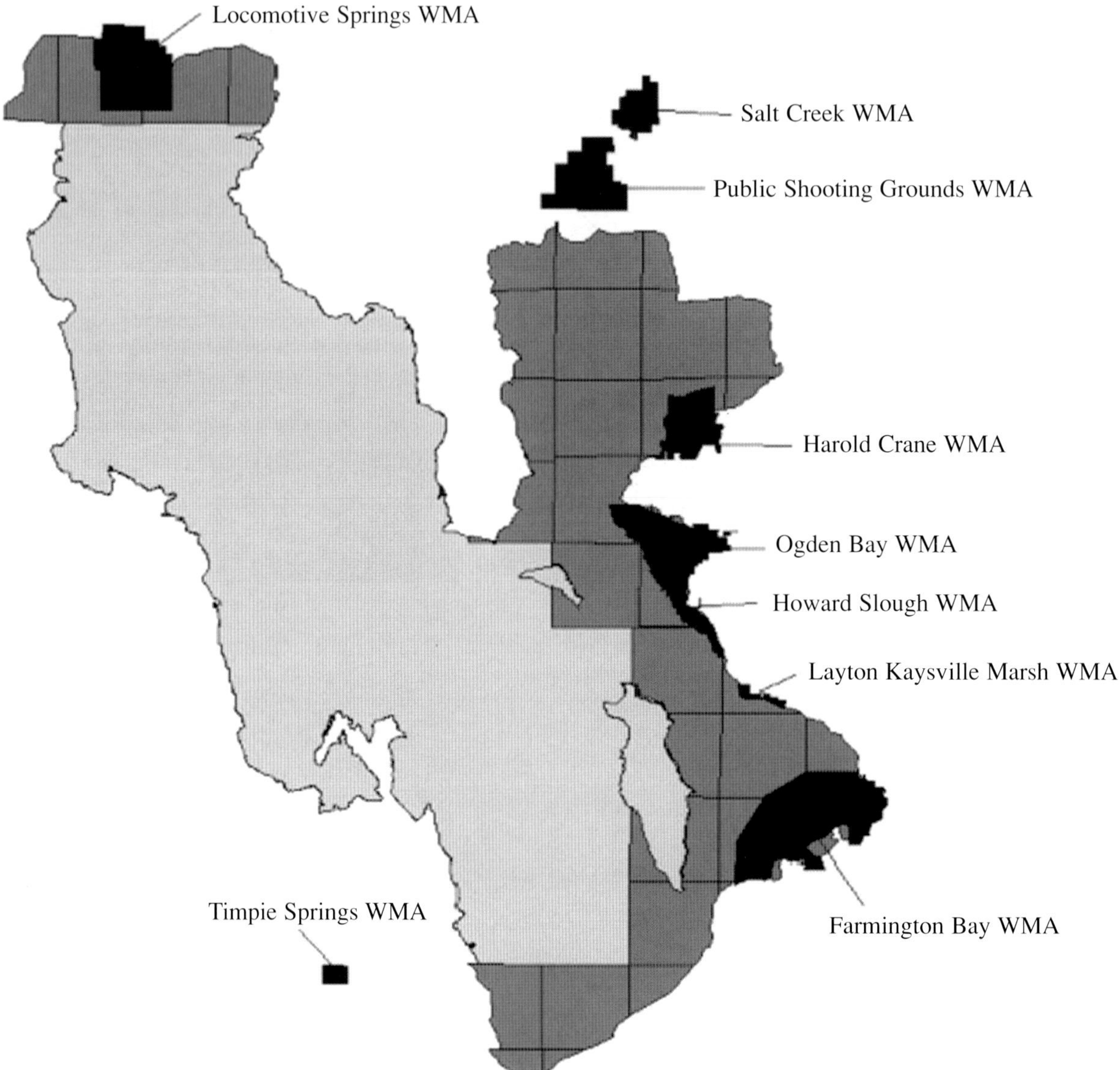

Figure 9. *State Waterfowl Management Areas (dark shading), and lands codified for wildlife management purposes (medium shading) on the Great Salt Lake, Utah.*

trol avian botulism. This refuge currently covers over 75,000 acres (30,353 hm^2) of land, and is the largest management block on the Bear River delta. The Bureau of Land Management also administers numerous sections of land around the north, west, and southwest shores of GSL, and actively manages approximately 11,000 acres (4,452 hm^2) of habitat within the Blue Springs and Salt Wells Wildlife Habitat Areas created during the 1980s and 1990s.

Federal presence on GSL was expanded in 1992 with the authorization of the Central Utah Project Completion Act, and the subsequent establishment of the Utah Reclamation Mitigation and Conservation Commission in 1994. This act provided funding to complete water development projects authorized in earlier legislation, but also provided funding to mitigate for the project's environmental damages. This act provided 14 million dollars for wetland acquisition, restoration, and enhancement in and around GSL during the 1990s and among other things, greatly facilitated the rebuilding of Farmington Bay and Ogden Bay WMAs.

Non-government organizations also have a significant presence on GSL, with the Nature Conservancy's 1,696 acre (686 hm^2) sanctuary (located within the Layton-Kaysville Marsh), initiated in 1983, and the National Audubon's 1,400 acre (567 hm^2) Gilmore Sanctuary on the south shore, initiated in 1992. Several managed mitigation projects protect wetlands created on the south shore, most notedly the 450 acre (182 hm^2), \$12 million dollar Salt Lake International Airport Wetland Mitigation area, and the 3,800 acre (1,538 hm^2) Inland Sea Shorebird Reserve established by Kennecott Utah Copper. Ducks Unlimited has been active in habitat conservation throughout most of the 20th century, and began funding waterfowl habitat projects in Utah with their Matching Aid to Restore States Habitat (MARSH) program, initiated in 1985. The MARSH program has provided \$226,000 for habitat protection on over 25,000 acres (61,774 hm^2) of habitat since 1986, most of which has been focused on GSL. Several other significant conservation actions impacted migratory bird management and protection in Utah during the 20th century (table 14).

Although wetland acquisition and preservation was a

Table 14. Some major conservation actions affecting migratory bird populations on the Great Salt Lake during the 20th century.

Year	Conservation Action	Purpose
1918	Migratory Bird Treaty Act	Regulated the take, propagation and possession of migratory birds.
1934	Migratory Bird Hunting Stamp Act	Established the federal duck stamp to fund habitat acquisitions.
1937	Federal Aid in Wildlife Restoration Act	Tax on firearms and ammunition to fund state wildlife restoration activities. Has provided more than $17 million for the GSL activities.
1977	Clean Water Act	Protects aquatic ecosystems. Sec. 404 requires permits to discharge fill materials into wetlands.
1985	Ducks Unlimited MARSH program	Provides private funding to states for protection and development of water fowl habitat. Has provided $226,000 on habitat projects in Utah.
1985-94	Utah Waterfowl Stamp	State stamp required to hunt waterfowl for hunters 16 and older. Generated 1.5 million in revenues for the restoration, preservation and development of waterfowl habitat.
1985, 1990, 1996	Farm Bill	Provides incentives and funding to landowners for certain conservation provisions including Wetland Reserve Program, Conservation Reserve Program, Swampbuster, Wildlife Habitat Incentives Program.
1986	North American Waterfowl Management Plan (NAWMP)	Coordinated plan between Canada, the U.S. and more recently Mexico to protect the continents wetlands and associated wildlife through creation of Joint Venture Plans.
1989	North American Wetlands Conservation Act (NAWCA)	Provides matching funds to State and private wetland protection activities. Is one of the main funding mechanisms for the NAWMP.
1991	Western Hemispheric Shorebird Reserve Network Designation (WHSRN)	The GSL was recognized as a Hemispheric Site within the WHSRN. A status only achieved by a few important ecosystems supporting shorebirds in the Western Hemisphere.
1992	Central Utah Project Completion Act	Provided funding for completing CUP water development in Utah. Provided $14 million in mitigation for GSL wetland acquisition, restoration and enhancement.
1995	Intermountain West Joint Venture Plan (IMWJV)	Formally recognized wetland values and established population and habitat goals for the Intermountain Region under the NAWMP.
1998	Great Salt Lake NAWCA Grant	Provided $1 million for GSL wetland acquisition, restoration and enhancement under the IMWJV of the NAWMP.

significant and key component of management throughout the 20th century on GSL, development and manipulation of those same lands share an equally important and informative history. Smith and Kadlec (1986), and Nelson (1954) detailed much of the history of agency actions on acquired lands through the early 1980s and is the basis for part of the following chronological summary.

In the early 1900s, most of the fresh water marshes in GSL existed on natural deltas of the Bear, Weber, and Jordan rivers of the east shore. There also were emergent wetlands associated with smaller streams and springs, but they undoubtedly were proportionately small. The vast majority of the GSL shoreline was likely a mud flat, remnants of Lake Bonneville, and too salty to support most plant species. Historical accounts suggest the marshes supported myriads of waterbirds, but were also plagued with enormous disease losses from avian botulism. Agency actions were subsequently focused on acquiring land and water in the delta areas, and capturing the flows through diversions and dike developments. The result was a vast expansion of fresh water marsh area, stabilization of water levels in many delta areas, and an apparent reduction in the magnitude of disease losses. Dramatic increases in the abundance of breeding birds and offspring were also noted during the early development period. Stabilized water levels, however, eventually converted diverse and desirable vegetative communities to dense, monotypic stands of cattail and other less-desirable plants. Waterfowl breeding populations and production declined in response to maturing marshes.

Concerns over declining marsh productivity led to a period of research and experimentation on management areas in the 1960s and 1970s by DWR wetland managers. These studies were directed toward identifying water needs and salinity tolerances of plants as well as management techniques useful in controlling plant succession. Initiation of additional research and implementation of successful management techniques was cut short in the early 1980s when back-to-back 100-year flood events caused a ten-foot (3.048 m) rise in GSL, inundating most managed marsh complexes. Wetland management on GSL essentially ceased for the large majority of public and private areas during the mid to late 1980s. Only the areas above flood levels maintained any kind of water level control which included Salt Creek, most of the Public Shooting Grounds WMAs, and some of the higher elevation duck clubs on the Jordan River delta.

Lake levels declined rapidly in the late 1980s and early 1990s. Management activities centered around the awesome challenge of restoring facilities on over 150,000 acres (60,705 hm^2) of managed marsh that had been developed through nearly a century of work, but destroyed in just a few short years. Restoration of state marshes and many private clubs essentially "chased" the receding lake shore, gambling that levels would continue to drop. By 1998, essentially all of the state areas and private clubs were restored and fully functional. Restoration activities on state management areas were funded largely through revenues of the Utah Duck Stamp program, grants from the Federal Emergency Management Agency, and Central Utah Project Completion Act of 1992 agreements. Restoration at BRMBR began in 1989, but remains incomplete as of this time.

Although the great flood of the 1980s had an enormous impact on wetland vegetation and dependent wildlife, there were some moderating management consequences. Rebuilding activities allowed new, and more efficient, design features to be incorporated into restoration efforts. The BRMBR in particular, included a water bypass system and subdivided many larger management units to enhance water control. Flooding also provided opportunities to acquire important adjacent land and water from willing sellers whose adjacent property was damaged by high water. The Nature Conservancy acquired areas in the Layton/Kaysville marsh. Expansions occurred on nearly all state areas during this period, including the 1,655 acre (670 hm^2) Weber Delta Unit at Ogden Bay WMA and the 1,200 acre (486 hm^2) Rainbow Unit of Harold Crane WMA. Wetland productivity ultimately was enhanced by flood-induced conversion of decadent and relatively low-productivity marsh to younger, more productive plant communities.

Management Challenges And Sustainable Resource Needs

Waterfowl

During the 20th century, an incredible amount of knowledge was gained about GSL biological resources through monitoring, exploration, and experimentation. Much of this information will help guide management strategies and preservation activities necessary to sustain migratory bird abundance and distribution during the new millennium. However, key management challenges in GSL basin remain, and for waterfowl, they include effective means of dealing with predators, disease, undesirable plant communities, and water needs.

Increasing impacts of predators on waterfowl populations have been well documented across North America. In GSL, predator abundance and impact appears to be increasing as well. Studies at BRMBR (Williams and Marshall, 1938), Ogden Bay (Nelson,1954) and Salt Creek (Hilliard,1974; Dalton,1976), suggest waterfowl nest success and abundance was much higher in the first half of the century than during more recent periods (Huener and Manes,1992a, b). Recent invasions of red fox, raccoons (*Procyon lotor*), apparent increases in other mammalian predators, and colonization of management area dikes by gulls may, in part, explain the change. Nest success rates reported by Huener and Manes (1992a, 1992b) were often below levels believed to be necessary for population maintenance. Future management actions focused on improving reproductive success of waterfowl, therefore, appear to be warranted to help sustain local breeding populations.

Waterfowl diseases represent a second mortality factor with significant consequences to populations in Utah. Disease, directly or indirectly, accounts for the largest proportion of non-hunting deaths in waterfowl (Bellrose, 1976), and is likely the main cause of death in un-hunted species. Estimated waterfowl losses to botulism on GSL in the summer of 1997 exceeded 500,000 birds, which was nearly twice the estimated statewide waterfowl harvest that year, and five times the likely number of ducks fledged on local marshes. This was one of the most significant botulism mortality events in the country for the decade. Although stabilization of marsh impoundments appears to have successfully reduced the frequency and distribution of severe avian botulism losses over the long term, low-level outbreaks continue to occur in most years (figure 10), and the potential for catastrophic losses remains. Research on the control of botulism on GSL has spanned the last 60 years, but new preventative solutions have yet to be discovered. Picking up carcasses has historically been and continues to be the main method of control, but the effectiveness of cleanup during epizootics has never been proven in large natural settings.

Whereas avian botulism was the major disease impacting waterbirds in the 20th century, avian cholera may be of equal concern in the next century. Cholera in the U.S. first appeared in wild birds in 1944, in Texas and California (Friend, 1987). Today, cholera occurs across North America, including arctic breeding areas, and losses can rival botulism in some years. In Utah, cholera made its debut in the fall and winter of 1994 to 95, when an estimated 30,000 eared grebes and ducks were lost along the Antelope Island causeway and

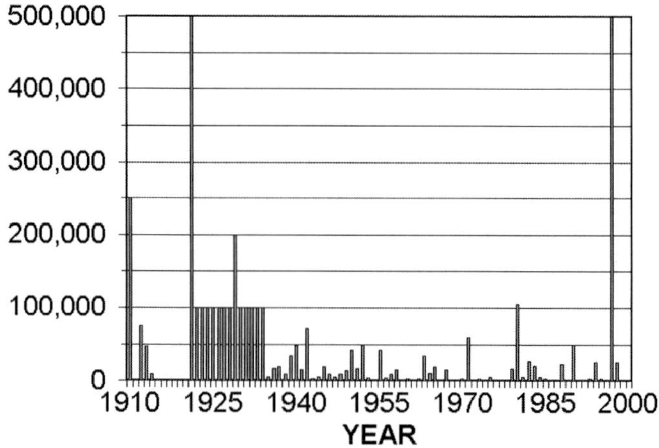

Figure 10. *Estimated annual waterfowl losses to avian botulism on the Great Salt Lake, Utah.*

south shore of GSL. In 1998, cholera reappeared killing over 35,000 grebes, gulls, ducks and shorebirds. Considering the rapid expansion of cholera in the U.S., it is likely that cholera on GSL will become a significant future management challenge. Although GSL bird populations likely evolved dealing with botulism mortality, cholera represents a new challenge of unknown consequence.

Control of undesirable plant communities around GSL will be an ongoing challenge for wetland managers. Most impounded areas are managed for stable water during spring and summer to enhance nest success and reduce disease losses of migratory birds. Unfortunately, stable fresh water in GSL basin also contributes to establishment of dense, and often monotypic, plant communities. This is particularly true where water depths are insufficient to prevent plant reproduction and establishment. In general, hemi-marshes (50:50 mix of emergent vegetation and open water) are believed to harbor the greatest diversity and abundance of birds (Weller and Spatcher, 1965) and are therefore preferred by area managers. Although management techniques exist to control plant densities, they are usually intensive, expensive, and often cannot be applied to broad landscapes.

Some species, including common reed, purple loosestrife (*Lythrum salicaria*), and tamarisk (*Tamarix chinensis*) are rapidly replacing plants with higher values to wildlife. All three species are hardy competitors and eventually exclude more desirable plants. In particular, common reed was among the first plants to colonize disturbed and unvegetated areas after GSL receded in the late 1980s. Efforts to date to control its spread have only been successful in local, intensively managed areas. Efforts to develop more cost-efficient means to deal with undesirable plants will be required to effectively deal with this problem on a broad scale. Using biological agents to control some of these plant species is currently being investigated and shows some promise of success.

Perhaps the greatest management challenge for protecting habitat and water bird populations on GSL is the procurement of adequate water. Water is needed not only to meet the needs of managed marshes, but also to supply unmanaged wetlands that exist below dikes and the myriads of small creeks and drains that intersect the GSL shoreline. These needs have long been recognized, but no attempt to quantify wetland requirements in Utah occurred until studies were initiated in the 1960s by DWR and Utah State University. Christensen and Low (1970) determined salinity tolerances of several wetland plants and generated a basic water model to predict water requirements of marshes in northern Utah. Jensen (1974) used these values to estimate water requirements of GSL wetlands and demonstrated that over 1.5 million acre-feet of water is annually needed for existing marshes (table 1). This represents approximately 80 percent of the 1.9 million acre-feet of surface inflow reaching GSL each year (Stauffer, 1980). The amount of water currently held by state, federal and private interests for wetland management has not been compiled, but is undoubtedly much less than the amount suggested by Jensen. Deficits between water needs and perfected rights likely exist in most marsh systems, but is likely greatest on the Jordan River delta, where approximately half the hunting clubs indicate they have insufficient rights to meet current and future needs (Dunstan and Martinson, 1995).

Although identifying water needs is an important step in securing wetlands, actually obtaining water rights for wetlands is problematic. Currently, Utah water law requires proof of beneficial use before a right can be perfected. By legal definition, beneficial use in Utah includes agricultural, industrial, domestic, and municipal uses. Although some western states now legally recognize wildlife and recreation as beneficial uses, Utah does not. The Utah State Water Engineer does recognize wildlife as a beneficial use by policy, but this could be legally challenged or changed under different administrations. In consideration of this condition, many private hunting clubs graze cattle, trap muskrats, and conduct haying operations to secure their water rights. Additionally, Utah law requires water to be diverted to be put to beneficial use. Protecting natural wetlands that exist on undiverted flows or ground water is therefore difficult without constructing dikes and other water management facilities. Future actions to quantify wetland water needs relative to perfected rights, evaluation of potential upstream depletions of GSL inflows, and development of a long-term comprehensive water plan for GSL wetlands will be needed to effectively protect GSL wetland resources in the future.

Shore and Colonial Birds

Management challenges for shore and colonial waterbirds, in many cases, do not differ from those of waterfowl. Similar challenges are coping with predation, disease, and water needs. Other concerns are the maintenance of shorebird habitats, issues concerning specialty foods such as halophiles, especially brine flies and brine shrimp, and the maintenance of fisheries and shallow water. Some of these concerns also play a role in waterfowl ecology and management.

Long-term wetland predators, such as the striped skunk (*Mephitis mephitis*), weasels (*Mustela* spp), northern harrier (*Circus cyaneus*), American bald eagle (*Haliaeetus leucocephalus*), peregrine falcon (*Falco peregrinus*), and California gull have played a role in the predation of shorebird and colonial waterbirds. Most studies, however, have demonstrated that the most significant wetland predator on the continent is the red fox. Sargeant and others (1984) reported an estimate of 900,000 adult ducks killed each year for the mid-continental North America. For GSL, the red fox is reported as the most important predator for snowy plover mortality (Paton, 1994). The raccoon has established itself as a primary predator on wetland bird nests within the GSL ecosystem. Both the raccoon and red fox are newcomers to GSL wetlands. These predators were rare in Utah wetlands and other habitat landscapes before the mid 1970s. In the 1980s, breeding populations of both predator species were well established. Currently, efforts through cooperative agreements between management agencies and Utah State University are in place to develop management strategies focused on preventing red fox and other predation on wetland birds. The BRMBR is experimenting with different predator exclosure devises to protect breeding wetland waterbirds (Vicki Roy, USFWS, personal communication, 2000).

Avian botulism is another cause of significant numbers of shorebird, gull, and ibis deaths. Records are incomplete, but subsampling indicates that they may represent 10 or more percent of waterfowl losses (author's field notes). There are at least 30 records of peregrine falcon mortalities or illnesses associated with GSL botulism outbreaks (White, 1963).

Fowl cholera is responsible for some shorebird and gull mortality. However, the winter timing of cholera outbreaks, to date, has precluded affecting most shorebirds due to their summer and early fall migratory habits. Eared grebes suffer the greatest loss from cholera as their numbers peak in winter when the disease is active. An outbreak of New Castle's disease in double-crested cormorant colonies of the BRMBR took place during the summers of 1997 through 1999, with significant loss of young (personal communication, Vicki Roy). The treatment and protocol for dealing with these, and other avian disease threats, is carried out by an interagency Disease Response Team. There is no shortage of opportunities to respond in large, rich, playa lake systems like GSL because of the immense concentration of waterbirds that often peak at periods favorable for major epizootics.

Water needs for shore and colonial waterbirds often are similar to those for waterfowl, but there are some exceptions. How water lies on the land, the iteration frequency at which it enters GSL, its abundance, salinity and depth are critical characteristics that dictate its value to shorebirds. Shorebirds require a variety of water patterns. For nesting, they often show preference for water that is nearby or surrounds the nesting substrate. For brood care, they require fresh water for drinking and shallow water with rich invertebrate resources for food. A gradient of wet mud to shallow water, one-half inch deep to seven inches deep, is a key foraging substrate for most shorebirds at all ages and activities. Shallow, brine fly-rich saltwater is perhaps the key water habitat to the immense fall migratory shorebird populations of GSL. The relationship between fresh water wetlands and saltwater is perhaps the most unique habitat characteristic of GSL.

Colonial and other waterbirds often need water that supports a fishery. This may be WMA impoundments, rivers, drainage ditches or large open bays like Bear River and the Willard Spur. The threats to these water conditions come from many fronts. Human water consumption in the burgeoning Wasatch Front communities is precluding the amount and distribution of water to wetlands and GSL. GSL is the terminus of our drainage systems and therefore receives the total sum of all chemicals that accumulate in passage through our towns and cities.

There is a constant threat of urban development to shorebirds and wetlands. Upland open-space buffers are converting from pastures to subdivisions. These shrinking open-space habitats are critical to white-faced ibis, Franklin's gulls, long-billed curlews, and Canada geese. On the lakeside there are discussions of inter-island diking, and attempts at lake stabilization which threaten the very pulse of the lake. As described elsewhere in detail, the transitory shoreline allows for habitat dynamics which provide critical shoreline changes for shorebirds, waterfowl, and colonial birds. Many irrigation companies are exploring the possibilities of converting from agricultural irrigation to pressurized water systems as the landscape changes from farms to housing.

Separation of the lake's brines with extensive dike systems has dramatically affected lake limnology and ecology (Jehl, 1994; Gwynn,1980). The consequences have played a profound role in the distribution and abundance of brine shrimp and brine flies around the lake. These keystone species, in turn, drive the occurrence and abundance of many waterbird species using the lake. Currently, the most viable lake regions for these macroinvertebrates and birds are Farmington Bay and Gilbert Bay. Any future diking that threatens to modify these brines needs serious contemplation. The consequences of changing brine densities through diking, or other developments, could seriously threaten the ecosystem and brine shrimp industry. Conversely, changes to the Bear River Bay fishery could threaten the conservation of one of North America's most stable, and important, white pelican colonies, as well as affect significant populations of breeding double-crested cormorants, western and Clark's grebes, Caspian, Forster's and black terns, and several species of herons and egrets.

CONCLUSIONS

Joseph R. Jehl, Jr., noted ornithologist of salt lake systems, when referring to GSL said, "The Great Salt Lake is ornithologically the most impressive salt lake on the continent." (Jehl, 1994). Lewis W. Oring and Larry Neel, authors of the Intermountain West Regional Report to the U.S. Shorebird Conservation Plan said, "The Great Salt Lake stands out as probably the most important inland shorebird site in North America ...". (Oring and Neel,1999)

GSL is recognized regionally, nationally and hemispherically as a wildlife habitat of significance. Many Utah citizens are learning that they virtually live near "something great" as is indicated in the title of an audio visual presentation by the Friends of the Great Salt Lake. Biological and ecological investigations continue to support these claims. Proper public and private stewardship of this richly diverse and abundant avian habitat will be the key to long-term sustainability.

REFERENCES

Ankney, C. D., and Afton, A. D., 1988, Bioenergetics of breeding Northern Shovelers - diet, nutrient reserves, clutch size, and incubation: Condor v. 90, p. 459-472.

Ankney, C. D., and MacInnes, C. D., 1978, Nutrient reserves and reproductive performance of female Lesser Snow Geese: Auk, v. 95, p. 459-471.

Behle, W. H., 1944, The pelican colony at Gunnison Island, Great Salt Lake in 1943: Condor, v. 46, p.198-200.

—1958, The bird life of Great Salt Lake: University of Utah Press, 203 p.

Bellrose, F. C., 1976, Ducks, geese and swans of North America: Stackpole Books, Harrisburg, Pa., 540 p.

Boyd, W. S., and Jehl,J. R., Jr., 1998, Estimating the abundance of eared grebes (*Podiceps nigricollis*) on Mono Lake, California by aerial photography: Colonial Waterbirds.

Capen, D. E., and Low, J. B., 1974, Recent studies of white-faced ibis in northern Utah: Proceedings Utah Acadamy Science, Arts, Letters, v. 51, p.198.

Christensen, J. E., and Low, J. B., 1970, Water requirements of waterfowl marshlands in northern Utah: Utah Department Fish and Game Publication, p. 69-123, 123p.

Collins, N., 1980, Population ecology of *Ephydra cinerea* (Jones), the only benthic metazoan of the Great Salt Lake, USA: Hydrobiologia, v. 68, n. 2, p. 99-112.

Colwell, M. A., and Jehl, J. R., Jr., 1994, Wilson's phalarope: The Birds of North America, no. 83.

Connell, J. H., 1975, Some mechanisms producing structure in natural communities - a model and evidence from field experiments, *in* Cody, M. L., and Diamond, J. M., editors, Ecology and evolution of communities: Harvard University Press, Cambridge, Mass., p. 460-90.

Cruff, R.W., 1986, Great Salt Lake hypsographic data: unpublished U.S Government memorandum, U.S. Geological Survey, Water Resources Division, Salt Lake City, Utah, 8 p.

Dalton, L. B., 1976, Livestock grazing and waterfowl utilization of marshlands: unpublished M.S. thesis, Utah State University, Logan, Utah, 66 pp.

Dole, D. A., 1986, Nesting and foraging behavior of American avocets: unpublished Master's thesis, University of Montana, Missoula.

Dunstan, F. M., and Martinson, W., 1995, South shore duck club study. A report to the Department of Interior: Utah Reclamation Mitigation and Conservation Commission, 12 p.

Fellows, S. and Edwards, T. C., Jr., 1991, Temporal and spatial distribution of shorebirds on the Great Salt Lake: Utah Department of Natural Resources, Division of Wildlife Resources, unpublished progress report.

Flannery, A. W., 1988, American white pelicans in northern Utah: Utah Birds v. 4, no. 1-4, p. 11.

Friend, Milton (editor), 1987, Field guide to wildlife diseases - general field procedures and disease of migratory birds: U.S. Department of the Interior, Fish and Wildlife Services, Resource Publication 167, p. 69.

Fuller, M., Yates, M., Schueck, L., Bates, K., Seegar, W. S., Henry, W., Young S., and Young G., 1998, Movements of American white pelicans from Nevada through the western United States: Wader Study Group Bulletin 85.

Gibson, F., 1971, The breeding biology of the American avocet (*Recurvirostra americana*) in central Oregon: Condor v. 73, p. 444-454.

Goodwin, S. H., 1904, Pelicans nesting at Utah Lake: Condor v. 6, p. 126-129.

Gwynn, J. W. (editor), 1980, Great Salt Lake - a scientific, historical, and economic overview: Utah Geographical and Mineral Survey Bulletin 116, 400 p.

Gwynn, J. W., and, Murphy, P. J., 1980, Recent sediments of the Great Salt Lake Basin *in* Gwynn, J.W., editor, Great Salt Lake - a scientific, historical, and economic overview: Utah Geological and Mineral Survey Bulletin 116, p. 83-96.

Haig, S. M., 1998, Wetland connectivity and waterbird conservation in the Western Great Basin of the United States: Wader Study Group Bulletin 85.

Halpin, M., and Paul,D. S., 1989, Survey of snowy plovers in northern Utah - 1988: Utah Birds v. 5, p. 21-32.

Hamilton, R. C., 1975, Comparative behavior of the American avocet and black-necked stilt (Recurvirostridae): American Ornithologist's Union, Monograph 17.

Hayward, C. L., Cottam, C., Woodbury, A. M., and Frost, H. H., 1976, Birds of Utah: Brigham Young University, Great Basin Naturalist Memoirs no. 1.

Hilliard, M. A., 1974, The effects of regulated livestock grazing on waterfowl production: unpublished M.S. thesis, Utah State University, Logan, Utah, 98 p.

Huener, J. D., and Manes, S., 1992a., Nesting study at Harold Crane Waterfowl Management Area: Utah Division of Wildlife Resources, unpublished report.

—1992b, Nesting study at Salt Creek and Public Shooting Grounds Waterfowl Management Areas: Utah Division of Wildlife Resources, unpublished report.

Jehl, J. R., Jr., 1988, Biology of the eared grebe and Wilson's phalarope in the non-breeding season - a study of adaptations to saline lakes: Studies in Avian Biology no. 12.

—1994, Changes in saline and alkaline lake avifaunas in western North America in the past 150 years: Studies in Avian Biology no. 15, p. 258-272.

—1998, Great Salt Lake eared grebe summary report: Utah Department Natural Resources, Division Wildlife Resources, unpublished Great Salt Lake Ecosystem Project contractual report.

Jehl, J. R., Jr., Henry, A. E., and Boyd S. I., 1998, Sexing eared grebes by bill measurements: Colonial Waterbirds, v. 21, no. 1, p. 98-100.

Jensen, F.C., 1974, Evaluation of existing wetland habitat in Utah: Utah State Division of Wildlife Resources, Resources Publication 74-17, 219 p.

Kelchin, E. P., 1998, White-faced ibis in the northwest Great Basin - relationship to surface water conditions and intercolony movements: Wetland connectivity and waterbird conservation in the Western Great Basin Proceedings, Bend, Oregon, February 18-19.

Knopf, F. L., 1975, Spatial and temporal aspects of colonial nesting of the white pelican (*Pelecanus erythrorhynchos*): Ph.D. dissertation, Utah State University, Logan, Utah.

Krapu, G. L., 1979, Nutrition of female dabbling ducks during reproduction, *in* Bookhout, T. A., editor, Waterfowl and wetlands-an integrated review: La Crosse Printing Co., La Crosse, Wis., p. 59-70.

Krohn, W. B., and Bizeau, E. G., 1979, Molt migration of the Rocky Mountain population of the Western Canada Goose., *in* Jarvis, R.L., and Bartonek, J.C., editors, Management and biology of Pacific Flyway geese: Oregon State University Book Stores, Inc., Corvallis, p. 130-140.

—1980, The Rocky Mountain population of the Western Canada Goose - its distribution, habitats, and management: U.S. Fish Wildlife Service, Special Science Report - Wildlife no. 229, 93 p.

MacDonald, G. H., 1980, The use of Atremia cysts as food by the flamingo (*Phoenicopterus ruber roseus*) and the shelduck (*Tadorna tadorna*), *in* Persoone,G., Sorgelloos, P., Roels, O., and Jaspers E., editors, The brine shrimp artemia, v. 3 Ecology, culturing, use in aquaculture: Universal Press, Wetteren, Belgium, 456 p.

Morrison, R. I. G., Ross, R. K., and Guzman, J., 1994, Aerial surveys of nearctic shorebirds wintering in Mexico - preliminary results of surveys of the southern half of the Pacific coast states of Chiapas and Sinaloa: Canadian Wildlife Service progress notes, no. 209.

Nagel, J., 1965, Field feeding of whistling swans in northern Utah: Condor v. 67, p. 446-447.

Nelson, N. F., 1954, Factors in the development and restoration of waterfowl habitat at Ogden Bay Refuge, Weber County, Utah: Utah Division Fish and Game Publication, 87 p.

—1966, Waterfowl hunting in Utah: Utah State Department Fish and Game Publication 66-10, 100 p.

Oldroyd, H., 1964, The natural history of flies: Weidenfeld and Nicolson, London, England, xiv, 324 p.

Oring, L. W., and Neel. L., 1999, U.S. Shorebird Conservation Plan, Intermountain West Regional Report: Manomit Center for Conservation Science Report.

Osmundson, B. C., 1990, Feeding of American avocets (*Recurvirostra americana*) during the breeding season: unpublished Master's thesis, Utah State University, Logan, Utah.

Owen, R. B., Jr., and Reinecke, K. J., 1979, Bioenergetics of breeding dabbling ducks, *in* Bookhout, T. A., editor, Waterfowl and wetlands-an integrated review: La Crosse Printing Co., La Crosse, Wis., p. 71-93.

Page, G. W., Bistrup, F. C., Ramer R. J., and L. E. Stenzel, 1986, Distribution of wintering snowy plovers in California and adjacent states: Western Birds v. 17, p. 145-170.

Page, G. W., Shuford, W. D., Kjelmyr, J. E., and Stenzel, L. E., 1992, Shorebird numbers in wetlands of the Pacific flyway - a summary of counts from April 1988 to January 1992: Point Reyes Bird Observatory Report.

Paton, P. W. C., 1994, Breeding ecology of snowy plovers at Great Salt Lake, Utah: Ph.D. dissertation, Utah State University, Logan, Utah.

—1997, Distribution, abundance, and habitat use patterns of the snowy plover (*Charadrius alexandrinus*) at Great Salt Lake, Utah: American Birding Association and National Fish and Wildlife Foundation, final report.

Paton, P. W. C., and Edwards, T. C., Jr., 1992, Nesting ecology of the snowy plover at Great Salt Lake, Utah - 1992 breeding season: Utah Department of Natural Resources, Division of Wildlife Resources, unpublished progress report.

Paton, P. W. C., Kneedy, C. C., and Sorensen, E., 1992, Chronology of shorebird and ibis use of selected marshes at Great Salt Lake: Utah Birds, v. 8, no. 1.

Paul, D. S., 1984, Current and historical breeding status of the California gull in the Great Salt Lake Regions: Utah Department of Natural Resources, Division of Wildlife Resources, Report from Los Angeles Department of Water and Power,

Paul, D. S., Jehl, J. R., Jr., and Yochem, P. K., 1990, California gull populations nesting at Great Salt Lake, Utah: Great Basin Naturalist v. 50, no. 3, p. 299-302.

Paul, D. S., Annand, E. M., and Flory, J., 1998a. Great Salt Lake eared grebe photo survey, 1997: Utah Department of Natural Resources, Division of Wildlife Resources, unpublished report.

—,1998b, Great Salt Lake Waterbird Survey, 1997 and 1998 seasons report: Utah Department of Natural Resources, Division of Wildlife Resources.

Raveling, D. G., 1979, The annual cycle of body composition of Canada geese with special reference to control of reproduction: Auk, v. 96, p. 234-252.

Rawley, E. V., 1976, Small islands of the Great Salt Lake: Utah Department of Natural Resources, Division of Wildlife Resources, unpublished internal report.

Rawley, E. V., 1980, Plant life of the Great Salt Lake study area, *in* Gwynn J.W., editor, Great Salt Lake - a scientific, historical, and economic overview: Utah Geological and Mineral Survey Bulletin 116, p. 337-350.

Sanderson, G. C., 1980, Conservation of waterfowl, *in* Bellrose, F. C., editor, Ducks, geese and swans of North America: Stackpole Books, Harrisburg, Pa., p. 43-58.

Sargeant, A. B., Allen, S. H., and Eberhardt, R. T., 1984, Red Fox predation on breeding chicks in mid-continent North America: Wildlife Monograph no. 89, 41 p.

Sharp, B., 1985, White-faced ibis management guidelines, Great Basin Population: US Fish and Wildlife Service, Portland, Oregon.

Shuford, W. D., Page, G. W., and Kjelmyr, J. E., 1998, Patterns and dynamics of shorebird use of California's Central Valley: Condor v. 100, no. 2, p. 227-244.

Shuford, W. D., Roy, V. L., Page, G. W., and Paul, D. S., 1995, Shorebird surveys in wetlands at Great Salt Lake, Utah: Point Reyes Bird Observatory Report no. 708.

Shuford, W. D., Hickey, C. M., Safran, R. J., and Page, G. W., 1996, A review of the status of white-faced ibis in winter in California: Western Birds, v. 27 no. 4.

Sidle, J. G., and Arnold, P. M.., 1982, Nesting of the American avocet in North Dakota: Prairie Naturalist, v. 14, p. 73-80.

Skagen, S. K., and Knopf, F. L., 1990, Towards conservation of mid-continental shorebird migrations: unpublished internal report, US Fish and Wildlife Service, National Ecology Center, Fort Collins, Colorado.

Smith, L. M., and Kadlec, J. D., 1986, Habitat management for wildlife in marshes of the Great Salt Lake: Transactions North American Wildlife Natural Resources Conference, v. 51, p. 222-231.

Stauffer, N. E., 1980, Computer Modeling of the Great Salt Lake *in* Gwynn, J.W., editor, Great Salt Lake - a scientific, historical, and economic overview: Utah Geological and Mineral Survey Bulletin 116, p. 265-272.

Steele, B. B., 1980, Reproductive success of the white-faced ibis - the effect of pesticides and colony characteristics: Master's thesis, Utah State University, Logan, Utah.

Stephens, D. W., 1977, Brine shrimp ecology in the Great Salt Lake, Utah for the period of July 1996 through June 1997, Administrative report: Utah Department Natural Resources, Division of Wildlife Resources Publication no. 76-19.

Utah Department of Natural Resources, 1999, Great Salt Lake Comprehensive management plan - planning document: Utah Department of Natural Resources, 303 p.

Waddell, K. M., and Fields, F. K., 1977, Model for evaluating the effects of dikes on the water and salt balance of Great Salt Lake, Utah: Utah Geological and Mineral Survey Water-Resource Bulletin 21, 54 p.

Weller, M. W., and Spatcher, C. E., 1965, Role of habitat in the distribution and abundance of marsh birds: Iowa State Agriculture Home and Economic Experiment Station Special Report no. 43, 31pp.

White, C. M., 1963, Botulism and myiasis as mortality factors in falcons: Condor v. 65, p. 442-443.

Williams, C. S., and Marshall, W. H., 1938, Duck nesting studies, Bear River Migratory Bird Refuge, Utah, 1937: Journal Wildlife Management, v. 2, p. 29-48.

Winkler, D. W., 1983, Ecological and behavioral determinants of clutch size - the California gull (*Larus californicus*) in the Great Basin: Ph.D. dissertation, University of California, Berkeley, California.

HISTORY OF THE BEAR RIVER MIGRATORY BIRD REFUGE BOX ELDER COUNTY, UTAH

by

J. Wallace Gwynn
Utah Geological Survey

ABSTRACT

The area occupied by the Bear River Migratory Bird Refuge has a long and interesting history, and serves as a habitat for millions of waterfowl and other wildlife. The area served as hunting and fishing grounds for the early Native American inhabitants as well as for the early settlers, and today, it serves as a source of both food and recreation. The diminishment of wetland areas due to increased land and water usage for agriculture, coupled with outbreaks of duck disease (botulism) and uncontrolled hunting, prompted the creation of the Bear River Migratory Bird Refuge in 1928. During the past 72 years, the refuge has served as a safe haven for wildlife, a duck-disease research center, and a popular tourism site. During 1986 and 1987, the physical facilities of the refuge were destroyed by flooding as Great Salt Lake rose nearly 12 feet to its historic high of 4,212.85 feet. As the lake receded, the U.S. Fish and Wildlife Service contemplated a number of alternatives, and chose to restore and expand the refuge. Restoration and expansion activities are taking place at the present time.

INTRODUCTION

The Bear River Migratory Bird Refuge is located in Box Elder County, 15 miles west of Brigham City, in northwestern Utah (figure 1). From its establishment in 1928 until the 1980s flooding of Great Salt Lake, the refuge comprised 64,500 acres of marsh, open water, and mud flats that were managed for use by migratory birds. Five shallow-water impoundments, each covering 5,000 acres, and having an extensive system of dikes and water-control structures, were originally developed to spread the water from the Bear River before it flowed into Great Salt Lake (U.S. Fish and Wildlife Service, 1998b) (figure 2 - upper). Since the 1980s flooding, the refuge is being expanded, bringing the total area of the refuge to 103,200 acres when expansion is complete (Trout, 1997), and increasing the number of ponds to about 29 (Hansen, 1991) (figure 2 - lower).

The refuge hosts over 200 species of birds and many millions of individual birds as they stop to rest and feed during their seasonal pilgrimages. Sixty species of birds nest and raise their young at the refuge. About 10,000 young gadwall, cinnamon teal, and redheads, the principal nesting species, are hatched each year (pre-flooding). The refuge is part of a major redhead nesting area along the shores of Great Salt Lake, and is considered to be one of the finest redhead production areas in the nation. In addition to the ducklings produced, approximately 2,000 Canada geese are fledged annually, along with countless thousands of various species of shorebirds, marsh birds, and songbirds. During the fall migration, up to 500,000 ducks and geese congregate on the refuge impoundments (U.S. Fish and Wildlife Service, 1998b). Large rafts of Wilson's phalaropes, American avocets, black-necked stilts, and other shorebirds can be observed as birds stage for the flight south. During most of the refuge's existence, scientific investigations have been conducted on avian botulism. Important discoveries have been made in controlling the disease, but the riddle has not been completely solved (U.S. Fish and Wildlife Service, 1998b).

The Bear River Migratory Bird Refuge is part of the National Wildlife Refuge System, the genesis of which was the establishment of the Pelican Island Federal Bird Reservation in 1903 by the Executive Order of President Theodore Roosevelt. By the end of his administration in 1909, Roosevelt had issued a total of 51 Executive Orders that established wildlife reservations in 17 states and 3 territories. The first two refuge acquisitions specifically for management of waterfowl came about with legislation that established the Upper Mississippi River Wildlife and Fish Refuge in 1924, and the Bear River Migratory Bird Refuge in 1928. The Migratory Bird Conservation Act, which became law in 1929, provided the primary authority under which the Refuge System grew in the years that followed. The Migra-

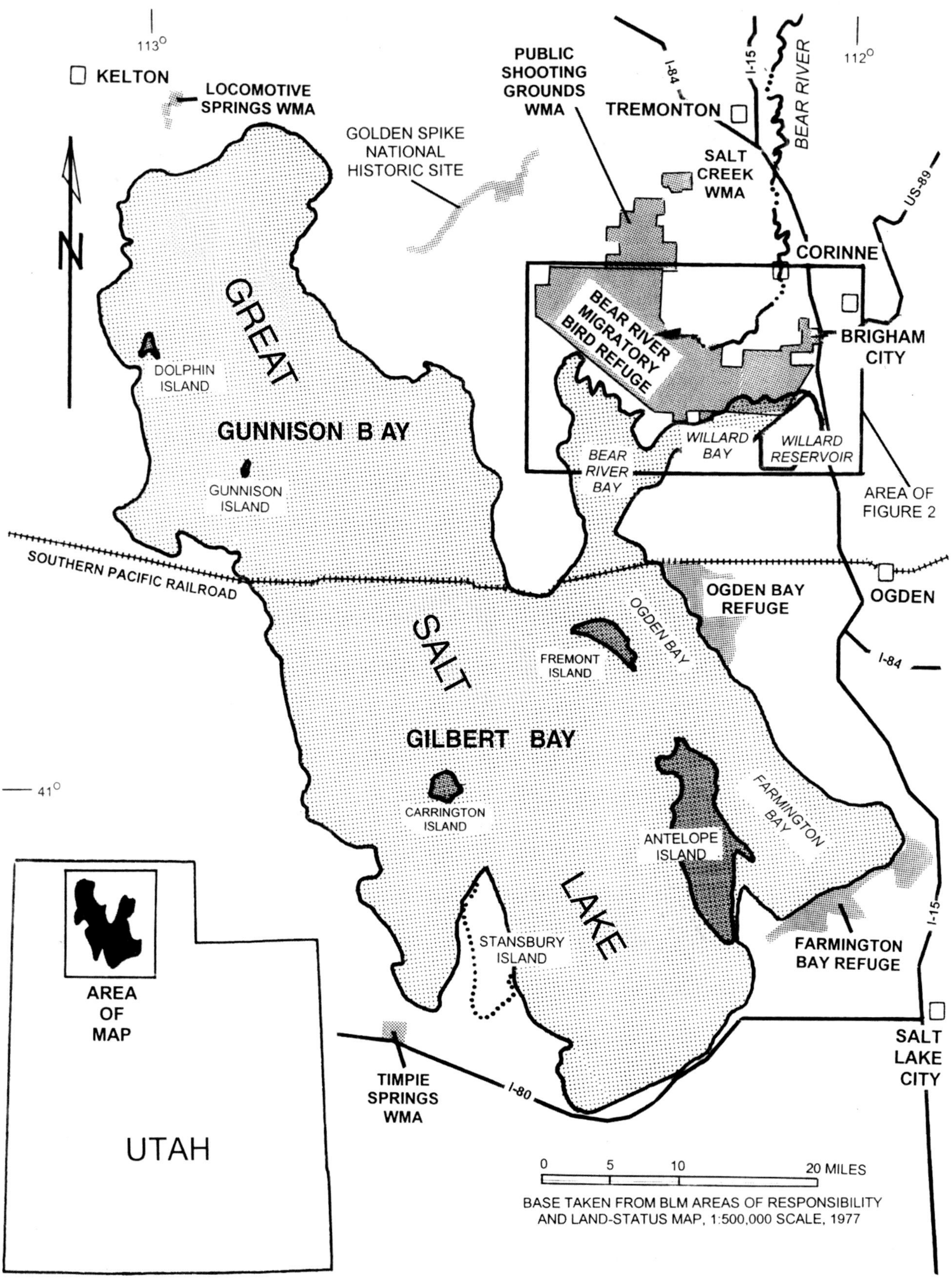

Figure 1. Map of Great Salt Lake and vicinity showing locations of the Bear River Migratory Bird Refuge and other waterfowl-management areas (WMA), the Golden Spike National Historic Site, the Southern Pacific Railroad, highways, and the area covered by figure 2.

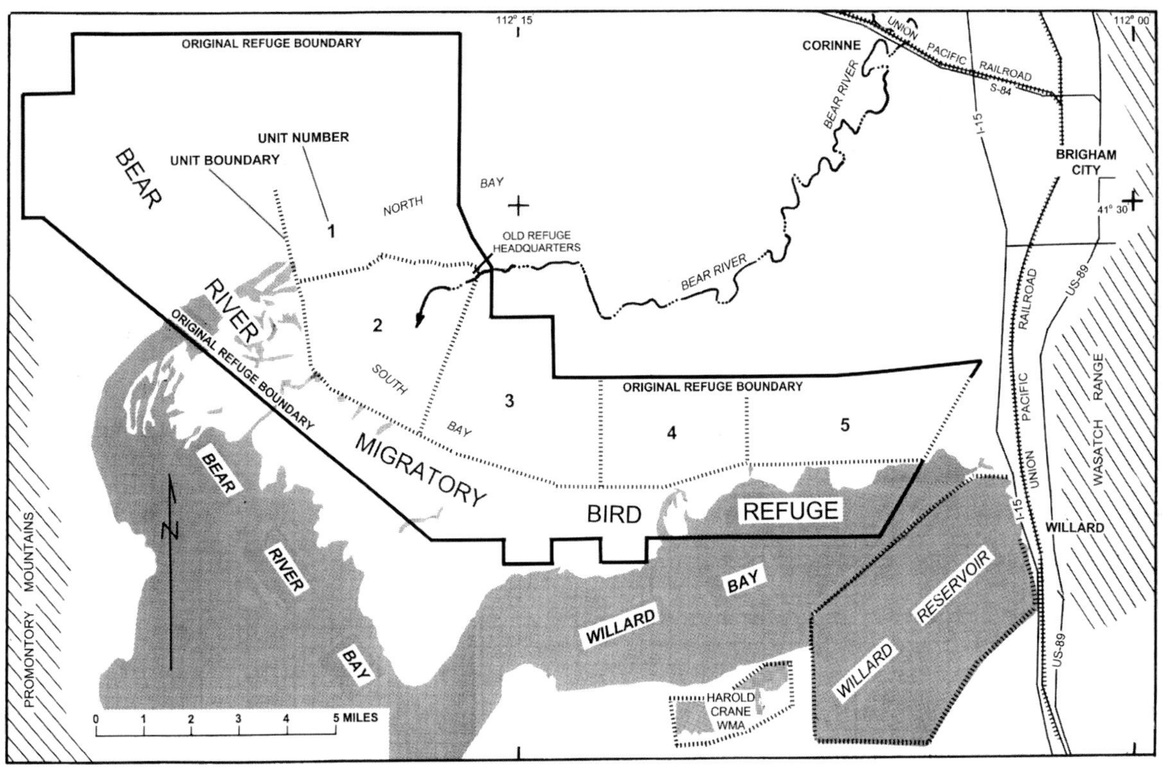

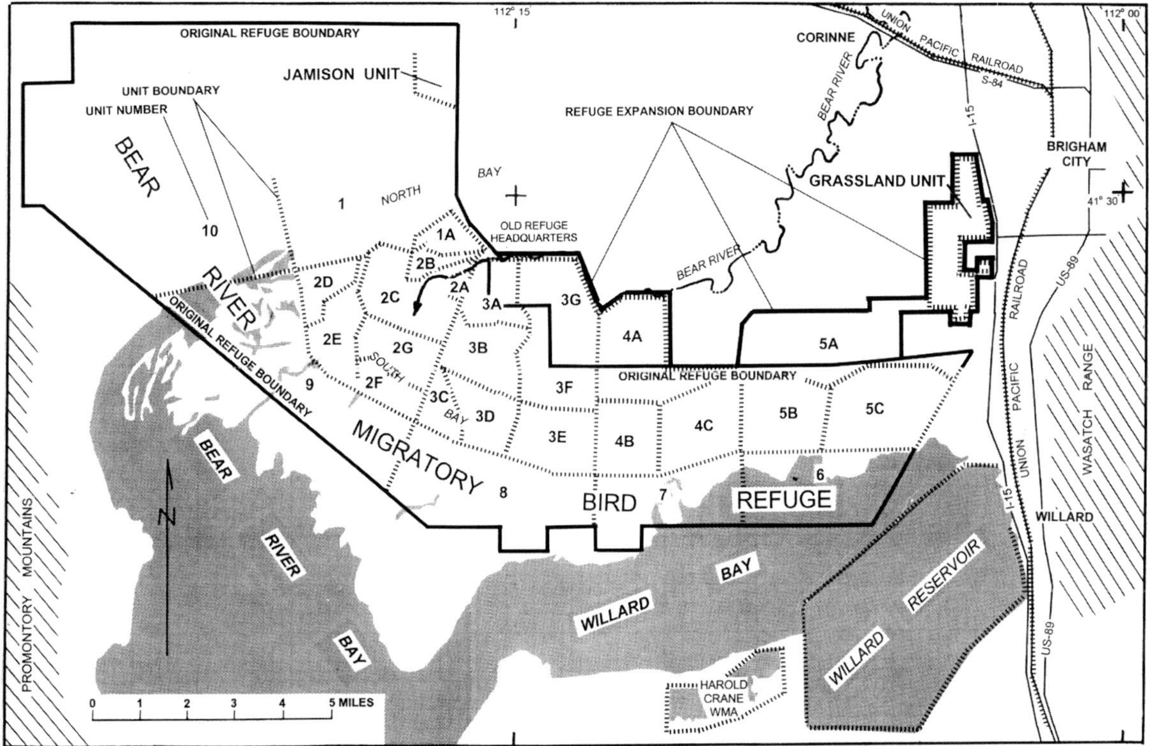

Figure 2. *(Above) Pre-flooding boundary and five-unit configuration of the Bear River Migratory Bird Refuge. (Below) Post-flooding, restoration and expansion-alternative boundary, and 29 -unit configuration of the refuge (after Hansen, 1991).*

tory Bird Hunting and Conservation Stamp Act (Duck Stamp Act) of 1934 provided a source of funds to buy migratory bird habitat. Other refuges were established on former Resettlement Administration lands and as overlays of Bureau of Reclamation water projects (U.S Fish and Wildlife Service, 1998a).

The refuge is a popular tourist attraction and educational center. The visitors' center and tour route records 30,000 visits yearly (unpublished data from Bear River Migratory Bird Refuge files).

EARLY HISTORY AND SETTLEMENT OF THE BEAR RIVER DELTA WETLANDS

Native Americans

The abundant wildlife resources of the Bear River delta were used by Native Americans for many years prior to the settlement of the area by white settlers. The Hukundüka or "porcupine grass-seed eaters" (Steward, 1938) group of the Northern Shoshoni Indians lived in an arc around Bear River Bay on the northeastern margin of Great Salt Lake. Their principal villages were located along the lower Bear and Malad Rivers, and west of Corinne near the area where the Bear River empties into the lake. Madsen (in preparation) notes that the single most significant resource that distinguished the Hukundüka from other groups both north and south of the lake was the presence of fish in significant numbers. Communal fish weirs along the lower Bear River were used to take chubs, suckers, and apparently some trout. A variety of marsh fauna was also collected, and communal duck drives were held in the Bear River marshes (Steward, 1938).

Explorers and Trappers

During the winter of 1824-25, Jim Bridger traveled from the encampment of William Henry Ashley's trappers near the present-day site of Franklin, Idaho, down the Bear River to determine its ultimate destination. Bridger followed the river to its point of discharge into Great Salt Lake and returned to report that he had reached an arm of the Pacific, an erroneous notion that was soon dispelled (Miller, 1980). Bridger was thus the first white man to have penetrated the huge marsh area, and reported millions of ducks and geese at its marshy mouth on the lake shore (Nelson, 1966).

It was not until 1843, however, that John C. Fremont first described the area. In a governmental report (Fremont, 1845), he recorded observations of September 3, 1843, as follows:

> The whole morass was animated with multitudes of water fowl, which appeared to be very wild - rising for the space of a mile around about at the sound of a gun, with a noise like distant thunder. Several of the people waded out into the marshes, and we had to-night a delicious supper of ducks, geese, and plover (in Behle, 1958).

During Howard Stansbury's governmental survey of Great Salt Lake (Stansbury, 1852), he naturally visited the marshes, and on October 22, 1849, he wrote:

> The Salt Lake, which lay about a half mile to the eastward, was covered by immense flocks of wild geese and ducks, among which many swans were seen, being distinguishable by their size and the whiteness of their plumage. I had seen large flocks of these birds before, in various parts of the country, and especially on the Potomac, but never did I behold any thing like the immense numbers here congregated together. Thousands of acres, as far as the eye could reach, seemed literally covered with them, presenting a scene of busy, animated cheerfulness, in most graceful contrast with the dreary, silent solitude by which we were immediately surrounded.

Early Mormon Settlers

Shortly after the Mormon leader, Brigham Young, and his first band of pioneers entered the valley of Great Salt Lake in July of 1847, parties of colonizers were sent out to explore the surrounding region. Those parties going north reported that they found great swamps at the mouth of the Bear River with millions of ducks and geese (Behle, 1958). Soon the Bear River delta region began to be settled and towns established such as Brigham City (starting in 1850), Corinne (1869), and Tremonton (first settled in 1888, with its "second colonization" around the turn of the twentieth century). An event that accelerated colonization of this area, and of the entire western U.S., was the completion of the transcontinental railroad; driving of the "Golden Spike" on May 10, 1869, at Promontory Summit occurred just northwest of the refuge (Fisher, 1998).

As settlements grew, the demands for water from the Bear River and its tributaries grew, and ambitious projects were undertaken to divert great quantities of river water for use by upstream settlements and farms. Because of these diversions, the marshes began to dry, and by the 1920s, only a few thousand acres of the original 45,000 acres of marshland were left. The loss of marshlands through drying seriously affected the survival of migrating birds, but attracted little attention or concern since the drying occurred slowly, (U.S. Fish and Wildlife Service, 1995b).

HUNTING

The vast waterfowl populations of the Bear River delta, as described by Bridger, Fremont, and Stansbury, no doubt supplied a readily available source of food for generations of Native Americans (Nelson, 1966). We can also assume that the early settlers relied heavily upon this same resource as a supply of fresh meat.

Following the period of exploration and early settlement came the market hunting era (1877 through 1900). There were no game laws then to limit the number of ducks killed, and hunters were permitted to sell game. As a result of these liberal policies, a number of men made duck hunting a business, since there was a good market for ducks and feathers (Nelson, 1966).

The center of market hunting activity was the Bear River marshes, though considerable activity occurred on the Weber River, Jordan River, and the Utah Lake marshes. All of these areas were located close to the ready markets in nearby population centers. In addition, large numbers of ducks were also transported by railroad as far away as Chicago, Kansas City, Denver, Butte, and San Francisco. Vincent F. Davis, a prominent early-day market hunter, estimated that 200,000 ducks were killed annually on the Bear River marshes during the market hunting era, just prior to 1900 (Nelson, 1966).

Around the turn of the century, sportsmen began to enter the picture and a number of duck clubs were established, some of which are still functioning. Club houses were erected, and some clubs acquired privately owned marshes (Behle, 1958). Nelson's (1966) report lists 32 duck clubs, along with year organized, acreage controlled, and location, that were located within Box Elder, Weber, Davis, and Salt Lake Counties. At the time of his report, six of the clubs were no longer in operation. As of 1992, there were 33 private gun/duck clubs, and possibly more, around the lake, as shown in table 1 (data from Utah Division of Wildlife resources). Information on the size and locations of the clubs is difficult to obtain as they are not regulated by the state, and their names frequently do not reflect their being duck or gun

Table 1. Duck clubs located around Great Salt Lake in 1992.

Bay View Duck Club*	Five Mile Club	Pintail Duck Club*
Bear River Club*	George East Gun Club*	Pioneer Duck Club*
Brown Duck Club*	Harrison Reclamation Co.*	Pothole
Burnham* (Utah Improvement Co.)	Hills' Folley	Rainbow Duck Club* (now a unit in Harold Crane WMA)
Chesapeake Duck Club*	Irvin Ranch & Petroleum Co.	Rudy Reclamation & Sportsman's Club*
Von's Club	Knutson (Knudson's) Club*	Sagebrush*
Club 41	Feather and Fin	Shotgun Springs
Black Hawk	Lakefront Gun, Fur & Reclamation Club*	Utah Airboats Inc.
Davis Duck Club	Jordan Fur & Reclamation Co. (New State)*	Utah Duck Club*
Duckville Club*	North Point Club*	Wasatch Duck Club
S.J. Duck Club	Ogden Duck Club*	Willard (Bay) Gun Club*

* Clubs also included in Nelson's (1966) list.

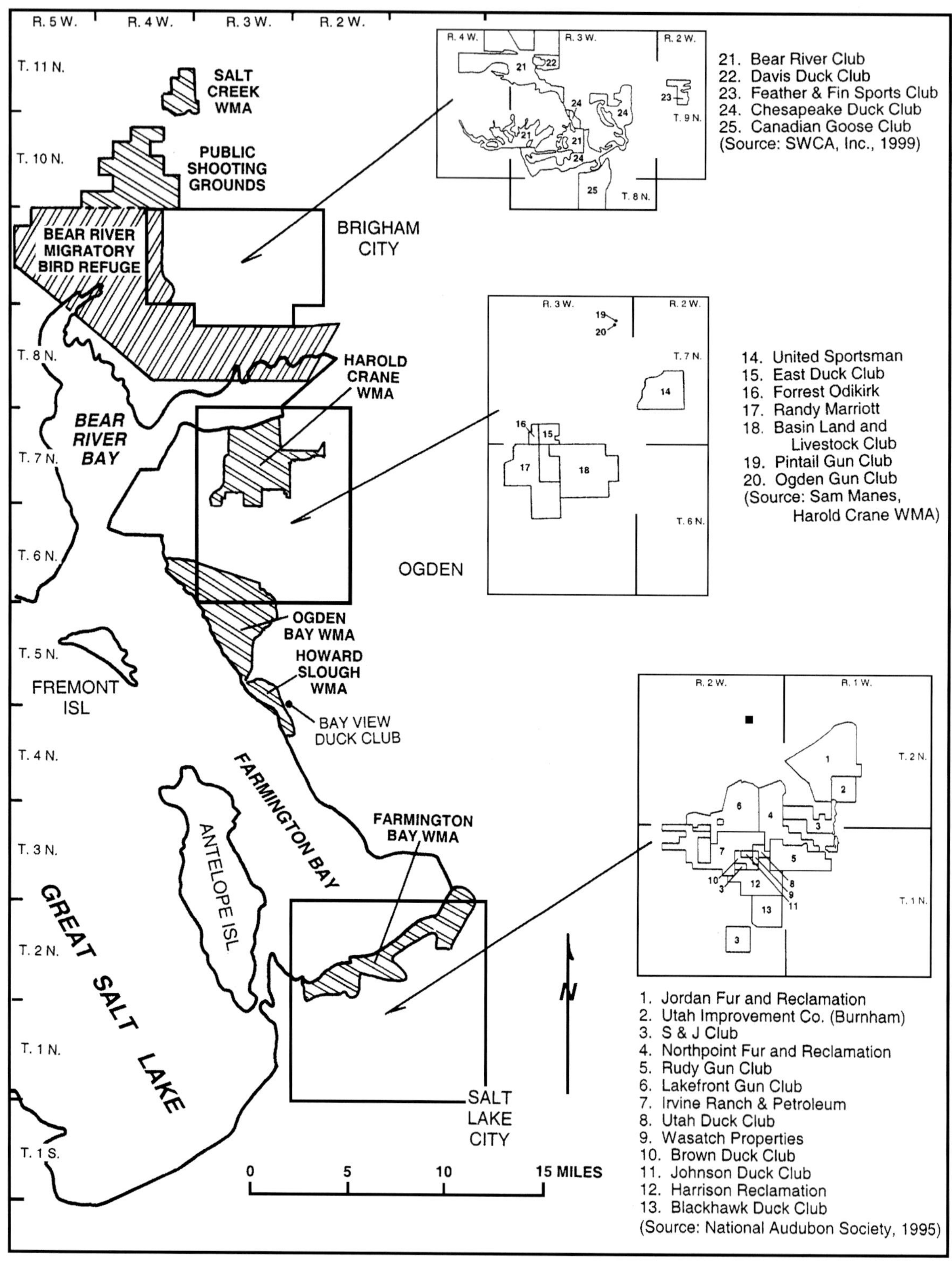

Figure 3. *Approximate locations and areas covered by State and Federal wildlife management areas, and prominent duck clubs, located near Great Salt Lake, Utah.*

clubs. Figure 3 shows the approximate locations and areas covered by the State and Federal wildlife management areas along the east side of the Great Salt Lake, and the locations of some of the clubs.

A few of the clubs were able to acquire established marshlands; however, the majority of the clubs were required to undertake extensive diking and water-control programs to maintain and develop their hunting areas. With reduced water supplies and waterfowl habitat, the clubs were a decisive factor in preserving and developing marshlands. The primary objective of the duck clubs was to develop better hunting areas. Yet, these clubs developed and maintained large tracts of marshland that serve as excellent nesting and producing areas for waterfowl. Some of these marshlands would have been lost without the efforts of these early clubs (Nelson, 1966).

DUCK SICKNESS

Botulism

Following the heyday of the market hunting period was a gradual decline of the Bear River marsh, due mainly to the development of irrigation systems, and a drought that lasted from about 1900 to 1906. By 1910, thousands of acres of marshland, once covered by heavy vegetation, lakes, ponds, and channels of fresh water, had become barren mud flats with stagnant pools of alkaline water. During that year, an outbreak of botulism began that killed over a half million waterfowl, including 85 percent of the ducks (Behle, 1958). Dead ducks dotted the marshes every few yards or were piled up in windrows. Between 1910 and 1925, an estimated seven million ducks died on the Bear River marshes, and Hansen (1991) reports that between 1932 and 1985, an estimated additional 514,300 birds were lost.

Alexander Wetmore's Research

By the early 1900s, the public was aroused by the serious botulism situation, especially since outbreaks of the malady had also been noted in several sites in the San Joaquin Valley and other areas in California. In 1914, Alexander Wetmore was sent to the Great Salt Lake area by the U.S. Biological Survey. He made field surveys at the mouths of the Jordan, Weber, and Bear Rivers, and other areas around the lake, over a period of three seasons (1914-1916) to study the mortality among the waterfowl of the area, especially the ducks (Behle, 1958).

Wetmore meticulously described the symptoms and the behavior of the affected ducks, and characterized duck sickness as a progressive paralysis of the nerve centers and muscles, so that in the early stages there is loss of the power of flight and the ability to dive. The ducks also experience difficulty in breathing, heart problems, and inflamation of the digestive tract before death occurs. Wetmore also noted that the birds became sickened in shallow water areas bordering mud flats that were characterized by poor drainage and high concentration of salt through evaporation. Thus he theorized that the poisoning was due to alkaline salts, principally magnesium chloride and calcium chloride. He found that fresh water was a remedy (Behle, 1958).

The actual cause of duck sickness was later found by Kalmbach (1930), and Giltner and Counch (1930), to be a toxin caused by the bacterium *Clostridium botulinum*, Type C, which apparently exists under moderate alkaline conditions such as those found and described by Wetmore within the Bear River marshes. It was also determined that certain types of vegetation, notably some species of the alga *Cladophora*, were found to be detrimental. Following their active growth period, masses of algal debris settle to the bottom where they decay, trap plankton animals, and contribute to the production of anaerobic conditions. This in turn encourages the growth of *Clostridium botulinum*. These conditions are especially serious in shallow water. In an effort to learn more about the disease, a research laboratory was established on the refuge by 1936 to study the cause of duck sickness and to work toward a cure. Development of an antitoxin for inoculation of sick ducks found in the refuge, along with keeping the ducks in fresh-water areas, provided a cure in a high percentage of cases (Behle, 1958).

BEAR RIVER MIGRATORY BIRD REFUGE

Refuge Establishment and Construction

There is little doubt that the work of Alexander Wetmore from 1914 through 1916 laid the groundwork for the ultimate creation of the refuge. His work called attention to the need for conversion of the area into a refuge with a managed, multipurpose program of: (1) controlling the duck sickness and raising water levels, and reducing the random spread of water over the alkali flats; (2) improving the habitat necessary for food and cover for breeding birds; and (3) providing a resting place for the thousands of water and shore birds passing through the region in spring and fall (Behle, 1958).

No matter how desirable it seemed to have a federal refuge, it still took years of planning and overcoming obstacles to create. It was necessary for the State of Utah to pass an enabling act that would allow the Federal Government to use state lands (Utah S.B. No. 77 by Mr. Young, 1929), and secure title for the use of Bear River water. About 15,860 acres of land for the refuge was initially purchased outright from private owners, and 36,632 acres of government land was set aside by executive order. Other parcels were acquired through land exchanges with various holding agencies, including the State of Utah (Behle, 1958).

Ultimately, however, all obstacles were overcome and on April 23, 1928, Congress authorized the establishment of the Bear River Migratory Bird Refuge (An Act to establish the Bear River Migratory Bird Refuge, Public Law No. 304 - 70th Congress), and an appropriation was made of $350,000. The first contract for construction was awarded on August 1, 1929; actual construction was started September 24, 1929, and was completed by August 1931. President Hoover signed a proclamation on September 26, 1932 (see appendix), declaring the area a refuge and establishing its boundaries. Approval of regulations for the administration of the refuge was given by the Secretary of Agriculture on September 28, 1932, and finally the area was established as a full-fledged refuge on October 1, 1932. The buildings for its headquarters were mostly completed by 1936 (Behle, 1958).

Refuge Mission and Hunting Provisions

The purpose of the Bear River Migratory Bird Refuge, as stated in the 1928 Act, was for the establishment and maintenance of a refuge for feeding, breeding, and resting habitat for migratory birds and other wildlife, while maintaining the natural diversity of plants and animals native to the Bear River Basin (Hansen, 1991). These conditions were included in the terms of the convention between the United States and Great Britain for the protection of migratory birds, that concluded August 16, 1916 (Lansing, 1916).

Because of the long history of hunting at the Bear River marshes, the provision was written into the Act "that at no time shall less than 60 per centum of the total acreage of the said refuge be maintained as an inviolate sanctuary for such migratory birds" (Behle, 1958; Hansen, 1991). Later, agreement with the State of Utah provided for hunting on 40 percent of the developed portion of the refuge. This agreement was set forth in regulation 3, Migratory Game Birds, of the Regulations for the Administration of the Bear River Migratory Bird Refuge in the State of Utah, effective October 1, 1932.

The 1980s Flood

In the early 1980s, after several extremely wet winters and cooler than normal summers, Great Salt Lake began to rise above its traditional elevation of about 4,200 feet, to 4,205 feet by the end of 1983, and to over 4,209 feet by the end of 1984. The dikes of the refuge were inundated as the lake reached the 4,209-foot elevation. By 1986-1987, the lake reached its historic high of nearly 4,212 feet, causing an estimated $8 million in damage to the capital improvements at the refuge (Bureau of Economic and Business Research, 1984). Although the flood water receded over the next six years, it destroyed everything on the refuge below the 4,212-foot level including the dikes, new visitors' center, machine shops, and housing facilities. All the geese nesting habitat and vegetation that once grew around the refuge headquarters were destroyed. Loss of habitat also meant that most of the once-common passerines (perching birds) that were associated with the trees are now seen only rarely in migration. With the refuge's history and value in mind, the U.S. Fish and Wildlife Service (FWS) evaluated options for reconstruction and/or expansion of the refuge (U.S. Fish and Wildlife Service, 1995a).

Reconstruction/Expansion Alternatives Evaluated

This section of the paper, and the Preferred Alternative - Expansion section that follows, are extracted from Hansen (1991).

A number of alternative reconstruction/expansion actions were considered to meet the mission of the FWS Refuge System. The following four alternatives were selected at the culmination of the 1991 project review.

No Action - The area would remain as it was after the flooding. The refuge would be allowed to revert to an appearance preceding development.

Restoration - Most of the refuge would be restored to the conditions existing prior to the damage caused by the flood. Dikes and water-control structures would be repaired, but no permanent buildings would be constructed.

Enhancement - Existing refuge lands would be more intensely managed for migratory birds. Additional dikes would be built to divide the units, each having water management capabilities.

Expansion - The refuge boundary would be expanded through land acquisition to allow for intensive wildlife and public use development and for protection of existing wetlands occurring outside the present boundary. Acquisition would occur on a willing-seller basis.

An alternative that was discussed, but rejected as having no merit, called for the FWS to divest itself of the refuge. State lands within the refuge would revert back to state ownership, and other lands would be handled according to federal laws dealing with the disposition of surplus lands. These actions would have required congressional action.

The Preferred Alternative - Expansion

The expansion alternative, detailed in the Environmental Assessment document (Hansen, 1991), was the FWS's preferred alternative. This preferred action included all items contained within the restoration and enhancement alternatives (summarized above) as well as the acquisition of wetlands identified as being important to wildlife.

The Preferred Alternative proposed expanding the refuge boundary through land acquisitions totaling 38,200 acres,

primarily north and east of the refuge. Water rights and mineral rights, where possible, would be acquired with surface rights to the land. Two types of land acquisitions were proposed: fee title - 16,891 acres, and long-term easements - 21,309 acres, bringing the total potential area of the refuge to 103,200 acres.

Six new impoundments would be created with 17.5 miles of diking. The refuge would be divided into approximately 29 units (figure 2, lower), each with individual water management capabilities. A canal and drain serving each unit would allow efficient water management, which would meet the needs of the entire marsh community. The water would be used several times as it moves through the refuge, and a variety of habitats would be created. With the canal system in place, excess spring flows would be by-passed through the refuge directly to the state-managed area to the south of the Bear River and Forest Street (not shown on figures 1 or 2, but runs west from the center of Brigham City). Wetlands north of Bear River and Forest Street would be protected through the purchase of perpetual easements. Few physical changes are anticipated in the natural marsh area on the east side of the fee-purchase area.

All land acquisitions, fee, and easement transactions, would be accomplished only on a willing-seller basis in accordance with the Department of Interior and FWS policies. Inclusion of lands within the boundary would not guarantee acquisition, but indicates that the FWS wishes to protect them as part of the refuge system. Some private lands in the fee-purchase area would be removed from tax rolls and placed in federal government ownership. The refuge would implement the preferred alternative and the master plan as funding become available. Acquisition of fee-title or easements would be funded with money from the Land and Water Conservation Fund or Migratory Bird Commission Fund.

Part of the purchased land would be above the established flood plain, allowing for the construction of new buildings and visitor facilities. Major developments would include additional impoundments, an administrative complex and visitor center near highway I-15, two auto tour routes (8 and 20 miles in length), nature trails, and an environmental education center. Environmental education efforts, to include workshops, would be implemented to assist educators and students in understanding the natural environment.

Under this alternative, the FWS would undertake intensive management activities for wildlife enhancement. Short-term disruptions of habitats for wetland-impoundment developments, or alterations of vegetation to favor particular species, would result in long-term improvements in desirable wildlife populations.

The ability to manage water within smaller marsh units would allow for the reduction and control of botulism outbreaks. It is anticipated that the number of birds lost to botulism would decrease by 80-85 percent (Hansen, 1991).

Mammalian and avian-predator populations would be limited by control practices, preventing their populations from growing to maximum levels. Species that prey on waterfowl nests, such as skunk, raccoon, red fox, ravens, and gulls, would be limited in favor of expanding waterfowl and other migratory bird species.

An additional 35,040 acres of freshwater marsh habitat would be placed under permanent protection as a component of the refuge system. As such, easement areas would be protected from drainage and wetland destruction to ensure that future generations would be able to enjoy this unique resource.

Refuge Restoration and Enhancement Accomplishments and Budget

Accomplishments at the refuge, from 1992 through 2000, include the following (U.S. Fish and Wildlife Service, 1999a, 1999b, 2000, 2001):

- Purchase of 8,535 acres of critical breeding habitat, and restoration of grasslands on 919 acres of new land.
- Approximately 45,364 acres of wetland habitat were restored.
- Cleaned up the old headquarters site.
- Constructed a new pavilion, 15,900-square-foot maintenance facility, five new ponds, restroom, kiosk, demonstration pond, boardwalk, two photo blinds, two observation platforms, and one hunting blind.
- Moved over 1.5 million cubic yards of dirt to restore or construct dikes.
- Restored and/or constructed of over 80 water-control structures.
- Protected 17 legal water rights associated with new lands.
- Restored 45 miles of dikes, and constructed 40 miles of new dikes and 12 miles of new canals. About 100 miles of dike work was completed during 2000 alone, including shaping, leveling, grading, and graveling.
- Some 100,000 cubic yards of dirt were moved to build islands, dikes, and canals.
- Major upgrades were made to the refuge's computer and communication systems.

The total budgeted cost for restoration and enhancement of the refuge is $10,134,000. Of this amount, $7,617,000 has been appropriated to date, with the remaining $2,517,000 outstanding (U.S. Fish and Wildlife Service, 2001).

Refuge Recovery, Education Opportunities, and Cooperative Groups

Since the 1980s flood waters have receded, waterfowl, particularly ducks, have not recovered to pre-flood numbers. They are beginning to recover slowly as habitat is being improved and predator problems are being solved. Shorebirds and waterbirds have recovered to pre-flood numbers or better (Karen Lindsey, Bear Rriver Migratory Bird Refuge, written communication, October 20, 1999).

In 1998, the refuge provided valuable opportunities for wildlife-oriented recreation and education. Over 500 people fished at the refuge, 6,000 hunters used the refuge, and photographers and film makers from as far away as Chile worked on the refuge. There were also 20,000 visitors who observed wildlife, over 2,500 who participated in staff-led education programs, and 10,000 who utilized the interpretive materials, and took the 12-mile auto tour route.

The refuge is working in cooperation with "Friends of the Bear River Refuge," a non-profit citizens' group made up of volunteers from all over the United States and many other countries. The main focus of the Friends is to raise money and support for an education center for the refuge. "Young Friends of Bear River Refuge (YFBR2)," an after-school club from the Adele Young Intermediate School in Brigham City, is also working with the refuge. Their mission is "a commitment to preserving wildlife, and showing local people the beauty that exists in their own backyard."

There is a current need for an education center at the refuge, to be designed for interactive education and the enlightenment of visitors. As envisioned, a new $8 million visitors' center and administrative complex would house exhibits, a 125-seat theater, classrooms, and laboratories. Congress has allocated initial funding for planning and voiced support for completion of the project, and construction of the new facility is scheduled to begin by 2002. The new visitors' center/administrative complex would be built on the refuge at a site near Interstate 15.

In the future, the FWS's "Partners for Fish and Wildlife" program, combined with other federal, state, and private programs, will be used to address off-refuge threats and provide opportunities to protect the FWS resources and interests. Tools that will be used to accomplish these goals include: (1) technical assistance from private landowners, the Natural Resources Conservation Service, and conservation groups; (2) riparian, wetland, and upland restoration and wetland creation; and (3) riparian, wetland, and upland habitat enhancement and improvement. Some of these goals have been achieved to-date.

ACKNOWLEDGMENTS

Thanks to Mr. Bryce Passey, Mount Logan Middle School, for permission to use the photographs used in this article, and to refuge staff members for their reviews of this article and helpful suggestions.

REFERENCES

Behle, W.H., 1958, The bird life of Great Salt Lake: Salt Lake City, University of Utah Press, 203 p.

Bureau of Economic and Business Research, 1984, Flooding and landslides in Utah - an economic impact analysis: Bureau of Economic and Business Research, University of Utah, 107 p.

Fisher, Denise, 1998, Bear River Migratory Bird Refuge, Utah (Est. April 23, 1928): Online, http://www.refugenet.com/beavr.htm.

Fremont, J.C., 1845, Report of the exploring expedition to the Rocky Mountains in the year 1842 and to Oregon and northern California in the years 1843-1844: Washington, D.C., Gales and Scaton Printers, 693 p.

Giltner, L.T., and Couch, J.F., 1930, Western duck sickness and botulism: Science, v. 72, p. 660.

Hansen, K.S., 1991, Restoration and expansion of Bear River Migratory Bird Refuge, Brigham City, Utah Environmental Assessment: U.S. Fish and Wildlife Service, U.S. Department of the Interior, variously paginated.

Kalmbach, E.R., 1930, Western duck sickness produced experimentally: Science, v. 72, p. 658.

Lansing, Robert, 1916, The Migratory Bird Treaty (1916): Online, http://www.icu.com/geese/mbt.html.

Madsen, D.B., in preparation, The Silver Island expedition: Anthropological archaeology in the Bonneville Basin: University of Utah Anthropological Papers.

Miller, D.E., 1980, Great Salt Lake - a historical sketch, *in* Gwynn, J.W., editor, Great Salt Lake, a scientific, historical and economic overview: Utah Geological and Mineral Survey Bulletin 116, p. 1-14.

National Audubon Society, 1995, South shore duck club study: National Audubon Society (cooperative agreement no. 4-WS-94-00302 between the Department of Interior and the National Audubon Society, Title III of the Central Utah Project Completion Act Public Law 102-575), various pagings.

Nelson, N.F., 1966, Waterfowl hunting in Utah: Utah State Department of Fish and Game, Publication no. 66-10, 100 p.

Stansbury, Howard, 1852, Exploration and survey of the valley of the Great Salt Lake of Utah: Philadelphia, Lippincott, Grambe, and Company, 487 p.

Steward, Julian, 1938, Basin-Plateau aboriginal sociopolitical groups: Bureau of American Ethnology Bulletin no. 120, 346 p.

SWCA, Inc., 1999, Box Elder County comprehensive wetlands management plan: SWCA, Inc., Environmental Consultants (submitted to Box Elder County Great Salt Lake Wetlands ecosystem plan steering committee), various pagings.

Trout, A.K., 1997, Bear River Migratory Bird Refuge comprehensive management plan approval: U.S. Fish and Wildlife Service, Region 6, 39 p.

U.S. Fish and Wildlife Service, 1995a, Bear River Migratory Bird Refuge - Birds: U.S. Fish and Wildlife Service information pamphlet, 1 p.

—1995b, Bear River Migratory Bird Refuge, Brigham City, Utah: U.S. Fish and Wildlife Service information Pamphlet, 1 p.

—1998a, History of the National Wildlife Refuge System: Online, <http://bluegoose.arw.r9.fws.gov/nwrsfiles/general/history.html.>, accessed August 3, 1998.

—1998b, Bear River Migratory Bird Refuge, Brigham City, Utah, annual narrative report, calendar year 1998: U.S. Fish

and Wild Life Service, 49 p.
—1999a, Fulfilling the promise at Bear River Migratory Bird Refuge: U.S. Fish and Wildlife Service, Bear River Migratory Bird Refuge, 27 p.
—1999b, U.S. Fish and Wild Life Service 1999 annual report: U.S. Fish and Wildlife Service, Bear River Migratory Bird Refuge, p 61-70.
—2001, History - 75 years of improvements for wildlife and people: U.S. Fish and Wildlife Service, Bear River Migratory Bird Refuge unpublished 2001 briefing document, 3 p.

APPENDIX - A PROCLAMATION

WHEREAS it is provided by section 2 of the act of Congress, approved April 23, 1928 (45 Stat. 448), entitled "AN ACT To establish the Bear River Migratory Bird Refuge," that lands acquired by the Secretary of Agriculture in accordance with said act "together with such lands of the United States as may be designated for the purpose by proclamations or Executive orders of the President, shall constitute the Bear River Migratory Bird Refuge;"

NOW, THEREFORE, I, HERBERT HOOVER, President of the United States, by virtue of the power in me vested by the aforesaid act of Congress, and otherwise, do hereby make known and proclaim that I do hereby reserve from settlement and entry and/or any other form of disposition under the public land laws, and do hereby set apart and designate for the purpose of the Bear River Migratory Bird Refuge, subject to existing valid rights in any parts or parcels thereof under the public land laws, the lands of the United States in Box Elder County, Utah, within the boundaries particularly described as follows, to wit:

....
................Legal Description of Refuge Area................
....

IN WITNESS WHEREOF, I have hereunto set my hand and caused the seal of the United States to be affixed.

DONE at the City of Washington this 26th day of September, in the year of our Lord nineteen hundred and thirty-two, and of the Independence of the United States of America the one hundred and fifty-seventh.

HERBERT HOOVER

ANAEROBIC MICROBIOLOGY AND SULFUR CYCLING IN HYPERSALINE SEDIMENTS WITH SPECIAL REFERENCE TO GREAT SALT LAKE

by

Kjeld Ingvorsen
and
Kristian Koefoed Brandt

Department of Microbial Ecology
Institute of Biological Sciences
University of Aarhus
Ny Munkegade, Building 540
DK-8000 Aarhus C, Denmark

E-mail: kjeld.ingvorsen@biology.au.dk
Phone: +45 8942 3245 (direct line)
Fax: +45 8612 7191

ABSTRACT

Hypersaline environments are widely distributed on earth and have existed for billions of years. Most microorganisms are not able to grow at the extreme salinities prevailing in these environments, and the microbiota, therefore, consists of specially adapted halophilic or halotolerant microbes. Many of the extant hypersaline environments are highly productive as evidenced by the development of microbial mats on the sediment surface, or blooms of microorganisms in the water column often imparting a dense, red coloration to the water.

Due to the low water depth of many hypersaline environments, and the low oxygen saturation value of concentrated brines, most of the organic material produced is ultimately degraded in the sediment by anaerobic microorganisms. Hypersaline water bodies often contain high concentrations of sulfate favoring bacterial sulfate reduction, which results in a high production of hydrogen sulfide - a highly toxic and ill-smelling gas.

This article addresses anaerobic microbial processes in general, and also presents some of our recent research on bacterial sulfate reduction in Great Salt Lake. Our research demonstrates the occurrence of very high rates of bacterial sulfate reduction (and thus hydrogen sulfide production) at sites in the moderately hypersaline southern arm of GSL. Some of the bacteria responsible for the sulfate reduction activity in the lake have been isolated and described in detail. All the isolated strains, including two newly discovered species, can be characterized as being merely halotolerant, actually preferring salinities of only 2-5%. In contrast to the high sulfate reduction activity in the southern arm of GSL, sulfate reduction is strongly inhibited at the high salinities prevailing in the extremely hypersaline northern arm of Great Salt Lake, indicating the existence of a severely stressed population of sulfate-reducing bacteria at this site. Our data strongly indicate that salinity is a major factor determining the intensity of sulfate reduction in organic-rich sediments of the Great Salt Lake.

INTRODUCTION

Microorganisms thriving in hypersaline environments have received much attention during the last decades, but most studies have focused on the diversity and physiology of aerobic or phototrophic halophiles (for reviews see: Rodriguez-Valera, 1988; Javor, 1989; Vreeland and Hochstein, 1993; Oren, 1994; Kamekura, 1998; Oren, 1999). The anaerobic microbiology of hypersaline ecosystems like the Great Salt Lake (GSL) and the Dead Sea has been studied only to a limited extent. Therefore, very little is known about the anaerobic degradation of organic carbon and the microorganisms catalyzing these reactions in sediments at extreme salinities. This is somewhat paradoxical because, owing to the low solubility of oxygen in brines and the high biological productivity of many hypersaline environments, a large part of the organic biomass produced within these ecosystems ultimately degrades under anoxic conditions.

Much of the microbiological research done on GSL has been referenced in publications by the following authors: Stephens and Gillespie, 1976; Post, 1977; Brock, 1979; Gwynn, 1980; Post, 1981; Rushforth and Felix, 1982; Zeikus, 1983; Post and Stube, 1988; Javor, 1989; Stephens, 1990. Results from other investigations of GSL, not mentioned in the above references, are discussed in due context in this report.

This article is not intended to be a comprehensive review on the anaerobic microbiology of hypersaline environments. Rather, it focuses on the microbial sulfur cycle of anoxic hypersaline sediments and, in particular, on the process of dissimilatory sulfate reduction as based on studies of GSL conducted recently at the University of Aarhus.

ANAEROBIC MICROBIAL BIODEGRADATIVE PROCESSES - A SHORT INTRODUCTION

The complete decomposition of organic matter under anoxic conditions requires the concerted action of various physiological groups of anaerobic bacteria which constitute the so-called anaerobic biodegradative microbial food chain (Zeikus, 1983; Oren, 1988; Zehnder, 1988; Lowe and others, 1993). Anaerobic decomposition of organic carbon is initiated by the enzymatic hydrolysis of complex biopolymers such as proteins, lipids and polysaccharides into oligomeric and monomeric components (figure 1). Polymer hydrolysis is carried out by hydrolytic fermentative bacteria and results in the production of low-molecular-weight compounds which in turn serve as substrates for a variety of bacterial fermentation processes. The major end-products excreted by fermentative bacteria are hydrogen, carbon dioxide, ammonia, and low molecular weight alcohols and organic acids. These organic end-products and hydrogen are finally consumed by sulfate-reducing bacteria (SRB), methane-producing archaebacteria, and homoacetogenic bacteria, all of which have been reported to be present in various hypersaline environments (Oren, 1988; Javor, 1989; Zhilina and Zavarzin, 1990; Ollivier and others, 1994). Other groups of prokaryotes known to be important in organic carbon mineralization in some fresh water and marine environments include S^o-, $S_2O_3^{2-}$-, Fe(III)- and Mn(IV)-reducing bacteria. Very little information is currently available on these processes in extremely hypersaline environments.

Whether the products of sulfate reduction (sulfide and carbon dioxide) or methanogenesis (methane and carbon dioxide) become the major end-products of anaerobic biodegradation in sediments depends on the sulfate concentration of the environment. In sulfate-rich marine environments, sulfate reduction is the dominating process (Jørgensen 1977; Jørgensen, 1982), whereas methanogenesis is the major terminal process in most freshwater sediments, where the sulfate concentration is limiting for sulfate reduction (Winfrey and Zeikus, 1977; Ingvorsen and Brock, 1982).

FERMENTATION - ANAEROBIC HALOPHILIC FERMENTATIVE BACTERIA

As mentioned in the previous section, many organic biomolecules are initially "attacked" by hydrolytic, fermentative bacteria during the anaerobic degradation process (Figure 1). This group of bacteria, therefore, is of great importance for anaerobic biodegradation.

Except for the halophilic Gram-positive *Clostridium halophilum* (Fendrich and others, 1990) and *Thermohalobacter berrensis* (Cayol and others, 2000), the majority of the known fermentative anaerobic bacteria isolated from hypersaline environments constitute a monophyletic group of bacteria. Recently a new order, the *Haloanaerobiales* was proposed to include these haloanaerobes as a result of phylogenetic and physiological analyses (Rainey and others, 1995). *Haloanaerobiales* currently contains 17 species of halophilic obligately anaerobic, Gram-negative fermentative bacteria grouped into nine genera. All species belonging to the *Haloanaerobiales* require at least 2 percent NaCl for growth, and generally show optimal growth between 10 and 18 percent NaCl. Species of *Haloanaerobiales* investigated so far osmoadapt by accumulating inorganic ions in their cytoplasm (Oren, 1999). Members of the *Haloanaerobiales* seem to be well adapted to hypersaline environments, as evidenced by their wide span of halotolerance and their relatively high salinity optima for growth (Cayol and others, 1995; Rainey and others, 1995).

The physiological diversity of the *Haloanaerobiales* is quite large. Most haloanaerobes ferment carbohydrates and some are able to use biopolymers such as peptides, starch, glycogen, pectin, chitin, and cellulose as a carbon and energy source. The major end-products formed by the carbohydrate-fermenting haloanaerobes are the same as those produced by their physiological counterparts in non-saline environments, namely short-chain fatty acids (format, acetate, propionate, butyrate), H_2, CO_2, ethanol and lactate.

The first species of *Haloanaerobiales* isolated from GSL was *Haloanaerobium praevalens* (Zeikus and others, 1983). *H. praevalens* is an interesting microorganism for several reasons. It has a wide salt range for growth (2 to 30 percent NaCl), and is capable of producing methylmercaptan, which is a methanogenic substrate (see next section), by fermentation of methionine. Because *H. praevalens* is found in both the northern and southern arms of GSL, and is present in fairly high numbers (> 10^8 cells/ml sediment), it could be an ecologically important microorganism in anaerobic carbon degradation in GSL.

More recently, Tsai and others (1995) isolated *Haloanaerobium alcaliphilum* from the northern arm of GSL. A distinctive characteristic of *H. alcaliphilum* is its ability to ferment glycine betaine to acetate and trimethylamine. Glycine betaine is an organic osmolyte produced by a wide variety of halophilic prokaryotes (Oren, 1990; Galinski, 1995). Relatively high concentrations of glycine betaine have been demonstrated in several hypersaline environments (King, 1988; Oren and others, 1994), and this compound is likely to be abundant in many hypersaline ecosystems.

METHANOGENESIS - THE HALOPHILIC METHANOGENIC ARCHAEBACTERIA

In hypersaline environments, methane formation seems to derive mainly from the methylotrophic substrates methanol, methylamines, and methylmercaptan (King, 1988; Giani and others, 1989; Oremland and King, 1989; Oren, 1990). Methylamines, for example, are likely to be abundant in many hypersaline environments resulting from the anaerobic degradation of the common microbial osmolyte glycine betaine (Fendrich, and others, 1990; Zhilina and Zavarzin, 1990; Tsai, and others, 1995).

Methanogenic activity in GSL sediments has been measured by several investigators (Phelps and Zeikus, 1980; Zeikus, 1983; Lupton and others, 1984). Experiments with ^{14}C labeled organic compounds suggest that methane is produced mainly from methylmercaptan, methanol, or the methyl group of methionine (Phelps and Zeikus, 1980; Zeikus, 1983; Lupton and others, 1984). Methylamines were not tested as substrates. The experiments also indicate that the common methanogenic substrates, acetate and H_2+CO_2, are not converted to methane at significant rates, and that methanogens which are able to use these substrates were present in very low numbers in the sediments of GSL (Zeikus, 1983).

In the hypersaline lagoons of the Arabat Spit (East Crimea, Russia), no methane is produced from acetate or formate. In this ecosystem, methanol, and mono- and dimethyl amines served as methanogenic precursors at salinities less than 20 percent, whereas trimethylamine is the main substrate for methanogenesis at higher salinities (Zhilina and Zavarzin, 1990). Furthermore, methanogens in the sediment decrease to very low numbers at the highest salinities (Zhilina and Zavarzin, 1990). Studies of a saltern in the Bretagne (France) indicate that methylated amines and methanol are the preferred substrates for methanogenesis at salinities up to 12 percent (Giani and others, 1989).

A number of methanogens have been isolated from hypersaline ecosystems (Zhilina, 1986; Zhilina and Zavarzin, 1990; Lowe and others, 1993; Ollivier and others, 1994; Ollivier and others, 2000). The moderately halophilic methanogen, *Halomethanococcus mahii*, strain SLP (now classified as *Methanohalophilus mahii*), was isolated from GSL (Paterek and Smith, 1985; Paterek and Smith, 1988). It grows at NaCl concentrations between 3 percent and 20 percent (optimum between 6 percent and 20 percent), and has a temperature optimum for growth at 37°C. Only one extremely halophilic methanogen, *Methanohalobium evestigatum*, has been described in the literature. *M. evestigatum* grows at salinities ranging from 15 to 30 percent NaCl, with an optimum of 25 percent. Its temperature optimum is rather broad at 40 to 55°C (Zhilina and Zavarzin, 1990).

Most methanogens isolated so far from hypersaline sediments are strictly methylotrophic and incapable of using acetate and H_2+CO_2, which are the principal methanogenic substrates in low-salt environments. *Methanocalculus halotolerans* is the most halotolerant hydrogenotrophic methanogen described so far (Ollivier and others, 1998). It produces methane from H_2+CO_2 and formate at NaCl concentrations ranging from zero up to 12.5 percent. Methanol and methylamines are not used for methanogenesis by *M. halotolerans*.

BACTERIAL SULFATE REDUCTION

Bacterial sulfate reduction is the main, terminal anaerobic oxidation process in marine and hypersaline sediments where sulfate is rarely a limiting factor (Jørgensen and Cohen, 1977; Jørgensen, 1982; Klug and others, 1985; Skyring, 1987; Canfield and Des Marais, 1993; Caumette and others, 1994). This process is carried out by SRB, which constitute a phylogenetically and physiologically diverse group of specialized anaerobic microorganisms that use sulfate as an electron acceptor for oxidation of organic compounds or molecular hydrogen (Widdel, 1988; Gibson, 1990; Hansen, 1993). Many species of sulfate-reducing bacteria (SRB) can also reduce sulfite or thiosulfate, and some also reduce elemental sulfur. SRB are ubiquitous in nature and can be isolated from almost any anoxic environment.

A generalized equation for energy production with sulfate by SRB is:

$$2(CH_2O) + SO_4^{2-} \rightarrow H_2S + 2\ HCO_3^-$$

Thus, sulfate is used as an oxidant for the degradation of organic material. Most of the sulfide produced by SRB is released into the environment, with less than a few percent used for synthesis of cell biomass. The substrate spectrum of the SRB described to date comprises over 125 different carbon compounds (Hansen, 1994). Examples of organic substrates used by many SRB (non-halophilic and halophilic) include hydrogen, formate, lactate, ethanol, pyruvate, and C_2 to C_{18} fatty acids. Most of these substrates are in fact end-products excreted by the anaerobic fermentative bacteria (see figure 1). Some SRB are able to grow on more exotic substrates, for example, crude oil components such as alkanes and toluene, as well as several types of substituted aromatics (Hansen, 1994). Some energy-yielding reactions carried out by SRB are shown in table 1.

Table 1. Examples of energy-yielding reactions of sulfate reducing bacteria (SRB). Reactions A and B show incomplete and complete oxidation of lactate, respectively.

	Reaction
A.	$2CH_3CHOHCOO^- + SO_4^{2-} \rightarrow 2CH_3COO^- + 2HCO_3^- + HS^- + H^+$
B.	$2CH_3CHOHCOO^- + 3SO_4^{2-} \rightarrow 6HCO_3^- + 3HS^- + H^+$
C.	$CH_3COO^- + SO_4^{2-} \rightarrow 2HCO_3^- + HS^-$
D.	$4H_2 + SO_4^{2-} + H^+ \rightarrow 4H_2O + HS^-$
E.	$S_2O_3^{2-} + H_2O \rightarrow SO_4^{2-} + H_2S$

Physiologically, SRB may be grouped into complete oxidizers and incomplete oxidizers, depending on whether they oxidize their organic substrates completely to CO_2 (HCO_3^-) (table 1, reactions B and C), or only partly to the level of acetate, which is excreted as an end-product (table 1, reaction A). The acetate excreted by incompletely oxidizing SRB (reaction A) may, in turn, serve as a substrate for completely oxidizing SRB (reaction C), thereby eventually leading to a complete oxidation of the original carbon substrate by sulfate reduction. Many species of SRB grow well with hydrogen as an energy source (table 1, reaction D). As mentioned above, hydrogen is a quantitatively important end-product of anaerobic fermentation reactions.

SRB are, however, not restricted to energy-production by reduction of sulfate. Many SRB are able to ferment organic compounds in the absence of sulfate and of disproportionation of inorganic sulfur compounds such as thiosulfate, sulfite, and elemental sulfur. For example, the halophilic SRB, *Desulfovibrio oxyclinae*, is able to grow by disproportionation of thiosulfate (Krekeler and others, 1997; table 1, reaction E). The physiology, biochemistry, and ecology of SRB has been reviewed extensively (Widdel, 1988, Gibson, 1990; Hansen, 1993; Hansen, 1994).

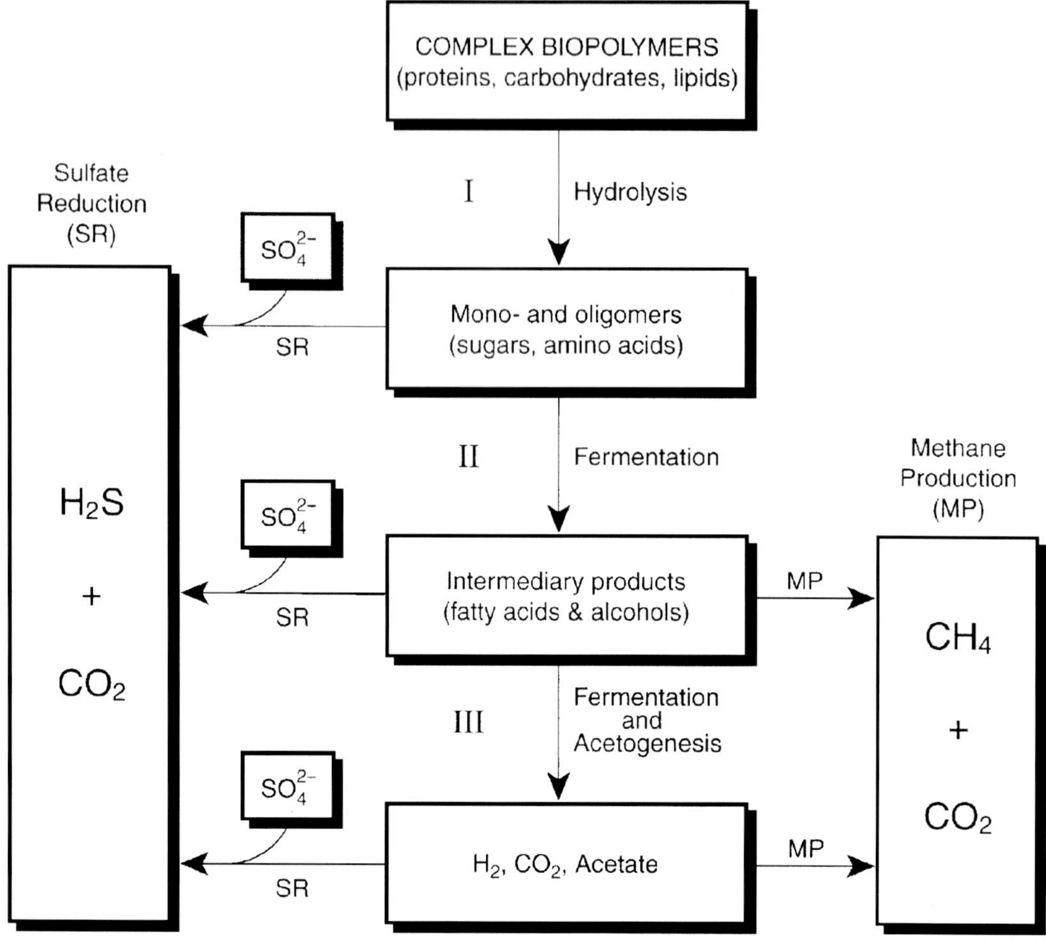

Figure 1. *Anaerobic degradation of organic compounds. The scheme is much simplified and does not depict all the complex interactions occurring between anaerobic bacteria of different trophic levels. Some anaerobes (for example members of the order* Haloanaerobiales*) are capable of carrying out both hydrolysis (step I) and fermentation (step II), while other anaerobes are only capable of fermentation (step II or III) or acetogenesis (step III).*

In marine and moderately hypersaline environments, sulfate-reducing bacteria usually outcompete methanogenic bacteria for acetate and hydrogen. Under these conditions, methanogenesis may occur from methylotrophic substrates such as methanol and methylamines (see text under "Methanogenesis") because most SRB are not capable of using these substrates.

Halotolerant and Halophilic SRB

To date, only ten species of SRB (representing six different genera) have been obtained in pure culture from hypersaline environments and validly described (for references see Tsu and others, 1998; Brandt and others, 1999).

Two SRB, *Desulfobacter halotolerans* and *Desulfocella halophila*, were recently isolated at the University of Aarhus from anaerobic sediments collected from the southern arm of GSL. *D. halotolerans* (strain GSL-Ac1) represents the only completely oxidizing sulfate-reducing bacterium isolated from a hypersaline environment, but resembles other *Desulfobacter* species by being metabolically very specialized and growing well only with acetate (table 1, reaction C). However, *D. halotolerans* only grows at salinities up to 13 percent NaCl, with an optimum at 1 to 2 percent NaCl, and hence appears to be poorly adapted to its natural habitat (Brandt and Ingvorsen, 1997). Attempts to isolate acetate-oxidizing SRB growing at higher salinities have been unsuccessful (Brandt and Ingvorsen, unpublished results). The other sulfate-reducing bacterium isolated from GSL, *Desulfocella halophila* (strain GSL-But2), is a representative of a novel genus of SRB, and is metabolically much more versatile and more halotolerant than *D. halotolerans*. *D. halophila* is an incomplete oxidizer being the first halotolerant/halophilic sulfate-reducing bacterium known that utilizes long-chain fatty acids (C4-C16). *Desulfocella halophila* grows at salt concentrations between 2 and 19 percent NaCl, with an optimum at 4 to 5 percent (Brandt, and others, 1999). An electron micrograph of *Desulfocella halophila* is shown in figure 2.

Extremely halophilic SRB have not yet been isolated in pure culture. Even though some of the known halophilic SRB were isolated from localities of high salinity, all of them show optimal growth below approximately 10 percent NaCl, and none grow beyond 25 percent salt in vitro. The most halotolerant species of SRB described are *Desulfohalobium retbaense*, with a salinity range of 3 to 25 percent NaCl and an optimum at around 10 percent (Ollivier and others, 1991; Ollivier and others, 1994), and *Desulfovibrio oxyclinae*, with a salinity range from 2.5 to 22.5 percent NaCl, and an optimum from 5 to 10 percent NaCl (Krekeler and others, 1997). It should be noted that several species of SRB isolated from

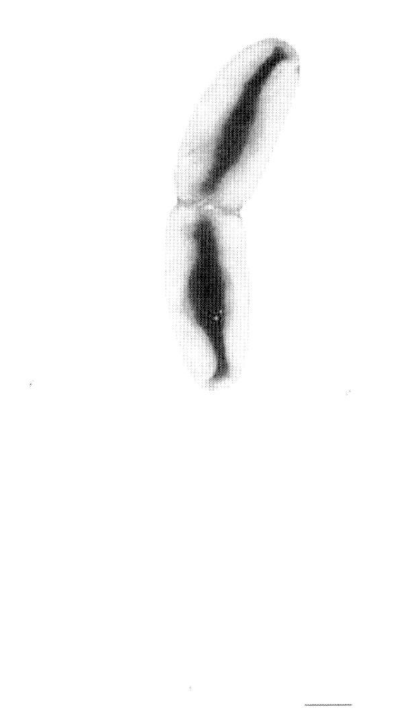

Figure 2. Transmission electron micrograph of a dividing cell of **Desulfocella halophila** *strain GSL-But2T after negative staining. Bar represents 0.5 μm. From Brandt and others, 1999. Reproduced with permission of the International Union of Microbial Societies.*

marine environments have a salinity range between 0.2 and 6 percent NaCl (or even up to 12 percent) and show optimal growth at salinities between 0.5 and 4 percent NaCl (Caumette, 1993).

Welsh and others (1996) recently reported that salt inhibition of *Desulfovibrio halophilus*, which is a moderate halophile, can be partially alleviated by the addition of glycine betaine to the growth medium. As demonstrated in this study, the osmoadaptation of *D. halophilus* further involved de novo synthesis of the carbohydrate trehalose and to a lesser extent the accumulation of potassium ions. Partial alleviation of salt-induced growth inhibition by betaine addition has also been demonstrated in *Desulfocella halophila*, *Desulfobacter halotolerans*, and a third GSL-isolate, *Desulfovibrio* sp., strain GSL-Lac3 (Brandt, 1998).

Glycine betaine, trehalose, and other compatible osmolytes are produced by a large variety of halophilic and halotolerant bacteria (Galinski and Trüper, 1994; Galinski, 1995), and the presence of such compounds in hypersaline sediments may extend the salinity range for growth, or prolong the survival of salt-stressed SRB.

Bacterial Sulfate Reduction in the Sediments of Great Salt Lake

In the sediments of the southern arm of GSL, high sulfate reduction rates have been measured by several investigators using in vitro incubations with ^{35}S-sulfate (Zeikus, 1983; Lupton, and others, 1984; Brandt and others, 2001).

Several studies have shown that acetate is a major substrate for sulfate reduction in marine and moderately hypersaline sediments (Skyring, 1988; Parkes and others, 1989). Thus, Skyring (1988) reported that acetate was the most important substrate for SRB in Lake Eliza (Australia) at a salinity of 12 percent. Other studies carried out with anoxic sediments from the GSL showed a significant conversion of ^{14}C-2 labeled acetate and ^{14}C-3 labeled lactate to $^{14}CO_2$, but not $^{14}CH_4$, at salinities of 24 to 29 percent (Phelps and Zeikus, 1980; Lupton, and others, 1984). These findings indicate the presence of unknown acetate-oxidizing anaerobes because known SRB do not oxidize acetate at these high salinities (see above).

Our laboratory recently investigated sulfate reduction in surface sediments at several localities in GSL (Brandt and others, 2001). The stations sampled were AS2 (Utah Geological Survey (UGS) station code), which is located in the deepest part of the southern arm, Station 20a, which is located in shallow water on the northern shore of the Antelope Island causeway, and Station 27 which is located outside Little Valley Harbour in the northern arm. The average sulfate reduction rates measured at the three sampling stations using ^{35}S-radiotracer are shown in figure 3. The sulfate reduction rates measured at Station 20a were extremely high, reflecting the high organic content of the sediment at this site. Sulfate reduction rates measured at 27 percent NaCl in the sediment outside Little Valley Harbour were one or two orders of magnitude lower than those determined at the two sampling stations in the southern arm (station AS2 and 20a), yet still far above the detection limit of the tracer method. In the same study (Brandt and others, 2001), high numbers of SRB were detected, especially in the sediments from the south arm. For

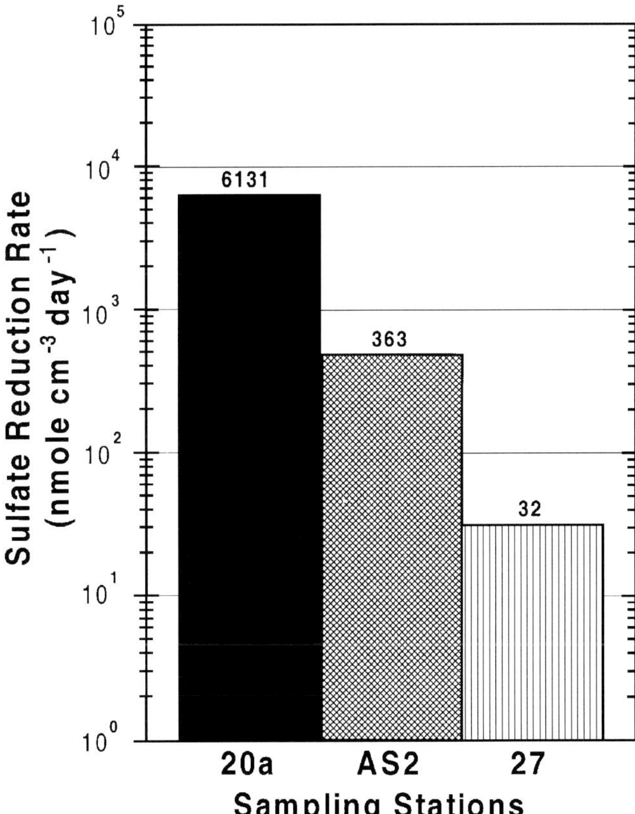

Figure 3. Sulfate reduction rates (average values) in surface sediment at the three study sites in GSL. The salinities of the pore waters at Stations 20a, AS2, and 27 were 11.6, 12.3, and 27.4 percent NaCl, respectively. Note the logarithmic scaling on the Y-axis.

example, by using a new tracer-based most-probable-number (MPN) procedure developed in our laboratory (Vester and Ingvorsen, 1998), sediment counts of SRB at Station AS2 yielded numbers ranging from 5 x 10^7 to 2 x 10^8 cells per ml of sediment (Brandt and others, 2001). These values are some of the highest MPN counts of SRB reported from anaerobic sediments (Jørgensen, 1978; Teske and others, 1998).

Investigations by Brandt and others (2001), and Ingvorsen (unpublished results) show that in vitro rates of bacterial sulfate reduction in surface sediment from the southern and the northern arms of GSL are greatly reduced at salinities above 20 percent NaCl. Strong evidence was also found that the sulfate-reducing populations in the sediments at all three sampling stations in GSL were not fully adapted to the prevailing salinities (Brandt and others, 2001; Ingvorsen unpublished results). Similar inhibitory effects of salinity on rates of methanogenesis, sulfate reduction, and various aerobic degradation processes in GSL have been reported previously (Ward and Brock, 1978; Lupton and others, 1984; Fendrich and Schink, 1988).

Although only few measurements of sulfate reduction and methanogenesis have been carried out in GSL, it seems justified to assume that sulfate reduction is the dominating terminal anaerobic process in sediment metabolism in the southern arm of the GSL (Brandt and others, in 2001; Lupton and others, 1984).

FATE OF SULFIDE PRODUCED IN HYPERSALINE SEDIMENTS

Biogenic Sulfide in Hypersaline Sediments.

Much of the sulfide produced in sulfate-rich anoxic sediments derives from bacterial dissimilatory sulfate reduction (Jørgensen, 1977; Widdel, 1988). However, biogenic H_2S is also formed by the fermentative breakdown of organic sulfur compounds (especially sulfur containing amino acids), and by the reduction of elemental sulfur and other oxidized inorganic sulfur compounds. Although quantitative data are lacking on the relative importance of these processes, sulfate reduction is likely to be the major source of sulfide in GSL, particularly in the sediments of the southern arm.

Part of the sulfide produced by dissimilatory sulfate reduction in anoxic sediment is rapidly precipitated and thus immobilized in the sediment by reactions with metal ions. Because iron oxides are usually the most abundant metals in sediments, various types of iron sulfides, such as amorphous ferrous sulfide (FeS), mackinawite ($FeS_{0.9}$), and greigite (Fe_3S_4) are formed. The iron sulfide compounds formed are subsequently converted into pyrite (FeS_2) by several reactions as described below. Studies of various sulfate-rich sediments have shown that only a fraction of the total sulfide produced by SRB is permanently buried in the sediment as reduced sulfur (Jørgensen and Cohen, 1977; Jørgensen, 1977, 1982, 1990; Skyring, 1987). The major reduced inorganic sulfur species accumulating in anoxic sediments are pyrite and iron sulfides.

The sulfide that does not react with metals diffuses towards the sediment surface where it is oxidized by a variety of chemical and microbiological reactions, though some sulfide may eventually escape to the atmosphere as H_2S. Before discussing these reactions, a brief description of the chemical and biological zonations of hypersaline surface sediments is given (see below).

Chemical and Biological Processes in Hypersaline Sediments

Surface sediments from various hypersaline environments have been described in detail (Jørgensen and Cohen, 1977; Javor and Castenholz, 1981; Canfield and Des Marais, 1991; Fründ and Cohen, 1992; Caumette, 1993; Caumette and others, 1994; Jørgensen, 1994; Ollivier and others, 1994; Teske and others, 1998). Sediment surfaces typically consist of multi-layered microbial mats with an oxygen-producing upper layer formed by cyanobacteria and microalgae covering layers of anoxygenic green and purple phototrophic bacteria (figure 4). In sediments with intense sulfate reduction, the oxygen-producing zone may constitute a thin surface layer only a few millimeters thick (Jørgensen, 1988; Teske and others, 1998; Canfield and Des Marais, 1991). At night, when photosynthesis is no longer active, the oxic zone becomes even thinner and the sediment may become anoxic right up to the surface because oxygen is now supplied only by downward diffusion into the sediment from the atmosphere or overlying water.

During the day, the activity of microorganisms carrying out oxygenic photosynthesis provides oxygen for oxidation of sulfide by chemical and biological processes (figure 4). Furthermore, the anoxygenic phototrophic bacteria situated beneath the algal/cyanobacterial layer act as an additional "sulfide trap" during daytime. During the night, however, these light-dependent microbial processes no longer serve as "sulfide-traps" and free sulfide may reach the sediment surface and escape to the atmosphere or the water column (figure 4).

Many colourless and phototrophic sulfide-oxidizing bacteria are motile and move actively up and down in the surface sediment in response to diurnal changes in profiles of oxygen, sulfide, and light; these bacteria often form dense and intensely coloured layers on the surface of organic-rich sediments (Hansen and others, 1978; Ingvorsen and Jørgensen, 1982; Jørgensen 1988; Caumette and others, 1994). Some of the transformations of sulfide occurring in sediments are outlined in figure 4 and described in more detail below.

Chemical Oxidation of Sulfide with Oxygen

(Reaction 1 in figure 4)

Chemical oxidation of sulfide in oxic waters (often referred to as spontaneous oxidation or autooxidation) is a complex process, the oxidation rate being dependent on numerous factors such as pH and the concentration of reactants. Furthermore, sulfide oxidation is significantly enhanced by low concentrations of metal ions. Half-lives of sulfide, ranging from minutes to several hours, have been reported under various conditions. Products of sulfide autooxidation include elemental sulfur (S^o), thiosulfate ($S_2O_3^{2-}$), tetrathionate ($S_4O_6^{2-}$), or more oxidized compounds such as sulfite (SO_3^{2-}) and sulfate (SO_4^{2-}).

Microbial Oxidation of Sulfide with Oxygen
(Reaction 2 in figure 4)

Microbial oxidation of sulfide is carried out by aerobic, colourless, sulfur bacteria like *Thiobacillus* and *Beggiatoa*. Representatives of these two genera have been reported from several hypersaline mats (Javor, 1989; Canfield and Des Marais, 1991; Wood and Kelly, 1991; Fründ and Cohen, 1992), and they could be important for sulfide oxidation in these environments. Aerobic bacterial sulfide oxidation is dependent on both sulfide and oxygen and, therefore, this process is restricted to a narrow sediment zone where these two compounds coexist (i.e., the interface between HI and HII in figure 4). As shown in figure 4, this zone where oxygen and sulfide meet moves up and down diurnally as a result of changing oxygen and sulfide gradients in the uppermost sediment. Whether present in the oxic waters or the sediment, aerobic sulfide oxidizing bacteria must compete with the chemical oxidation processes for both sulfide and oxygen. The colourless bacteria may also obtain metabolic energy by oxidizing sulfur compounds of intermediate oxidation states such as elemental sulfur and thiosulfate. The final product of microbial sulfide oxidation with oxygen is sulfate.

Microbial Oxidation of Sulfide Without Oxygen
(Reaction 3 in figure 4)

Microbial oxidation of sulfide may also occur in the absence of oxygen. For example, this process is carried out by the green and purple sulfur bacteria of the genera *Chlorobium*, *Chromatium*, and *Thiocapsa*, respectively, all of which use sulfide as an electron donor for anoxygenic photosynthesis (Caumette, 1993; Caumette and others, 1994). The final product of anoxygenic photosynthesis is sulfate with elemental sulfur occurring as an intra- or extra-cellular intermediate. Green and purple sulfur bacteria are anaerobes and, therefore, have to position themselves in a rather narrow layer within the anoxic sulfide-containing zone where light is still available (zone HII in figure 4). In many sediments, phototrophic bacteria may reach depths of 2 to 8 mm. Sulfide may also be oxidized by cyanobacteria and by non-sulfur purple bacteria belonging to the genus *Rhodospirillum*. The most extremely halophilic purple bacteria belong to the genera *Ectothiorhodospira* and *Halorhodospira*.

The complete oxidation of sulfide to sulfate in the absence of oxygen and light has been observed in marine sediments. This process, which is stimulated by Fe(III) and Mn(IV) oxides, probably occurs by a number of interacting chemical and biological reactions, some of which have been demonstrated experimentally in the laboratory. Elemental sulfur and thiosulfate are probably important intermediates in these processes. It is likely that sulfate-reducing bacteria, which can disproportionate elemental sulfur, thiosulfate (table 1, reaction E) and sulfite, contribute significantly to the light-independent anaerobic oxidation of sulfide to sulfate (Luther and others, 1986; Bak and Cypionka, 1987; Jørgensen, 1990, 1994; Elsgaard and Jørgensen, 1992).

Immobilization of Hydrogen Sulfide by Reactions with Metals
(Reaction 4 in figure 4)

As mentioned above, part of the sulfide produced within the anoxic zone of sediments (HII and HIII in figure 4) reacts rapidly with metal ions, in particular iron. The products of these reactions include solid-phase sulfur (iron sulfides, elemental sulfur, pyrite) and more oxidized soluble sulfur species (figure 4). Iron is introduced into the sediment as oxides or hydroxides on mineral particles, and as a constituent of organic material. Once in the anoxic sediment, ferric iron is reduced, either biologically or by chemical reaction with free sulfide present in the pore water, according to the following reaction:

$$2\,FeOOH + H_2S \rightarrow S^o + 2\,Fe^{2+} + 4\,OH^-$$

Ferrous iron reacts rapidly with free sulfide forming amorphous iron:

$$Fe^{2+} + H_2S \rightarrow FeS + 2H^+$$

which in turn reacts with the produced elemental sulfur to form pyrite:

$$FeS + S^o \rightarrow FeS_2$$

The mechanisms of pyrite formation in sediments are not completely understood, and the extent and rate of pyrite accumulation seems very dependent on the type of sediment (Howarth and Teal, 1979; Jørgensen, 1983; Skyring, 1987; Thode-Andersen and Jørgensen, 1989). Reduced sediment layers with intense sulfate reduction are usually black due to their high content of iron sulfide (FeS); deeper down, the sediment may turn grey as a result of pyrite accumulation.

If the reactive iron pool in the aphotic anoxic sediment (H III in figure 4) becomes exhausted, and the concentration of dissolved sulfide in the pore water increases, sulfide diffuses towards the sediment surface where it is oxidized biologically (reaction 2, and 3) or chemically (reaction 1), or "trapped" as iron sulfides (reaction 4). If the uppermost sediment zone turns completely anoxic in the night (this scenario is not depicted in figure 4), a high iron content in this layer may still prevent sulfide from diffusing out of the sediment. In that case, the iron sulfides formed during the night may be reoxidized during the following day as a result of photosynthetic oxygen production in the uppermost sediment. A dynamic, repetitious, diurnal redox cycling of iron between the Fe(II) and Fe(III) oxidation states may thus take place in the uppermost sediment layer.

Usually only a minor fraction of the total sulfide produced by sulfate reduction is permanently buried in the sediment as pyrite and iron sulfides (see above). However, some biogenic sulfide is also incorporated into organic material during diagenesis. Biomass contains less than 2 percent of its dry weight as sulfur, and it is believed that much of the organically bound sulfur in coal (up to 12 percent) and oil derive from biogenic sulfide (Sinninghe Damstè and others, 1989).

GSL - A Potential Source Area for Atmospheric H_2S

Due to the high intensity of bacterial sulfate reduction (see above), steep vertical gradients of oxygen and sulfide are likely to exist at the sediment surface at many localities in GSL. Therefore, the shallow organic-rich sediments present in the southern arm of GSL are potential source areas for the emission of biogenically produced hydrogen sulfide. Although no ecological studies on bacterial sulfide oxidation

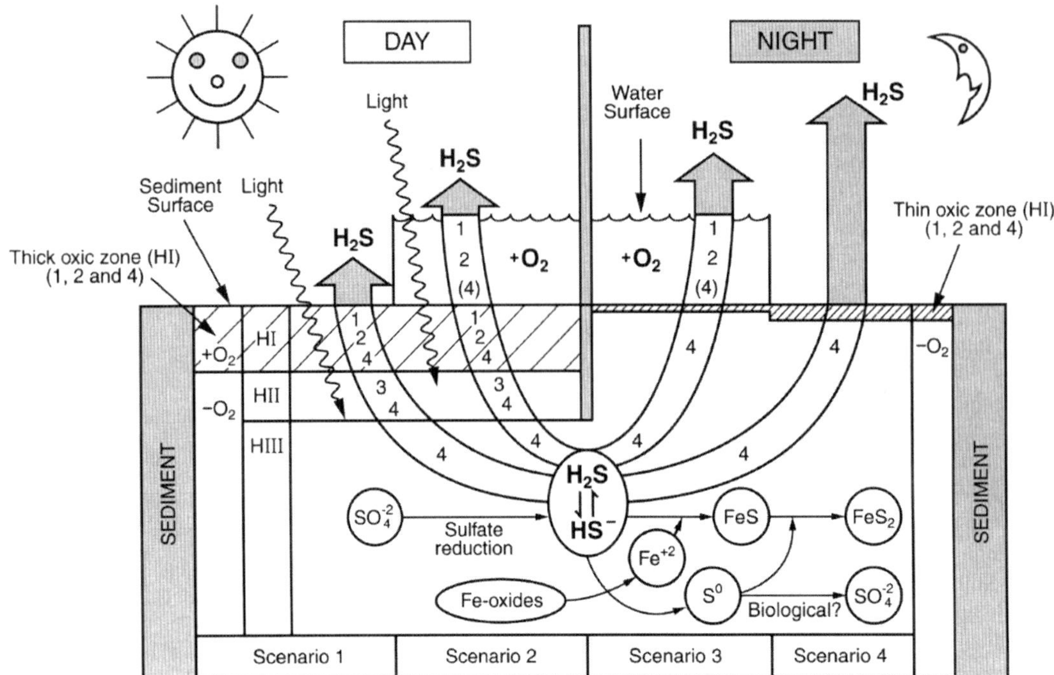

Figure 4. *Fate of hydrogen sulfide produced by sulfate reduction* in hypersaline sediments. Hydrogen sulfide produced in the anoxic zone of GSL sediments may undergo a variety of chemical and microbial transformations, all of which act to reduce H_2S emission to the atmosphere. The figure, which is highly schematic, shows the vertical zonations of sulfide transformations and different scenarios in relation to light and water covering conditions.*

During the day, the sediment surface is stratified into three distinct zones (not drawn to scale): the upper oxic (= oxygen containing) photic zone (HI), the site of oxygenic photosynthesis carried out by microalgae and cyanobacteria; an intermediate zone (HII) which constitutes the illuminated upper part of the anoxic (= devoid of oxygen) zone, this is the site of anoxygenic photosynthesis; and the lowermost aphotic, permanently anoxic zone (HIII) constituting the bulk of the sediment. Zone HIII usually has a black colour due to its content of insoluble iron sulfides (e.g., FeS).

At night, both oxygenic and anoxygenic photosynthetic processes are "turned off" in the uppermost sediment zones (HI and HII), respectively, and the oxic zone (HI) becomes very thin (scenarios 3 and 4) or non-existent (not shown).
The following scenarios are shown in figure 4. Scenario 1: daylight, no water covering of sediment. Scenario 2: day, sediment covered by oxic water layer. Scenario 3: night, sediment covered by oxic water layer. Scenario 4: night, no water covering of sediment.

The sulfur transformation reactions, Rx 1-4, are as follows:
Rx 1: autooxidation of sulfide by oxygen (chemical sulfide oxidation).
Rx 2: sulfide oxidation by colourless sulfur bacteria.
Rx 3: sulfide oxidation by anoxygenic phototrophic bacteria.
Rx 4: Immobilization and/or oxidation of sulfide by reactions involving metals.

Numbers within arrows show the types of sulfide transformations (Rx 1 thru Rx 4) which potentially could take place within the sediment/water layers indicated.
The term "sulfide" denotes both hydrogen sulfide and the bisulfide ion because at circum-neutral pH sulfide is present as almost equimolar concentrations of H_2S and HS^- ($pK1 \approx 7$).

**Note that elemental sulfur may also be reduced to hydrogen sulfide by a variety of microorganisms. This reaction is not shown in figure 4 because nothing is known about the importance of this process in hypersaline sediments.*

processes have been conducted in GSL, it seems likely that conditions observed in other hypersaline habitats may also exist in GSL. Halophilic phototrophic bacteria have been described from a large variety of hypersaline environments and sulfide oxidation mediated by phototrophic bacteria may be a quantitative important process in regulating sulfide emission from exposed sediments to the atmosphere during daytime (Fründ and Cohen, 1992; Caumette, 1993; Caumette, and others, 1994). For example, a strong odour of sulfide often develops around shallow-water, eutrophic sediments immediately after sundown, when the photosynthetic sulfide filters are "switched-off" (Hansen and others, 1978; Ingvorsen and Jørgensen, 1979; Ingvorsen and Jørgensen, 1982).

In hypersaline microbial mats, sulfide production may occur very close to the sediment surface during daytime. Several investigators have reported on high rates of sulfate reduction within highly oxygenated algal mats of hypersaline sediments (Jørgensen and Cohen, 1977; Canfield and Des Marais, 1991; Fründ and Cohen, 1992; Canfield and Des Marais, 1993; Teske, and others, 1998), indicating that such sediments could be a source of atmospheric H_2S, especially at night when the uppermost sediment turns anoxic. Furthermore, in hypersaline microbial mats with high sulfate-reduction activity, significant amounts of elemental sulfur could accumulate in the surface sediment during the day as a result of both chemical and biological oxidation of sulfide. At night, the elemental sulfur may be reduced back to H_2S

due to the respiratory activity of the indigenous microbial population.

The sulfate reduction rates measured in shallow-water sediments at approximately 12 percent salinity in the vicinity of the Antelope Island causeway (figure 3) are among the highest ever reported from a natural environment. The organic-rich sediments in this area are very soft and smell strongly of sulfide. Sulfate reduction in the sediments of nearby Farmington Bay are probably also high due to the low salinity and the (pre-1950s) discharge of untreated waste waters into this lagoonal bay. Other localities that might serve as sources of atmospheric H_2S at night include the sediments in the south and southwestern part of GSL, as well as salt marshes and areas around river outlets. It is well known that strong smells of hydrogen sulfide occasionally develop around Salt Lake City during north winds or after sundown (J.W. Gwynn, Utah Geological Survey, written communication, July, 1999).

The intensity of H_2S emissions from GSL sediments during the night depends on many factors - the primary factor being the intensity of sediment sulfate reduction. Sediment sulfate reduction on the other hand is highly influenced by salinity, temperature, and the organic content of the sediment. The iron content and the microbial populations of the surface sediment are, likewise, important factors that control the emission of H_2S.

SUMMARY - CONCLUDING REMARKS

Acetate and hydrogen are key intermediates of anaerobic carbon degradation (figure 1). Both compounds are major substrates of SRB and methanogens, but in marine and moderately hypersaline sediments, where sulfate is not a limiting factor, SRB seem to outcompete the methanogens for both acetate and molecular hydrogen. Due to their rapid consumption by the anaerobic microorganisms, concentrations of acetate and hydrogen normally do not build up in the pore waters of anoxic marine or freshwater sediments. However, studies by Lupton and others (1984), and Klug and others (1985) showed that both hydrogen and volatile fatty acids (in particular acetate) accumulated in anoxic porewaters as a function of increasing salinity. This indicates that sulfate reduction and methanogenesis, which are terminal steps in the anoxic mineralization of organic matter (see figure 1), are not functioning with these substrates, or at least are attenuated at extreme salt concentrations. Accordingly, it has been suggested by other researchers that acetate, and presumably also hydrogen, are terminal products in anoxic environments at extreme salinities. However, when reaching an oxic environment, e.g., by diffusion, both hydrogen and acetate are readily metabolized by aerobic halophilic microorganisms.

Presently, no extremely halophilic SRB have been isolated in pure culture from hypersaline habitats. The ecological importance of the extant halophilic SRB at salinities above 20 percent NaCl is unknown because in vitro studies indicate that they are poorly adapted to such high salinities. Additional studies are needed in order to identify the SRB responsible for the sulfate reduction rates that have been measured in extremely hypersaline sediments. Is sulfate reduction in these sediments carried out by unknown SRB and/or by salt-stressed populations of the currently described SRB which "hang on" due to their ability to produce or scavenge osmolytes?

It is noteworthy that known cultured species of halophilic fermentative bacteria, many of which produce hydrogen and acetate as end-products (see section on Fermentation), exhibit significantly higher salinity optima for growth in vitro than any of the SRB so far isolated from hypersaline environments. In a recent review, Oren (1999) used bioenergetic reasoning to explain why the known species of halophilic fermentative bacteria apparently are more salt-tolerant than the extant halophilic SRB.

Much remains to be understood concerning the microbial communities of anaerobic hypersaline sediments and the metabolic couplings between the different trophic groups of bacteria. The information available, though very limited, indicates that the terminal anaerobic processes of sulfate reduction and methanogenesis are increasingly inhibited by high salt concentrations (Lupton and others, 1984; Klug and others, 1985; Brandt and others, 2001). Future studies of hypersaline environments, no doubt, will reveal the presence of a large variety of novel types of anaerobic bacteria and yet unrecognized metabolisms. Isolation and characterization of such bacteria may help to show whether the anoxic biodegradative process of hypersaline sediments is significantly different from that of low-salinity environments.

As a case in point, recent work in our department using DNA-based molecular techniques clearly shows that we have a very incomplete knowledge of the diversity of SRB in GSL because these studies revealed the presence of deep-branching groups of SRB having no known representatives among neither pure cultures, nor "molecular isolates" from other environments (Thomsen, T.R., and others, personal communication, 2000).

ACKNOWLEDGMENTS

We would like to thank J. Wallace Gwynn and Jeffrey C. Quick, Utah Geological Survey, for useful suggestions and comments.

REFERENCES

Bak, F., and Cypionka, H., 1987, A novel type of energy metabolism involving fermentation of inorganic sulphur compounds: Nature, v. 326, p. 891-892.

Brandt, K.K., 1998, Sulfate-reducing bacteria from hypersaline environments - ecology, physiology and phylogeny; Ph.D. dissertation, University of Aarhus, DK, 174 p.

Brandt, K.K., and Ingvorsen, K., 1997, *Desulfobacter halotolerans* sp. nov., a halotolerant acetate-oxidizing sulfate-reducing bacterium isolated from sediments of Great Salt Lake, Utah: Systematic and Applied Microbiology, v. 20, p. 366-373.

Brandt, K.K., Patel, B.K.C., and Ingvorsen, K., 1999, *Desulfocella halophila* gen. nov., sp. nov., a halophilic, fatty-acid-oxidizing, sulfate-reducing bacterium isolated from sediments of the Great Salt Lake: International Journal of Systematic Bacteriology, v. 49, p. 193-200.

Brandt, K.K., Vester, F., Jensen, A.N., and Ingvorsen, K., 2001, Sulfate reduction dynamics and enumeration of sulfate-reducing bacteria in hypersaline sediments of the Great Salt Lake (Utah, USA): Microbial Ecology, v. 41, p. 1-11.

Brock, T.D., 1979, Ecology of saline lakes, *in* M. Shilo, editor, Strategies of Microbial Life in Extreme Environments: Dahlem Konferenzen, p. 29-47.

Canfield, D.E., and Des Marais, D.J., 1991, Aerobic sulfate reduction in microbial mats: Science, v. 251, p.1471-1473.

—1993, Biogeochemical cycles of carbon, sulfur, and free oxygen in a microbial mats: Geochimica et Cosmochimica Acta, v. 57, p. 3971-3984.

Caumette, P., 1993, Ecology and physiology of phototrophic bacteria and sulfate-reducing bacteria in marine salterns: Experientia, v. 49, p. 473-481.

Caumette, P., Matheron, R., Raymond, N., and Relexans, J.-C., 1994, Microbial mats in the hypersaline ponds of Mediterranean salterns (Salins-de-Giraud, France): FEMS Microbiology Ecology, v. 13, p. 273-286.

Cayol, J.L., Ducerf, S., Patel, B.K.C., Garcia, J.L., Thomas, P., and Ollivier, B., 2000, *Thermohalobacter berrensis* gen. nov., sp. nov., a thermophilic, strictly halophilic bacterium from a solar saltern: International Journal of Systematic and Evolutionary Microbiology, v. 50, 559-564.

Cayol, J.L., Ollivier, B., Patel, B.K.C., Ageron, E., Grimont, P.A., Prensier, G., and Garcia, J.L., 1995, *Haloanaerobium lacusroseus* sp. nov., an extremely halophilic fermentative bacterium from the sediments of a hypersaline lake: International Journal of Systematic Microbiology, v. 45, p. 790-797.

Elsgaard, L., and Jørgensen, B.B., 1992, Anoxic transformations of radiolabeled hydrogen sulfide in marine and freshwater sediments: Geochimica et Cosmochimica Acta, v. 56, p. 2425-2435.

Fendrich, C., Hippe, H., and Gottschalk, G., 1990, *Clostridium halophilum* sp. nov. and *C. litorale* sp. nov., an obligate halophilic and a marine species degrading betaine in the Stickland reaction: Archives of Microbiology, v. 154, p. 127-132.

Fendrich, C., and Schink. B., 1988, Degradation of glucose, glycerol, and acetate by aerobic bacteria in surface waters of Great Salt Lake, Utah, USA: Systematic and Applied Microbiology, v. 11, p. 94-96.

Fründ, C., and Cohen, Y., 1992, Diurnal cycles of sulfate reduction under oxic conditions in cyanobacterial mats: Applied and Environmental Microbiology, v. 58, p. 70-77.

Galinski, E.A., 1995, Osmoadaptation in Bacteria: Advances in Microbial Physiology, v. 37, p. 273-328, Academic Press.

Galinski, E.A., and Trüper, H.G., 1994, Microbial behavior in salt-stressed ecosystems: FEMS Microbiology Reviews, v. 15, p. 95-108.

Giani, D., Jannsen, D., Schostak, V., and Krumbein, W.E., 1989, Methanogenesis in a saltern in the Bretagne (France): FEMS Microbiology Ecology, v. 62, p. 143-150.

Gibson, G.R., 1990, Physiology and ecology of the sulphate-reducing bacteria: Journal of Applied Bacteriology, v. 69, p. 769-797.

Gwynn, J.W. (editor), 1980, Great Salt Lake - a scientific, historical and economic overview: Utah Geological and Mineral Survey, Bulletin 116, 400 p.

Hansen, M.H., Ingvorsen, K., and Jørgensen, B. B., 1978, Mechanisms of hydrogen sulfide release from coastal marine sediments to the atmosphere: Limnology and Oceanography, v. 23, p. 68-76.

Hansen, T.A., 1993, Carbon metabolism of sulfate-reducing bacteria, *in* J. M. Odom and R. Singleton, editors, The sulfate-reducing bacteria - contemporary perspectives: New York, Springer-Verlag, p. 21-40.

—1994, Metabolism of sulfate-reducing prokaryotes: Antonie van Leeuwenhoek, v. 66, p. 165-185.

Howarth, R.W., and Teal, J. M., 1979, Sulfate reduction in a New England salt marsh: Limnology and Oceanography, v. 24, p. 999-1013.

Ingvorsen, K., and Brock, T.D., 1982, Electron flow via sulfate reduction and methanogenesis in the anaerobic hypolimnion of Lake Mendota: Limnology and Oceanography, v. 27, p. 559-564.

Ingvorsen, K., and Jørgensen, B.B., 1979, Combined measurement of oxygen and sulfide in water samples: Limnology and Oceanography, v. 24, p. 390-393.

—1982, Seasonal variation in H_2S emission to the atmosphere from intertidal sediments in Denmark: Atmospheric Environment, v. 16, p. 855-865.

Javor, B., 1989, Hypersaline Environments - Microbiology and Biogeochemistry, 1st. edition: Springer-Verlag, Berlin, Heidelberg.

Javor, B., and Castenholz, R.W., 1981, Laminated microbial mats, Laguna Guerrero Negro, Mexico: Geomicrobiology Journal, v. 3, p. 237-273.

Jørgensen, B.B., 1977, The sulfur cycle of a coastal marine sediment (Limfjorden, Denmark): Limnology and Oceanography, v. 22, p. 814-832.

Jørgensen, B.B., 1978, A comparison of methods for the quantification of bacterial sulfate reduction in coastal marine sediments, III - Estimation from chemical and bacteriological field data: Geomicrobiology Journal, v. 1, p. 49-64.

—1982, Mineralization of organic matter in the sea bed - the role of sulphate reduction. Nature, 296, 643-645.

—1983, The microbial sulphur cycle, *in* Krumbein, W. E., editor, Microbial Geochemistry: Oxford, Blackwell Scientific Publications, p. 91-118.

—1988, Ecology of the sulphur cycle: Oxidative pathways in sediments, *in* Cole, J.A., and Ferguson, S.J., editors, The Nitrogen and Sulphur Cycles: Cambridge University Press, Cambridge, UK. p. 31-63.

—1990, A thiosulfate shunt in the sulfur cycle of marine sediments: Science, v. 249, p. 152-154.

—1994, Sulfate reduction and thiosulfate transformations in a cyanobacterial mat during a duel oxygen cycle: FEMS Microbiol Ecol, v. 13, p. 303-312.

Jørgensen, B.B., and Cohen, Y., 1977, Solar Lake (Sinai). 5, The sulfur cycle of the benthic cyanobacterial mats: Limnology and Oceanography, v. 22, p. 657-666.

Kamekura, M., 1998, Diversity of extremely halophilic bacteria: Extremophiles, v. 2, p. 289-295.

King, G.M., 1988, Methanogenesis from methylated amines in a hypersaline algal mat: Applied and Environmental Microbiology, v. 54, p. 130-136.

Klug, M., Boston, P., Francois, R., Gyure, R., Javor, B., Tribble, G., and Vairavamurthy, A., 1985, Sulfur reduction in sediments of marine and evaporite environments, *in* Sagan D., editor, The Global Sulfur Cycle: NASA Technical Memorandum 87570, National Aeronautics and Space Administration. p. 128-157.

Krekeler, D., Sigalevich, P., Teske, A., Cypionca, H., and Cohen, Y., 1997, A sulfate-reducing bacterium from the oxic layer of a microbial mat from Solar Lake (Sinai), *Desulfovibrio oxyclinae*, sp. nov.: Archives of Microbiology, v. 167, p. 369-375.

Lowe, S.E., Jain, M.K,. and Zeikus, J.G., 1993, Biology, Ecology, and Biotechnological Applications of Anaerobic Bacteria Adapted to Environmental Stresses in Temperature, pH, Salinity, or Substrates: Microbiological Reviews, v. 57, p. 451-509.

Lupton, F.S., Phelps, T J., and Zeikus, J.G., 1984, Methanogenesis, sulphate reduction and hydrogen metabolism in hypersaline anoxic sediments of the Great Salt Lake, Utah.: Baas Becking Geobiological Laboratory, Bureau of Mineral Resources, Canberra, Australia, p. 42-48.

Luther, G.W. III, Church, T. M., Scudlark, J.R., and Cosman, M., 1986, Inorganic and organic sulfur cycling in salt-marsh pore waters: Science, v. 232, p. 746-749.

Ollivier, B., Caumette, P., Garcia, J.L., and Mah, R. A., 1994, Anaerobic bacteria from hypersaline environments: Microbiological Reviews, v. 58, p. 27-38.

Ollivier, B., Fardeau, M..L., Cayol, J.-L., Magot, M., Patel, B.K.C., Prensier, G., and Garcia, J.-L., 1998, *Methanocalculus halotolerans* gen. nov., sp. nov., isolated from an oil-producing well: International Journal of Systematic Bacteriology, v. 48, p. 821-828.

Ollivier, B., Hatchikian, C.E., Prensier, G., Guezennec, J., and Garcia, J.-L., 1991, *Desulfohalobium retbaense* gen. nov., sp. nov., a halophilic sulfate-reducing bacterium from sediments of a hypersaline lake in Senegal: International Journal of Systematic Bacteriology, v. 41, p. 74-81.

Ollivier, B., Patel, B.K.C., and Garcia, J.L., 2000, Anaerobes from extreme environments, *in* J. Seckbach, editor, Journey to Diverse Microbial Worlds, Adaption to Exotic Environments: Kluiver Academic Publishers, Dordrecht, the Netherlands, p. 73-90.

Oremland, R.S., and King, G.M., 1989, Methanogenesis in hypersaline environments, *in* Cohen, Y., and Rosenberg, E., editors, Microbial mats - physiological ecology of benthic microbial communities: American Society for Microbiology, Washington D.C., p. 180-190

Oren, A., 1988, Anaerobic degradation of organic compounds at high salt concentrations: Antonie van Leeuwenhoek, v. 54, p. 267-277.

—1990, Formation and breakdown of glycine betaine and trimethylamine in hypersaline environments: Antonie van Leeuwenhoek, v. 58, p. 291-298.

—1994, The ecology of the extremely halophilic archaea: FEMS Microbiology Reviews, v. 13, p. 415-440.

—1999, Bioenergetic aspects of halophilism: Microbiology and Molecular Biology Reviews, v. 63, p. 334-348.

Oren, A., Fischel, U., Aizenshtat, Z., and others, 1994, Osmotic adaptation of microbial communities in hypersaline microbial mats, *in* Stal, L. J., and Caumette, P., editors, Microbial mats: Springer-Verlag, Berlin, p. 125-130.

Parkes, R.J., Gibson, G.R., Mueller-Harvey, I., Buckingham, W. J., and Herbert, R. A., 1989, Determination of the substrates for sulphate-reducing bacteria within marine and estuarine sediments with different rates of sulphate reduction: Journal of General Microbiology, v. 135, p. 175-187.

Paterek, J.R., and Smith, P.H., 1985, Isolation and characterization of a halophilic methanogen from Great Salt Lake: Applied and Environmental Microbiology, v. 50, p. 877-881.

—1988, *Methanohalophilus mahii* gen. nov., sp. nov., a methylotrophic halophilic methanogen: International Journal of Systematic Bacteriology, v. 38, p. 122-123.

Phelps, T., and Zeikus, J.G., 1980, Microbial ecology of anaerobic decomposition in Great Salt Lake: Annual Meeting of the American Society for Microbiology, Abstract 14, p. 89.

Post, F.J., 1977, The Microbial Ecology of the Great Salt Lake: Microbial Ecology, v. 3, p. 143-165.

—1981, Microbiology of the Great Salt Lake north arm: Hydrobiologia, v. 81, p. 59-69.

Post, F.J., and Stube, J.C., 1988, A microcosm study of nitrogen utilization in the Great Salt Lake, Utah: Hydrobiologia, v. 158, p. 89-100.

Rainey, F.A., Zhilina, T.N., Boulygina, E.S., Stackebrandt, E., Tourova, T.P., and Zavarzin, G.A., 1995, The taxonomic status of the fermentative halophilic anaerobic bacteria - Description of *Haloanaerobiales* ord. nov., Halobacteroidaceae fam. nov., *Orenia* gen. nov. and further taxonomic rearrangements at the genus and species level: Anaerobe, v. 1, p. 185-199.

Rodriguez-Valera, F., editor, 1988, Halophilic bacteria: CRC Press, Boca Raton, Florida.

Rushforth, S.R., and Felix, E.A., 1982, Biotic adjustments to changing salinities in the Great Salt Lake, Utah, USA: Microbial Ecology, v. 8, p. 157-161.

Sinninghe Damstè, J.S., Rijpstra, W.I.C., De Leeuw, J.W., and Schenck, P.A., 1989, The occurrence and identification of series of organic sulphur compounds in oils and sediment extracts, II. Their presence in samples from hypersaline and non-hypersaline palaeoenvironments and possible applications as source, palaeoenvironmental and maturity indicators: Geochimica et Cosmochimica Acta, v. 53, p. 1323-1341.

Skyring, G.W., 1987, Sulfate reduction in coastal ecosystems: Geomicrobiology Journal, v. 5, p. 295-374.

—1988, Acetate as the main energy substrate for the sulfate-reducing bacteria in Lake Eliza (South Australia) hypersaline sediments: FEMS Microbiology Ecology, v. 53, p. 87-94.

Stephens, D.W., 1990, Changes in lake levels, salinity and the biological community of Great Salt Lake (Utah, USA), 1847-1987: Hydrobiologia, v. 197, p. 139-146.

Stephens, D.W., and Gillespie, D.M., 1976, Phytoplankton production in the Great Salt Lake, Utah, and a laboratory study of algal response to enrichment: Limnology and Oceanography, v. 21, p. 74-87.

Teske, A., Ramsing, N.B., Habicht, K., Fukui, M., Kyver, J.,

Jørgensen, B.B., and Cohen, Y., 1998, Sulfate-reducing bacteria and their activities in cyanobacterial mats of Solar Lake (Sinai, Egypt): Applied and Environmental Microbiology, v. 64, p. 2943-2951.

Thode-Andersen, S., and Jørgensen, B.B., 1989, Sulfate reduction and the formation of 35S-labeled FeS, FeS_2, and S^o in coastal marine sediments: Limnology and Oceanography, v. 34, p. 793-806.

Tsai, C.-R., Garcia, J.-L., Patel, B.K.C., Cayol, J.-L., Baresi, L., and Mah, R.A., 1995, *Haloanaerobium alcaliphilum* sp. nov., an anaerobic moderate halophile from the sediments of Great Salt Lake, Utah: International Journal of Systematic Bacteriology, v. 45, p. 301-307.

Tsu, I. H., Huang, C.Y., Garcia J.L., Patel, B.K.C., Cayol, J.L., Baresi, L., and Mah R. A., 1998, Isolation and characterization of *Desulfovibrio senezii* sp. nov., a halotolerant sulfate reducer from a solar saltern and phylogenetic confirmation of *Desulfovibrio fructosovorans* as a new species: Archives of Microbiology, v. 170, p. 313-317

Vester, F., and Ingvorsen, K., 1998, Improved most-probable-number method to detect sulfate-reducing bacteria with natural media and a radiotracer: Applied and Environmental Microbiology, v. 64, p. 1700-1707.

Vreeland R.H., and Hochstein L.I., 1993, The biology of halophilic bacteria: CRC Press, Boca Raton, Florida, 320 p.

Ward, D.M., and Brock, T.D., 1978, Hydrocarbon biodegradation in hypersaline environments: Applied and Environmental Microbiology, v. 35, p. 353-359.

Welsh, D.T., Lindsay, Y.E., Caumette, P., Herbert, R.A., and Hannan, J., 1996, Identification of trehalose and glycine betaine as compatible solutes in the moderately halophilic sulfate reducing bacterium, *Desulfovibrio halophilus*: FEMS Microbiology Letters, v. 140, p. 203-207.

Widdel, F., 1988, Microbiology and ecology of sulfate- and sulfur-reducing bacteria, *in* Zehnder, A. J. B., editor, Biology of Anaerobic Microorganisms: Wiley Interscience, New York, p. 469-585.

Winfrey, M. R., and Zeikus, J. G., 1977, Effect of sulfate on carbon and electron flow during microbial methanogenesis in freshwater sediments: Applied and Environmental Microbiology, v. 33, p. 275-281.

Wood, A.P., and Kelly, D.P., 1991, Isolation and characterisation of *Thiobacillus halophilus* sp. nov., a sulphur-oxidising autotrophic eubacterium from a Western Australian hypersaline lake: Archives of Microbiology, v. 156, p. 277-280.

Zehnder, A.J.B., 1988, Biology of Anaerobic Microorganisms: Wiley Interscience, New York, New York, 872 p.

Zeikus, J.G., 1983, Metabolic communication between biodegradative populations in nature, *in* Slater H., Wittenbury E., and Wimpenny J., editors, Microbes in their natural environments, Society of General Microbiology Symposium 34: Cambridge University Press, London, p. 423-462.

Zeikus, J.G., Hegge, P.W., Thompson, T.E., Phelps, T.J., and Langworthy, T. A., 1983, Isolation and description of *Haloanaerobium prevalens* gen. nov. and sp. nov., an obligately anaerobic halophile common to Great Salt Lake sediments: Current Microbiology, v. 9, p. 225-234.

Zhilina, T.N., 1986, Methanogenic bacteria from hypersaline environments: Systematic and Applied Microbiology, v. 7, p. 216-222.

Zhilina, T.N., and Zavarzin, G.A., 1990, Extremely halophilic, methylotrophic, anaerobic bacteria: FEMS Microbiology Reviews, v. 87, p. 315-322.

- **ANTELOPE ISLAND, GREAT SALT LAKE, UTAH** - *by Tim Smith, Brock Cheney and Alan Millard*
- **GREAT SALT LAKE, A BOATING JEWEL OF THE WEST** - *by Steve Ingram*

ANTELOPE ISLAND, GREAT SALT LAKE, UTAH

by

Tim Smith, Brock Cheney and Alan Millard
Antelope Island State Park

INTRODUCTION

Within the human imagination, islands are special places that offer an escape from mainland demands and pressures. Great Salt Lake has 11 islands ranging from small, barren piles of rock protruding from the water, to those large enough to sustain complex ecosystems. With an area of 42 square miles, Antelope Island is the largest island in Great Salt Lake (figure 1). The island's size, fresh-water sources, and proximity to the mainland have allowed an interface with humans which spans thousands of years. This paper will discuss the island's history, purchase by the state, geology, vegetation, wildlife, recreation, and future. Information on park hours, the various fees, activities, special events and amenities are found in the appendix.

Figure 1. *Aerial view (looking north) of 42-square mile Antelope Island. The southern causeway can be seen extending to the right from the island (photo courtesy of John George).*

EARLY HISTORY OF THE ISLAND

Pre-1847 Period

Before Anglo settlement in the valley of Great Salt Lake, indigenous people used the island's resources for thousands of years. Although historical human activity has obscured the record of these earlier cultures, evidence of their existence continues to surface. Recently, several archaeological surveys revealed human activity at multiple sites on Antelope Island. Specifically, these surveys uncovered artifacts dating to more than 6,000 years ago. However, it is unclear to what extent these people utilized Antelope Island. Park management plans include additional archaeological surveys and research.

Recent archeological digs have revealed evidence that Antelope Island served as a temporary hunting site amongst the prehistoric people. In 1999, during preparations to install a water system at Mushroom Springs for the Garr Ranch facility, an archeological site was discovered, presumed to be a Fremont camp site dating back about 1,000 years (Ron Rood, Utah Division of State History, Antiquities Section, personal communication, August 2000). The most valuable part of this discovery was its distinct stratification, marking the times the camp was occupied, which revealed recurring visits over hundreds of years. Items found in the dig included deer bones, bowls, shards, arrow points, manos, metates, and charcoal (evidence of fires).

Native American use of the island is also documented in historical records of the nineteenth century. In October 1845, the Fremont expedition encountered Ute Indians in the Salt Lake Valley. Chief Wanship informed John C. Fremont that it was possible to ride a horse to the largest of Great Salt Lake's islands. Following Wanship's directions, Fremont led his expedition across a sand bar to the south end of Antelope Island. Kit Carson, who served as scout for the survey party, shot two antelope (pronghorn), whereafter Fremont wrote in his diary, "grateful for the supply of meat they furnish, I give their name to the island."

Mormon Pioneer Period

Soon after Fremont's expedition, Mormon pioneers arrived in the Salt Lake Valley. As part of their extensive surveys of the vicinity, Antelope Island was scouted to determine potential uses. Leaders of the Church of Jesus Christ of Latter Day Saints (Mormon Church) concluded that Antelope Island would provide both good forage and protection for the church's livestock herds. At this time, hard cash or currency was scarce, and most of the church wealth was in the form of livestock. Fielding Garr began the initial construction of the ranch buildings and corrals on what is now known as the Fielding Garr Ranch.

Fielding Garr and the Garr Ranch

Fielding C. Garr, a widower from Wayne County, Indiana, and his eight children were among the early Mormon pioneers to cross the prairies and mountains seeking domicile in the Salt Lake Valley. In the fall of 1848, Fielding Garr, Thomas Thurston, and Joseph B. Noble traveled to Antelope Island to determine if, in fact, the island would be suitable as a herd ground. The island was covered with good abundant grasses, starch root, sunflowers, wild rose bushes, sage, berry bushes, and a few willows and shrubbery in the ravines. There were fresh-water springs and excellent grazing fields on the eastern slopes of the island. They found Antelope Island to be unrivaled as a herd ground, and most appealing was the isolation and protection offered from theft and disease.

Garr, Thurston, and Noble drove the first herd of cattle 18 miles to their new grazing land on Antelope Island during the winter of 1848-49. The drive took three days. The lakebed was dry most of the way where they crossed from the mainland to the south end of the island. Garr's family came over later, crossing the lake by horse and wagon over sand bars covered with coarse salt.

A small, one-room cabin was built as temporary living quarters for the family until a permanent dwelling could be constructed. It was located very close to what became known as Garr Springs. The ranch house was built farther east of the springs.

A five-room adobe ranch house was built in the early pioneer architectural design. Two square rooms facing east had window-door-door-window facade, with a large fireplace in each room on the gabled end. A lean-to extension faced west and included three small bedrooms, each with a window. Two bedrooms opened into the livingroom area, the third opened into the kitchen where the hired hands ate their meals. Natural material from the island was used to form the adobe bricks; pieces of straw were used as binding material. The bricks were then sun dried. The walls were one foot thick. Fielding Garr, a skilled mason, built the house so well it has withstood the ravages of time over 150 years.

The original Garr Ranch complex covered 25 acres. The oldest and most historic buildings besides the ranch house are: the bunkhouse, a spring house, barn/stable, and a blacksmith shop; a large barn and corrals were built to the north of these buildings. The complex also included a garden and fruit trees, a pole fence in front of the house, and a landing wharf on the east shore of the island. Locust trees were planted on the east side of the house. A recent photograph of the ranch is shown in figure 2.

Fielding Garr was a bonded herdsman, and became the first manager of the Mormon Church's herds on the island. He was also asked to care for the herds belonging to the Stansbury Expedition in 1849-50. In September of 1850, the legislature of the Territory of Utah designated Stansbury and Antelope Islands for exclusive use by stock of the Perpetual Emigration Company (Morgan, 1947), which was established to generate money for Mormons to immigrate to Utah. In practice, the Mormon pioneers would pay this money back once they became established. At this time, Thomas Thurston, Joseph Noble, and others removed their cattle from Antelope Island, which was still managed by Garr. In addition to cattle, sheep were raised, and 600 horses were turned loose on the island.

Figure 2. *Aerial view (looking east) of Fielding Garr Ranch (photo courtesy of John George).*

By 1854, conditions on the island for grazing cattle began to deteriorate as feed on the island became scarce due to overgrazing. This necessitated cattle removal, and much of the church's herd was moved to the Elk Horn Ranch in Cache Valley. On June 15, 1855, Fielding Garr died on the island.

The Mormon church ran livestock on Antelope Island until the mid-1870s. At that time, the federal government had just completed the first extensive survey of the territory. This survey facilitated the introduction of homesteading to the island. The Homesteading Act didn't recognize squatters' rights, but only those homesteads that had made physical improvements on the land. As a result, the Garr Ranch could only claim the section of land the ranch was located on, while the rest of the island was opened for homesteading.

Little remains of the three homesteads that were established on the island during the late 1880s. One exception is the homestead of George and Alice Frary, established in 1891. George Frary was one of the first to sail the lake, carrying cattle from the ranch to the mainland. George also ferried the first ancestors of today's bison herd to the island. The Frarys lived on the island for five years until Alice's death in 1896. George buried Alice at the homestead (figure 3), and then moved his family to the mainland, but his life

Figure 3. *Fenced grave site of Alice Frary who was buried on the Frary homestead site in 1896 (photo courtesy of Bob Middlemas).*

Table 1. A chronology of events and occupations of the ranch and the island.

Year	Event
1843	John Fremont named Antelope Island for the antelope he observed.
1848	Fielding Garr built the ranch house.
1849	Fielding Garr became foreman of the Perpetual Emigration Company (PEC).
1855	Fielding Garr died.
1856	Briant Stringham replaced Fielding Garr as foreman of the PEC.
1871	Briant Stringham died.
1872	Christopher Layton became ranch caretaker.
1875	The Mormon Church lost interest in the island due to homesteading and patents to the Union Pacific Railroad for every odd section.
1884	John Dooly, Sr. and Frederick Meyer purchased a large part of the island. It became known as the Island Improvement Company and was later leased to John White and sons. The wild horses introduced and cultured earlier by Brigham Young were all shot to reduce grazing competition with other livestock.
1885-1903	Will Walker was foreman of the Island Improvement Company.
1891	George Frary established a homestead just north of Garden Creek.
1903	Ernest Bamberger became co-owner of the island.
1911	John Dooly, Jr. inherited the Dooly empire from his father, John Dooly, Sr., which included The Island Improvement Company.
1938-1942	Jabez Broadhead Harward became foreman of the Island Improvement Company. He lived on the island with his family.
1939	Glen Phipps was hired on as foreman number two.
1942-1951	J.B. Harward became general foreman of the Island Improvement Company ranches. He and his family lived alternately between the island and the summer range at Castle Rock, Utah near the Wyoming-Utah border.
1942	William H. Olwell was appointed manager of the Island Improvement Company.
1942-1947	Arnold Haskell became foreman. He lived on the island with his wife, Ida, and their family.
1945-1972	William H. Olwell was appointed president of the Island Ranching Company. The Island Improvement Company's name changed to Island Ranching Company in 1955.
1948-1950	Orren Hale became forman of the Island Improvement Company. He lived on the island with his wife LaVern and their family.
1950-1955	Arnold Haskell returned as foreman of the island ranch after the death of J.B. Harward in 1950.
1951-1972	Gene Phipps became general foreman of Island Improvement Company.
1951-1952	The south causeway was built over an existing road from the south end of the island to the mainland to the east.
1952	Island Improvement Company sold all their sheep and moved their cattle from the island to Castle Rock and Skull Valley ranches.
1955-1972	Island Improvement Company's name was changed to Island Ranching Company.
1955-1971	Alton and Janet Sorenson were foremen of the Island Ranching Company.
1967	State of Utah purchased 2,000 acres on the north end to establish a park.
1969	The north causeway from Syracuse to the north end of the island was finished.
1971-1972	Nelson Witt was foreman of the Island Ranching Company.
1972-1981	Anschutz Corporation of Denver, Colorado, purchased the Island Ranching Company including the island, Castle Rock, and Skull Valley. They also purchased Hatch Bros. Livestock Company. These were two of the three largest livestock companies in the state at that time, Deseret Livestock Company being the largest.
1981	State of Utah purchased the rest of the island from the Anschutz Corporation.
1983	Rising flood water inundated the causeway and forced the park to close.
1993	Park reopened after the flood waters receded and the road was rebuilt.
1995	A new 5,600 square foot Antelope Island visitors center is completed.

continued to intertwine with the lake. He spent his final years working an asphalt mine on the north arm of the lake near the Promontory Mountains. Today, the foundation of Frary's cabin and a few stone walls are all that remain.

Frary's activity on the lake coincided with the introduction of large-scale ranching on Antelope Island. In 1884, John Dooly purchased a large part of the island and formed the Island Improvement Company. Dooly's ranching efforts prospered to become one of the largest ranching operations in Utah, and the Garr Ranch continued to be the headquarters for these operations. A chronology of events and occupants of the ranch and the island, given in table 1, are taken from Harward (1996).

PURCHASE OF THE PARK PROPERTY AND THEN THE WHOLE ISLAND

Through the years, several attempts were made to purchase the ranch and Antelope Island for the public. On three separate occasions, Utah's Senator Frank Moss sponsored federal legislation to purchase the island for inclusion in the National Park system. Finally, in 1967, the State of Utah purchased 2,000 acres on the north end of the island that became Antelope Island State Park in1969. In an attempt to acquire all of Antelope Island, the State of Utah initiated a court battle with the island's owner, the Anschutz Corporation, to establish the right to eminent domain. However, after several years of negotiation, an out-of-court settlement was finally reached in 1981, which allowed the State to buy the remaining southern portion of Antelope Island from the Anschutz Corporation for $4.7 million.

GEOLOGY AND HYDROLOGY

Antelope Island is the largest island in Great Salt Lake. It is approximately 15 miles long and 4.5 miles across at the widest point, with an area of about 42 square miles. The highest peak (Frary Peak) is 6,597 feet, about 2,400 feet above the historical average (4,200-foot) level of the lake. Recent history pales when contrasted against geologic time. The following brief description of the geology and mineral resources of the island is taken from Doelling and others (1988).

Archean Rocks

Antelope Island is comprised of some of the oldest rocks exposed in Utah. The Farmington Canyon Complex - more than 2.5 billion years old - dominates the southern two-thirds of the island, and is classified as gneiss. A gneiss is a coarse-grained, irregularly banded metamorphic rock. There are banded gneisses, granite gneisses (pink or salmon-colored), amphibolite gneisses (dark), and quartz gneisses (almost white or translucent) on the island.

Proterozoic and Cambrian Rocks

The rocks immediately overlying the Farmington Canyon Complex on Antelope Island were deposited sometime between 1,600 and 700 million years ago. They are assigned to the Perry Canyon Formation, and are divisible into three parts. The lowermost part is diamictite (pebbles, cobbles and boulders mixed in a dark matrix of sand and granules). Overlying the diamictite is a distinctive layer of light tan to pink dolomite about 25 feet thick. The dolomite is especially easy to recognize on the east part of Elephant Head, the island's second highest peak, where it forms light-colored cliffs and ledges. More than 200 feet of gray, purple, brown, green, and red slate rest above the dolomite.

The Tintic Quartzite overlies the Perry Canyon Formation, and forms the pale-orange outcrops and rounded ridges on the northern third of the Antelope Island. The Tintic Quartzite is pale orange to light-gray quartzite with abundant layers of pebble conglomerate. These rocks were deposited about 570 to 540 million years ago in a beach, or shallow marine environment.

Tertiary Rocks

The earliest of the Tertiary rocks deposited on Antelope Island are known as the Wasatch Group and consist of gray and red conglomerate and minor reddish-orange silt, sand, and clay. These rocks were deposited at the front of a mountain range as coalescing alluvial fans, 60 to 45 million years ago. Younger Tertiary rocks, also present on Antelope Island, were deposited between 19 and 1.6 million years ago. These are called the Salt Lake Group and consist of tufaceous sandstone, volcanic tuffs, white to gray conglomerate, friable sandstone, and multi-colored bentonitic clay. Tertiary rocks crop out only on the east side of Antelope Island south of Sea Gull Point, and on the tip of Ladyfinger Point, on the island's extreme northern end.

Quaternary Lake Bonneville Deposits

Tufa deposits, formed on the island as a result of wave and algal action from ancient Lake Bonneville, blanket the much older Farmington Canyon Complex. These deposits are particularly prevalent along the Provo level terrace of Lake Bonneville. Perhaps nowhere else are the Lake Bonneville shorelines etched so dramatically into the landscape as on Antelope Island.

Rock and Mineral Resources

Sand, gravel, flag stone (slate), metals (copper and gold), and quartzite suitable for concrete aggregate are the chief resources of Antelope Island. Sand and gravel deposits are present in many places around the perimeter of the island and are the most important resource. In 1979 and 1980, about 16 million cubic yards of sand and gravel were excavated from alluvial fans and beach deposits on the southeastern side of the island for use in the construction of Interstate Highway 80 between Saltair and Salt Lake City. The material was conveyed from the island to the highway construction stockpile via a 13-mile long conveyor belt.

Small occurrences of metal-bearing minerals, such as chalcopyrite, chrysocolla, malachite, and limonite, have been discovered on Antelope Island disseminated in white quartz veins, especially in the central part and on the higher slopes of the island. In early times, a little slate was quarried for flagstone and crushed for roofing stone. Noteworthy deposits of oolitic sand (calcium carbonate) are abundant on the northern and western beaches of the island. Non-economic minerals present on the island include quartz, feldspar, hornblende, mica, red and brown garnet, staurolite, and sillimanite.

Springs and Groundwater

Although the fresh water of Lake Bonneville has receded to become the saline lake of today, fresh water gushes year round from more than 40 springs on Antelope Island. Precipitation falling in the higher elevations on the island seeps into the lacustrine and alluvial deposits, and deep into the geological formations mentioned above, forming and maintaining a saturated zone of groundwater. This groundwater then emerges as springs, typically below the Lake Bonneville Terrace, where it flows to lower elevations before disappearing into the alluvium (figure 4). Some springs re-emerge just above the Great Salt Lake shoreline, giving rise to the east-shore wetlands.

Figure 4. *Antelope Island has 40 major springs, some of which give rise to lush riparian zones and wetlands (photo courtesy of Garth Taylor).*

Figure 5. *Antelope Island has been called the last stronghold for Great Salt lake's long-billed curlews (photo courtesy of Robert Erwin).*

VEGETATION

Spring-fed wetlands in the eastern lowland areas are the lush, manifestation of Antelope Island's fresh-water resources. Upland habitats on the island are dominated by a sagebrush-grassland community. Historically, the early ranchers on Antelope Island recognized the value of these native grasslands as pastures for their livestock.

While exotic grasses are common today on the lower elevations of Antelope Island's rangelands, historical records indicate the presence of a more diverse browse community of native perennials. Disturbances from large-scale livestock grazing, and subsequent planting of exotic annual grasses, have increased the frequency and intensity of wildfires. Restoring island rangelands, by reducing fire frequency and seeding native perennial grasses, is an important goal of Antelope Island State Park. These native perennials are more nutritious for wildlife and stay green longer into the growing season, reducing the fire hazard.

WILDLIFE AND DOMESTIC HERDS

Introduction

Pronghorn, bighorn sheep, mule deer, and bison populations are all carefully balanced by park management to remain within the carrying capacity of the island's ranges. Bison herd size is managed through an annual roundup and sale. Bison sales are based on population models to ensure a healthy balance of bulls and cows across all age groups. The funds generated from bison sales are tied to park wildlife-management programs. This funding source has allowed the park to fund major range restoration projects and outside research.

In addition to range-carrying capacity, another wildlife management goal on Antelope Island is to provide recreation through "Watchable Wildlife" opportunities. Diverse populations of wild species combine to augment the austere beauty of Antelope Island. Visitors are drawn by an unusual array of wildlife which juxtaposes abundant large mammals with one of the hemisphere's largest concentrations of water birds (figure 5). The salinity of Great Salt Lake, which precludes most aquatic life, nurtures intense populations of brine shrimp and brine flies. These in turn provide protein-rich food sources for the millions of water birds and other upland birds that depend on the island ecosystem. As with most lands around Great Salt lake, Antelope Island is an outstanding birding area. The drive across the Antelope-Syracuse, or northern causeway, provides the birder an opportunity to observe pelagic species not readily visible from mainland marshes. Gull rookeries on Egg and White Rock Islands can be seen from the north end of Antelope Island. Antelope Island also provides rare undeveloped upland habitat above productive Great Salt Lake wetlands.

Bison

By the late 1800s it was estimated that only 800 bison were left on the continent of North America. At this time, conservationists around the country began to take steps to save bison from extinction. Two Utahns were instrumental in this effort. William Glassman and John Dooly brought bison to Antelope Island. Glassman was developing Garfield City on the south shore of the Great Salt Lake in Tooele County, which was to include a planned zoological garden with a "Buffalo Park." The venture was not successful and Glassman began to look for buyers for his bison. John Dooly, who owned most of Antelope Island, purchased some of the bison for the island. On February 15, 1893, 4 bulls, 4 cows and 4 calves were boated to the island. These twelve animals provided the foundation for what has grown into one of the oldest and largest publicly owned herds of bison in the nation.

The following two anecdotes are related by park personnel, and on the park's visitor-center video clips, about the island's buffalo herd.

In 1922, a portion of the silent film, "The Covered Wagon," directed by James Cruze, was filmed on the island in Camera Flat. The bison herd was stampeded through the wagons for the movie. A camera man was almost killed during the stunt, but a quick cowboy saved the day, downing an errant bison just before he reached the camera wagon. Buffalo steaks were enjoyed that evening by everyone, with the camera man refusing to partake.

In 1926, a large bison hunt took place on the island. This was advertised as "The Last Great Buffalo Hunt." The hunters were charged $300.00 a head for each animal taken. The famous boxer Jack Dempsy participated in the hunt. One source claimed that fewer than 100 of the 400 bison were taken and another claimed only 25 were spared.

In addition to bison, the island was home to the largest (figures 6 and 7) sheep operation west of the Mississippi River during the 1920s and 1930s. Over 10,000 sheep were kept on the island at that time. Since the mid-1800s the island has been over-grazed by domestic livestock.

Figure 6. Sheep entering the Fielding Garr shearing barn (unknown, circa 1926).

Figure 7. View inside the Fielding Garr shearing barn (unknown, circa 1926).

From 1893 to 1986 the only management of bison on the island was hunting to keep the herd size down to limit competition with livestock. In 1987, the Utah Division of Parks and Recreation prepared and initiated a bison-management program. Rangeland studies suggest that to maintain adequate forage for wildlife species other than bison, the bison herd should be maintained at between 550 to 700 animals. On November 7, 1987, the wild herd was captured for the first time, and blood samples were taken. Prior to the roundup, there were eight known gene alleles for the American Plains Bison (*Bison bison bison*). However, a ninth allele unique to the Antelope Island herd was found.

Each year since 1987, the bison are rounded up at the end of October as part of the park's range and wildlife management program. The annual bison round-up is the primary tool to control bison herd size on the island. The herd is managed to maintain a stock population of 550, which increases to more than 700 as the new calves are born. During the round-up the bison are also weighed, vaccinated, and checked for pregnancy. The excess animals are sold at auction, and a small number are designated for the annual bison hunt conducted by the Utah Division of Parks and Recreation in December. On Antelope Island most bison breed in July and August, and have a gestation period of nine months. A majority of the calves are born in April and May. The bison (figure 8) thrive on the island due to the fact that most of the island is a grassland and because of the 40 major fresh-water springs found on the island.

Figure 8. Antelope Island is home to one of the largest publicly owned bison herds in the nation (photo courtesy of Don Paul).

Pronghorn Antelope and Other Animals

In 1845, John C. Fremont named Antelope Island after the antelope he found there. By the 1930s the island's namesake animal had been extirpated from the island to reduce competition with sheep numbering in the thousands. In 1993, a cooperative effort between the Utah Divisions of Wildlife Resources and State Parks and Recreation resulted in the reintroduction of 24 pronghorn antelope (figure 9). By the 1995 fawning season, the population had nearly doubled in size, and has since leveled off at about 50 to 60. Long-term research by Weber State University monitors the population, helps determine critical habitat, and studies behavioral traits of the species. In 1997, a similar joint effort resulted in the introduction of bighorn sheep (figure 10). Other mammals found on the island include mule deer, coyotes, bobcats, badgers, porcupines, jackrabbits, and several species of rodents. A variety of snakes and other reptiles is also found.

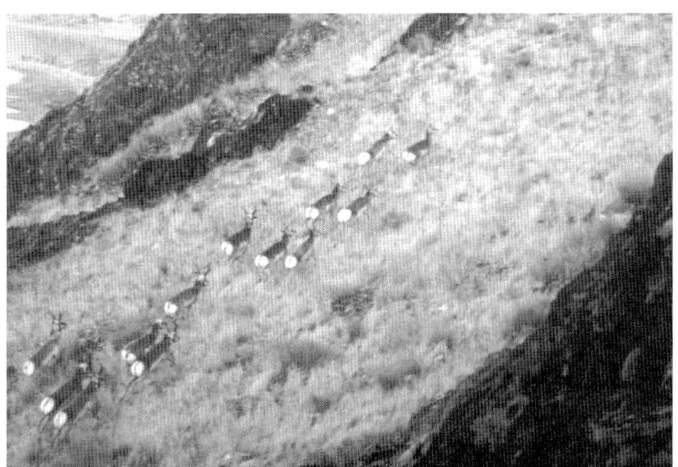

Figure 9. *Distant view of Pronghorn antelope (photo courtesy of Kevin Sherman).*

Figure 10. *Bighorn sheep inhabit Antelope Island's steep, rocky slopes (photo courtesy of Don Paul).*

Birds

Great Salt Lake is one of the most important avian breeding and migratory staging areas in the United States. The lake is of particular importance to long-distance migrants. This importance has been officially recognized with Great Salt Lake's designation as a Hemispheric Shorebird Reserve. A comprehensive review of the avian ecology of Great Salt Lake by Don Paul and Tom Aldrich is found within this volume.

Antelope Island and Great Salt Lake attract numerous migrating and nesting birds. Along the shoreline, avocets, black-necked stilts, willets and sanderlings can be observed. The island grasslands provide habitat for long-billed curlews, burrowing owls, chuckars, and several species of raptors, to include Golden Eagle and Peregrine Falcon. Great Salt Lake attracts incredible numbers of eared greebs, Wilson's phalaropes and California gulls, and is one of the most important natural features in the country for migrating birds. Other birds seen on the island in recent years include: Western Meadowlark, Baird's Sandpiper, Ruddy Turnstone, Sabine's Gull, Hudsonian Godwit, White-winged Scoter, Black Scoter, Surf Scoter, and Oldsquaw (Evans, 1998; Utah Division of Parks and recreation, 2000a).

RECREATION AND TOURISM

Introduction

Antelope Island has often been called the best place to see and experience Great Salt Lake; it annually and draws over 300,000 visitors. As early as 1856, Brigham Young used Antelope Island as a retreat for his recreation on the lake. Two miles of white oolitic sand beaches in Bridger Bay provide an opportunity for traditional waterfront recreation. The saline water in Bridger Bay allows bathers to enjoy the unsinkable swimming for which Great Salt Lake is most famous (figure 11). Stark mountains and islands combine with vast vistas of azure water to present a dramatically xeric landscape. Such naked beauty is visible at Buffalo Point on the north tip of the island. From this vantage point 800 feet above the lake, splashes of color from sunsets to the west paint the jagged peaks of the Wasatch Range to the east. The park's back-country trail system provides dramatic views of the lake and Great Basin mountain ranges, primarily from the island's isolated west side.

Figure 11. *Antelope Island's Bridger Bay is a great place to experience Great Salt Lake and view the shoreline terraces formed by Lake Bonneville (photo courtesy of Bob Middlemas).*

Figure 12. The Antelope Island State Park's marina is an idyllic place to begin a voyage on Great Salt Lake (photo courtesy of Bob Middlemas).

Today, Antelope Island State Park is one of the largest state parks in Utah. The proximity of the island's undeveloped open spaces to the dense populations of the Wasatch Front has drawn an increased interest in recent years. Further, the island's array of natural and cultural resources demands management solutions which balance accommodation of increasing interest and visitation with protection. Information on park hours, the various fees, activities, special events and amenities are found in the appendix.

Trails

Antelope Island State Park offers 35 miles of non-motorized trails allowing a visitor to discover the island, its wildlife, and wonders of Great Salt Lake (figure 13). The beginner and the advanced trail rider will both find trails to fit their individual skill level. These are biking, horseback riding (figure 14), and hiking trails. The trail system is the best way to explore the island and enjoy watchable wildlife. Most trails are moderate with some strenuous uphill and downhill segments. Weather can be extreme, so plan accordingly. Spring and fall offer more pleasant temperatures for trail use. Early morning and late afternoon are best in summer. The park's trail regulations are designed for visitor safety, and to protect the wildlife and unique natural resources of the island. A volunteer trail patrol helps park staff, provides information, and assists trail users. Trail volunteers wear Bike Patrol T-shirts or flourescent vests.

The Lakeside trail is an easy trail with some difficulty for bikes through the rocks. Trail head access is at Bridger Bay campground and White Rock Bay. The trail meanders along the lake shore and provides views of some of the island's unique geology, shorebirds, and wetlands.

The White Rock Loop trail is moderate and has long ascents and descents. It is sandy in places. This is the main trail that provides access to Elephant Head and Split Rock trails. It traverses from the west shore to a high ridge on the north end of the island, and provides good wildlife viewing opportunities.

The Elephant Head trail can be difficult, but it is a great

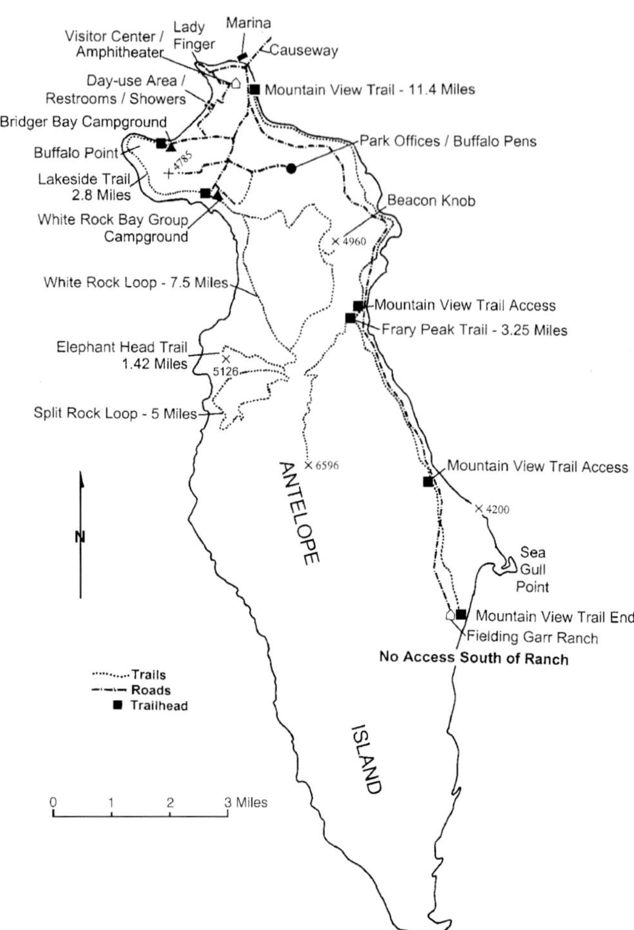

Figure 13. Map of Antelope Island showing the locations of trails, campgrounds, the marina, and other features.

Figure 14. The park's back country trail system takes visitors to the isolated interior of the island.

place to take in the island's ruggedness. The trail is accessed from the White Rock loop at the Split Rock loop overlook. A single-track trail leads to a spectacular overlook 900 feet above the Great Salt Lake and offers views of the island's west side.

The Split Rock Loop trail is moderate to difficult. The loop is accessed from White Rock loop at the Split Rock overlook. Island views and geology make this trail worth the effort. Bighorn sheep and mule deer are seen on occasion. The trail passes a historic horse corral built in the late 1800s; the trail to this site is moderately strenuous. South of the corral, the trail makes a dramatic descent into Split Rock Bay and loops back up to the overlook. Steep switchbacks and a long climb make this part of the trail difficult and challenging.

The Mountain View trail (closed May 15 through June 15 for antelope fawning season) is an easy but long trail with several access points. The trail head is .25 miles south of the marina. The trail winds along the east side of the island through wetlands and geological formations, stopping at the historical Fielding Garr Ranch. This is a great trail to discover the diversity of the island and view wildlife. It can also be accessed from the ranch. The trail crosses the road twice. Use caution when crossing.

The Frary Peak trail (closed April 1 through June 1 for bighorn sheep lambing season) is a hiking trail only. No dogs, horses, or bikes are allowed on the trail. It is moderate to difficult due to steepness and terrain. The trail head is accessed from the east side road. Climbing to an elevation of 6,596 feet, the trail reaches the highest peak on the island. It provides a bird's-eye view of the lake, other islands, and the Wasatch Range.

Other trails, not shown on figure 14, include the Buffalo Point trail, Egg Island Overlook/Ladyfinger Point trail, and Beacon Knob trails. The Buffalo Point trail trends westward from about the Buffalo Point concession stand, and reaches Buffalo Point, the highest point on the north end of the island. The Egg Island/Ladyfinger Point .25-mile-long trail winds across a rocky point to overlook both Egg Island and Bridger Bay, and is located on the northernmost extent of the island. The Beacon Knob trail is a short trail that spurs off the White Rock loop reaching the 4,960-foot Beacon Knob.

LOOKING TOWARD THE FUTURE

The primary challenge of the future will be to protect the island from its own popularity. The presence of diverse wildlife populations and readily accessible open spaces and amenities combine to make this park an irresistible recreation opportunity. The public's dual mandate of greater access and protecting the park's natural and cultural resources, requires a balancing of issues that have been the focus of several planning and research projects. Critical wildlife habitat has been identified, and development has been steered away from these sensitive zones. Range conditions are monitored and restoration projects prioritized.

A number of restrictions have been implemented to reduce impacts to wildlife from recreation. Visitors are restricted to roads and trails to limit impacts to defined corridors. In locations where trails impinge on critical habitat, trails are closed during sensitive times, such as lambing and fawning seasons.

The impacts of increasing visitation on the physical resources and social experience can be subjective and hard to quantify. An ongoing monitoring program is in place to measure visitors' dissatisfaction with crowding at specific sites on the island. Safeguarding the island's resources while maintaining the social experience promises to be an ongoing vexing issue with potentially controversial solutions. As life along the Wasatch Front becomes more hectic and crowded, Antelope Island's provision of refuge for wildlife and people will grow in value. Currently the park is managed with a resource management plan developed in 1994 with public input completed. An updated resource management plan was adopted in July 2001 and focuses on three primary issues: Antelope Island trails - use, management, and development; a potential public hunt of the park's mule deer population; and management of increasing visitor use on the park's east side (Utah Division of Parks and Recreation, 2000b).

SUMMARY

Before Anglo settlement in the valley of Great Salt Lake, ancient people used the island's resources for thousands of years. Shortly after the arrival of the Mormon pioneers in 1847, the Mormon church began to use the island as a home for its cattle, with Fielding Garr as the first manager of the church's herds. The island's ownership changed hands a number of times until the State of Utah purchased the northern 2,000 acres in 1967 to establish the Antelope Island State Park, and then purchased the remainder of the island from the Anschutz Corporation in 1981.

Antelope Island, the largest island in the Great Salt Lake, is reached via a 7.5-mile causeway from the mainland to the east. Island activities include saltwater bathing, bird watching, camping, hiking, biking, horseback riding, picnicking, sunbathing, exploring historical sites, photography, and viewing wildlife in its natural habitat. Antelope Island State Park offers 35 miles of non-motorized trails allowing a visitor to discover the island and its wildlife. Watchable wildlife viewing opportunities abound with a herd of 600 bison, many deer, coyotes, antelope, and waterfowl. Facilities include modern rest rooms, hot showers, picnic shelters, a group-use pavilion, a boat-launching ramp, marina, and a visitor center.

Antelope Island is comprised of some of the oldest rocks exposed in Utah. The Farmington Canyon Complex, more than 2.5 billion years old, dominates the southern two-thirds of the island. The rocks immediately overlying the Farmington Canyon Complex on Antelope Island were deposited sometime between 1,600 and 700 million years ago, and are assigned to the Perry Canyon Formation. The earliest of the Tertiary rocks, deposited on Antelope Island 60 to 45 million years ago, are known as the Wasatch Group. These rocks were deposited at the front of a mountain range as coalescing alluvial fans. Younger Tertiary rocks, called the Salt Lake Group, were deposited between 19 and 1.6 million years ago, and crop out only on the east side of Antelope Island south of Sea Gull Point, and on the tip of Ladyfinger Point on the island's extreme northern end. Quaternary tufa deposits, formed on the island as a result of wave and algal action from ancient Lake Bonneville, blanket the much older Farmington Canyon Complex. Sand, gravel, flag stone (slate), metals (copper and gold), and quartzite suitable for concrete aggregate are the chief mineral resources of Antelope Island. Fresh water gushes year round from more than 40 springs.

The primary challenge of the future will be to protect the island from its own popularity. Safeguarding the island's natural resources and the recreational opportunities promises to be an on going issue with potentially controversial solutions. Through park management's proactive solutions and community involvement, however, the timeless nature of the island's resources will be safeguarded into the future.

SELECTED BIBLIOGRAPHY

Doelling, H.H., Willis, G.C., Jensen, M.E., Davis, F.D., Gwynn, J.W., Case, W.F., Hecker, Suzanne, Atwood, Genevieve, and Klauk, R.H., 1988, Geology & Antelope Island State Park, Utah: Utah Geological and Mineral Survey in cooperation with Utah Division of Parks and Recreation, MP-88-2, 20 p.

Evans, Keith, 1998, Antelope Island & causeway - area description: Online, http://wasatchaudubon.org/tr_antelope_island.htm

Harward, Max, 1996, Where the buffalo roam - life on Antelope Island: [publisher unknown], 87 p.

Holt, C.J., 1994, History of Antelope Island: Syracuse, The Syracuse Historical Commission, 35 p.

Miller, D.E., 1997, Great Salt Lake - past and present: Salt Lake City, Publishers Press, 50 p.

Morgan, D.L., 1947, The Great Salt Lake: Albuquerque, University of New Mexico Press, 432 p.

Stokes, W.L., 1984, The Great Salt Lake: Salt Lake City, Starstone Publishing Company, 32 p.

Utah Division of Parks and Recreation, 2000a, Antelope Island State Park: Online, http://parks.state.ut.us/parks/www1/ante.htm.

—2000b, Antelope Island Resource Management Plan: Online, http://parks.state.ut.us/antelopeplan.htm.

APPENDIX - PARK HOURS, FEES, ACTIVITIES, SPECIAL EVENTS, AND AMENITIES

Hours: Antelope Island is open year round, as per the following current schedule:

May - August	7:00 a.m. to 10:00 p.m.
September	7:00 a.m. to 9:00 p.m.
October	7:00 a.m. to 7:00 p.m.
November - February	7:00 a.m. to 6:00 p.m.
March	7:00 a.m. to 7:00 p.m.
April	7:00 a.m. to 9:00 p.m.

The visitor center, completed in 1996, is open during the summer from 10:00 a.m. to 5:00 p.m., Monday through Friday, and during the winter, Monday through Friday, from 10:00 a.m. to 4:00 p.m. It is closed on Thanksgiving, Christmas, and New Year's Day. The center houses displays, exhibits, an art gallery, gift shop, and conference rooms.

Activities: Visitors to the island can enjoy numerous activities, including camping, horseback riding, picnicing, boat tours, sailing (figure 12 - Smith 13), biking, saltwater bathing, bird watching, photography, wildlife and geology viewing, and touring the Fielding Garr Ranch historical site. Amenities for the physically challenged include parking, restrooms, picnic shelters, and camp sites.

Special Events: Each year, the annual Bison Roundup takes place during the last weekend in October and first week in November. Visitors can also tour the Fielding Garr Ranch, attend star-gazing parties, and enjoy mountain biking and running events.

Concessions: Buffalo Point offers souvenirs, food, and drinks. Horse rental and rides, bike rentals, and boat tours are also available.

Boat-mooring: The marina at the north end of the island has 92 slips that will accommodate various sized boats. Slip rentals currently range from $100 to $140 per month, depending on slip length. The general $7.00 day-use fee covers the cost of using the launching ramp facilities.

General Fees: The current day-use fee for using the park is $7.00 per motorized vehicle; the walk-in fee is $4.00. Both fees include a Davis County Causeway fee of $2.00.

Camping facilities and fees: Camping areas have been established at both Bridger Bay and White Rock Bay; reservations are required to insure accommodations. For camping on Bridger Bay, call 1-800-322-3770, and for groups 25 or more at White Rock Bay, call 1-801-773-2941. At Bridger bay there are 26 RV/tent sites. Within a mile of Bridger Bay, each of the following are available at Bridger Bay beach: drinking water, modern rest rooms, hot-water showers, sewage disposal station, covered picnic tables/shelters, barbecue grills and fire pits. At White Rock Bay there are 5 group-camp sites that, combined, can accommodate up to 300 people, with the following facilities: picnic tables, vault toilets and fire rings.

The current camping fee at Bridger Bay is $10.00 per night. The maximum trailer length for back-in-sites is 35 feet, and for pull-through sites, the maximum trailer length is 60 feet. At the White Rock Bay group camp sites, there is currently a $25.00 minimum fee, with an additional $1.00 fee for each additional person over 25.

GREAT SALT LAKE, A BOATING JEWEL OF THE WEST

by

Steve Ingram

ABSTRACT

The Great Salt Lake has lured people to it, first to its shores, then upon its waters, and finally to its islands. Boating on the lake has spanned the time from when Fremont Indians first paddled the lake in crude canoes to the present time. During this time, bull boats, rafts, sail powered, steam-powered, and diesel-powered boats have been used to move people, animals, lumber, and ore about the lake for commercial or recreational purposes. The Great Salt Lake Yacht Club, founded in 1877, has promoted pleasure sailing and racing on the lake intermittently for 124 years. The greatest challenge to boating on the lake has been its constant change in surface elevation, which, during low-level periods, can affect boating activities by restricting access to its shores and harbors. Two harbors provide access to the lake. The largest of the two harbors is located at the south end of the lake at the Great Salt Lake State Marina. The smaller harbor is located at the north end of Antelope Island, as part of the Antelope Island State Park. Boaters on the Great Salt Lake must be aware of, and prepared for, severe storms that can occur suddenly.

MY INTRODUCTION TO GREAT SALT LAKE

The lure of a shimmering body of water has always acted as a magnet to a curious soul. History recounts many episodes of man's fascination with the world's vast oceans, seas, and lakes. Large lakes, in particular, attract inland explorers no matter how old or young they might be.

Great Salt Lake captured my attention in a vastly different way than, for instance, Jim Bridger and John C. Fremont, who explored the lake using an India-rubber boat in 1843 (Miller, 1980), or even Captain Howard Stansbury who used a crude sail boat and a flat-bottomed skiff to explore the lake (Morgan, 1947). My first introduction to the lake was at the very innocent age of eight while on a family outing to the then-popular Silver Sands area on the lake's south shore. Since then, 30 years of fluctuating lake levels, development, and changing land leases have vastly altered the overall look of the shore, but my memory of the water, sky, and islands remains unchanged. The crystal-clear water, warm with the solar heating of early summer, splashed against my legs as gentle waves, kicked up by a 10-knot (11.5 miles per hour) "sea" breeze. Many different types of sailing craft plyed the water against a backdrop of mysterious islands. Gazing across the water, I could only imagine sailing to remote places on this backyard sea, and bringing tales of high adventure to my landlubber friends back home!

The fascination of this beautiful lake also captured my father's imagination, and he chose to buy a small boat instead of a mountain cabin in 1970. Less than four years later, we were piloting our 22-foot sailboat over the lake's waters. Within a month of our learning to sail, we were sailing beyond sight of the safe waters of the marina. After two months, we began campaigning our little sloop, named Renegade, in the Great Salt Lake Yacht Club's regattas. Not only had the lure of the lake captured us, but its recreational possibilities had completely consumed our every thought. I personally was hooked, not just to sailing, but to existing in harmony with this huge new friend I had found so close to home. Few things in my life came as naturally to me as the sport of sailing.

My admiration of Great Salt Lake grew out of a natural love for the water. However, the enthusiasm that I held for my new friend, was shared by very few others. In fact, the lake, as close as it is to a major population area, was relatively untouched as a recreational destination. Over the last 30 years, I have found myself extolling the virtues of the lake to local residents, some of whom refuse to visit it and hold common misconceptions about it, which include: it stinks, it's dead, it's polluted, and it's too shallow to sail. The late Thomas C. Adams (figure 1) must have felt a similar frustration when faced with opposition as he lobbied the local Chamber of Commerce, and the State Parks' Commission to save the aging Saltair resort in the early 1960s.

VESSELS ON THE LAKE

Throughout time, the mystique of Great Salt Lake (GSL) has drawn the curious and adventurous to its shores. Within days of the Mormon pioneers' arrival in the Salt Lake Valley in 1847, the new settlers visited the lake, and were soon drawing plans for small working sailboats. In fact, Brigham Young commissioned the first vessel to be used in exploration of the eastern part of the lake in the fall of 1847. Although small, the Mud Hen was launched at the mouth of the Jordan River and ushered in the era of human commercial and recreational use of the lake (Van Alfen, 1995). Later, an ingenious launching/docking rail-track system was constructed at the south end of Antelope Island. Even today, more than 150 years later, the remains of this launching system are visible (figure 2).

Sail-Powered Vessels

For approximately the next 60 years after the pioneers arrived in the valley, the lake was home to many different

Figure 1. Thomas C. Adams looking at the Carson Cross, reportedly carved by Kit Carson in 1843, near the top of Fremont Island. Adams reinstated the Great Salt Lake Yacht Club, and created its bylaws in 1928 (photo courtesy of Utah State Historical Society).

types of vessels using the lake for commercial transportation of goods, ore, lumber, cattle, and human passengers. Until 1868, these vessels were exclusively sail powered. The size of the lake, and the remoteness of its islands, necessitated the use of large, seaworthy vessels, capable of carrying several hundred sheep or cattle at once. At least one of these, the Lady of the Lake (figure 3), built by the Miller brothers who owned Fremont Island, was of a very unusual design. It had high bows and two decks, making it suitable for transporting animals. This 50-foot vessel was built in 1876, and was well constructed for its day. The Lady of the Lake was later purchased by Judge Wenner and renamed the Argo. From 1886 until 1891, Judge Wenner homesteaded on Fremont Island and used the Argo to transport mail and supplies to his family over the six-mile stretch of water from the mainland to the island at least twice a month (Van Alfen, 1995).

Steam-Powered Vessels

Steam power was introduced on the lake in 1868 with an unusual side-wheel steamer called the Kate Conner (figure 4). Its poor design and light construction, however, caused it to have a short life on the rough waters of the lake. The largest vessel ever to navigate on the lake, the 150-foot City of Corinne, was launched on the Bear River near Corinne in 1871. The City of Corinne was designed and built much sturdier than her predecessor, the Kate Connor. Both vessels were used initially to haul lumber, ore, and other goods between Corinne, on the north end of the lake, and Lake Point, on the south end (Morgan, 1974; Van Alfen, 1995). This passage of nearly 60 miles was a challenge to the Kate Conner because of her side-wheel design and her limited hold capacity The stern-wheel design of the City of Corinne proved to be much more capable, and it operated for several years, providing commercial transportation on the lake until the fledgling railroads put her out of business in the late 1870s.

Figure 2. The Mud Hen's ingenious launching/docking rail-track system extending into the lake off the south end of Antelope Island (photo by Steve Ingram).

Figure 3. The Lady of the Lake, built by the Miller brothers, owners of Fremont Island, was later purchased by Judge Wenner and renamed the Argo (photo courtesy of Daughters of the Utah Pioneers).

Figure 4. The Kate Conner, an unusual side-wheel steamer, introduced steam power on the lake in 1868 (photo courtesy of Utah State Historical Society).

With the explosion of popular beach resorts beginning around 1870, the general public demanded to have access to, and beyond, the lake's vast shoreline. This demand caused the City of Corinne to be refitted into an excursion vessel and renamed the General Garfield in about 1877. The new name was based on a rumor that this boat carried future U.S. President Garfield on a private cruise (Morgan, 1974). For years, the General Garfield (figure 5) could be seen with its oil lights ablaze on nightly dining and dancing cruises, plying the water between the south shore and Antelope Island. Many smaller excursion boats were built and operated out of the larger lake resorts such as Saltair, Lake Park, and Garfield Beach, including the lake's first propeller driven vessel, the Whirlwind (figure 6).

Figure 5. The General Garfield moored aside the pier at Garfield Beach (photo courtesy of Utah State Historical Society).

Figure 6. The Whirlwind, the lake's first propeller-driven vessel, shown among the bathers at Garfield Beach (photo courtesy of Utah State Historical Society).

Steam power on the lake presented many special challenges because of the high salinity of the water. Heat exchangers were fabricated to catch and recirculate the condensed water off of the boilers to conserve as much fresh water as possible. Even then, these steam-driven vessels had to replenish their fresh-water supply at Antelope Island on their long, north-south runs.

Gas and Diesel Powered Boats

Gasoline-engine powered boats operated on GSL as early as 1902, when the Tiddley Addley became the pioneer of the Southern Pacific Railroad's GSL fleet during the construction of the Lucin Cutoff across the lake (Peterson,

2001). Gasoline-engine-powered, and later diesel-engine-powered boats have been used since that time. They were used during the construction of the Southern Pacific's causeway, during AMOCO's search for oil in the north arm of the lake during the late-1970s, and they are used today by both State and Federal agencies conducting research on the lake, by brine-shrimp cyst harvesters, and by others.

So many misconceptions abound when talking about boating on GSL. As captain of the Island Serenade, I hear my share. Among the top five would have to be the unnatural corrosive properties of the lake's saline waters. Even local residents have 'heard' that no one should ever attempt to operate an engine-powered vessel on the lake. My quick response to passengers' questions alludes to the fact that they indeed are a passenger aboard a metal-hulled vessel with diesel engines using lake water as a coolant.

The trick to avoiding problems, I explain, is routine maintenance and a healthy dose of fresh water when returning to the dock. Any vessel that can be operated in ocean water will do just fine in the GSL. A closed cooling system, utilizing a fresh water/coolant mixture is quite common for marine engines. This protects the engine block from the corrosive effects of salt water. Changing zinc sacrificial anodes within the heat exchanger and exhaust system, as well as on any underwater metals, extends the life of mechanical-propulsion systems greatly.

Figure 7. *The Island Serenade, shown with its captain, is a 65-foot, diesel-powered, steel-hull vessel. It was designed and built for the waters of Great Salt Lake, and was launched in February 1996 (photo courtesy of Wally Gwynn).*

Ironically, the 300-plus sailboats are the largest population of motor-driven vessels on the lake. Most of the sailors use outboard engines, especially on the smaller craft, and sailboats over 27 feet use inboard gas or diesel engines, with and without closed cooling systems. Some of the boats operating on the lake have been in service for years and have never experienced engine failures. Of course, the best chance for long-term service requires a closed cooling system, proper zinc anode placement, and routine anode replacement.

The bottom line is that any vessel can operate trouble free on the GSL, just as it can on any other body of salty or fresh water. If you perform proper maintenance, and rinse off excess salt, and there is no reason why a boating experience on this unique lake would harm the vessel or its propulsion systems in any way.

GREAT SALT LAKE YACHT CLUB

Yachting, or pleasure sailing, began on May 10, 1877, when a lake lover named David L. Davis founded the Great Salt Lake Yacht Club (GSLYC) and became its first commodore. Members included all crew members of the vessels Waterwitch, Petrel, Mary Askey, and an unnamed boat belonging to a Mr. Hudson. In 1879, two regattas, complete with gold and silver cups were held. Other cruises and regattas were held throughout the 1880s and 1890s (Great Salt Lake Yacht Club, 2001). The GSLYC actively sponsored sailing craft and rowing-vessel races until 1926 when the lake's elevation began to drop rapidly. This left many sailors high and dry with no deep-water moorage anywhere on the lake.

With the drop in the lake's surface elevation, the GSLYC virtually disbanded until 1928, when Thomas C. Adams headed an effort to build a boat harbor on the lake's southern shore with a deep-water channel leading to the main body of the lake. Once the harbor was completed in 1929 (figure 8), the GSLYC charter was reinstated by Adams using its original 1877 charter. In 1932, a $200 club house was opened beneath Saltair's south pier, and the GSLYC was legally incorporated (Great Salt Lake Yacht Club, 2001). As of the end of 2000, the operation of the Yacht Club under its 1932 articles of incorporation and its 1877 charter made it among the oldest yacht clubs in the nation. There was not a lot of sailing during the early 1960s due to the record low lake level, but the lake began to rise in 1964, and by 1970 the GSLYC was able to resume activities (Great Salt Lake Yacht Club, 2001). In 2000, the GSLYC's bylaws were reviewed and rewritten to make them both modern and consistent with current laws. In 2001, the GSLYC reincorporated under the new articles of incorporation (Darin Christensen, GSLYC, written communication, 2001).

A recent upswing in interest, related to cruising and junior-class events, has brought out a more varied population to GSLYC-sponsored events. Many GSLYC members and nonmembers are enjoying the social feeling of sharing weekends with other boaters who have a common attraction to GSL and its islands. Even though membership is down from its high mark of about 100 in the 1970s, the club has been able to retain the interest of the new boaters, and the ol' salts alike.

Figure 8. *View of the old Salt Lake County boat harbor (1939 to about 1980) (photo courtesy of Utah State Historical Society).*

CHANGING LAKE LEVELS

Closely woven into the pattern of recreation on the lake, is the lake's fickle manner. A small change in water inflow, or in water loss through evaporation, can cause a large and relatively sudden surface-elevation change in the lake. This change in elevation directly relates to many aspects of boating on the lake. For instance, if the lake's elevation were to fall only two feet below its mean level of 4,200 feet above sea level, 75 percent of its shoreline changes dramatically. Sailors would experience greater difficulty in approaching these shores, as well as coming much closer to submerged sand spits that extend outward miles from shore. Some sand spits are over three miles in length, and at a lake elevation of 4,200 feet, are only three to four feet under the surface. One can be sailing along, perceiving he is near the center of the lake in 28 feet of water, when suddenly, the boat skids to a halt in sand. A look over the side reveals only three feet of water. Even keeping a close eye on the depth sounder would not be enough in some areas to avoid grounding. On GSL, there is no substitute for experience, and a good navigation chart (Great Salt Lake Yacht Club, 1992). If the lake's elevation were to rise two feet, the boater would have about 400 more square miles of surrounding water. This change in area also affects the lake's daily wind patterns, seasonal precipitation, the amount of water available for evaporation, and the depth under a vessel's keel when approaching the shorelines.

Lake elevation frequently drops two feet between spring and fall of the same year, a small time frame for such a dramatic change. During the past 150 years, the lake's elevation has often changed more than two feet annually, especially in the south arm since the construction of the Southern Pacific Railroad causeway in 1959. For example, in 1969, 1971, and 1986, seasonal fluctuations of about three feet occurred, and in 1986 and 1987, the lake varied by over 4.5 feet. Prolonged lake-level fluctuations can affect boating activities tremendously, especially drops in the lake's elevation. From 1976 through 1979, for example, the lake dropped over 4.5 feet, significantly restricting boat access to the lake's shores and harbors.

PUBLIC SAFE HARBORS

The GSL has only two public safe harbors, which provide boaters with access to the lake. One is on the north end of Antelope Island as part of the Antelope Island State Park, and the other is at the south end of the lake, as part of the Great Salt Lake State Marina. Little Valley harbor, located on the north arm of the lake, was built by the Southern Pacific Railroad during the construction of the railroad causeway across the lake in 1959. This harbor is no longer maintained; additionally it rests on private ground rendering it inaccessible to the public.

The Antelope Island Marina (figure 9) offers solitude, good sailing, and starlit nights. It is a short drive from many northern Wasatch Front communities; only about 37 miles north from Salt Lake City. The marina has 92 year-round slips, intermittent concession services, and undrinkable water at the docks. Dry storage is allowed. The marina's breakwaters are low, and the strong winds that blow from the northwest towards Antelope Island can kick up nearly two foot-high waves inside the marina's breakwaters. Close to the marina, park facilities on the island include modern rest rooms, hot showers, picnic shelters, a group-use pavilion, concessions, and a visitor center.

The Great Salt Lake State Marina (figure 10) is located 18 miles west of Salt Lake City. It is almost a full-service coastal marina with a ship's store, visitor information center, ice, boat excursions, sailing lessons, boat rentals, and charters. This marina has 300 slips, all of which have fresh, potable water, and 80 percent of them have power available. Live-aboards, limited to 12 boats, are welcome in the marina (after application and acceptance) for an additional fee of

Figure 9. *View looking northeast of the Antelope Island Marina (photo courtesy of Wally Gwynn).*

Figure 10. View looking southwest of sail boats in the Great Salt Lake State Marina, located at the south end of Great Salt Lake (photo courtesy of Wally Gwynn).

$50. The marina also has a group pavilion, modern rest rooms, and areas for picnicking and swimming. Unlike the Antelope Island Marina, the Great Salt Lake Marina's breakwaters are nearly 20 feet tall. Boaters sometimes complain because the lack of wind inside the harbor makes it difficult for the purists to sail in or out of their slip. In years past, millions of dollars have been spent at the Great Salt Lake State Marina both to construct and maintain strong breakwaters for high lake levels, (figure 11) and to dredge deep-water channels for low lake levels.

Figure 11. Construction of the outer breakwater at the Great Salt Lake State Marina during the late1970s (photo from Utah Geological Survey photo archives).

STORMS

Because of its location within a large, wide basin, GSL is prone to sudden high winds. South winds may blow in excess of 45 knots (53 miles per hour) over open stretches of water for two to three days in advance of a storm front. These winds can create treacherous waves in a very short time. It is not uncommon to see short, steep-breaking waves up to 10 feet high in the deeper parts of the lake. Combine the lake's shallowness, having an average depth of 18 feet or less, its high water density, its open expanse, and heavy weather, and you have the formula for potentially dangerous conditions for vessels of any size or type. The waves build quickly with a steep face and deep troughs due to the shallow water and, unlike an ocean swell, these waves are spaced close together. The wary boater can sometimes outrun approaching weather (figure 12), but not always. If a storm hits, the boater may anchor to ride it out, but even so, the crew and their vessel might be battered and injured due to the violent motions of the boat in 8- to 10-foot seas.

Figure 12. Racing the storm (photo courtesy of Eugene Morgan, Great Salt Lake Yacht Club).

Operating a vessel, either sail or power, in rough conditions can be challenging. Boats have sunk and lives have been lost because of violent storms on the lake. Salt spray can often be thickly caked on rigging and lines, rendering them nearly useless unless washed often with fresh water. Because of this, sailors venturing out more than a few days need to carry plenty of water, as well as communication equipment, and a good anchor. One positive note about weathering storms on GSL, is that anchoring is possible nearly everywhere. With the proper anchor and enough rope and chain, even a large boat can ride out a storm "on the hook."

To venture out for more than five miles from harbor or for more than one day, the GSL boater needs to be as prepared as taking a coastal cruise on the ocean. The remoteness of the lake and its islands means that help is far away, or not available at all, making storms frightening for even experienced boaters. Night time can be especially scary in a storm, since one slip on a wet deck can send a crew member overboard in conditions that might make a rescue nearly impossible. Self-sufficiency is, therefore, a necessity for those who wish to visit and experience the tremendous beauty of the lake for long excursions.

ISLANDS

The islands of GSL are often the destination of an afternoon excursion, or an over night adventure. They can provide the boater with rocky headlands to explore, wildlife viewing, and clean white beaches. To the explorer and historian, the islands are outdoor museums full of untouched remnants and artifacts hundreds of years old, just begging for the camera lens of a visiting sailor. From the islands, or at anchor nearby, the cruising sailor can a enjoy sunsets that are truly world-class.

The seven major GSL islands truly have individual characteristics of their own. In the south arm, Antelope Island, the lake's largest at 28,000 acres, contains some of the most beautiful headlands, and exposed geologic features on its west shore. The cruising sailor can easily anchor in half dozen or more small bays with pristine, white sandy beaches. For wildlife viewing, this island is probably the best. The island is home to over 700 American bison, well over 200 mule deer, as well as about 60 prong horn antelope from which the island takes its name.

Sixteen miles to the west of Antelope Island lies the second largest island named Stansbury. This island is actually a peninsula at lake levels below 4,202 feet, and covers 22,000 acres. It is named for Captain Howard Stansbury who headed the comprehensive, government funded survey from 1849 to1850 (Morgan, 1947). Perhaps one of the most visited places on the lake for the overnight sailor would be Crystal Bay, located of the extreme northern tip of Stansbury Island.

Fremont Island lies almost 18 miles due north of Stansbury Island. This 2,900-acre island is named after John C. Fremont, who actually visited this island in 1843 after making a treacherous passage in his inflatable rubber boat with eight other men. A small cross carved by one of the expedition members, Kit Carson, can still be seen quite clearly on a rock face at the island's summit (see figure 1). This island provides boaters some of the most historic points of interest of any of the islands on the lake, such as the remnants of Judge Wenner's homestead on the south end, and the graves of Judge and Kate Wenner located nearby.

Carrington Island, the fourth largest at 1,700 acres, is located 5 miles northwest of Stansbury Island. The island was used as a precision bombing target for the Army Air Corps during World War II, and is still littered with crators and shrapnel. Boaters brave enough to approach this island's shores may then actually find unexploded bomb shells on their way to its summit, 500 feet above the lake's surface. From the island's summit, boaters have some of the finest views of any place on the lake. A small 22-acre rocky island, named Hat, lies 4 miles northwest of Carrington Island, and is one of two federally-designated habitats for nesting birds.

The islands in the north arm of the lake, Gunnison and Dolphin, are largely inaccessible to any type of vessel because of the Southern Pacific Railroad causeway. These two smaller islands, both less than 200 acres in size, provide valuable nesting habitat for migratory birds such as the Great White pelican, and California gulls. Gunnison Island's remoteness, and distance from the nearest shore, make it impossible for predators and very difficult for humans to reach.

CONCLUDING THOUGHTS

The weather and the mood of the lake are constantly changing. It is through these changes, both daily and seasonal, that those who truly enjoy and love the lake find their greatest rewards. Die-hard boaters have never been discouraged by the fickle nature of the lake and its climatology. They have always found ways to adapt, and remain true believers in the idea that their rewarding experiences on the lake cannot be compromised by weather, or the rise and fall of its waters.

All the qualities that make a safe boater, seaman, and the patient, adaptable outdoors lover must come together for the long-term enjoyment of boating and recreating on GSL. For 30 years, I have seen the human response to hot, sticky, calm, stormy, and brine-fly filled days. I have also seen the terror in the eyes of even experienced sailors during severe storms. I have seen the many difficult and unique situations that have been experienced, and then must be overcome in the minds of boaters, for them to return to the lake. For those that do return day after day, and season after season, the lure of the lake and its beauty keep drawing them back. The adversities, and variability of the lake are all accepted by the understanding boater who has a simple, total enjoyment of a very unique and beautiful place so close to home.

As the owner of the largest excursion-vessel company to operate on the lake's difficult waters, and as captain of the diesel-powered Island Serenade (figure 7), I have tremendous respect for my predecessors who were successful in their similar endeavors years ago. Dealing with the lake's salty, dense water, its dangerous storms and steep waves, and its reefs would be enough of a stumbling block for any enterprise. But, factor in the common misconceptions about the lake, and a considerable challenge is created. I am sure the captains of the early steamers were up against similar odds. The rewards of a successful cruise on the lake when one more passenger comes away with a deeper appreciation of GSL far outweigh the inconveniences of bad weather and salty water.

REFERENCES

Great Salt Lake Yacht Club, 1992, Great Salt Lake south arm - soundings in feet at datum lake level 4,200 feet: Great Salt Lake Yacht Club, map scale of 1:90,000.

Great Salt Lake Yacht Club, (October 2001), GSL Yacht Club History: Great Salt Yacht Club, Website, http://www.gslyc.org/history.html.

Miller, D.E., 1980, Great Salt Lake, a historical sketch, *in* Gwynn, J.W., editor, Great Salt Lake, a scientific, historical, and economic overview: Utah Geological and Mineral Survey Bulletin 116, p. 1-14.

Morgan, D.L., 1947, The Great Salt Lake: Albuquerque, New Mexico, University of New Mexico Press, 432 p.

Peterson, David, 2001, Tale of the Lucine - a boat, a railroad and the Great Salt Lake: Trinidad, California, Old Waterfront Publishing, 158 pages.

Van Alfen, P.G., 1995, Sail and steam - Great Salt Lake's boats and boatbuilders, 1847-1901: Utah Historical Quarterly, v. 63, no. 3, p. 196-221.

BONNEVILLE SALT FLATS

HISTORY OF POTASH PRODUCTION FROM THE SALDURO SALT MARSH (BONNEVILLE SALT FLATS), TOOELE COUNTY

by
J. Wallace Gwynn

Reprinted from Survey Notes, V. 28, no. 2, April 1996

The first 10 to 12 miles of Interstate 80 east of Wendover, Utah traverse the seemingly endless, flat, white, salt-covered expanse of the Bonneville Salt Flats, known in the early 1900s as the Salduro Salt Marsh. The salt flats and the surrounding Great Salt Lake Desert are remnants of the bed of an ancient, large, cyclic lake whose latest cycle, called the Lake Bonneville cycle, occurred from about 32,000 to 14,000 years ago. Lake Bonneville had a depth of more than 1,000 feet, and covered an area of 20,000 square miles in western Utah plus small portions of southern Idaho and eastern Nevada. Though the water of Lake Bonneville was relatively fresh, it contained small amounts of dissolved salt, including chlorides and sulfates of sodium, potassium, and magnesium. These dissolved salts precipitated on the surface of the Salduro Salt Marsh as the lake evaporated and are the source material from which potash is produced.

The Salduro Salt Marsh in particular played an important role in the early development and production of potash or potassium chloride in the United States. The major use of potash is as a fertilizer; lesser amounts are used in the chemical industry. Prior to World War I (1914-1918), Germany supplied nearly all of the growing demands for agricultural and industrial potash in the U.S. With the blockade of Germany by Great Britain during the war, however, these supplies were cut off, and the U.S. was forced to find alternative sources of potash. By the end of the war, at least 128 U.S. plants were producing potassium compounds from kelp, wood ashes, lake brines, alunite, cement dust, sugar-beet waste, blast-furnace dust, and other sources.

As early as 1906 or 1907, the existence of the salt beds of the Salduro Salt Marsh was brought to national attention by the engineers building the Western Pacific Railroad across the western Utah desert. The salt beds were soon covered by mining claims, and almost immediately the claim owners organized the Montello Salt Company, headquartered in Ogden, Utah. After several years of unprofitable attempts to produce salt, the claims were leased to the Capell Salt Company of Salt Lake City, Utah. Capell erected a small mill near Salduro, about 10 miles east of Wendover, and produced and sold common salt for a short time.

In about 1916, the Capell Salt Company merged into or was transferred to the potash enterprise of the Solvay Process Company (headquarters unknown). During the war years, the Solvay Process Company investigated many saline deposits and in 1916 began to extract potash from the subsurface brines of the Salduro Salt Marsh. The operation was constructed on the south side of the Western Pacific Railroad at Salduro station. There, in the center of the Salduro Marsh salt beds, concentric, circular canals were dug into the salt

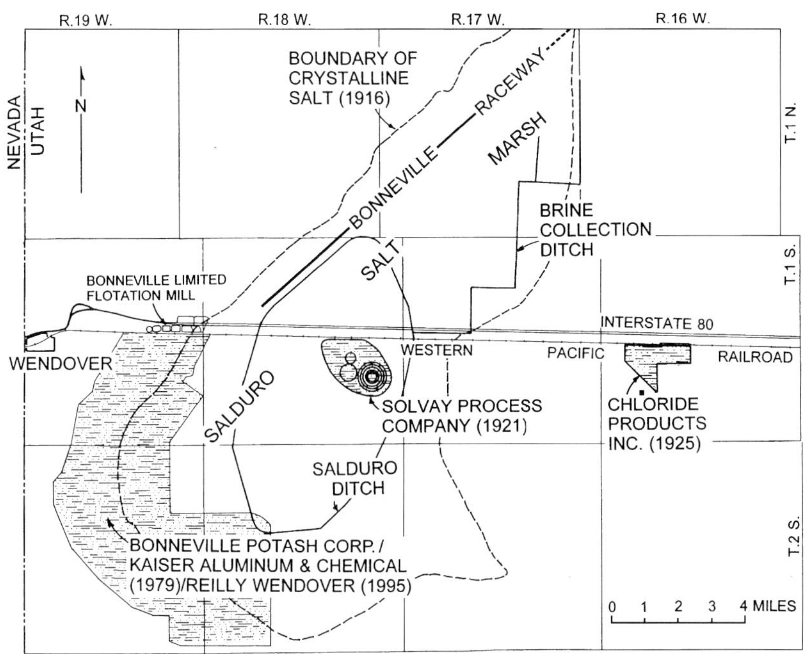

Figure 1. *Location of Salduro Salt Marsh.*

Figure 2. *Solar evaporation pond covered by precipitated salt, Salduro Salt Marsh (1921-1925, Chloride Products, Inc.).*

Figure 3. Wide-tired car used for transportation over the salt marsh (1921-1925, Chloride Products, Inc.).

Figure 4. Pump (elevator) used to move brine from ditch into solar-evaporation pond, (1921-1925, Chloride Products, Inc.).

Figure 5. Trenching machine building solar-evaporation-pond dikes (1921-1925, Chloride Products, Inc.).

and underlying muds. Salty water, or brine, flowed into these canals, where it concentrated through the process of solar evaporation. The most concentrated brines were continually pumped inwards, over the dikes separating the outer from the inner concentric canals. Potassium-bearing salts precipitated from the highly concentrated brine within the innermost canal. From there, the salts were harvested and processed to produce potassium chloride.

Production of potassium chloride began in 1917 and, at the end of 1918, Solvay transferred its interest in the potash operation to the Utah-Salduro Potash Company (USPC). By 1920, USPC was the largest single producer of potash in the United States. In 1921, the plant was suddenly closed, due mainly to the reduction of high war-time potash prices and the reorganization of the activities of the Solvay Process Company. After that time, USPC restricted its operations to the production of common salt.

In 1919, the Bonneville Potash Corporation (BPC), formed by J.L. Silsbee of Salt Lake City, erected a potash plant at Wendover, Utah, near the Utah-Nevada border. During the period 1920 to 1936, BPC unsuccessfully attempted to commercially produce potash through the solar evaporation of brines. In 1936, a new operating company, Bonneville Limited, was formed and it built a new plant to recover potassium chloride by flotation-recovery from solar-precipitated salts. The first potash from this new plant was shipped in 1938. By 1939, Bonneville Limited was successfully producing potash and went on to become a significant, long-term potash supplier. Since that time, the operation has survived several ownership changes, and now operates under the name of Reilly Wendover.

A third company, Chloride Products Incorporated (CPI), formed in 1921 by Frank Cook and a group of California capitalists, also attempted to produce potash from the brines of the Salduro Salt Marsh. CPI constructed canals, evaporation ponds, and a small processing plant near Arinosa, a few miles east of the Utah-Salduro operations. CPI's developmental work is recorded through at least 1925, after which no further information is available concerning their activities.

The growth and development of the potash industry on the Salduro Salt Marsh faced many challenges. The post-war decrease in potash prices made it increasingly difficult for the Utah potash companies to compete with domestic and foreign suppliers. The surface conditions on the marsh were another critical factor. During spring, the surface was normally covered with water, hindering development work, and wave action frequently destroyed dikes and filled the brine-collection ditches with sediment. Also, heavy equipment frequently broke through the salt crust and sank into the underlying mud, necessitating the invention and use of special wide metal wheels on the equipment. Hot, dry summers and cold winters, accentuated by the ever-present wind, made working conditions on the Salduro Salt Marsh unpleasant.

The early production of potash from the brines of the Salduro Salt Marsh by the Solvay Process/Utah-Salduro Potash Company played an important and sometimes singular role in supplying the U.S. with fertilizer during the latter part of World War I. In spite of other U.S. suppliers, international competition, and monumental economical, logistical, and climatological obstacles, the potash industry on the Bonneville Salt Flats has survived. Today, Reilly Wendover, the potash industry's lone survivor on Utah's West Desert, is an important contributor to potash production in the United States, and to the economic base of both Tooele County and the State of Utah.

INVESTIGATION OF HYDROLOGY AND SOLUTE TRANSPORT, BONNEVILLE SALT FLATS, NORTHWESTERN UTAH

by

James L. Mason and Kenneth L. Kipp, Jr.
U.S. Geological Survey

ABSTRACT

The Bonneville Salt Flats, located in the western part of the Great Salt Lake Desert in northwestern Utah, appeals to land users of conflicting interests. The mostly smooth, hard surface of the salt crust has been used for high-speed automobile racing, whereas, brine has been removed from the underlying aquifer for mineral production. A decrease in thickness and overall extent of the salt has been documented from 1960 to 1988. A study to determine the natural and human-induced processes causing the loss of salt from the crust was completed by the U.S. Geological Survey in cooperation with the Bureau of Land Management. Data collected during the study were used to define the hydrologic system and determine potential causes for salt loss. The water-table configuration for the shallow-brine aquifer indicates that the primary source of fluid recharge is infiltration of precipitation on the playa surface, which dissolves salt as it seeps into the subsurface. Discharge from the shallow-brine aquifer is primarily through evaporation and through ground-water seepage to the brine-collection ditches to the east and south of the salt crust. Extensive flooding of the Bonneville Salt Flats during the winter of 1992-93 dramatically changed the salt surface. An estimated 10 to 14 million tons of salt was dissolved and redeposited on the playa surface or infiltrated into the shallow-brine aquifer. Variable-density ground-water flow simulations were used to determine that brine withdrawal is a major cause of salt loss from the crust.

INTRODUCTION

The Bonneville Salt Flats study area is located in the western part of the Great Salt Lake Desert in northwestern Utah, about 110 miles west of Salt Lake City (figure 1). The study area lies within the Great Basin part of the Basin and Range physiographic province (Fenneman, 1931, pl. 1). The salt crust is unique for its mostly smooth, hard surface, size, and accessibility. To protect these characteristics, the Bureau of Land Management designated the Bonneville Salt Flats as an Area of Critical Environmental Concern in 1985.

The salt crust has been used for high-speed automobile racing since 1914. Members of the racing community have been concerned about the progressive deterioration of the racing surface. They have suggested that brine withdrawal for mineral production has caused the decreasing thickness and extent of the salt crust.

Mining of halite from the Bonneville Salt Flats began in the early 1900s. Extraction of potash from brine beneath the Bonneville Salt Flats began in 1917, when supplies of potassium salts from Germany were interrupted during World War I. The area from which brine is withdrawn for potash production was expanded in 1963, when mineral leases on 25,000 acres of Federal land were issued.

A decrease in thickness and overall extent of the salt crust on the Bonneville Salt Flats during 1960-88 has been documented (S. Brooks, Bureau of Land Management, written communication, 1989). Maximum salt-crust thickness was 7 feet in 1960 and 5.5 feet in 1988. No definitive data were available to identify and quantify the processes that cause salt loss. More than 55 million tons of salt are estimated to have been lost from the salt crust during the 28-year period.

The Bureau of Land Management needed to know the causes of salt loss to make appropriate management decisions. To meet this need, the U.S. Geological Survey (USGS), in cooperation with the Bureau of Land Management, completed a study designed to define the hydrology of the ground-water system and its relation to natural and human-induced processes that cause the loss of salt from the crust. Because brine is withdrawn for mineral production from the shallow-brine aquifer, and the near-surface hydrology has the greatest effect on the salt crust, the shallow-brine aquifer was studied in detail. Specific areas of study included solute transport by ground-water flow and movement of ponded surface water. The term solute, as used here, refers to salt dissolved in water.

Data collected by the USGS (Mason and others, 1995) during 1991-93 were used to make hydrologic interpretations and to develop computer simulations to achieve a better understanding of fluid flow and salt-transport processes in the shallow-brine aquifer. Locations of hydrologic data collection sites are shown in figure 2. A complete description of the study, interpretations of hydrology and brine chemistry, and results of model simulations are presented in Mason and Kipp (1998a). A synopsis of this interpretative report is presented by Mason and Kipp (1998b). All data, figures, tables, and interpretations presented in this manuscript have been published in these two reports unless cited otherwise.

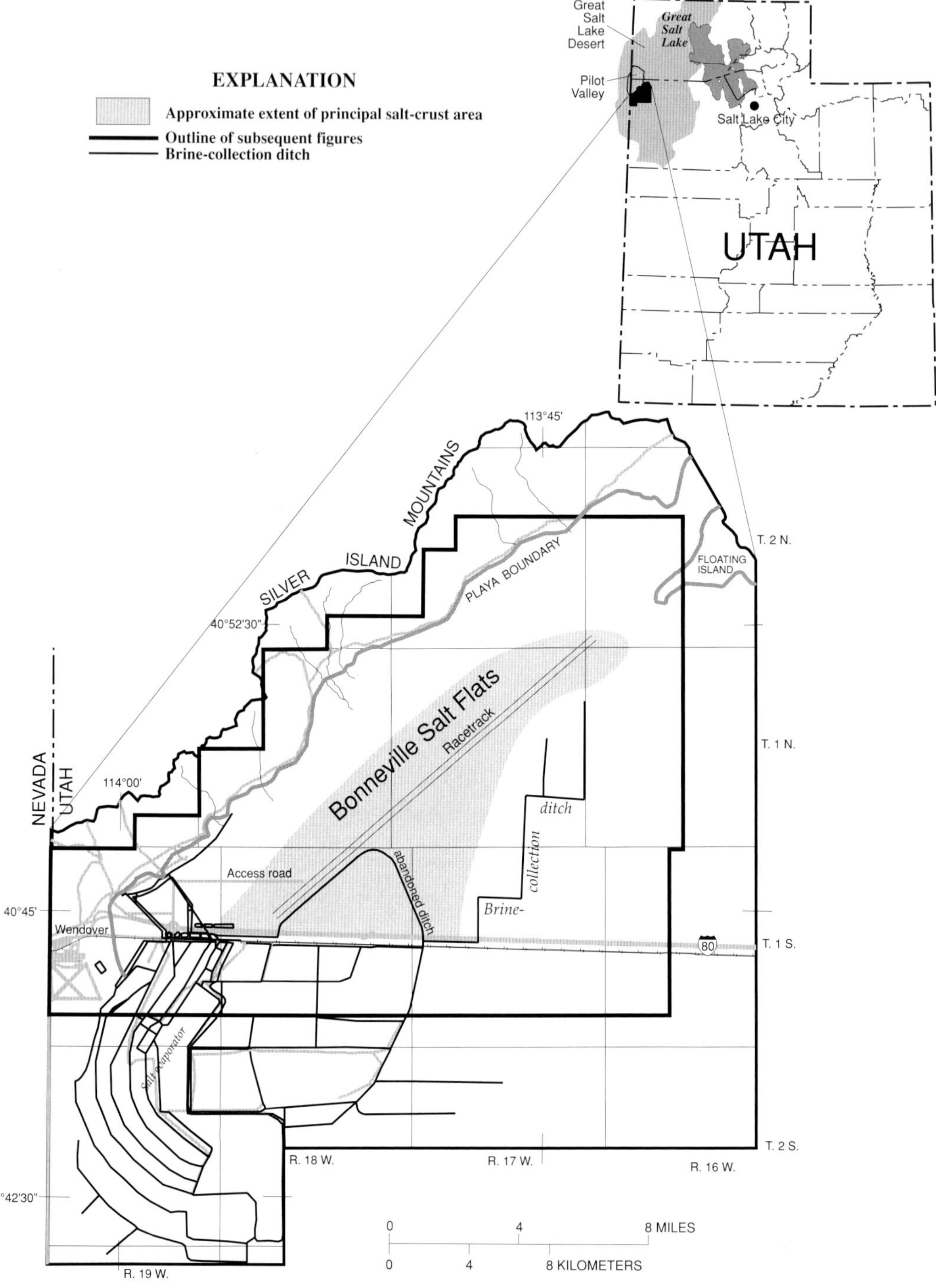

Figure 1. *Location of the Bonneville Salt Flats study area, northwestern Utah.*

Previous studies of the hydrology of the Bonneville Salt Flats were done by Turk (1969, 1978) and Lines (1979). Data used by Turk (1969) were limited to areas adjacent to brine collection ditches and evaporation ponds. Turk (1978) was the first to use satellite imagery to delineate the seasonal variation in salt-crust extent and to develop a simple, two-dimensional, ground-water flow model of the shallow-brine aquifer. Lines (1979) estimated the volume of brine and amount of dissolved salt that flows into and out of the shallow-brine aquifer.

PLAYA MORPHOLOGY

The Bonneville Salt Flats playa is a topographically low, flat area where evaporation is the only form of water loss from this arid region. The playa is flanked by the Silver Island Mountains on the northwest, and the land surface gradually slopes upward for many miles in all other directions. The salt crust and associated near-surface deposits in the Bonneville Salt Flats playa are the result of desiccation of late Pleistocene-age Lake Bonneville. Lake desiccation can

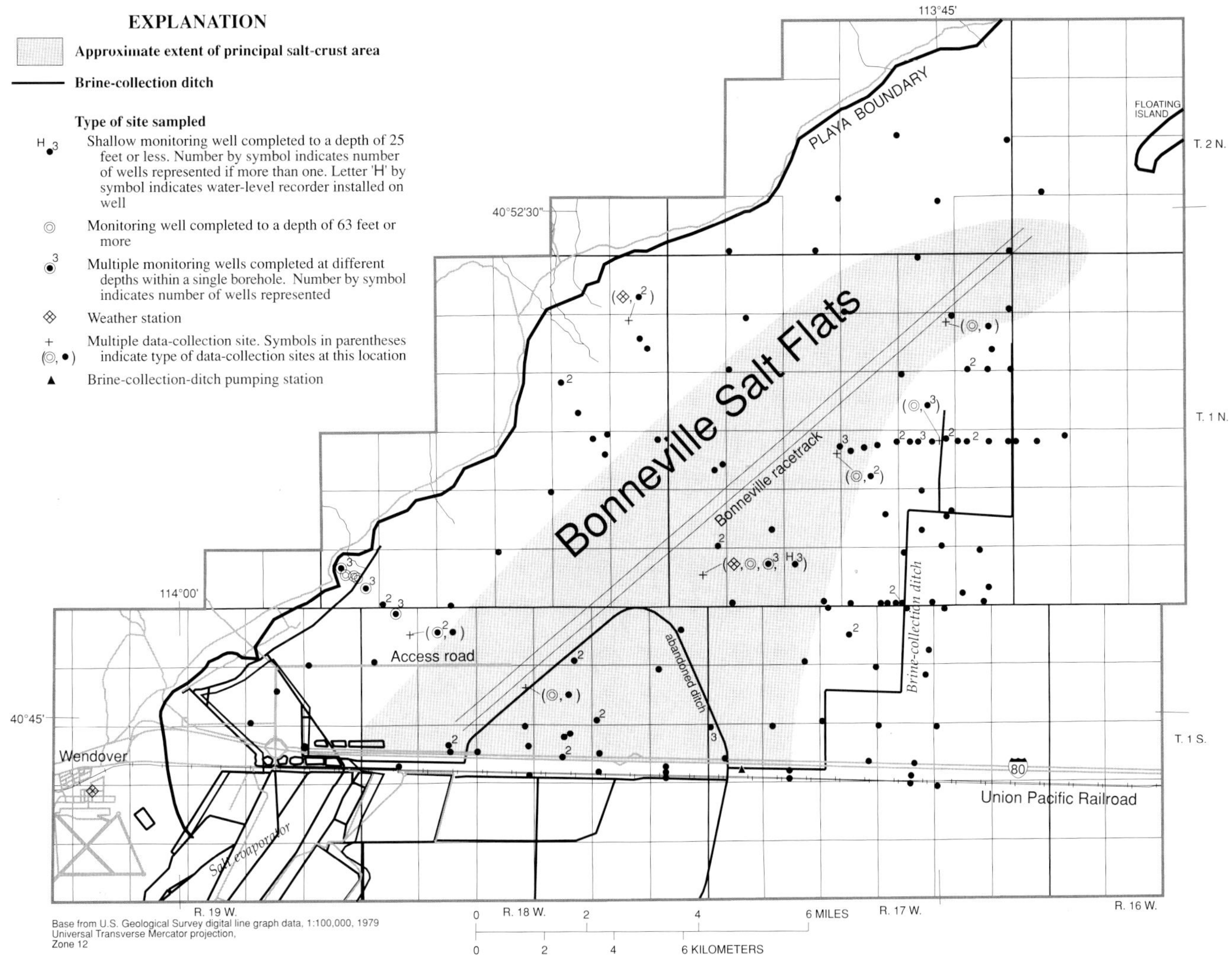

Figure 2. *Location of hydrologic data-collection sites, Bonneville Salt Flats, northwestern Utah.*

result in evaporite mineral zonation as described by Hunt and others (1966, p. B46-B48). Mineral zonation in the Bonneville Salt Flats is divided into three zones; the carbonate, sulfate, and chloride zones. Carbonate minerals, primarily aragonite, are the least soluble and precipitate first, followed by sulfate minerals (primarily gypsum) and finally chloride minerals (primarily halite). Because halite is the most soluble mineral, the extent of the salt crust, which covers about 50 square miles, can vary yearly as a result of salt dissolution during surface-pond formation and movement in the winter, and salt deposition during evaporation in the summer.

Situated at the lowest altitude in the study area, the salt crust generally is the final destination for surface runoff. Runoff from the flanks of the adjacent Silver Island Mountains generally occurs during intense summer thunderstorms and flows in distinct channels from the mountains toward the playa. These flows cross the alluvial fans and seep into the subsurface before reaching the playa, either rapidly through fractures or more slowly when ponded. Surface flow from the northeast, east, and south during wetter-than-normal conditions can migrate from several miles to the salt-crust area. Slight changes in altitude of the playa surface (less than 1 foot), however, can create small depressions where saline water can accumulate. Surface flow toward the salt crust is sometimes temporarily impeded by these storage depressions, but they are easily breached when the water level rises as a result of additional precipitation or when strong winds move sheets of water across the playa surface. Some of the runoff toward the salt crust is impeded by human-made structures such as the brine-collection ditch and associated berm located east of the salt crust, and the embankments created for Interstate Highway 80 and the railroad tracks to the south.

GROUND-WATER HYDROLOGY

Ground water in the Bonneville Salt Flats study area is present in an alluvial-fan aquifer along the northwestern margin of the playa, in a basin-fill aquifer, and in a shallow-brine aquifer. The spatial relation of the three aquifers is shown in figure 3. Brine of economic value is present primarily in the shallow-brine aquifer, but brine also has been withdrawn by wells completed in the deep part of the basin-fill aquifer. Ground-water density in the three aquifers varies from less dense brackish water in the alluvial-fan aquifer to brines in

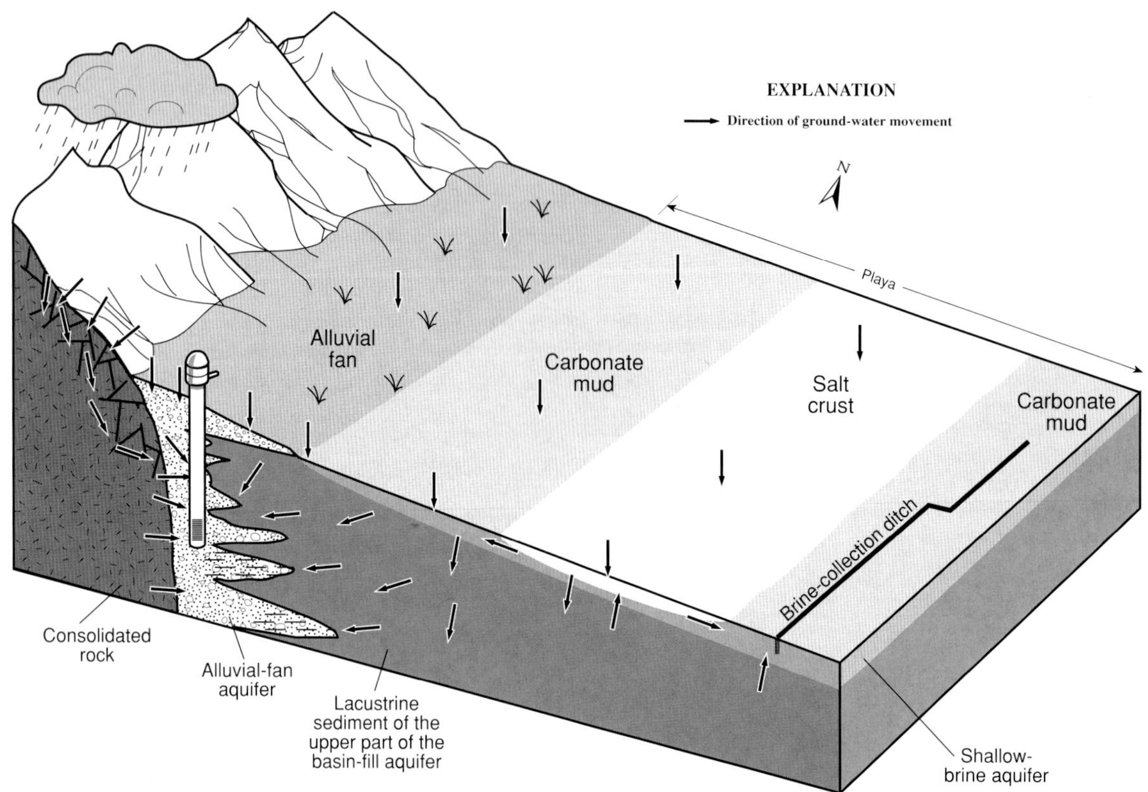

Figure 3. *Generalized block diagram showing the relation of aquifers and direction of ground-water flow in the Bonneville Salt Flats, northwestern Utah.*

the basin-fill and shallow-brine aquifers. The density of water in the alluvial-fan aquifer varies from about 1.001 grams per cubic centimeter (g/cm^3) near the Silver Island Mountains to about 1.080 g/cm^3 closer to the salt crust. The alluvial-fan aquifer readily yields water to wells and is used as process water for mineral production. The density of brine in underlying Lake Bonneville sediment of the upper part of the basin-fill aquifer is generally about 1.110 g/cm^3. The density of brine in the shallow-brine aquifer varies from a maximum of 1.202 g/cm^3 in and immediately below the salt crust to a minimum of 1.130 g/cm^3 farther from the salt crust. The density of brine in the shallow-brine aquifer generally decreases by dilution with infiltration of fresher water during winter and spring and increases with the concentrating effect of evaporation during summer.

The shallow-brine aquifer is composed of a halite and gypsum surface crust in the lowest part of the playa and is surrounded by carbonate mud, both on the surface adjacent to the crust and immediately beneath the crust. The salt crust can vary in thickness from about 1 inch to more than 5 feet. Vertical fractures in the shallow-brine aquifer are exposed along the sides of brine collection ditches and appear as salt ridges in a polygonal pattern on the salt-crust surface. These fractures, which can be up to an inch wide, and the more porous salt crust are probably the principal conduits through which brine moves. The spacing between fractures varies; a 1-foot interval is common.

The sand and gravel of the alluvial-fan aquifer are interbedded with the carbonate mud of the shallow-brine aquifer and lake sediment of the upper part of the basin-fill aquifer. As these interbedded layers extend toward the playa, the coarser material becomes intermixed with fine-grained sediment. The horizontal and vertical extent of these interbedded layers, and thus, the hydraulic connection with the shallow-brine aquifer and the upper part of the basin-fill aquifer are unknown.

Water-level and brine-density data were used to map the seasonal water-table surfaces of the shallow-brine aquifer. Because of varying density of brine in the shallow-brine aquifer, water level measurements must be normalized to an average density for the two-dimensional representation of the water-table. A more detailed explanation of variable-density flow is presented in Mason and Kipp (1998a, p. 19-22). The water-table configuration (figure 4) represents conditions near the end of the evaporation season, after most of the brine used for mineral production has been withdrawn through the brine-collection ditches. The general configuration of the water table is very similar to that mapped by Lines (1979, figure 35) for the same time of year. This comparison indicates that there has been no substantial long-term change in water levels within the shallow-brine aquifer despite seasonal and yearly fluctuations in fluid inflows and outflows. The horizontal movement of brine through the shallow-brine aquifer is from the center of the playa, where ground-water levels are high, toward the northwestern margin of the playa and toward the brine-collection ditches on the east and south, where water levels are low. Brine that moves in the shallow-brine aquifer from the east and southeast is intercepted by the brine-collection ditch before reaching the salt-crust area.

Water-level and density data collected at five locations during 1993, representative of wetter-than-normal conditions, indicate that a potential for downward movement

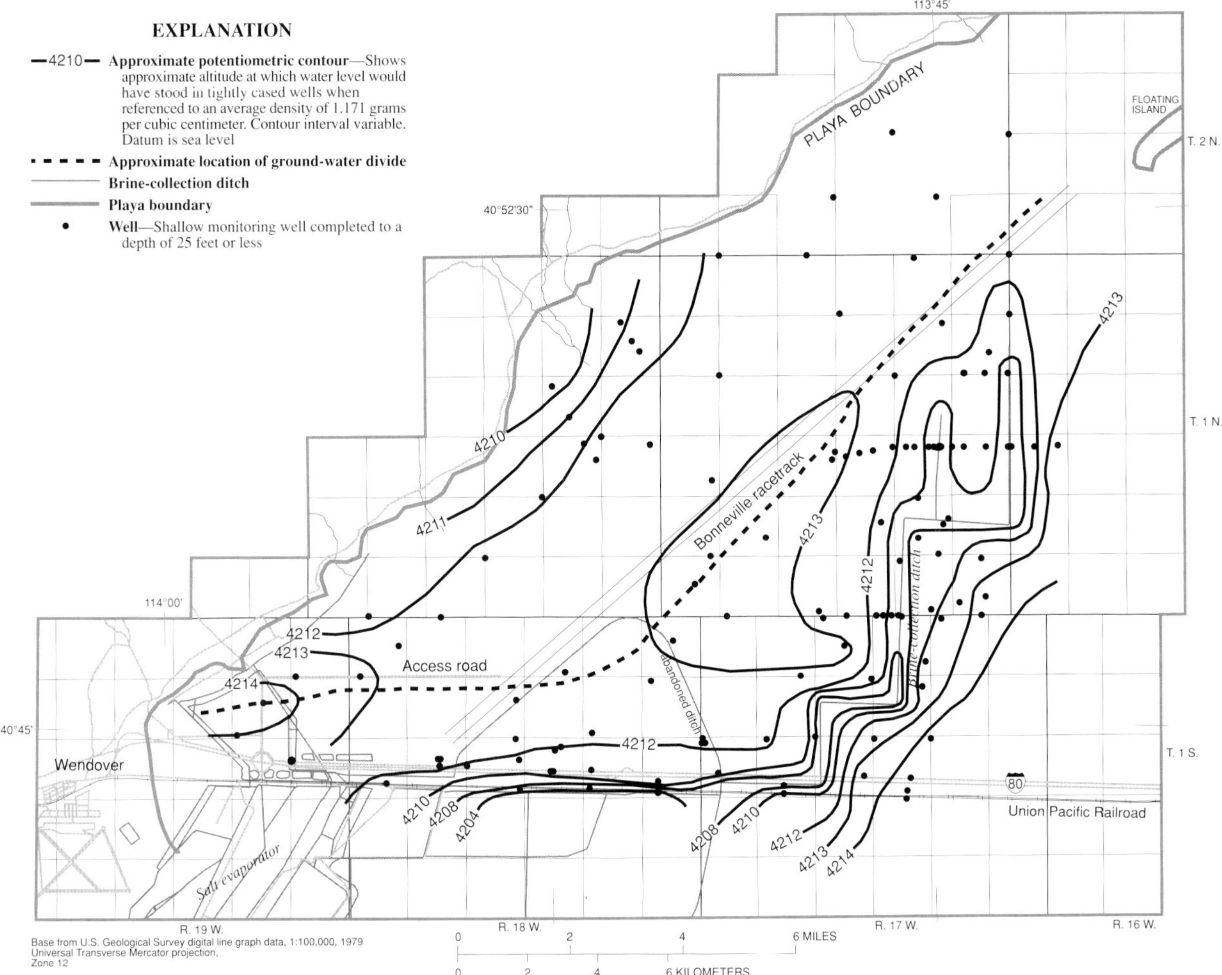

Figure 4. *Approximate potentiometric surface, referenced to an average density, in the shallow-brine aquifer of the Bonneville Salt Flats, northwestern Utah, September 1992.*

existed between the shallow-brine aquifer and the upper part of the basin-fill aquifer. Despite measurements that indicate potential downward movement, contradictory geochemical evidence indicates that brine may move upward near the center of the salt crust.

Tritium values were determined for brine collected from several wells completed in the shallow-brine aquifer and one well completed in the lake sediment of the upper basin-fill aquifer. Tritium is an unstable isotope of hydrogen that decays with a half-life of 12.3 years. (A tritium value for ground water that is isolated from mixing with other water will decrease by one-half of its original value after 12.3 years.) Tritium occurs naturally in the atmosphere, but the largest source was generated during atmospheric nuclear testing from 1952 to 1969. Atmospheric tritium values during sampling (1992-93) are almost the same as those measured prior to atmospheric nuclear testing, about 10 tritium units. Tritium values measured in brine collected from the shallow-brine aquifer during this study range from 0 to 24 tritium units, whereas a tritium value measured in brine from the upper part of the basin-fill aquifer was 0. Tritium values of about 10 tritium units measured in brine collected from the shallow-brine aquifer are indicative of water that seeped into the subsurface within a few years prior to sampling. Tritium values of greater than about 10 tritium units are indicative of water that has seeped into the subsurface since atmospheric nuclear testing and several years prior to sampling. Tritium values greater than 10 tritium units were in brine collected from sampled wells located horizontally farthest from the center of the salt crust. Of greatest importance are tritium values of almost 0 tritium units measured in brine collected from wells completed in the shallow-brine aquifer beneath the center of the salt crust. These values indicate mixing with brine that has a tritium value of 0 tritium units. The only possible source for this brine is upward movement from the underlying lake sediment in the upper part of the basin-fill aquifer.

Recharge to the shallow-brine aquifer includes infiltration of precipitation that falls on the playa surface, subsurface inflow from adjacent areas to the north and east, and possibly upward leakage. Fresher water from the playa surface dissolves and transports salt from the crust as it infiltrates into the shallow-brine aquifer. Subsurface inflow transports dissolved salt from the north and east along with upward movement from the underlying lake sediment in the upper part of the basin-fill aquifer, if it occurs, into the shallow-brine aquifer.

The amounts for recharge by infiltration and subsurface inflow during 1992 are presented in table 1. Estimates for associated salt transport were determined by multiplying the

Table 1. Estimated fluid flow and salt transport into and out of the shallow-brine aquifer, Bonneville Salt Falts, northwestern Utah.

	1976 (Lines, 1979)		1992 (this study)	
Component	**Acre-feet**	**Tons of solute**	**Acre-feet**	**Tons of solute**
Recharge				
Subsurface inflow from east and northeast	40	14,600	60	21,900
Infiltration from playa surface	11,740	--	8,300 - 12,800	--
Total (rounded)	11,800	14,600	8,400 - 12,900	21,900
Discharge				
Subsurface outflow to south	2,000	780,000	1,800 - 2,300	700,000 - 895,000
Subsurface outflow to brine-collection ditch	680	264,000	1,500	580,000
Subsurface outflow to northwest	70	27,200	35	13,600
Downward leakage	--	--	58 - 580	22,600 - 226,000
Evaporation	10,900	0	15,000	0
Total (rounded)	13,600	1,071,000	18,400 - 19,400	1,316,000 - 1,715,000

[-- No estimate]

volume of flow by an average density for that flow. These estimates compare favorably with those made by Lines (1979). This is reasonable because the ground-water system appears to be unchanged over the long term despite seasonal and yearly variations.

Discharge from the shallow-brine aquifer is primarily by evaporation, which causes crystalline salt to precipitate on the playa surface. Additional discharge from the shallow-brine aquifer is by ground-water outflow to the south, primarily to a brine-collection ditch, and ground water seepage to the brine-collection ditch east of the salt crust. Brine obtained from collection ditches is used for mineral production. Ground-water outflow to the adjacent alluvial-fan aquifer along the northwestern margin of the playa and downward movement to the underlying lake sediment in the upper part of the basin-fill aquifer are relatively minor components. All subsurface outflows transport salt out of the shallow-brine aquifer in the study area. Evaporation from the shallow-brine aquifer in 1992 was estimated to be about 15,000 acre-feet, considerably more than that estimated by Lines (1979). Estimated subsurface outflow to the south ranged from 1,800 to 2,300 acre-feet. Seepage to the brine-collection ditch east of the salt crust was estimated to be 1,500 acre-feet from data collected at a pumping station south of Interstate Highway 80. This volume of brine was more than double the 680 acre-feet reported by Lines (1979, p. 85). Estimated downward leakage into the underlying lacustrine sediment ranged from 58 to 580 acre feet, and subsurface outflow to the northwestern margin was about 35 acre-feet.

Estimated discharge from (18,400 to 19,400 acre-feet) and recharge (8,400 to 12,900 acre-feet) to the shallow-brine aquifer for 1992 were not equal as would be expected for a system under equilibrium conditions. The difference is partly the result of uncertainty in values estimated for the hydrologic properties that describe the shallow-brine aquifer. Subsurface outflow to the south was estimated using water-level data collected in 1992 and transmissivity values determined prior to construction of Interstate Highway 80. Transmissivity is a measure of the capacity for water to pass through an aquifer. Removal of salt crust and compaction of underlying carbonate mud during highway construction probably resulted in a decrease in aquifer transmissivity, thereby decreasing discharge. Also, estimated recharge could be greater if the estimated storage capacity of the shallow-brine aquifer were larger and the amount of upward movement from the underlying lacustrine sediment were known.

Comparison of water-chemistry data indicates that the dissolved-solids and potassium concentrations in brine from the shallow-brine aquifer have not changed appreciably from 1976 to 1992. Despite the loss of salt by subsurface outflow to the brine-collection ditches, the high dissolved-solids concentration has been maintained primarily by dissolution of the salt crust and possibly by the addition of salt through upward movement of brine from the underlying lake sediment in the upper part of the basin-fill aquifer. Similarly, the potassium concentration probably is maintained by diffusion from pore fluid within the carbonate mud and clays into the brine that migrates through fractures and possibly by upward movement of brine from the underlying basin-fill aquifer.

POTENTIAL FOR SALT LOSS BY SEASONAL PONDING

Extensive flooding on the playa surface during the winter and early spring of 1992-93 dramatically changed the salt surface. The flooding was the result of greater-than-normal precipitation during January and below-normal temperatures during January and February. Because of these climatic conditions, a large amount of surface water, primarily from the south and east of the study area, moved toward the playa. The density of this inflow ranged from 1.014 to 1.030 g/cm^3.

The amount of this inflow could not be estimated and, therefore, the amount of salt added to the salt-crust area is unknown. Much of the salt in the upper crust was dissolved by this inflow and transported north beyond the areal extent of the crust as it existed during 1992. Salt eventually precipitated on the playa surface as the pond evaporated in the summer of 1993. Satellite imagery was used to determine the extent of the salt crust, which was 43 square miles in September 1992 and 58 square miles in August 1993 (figure 5). The newly redeposited salt extended about 3 miles beyond the brine-collection ditch to the northeast. The amount of salt that was dissolved and subsequently redeposited on the playa surface or that was transported by infiltration into the shallow-brine aquifer was estimated to be about 10 to 14 million tons. Some salt might have been deposited beyond topographic divides where it was not able to coalesce with the salt crust during subsequent dissolution and overland flow. The amount of salt that might have been permanently lost from the crust by this type of transport could not be determined; however, extensive flooding, which also occurred during 1983-84, has the potential to transport millions of tons of salt beyond the salt-crust area.

SIMULATION OF VARIABLE-DENSITY GROUND-WATER FLOW

In a ground-water flow system in which the density of the water varies spatially, the movement of ground water and the spatial variations in the solute concentration are closely related. As the amount of solute increases, the density of the water also increases. Changes in density will cause changes in ground-water flow velocities, both speed and direction. The only practical way to compute the flow of ground water with variable density is by computer simulation. The three dimensional Heat and Solute Transport (HST3D) computer code (Kipp, 1987, 1997) of the U.S. Geological Survey was the simulator used to model brine movement through the shallow-brine aquifer. This simulator was chosen because of its ability to model variable-density flow and solute transport in three dimensions. Effects of temperature variation on the density of the brine were assumed to be negligible for the purposes of this study and therefore, only ground-water flow and associated solute transport were simulated.

The goals of model simulations were to (1) develop a fluid and solute balance for the shallow-brine aquifer, (2)

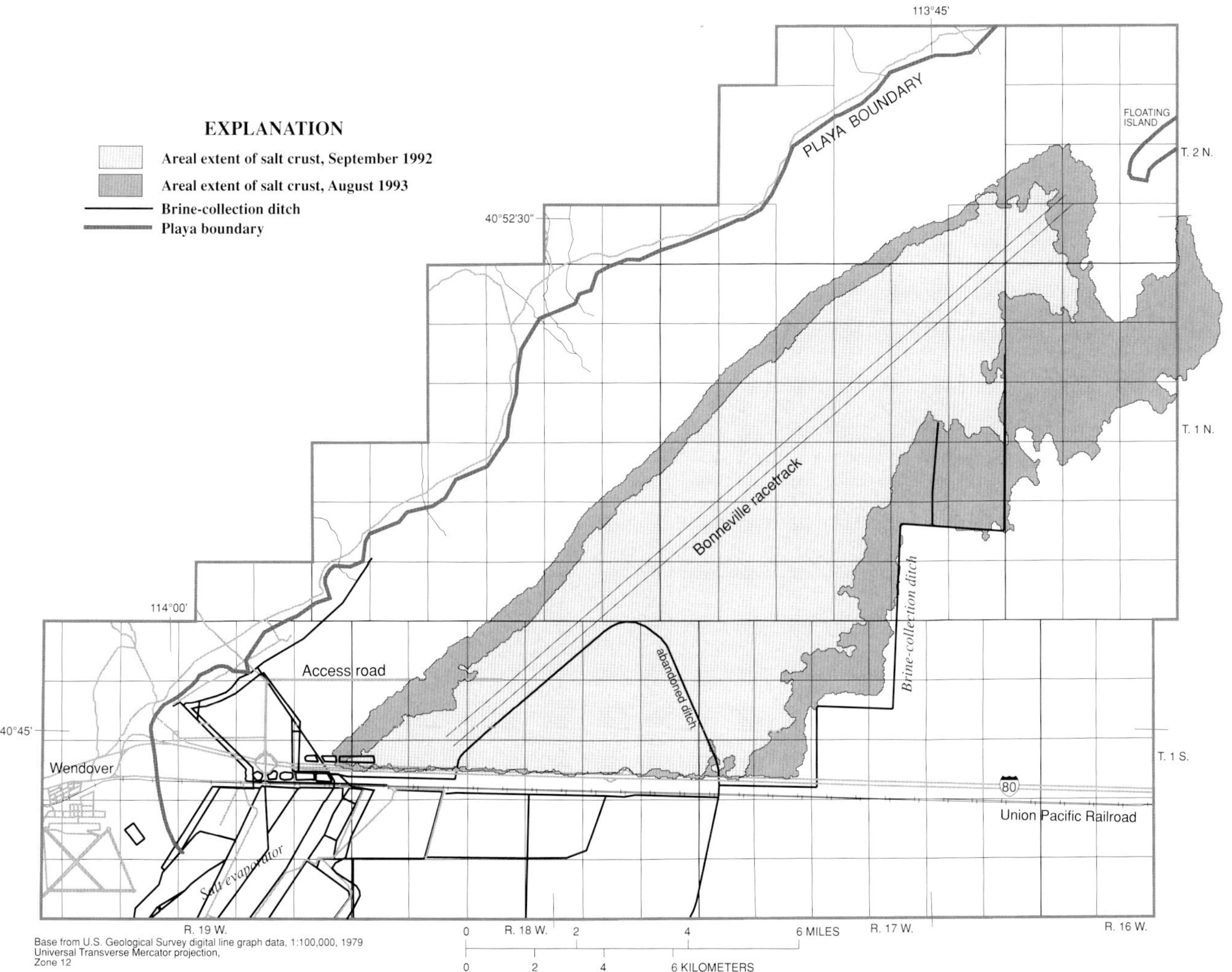

Figure 5. *Increase in areal extent of the salt crust from September 1992 to August 1993, Bonneville Salt Flats, northwestern Utah.*

evaluate the effect of brine production from the ditches on the salt crust, and (3) identify brine movement to and from the shallow-brine aquifer. These model simulations aided in understanding how brine production affects the gradual loss of salt from the crust.

The simulation region (figure 6) extends east of the brine-collection ditch located east of the salt crust and is bounded to the south by the brine-collection ditch south of Interstate Highway 80 and to the northwest by the alluvial-fan aquifer. The vertical extent of the simulation region represents the thickness of the shallow-brine aquifer from the water at or near land surface down to the contact with the lake sediment of the upper part of the basin-fill aquifer. The seasonal surface pond that forms during the winter was incorporated into the simulation region as an area that approximately corresponds with a salt-crust thickness of at least 1 foot. Although this region was simulated as a highly porous medium with an assigned porosity of unity and permeability up to four orders of magnitude greater than the shallow-brine aquifer, the result was similar to simulating an actual free-surface water body connected to the ground-water system.

Inflow to the simulation region included infiltration of precipitation, and ground-water flow primarily from the east and north. The simulator does not have the ability to model salt dissolution and precipitation and therefore, infiltration from land surface was assigned an appropriate density depending on whether it occurred through the salt crust or carbonate mud. Brine outflow from the simulation region included evaporation, which precipitates crystalline salt on the playa surface; seepage to brine-collection ditches east of the salt crust and along the south boundary; outflow to the alluvial-fan aquifer along the northwest margin; and downward leakage to the underlying basin fill aquifer. The amount of salt precipitated at land surface was calculated from the volume of fluid evaporated and brine density. Although tritium data suggest the possibility of upward leakage from the underlying basin-fill aquifer, water-level data collected during the study indicate the potential for downward leakage. The water-level data were used for model calibration. More data are needed to determine whether upward or downward leakage is more predominant in the long term.

Insufficient data and yearly variation in climatic conditions prevented the simulation of a specific time period. The approach, therefore, was to simulate a typical climatic year with seasonal average values for boundary pressures, recharge, leakage, and evaporation. Brine production from the brine-collection ditch east of the salt crust was set equal

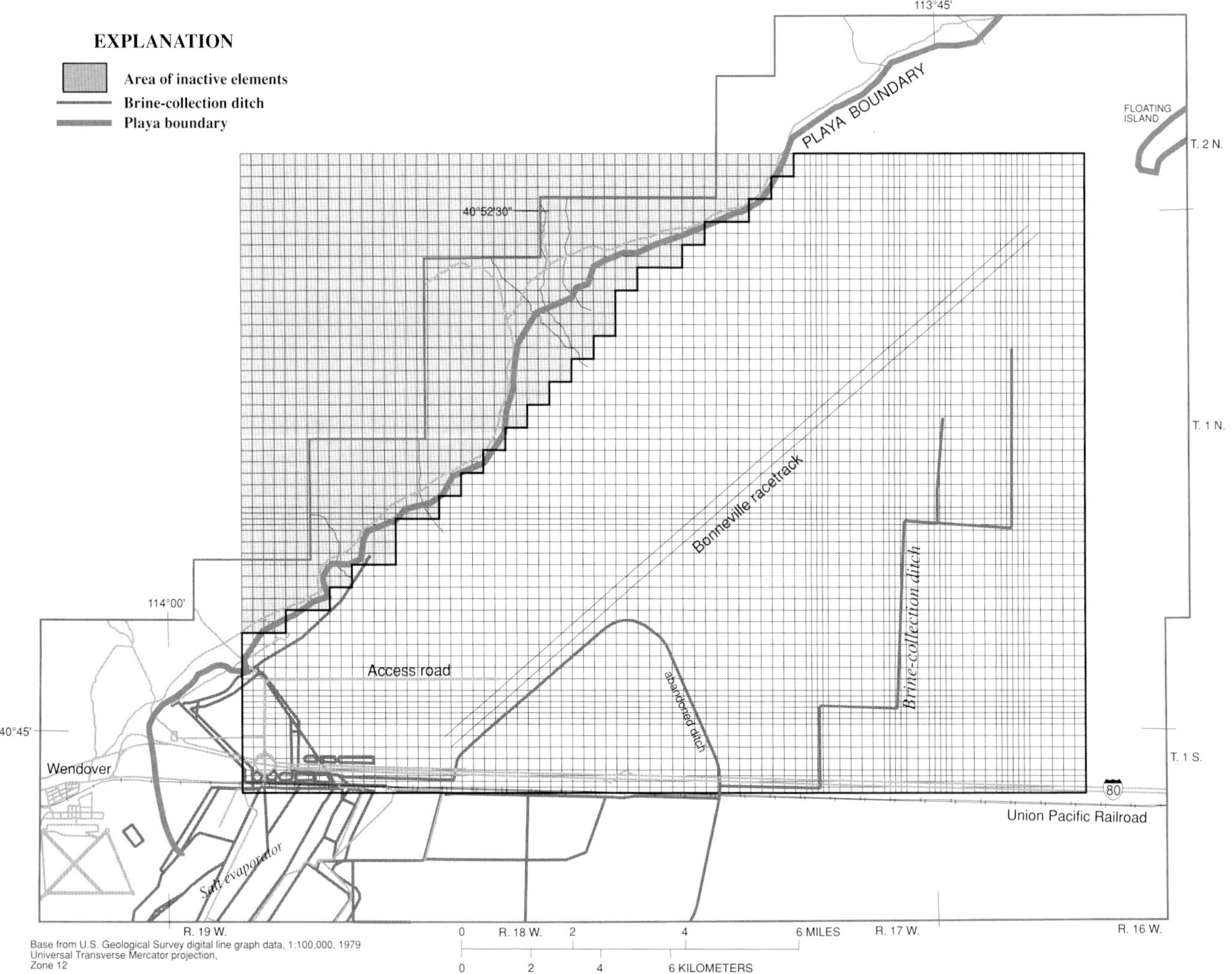

***Figure 6.** Location of model grid and inactive elements for the model of the shallow-brine aquifer, Bonneville Salt Flats, northwestern Utah.*

to the amount measured in 1992. Simulations were carried out to a periodic steady state with a repeated seasonal sequence representing a typical 6-month summer-fall season of brine production and evaporation followed by a 6-month winter-spring season of aquifer recovery and recharge. By simulating a periodic steady state, the results are removed from dependence on inadequately defined initial conditions that may not be truly representative of a typical year.

To reasonably match measured water levels near Interstate Highway 80 and to balance fluid recharge with discharge, transmissivity values were reduced to one-sixth of the previously estimated values reported by Lines (1979, figure 33) through the model calibration process. Simulated subsurface outflow to the south was about 700 acre-feet of fluid and about 285,000 tons of solute, about one-third of the estimated amount for 1992 as shown in table 2.

Model simulations indicate that brine withdrawal is a major cause of salt loss from the crust. Other than the cycling of fluid and solute through the playa surface each year as a result of infiltration of precipitation and evaporation, subsurface brine flow and solute transport to the brine collection ditches east and south of the salt crust are the largest contributors to salt removal from the shallow-brine aquifer. Model simulations, however, do not account for the occasional loss of salt from the crust by extensive flooding described previously.

On the basis of model simulations, the loss of crystalline salt from the playa surface is estimated to be about 975,000 tons per year. The concurrent subsurface loss of salt in solution was computed to be 850,000 tons per year. Uncertainties exist in the simulations because they were based on sparse data. The difference in computed loss of crystalline salt and salt transported through subsurface outflow of brine is within the accepted simulation error of 15 percent.

The sensitivity of model simulations to uncertainties in the data was tested. Subsurface outflow of solute and the annual loss of crystalline salt were most sensitive to changes in the capacity of the shallow-brine aquifer to transmit brine near Interstate Highway 80 (figure 7). The annual loss of crystalline salt also was sensitive to the density of brine infiltrating from the playa surface. All sensitivity simulations resulted in a loss of crystalline salt during a typical year.

Some limitations of this model include: (1) simulating the pond as a fixed region of very high permeability and 100-percent porosity, rather than as a surface-water feature coupled to the ground-water system; (2) having no mechanism for salt dissolution and precipitation at the land surface and above the water table; (3) representing the brine-collection ditches as a specified pressure boundary rather than a surface-water feature coupled to the ground-water system; (4) approximating flow in fractured sediment and salt crust as equivalent porous media; (5) extrapolating and interpolating the distribution of model parameters from sparse data in some areas; and (6) simplifying the transient system into two representative seasons, thus excluding changes on a daily or monthly time scale.

CONCLUSIONS

Interpretations made from data collected during the study indicate that there are no long-term changes in the fluid balance and dissolved-solids and potassium concentrations in the shallow-brine aquifer of the Bonneville Salt Flats despite yearly and seasonal variations. The water-table configuration for the shallow-brine aquifer indicates that the primary source of fluid recharge is by infiltration of precipitation on the playa surface, which dissolves salt as it seeps into the subsurface. Geochemical data indicate that upward leakage is possible from the upper part of the basin-fill aquifer.

Table 2. Estimated and simulated annual fluid flow and solute transport into and out of the shallow-brine aquifer, Bonneville Salt Falts, northwestern Utah.

Component	1992		Calibrated simulation	
	Acre-feet	Tons of solute	Acre-feet	Tons of solute
Recharge				
Subsurface inflow from east and northeast	60	21,900	88	27,750
Subsurface inflow from west	--	--	9	-750
Infiltration from playa surface	8,300-12,800	NA	11,830	(4,712,000)
Total (rounded)	8,400-12,900	21,900	11,930	27,000
Discharge				
Subsurface outflow to south	1,800-2,300	700,000-895,000	705	285,100
Subsurface outflow to brine collection ditch	1,500	580,000	1,450	525,000
Subsurface outflow to northwest	35	13,600	315	50,590
Downward leakage	58-580	22,600-226,000	27	11,000
Evaporation	15,000	NA	10,050	(3,742,000)
Total (rounded)	18,400-19,400	1,316,000 1,715,000	12,550	872,000

[--, no estimate; -, indicates direction of movement is opposite of designated category; NA not applicable because salt not leaving shallow-brine aquifer; () indicates value not added to total]

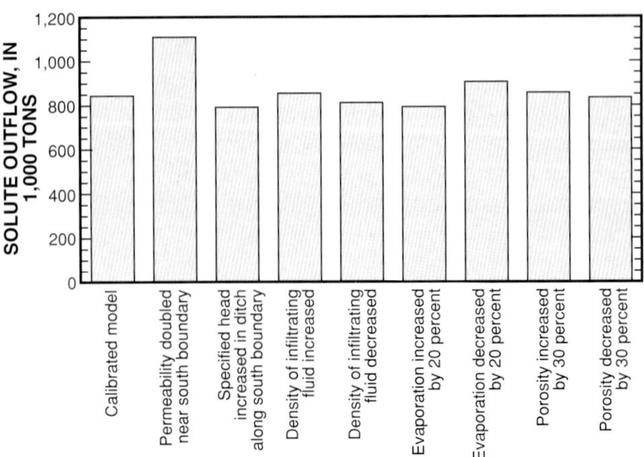

Figure 7. *Total subsurface outflow for the calibrated model of the shallow-brine aquifer and for each sensitivity simulation, Bonneville Salt Flats, northwestern Utah.*

Discharge from the shallow-brine aquifer is primarily through evaporation, which results in the precipitation of crystalline salt on the playa surface, and through ground-water seepage to the brine-collection ditches to the east and south of the salt crust, which results in the transport of solute out of the shallow-brine aquifer.

Extensive flooding of the Bonneville Salt Flats during the winter of 1992-93 dramatically changed the salt surface. Much of the near-surface salt was dissolved and remained in solution into the summer of 1993, when the salt precipitated on the playa surface as the pond evaporated. Because of the flooding, salt was redeposited beyond the extent of the crust mapped in 1992. An estimated 10 to 14 million tons of salt were dissolved and redeposited on the playa surface or infiltrated into the shallow-brine aquifer. An unknown amount of salt might have been deposited in an area where the salt would not be able to coalesce with the main salt body or may have infiltrated into the shallow-brine aquifer and transported by ground-water flow to the brine collection ditch.

A three-dimensional ground-water flow and solute-transport model was developed for the shallow-brine aquifer in the Bonneville Salt Flats. The model simulated a typical year representing average climatic and hydrologic conditions for a summer production season and a winter recovery season. Simulation results indicated a net loss of salt of about 850,000 tons during a typical year through subsurface flow from the simulation region primarily through seepage to brine-collection ditches. Model simulations, in which the 1992 rate of withdrawal from the brine- collection ditch east of the salt crust and average climatic conditions were used, indicate that brine withdrawal is a major cause of salt loss from the crust.

REFERENCES

Fenneman, N.M., 1931, Physiography of the Western United States: New York, McGraw-Hill, 534 p.

Hunt, C.B., Robinson, T.W., Bowles, W.A., and Washburn, A.L., 1966, Hydrologic basin, Death Valley, California: U.S. Geological Survey Professional Paper 494-B, 138 p.

Kipp, Jr., K.L., 1987, HST3D: A computer code for simulation of heat and solute transport in three-dimensional ground-water flow systems: U.S.Geological Survey Water-Resources Investigations Report 86-4095, 517 p.

—1997, Guide to the revised heat and solute transport in three-dimensional ground-water flow systems: U.S. Geological Survey Water-Resources Investigations Report 97-4157, 149 p.

Lines, G.C., 1979, Hydrology and surface morphology of the Bonneville Salt Flats and Pilot Valley playa, Utah: U.S. Geological Survey Water-Supply Paper 2057, 107 p.

Mason, J.L., Brothers, W.C., Gerner, L.J., and Muir, P. S., 1995, Selected hydrologic data for the Bonneville Salt Flats and Pilot Valley, western Utah, 1991-93: U.S. Geological Survey Open-File Report 95-104, 56 p.

Mason, J.L., and Kipp, Jr., K.L., 1998a, Hydrology of the Bonneville Salt Flats, northwestern Utah, and the simulation of ground-water flow and solute transport in the shallow-brine aquifer: U.S. Geological Survey Professional Paper 1585, 108 p.

—1998b, Investigation of salt loss from the Bonneville Salt Flats, northwestern Utah: U.S. Geological Survey Fact Sheet FS-135-97, 4 p.

Turk, L.J., 1969, Hydrology of the Bonneville Salt Flats, Utah: Unpublished PhD dissertation, Stanford University, 307 p.

—1978, Bonneville salt crust study: Kaiser Aluminum and Chemical Corporation (as of 1988, Reilly Industries), Wendover, Utah, 103 p.

SALT LAYDOWN PROJECT - REPLENISHMENT OF SALT TO THE BONNEVILLE SALT FLATS

by

W. W. White III, P.G.
U.S. Bureau of Land Management - Salt Lake Field Office
2370 South 2300 West, Salt Lake City, UT 84119

ABSTRACT

Reilly Industries, Inc. and U.S. Bureau of Land Management have been conducting a Salt Laydown Project to increase the salt-crust thickness of the Bonneville Salt Flats. The project began delivering sodium chloride brine to Bonneville Salt Flats in November 1997. The objective of the five-year experimental Salt-Laydown Project was to add up to 1.5 million tons of salt to Bonneville Salt Flats each year of the experiment. Three years of Laydown-Project operation have demonstrated that annual salt-tonnage loss from Bonneville Salt Flats can be replenished by the laydown facility. During this period, a sodium chloride salt mass of 4.6 million tons was delivered to Bonneville Salt Flats. The average annual addition of 1.5 million tons exceeded an estimated annual salt loss of 0.85 million tons. The salt addition appears to be distributed between the existing salt-crust, a new salt-crust area, and the shallow-brine aquifer.

Monitoring during the project included measurement of changes in salt-crust thickness, areal extent, and mass. Ten test pits showed that at least five different strata comprise the salt crust. Thickness measurements of the dense-cemented halite stratum (surface stratum of the salt crust), and the cemented-coarse-porous halite stratum (sub-surface stratum), showed substantial thickness changes from year to year, while little change in total salt-crust thickness was observed at nine locations during a 10-year period. Based on thickness comparisons from 1994 through 2000, the dense-cemented halite stratum thickness increased at one location and decreased at three other locations. However, cemented-coarse-porous halite stratum thicknesses increased when the dense-cemented halite stratum thicknesses decreased at the same locations. An unusually wet (or dry) year could measurably decrease (or increase) the thickness of the dense-cemented halite stratum from year to year. This yearly variation in thickness could occur despite the increased salt tonnage added to Bonneville Salt Flats by the Laydown Project. Although the yearly added tonnage would maintain or increase the current mass of sodium chloride in the salt-crust deposit and shallow-brine aquifer, annual variation in dense-cemented halite stratum could easily mask the annual 0.4-inch increase in salt-crust thickness predicted by the 1991 Laydown Project feasibility study.

An estimated 5-square mile increase in salt-crust areal extent was observed between September 1997 and October 1999. Based on assumed changes in thickness of 0.25 to 1 inch, and bulk salt densities of 79.4 and 109.8 pounds per cubic foot, as much as 0.6 million tons of salt were added to the new salt-crust area by the Laydown Project. The remaining 4 million tons were apparently distributed to the main body of the salt crust (26 square miles in 1997) and the shallow brine aquifer (the tonnage actually distributed to each is currently unknown).

Geochemical modeling of brine compositions, determined during the project, produced several observations regarding salt additions to the salt crust and shallow-brine aquifer. The TEQUIL model accurately predicted brine chemistries, based on agreement between modeled output and solar-pond brine compositions from a commercial potash operation. The model also showed that the shallow-brine aquifer has the capacity to assimilate an additional 17 to 25 million tons of sodium chloride. This tonnage is four to five times the 4.6 million tons of laydown salt delivered to Bonneville Salt Flats during the first three years of the Laydown Project. The ability of the shallow brine aquifer to assimilate additional salt may help account for the laydown-delivered salt tonnage that was in excess of the amount estimated in the 5 square miles of new salt crust.

The laydown sodium chloride mass assimilated into the shallow-brine aquifer is expected to eventually be redistributed in the salt crust as part of new surface and additional subsurface halite crystal growth. TEQUIL also predicted that the addition of laydown brine to the shallow-brine aquifer does not change the salt-crust mineral assemblages. Anhydrite and halite are the only minerals predicted to precipitate from two different mixing ratios of laydown brine and shallow aquifer brine in an open system such as Bonneville Salt Flats. Potassium and magnesium salts do not precipitate in the open system of the Bonneville Salt Flats salt crust because the addition of rain water or the shallow-aquifer brine dissolves them immediately. All TEQUIL simulations showed that more than 90 percent of the water would have to be evaporated, and 96 to 98 percent of the sodium chloride would have to be precipitated before potassium and magnesium minerals precipitated from the brine. This condition could only be achieved if a mixture of laydown brine and shallow-aquifer brine were isolated from the shallow-brine aquifer and subjected to conditions similar to a commercial solar-pond process.

INTRODUCTION

The Bonneville Salt Flats (BSF) is located in the western part of the Great Salt Lake Desert of northwestern Utah. BSF is part of a large playa that occupies one of several enclosed sub basins that comprise the Great Salt Lake Desert (figure 1). These sub-basins include the Bonneville Salt Flats, Pilot Valley, and the Newfoundland Basin. BSF is roughly divided into a north and south half by the east-west-trending Interstate Highway 80 (I-80) and the adjacent Western Pacific Railroad right-of-way (now Union Pacific) which is parallel to, and 1,400 feet south of I-80. The north half of BSF is dominated by public land, while the south half is mainly private. The twin cities of Wendover, UT and Wendover, NV are adjacent to I-80 and about 4 miles west of BSF's western margin. The economic potential of potash-bearing brines beneath the surface of BSF was recognized as early as 1914, and commercial potash (KCl) production from these brines has been continuous since 1939 (Bingham, 1980, p. 230-231).

BSF is listed on the National Register of Historic Places as site of world land-speed records. Because of its unique geologic characteristics, BSF was designated by the U.S. Bureau of Land Management (BLM) as an Area of Critical Environmental Concern (ACEC). Historic activities and unique physical characteristics have made BSF an area of national and international interest.

The BSF sub-basin is geologically and hydrologically complex, and this complexity is reflected by both seasonal and yearly variation in salt crust areal extent and thickness. Variation of salt-crust area and thickness has been documented since 1927. Early measurements made by Nolan (1927, p. 34) in 1925 indicated that the salt crust (north and south halves) covered about 150 square miles, and had a measured thickness of 3.5 feet near Salduro Station (9 miles east of Wendover on the Western Pacific Railroad). Using Nolan's map of "crystalline salt" (1927, Plate 3), Turk (1978, p. 9) estimated that the 1925 salt-crust area north of the Western Pacific Railroad was 68 square miles. For comparison, Turk (1978, p. 9-11) also documented 22 salt-crust area measurements that were taken during 1972-1976 from BSF north of the Western Pacific Railroad. The average area from the 22 measurements was 47 square miles, while minimum and maximum areas were 38 and 56 square miles respectively. Lines (1979, p. 2) measured a salt-crust area of about 40 square miles during fall 1976, of which 30 square miles were perennial salt crust with thicknesses of 1 to 3 feet (estimated from Lines, 1979, p. 30, 45-56). Mason and Kipp (1998, p. 1) reported a measured salt-crust area of 43 square miles in late summer 1992; of the 43 square miles, about 34 square miles were perennial salt crust with thicknesses of 1 foot or more. Lines (1979, figure 11) and Mason and Kipp (1998, figure 26) show these areas were based on measurements of salt crust exposed north of the Western Pacific Railroad and I-80.

Reported depletion of salt-crust thickness has been a concern to the public and land-managing agencies for at least 26 years (McMillan, 1974, p.1; Lines, 1979, p. 4). This concern is based on salt-crust area and volume changes reported between 1960 and 1988, that were measured north of the Western Pacific Railroad and I-80. McMillan (1974, p. 3) reported a 9 and 15 percent respective decrease in salt-crust area and volume from 1960 through 1974. Brooks (1991, p. 8) calculated a 20 and 30.6 percent respective decrease in salt-crust area and volume from 1960 through 1988.

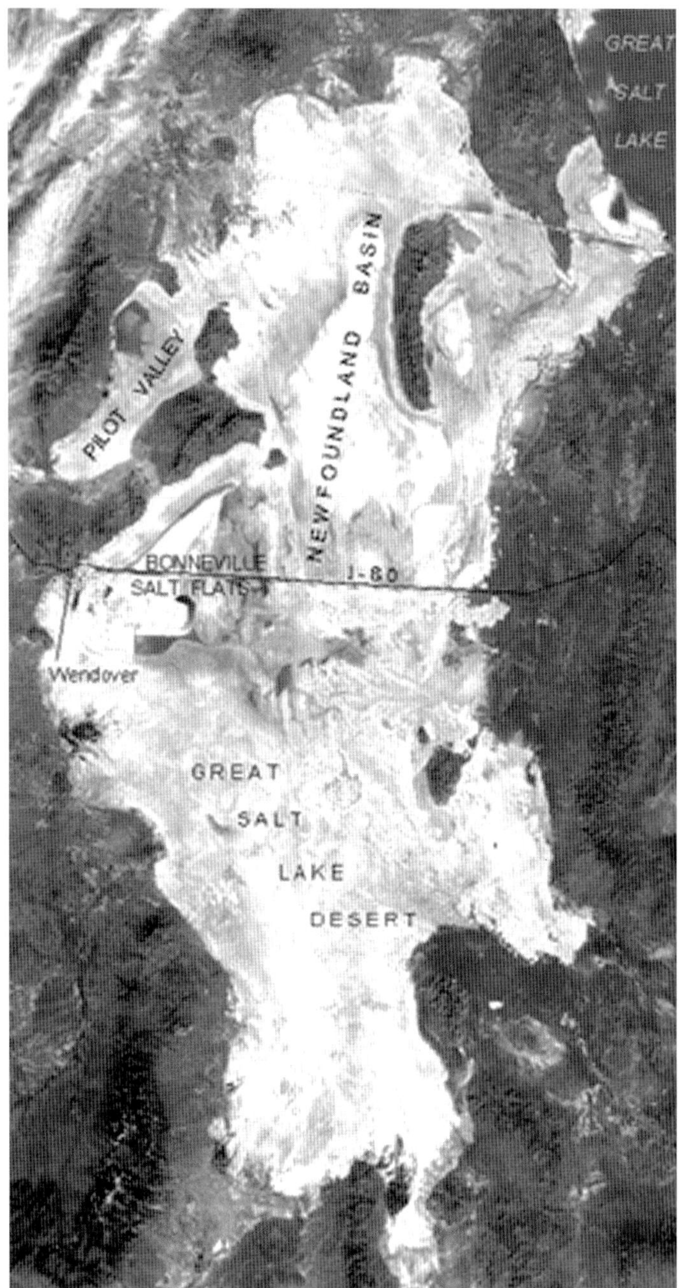

Figure 1. Great Salt Lake Desert showing sub-basin locations (Bonneville Salt Flats, Pilot Valley, and Newfoundland Basin), western Utah.

A more recent study by the U.S. Geological Survey (USGS) suggested that an estimated 850,000 tons of salt could be lost annually from the north half of BSF (Mason and Kipp, 1998, p. 106). The USGS estimate was based on an annual brine withdrawal of 1,500 acre feet from a federal-lease collection ditch located east of the salt crust. Salt tonnage was derived from computer simulations that used data collected from BSF from 1991 through 1993. This estimate is in reasonable agreement with Reilly Industries, Inc.'s (Reilly) annual production from the same lease collection ditch; a range of 0.49 to 1.03 million tons per year of salt were removed during the period from 1995 through 1998 (A. Frye, Chief of Counsel, Reilly Industries, Inc., personal communication, May 19, 1999).

Because the BLM, Reilly, and the racing community (represented by "Save the Salt" [STS]) are concerned about the reported deterioration of the BSF, they are attempting to replenish salt to the BSF through cooperative agreements. In 1991, Reilly and STS jointly funded a salt-replenishment feasibility study that resulted in a salt-laydown facility plan (Bingham, 1991). According to the plan, sodium chloride brine would be pumped out onto the BSF at a rate of 6,000 gallons per minute (gpm), 24-hours per day, for six months (November - April) during each year of the program. This experimental program was anticipated to have an initial life of at least five years. Based on the engineering design, up to 7.5 million tons of salt could be deposited during a five-year period over a 28-square mile area. According to Bingham (1991, p. 2), this would result in an additional salt-crust thickness of about 0.4 inches per year.

Based on the 1991 salt-replenishment feasibility study, BLM and Reilly entered into a salt-laydown agreement in 1995. Under the laydown agreement, Reilly financed the installation and operation of a $1,000,000 salt-laydown facility, and BLM and Reilly initiated a cooperative monitoring agreement to measure the amount of salt delivered to BSF each year of the program. To ensure the pumped brine meets salt-laydown-design specifications, Reilly and BLM independently sample and analyze the brine being pumped onto BSF. The Laydown Project began delivering brine to BSF on November 1, 1997.

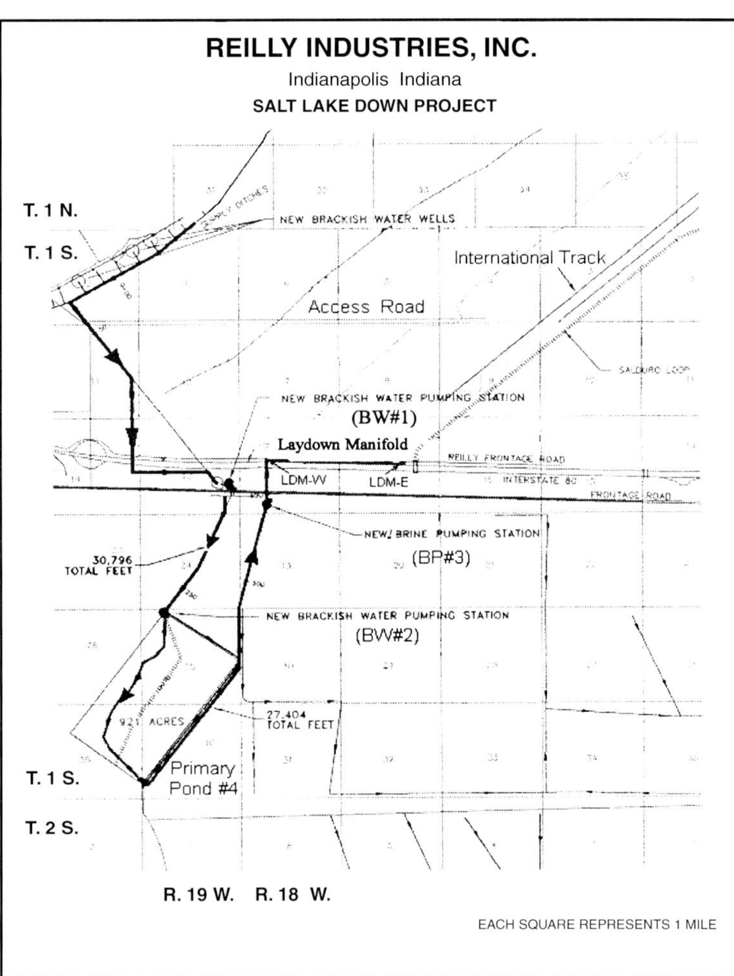

Figure 2. Index map showing location of Reilly Laydown facilities with respect to Interstate 80 and Bonneville Salt Flat features (for example, Salduro Loop, International Track, and county access road. Arrows on ditches show water-flow direction (after Reilly Industries, 1997).

OBJECTIVES

This paper describes the first three years of the Laydown Project, and includes the November through April pumping periods of 1997 to 1998, 1998 to1999, and 1999-2000. The objectives of this report are to: 1) describe components and operation of the salt laydown facility, 2) summarize BSF and laydown monitoring methods, 3) explain and quantify results of the Laydown Project's first three years, and compare them with BSF baseline data, 4) compare pre- and post-laydown salt-crust thicknesses and areal extents, and 5) evaluate the effects of mixing laydown brine with shallow-aquifer brine using the TEQUIL model.

LAYDOWN FACILITY DESCRIPTION

The laydown facility (figure 2) is composed of brackish-water supply wells, about 11 miles of transfer ditches and associated pumps, a 921-acre bedded salt deposit (primary pond no. 4 [PP no. 4]), and a brine-distribution manifold. Because the ditch system and the brine-distribution manifold traverse a 13-mile distance over flat terrain, the biggest challenge is to move water from one end of this flat area to the other. The following description of the laydown facility shows how this water movement is accomplished.

Brackish water is obtained from seven alluvial-fan wells on the south flank of the Silver Island Range (water contains from 6.2 to 8 g/L TDS [Mason and Kipp, 1998, p. 49] compared with shallow-brine aquifer which averaged 244 to 297 g/L TDS during 1994-2000). From these wells the brackish water is pumped into a series of transfer ditches that move water south and under I-80, a distance of about 4.5 miles. To keep the brackish water flowing, brackish-water pump no. 1 (BW no. 1) was installed in the transfer ditch south of I-80. A second pump (brackish-water pump no. 2 [BW no. 2]) was installed 1.5 miles south of BW no. 1. With the aid of BW no. 1 and BW no. 2, brackish water is ultimately delivered through a 24-inch diameter discharge pipe into the northwest corner of primary pond no. 4 (PP no. 4) where the salt dissolution process is initiated.

PP no. 4 is an old solar-evaporation pond that was used to precipitate NaCl (as the mineral halite) from potassium-bearing brine during the economic mineral-recovery process. NaCl in the 921-acre pond is approximately 3 to 4 feet thick, and totals about 8 million tons. During its residence time in PP no. 4 (about 15 to 30 days), the brackish water dissolves as much as 2 lbs of salt per gallon, and the density of the resulting brine approaches 1.2 g/mL (sodium chloride saturation).

From the southeast corner of PP no. 4, the brine flows

through a north-trending, 3.3-mile transfer ditch that terminates at brine pump site no. 3 (BP no. 3, figure 3). Brine is lifted approximately 11 feet vertically above the collection ditch by the active pump and discharged into a 5-foot wide, 7-foot long, 7-foot high concrete vault. Pumping the brine into the concrete vault provides sufficient hydraulic head for the brine to flow north for a distance of about 0.4 miles through a 24-inch diameter discharge pipe. The north-trending discharge pipe passes under I-80 and discharges brine into the west end of a brine-laydown manifold. The laydown manifold is an east-west-trending, 1.5-mile long, 30-foot wide area between west-bound I-80 and a parallel frontage road. Using a brine-density range of 1.14 to 1.18 g/mL, an average brine flow rate of 6,688 gpm would deliver about 350 to 485 tons per hour (tph) of NaCl to the brine-laydown manifold. Brine flows north from the manifold onto BSF through twelve culverts that are uniformly spaced along the manifold's length. Values for tph were calculated using equations 1-3 in White and Wadsworth, 1999, appendix 3, p. A3-1; densities of 1.14 and 1.18 g/mL correspond with NaCl wt percent values of 18 and 24, and were determined from linear regression of daily laydown-brine density and NaCl wt percent values (n=70, r^2=0.976; White and Wadsworth, 1999, appendix 3, tables A3.1, A3.5, and A3.6).

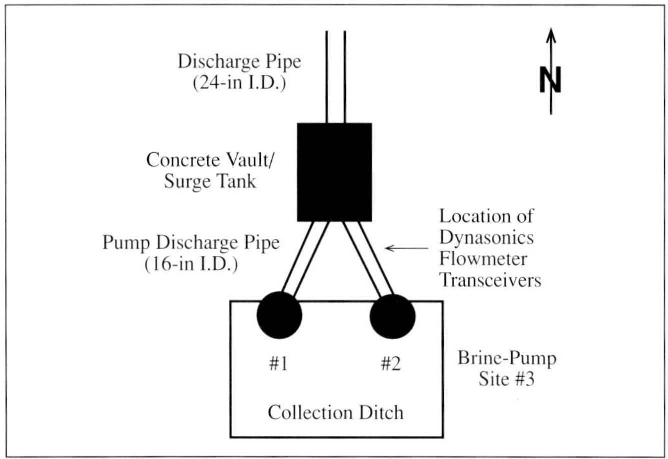

Figure 3. *Plan view of brine-pump site no. 3; pump no. 2 is active, and pump no. 1 is standby.*

METHODS

Pre-1994 Monitoring Well Database

Three USGS studies of BSF were conducted from 1976 through 1993: 1976 through 1978 (Lines, 1978; 1979), 1981, and 1991 through 1993 (Mason and others, 1995; Mason and Kipp, 1998). Chemical analyses of 186 monitoring-well samples from these studies (appendix A1) are used in this paper as an historical comparison with 1994 through 2000 BLM monitoring-well sample analyses. Four different well-numbering systems were used to identify four groups composing the 186 samples (well numbers prefixed by "USGS/BLM", "B", "K", and "L"). These systems are described below and listed in table 10.

The system of numbering wells in Utah is based on the cadastral land-survey system of the U.S. Government, and is described by Lines (1978, p.1-2) and Mason and Kipp (1998, p. 6). Although USGS uses this method as their primary well-numbering system, the primary system for BSF is commonly linked with a secondary alpha-numeric system characterized by a "USGS" and/or "BLM" prefix (that is, the primary well number [B-1-17] 11aac-1 is equivalent to USGS-2/BLM-60). To simplify well labeling for figures and tables presented in this paper, the "USGS" and/or "BLM" alpha-numeric system is used instead of the primary numbering system. The 186 samples identified by the "USGS" and/or "BLM" alpha-numeric identification system, are listed in appendix A1.

Monitoring-well sample data from Lines (1978) are numbered with "USGS" and "K" prefixes (figure 4). Conservation Division sample data are numbered with "B", "K" and "L" prefixes (figure 5). Sample data from Mason and others (1995) are numbered with "USGS and/or BLM" prefixes (figure 6). Monitoring-well numbers starting with "K" represent shallow wells (19- to 23-feet deep) that were hand augered during 1965-1967 (Turk, 1969, p. 64-65), and subsequently sampled during the Lines and Conservation Division studies. Conservation Division well numbers prefixed with "B" or "L" were shallow auger holes drilled to depths of 10 feet or less during 1981 (J. F. Kohler, Geologist, BLM Utah State Office, personal communication, June 29, 2000).

Chemical analyses from the 186 monitoring wells that were 25-feet deep, or less, were compared with the 1994 through 2000 BLM monitoring-well sample analyses. Molar concentrations were calculated for each of the major ions (Na, Mg, K, Ca, Cl, and SO_4) listed in the analyses of 186 samples. The resulting molar concentrations were averaged, and the results used to determine average mole ratios of K/Mg, Cl/Na, and SO_4/Ca for each of the four USGS well-numbering systems listed in table 10.

1994-2000 Monitoring Well Sampling and Analyses

BLM started collecting annual BSF shallow-brine aquifer samples in 1994. The purpose of the sampling program was to identify historical trends, and make yearly comparisons between BSF-brine and current laydown-brine chemistry, so that future effects of the project on the natural system could be measured. The brine samples were taken during an August-September period from a suite of 27 monitoring wells in the shallow-brine aquifer. This period was selected because effects of evaporation on BSF are historically greatest at this time of year, and maximum yearly brine concentrations would be expected; because most of the pre-1994 samples were collected during the same period, comparisons could be made. Because the depth of the shallow-brine aquifer has been reported as ranging from 15 to 25 feet (Turk, 1973, p. 8; Lines, 1979, p. 65; Mason and Kipp, 1998, p. 22), wells included in the 27-well suite have depths of 25 feet or less (BLM-27 is the only exception with a measured depth of 27.7 feet). These 27 wells were also selected because of their proximity to the International Track and the federal-lease collection ditch and are shown in figure 7. Samples have been collected during Fiscal Years 1994, 1995, 1996, 1997, and 2000 (the period October 1 through September 30 of the next year - FY94, FY95, FY96, FY97, and FY2K). Although BLM's original intent was to sample the same wells each year, changes in accessibility and loss of some surface casings (for example, USGS-9 and BLM-60)

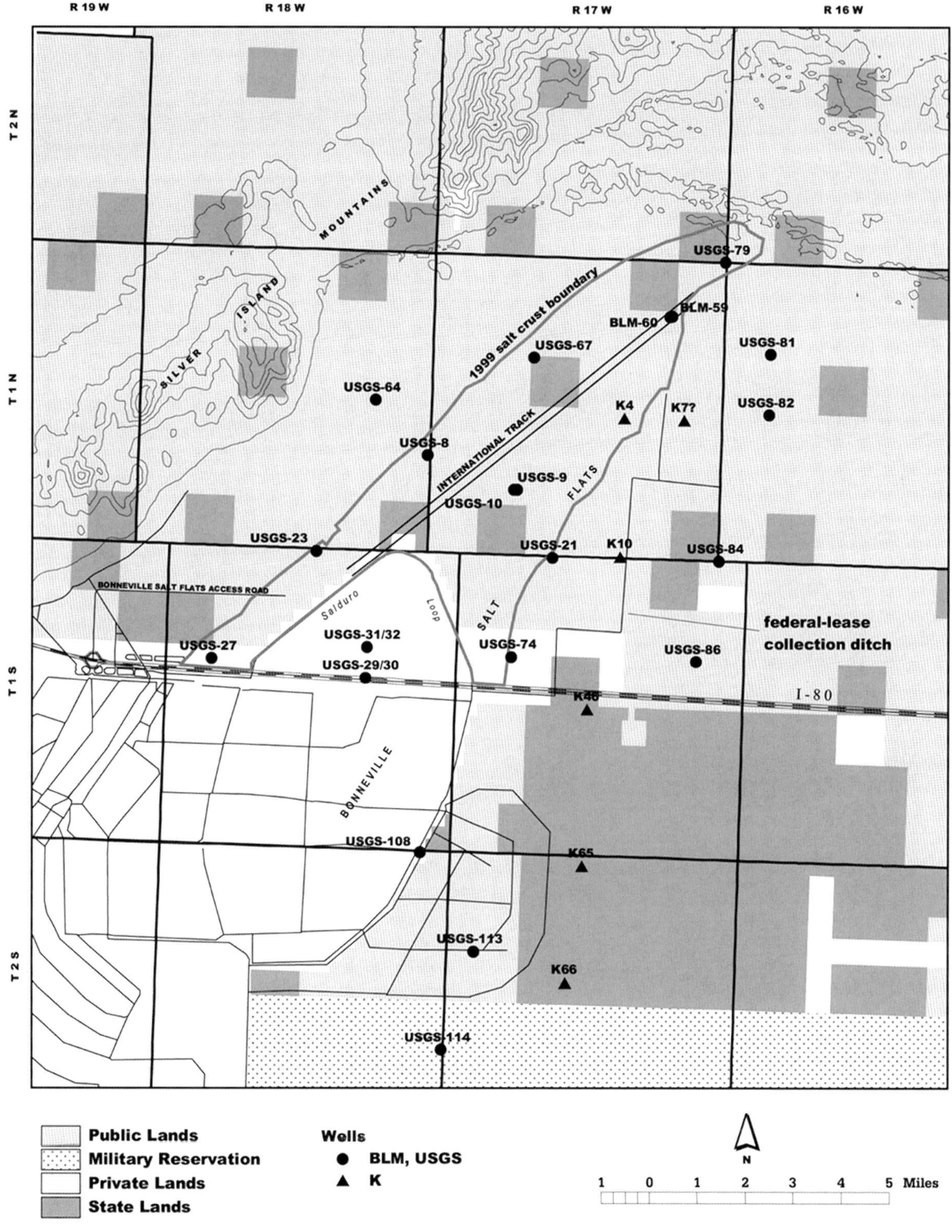

Figure 4. *Location of 1976-78 USGS monitoring-well samples (Lines, 1978-1979). Well sites are identified by "USGS" and "K" prefixes.*

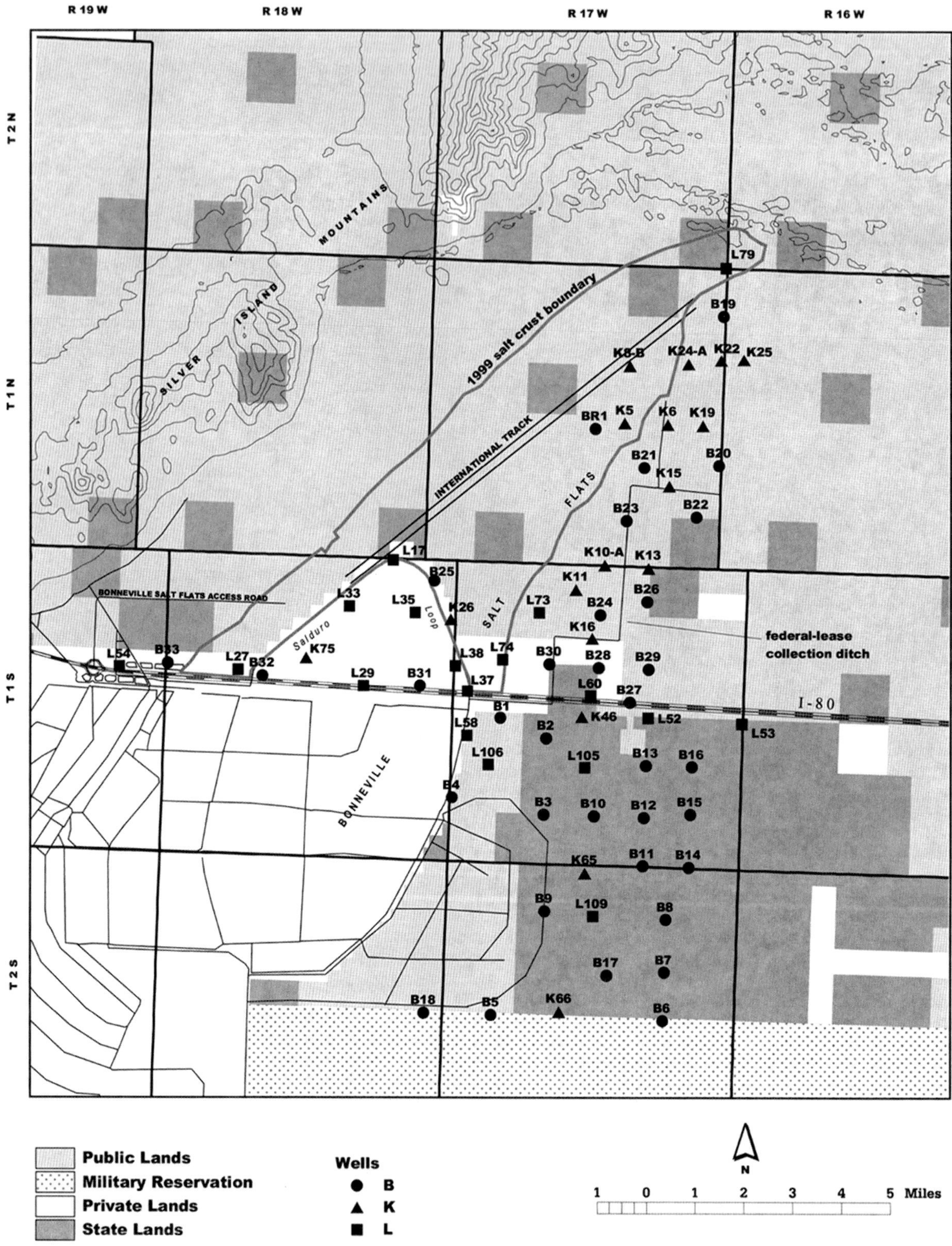

Figure 5. Locations of 1981 USGS (Conservation Division) monitoring-well and auger-hole samples. Well sites are identified by "K" prefixes, and auger holes are identified by "B" and "L" prefixes.

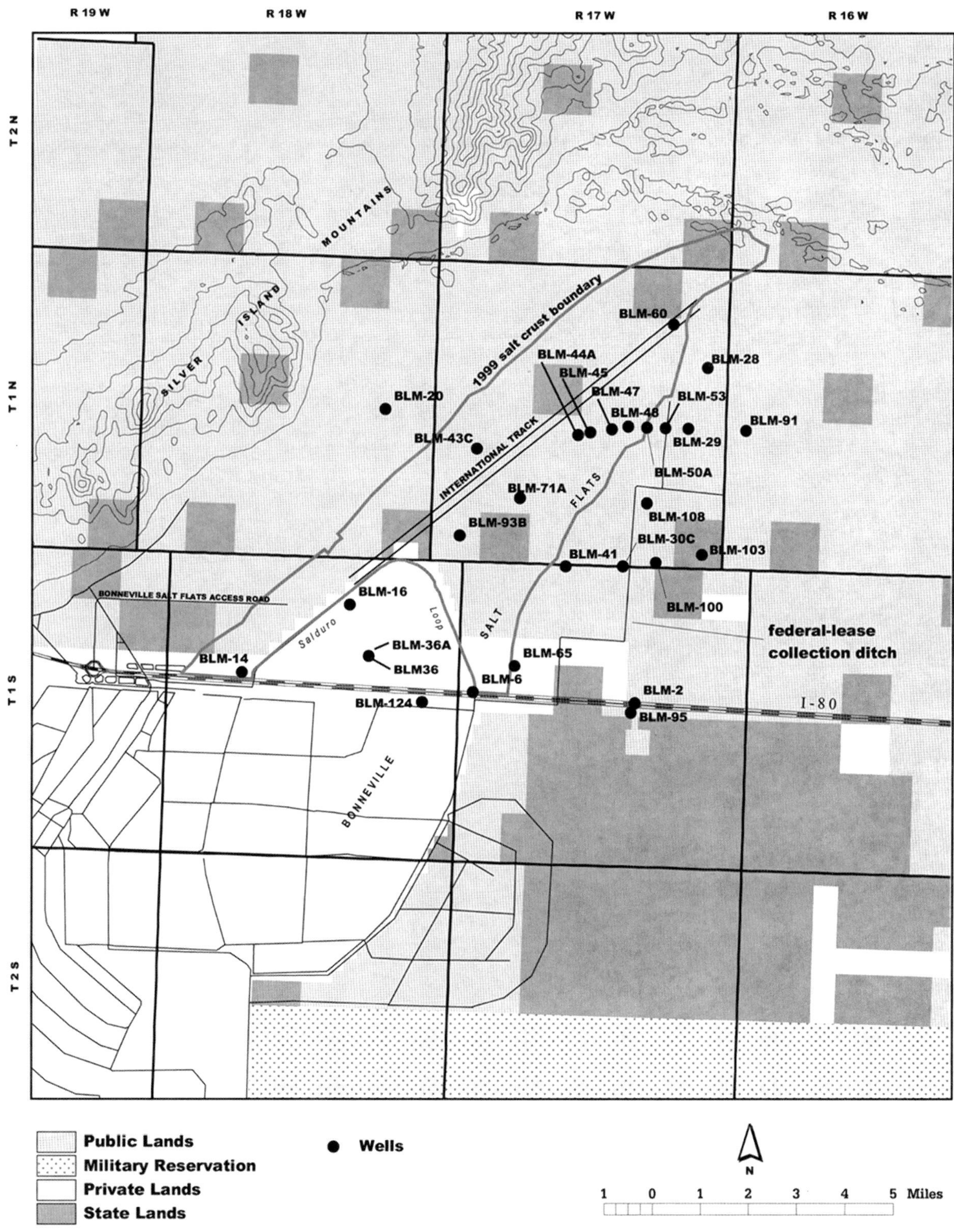

Figure 6. Monitoring-well sample locations for 26 wells sampled by USGS in 1992-93 (Mason and others, 1995, p. 46-50). Nine of these monitoring wells were also sampled by BLM during its 1994-97 sampling period.

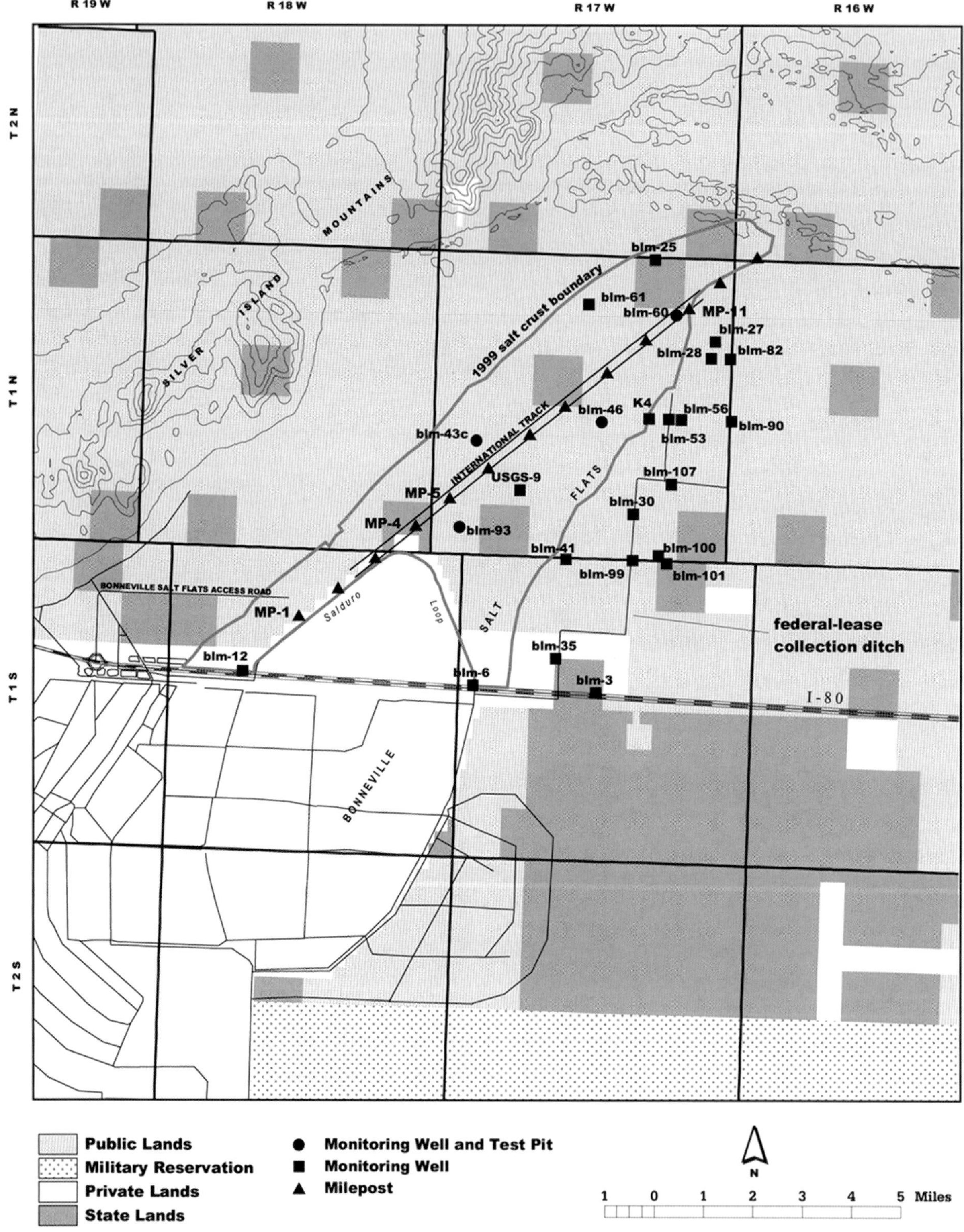

Figure 7. Location map showing positions of 27 monitoring-well and test-pit locations with respect to the 1999 salt-crust boundary.

resulted in yearly variation of the sample suite (table 1).

BSF shallow-brine aquifer samples were collected from the selected monitoring wells with a 3-foot-long, 1.6-inch O.D. polyethylene bailer, and stored in 8-ounce (237 mL), low-density polyethylene sample bottles. Density measurements and chemical analysis for specific ions were determined by Chemical and Mineralogical Services (CMS), Salt Lake City, UT. Sodium and potassium concentrations were determined using an Instrumentation Laboratories Model 343 flame photometer. Calcium and magnesium were determined by disodium ethylenediaminetetraacetie acid (EDTA) titration (EPA methods 311C and 314B). Sulfate (SO_4) was determined gravimetrically by barium chloride precipitation (EPA method 426A), and chloride was determined by silver nitrate titration using the Mohr method. Sample densities were determined with a density meter and verified with a hydrometer (EPA method 210B). Total dissolved solids (TDS) were calculated by summation of cation and anion concentrations. To check the accuracy of the analyses, the mole-balance method described by Sturm and others (1980, p. 175) was used.

Chemical analyses of monitoring well brines, sampled during FY94, FY95, FY96, FY97, and FY2K, are listed in appendix A2. Molar concentrations were calculated from

Table 1. Summary of well depths and sampling history for 27 BSF monitoring wells.

Monitoring Well no.[1]	**Well Depth feet**[2]	**Sample Collected and Analyzed (X)**				
		FY94	**FY95**	**FY96**	**FY97**	**FY2K**
BLM-3/USGS-60	8.0	X	X	X	X	X
BLM-6/USGS-37	19.0	X	X		X	
BLM-12/USGS-28	19.0	X	X	X	X	
BLM-19/USGS-68	9.0		X			
BLM-25/USGS-1	19.0	X	X	X	X	
BLM-27	27.7*				X	X
BLM-28	25.0	X	X			
BLM-30	14.0	X		X	X	X
BLM-35/B-30	6.1*	X		X		X
BLM-41/USGS-21	18.0	X		X	X	X
BLM-43C/USGS-5	16.0	X	X	X	X	
BLM-46	9.0	X	X	X	X	
BLM-50A/K-4A	23.0	X	X	X		
BLM-50B	8.3				X	X
BLM-53/K-3	25.0	X	X	X	X	X
BLM-56	9.0					X
BLM-60/USGS-2	19.0	X	X	X	X	
BLM-61/USGS-66	5.0	X	X	X	X	
BLM-71A/USGS-9	16.0	X				
BLM-82/K-22	25.0	X	X		X	
BLM-90	10.4				X	X
BLM-93	12.0			X	X	
BLM-99	9.7	X		X	X	
BLM-100	9.9				X	
BLM-101/K-14	11.0*	X		X		
BLM-107	10.3	X		X	X	X
BLM-107A	NL				X	
n =		19	13	16	20	10

NL Not listed
* Measured
[1]From Mason and others, 1995, table 1; BLM survey number with matching alternate number (if available).
[2]Mason and others, 1995, table 2; all depths are reported unless otherwise indicated.

each of the major-ion concentrations listed for each sample (Na, Mg, K, Ca, Cl, and SO_4), and then averaged by sample year. Molar ratios of K/Mg, Cl/Na, and SO_4/Ca were then calculated from the average molar concentrations for each sample year.

Laydown Brine Sampling and Analyses

During the November-April pumping periods of FY98, FY99, and FY2K, Reilly collected brine samples daily from BP no. 3. Brine-pump hour-meter readings were recorded concurrently with sample collection. Brine density and concentration of specific ions were determined, for each sample, in Reilly's Wendover laboratory. Sodium and potassium concentrations were determined by flame photometry using an Automatic FP 1L943. Calcium plus magnesium concentrations were determined by EDTA titration using Calmagite as an endpoint indicator. Sulfate (SO_4) was determined gravimetrically using barium chloride precipitation. Daily sample sodium, magnesium, and potassium concentrations (g/L) were converted to their respective chloride salts, in units of weight percent and tons, based on mole balance and density. Analytical results for the FY98 and FY99 daily samples are listed in White and Wadsworth (1999, appendix 3, tables A3.1-A3.13). FY2K daily samples are in White and Wadsworth (2001).

To provide independent verification of laydown-brine quality, BLM collected triplicate samples of laydown brine from sample sites at key facility locations. These samples were collected during the November-April pumping periods of FY98, FY99, and FY2K. A total of 255 sample triplicates, comprising 85 samples, were collected during the 3-year period. Sample triplicates were identified as "A", "B", and "C", and attached to a sequential number (for example, A-1, B-1, C-1; A-2, B-2, C-2; etc.). The "A" triplicates were submitted to Reilly, "B" triplicates were retained by BLM to be analyzed by CMS (BLM's contracted analytical laboratory), and the "C" triplicates were submitted to University of Arizona, Department of Chemistry [(U of Az), STS's contracted analytical laboratory]. Laydown-sample analyses are listed in appendix A3.

The key-facility sample sites include the concrete vault at BP no. 3, and the west and east end of the laydown manifold (LDM-W and LDM-E); see figures 2 and 3. These sample sites were selected because they comprise the primary laydown-brine distribution system which directly affects the BSF. Additionally, a lag in brine travel time exists between BP no. 3 and LDM-E, and these sample sites provided easily accessible check points to more closely monitor changes in brine quality. Brine-density measurements using a CL-USA no. 458438 hydrometer were determined in the field for each sample.

CMS chemical analyses for "B" triplicates followed the protocol summarized in "Monitoring Well Sampling and Analyses." U of Az chemical analyses for "C" triplicates used the following protocol: after dilution, the samples were analyzed for sodium (Na), calcium (Ca), potassium (K), and magnesium (Mg), using a Thermo Jarrell Ash IRIS-HR inductively coupled plasma (ICP) emission spectrometer; the spectrometer was calibrated using solutions made from pure salts. Because U of Az did not report chloride and sulfate (Cl and SO_4) concentrations, CMS values for Cl and SO_4 from matching "B" triplicates were used to estimate mole balance for each "C" triplicate sample.

Laydown Brine Flow-Rate Measurement

Laydown-brine flow rates were measured from two locations at BP no. 3: 1) a hand-held Swaffer Instruments Inc. Current/Flow Velocity flowmeter 2100-LX was used to measure flow where the active-pump discharges inside the concrete brine-storage vault, and 2) a Dynasonics Flowmeter model M3-901 connected to a Campbell Scientific CR10X datalogger continuously monitored flow from the active-pump discharge pipe at a point midway between the pump and the concrete vault (see figure 3).

A flow-rate range was measured with the 2100-LX in April 1998, and averaged 6,688 gpm. During the FY99 and FY2K pumping periods, an additional 23 measurements with the 2100-LX flowmeter averaged 6,685 gpm, which is in good agreement with the initial April 1998 average.

A recent compilation of Dynasonics M3-901 flow-rate measurements, taken over a 77-day period (December 15, 1999 through 29 February 2000), produced an average flow rate of 6,600 gpm, which is in reasonable agreement with the 2100-LX average of 6,685 gpm. However, Dynasonics flow rates were typically lower than 2100-LX flow rates. Because the Dynasonics was calibrated with 2100-LX-measured flow rates, the Dynasonics average will be considered an estimate until differences between both flow-meter values can be reconciled. Consequently, in this paper, the average 2100-LX flow-meter measurements are being used to calculate NaCl tonnages. For conversion of gallons of laydown brine to tons of NaCl, see White and Wadsworth (1999, appendix 3, equations 1-3).

Geochemical Modeling of Laydown and BSF Shallow-Aquifer Brine Mixing

A transient pond usually forms in low areas of BSF from November through March or April (Lines, 1979, p. 11, 51, 85-86). This pond is a combination of seasonal meteoric precipitation and shallow-brine aquifer ponding on the BSF surface during the winter months (Mason and Kipp, 1998, p. 65-66). Because the laydown-pumping cycle is from November through April, laydown brines mix with the transient pond, and ultimately with the shallow-brine aquifer.

Geochemical modeling was performed to determine if this mixing has any effect on the chemistry of shallow-brine aquifer and salt-crust composition, and to examine the fate of laydown-brine NaCl after being mixed with the water of the shallow-brine aquifer. A temperature-dependent chemical equilibrium model (TEQUIL) was used to: 1) calculate the equilibrium composition of six years of shallow-brine aquifer samples and six brine-mixing simulations, and 2) predict which salts would precipitate from these simulations at each step of a 45-step evaporation routine.

To assess the impact of mixing laydown brine with the shallow-brine aquifer, TEQUIL was used in the following applications:

- Determine if chemical differences exist among six different years of shallow-brine aquifer samples collected prior to the Laydown Project, and classify the annual sample suites by their predicted evaporative-mineral assemblages.

• Use the TEQUIL-predicted mineral assemblages to help classify any differences between the shallow-aquifer brine and laydown brine.

• Predict the mineral assemblages that would result from mixing laydown brine with shallow-aquifer brine (four mixing simulations).

• Predict the mineral assemblages that would result from mixing transient-pond brine with rainwater, and with shallow-aquifer brine (two mixing simulations).

• Identify the fate of laydown NaCl after mixing with the shallow-brine aquifer.

• Evaluate TEQUIL's integrity by simulating the solar-pond conditions used in potash production, and comparing modeled output with published data.

Salt-Crust Thickness Measurements

Since 1994, BLM has excavated 12 test pits and drilled 11 auger holes in the BSF salt crust. The purpose of the test pits was to: 1) identify the stratigraphic components of the salt crust, and 2) determine if a recognizable geologic datum exists that could be used to measure potential changes in salt-crust thickness after each 6-month pumping period. The auger holes were drilled to record the salt-crust thickness at specific locations. Two previous salt-crust thickness studies were conducted by the Utah State Department of Highways (UDOT) and BLM between 1960 and 1988 (McMillan, 1974; Brooks, 1991). UDOT and BLM each augered more than 100 holes in the salt crust during their respective studies. Salt-crust thickness differences between the UDOT and 1988 BLM studies are summarized in White and Wadsworth (1999, p. 36-37). Because the 1988 BLM auger-hole locations were surveyed (Brooks, 1991, p. 4), and their salt-crust thicknesses were the most recent historical measurements, 1988 thicknesses were compared with 1998 and 2000 BLM measurements to see if thickness changes could be observed. The locations of 1998 and 2000 auger holes were plotted on a computer-generated isopach map of the 1988 salt-crust thickness measurements. The 1988 isopach map was generated by the Radian Corporation program CPS/PC v. 4.2. Salt-crust thicknesses, corresponding to the 1998 and 2,000 auger-hole locations, were read directly from the 1988 isopach map through the computer program (J. F. Kohler, Geologist, BLM Utah office, personal communication, December 2000).

During a three-year period prior to the start of the Laydown Project (September 1994 and 1995, and October 1997), BLM excavated a total of 10 test pits along a 7-mile segment of the International Track, within 1 mile of the track centerline (see figure 7). This segment extended from mile post no. 4 to mile post no. 11 (original 1960 and 1974 UDOT locations). The 10 pits were located adjacent to six, reference BSF monitoring wells (BLM-42a, BLM-43c, BLM-46, BLM-60/USGS-2, BLM-93, and BLM-71A/USGS-9). The first few pits were excavated with a variety of hand tools (screwdriver and geologic pick, J. F. Kohler, Geologist, BLM Utah office, personal communication, September 1997). A chain saw was much more efficient in excavating the later pits. Successive salt and gypsum strata comprising the salt crust exposed at each pit site were described and their respective thicknesses measured.

Since the start of the Laydown Project (November 1997), BLM excavated two additional test pits and drilled nine auger holes to determine if measurable changes in salt-crust strata thickness had occurred. The two additional pits were located adjacent to the BLM-93 monitoring well. A chain saw was used to cut pits an average of 24 inches square and about 14 inches deep. The first test pit was excavated in August 1998, approximately 95 feet NW of BLM-93 (see figure 7). The second test pit was excavated in July 2000, within 10 feet of the August 1998 pit. Total salt-crust thickness at the July 2000 test-pit site was determined by drilling an auger hole adjacent to the pit. Salt and gypsum strata exposed in the two pits were described and their thicknesses measured.

During 1998-1999, five auger holes were drilled along the International Track with a 4.0-inch outside diameter soil auger. Auger-hole locations were determined by a Trimble ScoutMaster Global Positioning System receiver. Salt-crust strata were described and their thicknesses recorded. Total salt-crust thickness for each hole was determined by identifying the salt-crust/carbonate-mud interface in the recovered auger core, and by matching it with the corresponding auger-drilling depth.

Four additional auger holes were drilled in July and October 2000, and located near BLM-43C, BLM-46, BLM-60, and BLM-93. Each hole was started with an "Earthquake" model 8900E, 4-inch diameter motorized auger, and completed with an AMS 4.625-inch O.D. mud auger. Respective drilling depths for each tool were recorded. Each hole was completed with the mud auger so that the salt-crust/carbonate-mud interface could be visually verified in the recovered core. Matching the confirmed salt-crust/carbonate-mud interface with recorded drilling depths provided more precise salt-crust thickness measurements. Successive salt-crust strata, exposed in the five auger holes, were described and their thicknesses measured. Because ponded water was present on BSF, test pits were not excavated.

Salt-Crust Areal Extent Measurements

Landsat 5 Thematic Mapper images (Jensen, 1996, p. 37-44) for September 8, 1997, August 28, 1998, and October 8, 1999 were used to estimate the areal extent of BSF salt crust one month prior to, and during the first two years of, the Laydown Project. Because the imagery was selected and purchased after its flight date, field checking to confirm the salt-crust boundaries on those dates was not possible. A qualitative estimate of the salt-crust/mud-flat boundary location for each of the three years was determined by examining images produced by a color composite and selected individual bands. The color composite was made up of bands 2, 3, and 4 (Jensen, 1996, p. 40), and most closely approximated the color of the natural ground surface. Bands 2, 3, and 4 were also examined individually.

All three years of imagery showed that the salt-crust's north-west boundary remained relatively static, while the east boundary, adjacent to the Federal lease collection ditch, showed visible change during the 1997-1999 period.

The color composite images for each of the three years were examined first to estimate the location of the east salt-crust boundary. Areas of known salt crust and mudflat were compared with what appeared to be the east salt-crust/mudflat boundary, and a tentative east boundary was drawn. The boundaries for each delimited area were then superimposed on the corresponding Landsat scene, displayed as a single band gray-scale image (either band 3 or 4). The tentative east boundary from the color composite was compared with the single-band gray-scale image. If an eastern boundary suggested by the single band gray-scale image was in reasonable agreement with the boundary suggested by the color-composite image, then that boundary was used in the area calculations. Reasonable agreement between eastern boundaries from the color-composite and single-band gray-scale images was achieved for all three years of imagery.

Because I-80 and the Salduro Loop berm formed a consistent southern boundary for the study area, and the northwest boundary remained relatively static, the tentative east boundary was connected with the southern and northwestern boundaries to delimit the salt-crust area for each year. The area inside the Salduro Loop was not included in the measurements.

RESULTS AND DISCUSSION

Laydown brine has been delivered to BSF for the past three years with the objective of adding salt to the salt crust. The laydown brine was discharged to the salt crust from November through April where it mixed with the winter transient pond. Depending upon yearly weather conditions, this pond can cover most of the BSF playa surface (Mason and Kipp, 1998, p. 33). The transient pond typically begins to form on the surface of the salt crust in October or November when temperatures cool, evaporation decreases, and discharge from the shallow-brine aquifer pools at the land surface (Lines, 1979, p. 11, 51, 85-86; Mason and Kipp, 1998, p. 65-66). Because of its source and location (mainly on the salt crust), the transient pond is hydrologically connected with the shallow-brine aquifer.

Meteoric precipitation also contributes to formation of the transient pond. Rainwater from fall and winter storms falls on the developing pond and playa surface where it dissolves halite from the salt crust until the fresh water is saturated, and ultimately mixes with the transient pond by wind action (Mason and Kipp, 1998, p. 55). The transient pond eventually dissipates as evaporation from the pond increases during late spring and early summer. As the transient pond begins to shrink from evaporation, it becomes saturated with respect to NaCl. Eventually, the pond disappears when evaporation reaches its peak during the summer. The extent of mineral precipitation and the type of mineral species precipitated are governed by the chemical composition of the transient pond and the amount and duration of evaporation.

Although the engineering design estimated that up to 0.4 inches of NaCl would be added to the salt crust each year (Bingham, 1991, p. 2), the effects of adding laydown brine to the transient pond were unknown. The results that are presented focus on the following questions regarding the influence of adding laydown brine to the transient pond:

- How much dissolved salt was delivered to BSF by the Laydown Project?
- How much salt was added to the salt crust?
- Does shallow-aquifer brine chemistry vary historically, and how does it compare with laydown brine chemistry?
- How does the composition of salt, precipitated as a result of the Laydown Project, compare with that of pre-laydown salt crust?
- How much dissolved laydown salt may be assimilated into the shallow-brine aquifer?
- Has the composition of the minerals precipitated from the shallow-brine aquifer changed as a result of the Laydown Project?

Progress of Salt-Crust Restoration

Tons of Salt Delivered to BSF

After three years of operation, the Laydown Project has delivered 4.5 billion gallons of brine containing 4.6 million tons of NaCl to BSF (table 2). This amount exceeds the expected tonnage at this stage of the project - the feasibility study (Bingham, 1991, p. 2) anticipated yearly tonnages of up to 1.5 million tons.

Because of start-up problems during the first year of the project, 825,206 tons of NaCl, or about half of the anticipated tonnage, was produced in FY98. However, the FY98 tonnage nearly matched the 850,000 tons of annual salt loss estimated by the USGS (Mason and Kipp, 1999, p. 93). The lower than expected FY98 tonnage was mainly due to the mechanical failure of BW no. 1 that caused a 37-day down period from November 15 to December 22, 1997.

An additional delay, totaling 27 days in 1998, occurred because of deteriorating brine quality in primary pond no. 4 (PP no. 4). Brine densities for December and January were 1.080 and 1.074 g/mL, respectively (table 3), and corresponding salt (NaCl) tonnages delivered to BSF were 30,739 and 80,359 tons (appendix A4). Brine pumping was discontinued from January 28 through February 16, 1998 while the system was modified to increase salt dissolution (see White and Wadsworth, 1999, p. 22-23). As a result, average brine densities increased from respective values of 1.080 and 1.074 g/mL in December and January, to respective values of 1.178 and 1.148 g/mL in late February and March (table 3). February and March NaCl tonnages increased to 129,792 and 278,446 tons, respectively (appendix A4). When brine quality dropped below acceptable limits in April, Reilly again closed the discharge gate at the southeast corner of PP no. 4 for a seven-day period (April 9-15, 1998) until brine quality improved.

During FY99, the second year of the Salt-Laydown Project produced 1.965 million tons of NaCl (table 2). This tonnage was 1.3 times the anticipated tonnage (1.5 million tons per year) and 2.3 times the USGS estimated annual salt loss. Additionally, the FY99 salt tonnage was 2.4 times the total salt tonnage delivered to BSF during FY98. The marked increase in FY99 salt tonnage was due to improvements made to the laydown facility during the summer and fall of

1998. Improvements included doubling the time required to fill PP no. 4 (from 30 to 60 days), and ripping the pond's salt-crust surface. FY99 brine-density values (table 3) also reflected effects of the facility improvements; average densities approached NaCl saturation and had a narrow range of 1.183 to 1.194 g/mL for the 6-month period.

Effects of the laydown facility improvements continued and are reflected in FY2K production figures. The third year of the project produced 1.833 million tons of NaCl, or 1.2 times anticipated tonnage and about 2.2 times the USGS estimated annual salt loss. Although brine density values decreased slightly in January and February, brine density was increased in March by bulldozer ripping the salt-crust surface of PP no. 4 (table 3).

Accuracy of Chemical Analyses

Sixty-nine of the 85 samples of laydown brine, collected during the first three years of the Laydown Project, were subjected to interlaboratory analyses by Reilly, CMS, and U of Az. Agreement for NaCl analyses among the laboratories was very good, with replicate analyses being within about 6 percent of the mean value (appendix A5).

The interlaboratory NaCl analyses of laydown-sample triplicates collected by BLM during FY98 through FY2K are summarized in table 4. Reilly and CMS average analyses of the FY2K samples were in excellent agreement (0 percent difference), as were their respective average analyses of the FY99 samples (2 percent difference from the mean). Reilly, CMS, and U of Az average analyses of the FY98A samples were also in excellent agreement (2.2 percent difference from the mean), while good agreement was observed between Reilly, CMS, and U of Az average analyses of the FY98B samples (5.7 percent difference from the mean).

Amount of Salt Added to the Salt Crust

Area - Based on measurements from Landsat 5 imagery, the areal extent of the salt crust increased by about 5 square miles or 17 percent during a two-year period (September 1997 to October 1999). BSF salt-crust areas measured from fall 1997, 1998, and 1999 Landsat 5 scenes were approximately 26, 29, and 31 square miles, respectively. Figure 8 shows the 1997, 1998, and 1999 salt-crust boundaries super-

Table 2. FY98-2K totals of laydown-NaCl tonnage, acre feet, gallons, and pump hours.

TOTALS	NaCl,	Pumped Brine		Hours Pumped Per Year (Nov.-Apr.)		
	Tons (dry)	Acre Feet	Gallons	Maximum Possible	Actually Pumped	% of Possible
FY98	825,206	3,161	1,030,008,750	4,344	2,567	59.1
FY99	1,965,068	5,335	1,738,215,000	4,344	4,332	99.7
FY2K	1,833,074	5,345	1,741,425,000	4,368	4,340	99.4
FY98-2K	4,623,348	13,841	4,509,648,750	13,056	11,239	86.1

Table 3. Average monthly laydown-brine-density values (expressed in g/mL) for November through April during FY98-FY2K[1].

MONTH	NOV	DEC	JAN	FEB	MAR	APR
FY98	1.161	1.080	1.074	1.178	1.148	1.133
FY99	1.194	1.192	1.183	1.187	1.186	1.189
FY2K	1.187	1.178	1.152	1.145	1.178	1.202

[1] Density values based on daily brine-sample measurements conducted in Reilly's Wendover laboratory.

Table 4. Three years of laydown-brine NaCl analyses; all samples collected by BLM and analyzed by Reilly, CMS, and U of Az; average Na and Cl concentrations converted to chloride-salt equivalents (expressed as wt percent in solution).

SAMPLES	n	Reilly	CMS	U of Az	mean
FY2K	29	21.62	21.62	ND	21.62
FY99	23	21.97	21.13	ND	21.55
FY98A	9	7.59	7.39	7.69	7.56
FY98B	8	17.49	16.05	16.12	16.55

ND Not determined

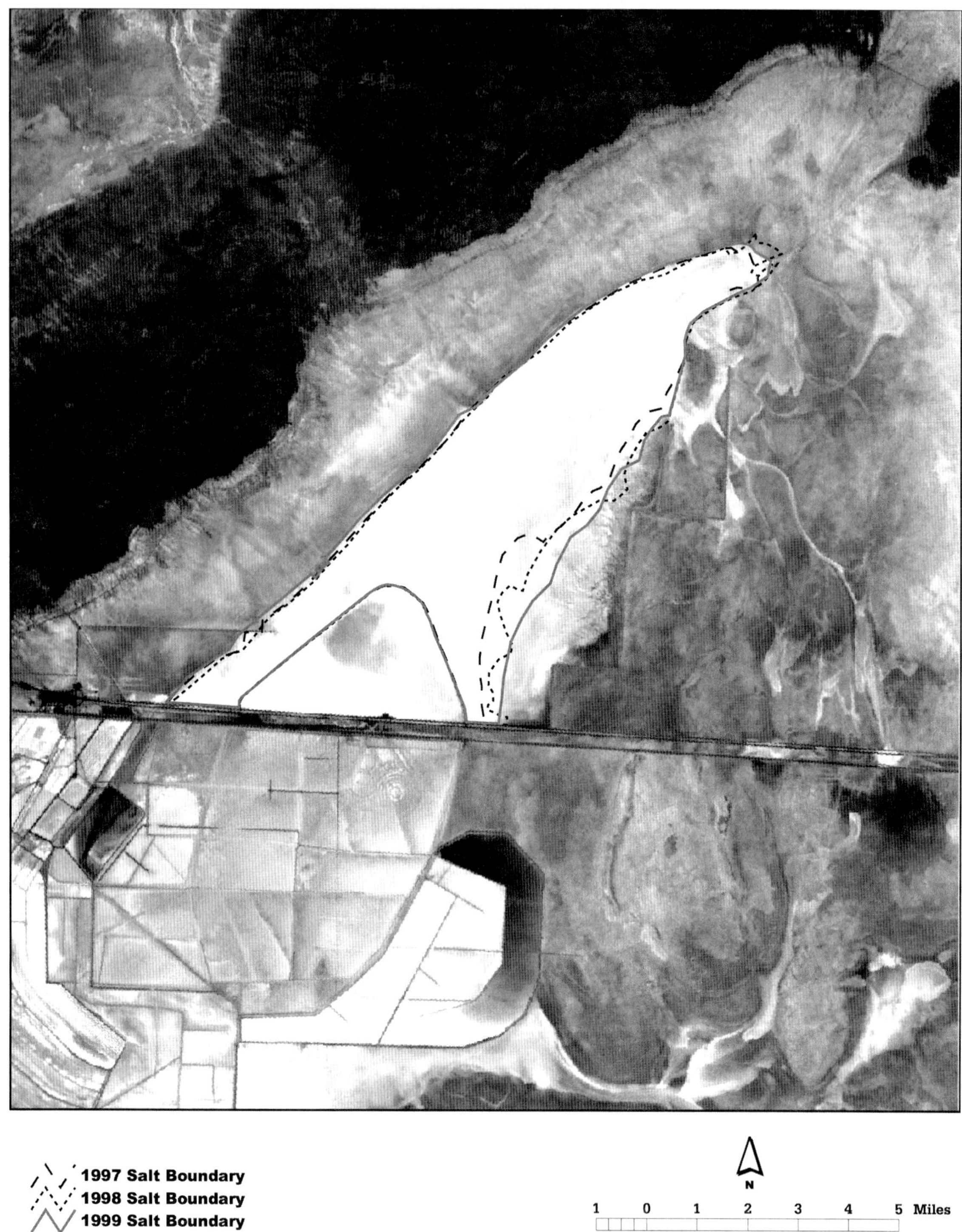

Figure 8. *1999 Landsat 5 image showing progressive increase in salt-crust area from September1997 through October 1999.*

imposed on the 1999 Landsat 5 image of BSF. The 1999 image was selected to show the compared areas because its salt-crust areal extent was the largest of the three years measured.

Because 2.8 million tons of NaCl had been delivered to BSF between November 1997 and October 1999 (FY98 and FY99 laydown tonnage - see table 2), estimates of NaCl tonnage contained in the 5-square mile area were made for comparison with the amount delivered during those two years. The estimated tonnage contained in the new area of salt crust was based on the following:

- Reported halite densities (dry) ranged from 79.4 to 109.8 pounds per cubic foot (Brent Bingham, 1998, Bingham Engineering, personal communication; Mason and Kipp, 1998, p. 54).
- New salt-crust thickness was estimated to range from 0.25 to 1.0 inches (from preliminary thickness measurements taken in June 1998).

Based on these parameters, new salt-crust tonnages for the 5-square mile area ranged from about 0.12 to 0.64 million tons. The 0.64 million tons represents about 23 percent of the 2.8 million tons of NaCl delivered by the Laydown Project during FY98 and FY99. To contain 2.8 million tons of salt, the 5-square mile area of new salt crust would have to average more than 4 inches thick. Salt-crust thickness studies, and geochemical modeling of the effects of mixing laydown brine with the shallow-aquifer brine (described in the following sections), suggest that the remainder of the laydown salt tonnage not included in the 5-square mile area was distributed between the main body of the salt crust (26 square miles in 1997) and the shallow-brine aquifer.

Surface Stratum Thickness - Based on the salt-replenishment feasibility study by Bingham (1991, p. 2), it was anticipated that about 0.4 inches per year of salt thickness would be added to the surface of the salt crust. To quantify this addition, BLM evaluated two methods for measuring changes in thickness. The first method was to measure elevation differences between the salt-crust surface and the surveyed top-of-casing elevations from selected monitoring wells located on the salt crust. However, the combination of uneven salt-crust surfaces and the cratering of the salt-crust surface surrounding each casing (caused by rainwater running down the sides of the casing) made thickness-change measurements highly subjective. For this reason, this method was not used.

The second method was to use a distinct salt crust stratum as a horizon from which to reference all thickness measurements. Such a stratum was identified, based on 10 test pits excavated by the BLM between 1994 and 1997. These pits were adjacent to six reference BSF monitoring wells (BLM-42a, BLM-43c, BLM-46, BLM-60/USGS-2, BLM-93, and USGS-9) located along the trend of the International Track (see figure 7). Examination of these pits indicated that five distinct strata comprised the salt crust and were consistent in sequence and composition from pit to pit (table 5; Kohler, 1995, and Kohler and White, 1997).

An uncemented gypsum stratum, ranging from 0.5 to 1.2 inches thick, was always the first stratum underneath the dense-cemented halite stratum (the bedded halite that makes up the surface of the salt crust). The upper surface of the gypsum stratum was used as the thickness-measuring datum for all salt-crust thickness measurements reported in this paper. Note that the sequence contains two different uncemented gypsum strata, and that the first (top) gypsum stratum is the measurement datum. More detailed descriptions of each salt-crust strata are in appendix 6.

Thickness measurements for several of the strata were complicated by the presence of undulating top and bottom surfaces. The dense-cemented halite stratum typically had an undulating bottom surface, and the cemented-coarse-porous halite stratum had undulating top and bottom surfaces. To account for the variable thickness caused by these surfaces, three measurements were usually taken of each stratum (one in the middle, and one at each end of the stratum length exposed in the test pit). Based on these three measurements, thicknesses for each stratum were usually presented as a range of values. However, to facilitate stratum thickness comparisons from year to year, the range of thickness values for each stratum were averaged.

Two additional test pits and nine auger holes were excavated during 1998-2000 to determine variations in the thickness of the dense-cemented halite stratum as well as other strata. These excavations provided salt-crust, stratum-thickness comparisons with four of the10 test pits excavated during the 1994-1997 period, and a limited comparison of total salt-crust thickness with 1988 thickness measurements by Brooks (1991). Tables 6 and 7 list stratum thickness comparisons at BLM-93, BLM-43c, BLM-46, and BLM-60. The following observations can be made from these comparisons:

- Dense-cemented halite stratum thickness increased at BLM-93 by 0.4 to 0.5 inches (an approximate 20-25 percent increase), and decreased by 0.9 to 1.3 inches (an average decrease of about 35 percent) at BLM-43c, BLM-46, and BLM-60.
- Cemented-coarse-porous halite stratum thicknesses increased by 0.4 and 2.2 inches (about a 15 and 120 percent increase) at BLM-43c and BLM-60 sites when dense-cemented halite stratum thicknesses decreased at the same locations.

Table 5. Salt-crust strata sequence and thickness ranges from 10 BLM test pits (Kohler, 1995, and Kohler and White, 1997).

STRATA	Thickness, inches
Dense-cemented halite (surface bed of the salt crust)	1.3 to 2.9
1st uncemented gypsum mixed with carbonate clay	0.5 to 1.2
Cemented-coarse-porous halite	1.4 to 7.7
2nd uncemented gypsum mixed with carbonate clay	0.4 to 1.8
Uncemented-coarse halite	18.0 to 24.0[1]
Carbonate clay	ND

ND Not determined

[1]Thickness range is based on one test pit which was excavated to the salt/mud interface.

It was originally thought that annual measurements of the surface stratum (dense-cemented halite) would show a gradual thickness increase from salt addition by the Laydown Project (0.4 inches per year). However, based on thickness comparisons in tables 6 and 7, dense-cemented halite stratum thicknesses increased at BLM-93, but decreased at BLM-43c, BLM-46, and BLM-60. These thickness decreases occurred, despite an addition of 4.6 million tons of NaCl salt to BSF during the first 3 years of the Laydown Project. Several factors may have contributed to the measured decrease in dense-cemented halite stratum thickness:

- Annual variations in weather.
- Laydown salt contributions to the shallow-brine aquifer.
- Laydown salt contributions to increased salt-crust area.
- Laydown salt contributions to other salt-crust strata.

Annual variations in weather can produce changes in salt-crust strata thicknesses. The thickness of the dense-cemented halite stratum would tend to decrease if an unusually wet winter were followed by a cool summer (when rain fall is high and evaporation is low due to colder temperatures). An unusually wet winter could increase dissolution of the dense-cemented halite stratum thickness because of increased rainfall on the surfaces of the salt-crust and the transient pond. For example, Mason and Kipp (1998, p. 55) estimated 10 to 14 million tons of salt were dissolved from the salt crust just north of I-80 during the winter of 1993. This was due to greater than normal precipitation in January, and unseasonably cool temperatures in January and February. During a cool summer, less evaporation could reduce replenishment of the dense-cemented halite stratum. Conversely, if a dry winter were followed by a hot dry summer the dense-cemented halite stratum thickness would tend to increase. A dry winter would lead to less salt-crust dissolution. A dry summer would tend to increase evaporation of the shrinking transient pond and discharging shallow-aquifer brine; consequently, salt precipitation, and salt-crust thickness would be increased. Annual thickness changes in the dense-cemented halite stratum, due to changes in annual weather conditions, would make it difficult to measure the thickness added to the surface stratum by the Laydown Project, regardless of the measurement method. For example, a decrease in thickness of 0.5 inches (or more) in the dense-cemented halite stratum from weather-related conditions could easily mask the annual 0.4-inch increase in salt-crust thickness predicted by the Laydown Project design. Depending upon the amount of halite dissolved from the surface stratum, an apparent thinning of the dense-cemented halite stratum could occur during unseasonably wet years in spite of increased salt tonnage added to BSF by the Laydown Project.

TEQUIL modeling calculations indicate that between 17 and 25 million tons of laydown-brine NaCl could be assimilated into the shallow brine aquifer (see "Fate of laydown-brine NaCl when mixed with the shallow-brine aquifer"). Whereas this salt would eventually be incorporated into the salt crust as halite, there may be a lag time between the period of laydown-brine delivery and this incorporation. Consequently the effects of this salt addition would not be measured until some time in the future.

As was previously discussed, laydown NaCl mass added to the salt crust can manifest itself through increases in thickness or increases in area. From September 1997 to October 1999 the area of the BSF increased by five square miles. This area could contain as much as 0.6 million tons of laydown NaCl.

Laydown NaCl mass added to the salt crust may also be added to the cemented-coarse-porous halite, and the uncemented-coarse halite strata (tables 6, 7). Data from test pits adjacent to monitoring well BLM-93, and those adjacent to monitoring wells BLM-43C and BLM-60, can be used to assess changes in the thickness of the cemented-coarse-porous halite stratum. All three sites indicate substantial increases in the thickness of the cemented-coarse-porous halite stratum measured in 2000 (tables 6, 7). It is conceivable that some of this increase is the result of the Laydown Project. Because the subsurface salt-crust strata (cemented-coarse-porous halite or uncemented-coarse halite strata) are saturated by the shallow-brine aquifer, their observed thickness changes may be due to alternating halite crystal growth and dissolution:

- Existing halite crystals could act as nucleating sites for additional crystal growth, especially during the summer when evaporation of shallow-brine aquifer could cause a concentration gradient in the brine.
- Conversely, a wetter than normal winter, such as 1993 (Mason and Kipp, 1998, pp 26, 55), would lower shallow-aquifer brine densities (that is, 1.186 gm/cm^3 for 1993 vs 1.194 gm/cm^3 for 1992 - see table 9) which could result in some partial halite-crystal dissolution.

Total Salt-Crust Thickness - During October 1998 and September 1999, five auger holes (MB-1 - MB-3, MB-6, and MB-7) were drilled by the BLM in the salt crust adjacent to the International Track between mile posts no. 1 and no. 5 (see figure 7). Total salt-crust thicknesses measured from these five auger holes, and from four holes drilled in 2000 (adjacent to BLM-43c, BLM-46, BLM-60, and BLM-93),

Table 6. Salt-crust strata sequence and thickness ranges from 5 BLM test pits adjacent to monitoring well BLM-93.

STRATA	Strata Thickness, inches				
	1994 (pit 1a)[1]	1995 (pit 1b)[1]	1997 (pit 1c)	1998 (pit 1d)	2000 (pit 1e)
Dense-cemented halite	1.8	1.9	1.9	2.25 to 2.5	1.5 to 3.0
1st uncemented gypsum mixed with carbonate clay	1.0	0.7	0.6	1.0 to 1.25	1.0
Cemented-coarse-porous halite	2.9	2.4	2.0 to 3.5	1.0 to 2.25	4.5 to 7.5
2nd uncemented gypsum mixed with carbonate clay	0.6	0.6	0.5 to 0.75	0.75 to 2.25	0.0 to 1.0
Uncemented-coarse halite	ND	ND	ND	ND	29.0[2]

ND Not determined
[1]Kohler, 1995 unpublished field notes.
[2]Total thickness of uncemented-coarse salt (depth to salt/mud interface measured).

Table 7. Strata-thickness comparisons between three October 2000 auger holes and three 1994 test pits located adjacent to BLM-43c, BLM-46 and BLM-60.

STRATA	STRATA THICKNESS, INCHES					
	BLM-43c		BLM-46		BLM-60	
	1994	2000	1994	2000	1994	2000
Dense-cemented halite	2.9	2.0	3.1	2.0 to 2.25	2.8	1.5
1st uncemented gypsum	0.8	1.0	0.8	0.5	1.0	1.0
Cemented-coarse-porous halite	2.6	3.0	ND	NP	1.8	4.0
2nd uncemented gypsum	ND	NP	-	NP	0.5	13.0[1]
Uncemented-coarse halite	-	25.8	-	21.25	ND	-

ND Not determined
NP Not present
[1]Total depth of salt crust measured.

were compared with 1988 total salt-crust thicknesses measured by Brooks (1991) near the same locations. A slight decrease in thickness from 1988 to 1998-2000 was observed for seven of the nine locations. One location (MB-7) showed a slight increase (+ 0.2 feet). With the exception of BLM-93 (which showed a decrease of 0.4 feet), thickness differences were from 0 to 0.2 feet, and were within the margin of measurement error reported by Kohler (1995).

Brooks (1991, p. 3) measured 1988 total salt-crust thicknesses from each auger hole using a method similar to that used by UDOT, and reported difficulty in replicating measurements within the same hole. A hook mounted at the base of a 1-inch diameter, 5-foot long pole was dragged up along the auger-hole wall from the bottom of the hole to the surface of the salt crust. The interface between the salt crust and the underlying carbonate mud was indentified by catching the hook on the underside of the more resistant of the two surfaces that comprise the interface. The distance from interface to salt-crust was measured by reading the footage increment on the pole where it intersected the salt-crust surface. Kohler (1995) also attempted to replicate total salt-crust thickness measurements within the same hole using the USDH method, and reported thickness-measurement differences of ± 0.1 foot.

Although table 8 summarizes thickness changes over a 10 to 12-year period, only the most recent three of the 12 years would have been affected by the Laydown Project. Additionally, this preliminary comparison is based on a small population of data from nine drill-hole locations. To more accurately assess effects of the Laydown Project on total salt-crust thickness, thickness measurements from a larger population of drill holes need to be collected on a yearly schedule.

This assessment could be accomplished by drilling the same number of holes each year for at least a five-year period. The 14 transect lines Brooks (1991, figure 4) established in 1988 could be used to locate the holes. These transect lines were oriented perpendicular to the trend of the International Track and spaced on 1-mile centers. From three to five holes per transect could be drilled close to the same locations each year. This would provide yearly thickness measurements from 42 to 70 holes. With this larger population of measurements being collected annually over a 5-year period, a trend in total salt-crust thickness change could be established, and the influence of the Laydown Project on that thickness change could be assessed with greater confidence.

Table 8. Comparison of salt-crust thickness measurements in nine auger holes drilled in 1998-2000 with correlative 1988 thicknesses.

Auger Hole no.	1988[1] Thickness, ft.	1998-2000[2] Thickness, ft.
MB-1	1.2	1.0
MB-2	1.0	1.0
MB-3	1.7	1.6
MB-6	1.9	1.8
MB-7	2.8	3.0
BLM-43c	2.8	2.7
BLM-46	2.2	2.0
BLM-60	1.7	1.6
BLM-93	3.9	3.5

[1]Brooks, 1991
[2]MB holes measured in 1998, and BLM-43c, etc., measured in 2000.

Brine Composition

Comparison of 1994-2000 Shallow-Aquifer Brine Samples with Historical Data

Five years of recent BLM samples of shallow-aquifer brine were compared with 186 USGS samples that were collected over a 17-year period (table 9). Sodium and chloride ions dominated both groups of samples. Sodium made up about 91 percent of cation content, and chloride made up 97 percent of anion content in both BLM and USGS samples. Magnesium and potassium jointly made up about 8 percent

Table 9. Comparison of recent (BLM) samples with historic (USGS) samples of the BSF shallow-aquifer brine (samples usually collected during August and September unless otherwise indicated).

SAMPLE SUITE	n	Density	Average Major Ion Concentrations, moles/L							Avg Mole Balance[1]
			Na	Mg	K	Ca	Cl	SO_4	TDS	
BLM:										
FY2K	10	1.173	3.74	0.102	0.110	0.033	4.09	0.051	244	0.003183
FY97	20	1.185	3.95	0.109	0.134	0.028	4.24	0.058	256	0.001462
FY96	16	1.182	4.11	0.140	0.155	0.021	4.56	0.051	282	0.003822
FY95	13	1.196	4.47	0.189	0.152	0.026	4.71	0.052	286	0.010115
FY94	19	1.192	4.47	0.224	0.162	0.022	5.00	0.055	297	0.002707
USGS:										
FY93	20	1.181	4.31	0.154	0.139	0.028	4.37	0.060	297	0.013832
FY92	26	1.185	4.44	0.162	0.170	0.028	4.50	0.070	309	0.014853
FY81- B[2]	34	1.185	4.22	0.120	0.095	0.031	4.70	0.053	276	-0.007897
FY81- K[2]	17	1.188	4.37	0.141	0.085	0.032	4.77	0.055	283	-0.003655
FY81- L[2]	18	1.189	4.23	0.134	0.114	0.030	4.80	0.053	281	-0.009904
FY76-78- K	10	1.167	3.88	0.114	0.120	0.030	4.23	0.055	253	0.002155
FY76-78- USGS	43	1.189	4.51	0.148	0.159	0.031	4.86	0.056	293	-0.002053
1925	18	ND	ND	0.126	0.131	ND	ND	ND	ND	ND

TDS Total Dissolved Solids, g/L
ND Not determined
[1]Calculated using method of Sturm and others, 1980, p. 175; acceptable limits are ± 0.0055 moles.
[2]Samples were collected in May.

of cation content in both sample groups, with magnesium and potassium averaging 3 and 5 percent, respectively. Although some yearly variability is evident between individual cation and anion concentrations, their respective percents of total cation and anion content remained relatively constant. Average potassium and magnesium concentrations from 18 samples of shallow-aquifer brine collected by USGS in 1925 (Nolan, 1927, Plate 3 - see appendix A1) were included for comparison with the more recent USGS and BLM samples. The 1925 potassium and magnesium concentrations are in reasonable agreement with the concentration range of the more recent samples. Although (TDS) reached a peak of 309 g/L in FY92, the TDS range from 1976 to 1981 was 253 to 293 g/L, which was similar to the 1993-2000 range of 244 to 297 g/L.

The use of mole ratios to help classify geologic processes and their products is a common geochemical practice. Ratios have been used to help distinguish between two different water sources (White, 1955, p. 148) and to help understand crystallization pathways in brines (Krauskopf, 1967, p. 341-343). More recently, Kohler (this volume) used K/Mg and Ca/SO_4 molar ratios to compare differences between brines from the Great Salt Lake and the Newfoundland Basin. Mole ratios are used in this paper to:

- Identify any subtle variability or trends in shallow-aquifer brine chemistry.
- Determine if shallow-aquifer brine chemistry is affected by laydown brine.

Mole ratios of K/Mg, Cl/Na, and SO_4/Ca were calculated from average molar concentrations of the BLM and USGS shallow-aquifer samples listed in table 9. The BLM and USGS shallow-aquifer samples comprise 13 suites of samples, and their respective mole ratios are compared in table 10.

Historically, the Cl/Na ratio was the most stable over time, ranging from 1.052 to 1.135. All reported values were within 4.2 percent of the mean (1.089) for 12 sample suites. The 13th sample suite (composed of 18 samples collected in 1925) was analyzed for potassium and magnesium only, so it is not included in the discussion of Cl/Na and SO_4/Ca ratios. This stability, and the ratio near 1:1, are consistent with the probability that the sodium and chloride concentrations of the shallow-aquifer brine are controlled by dissolution and precipitation of halite.

The SO_4/Ca was more variable, with values ranging from 1.532 to 2.475. The variation from the mean value of 2.002 was as much as 24 percent. The K/Mg ratio was the most variable of the three ratios, with values ranging from 0.604 to 1.234. Values varied from the mean (0.952) by as much as 37 percent.

Why K/Mg ratios for 5 of 13 suites (three sample suites collected by USGS in 1981 and two sample suites collected by BLM in 1994 and 1995) were 37 percent less than a ratio of about 1.0 is unknown. The 1981 USGS samples (which had the lowest ratios) were the only monitoring-well samples collected during the spring (May). This time of collection may have had some effect on the K/Mg ratio, although dilution alone should not have changed the ratio.

The BLM's FY94 and FY95 samples (with ratios of 0.721 and 0.805) were collected in the fall (September) when their concentrations were at a maximum due to effects of summer evaporation. Mason and Kipp (1998, p. 26, 55) reported wetter than normal conditions on BSF during the winter and summer months of 1993, which resulted in extensive flooding of the salt crust. Lower K/Mg ratios for the FY94 and FY95 samples could possibly be a delayed effect

Table 10. Comparison of average mole ratios from 13 suites of shallow-aquifer brine samples (from monitoring wells - MW, and auger holes - AH).

SAMPLE SUITES	n	Average Mole Ratio		
		K/Mg	Cl/Na	SO_4/Ca
BLM-MW: 1994-2000				
FY2K	10	1.080	1.093	1.532
FY97	20	1.234	1.072	2.085
FY96	16	1.108	1.109	2.403
FY95	13	0.805	1.054	2.000
FY94	19	0.721	1.118	2.468
USGS-MW (Mason et al., 1995): 1992-1993				
FY93	20	0.900	1.013	2.119
FY92	26	1.054	1.013	2.486
USGS (CD[1]): 1981				
B (AH)	34	0.791	1.114	1.699
K (MW)	17	0.604	1.093	1.732
L (AH)	18	0.855	1.135	1.752
USGS-MW (Lines, 1978): 1976-1978				
K	10	1.054	1.089	1.812
USGS	43	1.078	1.079	1.817
USGS-AH (Nolan, 1927): 1925	18	1.040	ND	ND

ND Not determined
[1]Conservation Division

of these conditions, if: 1) potassium and magnesium were leached from lacustrine clays or marine sedimentary rocks adjacent to BSF, and 2) slow, lateral subsurface inflow transported the additional potassium and magnesium towards BSF (Turk, 1969, p. 156-158). However, the average K/Mg ratio for nine of the monitoring-well samples collected by USGS during late June 1993 was 1.02.

The FY2K monitoring-well samples had the potential to be affected by two years of laydown brine delivery. However, the average FY2K K/Mg mole ratio (1.080) was generally in good agreement with sample suites that pre-dated the start of the Laydown Project (the exceptions were the three 1981 sample suites that were collected in the spring, and the FY94 and FY95 sample suites). In contrast, the FY2K SO_4/Ca mole ratio (1.532) was significantly lower and differed from the 12 sample-suite mean by 23 percent (there were no sulfate and calcium data for the 1925 sample suite). The lower ratio was due to a lower average sulfate concentration and a higher average calcium concentration. The cause of this difference is unknown. Additional, yearly shallow-aquifer brine samples will be required to see if a trend exists or if this ratio is part of the natural variability.

Comparison of Laydown Brine with Shallow-Aquifer Brine

Although the FY98 laydown brine contained about 15 to 64 percent less salt (as indicated by solution density) than the FY2K BSF shallow-aquifer brine, the FY99 and FY2K laydown-brine values were close in density to those of the shallow-aquifer brine (table 11). Overall composition of the laydown brine was generally similar to that of the BSF shallow-aquifer brine in that sodium and chloride were the dominant ions in both waters. On average, sodium made up 98 percent of the cation content in the laydown brine and 90 percent of that in the shallow-aquifer brine. Chloride made up nearly 98 percent of the anions in both waters. Concentrations of sodium and chloride in the FY2K laydown brine were 16 and 8 percent higher than those in the FY2K shallow-aquifer brine. The slightly elevated values are not unexpected because NaCl is the main salt present in PP no. 4.

Compared to sodium and chloride, magnesium, potassium, calcium, and sulfate were present in relatively low concentrations in both shallow-aquifer and laydown brines. Potassium and magnesium made up 5 and 4 percent of the average cation content for the FY94-2K shallow-aquifer brines. However, magnesium and potassium concentrations in the laydown brines were an order of magnitude less than those in the shallow-aquifer brine. Traces of potassium and magnesium in the laydown brine may have been contributed from microscopic volumes of shallow-aquifer brine entrapped as fluid inclusions in the precipitated halite in PP no. 4, and possibly entrained in the pore spaces between halite crystals. Recent microscopic examination of similar halite salt-crust samples from an adjacent Great Salt Lake Desert basin revealed presence of fluid inclusions in the halite (B. F. Jones, Research Geochemist, USGS, personal communication, September 5, 2000). Calcium concentrations in the two waters were similar, while sulfate concentrations in the laydown brine were about half those in the shallow-aquifer brine (table 11).

The mineral species that precipitate from a given solution upon evaporation, and the mass of minerals precipitated,

Table 11. Five years of shallow-aquifer brine analyses from selected BSF monitoring wells (MW) compared with 3 years of laydown-brine analyses; samples collected by BLM and analyzed by CMS.

SAMPLE SUITE	n	Density	Average Major Ion Concentrations, moles/L							Avg Mole Balance[1]
			Na	Mg	K	Ca	Cl	SO_4	TDS	
BSF MW Brine[2]										
FY2K	10	1.173	3.74	0.102	0.110	0.033	4.09	0.051	244	0.003183
FY97	20	1.185	3.95	0.109	0.134	0.028`	4.24	0.058	256	0.001462
FY96	16	1.182	4.11	0.140	0.155	0.021	4.56	0.051	282	0.003822
FY95	13	1.196	4.47	0.189	0.152	0.026	4.71	0.052	286	0.010115
FY94	19	1.192	4.47	0.224	0.162	0.022	5.00	0.055	297	0.002707
Laydown Brine[3]										
FY2K	30	1.164	4.33	0.012	0.014	0.033	4.43	0.031	262	0.004198
FY99	27	1.185	4.29	0.013	0.009	0.034	4.45	0.031	261	0.004801
FY98A	9	1.058	1.34	0.005	0.005	0.016	1.32	0.016	80	0.001891
FY98B	8	1.130	3.16	0.008	0.008	0.025	3.33	0.026	195	0.006570

TDS Total Dissolved Solids, g/L
[1] Calculated using method of Sturm and others (1980, p. 175); acceptable limits are ± 0.0055 moles.
[2] Samples collected in August and/or September from selected BSF monitoring wells in a 27-well suite.
[3] With exception of FY98A and B, average sampling frequency was two times per month (Nov through April); FY98A and B were collected in January and March-April 1998, respectively.

are dependent on the solution composition. The elevated concentrations of sodium and chloride in the laydown brine indicate that more halite will precipitate from a unit volume of laydown brine than from the same volume of shallow-aquifer brine. Conversely, the comparatively low concentrations of magnesium, potassium, and sulfate in the laydown brine indicate that minerals containing these components will be less likely to precipitate from the laydown brine than from the shallow-aquifer brine. These statements are generalizations based on evaporating fixed volumes of both brines in a closed system, contrasted with the open system of the BSF.

Predicted Mineral Precipitation from Transient-Pond Brine

The BSF transient pond currently receives input from the shallow-brine aquifer, meteoric precipitation, and the laydown Project. To identify which minerals could precipitate from the transient-pond brine, its chemistry was simulated using a range of compositions as input for TEQUIL modeling. The compositions used were those of the pre-laydown shallow-aquifer brine, the laydown brine, a 90 percent shallow-aquifer brine + 10 percent laydown mixture, a 50 percent shallow-aquifer brine + 50 percent laydown mixture, and the post-laydown shallow-aquifer brine. Effects of rainwater and shallow-aquifer brine on the minerals predicted to precipitate were also examined.

Simulation of Mixing Laydown Brine with the Transient-Pond Brine

The shallow-aquifer brine and the laydown brine represent the two compositional extremes possible for mixtures used to simulate effects of mixing laydown brine with the transient pond. Simulated transient pond compositions were based solely on contributions from shallow-aquifer and laydown brines (input from rainfall was ignored). Two different mixtures of the two components were simulated for modeling purposes and their mixing ratios were selected based on assumptions described below.

The 90 percent shallow-aquifer brine + 10 percent laydown brine mixture simulates a condition in which the transient pond interacts with the shallow-brine aquifer as a well-mixed system. The brine volume in this system was then mixed with the laydown-brine volume delivered to BSF during FY98 through FY2K (4.5 billion gallons; table 2) (complete mixing of laydown brine with the 52.5 billion-gallon volume of the shallow-brine aquifer probably does not occur because the increased hydraulic head from addition of laydown volume reduces the upward gradient of the discharging shallow brine aquifer; however, to compute the mixing ratio, complete mixing was assumed). The volume of the BSF shallow-brine aquifer affected by the Laydown Project was estimated as 52.5 billion gallons (the transient-pond volume was assumed to be contained in this volume). This estimate was based on an average shallow-brine aquifer thickness of 20 feet, an average aquifer porosity of 0.45 percent (Mason and Kipp, 1998, p. 54), and an area of 28 square miles (identified as the affected area in the salt-replenishment feasibility study [Bingham, 1991, p. 1]). The shallow-brine aquifer and laydown brine volumes represent 92 percent and 8 percent of the total volume, respectively ($100 \times 52.5/[52.5 + 4.5] = 0.92$; $100 \times 4.5/[52.5 + 4.5] = 8$). These values were rounded to 90 percent and 10 percent for modeling.

The 50 percent + 50 percent mixing ratio simulates a condition in which the transient-pond volume does not mix with the shallow-brine aquifer, and only the transient-pond volume is considered. Although it is known that the transient pond and shallow-brine aquifer are hydrologically connected, this simulation provides a case in which the laydown-brine component is larger than in the previous case. The volume of the pond was determined and mixed with the volume of laydown brine added in one year. Transient-pond volume was estimated to be about 500 million gallons per 1 inch of pond depth over an area of 28 square miles. Because transient-pond depth was reported to be from 3 inches to more than 12 inches deep in some areas (White, 1998, p. 37; Mason and Kipp, 1998, p. 55), a pond depth of 3 inches was chosen to calculate the pond volume. Monthly laydown volume averaged 290 million gallons during FY99 and FY2K. Using these conditions, the transient-pond volume could be as much as 1.5 billion gallons and total laydown volume for six months could be about 1.7 billion gallons, or nearly a 50-50 ratio.

Chemical analyses from FY94 and FY97 samples of the shallow-brine aquifer were selected as the end members that represent the compositional range of the shallow-brine aquifer. Pre-laydown transient pond composition was assumed to be similar to that of the shallow-brine aquifer. To simulate mixing the laydown brine with the transient pond, the two end members were combined with the FY2K laydown-brine composition to make two different brine mixtures (FY94 shallow-aquifer + FY2K laydown brine, and FY97 shallow-aquifer + FY2K laydown brine). Each of the brine mixtures was subdivided into a 90 percent + 10 percent and a 50 percent + 50 percent mixing ratio to make four different simulations. To calculate new cation and anion molar concentrations for the four simulations, the FY94 and FY97 shallow-aquifer and FY2K laydown brine analyses were run through the mixing function of AquaChem v. 3.7 software (Waterloo Hydrogeologic, Inc., 1998). Calculated molar concentrations from these four simulations, and an excess amount of halite, were used as inputs to the TEQUIL 25°C version. Ten moles of halite were input to simulate placing each brine mixture on the surface of halite-dominated salt crust.

TEQUIL Modeling

Description - TEQUIL is based on Pitzer electrolyte equations and calculates liquid-solid-gas equilibria in complex brine systems (Moller and others, 1997). The TEQUIL 25°C version for the Na-K-Ca-Mg-H-Cl-OH-SO_4-HCO_3-CO_2-H_2O system (Harvie and others, 1984) was selected for use in this study.

Parameters and Output - To simulate brine evaporation in a closed system, TEQUIL was constrained to reduce the original water mass in the brine by 10 percent in the first evaporation step, and then reduce the resulting new water mass by 10 percent in the second evaporation step, and so on. After 45 evaporation steps, more than 99 percent of the original water mass was depleted. The original water mass was 1,000 grams or 55.508 moles.

Output from the model lists the new brine composition from each evaporation step, and identifies the mineral species that precipitate at each step. Brine composition from each evaporation step is expressed as moles of water, major ions, and precipitated mineral species. TEQUIL uses the brine composition from the previous evaporation step as input to calculate the brine composition for the next evaporation step. This step-wise process continues until brine compositions have been calculated for all 45 evaporation steps. In a separate output, the model also calculates cation and anion molar concentrations for the brine at each evaporation step. New molar concentrations of the brine, at any step in the simulated evaporation sequence, can be obtained for additional modeling simulations (see "Model Integrity", and "Impact of Rain and Shallow-Brine Aquifer on Predicted Potassium and Magnesium Salts").

TEQUIL-generated plots from the simulations in this paper are presented as mineral mass precipitated versus percent water remaining as evaporation progresses. Because halite mole values far exceed those of all other predicted mineral precipitates, halite is plotted on the primary Y axis, and the other predicted mineral precipitates are plotted on the secondary Y axis.

Model Integrity - To test the goodness or validity of the model, TEQUIL-predicted brine concentrations from simulated evaporation of FY96 and FY97 shallow-aquifer brine samples were compared with solar-pond brine concentrations from Reilly's potash operation at Wendover, Utah. FY96 and FY97 samples were selected because they pre-date the start of the Laydown Project. Halite and minor gypsum are the first precipitates in the commercial fractional-crystallization process, and are followed by sylvite and carnallite when concentrated brines from the primary-pond system are moved to the harvest and carnallite ponds (Bingham, 1980, p. 237-239). Brine concentrations, from the point where sylvite and carnallite first appeared in the solar ponds, were compared with modeled FY96 and FY97 brine concentrations.

Modeling showed that the first occurrences of precipitated sylvite and carnallite in the FY96 and FY97 brines usually appeared when about 93 and 97 percent of the original water mass was evaporated. At this stage of modeled evaporation, molar concentrations of the six major ions in the TEQUIL output were converted to chloride salts of sodium, potassium, and magnesium (appendix 7) and compared with solar-pond concentrations (table 12).

Table 12. Typical brine concentration as sylvite and carnallite begin to precipitate - comparison of solar-pond analyses (Bingham, 1980, p. 237-239) with TEQUIL output (concentration as wt percent in solution).

Salt	Bingham, 1980	TEQUIL - FY96	TEQUIL - FY97
First Sylvite:			
NaCl	12.5	11.6	11.9
KCl	7.5	8.7	8.8
$MgCl_2$	9.8	13.6	13.2
Density (g/mL)	1.245	1.274	1.275
First Carnallite:			
NaCl	4.0	3.2	2.9
KCl	5.0	4.1	3.8
$MgCl_2$	18.0	27.9	28.5
Density (g/mL)	1.257	1.286	1.284

Based on the TEQUIL output, the following was observed:

- Modeled concentrations were in reasonable agreement with solar-pond concentrations.

- Closer agreement was achieved between modeled and solar-pond sodium and potassium chloride salts, while modeled magnesium chloride values were about 50 percent more than solar-pond values.

Modeled magnesium values are believed to be greater than the solar-pond values because depletion of magnesium due to its adsorption on smectite clays was not accounted for in the modeling (B. F. Jones, Research Geochemist, USGS, personal communication, January 19, 2001). It is unclear why modeled density values are about 2 percent greater than the solar-pond values, because modeled initial-density values for both FY96 and FY97 monitoring well samples were in good agreement with their actual density values (for example, actual FY96 density versus modeled FY96 density was 1.182 and 1.184 g/mL, respectively).

Mineral Precipitation from Pre-Laydown Transient Pond

TEQUIL modeling was used to predict mineral precipitation from the transient pond. In the first modeling runs, chemical composition of the transient pond was assumed to be similar to that of the BSF shallow-brine aquifer. Pre-laydown shallow-brine aquifer composition was represented by chemical analyses from six years of monitoring-well samples (FY92 - FY97). Average molar concentrations of six major ions from six years of monitoring-well sample analyses (table 13) were modeled.

The minerals predicted to precipitate from the FY92-93 & FY96-97 brine compositions were nearly identical, while the FY94 and FY95 results were slightly different. Predicted mineral-species plots for FY92, FY94, and FY97 are shown in figures 9-11, and plots for FY93, FY95, and FY96 are in appendix A8, figures A8.1 -A8.3. The order of mineral precipitation (listed in order of predicted precipitation; see appendix 9) for the various years was as follows:

- FY92-93 & FY96-97: anhydrite-halite-syngenite-sylvite-polyhalite-carnallite-kainite.
- FY94: anhydrite-halite-polyhalite-sylvite-carnallite-kainite-kieserite-bischofite.
- FY95: anhydrite-halite-polyhalite-sylvite-carnallite-kieserite.

The predicted potassium and magnesium mineral precipitates only occurred after 85 to 90 percent of the water was evaporated.

While the FY92-93 and FY96-97 mineral precipitates were identical, there was some slight variation in their respective sylvite and carnallite peak values; these were due to slight differences in potassium and magnesium concentrations in the FY92-93 and FY96-97 sample analyses. Unlike FY92-93 and FY96-97, the FY94-95 predicted mineral precipitates included kieserite, but were missing syngenite. These differences are in response to a marked increase in average FY94 and FY95 magnesium concentrations, compared to those in the FY92-93 and FY96-97 samples.

Based on differences observed from plots of TEQUIL-predicted mineral precipitates, analyses from the FY92-93 and FY96-97 samples, and the FY94-95 samples, represented the two compositional extremes of the transient pond during the sampling period. Predicted mineral-precipitate plots from the two compositional extremes were compared with those from four actual transient pond samples collected by USGS in 1993 (see table 17). The predicted mineral-precipitate plots from the June and July 1993 pond samples (appendix A10, figures A10.1 and A10.2), which were collected during peak evaporation, were nearly identical to plots from the FY92-93 and FY96-97 monitoring-well samples (see figures 9 - 11, and appendix A8, figures A8.1 - A8.3).

Table 13. Pre-laydown molar concentrations of the shallow-brine aquifer from four years of BLM monitoring-well samples and two years of USGS monitoring-well samples

Sample	n	Density	Na	Mg	K	Ca	Cl	SO_4
BLM Samples:								
FY97	20	1.185	3.95	0.109	0.134	0.028	4.24	0.058
FY96	16	1.182	4.11	0.140	0.155	0.021	4.56	0.051
FY95	13	1.196	4.47	0.189	0.152	0.026	4.71	0.052
FY94	19	1.192	4.47	0.224	0.162	0.022	5.00	0.055
USGS Samples:								
FY93[1]	9	1.186	4.38	0.160	0.163	0.029	4.65	0.063
FY92[1]	9	1.194	4.44	0.195	0.193	0.028	4.67	0.068

[1]These nine samples are from 9 of 26 BSF monitoring wells sampled by Mason and others (1995, p. 50, table 4) in 1992-93; these same nine wells were also sampled by BLM during 1994-97 (see appendix A1, tables A1.9-A1.10).

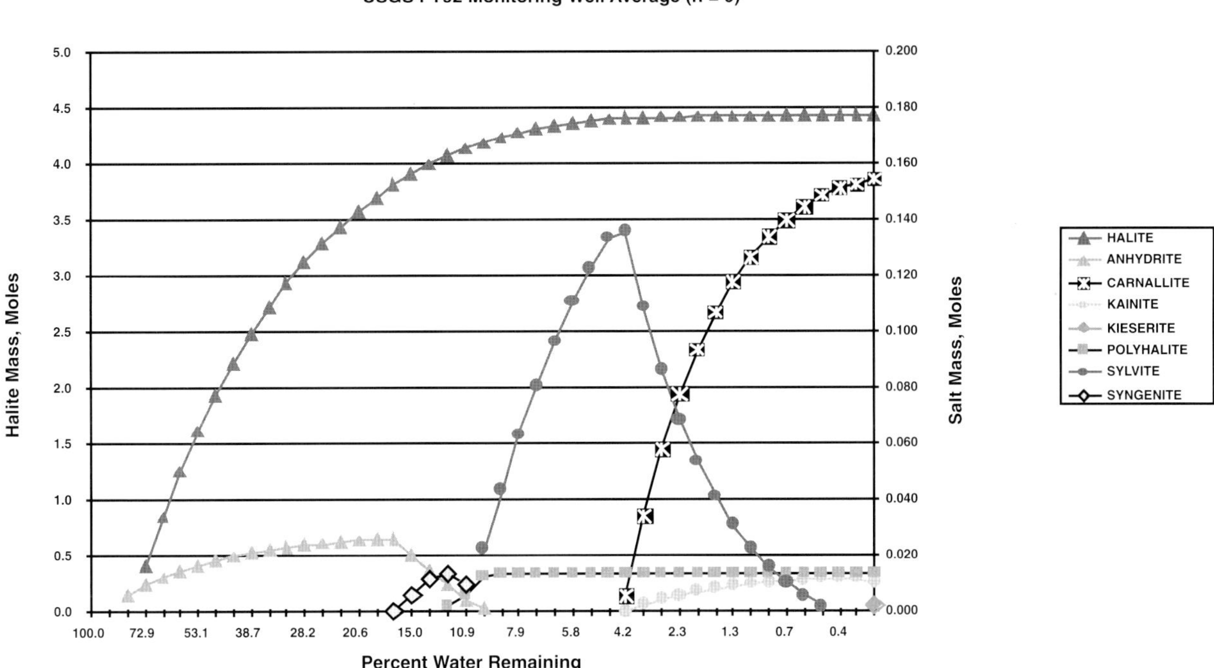

Figure 9. *TEQUIL-predicted mineral precipitation plots from average analyses of USGS FY92 monitoring-well samples.*

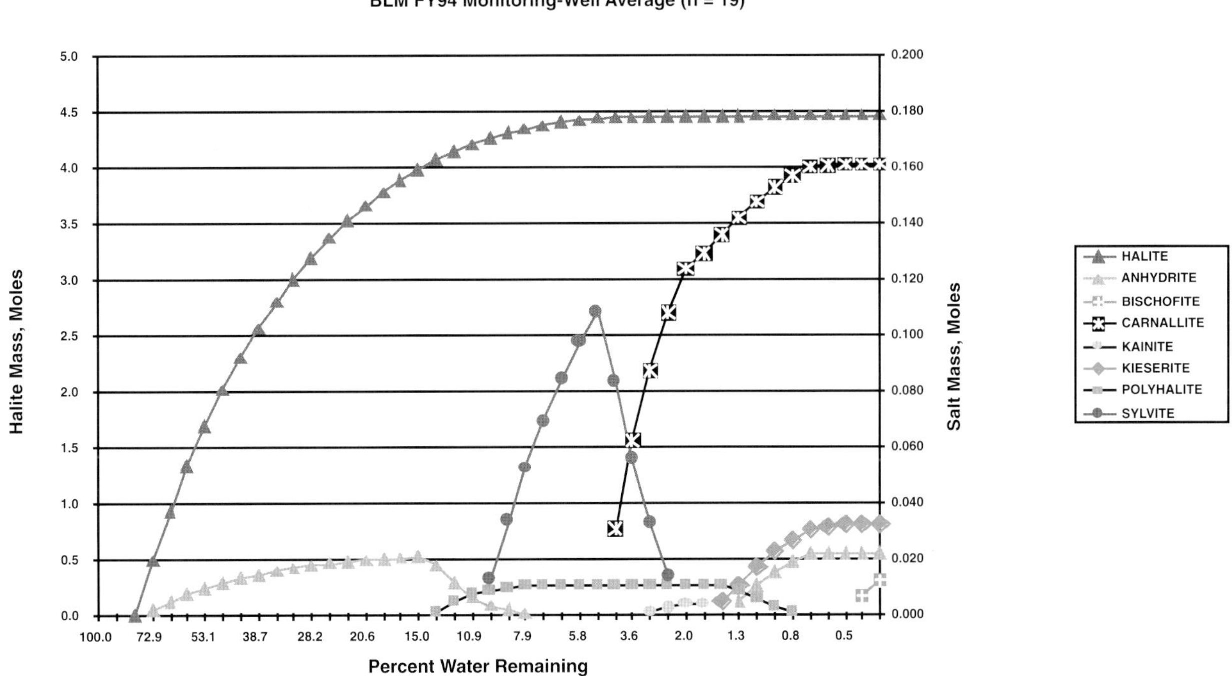

Figure 10. *TEQUIL-predicted mineral precipitation plots from average analyses of BLM FY94 monitoring-well samples.*

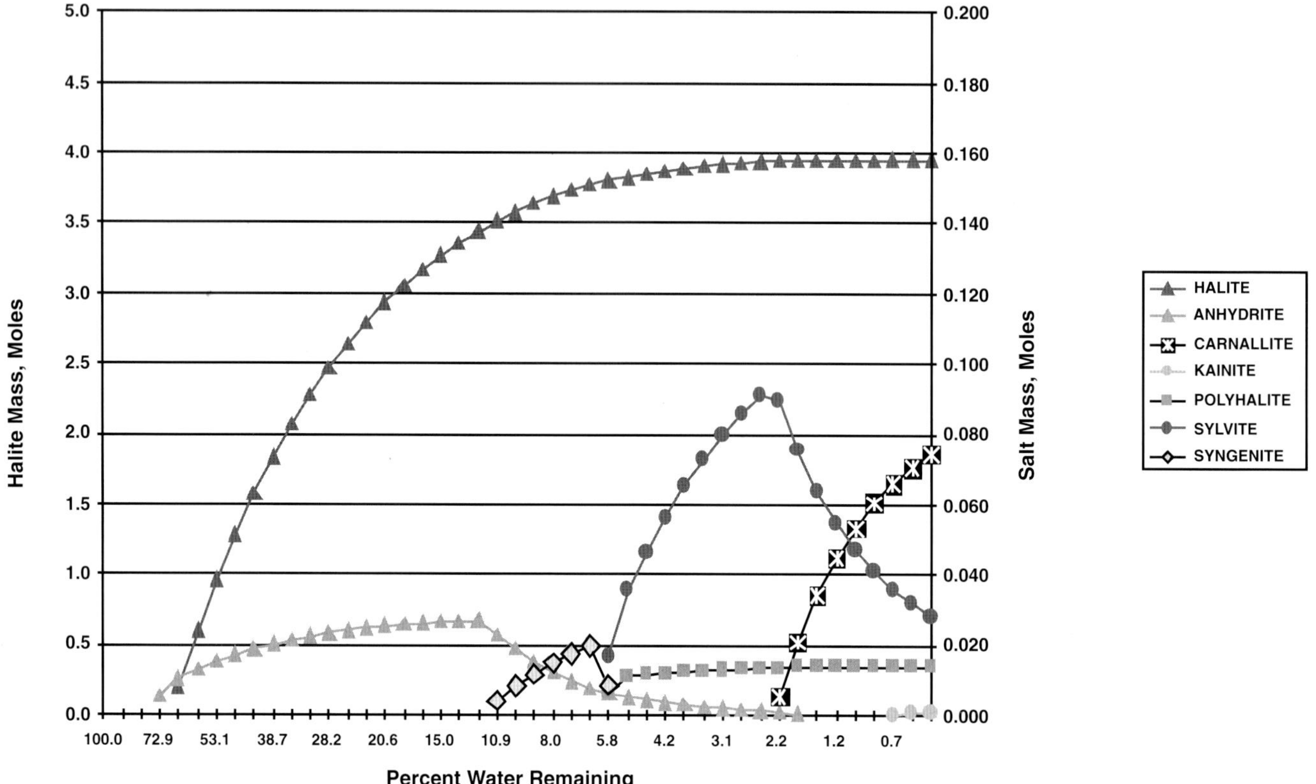

Figure 11. *TEQUIL-predicted mineral precipitation plots from average analyses of BLM FY97 monitoring-well samples.*

To simplify the modeling of mixing the transient pond with laydown brine, FY94 and FY97 monitoring-well-sample analyses were chosen to represent the compositional range of the transient pond. FY94 was selected because its predicted mineral precipitates differed the most from the 6 years of monitoring-well samples examined. FY97 was selected because its average chemical composition was closest to shallow-brine aquifer and subsequent transient pond compositions just prior to the first delivery of laydown brine to BSF on November 1, 1997.

Mineral Precipitation from Laydown Brine

In contrast with the seven or eight minerals predicted to precipitate from the two transient-pond end members, only anhydrite and halite were predicted to precipitate from the FY99 and FY2K laydown brines. Because of their very low concentrations, all potassium and magnesium contained in these brines remained in solution after 99 percent of the water was removed (figure 12).

Mineral Precipitation from Mixtures of Transient Pond and Laydown Brines

The TEQUIL-predicted mineral precipitation from the four brine-mix simulations were essentially the same as those predicted for their end-member components (FY94 and FY97 monitoring-well brines [see figures 10 and 11]). However, as the mixing ratios were changed from 90 percent shallow-aquifer brine + 10 percent laydown brine to 50 percent + 50 percent, the first appearances of most mineral-precipitate plots were shifted to later evaporation steps, and correspondingly mineral masses precipitated were reduced proportionately depending upon the mixing ratio.

Although both sets of mixing ratios approached halite saturation, they were able to dissolve additional halite (haltite saturation is equivalent to a brine density of 1.2 g/mL; average densities for FY94 and FY97 monitoring-well samples and FY2K laydown samples were 1.192, 1.185 and 1.164 g/L, respectively). TEQUIL output for both mixing ratios showed that from 1.2 to 1.9 moles (or 12 to 19 percent) of halite from the 10 moles used to simulate contact with the halite-dominated salt crust were dissolved during the first evaporation step (figures 13 - 16). The model predicted that the portion of the 10 moles dissolved in the first evaporation step would be reprecipitated by the 4th or 5th evaporation step. After the reprecipitation restored the halite 10-mole mass, halite in excess of 10 moles was contributed by Na^+ and Cl^- originally contained in the 90 percent + 10 percent and 50 percent + 50 percent mixing simulations prior to their contact with 10 moles of halite.

As was previously mentioned, the FY2K laydown brine was enriched in sodium and chloride, and depleted in potassium and magnesium. TEQUIL output at evaporation step one showed that mixing laydown brine with the two transient-pond end members generally resulted in increased con-

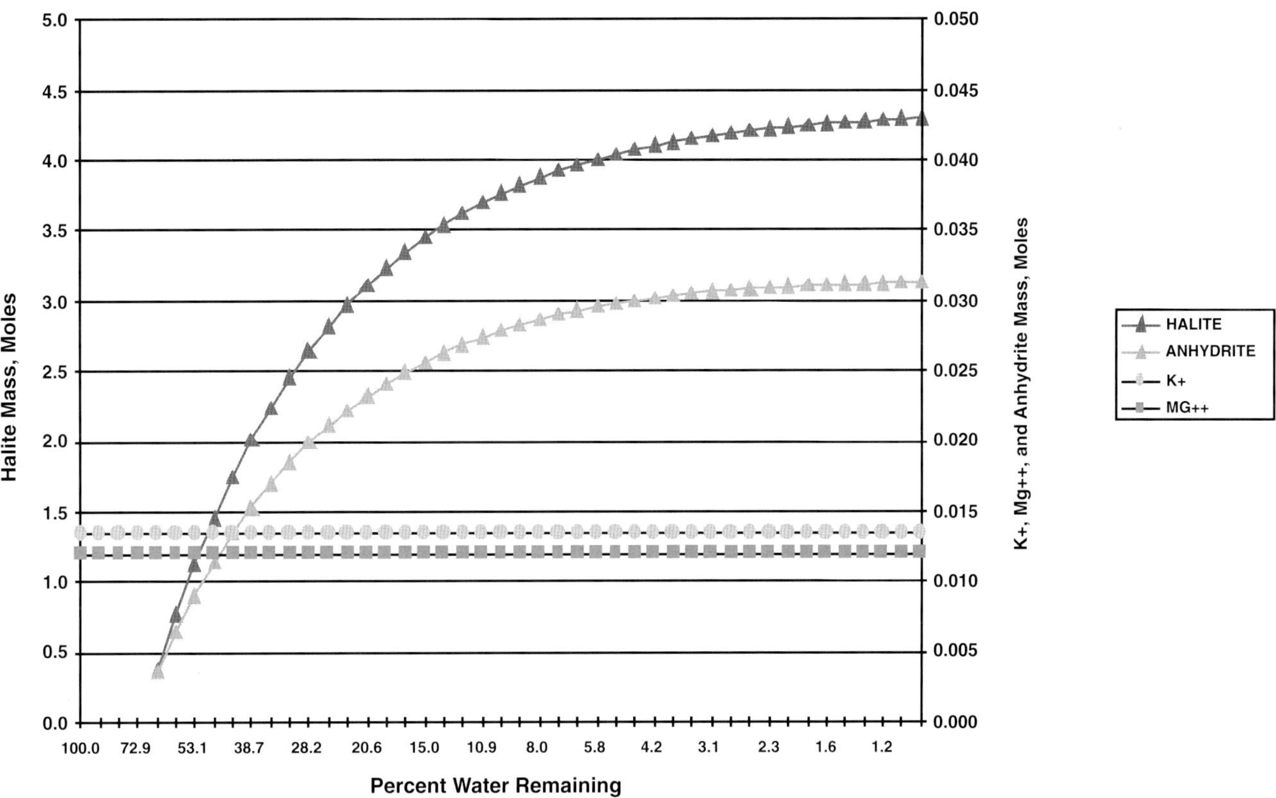

Figure 12. *TEQUIL-predicted mineral precipitation plots from average analyses of BLM FY2K laydown-brine samples.*

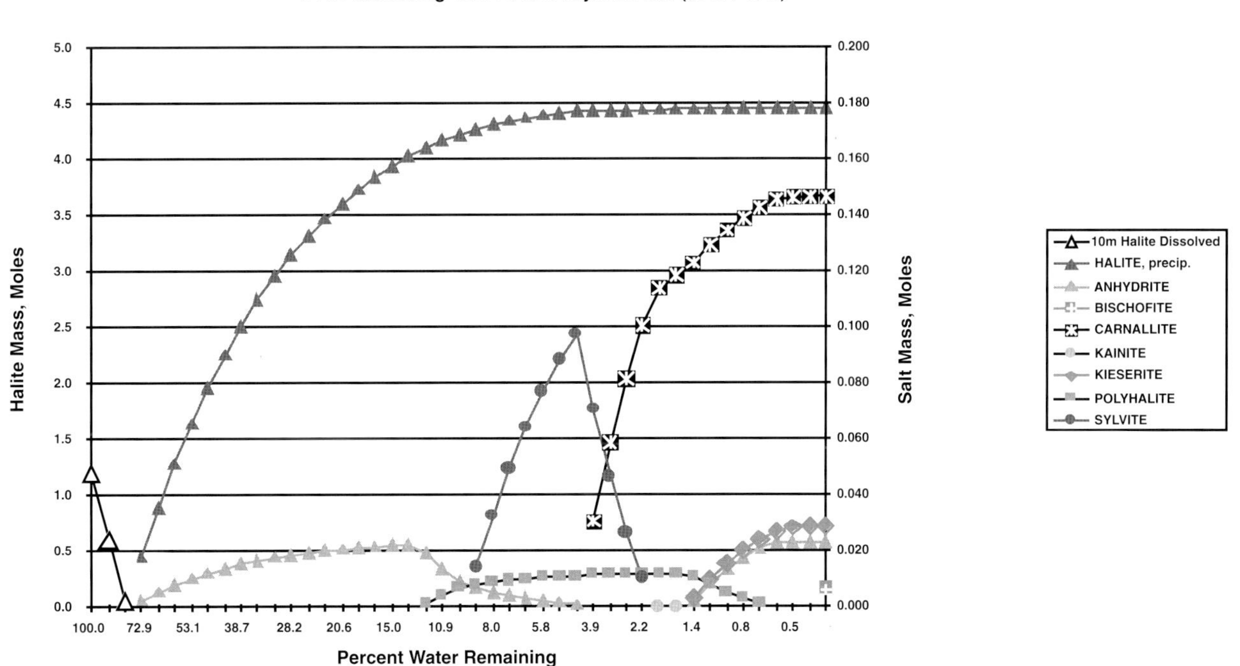

Figure 13. *TEQUIL-predicted mineral precipitation plots from a mixing simulation of FY94 monitoring-well + FY2K laydown brines (90% + 10% mixing ratio).*

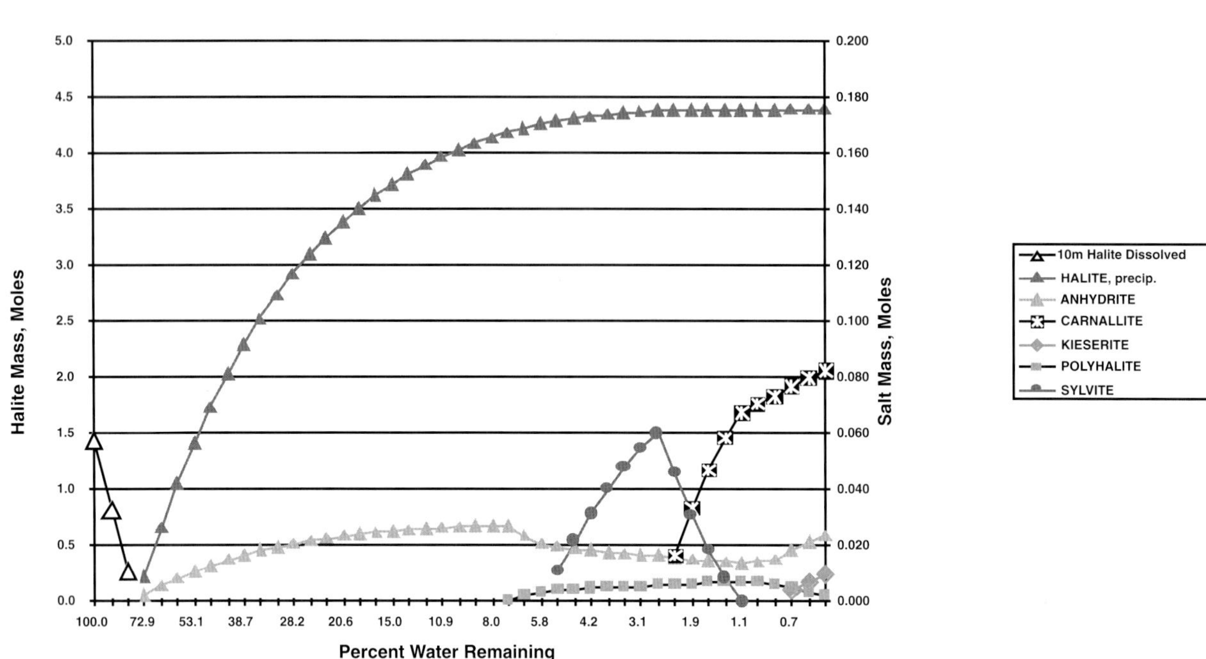

Figure 14. *TEQUIL-predicted mineral precipitation plots from a mixing simulation of FY94 monitoring-well + FY2K laydown brines (50% + 50% mixing ratio).*

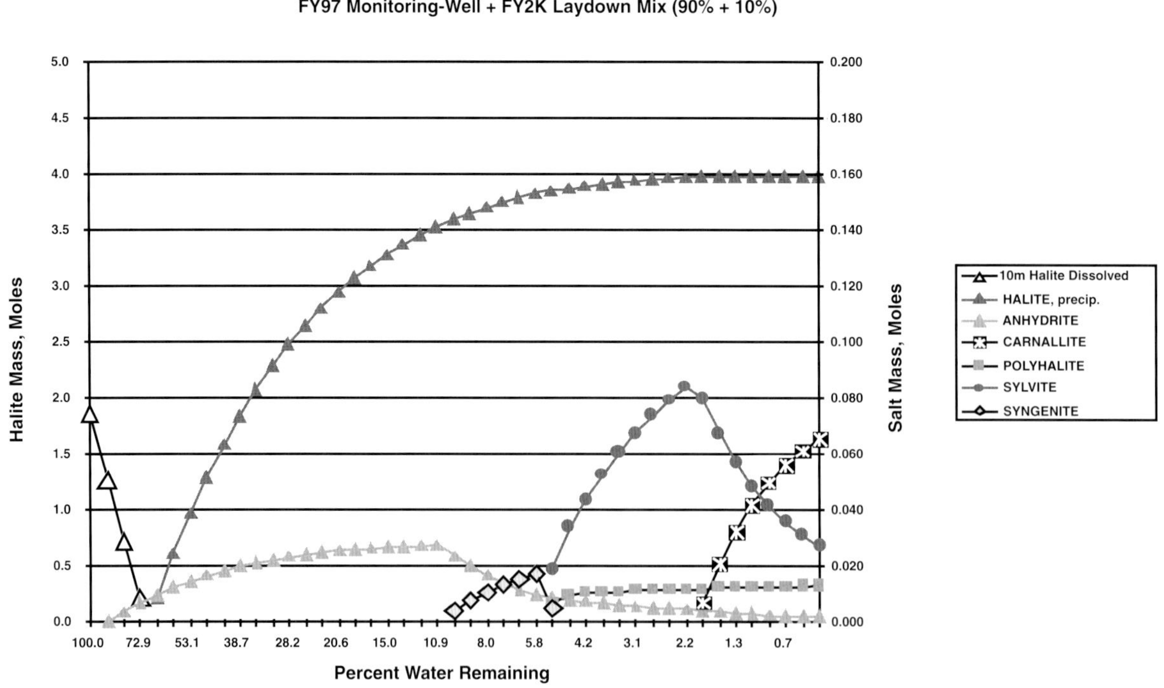

Figure 15 *TEQUIL-predicted mineral precipitation plots from a mixing simulation of FY97 monitoring-well + FY2K laydown brines (90% + 10% mixing ratio).*

centrations of NaCl (see "Fate of laydown Brine NaCl..."), and a dilution of potassium and magnesium concentrations. Consequently, a larger reduction of original water mass was required to initiate precipitation of carnallite and sylvite. For example, first occurrence of sylvite in the FY94 monitoring-well plot (figure 10) coincided with a 90 percent reduction in original water mass; however, first occurrence of sylvite in the FY94 monitoring-well + FY2K laydown plot (50 percent + 50 percent mixing ratio) coincided with a 95 percent reduction in water mass (figure 14).

In summary, TEQUIL modeling of mineral precipitation from the four brine-mixing simulations showed the following:

- No new mineral species were introduced as a result of mixing laydown brine with transient-pond brine.
- Anhydrite and halite were the first mineral species to precipitate and were the only minerals present until the original water mass was reduced by 87 and 93 percent, respectively (figures 13 - 16).
- Potassium and magnesium mineral precipitates (sylvite and carnallite) did not appear until the original water mass was reduced by 91 to 97 percent, and 96 to 98 percent of the halite was precipitated.
- When compared with the two transient-pond end members (FY94 and FY97), the 90 percent + 10 percent and 50 percent + 50 percent mixing ratios usually had their plot positions shifted to later evaporation steps, and the maximum predicted mass of carnallite and sylvite precipitated decreased by about 10 to 70 percent depending upon the mixing ratio (compare figures 10 & 11 with figures 13 - 16).

Mineral Precipitation from Shallow-Brine Aquifer: Pre- and Post-Laydown

TEQUIL model simulations were also used to compare predicted mineral precipitation between pre- and post-laydown samples from the BSF shallow-brine aquifer. FY97 monitoring-well samples collected one month prior to the start of laydown brine delivery were selected for comparison with FY2K monitoring-well samples. FY2K samples were collected 1 month prior to the start of the third-year of laydown brine delivery to BSF. Predicted mineral-precipitate plots for FY97 and FY2K monitoring-well samples were nearly identical (figures 11 and 17):

- Predicted mineral precipitates were the same for both monitoring-well samples; halite and gypsum were the first minerals to precipitate and were the only minerals present until 90 percent of the original water mass was evaporated.
- Relative positions of the sylvite and carnallite plots were the same for both samples; sylvite and carnallite first occurrences appeared when evaporation reduced the original water mass by 94 and 98 percent, respectively, and 96 percent of the halite precipitated.
- Maximum predicted masses for carnallite were the same for FY97 and FY2K (0.075 moles).
- Maximum predicted masses for sylvite were nearly the same for FY97 and FY2K (0.091 and 0.079 moles, respectively).

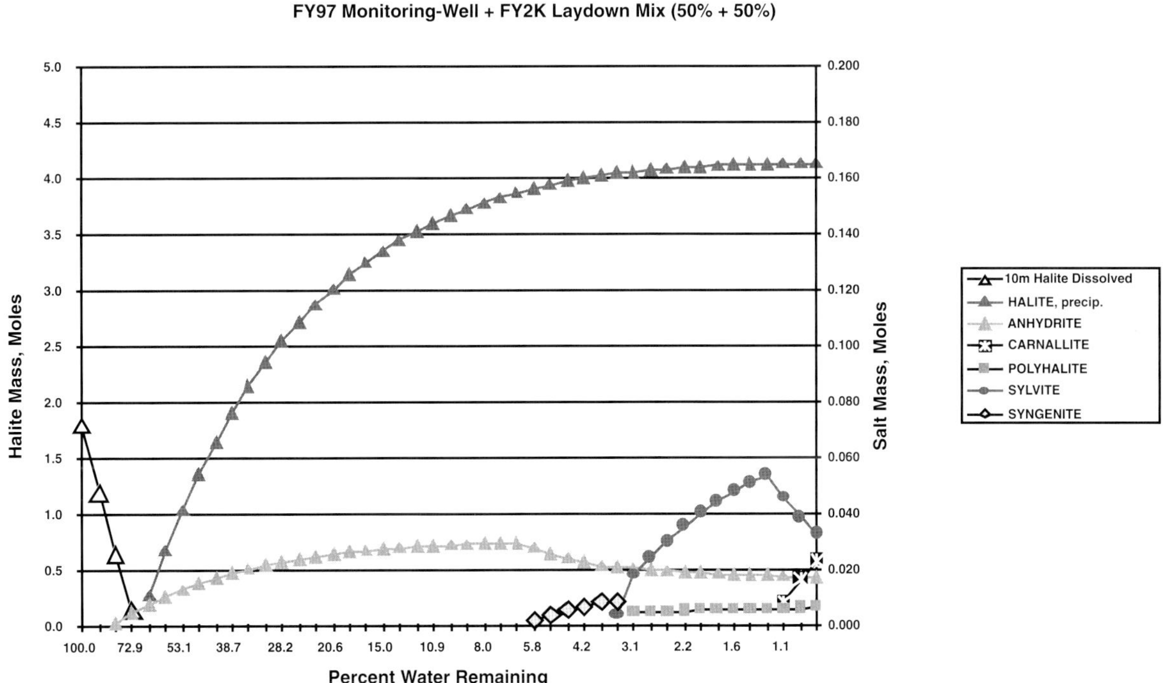

Figure 16. TEQUIL-predicted mineral precipitation plots from a mixing simulation of FY97 monitoring-well + FY2K laydown brines (50% + 50% mixing ratio).

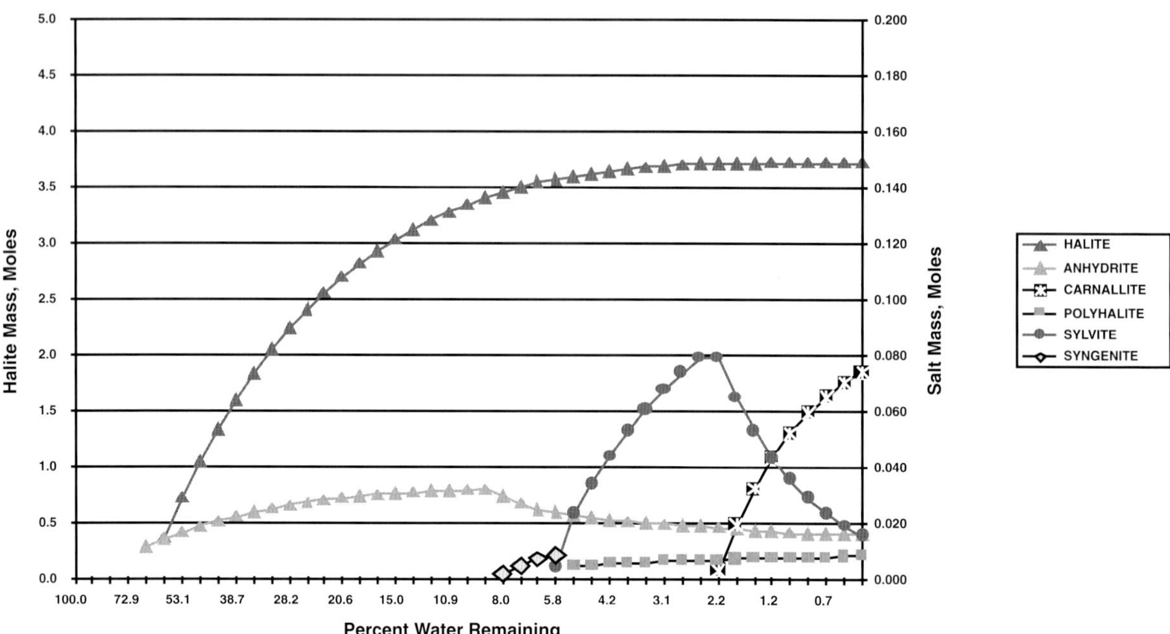

Figure 17. TEQUIL-predicted mineral precipitation plots from average analyses of BLM FY2K monitoring-well samples.

Impact of Rain and Shallow-Brine Aquifer on Predicted Potassium and Magnesium Salts

Potassium and magnesium mineral species predicted in the previous closed-system simulations resulted because TEQUIL was constrained to evaporate a fixed volume of brine to dryness without its being replenished by rainfall or the shallow-brine aquifer. However, in a recent mineralogic examination of salt-crust samples from a natural open system, neither potassium or magnesium minerals were observed with the halite and gypsum.

To understand why potassium and magnesium minerals were absent from these salt-crust samples, and to determine the efficacy of potassium and magnesium mineral formation in an open system, two additional mixing simulations were modeled with TEQUIL:

- 50 percent transient pond + 50 percent rainwater, and
- 50 percent transient pond + 50 percent shallow-aquifer brine.

These simulations were considered to be a more realistic representation of open-system effects from addition of rainwater and shallow-aquifer brine to an evaporating transient pond. Prior to being mixed either with rainwater, or with shallow-aquifer brine, it was assumed that the transient pond's water volume was reduced by 95 percent through evaporation. This extreme volume reduction was required to produce potassium and magnesium mineral precipitates whose fate could then be determined by mixing with rainwater or shallow-aquifer brine. Because of the variety of potassium and magnesium salts predicted to precipitate by TEQUIL, in the closed-system modeling of FY94 monitoring-well brine, FY94 chemical analyses were used to approximate the transient pond and shallow-aquifer brine compositions in the simulations.

Evaporated transient-pond brine composition was simulated by using the TEQUIL-predicted concentrations of the FY94 monitoring-well brine at the point where its water volume was reduced by 95 percent (table 14). At this point, sylvite and polyhalite precipitation approached peak values and carnallite precipitation first occurred (see figure 10).

Shallow-aquifer brine composition was represented by average chemical analyses of the FY94 monitoring-well brine (see table 13). Rainwater composition was taken from Hem (1989) (table 15). The mixing function of AquaChem v. 3.7 (Waterloo Hydrogeologic, Inc., 1998) was used to calculate new cation and anion molar concentrations from mixing evaporated transient pond with rainwater and with shallow-aquifer brine. These new cation and anion concentrations were the aqueous-phase input to TEQUIL for modeling both transient-pond mixing simulations (table 16). Because the precipitated minerals listed in table 14 were in equilibrium with the evaporated transient-pond composition, their mole values were also included as solid-phase input to the modeling.

TEQUIL-generated plots from the two simulations showed that potassium and magnesium minerals from the evaporated transient pond were re-dissolved when mixed with rainwater or shallow-aquifer brine. Halite and anhydrite were the only salts to precipitate during the first four evaporation steps of the TEQUIL output (figures 18 and 19).

To illustrate the difference between open-system conditions and those required for potassium and magnesium mineral precipitation, brine analyses from four transient-pond samples collected by USGS in 1993 were compared with TEQUIL-predicted brine concentrations from the two transient-pond mixing simulations (table 17). The simulation

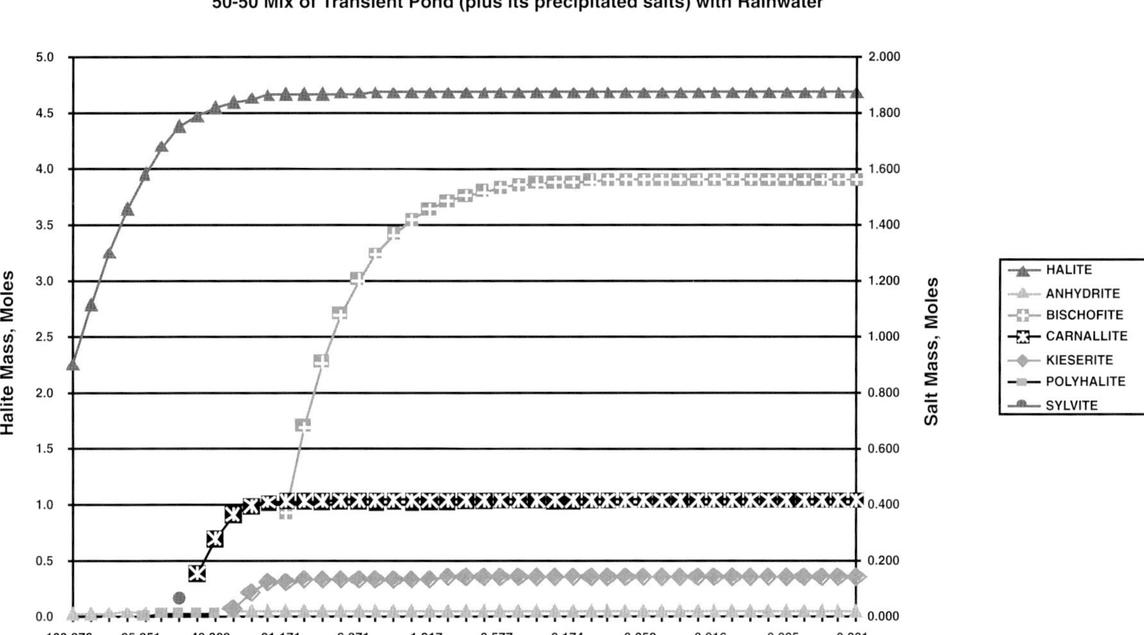

__Figure 18.__ TEQUIL-predicted mineral precipitation plots from a mixing simulation of evaporated FY94 monitoring-well brine + rainwater (50% + 50% mixing ratio. Note: original water mass is greater than 100% because model accounted for water of hydration contained in salts that were used as solid-phase input to the model (see table 14).

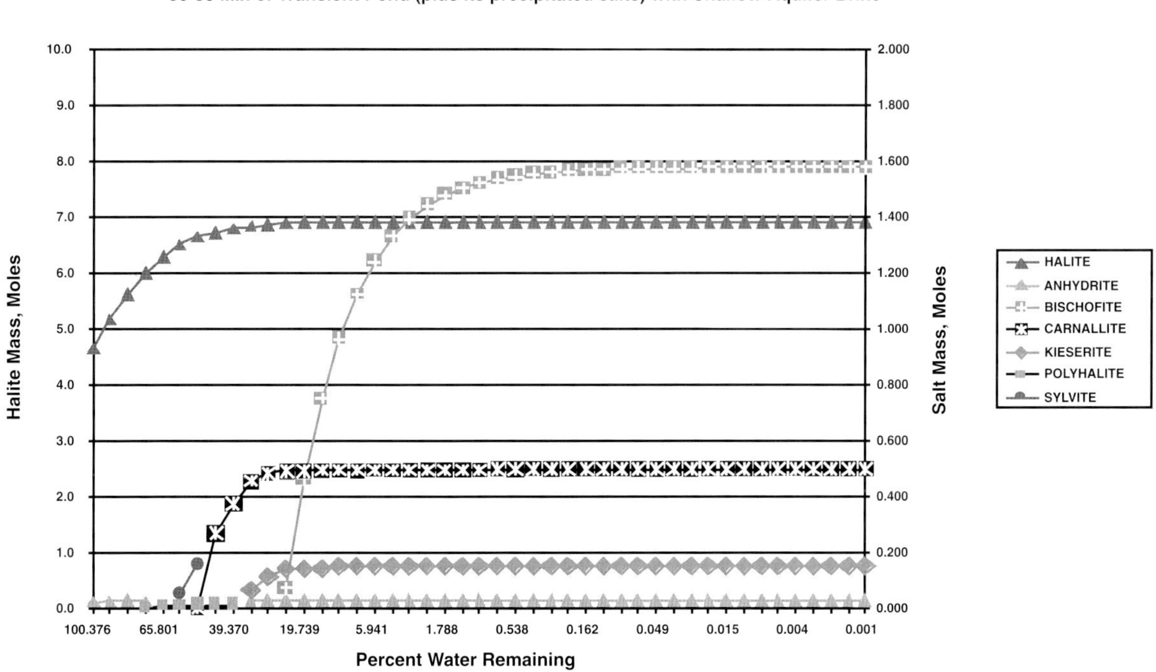

__Figure 19.__ TEQUIL-predicted mineral precipitation plots from a mixing simulation of evaporated FY94 monitoring-well brine + shallow-aquifer brine (50% +50% mixing ratio). Note: original water mass is greater than 100% because model accounted for water of hydration contained in salts that were used as solid-phase input to the model (see table 14).

Table 14. Simulated transient pond composition from TEQUIL-predicted concentrations of FY94 monitoring-well brines after water volume was reduced by 95 percent.

Aqueous-phase Species, moles/L		Solid-phase Species, moles	
Na^{+}	0.483	Carnallite	0.031
K^{+}	0.566	Halite	4.452
Ca^{2+}	0.001	Polyhalite	0.011
Mg^{2+}	4.174	Sylvite	0.084
Cl^{-}	8.916		
$SO4^{2-}$	0.242		

Table 15. Rainwater composition (from Hem, 1989, p. 36, table 6, item 2).

Species	mg/L	moles/L
Na^{+}	0.56	0.000024
K^{+}	0.11	0.000003
Ca^{2+}	0.65	0.000016
Mg^{2+}	0.14	0.000006
Cl^{-}	0.57	0.000016
SO_4^{2-}	2.18	0.000023
HCO_3^{-}	0.42	0.000007

[1]Bicarbonate was added to achieve charge balance required by TEQUIL.

Table 16. Molar concentrations used as input for TEQUIL simulations of evaporated transient pond mixed with rainwater and with shallow-aquifer brine.

Sample	Na	Mg	K	Ca	Cl	SO_4
Transient Pond[1] + rainwater	0.241	2.087	0.283	0.001	4.458	0.121
Transient Pond[1] + Shallow-aquifer brine	2.487	2.199	0.364	0.012	6.958	0.148

[1]Original volume reduced by 95%

Table 17. Molar concentrations from four USGS samples of the 1993 transient pond and from two TEQUIL simulations of evaporated transient pond mixed with rainwater and with shallow-aquifer brine.

Sample	Na	Mg	K	Ca	Cl	SO_4
1993 Transient Pond Near BLM-93:[1]						
January 21	4.349	0.005	0.010	0.027	4.513	0.029
May 13	4.785	0.023	0.041	0.037	4.231	0.046
June 16	5.219	0.053	0.072	0.042	4.231	0.044
July 7	4.872	0.132	0.207	0.035	5.416	0.054
TEQUIL Simulations:						
Transient Pond[2] + rainwater	0.308	2.118	0.334	0.001	4.637	0.121
Transient Pond[2] + Shallow-aquifer brine	0.412	2.225	0.412	0.001	5.022	0.127

[1]From Mason and others (1995, p. 50, table 4).
[2]Original volume reduced by 95 percent

concentrations were selected from the point where sylvite first occurred in the modeled output (or the point where nearly 50 percent of the original water mass was evaporated).

Two marked differences were observed:

- Sodium concentrations from the four 1993 transient-pond samples were more than 10 times the sodium concentrations predicted for the two transient-pond mixing simulations.
- Potassium and magnesium concentrations from July 1993 transient-pond samples were about one-half and one-seventeenth of corresponding mixing-simulation concentrations.

Despite the occurrence of peak evaporation on the salt crust during June 1993 (Mason and Kipp, 1998, p. 34), the sodium concentration of the June pond sample was still 10 times that of the two simulations. Before potassium and magnesium minerals could precipitate from the 1993 transient pond, the pond would have to be evaporated to the point where 99.5 percent of its sodium was precipitated as halite, and it remained in the solid phase. This is unlikely in an open system where the transient pond receives seasonal rainwater and is connected with the shallow-brine aquifer. The 1993 transient-pond brine would have to subjected to closed-system conditions similar to the commercial solar-pond process (Bingham, 1980, p. 237-239) in order for its composition to approach that of the two transient-pond mixing simulations shown in table 17.

Fate of Laydown-Brine NaCl When Mixed With the Shallow-Brine Aquifer

In the first three years of the Laydown Project, 4.6 million tons of NaCl were delivered to BSF. However, the amount of laydown NaCl precipitated as salt crust versus the amount of NaCl that remained in solution in the shallow-brine aquifer was unknown. TEQUIL outputs from previously described simulations were examined to see if the fate of laydown NaCl could be quantitatively described.

TEQUIL output from 90 percent shallow brine aquifer brine + 10 percent laydown brine and 50 percent + 50 percent mixing ratios, that were contacted with an excess of halite (10 moles), showed that the simulated solutions could dissolve additional NaCl. Based on this output, TEQUIL was used to determine how much of the 4.6 million tons could be assimilated by the shallow-brine aquifer within the 28 square mile area of BSF affected by the Laydown Project. Model output from each of the following brines and brine-mixture simulations was examined: 1) pre-laydown transient pond end members (FY94 and FY97 MW samples), 2) laydown brine (FY2K LD), and 3) 90 percent shallow-brine aquifer brine + 10 percent laydown brine and 50 percent + 50 percent mixing ratios of the FY94 MW + FY2K LD and FY97 MW + FY2K LD simulations. TEQUIL-calculated moles of sodium from the first evaporation step of these simulations were converted to pounds and tons of NaCl dissolved per gallon (table 18).

Based on the TEQUIL output at the first evaporation step (table 18), the transient-pond end members (FY94 MW and FY97 MW) contained 2.179 lbs/gal (1.089 X 10^{-3} tons/gal) and 1.926 lbs/gal (0.963 X 10^{-3} tons/gal) of dissolved NaCl, respectively. When FY94 MW and FY97 MW were compared to their corresponding 50-50 mixtures, the dissolved NaCl weight increased to 2.842 lbs/gal (1.421 X 10^{-3} tons/gal) and 2.896 lbs/gal (1.448 X 10^{-3} tons/gal), respectively. When the weight of dissolved NaCl in the 50-50 mix is subtracted from the weight of dissolved NaCl in the transient-pond end member, this results in an increased capacity of 0.663 lbs/gal (0.332 X 10^{-3} tons/gal) and 0.97 lbs/gal (0.485 X 10^{-3} tons/gal) for FY94 MW and FY97 MW.

When the increased capacity is multiplied by the 52.5 billion-gallon volume of the shallow-brine aquifer affected by the Laydown Project, the FY94 MW and FY97 MW end members could accommodate 17 and 25 million tons of additional NaCl, respectively. This is about four to five times the 4.6 million tons delivered to BSF by the Laydown Project. This suggests that the 4.6 million tons could initially be assimilated into the shallow-brine aquifer as NaCl in solution.

The capability of the shallow-brine aquifer to accept additional NaCl is significant, because it is the shallow-brine aquifer that regulates the distribution of NaCl mass to the BSF salt crust. If the laydown NaCl mass is assimilated by the shallow-brine aquifer, then: 1) one would anticipate an increase of NaCl concentration in the affected area of the shallow-brine aquifer, and 2) it would not be unreasonable to expect that more halite mass would be added to the existing salt crust as a result of this increased NaCl concentration.

Based on salt-crust monitoring using Landsat 5 imagery, thickness measurements, and geochemical modeling, new

Table 18. TEQUIL-calculated moles of Na at the first evaporation step converted to pounds and tons of NaCl dissolved per gallon of brine.

Sample	Na, moles	NaCl, lbs/gal	NaCl, tons/gal ($*10^{-3}$)
FY94 MW (n=19)	4.473	2.179	1.089
FY97 MW (n=20)	3.954	1.926	0.963
FY2K LD (n=30)	4.328	2.108	1.054
90% + 10% 94 + 2K[1]	5.666	2.760	1.380
50% + 50% 94 + 2K[1]	5.835	2.842	1.421
90% + 10% 97 + 2K[1]	5.866	2.857	1.428
50% + 50% 97 + 2K[1]	5.947	2.896	1.448

[1]Includes 10 moles of halite to simulate contact with salt crust.

halite deposition from 3 years of brine delivery by the Laydown Project is probably distributed into:

- Increased salt-crust area (5-square mile increase since 1997),
- Increased dissolved solids in the shallow-brine aquifer (with NaCl assimilation capacity of 17 to 25 million tons), and
- Thickened halite strata (dense-cemented halite, cemented-coarse-porous halite, and uncemented-coarse halite strata) that comprise the 26-square mile area of salt crust existing as of 1997.

CONCLUSIONS

Three years of Salt-Laydown Project operation have demonstrated that annual salt-tonnage loss from BSF can be replenished by the laydown facility. During this period, a sodium chloride salt mass of 4.6 million tons was delivered to BSF. The average annual addition of 1.5 million tons exceeded the estimated annual salt loss of 0.85 million tons.

The salt addition appears to be distributed between new salt-crust area, various salt-crust strata, and the shallow-brine aquifer.

Monitoring during the project produced the following conclusions regarding changes in salt crust thickness, areal extent and mass:

- At least five different strata comprise the salt-crust; thickness measurements of the dense-cemented halite stratum and the cemented-coarse-porous halite stratum showed substantial thickness changes from year to year, while little change in total salt-crust thickness was observed at nine locations during a 10-year period.
- Based on salt-thickness comparisons from 1994 to 2000, the dense-cemented halite stratum thickness increased at one location and decreased at three other locations; however, cemented-coarse-porous halite stratum thicknesses increased when the dense-cemented halite stratum thicknesses decreased at the same locations.
- An unusually wet (or dry) year could measurably decrease (or increase) the thickness of the dense-cemented halite stratum from year to year.
- Although the yearly laydown tonnage would maintain or increase the current mass of NaCl in the salt-crust deposit and shallow-brine aquifer, annual variation in dense-cemented halite stratum could easily mask the annual 0.4-inch increase in salt-crust thickness predicted by the laydown feasibility study.
- Depending upon the amount of halite dissolved from the dense-cemented halite stratum, an apparent thinning of the dense-cemented halite stratum could occur during unseasonably wet years in spite of increased salt tonnage added to BSF by the Laydown Project.
- An estimated 17 percent (5-square mile) increase in salt-crust areal extent was observed between September 1997 and October 1999 (the first laydown brine was delivered to BSF November 1, 1997).
- Using the new 5-square mile area of salt crust, assumed thickness of 0.25 to 1 inch, and bulk salt densities of 79.4 and 109.8 pounds per cubic foot, an estimated 0.1 to 0.6 millions tons of salt was added to the salt crust as new area by the Laydown Project; the remainder was apparently distributed to the main body of the salt crust (26 square miles in 1997), and to the shallow-brine aquifer.

Geochemical modeling of brine compositions, determined during the project, produced the following conclusions regarding salt addition to the shallow-brine aquifer:

- The TEQUIL model accurately predicted chemistries, based on agreement between modeled output and solar pond brine compositions from a commercial potash operation.
- The model also showed that the shallow brine aquifer has the capacity to accept 17 to 25 million tons of NaCl; this tonnage is four to five times the 4.6 million tons of laydown salt delivered to BSF during the first three years of the Laydown Project.
- The ability of the shallow-brine aquifer to assimilate additional salt may help account for the laydown-delivered salt tonnage that was in excess of the amount estimated in the 5 square miles of new salt crust.
- The laydown NaCl mass assimilated into the shallow-brine aquifer is eventually redistributed in the salt crust as part of new surface and additional subsurface halite crystal growth.
- The addition of laydown brine to the shallow-brine aquifer does not change the salt-crust mineral assemblages; anhydrite and halite were the only minerals predicted to precipitate from two different mixing ratios of laydown brine and shallow-aquifer brine in an open system such as BSF.
- Potassium and magnesium salts do not precipitate in an open system such as BSF; the addition of rainwater or the shallow-aquifer brine dissolves them immediately. All TEQUIL simulations showed that more than 90 percent of the water would have to be evaporated, and 96 to 98 percent of the NaCl would have to be precipitated before potassium and magnesium minerals precipitated from the brine; this condition could only be achieved if a mixture of laydown brine and shallow-aquifer brine were isolated from the shallow-brine aquifer and subjected to conditions similar to a commercial solar-pond process.

CONSIDERATIONS FOR CONTINUED ASSESSMENT

To continue the assessment of the Laydown Project's effectiveness, thickness measurements from a larger population of drill holes are needed, and an additional 42 to 70 auger holes per year are recommended:

The number of holes is based on drilling three to five holes on each of Brooks' (1991, figure 4) 14 drill-hole transects across the International Track.

The same number of holes would be drilled each year close to the same points on each transect, and total salt-crust thickness measurements would be recorded.

Additionally, each stratum thickness should be measured and compared with total salt-crust thickness at each specified sample site; using the thickness changes in the dense-cemented halite stratum alone may be misleading.

Purchase and use of satellite imagery should continue so that changes in salt-crust areal extent can be monitored; salt-crust thickness measurements should be made adjacent to monitoring wells that are located near the east margin of the salt crust so that new salt-crust area tonnages can be calculated.

ACKNOWLEDGMENTS

The author gratefully acknowledges technical peer reviews that were conducted by the following: Blair F. Jones, USGS, National Center, Reston, VA; Kim A. Lapakko, Principal Engineer, Division of Lands and Minerals, Minnesota Department of Natural Resources, St Paul, MN; James L. Mason, USGS, Water Resources Division, Salt Lake City, UT. Comments and suggestions from these peer reviews helped enhance the quality and organization of the report.

Nancy Moller Weare, Research Chemist, University of California at San Diego, provided guidance in the use of the TEQUIL model. Fred C. Kelsey, Geologist, formerly with ARCO Coal Company, Denver, CO., wrote the Output Formatter program which transfers TEQUIL output to a spreadsheet for graphing purposes. Brent Bingham, Bingham Engineering, refined our brine travel-time calculations for collection ditches and the brine-distribution manifold.

Cheryl Martinez and Michael Johnson, BLM Salt Lake Field Office geologists, assisted with BLM's twice-monthly brine-sample collection program; Michael Johnson assisted with excavation of BLM 1997-98 test pits. Cheryl Johnson, Geographic Information Systems Coordinator, BLM Salt Lake Field Office, determined salt-crust areas from Landsat 5 images, and prepared monitoring-well location maps and satellite maps. Landsat 5 imagery was provided through a 1999 grant from the BLM Director's Field Incentive Program.

The author also thanks BLM geologists James Kohler and Philip Allard for providing unpublished field notes, historical data files, project guidance, and assistance in the field. Save the Salt funded the independent brine analyses (University of Arizona, Department of Chemistry) and shared the results with BLM. Michelle Stalter and John Norris, Analytical Chemists, University of Arizona, Department of Chemistry, analyzed the "C"-triplicate brine samples.

The author especially thanks Glenn D. Wadsworth, former Reilly-Wendover plant manager, Reilly Industries, Inc., and the Bureau of Land Management; without their support this study would not have been possible.

REFERENCES

Bingham, B. S., 1991, Bonneville International Raceway Salt Laydown Project Feasibility: Prepared for Save the Salt and Reilly Industries: Bingham Engineering, Salt Lake City, UT, 25 p.

Bingham, C. P., 1980, Solar production of potash from the brines of the Bonneville Salt Flats, p.229-242, *in* Gwynn, J. W., editor, Great Salt Lake- a scientific, historical and economic overview: Utah Geological and Mineral Survey Bulletin 116, 400 p.

Braitsch, O., 1971, Salt Deposits - Their origin and composition: Springer Verlag, New York, 297 p.

Brooks, S. J., 1991, A Comparison of Salt Thicknesses on the Bonneville Salt Flats, Tooele County, Utah during July 1980, October 1974, and October 1988: Unpublished Technical Memorandum, 15 p. Available upon request from BLM Salt Lake Field Office.

Harvie, C. E., Moller, N., and Weare J. H., 1984, The prediction of mineral solubilities in natural waters - The Na-K-Mg-Ca-H-Cl-SO_4-OH-HCO_3-CO_3-CO_2-H_2O system to high ionic strengths,: Geochimica et Cosmochimica Acta v. 52., p. 821-837

Hem, J. D., 1989, Study and interpretation of the chemical characteristics of natural water, 3rd edition, U.S. Geological Survey Water-Supply Paper 2254, 263 p.

Jensen, R. J., 1996, Introductory digital image processing - A remote sensing perspective, 2nd edition: Prentice-Hall, Inc., 316 p.

Kohler, J. F., 1995, Salt Flats Investigation - Measured sections of salt crust exposed in test pits: Unpublished field notes, 8 p. Available upon request from BLM Salt Lake Field Office.

Kohler, J. F. and White.,W. W., III, 1997, Measured sections of salt crust exposed in test pits on the Bonneville Salt Flats: Unpublished field notes, 2 p. Available upon request from BLM Salt Lake Field Office.

Krauskopf, K. B., 1967, Introduction to Geochemistry: McGraw-Hill Book Company, 721 p.

Lines, G. C., 1978, Selected ground-water data - Bonneville Salt Flats and Pilot Valley, western Utah: U.S. Geological Survey Utah Basic-Data Release no. 30, 14 p.

—1979, Hydrology and surface morphology of the Bonneville Salt Flats and Pilot Valley Playa, Utah: U.S. Geological Survey Water-Supply Paper 2057, 107 p.

Mason, J. L., Brothers, W. C., Gerner, L. J., and Muir, P. S., 1995, Selected hydrologic data for the Bonneville Salt Flats and Pilot Valley, western Utah, 1991-1993: U.S. Geological Survey Open-File Report 95-104, 56 p.

Mason, J. L. and Kipp K. L., 1998, Hydrology of the Bonneville Salt Flats, northwestern Utah, and simulation of ground-water flow and solute transport in the shallow-brine aquifer: U.S. Geological Survey Professional Paper 1585, 108 p.

McMillan, D. T., 1974, Bonneville Salt Flats - a comparison of salt thickness in July, 1960 and October, 1974: Utah Geological and Mineral Survey Report of Investigation no. 91, 6 p.

Moller, N., Weare, J. H., Duan, Z., and Greenberg, J. P., 1997, Chemical models for optimizing geothermal energy production: Online, http://geotherm.ucsd.edu/tequil.

Nielson, D., Watkins, R. K., Christiansen, J. E., and Sabnis, P., 1960, Project EC 35 progress report, Salt Flat investigations - Part V, physical tests on salt samples: Utah State University, College of Engineering, Engineering Experiment Station, Logan, Utah, 19 p.

Nolan, T. B., 1927, Potash brines in the Great Salt Lake Desert, Utah: U.S. Geological Survey Bulletin 795-B, p. 25-44.

Reilly Industries, 1997, Index Map - Salt Laydown Project: Reilly Industries, Inc., information folder, Laydown Partnership Celebration, November 13, 1997, Wendover, UT.

Sturm, P. A., McLaughlin, J. C.. and Broadhead, Ramond, 1980, Analytical procedures for Great Salt Lake Brine, *in* Gwynn,

J. W., editor, Great Salt Lake - a scientific, historical and economic overview: Utah Geological and Mineral Survey Bulletin 116, p. 175-193.

Turk, L. J., 1969, Hydrogeology of the Bonneville Salt Flats, Utah: Ph. D. dissertation, Stanford University, 307 p.

Turk, L. J., 1973, Hydrogeology of the Bonneville Salt Flats, Utah: Utah Geological and Mineral Survey Water-Resources Bulletin 19, 81 p.

Turk, L. J., 1978, Bonneville salt crust study: Turk, Kehle & Associates, Austin, Texas. v. 1. 103 p.

White, D. E., 1955, Thermal springs and epithermal ore deposits, *in* Bateman, A. M., editor: Economic Geology, 50th Anniversary Volume, Part 1, p. 99-154.

White, W. W., III, 1998, Depth of ponded brine at shear-strength test hole MB-1, Bonneville Salt Flats: Unpublished field notes, p. 37. Available upon request from BLM Salt lake Field Office.

White, W. W., III, and Wadsworth, G. D., 1999, Salt Laydown Project - Bonneville Salt Flats - 1997-99 progress report: Online, Http://www.ut.blm.gov/wh3bsfsalt.html, 57 p., 10 appendices.

White, W. W., III, and Wadsworth, G. D., 2001, Salt Laydown Project - Bonneville Salt Flats - FY2000 Laydown brine production and analytical data: Online Http://www.ut.blm.gov/wh3bsfsalt.html, 15 p.

Note: The use of brand names in this paper does not imply endorsement of any product by Reilly, BLM, or Department of Natural Resources.

APPENDIX A1

PRE-1994 BSF MONITORING-WELL SAMPLE ANALYSES BY USGS

Table A1.1. Summary of 18 auger-hole brine samples collected on and adjacent to BSF (Nolan 1927, Plate 3); concentrations expressed in g/L and moles/L.

Sample	g/L		moles/L	
	Mg	K	Mg	K
1	0.4	1	0.016	0.026
2	1.3	2.8	0.053	0.072
3	3.4	4.6	0.140	0.118
5	6.4	10.2	0.263	0.261
8	7.3	12.5	0.300	0.320
9	4.8	8.7	0.197	0.223
10	7.2	12.2	0.296	0.312
11	5.9	9.2	0.243	0.235
12	3.9	7.4	0.160`	0.189
13	2.3	4.2	0.095	0.107
14	2.2	4	0.091	0.102
15	1.6	2.9	0.066	0.074
16	2.8	3.7	0.115	0.095
17	2.1	3.2	0.086	0.082
23	0.3	0.7	0.012	0.018
255	0.5	1.1	0.021	0.028
260	0.5	0.9	0.021	0.023
265	2.2	3.2	0.091	0.082
Mean	3.1	5.1	0.126	0.131
Minimum	0.3	0.7	0.012	0.018
Maximum	7.3	12.5	0.300	0.320

Table A1.2. Summary of 43 "USGS" BSF monitoring-well brine samples (Lines, 1978; Lines, 1979); cation/anion values expressed as g/L; density expressed as g/mL.

Sample	Date	Density	Na	Mg	K	Ca	Cl	SO_4	TDS
USGS-29/30	10/02/78	1.202	120.0	3.20	7.10	1.20	180.0	5.60	317.0
USGS-29/30	10/12/77	1.198	110.0	4.10	6.20	1.20	180.0	5.50	307.0
USGS-29/30	10/02/78	1.200	120.0	1.20	2.70	1.30	190.0	4.70	320.0
USGS-29/30	10/26/77	1.200	120.0	1.40	3.10	1.40	180.0	5.20	311.0
USGS-74	09/23/76	1.196	100.0	5.70	9.20	1.10	180.0	5.30	301.0
USGS-31/32	09/27/76	1.189	110.0	0.96	2.40	1.50	180.0	4.10	299.0
USGS-31/32	09/27/76	1.202	120.0	1.40	2.90	1.40	190.0	4.00	320.0
USGS-23	09/27/76	1.204	110.0	3.00	7.00	1.30	190.0	4.30	316.0
USGS-23	09/27/76	1.198	100.0	4.60	8.40	1.20	180.0	5.00	299.0
USGS-8	09/24/76	1.202	120.0	1.70	4.00	1.80	190.0	3.80	321.0
USGS-8	09/24/76	1.198	110.0	2.80	4.90	1.30	190.0	4.40	313.0
USGS-9	09/24/76	1.197	100.0	5.50	9.60	1.00	180.0	6.20	302.0
USGS-9	10/02/78	1.200	100.0	5.70	11.00	0.92	180.0	7.10	305.0
USGS-9	10/11/77	1.198	110.0	5.80	9.20	1.00	180.0	6.60	313.0
USGS-10	10/02/78	1.205	110.0	2.90	6.10	1.10	180.0	5.10	305.0
USGS-10	09/24/76	1.201	110.0	5.50	9.30	0.96	200.0	5.90	332.0
USGS-10	10/26/77	1.203	110.0	3.30	5.50	1.20	170.0	4.90	295.0
USGS-82	09/23/76	1.145	85.0	1.90	3.10	1.50	130.0	4.60	226.0
USGS-82	10/03/78	1.185	100.0	2.90	4.80	1.20	160.0	6.10	275.0
USGS-82	10/12/77	1.157	89.0	2.30	3.80	1.50	140.0	5.60	242.0
K7?	09/23/76	1.129	71.0	1.20	2.60	1.80	120.0	4.60	201.0
K4	10/02/78	1.195	110.0	4.80	9.30	0.99	170.0	6.70	302.0
K4	09/23/76	1.194	100.0	4.70	7.90	1.10	180.0	5.70	299.0
K4	10/11/77	1.195	100.0	5.00	8.00	1.10	170.0	6.30	290.0
USGS-81	10/12/77	1.154	75.0	1.60	2.80	1.50	130.0	4.20	215.0
USGS-81	10/03/78	1.190	110.0	2.00	0.36	1.30	170.0	5.00	289.0
USGS-81	09/23/76	1.186	100.0	1.90	3.80	1.30	170.0	4.10	281.0
USGS-2/BLM-60	09/24/76	1.187	100.0	4.90	7.50	1.10	170.0	5.70	289.0
USGS-96/BLM-59	09/24/76	1.204	110.0	3.00	6.40	1.30	190.0	4.30	315.0
USGS-79	09/21/76	1.161	110.0	4.00	6.60	1.20	150.0	4.10	276.0
USGS-108	09/28/76	1.194	93.0	8.50	12.00	0.91	180.0	7.10	302.0
USGS-113	09/28/76	1.180	88.0	6.90	11.00	1.10	160.0	6.50	274.0
USGS-114	09/28/76	1.184	100.0	4.00	6.30	1.20	170.0	5.70	287.0
USGS-21	09/24/76	1.192	100.0	5.30	8.00	1.10	180.0	5.80	300.0
USGS-27	09/27/76	1.195	110.0	4.10	7.60	1.10	180.0	5.10	308.0
USGS-27	10/12/77	1.196	110.0	4.40	7.30	1.10	180.0	5.60	308.0
USGS-27	10/02/78	1.199	110.0	4.20	7.90	1.10	180.0	5.90	309.0
USGS-64	09/22/76	1.156	88.0	1.90	3.60	1.20	150.0	4.50	249.0
USGS-67	09/22/76	1.195	110.0	4.40	8.00	1.10	180.0	5.40	309.0
USGS-84	09/23/76	1.186	100.0	2.90	5.10	1.30	170.0	5.20	285.0
USGS-84	10/12/77	1.183	100.0	2.80	4.80	1.30	160.0	6.10	275.0
USGS-84	10/03/78	1.184	96.0	2.80	5.10	1.30	160.0	6.30	270.0
USGS-86	09/23/76	1.205	110.0	3.40	5.70	1.10	190.0	5.50	316.0
Mean		1.189	103.6	3.59	6.23	1.23	172.3	5.33	292.5
Minimum		1.129	71.0	0.96	0.36	0.91	120.0	3.80	201.0
Maximum		1.205	120.0	8.50	12.00	1.80	200.0	7.10	332.0

Table A1.3. Summary of 10 "K" BSF monitoring-well brine samples (Lines, 1978; Lines, 1979); cation/anion values expressed as g/L; density expressed as g/mL.

Sample	Date	Density	Na	Mg	K	Ca	Cl	SO_4	TDS
K10	09/23/76	1.194	100.0	5.20	7.90	1.00	180.0	6.30	300.0
K10	10/03/78	1.188	94.0	4.80	7.50	1.00	160.0	7.00	274.0
K10	10/12/77	1.190	93.0	5.40	7.50	0.99	160.0	7.10	274.0
K46	09/28/76	1.175	95.0	2.10	4.20	1.40	160.0	4.50	267.0
K65	10/13/77	1.127	72.0	0.86	1.80	1.30	120.0	3.70	200.0
K65	09/28/76	1.125	70.0	0.98	1.90	1.20	110.0	3.60	188.0
K65	10/04/78	1.128	70.0	0.82	2.10	1.20	110.0	4.10	188.0
K66	10/04/78	1.178	99.0	2.20	4.90	1.30	160.0	5.70	273.0
K66	09/28/76	1.182	100.0	2.60	4.70	1.30	170.0	5.00	284.0
K66	10/13/77	1.181	100.0	2.70	4.40	1.40	170.0	5.50	284.0
Mean		1.167	89.3	2.77	4.69	1.21	150.0	5.25	253.2
Minimum		1.125	70.0	0.82	1.80	0.99	110.0	3.60	188.0
Maximum		1.194	100.0	5.40	7.90	1.40	180.0	7.10	300.0

Table A1.4. Summary of 34 "B" BSF auger-hole sample analyses (data file provided by J.L. Mason, Hydrologist, USGS, written communication, March 7, 2001); cation/anion values expressed as g/L; density expressed as g/mL.

Sample	Date	Density	Na	Mg	K	Ca	Cl	SO_4	TDS
B1	05/11/81	1.195	93.0	4.40	7.60	1.20	180.0	5.80	292.0
B10	05/14/81	1.115	80.0	0.93	1.00	1.10	110.0	4.30	197.0
B11	05/14/81	1.147	90.0	1.80	1.10	1.10	130.0	5.90	230.0
B12	05/14/81	1.149	70.0	1.20	2.30	1.20	130.0	5.10	210.0
B13	05/14/81	1.167	85.0	1.40	2.50	1.40	150.0	5.10	245.0
B14	05/14/8	1.164	80.0	1.60	1.70	1.20	140.0	5.40	230.0
B15	05/14/81	1.173	90.0	1.40	1.70	1.50	140.0	6.90	242.0
B16	05/14/81	1.169	90.0	1.90	1.90	1.60	150.0	5.30	251.0
B17	05/15/81	1.177	87.0	1.70	2.90	1.40	150.0	5.20	248.0
B18	05/15/81	1.210	120.0	8.10	9.00	0.90	200.0	5.30	343.0
B19	05/19/81	1.188	100.0	3.20	2.80	1.30	150.0	4.80	262.0
B2	05/11/81	1.209	100.0	2.70	4.10	1.30	180.0	5.20	293.0
B20	05/19/81	1.178	96.0	2.30	1.60	1.40	160.0	5.20	267.0
B21	05/19/81	1.204	100.0	4.00	6.30	1.20	190.0	4.50	306.0
B22	05/20/81	1.197	110.0	2.90	3.40	1.40	190.0	4.90	313.0
B23	05/20/81	1.202	110.0	3.80	2.20	1.20	180.0	5.70	303.0
B24	05/20/81	1.206	100.0	4.80	7.60	1.20	190.0	5.60	309.0
B25	05/20/81	1.221	110.0	3.40	2.10	1.30	200.0	0.43	317.0
B26	05/21/81	1.190	110.0	3.20	3.90	1.30	180.0	5.10	304.0
B27	05/21/81	1.194	97.0	2.30	4.60	1.30	170.0	5.30	281.0
B28	05/21/81	1.202	110.0	3.80	2.40	1.20	180.0	2.90	300.0
B29	05/21/81	1.196	100.0	3.00	4.80	1.40	180.0	5.20	294.0
B3	05/12/81	1.169	87.0	1.50	2.90	1.40	150.0	4.30	247.0
B30	05/22/81	1.190	89.0	4.10	6.60	1.00	150.0	5.30	256.0
B31	05/22/81	1.216	110.0	1.10	2.80	1.20	200.0	3.80	319.0
B32	05/28/81	1.214	160.0	1.50	2.00	1.30	240.0	3.60	408.0
B33	05/28/81	1.192	90.0	3.70	5.60	1.20	160.0	6.00	267.0
B4	05/12/81	1.196	91.0	7.40	11.00	0.85	180.0	7.10	297.0
B5	05/13/81	1.196	100.0	5.50	5.30	0.91	180.0	7.30	299.0
B6	05/13/81	1.177	90.0	2.00	2.10	1.50	160.0	5.40	261.0
B7	05/13/81	1.166	90.0	2.70	2.00	1.30	150.0	6.90	253.0
B8	05/13/81	1.161	81.0	1.70	2.30	1.30	140.0	5.80	232.0
B9	05/14/81	1.128	70.0	1.00	1.50	1.50	120.0	4.50	199.0
BR1	05/20/81	1.216	110.0	3.30	4.80	1.20	200.0	5.00	324.0
Mean		1.185	96.9	2.92	3.72	1.26	166.5	5.12	276.4
Minimum		1.115	70.0	0.93	1.00	0.85	110.0	0.43	197.0
Maximum		1.221	160.0	8.10	11.00	1.60	240.0	7.30	408.0

Table A1.5. Summary of 17 "K" BSF monitoring-well sample analyses (data file provided by J.L. Mason, Hydrologist, USGS, written communication, March 7, 2001); cation/anion values expressed as g/L; density expressed as g/mL.

Sample	Date	Density	Na	Mg	K	Ca	Cl	SO_4	TDS
K10-A	05/20/81	1.213	100.0	5.50	4.00	1.10	180.0	6.90	298.0
K11	05/20/81	1.215	100.0	4.70	3.60	1.10	180.0	5.40	295.0
K13	05/21/81	1.114	50.0	2.40	3.50	1.50	96.0	5.80	159.0
K15	05/19/81	1.183	100.0	3.70	3.30	1.30	160.0	4.90	273.0
K16	05/21/81	1.177	90.0	4.70	4.00	1.20	150.0	5.80	256.0
K19	05/19/81	1.197	110.0	2.20	2.40	1.30	190.0	4.50	310.0
K22	05/19/81	1.194	110.0	2.60	3.10	1.40	170.0	4.30	291.0
K24-A	05/19/81	1.194	100.0	4.80	5.00	1.10	180.0	6.10	297.0
K25	05/19/81	1.201	110.0	2.30	2.60	1.40	180.0	4.10	300.0
K26	05/20/81	1.199	110.0	6.10	4.10	0.96	190.0	6.10	317.0
K46	05/12/81	1.192	96.0	2.20	3.70	1.70	170.0	4.70	278.0
K5	05/19/81	1.212	100.0	5.00	2.00	1.00	180.0	6.80	295.0
K6	05/19/81	1.167	100.0	2.80	2.90	1.50	150.0	5.30	263.0
K65	05/14/81	1.132	100.0	0.86	1.00	1.10	140.0	3.80	247.0
K66	05/13/81	1.189	100.0	2.50	3.30	1.30	170.0	5.20	282.0
K75	05/28/81	1.205	110.0	3.60	6.40	1.20	190.0	5.10	316.0
K8-B	05/19/81	1.207	120.0	2.10	1.50	1.30	200.0	4.30	329.0
Mean		1.188	100.4	3.42	3.32	1.26	169.2	5.24	282.7
Minimum		1.114	50.0	0.86	1.00	0.96	96.0	3.80	159.0
Maximum		1.215	120.0	6.10	6.40	1.70	200.0	6.90	329.0

Table A1.6. Summary of 18 "L" BSF auger-hole sample analyses (data file provided by J.L. Mason, Hydrologist, USGS, written communication, March 7, 2001); cation/anion values expressed as g/L; density expressed as g/mL.

Sample	Date	Density	Na	Mg	K	Ca	Cl	SO_4	TDS
L105	05/12/81	1.173	100.0	1.50	2.50	1.40	150.0	3.90	259.0
L106	05/12/81	1.194	94.0	3.70	8.10	1.20	170.0	5.40	282.0
L109	05/14/81	1.157	78.0	1.10	2.10	1.30	140.0	4.80	227.0
L17	05/20/81	1.204	120.0	3.20	2.80	1.20	200.0	4.60	332.0
L27	05/28/81	1.213	110.0	3.90	6.60	1.20	190.0	4.80	317.0
L29	05/28/81	1.202	110.0	0.85	1.90	1.20	200.0	3.80	318.0
L33	05/22/81	1.212	100.0	5.20	2.30	1.10	190.0	5.60	304.0
L35	05/22/81	1.206	110.0	5.30	4.90	1.20	190.0	5.30	317.0
L37	05/22/81	1.205	100.0	7.10	7.70	1.10	180.0	5.90	302.0
L38	05/22/81	1.204	100.0	3.70	7.50	1.10	190.0	5.10	307.0
L52	05/12/81	1.166	79.0	1.70	3.10	1.40	140.0	5.40	231.0
L53	05/15/81	1.191	100.0	2.30	2.10	1.30	170.0	4.90	281.0
L54	05/28/81	1.092	50.0	1.70	1.80	0.87	82.0	4.00	140.0
L58	05/12/81	1.191	100.0	4.80	6.70	1.20	170.0	5.50	288.0
L60	05/21/81	1.180	88.0	2.50	4.20	1.30	150.0	5.30	251.0
L73	05/20/81	1.201	110.0	3.80	3.70	1.20	190.0	5.50	314.0
L74	05/22/81	1.205	100.0	1.70	8.00	1.30	180.0	6.00	297.0
L79	05/19/81	1.203	100.0	4.40	4.40	1.20	180.0	5.60	296.0
Mean		1.189	97.2	3.25	4.47	1.21	170.1	5.08	281.3
Minimum		1.092	50.0	0.85	1.80	0.87	82.0	3.80	140.0
Maximum		1.213	120.0	7.10	8.10	1.40	200.0	6.00	332.0

Table A1.7. Summary of 26 BSF monitoring-well samples collected in 1992 by USGS (Mason and others, 1995) - see text table 9; cation/anion values expressed as g/L; density expressed as g/mL.

Sample	Date	Density	Na	Mg	K	Ca	Cl	SO_4	TDS
BLM-91	10/02/92	1.179	110.0	3.00	4.70	1.30	180.0	6.20	321
BLM-60/USGS-2	08/12/92	1.190	100.0	5.00	7.30	1.20	160.0	6.40	301
BLM-28	08/12/92	1.190	100.0	4.70	7.10	1.20	160.0	6.20	307
BLM-43C	08/12/92	1.200	110.0	3.70	7.10	1.10	160.0	5.40	318
BLM-44A	08/11/92	1.202	110.0	3.00	4.70	1.20	160.0	8.00	336
BLM-48/K-5	08/11/92	1.190	110.0	4.10	7.10	1.10	180.0	6.20	313
BLM-47	08/11/92	1.200	99.0	5.40	8.20	1.00	160.0	9.50	327
BLM-45	08/11/92	1.200	100.0	4.50	7.10	0.92	160.0	9.40	329
BLM-53	08/11/92	1.198	110.0	3.80	4.70	1.30	170.0	5.70	309
BLM-29/K-7	08/12/92	1.190	100.0	3.20	6.80	1.20	160.0	5.60	303
BLM-108	10/02/92	1.182	95.0	4.70	5.80	1.10	160.0	6.50	305
BLM-71A/USGS-9	08/25/92	1.190	98.0	5.40	10.00	0.94	170.0	9.30	329
BLM-93B	08/25/92	1.176	94.0	4.60	8.10	1.00	150.0	8.00	274
BLM-30C	08/25/92	1.186	100.0	4.40	7.60	1.10	160.0	6.70	292
BLM-103	10/02/92	1.180	92.0	4.80	3.50	1.10	150.0	6.10	302
BLM-20/USGS-64	08/25/92	1.160	89.0	1.90	3.50	1.20	140.0	7.50	262
BLM-41	08/25/92	1.192	100.0	4.30	7.30	0.94	170.0	6.40	323
BLM-2/B-27	08/26/92	1.170	100.0	1.80	4.30	1.10	150.0	5.50	285
BLM-95	10/02/92	1.148	79.0	1.40	2.30	1.60	120.0	5.60	236
BLM-65/USGS-74	08/25/92	1.190	100.0	5.20	9.60	1.00	160.0	6.20	312
BLM-6/USGS-37	08/26/92	1.190	100.0	5.60	9.50	1.00	160.0	6.40	315
BLM-16/USGS-34	08/26/92	1.194	100.0	4.80	9.10	1.00	170.0	8.50	329
BLM-36A/USGS-31	08/26/92	1.190	120.0	1.80	4.60	1.20	170.0	4.50	340
BLM-36/USGS-32	08/26/92	1.200	120.0	3.40	7.90	1.20	170.0	4.90	325
BLM-124	10/02/92	1.120	110.0	4.60	8.50	1.10	160.0	5.40	310
BLM-14/USGS-27	08/26/92	1.200	110.0	3.10	6.80	1.10	140.0	7.90	336
Mean		1.185	102.2	3.93	6.66	1.12	159.6	6.69	309
Minimum		1.120	79.0	1.40	2.30	0.92	120.0	4.50	236
Maximum		1.202	120.0	5.60	10.00	1.60	180.0	9.50	340

Table A1.8. Summary of 20 BSF monitoring-well samples collected in 1993 by USGS (Mason and others, 1995) - see text table 9; cation/anion values expressed as g/L; density expressed as g/mL.

Sample	Date	Density	Na	Mg	K	Ca	Cl	SO_4	TDS
BLM-60/USGS-2	07/13/93	1.180	99.0	3.80	6.20	0.97	172.0	6.00	300
BLM-28	07/01/93	1.180	98.0	2.20	5.70	2.10	170.0	5.90	283
BLM-44A	06/23/93	1.200	110.0	2.50	4.20	1.10	160.0	4.20	288
BLM-47	06/23/93	1.190	100.0	4.30	6.70	0.97	160.0	6.30	319
BLM-53	06/23/93	1.170	97.0	3.20	5.00	1.30	150.0	4.70	294
BLM-29/K-7	06/23/93	1.180	100.0	3.20	4.30	1.20	160.0	4.80	298
BLM-108	06/23/93	1.180	100.0	3.20	4.30	1.20	160.0	4.80	298
BLM-71A/USGS-9	07/13/93	1.190	103.0	4.80	8.30	0.83	186.0	6.90	318
BLM-93	09/01/93	1.180	100.0	5.80	8.10	1.10	170.0	7.50	292
BLM-93B	09/01/93	1.170	81.0	4.50	4.10	1.00	140.0	7.80	261
BLM-93C	09/01/93	1.190	110.0	3.90	3.80	1.10	180.0	5.10	320
BLM-30C	06/24/93	1.170	98.0	4.20	6.10	1.00	150.0	6.30	297
BLM-103	06/23/93	1.170	100.0	3.00	4.30	1.10	150.0	6.80	284
BLM-20/USGS-64	06/03/93	1.160	81.0	1.80	3.90	1.10	120.0	3.30	244
BLM-41	06/24/93	1.180	98.0	4.70	7.20	1.00	140.0	6.40	309
BLM-2/B-27	06/08/93	1.170	100.0	2.20	4.40	1.40	150.0	4.80	278
BLM-65/USGS-74	07/07/93	1.180	92.0	5.00	0.72	0.97	180.0	5.60	306
BLM-6/USGS-37	07/13/93	1.190	102.0	5.10	8.50	0.96	185.0	6.40	313
BLM-36/USGS-32	07/14/93	1.190	104.0	4.30	7.40	1.10	196.0	6.20	323
BLM-14/USGS-27	07/14/93	1.190	110.0	3.30	5.40	1.10	20.0	5.00	322
Mean		1.181	99.2	3.75	5.43	1.13	155.01	5.74	297
Minimum		1.160	81.0	1.80	0.72	0.83	20.0	3.30	244
Maximum		1.200	110.0	5.80	8.50	2.10	196.0	7.80	323

Table A1.9. Summary of 9 samples selected from the 26 BSF monitoring-well samples collected in 1992 by USGS (Mason and others, 1995); these selected samples were compared with BLM samples collected from the same wells during 1994-1997 (see text table 13); cation/anion values expressed as g/L; density expressed as g/mL.

Sample	Density	Na	Mg	K	Ca	Cl	SO_4	TDS
BLM-6	1.196	100.00	5.60	9.50	1.00	160.00	6.40	315.00
BLM-28	1.194	100.00	4.70	7.10	1.20	160.00	6.20	307.00
BLM-41	1.192	100.00	4.30	7.30	0.94	170.00	6.40	323.00
BLM-43C	1.200	110.00	3.70	7.10	1.10	160.00	5.40	318.00
BLM-50A	1.194	100.00	4.30	6.80	1.20	170.00	5.90	309.00
BLM-53	1.198	110.00	3.80	4.70	1.30	170.00	5.70	309.00
BLM-60	1.192	100.00	5.00	7.30	1.20	160.00	6.40	301.00
BLM-71A	1.198	98.00	5.40	10.00	0.94	170.00	9.30	329.00
BLM-93	1.182	100.00	5.80	8.10	1.10	170.00	7.50	292.00
Mean	1.194	102.00	4.73	7.54	1.11	165.56	6.58	311.44
Miniumum	1.182	98.00	3.70	4.70	0.94	160.00	5.40	292.00
Maximum	1.200	110.00	5.80	10.00	1.30	170.00	9.30	329.00

Table A1.10. Summary of 9 samples selected from the 20 BSF monitoring-well samples collected in 1993 by USGS (Mason and others, 1995); these selected samples were compared with BLM samples collected from the same wells during 1994-1997 (see text table 13); cation/anion values expressed as g/L; density expressed as g/mL.

Sample	Density	Na	Mg	K	Ca	Cl	SO_4	TDS
BLM-6	1.192	102.00	5.10	8.50	0.96	185.00	6.40	313.00
BLM-28	1.188	98.00	2.20	5.70	2.10	170.00	5.90	283.90
BLM-41	1.188	98.00	4.70	7.20	1.00	140.00	6.40	309.00
BLM-43C	1.200	110.00	2.50	4.20	1.10	160.00	4.20	288.00
BLM-53	1.174	97.00	3.20	5.00	1.30	150.00	4.70	294.00
BLM-60	1.184	99.00	3.80	6.20	0.97	172.00	6.00	300.00
BLM-71A	1.192	103.00	4.80	8.30	0.83	186.00	6.90	318.00
BLM-93	1.182	100.00	5.80	8.10	1.10	170.00	7.50	292.50
BLM-100	1.174	100.00	3.00	4.30	1.10	150.00	6.80	284.00
Mean	1.186	100.78	3.90	6.39	1.16	164.78	6.09	298.04
Minimum	1.174	97.00	2.20	4.20	0.83	140.00	4.20	283.90
Maximum	1.200	110.00	5.80	8.50	2.10	186.00	7.50	318.00

APPENDIX A2

1994-2000 BSF MONITORING-WELL SAMPLE ANALYSES BY BLM

Table A2.1. Summary of 19 FY94 BSF monitoring-well brine samples collected by BLM September 19-20, 1994 (concentrations expressed as g/L; density expressed as g/mL)

Sample	Density	Na	Mg	K	Ca[1]	Cl	SO4	TDS	mole bal.[2]
94BLM-3	1.174	95.2	2.70	3.80	1.10	156.7	5.02	264.5	-0.000498
94BLM-6	1.200	99.3	8.50	9.40	0.86	185.4	5.79	309.3	-0.002045
94BLM-12	1.205	116.9	4.20	5.70	0.90	192.5	4.34	324.5	0.003956
94BLM-25	1.193	98.3	5.60	6.20	0.80	173.3	5.59	289.8	-0.002644
94BLM-30	1.177	96.1	3.30	3.80	1.10	160.3	5.11	269.7	-0.001187
94BLM-35	1.177	95.8	3.00	4.30	1.10	159.5	5.02	268.7	-0.001196
94BLM-41	1.198	100.5	6.80	7.30	0.70	181.2	5.18	301.7	-0.002410
94BLM-43C	1.202	111.7	5.30	6.40	0.70	186.8	4.83	315.7	0.005418
94BLM-46	1.204	112.6	7.60	8.30	0.60	195.8	5.31	330.2	0.005817
94BLM-50A	1.199	110.6	4.80	5.50	0.80	186.8	5.02	313.5	0.000596
94BLM-53	1.192	111.2	5.10	5.80	0.80	184.7	4.83	312.4	0.005734
94BLM-61	1.201	111.6	6.00	7.70	0.70	189.8	5.12	320.9	0.005203
94BLM-82	1.195	111.3	4.10	5.50	0.86	185.4	4.92	311.9	0.001223
94BLM-99	1.166	88.9	6.10	5.80	0.60	152.6	6.01	260.0	0.006078
94BLM-101	1.188	97.2	5.60	5.40	1.40	169.9	5.70	285.2	-0.001537
94BLM-107	1.188	97.4	5.30	5.90	0.90	169.5	5.58	284.6	-0.001086
94BLM-60	1.193	98.5	5.90	7.30	0.70	174.8	5.59	292.8	-0.001862
94BLM-71/USGS-9	1.200	100.4	7.50	9.30	0.86	184.4	6.08	308.4	-0.002647
94BLM-28	1.196	100.5	6.10	6.60	0.86	178.5	5.49	297.9	-0.002615
Mean	1.192	102.8	5.45	6.32	0.89	177.3	5.29	298.1	0.002830
Minimum	1.166	88.9	2.70	3.80	0.76	152.6	4.34	260.5	-0.002647
Maximum	1.205	116.9	8.50	9.40	1.08	195.8	6.08	330.4	0.006078

[1]Because Ca was not included in 1994 analyses, either matching or average (0.86) 1996 Ca values were used.
[2]Because of the occurrence of negative values, the mole-balance mean was calculated by converting sample mole balances to absolute values; this was done to avoid the possibility of a zero mean value.

Table A2.2. Summary of 13 FY95 BSF monitoring-well brine samples collected by BLM in September 1995 (concentrations expressed as g/L; density expressed as g/mL).

Sample ID	Density	Na	Mg	K	Ca1	Cl	SO4	TDS	mole bal.[2]
95BLM-3	1.170	92.7	2.90	3.70	1.05	144.2	5.47	249.0	0.010124
95BLM-6	1.201	97.4	7.25	9.00	1.05	170.3	5.78	289.7	0.008002
95BLM-12	1.205	106.6	3.89	5.40	1.05	169.8	4.50	290.2	0.010979
95BLM-19	1.198	107.2	2.95	3.60	1.05	170.3	3.99	288.0	0.006837
95BLM-25	1.195	99.5	4.97	5.90	1.05	165.8	5.38	281.6	0.006356
95BLM-28	1.195	100.4	4.97	5.90	1.05	163.7	5.13	280.1	0.010750
95BLM-43C	1.200	105.9	4.89	6.00	1.05	172.3	4.23	293.3	0.011120
95BLM-46	1.206	107.3	6.30	7.70	1.05	179.6	5.30	306.2	0.010742
95BLM-50A	1.194	102.9	3.93	5.50	1.05	164.5	4.81	281.6	0.010575
95BLM-53	1.194	104.6	4.25	5.60	1.05	168.5	5.01	288.0	0.009983
95BLM-60	1.193	98.7	4.83	7.20	1.05	161.3	5.77	277.8	0.010799
95BLM-61	1.205	107.4	4.80	6.60	1.05	174.6	4.90	298.3	0.010840
95BLM-82	1.196	104.3	3.85	5.30	1.05	165.9	4.84	284.2	0.010938
Mean	1.196	102.7	4.60	5.95		167.0	5.01	285.2	0.009850
Minimum	1.170	92.7	2.90	3.60		144.2	3.99	249.0	0.006356
Maximum	1.206	107.4	7.25	9.00		179.6	5.78	306.2	0.011120

[1]Calcium was not analyzed; values are averages of USGS FY92-FY93, and BLM FY96-FY97 calcium analyses.

Table A2.3. Summary of 16 FY96 BSF monitoring-well brine samples collected by BLM September 24, 1996 (concentrations expressed as g/L; density expressed as g/mL)

Sample	Density	Na	Mg	K	Ca	Cl	SO4	TDS	mole bal.[2]
96BLM-3	1.171	93.1	1.90	4.00	1.10	149.4	4.85	254.3	0.002029
96BLM-12	1.204	116.5	2.80	6.00	0.90	190.8	4.20	321.2	0.001129
96BLM-25	1.187	95.4	3.40	5.90	0.80	163.8	5.10	274.4	-0.004556
96BLM-30	1.174	94.9	1.60	3.55	1.10	157.0	4.45	262.5	-0.004958
96BLM-35	1.179	96.6	1.80	3.90	1.10	156.6	4.10	264.1	0.000157
96BLM-41	1.196	98.6	4.70	8.10	0.70	176.2	5.15	293.4	-0.006752
96BLM-43C	1.201	103.3	3.60	6.90	0.70	178.2	4.45	297.1	-0.005000
96BLM-46	1.206	99.6	5.20	8.45	0.60	179.8	4.80	298.4	-0.006930
96BLM-50A	1.194	97.3	3.60	5.90	0.80	168.2	4.70	280.5	-0.005077
96BLM-53	1.194	101.5	3.55	5.55	0.80	175.2	4.45	291.1	-0.006083
96BLM-61	1.208	101.9	4.80	8.45	0.70	180.8	4.45	301.1	-0.004799
96BLM-93	1.189	92.8	3.80	8.10	0.80	157.0	6.45	269.0	0.001421
96BLM-99	1.159	91.6	4.90	5.35	0.60	158.7	6.60	267.8	-0.002557
96BLM-101	1.076	37.8	0.90	2.20	1.40	60.9	4.85	108.1	0.001194
96BLM-107	1.184	93.9	3.50	6.10	0.90	160.8	5.00	270.2	-0.002887
96BLM-60	1.193	96.2	4.30	8.45	0.70	170.7	5.30	285.6	-0.005627
Mean	1.182	94.4	3.40	6.06	0.86	161.5	4.93	271.1	0.003820
Minimum	1.076	37.8	0.90	2.20	0.60	60.9	4.10	108.1	-0.006930
Maximum	1.208	116.5	5.20	8.45	1.40	190.8	6.60	321.2	0.002029

[2]Because of the occurrence of negative values, the mole-balance mean was calculated by converting sample mole balances to absolute values; this was done to avoid the possibility of a zero mean value.

Table A2.4. Summary of 20 BSF FY97 monitoring-well brine samples collected by BLM September 23-24, 1997 (concentrations expressed as g/L; density expressed as g/mL).

Sample	Density	Na	Mg	K	Ca	Cl	SO_4	TDS	mole bal.[2]
97BLM-6	1.197	93.9	4.70	9.00	0.88	162.9	5.70	277.1	0.001333
97BLM-3	1.162	84.4	1.45	3.55	1.50	137.0	5.25	233.2	-0.000744
97BLM-12	1.202	97.0	1.25	5.90	1.04	156.9	4.80	266.9	-0.000019
97BLM-25	1.188	93.1	3.15	6.05	1.02	155.2	5.95	264.5	0.000551
97BLM-27	1.112	48.4	0.55	1.90	1.47	76.0	4.95	133.3	0.001161
97BLM-30	1.173	92.7	1.00	3.30	1.54	149.9	4.90	253.3	-0.002313
97BLM-41	1.197	94.3	2.85	6.25	0.90	156.6	5.95	266.9	0.000010
97BLM-43C	1.201	97.6	3.00	6.00	1.05	162.9	4.80	275.4	0.000144
97BLM-46	1.207	96.4	2.80	6.50	0.85	158.2	5.80	270.6	0.002045
97BLM-50B	1.197	98.7	3.20	5.55	1.10	166.4	5.30	280.3	-0.002104
97BLM-53	1.196	96.8	2.45	5.00	1.15	156.3	5.15	266.9	0.003416
97BLM-60	1.195	94.5	4.05	6.55	1.05	162.2	5.90	274.3	-0.001425
97BLM-61	1.204	94.6	3.70	6.80	0.95	158.2	5.85	270.1	0.002357
97BLM-82	1.198	97.6	2.55	4.90	1.15	162.5	4.85	273.6	-0.001939
97BLM-90	1.17	86.8	1.45	2.65	1.15	138.8	4.15	235.0	0.000801
97BLM-93	1.191	94.5	4.55	7.00	0.95	162.8	7.86	277.7	0.001853
97BLM-99	1.164	76.0	2.25	4.20	1.05	122.6	7.10	213.2	0.001931
97BLM-100	1.189	96.3	3.85	5.55	1.05	159.8	6.60	273.2	0.002326
97BLM-107	1.183	90.1	2.45	4.50	1.10	147.0	6.00	251.2	0.000828
97BLM-107A	1.177	94.5	1.60	3.80	1.50	154.2	5.30	260.9	-0.001929
Mean	1.185	90.9	2.64	5.25	1.12	150.3	5.61	255.9	0.001460
Minimum	1.112	48.4	0.55	1.90	0.85	76.0	4.15	133.3	-0.002310
Maximum	1.207	98.7	4.70	9.00	1.54	166.4	7.86	280.3	0.003416

Table A2.5. Summary of 10 FY2K BSF monitoring-well brine samples collected by BLM October 1, 1999 (cation/anion values expressed as g/L; density expressed as g/mL).

Sample	Density	Na	Mg	K	Ca	Cl	SO_4	TDS	mole bal.[2]
2KBLM-3	1.165	82.2	2.49	3.50	1.51	139.4	4.77	233.8	-0.003756
2KBLM-27	1.191	93.6	2.94	5.34	1.24	159.4	4.77	267.3	-0.003524
2KBLM-30	1.147	76.0	1.31	1.89	1.68	125.6	4.44	210.9	-0.003962
2KBLM-35	1.173	87.4	1.54	3.19	1.47	143.2	4.33	241.1	-0.001966
2KBLM-41	1.196	93.1	4.12	6.74	1.11	162.0	5.23	272.3	-0.002660
2KBLM-50B	1.197	96.2	2.94	5.70	1.08	164.0	6.07	276.0	-0.005269
2KBLM-53	1.186	91.6	2.77	5.06	1.18	155.6	4.71	260.9	-0.003653
2KBLM-56	1.186	93.1	2.12	4.59	1.30	155.2	4.74	261.0	-0.003023
2KBLM-90	1.121	61.8	1.71	1.89	1.50	100.7	4.46	172.1	0.000851
2KBLM-107	1.171	84.4	2.79	5.06	1.23	143.8	5.32	242.6	-0.003165
Mean	1.173	85.9	2.47	4.30	1.33	144.9	4.88	243.8	0.003180
Minimum	1.121	61.8	1.31	1.89	1.08	100.7	4.33	172.1	-0.005270
Maximum	1.197	96.2	4.12	6.74	1.68	164.0	6.07	276.0	0.000851

[2]Because of the occurrence of negative values, the mole-balance mean was calculated by converting sample mole balances to absolute values; this was done to avoid the possibility of a zero mean value.

APPENDIX A3

FY98-FY2K LAYDOWN-BRINE SAMPLE ANALYSES BY BLM

Table A3.1. Summary of 9 BLM laydown-brine analyses from FY98A sampling (concentration expressed as g/L; density expressed as g/mL).

Sample	Date	Location	Density	Na	Mg	K	Ca	Cl	SO_4	TDS	mole bal[1]
FY98A B-1	15Jan98	BP#3	1.064	34.4	0.12	0.20	0.59	52.8	1.40	89.5	0.001050
FY98A B-2	15Jan98	LDM-W	1.065	33.5	0.09	0.20	0.48	51.1	1.20	86.6	0.001285
FY98A B-3	15Jan98	LDM-E	1.057	33.5	0.14	0.20	0.71	48.9	1.75	85.2	0.004492
FY98A B-4	23Jan98	BP#3	1.082	43.6	0.11	0.25	0.64	67.8	1.60	114.0	-0.000082
FY98A B-5	23Jan98	PP#4	1.051	28.2	0.11	0.20	0.57	41.7	1.40	72.1	0.003108
FY98A B-6	23Jan98	LDM-W	1.080	41.5	0.11	0.25	0.63	65.1	1.50	109.0	-0.000646
FY98A B-7	23Jan98	LDM-E	1.043	23.3	0.09	0.20	0.59	35.2	1.40	60.7	0.001672
FY98A B-8	26Jan98	LDM-E	1.041	19.5	0.11	0.20	0.64	28.2	1.40	50.0	0.003419
FY98A B-9	26Jan98	LDM-W	1.041	20.2	0.18	0.20	0.80	30.9	1.95	54.2	0.001266
Mean			1.058	30.9	0.12	0.21	0.63	46.8	1.51	80.2	0.001891
Minimum			1.041	19.5	0.09	0.20	0.48	28.2	1.20	50.0	-0.000650
Maximum			1.082	43.6	0.18	0.25	0.80	67.8	1.95	114.0	0.004492

Table A3.2. Summary of 8 BLM laydown-brine analyses from FY98B sampling (concentration expressed as g/L; density expressed as g/mL).

Sample	Date	Location	Density	Na	Mg	K	Ca	Cl	SO_4	TDS	mole bal[1]
FY98B B-12	2Mar98	LDM-E	1.179	102.9	0.25	0.29	1.35	167.7	2.35	274.8	-0.008809
FY98B B-13	10Mar98	LDM-W	1.159	91.5	0.20	0.29	1.07	148.4	2.49	244.0	-0.007775
FY98B B-15	10Mar98	BP#3	1.158	92.0	0.20	0.29	0.95	149.4	2.57	245.4	-0.008485
FY98B B-16	27Mar98	BP#3	1.156	89.0	0.20	0.29	1.13	142.1	2.93	235.6	-0.005174
FY98B B-17	27Mar98	LDM-W	1.159	87.5	0.17	0.29	1.14	141.5	3.23	233.8	-0.007506
FY98B B-19	9Apr98	LDM-W	1.041	19.2	0.16	0.29	0.59	32.9	2.10	55.2	-0.004155
FY98B B-21	9Apr98	BP#3	1.040	21.4	0.17	0.35	0.77	35.5	2.17	60.4	-0.002609
FY98B B-22	22Apr98	BP#3	1.149	78.1	0.25	0.45	0.90	128.0	2.24	209.9	-0.008049
Mean			1.130	72.7	0.20	0.32	0.99	118.2	2.51	194.9	-0.006570
Minimum			1.040	19.2	0.16	0.29	0.59	32.9	2.10	55.2	-0.008810
Maximum			1.179	102.9	0.25	0.45	1.35	167.7	3.23	274.8	-0.002610

[1]Because of the occurrence of negative values, the mole-balance mean was calculated by converting sample mole balances to absolute values; this was done to avoid the possibility of a zero mean value.

Table A3.3. Summary of 29 BLM Laydown-brine analyses from FY99 sampling (concentration expressed as g/L; density expressed as g/mL).

Sample	Date	Location	Density	Na	Mg	K	Ca	Cl	SO_4	TDS	mole bal[1]
FY99 B-25	6Nov98	BP#3	1.180	97.5	0.24	0.25	1.25	156.7	2.86	258.8	-0.006348
FY99 B-26	6Nov98	LDM-W	1.180	97.1	0.23	0.20	1.25	155.3	2.93	257.0	-0.005563
FY99 B-27	6Nov98	LDM-E	1.198	100.9	0.29	0.30	1.40	161.4	3.30	267.6	-0.005458
FY99 B-28	17Nov98	LDM-W	1.196	102.6	0.25	0.25	1.36	162.3	2.98	269.7	-0.003433
FY99 B-29	17Nov98	LDM-E	1.198	102.7	0.24	0.20	1.37	162.8	3.10	270.4	-0.004005
FY99 B-30	17Nov98	BP#3	1.193	100.9	0.25	0.30	1.36	162.2	2.93	267.9	-0.006325
FY99 B-31	3Dec98	BP#3	1.193	99.6	0.28	0.35	1.34	160.5	2.95	265.0	-0.006588
FY99 B-32	3Dec98	LDM-W	1.194	100.1	0.28	0.25	1.34	161.4	2.85	266.2	-0.006754
FY99 B-33	3Dec98	LDM-E	1.194	101.0	0.27	0.35	1.35	162.1	2.82	267.9	-0.005822
FY99 B-34	14Jan99	LDM-W	1.183	99.8	0.29	0.30	1.31	158.2	2.90	262.8	-0.003572
FY99 B-35	14Jan99	LDM-E	1.182	98.9	0.31	0.30	1.30	157.1	3.02	260.9	-0.003976
FY99 B-36	14Jan99	BP#3	1.182	99.2	0.29	0.25	1.27	156.5	2.86	260.4	-0.002754
FY99 B-37	22Jan99	LDM-W	1.155	86.8	0.32	0.40	1.18	136.0	2.66	227.4	-0.000878
FY99 B-38	22Jan99	LDM-E	1.182	199.6	0.33	0.40	1.34	160.4	2.84	264.9	-0.006205
FY99 B-39	22Jan99	BP#3	1.157	87.8	0.33	0.35	1.19	140.9	2.63	233.2	-0.004941
FY99 B-40	19Feb99	LDM-W	1.184	99.5	0.36	0.45	1.39	159.7	2.95	264.4	-0.005377
FY99 B-41	19Feb99	LDM-E	1.187	100.7	0.39	0.50	1.42	162.6	3.12	268.7	-0.006538
FY99 B-42	19Feb99	BP#3	1.183	100.6	0.39	0.45	1.39	162.2	3.05	268.1	-0.006323
FY99 B-43	25Feb99	BP#3	1.189	100.8	0.39	0.50	1.40	162.4	2.92	268.4	-0.005974
FY99 B-44	25Feb99	LDM-W	1.190	101.0	0.39	0.45	1.40	162.5	2.93	268.7	-0.005785
FY99 B-45	25Feb99	LDM-E	1.190	101.9	0.35	0.40	1.43	162.8	3.04	269.9	-0.004721
FY99 B-49	12Mar99	LDM-E	1.179	97.8	0.34	0.45	1.38	155.7	3.05	258.7	-0.003930
FY99 B-50	12Mar99	LDM-W	1.166	92.7	0.33	0.40	1.30	145.0	2.99	242.7	-0.000753
FY99 B-51	12Mar99	BP#3	1.164	90.5	0.32	0.35	1.29	143.4	2.91	238.8	-0.002967
FY99 B-52	19Apr99	BP#3	1.195	100.9	0.35	0.40	1.45	162.0	3.16	268.3	-0.005639
FY99 B-53	19Apr99	LDM-W	1.195	101.5	0.35	0.35	1.45	161.5	3.25	268.4	-0.004089
FY99 B-55	20Apr99	LDM-E	1.197	101.8	0.32	0.40	1.46	162.8	3.05	269.8	-0.004924
Mean			1.185	98.7	0.31	0.35	1.35	157.6	2.96	261.3	-0.004801
Minimum			1.155	86.8	0.23	0.20	1.18	136.0	2.63	227.4	-0.006750
Maximum			1.198	102.7	0.39	0.50	1.46	162.8	3.30	270.4	-0.000750

[1]Because of the occurrence of negative values, the mole-balance mean was calculated by converting sample mole balances to absolute values; this was done to avoid the possibility of a zero mean value.

Table A3.4. Summary of 30 BLM Laydown-brine analyses from FY2K sampling (concentration expressed as g/L; density expressed as g/mL).

Sample	Date	Location	Density	Na	Mg	K	Ca	Cl	SO_4	TDS	mole bal[1]
FY2KB-56	9Nov99	BP#3	1.185	110.0	0.25	0.46	1.41	172.5	2.85	287.5	-0.001502
FY2KB-57	9Nov99	LDM-W	1.187	114.1	0.04	0.46	1.47	175.3	2.84	294.2	0.002067
FY2KB-58	9Nov99	LDM-E	1.199	116.9	0.46	0.77	1.43	182.2	3.41	305.1	0.000203
FY2KB-59	30Nov99	BP#3	1.167	99.5	0.04	0.31	1.50	155.9	3.11	260.4	-0.002101
FY2KB-60	30Nov99	LDM-W	1.168	101.8	0.14	0.31	1.28	154.6	3.62	261.7	0.003278
FY2KB-61	30Nov99	LDM-E	1.174	101.4	0.12	0.31	1.33	158.7	2.87	264.7	-0.001752
FY2KB-62	14Dec99	BP#3	1.163	98.2	0.47	0.38	1.31	157.8	2.84	261.0	-0.005411
FY2KB-63	14Dec99	LDM-W	1.163	96.8	0.68	0.38	1.41	146.3	3.29	248.8	0.006544
FY2KB-64	14Dec99	LDM-E	1.168	98.4	0.51	0.38	1.34	149.0	2.93	252.6	0.005735
FY2KB-65	29Dec99	BP#3	1.167	98.4	0.16	0.38	1.31	149.0	3.09	252.4	0.004300
FY2KB-66	29Dec99	LDM-W	1.167	100.1	0.05	0.38	1.31	151.8	3.09	256.7	0.003745
FY2KB-67	29Dec99	LDM-E	1.168	100.1	0.09	0.38	1.37	156.7	2.84	261.5	-0.001683
FY2KB-68	18Jan2K	BP#3	1.148	92.8	0.30	0.56	1.27	148.8	3.10	246.8	-0.005338
FY2KB-69	18Jan2K	LDM-W	1.146	92.8	0.30	0.56	1.27	149.0	2.98	247.0	-0.005534
FY2KB-70	18Jan2K	LDM-E	1.144	90.3	0.22	0.56	1.27	144.9	2.81	240.1	-0.005325
FY2KB-71	07Feb2K	LDM-E	1.117	71.3	0.41	0.65	1.07	117.3	2.20	192.9	-0.006677
FY2KB-72	07Feb2K	LDM-W	1.117	70.4	0.32	0.65	1.01	115.7	2.05	190.1	-0.006710
FY2KB-73	07Feb2K	BP#3	1.112	71.9	0.39	0.65	0.98	116.7	1.95	192.6	-0.004789
FY2KB-74	01Mar2K	LDM-E	1.141	89.6	0.42	0.74	1.27	146.0	2.95	241.0	-0.007288
FY2KB-75	01Mar2K	LDM-W	1.141	84.2	0.42	0.74	1.27	133.9	2.71	223.2	-0.002307
FY2KB-76	01Mar2K	BP#3	1.137	85.5	0.42	0.65	1.22	136.0	2.72	226.5	-0.002703
FY2KB-77	17Mar2K	BP#3	1.157	97.2	0.34	0.65	1.38	154.3	3.46	257.4	-0.003599
FY2KB-78	17Mar2K	LDM-W	1.156	97.2	0.24	0.56	1.42	153.2	3.15	255.8	-0.002301
FY2KB-79	17Mar2K	LDM-E	1.161	96.2	0.48	0.83	1.41	153.2	3.55	255.7	-0.003397
FY2KB-80	31Mar2K	LDM-E	1.197	120.0	0.18	0.56	1.52	189.6	3.24	315.1	-0.003757
FY2KB-81	31Mar2K	LDM-W	1.189	118.8	0.31	0.46	1.32	189.3	3.57	313.8	-0.006059
FY2KB-82	31Mar2K	BP#3	1.189	117.8	0.24	0.46	1.44	187.7	3.44	311.1	-0.005796
FY2KB-83	27Apr2K	LDM-E	1.194	117.8	0.22	0.46	1.40	188.4	2.85	311.2	-0.006308
FY2KB-84	27Apr2K	LDM-W	1.195	117.8	0.26	0.56	1.46	188.1	3.40	311.5	-0.005963
FY2KB-85	27Apr2K	BP#3	1.188	117.8	0.29	0.65	1.47	186.3	3.50	310.0	-0.003775
Mean			1.164	99.5	0.29	0.53	1.33	156.9	3.01	261.6	0.004198
Minimum			1.112	70.4	0.04	0.31	0.98	115.7	1.95	190.1	-0.000729
Maximum			1.199	120.0	0.68	0.83	1.52	189.6	3.62	315.1	0.006544

[1]Because of the occurrence of negative values, the mole-balance mean was calculated by converting sample mole balances to absolute values; this was done to avoid the possibility of a zero mean value.

APPENDIX A4

FY98-FY2K LAYDOWN-BRINE MONTHLY PRODUCTION SUMMARY

Table A4.1. FY98-FY2K monthly Laydown NaCl tonnage, pump hours, acre feet, and gallons.

Sample Months	NaCl, Tons (dry)	Pumped Brine		Hours Pumped Per Month		
		Acre Feet	Gallons	Maximum Possible	Actually Pumped	% of Possible
NOV 97	118,982	368	119,973,750	720	299	41.5
DEC 97	30,739	208	67,811,250	744	169	22.7
JAN 98	80,359	616	200,625,000	744	500	67.2
FEB 98	129,792	356	115,961,250	672	289	43.0
MAR 98	278,446	911	296,925,000	744	740	99.5
APR 98	186,888	702	228,712,500	720	570	79.2
NOV 98	336,510	872	284,085,000	720	708	98.3
DEC 98	347,383	916	298,530,000	744	744	100.0
JAN 99	322,864	916	298,530,000	744	744	100.0
FEB 99	286,451	828	269,640,000	744	744	100.0
MAR 99	332,375	916	298,530,000	744	744	100.0
APR 99	339,486	887	288,900,000	720	720	100.0
NOV 99	315,518	883	287,696,250	720	717	99.6
DEC 99	322,171	916	298,530,000	744	744	100.0
JAN 2K	269,879	916	298,530,000	744	744	100.0
FEB 2K	242,421	857	279,270,000	696	696	100.0
MAR 2K	332,938	916	298,530,000	744	744	100.0
APR 2K	350,147	856	278,868,750	720	695	96.5

APPENDIX A5

FY98-FY2K LAYDOWN-BRINE Na & Cl ANALYSES EXPRESSED AS PERCENT NaCl

Table A5.1. Summary of 9 BLM Laydown-brine analyses from FY98A sampling; Na and Cl concentrations from Appendix A3 converted to chloride salt equivalents (concentration expressed as Wt% NaCl in solution).

Sample	Location	Reilly	CMS	U of Az
FY98A A,B,C-1	BP#3	8.40	8.22	8.49
FY98A A,B,C-2	LDM-W	8.60	8.00	8.39
FY98A A,B,C-3	LDM-E	7.30	8.06	7.45
FY98A A,B,C-4	BP#3	10.60	10.24	10.78
FY98A A,B,C-5	PP#4	6.80	6.82	6.98
FY98A A,B,C-6	LDM-W	10.60	9.77	10.47
FY98A A,B,C-7	LDM-E	5.70	5.68	5.90
FY98A A,B,C-8	LDM-E	5.20	4.76	5.35
FY98A A,B,C-9	LDM-W	5.10	4.93	5.37
Mean		7.59	7.39	7.69
Minimum		5.10	4.76	5.35
Maximum		10.60	10.24	10.78

Table A5.2. Summary of 8 BLM Laydown-brine analyses from FY98B sampling; Na and Cl concentrations from Appendix A3 converted to chloride salt equivalents (concentration expressed as Wt% NaCl in solution).

Sample	Location	Reilly	CMS	U of Az
FY98B A,B,C-12	LDM-E	24.60	22.19	22.56
FY98B A,B,C-13	LDM-W	21.20	20.07	19.39
FY98B A,B,C-15	BP#3	21.20	20.19	19.83
FY98B A,B,C-16	BP#3	21.80	19.56	19.74
FY98B A,B,C-17	LDM-W	21.80	19.19	19.48
FY98B A,B,C-19	LDM-W	4.70	4.69	4.85
FY98B A,B,C-21	BP#3	4.80	5.23	5.05
FY98B A,B,C-22	BP#3	19.80	17.27	18.09
Mean		17.49	16.05	6.12
Minimum		4.70	4.69	4.85
Maximum		24.60	22.19	22.56

Table A5.3. Summary of 23 BLM Laydown-brine analyses from FY99 sampling; Na and Cl concentrations from Appendix A3 converted to chloride salt equivalents (concentration expressed as Wt% NaCl in solution).

Sample	Location	Reilly	CMS
FY99 A,B-25	BP#3	22.40	21.01
FY99 A,B-30	BP#3	22.10	21.50
FY99 A,B-31	BP#3	23.90	21.22
FY99 A,B-36	BP#3	22.70	21.34
FY99 A,B-39	BP#3	19.70	19.29
FY99 A,B-42	BP#3	22.20	21.62
FY99 A,B-43	BP#3	20.20	21.55
FY99 A,B-51	BP#3	20.80	19.77
FY99 A,B-52	BP#3	24.70	21.46
FY99 A,B-29	LDM-E	22.20	21.79
FY99 A,B-33	LDM-E	23.80	21.50
FY99 A,B-35	LDM-E	22.70	21.27
FY99 A,B-38	LDM-E	20.10	21.42
FY99 A,B-41	LDM-E	22.30	21.58
FY99 A,B-45	LDM-E	23.50	21.77
FY99 A,B-49	LDM-E	20.30	21.09
FY99 A,B-28	LDM-W	22.30	21.81
FY99 A,B-32	LDM-W	24.60	21.31
FY99 A,B-34	LDM-W	22.80	21.45
FY99 A,B-37	LDM-W	19.10	19.10
FY99 A,B-40	LDM-W	22.40	21.36
FY99 A,B-44	LDM-W	19.80	21.58
FY99 A,B-50	LDM-W	20.70	20.21
Mean		21.97	21.13
Minimum		19.10	19.10
Maximum		24.70	21.81

Table A5.4. Summary of 29 BLM Laydown-brine analyses from FY2K sampling; Na and Cl concentrations from Appendix A3 converted to chloride salt equivalents (concentration expressed as Wt% NaCl in solution).

Sample	Location	Reilly	CMS
FY2K A,B-59	BP#3	21.90	21.67
FY2K A,B-62	BP#3	21.10	21.47
FY2K A,B-65	BP#3	21.60	21.44
FY2K A,B-68	BP#3	19.80	20.55
FY2K A,B-73	BP#3	16.10	16.44
FY2K A,B-76	BP#3	19.00	19.12
FY2K A,B-77	BP#3	22.20	21.36
FY2K A,B-82	BP#3	25.50	25.19
FY2K A,B-85	BP#3	26.00	25.21
FY2K A,B-58	LDM-E	23.10	24.79
FY2K A,B-61	LDM-E	22.50	21.96
FY2K A,B-64	LDM-E	21.60	21.42
FY2K A,B-67	LDM-E	21.90	21.79
FY2K A,B-70	LDM-E	19.20	20.07
FY2K A,B-71	LDM-E	16.30	16.23
FY2K A,B-74	LDM-E	19.50	19.96
FY2K A,B-79	LDM-E	22.90	21.06
FY2K A,B-80	LDM-E	25.70	25.49
FY2K A,B-83	LDM-E	26.00	25.08
FY2K A,B-57	LDM-W	22.10	24.44
FY2K A,B-60	LDM-W	22.00	22.16
FY2K A,B-63	LDM-W	21.10	21.16
FY2K A,B-66	LDM-W	21.50	21.81
FY2K A,B-69	LDM-W	19.20	20.59
FY2K A,B-72	LDM-W	16.10	16.02
FY2K A,B-75	LDM-W	19.10	18.76
FY2K A,B-78	LDM-W	22.50	21.38
FY2K A,B-81	LDM-W	25.40	25.40
FY2K A,B-84	LDM-W	26.00	25.06
Mean		21.62	21.62
MinimuM		16.10	16.02
Maximum		26.00	25.49

APPENDIX A6

SALT-CRUST STRATA DESCRIPTIONS

Distinguishing characteristics of the strata comprising the salt crust are as follows:

- **Dense-cemented salt** - dense, stratified stratum of interlocking halite crystals that range from 0.125 to 0.25 inches on exposed crystal faces. Stratification results from segregation of halite crystals into layers of smaller and larger crystals. The dense, cemented-salt stratum occasionally exhibits an undulating bottom surface, which may be a response to the undulating top surface of the underlying gypsum stratum.

- **Uncemented gypsum** - uniformly small, loose, doubly-terminated crystals of gypsum, with some minor amounts of clay-sized material. The small, uniform gypsum-crystal size of approximately 0.063 by 0.125 inches suggests that their initial source may be wind-transported gypsum from the gypsum-mud deposit that bounds the periphery of BSF (see Lines, 1979, pp. 28-32).

- **Cemented-coarse-porous salt** - porous, but cemented coarse-halite crystal structure. Halite crystals are relatively uniform in size and commonly measure 0.25 by 0.25 inches. Presence of solution cavities give this stratum a "sponge"-like appearance. Commonly has an undulating top and bottom surface (undulation is greater on the bottom surface).

- **Uncemented-coarse salt** - large, coarse, commonly uncemented halite crystals that range in size from 0.25 to 0.5 inches. Typically, the halite crystals increase in size with depth. Occasionally, cemented clumps of halite crystals that measure 2 by 3 inches are observed.

During pit excavation, the dense-cemented salt and the cemented-coarse-porous salt strata were usually removed from the test pit as intact slabs. In contrast, the uncemented-gypsum strata and the uncemented-coarse salt stratum were typically removed with a shovel as loose crystals and as loosely aggregated masses.

The stratum sequence described appears to have remained relatively consistent since at least 1960. This is supported by the presence of a similar sequence shown in two photographs in a joint Utah State University - USDH engineering study of the proposed I-80 route through BSF (Nielson and others, 1960, pp. 10-11, figures 1 and 4). While individual descriptions of each stratum are not included in the joint study, number, orientation and relative thicknesses of each stratum shown in the two photographs appear to be consistent with correlative stratum described in BLM's 1994 1997 field investigations.

APPENDIX A7

Table A7.1. TEQUIL-predicted concentrations from simulated brine evaporation to the point where precipitation of sylvite and carnallite first occurred; average chemical analyses from FY96 and FY97 BSF monitoring-well samples were used as input to TEQUIL (see Appendix A2, tables A2.3 and A2.4).

Aqueous Species	FY96		FY97	
	1st sylvite	1st carnallite	1st sylvite	1st carnallite
Na	58.06	15.97	59.61	14.56
K	58.24	27.33	59.14	25.61
Ca	0.41	0.06	0.41	0.14
Mg	44.28	91.76	43.01	93.26
Cl	260.46	299.14	259.91	307.19
SO4	15.94	24.46	16.05	14.64
% Water Remaining	7.2	3.4	5.8	2.5

Note: sodium, magnesium, and potassium concentrations were converted to their corresponding chloride salts using equations 2-7 in White and Wadsworth, 1999, Appendix 1, pp. A1-6-A1-7.

APPENDIX A8

TEQUIL-PREDICTED MINERAL PRECIPITATION PLOTS: Monitoring-Well Samples

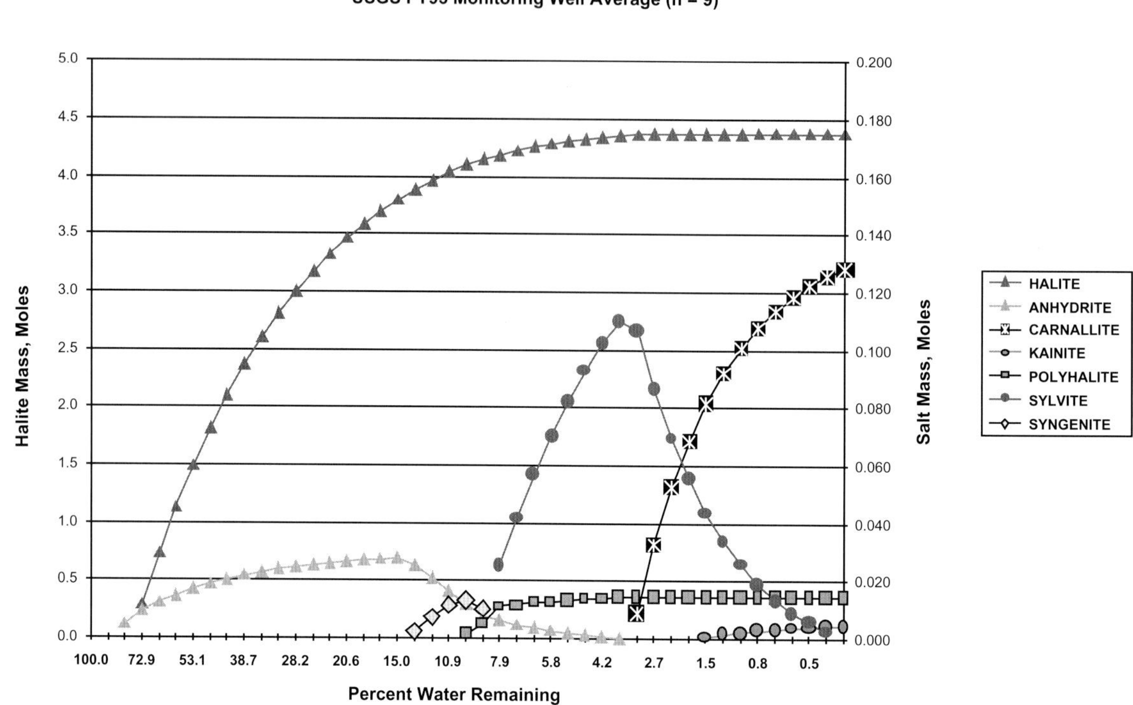

Figure A8.1. *TEQUIL-predicted mineral precipitation plots from average analyses of USGS FY93 monitoring-well samples.*

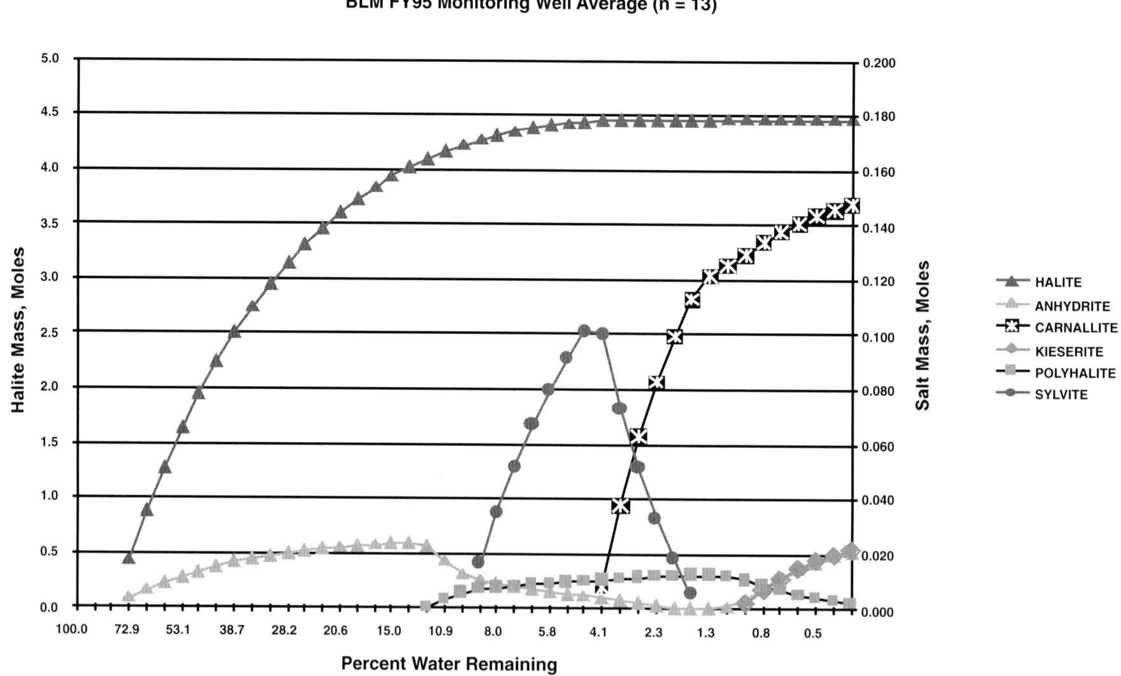

Figure A8.2. *TEQUIL-predicted mineral precipitation plots from average analyses of BLM FY95 monitoring-well samples.*

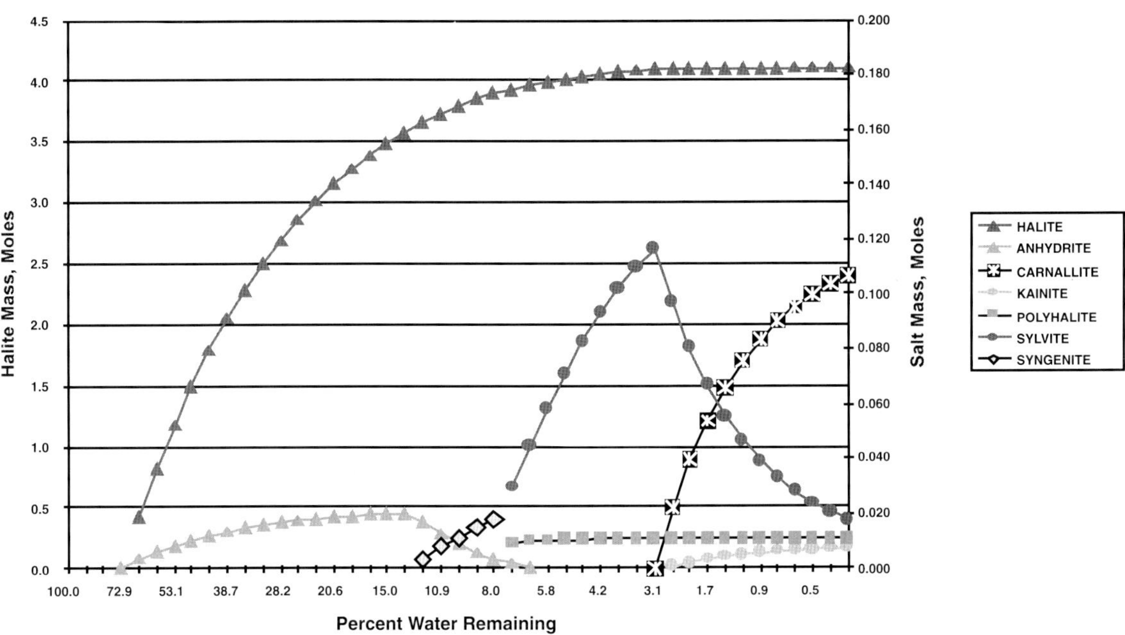

Figure A8.3. *TEQUIL-predicted mineral precipitation plots from average analyses of BLM FY96 monitoring-well samples.*

APPENDIX A9

Table A9.1. Salt-mineral species (*common minerals italicized*); after Braitsch, 1971, table 4.

Name	Formula	Molecular Weight, g/mole
Anhydrite	$CaSO_4$	136.15
Bischofite	$MgCl_2 \cdot 6H_2O$	203.33
Carnallite	$KMgCl_3 \cdot 6H_2O$	277.88
Gypsum	$CaSO_4 \cdot 2H_2O$	172.18
Kainite	$KMgClSO_4 \cdot 11/4H_2O$[1]	244.48
Kieserite	$MgSO_4 \cdot H_2O$	138.41
Polyhalite	$Ca_2K_2Mg(SO_4)_4 \cdot 2H_2O$	602.98
Rock salt (*Halite*)	NaCl	58.45
Sylvite	KCl	74.55
Syngenite	$K_2Ca(SO_4)_2 \cdot H_2O$	328.43

[1]Water content is expressed as a fraction so that resulting molecular weight is comparable with other salt minerals (required for subsequent calculations of salt precipitation).

APPENDIX A10

TEQUIL-PREDICTED MINERAL PRECIPITATION PLOTS: Surface-Pond Samples

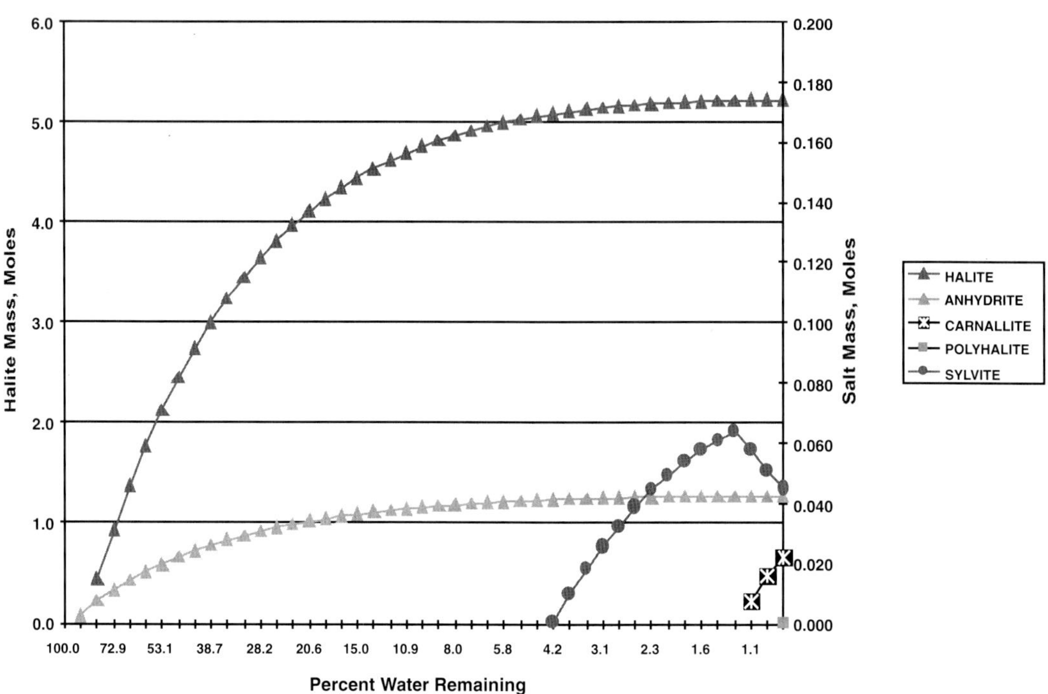

Figure A10.1. TEQUIL-predicted mineral precipitation plots from analyses of USGS surface pond sample collected near monitoring well BLM-93 on June 15, 1993.

July 7, 1993 USGS Surface-Pond Sample Near BLM-93

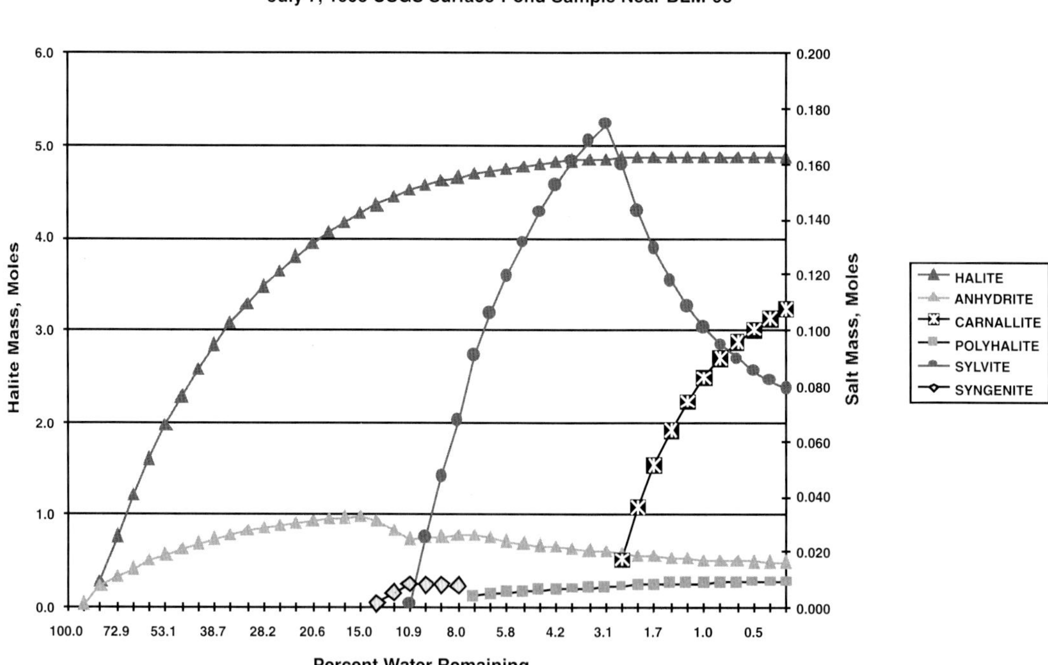

Figure A10.2. TEQUIL-predicted mineral precipitation plots from analyses of USGS surface pond sample collected near monitoring well BLM-93 on July 7, 1993.

EFFECTS OF THE WEST DESERT PUMPING PROJECT ON THE NEAR-SURFACE BRINES IN A PORTION OF THE GREAT SALT LAKE DESERT, TOOELE AND BOX ELDER COUNTIES, UTAH

by

James F. Kohler
U.S. Bureau of Land Management

Introduction

In the early 1980s, the elevation of the Great Salt Lake increased significantly, which resulted in property damage due to flooding of lands adjacent to the lake. The State of Utah investigated various solutions to mitigate the damages caused by the rise in the lake, and in 1986 began construction of the West Desert Pumping Project (WDPP). The project was designed to pump Great Salt Lake waters into an area of the Great Salt Lake Desert and allow the volume of the pumped waters to be reduced through evaporation. Figure 1 shows the area of the Great Salt Lake Desert that was inundated by the WDPP, often referred to as the West Pond. West Pond was confined on the south and southwest by a dike constructed to keep the pond from impinging upon Interstate 80 and the Bonneville Salt Flats. West Pond is bounded on the east by the Newfoundland Mountains and on the west by Floating Island, Silver Island Mountains, and Crater Island. At the south end of the Newfoundland Mountains, an additional dike, the Newfoundland Weir, was constructed with outlet structures to control the level of the water in West Pond and allow the overflow to drain back to the Great Salt Lake. To avoid filling shallow basins in the Great Salt Lake Desert with salt and depleting the mineral resources of the Great Salt Lake, provisions were made in the original design of the project to use less saline brine from the south arm of the Great Salt Lake and convey the brine concentrated in West Pond back to Great Salt Lake by maintaining a return flow averaging 500 cubic feet per second (cfs) (EWP, 1983).

In May, 1986, the Utah State Legislature authorized construction of a "bare-bones" project which departed from the original design by withdrawing water from the more saline north arm of the Great Salt Lake and eliminating some of the features east of the Newfoundland Weir designed to facilitate the flow of the concentrated brines back into the Great Salt Lake (Bureau of Land Management, 1986). The pumping project started operation on April 10, 1987 and continued through June, 30, 1989. During this period, the United States Geological Survey (USGS) estimated that Great Salt Lake brines containing 695 million tons of salt were pumped into the West Desert (Wold and Waddell, 1993). At the end of pumping, on June 30, 1989, they estimated that 315 million tons remained in the evaporation pond in the West Desert, 71 million tons were contained in a salt crust on the floor of the pond, 10 million tons infiltrated the subsurface of the inundated area, 88 million tons were withdrawn as part of a mineral extraction operation, and 123 million tons were discharged from the evaporation pond back towards the Great Salt Lake. In this study, the USGS was unable to account for about 88 million tons of the 695 million tons they estimated had been removed from the Great Salt Lake.

The area of the Great Salt Lake Desert inundated by the WDPP was known to contain a shallow subsurface brine prior to being flooded. This near-surface brine resource in the Great Salt Lake Desert was also known to differ chemically from Great Salt Lake. The purpose of this investigation is to document what was known about these shallow brine resources before the area was flooded by the WDPP and to assess any impacts the introduction of Great Salt Lake brine to the West Pond may have had on these resources.

GEOLOGIC SETTING

The Great Salt Lake Desert is a broad valley in the Great Basin portion of the Basin and Range physiographic province. The Great Salt Lake Desert is a closed basin with all drainage towards the center of the basin. This broad basin consists of a number of smaller subbasins that are separated by very low topographic divides. The principle subbasins within the Great Salt Lake Desert include the Bonneville Salt Flats and Pilot Valley. Two other basins, which have not been formally named, include the areas to the east and west of the Newfoundland Mountains, which will be referred to in this report as the East Pond and Newfoundland Basin. These subbasins are designated on a satellite image of the West Desert shown on figure 2. In this report, the term Great Salt Lake Desert will be used in a restricted sense to include the areas of mud flats or salt flats within the basin.

The sediments comprising the mudflats consist primarily of carbonate muds, clay, gypsum and salt. Minor occurrences of oolitic sand and thin carbonate layers of probable algal origin are also present in the surface sediments within the Great Salt Lake Desert. Salt crusts cover the playa surfaces of Pilot Valley and Bonneville Salt Flats. Since cessation of the WDPP in 1989, a salt crust has also been deposited on the playa in the deeper part of the Newfoundland Basin.

The area of the mud flats exhibits very low relief with elevations generally ranging from about 4,210 to 4,225 feet

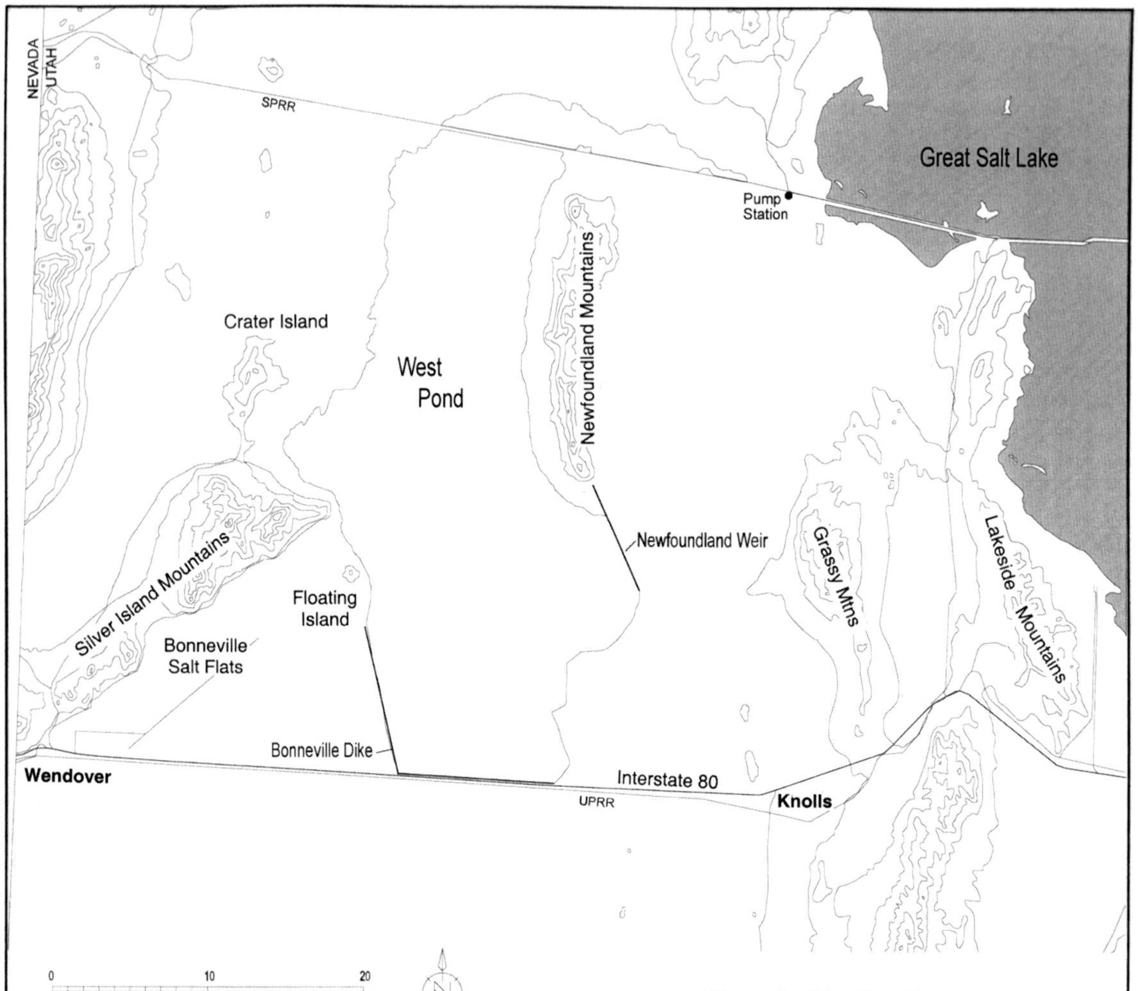

NEVADA
UTAH
SPRR
Great Salt Lake
Pump Station
Crater Island
West Pond
Newfoundland Mountains
Silver Island Mountains
Floating Island
Bonneville Salt Flats
Newfoundland Weir
Grassy Mtns
Lakeside Mountains
Bonneville Dike
Wendover
Interstate 80
UPRR
Knolls
0
10
20
Scale (Miles)

Figure 1. *West Pond location map.*

Pilot Valley
Newfoundland Basin
East Pond
Bonneville Salt Flats

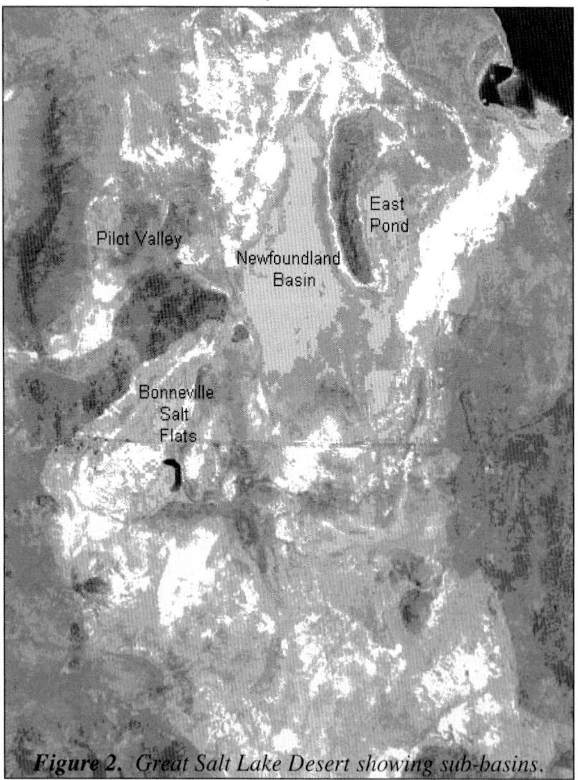

Figure 2. *Great Salt Lake Desert showing sub-basins.*

above sea level. Locally, small dunes of wind-blown gypsum sand provide a slight amount of relief to the otherwise flat surface. Extensive dunes of gypsum sand are present along the eastern edge of the Newfoundland Basin playa.

The hydrologic system within West Pond generally consists of three types of aquifers: an alluvial fan aquifer adjacent to the mountain ranges around the perimeter of a playa, a shallow brine aquifer that occurs in the upper 20 feet of a playa, and a deep basin-fill aquifer. The shallow brine aquifer is the only part of the ground-water system that was affected by the WDPP.

PREVIOUS WORK

The shallow brine resources of the West Pond were investigated at various times in the past. The sampling points from these investigations are shown on figure 3. Thomas Nolan of the USGS conducted an initial assessment of the area in the 1920's (Nolan, 1927). Nolan completed 405 shallow test holes distributed over the entire Great Salt Lake Desert. Brine samples were collected and analyzed from these drill holes, but complete analyses were not included in Nolan's published report. However, the report does include a composite analysis of 126 samples taken during his field

Table 1. Summary of pre-WDPP shallow brine samples from the West Pond area. Ion concentrations reported in moles per liter.

Data Source	Number of Samples	Na^{+}	K^{+}	Ca^{++}	Mg^{++}	Cl^{--}	SO_4^{--}
Nolan	12	2.492	0.075	0.038	0.079	2.712	0.042
Lindenburg	13	1.875	0.072	0.045	0.060	2.057	0.071
Reynolds N	72	3.128	0.099	0.045	0.129	3.354	0.061
Reynolds S	20	2.537	0.148	0.048	0.107	2.728	0.060
Dames and Moore	7	2.187	0.084	0.036	0.077	2.401	0.024

Table 2. Summary of brine samples representative of Great Salt Lake brine pumped into the West Pond. Ion concentrations reported in moles per liter.

Data Source	Number of Samples	Na^{++}	K^{++}	Ca^{++}	Mg^{++}	Cl^{--}	SO_4^{--}
West Desert Pump	1	2.175	0.066	0.005	0.160	2.454	0.115
Great Salt Lake Average, April 1987 to June 1989	13	2.301	0.075	0.006	0.217	2.591	0.101
West Pond June, 1989	24	4.789	0.153	0.010	0.342	5.308	0.188

When pumping stopped in June 1989, water remaining in the West Pond continued to evaporate and the pond decreased in size and retreated to the deepest part of the basin where it eventually dried up. Satellite imagery of the West Pond suggests that it disappeared by July 22, 1991, leaving behind an extensive salt crust. Since 1991, a surface pond has formed in the deeper part of the West Pond during periods of low evaporation in the winter and following heavy precipitation events.

PRESENT SHALLOW BRINE AQUIFER IN THE WEST POND

In order to make a preliminary assessment of the brine presently in the shallow brine aquifer in the West Pond, a field investigation was undertaken by the BLM during the summer of 1999. As part of this investigation, brine samples were recovered from shallow (6 to 8 feet) boreholes drilled in various locations throughout the West Pond area. These boreholes were drilled with a standard 4-inch bucket soil auger and the sedimentary sequence and the depth to brine was recorded. The brine was then pumped from the borehole with a portable centrifugal pump at a rate of 1 to 2 gallons per minute until approximately 50 gallons of brine had been removed from the hole before samples were collected for analysis. In a typical borehole, the pumping rate of 1 to 2 gallons per minute lowered the water level a few inches, where a sustainable level was maintained. When pumping stopped, the water level in the borehole returned to the original static level. In the boreholes away from the center of the basin, the brine was typically under confined conditions and the water level would rise a few inches above the level where it was first encountered. In the deepest part of the basin, no such confined conditions were observed. Three of the samples collected during this investigation, obtained in the deepest portion of the basin on or near locations sampled by Reynolds Metals in the 1960s, are of particular interest. They provide a basis for identifying changes to the shallow brine aquifer chemistry in the area last occupied by the waters from the WDPP. The analyses from these three samples are summarized in table 3.

Not surprisingly, the marked increases in concentrations of potassium and magnesium reflect brine that has been concentrated through evaporation.

COMPARISON OF THE BRINES USING IONIC RATIOS

Because brines from the West Desert and the Great Salt Lake are dominated by sodium chloride, subtle differences in the brines are not always evident from comparison of chemical analyses of the brines. In addition, the shallow brine aquifer receives recharge from direct precipitation that may dilute the brine during wet years. A similar situation occurs with the Great Salt Lake in that, during wet weather cycles, the level of the lake rises and the lake brine becomes diluted. The variable composition of the brines resulting from this dilution further masks subtle difference in the brine chemistry.

Initially, comparisons of Great Salt Lake and West Desert brines were made with ratios of major ions. Previous work had indicated that brine from the Great Salt Lake had a higher magnesium and sulfate content than Great Salt Lake Desert shallow brine aquifer brine. Nolan (1927) noted that ground-water brine in the Great Salt Lake Desert had a potassium to magnesium ratio approaching one and a molar com-

Table 3. Brine samples from the shallow brine aquifer collected in the deep part of the West Pond, 1999. Ion concentrations reported in moles per liter

Well Designation	Na^{++}	K^{++}	Ca^{++}	Mg^{++}	Cl^{--}	SO_4^{--}
NB-8	3.454	0.425	0.004	0.823	4.727	0.365
NB-9	3.341	0.527	0.008	0.848	5.012	0.373
NB-10	3.310	0.486	0.008	0.859	5.012	0.318
Average	3.367	0.479	0.007	0.843	4.917	0.352

Table 4. Comparison of K/Mg and Ca/SO_4 for brine samples from the Great Salt Lake, pre-WDPP shallow brine aquifer in the Newfoundland Basin, and present (1999) shallow brine aquifer in the Newfoundland Basin. Ratios are calculated on a molar basis.

Source of Brine	K/Mg	Ca/SO_4
Average of Great Salt Lake (1987 to 1989)	0.34	0.059
Average brine in West Pond (June 1989)	0.45	0.053
Average Reynolds N brine samples (1960)	0.77	0.738
Average of NB-8, NB-9, and NB-10 (1999)	0.57	0.020

position of calcium and sulfate close to being in balance with precipitation of gypsum. Based on these observations, a comparison was made of the potassium to magnesium and calcium to sulfate ratios for ground water from the Great Salt Lake Desert prior to being inundated by the West Desert Pumping Project, brine from the Great Salt Lake during the period that the pumping project was being operated, brine in the West Pond at the end of pumping in June, 1989, and shallow ground-water brine presently in the area formerly covered by the West Pond. The results of this comparison are shown in table 4.

The K/Mg ratio for the pre-WDPP brine in the Newfoundland Basin (i.e., Reynolds N) is more than twice the value for the Great Salt Lake during the period of the pumping project. The K/Mg ratio for the brine presently in the shallow brine aquifer in the basin lies between the values for the Great Salt Lake and the pre-WDPP ground water. However, comparison of the Ca/SO_4 ratio does not yield the same results. Rather than a Ca/SO_4 ratio that lies between the Great Salt Lake and the pre-WDPP shallow brine aquifer, the ratio for the present shallow brine aquifer is actually lower. This could be the result of the excess sulfate from the Great Salt Lake brine combining with the calcium in the shallow brine aquifer brine resulting in the precipitation of gypsum. The removal of the calcium in this manner results in a lower Ca/SO_4 ratio.

This comparison of the ionic ratios suggests that the chemistry of the present shallow brine in the Newfoundland Basin could result from mixing Great Salt Lake brine with the pre-WDPP ground water. To better understand this relationship, a model was used to simulate evaporation of the brines and then make a comparison based on the various precipitated minerals.

Geochemical Modeling of West Pond Brines

The TEQUIL program was developed by the University of California at San Diego for the geothermal industry to predict the chemical behavior of the natural brines from which the thermal energy is extracted. Although the program was developed initially for the geothermal industry, is has been found to have wide applications in other industries including solar pond engineering (Moller and others, 1997). The program is based on a model which uses semi-empirical equations of Pitzer (1973) that were used by Harvie and Weare (1980) to show that the free energy calculations on electrolyte solutions could be used to accurately predict complex solubility relationship in the $Na\text{-}K\text{-}Ca\text{-}Mg\text{-}SO_4$ system (Moller and others, 1997). Various versions of this program are available, but for the purposes of this investigation, the 25°C model for the $Na\text{-}K\text{-}Ca\text{-}Mg\text{-}H\text{-}Cl\text{-}OH\text{-}SO_4\text{-}HCO_3\text{-}CO_3\text{-}CO_2\text{-}H_2O$ system (Harvie and others, 1984) was used.

TEQUIL provides a user interface to input the composition of the initial solution in moles. The model then simulates removal of 10 percent of the water through an iterative process to simulate evaporation. For each iteration, the program calculates the solubility for the various minerals that could be expected to precipitate from the brine. When the solution becomes saturated with a given mineral, the mineral is allowed to precipitate. The precipitated salt remains in contact with and is allowed to react with the brine. The results of the evaporation simulation are saved by the program in a text file that lists the molar composition of the concentrated brine and any minerals that precipitate from the brine.

In order to better understand how the brines introduced to the West Pond from the Great Salt Lake may have interacted with the subsurface brines contained within the shallow brine aquifer, the TEQUIL program was used to identify the minerals that would precipitate from the brine and the sequence of precipitation. The model shows that evaporation of samples of brine from the Great Salt Lake produced a very consistent sequence of minerals even though the brine chemistries appear to be different. The mineral sequence projected by the model for the Great Salt Lake is consistent with actual results reported from the evaporation of Great Salt Lake brine (Butts, 1993; Jones and others, 1997).

The model also shows that when brines from the shallow brine aquifer from the Great Salt Lake Desert before the WDPP are evaporated, an equally consistent sequence of minerals results which is significantly different than the minerals which precipitate from the Great Salt Lake. The mineral sequence projected by the model for brine from the shallow brine aquifer in the Great Salt Lake Desert is consistent with the sequence reported for Reilly-Wendover's potash extraction operation near the Bonneville Salt Flats (Bingham, 1980).

The model was applied to the following brines: (1) subsurface brine from the deepest part of the West Pond before the area was inundated by the WDPP, (2) brine from Great Salt Lake during the WDPP pumping period, (3) brine contained within the West Pond when pumping ceased in 1989, and (4) subsurface brine presently found in the deepest part of the basin. The program was run for 45 iterations to simulate the removal of over 99 percent of the water through evaporation. The model results were then graphed to show the sequence of minerals that precipitated from the brine. The primary minerals expected to precipitate from the brines used in this exercise are as shown in table 5. The model projected precipitation of all of the minerals in table 5 except for mirabilite that is known to precipitate from Great Salt Lake brine during the winter. Because the model was run at a constant 25°C, the temperature-dependent precipitation of mirabilite was not reflected in the model.

The results of using the TEQUIL 25°C model to simulate evaporation of a brine with a composition of the average of the Great Salt Lake during the period the WDPP was in operation (1987 through 1989) are shown on figure 6. The y-axis values on the left side of the graph show the percent of the original water in the brine. The y-axis values on the right side of the graph show the amount of the various salts precipitated in moles.

On the graphical outputs from the model, halite is excluded because its greater abundance would mask the other minerals. For Great Salt Lake brine, the model indicates that the non-halite minerals precipitate out in the following sequence: anhydrite - glauberite - bloedite - polyhalite - epsomite - hexahydrite - kainite - kieserite - carnallite.

The model's projection of minerals that would precipitate from the average subsurface brines from the deep part of the West Pond basin before the area was flooded by the

Table 5. Evaporite minerals derived from West Desert brines (adapted from Braitsch, 1971).

Mineral	Formula	Molecular Weight
Anhydrite	$CaSO_4$	136.15
Bischofite	$MgCl_2 . 6 H_2O$	203.33
Bloedite	$Na_2Mg(SO_4)_2 . 4 H_2O$	334.51
Carnallite	$KMgCl_3 . 6 H_2O$	277.88
Epsomite	$MgSO_4 . 7 H_2O$	246.50
Gypsum	$CaSO_4 . 2 H_2O$	172.18
Glauberite	$Na_2Ca(SO_4)_2$	278.21
Halite (common salt)	NaCl	58.454
Hexahydrite	$MgSO_4 . 6 H_2O$	228.49
Kainite	$KMgClSO_4 . 3 x H_2O$	244.48
Kieserite	$MgSO_4 . H_2O$	138.41
Leonite	$K_2Mg(SO_4)_2 . 2 H_2O$	366.71
Mirabilite	$Na_2SO_4 . 10 H_2O$	322.22
Polyhalite	$Ca_2K_2Mg(SO_4)_4 . 2 H_2O$	602.98
Sylvite	KCl	74.553
Syngenite	$K_2Ca(SO_4)_2 . H_2O$	328.43

WDPP is shown on figure 7.

The minerals precipitated from the subsurface West Desert brine as projected by the model are quite different than minerals derived from Great Salt Lake brine. The non-halite minerals precipitate in the following order: anhydrite - polyhalite - sylvite - carnallite - kieserite.

The model's projection of minerals precipitating from the final brine in the West Pond in June 1989 is shown on figure 8.

When figure 8 is compared with the original Great Salt Lake brine on figure 6, the results appear fairly similar except for the greater amount of precipitated salts. This is most likely due to the fact that the brine in the West Pond represents brine that had already been concentrated by evaporation. The modeled output differs slightly from the Great Salt Lake brine in the relative abundance of the sodium sulfate minerals bloedite, epsomite, and hexahydrite. Instead, a different mineral, leonite ($K_2Mg(SO_4)_2 . 4 H_2O$), precipitates. The non-halite minerals precipitate as follows: glauberite - polyhalite - bloedite - leonite - epsomite - kainite - hexahydrite - kieserite - carnallite.

The TEQUIL model was then applied to the average shallow brine aquifer in the deepest part of the West Pond basin during the summer of 1999. The results, as shown on figure 9, appear to be significantly different than the three previous brines. First of all, the brine has been significantly concentrated by evaporation. Because of this, the maximum value on the y-axis for the precipitated salts had to be in-

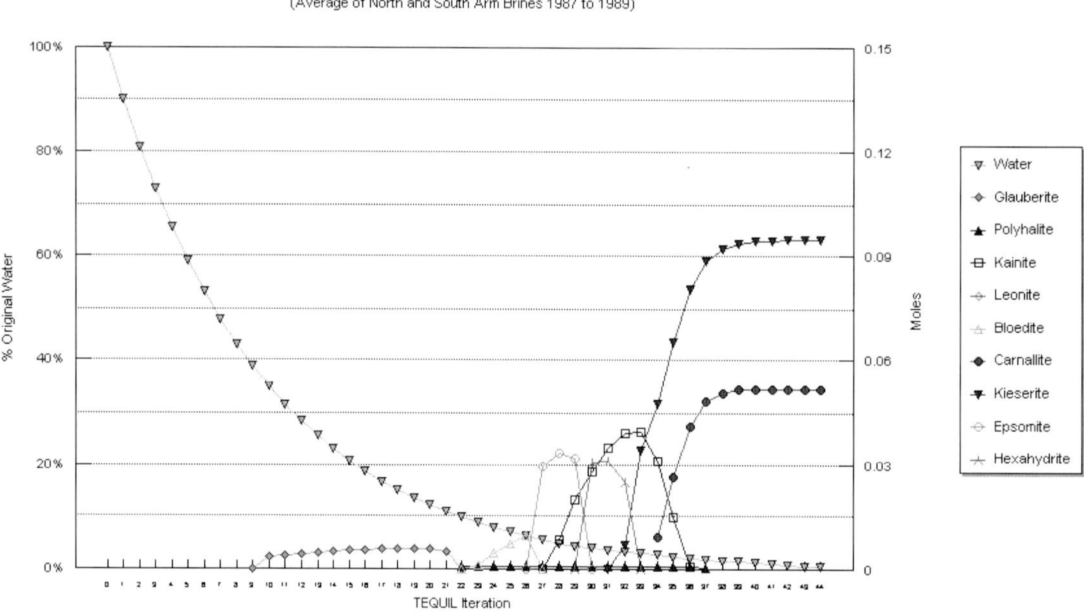

Figure 6. *TEQUIL model - salts precipitated from Great Salt Lake brine.*

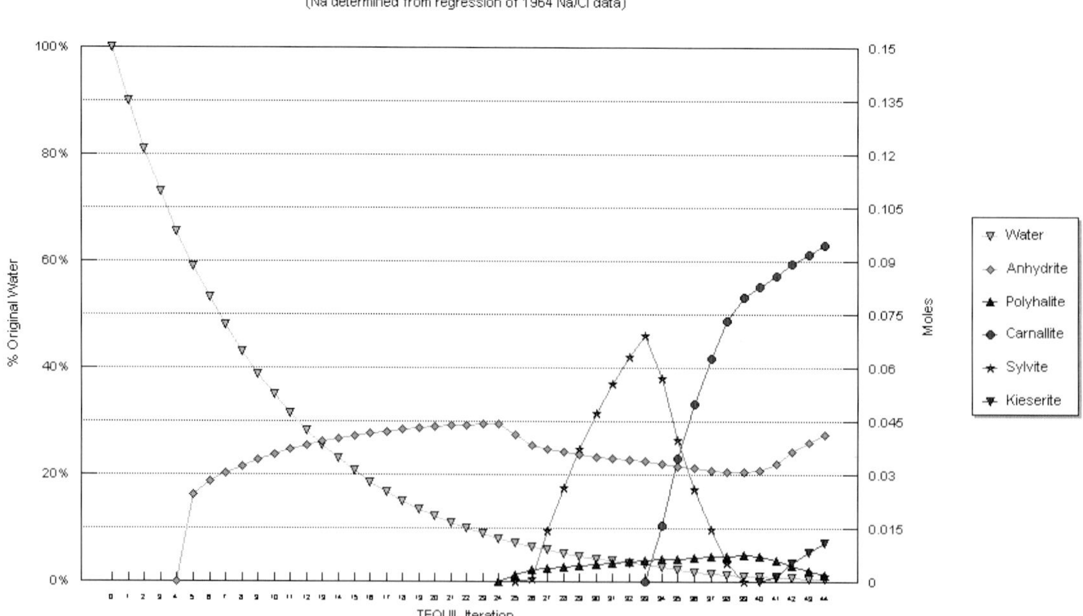

Figure 7. *TEQUIL model - salts precipitated from Reynolds north data.*

creased to 0.5 moles from the 0.15 moles used on the other graphs to adequately show the distribution of the precipitated minerals. For the present ground water brine in the deepest part of the basin, the non-halite minerals precipitated out as follows: glauberite - polyhalite - leonite - sylvite - kainite - carnallite - kieserite.

The mineral suite projected by the model for the shallow ground water brine in the deepest part of the West Pond is different than that projected for the Great Salt Lake and the original ground water. However, it contains components of both which, suggests that the brine presently contained in the shallow brine aquifer in the deepest part of the basin represents a mixture of the original ground water brine and brine from the Great Salt Lake.

SUMMARY AND CONCLUSION

Subtle differences in brine chemistry are not always readily apparent in brine from the Great Salt Lake and the Great Salt Lake Desert when these brines are compared using analyses of the major ions contained within the brine. This may be due to dilution of the brine by meteoric water, or simply due to the fact that the brines are dominated by sodium and chloride, which tend to mask differences in the other ions. By evaporating the brines and identifying the minerals that precipitate out of the solution, these subtle differences become more evident. The TEQUIL model at 25°C seems to produce a reasonable projection of the simple Na-K-Ca-Mg-Cl-SO_4-H_2O system to which these brines belong. Application of the model shows that the mineral suite precipitating

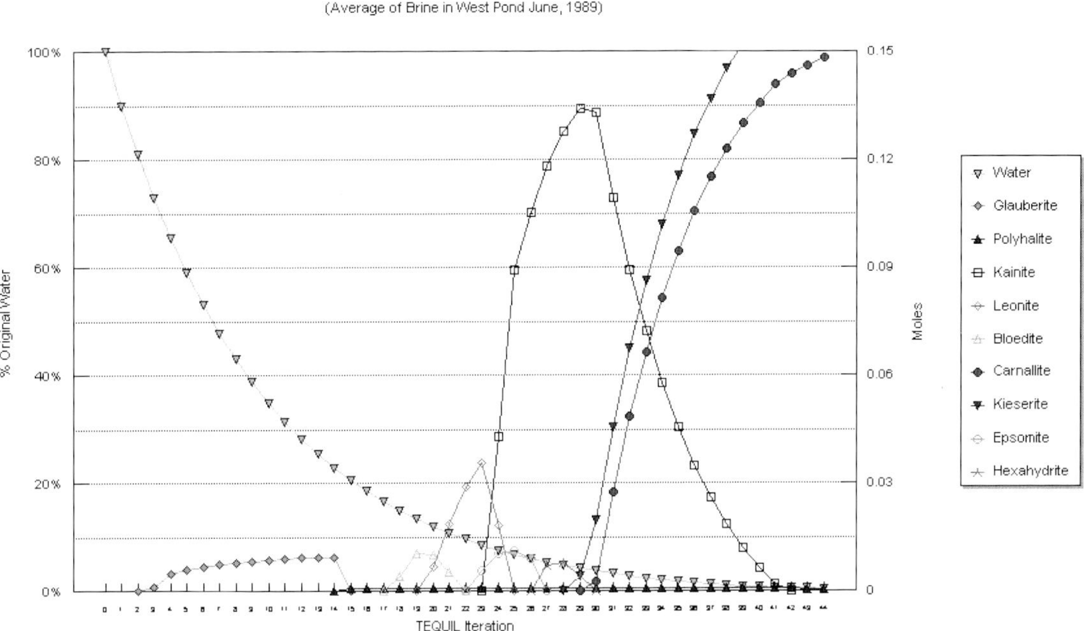

Figure 8. *TEQUIL model-salts precipitated from West Pond brine.*

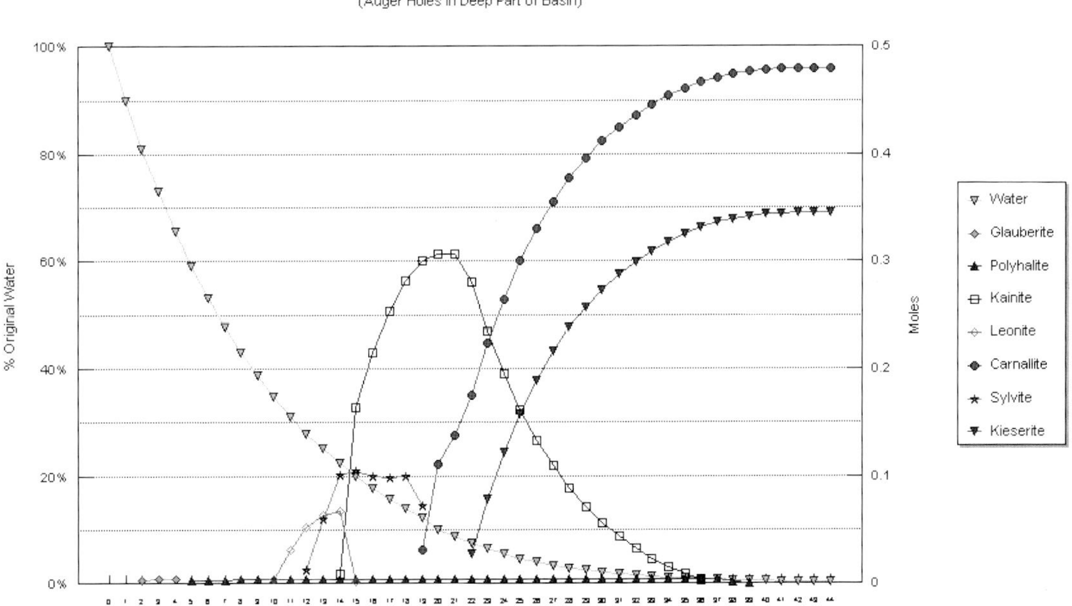

Figure 9. *TEQUIL model-salts precipitated from shallow brine aquifer.*

from Great Salt Lake brine is very different than the minerals precipitating from ground-water brines in the Great Salt Lake Desert. The model also shows that the brines presently in the shallow brine aquifer in the portion of the Great Salt Lake Desert inundated by Great Salt Lake brine by the WDPP contain components of both brines and could be considered a mixed brine. These conclusions are based on a limited field investigation, and more sampling should be undertaken to confirm and refine the interpretations. However, these data suggest that the subsurface brine resource that existed in the West Pond has been impacted by the WDPP in two significant ways. First, the brine from the Great Salt Lake mixed with the existing ground-water brine, which changed the chemical character of the brine. Instead of a simple halite - sylvite - carnallite system, the higher levels of sulfate from the Great Salt Lake have resulted in a system with a more complex mineralogy. Secondly, the brine left in the basin was concentrated through evaporation, which significantly increased the amount of potassium and magnesium in the brine in the deepest part of the basin.

REFERENCES

Bingham, C. P., 1980, Solar production of potash from the brines of the Bonneville Salt Flats: *in* Gwynn, J.W., editor, Great Salt Lake, a scientific, historical and economic overview: Utah Geological and Mineral Survey Bulletin 116, p. 229-242.

Bureau of Land Management, 1986, West Desert Pumping Project final environmental impact statement: Salt Lake District Office, Salt Lake City, Utah, 98 p.

Braitsch, O., 1971, Salt deposits, their origin and composition: Springer-Verlag, 1971, 288 p.

Butts, D., 1993, Chemicals from Brine: Kirk-Othmer Encyclopedia of Chemical Technology Fourth Edition, Volume No. 5, John Wiley & Sons, Inc, p. 817-837.

Dames & Moore, 1983, Report: Geotechnical and ground water investigation, West Desert Pumping Project, Great Salt Lake Desert, Utah: unpublished report submitted to the Utah Division of Water Resources, Salt Lake City, Utah, November, 1983, various pagings.

EWP (Eckhoff, Watson, and Preator, Dames & Moore, International Engineering Co.), 1983, West Desert Pumping Alternative - Great Salt Lake: unpublished report submitted to the Utah Division of Water Resources, Salt Lake City, Utah, December, 1983, various pagings.

Harvie, C.E., and Weare, J.H., 1980, The prediction of mineral solubilities in natural waters - the Na-K-Mg-Ca-Cl-SO_4-H_2O system from zero to high concentration at 25°C: Geochimica Cosmochimica Acta 44, pp. 981 - 987.

Harvie, C.E., Moller, N., and Weare, J.H., 1984, The prediction of mineral solubilities in natural waters: The Na-K-Mg-Ca-H-Cl-SO_4-OH-HCO_3-CO_3-CO_2-H_2O system to high ionic strengths: Geochimica et Cosmochimica Acta 52, pp.821-837.

Jones, B. F., Carmody, R., and Frape, S. K., 1997, Variations in principal solutes and stable isotopes of Cl and S on evaporation of brines from the Great Salt Lake, Utah, USA: Geological Society of America, Abstracts with Programs, 1997 annual meeting, Salt Lake City, Utah, p. A-261.

Lindenburg, G.J., 1974, Factors contributing to the variance in the brines of the Great Salt Lake Desert and the Great Salt Lake: unpublished master's thesis, University of Utah, Salt Lake City, Utah, 70 p.

Moller, N., Weare, J.H., Duan, Zhenhao, and Greenberg, J.P., 1997, Chemical models for optimizing geothermal energy production: unpublished internet document, U.S. Department of Energy Technical Site, Research Summaries - Reservoir Technology.

Nolan, T.B., 1927, Potash brines in the Great Salt Lake Desert, Utah: U.S. Geological Survey Bulletin 795-B, p. 25-44.

Pitzer, K.S., 1973, Thermodynamics of electrolytes- I. Theroetical basis and general equations: Journal of Physical Chemistry, v. 77, pp. 97 - 142.

Reynolds Metals Company, 1966, unpublished letter to Regional Mining Supervisor, U. S. Geological Survey: Reynolds Metals Company, Richmond, VA, December 30, 1966.

Wold, S. R., and Waddell, K. M., 1993, Salt budget for West Pond, Utah, April 1987 to June 1989: U. S. Geological Survey Water-Resources Investigations Report 93-4028, 20 p.

APPENDIX 1. Brine analyses from the shallow brine aquifer in the Great Salt Lake Desert, 1966 (data from Reynolds Metals Co., 1966).

Sample ID	UTM Northing	UTM Easting	Na * *mg/l*	K *mg/l*	Mg *mg/l*	Ca ** *mg/l*	SO_4 *mg/l*	TDS *mg/l*	Density
60135	4521304	292806	45593	3100	2100	1742	4900	133336	1.083
60136	4521349	291198	48042	3000	2000	1992	4800	139734	1.087
60137	4522823	294451	52693	5900	2200	2112	5300	155706	1.096
60138	4524486	296111	58447	5700	2500	2250	6100	171896	1.106
60139	4524531	294501	74544	7200	3800	1359	7100	217202	1.134
60140	4524219	294404	75010	8000	3700	1402	6700	218812	1.135
60141	4522868	292842	55325	5400	2400	2044	5600	162569	1.101
60142	4522913	291234	51102	3800	2200	1779	4700	148481	1.092
60143	4524575	292892	78216	7600	3200	2371	6700	227287	1.141
60144	4524620	291282	56121	6100	2300	2292	5700	165613	1.102
60145	4521215	296021	44308	4000	2000	2216	5800	132124	1.082
60146	4521171	297628	64751	7600	3100	1087	4900	188638	1.116
60147	4521126	299236	58263	5900	2700	1619	5400	170481	1.105
60148	4522733	297664	54346	4800	2300	2158	5600	159404	1.099
60150	4522688	299273	56978	5900	2400	2005	5300	167082	1.103
60151	4524397	299330	56672	5600	2400	2072	5500	166244	1.103
60152	4524326	301142	62241	6600	2900	1996	6600	183437	1.113
60153	4524352	300938	62792	8900	3000	1859	6600	187152	1.115
60154	4524307	302547	56243	4600	2500	2261	6400	165304	1.103
60155	4522778	296056	54958	6000	2500	1860	5600	162118	1.100
60201	4556536	294254	73564	4000	3000	2293	6300	210757	1.132
60202	4551854	293643	80174	5100	3700	1440	6200	229014	1.143
60203	4551964	292058	68117	4000	3000	1802	5800	195419	1.122
60204	4552064	290798	62853	3800	2700	1982	5700	181135	1.113
60205	4550422	292015	75829	4000	3600	1421	6300	216450	1.135
60206	4550454	290752	68117	3800	3100	1804	6200	195721	1.122
60207	4548824	290730	80664	5100	4100	681	5900	229645	1.143
60208	4548809	291969	76074	3000	2800	2544	5800	215918	1.135
60209	4547199	291924	80848	5100	3600	1473	5800	230321	1.143
60210	4547225	290867	80419	5300	3700	827	4700	227746	1.142
60211	4545591	291878	79256	4700	3300	1968	6000	226125	1.141
60212	4544923	291233	76013	4700	3700	473	4400	214886	1.134
60213	4547153	293509	79746	4400	3600	1833	6800	228079	1.142
60214	4558193	292692	79746	4400	4500	0	5200	224986	1.140
60215	4558239	291086	66954	3500	4100	1013	8400	194767	1.122
60216	4556630	291039	58997	3000	2700	1239	4400	168137	1.105
60217	4556584	292645	74482	3700	3900	606	5700	211489	1.132
60218	4554974	292598	71055	4000	3300	1126	5000	201981	1.126
60219	4554926	294207	79318	4200	3100	1884	5000	224502	1.140
60220	4553317	294160	75156	4200	3000	2334	6200	215089	1.134
60221	4559754	294346	85683	4400	3800	1185	5300	241768	1.151
60222	4561363	294394	77910	4200	2800	2139	4600	220349	1.138
60223	4561410	292786	79269	4200	3700	1852	7300	227241	1.142
60224	4559802	292739	80174	4500	4000	1237	6900	229212	1.143
60225	4551452	287883	75400	4500	3800	1319	6900	216520	1.135
60226	4551499	286273	78155	4000	3200	1867	5500	221822	1.138
60227	4552570	284692	39534	1600	1400	1622	2600	112756	1.071
60228	4551594	283052	72401	2600	2800	2270	5600	205372	1.129
60229	4552665	281471	82255	2700	3600	544	3400	228300	1.143
60230	4552618	283082	70626	2800	2600	1548	3300	197675	1.124
60231	4552427	289523	64445	3200	2800	1816	5500	184461	1.115
60232	4552475	287913	68545	3100	2500	3191	7100	197836	1.124
60233	4552523	286302	92538	4500	3700	2081	6200	261619	1.163
60234	4549938	284617	71055	4500	2800	2409	6100	204364	1.127
60235	4549986	283006	67383	3200	3000	2181	6800	194064	1.121
60236	4548377	282961	68974	3600	2800	2426	6400	198300	1.124
60237	4548330	284571	70626	3800	3400	1481	6300	202408	1.126
60238	4546720	284520	65118	3800	3600	1200	7100	188617	1.118
60239	4546673	286130	71483	4100	4500	964	9300	208547	1.130
60240	4546625	287741	80664	5000	4300	601	6500	230265	1.143
60241	4551404	289494	66097	3700	3400	1747	7500	191844	1.120
60242	4549796	289448	67627	4000	3100	1904	6500	195031	1.122
60243	4549843	287838	68790	4000	2800	2584	6800	198774	1.124

(Appendix 1 continued)

Sample ID	UTM Northing	UTM Easting	Na * mg/l	K mg/l	Mg mg/l	Ca ** mg/l	SO_4 mg/l	TDS mg/l	Density
60244	4549891	286227	77420	4200	3100	1493	4300	218414	1.136
60245	4548282	286182	62853	3700	3000	2196	7400	183250	1.114
60246	4548235	287792	70810	3800	3100	2236	6900	203946	1.127
60247	4548187	289403	81337	5300	4300	594	6400	232232	1.144
60248	4553412	290946	63282	3500	3000	1801	6400	182783	1.114
60249	4555020	290993	63282	3500	2700	2296	6400	182978	1.114
60250	4555066	289388	77910	4600	3700	1740	7200	223850	1.140
60251	4556675	289434	75400	4800	4100	950	7200	217051	1.135
60253	4559893	289529	69096	4000	4100	1166	8500	201162	1.125
60254	4559848	291133	69953	4000	4600	1387	10900	206540	1.129
60255	4561457	291179	71055	3700	4500	1609	10900	209264	1.131
60256	4561501	289574	78889	4400	4000	1797	8400	227786	1.142
60257	4553494	286121	80174	4000	2900	2467	5500	227441	1.142
60258	4555155	286169	77787	3700	2500	2585	4500	219573	1.137
60259	4556766	286216	74299	3700	2300	2734	4500	210333	1.131
60260	4558376	286263	71300	3500	2100	2783	4200	201783	1.126
60261	4559985	286310	62792	3200	1600	2832	3400	177825	1.111
60262	4559939	287920	59303	2500	2100	2453	4900	169557	1.106
60263	4558330	287873	61935	2700	2200	1674	3100	174209	1.109
60264	4556720	287826	68913	3500	2500	2000	4200	195113	1.122
60265	4555110	287779	83235	4800	3400	1676	5200	235711	1.147
60266	4553500	287733	66036	4200	2700	1855	5000	189091	1.118
60268	4558468	283044	56733	3100	1200	3470	4100	162703	1.102
60269	4556857	282997	55631	2600	1500	2084	2100	156215	1.098
60270	4556813	284606	69341	3200	1800	2675	3000	194717	1.122
60271	4555245	282950	52571	3000	1300	2755	3300	150226	1.094
60272	4555201	284559	63037	3000	1800	2474	3300	178011	1.111
60273	4553537	284511	79318	3600	2600	3042	5800	225360	1.141

* Na determined from linear regression of Na/Cl for 80 brine samples
** Ca estimated from value needed to achieve mole balance for the sample

THE U.S. BUREAU OF LAND MANAGEMENT'S ROLE IN RESOURCE MANAGEMENT OF THE BONNEVILLE SALT FLATS

by

Glenn A. Carpenter, Field Office Manager
Bureau of Land Management - Salt Lake Field Office
2370 South 2300 West, Salt Lake City, UT 84119

ABSTRACT

The Bonneville Salt Flats, which is famous for automobile land-speed records, unusual geology, stark contrasts, and unique scenery, is managed mainly by the U.S. Bureau of Land Management (BLM). BLM balances the administration of valid existing rights, such as mineral extraction from federal leases, with recreational and commercial uses that include: 1) land-speed racing, 2) competitive archery meets and rocket launches, and 3) commercial filming for motion pictures and magazine and television commercials.

To ensure the health of the land and help pay for administrative costs, BLM collects royalties, and usage and cost-recovery fees from these activities. Lease royalties and cost-recovery fees go to the General Treasury. However, as much as 85 percent of usage fees collected by BLM for land-speed racing, competitive archery meets, rocket launches, and commercial filming are returned to the BLM Salt Lake Field Office to help defray the costs of permit processing and resource monitoring for these activities. In the 2000 calendar year, Special Recreation Permit usage fees totaled $8,775, and filming use fees totaled $14,775. During the same time, an estimated 125,000 people visited the Salt Flats either as participants in some of the activities, or for purposes of general recreation. No fees are currently charged for casual or non-competitive, non-commercial uses such as sightseeing and general recreation.

In addition to managing and permitting a wide range of activities on the Bonneville Salt Flats, BLM is also responsible for implementing federal resource protections to preserve the unique character of the salt flats. Examples of these protections include two federal mineral withdrawals that protect specific areas of the salt flats from surface entry and mining, and two federal management plans (Bonneville Salt Flats Special Recreation Management Plan and Bonneville Salt Flats Area of Critical Environmental Concern). These management plans identify mineral management restrictions and compatible uses within 30,203 acres of the Salt Flats.

To help promote public and scientific knowledge of this unique area, BLM also participated in funding three major scientific studies of the Bonneville Salt Flats. These studies included two hydrologic investigations by the U.S. Geological Survey during the 1970s and 1990s, and an ongoing meteorologic study by Utah State University. In a cooperative effort to preserve the character of the Bonneville Salt Flats, Reilly Industries, Inc. and BLM entered into an agreement to help replenish salt to the Salt Flats called the Salt Laydown Project. This five-year experimental program began delivering salt to the salt flats in 1997, and as of April 30, 2000 added 4.6 million tons of salt to the salt flats north of Interstate 80.

INTRODUCTION

The Bonneville Salt Flats is located in the western part of the Great Salt Lake Desert of northwestern Utah, and its western margin is approximately 4 miles east of the twin cities of Wendover, Utah and Wendover, Nevada. The salt flats are roughly divided into a north and south half by the east-west-trending Interstate Highway 80 (I-80). The north half, which includes the sites of the historic circular race track and 10- to 12-mile long International Track, is dominated by public land, while the south half is mainly private. Based on measurements from Landsat 5 imagery, areal extent of the salt flats north of I-80 was approximately 31 square miles (19,840 acres) in October 1999 (White, this volume). As the site of automobile land-speed records (figure 1), unusual geology, stark contrasts, and unique scenery, the Bonneville Salt Flats has become famous as an area of national and international interest.

Since the early 1900s, the federal government's management role in the Bonneville Salt Flats was custodial. Up until 1946, lands actions affecting the salt flats were performed by the General Land Office. When the BLM was established in 1946, its land management practices were also custodial and included issuing mineral patents, leases, and rights-of-way.

BLM's involvement in the management of the Bonneville Salt Flats was expanded with establishment of the first "Speed Week" which was held on the salt flats in 1949 and sponsored by Bonneville Nationals Inc. Subsequently, BLM was contacted by the Bonneville Speedway Association about continuing land-speed racing on the salt flats. As a result, BLM issued a special land-use permit to the Bonneville Speedway Association in 1953 for automobile racing. The special land-use permit was continued and reissued through 1969, technically making the Bonneville Speedway Association the manager of the salt flats. Their management

Figure 1. *Record-setting 459-mph run of the wheel-driven "Team Vesco Turbinator" during Speed Week 2001 (photography by Ron Christensen - used with permission).*

objective was to maintain the Bonneville Salt Flats for all types of racing events and protect the vast expanse of open space and hard salt. The Association was responsible for granting permission for all racing activities, and managed all promotion, publicity, and track maintenance. Additionally, they gave permission for commercial movie filming and photography (Morgan, 1985, p. 6).

Bonneville Speedway Association disbanded in 1971, and the Utah Division of Parks and Recreation obtained the Association's special land-use permit from the BLM in 1971. The special land-use permit allowed Parks and Recreation to manage speed trials only. BLM managed all other activities on the Salt Flats. The purpose of the Utah Division of Parks and Recreation's involvement in the Salt Flats was to perpetuate land-speed racing and promote this famous Utah tourist attraction. In 1972, the acreage contained in the special land-use permit, along with 640 acres of private land purchased by the State of Utah became the Bonneville Salt Flats Outdoor Recreation Reserve. As part of its involvement in the Reserve, the State of Utah nominated the Bonneville Salt Flats Race Track to be listed on the National Register of Historic Places in 1973. However, by 1974, the Division of Parks and Recreation was unable to obtain the necessary funding to continue managing the Reserve, and they relinquished the special land-use permit to BLM. BLM has actively managed the salt flats since then (Morgan, 1985, p. 6-7).

With the passage of the National Environmental Policy Act in 1969 and the Federal Land Policy and Management Act of 1976, BLM's land-management role has focused on multiple-use planning and management which provided environmental protection. Since 1974, BLM has completed four management plans related to the Bonneville Salt Flats. The first management plan was completed in 1977 and provided day-to-day guidance for activities affecting the salt flats. BLM's Tooele County Management Framework Plan was completed in 1984, and identified additional measures that were to be used to preserve the unique salt resource (Morgan, 1985, p. 7). In October 1985, BLM completed the Bonneville Salt Flats Recreation Area Management Plan which established criteria and standards for management of a Special Recreation Management Area (SRMA) and an Area of Critical Environmental Concern (ACEC) (Morgan, 1985). The Pony Express Resource Management Plan and Final Environmental Impact Statement was completed in September 1988, and carried forward decisions to continue managing 30,203 acres of the salt flats as an ACEC, and to continue the closure of 104,814 acres within the Bonneville Salt Flats Recreation Area to non-energy leasable minerals (U.S. Bureau of Land Management, 1988, p. 22, 35). Based on BLM recommendations derived from these management plans, the Secretary of Interior, through 43 CFR Public Land Order 6941, withdrew the 30,203 acres of the salt flats covered by the ACEC from surface entry and mining to protect the unique resources of the Bonneville Salt Flats.

In addition to formulating and implementing management plans that help protect and preserve the Bonneville Salt Flats, BLM joined with other federal and state agencies, as well as private industry, in cooperative efforts to help understand and preserve the Bonneville Salt Flats. During the 1970s and 1990s, BLM participated in the funding of two U.S. Geological Survey (USGS) hydrologic studies of the salt flats (Lines, 1979; Mason and Kipp, 1998). BLM also continues to participate in a cooperative agreement with Utah State University's Office of the State Climatologist to conduct weather studies of the Bonneville Salt Flats. As part of the agreement, BLM helped fund the installation and maintenance of two weather stations located on, and adjacent to, the salt flats. In a unique cooperative effort between private industry and a federal government agency, Reilly Industries, Inc. and BLM entered into a salt laydown agreement to help replenish salt to the Bonneville Salt Flats. This experimental Salt Laydown Project began delivering salt to the salt flats in 1997 (White, this volume), and will continue through April 2002. During 2002, Reilly and BLM will review the progress made by the Salt Laydown Project, and mutually decide upon the feasibility and scheduling of an extension of the project.

The following discussion summarizes the major federal resource protections instituted for the Bonneville Salt Flats, both through BLM planning efforts, and from cooperative efforts of other state and federal agencies. Also summarized are some of the unique activities that are conducted on the salt flats through BLM's permitting process.

FEDERAL RESOURCE PROTECTION

Although the Bonneville Salt Flats has been an area of scientific and commercial interest since the early 1900s, no formal protective measures were instituted until the 1950s. Beginning in1952, the following federal resource protections for public lands contained within the north half of the Bonneville Salt Flats (the portion north of I-80) were established to help preserve the unique character of the salt flats (figure 2):

- 1952 - Public Land Order 852: Circular-track portion (8,927 acres) of the Bonneville Salt Flats designated as an automobile racing and testing ground.

- 1975 - National Register of Historic Places: 36,650 acres of Bonneville Salt Flats listed on the Register.

- 1985 - Bonneville Salt Flats Special Recreation Management Area: 30,203 acres of Bonneville Salt Flats identified as the most significant area of the salt flats to be managed.

- 1985 - Bonneville Salt Flats Area of Critical Environmental Concern: 30,203 acres of Bonneville Salt Flats designated.

- 1992 - Public Land Order 6941: 30,203 acres of Bonneville Salt Flats withdrawn from surface entry and mining location for a period of 20 years.

In addition to the traditional federal protective measures listed above, private industry and BLM entered into an historic cooperative agreement in 1995 to replenish salt to the Bonneville Salt Flats. Through the 1995 Salt-Laydown Agreement, Reilly Industries, Inc. and BLM initiated a joint 5-year experimental Salt Laydown Project which began in November of 1997.

Public Land Order 852

On May 26, 1952, 8,927 acres of public land on that portion of the Bonneville Salt Flats containing the historical circular track was withdrawn from mineral location, and reserved for administration as an automobile racing and testing ground. The withdrawal, which was subject to valid existing rights, prohibited mineral prospecting, location, or purchase under the mining laws.

Prior to 1949, the circular track was the most popular track on the salt; however, when the Bonneville Nationals annual racing event began in 1949, the straight or International Track became the track of choice. It is unclear as to why the International Track was not included in the PLO 852 withdrawal. The withdrawal expired in 1982 (Morgan, 1985, p. 6).

National Register of Historic Places

The Bonneville Salt Flats Race Track was nominated to be listed on the National Register of Historic Places by the Utah Governor's Historic and Cultural Sites Review Committee and the Utah State Preservation Officer in October 1973. At the time of its nomination, the Bonneville Salt Flats had earned international recognition as a raceway and site of all the major world land speed records. Additionally, because of its ideal racing conditions, the Bonneville Salt Flats was important to the history of automobile racing because: 1) the sport developed more rapidly than otherwise possible, and 2) the American automobile industry benefitted from mechanical innovations related to the salt flats racing (National Park Service, 1975).

On December 18, 1975, 36,650 acres of Bonneville Salt Flats were officially listed on the National Register of Historic Places. By virtue of this listing, the designated acreage was afforded two protections under the National Historic Preservation Act of 1966 (P.L. 89-665).

- An assessment of the effects of such a project on the listed site will be made, and

- The head of the involved federal agency or department shall afford the Advisory Council on Historic Preservation a reasonable opportunity to comment on the proposed project.

These two protections must be implemented prior to expending federal funds for a proposed federal or federally assisted project that could have potential to affect any district, site, building, structure, or object that is included in the National Register:

BLM Recreation Area Management Plan

The BLM Recreation Area Management Plan, completed in October 1985 (Morgan, 1985), established two entities, the SRMA and the ACEC. Both the SRMA and the ACEC address the protective management of the same 30,203 acres of "hard crusted salt" north of I-80 (Morgan, 1985, p. 18-27).

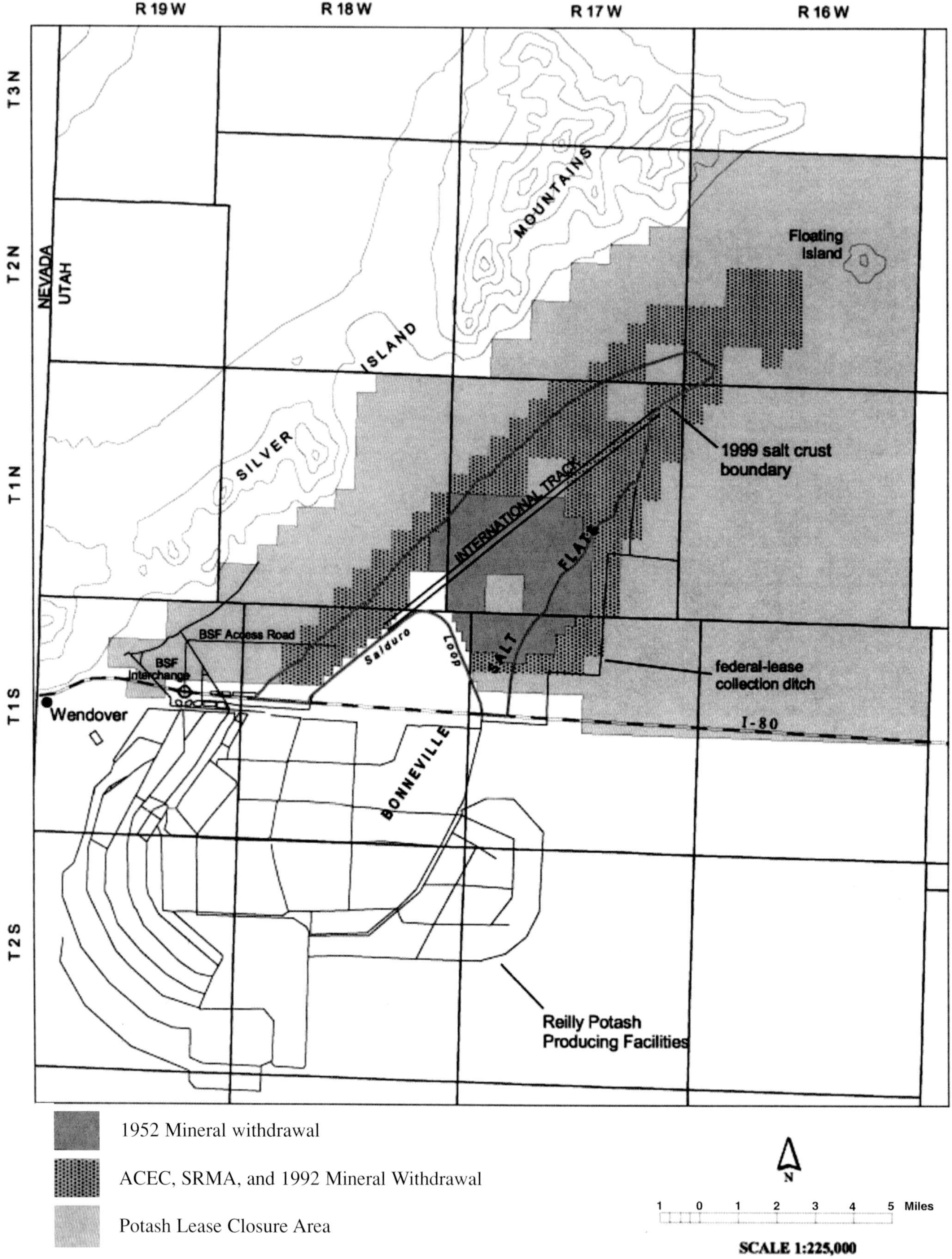

Figure 2. *Index map of Bonneville Salt Flats and vicinity showing locations of five federal-resource protection actions. Note that the Area of Critical Environmental Concern (ACEC), the Special Recreation Management Area (SRMA), and the 1992 Mineral withdrawal each affect the same 30,203 acres, and are represented on the map by the same stippled pattern.*

Bonneville Salt Flats Special Recreation Management Area (SRMA)

Some of the objectives in the SMRA plan are:

- The Bonneville Salt Flats will be managed and retained in public ownership by the BLM as a recreation resource of national and international significance.
- Recreation will be managed as a year-round activity, weather permitting.
- Land speed racing or filming groups which use the Bonneville Salt Flats speedway will contribute funds for track preparation and maintenance.
- Mineral extraction from the Bonneville Salt Flats will be managed to allow continued recreational use of the SRMA; to this end, several management decisions (subject to valid existing rights) were made to cover lands within, or adjacent to, the 30,203-acre SRMA.
- 18,529 acres of federal mineral estate shall be closed to further oil and gas leasing.
- 12,153 acres of federal mineral estate shall be open to oil and gas leasing, but with a provision to prohibit use of the surface.
- 104,814 acres of federal mineral estate surrounding the 30,203-acre SRMA/ACEC shall be closed to further mineral leasing for potash, salts, and other similar salines.

Bonneville Salt Flats Area of Critical Environmental Concern (ACEC)

Because of its unique racing history, stark and unusual landforms, and its complex hydrologic and geologic setting, the Bonneville Salt Flats qualified to be designated as an ACEC. Under the special-management protective umbrella of the ACEC, the following restrictions and compatible uses were identified:

Restrictions:

- Designate areas of the ACEC that will be closed or have specified limitations on locatable mineral exploration and development, oil and gas and geothermal leasing, and non-energy mineral leasing (for example, potash, halite and other similar saline minerals).
- Restrict placement of any permanent structure or facility on the salt crust.

Compatible (Non-Surface Disturbing) uses:

- Recreational uses include automobile and motorcycle racing, off-road vehicle exploration, photography, use monitoring, and site interpretation.
- Scientific uses include automobile testing, environmental study, and experimentation.
- Commercial uses include filming (still photography, commercials, and movies).

Public Land Order 6941

In August 1992, the 30,203 acres of public land contained in the Bonneville Salt Flats SRMA/ACEC were withdrawn from settlement, sale, location, or entry under the general land laws including the United States mining laws. However, the withdrawal did not affect valid existing rights, or leasing under the mineral leasing laws. The objective of the withdrawal was to protect the unique geologic, recreational, and visual resources of the Bonneville Salt Flats. The withdrawal is scheduled to expire 20 years from the date of the order (August 6, 2012) unless extended by the Secretary of Interior.

Salt Laydown Project - Restoration of Salt to the Salt Flats

Reilly Industries, Inc. (Reilly) and BLM have been conducting a Salt Laydown Project to increase the salt-crust thickness of the Bonneville Salt Flats. Reilly funds the Salt Laydown Project's operation which includes capital costs of at least $1,000,000, and operating costs of $80,000 per year. In 1995, BLM joined Reilly in a Salt Laydown Agreement where Reilly and BLM agreed to jointly monitor the Salt Laydown Project's daily and monthly brine chemistry and flow rates.

The project began delivering sodium chloride (NaCl) brine to Bonneville Salt Flats in November 1997. The objective of the five-year experimental Salt-Laydown Project was to add up to 1.5 million tons of salt to the Bonneville Salt Flats during each year of the experiment. Three years of operation have demonstrated that annual salt-tonnage loss from the Bonneville Salt Flats can be replenished by the Laydown Project. From November 1997 through April 2000, a sodium chloride salt mass of about 4.6 million tons was delivered as a brine, through large-diameter pipes and ditches, to the Bonneville Salt Flats. The average annual addition of 1.5 million tons exceeded an estimated annual salt loss of 0.85 tons. The salt addition appears to be distributed between existing salt-crust, an expanded salt-crust area, and the shallow-brine aquifer. For technical details, see White this volume.

ACTIVITIES MANAGEMENT

An estimated 125,000 people visited the Salt Flats in 2000, either as participants and observers in organized activities, or for purposes of general recreation. Typical activities administered by the BLM for public use of the Bonneville Salt Flats fall into three general categories: commercial, competitive, and casual use (that is, general recreation). Examples of commercial uses include potash mineral extraction and filming. Competitive uses that require a Special Recreation Permit include land-speed racing, archery meets, and rocket launch contests. Examples of casual use or general recreation include sightseeing and off-road vehicle exploration. No use fees are currently charged for casual use. Use and cost-recovery fees are charged for commercial and competitive uses, and production royalties are collected from potash mineral leases.

Because potash mineral leases, Special Recreation Permits, filming permits, and casual use (i.e., general recreation) constitute the major Salt Flats activities administered by BLM, they are described in more detail in the following sections.

Potash Mineral Leases

Potash extraction was the first commercial use of the Bonneville Salt Flats, and development of potash processing from Salt Flat brines was started by the Utah Salduro Company in 1917. Most of the Salt Flats south of I-80 were patented between 1917 and 1927 when potash was a locatable mineral. Passage of the Potassium Act of 1927 initiated potassium leasing (Stan Perkes, Mining Engineer, BLM Utah State Office, personal communication September 19, 2001). Utah Salduro Company closed its plant in 1921, and the plant was idle until 1936, when Bonneville Limited acquired the plant and patented property. Bonneville Limited began commercial potash operations in 1939, and production has continued to the present time. During 1963, 10 federal leases (including the collection ditches on the eastern margin of the Salt Flats north of I-80) were issued by BLM to Bonneville Limited. In 1964, Kaiser Aluminum and Chemical Corporation (Kaiser) acquired Bonneville Limited and its Bonneville Salt Flat holdings (Bingham, 1980, p. 230-231). Kaiser operated the privately held potash operations and federal leases until they were sold in 1988 to Reilly Industries, Inc., the current operator.

Federal potash leases are issued for an indeterminate time period, and are subject to readjustment at the end of each 20-year period. The federal leases held by Reilly are due for readjustment in January 2003, and BLM is responsible for performing the readjustment. Although two protective mineral withdrawals and several closures to specific mineral-leasing activities have been initiated within the Bonneville Salt Flats ACEC (see previous section on Resource Protection), Reilly's federal leases are pre-existing rights and are exempted from the restrictions contained in the withdrawals and lease closures. However, during the 2003 lease adjustment process, BLM will determine any changes or additions to the current lease stipulations.

Federal Mineral-Lease Royalties

Lessees of non-energy federal minerals such as potash pay the federal government royalties based on production. For example, the royalty on potash is 3 percent of sales value. Rather than being paid to BLM, these royalty payments are collected by the Minerals Management Service (MMS). MMS distributes the royalty payments as follows: 50 percent of the funds are distributed to the state of origin, 40 percent are placed in the Reclamation Fund, and 10 percent are directed to the General Fund of the U.S. Treasury (Minerals Management Service, 1999, p. 2).

Special Recreation Permits

Each year, BLM issues an average of five Special Recreation Permits for scheduled activities/events on the Bonneville Salt Flats. Depending upon weather conditions, these activities/events are typically scheduled from the beginning of August through the end of October. Land-speed racing and competitive archery matches and rocket launches are examples of these permitted activities.

Land-Speed Racing

Permitted land-speed racing events include Speed Week (August), World of Speed (September), and World Finals (October). Land-speed racers do not compete against each other in a multiple-vehicle race, but rather each competitor races individually over a specified distance against the clock in an attempt to set the land-speed record for his or her particular vehicle category (figure 3). Speed Week, for example, has four different vehicle categories: Special Construction, Vintage (pre-1948 automobiles), Modified, and Diesel Truck. The events all use straight courses which allow ample space for acceleration as well as deceleration. The longest course in 2001 was about 11 miles. As salt is added from the Salt Laydown project, it may be possible to have an even longer course in the future. The course is seasonably obliterated each winter by rising ground waters.

Speed Week, World of Speed, and World Finals are sponsored respectively by Southern California Timing Association (SCTA) and Bonneville Nationals Inc. (BNI), Utah Salt Flats Racing Association (USFRA), and SCTA and BNI. All three events are well attended by U.S. and international racing enthusiasts. For example, the number of people attending the 2000 racing events were:

- Speed Week - 6,316
- World of Speed - 864
- World Finals - cancelled due to rain

Competitive Archery Matches and Rocket Launches

Non-racing events include annual activities such as competitive archery matches and rocket launches sponsored by the United States National Archery Association and the Utah

Figure 3. *Preparation for land-speed racing on the Bonneville Salt Flats during the 2001 "World of Speed" sponsored by the Utah Salt Flats Racing Association (photography by Larry Volk, chairman, Save the Salt - used with permission of the photographer).*

Rocket Club. The National Archery Association has been conducting archery matches at the Bonneville Salt Flats since 1961 (figure 4); participants include U.S. archers, as well as international archers from Federation International' de Tir à l'Arc (FITA) (Rulon Hancock, U.S. National Archery Association, personal communication, September 18, 2001). Utah Rocket Club, which is the local section of the National Association of Rocketry has been sponsoring rocket launches at Bonneville Salt Flats since 1992 (figure 5); participants are typically from U.S., Canada, Great Britain, and Australia (Neal Baker, President, Utah Rocket Club, personal communication, October 19, 2001).

Respective objectives of the archery and rocket events are to set distance and altitude records by individual classes. For example, archery classes include conventional, compound, recurve field and target, modern longbow, and primitive (that is, wooden arrow shafts with sinew-attached points) categories. Rocketry types are divided into two classes, model rockets and high-power rockets. Model rockets are typically under 3 pounds and may reach heights of 50 to 3,000 feet above ground level (AGL). High-power rockets range from 3 to 100 pounds and may achieve heights of 500 to 25,000 feet AGL. High-power rocket flights are subject to Federal Aviation Administration (FAA) air-traffic regulations.

Figure 4. Shooting for distance records during a competitive archery match sponsored by the U.S. National Archery Association (photography by Rulon I. Hancock, U.S. National Archery Association - used with permission of the photographer).

Figure 5. Rocket launching during the Utah Rocket Club's "Hellfire 6" activity in September 2000 (photography by mark Hamilton, Utah Rocket Club - used with permission from the Utah Rocket Club).

The recorded number of people attending the 2000 archery match and rocket launch competitions were:

- United States National Archery Association - 20;
- Utah Rocket Club - 666.

Special Recreation Usage Fees

Competitive events such as land-speed racing, archery matches, and rocket launches are subject to BLM competitive-use fees. Competitive-use fees are $4.00 per person per day for participants and spectators, or 3 percent of gross receipts, whichever is greater. The fees generated by these events have been distributed as follows: 15 percent - General Treasury, and 85 percent - BLM Salt Lake Field Office recreation program. The portion of the competitive use fee returned to BLM is used to help defray costs of staff time spent on processing event/activity permits, preparing associated environmental documentation, and providing Salt Flat interpretative studies. Competitive use fees collected by BLM during FY 2000 totaled $8,775.

Filming Permits

The unique scenery of the Bonneville Salt Flats makes an ideal backdrop for movies, television commercials, artistic videos, and still photography. The film industry is attracted to the Salt Flats because the unusual nature of its scenery is accentuated by seasonal change. For example, during the summer months, stark white expanses of salt crust stretch to a horizon made up of rugged desert mountain ranges. Conversely, during the winter and spring months, a shallow inland sea covers the salt crust and provides mirror-like reflections of the distant mountains. This unusual scenery has been the backdrop for segments of major motion pictures such as *Independence Day, Mulholland Falls*, and *Con-Air*. Additionally, the Salt Flats have been the site of eye-catching television and magazine commercials for Nissan, Daimler/Chrysler, and Harpers Magazine (figures 6 and 7).

BLM usually allows commercial filming on the Bonneville Salt Flats from late May into late October. This is accomplished through a formal permitting process that includes use and cost-recovery fees, and reclamation bonds. Use fees are retained by the BLM Salt Lake Field Office, and the cost recovery fees are deposited with the General Treasury. The amount charged for use fees varies between video and still photography. Use fees depend on size of the crew, and range from $250/day to $600 a day for video, and $100/day to $500/day for still. Cost-recovery fees range from $375 to $1175 depending on complexity of production and monitoring requirements by BLM personnel. Bonding for aerial filming is $10,000. Reclamation cash bonds are also required at BLM discretion. These bonds are all refundable if the

Figure 6. *Preparation for filming an automobile commercial on the Bonneville Salt Flats (photography by Anita Jones, Land Law Examiner, BLM Salt Lake Field Office).*

Figure 7. *Filming of an automobile commercial on the Bonneville Salt Flats (photography by Anita Jones, Land Law Examiner, BLM Salt Lake Field Office.*

permitted activity results in no damage to the Bonneville Salt Flats. Film-related use and cost-recovery fees collected by BLM during FY 2000 totaled $14,775.

Casual Use - General Recreation

Casual use or general recreation on the Bonneville Salt Flats includes activities such as sightseeing and the pursuit of solitude in an area of stark contrasts and extensive vistas. The Bonneville Salt Flats north of I-80 is one of the few land forms that exhibits a naturally flat surface covering an area of more than 30 square miles. Because of its smooth surface and large areal extent, the curvature of the earth is actually observable from the salt flats. An example of this can be seen when viewing down the axis of one of several telephone pole lines that extend across portions of the salt flats; the pole line gradually grows smaller and eventually disappears below the horizon at distances of about 10 miles from the observer. Additionally, optical illusions of "floating" mountain ranges result from the mirage effect that is accentuated by the extensive flat surface of the salt flats. This mirage effect is most notable during the summer heat and makes landmarks such as Floating Island (see figure 2) appear to be floating just above the horizon of the salt flats on a thin, shimmering cushion of air.

BLM encourages the public to enjoy the Bonneville Salt Flats as part of their public lands heritage. Additionally, when the public is engaged in casual use of the Salt Flats, no use fees are charged. Although visitation to the area is welcomed during most of the year, some precautions are necessary, and the public should be prepared for the extreme conditions that are unique to the salt flats. The following are examples: 1) temperatures can range from freezing during the winter to more than 100 degrees F during the summer; 2) while the salt crust is thickest along the axis of the International Track (see figure 2), it thins to a feather edge on its western and eastern margins; the unwary who attempt to drive their vehicles to the edge of the salt crust run the risk of being entrapped in the mud flat that surrounds the salt flats; 3) distances are deceiving - it is 5 miles to the nearest telephone and help from the end of the Bonneville Salt Flats access road to the truck stop at the junction of the access-road off ramp at I-80 (see figure 2).

Visitors are discouraged from driving off the access road onto the salt flats during the winter months for their own protection, and for protection of the Bonneville Salt Flats. This is because a shallow transient pond of salt water covers most of the salt crust area during the winter. Additionally, the Salt Laydown Project, which operates from November 1 through April 30, also contributes salt water to the transient pond as part of its replenishment of salt to the salt flats. Visitors who attempt to drive onto the flooded salt flats may accidentally spray salt water on critical electrical components of their engines which could result in electrical shorting and consequent disabling of their vehicle.

While the public should be aware of the potentially harsh nature of the Bonneville Salt Flats environment, the rewards from visiting this unique resource (even if one only goes to the end of the access road) include unusual geology, stark contrasts, and surreal scenery.

SUMMARY

- BLM is responsible for implementing formal protections that have been instituted to preserve the unique character of the salt flats. Examples of these protections include:
  - Declaration of two federal mineral withdrawals that protect specific areas of the salt flats from surface entry and mining.
  - Placement of 36,650 acres of Bonneville Salt Flats on the National Register of Historic Places.
  - Preparation of a plan for the SRMA and ACEC

that identifies mineral-leasing restrictions and compatible uses within 30,203 acres of the salt flats.

- Potash mineral leases, Special Recreation Permits, and filming permits are the main categories of activity administered by the BLM for public use of the Bonneville Salt Flats:
  - The royalty paid by Reilly on its federal potash leases is 3 percent of sales value; 50 percent of the royalties are distributed to the state of origin, 40 percent are placed in the Reclamation Fund, and 10 percent are directed to the General Fund of the U.S. Treasury.
  - As much as 85 percent of usage fees collected by BLM for land-speed racing, competitive archery meets and rocket launches, and commercial filming are returned to the BLM to help defray the costs of permit processing and resource monitoring for these activities.
  - During the 2000 calender year, special recreation usage fees totaled $8,775, and filming use fees totaled $14,775.
- To help promote scientific and public knowledge of this unique area, BLM has also participated in funding three major scientific studies of the Bonneville Salt Flats. These studies included:
  - Two hydrologic investigations by the U.S. Geological Survey during the 1970s and 1990s.
  - An ongoing meteorologic study by Utah State University.
- In a cooperative effort to preserve the character of the Bonneville Salt Flats, Reilly Industries Inc, and BLM entered into a Salt Laydown agreement to help replenish salt to the salt flats. This five-year experimental program began delivering salt to the salt flats in 1997 and, as of April 30, 2000, has added about 4.6 million tons of salt.
- The public is encouraged to enjoy this unique resource but should be prepared for the extreme conditions that are unique to the salt flats. No fees are currently charged for casual or non-competitive, non-commercial uses such as sightseeing and general recreation.

ACKNOWLEDGMENTS

The following Salt Lake Field Office staff contributed significantly to the preparation and review of this paper: Grace Jensen, Realty Specialist; Cheryl Johnson, GIS Coordinator; Britta Laub, Outdoor Recreation Planner; Michael Nelson, Realty Specialist; and William White, Physical Scientist. Additionally, Larry Volk, Chairman, Save-The-Salt, Ron Christensen, Utah Salt Flats Racing Association, Rulon Hancock, U.S. National Archery Association, Neal Baker, President, Utah Rocket Club, and Mark Hamilton, Utah Rocket Club, contributed to descriptions of their respective activities and provided some of the photographs. Editorial and peer reviews were performed by Stan Perkes, Mining Engineer, and James Kohler, Branch Chief Solid Minerals, BLM Utah State Office, and J. Wallace Gwynn, Ph.D. and staff from the Utah Geological Survey; their comments and suggestions helped enhance the quality and organization of the paper.

DEDICATION

This paper is dedicated to the memory of Lewis William Kirkman, Jr., Outdoor Recreation Planner, BLM Salt Lake Field Office. Lew will be remembered for his uncompromising devotion to the protection of the public lands, and his service and friendship to the organizers and participants of the many activities conducted on the Bonneville Salt Flats.

REFERENCES

Bingham, C.P., 1980, Solar production of potash from the brines of the Bonneville Salt Flats, *in* Gwynn, J.W., editor, Great Salt Lake - a scientific, historical and economic overview: Utah Geological and Mineral Survey Bulletin 116, p. 229-242

Lines, G.C., 1979, Hydrology and surface morphology of the Bonneville Salt Flats and Pilot Valley Playa, Utah: U.S. Geological Survey Water-Supply Paper 2057, 107 p.

Mason, J.L. and Kipp, K.L., 1998, Hydrology of the Bonneville Salt Flats, northwestern Utah, and simulation of ground-water flow and solute transport in the shallow-brine aquifer: U.S. Geological Survey Professional Paper 1585, 108 p.

Minerals Management Service, 1999, Mineral Revenue Distributions, Fiscal Year 1999, Royalty Management Program: Minerals Management Service, U.S. Department of the Interior, 23 p., Online: http://www.rmp.mms.gov/stats/stat-srm.htm.

Morgan, G.B., 1985, Recreation Management Plan for the Bonneville Salt Flats Special Recreation Management Area and Area of Critical Environmental Concern, Utah: U.S. Department of the Interior, Bureau of Land Management, Salt Lake District - Utah, 46 p., 21 Appendixes.

National Park Service, 1975, National Register of Historic Places Inventory - Nomination Form, Bonneville Salt Flats Race Track: U.S. Department of the Interior Form 10-300, 8 p. Copies available upon request from BLM, Salt Lake Field Office.

U.S. Bureau of Land Management, 1988, Proposed Pony Express Resource Management Plan and Final Environmental Impact Statement, September, 1988: U.S. Department of the Interior, Bureau of Land Management, Salt Lake District - Utah, 144 p.

STATE AND FEDERAL GSL ISSUES

GREAT SALT LAKE - A NEW TEACHING TOOL THAT REFLECTS SOCIETY CONCERNS

by

Sandra N. Eldredge
Utah Geological Survey

ABSTRACT

Reacting to growing national and world problems, the 1990s have been a decade of changes in how we see and interact with our environment. The technologically oriented decade has also changed the ways we seek to solve problems. Partly as a result of these concerns, new national science education standards have evolved to create a science-literate population for the 21st century - a population that can assess the pros and cons of science and technology solutions to these problems. Interestingly, Great Salt Lake has provided a unique conduit for putting these new standards into effect. In addition to curricula and courses developed with a Great Salt Lake theme, public and private groups have formed that are concerned about the lake and its future. The outcome is that the lake is no longer ignored or perceived as an isolated place, but a place with a dynamic complex environment that affects us all.

GREAT SALT LAKE AND NEW SCIENCE EDUCATION STANDARDS

Although Great Salt Lake has been here for thousands of years, only recently has it become a unique teaching tool for "new" ways of teaching science. Up until the 1990s, grades Kindergarten - 12th (K-12) students typically learned a few sporadic facts about the lake, such as: it is located in northern Utah, it's large, it's saltier than the ocean, and you can float in it. These facts are okay, maybe even interesting. But the lesson usually ended there - no excitement, no involvement, no relevancy. This decade brought a change, however, and now our school children get excited about the subject of Great Salt Lake. They experiment with lake water, roll oolitic sand grains between their fingers and look at it with hand lenses, grow brine shrimp, investigate how the lake came to be here, research the effects of salinity on the various lake industries, or debate the issues involved with the proposed Legacy-West Davis highway and its effects on wetlands. Why the change in instruction? The reason is new education standards, and Great Salt Lake is an ideal topic to put them into effect.

The new standards in science education result from our nation's students ranking low in science and math performance, and from alarming national and global problems such as unchecked population growth, acid rain, pollution, inequities of resource distribution ... the list goes on. In a world shaped by science and technology, the 1990s cry is for a science-literate population that has the ability to maintain a balanced well-being of the United States, as well as the world (National Research Council, 1996; National Science Foundation, 1997).

The new teaching methods and geoscience concepts advocated have resulted in a much more relevant way of learning for students: students become active learners, where they actually do science (hands-on science), not just read about it. Utah is one of many states that has adopted the national science education standards and revised its science core curriculum in the mid-1990s (Utah State Office of Education, 1994, 1995). The new core includes an increase in geoscience concepts taught at more grade levels between Kindergarten and 12th grade, a multidisciplinary approach, active learning, and a focus on meanings, connections, and contexts rather than fragmented bits and pieces of information/facts. Also emphasized is personal relevance so that students' understanding of science should enable them to make informed and responsible decisions. To this end, relationships among the sciences, and science, technology, and society permeate the core.

Great Salt Lake is a perfect topic for students that meets both the new national standards and Utah's science core curriculum. The lake and its environs offer not only multidisciplinary topics (chemistry, biology, geography, history, geology, hydrology, etc.), but interrelate the topics. The lake can be looked at as a system, a lab, or a discussion point of current issues. Students out of state can study the lake from afar, through activities in labs or classrooms. Utah students can use the lake as a local model for a global view of systems, cycles, interactions, and issues. In both cases, the content portion (the scientific concepts learned at specific grade levels) of the *National Science Education Standards* is met: properties of earth materials and changes in the earth (K-4); structure of the earth system and earth's history (5-8); and energy in the earth system, geochemical cycles, origin and evolution of the earth system (9-12).

Utah students have the extra benefit of applying their learning personally. This personal relevance, or how the lake affects them and vice versa, is an extremely important part of learning science. In addition, Great Salt Lake has played an interesting role in Utah history, so it is a pertinent topic in both 4th-grade and 7th-grade Utah Studies classes. Classes in Utah with varying Great Salt Lake themes are now offered for both teachers and students.

GREAT SALT LAKE DYNAMIC SYSTEM

(What Can Be Learned)

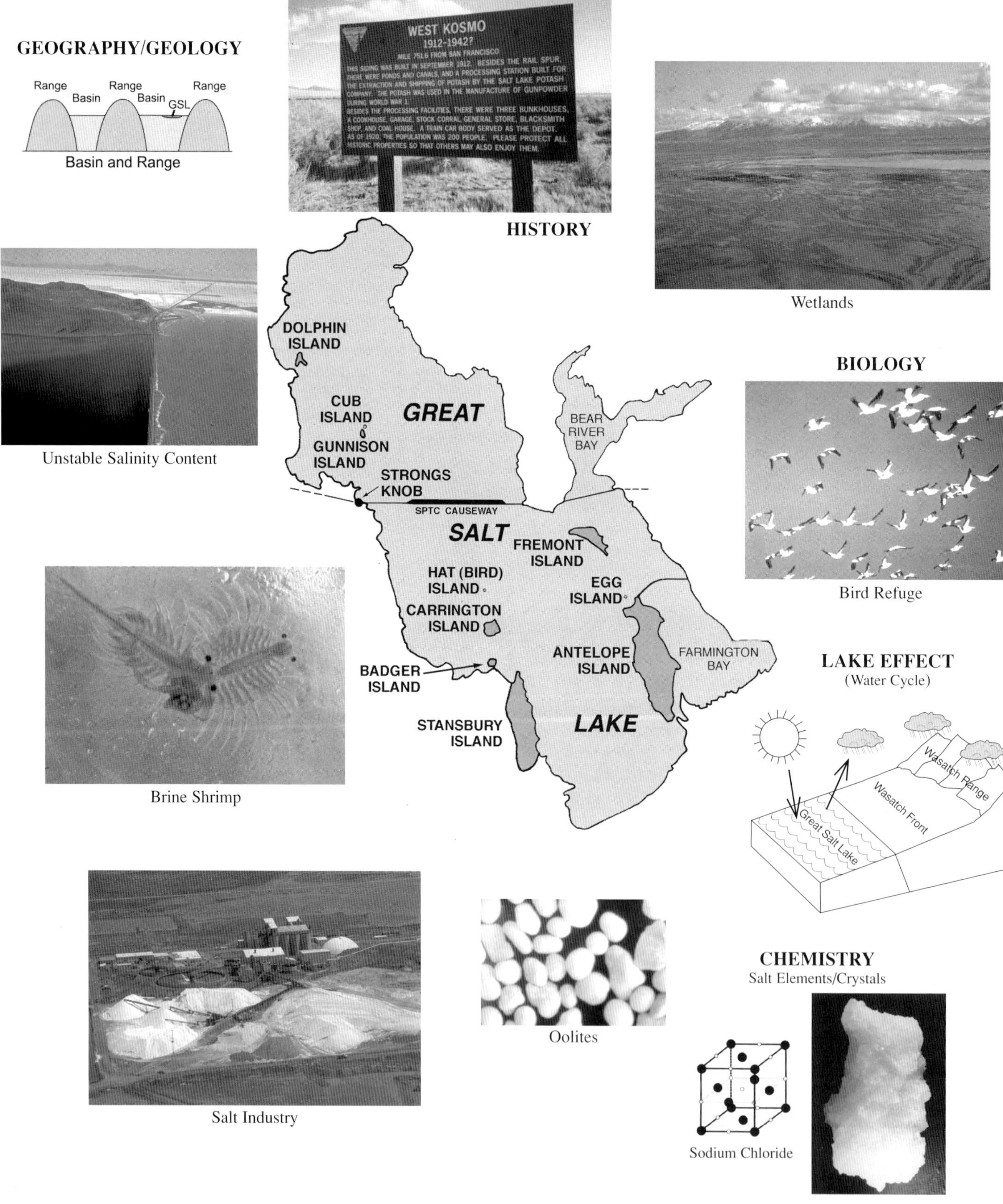

Wetlands

Unstable Salinity Content

Bird Refuge

Brine Shrimp

Salt Industry

Oolites

Sodium Chloride

Examples of Curricula and Courses

The Great Salt Lake Story, an Interdisciplinary Activity Guide (Utah Museum of Natural History, 1995) is used by many teachers of various grade levels, especially grades 3-8. Developed in the early 1990s, with numerous activities, different science topics, and interrelated topics, the two goals of this activity guide are: "to increase student awareness of the lake as a valuable natural resource that needs our understanding, appreciation, and protection," and "to help (the teacher) use the lake as a unifying theme to make general science, social studies, and environmental concepts more relevant to your students." The activities range from basic classroom science observations to evaluating the effects of developing Antelope Island or creating a freshwater lake in one section of Great Salt Lake.

Earth Systems One: Content and Inquiry (University of Utah Center for Integrated Science, 1998), developed by the University of Utah entities of Educational Studies, Geology and Geophysics, and the Center of Integrated Science and the Utah Geological Survey, addresses the 9th-grade core called "Earth Systems." The core intends that students will "develop an understanding of interactions and interdependence within and between Earth systems and changes in Earth systems over time." The systems outlined are biologic, geologic, hydrologic, atmospheric, and energy. What better place to learn about systems than Great Salt Lake! The course centers on the lake to help teachers become familiar with the systems approach (matter cycles, energy flows, and life webs) to the study of the earth by developing knowledge of specific earth systems, and how these systems all interact and are interdependent. For example, just one simple question such as "Why is the lake here?" leads into a myriad of connections among earth's systems: topography, which relates to earthquakes, faulting, and internal energy (geologic system); changes through time, including climate (atmospheric, hydrologic, biologic, and energy systems); interactions of atmosphere and topography - including the lake effect (geologic, hydrologic, atmospheric, and energy systems); and so on.

Exploring the Geologic System: Earth System Interactions Involving Earthquakes (University of Utah Earthquake Education Services, 1997) also addresses the 9th-grade core. Part of this curriculum explores relationships between Great Salt Lake and earthquakes. One activity, for example, explores earthquake-generated seiche waves and their effects.

A variety of Utah Museum of Natural History classes that investigate different aspects of the lake are offered virtually every year to the public and teachers and students. Not surprisingly, Antelope Island has become a popular place for field trips and classes to study numerous topics, including many relating to the lake. Many people, both Utahns and out-of-staters, visit the lake and/or the island and walk away with some gained knowledge.

PARALLEL SOCIETAL CONCERNS

Paralleling education interests in Great Salt Lake, is of course public (individual and society) interest in the lake and its resources and environs. The lake was recently designated as part of the Western Hemispheric Shorebird Reserve Network - a recognition of its international importance to the survival of bird species. Its shorelines, wetlands, and formalized preserves draw hundreds of birdwatchers every year. Nature writers and photographers abound who are attracted to the lake's beauty and magic. Conflicts due to multiple uses of the lake and burgeoning populations mean Great Salt Lake issues are constantly arising. Concerned citizens formed "Friends of the Great Salt Lake" to increase public knowledge of the lake's issues. The Utah Department of Natural Resources' Great Salt Lake Planning Project Team has developed a comprehensive resource management plan for the lake (see in this book). These are just some of the groups and individuals who appreciate the lake and are concerned about the future of Great Salt Lake and interacting resources. This brings us full circle to new education standards resulting, in part, from such societal concerns. The lake is no longer perceived as an isolated spot on the map, but a dynamic complex environment that affects us all.

ACKNOWLEDGMENTS

Reviewers Deedee O'Brien, Wally Gwynn, and Paula Wilson helped me focus my viewpoint (Paula), expand my viewpoint (Deedee), and enhance the outcome (Wally).

REFERENCES

National Research Council, 1996, National science education standards: National Academy of Sciences, National Academy Press, 262 p.

National Science Foundation, 1997, Geoscience education - a recommended strategy: National Science Foundation, NSF 97-171, 32 p.

University of Utah Center for Integrated Science, 1998, Earth systems one - content and inquiry: unpublished materials, unpaginated.

University of Utah Earthquake Education Services, 1997, Exploring the geologic system - earth system interactions involving earthquakes: unpublished materials, 86 p.

Utah Museum of Natural History, 1995, The Great Salt Lake story, an interdisciplinary activity guide: Utah Museum of Natural History, University of Utah, 249 p.

Utah State Office of Education, 1994, Science core curriculum, grades K-6: Utah State Board of Education, 34 p.

Utah State Office of Education, 1995, Science core curriculum, grades 7-12: Utah State Board of Education, 64 p.

OWNERSHIP AND MANAGEMENT OF GREAT SALT LAKE, UTAH, FROM STATEHOOD THROUGH 1999

by

James W. Carter
Utah Division of Oil, Gas and Mining, now with Bear West, Salt Lake City, UT

J. Wallace Gwynn
Utah Geological Survey

and

Karl Kappe
Utah Division of Forestry, Fire and State Lands

ABSTRACT

Since statehood, Utah's leadership has recognized the need for planned management of Great Salt Lake. The U.S. Bureau of Land Management challenged the state's claim to land lying between the lake waterline and adjacent federally owned property in 1959. After years of administrative decisions unfavorable to the state, the Federal Government's claim to the lakeshore was challenged in federal court by the state in 1967. In 1976, Utah prevailed, and the ownership of all lands, brines, and other minerals within the waters of the lake, and the bed, and all shore lands located within the official surveyed meander line, was conveyed to the state. Even before the 1975 ruling of the court, Utah had assumed responsibility for managing the lake and its resources. Lake management and planning responsibilities have rested with the Great Salt Lake Authority (1963-1967), the Division of Parks and Recreation (1967- 1975), the Division of Great Salt Lake (1975-1979), the Department of Natural Resources (DNR) (1979-1988), and finally the Division of State Lands and Forestry (DSL&F), now called the Division of Forestry, Fire and State Lands (DFF&SL) (1988-present). Since 1963, these agencies have developed a number of management plans, framed to provide guidance for the management and development of the lake and its resources by governmental and private entities.

In 1995, the Great Salt Lake Technical Team, a multi-interest advisory group convened by DSL&F, published The Great Salt Lake Comprehensive Management Plan- Planning Process and Matrix which "...provided needed information and guidance in the form of recommendations to federal, state and local governments, ...to facilitate and enhance management of Great Salt Lake and its environs to assure protection of the unique ecosystem of the lake while promoting balanced multiple resource uses (Flandro, 1995)." Those recommendations provided the starting point for the Great Salt Lake Planning Project, initiated in 1997 with the joint sponsorship of DNR and DFF&SL. The purposes of the Great Salt Lake Planning Project are: (1) to establish unifying DNR management objectives and policies for the trust resources of Great Salt Lake, (2) to coordinate the management, planning and research activities of the department's divisions on Great Salt Lake, (3) to coordinate management with the actions of land and resource owners and managers on, and adjacent to, Great Salt Lake, (4) to develop a sovereign lands and resources management plan, and (5) to establish processes for plan implementation, monitoring, evaluation and amendment. At the date of publication of this volume, the Great Salt Lake Planning Project is completed and the plan is being implemented.

INTRODUCTION

Federal court actions gave Utah title to the lands, brines, and minerals of Great Salt Lake within the surveyed meander line. Through legislative and administrative actions, management responsibility for Great Salt Lake has shifted over time from state agency to state agency. Management plans prepared by the different agencies have reflected the different philosophies of resource preservation and resource development prevalent at the time. The purpose of this paper is to summarize the historical and current management of the lake and provide an understanding of the state's responsibility under the Public Trust Doctrine and other mandates to manage the lake.

OWNERSHIP OF THE LAKE

From the time of statehood in 1896, Utah believed that it owned Great Salt Lake, the water-covered bed, and the shore lands located within the meander line officially surveyed and approved by the United States Government. This claim was based on the Equal Footing Doctrine, which holds, among other things, that at the date of statehood all states received rights of ownership and management to all tidelands and the beds of navigable lakes and streams within their borders. Utah's claim to the lake was further asserted in Chapter 9 of the Laws of Utah, 1927 (Senate Bill No. 12), which reports the enactment by the Utah Legislature of a statute that reaffirms Utah's title to the beds of all navigable lakes and streams within the state of Utah. Since statehood Utah has assumed ownership and control of Great Salt Lake and its shore lands, and has managed them for various uses, including recreation, wildlife, mineral development, grazing and a variety of other purposes (Mahoney, 1966; and Dewsnup and Jensen, 1980).

In 1959, the U.S. Bureau of Land Management (BLM) notified the Utah State Land Board that it intended to survey a new State/Federal ownership boundary line along the shore of Great Salt Lake at an elevation of 4201.8 feet, the lake's elevation at statehood. That notification disallowed Utah's claim to the bed of the lake above the 4,201.8 elevation. "The BLM held that, where the United States was an owner of lands bordering on the lake, it possessed, by virtue of the common law right of accretion and reliction, a vested right to land which emerged from the lake and bordered land already owned by the Federal Government" (Moss, 1965). After a number of administrative reviews that were unfavorable to the state, Utah filed an original action in federal court on March 1, 1967, to obtain ownership of the bed of Great Salt Lake up to the officially surveyed meander line. In 1976, after a lengthy period of multi-phase litigation, Special Master Judge Charles Fahy ruled against the United States and in favor of the state of Utah. The court's acceptance of the Special Master's report confirmed ownership of all lands, brines, and other minerals within the waters of the lake and within the bed, and all shore lands located within the official surveyed meander line, in the state of Utah (Burnham, 1980; Dewsnup and Jensen, 1980).

MANAGEMENT RESPONSIBILITIES AND LAKE PLANNING

Great Salt Lake Authority (1963)

The 1963 Utah Legislature created the Great Salt Lake Authority and an advisory council to the Authority (Laws of [Utah] 1963, chapter 161 [H.B. no. 33]). The Authority was empowered to:

> (1) determine the policies and develop the program best designated to accomplish the objectives and purposes set out in the act,
> (2) construct facilities and to acquire real and personal property in the name of the Authority by all legal and proper means,
> (3) protect property controlled or administered by the Authority,
> (4) coordinate multiple use of property for such purposes as grazing, fish and game, mining and mineral removal, development and utilization of water and other natural resources, industrial, and other uses in addition to recreational development, and adopt such reasonable rules and regulations as the Authority may deem advisable to insure the accomplishment of the objectives and purposes of the act" (Laws of 1963).

The Authority was empowered and instructed to proceed without delay to gather all studies, investigations, and reports regarding Great Salt Lake concerning the development of any part of Antelope Island for tourists and recreational uses. The Authority was also empowered to take all steps necessary to secure, by donation, purchase agreement, lease, or other lawful means, such part of Antelope Island as was deemed necessary for recreational use (Laws of 1963). In the same bill, it was specified that both the State Department of Fish and Game, and the State Land Board would retain the power and jurisdiction conferred upon them within the boundaries of jurisdiction of the Authority with reference to their functions, subject to such reasonable rules and regulations as the Authority may make to ensure the accomplishment of the objectives and purposes of the act. The bill also provided that, within the limitations of available funds, the State Road Commission was authorized to construct a road from the town of Syracuse to the northern end of Antelope Island, along with the necessary roads on the island, to serve the recreational development.

Great Salt Lake Authority Management Plans (1964-1965)

A preliminary feasibility study for the recreational development of the north end of Antelope Island was prepared by Snedaker & Budd and Allred & Associates for the Great Salt Lake Authority and submitted on June 26, 1964. Then in 1965, a document entitled "A Preliminary Master Plan for the Development of Great Salt Lake - Over a Period of the Next 75 Years" (Richards, 1965), was prepared for the Great Salt Lake Authority. This plan envisioned the use of surplus waters from the Bear River, Weber River, and Jordan River drainage areas, as well as the use of Kennecott tailings material for the construction of dikes, highways, and land reclamation within Farmington Bay.

The Authority was not effective in carrying out its duties, and in 1966, the Utah Supreme Court declared that the act creating the Authority was unconstitutional, as it failed to define the Authority's geographical jurisdiction.

Re-establishment of Authority (1967)

In 1967, the Legislature cured the jurisdictional defect when it recreated the Great Salt Lake Authority (Laws of [Utah] 1967, ch. 187 [S.B. no. 60]). Within this legislation, the Authority's geographical jurisdiction was defined, and included the mainland, peninsulas, islands, and water within the Great Salt Lake meander line established by the United States Surveyor General. The purpose of the recreated Authority was to establish and coordinate programs for the development of recreational areas and water conservation within Great Salt Lake and its environs, and to provide:

> (1) the development of such area of Antelope Island as the Authority may determine to be suitable and desirable for recreational usage,
> (2) testing the feasibility of using [Kennecott Utah Copper] tailings in the development of Great Salt Lake and its environs,
> (3) the restoration and preservation of points of historical interest on Antelope Island.

The Authority was also given the following powers and duties to implement:

(1) to acquire real and personal property in the name of the Authority,
(2) to construct and maintain those facilities consistent with the purposes of the act,
(3) to operate under its own administration,
(4) to hire officers and employees as required,
(5) to deputize, or obtain from local authorities by contract, such peace officers as may be necessary to protect Authority property,
(6) to seek assistance from federal and local governments in planning and developing uses of Great Salt Lake,
(7) to proceed without delay to gather all studies, investigations, and reports regarding Great Salt Lake and concerning the development of Antelope Island.

Abolition of Great Salt Lake Authority (1967)

After the creation of the Utah Department of Natural Resources (DNR) in 1967, the Great Salt Lake Authority, operating with a half-time director, was abolished. The functions of the Authority were later merged into the Utah Division of Parks & Recreation (Flandro, 1995).

Division and Board of Great Salt Lake (1975)

In 1975, the Utah Legislature enacted House Bill Number 23, which established a Board and a Division of Great Salt Lake within DNR for the purpose of establishing and coordinating programs for the development of recreation areas, flood control, wildlife resources, industrial uses, and conservation of Great Salt Lake. The Great Salt Lake Division was given the responsibility to determine the direction and implementation of activities through existing agencies, and was given the following powers and duties:

(1) to direct the preparation of and adopt a comprehensive plan for the lake in a manner which will assure the maximum interchange of information, ideas, and programs with affected state, federal and local agencies, private concerns, and the general public,
(2) to implement the provisions of the plan by utilizing the existing authority of the various state and local entities or agencies concerned,
(3) to weigh the policies and programs of agencies that affect the lake to ensure their compatibility with the adopted comprehensive plan,
(4) to revise and update the plan at periodic intervals
(5) to employ assistants and advisors deemed necessary for the purposes of the act,
(6) to initiate studies of the lake and its related resources
(7) to publish or authorize the publication of scientific information,
(8) to define the lake's flood plain,
(9) to qualify for, accept and administer loan payments, grants, gifts, loans or other funds for carrying out any functions under the act,
(10) to determine the need for and desirability of public works and utilities for the lake area,
(11) to cooperate with the State Engineer and all upstream entities in considering the water relationship between the lake and its tributaries,
(12) to perform all other acts reasonably necessary to carry out the purposes and provisions of the act.

Comprehensive Management Plan (1976)

Under the direction of House Bill Number 23, the Division of the Great Salt Lake began preparation of a comprehensive management plan in July of 1975. The plan was developed through the efforts of an Inter-Agency Technical Team that was established under the terms of the 1975 enabling legislation. The Inter-Agency Technical Team was made up of representatives of a variety of interests, public and private, and included representatives from several divisions of DNR, the Department of Transportation, county commissioners from the five counties surrounding the lake, and other representatives (primarily industry)(Burnham, 1980).

The Comprehensive Management Plan for Great Salt Lake (Burnham, 1976) was intended to serve as a general statement of objectives for use and management of the lake. Goals and policies based on the concepts set forth in the legislation, and as adopted by the Great Salt Lake Board, served as a guide for preparation of the plan. The plan covers six major elements: minerals, recreation, tourism, wildlife, transportation, and hydrology.

The minerals element focuses primarily on the removal of minerals from the lake brines, but also addresses mineral development's interrelationships with other planning considerations.

The wildlife element consists of recommendations regarding commercial brine shrimping, waterfowl management areas, and Great Salt Lake islands. The recreation element of the plan recommends that Antelope and Fremont Islands be acquired for public recreation, and that the South Shore, Promontory Point marina, and Little Valley harbor be publicly owned, and that they be developed by public or private enterprise. It is also recommended that Stansbury Island, Black Rock Canyon, Lake Point, and the Farnsworth Peak tramway areas be committed to public uses through private or public ownership and development. The tourism element of the plan recommends that development of facilities for the recreational, educational, historical, and view sites of the lake be oriented towards the needs of tourists, and that the lake be managed as a preferred tourist destination. The transportation element designates a network of different modes of transportation for both lake visitors and commerce. Included were air, boat, highway, pipeline, and rail terminals. The hydrology element addresses the hydrologic system of Great Salt Lake, which is composed of a network of streams and rivers that collect runoff from throughout a closed drainage basin and deliver it to the lake.

The management plan for each of the elements was developed after careful consideration of the interrelationship of each plan element with the others. The plan was recommended as a general guide for development rather than a detailed development plan for private agencies, or for divisions of local, state, or federal government (Burnham, 1976).

Great Salt Lake Environs Report (1976)

A companion report to the 1976 Comprehensive Management Plan, the Great Salt Lake Environs Report (Millard, 1976), was prepared at the same time. The purpose of the report was to summarize and graphically portray the most current, and reliable, data available concerning land use, land ownership, soils, vegetation, man-made structures, access ways, fresh water, and the utilities lying between the water's edge, as of January 1, 1976, and the upper limits study line established at approximately the 4,212-foot level.

Division of State Lands and Forestry (1979)

In 1979, the Division of Great Salt Lake was eliminated by the legislature, and the staff functions for the management of Great Salt Lake were transferred to DNR. Later, management responsibilities were administratively delegated to the Division of State Lands and Forestry (Flandro, 1995).

Great Salt Lake Contingency Plan (1983)

In 1983, the water level of Great Salt Lake began a rapid rise which prompted the Division of State Lands and Forestry to draft the Great Salt Lake Contingency Plan (Elmer, 1983). This plan was designed to meet a legislative mandate to maintain the level of Great Salt Lake below an elevation of 4,202 feet. It sets forth background, analysis and recommendations for influencing both the high and low water levels of Great Salt Lake. The conclusions of the Plan state: "It is anticipated that lake levels will peak at approximately 4,203 feet in 1983 with potential resultant damages of $20 to $30 million." Ironically, the lake level peaked at approximately 4,205 feet in 1983, and continued rising to an elevation of nearly 4,212 feet in 1987; estimated capital damages exceeded $250 million (Bureau of Economic and Business Research, 1983).

Beneficial Development Area (1985)

During the 1980s flooding, the concept of a Beneficial Development Area (BDA) was formed. The Great Salt Lake Technical Team, in its April 18, 1985 meeting, recommended that (1) the elevation 4,217 feet should be the basis for planning by the state and other entities in relationship to activities around the lake, and (2) the state should work with counties to define the flood plain and to work out strategies for managing it as a hazard zone. In the1985 Hazard Mitigation Plan for Utah (Utah Division of Comprehensive Emergency Management, 1985), the concept of the BDA was introduced, "...as a method for developing hazardous areas around lake shores (to an elevation of 4,217 feet for Great Salt Lake) to the maximum prudent use for the people of the state of Utah, while avoiding unnecessary disaster losses." Of the five counties bordering Great Salt Lake, only Salt Lake and Davis Counties have adopted the BDA concept as part of their master plans (Fred May, Division of Comprehensive Management, verbal communication, November, 1999).

General Management Plan (1987)

As Great Salt Lake reached its historic high water level of 4,211.85 feet in 1987, a five-year General Management Plan, Great Salt Lake (1987-1992) was prepared for the Great Salt Lake Advisory Council. The General Management Plan was "designed to assist decision makers by providing analyses of situations and alternatives, and by making recommendations to meet identified needs, solve problems, resolve issues, achieve goals, and otherwise prepare for the foreseeable future" (Elmer, 1987). The General Management Plan, and the BDA concept, were cooperative outlines of the best strategies for land uses near the lake under 1987 and anticipated future conditions.

Great Salt Lake Advisory Council (1988)

In 1988, the Great Salt Lake Advisory Council (GSLAC) was created by the Legislature to advise the Board of the Division of State Lands and Forestry, which was the division-designated manager of the lake at that time. The Great Salt Lake Technical Team (GSLTT), a successor in composition and function to the Inter-Agency Technical Team, was designated by the 1988 legislation to the support of the division in pursuit of its lake-management activities.

Division of Sovereign Lands and Forestry, Sovereign Lands Advisory Council (1994)

In 1994, the Utah Legislature removed management of state school trust lands from the purview of DNR, and created a separate management entity named the School and Institutional Trust Lands Administration. In the same act, the Board of State Lands and Forestry and the GSLAC were legislatively replaced by the Sovereign Lands Advisory Council (SLAC), which was created to advise the newly named Division of Sovereign Lands and Forestry (DSLF) of DNR.

With these changes, DSLF was able to concentrate its planning and management activities on public-trust lands, leaving the management of institutional trust lands to the newly created agency. The broader public-trust management responsibilities brought renewed emphasis on how the lake's non-economic resources were interrelated with its mineral and economic resources. In 1996, the name of the division was changed to the Division of Forestry, Fire and State Lands (DFF&SL).

Great Salt Lake Comprehensive Management Plan (1995)

The Great Salt Lake Technical Team (GSLTT) prepared the Great Salt Lake Comprehensive Management Plan - Planning Process and Matrix (Flandro, 1995) for the division and board of State Lands and Forestry, and DNR. The plan's purposes were "to provide needed information and guidance in the form of recommendations to federal, state and local governments, and recommended legislation to the state legislature to facilitate and enhance management of Great Salt Lake and its environs to assure protection of the unique ecosystem of the lake while promoting balanced multiple-resource uses." The plan recites the legislative policies set forth in Section 65A 10-8 of the Utah Code as follows:

> (1) develop strategies to deal with a fluctuating lake level,
> (2) encourage development of the lake in a manner which will preserve the lake, encourage availability of brines to lake extraction industries, protect wildlife, and protect recreation facilities,
> (3) maintain the lake's flood plain as a hazard zone,

(4) promote water quality management for the lake and its tributary streams,
(5) promote the development of lake brines, minerals, chemicals, and petro-chemicals to aid the state's economy,
(6) encourage the use of appropriate areas for the extraction of brines, minerals, chemicals, and petro-chemicals
(7) maintain the lake and the marshes as important to the waterfowl flyway system,
(8) encourage the development of an integrated industrial complex,
(9) promote and maintain recreation areas on and surrounding the lake,
(10) encourage safe boating use of the lake,
(11) maintain and protect state, federal, and private marshlands, rookeries, and wildlife refuges, and
(12) provide public access to the lake for recreation, hunting and fishing.

The plan contained a number of recommendations formulated by the GSLTT to address current and anticipated management problems and issues, and directed to federal, local, and state agencies including DNR and its divisions.

Mineral Leasing Plan for Great Salt Lake (1996)

In 1995, the DSL&F announced the withdrawal of all sovereign lands from mineral leasing, including Great Salt Lake, as part of a comprehensive planning process for the management of minerals on public trust lands. To accomplish its planning and management mandates, DSL&F (now DFF&SL) is creating mineral leasing plans for each public trust resource area. The Mineral Leasing Plan for Great Salt Lake (Utah Division of Sovereign Lands & Forestry, 1996) is the first of these plans completed. The plan reviews the history of mineral ownership and leasing, inventories mineral resources, and examines potential and existing conflicts among resource management and uses on the lake. As the result of this analysis, the plan established mineral-production zones with limitations and special conditions appropriate to protect the other trust resources, and establishes new mineral leasing procedures.

OTHER LAKE MANAGEMENT INTERESTS AND AUTHORITIES

The responsibility for overall planning and management for Great Salt Lake currently falls to DFF&SL pursuant to Section 65A-2-2 of the Utah Code. Other state and federal government entities also have regulatory and non-regulatory interests on the lake, which are described in the following sections.

U.S. Army Corps of Engineers

The U.S. Army Corps of Engineers is responsible under Section 404 of the Clean Water Act for regulating placement of fill material and excavation in the nations's waters, including Great Salt Lake. The purpose of the act is to protect the nation's aquatic resources from unnecessary adverse impacts (Dennis Blinkhorn, U.S. Army Corps of Engineers, written communication, 1998).

U.S. Fish and Wildlife Service

The U.S. Fish and Wildlife Service (USFWS) manages the Bear River Migratory Bird Refuge west of Brigham City at the mouth of the Bear River. USFWS is responsible for the protection of threatened and endangered species present in Great Salt Lake environs, and the protection of migratory birds, as specified by treaty.

Utah Division of Oil, Gas and Mining

The Division of Oil, Gas and Mining is the regulatory agency for mineral exploration, development, and reclamation on Great Salt Lake. The division approves development and reclamation plans, issues permits, and inspects mineral development activities for compliance with Utah laws and regulations. This regulatory role is conducted in close coordination with DFF&SL.

Utah Division of Parks and Recreation

The Division of Parks and Recreation manages Antelope Island State Park, and is directly responsible for boating enforcement on Great Salt Lake. Parks personnel also work closely with five county Sheriffs' offices to participate in search and rescue activities on the lake. Search and rescue along the eastern shore is handled in a cooperative effort with Davis and Weber County Sheriffs' offices.

Utah Geological Survey

The Utah Geological Survey (UGS) is responsible for surveying, collecting, preserving, publishing, and distributing reliable information on geology, brine and mineral resources, and geologic hazards for the entire state, including Great Salt Lake. The UGS is also responsible for assisting, advising, and cooperating with state and local agencies and state educational institutions on all subjects related to geology.

Utah Division of Water Quality

The Utah Water Quality Board and the Division of Water Quality are charged to maintain, protect, and enhance the quality of Utah's surface and ground water resources, and to develop programs for prevention and abatement of water pollution. The division's mission is to protect public health and all beneficial uses of water by maintaining and enhancing the chemical, physical, and biological integrity of Utah's waters. All discharges to the lake and its tributaries are regulated and inspected by the division.

Utah Division of Water Resources

The mission of the Utah Board and Division of Water Resources is to direct the orderly and timely planning, conservation, development, protection, and preservation of Utah's water resources to best meet the beneficial needs of the citizens of the state of Utah. Although the division has not been given any direct regulatory responsibility on Great Salt Lake, performance of their mission may affect the lake.

The division collects and analyzes hydrologic information on the lake, and is also responsible for the maintenance and operation of the West Desert Pumping Project.

Utah Division of Water Rights

The Division of Water Rights regulates the appropriation and distribution of water in the state of Utah. The State Engineer, who is the director of this division, must give approval for the diversion and use of any water, regulates the alteration of natural streams, and has the authority to regulate dams for the purpose of protecting public safety. All diversions of lake brines for mineral extraction and other purposes require prior approval by the State Engineer. Any dam or dike placed in or around Great Salt Lake also requires prior approval from the division or State Engineer.

Utah Division of Wildlife Resources

The division's duties are to protect, propagate, manage, conserve and distribute protected wildlife throughout the state, including regulating the harvesting of brine shrimp from Great Salt Lake. The legislature has authorized the division to use all or parts of 38 townships below the surveyed meander line for the "creation, operation, maintenance and management of wildlife management areas, fishing waters, and other recreational activities." The division manages wildlife and waterfowl refuges and management areas on the lake, and regulates hunting of terrestrial wildlife in the lake environs.

THE GREAT SALT LAKE PLANNING PROJECT (1997)

Great Salt Lake is a navigable body of water. Its bed is sovereign land, the title to which passed to the state from the Federal Government upon statehood. Sovereign lands are managed under the Public Trust Doctrine, which holds that navigable waters are owned by the public, but held in trust by the state for the benefit of the public. The purposes of the trust were originally characterized as commerce, navigation and fishery, in recognition of the use of major waterways as highways of commerce. As other public values increased in importance, courts and legislatures expanded the scope of the trust to include benefits such as economic development, recreation, and environmental protection.

In August 1997, DNR and DFF&SL initiated the Great Salt Lake Planning Project to articulate the department's management objectives for Great Salt Lake and to reconcile the diverse resource management mandates of the divisions of the department. The purposes of the Great Salt Lake Planning Project are:

(1) to establish unifying DNR management objectives and policies for Great Salt Lake public trust resources,
(2) to coordinate the management, planning and research activities of department divisions on Great Salt Lake,
(3) to coordinate management with the actions of land and resource owners and managers on and adjacent to Great Salt Lake,
(4) to develop a sovereign lands and resources management plan, and
(5) to establish processes for plan implementation, monitoring, evaluation and amendment.

The planning project began with internal and external scoping to determine the subjects and issues that the plan should address. Planning team members (1) interviewed state, federal, and local government agency personnel, and (2) conducted public meetings, presentations, and meetings with companies, associations, clubs, and individuals interested in Great Salt Lake resources and management. The team then gathered available information relevant to the subject matter and issues raised during scoping and prepared a draft "Statement of Current Conditions and Trends" as a background for comment and suggestions by interested parties. The team then developed a series of alternative management scenarios for public review and comment. A second series of public meetings was held in the five counties around the lake to present the alternative management scenarios and invite public comment on the alternatives. Due to intensive public interest, and the potential effects on economic interests resulting from management actions on the lake, DNR sought verification and validation of the scientific information used by the planning team. A scientific review panel from the academic sector, independent of the department, was assembled for this purpose. The planning team considered the comments received from the public, DNR leadership, and the science review panel to develop the preferred alternatives presented in a draft plan for public review and comment.

With approval of the new plan, division-specific implementation plans are being developed to accomplish the overall plan objectives. DNR has established an administrative structure to review proposals for consistency with the plan, to consider plan amendments, and to accomplish the intra- and inter-agency coordination elements of the plan.

This planning project served as a possible model for future departmental collaboration on other sovereign lands. The process included several steps not required by statute or rule, and not used in previous plans. Additional steps included the distribution of the draft Statement of Current Conditions and Trends, distribution of the array of alternatives, and the second round of public meetings. These additional steps were intended to increase public knowledge about the plan and stimulate public involvement.

CONCLUSIONS

Through legislative and administrative actions, management responsibility for Great Salt Lake has shifted over time from agency to agency. Management plans prepared by the different agencies have reflected the different philosophies of resource preservation and resource development prevalent at the time. As understanding of the implications of the state's responsibility under the Public Trust Doctrine has increased, the management plans have reflected greater appreciation of natural systems. Development-oriented plans have given way to more conservation-oriented management, and less irretrievable commitment of resources. The DNR's attention and commitment to the planning process, as a collaborative priority within the department, demonstrates this continuing evolution.

REFERENCES

Bureau of Economic and Business Research, 1983, Actual and potential damages incurred by industries, governments, and other lake users due to the rising level of the Great Salt Lake: Bureau of Economic and Business Research, University of Utah, 118 p. (variously paginated).

Burnham, O.W., 1976, Great Salt Lake Comprehensive Plan: Utah Department of Natural Resources, Division of Great Salt Lake, 69 p.

—1980, The Great Salt Lake comprehensive plan, *in* Gwynn, J.W., editor, Great Salt Lake - a scientific, historical and economic overview: Utah Geological and Mineral Survey Bulletin 116, p. 47-51.

Dewsnup, R.L., and Jensen, D.W., 1980, Legal battle over ownership of the Great Salt Lake, *in* Gwynn, J.W., editor, Great Salt Lake - a scientific, historical and economic overview: Utah Geological and Mineral Survey Bulletin 116, p. 15-18.

Elmer, Stan, 1983, Great Salt Lake Contingency Plan: Utah Department of Natural Resources, Division of State Lands and Forestry, 47 p.

—1987, General Management Plan - Great Salt Lake, 1987-1992: Utah Department of Natural Resources, Division and Board of State Lands and Forestry, 55 p.

Flandro, S.C., 1995, Great Salt Lake Comprehensive Management Plan - planning process and matrix: Utah Department of Natural Resources, Division of Forestry, Fire and State Lands, 118 p.

Mahoney, J.R., 1966, The rights of states to beds of their navigable lakes and streams, and some special problems of determining boundaries establishing Utah's claim to the Great Salt Lake, *in* Stokes, W.L., editor, Guidebook to the geology of Utah: Utah Geological Society, p. 165-173.

Millard, K.R., 1976, Great Salt Lake environs report: Millard Consultants, 2200 South 9th East, Salt Lake City, Utah, prepared for the [Utah] Division of Great Salt Lake, 30 p.

Moss, Frank E., 1965, "Testimony" at Hearings before the subcommittee on public lands of the committee on interior and insular affairs, United States Senate, eighty-ninth congress, first session on S. 265: U.S. Government Printing Office, Washington, D.C., 146 p.

Richards, A.Z., 1965, A preliminary master plan for the development of the Great Salt Lake over a period of the next 75 years: Caldwell, Richards & Sorensen, Inc., Salt Lake City, Utah, prepared for the Great Salt Lake Authority, 32 p.

Utah Division of Comprehensive Emergency Management, 1985, Hazard Mitigation Plan, Utah - 1985: Utah Division of Comprehensive Emergency Management, 20 p.

Utah Division of Sovereign Lands & Forestry, 1996, Mineral Leasing Plan, Great Salt Lake, Final: Division of Sovereign Lands & Forestry, 49 p.

HISTORY OF STATE OWNERSHIP, RESOURCE DEVELOPMENT, AND MANAGEMENT OF GREAT SALT LAKE

*Edie Trimmer and Karl Kappe
Division of Forestry, Fire and State Lands
Utah Department of Natural Resources
1594 West North Temple
Salt Lake City, UT 84114-5703
*etrimmer@state.ut.us

ABSTRACT

Utah statute defines sovereign lands as "those lands lying below the ordinary high water mark of navigable bodies of water at the date of statehood, and owned by the state by virtue of its sovereignty." The lands within the bed of Great Salt Lake (GSL) are, by this definition, sovereign lands, acquired at statehood in 1896 in accordance with the "equal footing" doctrine, granting each state control and ownership of navigable waters and the lands underneath those waters within its borders. Under public trust doctrine, the State, as trustee for the people, bears responsibility for preserving and protecting the right of the public to use of the waters for navigation, commerce, fishing, recreation, and wildlife habitat.

Also by statute, sovereign lands are defined as "state lands," to be managed by "multiple-use sustained-yield principles." The Division of Forestry, Fire and State Lands is given management authority for sovereign lands and, as manager, has responsibility to prepare comprehensive plans, initiate studies of the lake and its resources, implement comprehensive plans through state and local entities, and coordinate the activities of various divisions within the Department of Natural Resources (DNR). The Division of Forestry, Fire and State Lands also has responsibility for management of mineral leasing on sovereign lands. The many resources on the lake--water, minerals, wildlife, recreation, archeological and historical values--are managed by as many state agencies which occasionally creates conflicts.

The brines of GSL contain several ions that crystalize into valuable minerals during evaporation. The major ions in the lake are, in order of relative abundance, chloride, sodium, sulfate, magnesium, and potassium. Mineral products which are currently extracted from lake brines are sodium chloride, magnesium chloride brine which can be sold as flake magnesium chloride or further processed into magnesium and chlorine gas, and potassium sulfate. Mineral products which have potential for extraction include gypsum, sodium sulfate, and trace amounts of lithium, boron, and bromine.

The GSL contained an estimated 4.3 billion short tons (st) (3.9 billion metric tons [mt]) of dissolved salts in 1998. Utah Geological Survey (UGS) estimates of the dissolved salt content in GSL have fluctuated from 4.0 to 5.5 billion st (3.6 to 5.0 billion mt) due to the dynamic conditions in the lake as salts are precipitated and redissolved, and due to the diversion of brines from GSL, such as the West Desert Pumping Project. The lake has four areas of varying salinity, separated by dikes or other man-made structures: north arm and Stansbury Bay brines at near saturation (25 to 27 percent total dissolved solids [TDS]); the main body of the south arm with concentrations ranging from 7 to 15 percent TDS as lake elevations fluctuate; the waters in Farmington Bay at approximately 3 to 5 percent TDS; and Bear River Bay at <1 to 7 percent TDS. The percent TDS in Bear River Bay fluctuates with lake level, and changes in Bear River inflow. The transfer of salts from the south arm to the north arm has raised questions about the viability of the mineral and brine shrimp industries. The UGS and the U.S. Geological Survey (USGS) continue to monitor salinities at designated sites on the lake to document changing lake salinity. A recurrent theme is that placement of dikes and diversions can have significant and rapid impacts on various conditions in the lake.

Hydrocarbon resources on the lake are significant, but presently undeveloped. The hydrocarbons are low gravity (4 to 9 degree API) and tar-like, contain high nitrogen concentrations, and up to 12 percent sulfur. The unusual characteristics of the oil have been the subject of studies by chemists at Weber State University and Université Louis Pasteur de Strasbourg. However, these resources are difficult, and at present, uneconomic to extract using current technology because of the nature of the hydrocarbons, and production in "an offshore, highly saline environment."

Oolitic sand deposits make up many of the beaches and shorelines around the lake. Because of their high calcium carbonate content, oolites have been used by Magnesium Corporation of America (Magcorp) and its predecessors for acid neutralization and dike construction. Oolites are also used in very minor amounts in flower drying. The Utah Division of Oil, Gas and Mining reports up to 130,000 st (118,000 mt) mined annually by Magcorp from U.S. Bureau of Land Management (BLM) lands adjacent to GSL.

Currently, there are twelve producing mineral leases which generated slightly more than $1,000,000 in royalties during calendar year 1998. IMC Kalium Ogden Corp. (IMC Kalium) produces potassium sulfate and magnesium chloride from brines concentrated through solar evaporation in Bear River Bay and Clyman Bay. By-product sodium chloride is transferred to IMC Salt, which packages and sells the salt. MagCorp produces magnesium metal from brines concentrated in Stansbury Bay. Cargill Salt produces sodium chloride from brines provided by Magcorp under a lease agreement. Morton Salt produces salt at the southeast end of Stansbury Island. Lastly, North Shore Limited produces concentrated brines for use in dietary and mineral/vitamin supplements near Spring Bay in the north arm of the lake.

Producers of magnesium, potash, and salt from GSL contribute significantly to the value of metals and industrial minerals in Utah. Together these companies contribute approximately $240 million in gross value, or 18 percent of the value of the state's nonfuel mineral production. Most of this production is exported.

INTRODUCTION

The U.S. government acquired lands that are now Utah from Mexico through treaty following cessation of war with Mexico on February 2, 1848, six months after the arrival of the first Mormon pioneers into Salt Lake Valley. Congress created the Utah Territory from those lands on February 21, 1855. As a territory, these lands were managed for disposal by the General Land Office which sold or granted lands for railroads, Native American allotments and reservations, forest lands designation, homesteading and mineral entries, and land grants to the territorial government.

The area had been explored by American, French, and British fur trappers starting in 1829. Captain John Charles Fremont led an exploration, authorized by the U.S. Congress in 1843, of the Great Basin, including GSL. In 1849-50, Captain Howard Stansbury, as part of the U.S. Army Corps of Topographical Engineers, was in charge of a "trigonometrical and nautical survey" of GSL.

Captain Stansbury's observations in his accounts of the survey range between fascination with the vastness of the lake and disappointment in its desolation. He gives this account soon after his arrival at Bear River Bay on October 22, 1849:

> Morning clear and calm. The Salt Lake, which lay about half a mile to the eastward, was covered by immense flocks of wild geese and ducks, among which many swans were seen, being distinguishable by their size and the whiteness of their plumage. I had seen large flocks of these birds before, in various parts of our country, and especially upon the Potomac, but never did I behold anything like the immense numbers here congregated together. Thousands of acres, as far as the eye could reach, seemed literally covered with them, presenting a scene of busy, animated cheerfulness, in most graceful contrast with the dreary, silent solitude by which we were immediately surrounded (Stansbury, 1852).

His observations from Promontory Point that evening describe his ambivalence about GSL:

> The evening was mild and bland, and the scene around us one of exciting interest. At our feet and on each side lay the waters of the Great Salt Lake, which we had so long and so ardently desired to see. They were clear and calm and stretched to the south and west...On the west appeared several dark spots, resembling..islands, but the dreamy haze hovering over this still and solitary sea threw its dim, uncertain veil over the more distant features of the landscape...The stillness of the grave seemed to pervade both air and water; and, excepting here and there a solitary wild-duck floating motionless on the bosom of the lake, not a living thing was to be seen. The night proved perfectly serene, and a young moon shed its tremulous light upon a sea of profound, unbroken silence....The bleak and naked shores, without a single tree to relieve the eye, presented a scene so different from what I had pictured in my imagination of the beauties of this far-famed spot, that my disappointment was extreme (Stansbury, 1852).

HISTORY OF STATE OWNERSHIP OF GREAT SALT LAKE

What Are Sovereign Lands?

At statehood, the State was not only granted school trust lands (four sections out of every township) for support of its schools and institutions, but also sovereign lands (lands under navigable waters). These sovereign lands were to be managed by the states for public trust purposes such as navigation, commerce, and fisheries. In southwestern states, there has been slow recognition of management responsibilities for sovereign lands, where water for diversion and irrigation were far more important than preserving a waterway for these public trust purposes. The first chapter of "The Great Salt Lake," describes the attitudes toward this lake: "Lake of paradoxes, in a country where water is life itself and land has little value without it, Great Salt Lake is an ironical joke of nature--water that is itself more desert than a desert" (Morgan, 1947).

Utah statute defines sovereign lands as "those lands lying below the ordinary high water mark of navigable bodies of water at the date of statehood and owned by the state by virtue of its sovereignty" (Utah Code Ann., Section 65A-1-1(5)). The lands within the bed of GSL are by this definition sovereign lands, acquired at statehood in 1896 in accordance with the "equal footing" doctrine granting each state control and ownership of navigable waters and the lands underneath those waters within its borders. These lands are managed according to the doctrine of public trust, a system of court-interpreted common law dating back to the sixth century Roman law as codified in Institutes and Digests of Justinian (Coastal States Organization, 1997).

The BLM and its predecessor the General Land Office were responsible for surveying the public domain so that land could be conveyed into various ownerships. Unappropriated lands remained under the management of the federal government. The BLM's 1973 Manual of Surveying Instructions notes: "Beds of navigable bodies of water are not public domain and are not subject to survey and disposal by the United States. Sovereignty is in the individual states. Under the laws of the United States the navigable waters have always been and shall forever remain common highways" (U.S. Bureau of Land Management, 1973).

Public trust principles guide the management of sovereign lands. This doctrine has evolved over several centuries. Black's Law Dictionary (1990) defines public trust doctrine in terms of the state's responsibility, as trustee for the people, to preserve and protect the submerged or submersible lands for public use in navigation, fishing, and recreation. There have been rulings by the courts which expand the definition of public trust uses on submerged lands to mean protection of visual, wildlife, and open-space values for the benefit of all the state's citizens. Coastal States Organization (1997) asserts the public trust doctrine "provides that public trust lands, waters, and living resources in a state are held by the state in trust for the benefit of all of the people, and establishes the right of the public to fully enjoy public trust lands, waters, and living resources for a wide variety of recognized public uses." The living resources (for example, the fish and aquatic plant and animal life) inhabiting these lands is also subject to the Public Trust Doctrine (Coastal States Organization, 1997).

Utah Statutes Defining Sovereign Lands and Public Trust

Each state, through its constitution and statutes, determines how the multiple resources within sovereign lands are to be managed for the public trust. There exists what the Coastal States Organization (1997) refers to as a "pyramid of authority over navigable waters." At the top, and operating within the narrow scope of 'improvements to navigation' is the federal navigational servitude. Next is the state authority, as trustee, to manage its trust lands, waters, and resources for the benefit of the public's various trust uses, including the authority to reasonably regulate riparian rights, or to deny them altogether. Finally, "riparian owners have certain rights..." In this context, the 'federal navigational servitude' is defined as the "dominant servitude over navigable waters. 'The right of the United States in the navigable waters within the several States is, however, limited to the control thereof for the purposes of navigation' (Coastal States Organization, 1997).

Utah's constitution does not address sovereign or public trust lands directly but rather defines the use of public lands as follows in Article XX, Section 1:

> All lands of the State that have been, or may thereafter be granted to the State by Congress, and all lands acquired by gift, grant, or devise, from any person or corporation, or that may otherwise be acquired, are hereby accepted, and, except as provided in Section 2 of this article, are declared to be public lands of the State; and shall be held in trust for the people, to be disposed of as may be provided by law, for the respective purposes for which they have been or may be granted, donated, devised, or otherwise acquired (Utah Constitution, Article XX, Section 1, 1896).

By statute, sovereign lands are defined as "state lands," to be managed by "multiple-use sustained yield principles" (Utah Code Ann., Section 65A-2-1). 'Multiple use' is defined as "management of various surface and subsurface resources in a manner that will best meet the needs of the people of this state"(Utah Code Ann., Section 65A-1-1(3)). 'Sustained yield' is defined as "the achievement and maintenance of high-level annual or periodic output of the various renewable resources of land without impairment of the productivity of the land" (Utah Code Ann., Section 65A-1-1(7)).

The Division of Forestry, Fire and State Lands (DFF&SL) has been given management authority for sovereign lands and, as manager, "may exchange, sell, or lease sovereign lands but only in the quantities and for the purposes as serve the public interest" (Utah Code Ann., Section 65A-10-1). Sales and exchanges have been few in number compared to other dispositions. Utah statute further defines powers and duties for management of sovereign lands within GSL. These duties include preparing comprehensive plans, initiating studies of the lake and its resources, defining the lake's flood plain, determining the need for public works and utilities, implementing comprehensive plans through state and local entities, coordinating the activities of various divisions within the DNR, and encouraging the continued activity of the GSL technical team.

Just as public trust doctrine has evolved over several centuries and has been interpreted by the laws and customs of several governments, the management directives for sovereign lands in Utah have also changed over the years since statehood. Nevertheless, a common thread is that, from statehood, the Utah State Land Board (Land Board) was intended to be a "conservation organization, empowered to protect vital watersheds " in its management of timber and rangelands (Smith, 1960).

Management and disposition of minerals initially was a low priority relative to agricultural uses and water rights on, or adjacent to, sovereign lands. The Land Board legislation in 1917 allowed the sale of submerged lands but only if lakes or waterways were dewatered "to reclaim the bed thereof for agricultural purposes..." (Utah State Legislature, 1917, Chapter 114). By 1925, submerged lands could also be sold if riparian landowners had made valuable improvements below the water's edge, but with mineral rights to be reserved to the state (Utah State Legislature, 1925, Chapter 31). In 1929, the Utah Legislature granted use of some sovereign land in Bear River Bay to the United States for use as a migratory bird refuge. In 1931, the Land Board was authorized to make surveys of lands for flood control. This charge to DFF&SL is still in effect, although flood control activities were given to the State Engineer in 1936 (Smith, 1960).

Development of Management Principles for Sovereign Land

In 1933, legislation was added to the Utah Code which began to define public trust goals for the state's sovereign lands. This legislation allowed sovereign lands to be sold for "public or quasi-public use or service" (Utah State Legislature, 1933) as long as such sales did not interfere with navigation. In 1956, sovereign land near Little Mountain in Weber County was sold to Marquardt Aircraft Company, presumably for national defense purposes. In 1957, land was sold to Weber Basin Water Conservancy District; Willard Bay Reservoir now covers this land. In 1973 and 1982, parcels were sold to the town of Perry for its sewage lagoons. Both of these sales include reversionary interest clauses through which title reverts to the state if the land is not used for a public purpose. In 1984 and 1991, land exchanges were consummated to resolve ownership disputes in Farmington Bay and Bear River Bay.

The 1982 sale to the town of Perry was the last sale of sovereign land. The administrative rule for the sale of land has since lapsed. Any future sale proposal will be closely scrutinized from a public trust perspective, and rulemaking will be required. Rules are in place for leases, exchanges, and other dispositions.

With exception of the Bear River Migratory Bird Refuge and state waterfowl management areas, most early non-mineral development on the bed of GSL occurred during the heyday of recreational resort development. Much of this development took place before there was an administrative framework for sovereign land management, and probably with some confusion over the nature of riparian rights or the lack thereof. It was not until the late 1950s that the Land Board required formal applications to request special-use leases for the kinds of development that had occurred earlier.

The earliest lease applications were for a boat harbor, public resorts, a fresh-water, artificial lake, and the use of mill tailings in land reclamation. Some of these proposed projects were studied, but not constructed. Others flourished briefly, but eventually succumbed to the changing level of GSL, fires, or the changing recreational pursuits of an increasingly mobile society. Today, the only remaining early developments are the GSL Boat Harbor, owned by the Utah Division of Parks and Recreation, and the privately owned Saltair Resort.

U.S. Supreme Court Decision

In the early 1960s, as interest grew in non-salt minerals in the lake, BLM served notice on the Land Board that it intended to survey a boundary line along the GSL to separate state and federal ownership, and that it would locate such boundary line at an elevation of 4201.8 feet (1,280.7 m) above mean sea level (m.s.l.), which was the same elevation as the water level on January 4, 1896, when Utah obtained statehood. Utah objected because it believed that the State owned the lake, the water-covered bed, and the shore lands located within the surveyed meander line as officially surveyed and approved by the U.S. Government during the 18 surveys performed from 1855 through 1966 (Dewsnup and Jensen, 1980). The lake level was at many different elevations during that 111-year period.

In 1975, after nearly 15 years of congressional and legal proceedings, the U.S. Supreme Court ruled to affirm "in Utah, ownership of all lands, brines, and other minerals within the waters of the lake and within the bed and all shore lands located within the official surveyed meander line as duly surveyed prior to or in accordance with Section 1 of the Act of June 3, 1966, 80 Stat. 192." (Utah vs. United States, 1975). The Supreme Court's final decree did not address the title to lands within the Bear River Refuge, the Weber Basin Federal Reclamation Project, and the Hill Air Force Range as bounded by the water's edge June 15, 1967.

Ownership of Islands in Great Salt Lake

Utah considers unsurveyed islands as being sovereign lands while surveyed islands are owned by the "upland" land owner. Surveyed islands in the lake are Antelope, Fremont, and Carrington Islands (see figure 1). Unsurveyed islands are Gunnison/Cub, Dolphin, Egg, Goose, Hat, and Badger Islands. At the time of the U.S. Supreme Court decision, significant parts of Gunnison/Cub and Hat Islands were in private ownership. Following legislation passed in 1977, lands in private ownership on Gunnison/Cub and Hat Islands were purchased by the state and subsequently designated as wildlife management areas for the protection of the American White Pelican.

MANAGEMENT OF GREAT SALT LAKE RESOURCES

The lake's many resources; water, minerals, wildlife, recreation, archeological, and historical values, are managed by several state agencies. Examples of conflicts involving one or more state agency include: conflicts in water requirements among mineral operations and wildlife; conflicts between recreational use (for example, access, water use, and visual impacts) and issuance of mineral leases; conflicts between recreational use and wildlife use, especially during breeding and nesting seasons; disturbances or hazards to wildlife by mineral development or commercial activities, especially in areas owned by the Division of Wildlife Resources (DWR) or authorized for use as wildlife management areas, fishing waters, and recreational activities by Utah Code Ann., Section 23-21-5.

Several islands have been acquired by state agencies to be dedicated to wildlife management or recreational use. The DWR purchased Gunnison/Cub and Hat Islands, and the Division of Parks and Recreation purchased Antelope Island. In addition, the Land Board withdrew areas around Hat and Dolphin Islands from mineral leasing (Utah State Land Board minutes, 1976) and dedicated sovereign lands below Harold Crane Wildlife Management Area (WMA) to wildlife use (Utah State Land Board minutes, 1982). Table 1 lists these areas and figure1 shows their location on GSL.

Current Management of Sovereign Lands

In 1988, the management direction for sovereign lands was changed by the state legislature to ".....sell or lease sovereign lands but only in the quantities and for the purposes as serve the public interest and do not interfere with the public trust." (Utah Code Ann., Section 65A-10-1). 'Public trust assets' were defined as lands and resources, including sovereign lands, administered by the division that are not part of the school or institutional trust lands. Specific management responsibilities are set forth in statute for sovereign lands within GSL. Statute allows the state as represented by DFF&SL to exchange, sell or lease sovereign lands, to set aside sovereign lands for recreational purposes, and instructs DFF&SL to "develop plans for the resolution of disputes over the location of sovereign land boundaries." In addition, DFF&SL may enter into agreement with state agencies and private parties to establish boundaries (Utah Code Ann., Section 65A-10-3(1-2)).

The DFF&SL was created in 1994 with the responsibility for management and planning of all mineral resource development on sovereign lands, in addition to responsibilities for comprehensive planning and coordination of activities of public and private entities. Management of many of these resources has been delegated to other natural resource agencies; recreation and boating to the Division of Parks and Recreation; the various aspects of water resources to the Divisions of Water Resources, Water Quality, and Water Rights; and wildlife to the DWR.

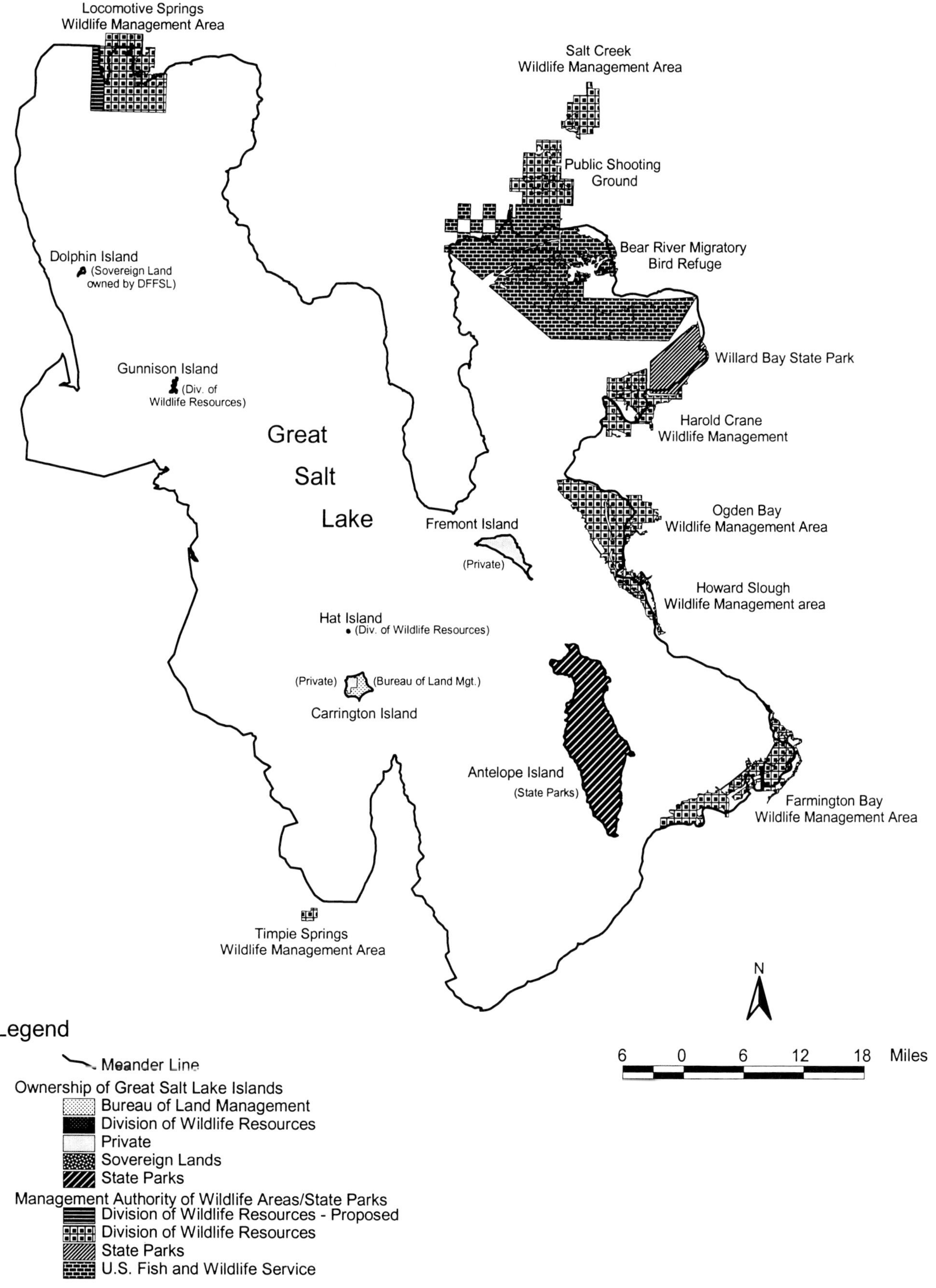

Figure 1. *Ownership or management authority for lands in the vicinity of Great Salt Lake.*

Table 1. Lands in the vicinity of Great Salt Lake owned or managed by state agencies other than the Division of Forestry, Fire, and State Lands. The column "Year" is the year that the Land Board delegated management responsibility.

Agency	Area	Acres	Year
Division of Wildlife Resources	Locomotive Springs	17,937 2,657	1931 Additional acreage proposed
Division of Wildlife Resources	Public Shooting Gnds	13,063	1923
Division of Wildlife Resources	Harold Crane	8,593	1963
Division of Wildlife Resources	Ogden Bay	18,395	1937
Division of Wildlife Resources	Howard Slough	3,300	1958
Division of Wildlife Resources	Layton-Kaysville	25,000	1975
Division of Wildlife Resources	Farmington Bay	10,772	1935
Division of Wildlife Resources	Bird (Hat) Island	22	1977
Division of Wildlife Resources	Gunnison Island	163	1977
Division of Wildlife Resources	Dolphin Island	624	1977
Division of Wildlife Resources	West Bear River Bay	>5	Proposed for public access
Division of Parks and Recreation	Antelope Island	28,022	1969/1981
Division of Parks and Recreation	South Shore	5,874	1977

Planning Efforts for Great Salt Lake

Sovereign lands within GSL are the largest contiguous area (approximately 1.35 million acres [0.547 million ha] within the surveyed meander line) to be managed by the State of Utah. These "lands" are part of a hypersaline lake rich in mineral resources, wildlife ("ornithologically the most impressive salt lake on the continent"- Jehl, 1994), recreational values, and vast expanses of view and water. Management plans were prepared for GSL in 1976 and 1987, however, planning for mineral resources was not fully incorporated into those plans because mineral leasing was administered by the Division of State Lands and Forestry while planning and coordination was done by the Division of Great Salt Lake (1976 to1979) and then by DNR (1980 to 1988).

Three levels of planning satisfy the statutory requirement for the management of sovereign lands. The most specific planning involves preparation of a record of decision and usually requires a site-specific plan. The next level is resource management planning, such as the 1996 Mineral Leasing Plan for Great Salt Lake. The broadest level of planning is a comprehensive management plan, such as the Great Salt Lake Planning Project initiated in 1997 and completed in 1999.

The DFF&SL, with the approval of the executive director of DNR and the governor, may set aside sovereign land for public or recreational use. Management authority for those lands may be delegated to any state agency. In 1977, the Land Board exercised this authority for the creation of Great Salt Lake State Park and delegated management authority to the Division of Parks and Recreation. The sovereign land included in this action extended one mile (1.6 km) lakeward from the Meander Line, generally between the Goggin Drain and Black Rock. The state park existed until 1997 when a partial recission of the delegation of management authority was executed to return management authority for most of the sovereign land to DFF&SL. This was done after the Division of Parks and Recreation reconsidered the suitability of the land as a state park and decided it was not suitable park land. This decision was based, in part, on the difficulty of dealing with the fluctuating level of GSL - the same difficulty encountered by resort owners decades earlier. In what was the state park, the Division of Parks and Recreation retained management authority for only the Great Salt Lake Marina. A similar delegation of management authority was executed for sovereign land around Antelope Island State Park in 1985. The Division of Parks and Recreation manages an irregular boundary, generally extending one mile (1.6 km) lakeward, around the island.

The DFF&SL has established general statewide sovereign land classifications through administrative rule:

Class 1. Lands managed to protect existing resource development uses,
Class 2. Lands managed to protect potential resource development options,
Class 3. Lands managed as open for consideration of any use,
Class 4. Lands managed for resource inventory and analysis,
Class 5. Lands managed to protect potential resource preservation options, and
Class 6. Lands managed to protect existing resource preservation uses.

These classifications were applied to GSL in 1995, and were reviewed under the current (2000) Great Salt Lake Planning Project. Lands under these classifications will change over time in response to changes in public demand, and changes in legislative and administrative policy.

In addition to the classification of lands, several easement corridors have been created to expedite commerce on sovereign lands within GSL. The main north-south easement corridor is along the east side of the lake, primarily for utility lines. The main east-west corridors are the two Union Pacific Railroad causeways. Major easements in the future will be directed to these corridors to attempt to lessen the cumulative impacts on sovereign lands.

Section 23-21-5 Lands

The Utah Legislature has authorized the DWR to use all or parts of 39 townships of GSL for the creation, operation, maintenance, and management of wildlife areas, fishing waters, and other recreational activities (Utah Code Ann., Section 23-21-5). This geographic area covers Bear River Bay, Ogden Bay, Farmington Bay, portions of the south shore area, and the north end of Spring Bay. This statutory authorization is interpreted as establishing wildlife management and wildlife-related recreation as the primary intended land uses, except for areas identified for other uses through a planning process. Land uses with significant adverse impacts on wildlife and recreation values may be prohibited, even though mitigation strategies are available. Within this area, the DWR established the Harold Crane WMA, Howard Slough WMA, Ogden Bay WMA, Farmington Bay WMA, and Locomotive Springs WMA (figure1).

Use of this area for wildlife is influenced by the level of GSL. At high lake levels, WMA dikes may be overtopped and fresh water impoundments lost to salt water inundation. At low lake levels, expansive mud flats are exposed lakeward of the WMA dikes. Wildlife species using the area change accordingly. Extensive non-wildlife developments such as IMC Kalium's evaporation ponds have been permitted. The record is not clear on the extent of consultation with the DWR when mineral leases were issued in the early 1960s, but by the late 1960s it is clear that the Land Board was concerned that consultation should occur before additional mineral leasing. IMC Kalium and the DWR and DFF&SL have recently cooperated as opportunities arise to relocate undeveloped lease acreage to relatively less sensitive areas of GSL, primarily its west side.

Mineral Leasing Plan for Great Salt Lake

The DFF&SLs' focus in the past has been on management and leasing of minerals and rights-of-way on sovereign lands. With regard to mineral extraction, DFF&SL has several directives under Utah Code, Section 65A-10-18. These directives are:

(1) Encourage development of the lake in a manner which will preserve the lake;
(2) Encourage availability of brines to lake extraction industries;
(3) Protect wildlife and recreation facilities;
(4) Promote the development of lake brines, minerals, chemicals, and petro-chemicals to aid the state's economy;
(5) Encourage the use of appropriate areas for the extraction of brines, minerals, chemicals, and petro-chemicals; and
(6) Encourage the development of an integrated industrial complex.

MINERAL AND ENERGY RESOURCES

Captain Howard Stansbury's accounts of his exploration of GSL, when the lake elevation was around 4,201 feet (1,281 m), make many references to the very saline character of the lake. The chemical analyses of the lake in the appendices of his report described lake brines as "perfectly clear, and had a specific gravity of 1.170, water being 1.00." Dissolved solids were reported as 22.42 percent, and sodium chloride as 20.196 percent (Stansbury, 1852). Stansbury describes his attempts to preserve meat at Black Rock:

> Before leaving Black Rock, I made an experiment upon the properties of the water of the lake for preserving meat. A large piece of fresh beef was suspended by a cord and immersed in the lake for rather more than twelve hours, when it was found to be tolerably well corned.

The method worked so well that the survey party preserved all their beef with lake brines. In fact, the brine had to be diluted with fresh water to avoid the meat from becoming "what the sailors call 'salt junk' "(Stansbury, 1852).

Besides the salt in the lake, Stansbury also noted other mineral resources in the course of his year-long survey, including the asphalt seeps at Rozel Point, and the presence of oolitic sands. At the beaches west of Rozel Point, he describes the sands under a magnifying glass as "rounded globules, chiefly of calcareous rocks, worn doubtless by attrition into their present form, not an angular particle being found among them. It is variegated by different and brilliant colors...A piece of bitumen was found buried in the sand, which had adhered to it when softened by the sun, and completely frosted it over, so that it very much resembled one of the small chocolate lozenges of the shops, covered with miniature sugar plums" (Stansbury, 1852).

Mineral Resources in Brines

The brines of GSL contain several ions that crystallize into valuable minerals during the evaporative process. The major ions in the lake are, in order of relative abundance, chloride, sodium, sulfate, magnesium, and potassium. The composition of GSL water is similar to the oceans although GSL brines are significantly more concentrated. Compared with other inland seas, GSL is high in sulfates, which aids in the production of schoenite (a hydrated potassium, magnesium sulfate). Schoenite is, in turn, a key ingredient used in producing potassium sulfate from high magnesium/potassium harvest salts.

Mineral products currently extracted from lake brines are sodium chloride, magnesium chloride brines (and from that, magnesium metal and chlorine gas), and potassium sulfate. Production of sodium sulfate (salt cake) recently ceased due to declining demand and competition with sodium sulfate produced as a waste product in the manufacture of nylon and other products.

Mineral products that have the potential for commercial production include gypsum, and, in trace amounts, lithium, boron and bromine. Of these, the extraction of lithium has attracted the most interest. MagCorp concentrates lithium in their electrolytic cells, but does not produce lithium chloride for sale at present. The company also extracts boron as part of the brine preparation, but discards it as a waste product. Lithium Corporation of America, predecessor to IMC Kalium, investigated the production of lithium and bromine, but did not pursue the idea.

Salt Resources

An estimated 4.3 billion short tons (3.9 billion mt) of dissolved salts were contained in GSL in 1998. UGS estimates of dissolved solids in GSL have fluctuated from 4.0 to 5.5 billion st (3.6 to 5.0 billion mt) (J.W. Gwynn, Utah Geological Survey, personal communication, 1999) due to the dynamic conditions in the lake as salts are precipitated and redissolved, and to the diversions of brines from GSL, such as the West Desert Pumping Project. About 2.0 million st (1.8 million mt) of dissolved solids flow into the lake each year from surface runoff. Flow-volume, weighted average calculations of these dissolved solids indicate the following chemical composition: sodium-23 percent; chloride-36.4 percent; bicarbonate-25.4 percent; sulfate-5.3 percent; calcium- 5.08 percent; magnesium-2.9 percent; and potassium-1.6 percent. Most of the bicarbonate, calcium, and sulfate will precipitate onto the lake bed as calcite and gypsum. The composition of dissolved solids in streams flowing into the lake is somewhat influenced by agricultural, municipal, and industrial use of tributary waters. Salts are also deposited in the lake by wind and ground water but there are no estimates of the composition and amounts from these sources (J.W. Gwynn, unpublished data, 1998).

Brine Concentration

Brine concentration and the precipitation of salts are dependent upon a number of factors. These factors include water elevation and fresh water flow into the lake, construction of causeways and other diking systems, pumping of lake brines for flood control, and seasonal variations in temperature. Precipitation of salts, whether on the lake bed, in the Newfoundland Evaporation Basin, or in solar evaporation ponds, removes salts from the lake water. Salts on the lake bed have historically precipitated from, and redissolved into, the lake brine system in response to changing lake levels. Mirabilite (sodium sulfate with ten waters of hydration) precipitates out during winter months and is redissolved as the water warms during the summer.

Impacts of Causeway, Diking, and Diversion Operations

Variations in brine concentrations throughout the lake are directly influenced by the causeway and other diking systems. Continuous monitoring of brine concentrations began in 1966, seven years after the construction of the northern Union Pacific causeway. Prior to that time, there is little information about brine concentrations beyond occasional historical references. These early records indicate the lake was a relatively homogenous saline body of water with somewhat higher concentrations of brine on the western side of the lake due to smaller inflows of fresh water and higher rates of evaporation. After construction of the northern Union Pacific causeway, the lake was divided into two bodies of water. There was limited interchange of brines through two culverts in the causeway and through the causeway itself; over time, however, the two arms developed distinct physical and chemical characteristics. In addition to higher concentrations along the west side of the lake, the north arm of the lake had concentrations nearly twice the concentrations of the south arm. The south arm was stratified into a shallow, less concentrated layer (to a depth of 23 feet [7 m]) and a deep layer (below 23 feet [7 m]) of dense, fetid brine (due to hydrogen sulfide and considerable organic matter) at the center of the lake. Concentrations of these deep brines were approximately two times that of the upper layer.

Flood management during the high water years of the early to mid-1980s had significant impacts on lake salinity. When the 300-foot (91 meter) breach on the western end of the northern Union Pacific causeway was opened, a very large volume of dilute south arm brine flowed through the opening at the surface into the north arm of the lake, and a large volume of more dense north arm brine moved through the same opening into the depths of the south arm as return flow. As a result, the north arm of the lake became temporarily stratified. By mid-1991, due to wave action, the north arm of the lake became totally mixed from top to bottom and all signs of stratification were gone. Due to the general lowering of the lake level from 1990 through 1995, the denser brine from the north arm stopped flowing through the breach to the south arm in 1990, when the brine depth at the breach decreased to approximately four feet (1.2 m). As the lake rose from 1996 through most of 1998, brines flowed through the breach from south to north, resulting in the transfer of large volumes of salt to the north arm. In the fall of 1998, as the elevation of the two arms of the lake were within a foot of each other, there was some flow of dense north arm brines into the south arm. There is significant concern that the settling of the existing causeway fill and the addition of extra fill along the length of the causeway has reduced the return flow from the north arm through the causeway fill.

Diking elsewhere in the lake, at the north end of Antelope Island and Stansbury Bay, causes differences in salinity. Farmington Bay is more dilute than the rest of the south arm of the lake while Stansbury Bay has concentrations approaching those of the north arm. As a result, the lake has four areas of salinity induced by diking or other engineered structures (figure 2):

(1) North arm and Stansbury Bay brines are at, or near, saturation (25 to 27 percent),
(2) The main body of the south arm has concentrations ranging from 7 to 15 percent as lake elevations fluctuate,
(3) The waters in Farmington Bay are at approximately 3 to 5 percent, and
(4) Bear River Bay, which fluctuates with the lake level and Bear River inflow, varies between fresh water and salinities from 7 to 10 percent. In comparison, the salinity of seawater ranges from 3 to 5 percent.

The transfer of salts from the south arm to the north arm gave rise to charges that the ability of south arm mineral industries to extract salts from brines is impaired and that the viability of brine shrimp, which thrive in a salinity range of 13 to 19 percent, is threatened. To help resolve these issues, the UGS continues to monitor salinities at designated sites on the lake.

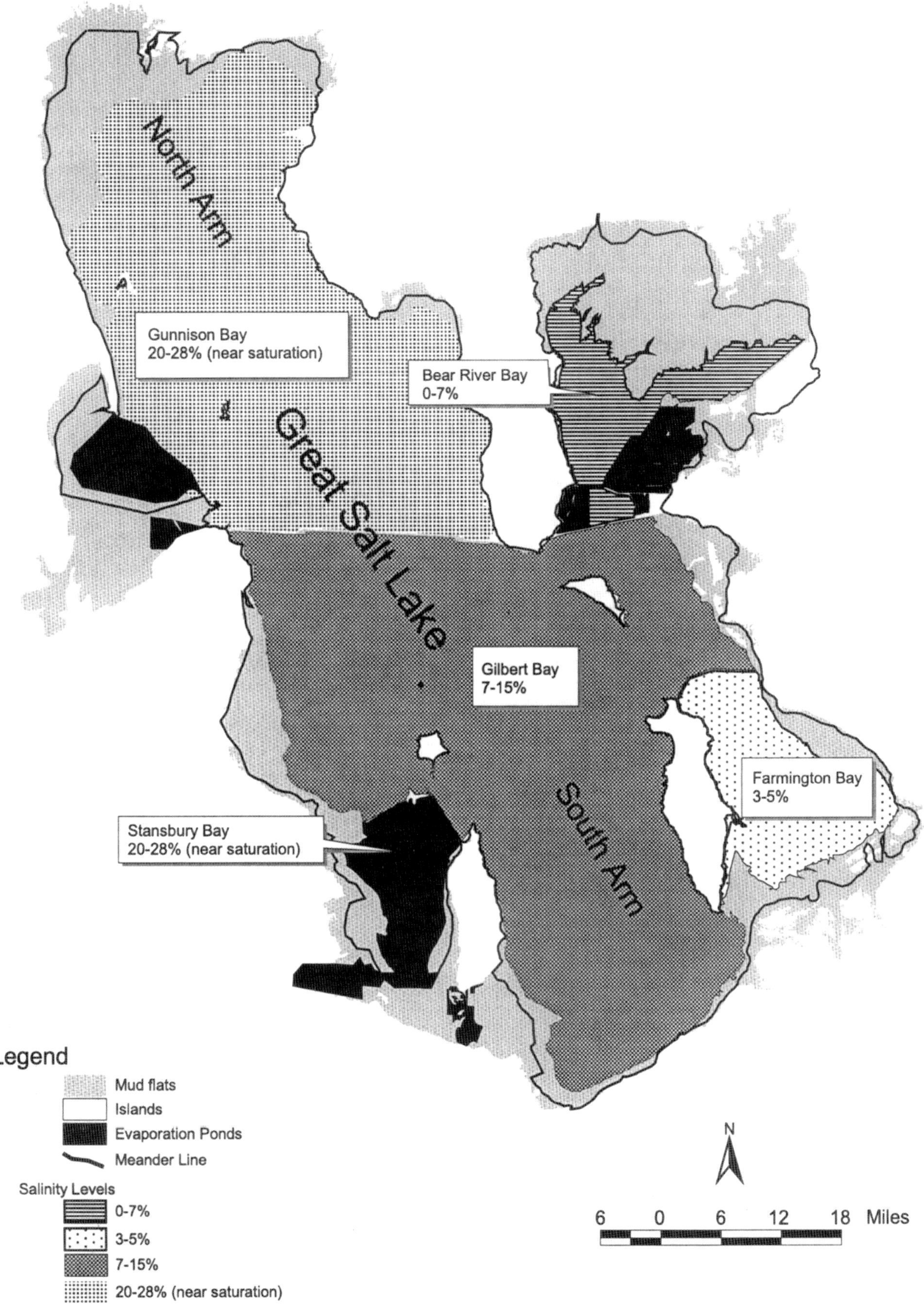

Figure 2. *Areas of salinity levels within Great Salt Lake.*

Natural Factors in Brine Concentrations

Salinity is also affected by natural factors. Water inflow and evaporation rates are the most important influences, natural or man-caused. In the south arm, the relationship between water elevation and salinity is an inverse one; as the lake goes up due to increased inflows, precipitation, and decreased rates of evaporation, the salinity goes down. In the north arm, for the period 1966 through 1983, salinity remained at or near saturation regardless of the north arm elevation. This was because evaporation was greater than the diluting effect of the south arm brines that flowed from south to north. There appears to be variation in salinity in the north arm, with higher concentrations to the north and west, due to inflows at the causeway breach and because of counterclockwise currents in the lake. During the high water years of 1983 to 1987, the salinity of the north arm decreased. The brines in the north arm again became concentrated enough to precipitate salts in the summers of 1995, 1996, and 1997, but were below the saturation point in 1998 (J.W. Gwynn, unpublished data, 1998). According to IMC Kalium's observations, precipitation of salts occurs mostly in the shallow areas in the north and west sides of the north arm.

Salt Resource Consumption and Loss in Great Salt Lake

Salts have been extracted in significant amounts, and at an increasing rate since 1965. In that year, salt companies produced about 0.3 million st (0.27 million mt) of sodium chloride. By 1994, production from GSL was over 2 million st (1.8 million mt) of sodium chloride and approximately 0.3 million st (0.27 million mt) of other salts. To produce salable tons of potassium, and magnesium salts, many more tons of salt (sodium chloride) are deposited in concentration ponds as waste. Some of these salts are returned to the lake if sufficient fresh water is available to flush the salts from the ponds.

A far more significant depletion of the resource was the loss of approximately 12 percent of the lake's dissolved salts as a result of the pumping of brines into the Newfoundland Evaporation Basin west of GSL. Even so, geologists and industry representatives estimate it will take at least 200 years to deplete 10 percent of the lake's remaining sodium chloride at current rates of extraction (or 1,000 years to deplete the sodium chloride in the lake to the point that further extraction is not economic), not accounting for the annual replenishment of salts from surface and ground water inflows. Some ions, notably the sulfate, magnesium, and potassium, are more limited in supply than sodium and chloride ions but their production cannot be separated from the production of salt. The DFF&SL has not monitored the amounts of salts which are deposited into evaporation ponds and how much of these salts are flushed back into GSL. These data could provide insight into the long-term resources contained in the lake.

Waste Salts

Industries extracting magnesium or potassium salts produce substantially more salts than they process for sale. Significant amounts of sodium chloride are used to form floors for the harvesting ponds. Remaining, unwanted salts either accumulate on the bed of the evaporation ponds, as happens in the Stansbury Bay portion of Magcorp's operations and in IMC Kalium's west pond, or are partially flushed with fresh water from the ponds each season, as is the case with IMC Kalium's ponds in Bear River Bay. At both these locations, significant amounts of sodium chloride are harvested by other salt companies under agreements with the original lessees. The returned salts are altered in chemical composition from lake brines because target ions have been removed. Millions of tons of salts are precipitated out annually in Stansbury Basin and Clyman Bay, or are used in harvesting ponds in all salt operations, and therefore not returned to lake brines. Some of these salts are harvested and sold, some are flushed back into the lake, and others remain on pond floors.

The accumulation of waste salts on pond floors becomes a significant problem for producers. So that production of target chemical salts can continue, higher and higher dikes must be built or considerable sums must be spent for the removal of these waste salts. Removal of salts into areas of lower salinity (as in Bear River Bay) can create environmental problems.

Salt Extraction Requirements

The most important factor in the salt extraction process is the original concentration of lake brines. The lower the concentration of brines, the greater the evaporation pond area required for a given volume of produced salt, or alternatively, a limitation on the amount of salt that can be produced. Similarly, those companies extracting potassium, sulfate, or magnesium ions require much larger ponding areas than companies extracting salt because those ions are far less abundant in lake brines.

Evaporation ponds require large areas with suitable soil conditions, access to transportation and utilities, availability of fresh water to flush excess salts from evaporation ponds, and a location conducive to high evaporation rates. These conditions place the greatest constraints for new or expanded operations on the lake. Currently, there are over 100,000 acres (40,000 ha) in evaporation ponds around GSL. Potential ponding sites that are currently unleased include Rozel Bay, Spring Bay, the northwest portion of Gunnison Bay, and the mudflats along the south shore between Stansbury Island and Lake Point.

Lake-bed Deposits

Sodium chloride precipitates on the lake bed as salinity increases in the north arm and in Stansbury Bay. If the north arm of the lake stabilizes at or near the saturation point for sodium chloride, as happened for the period from 1966 to 1983, these lake-bed deposits will accumulate.

Sodium sulfate (mirabilite) also precipitates on the lake bed during winter months in response to cooler temperatures and higher salinity. Mirabilite deposits are found at the southwestern tip of Promontory Point and throughout the lake during winter months if brines become cold and concentrated enough to precipitate. A substantial portion of these deposits redissolve as temperatures warm again. "Permanent" deposits probably occur around much of the perimeter of the lake. These are mirabilite-cemented sands which were probably formed by mirabilite being blown upon the beaches, dissolving, then resolidifying, cementing the sands at depth (Wilson and Wideman, 1957). These cemented sands have been found at Saltair and the south shore marina, where they had to be blasted to deepen the marina, and in other areas around the lake. There is also a very thick layer of mirabilite westward from the southern tip of Promontory Point (Eardley, 1962). The potential for mineral extraction of these deposits is small due their low value, limited or declining markets, and high extraction costs.

Hydrocarbon Resources

Hydrocarbon resources underlying the lake are significant, but presently undeveloped. The hydrocarbons are low gravity (4 to 9 degree API) and tar-like. They contain high nitrogen concentrations and up to 12 percent sulfur. The UGS reports: "The oil is chemically similar to ichtyol, a rare substance used for medicinal purposes, and thus has the potential to be an extremely valuable commodity. Higher molecular weight fractions, when added to oil, are known to increase the lubricity of the oil" (Chidsey, 1995). The unusual characteristics of the oil have been the subject of studies by chemists at Weber State University and Université Louis Pasteur de Strasbourg. However, these resources are expensive to process and, at present, uneconomic to extract using current technology because of the nature of the hydrocarbons and production in "an offshore, highly saline environment" (Kendall, 1993).

Two oil fields have been discovered on GSL, Rozel Point and West Rozel. The Rozel Point field is located in T. 8 N., R. 7 W., Salt Lake Base Line, along the north shore of the lake. West Rozel field is located in T. 8 N., R. 8. W., three miles (5 km) from the shoreline of the Rozel Point field. The estimated area of the Rozel Point field is about ten acres (4 ha). The field has a low reserve estimate because of the poor reservoir seal. Small amounts of hydrocarbons or asphaltum have been recovered from natural seeps and shallow wells at Rozel Point since the turn of the century. Earliest use was as a lubricant. More recent uses have been to resurface roads and for impregnating tire cords (Chidsey, 1995). In the early 1960s, several wells were drilled on a one-acre (0.4 ha) spacing order from the Division of Oil, Gas and Mining (DOGM). One 40-acre (16.2 ha) lease in this field expires in 2002.

The West Rozel field was discovered as a part of Amoco Exploration Company's exploration program on GSL in the late 1970s. Thirteen "offshore" wells were drilled between June 1978 and December 1980, resulting in the discovery of the West Rozel field, and with oil shows reported in eight wells. Two development wells were drilled at West Rozel, identifying a field covering 2,300 acres (930 ha) with heavy oil similar to that at Rozel Point. Reserve estimates for the field are high, in contrast to Rozel Point, with a primary recovery of 1 to 10 million barrels (0.16 to 1.6 million m3) of oil. However, Amoco did not develop the field "because of the high water cut and the high cost of operating an 'offshore' field" (Bortz, 1987). The unusual character of the oil also contributed to this decision. This field is considered by UGS to have "low development potential" because the oil cannot be economically produced. Changing technology may make the field viable. Oil characteristics and reserve estimates for the West Rozel and Rozel Point fields are summarized in table 2.

Condemnation proceedings against Anschutz Ranch to establish a purchase price for Antelope Island documented the low potential for oil and gas in the vicinity of the island (Howard Ritzma, private consultant, personal communication, 1995).

World-wide, there appears to be low prospects for tar sands development and even less for "super" heavy oil extraction under such difficult conditions. Only Canada and Venezuela, with lead time, infrastructure, and experience, have been successful in developing their heavy oil and tar sands industries (Howard Ritzma, personal communication, 1995).

Other Mineral Resources

Oolitic sand deposits make up many of the beaches along the shoreline around the lake, with higher concentrations along northwestern Antelope Island, the east side of Spring Bay, the western side of Stansbury Island, and on Carrington Island. Because of their high calcium carbonate content, oolites have been used by Magcorp, and its predecessors, for acid neutralization and dike construction. Oolites are also used in very minor amounts in flower drying. Until recently, the DOGM reported up to 130,000 st (118,000 mt) mined annually by Magcorp on BLM lands adjacent to GSL.

Table 2. Oil characteristics and reserve estimates of Great Salt Lake oil fields (from Chidsey 1995).

Field	Oil Characteristics		Reserve Estimates
West Rozel	Gravity	4° API	Proved area, 500 acres
	Color	Dark Brown	Primary Recovery, 1-10 mmbo
	Sulfur	12.5%	
	Pour Point	75°	
	Viscosity	3000-4000 cp @ 140°F	
Rozel Point	Gravity	5° API	Proved area, 10 acres
	Color	Black, tar-like	Primary Recovery, 2,665 bo
	Sulfur	12%	

Explanation: mmbo=million barrels of oil bo=barrels of oil

HISTORY OF MINERAL LEASING ON GREAT SALT LAKE

Sodium Chloride Leases

Sodium chloride is the mineral with the longest history of successful extraction on GSL. Native Americans and early explorers extracted small amounts of salt. Permanent salt production facilities began in 1850, using boilers to evaporate salt. In 1888, Inland Salt Company developed a process they called fractional crystallization, which used a series of evaporation ponds to produce salt free of contaminating chlorides and sulfates of magnesium, calcium, and potassium. Inland Salt Company was the predecessor of a succession of salt companies that dominated the salt industry in Utah. Inland Salt Company was bought by Morton Salt in 1923, and Morton continued to dominate the salt market until the 1950s, operating from sites near Saltair and Burmester. In a highly competitive market, several salt companies came and went as the waters of the lake fell and then rose, and as industrial markets for salt followed the rise and fall of silver mining (Clark and Helgren, 1980).

Leases for the extraction of salt and other minerals have been issued by the Land Board for various products "under the waters of Great Salt Lake" since 1919, after authority for management of sovereign lands was given to the Land Board. Leased minerals included sodium sulfate, salt, magnesium, and oil and gas, sometimes all in one lease. In 1935, the State legislature made reservation of coal and other minerals on state lands mandatory and "reserved from sale, except on a rental and royalty basis..."(Utah State Legislature, 1935).

In 1940, despite this long history of leasing, Deseret Land and Livestock filed for a diversion from GSL under the State's water appropriation laws for the purpose of extracting salt. Deseret claimed the State did not own the salts contained in brines and that no royalty or lease agreement was needed. In 1941, the legislature added a section to Utah statute that included "salts and other minerals in the waters of navigable lakes and streams" as minerals to be managed by the Land Board. A 1946 Utah Supreme Court decision affirmed State ownership of minerals in the waters of the lake. Royalty terms on these leases were variously 35 cents and 50 cents per short ton (38 cents to 55 cents per mt), but this rate was disputed by Deseret Land and Livestock Company and Morton Salt, in particular. After eight years of negotiation with the Land Board, Morton negotiated a 15-year lease to extract sodium chloride from brines, and the lease did not provide any acreage within the meander line. The royalty rate was 10 cents per short ton (11 cents per mt). These lease terms were subsequently offered to all producers and new lessees on the lake. Many lessees negotiated mineral leases in connection with royalty agreements which allowed sovereign lands to be used for evaporation ponds or for the lessee to extract salt precipitated on the bed of the lake. As the royalty agreements reached the end of their terms in the late 1960s, the Land Board added language that allowed leases to be held by production. Royalty rates were left at 10 cents per st (11 cents per mt).

In 1997, more than 40 years after the 10 cents per st (11 cents per mt) rate was agreed to by the Land Board, DFF&SL enacted new rules to increase the royalty rate to 50 cents per short ton (55 cents per mt), that will be phased in over a five to ten year period, depending on each salt producer's circumstances. After the royalty rate reaches 50 cents per st (55 cents per mt) for all producers, it will be adjusted annually for inflation by an index tied to the Producer Price Index for industrial commodities.

Non-Halite Salt Leases

During the 1960s, the Land Board entered into agreements with three companies, two interested in extracting magnesium chloride to be refined into magnesium (H-K, Inc. and Bonneville-on-the-Hill, now MagCorp), and the third interested in lithium and potassium products (Lithium Corporation, now IMC Kalium). These agreements had 49-year terms and an ad valorem royalty rate beginning at 1.5 percent applied against "dry" products and escalating to 5 percent over a 25 year period. The magnesium chloride producer was granted an exclusive right to produce that product from 1961 to 1969. At the end of that period, salt lessees were offered an opportunity to convert their royalty agreements, which allowed the extraction of sodium chloride only, to an agreement which allowed extraction of all minerals, including magnesium chloride, contained in brines. Also at the end of that period, a royalty rate for magnesium was added to the royalty schedule beginning at 0.1259 percent and escalating to 0.4196 percent. This rate was meant to produce equivalent royalty revenues when applied to the value of magnesium metal that the 1.5 percent to 5 percent rate would have generated if applied against the value of anhydrous magnesium chloride. However, the new schedule fell significantly short of accomplishing this goal. All royalty agreements contained a provision which entitled lessees to the lowest royalty rate granted to any other lessee on the lake. This provision was applied primarily to royalties on sodium chloride in those agreements which contain that clause.

Lessees under these royalty agreements began production in the mid 1970s. Ten years later, as GSL was approaching its historic high elevation of 4,211.60 feet (1,283.7 m) above m.s.l., and most producers on the lake were experiencing major damage to their dikes, the Land Board granted both companies royalty relief by starting the clock at year one of the royalty schedule.

An Attorney General's Opinion dated June 9, 1966, in response to a request for an opinion by the Director of State Lands, stated that "salt and salt derivative leases on the bed of GSL would be subject to the simultaneous filing provisions of the statute" if the lands were under a lease which was terminated by board action or released from a withdrawal from mineral leasing (State Land Board minutes, 1966). However, salt leases on GSL have never been offered as part of a simultaneous filing.

Current Salt Operations

Although the two non-halite salt companies (Magcorp and IMC Kalium) have maintained their leases since the 1960s, sodium chloride leases have dwindled from 8 to 10 small operations on the lake during the 1940s, to three significant operations today. Following the high lake levels in the mid-1980s, Morton Salt Company relinquished its original royalty agreement negotiated with the Land Board in 1954, to assume control of a site southeast of Stansbury Island, formerly operated by American Salt Company. SolAire Salt (now owned by Cargill Salt) relinquished its operations near Lake Point, the original lease site for Deseret Livestock Company and its many successors, and now operates from a royalty agreement by acquiring concentrated brines from Magcorp. American Salt (now IMC Salt) moved its operations to Bear River Bay and operates on a sublease agreement, purchasing crude salt from IMC Kalium Ogden Corp. Mineral extraction from GSL brines currently provides the largest source of mineral royalties on sovereign lands with between 1.5 and 2.0 million st (1.4 million and 1.8 million mt) of sodium chloride and close to 300,000 st (272,000 mt) of other minerals salts extracted each year.

Currently, there are 12 producing mineral leases totaling 164,950 acres (66,780 ha) that generated almost $1,900,000 in royalties during calendar year 2000. IMC Kalium Ogden Corp. produces potassium sulfate and magnesium chloride from 89,257 leased acres (36,069 ha) in Bear River Bay and Clyman Bay. Sodium chloride, produced as a byproduct, is transferred to IMC Salt, which packages and sells the salt. MagCorp produces magnesium metal from its 75,610 leased acres (30,611 ha) in Stansbury Bay. Cargill Salt Company produces sodium chloride from brines provided by Magcorp under a lease agreement with DFF&SL. Morton Salt Company produces salt on ponds above the meander line with an 83 acre (33.6 ha) lease from DFFSL which serves as a right of way to GSL brines from the southeast end of Stansbury Island. Finally, North Shore Limited Partnership produces high magnesium brines for use in dietary and mineral/vitamin supplements. Lake brine is brought from the lake through a canal, located on a state right of way, onto its ponds on private lands near Spring Bay in the north arm of the lake. Lessees, acres, producing status, and expiration dates for active leases on GSL are summarized in table 3.

Oil, Gas, and Hydrocarbon Leases and Development

Interest in oil and gas leasing on the bed of GSL is long standing. Leases have been issued with the standard 10 year primary term and 12 percent royalty rate. Natural seeps at Rozel Point have attracted oil industry interests since the turn of the century, and the area has been under nearly continuous lease with a number of different lessees.

At present there is a single 40-acre (16.2 ha) lease covering the majority of wells in the Rozel Point area. Despite a long history of leasing and some efforts to stimulate production by electric heaters and steam injection, there has been minimal production and no payment of royalties. Including adjacent private land, the Rozel Point site has abandoned wells, the remains of drilling and production piping, buildings, tanks, abandoned vehicles, tires and other debris, and seeps spontaneously ooze oil, especially on warm days. All wells except one were drilled before the implementation of Division of Oil, Gas and Mining's (DOGM's) current regulations. Several abandoned wells were capped in 1996, in cooperation with DOGM, UGS, Environmental Protection Agency, and the U.S. Fish and Wildlife Service.

In 1972, the Land Board held public hearings regarding plans for large-scale drilling on the lake in response to lease applications for oil and gas exploration by Marvin Wolf in

Table 3. Current mineral, oil, gas, and hydrocarbon leases on Great Salt Lake (from Division of Forestry, Fire and State Lands lease files).

Lease No.	Lessee	Lease Type	Acres	Lease Status	Expiration Date
ML 9300-SV	Morton	Chemical Salts	83	Producing	Held by production
ML 18779-SV	Magnesium Corp	Chemical Salts	75,610	Producing	Held by production
ML 19024-SV	IMC Kalium	Chemical Salts	10,413	Producing	Held by production
ML 19059-SV	IMC Kalium	Chemical Salts	1,282	Producing	Held by production
ML 21708-SV	IMC Kalium	Chemical Salts	20,860	Producing	Held by production
ML 22782-SV	IMC Kalium	Chemical Salts	7,580	Producing	Held by production
ML 23023-SV	IMC Kalium	Lake-bed Salts	14,381	No production	Past primary term
ML 24631-SV	IMC Kalium	Chemical Salts	1,911	No production	Termination of Royalty Agreement
ML 25859-SV	IMC Kalium	Chemical Salts	10,583	Producing	Held by production
ML 29864-SV	Willam J. Coleman	Chemical Salts	0	No production	12/31/2003
ML 43388-SV	IMC Kalium	Chemical Salts	709	Producing	Held by production
ML 44455-SV	Nova Natural Res.	Oil Gas & Hydro.	640	No production	8/31/1999
ML 44456-SV	Nova Natural Res.	Oil Gas & Hydro.	960	No production	8/31/1999
ML 44607-SV	IMC Kalium	Chemical Salts	37,830	Producing	Held by production
ML 45646-SV	Kenneth Pixley	Oil Gas & Hydro.	49	No production	6/30/2002
ML 45741-SV	Coleman Morton	Oil Gas & Hydro.	765	No production	9/30/2002
ML 45772-SV	Coleman Morton	Oil Gas & Hydro.	647	No production	12/31/2002
200-00001	North Shore Limited	Chemical Salts	0	Producing	Held by production
200-00002	Cargill Salt	Salt	0	Producing	Held by production

1972 for approximately 180,000 acres (73,000 ha) along the east shore of the lake and by Amoco Production Company for over 600,000 acres (240,000 ha) in the main body of the lake. Ultimately, drilling rules were approved by DOGM and a lease form was approved by the Division of State Lands in the summer of 1973. Both DOGM rules, and leases issued by the Land Board, placed timing and location restrictions on drilling unless permission to drill was granted by both the Board of Oil, Gas and Mining and the Land Board. Leases were issued to Amoco in 1973. At that time the Land Board decided to take no action on the Marvin Wolf lease applications along the east shore because of concerns about "the ecology factor."

From 1973 to 1985, Amoco Production Company conducted its exploration program drilling 13 exploration wells and two development wells. Amoco established five units on the lake, the most promising of which was the West Rozel unit in T. 8 N., R. 8 W., Salt Lake Base Line. All units were abandoned in the early 1980s, and leases were terminated in 1985. In 1978, the Land Board reversed its original decision to lease lands along the east shore and issued leases which were ultimately acquired by Phillips Petroleum, Sun Exploration, and other oil and gas companies. Most of these leases were relinquished in 1986.

SIGNIFICANCE OF MINERALS EXTRACTED FROM GREAT SALT LAKE

Producers of magnesium, potash, and salt from GSL have contributed significantly to the growth in metals and industrial minerals production in Utah. Production of magnesium and potash began in the early 1970s and by 1980 had reached over 25,000 st (23,000 mt) of magnesium metal and nearly 100,000 st (90,000 mt) of potassium oxide equivalent. There was a lapse in production during the lake's high-water years for all products followed by a second surge in production to current levels. Production of salt also increased significantly during this same twenty-year period. Together these companies contribute approximately $240,000,000 in gross value, or 18 percent of the value of the state's non-fuel mineral production. Most of this production is exported. These companies together employ more than 1,000 people; approximately 250 by salt producers, 550 in magnesium production, and 220 in the production of potassium sulfate and magnesium chloride.

The importance of the mining industry in Utah has declined as a percentage of the state's gross product, falling from 6.3 percent in 1965 to 3.2 percent in 1996. However, this statistic alone understates the importance and continued growth in the mining sector, especially for non-fuel mineral products (metals and industrial minerals). In constant dollars, the value of industrial minerals and metals has more than tripled from 1980 through 1996 according to the U.S. Bureau of Economic Analysis' estimate of gross state product (Utah Office of Planning and Budget, 2000). The state's economy has instead become increasingly diverse so that mining plays a much smaller role than in the past. In addition, the value of industrial minerals may be significantly under-reported in these estimates, possibly because much of the value of industrial minerals is attributed to the manufacture of end products rather than mining of the raw material. The UGS, for example, reports the value of industrial minerals in 1998 as $533 million (Utah Office of Planning and Budget, 2000), while the value of non-metallic minerals is reported in gross state product estimates as only $30 million (Utah Office of Planning and Budget, 2000).

MARKETS FOR MINERALS EXTRACTED FROM GREAT SALT LAKE

Halite

In 1998, approximately 1.6 million st (1.4 million mt) of salable solar salt were produced from GSL (Division of Forestry, Fire and State Lands, unpublished data, 1998). Most of the solar salt production from GSL (75 to 80 percent) is exported, primarily to other Rocky Mountain and Midwest states where it is used for agriculture, water conditioning, industrial/chemical use, and highway deicing. Most of the salt not exported is used for highway deicing.

The USGS's 1999 Minerals Yearbook reports 1997 prices of $16.21 per st ($17.83 per mt) for dry bulk solar salt (Kostick, 2000a). The USGS's 1999 Mineral Commodity Survey for salt reports an average 1998 price for all solar salt of $29 per st ($32 per mt) (Kostick, 2000b). The Salt Institute reports 1997 average salt prices of $52 per st ($57 per mt) for agricultural uses and $91 per st ($100 per mt) for water conditioning. Since 1981, these prices for solar salt have increased by approximately 4 percent per year. In contrast, prices for vacuum pan and rock salt have increased at rates of 2 to 3 percent per year (Kostick, 2000a).

Bulk solar salt can be produced for prices ranging from $7 to $10 per st ($8 to $11 per mt) so Utah producers appear to be in a competitive position to maintain markets in the Midwest. Total sales revenues from solar salt production on GSL range between $30 to $50 million per year (Division of Forestry, Fire and State Lands, unpublished data, 1998).

Solar salt produced from GSL represents a significant, and increasing, share of total domestic solar salt production. Figure 3 shows this relationship over the past 15 years. The

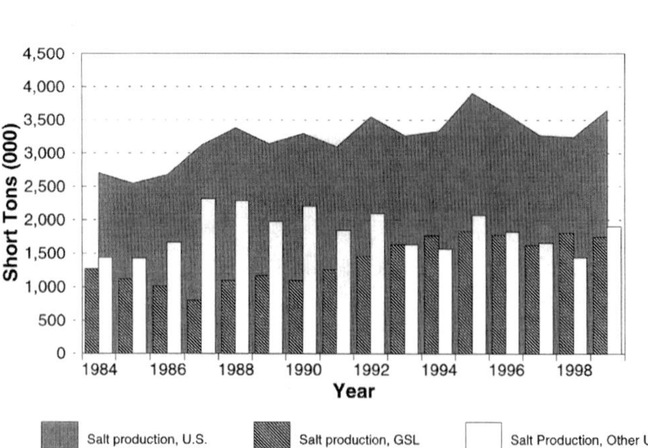

Figure 3. Comparison of solar salt production from Great Salt Lake (GSL) to total U.S. solar salt production and to solar salt production from other U.S. states from 1984 through 1998 (from Kostick, D.S., 2000a)

remainder of solar salt produced in the U.S. is primarily from California with some production from New Mexico.

Solar salt competes in regional markets with rock salt and vacuum pan salt for chemical and industrial, water conditioning, and agricultural uses. Nationwide, the consumption of rock salt is four times that of solar salt. However, in markets other than road salt (primarily rock salt) and food processing (primarily vacuum pan salt), the three types of salt are in closer competition. Table 4 shows the historic pattern of consumption for Utah solar salt where its uses compete with vacuum pan and rock salt. These include chemical, industrial, water treatment, and agricultural end uses, but exclude uses of salt for deicing and food processing. Since 1987, solar salt consumption has grown annually by about 6 percent while rates of growth in consumption of vacuum pan and rock salt have grown at a much lower rate. USGS data show that markets for most salt products are regional, but the market for road salt is local. In Pacific Coast states (California, Oregon, and Washington), consumption of rock salt is negligible while consumption of solar salt, stable for several years at 1.2 million st (1.1 million mt), has recently increased to about 1.5 million st (1.4 million mt) (figure 4). California produces solar salt but Pacific Coast states import between 0.5 and 1.0 million st (0.45 million and 0.9 million mt) of solar salt each year. Imports of solar salt have recently declined from a previously stable level of 1 million st (0.9 million mt). In the past, Pacific Coast states were a difficult market for Utah solar salt to penetrate. However, solar salt from Utah may be finding its way to these markets due to increases in consumption and a decline in exports of solar salt by California.

Solar salt consumption in Rocky Mountain states (Utah, Colorado, Nevada, Wyoming, Idaho, and Montana) is just over 1 million st (0.9 million mt) per year. Consumption of solar salt shows some growth in these states for uses such as road salt, where demand is stable, or chemical uses, where demand could decline.

In Midwestern states (Illinois, Indiana, Iowa, Kansas, Minnesota, Nebraska, North Dakota, South Dakota, and Wisconsin), solar salt markets have grown from 0.43 million st (0.38 million mt) to 0.9 million st (0.8 million mt) in the past nine years. As with the Pacific Coast markets, these markets have been difficult to penetrate. Despite this, as figure 5 illustrates, production of Utah solar salt closely parallels the consumption of solar salt in Midwestern and Rocky Mountain states, where combined consumption of solar salt has grown from over 1 million st (0.9 million mt) to almost 2 million st (1.8 million mt). Since the only other major sources of solar salt for these areas would be from Oklahoma, the Great Lakes region, or Mexico, via the Mississippi River, the data suggest that Utah producers are making steady inroads into Midwestern markets.

Table 4. Annual production of halite by type of recovery process (in thousands of metric tons), 1987 through 1997. Competing end uses used in this table are: chemical, industrial, water treatment, and agricultural (From Kostick, 2000a).

Year	Solar	Vacuum	Rock	Total	% Solar
1987	2,188	2,347	3,949	8,484	25.79
1988	2,674	2,446	4,173	9,293	28.77
1989	2,781	2,881	4,415	10,077	27.60
1990	3,028	2,926	4,438	10,392	29.14
1991	2,956	2,704	3,941	9,601	30.79
1992	2,904	2,584	4,170	9,658	30.07
1993	3,309	2,529	4,482	10,320	32.06
1994	3,220	2,530	4,682	10,432	30.87
1995	3,460	2,446	4,031	9,937	34.82
1996	3,840	2,439	4,360	10,639	36.09
1997	3,779	2,239	4,037	10,055	37.58

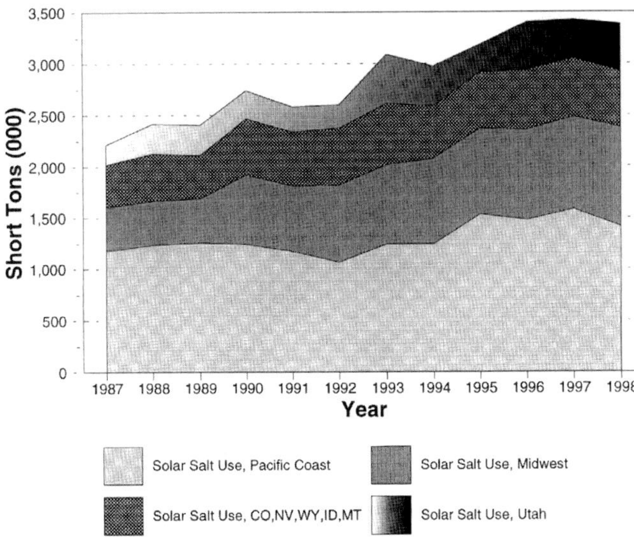

Figure 4. Growth in solar salt consumption in regional markets for chemical, industrial, water treatment, and agricultrual end uses from 1987 through 1997 (from Kostick, D.S., 2000a).

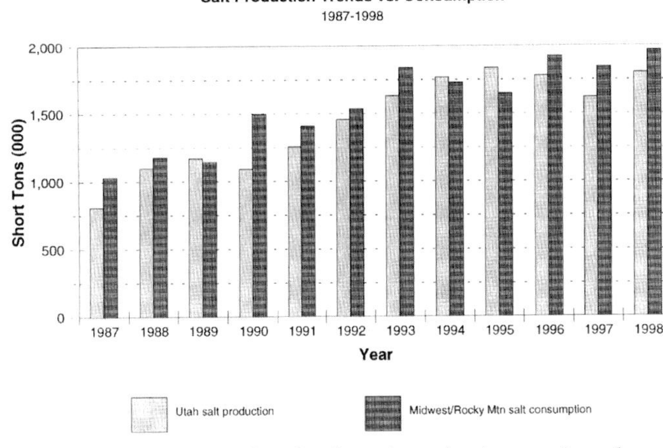

Figure 5. Comparison of Utah solar salt production to solar salt consumption in the Rocky Mountain/Midwest regions from 1987 through 1998 (from Kostick, D.S., 2000a).

In summary, the production data since 1987 indicate:

- Utah solar salt companies produce a significant and increasing share (40 to 55 percent) of solar salt produced domestically.
- Solar salt competes closely with rock salt and vacuum pan salt for end uses such as chemical/industrial, water treatment, and agriculture; solar salt is

the dominant source for agricultural and industrial uses.

- In total, consumption of salt for these uses is not growing steadily, but solar salt consumption is increasing at a stable and significant rate (approximately 6 percent/year), while rock salt shows uneven growth, and vacuum pan salt shows a decline in tons sold. Likewise, prices for solar salt have increased more rapidly than those for vacuum pan and rock salt.
- Markets for solar salt in Pacific Coast states have grown in the past five years, while imports of solar salt have declined.
- Solar salt consumption is increasing in Midwestern and Rocky Mountain states at a level which parallels Utah solar salt production; Midwestern markets consume 40 to 45 percent of Utah's production and have the highest rate of growth in tons consumed.
- Based on these observations, DFF&SL expects the production of solar salt to increase at an average of 3 percent per year for the next several years. Fluctuations in production can be expected due to weather, lake levels, and precipitation.

Potash (Potassium Sulfate)

IMC Kalium Ogden, Inc. produces potassium sulfate and magnesium chloride brines from its facilities at Bear River Bay. The company has the capacity to produce 0.34 million st (0.3 million mt) of potassium sulfate from its 37,000 acres (15,000 ha) of evaporation ponds in Bear River Bay and Clyman Bay (on GSL's west side) (Odgen Standard-Examiner, 1992). IMC Kalium also produces magnesium chloride brines as a byproduct and sells salt to a related company, IMC Salt, to process and market.

Potassium sulfate is preferred for most agricultural applications because the sulfate base is better for plants than the chloride in muriate of potash (KCl). However, because of its much higher cost to produce, and resulting higher selling price, potassium sulfate is used primarily as a specialty fertilizer for certain crops. In 1997, the K_2O equivalent price for potassium sulfate was $350 per st ($385 per mt). In comparison, the K_2O equivalent price for muriate of potash was about $130 per st ($140 per mt). Prices for potassium sulfate have fluctuated somewhat since 1991, when K_2O equivalent sold for $289 per st ($318 per mt) of K_2O equivalent. Over all, prices for potassium sulfate have increased about 3 percent per year.

About 20 to 25 percent of all potassium fertilizer produced in the U.S. is potassium sulfate. Potassium sulfate comprises 30 percent of all potash shipped to overseas markets such as the Pacific Rim and South America. The tobacco-producing states also buy significant quantities of potassium sulfate (Searls, 1999).

The U.S. potash industry operates in the shadow of Canadian producers. Canadian potash reserves and production are far greater than those in the U.S., producing close to 10 million st (9 million mt) of K_2O equivalent, compared to U.S. production of 1.5 million st (1.4 million mt) per year. Markets for potassium sulfate are much smaller, because of the high price of the product. However, U.S. production of potassium sulfate is more competitive with foreign producers, principally in Canada and Germany. While the U.S. is dependent on imports to satisfy agricultural grades of potassium chloride, U.S. producers of potassium sulfate supply most U.S. consumption and export most of their remaining production to other countries. Total U.S. potassium sulfate production is approximately 0.3 million st (0.27 million mt) of K_2O equivalent.

IMC Kalium Ogden Corp., which purchased Great Salt Lake Minerals Corporation from Harris Chemical in 1998, is now the sole producer of potassium sulfate in the U.S. In North America, IMC Kalium is in competition with only one other producer, Potash Corporation of Saskatchewan. IMC Kalium Ogden Corp.'s plant on GSL enjoys the advantage of a unique resource in lake brines which are rich in sulfate ions as well as potassium. Other producers must manufacture potassium sulfate from potassium chloride; IMC Kalium extracts potassium minerals from the lake brines from which potassium sulfate is produced through a leaching process. IMC also supplements production by combining excess sulfate ions with potassium chloride to convert it to potassium sulfate.

Markets for potassium sulfate, although small, grew from 0.25 million st (0.22 million mt) to more than 0.3 million st (0.27 million mt) from 1991 to 1995. Production will likely be stable at these levels and IMC Kalium will continue to be a major supplier.

Magnesium Chloride

IMC Kalium Ogden Corp. also produces magnesium chloride brines. Its annual production capacity is reported to be 0.12 million st (0.11 million mt) of MgO equivalent, or about 10 percent of U.S. production capacity; however, its actual production of magnesium chloride brines is far below this capacity. Much of this brine is sold for dust control for $50 per st ($55 per mt) or less. IMC Kalium Ogden Corp. sells magnesium chloride hexahydrate as well. Because of the additional processing, this product sells for close to $300 per st ($330 per mt), and is used as a chemical intermediate or in refractories (Kramer, 1999).

Magnesium Metal

Magcorp produces magnesium chloride brines from its operations on Stansbury Bay and Knolls ponds. The company refines these brines and produces magnesium metal through an electro-winning process. The plant has a capacity of 45,000 st (41,000 mt) per year of magnesium metal, which is half of U.S. magnesium metal production capacity, and approximately 10 percent of current world capacity. Until November 1998, there were two other U.S. producers, Northwest Alloys which produces magnesium from dolomite in Washington, and Dow Chemical which produced magnesium from seawater in Texas. In November 1998, Dow Chemical closed its 71,000 st (65,000 mt) per year capacity plant due, in part, to damage sustained from a lightning strike in June of that year and damage from floods during Hurricane Francis in September. Northwest Alloys is the only other U.S. producer with an annual production similar to Magcorp. Magcorp is undergoing an electrolytical cell upgrade which will temporarily lower its production capacity for the next two years.

U.S. magnesium metal production in 1997 was at 86 percent of plant capacity, down from close to 100 percent capacity in 1995 and 1996. These high levels of production were in response to high price levels, which peaked in 1995 at $2.09 per lb ($4.60 per kg) and dropped to $1.55 per lb ($3.41 per kg) in 1998 (Kramer, 1999). Of the minerals extracted from GSL, only magnesium metal has this price volatility.

In contrast to salt (sodium chloride), but similar to potassium sulfate, magnesium metal enjoys both a national and international market. Magcorp faces great competition from other domestic and foreign producers. In addition, much of the demand for magnesium metal is met by secondary sources through recycling of magnesium products. The U.S. produces the largest tonnages of both primary and secondary magnesium, but with new applications in the automotive industry, new production capacity is being brought on line in Canada, Australia, Israel, China, and possibly the Congo. These additions will have a significant impact on a relatively small industry. World-wide magnesium production is approximately 0.4 million st (0.36 million mt) per year, but is one or two orders of magnitude less than aluminum, at 4.0 million st (3.6 million mt) per year, or steel, at 100 million st (90 million mt) per year. Magnesium is used as an alloying agent for these metals, as well as competing with aluminum in diecasting applications for the automotive industry (Kramer, 1999).

The U.S. magnesium industry has been concerned enough about foreign imports of magnesium in recent years to file anti-dumping investigations against magnesium producers in Canada and the Ukraine. As a result of these investigations, antidumping duties were set at varying rates against a Canadian firm from 1991 through 1995, and against imports from the Ukraine, although, this latter decision is under appeal before the Court of International Trade (Kramer, 1999).

Because of the competitiveness and secrecy associated with each company's production processes, there has been little opportunity for technology transfer within the industry and production costs have remained high (Howard-Smith, 1998). The cost of producing magnesium for a new producer in Tasmania and Australia ranges between $0.65 and $0.90 per pound ($1.43 and $1.98 per kg) (Kramer, 1999).

ACKNOWLEDGMENT

Dr. J. Wallace Gwynn of the Utah Geological Survey contributed most of the data and history for the section on mineral resources of GSL. His contributions to and review of that section are greatly appreciated.

REFERENCES

Black's Law Dictionary (6th edition), 1990, St. Paul, Minnesota, West Publishing Group.

Bortz, L.C., 1987, Heavy-oil deposit(s), Great Salt Lake, Utah, *in* Meyer, R.F., editor, Exploration for heavy crude oil and natural bitumen: Tulsa, Oklahoma, American Association of Petroleum Geologists Studies in Geology No. 25, p. 555-562.

Chidsey, T.C., Jr., 1995, Rozel Point: harbinger of petroleum potential: Utah Geological Survey Survey Notes, v. 27, no. 3, p. 10-11.

Clark, J.L., and Helgren, Norman, 1980, History and technology of salt produced from Great Salt Lake, *in* Gwynn, J.W., editor, Great Salt Lake - a scientific, historical, and economic overview: Utah Geological and Mineral Survey Bulletin 116, p. 206-210.

Coastal States Organization, Inc., 1997, Putting the public trust doctrine to work: Boston, University of Massachusetts Press, p. 289-295.

Dewsnup, R.L., and Jensen, D.W., 1980, Legal battles over ownership of the Great Salt Lake, *in* Gwynn, J.W., editor, Great Salt Lake, a scientific, historical, and economic overview: Utah Geological and Mineral Survey Bulletin 116, p. 15-18.

Eardley, A.J., 1962, Glaubers salt bed west of Promontory Point, Great Salt Lake: Utah Geological and Mineral Survey Special Study No. 1, 12 p.

Howard-Smith, Ian, Roskill Metals Analysis, November 27, 1998.

Jehl, J.R., Jr., 1994, Changes in saline and alkaline lake avifaunas in western North America in the past 150 years, *in* Johnson, N.K. and Jehl, J.R., Jr., editors, A century of avifaunal change in western North America, proceedings of an international symposium at the centennial meeting of the Cooper Ornithological Society, Sacramento, California, April 17, 1993: Camarillo, California, Cooper Ornithological Society Studies in Avian Biology No. 15, p. 264.

Kendall, C.F., 1993, West Rozel oil field, Box Elder County Utah, *in* Hill, B.G. and Bereskin, S.R., editors, Oil and gas fields of Utah: Utah Geological Association Publication 22, unpaginated.

Kostick, D.S., 2000a, Salt: U.S. Geological Survey, 1998 minerals yearbook, 19 p.

—2000b, Salt: U.S. Geological Survey, 2000 mineral commodity summary, 2 p.

Kramer, D.A., 1999, Magnesium: U.S. Geological Survey, 1998 minerals yearbook, 9 p.

Morgan, D.L., 1947, The Great Salt Lake: Salt Lake City, University of Utah Press, p. 17.

Ogden Standard-Examiner, Chemical Firm Expects to Double Potash Output, March 2, 1992.

Searls, J.P., 1999, Potash, U.S. Geological Survey, 1999 mineral industry surveys, 4 p.

Smith, A.D., 1960, Utah's federal grant lands: Logan, Utah State University, p. 19-20.

Stansbury, Howard, 1852, Exploration and survey of the valley of the Great Salt Lake of Utah: Philadelphia, Lippincott, Grambe and Company, p. 101-178.

U.S. Bureau of Land Management, 1973, Manual of surveying instructions: Washington, D.C., U.S. Government Printing Office, Sections 1-12.

Utah Code Annotated, Section 23-21-5

—Section 65A-1-1(3).

—Section 65A-1-1(5).

—Section 65A-1-1(7).

—Section 65A-2-1

—Section 65A-10-1

—Section 65A-10-3 (1,2)

—Section 65A-10-18.

Utah Constitution, Article XX, Section 1, 1896, as amended 1997.

Utah Office of Planning and Budget, 2000, 1999 Economic Report To The Governor: Salt Lake City, 212 p.

Utah State Land Board minutes, June 14, 1966: Salt Lake City, Utah.

—August 17, 1976.: Salt Lake City, Utah.

—January 20, 1982: Salt Lake City, Utah.

Utah State Legislature, 1917, Laws of Utah, Chapter 114.

—1925, Laws of Utah, Chapter 31.

—1933, Laws of Utah, Chapter 46.

—1935, Laws of Utah, Chapter 97.

Utah vs. United States, 427 U.S. 461 (1975), U.S. Supreme Court, June 28, 1976.

Wilson, S.R., and Wideman, F.L., 1957, Sodium sulfate deposits along the southeast shore of Great Salt Lake, Salt Lake and Tooele Counties, Utah: U.S. Bureau of Mines Information Circular 7773, 10 p.

RECENT LEGAL ISSUES RELATING TO GREAT SALT LAKE

by

Michael M. Quealy
Assistant Utah Attorney General

ABSTRACT

This chapter discusses legal issues which have developed on Great Salt Lake after the U.S. Supreme Court quieted title in the State of Utah. These issues deal with private claims related to lands below the meander line, and the breaching of the railroad causeway during the 1980s flooding years.

INTRODUCTION

The authors of a similar article (Dewsnup and Jensen, 1980) in a preceding volume on Great Salt Lake, predicted prospective legal controversies between the state and private landowners over private claims to land below the Great Salt Lake meander line. Since the publication of that work in 1980, Dewsnup and Jensen's prediction has proven accurate, as several controversies along those lines have indeed arisen. Further, the dramatic rise in the level of Great Salt Lake in the mid-1980s gave rise to a whole new set of legal issues which were as unforeseen as the elevated lake levels. Each of these new issues will be discussed here.

PRIVATE LAND CLAIMS BELOW THE MEANDER LINE

As predicted, a number of private persons and entities have asserted ownership claims on lands below the meander line of Great Salt Lake. Such claims have usually been asserted by private landowners with patented lands adjacent to the meander line, although without exception, none of those landowners had patents which granted lands below meander. Usually such claims are based on a theory that a patent adjacent to the meander line somehow grants private rights below meander, or on the theory of reliction. However, the case law is clear that a patent above a meander line does not, and cannot, grant or convey a private interest to sovereign lands below meander, since the state owns those lands in its sovereign capacity by its admission to the Union under the Equal Footing Doctrine, and the United States had no power to convey such lands either before, or after, Utah's statehood. Moreover, any claims based on reliction are fairly weak. The dramatic rise in lake levels during the 1980s clearly shows the wisdom of the United States Supreme Court in adopting the meander line as the boundary of the state-owned lake bed given the wild fluctuations in lake levels over a relatively short period of time. For example, when the lake reached its highest levels in 1986 to 87, it actually exceeded the meander line in many places.

The state has, without exception, strongly defended its ownership below meander when challenged, and will continue to do so. Nevertheless, most private claims below meander have been settled through negotiation rather than litigation.

One of the major claims below the meander line was asserted by a group of ranchers in 1980 for lands in the vicinity of Locomotive Springs on the extreme northern end of the lake. The ranchers owned large tracts of patented land above the meander line, but they also laid claim to approximately 12,000 acres of adjacent land below meander. Since the early 1900s, the ranchers irrigated much of the land with water from Locomotive Springs, cutting wild hay in the summer, and grazing cattle in the winter. To further complicate matters, the ranchers (who were all related) traded and conveyed various tracts to each other over the years, creating a lengthy chain of title--but with no root patents. In addition, they paid county taxes on the property below meander for many years.

In 1980, the ranchers filed suit against the state to quiet titles to tracts below the meander line. After a year of legal wrangling, the parties negotiated a settlement of the suit. The ranchers agreed to recognize state ownership of all lands below meander in exchange for a perpetual easement on those lands to cut wild hay, and graze a limited number of cattle subject to certain limitations. Public access to the lake bed lands was preserved. The ranchers also granted to the state a perpetual right to maintain several waterfowl impoundments on private lands above meander, preserving habitat, public hunting access to those ponds, and a large state investment. The ranching operation has since gone out of business.

The second major claim below meander was made by the Lake Front Duck Club in the Farmington Bay area of Davis County. The Duck Club claimed several hundred acres of waterfowl marshes below the meander line. The Duck Club had occupied the land for many years and had constructed several dikes. The meander line cut an "S" curve through the Duck Club area and did not tie into the system of dikes which were connected to other adjoining state and private waterfowl management areas. A lawsuit was not brought, but again the parties negotiated a settlement in 1981. The Duck Club acknowledged the state's ownership below the meander line, and traded certain acreage above meander for limited acreage below. Also, the Duck Club entered into a 40-year lease with the state for additional land below meander in order to amortize improvements constructed by the club.

Other claims have arisen in the Willard/Bear River Bay areas, and these, too, were settled by land trades and/or leases. Throughout all the claims, the state has remained steadfast in its defense of ownership below the meander line. Nevertheless, negotiation of such claims remains the preferable resolution--with litigation as a last resort.

THE FLOODING YEARS -- LEGAL ISSUES

Due to extraordinary precipitation levels in the early to mid-1980s, the Great Salt Lake rose dramatically. Due to the shallow nature of the lake, the rising levels inundated thousands of acres surrounding the lake. The majority of the inflow to the lake comes in south of the railroad causeway, and the causeway acted as a dam, further exacerbating flooding on the south portion of the lake. At its highest, the level on the south arm was nearly four feet higher than the north arm. Interstate highways, the railroad causeway, sewage treatment plants, the Salt Lake City Airport, waterfowl management areas, and marinas were being seriously threatened or inundated.

In response to these serious public safety threats, the Utah Legislature passed the Great Salt Lake Causeway Act in its 1984 session, directing that the state coordinate with Southern Pacific Transportation Company to breach the railroad causeway in order to lower the level of the south arm of the lake. During the same session, the legislature amended the Utah Governmental Immunity Act to limit the liability of governmental entitles for the management of flood waters.

The Great Salt Lake Causeway Act authorized the state's Divisions of State Lands, and Water Resources, to work with the Southern Pacific Transportation Company to construct a breach of the railroad causeway near Lakeside, on the far west shore of the lake. An emergency permit was obtained from the U.S. Army Corps of Engineers to accomplish the breach, and construction of the breach and a trestle was begun almost immediately.

In spite of the rapidly rising lake levels, the Great Salt Lake Mineral & Chemical Company actively opposed breaching the causeway. The company drew its brines from the north arm of the Great Salt Lake, and they felt the breach would dilute their brine source. In March of 1984, the company filed suit in federal court against the state and the U.S. Army Corps of Engineers, challenging the Corps' issuance of the emergency dredge-and-fill permit for the breach. Soon after the lawsuit was filed, William Colman intervened, claiming his interests would also be harmed by the breach. Colman alleged that he operated and maintained an underwater brine-intake canal running parallel to, and approximately 1,300 feet north of the causeway under a lease from the state. Although Colman was not then in active salt production, he claimed the breach would destroy his canal and would impair his ability to withdraw deep lake brines from the north arm.

In late May of 1984, federal Judge Aldon J. Anderson heard evidence and oral arguments on whether an injunction should be issued to halt the causeway breach. The only issue before the court was whether issuance of the emergency permit by the U.S. Army Corps of Engineers had been arbitrary or an abuse of its discretion, for failing to adequately address the alleged adverse impacts of the breach on north arm resources.

On June 1, 1984, Judge Anderson issued a ruling denying plaintiff's requested injunction, thus allowing construction of the breach to continue. The judge reasoned that the benefits in alleviating flooding on the south arm of the lake outweighed any damage to north arm interests, and that the Corps of Engineers had adequately addressed environmental and other issues prior to issuing the permit. Further, the court held the breach to be in the "public interest" to alleviate the dramatic flooding on the lake.

Once the injunction was denied, the state and Southern Pacific pushed ahead with all dispatch to complete the breach. The target date for completion was August 1, 1984. But the court battles were not quite over.

On July 20, 1984 (10 days before the breach was to occur), William Colman filed a second lawsuit to halt the breach. This time, the action was filed in State District Court. The lawsuit alleged that the breach would be an unconstitutional taking of Colman's property. The court denied Colman's motion for an injunction after an evidentiary hearing on July 31, 1984, and the causeway was breached the next day.

Even though the causeway had been breached, Colman pushed his claim for damages to his brine ditch. The state moved to dismiss the action on the grounds of governmental immunity, exercise of police powers to alleviate a flood emergency, the exercise of the state's public trust duties on a navigable body of water, and that there was no compensable taking of a property interest. On May 2, 1986, the court granted the State's motion to dismiss.

Colman appealed the dismissal to the Utah Supreme Court. In April of 1990, the Utah Supreme Court reversed and remanded the matter back to the District Court for trial. The Supreme Court ruled that there were factual issues on which Colman should have been allowed to present evidence. The court reasoned that Colman may well have had a compensable property interest which may have been damaged by the breach. The court further ruled that the flooding on the lake was not a true "flood emergency" since the lake had been rising for several years. The court also ruled that since an unconstitutional taking of Colman's property was alleged, the doctrine of sovereign immunity could not protect the state from such a claim. Finally, the court expressed doubts as to whether the state could damage Colman's brine ditch in furtherance of the Public Trust Doctrine.

Once the case was remanded for trial, negotiations resulted in settlement of the litigation, and the case was ultimately dismissed. However, the Utah Supreme Court's decision presents the state with future uncertainty as to how far it can go in managing the Great Salt Lake and its numerous resources without being subject to litigation and claims for damages. There are numerous competing business and recreation interests on the Great Salt Lake such as salt extraction, brine shrimp harvesting, and waterfowl management. How the state balances these competing and future interests, while preserving the lake under the Public Trust Doctrine, will indeed be a challenge for the future.

The finale of the 1980s flooding years was the West Desert Pumping Project. The details of that ambitious project will undoubtedly be discussed elsewhere in this volume. From a legal standpoint, the project fortunately did not result in any litigation. Nevertheless, that massive undertaking kept state attorneys busy for months negotiating contracts and permits with the Army Corps of Engineers, the U.S. Bureau of Land Management, private landowners, the railroad, natural gas suppliers, and the manufacturers of the giant pumps and engines.

REFERENCE

Dewsnup, R.L., and Jensen, D.W., 1980, Legal Battle over ownership of the Great Salt Lake, *in* Gwynn, J.W., editor, Great Salt Lake - a scientific, historical and economic overview: Utah Geological and Minerals Survey Bulletin 116, p. 15-18.

THE ARMY CORPS OF ENGINEERS' REGULATORY PROGRAM AND THE GREAT SALT LAKE, UTAH

by
Dennis Blinkhorn and Mike Schwinn
Army Corps of Engineers
Utah Regulatory Office
1403 South 600 West, Suite A
Bountiful, Utah 84010

ABSTRACT

The U.S. Army Corps of Engineers has responsibility for regulating the placement of fill in the nation's waters, including the Great Salt Lake, under Section 404 of the Clean Water Act. The purpose of the act is to protect the nation's aquatic resources from unnecessary adverse impacts. Anyone who places dredged or fill material in a stream, lake, pond, playa or wetland must have a Section 404 permit from the Corps of Engineers. Permits can only be issued for the least damaging practicable alternative. This requires a review process intended to avoid and minimize adverse project impacts to the aquatic environment. Typical regulated activities include fill work associated with the installation of utility lines, road, driveway and bridge construction, and bank stabilization projects. Once it is determined that a Section 404 permit is required, a number of other Federal laws come into play. These include the Endangered Species Act, the Fish and Wildlife Coordination Act and the National Environmental Policy Act. A review of several important permit actions on the Great Salt Lake is included.

INTRODUCTION

The Great Salt Lake and the streams and wetlands surrounding it constitute a large, complex and unique ecosystem. Nowhere else in the country is there such an expanse of inland saline water and wetlands. Anything that affects any component of this ecosystem affects the entire ecosystem. The U.S. Army Corps of Engineers' (Corps) regulatory program attempts to protect the public interest in this resource while recognizing the rights and needs of property owners.

The Corps regulates certain types of activity on the Great Salt Lake and adjacent waters under Section 404 of the Clean Water Act. Originally named the Federal Water Pollution Control Act of 1972 and renamed in 1977, the objective of this law, as stated in its opening paragraph, is to restore and maintain the chemical, physical and biological integrity of the nation's waters. Section 404 of this act, administrative responsibility for which was delegated by Congress to the Secretary of the Army, regulates the discharge, or placement, of dredged and fill material in waters of the United States. Under Section 404, the Corps functions not as a military organization but as a regulatory agency, not unlike the Environmental Protection Agency, to assure that the mandate of Congress is carried out. In fact, the Corps and the Environmental Protection Agency share Clean Water Act responsibility and cooperate in a number of ways to carry out the law.

The Corps has been involved in regulating activities on the nation's waters since the passage of the Rivers and Harbors Act in 1899. Until 1972, the focus of the Corps' regulatory program was protecting and maintaining commercial navigation on major streams and in harbors. With the passage of the Clean Water Act, the Corps' regulatory program began to evolve and it has continued to evolve, as the result of several new laws and a number of judicial decisions, into one that focuses not only on commercial navigation but also on protecting the nation's aquatic resources from unnecessary adverse impacts. To accomplish this, each permit application is reviewed in light of a number of public interest factors and the decision to issue or deny the requested permit is made by balancing favorable project impacts against anticipated detrimental impacts. To implement the Section 404 program in the manner intended by Congress, the Corps now asserts Section 404 jurisdiction over essentially all streams, lakes, wetlands and playas. The Great Salt Lake and its adjacent wetlands and playas, as well as the streams leading into the lake, are all subject to Section 404.

Section 404 gives special protection to certain types of waters, such as wetlands, in light of their unique benefits and diminishing numbers. Wetlands provide important benefits such as wildlife habitat, water quality enhancement, flood water retention, ground water recharge and erosion control buffers. For this reason, permit applications for projects that would adversely affect wetlands are examined very closely and project proponents are required to prove that there is no reasonable alternative that would be less damaging to the wetland. Permits that allow wetlands to be filled usually require the permittee to mitigate, or replace, wetlands lost as a result of the activity by creating new wetlands at another location.

TYPES OF ACTIVITIES REQUIRING PERMITS

Under Section 404, the Corps regulates the placement of dredged and fill material in jurisdictional waters. Fill is defined as any material which changes the bottom contours of the waterbody. Earth, rock, sand, gravel and trees are examples of material that can be used as fill. However, it is important to remember that, in identifying fill material, it is not so much the type of material that is important but rather the effect on the bottom contours of the waterbody. Flood protection dikes, road embankments, bridge abutments, rock placement for erosion control and fills for buildings, driveways and parking lots are a few examples of the types of fills regulated under Section 404. A Department of the Army Section 404 permit is required for all of the above activities, or for any similar activity, if placing fill in a jurisdictional water is part of the project.

There are three types of Department of the Army Section 404 permits: nationwide permits, general permits and individual permits. Nationwide permits are the most commonly used form of authorization. These were developed at the national level for certain types of fill activities which normally result in minimal environmental impact. The intent of the nationwide permit program is to limit paperwork and reduce application processing time. Some nationwide permits require that the Corps be notified before work is initiated, while others require no notification. Each nationwide permit requires adherence to limitations and conditions developed specifically for that permit and all nationwide permits require that best management practices be followed in order to further reduce adverse impacts. Currently there are 39 nationwide permits. Commonly used nationwide permits are those which apply to fills associated with the installation of utility lines, such as water and sewer lines, the stabilization of eroding shorelines and stream banks, the repair or maintenance of previously authorized structures or facilities, the installation of fish and wildlife enhancement structures and boat ramp construction. Another nationwide permit, and one that is sometimes used to authorize projects in wetlands adjacent to the Great Salt Lake, is Nationwide Permit No. 14. This permit authorizes discharges of fill material associated with minor road crossings. It is often used to authorize the construction of driveways across wetlands. Local governments sometimes receive Nationwide Permit 14 authorization for minor fills associated with road construction or repair work.

The nationwide permit program changes periodically. For this reason and to assure that a proposed project satisfies the terms and conditions of a nationwide permit, it is recommended that anyone proposing to complete work under a nationwide permit first contact the Corps for verification that the work is authorized.

General permits are a form of authorization developed by local Corps offices specifically for certain types of frequently occurring discharges that are not covered by any of the nationwide permits. General permit limitations are well defined and adverse impacts associated with authorized projects are further limited by conditions attached to each individual authorization. General permits currently in use by the Corps' Utah Regulatory Office include General Permit 40 for those projects which have already received a stream alteration permit from the state engineer and General Permit 59 for fills associated with emergency flood protection work or the repair of flood damage. In addition, General Permit 44 is available to authorize certain fills associated with work in waterfowl management areas. Under this general permit, managers of refuges adjacent to the Great Salt Lake can apply for and receive timely Section 404 authorization to construct or repair dikes, canals, nesting islands and other such facilities on their refuges. Anyone contemplating doing work under a general permit must first contact the Corps for confirmation that the work is authorized by the general permit.

Any proposed discharge of dredged or fill material that cannot be authorized by either a nationwide permit or a general permit must be authorized by an individual permit. An individual permit requires that a project proponent complete an application form and submit it to the Corps. This starts a public-interest review process which begins with the issuance of a public notice inviting comment from interested Federal, state and local agencies, organizations and individuals. After the close of the comment period, the Corps evaluates all available information and decides if issuing the requested permit would be counter to the public interest. A central part of this public interest review are the Section 404(b)(1) Guidelines, which are an evaluation of the impacts of the proposed work on the aquatic environment. The 404(b)(1) Guidelines only permit selection of the alternative which will result in the least damage to the aquatic environment, while still achieving project goals. For example, an applicant who proposes to construct a commercial development in a wetland might be required to explain why the wetland could not be avoided entirely. If this point were satisfactorily addres-sed, the applicant might be asked to minimize wetland im-pacts by relocating a portion of the project or reducing the size of the fill area. Finally, for those adverse impacts that cannot reasonably be avoided, mitigation is required. Usual-ly this means that someone who is issued a Department of the Army permit to fill a wetland must create a new wetland to offset the loss of wetland functions associated with the fill activity. Only after this review has been completed can an individual permit be issued. Several important individual permit actions are examined in detail below.

OTHER APPLICABLE LAWS

Although Section 404 is the key to the Corps' involvement on the Great Salt Lake, once it is determined that Section 404 is applicable, the Corps has obligations under several other Federal laws. Two such laws are the Fish and Wildlife Coordination Act and the Endangered Species Act, which require coordination with the U.S. Fish and Wildlife Service to assess project impacts to fish, wildlife and plant resources. The National Historic Preservation Act requires coordination with the State Historic Preservation Office to identify possible impacts to cultural resources which might be listed on or eligible for the National Register of Historic Places. Sometimes previously unknown archaeological resources, such as Native American burial sites or camp sites, are discovered during the construction of authorized projects. In such cases, Section 404 permits require that work stop until the National Register eligibility of the site can be determined. In addition, the National Environmental Policy Act applies to all Corps permit actions. This act requires that the Corps complete either an environmental assessment or an environmental impact statement for each permit action. These documents must include an examination of all anticipated impacts to the human environment as a result of the proposal, not just those associated with the aquatic resource. It is because of these laws that the Corps must consider non-water-related impacts when evaluating permit applications.

One additional consideration is Section 401 of the Clean Water Act. Section 401 review assures that applicable water quality standards will not be violated by work completed under authority of a Section 404 permit. In Utah, Section 401 certification responsibility has been delegated to the Utah Department of Environmental Quality, Division of Water Quality. While nationwide permits and general permits normally receive blanket Section 401 certification when they are issued, each application for an individual Section 404 permit must be reviewed and receive Section 401 certification from the Division of Water Quality before the Corps can issue a permit.

SIGNIFICANT INDIVIDUAL SECTION 404 PERMIT ACTIONS ON THE GREAT SALT LAKE

Everyone, individuals, corporations and Federal, state and local government agencies, must get a Section 404 permit for regulated activities. While the objectives of the different projects undertaken by the various entities may vary, each must go through the same review process. This section summarizes some of the more visible public and private Section 404 permit actions on the Great Salt Lake in recent years.

Kennecott Utah Copper Corporation

Kennecott Utah Copper Corporation operates the Bingham Canyon Mine 30 miles southwest of Salt Lake City. The mine produces 152,000 tons of tailings each day. These tailings are stored in an impoundment near Magna. In 1985 the company identified a need to upgrade and expand the impoundment in order to accommodate the anticipated 1.9 billion tons of tailings to be generated during the remaining 25 to 30 years the mine is in operation. As the lead Federal agency under the National Environmental Policy Act, the Corps played an expanded role that included not only considering the company's request for a Section 404 permit to impact wetlands and playas but also overseeing the preparation of an environmental impact statement. The responsibility of the lead Federal agency in the preparation of an environmental impact statement is to assure that all reasonable alternatives to the proposed work and the anticipated impacts associated with the alternatives are identified, disclosed to the public, and considered in reaching an agency decision on the applicant's permit request. In the case of Kennecott, the no-action alternative would have resulted in the exhaustion of impoundment storage space and closing of the mine by the year 2004. Several off-site alternatives were considered and determined to be impracticable. Two on-site alternatives were also considered, one which would have affected 1,705 acres of wetlands, playas and other Section 404 waters and another which would have affected 1,055 acres of Section 404 waters. Other possible impacts to public use areas, aquatic life, the quality of both surface and ground water, and cultural resources were also identified. By considering the data contained in the environmental impact statement and the information gathered during the public interest review, the Corps was able to identify and permit the alternative that would result in the least damage to the human environment, including waters of the United States, and issue a permit for Kennecott to impact jurisdictional wetlands so that the impoundment expansion could take place and the mine could continue operating. To compensate for unavoidable wetland impacts, a condition was attached to the permit that required Kennecott to design and implement a mitigation plan for the creation of replacement wetlands. The approved plan, which called for the creation of a 2,500-acre wetland complex, was implemented on land near the tailings impoundment. Initial monitoring reports indicate that the mitigation wetlands are developing quickly.

Salt Lake City International Airport Expansion

In 1988 the Salt Lake City Airport Authority identified the need to expand the facilities at the airport in order to meet anticipated future demand. Since such an expansion would involve a significant impact on the human environment, the environmental impact statement process was initiated. In this case, the Corps participated not as the lead Federal agency but as a cooperating agency in development of the impact statement, in addition to its role under Section 404.

The Corps' Section 404 public interest review coincided with development of the final environmental impact statement. Using the information published in the environmental impact statement, that review identified eleven different alternatives. These alternatives ranged from the no-action alternative to four off-site alternatives. Six on-site alternatives were also identified and considered. The selected alternative, that is, the one identified as the least damaging practicable alternative, was determined to impact 336 acres of wetland. A permit to fill these wetlands was issued in 1992, conditioned to require the creation of mitigation wetlands to replace wetland functions lost as a result of the project. These wetlands were successfully created on land near the airport and today support a thriving population of waterfowl, shorebirds and birds of prey, in addition to a healthy mammal population (figures 1 and 2).

Great Salt Lake Minerals and Chemicals Corporation

Great Salt Lake Minerals and Chemicals Corporation, now known as IMC Kalium Ogden Corporation, is a company that produces potassium sulfate and other minerals. These minerals are extracted from the waters of Great Salt Lake through a solar evaporation process. When the heavy runoff from snow melt and precipitation effectively reduced the salt content of the lake's north arm waters in the 1980s, the company's production costs significantly increased. This problem was magnified by several breaches in the Southern Pacific Railroad causeway, which allowed the less salty water from the lake's south arm to flow more freely into the lake's north arm.

To solve this problem, in 1991 the company proposed to build a new pump station on the west shore of the lake.

Figures 1 and 2. *Mitigation wetlands created during the expansion of the Salt Lake City International Airport (photos by Dennis Blinkhorn).*

Associated with this pump station was to be a solar evaporation pond 18,000 acres in size. This pond was to be created by using earth and rock fill material for the construction of an 8.5-mile-long containment dike across the lake bed between Strong's Knob and Finger Point. To carry the concentrated brine from the evaporation pond to an existing pump station on Promontory Point, the company proposed dredging a 20-mile-long submerged channel in the lake bed. The channel was to be up to 110 feet wide and as much as 34 feet deep. Its construction would require the dredging of 3,000,000 cubic yards of lake bed material, which were to be discharged back to the lake a short distance from the channel. The channel was designed to slope to the east, so that gravity would pull the heavier brine to the Promontory Point pump station, from where it would be pumped to the company's evaporation ponds on the east shore for final processing.

During the public interest review, a concern was raised about possible project impacts to lake salinity levels and how this might affect the brine shrimp industry and other mineral producing companies that also rely on the Great Salt Lake's waters. Through a series of meetings between the Corps, the company and the concerned parties, the proposed work was modified and these issues were resolved. The necessary Section 404 permit was issued in 1991 and work began on the evaporation pond and the channel, which is sometimes called the Behrens' Trench.

West Desert Pumping Project

Historically the level of the Great Salt Lake is relatively stable, aside from seasonal fluctuations. However, between 1982 and 1985 the lake had risen to nearly 10 feet above normal, resulting in hundreds of millions of dollars of property damage. With predictions of even higher water, the state decided to take action.

After studying a number of alternatives, the state decided that the answer would be a series of dikes to protect property on the east shore and a pump station on the west shore to pump water from the north arm of the lake to two evaporation ponds. These two ponds would total 463,000 acres and would provide for the evaporation of approximately 1,000,000 acre-feet of water annually. Engineers estimated that the project would lower the lake level by two feet.

In order to implement the plan, additional work had to be completed. An intake canal, 1,600 feet long and 13 feet deep, had to be dredged in the lake bed to carry lake water to the pump station. It was also necessary to breach the Southern Pacific Railroad causeway and construct trestles at two locations. One breach near Hogup Ridge would allow the intake canal to pass through the causeway. A second breach near Lakeside would allow the construction of a five-mile-long return canal to carry concentrated brine from the evaporation ponds back to the north arm of the lake. To carry trains around these breaches during trestle construction, additional fills were necessary for the construction of temporary bypass causeways, each 1,000 feet long.

This large project with such significant impacts required an environmental impact statement. The Bureau of Land Management acted as the lead Federal agency with the Corps participating as a cooperating agency to provide input into the planning process and alternatives analysis to assure that the environmental impact statement was complete and to expedite the processing of the Section 404 permit application. After considering all reasonable alternatives, the Corps issued the necessary Section 404 permit to the state on July 21, 1986, and construction began soon thereafter.

CONCLUSION

The Corps' regulatory program has been successful in protecting and enhancing the quality of the nation's aquatic resources. The regulatory program, along with efforts by the Environmental Protection Agency and other Federal and state agencies, has greatly reduced the unnecessary loss of wetland acreage around the Great Salt Lake. This success assures that the Great Salt Lake and associated streams and wetlands will continue to function as a complete ecosystem to the benefit of the public and wildlife.

HUMAN HABITATION AND ART

ROBERT SMITHSON'S *SPIRAL JETTY*

by

Hikmet Sidney Loe
Salt Lake City Public Library

INTRODUCTION

"Through the vaporous abstraction of Box Elder County, Utah, I beheld a wide expanse of lake, whose waters were so bloody a hue as to bring to my mind a geography of unspeakable carnage. Yet at the same time, a voluptuous color prevailed. A luminous languor coupled with a foreboding sense of menace produced a gyratory dimension. Innumerable spirals flashed from the lake, assuming no distinct or definite existence, but instead whirled off into burning distances. There are times when the water glows like scarlet syrup, and times when it fades into colorlessness (Robert Smithson Papers, Archives of American Art, Smithsonian Institution, roll 3834, frame 130)."

It was in 1970 that Robert Smithson (1938-1973) visited northern Utah to select a site to construct an Earthwork. He chose Rozel Point, located in a remote section of the north end of Great Salt Lake in Box Elder County. The site is surrounded by nothing much except sparse hills leading to the shores of the water, vistas of surrounding mountains, and the flat expanse of the lake. This barren area was to work its way into Smithson's imagination in just a few months to yield his Earthwork, *Spiral Jetty*, a 1,500-foot spiral of earth and rocks jutting from the shore into Great Salt Lake (figure 1), to be followed by a film and several essays of the same name.

In a short list of twentieth century American art works, the *Spiral Jetty* would be among the few that serve as a turning point in our collective consciousness, joining the pantheon of all that we call sculpture and Earthworks. It acts as signifier of all that is American in the grandness of art produced in the vast expanse of the country's West. It is an artwork of paradoxes and multiple dimensions, constructed by bulldozers and trucks from earth and rocks. It is a permanent installation in a lake that continually covers it up, only to reveal it every few years to travelers seeking art in a desert.

Smithson developed permanent records of his Earthwork in a variety of forms. He produced a multitude of drawings of the *Spiral Jetty* and its site before, during, and after building the work. Smithson filmed the *Spiral Jetty* (Smithson, 1970a) during its construction and after it was completed, lending more permanence to a shifting image, although the film is not easily accessible to viewers. Shortly thereafter, Smithson (1972) wrote an essay of the same name perfecting the self-generating history of the *Spiral Jetty*. The article both romanticizes the Earthwork and the film while providing factual information on the Jetty's construction.

Figure 1. Spiral Jetty, *April 1970, Great Salt Lake. Photograph by Gianfranco Gorgoni. Art © Estate of Robert Smithson/VAGA, NY, NY.*

Due to its remote location and frequent inundation by the lake, the *Spiral Jetty*, more than the other large-scale Earthworks created by Smithson, relied on mass media for the dissemination of its existence and history. Within Smithson's body of work, this sculpture was the most elaborately documented by Smithson himself and other artists. Since its construction, it has been documented in almost every treatise on the movement entitled, alternately, as Earth Art, Earthworks, Land Art, and Environmental Art, as well as in other publications covering film, sculpture, photography, and twentieth century art. This has resulted in an ongoing mythology of the work.

At the time of its construction in 1970, there were few references associating the *Spiral Jetty* to its location in Utah. While the national art world wrote extensively on the *Spiral Jetty*, local attention to the piece was scarce; it was to take over one year for an article to appear about the site in any publication in Utah. The 1990s saw a widening interest in the *Spiral Jetty* while the national art world paid continued homage to it as an Earthwork. Local interest by Utah's artistic and media communities in the *Spiral Jetty* has increased, however, since its sporadic re-emergence from the waters of the lake in 1993.

ROBERT SMITHSON AND THE *SPIRAL JETTY*: ET IN UTAH EGO

Robert Smithson came to Great Salt Lake through circuitous means. His original intent was to construct an Earthwork related to an inland body of red saline water. In 1969, he read a book titled *Vanishing Trails of Atacama* (Rudolph, 1963), which described the process of mineralization in isolated bodies of water and the strange colors produced through this process. Smithson considered looking into the lakes of Bolivia, which contain such bodies of water, but decided that their geographical remoteness would make construction prohibitive. After further investigation, he heard of Great Salt Lake and the bacteria that dwell in its northern part, which often give it a reddish tinge. In his essay, *The Spiral Jetty*, Smithson (1972) recounts hearing of this location:

> From New York City I called the Utah Parks Development and spoke with Ted Tuttle, who told me that water in the Great Salt Lake, north of the Lucin Cutoff which cuts the lake in two, was the color of tomato soup. That was enough of a reason to go out there and have a look (Smithson, 1972).

This brief statement does not hint at the extensive amount of research Smithson was to conduct in creating *Spiral Jetty*. It is apparent from reviewing Smithson's project files on the *Spiral Jetty* (Robert Smithson Papers, Archives of American Art, roll 3,835, frames 798-1,045) that he spent considerable time researching this project. He collected articles, maps, brochures, and clippings on Rozel Point and the surrounding oil fields, the Great Salt Lake (its contents, history, and fluctuating levels), and information on Utah culture and sites of interest.

According to Smithson, the red color of the lake was one preliminary factor that attracted him to the site. Along with the bacteria, several other factors give the lake its unearthly color. The north arm of Great Salt Lake, and the location of Rozel Point, contains a higher level of salt than the southern half, creating a rather harsh environment in which few organisms can live. Of these few organisms, red algae and bacteria are able to thrive. The brine shrimp are a brilliant red color, and during the fall molting season they deposit their shells in the lake. While the shrimp contribute to the lake's red color, the bacteria provide almost a constant source of color, ranging from blood red to rust orange and purple pink. The bacteria live on the remains of the brine flies and shrimp in close proximity to the algae (Post, 1975). It is the bacteria that gives the lake its constant red color, and, combined with the algae and brine shrimp, provides a spectacular color to the lake as observed from the air as well as from the shore.

In the 1970s, Rozel Point contained dilapidated structures related to oil drilling, reminding Smithson of more familiar urban sprawls back east. His interest in sites that contained vestiges of human waste or, in his words, "sites that had been in some way disrupted or pulverized" (Alloway, 1972), was fulfilled at Rozel Point, an area that contained no other signs of human interaction except for stray cattle from nearby Promontory Ranch. In other words, this landscape, equipped with the abandoned remains of human endeavors, was the perfect "entropic landscape" for Smithson.

There was a rumor from the nineteenth century that Great Salt Lake contained whirlpools connected to subterranean channels leading to the Pacific Ocean. This appealed to Smithson's sense of the romantic, so much so that he mentions this legend in his film. Early trappers in Utah, including Jim Bridger, who was one of the first explorers to see Great Salt Lake, mistook the lake for the Pacific Ocean due to its high salt content. Also, cartographers in the early 1800s began to connect the as-yet-unnamed (at least by European immigrants) lake to the Pacific Ocean by means of a river called Rio Buenaventura. Even as the shores of the lake became more clearly defined through subsequent exploration, myths still continued to connect the lake to the ocean due to its salt content. Years later, in 1845, the explorer John Charles Frémont reported that: "It was generally supposed that it has no visible outlet [Great Salt Lake]; but among the trappers, including those in my own camp, were many who believed that somewhere on its surface was a terrible whirlpool, through which its waters found their way to the ocean by some subterranean communication" (Fremont, 1845).

While Smithson may not have known about these legends when he originally went in search of a site to build an Earthwork, he knew intuitively what he was searching for, which, in the following passage, describes the whirlpool effect the land was to make on him:

> About one mile north of the oil seeps I selected my site. Irregular beds of limestone dip gently eastward, massive deposits of black basalt are broken over the peninsula, giving the region a shattered appearance. It is one of the few places on the lake where the water comes right up to the mainland. Under shallow pinkish water is a network of mud cracks supporting the jigsaw puzzle that composes the salt flats. As I looked at the site, it reverberated out to the horizons only to suggest an immobile cyclone while flickering light made the entire land-

> scape appear to quake. A dormant earthquake spread into the fluttering stillness, into a spinning sensation without movement. This site was a rotary that enclosed itself in an immense roundness. From that gyrating space emerged the possibility of the *Spiral Jetty* (Smithson, 1972).

Finally, Smithson was familiar with the West, having traveled from the East Coast with his parents on family vacations, then later hitchhiking around the country, traveling west again on several occasions. As an artist working with the earth, he traveled west to collect specimens and images. Another great appeal for Smithson in choosing Great Salt Lake as his site was the mirror-like quality of the lake, especially in this sheltered region in the lake's north arm. Not only could he create a spiral, but the water would act like an inverse spiral and a mirror at the same time. The lake could mirror the spiral, but also mirror the surrounding land and sun, turning landscapes into themselves.

Smithson began his preliminary work in Utah on *Spiral Jetty* by February 1970. The finished documents of the *Spiral Jetty* give the impression that it was constructed as originally conceived, while in fact, Smithson's plans were to undergo a number of revisions before the resulting *Spiral Jetty* was complete. The concept was originally of a jetty with an island in the center of the spiral. A human-made island in Great Salt Lake would have blended in with the lake's topography, adding one more rounded land mass to the 10 islands already in existence in the lake. But, while the idea of the island was to remain throughout Smithson's many revisions of the work until almost the end, the spiral was to always dominate.

Robert Smithson spent several weeks in Utah in March 1970, looking for a site on which to build his Earthwork. On March 10th, he submitted a Special Use Lease Application to the Division of State Lands at the Capitol Building in Salt Lake City. His application letter mixes tones of factual data: "...I am respectfully submitting our Special Use Lease Application for a period of twenty (20) years, and the total area will be approximately ten (10) acres. I offer $10.00 per-acre-per-annum rental totaling $100.00 per year," with artistic intent:

> This project will resemble a jetty in the shape of a spiral. The structure will be made of rock and gravel. The purpose of placing the rock on the mud flat area will be to induce salt crystals on the rock and gravel as incrustations that will develop over a period of time. These will contrast with the red color of the water. Its purpose is purely esthetic and it can be viewed from airplane or from the road (Robert Smithson Papers, Archives of American Art, roll 3833, frame 90).

After visiting Rozel Point, Smithson began to conceptualize his artistic piece using maps of Great Salt Lake to fix the site. With lease in hand, he contacted one construction company to build his Earthwork, which lead him to a second company, Parson Asphalt Products, Inc., and to a close association with the project manager, Bob Phillips of Ogden, Utah. Smithson briefly recounts this initial interaction in his essay:

> After securing a twenty year lease on the meandering zone, and finding a contractor in Ogden, I began building the jetty in April, 1970. Bob Phillips, the foreman, sent two dump trucks, a tractor, and a large front loader out to the site. The tail of the spiral began as a diagonal line of stakes that extended into the meandering zone...(Smithson, 1972).

As with any art work, Smithson moved through a progression of ideas for the *Spiral Jetty*. His original intent, and the first impression that Phillips had of the work, was to construct a site shaped in a clockwise "J" with an island in the center. From there, the site quickly turned into a counterclockwise J-shaped spiral. The spiral then opened up, with the island remaining in the center. It is apparent in viewing Smithson's working drawings that the concept of the island remained with him through the process. It is also apparent from the dimensions included on the drawings that Smithson had a much smaller spiral in mind, one that would extend 300 feet into the water, rather than the lengthy spiral that was ultimately built.

Phillips drew up a work order for building the Earthwork in the beginning of April for $6,000. It was promptly financed by New York City gallery owner Virginia Dwan, and the construction began. The counterclockwise spiral that was constructed used approximately 6,650 tons of earth and basalt from the adjacent shoreline. The *Spiral Jetty*'s construction, completed in approximately six days in April, took "292 truck-hours (taking 30 to 60 minutes per load) and 625 man-hours (adding up to more than 10 tons of material per hour)" to complete (Hobbs, 1981).

The use of massive caterpillars and huge dump trucks, while standard equipment for the time, lent an air of antiquity to the project. In the film *Spiral Jetty*, Smithson romanticizes the monstrous quality of the machinery by comparing them to dinosaurs such as the Tyrannosaurus rex. Virginia Dwan also made the same comparison when she discussed the construction as "the sound of the machinery and the machinery itself and its grossness and massiveness and voraciousness of the backhoe and the dump truck, biting in and gobbling up...it's like another dinosaur at a later date, and placed in the context of an inland sea" (Dwan, 1984).

Visiting the site today yields no answers as to where the 6,650 tons of rocks and earth were obtained to build the *Spiral Jetty*, but according to Phillips, the rocks and dirt were taken from the immediate shoreline and nearby hill. After constructing the work, the crew was ordered by Smithson to cover up the area of land which surrendered the rocks to make it look less disturbed.

When the project was completed, the crew and Smithson left. Upon inspecting the site the next day, Smithson determined that the central island in the project was all wrong and ordered it removed. Another work order was drawn up for $3,000 to remove the island. The crew returned the following day to take out the island and round out the end of the spiral as can be seen in this drawing (figure 2). Another of Smithson's working drawings (dated April 1970) shows the spiral without the island, now approximately 800 feet from the shore, with a temporary truck crossing, which, according to Phillips, was never used. In all of the construction, and then reconstruction of the *Spiral Jetty*, the crew drove into and backed out of the fifteen-foot-wide, 1,500 foot-long strip of earth each day.

Smithson (1972) captured the mood of the existing landscape when he wrote:

Figure 2. Spiraling Jetty in Red Salt Water, Ogden, Utah, *1970. Pencil on paper by Robert Smithson. (Sarah-Ann and Werner H. Kramarsky owners). Courtesy James Cohan Gallery, NY. Art © Estate of Robert Smithson/VAGA, NY, NY.*

> An expanse of salt flats bordered the lake, and caught in its sediments were countless bits of wreckage. Old piers were left high and dry. The mere sight of the trapped fragments of junk and waste transported one into a world of modern prehistory. The products of a Devonian industry, the remains of a Silurian technology, all the machines of the Upper Carboniferous Period were lost in those expansive deposits of sand and mud...This site gave evidence of a succession of man-made systems mired in abandoned hopes (Smithson, 1972).

The 16 mm color film Smithson created of the project was shot mostly in Utah, with interior shots filmed in New York City at Smithson's favorite museum, the Museum of Natural History. The completed film, which runs approximately thirty-five minutes, interweaves several of Smithson's favorite topics - geological time, history, science, entropy, and dinosaurs - with images of the *Spiral Jetty*, in two distinct parts. The first part of the film documents the surrounding area of the *Spiral Jetty*, shots of dusty roads and the construction work, interspersed with footage of foaming red water. The latter part of the film is more contemplative as it shows Smithson running the length of the jetty with the helicopter (which films him) flying overhead around him in endless spirals. Allusions fill the film, from echoes of vast spans of time to geographic changes, from the extinction of species to the hidden mysteries of moving earth in order to form a spiral. These allusions are noted in the following article:

> The film begins with a series of contrasts and variations of the idea of the spiral. It is a progression from spirals of dust behind the car on the road to the Jurassic and dinosaurs to trucks and bulldozers, with peaceful interludes of pink water on which float patches of silvery foam. The reflection of the sun dances on the water in the curve of the spiral. The film reaches a hypnotic climax a long way from the prosaic road and the mundane maps and lake charts of the beginning (Utah Geological and Mineralogical Survey, 1971).

After the *Spiral Jetty* was built and the film was completed, Smithson turned to writing several essays on the *Spiral Jetty*. In 1970, he published "From the Center of the Spiral Jetty" (Smithson, 1970b), consisting of one of Gianfranco Gorgoni's stunning color photographs of the *Spiral Jetty* and Smithson's directional essay on the Earthwork. This essay was chanted by Smithson during the second half of the film while the helicopter circled the work. It also appeared in his major essay, "The *Spiral Jetty*." (Smithson, 1972) in Gyorgy Kepes' book, *Arts of the Environment* (the essay has since been reprinted several times in various publications). As one can tell by the quotes reprinted here from Smithson's essay, his writing weaves fact and fantasy together in impassioned prose, coupling Smithson's work on the sculpture and the film with a visceral account of Smithson's emotions while viewing the land through the production of these pieces.

Through Smithson's personal history with the *Spiral Jetty*, and particularly in this passage within his essay, Smithson addressed the topic of continuation and mortality while questioning his existence within the construct of the work. "All existence seemed tentative and stagnant. The sound of the helicopter motor became a primal groan echoing into tenuous aerial views. Et in Utah Ego. I was slipping out of myself again, dissolving into a unicellular beginning, trying to locate the nucleus at the end of the spiral" (Smithson, 1972). The statement "Et in Utah Ego" is an art-historical play on the painting *Arcadian Shepherds* (Nicolas Poussin, 1636) in which a group of shepherds find a tomb with the inscription "Et in Arcadia Ego," translated as "I, too, dwelt in Arcadia." Set within an idealized pastoral landscape, Poussin's goal was to point us towards the unattainable within a specific site.

In stating "Et in Utah Ego," Smithson reminds us that he,

too, was in Utah, but that it could be interpreted as only a state of mind, not a physical location in the here and now. By choosing to locate his piece at Great Salt Lake, Smithson pointed not to a pastoral landscape from which to gain joy, but to a site of ancient times and lakes that have left their marks etched on neighboring mountains from dead seas that are inhabited by few creatures. A place, in other words, marked by immortality on a grander scale than humanity. This landscape was to enter the psyche of the art world by the fall of 1970, beginning the canonization of the *Spiral Jetty*.

THE *SPIRAL JETTY'S* CONTINUING HISTORY

The *Spiral Jetty*, in all of its formats, received immediate attention within the media and the art world. First and foremost, the work's intrinsic importance was recognized along with its place in the establishment of a new paradigm of art. Then there was Smithson's passion for, and multimedia presentation of the *Spiral Jetty*, which reinforced the great attention afforded to it. Initial reaction to the *Spiral Jetty* seemed synchronized with the rise and fall of the lake, ebbing and flowing through the 1970s, then coming almost to a standstill in the early 1980s. After Smithson built his work, he had no idea what the surface elevation of the lake would be in the future. When the *Spiral Jetty* was built in 1970, the lake was at a low point. In May of 1971, the work was visible, according to Smithson, but by June it was underwater and was to stay underwater through July. In August he noted that "the water evaporated leaving the jetty totally encrusted with crystals. The lake was at its highest water level in 17 years, but the jetty showed no signs of erosion since it is almost solid rock..." (Robert Smithson Papers, Archives of American Art, roll 3833, frame 376).

The lake continued to rise steadily in the early 1970s: the rise in the lake was due in part to normal spring runoff from the mountains and increasing precipitation in what is usually an arid, desert climate. When Smithson built the *Spiral Jetty* in April 1970, the level of the north arm of the lake was approximately 4,195.15 feet above sea level. By August 1974 it was calculated at around 4,198.35 feet - some 3.2 feet higher than 1970. The lake would rise to unprecedented heights by 1975 (4,199.05 feet), then rise even higher in the 1980s (4,210.95 feet).

While the international art world knew immediately of the creation of the *Spiral Jetty*, it took much longer for local attention to be directed towards what would arguably become one of Utah's most popular art works, even during the work's brief periods of visibility. Only three articles were published locally on the Earthwork by 1973. The first article that featured the *Spiral Jetty* (Utah Geological and Mineralogical Survey, 1971) gave it front-page prominence (figure 3). It differs markedly from national articles on the *Spiral Jetty* in that it informs the reader of factual information concerning the work, such as the terms of the 20-year lease with the Utah State Land Board for the 10-acre site at Rozel Point, and the joint financing of the project for $25,000 by the Dwan Gallery of New York and the Ace Gallery of Vancouver, B.C.

Local reaction to the *Spiral Jetty* has reflected the visible state of the site only within the past few years. In November of 1992, the lake had reached a low of 4,197.1 feet, only 2 feet higher than when Smithson built the *Spiral Jetty* 22 years before. It was not until the summer of 1993, when the lake's level would drop to 4,196.9 feet in July, an all-time low for July since 1971, that visitors would venture to the north end of the lake, past the Golden Spike National Historic Site, to seek out the *Spiral Jetty*. Interest began with a news broadcast on KSL TV, Salt Lake City, which ran a story on the evening news on August 25th, 1971, about the *Spiral Jetty*. After seeing this broadcast, anyone previously interested in the *Spiral Jetty*, or in Earth Art, had license to travel almost three hours north of Salt Lake City in search of the work, knowing that they would finally find it.

By 1995, local residents were informed of the *Spiral Jetty*'s existence through some of the above activities, and through the publication of aerial photographs of the *Spiral Jetty*, which appeared in newspapers such as The Salt Lake Tribune. By this time, Rick Wilson, chief ranger at the Golden Spike National Historic Site had become accustomed to visitors stopping off at the site asking directions for the last leg of the journey to the *Spiral Jetty*. The site's office compiled a notebook of information regarding the *Spiral Jetty* (as well as on the *Sun Tunnels*, located west and south of the site near Lucin, close to the Utah-Nevada border; see article by Holt and Loe, this volume). Visitors can use the information to learn about the work, as well as to obtain a detailed map with directions to the *Spiral Jetty*. See appendix for directions to the *Spiral Jetty*.

In November 1995, Nancy Holt (wife of Robert Smithson) and two friends visited the *Spiral Jetty*, which was about 70 percent above water. They walked the length of it and flew over it - pleased to find that it was still visible, encrusted with salt crystals, and surrounded by the lake's red waters.

CONCLUSION

Robert Smithson (figure 4) would be impressed that this work has continued to rebuild itself through sporadic appearances, continuing documentation, and international publicity. While the art world was quick to give the *Spiral Jetty* an elevated status within the historical context of twentieth century art, Utahns have only recently allotted the *Spiral Jetty* its rightful place within the artistic culture and history of the State.

In the fall of 1999, the Dia Center for the Arts acquired the *Spiral Jetty* through a donation from the Estate of Robert Smithson. Although based in New York City, the Dia, an internationally renowned center for interdisciplinary art and criticism, has undertaken and managed permanent-sited works of art, including Earthworks such as Walter de Maria's *Lightning Field* in New Mexico. With the *Spiral Jetty* under its care, Dia intends to facilitate easier site access, and may raise the *Spiral Jetty*'s elevation higher so that it is above water more frequently. As the *Spiral Jetty* has not passed its 30th anniversary, it will be interesting to see what the future brings to this work in terms of access and continual visibility.

Over thirty years ago, at the end of the film "*Spiral Jetty*," Smithson chanted his verse "From the Center of the *Spiral Jetty*" as the helicopter flew in dizzying spirals around the *Spiral Jetty* - his voice and the images echoing the musi-

UTAH GEOLOGICAL AND MINERALOGICAL SURVEY

QUARTERLY REVIEW

Vol. 5, No. 2 *Geologic Investigation in the State of Utah* May 1971

EARTH ART IN GREAT SALT LAKE

An overwhelming work of art has been constructed in Great Salt Lake. In April of 1970 Robert Smithson, builder of earthworks as art objects, supervised the construction of a spiral jetty of earth and black basalt rocks. It is 1,500 feet long and 15 feet wide.

The project was financed by $25,000 from the Dwan Gallery of New York and the Ace Gallery of Vancouver, B. C.

Smithson, seemingly intent like the ancient Egyptians on creating landmarks on a grand scale, negotiated a 20-year lease with the Utah State Land Board for a 10-acre site at the north end of the Lake. The dark coil of the jetty lies in this barren area in water colored red by algae. The jetty itself is outlined by the white of precipitated salt.

To make this piece accessible to earthbound viewers, Smithson directed a 16 mm film of it, starting with the actual construction. While the film is called a documentary, it is a work of art in itself. It is being shown and marketed in an edition of 300, at $300 a print.

The film begins with a series of contrasts and variations of the idea of the spiral. It is a progression from spirals of dust behind the car on the road to the Jurassic and dinosaurs to trucks and bulldozers, with peaceful interludes of pink water on which float patches of silvery foam. The reflection of the sun dances on the water in the curve of the spiral.

The film reaches a hypnotic climax a long way from the prosaic road and the mundane maps and lake charts of the beginning. Smithson holds his audience transfixed for 35 minutes, longer than some people spend at an entire art exhibition.

Spiral jetty, by Robert Smithson. This work of art, 1,500 feet long and 15 feet wide, was constructed in the Great Salt Lake in April 1970. Photo courtesy of Dwan Gallery, New York.

Figure 3. Spiral Jetty *photograph, view to the north. Photo courtesy of Dwan Gallery, originally published Utah Geological and Mineralogical Survey (1971).*

Figure 4. Robert Smithson on the Spiral Jetty*, 1970. Photograph by Gianfranco Gorgoni. Art © Estate of Robert Smithson/VAGA, NY, NY.*

cality of the rounded spiral that has become imbedded into the roundness of the lake, the neighboring hills, and the islands in the distance. This was his mythology, the creation of a whirlpool within an ancient inland sea, recounting old legends, bringing past histories into the future. The seasons roll by in Utah, and according to Rick Wilson, visitors to the Golden Spike National Historic Site continue to ask for directions to Rozel Point, hoping to catch a glimpse of Utah's most famous intermittent illusion, the *Spiral Jetty*.

ACKNOWLEDGMENTS

This article is a condensed version of my art history master's thesis (Dogu, 1996). The many people who were instrumental in the development of that thesis were acknowledged in its pages. All excerpts from Robert Smithson's writings and images have been reproduced by permission of the Estate of Robert Smithson.

SELECTED BIBLIOGRAPHY

Alloway, Lawrence, 1972, Robert Smithson's development: Artforum vol. 11, no. 3 (November), p. 54 (includes previously written article - Discussions with Heifer, Oppenheim, Smithson: Avalanche (Fall 1970): p. 48-71 - from which this quote is derived)

Beck, D.L., 1973, A massive art work lies under lake surface: The Salt Lake Tribune (5 August): p. H5.

Bergeson, Kevin, 1994, The return of *Spiral Jetty*: Salt Lake City Magazine, vol. 5, no. 1 (January/February), p. 50.

Box Elder Journal, 1972, Spectacular jetty: it's under water: Box Elder Journal vol. 65, no. 10 (9 March): p. 1, [3].

Deseret News, 1996, *'Spiral Jetty'* makes bold statement in '90s: Deseret News (6 July), p. B4.

Dogu, Hikmet, 1996, An intermittent illusion - local reaction to Robert Smithson's *Spiral Jetty*: Thesis, Hunter College of the City, University of New York, 208 p.

Dwan, Virginia, 1984, Virginia Dwan interviews: Archives of American Art, 6 June, p. 35-36.

Egan, Dan, 1999, Coming around - first a joke, then a jewel for the guys who built *Spiral Jetty*: The Salt Lake Tribune (28 November), p. J1, J3.

Forsberg, Helen, 1996, Dancing on the Jetty: The Salt Lake Tribune (24 March), p. E10.

Fremont, J.C., 1845, Report of the exploring expedition to the Rocky Mountains in the year 1842, and to Oregon and North California in the years 1843-44. 28 Congress, 2 Session, House Executive Document 166 (1845). Quoted in Gloria Griffin's Exploring the Great Basin: Norman, OK, University of Oklahoma Press, p. 6.

Fulton, Ben, 1995, Seven weird wonders: Private Eye Weekly (1 June), p. 21-23.

Hartmann, Al, 1995, Floating nine: The Salt Lake Tribune (27 September), p. D1.

Hassett, Rick, 1976, The elusive Jetty: Utah Holiday (18 October), p. 11.

Hobbs, Robert, 1981, Robert Smithson - sculpture: Ithaca, NY, Cornell University press, p. 191.

Howe, John, 1993, A Desert Sea, a 58 minute film produced by John Howe, hosted by Terry Tempest Williams, produced by KUED Television, Salt Lake City, Utah, (November).

Karren, Melissa, 1996, Monuments - Eclectic Utah: Deseret News (8 August), p. A19, A20.

Post, F.J., 1975, Life in the Great Salt Lake: Utah Science, vol. 36, p. 45.

Rudolph, W.E., 1963, Vanishing trails of Atacama: New York, American Geographical Society, Research Series no. 24, 86 p.

Saal, Mark, 1996, The *Spiral Jetty*: Ogden Standard Examiner (23 June), p. D1, D6, D7.

Smithson, Robert, 1970a, *Spiral Jetty*, Thirty-five minute 16mm color film directed and produced by Robert Smithson (available for purchase from the Museum of Modern Art, New York).

—1970b, From the center of the *Spiral Jetty*: Art Now, New York 2, no. 9 (November), n.p.

—1972, The *Spiral Jetty*, in Gyorgy Kepes, ed., Arts of the Environment: New York, Braziller, 1972. p. 222-232.

South, Will, and May, D.L., 1996, Images of the Great Salt Lake: Salt Lake City, Utah Museum of Fine Arts, University of Utah, np.

Utah Geological and Mineralogical Survey, 1971, Earth art in Great Salt Lake: Utah Geological and Mineralogical Survey Quarterly Review, vol. 5, no. 2 (May), p. 1.

Van Wagoneer, Carol, 1993, Journey to the Jetty: The Salt Lake Tribune (30 November), p. D1, D8.

APPENDIX - DIRECTIONS TO THE *SPIRAL JETTY*

Go to the Golden Spike National Historic Site, 30 miles west of Brigham City, UT (which is one hour north of Salt Lake City on Route I-15). From the Site's visitor center, drive 5.6 miles west on the main dirt road (county dirt road - not the railroad grade). At the intersection, take the south route. You immediately cross a cattle guard (no.1) (you cross four cattle guards before reaching the *Spiral Jetty*). Drive 1.3 miles south. Take the southwest fork, just north of the corral. After turning here, go 1.7 miles to cattle guard no. 2. Continue 1.2 miles to cattle guard no. 3, a fence, a gate, and a sign which reads "Promontory Ranch." Continue 2.3 miles south/southwest to the combination fence, cattle guard no. 4, where there is a "No Trespassing" sign. At this gate, the class D road designation ends and the road gets rough. If you choose to continue south another 2.3 miles, around the east side of Rozel Point, you will see the lake and an abandoned oil jetty. As you approach the lake, there is an abandoned pink and white trailer, an old amphibious landing craft, and an old Dodge truck. Drive slowly past the trailer, turn immediately from the southwest to the west, onto a two-track trail that contours above the oil-drilling debris. This is not much of a road! Only high-profile vehicles should advance beyond the trailer! Don't hesitate to park and walk, the *Spiral Jetty* is just around the corner. Drive or walk .6 miles west/northwest around Rozel Point and look toward the lake. The *Spiral Jetty* may be in sight. (Condensed and adapted from a map developed by the Golden Spike National Historic Site, 1994).

HISTORY OF THE SUN TUNNELS NEAR LUCIN, UTAH

by

Nancy Holt and Hikmet Sidney Loe[1]

[1]*Salt Lake City Public Library*
210 East 400 South
Salt Lake City, UT 84111
hloe@mail.slcpl.lib.ut.us

INTRODUCTION

The following introduction was excerpted, with permission, from Holt (1977).

Sun Tunnels (1973-76) is built on 40 acres which I bought in 1974 specifically as a site for the work. The land is in the Great Salt Lake Desert in northwestern Utah, about 4 miles southeast of Lucin and 9 miles east of the Nevada border (figure 1).

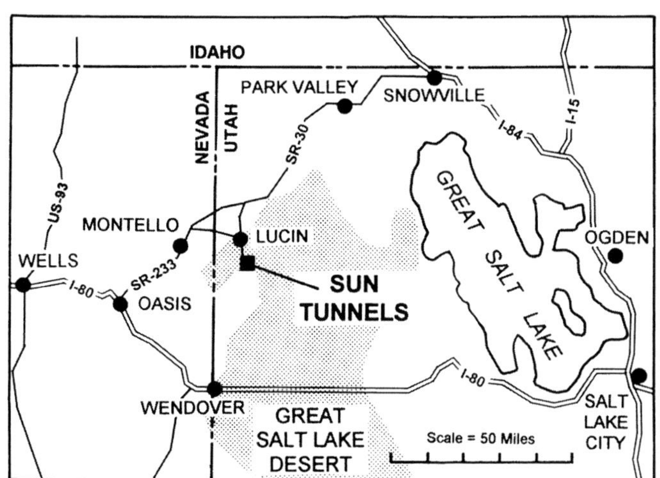

Figure 1. *Index map showing route from Ogden/Salt Lake City to the* Sun Tunnels *near Lucin.*

Figure 2. Sun Tunnels, *1973-76, by Nancy Holt. Total length: 86 feet, length of tunnels, 18 feet each. Great Basin Desert, Lucin, Utah. Sunset on the summer solstice.*

Figure 3. Sun Tunnels, *1973-76, by Nancy Holt. Detail of sunset on the summer solstice.*

Sun Tunnels marks the yearly extreme positions of the sun on the horizon - the tunnels being aligned with the angles of the rising and the setting of the sun on the days of the solstices, around June 21st (figures 2 and 3) and December 21st. On those days, the sun is centered through the tunnels and is nearly centered for about 10 days before and after the solstices.

The four concrete tunnels are laid out on the desert in an open "X" configuration 86 ft. long on the diagonal. Each tunnel is 18 ft. long and has an outside diameter of 9 ft.- 2 1/2 in. and an inside diameter of 8 ft. with a wall thickness of 7 - 1/4 in. (figure 4). A rectangle drawn around the outside of the tunnels would measure 68 - 1/2 ft. by 53 ft. Each tunnel weights 22 tons and rests on a buried concrete foundation.

Four different-sized holes (7, 8, 9, and 10 in. in diameter) are cut through the wall in the upper half of each tunnel. Each tunnel has a different configuration of holes corresponding to stars in four different constellations - Draco, Perseus, Columba, and Capricorn. The sizes of the holes vary relative to the magnitude of the stars to which they correspond. During the day, the sun shines through the holes, casting a changing pattern of pointed eclipses and circles of light on the bottom half of each tunnel (figure 5). On nights when the moon is more than a quarter full, moonlight shines through the holes casting its own paler pattern. The shapes and positions of the cast light differ from hour to hour, day to day, and season to season, relative to the positions of the sun and moon in the sky.

Due to the density, shape, and thickness of the concrete, the temperature is 15 to 20 degrees cooler inside the tunnels in the heat of day. There is also a considerable echo inside the tunnels.

Figure 4. Sun Tunnels, *1973-76, by Nancy Holt.* Sun Tunnels, *shown with an onlooker, who imparts a sense of scale to the environs.*

Figure 5. Sun Tunnels, *1973-76, by Nancy Holt. Interior of a tunnel with the constellation Draco. Spots of sunlight, cast through the holes on the top of the tunnel, mirror the constellation on the bottom half of the tunnel's interior.*

OVERVIEW OF NANCY HOLT AND HER WORK

Nancy Holt is an American sculptor who has created large-scale, environmental, site-specific works both nationally and internationally. Her works can be found in very urban, accessible locations, as well as in remote locations such as Utah's Great Salt Lake Desert. Originally from the East Coast and New York City, she has relocated to New Mexico where she continues to create site-specific works.

Nancy wrote the preceding excerpted description on *Sun Tunnels* (Holt, 1977) 10 months after its completion. *Sun Tunnels* was one of Holt's first major works - a lyrical statement in a barren desert, a work placed within the history of art that has become know as Earthworks. As a form of art, Earthworks first emerged in the late 1960s as artists became more aware of the outside environment, while reacting against the confines of showing their art in gallery and museum spaces. Whereas those spaces were limited in size and shape, the vast landscapes of the outdoors held an appeal to artists who wanted to remove themselves from structured spaces to create individual, monumental works with no preconceived limitations. By creating Earthworks, artists also establish a unique, paradoxically intimate relationship between the viewer, the artwork, and the environment that cannot exist within a gallery/museum setting. *Sun Tunnels* has a distinctive place within the history of Earthworks. It is an icon of art from the 1970s, just as Utah's other famous Earthwork is, *Spiral Jetty*, by her late husband Robert Smithson. More than any other art that has emerged from Utah, *Sun Tunnels* and *Spiral Jetty* place us [Utah] on the international map of art. But, unlike the *Spiral Jetty*, whose placement in the ever-changing Great Salt Lake determines its visibility and accessibility, *Sun Tunnels* is always visually accessible.

While Earthworks are usually considered as artworks to be experienced on a solitary basis, the relationship that Holt has established between the heavens and those viewing *Sun Tunnels* on the summer and winter solstices is unique, since groups of people often gather to celebrate the solstice together at her site. It is left for time present and future to be viewed by individuals, to be understood and appreciated within each individual's system of knowledge.

CONSTRUCTION AND SIGNIFICANCE OF *SUN TUNNELS*

The following interview by Hikmet Sidney Loe (HSL) of Nancy Holt (NH), was conducted in November 1998, and details Holt's thoughts about the construction and the significance of *Sun Tunnels*.

HSL: You wrote earlier that you looked for sites in Utah, Arizona, and New Mexico to create an Earthwork - how did you decide on Utah?

NH: In 1974, I saw some land, first in New Mexico, near Deming, then in Arizona, before going up to Utah. I had already known this area in the northwest part of Utah because Bob [Robert Smithson] and I had spent some time there [1969-70], and we had bought a piece of land there while he was alive. It was 40 acres, a rocky outcrop right on the Salt Flats, about 9 miles south of the site that I eventually bought for *Sun Tunnels*, so I was aware that land was for sale there. There was a window-of-time when you could buy 40-acre parcels. A year or two after I bought my land, they changed the law so you could only buy 160-acre parcels. I was able to contact the original company that I bought the land from in 1970 and see what else they had there. I knew it was the right kind of site for *Sun Tunnels*, because it's flat and barren and had no other use - you couldn't really grow anything there. The soil was alkaline, it had been under salt water at one time, when the Great Salt Lake was a sea. In the mountains surrounding the area you can see the lines where the sea bit into the mountains as it was going down, making striations in the rock. I liked that, it was a way of seeing time physically inscribed on the rocks. And since *Sun Tunnels* was about time, the cyclical time of the sun, a more universal kind of time, I thought it was a very fitting site, since you had this flat plane and the mountains, and you could see time written on the rocks.

And, it was affordable, within my budget, and the land was accessible because there was a gravel road that went to within 3/4s of a mile of it. This gravel road wasn't used much anymore, it had been built by an oil exploratory company to get to oil on the edge of the Great Salt Lake. If you go straight on that road, and don't turn towards *Sun Tunnels*, in about 10 miles you eventually get to an abandoned oil rig. From this existing gravel road I built a 3/4-mile gravel road to *Sun Tunnels*. That was one of the first things I did.

HSL: Was the crew you worked with in the actual construction from Salt Lake City, or were they people more local to the area?

NH: In building the 3/4-mile road I employed a contractor from Wendover. He had a truck and a front end loader. The gravel was easy to get because there were abandoned gravel pits in my area. The gravel pit he used was about 2 miles away. So after he cleared a road out to my site he was able to get the gravel with his front loader, put it in his dump truck, and dump it on my road. It took a few days to complete the work.

Carl Gorder from Salt Lake helped me make the footings. After the holes were dug by Carl and Chris Lund (who had assisted in archeological digs), the forms which Carl made in Salt Lake City were placed in the holes, and we were ready to pour the footings. I couldn't bring the concrete in from Salt Lake City because it was too far, it would have solidified before it got to my site, so I was in a dilemma about how to get concrete. I knew that they were building a road out by Oasis [Nevada] because I would pass it frequently, so I found out who was the owner of the construction company that was building this road. I heard that he liked to gamble and stay in fancy hotels in Nevada, a real man about town (laughs). I called the different hotels he was staying in, trying to track him down. It took me a couple of weeks to finally get to him, and I asked him if he could send one of his concrete trucks out to my site.

I tried to get it cheap (laughs), I was always making deals, you know, I was living out there, doing this unusual project. I was using a lot of my own money, and I didn't have that much money. I had a couple of grants, an NEA [National Endowment for the Arts] and a New York State Council Grant, which helped a lot, but, it still wasn't enough. It was one of my first experiences trying to get a large project done inexpensively, trying to interest people, getting them involved in the project, so that they would want to do things at cost. I was sometimes successful. I can't remember, I might have paid the full price this time, I was just so happy that he would agree to get some concrete out there.

So it was all set up, and I went to Montello, NV to meet the concrete truck. I was sitting at the bar at the Cowboy Bar, when the foreman of the road crew came in. I had met him before, so I said to him that I was waiting for one of his concrete trucks, and he said that that was impossible. "I don't know anything about that" he said, "I know all the scheduling, and that can't be." And I said, "Yes, I spoke to the owner and he said he was sending it." "I don't believe it" he said. So, I was worried that it wouldn't come (laughs). But it did come, and the foreman was amazed that he hadn't known about this! He felt badly that he had misled me, so that might have been one reason he jumped right on the truck, and said he would go out there and help us. As it turned out, we really needed the help. Carl was there alone waiting for the concrete, it was a big job, forming those footings.

HSL: What time of year was this? Were you working against the heat of the desert?

NH: No, it was April, I think. You have to work so fast, once the concrete is poured, the concrete has to be stirred with a bar so that it settles well and air pockets don't form. It's a very strenuous activity. The concrete truck had gone about its maximum distance, it had to dump its concrete fast, it didn't have a lot of time to hang around. We had to get the job done. So, if this foreman hadn't gone out and done a lot of work for nothing (laughs), I don't know, maybe we could have done it ourselves, but it was an iffy kind of thing.

So, those people were from the area, at least they were local people at that time. A road was being built out near Oasis, so a lot of these construction people were living in either Montello, or Oasis, or Wendover. The nearest town, Montello, Nevada, is 26 miles away (laughs), so when you say local people, the local area is pretty far reaching.

HSL: Were these tools that you worked with similar to the tools you had used in other works you've created?

NH: Yes. When I build *Hydra's Head* along the Niagara River up in New York State, we had to clear the site with a front loader, and make the holes in the ground with a back hoe. I used concrete pipes there too, and they had to be dropped into the holes with a front loader. Then I had a concrete truck come to pour the bottoms of the pools. So, I had used similar equipment before, only on a smaller scale.

HSL: Can you describe the process of work on the *Sun Tunnels*, the continuum of interest, research, knowledge, art, science, technology...

NH: In 1968, I went to the western desert for the first time and had an overwhelming experience. It was as if my inner landscape and the outer landscape were identical, there was a pervasive sense of oneness. It was like a nirvana experience. For three days I didn't sleep, I was euphoric. It was really the beginning of my art. Everything evolved out of that, to this day that experience is the root of everything I do.

Around 1970, I was working with light indoors. I was doing room installations where I would build walls with holes and cast light through the holes (figure 6). The light that was cast through the holes would end up shaped into circles of light on the opposing wall. I was also casting light off of mirrors and having a similar thing happen, circles or ovals of light were cast around the room. My concern with artificial light lead me to think about sunlight. In New York I began doing a few models up on my roof to test the light and shadows. I also put some cutouts in my windows.

Figure 6. Holes of Light, *1972, by Nancy Holt. Tungsten halogen lights on each wall alternately go on and off every 1.5 minutes, causing circles of light to be cast through the holes onto the opposing wall. These light circles are drawn in pencil, leaving outlines on the wall when the light is absent.*

So, it only seemed natural that the next step would be to go out into the landscape and actually work with the sun. I had already begun making works in the landscape. I made *Views through a Sand Dune* in 1972 with an 8 inch diameter by 66-inch-long pipe through a dune, which created a tunnel that you could look through, but not one that you could walk through (figures 7 and 8). In 1972, I also made *Missoula Ranch Locators* in a western location in Montana. So, my various concerns were coming together. In 1973, I began working on *Sun Tunnels* by building a model. It was not until 1974, that I had the time to look for the land and buy it. In 1975, I went back to Utah and lived there for more than a year, mainly in Salt Lake City. I had a camper van, so sometimes I would just camp out at the site, or stay in Montello.

For my [project] research, I went to the planetarium in Salt Lake City and met with the astronomer, John Mosley, to discuss the solstice angles with him. I also got an engineering firm, Edmund Allen & Associates in Salt Lake City, and, through them, a surveyor, Harold Stiles. I did a lot of photography from different centers on my site to see what the views were, so I could select views of the landscape to be framed; I was trying to juggle both the angles of the sunrises and the sunsets and the views from the site. Eventually I decided where I should position the work, which ended up right in the middle of the 40 acres, and Harold did the surveying out on the site.

Figure 7. Views Through a Sand Dune, Narragansett Beach, Rhode Island, *1972, by Nancy Holt. Early work which incorporates the elements of light, landscape, and visual perception.*

Figure 8. Views Through a Sand Dune, Narragansett Beach, Rhode Island, *1972, by Nancy Holt. Sky, earth, and the Narragansett River as seen through the hole in the dune.*

Although I'm really interested in looking at the sky and the stars and the sun, I don't really know much about astronomy, so I did have to engage an astrophysicist, Leslie Fishbone, who was then at the University of Utah. He did all the calculations. There are lots of things you have to be aware of to calculate where the sun sets - there's a refraction of light so the sun isn't really setting where you see it setting, and the mountains on the horizon change the angle of the rising and setting of the sun, so they have to be taken into consideration. So, it was complicated. After I got the final calculations for the solstice angles, I didn't trust the numbers somehow, I just couldn't believe that anyone could figure that out (laughs)! I stopped the process of building *Sun Tunnels*, so that we could go out there on December 22 of 1975 and check to make sure the astrophysicist was correct, which he was. I just couldn't bear to build the whole thing, and not have seen the sun rising on the solstice myself.

HSL: When creating *Sun Tunnels*, did you create their scale to suit their geographical environment, or to suit the scale of visitors who would interact with them?

NH: Actually, I build it according to human scale, I wasn't interested in building a megalithic monument that dwarfs humankind. I don't think there is a way to build to the geographical scale of the environment, because it's so vast, so huge, I don't think you can compete with desert scale, even if you wanted to (figure 9)! I like the fact that as I approach *Sun Tunnels* from a couple of miles away, I think, oh, they must be pretty big because they're far away and I can see them. Which is true, they are big, but it depends on relative to what, you know. As you're approaching them in a car, you're not quite sure (if you haven't seen a picture before) what the configuration of the tunnels is, because sometimes two tunnels will line up, and when they do, from far away, they seem to disappear. So you're questioning how many are there and what their configuration is. It's a mystery as you're approaching them. You just can't read them until you get out of the car and go over to them, then you start to understand the scale. Once you get inside the tunnels, they take on another scale altogether.

When I was doing the photography, which is a very important aspect of doing a remote work like this because you need to let others know of the existence of the work through photographs, I used a local person to be the scale model. I always put a person in photographs for scale, but, being in the middle of the desert, it was hard to find somebody to do that.

I am interested in the perceptual questions that arise. The work isn't just about the sun. It's important to keep in mind that the work comes out of my concerns with perception. Whether the sun is out or not, you can go there and have a sculpture experience. I was very interested in what the views were through the tunnels which frame the landscape. Also, when you're out in the middle of the desert, there's nothing to refer to, it's very vast and its very overwhelming and you get dislocated. The sculpture tends to focus your attention. Framing the landscape makes you more aware of the landscape around you and its distinctive elements. There is also the perceptual questioning I spoke of before. As you are driving in, you're not quite sure of the configuration, and their scale is questionable. Also, often in the heat of the summer the work seems to be hovering above the earth because of the mirages which are frequent out there. Sometimes the tunnels are actually reflecting upside down in the mirage.

HSL: So it's almost like having a pool of water, where you're seeing them upside down in the landscape, when there's no water out there?

NH: Yes. In the film I managed to capture a little of that, but not to the extent that you do with your naked eye.

HSL: When did you create the film about *Sun Tunnels*?

NH: The film was finished in 1978 (Holt, 1978). The article about *Sun Tunnels* came out in April, 1977 (Holt, 1977). It took me awhile to write the article because I had to get some distance from the work, and also, I wasn't used to writing. I just finished a big project, *Up and Under*, in Finland this year and I haven't found a way to present it to the world yet. It probably won't get out into the public until next year. It's the same thing with all these big projects, they last for a long time, they generate their own history, they don't have the same time frame as an art exhibition. If you have an exhibition, all the publicity has to go out while the show is on, it all has to happen then - it has to be reviewed and get out in the art magazines within the next two months. But, with these

Figure 9. Sun Tunnels, *1973-76, by Nancy Holt. Aerial view of* Sun Tunnels. *The tunnels are aligned with the sunrises and sunsets in the winter and summer solstices.*

permanent, site-specific projects, there's a whole different sense of time. I mean, here we are doing an interview about this project, twenty-two years later (laughs)!

I was the producer, the director, and did most of the shooting for the film, which is titled *Sun Tunnels*. On the big day when the tunnels were trucked in and they were lowered onto their foundations with a 60-ton crane, camera people came out from the local TV station [KUTV, Channel 2]. I had a friend, Judy Hallet, who worked at KUTV making documentaries. They wanted footage of the construction for the news, and maybe for a future documentary. All the time that we were doing the different processes, I was filming them on 16mm color film. Even the smallest details, like cutting the wood for the foundation forms, or the different stages of making the concrete pipe at the U.S. Pipe Company yard down in Pleasant Grove (south of Salt Lake), I captured on film. My tunnels were custom made, they weren't just ordinary culvert pipes. The reinforcing steel rebar had circles cut out and rings welded in wherever there was going to be a star hole. Pipes usually have male or female ends, but these were made flat, and exactly the length I wanted. Then the holes were core-drilled. After the concrete went on top of the steel we no longer knew where the holes were without an accurate template. That was a tough part for me. So I filmed the core drilling, the hoisting of the pipes onto each of four huge lowboy trucks, and then the 60-ton crane, slowly traveling the 200 miles to the site.

I was shooting constantly, getting all of the details of how the work got put together. It's all in the film!

HSL: Is that typical of the process of how you work on these major projects?

NH: Yes, I have filmed other projects of mine. This was the first one that I did, though.

HSL: When was the last time you visited *Sun Tunnels*?

NH: It was November 20th, 1995, three years ago.

HSL: Did you notice any structural changes that have taken place at the site or with the Tunnels?

NH: No structural changes, the structure is still there, exactly the way that I did it. Most changes were not environmental, they were man-made. There are spiraling lines inside the tunnels. These spiraling lines may be the trajectories of bullets, I'm not absolutely sure, but there were a lot of bullet casings on the ground there. Anyway, somehow the lines were formed around the inside of the tunnel. Somebody might have fired the bullet on the side of the tunnel and it just took off, because it had a lot of velocity. It's not done with marking equipment, pencils or pens, it's something physical that caused this line to spiral around the inside. It's not that bad, the lines have a certain energy. I can accept that, it's just part of having something out there in the world.

The only other change is a small amount of erosion around the footings. The footings hug the base of the tunnels, and the earth is piled, covering the footings. Natural erosion, wind and rain, cause the earth to settle. A few years after I built the work there was some erosion, which you would expect, because the earth was fresh; it hadn't settled. I went out with some art students from the University of Utah. They dug earth for me and covered up the foundations. After that was done a couple of times, the earth settled, so now when I go out there, there's not that much erosion.

Also, every so often, maybe every four or five years, a rancher in the area lets his cattle come over to my land, leaving cow flaps all over the place (laughs).

HSL: Where they were grazing?

NH: Yeah, and they like to be by the tunnels because they make shade for them, and they leave their cow flaps near them. That was another thing I had these students do, get rid of the cow flaps. There still are a few, usually, but it is not so bad, just a necessary aspect of the environment.

There are hairline cracks in the tunnels. Most of them were actually there from the beginning. In the transportation and in picking them up with the crane, little hairline cracks formed. My worry was that ice and snow was going to get into those cracks and make them bigger, and that it was going to be a problem, and I'd have to keep fixing them all the time, but that hasn't occurred. The hairline cracks are there, but they don't seem too bad yet, I haven't felt the need to do anything about them. The last time I was out there was in 1995, so it's been almost twenty years with little change. That's not bad.

There is a concrete core in the ground, between the four tunnels, which indicates the center of the work. It's actually three of the largest cores (10 inch diameter), from the star holes, that I epoxied together, and then set in the ground, so that it was flush with the surface of the earth. I imagined the core would get covered up with soil, that it would not be visible, and that every time I went out there I'd have to dig down and find it. But, that didn't happen! The wind there blows on top of the earth, and keeps that circle flush with the ground. You never have to try to find it, you can always see it. Well, that was a surprise!

HSL: At this point do you receive much feedback from visitors who go out to *Sun Tunnels*?

NH: Once in awhile I get some feedback, especially from art students or curators who have contacted me beforehand for directions to the site. I ask them to report back to me, which they usually do. What I hear is that quite a few people go out there now for the summer solstice (figure 7). A few years ago a woman went out there, thinking she'd be alone in the quiet silence of the desert. There ended up being thirty people the first day, and forty people the second day (laughs)!

HSL: Can you talk about the other pieces of land you own out there?

NH: All together I own four pieces of land in that area. I mentioned the 40-acre parcel that Bob (Robert Smithson) and I bought together, which is nine miles south of the *Sun Tunnels* site. In 1976, I bought two 160-acre parcels, one is five miles to the east [of *Sun Tunnels*]. If you continue straight down the old gravel road towards *Sun Tunnels*, but don't turn down my gravel road, and go another five miles you come to the base of Pigeon Mountain. There's an old gravel pit on the property which is kind of interesting, I call it my miniature Grand Canyon. And there's the only water hole in the area. The last time I was there (in November of 1995) it was dry, but usually there's water in it. It may have silted up after all these years, and I may have to dig it out. There's a crescent-shaped earth mound surrounding this pond. You can stand there and see all the animals that go there, leaving their footprints in the earth along the edge of the pond. Since it's the only water hole, you don't have to wait very long to see creatures come to drink there. So, to me it's a fascinating place. You have the base of Pigeon Mountain, the miniature Grand Canyon, and the water hole! It's 160 acres, when I bought it in 1976, you couldn't buy 40 acre parcels anymore.

I bought another 160 acres, five miles to the west of *Sun Tunnels*, that includes the top of a butte. Once again it's accessible. There's a jeep trail that goes practically to the top of it. You look out and see this 360-degree view, it's just breathtaking! It's off the dirt road from Lucin to Montello. So, both sites are accessible and both sites are very interesting. I've owned them for 22 years with the idea of going back there and building more works, but I haven't done that. Mostly it's a problem with money, because public agencies are not too inclined to give you money to build on your own land! So, I've been trying to raise the money from other sources. Altogether I own 400 acres in that area, so I have Utah roots, and I pay taxes on that land every year.

HSL: Now that it's been 22 years since you completed *Sun Tunnels*, how would you place it within the artistic body of the work you've created?

NH: Well, I think it's one of my three or four most major works. Certainly, it's the one that's sustained a lot of interest for a long, long time. Over the last four or five years it has been reproduced in several European books, which cover works of this nature. Mainly, these books deal with art in the late 60s or early 70s, so the books never go beyond the time that I did *Sun Tunnels*. That is one reason *Sun Tunnels* gets reproduced a lot, because there are so many people concerned with that era of art history. I think a whole new paradigm was established then, there hasn't been anything that radical since, so people are very interested and want to know more about my contribution to that era.

HSL: Can you describe how viewing *Sun Tunnels* might result in a joining of self and nature?

NH: *Sun Tunnels* brings people to a site they would not have gone to otherwise, even the local people there don't go out to this particular area. They were amazed that I would find anything special about it, that I would want to spend time there and build a sculpture there. As soon as *Sun Tunnels* was finished, I had a summer solstice celebration. I had a tent and food, and I invited the local people, many of whom came. But most of them had never been out there before (laughs)!

As I was saying before, *Sun Tunnels* isn't just about the sun, it's about perception. Also, it echos the landscape, or I should say, the skyscape. Out in the desert you see a vault of stars over your head, so *Sun Tunnels* repeats that, you have a vault of star holes in each tunnel. The sun itself, being one star among many, is shining through the star holes, casting spots of light in the bottoms of the tunnels in the configuration of stars. In the tunnels you can actually walk on stars. So you have the sky being brought down to earth, which is a phenomena that's in a lot of my works. People who camp out there at night, and have the experience of seeing the vault of the stars and watching the moon shining through the star holes, are able to bring a greater awareness to the work.

Also, there's the silence of the site. It's so silent out there, unless the wind is blowing. Then, the sound of the wind whistling through the tunnels is heard.

You also have a sense of being on the planet Earth out there, because it's so vast. You're probably walking on land that's never been walked on before, a sensation that is really thrilling, like the astronauts taking their first steps on the Moon.

HSL: Does *Sun Tunnels* relate to prehistorical structures that combine scale and astronomy?

NH: This work was not directly inspired by any prehistoric structures. That doesn't mean there wasn't an unconscious inspiration...I had been to Stonehenge and I did know about some other ancient structures. Many people bring this up. They think I must have gotten the idea for *Sun Tunnels* from Stonehenge (laughs)! But I didn't. It really came out of my own work, from a grappling with my inner structures, and from following some of the impulses I've had doing my art. It's okay if people make those connections with prehistoric monuments themselves, but it certainly wasn't one that I made in any conscious way.

HSL: How important do you believe sites will be to your future work?

NH: *Sun Tunnels* is a very site-specific work, but then almost everything I do is site-specific. When I make indoor installations, I do them to fit exactly into a room. I think everything I do is related to the site, to the environment. Like the work I just completed, *Up and Under*, in Finland. It is in a sand quarry and it's very much part of its site. The whole idea for it came from its unique site. It's unusual for me to do artwork that I feel could be moved to another site.

REFERENCES

Holt, Nancy, 1977, *Sun Tunnels*: Artforum, vol. XV, no. 8 (April): p. 32-37.

—1978, *Sun Tunnels*, 26.31-minute 16 mm color film directed and produced by Nancy Holt.

—1996, "*Sun Tunnels*", in Baile Oakes, editor, Sculpting with the Environment: A Natural Dialogue: New York, NY, Van Nostrand Reinhold.

Selected Additional Articles, Books, and Presentations

Anonymous, 1993, Utah, State of the Arts: By an Assembly of Experts: Ogden, UT, Meridian International, (front/back cover illustration is of *Sun Tunnels*, film still from a student film project).

Brewer, Ted, 1995, Utah-Off the Beaten Path: Old Saybrook, CT, Globe Pequot Press. (travel book that includes *Sun Tunnels*).

Castle, Ted, 1982, Nancy Holt, siteseer: Art in America, vol. 70, no. 3 (March): p. 84-91 (author discusses his travels and experience in viewing *Sun Tunnels* with Nancy Holt).

Howe, John, producer, 1993, A Desert Sea: a 58-minute film produced by John Howe, hosted by Terry Tempest Williams, produced by KUED Television, Salt Lake City, Utah, November 1993 (includes aerial footage of the *Sun Tunnels* with narration).

Fulton, Ben, 1995, Seven weird wonders: Private Eye Weekly (June 1): p. 21-23, (The *Sun Tunnels* are Wonder number three, with a description).

Gertsch, Guy, 1992, Desert Oddities: Salt Lake City Magazine, vol. 3, no. 5 (July/August) p. 66 (part of feature article,

Great Salt Lake Desert: the last resort, which includes *Sun Tunnels, The Tree of Utah,* and Bangerter pumps).

Holt, Nancy, 1977. "On this planet, rotating in space in universal time." Deseret News (3 December): p. W1.

Karren, Melissa, 1996, *Sun Tunnels'* Frames Space in a Human Scale; Deseret News (August 8-9): p. A20, (Article is one in this section entitled "Monuments: Eclectic Utah).

Lippard, L.R., 1983, Overlay: Contemporary Art and the Art of Prehistory: New York, NY, Pantheon Books (cover illustration is color detail of *Sun Tunnels,* also includes text).

Peterson, Randy, 1993, Desert Sea: The Salt Lake Tribune (November 30), p. D7, (review of video broadcast on KUED TV, which features the *Sun Tunnels*).

Poore, Alan, 1992), *Sun Tunnels* at Solstice: The Salt Lake Tribune (June 14), p. E1, E3.

Smart, W.B, 1995, Utah: A portrait in the western desert: Deseret News (October 01), p. S1-S2, (description of *Sun Tunnels* and writings by Nancy Holt).

Smith, Christopher, 1996, In a barracks on an abandoned Wendover military base, some awfully off-beat exhibits combine to form...a monument to the obscure. Museum celebrates obscure: The Salt Lake Tribune (August 19), p. A1 (review of photo exhibit at the Center for Land Use Development's Wendover Exhibit Hall, which features the *Sun Tunnels*).

Williams, T.T., 1991, Refuge: an unnatural history of family and place: New York, NY, Pantheon Books, p. 266-270, (Great Blue Heron, describes a visit to *Sun Tunnels* and includes excerpts from Holt's original article on the work).

Woodward, D.C., 1977, *Sun Tunnels'* mark time in wasteland: Deseret News (December 3), p. W1, (article on *Sun Tunnels* accompanies article by Holt on same page).

APPENDIX - DIRECTIONS TO *SUN TUNNELS*

SOUTHERN ROUTE - 200 miles (recommended). Take Route 80 west from Salt Lake City through Wendover (a wild border town with casinos) to Oasis, Nevada. At Oasis take Routes 233/30 NE through Montello, Nevada (Pilot's Motel, Cowboy Bar and Cafe where they have *Sun Tunnels* maps and information, last gas and water) back into Utah. About 10 miles past the border is a sign saying "to Lucin." Turn right onto the gravel road for 5 miles to Lucin (the town population has been 0 since 1992). Bear right at the unmarked fork in the road, cross over the railroad tracks and continue south of Lucin for about $2^1/_2$ miles (you should be able to see *Sun Tunnels* in the distance on your lower left - SE) for about $2^1/_2$ miles. Turn left at the steel post with two orange reflectors onto another gravel road (when this road is overgrown it is sometimes difficult to see the turn off) for 2 miles and then right $^3/_4$ of a mile to *Sun Tunnels*. Even thought there are car tracks already there in the earth, please park on the graveled area and walk to *Sun Tunnels*, the desert ecology is fragile. You may camp on my land, but please leave everything the way you found it.

NORTHERN ROUTE - also 200 miles, but a little slower. Take Route 15 north from Salt Lake City to Route 84. Take Route 84 to Snowville (Outsider's Motel, Mollie's Cafe, gas) just after which you turn onto Route 30 going southwest. You will go through Park Valley (Overland Trail Motel, last gas, water). After about 100 miles on Route 30 you will see a sign on the right saying "to Lucin." Turn left to Lucin and follow directions above.

THE TREE OF UTAH - ONE MAN'S DREAM

by

Herman du Toit Ph.D.
Brigham Young University Museum of Art

INTRODUCTION

On a wintry Saturday morning in January 1986, Utah became the beneficiary of one of the most controversial public sculptures. The *Tree of Utah* was constructed, at enormous personal expense, by European sculptor Karl Momen. The Tree is situated on the flat and barren Great Salt Lake Desert (figure 1). At the dedication ceremony of this singularly unique structure, over 1,000 people gathered to hear outgoing Utah Governor, Norman Bangerter, accept the Tree from its maker on behalf of the people of Utah. Notwithstanding the cultural conservatism of the state, and the reticence of local authorities to come to terms with such an extraordinary manifestation of a single artist's vision and resourcefulness, the Tree has become a permanent feature of this arid terrain. The previous art work of these proportions to change the Utah landscape was Robert Smithson's *Spiral Jetty*, at times submerged beneath the rising waters of the Great Salt Lake.

A BOLD ARTISTIC GESTURE FOR A CONSERVATIVE STATE

The physical dimensions and mode of construction of the Tree are impressive enough. As a manifestation of engineering virtuosity, the Tree stands as a significant example of the finest practices in steel frame and pre-cast concrete construction. It is located adjacent to the westbound lane of Interstate 80, where this highway cuts across the white desert floor, approximately 26 miles east of Wendover and 75 miles west of Salt Lake City. Rising 87 feet above the salt flats (figure 2), the structure was built to withstand desert winds gusting over 130 miles per hour, and earthquakes of 7.5 magnitude. It is estimated by the local Highway Patrol that two million cars travel past the Tree annually, and that five to seven cars an hour stop in violation of the law along this highway so occupants can study the Tree at close range. No provision has yet been made for a pullout or rest area for motorists at the site, although proposals have been submitted to local authorities.

On a clear day the Tree is visible to travelers on the highway at a distance of 17 miles. Motorists first see the multicolored spheres, as though they are suspended by seemingly invisible means above the desert. In warm weather, the trunk is lost in the convection currents of hot air rising from the blanched desert floor. Only the spheres shimmer mysteriously and silently in the arid atmosphere high above the desert surface, changing hue and intensity with prevailing conditions of light and weather. On nearing the Tree, the trunk slowly becomes visible and the enormous scale of the sculpture becomes evident. The Tree has been described by some as one of the boldest pieces of visual art to be conceived in the state.

Figure 1. Tree of Utah, *shown standing on the white desert floor approximately 26 miles east of Wendover, Utah during the January 1986 dedication ceremony.*

Figure 2. The Tree of Utah, *rising 87 feet above the salt flats, stands as a significant example of the finest practices in steel frame and precast concrete construction. The structure was built to withstand desert winds gusting over 130 miles and hour, and earthquakes in the order of magnitude 7.5.*

KARL MOMEN - ARCHITECT, PAINTER, SCULPTOR.

The curious onlooker may well ask what motivated the erection of such an apparently incongruous structure in this desolate location? The answer lies in the dream of its maker, the architect, painter and sculptor, Karl Momen who has carved out a significant reputation for himself in Europe, the U.S. and Japan. Momen has had important solo exhibitions of his paintings and sculpture in New York, Stockholm, Tokyo, Berlin and Salt Lake City. He has also gained a reputation for being somewhat of a renegade in the art world, choosing to go his own way rather than abiding the wisdom of his peers or the art cognoscenti of the day. But how did this international artist gain such an abiding interest in Utah? The answer to this question lies in Momen's maverick personality and his idiosyncratic approach to the visual arts evinced throughout his rich and colorful career.

Momen was born in 1934, near the Russian border in Iran. He trained in art and architecture, first in Germany and then in Sweden. He studied briefly under the tutelage of Max Ernst and worked on projects with the famed architect Le Corbusier. In 1961, he emigrated to Sweden with the intention of working on a six-month architectural project. He stayed and became a highly respected architect, working in Stockholm.

By the late 1980s, Momen was kept busy supplying a waiting list of over 40 European companies with paintings and sculptures produced under his hand. His bronze sculptures were purchased predominantly by corporate customers in Europe, but he also had buyers in the U.S. and Japan. Although a Swedish citizen, Momen acquired a residence and studio in Sausalito, California, where he would spend his winters. By the early 1990s, his sculptures and paintings could be found in such disparate venues as Monte Carlo and Brigham Young University.

Significantly, almost all of Momen's sculptures, about 800 of them, have been cast at a bronze foundry in Lehi, Utah, under the supervision of Neil Hadlock, a former professor of sculpture at Brigham Young University and a sculptor in his own right. Momen found that Hadlock's casting was equal to the finest work he could have commissioned in Europe.

THE TREE SYMBOL EMERGES

The image of a stylized tree in Momen's work began as early as 1962, and can be traced to design elements resembling the *Tree of Utah* in his architectural drawings at the time. It is common practice for architects to include abstracted impressions of trees as landscape elements in their renderings of proposed buildings in order to more fully represent the natural setting for proposed structures. Momen developed a stylized interpretation of a tree that used a trunk supporting spheres of differing sizes that would complement and contrast the rectilinear elements of his international style architectural designs. These spheres would often gently impinge on each other like under-inflated beach balls that flatten under pressure at their points of contact. This particular stylization recurred in Momen's architectural renderings at his practice in Sweden throughout the 1960s. Momen's characteristic tree symbol made its first appearance in some of his major paintings in the early 1970s. *Structural Change* (1973) and *Detonation* (1974) both include this now familiar symbol. *Detonation* was exhibited in various international galleries in 1978 to 1982, including the Striped House Museum in Tokyo in 1978, and the Cultural Center in Berlin in 1982. It was shown at the Springville Art Museum in Utah in 1984, and the following year in the Salt Lake Art Center.

Initially, the recurring image of the tree was a veiled reference by Momen to concerns with environmental issues that were prevalent during the 1970s. Later in Momen's iconography, the tree became a symbol that represented the natural order of all living things - an order that was under assault by rampant industrialization and urbanization. This tension between nature and technocracy is the subject of his painting *Structural Change*. Momen also expressed his concern with the dangers of the nuclear arms race at the time. His use of spherical forms in the work *Detonation* evokes the awesome power of nuclear fission, and is rendered as an injunction to the awesome destructive power of the atomic bomb. In later years Momen was to interpret his tree as a symbol of preservation and survival that also represented the essential beauty of the American nation, and even as a metaphor for the dynamic balance and order of the universe.

A DESERT VISION

Momen's affiliation with Utah began accidentally. He once exclaimed: "I did not pick Utah. Utah picked me!" The story of the *Tree of Utah* began with Momen's journey across the breadth of the United States during the summer of 1981. Momen had decided to travel by car from Washington, D.C. to the west coast in order to become better acquainted with the vast American landscape. After a week of cross country driving, he reached Salt Lake City in the sweltering heat of an August afternoon. With no local knowledge, and no experience of the city's road numbering system, he was soon frustrated in his attempts to find a hotel where he could stay the night. After giving up on his road map, he found himself on the road to the airport where he conveniently checked in at the local Hilton hotel.

Momen found the hotel to be well appointed, and after the wearisome drive, hospitable. However, he was soon to be introduced to an important aspect of the local Mormon culture. When he tried to order a glass of wine, he was politely told by a courteous hotel attendant that he could procure the wine for himself at the local newsstand, which he proceeded to do! The following morning he was faced with two choices about the route he could take. He could either take Interstate 15 and travel to Los Angeles via Las Vegas, or he could travel on Interstate 80 via Reno and Lake Tahoe to San Francisco. Notwithstanding his interest in visiting Las Vegas, he chose to proceed on to San Francisco. Although it was only nine o'clock in the morning, it was already hot and he stopped briefly to buy some light refreshments for the long car journey ahead.

After a couple of hours Momen had left Salt Lake City far behind, and had become aware of the almost surreal desert landscape along the dead straight highway leading to his first rest stop at the border town of Wendover. The desert heat and blazing sun would have been unbearable but for the air-conditioned environment of the car. The thought occurred to him that it was as if he was driving over a gigantic, blanched canvas. The straight road disappeared toward the horizon and the heat eddies in the air obliterated the distant mountains. He felt as if he was driving through unlimited white space, almost as if he had left the planet. After a while Momen reached into the back for one of the tomatoes and some lettuce he had brought with him when it occurred to him that he did not have salt for his tomato. The thought crossed his mind that he was traversing one of the largest and most famous dried salt flats in the world with what appeared to be endless expanses of the white substance in every direction!

Momen stopped the car on the shoulder of the road, got out, and was immediately blasted by the searing hot air. It was as if he had opened the door of an oven. He walked over the white crunchy salt and reflected on the fact that it was not unlike powder snow, but for the 120 degree (F) temperature! He stooped down, took a pinch of the white power and ate the tomato with some haste, fearing that the tomato might get scorched in the inferno of hot air and reflected heat that rose from the white terrain. After taking in the alien panorama, Momen returned to the welcome air-conditioned comfort of his vehicle and resumed the journey. As he progressed he watched as tiny black dots in the distance grew into large vehicles, only to disappear behind him as they passed on the dead straight east-bound lane of the highway.

The thought struck him that the monotony of this seemingly endless journey could, and perhaps should, be relieved by some reference point in the desert - some focus of color that would arrest the eye amid this expansive, yet featureless terrain. Momen's thoughts raced ahead. He felt a surge of inspiration as he contemplated the possibility of superimposing some element of color upon that sterile environment as a reference point for the eye, and perhaps for the entire soul. Thoughts and images flashed through his mind as he realized he was on the verge of a unique and powerful new discovery. By the time he left Wendover he already had the concept of what he would create.

Momen envisioned a large structure, symbolic of a tree, rising from the desert and visible for miles around. Although the image was peculiarly similar to the tree symbol that he had so often incorporated in his designs, he had never contemplated locating it in such a desolate location. The tree would hold his characteristic spheres like fantastical fruit in full bloom, high above the flat desert. On reaching San Francisco, he pulled out one of his earlier lithographs that contained an image of his tree symbol. He cut out the image and used it to make a three-dimensional montage on a topographical map of the desert, depicting the tree rising adjacent to the highway.

LOCAL AGENCIES AND BENEFACTORS TO THE PROJECT

Momen became increasingly excited about his concept, but had no idea to whom he should communicate his idea, or who the responsible authorities might be who would have say over such a project. After some inquiries he was referred to the Utah Arts Council. He telephoned the Council in early September 1981, and spoke with Ruth Draper and Arley Curtz, the director and assistant director at the time. They asked Momen to mail them a written proposal for consideration by the Council. He immediately went to work and produced a small maquette and perspectival drawings of the tree as he envisioned it adjacent to the highway near Wendover. On completion he packed his travel bag, and with the model under his arm, boarded the first flight to Salt Lake City. Within hours he was speaking face-to-face with members of the Arts Council, only two days after his initial telephone call to them from San Francisco. Somewhat incredulously they ushered Momen into the first floor conference room of the Arts Council where he proceeded to present his idea with the aid of the maquette he had brought with him. He was to revisit this room many times over the next several years as the project evolved. However, at this initial meeting, Draper and Curtz could hardly believe that this stranger, whom they had only spoken with briefly on the telephone, was seriously committed to such an unusual undertaking in an alien environment so far removed from his home.

The Utah Arts Council was excited about the project, and Arley Curtz sent Momen an encouraging letter offering the Council's assistance. After this initial meeting, Momen returned to Sweden to give the project more serious consideration and to consult with engineering colleagues whom he had worked with in his former architectural practice. In the initial architectural design of the project he envisioned a structure between 75 and 85 feet high.

The news media wasted no time in publicizing Momen's proposal for a large sculpture in the western American desert. The project received immediate coverage in Swedish television news broadcasts, and photographs of Momen's model for the project were publicized far and wide. In April 1982, a public announcement about the initiation of the project was made at the Cultural Center in Berlin. At a special reception, the director of the Center presented Momen with a 10-foot-high cake shaped in the form of the tree. The following day the event made national and international headlines and Momen knew that, for better or worse, he was unconditionally committed to the project.

In the fall of 1982, not long after this enthusiastic European reception, Momen received a chilling letter from the Utah Arts Council informing him that they were distancing themselves from his project, and although they would not stand in his way, they would also not be able to give him any direct assistance. Upon receiving this disconcerting news, Momen immediately returned to Salt Lake City to discuss the matter with Ruth Draper, Arley Curtz, and Dennis Smith who was the Chairman of the Visual Art Committee of the Arts Council at the time. Momen respected Dennis Smith as a sculptor and for the good work that he was doing in the promotion of the arts. Momen was to establish an abiding friendship with Smith. Although Smith resisted Momen's ideas initially, he soon warmed to Momen's enthusiasm. After a subsequent meeting, Smith agreed that, although the Council would not be able to give any material assistance, they would act in a consultative capacity offering advice and general guidance as the project developed. Undaunted by this lukewarm response from the Utah Arts Council, Momen turned to the task of securing a site for his sculpture in the desert along the I-80 highway. Dennis Smith accompanied Momen to a meeting with representatives of the Utah Land Board. At this meeting Momen learned that since he was not an American citizen, he would be unable to purchase land in what was considered to be a strategic area near two U.S. Air Force bases. The specious nature of this argument was brought into sharp focus by Momen's observation that he had recently purchased a condominium in Washington D.C. not far from the White House!

The Land Board representatives did, however, refer Momen to the owner of large tracts of land in this arid region, Khosrow Semnani, a local entrepreneur and president of Envirocare, a waste disposal company in Salt Lake City. Although Momen was scheduled to fly out that evening, he was able to arrange a meeting with Semnani that same afternoon. He was graciously received, and after lengthy and wide-ranging conversation, Momen showed Semnani his model and drawings of his idea for the *Tree of Utah*. Semnani was intrigued by the idea, and after further discussion, he generously offered to sponsor the project to the extent of providing the land upon which the Tree would be located. Although the particular site in Tooele County that Momen had earmarked for the sculpture belonged to the State Board of Education, Semnani was able to arrange a trade involving some of his own land in the area. At the conclusion of their meeting, Momen and Semnani discovered to their mutual astonishment, that they were both of Iranian parentage and both were fluent in Farsi, their mother tongue. Upon realizing this most unlikely coincidence, Semnani pledged even greater assistance to Momen for his project. Semnani would ultimately become the project's chief patron. He greatly assisted Momen in preparing the numerous applications and submissions to the various authorities to obtain the necessary approvals and permits for the structure. He pushed the project through local authorities such as the Utah State Lands and Forestry Division, the Tooele County Planning Commission, and the Federal Aviation Administration - the latter because of the Tree's height of 87 feet.

DON REIMANN COMMENCES CONSTRUCTION

By October, 1983, Momen had called for tenders from several local construction companies for the building of the structure. He received three or four estimates which ranged from $300,000 to $500,000, however he could not obtain a firm estimate as no one at this stage new the condition of the subsoil at the site. In working with the various local authorities, Momen soon discovered that there were mixed feelings toward the project. Some people were very excited and very helpful, while others were quite negative and even resentful about the plans that were being made. In October 1983, Momen received a construction proposal from Don Reimann, a Salt Lake contractor with a company by the name of Style Crete. Reimann had a reputation as the finest concrete caster in the state. The Reimann family had been in the stone casting business since the pioneers entered the valley. They had worked on a variety of imposing structures including the ZCMI Center in downtown Salt Lake City and the Ogden and Provo temples for the Church of Jesus Christ of Latter-day Saints. Moreover, Style Crete was one of the few companies in the world that would attempt to cast concrete spheres over 13 feet in diameter. One of Reimann's sons, John, was completing an architecture degree at the University of Utah, and Reimann thought that this would be an interesting and unusual project to introduce his son to the intricacies of his profession during the final year of his studies. Reimann recommended Devon Stone of Midvale as the structural engineer for the project. A building permit for the Tree was granted at the end of January 1984, and Don Reimann's company was contracted. Momen took a permanent suite at the Hilton Hotel in Salt Lake City, and hired Frank Harris, a cameraman from Los Alamos, to document the construction process. Harris and his wife had recently settled in Utah and both were soon engaged on a daily basis, filming the preparations for the Tree at Reimann's warehouse in Salt Lake City.

ASSEMBLING MATERIALS AND TECHNIQUES

One of the most important considerations at the outset of the project was the kind of material finishes that were to be employed on the large concrete spheres. In the absence of any better material, Momen and Reimann contemplated using some form of ceramic tile, but many tiles were not satisfactory since they could crack and break over time. During the month of March 1984, after lengthy discussions and numerous inquiries about various materials, Momen and Reimann discovered that chrysocolla, a hydrous copper silicate, was ideal for their purpose, and that these unique materials were native to Utah! The blue and green rocks were quarried northwest of Milford, about 200 miles south of Salt Lake City. Don Reimann contacted the owners of the quarry who agreed to donate one 100 tons of the stone to the project. Reimann and sons, however, had to collect the material themselves. In mid-March 1984, Reimann and his eight sons spent three days hauling the stone from southern Utah with the aid of two large trucks.

A stone cutter was immediately contracted to cut the stone pieces to size with a diamond saw, and the Reimann family set to work with the casting of the enormous concrete spheres. Each sphere required a mold. In order to make the molds, wooden jigs were constructed and plaster of paris was carefully applied over the jigs using a hemispherical screed that pivoted on the north and south poles of a convex former. Once the concrete had set, the surface was waxed and new concrete was cast over the convex forms. The mold was then separated from the convex form, and it, in turn, became the concave mold for the final sphere. The molds were produced in quarter segments, allowing four successive castings to be welded and bolted together to make a complete sphere. Each of the six spheres was constructed in this fashion. The trunk of the Tree was cast in like fashion. A tubular former made of wood and plaster of paris was used to make a mold that would cast three longitudinal sections. The sections were welded and bolted together on site to create the central trunk of the Tree.

TECHNICAL AND FINANCIAL CHALLENGES

Good fortune and propitious circumstances were soon to be overshadowed by some of the most daunting challenges that Momen had ever faced in his career. A few months into the project, it was discovered that the amount of steel reinforcing required by the structure had been underestimated. New calculations by Devon Stone now indicated that the final weight of the piece would far exceed the 100 tons originally estimated. It was also at this critical time that the media again became interested in the project. Local TV stations KUTV and KSL sent crews out to interview Momen and to obtain footage of the construction site near Wendover. Public interest was heightened by a flood of articles and television reports. The exposure received by the project also brought with it controversy and debate about the artistic merits and the appropriateness of the project as a public structure. Due to technical difficulties with the steel reinforcement of the structure, Momen moved the completion date of the Tree back from June to October 1984. He sent invitations to numerous civic and national dignitaries, including Ronald Reagan, then President of the United States. Reagan graciously declined the invitation, but nevertheless sent a message wishing Momen great success with his project.

At the height of this intense focus of media attention, Momen received the blood-chilling news that the soil tests had indicated that the selected site was wholly unsuited to supporting the finished weight of the projected construction. Whereas original plans had indicated a modest concrete platform at the base of the structure, new calculations called for a 90-foot-deep foundation comprised of 25 reinforced concrete piles of one foot diameter each supporting an eight foot thick concrete slab! This new revelation would result in a delay of at least six months and would more than double the cost of the project. Momen weighed the options before him. He could possibly redesign the Tree, scaling down the size and mass of its main trunk, its branches, and the spheres. Thoughts of canceling the project crossed his mind. Momen took a few days off to ponder his predicament and finally decided that there was no going back or opting out of the project to which he had committed himself and his friends. He decided to sell the condominiums he had purchased a couple of years previously. He also sold his cabin in northern Sweden as well as some prize pieces of his personal art collection. With the funds raised by liquidating his assets, Momen was able to instruct the construction team to proceed after making the necessary changes to the original design.

Word got out about Momen's unexpected obstacles and shortfalls, and suddenly suppliers refused to deliver materials on account. Some suppliers demanded pre-payment before even considering an order for the project. However, Momen's additional funds soon became available and he set about ordering the 12-inch steel pipes for the piles. The pipe was supplied in 30-foot lengths and three sections had to be welded together to make up the full length of each pile. Before they could be driven into the desert, they had to be sand blasted and treated for corrosion from the hostile saline environment. The excessive length of the piles meant there were not many companies who were equipped to successfully drive these long pipes into the ground. A large crane had to be brought in from out of state by Acme Crane. Further, the unknown condition of the subterranean soil at the site kept Momen from getting a fixed estimate for the work from the pile- driving company; he proceeded on blind faith. It turned out that the piles sank easily and quickly through the first 30 feet of the soft residue of the pre-historic lake bed that had covered this region. Reinforcing steel was placed inside each pipe and they were filled with concrete. This operation occupied Don Reimann and his sons for three continuous weeks. By the end of the fall they had managed to pour the enormous foundation base that rested on the subterranean piles. At the onset of winter, the crew withdrew to Reimann's Salt Lake factory to work on the large concrete spheres.

The next important decision was to determine how the color striations and markings on the large spheres were to be accomplished. There were no natural materials available locally that matched the colors called for by Momen's design. After considerable deliberation, Momen decided to import the finest quality, weather-resistant ceramic tile from Italy, again at great personal expense. Because of the scarcity of this high-quality tile, the order had to be pre-paid before it was shipped. The best means to attach the tiles to the concrete spheres also had to be determined. A special brand of epoxy was identified by Reimann as the most suitable adhering material. It also turned out to be the most expensive product of its kind. For three months after Christmas 1984,

Don Reimann and his workers occupied themselves with gluing, first ceramic material, and then the blue and green chrysacolla and anhydrous copper silicate, piece by piece, to the surface of each of the concrete spheres. As soon as the spheres were completed, work commenced on the trunk and supporting limbs of the Tree. However, Momen's cash reserves had run dangerously low by this time and he was compelled to seek various loans. A loan raised in Sweden was not enough and he turned to his local artist friends, Dennis Smith and Gary Smith who agreed to co-sign for a $30,000 loan from Copper State Saving and Loan Company.

The project lurched precariously forward, but Momen was soon forced to turn to Khosrow Semnani for yet another loan of $150,000. The end was not yet in sight, and by September 1995 Momen had exhausted these funds also. At this late stage in the project, he was compelled to call a halt to the operation and Don Reimann and his team were called off the job. Work came to a halt and Reimann moved on to other contracts. Eventually, Momen was successful in raising yet another loan of $30,000 in Sweden, and work on the Tree resumed. The project was now very near to completion and plans were being made to erect the structure on its base. However, just before this final operation, Reimann discovered that the Tree would need cathodic protection to stop the steel reinforcing materials from being corroded by the salt environment. They would have to contract a company that specialized in this work. Momen was barely able to find the money for this unexpected expense. Then, in what seemed like a cruel twist of fate, just when Momen and his team thought that they had accounted for every outstanding contingency, it was determined that in order to comply with building code, the structure would also have to be equipped with lightning protection. Miraculously, Momen was able to come up with sufficient funds to pay for this. The construction was completed in September 1985, 17 months after construction had commenced.

A STRUCTURE OF SIGNIFICANT STATURE

The Tree consists of a central steel-reinforced concrete trunk supporting six large pre-cast concrete spheres, the largest of which measures 13 feet, 6 inches in diameter and weighs 60 tons (figure 3). The spheres are supported by thick-walled diagonal steel pipes, or limbs, that penetrate the spheres and continue through the hollow cores to their opposite surfaces. The steel limbs are secured to the vertical surface of the trunk by the dual means of bolting and welding to steel plates on the trunk's surface. The steel plates are in turn secured by horizontal rebar embedded within the concrete trunk. Each weld was heat treated several times to relieve stresses in the metal to ensure that the joints would not fail. The combined weight of the spheres, the limbs and the main trunk is supported on an eight-foot-thick concrete found-

Figure 3. *The* Tree of Utah *consists of a central steel-reinforced concrete trunk supporting six large pre-cast concrete spheres, the largest of which measures 13 feet, 6 inches in diameter and weighs 60 tons. The Tree's European sculptor, Karl Momen, is shown standing next to one of the giant concrete spheres during its fabrication (photo by George Janecek).*

ation. The foundation is supported, in turn, by 25 steel-encased concrete piles that were driven 90 feet into the loosely compacted desert clay to ensure permanent stability of the structure. Allowance also had to be made for the expansion and contraction of the materials during the wide temperature variations in the desert. On completion, Devon Stone who had been responsible for structural specifications throughout the construction, was satisfied that the Tree complied "300%" with all safety criteria.

The original estimate of materials called for 200 tons of concrete, 100 tons of rock and 100 tons of steel. However, by the conclusion of the construction the project had consumed 100 tons of chrysacolla rock, 4 tons of epoxy, 160 tons of steel, 15 tons of colored cement and sand, 18,000 imported ceramic tiles, 5 tons of welding rods, 7 tons of timber for mold formers, and 20 tons of plaster. All in all, the structure ended up weighing 875 tons and had consumed 21,000 man-hours of labor. The project would also cost Momen $2.2 million of his personal funds, many times more than the original estimates at the outset of the project.

CRITICAL RESPONSE TO THE TREE OF UTAH

Even before ground had been broken for the foundation, media interest descended on the project. Newspaper headlines such As: Tree Artist Wants to Put More Flavor in the Salt, and Utah Mulls over Karl Momen Monumental Meatballs proliferated. Not to be outdone the Washington Journal proclaimed: Sure, the Redwoods Grow Taller, But They Don't Have Coconuts. Other writers referred to the Tree as "a coagulated mushroom cloud" or as "giant scoops of ice-cream." These statements by the media were indicative of the derisive attitude adopted by many critics. Many self-appointed critics voiced their condemnation of something that had suddenly sprung up within their midst and which was contrary to anything that they were familiar with. Although the Tree stood in a swirl of local controversy, world-wide interest was aroused and more than 600 individual publications have carried news of the Tree world-wide since its installation. The Swedish press gave the Tree prominent coverage because of Momen's nationality. The Japanese and German media also showed interest. Numerous articles appeared in art and architecture magazines. Overseas interest was more informed and more positive than local reports of this unique manifestation in the Utah desert. Elaine Jarvis, a local critic, noted the *Tree of Utah* is arguably Utah's most visible and internationally well known piece of art. Significantly, the *Tree of Utah* may have raised more critical awareness overseas than in Utah itself.

Opinions of the work varied from critics who saw the Tree as an unfortunate invasion of the desert, to others who heralded the work as a sophisticated piece of environmental art. Some local reports questioned Momen's generosity and his motives. Many were cynical about Momen embarking on such a costly endeavor and then making a magnanimous donation of the Tree to the people of Utah. Proposals to construct a pullout for motorists at the site were condemned by some for encroaching on Utah's skimpy road finances. However, many applauded Momen for his innovation, his generous spirit, and his artistic sensibility. Katherine Metcalf, an independent art critic writing for Utah Holiday magazine, proclaimed the work to be the product of a sterile Bauhaus aesthetic, reflecting a "pre-ecological, monument-making consciousness that is very far from the 'site-determined' works that are created in the western desert by younger artists such as Nancy Holt, James Turrell, and the late Robert Smithson." Other critics adopted opposing views of the Tree, namely that the Tree is in fact site-determined and that the whole impetus for the work was prompted by its unique and very specific environment. There is no doubt that Momen is a product of the abstract constructivist movement and that he was influenced during his formative years by the work of Russian Constructivist artists Naum Gabo, Anton Pevsner, and Kasimir Malevich. The fact that the aesthetic of pure form still survives in Momen's work is not an anachronism as much as it bears witness to the tenacity with which Momen embraced the tenets of significant form. The fact that the Tree did not please everyone did not detract from its function as significant public sculpture in the eyes of many. Carol Nixon, director of the Utah Arts Council, noted, "the artist's job is not to create something that will satisfy everyone, it is to create dialogue and present new ways of viewing the world."

Utah environmentalists did not seem particularly concerned about the Tree. Utah Wilderness Association coordinator, Dick Carter proclaimed, "With all the important environmental issues facing the state I'm stunned that anyone is making a fuss about it." Environmental activist Alex Kelner said that he was not offended by the sculpture, "I'm much more offended by high rises in the canyons – they're much more permanent." Kelner expected vandals to mar the surface and thought that it was just a matter of time before the marshy terrain eventually swallowed up the Tree. Momen was unmoved by his environmental critics: "What about Mount Rushmore?" he countered at the time.

An avid supporter of the project, Scott L. Beesley wrote to the editor of the Deseret News that he thought that Momen should be made an honorary prophet of his adopted state of Utah. Springville Museum Director, Vern Swanson appreciated Momen's work as architectonic, and thought that it was Momen's intention to "enlarge the dot on the Utah map of art history that Smithson had created." Other more utilitarian views were expressed by those who saw no value in the expenditure of money on anything that did not result in materialistic advantage to the poor and needy. Olle Granath director of the Museum of Modern Art in Stockholm stated, "There is a kind of poetic craziness in the project which ought to get encouragement." As far as she knew, there was no other Swedish artist who had made such a significant international creation. Stig Johansson, a prominent Swedish art critic and writer, saw the Tree as a sparkling mosaic giving life to something generally considered sterile – the desert. On the other hand, Janet Koplos, and independent art critic, writing in the Asahi Evening News, condemned the Tree as something drained of life. She saw the Tree as something "petrified, consisting of spheres attached to a column without the least intimation of growth, movement or even grace."

Momen has reserved judgment on his creation, allowing people to form their own opinions. He knew that people would have their own interpretations and it was for this reason that he had at an early stage contemplated naming the piece "Metaphor." He proclaimed that the tree was the most elemental symbol of life, and that in this work he had brought together the disparate elements of space, nature, myth and technology. He also maintained that we live in a world in which technology has taken over, and that the Tree represents the tension that exists between the natural world and the world of technology. Also, Momen believes the Tree could be seen as a metaphor for the confrontation between life and death, hope and despair. Momen has tried to avoid doctrinaire statements about the critical significance of the Tree, or of its many interpretations – "I don't want to be a messenger. I'm just an artist."

The future significance of the *Tree of Utah* is difficult to foretell. At the time of writing, Momen is engaged in plans for the creation on a large sculpture park surrounding the base of the Tree. It is his hope that many significant international artists would contribute their artworks to this project, and that it would become a major attraction to the multitude of travelers on this lonely highway, as well as to the many visitors and tourists that such a venture would attract. Whatever the outcome of these plans, it is certain that the Tree is here to stay and that its manifestation will continue to intrigue and prompt the imaginations of successive generations of travelers along I-80 through the Great Salt Lake Desert.

Editor's Note: Herman du Toit, Ph.D., curated the exhibition of Karl Momen's paintings and sculpture at the Brigham Young University Museum of Art in the summer of 1995, and is the author of the book "Vision in the Desert - *Tree of Utah*" which further explores the subject of Momen's work in the Great Salt Lake Desert.

NATIVE AMERICAN ROCK ART AND THE GREAT SALT LAKE - AN ANCIENT TRADITION AND UNIQUE HERITAGE

by

Kenneth Sassen

Department of Meteorology
University of Utah
Salt Lake City, Utah

INTRODUCTION

Following the end of the last ice age, humans witnessed the recession of Lake Bonneville and adapted new survival strategies as the changing climate brought an end to the big-game-hunter way of life. The archeological evidence from the classic cave excavations conducted in Danger Cave, Hogup Cave, and other sites surrounding Great Salt Lake (Jennings, 1978), suggests that indigenous peoples occupied the shorelines as the waters finally receded, and exploited this environment for a variety of foodstuffs as early as about 11,000 years ago. Not only were new technologies needed, but it can be surmised that a cultural and spiritual reorientation took place. This was necessitated by the changing life-way, as the hunting rituals of the Pleistocene gave way to the vastly different concerns of the hunter-gatherers of the Archaic Period, and the cultivators that followed them. Judging by the cultural debris preserved in arid caves, mountain sheep and rabbit became primary foodstuffs, pottery and the bow-and-arrow technologies were gradually incorporated, and eventually, at least a partial reliance on agriculture eventually made a more sedentary lifestyle possible (Jennings, 1978). Paralleling these adaptations, there must have been major changes in religious beliefs (Eliade, 1963). While these changes were occurring in the eastern Great Basin, there is also evidence for interactions with the surrounding regions (Schaafsma, 1971). Communication lines were open in what is now northern Utah, with the peoples of the northern Great Plains, the western Great Basin, the Northwest, and the Southwest.

Although it is unclear how far back in time the native American traditions of pecking and painting images on stone extend (termed petroglyphs and pictographs, respectively), an evolving rock-art legacy is preserved along the margins of Great Salt Lake. About 1,500 years ago, the predominantly abstract Desert Archaic rock-art style, based on circular, curvilinear meander, and stick-figure designs, gradually gave way to the more representational Fremont style. This new style blended abstract figures with realistic mountain sheep and other zoomorphs (animal forms), rainbows (Sassen, 1991), snakes, and broad-shouldered anthropomorphs (human forms), often holding shields or adorned with horned headdresses. It has been suggested that external stylistic influences, particularly from the Great Plains and the Anasazi culture that extended into south-central Utah, are also represented in this area's rock art (Schaafsma, 1971, 1980). Finally, a few modern horse figures establish that nomadic Ute latecomers added to the rock legacy during the historic period, which began after contact was established with the Spanish well to the south.

Before describing the rock art, it should be stressed that my interpretations of the meaning of prehistoric native American symbols are speculative. Thus, I tried to divest myself of modern artistic and religious preconceptions, and rely, as others researchers do, on ethnographic evidence from recorded interviews with the modern-day Native American inhabitants. However, these interviews may or may not represent culture-wide accepted beliefs, or more fundamentally, those of the extinct local peoples. On the other hand, in the study of comparative religion, it is suggested that the human mind conceptualizes our environment in a manner that depends on the state of cultural progress (Eliade, 1963). Although the following interpretations are undoubtedly biased by a modern view of the world, they reflect the collective results of comparative rock art studies from many locations.

In my view, the rock-art sites surrounding Great Salt Lake represent an unusual archeological heritage compared to most of the rock art sites in the Great Basin/Colorado Plateau region. The diversity of styles indicates this was an area of mixing cultures, and that the apparent great age of some sites, inferred by weathering, suggests unusual antiquity for North American rock art. At some sites, the regrowth of desert varnish surface staining (repatination) on the figures, pecked into the dark desert-varnished rocks, is essentially complete, and the rocks themselves are disintegrating, indicating the great age of these petroglyphs, or an unusual weakness to the rocks. The original formation of desert varnish and repatination is caused by the surface deposition of metal oxides that leach out of the rocks following rain showers. Moreover, as hinted here photographically, there are a number of exceptional sites involving seemingly mythical panel illustrations that are perhaps not well studied by the scientific community because of their isolation in the Great Salt Lake Desert.

ROCK ART LOCATIONS

Areas of Human Activity

Rock-art sites are found in a variety of locales that reflect the semi-nomadic culture of the ancient northern Utahans who created the rock art. Although a subject of continuing research, it can be said that the sites are frequently, but not always, located where a variety of human activities would be expected, provided that boulders, cliffs, and caves, suitable for art work, occur locally. Since the late Fremont peoples in northern Utah probably remained semi-nomadic, in order to tap seasonally available foodstuffs in this extensive region of deserts and mountain ranges, the associated rock-art sites undoubtedly trace subsistence strategies. Key to these subsistence strategies, and thus the locations of rock art, were the locations of foodstuffs. The highest and wettest mountain ranges yielded a variety of game, pine nuts, and other vegetable matter used for food and tools. Opportunities for the cultivation of corn and other crops occur where dependable water flows out of the higher mountains, such as along the Wasatch Range. The minor, drier ranges were probably a focus for hunting desert bighorn sheep, and in the surrounding grasslands, antelope. Springs and seeps naturally attract both humans and animals, and the trails leading to them and other resources, become culturally significant in their own right. Finally, the margins of the Great Salt Lake, with its fresh water tributaries, offer a rich source of small animals, bird life, and materials for tools and medicines. Excavations at Danger and Hogup Caves, for example, show how effectively the Great Salt Lake area resources were utilized in support of ancient cultures (Jennings 1978). Well-crafted nets for rabbits, and duck decoys made of local materials, have been found here excellently preserved.

Ceremonial Sites

Rock-art locations also reflect human ceremonial activities. These sites may be associated with every-day activities, such as cultivation, gathering food and medicine plants, hunting, and seasonal migrations. The distinctions between the profane and sacred worlds, however, become blurred in religious societies who use rites to ensure harmony with their environment. Human and animal fecundity, puberty, healing, and weather-control are but a few of the universal concerns of humans needing divine intervention (Eliade, 1963). The model of rock art as hunting "magic" surely is an oversimplification, but it may apply in some cases. One such example is found in the Coso Range in the California desert, where an enormous amount of activity went into the pecking of images of mountain sheep and their hunters onto volcanic cliffs (Grant and others, 1968). There is evidence, on the other hand, that some forms of rock art were created purely for the shamanistic purposes of ensuring harmony (Whitley, 2000). These forms of rock art are tucked away in places remote from normal human activity, then or now. The Barrier Canyon style of unworldly pictographs and petroglyphs, for example, is widespread in central Utah (Schaafsma, 1980). It appears that their remote locations, often high in alcoves in desolate canyons, were chosen as an integral part of the ceremonial process that removed them from the profane world. Unlike many readily accessible rock-art sites which are found along ancient trails that we have adopted for roadways, Barrier Canyon grottos are still being discovered, because of their very remoteness.

Sites Encircling the Great Salt Lake

Wasatch Front

The locations of rock-art sites along the Wasatch Front, and encircling Great Salt Lake, will be described in a counter clock-wise order, starting with the Lake Mountains site (figure 1). Although the emphasis is on rock art sites that lie more or less in view of Great Salt Lake, this discussion will include a few areas away from the lake, because this inland sea is part of a greater drainage basin. Numerous petroglyph sites are found around Utah Lake, which drains northward through the Salt Lake Valley into Great Salt Lake. These sometimes extensive sites are found mostly along the east side of the Lake Mountains, and appear to be of great antiquity based on degree of weathering, and almost exclusively portray the curvilinear Great Basin Archaic style. The petroglyphs cover low cliff bands and boulders, some flush with the ground or partially buried, with circles, spirals, meandering lines, and stick-figure zoomorphs occurring in abundance. One apparently unique petroglyph (figure 2) has been interpreted as a realistic portrayal of a halo/arc display in a cirrus cloud (Sassen, 1994).

In the Salt Lake Valley, and moving northward along the Wasatch Front, are a handful of pictograph sites in caves, boulder alcoves, and rock shelters near the Lake Bonneville bench level (about 5,200 feet above sea level). These are classic Fremont ceremonial sites containing a variety of anthropomorphic styles. A cave site near Holladay is highlighted by a large painted lizard and sun/halo symbols (Sassen, 1994). A line of hand-holding figures occurs in Parrish Canyon above Centerville (Color Section, Photo 1). The east Ogden "boulder" bench sports alcoves containing red petroglyphs. Several small triangular-shaped torsos, some adorned with horned headdresses, are located in smoke-blackened caves overlooking Willard Bay (figure 3). There have been numerous excavations of Fremont camps along the Wasatch Front along the recently varying shoreline of the Great Salt Lake, showing many visitations to this area (see Simms and Stuart this book).

Promontory Range

In the Promontory Mountains, and in the Blue Springs Hills and Little Mountain areas to the northeast, there are several unusual sites. Several caves contain small nondescript painted abstract designs and mountain sheep, of uncertain cultural affiliations. A cave near Narrow Springs on the Promontory Mountains appears to hold figures from the historic period. Overlooking Connor Spring is a unique site comprised of petroglyph-covered boulders portraying a wide variety of figures that reflect various cultural affiliations. These figures include Archaic abstract and representational Fremont anthropomorphs and zoomorphs (figure 4), vegetative motifs (figure 5), and a bizarre, mythical panel showing a superimposed historic Ute horseman (figure 6). The latter is most interesting. One interpretation is of a birth scene with

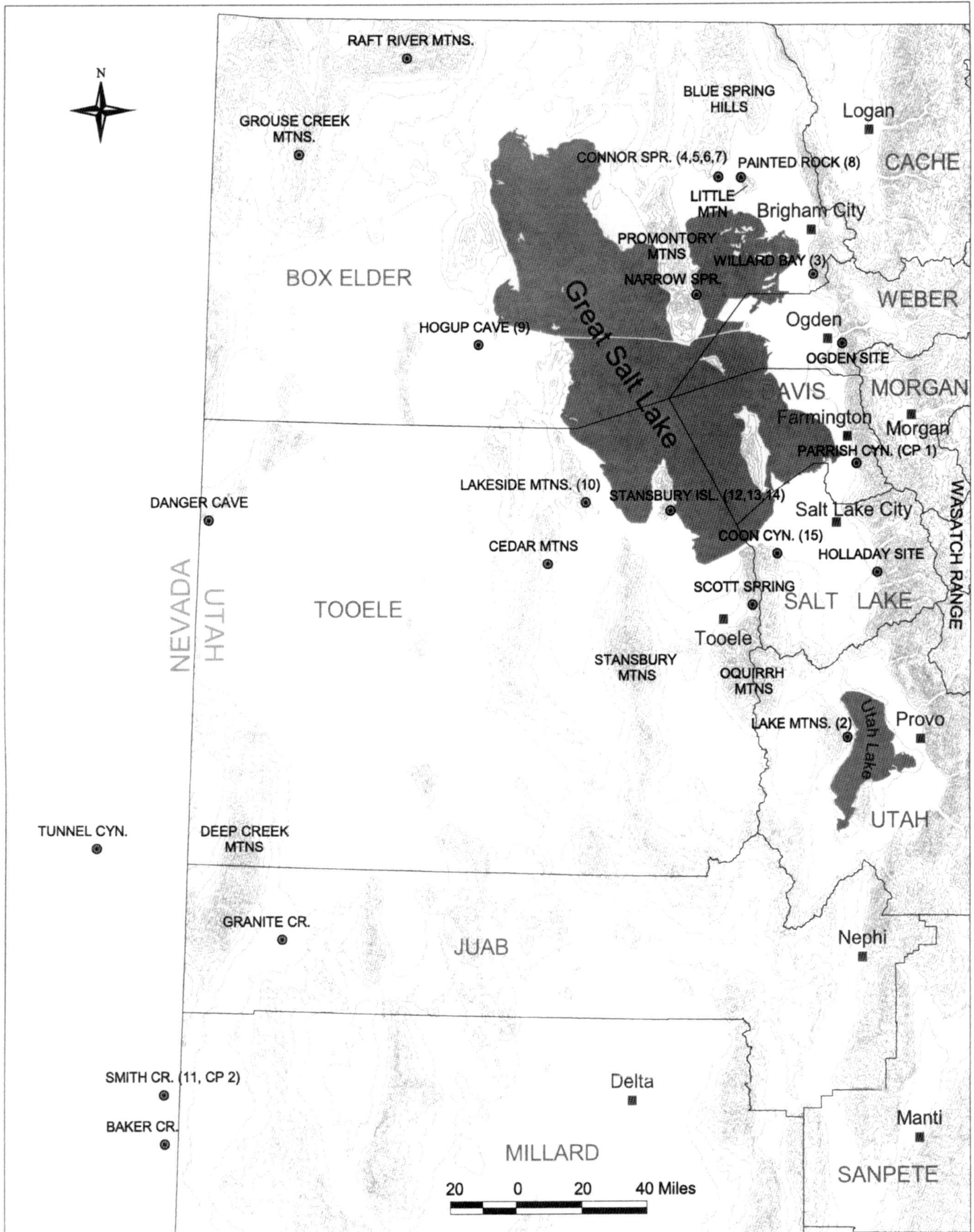

Figure 1. Generalized locations of rock-art sites in the region around Great Salt Lake, Utah and Nevada. Solid squares represent towns, and circles with a center dot represent rock-art sites. Numbers in parentheses (7, 8, etc.), indicate figures, and (CP 1, and CP 2) indicate photographs located in the color section of this volume.

Figure 2. *A cliff band in the Lake Mountains near Utah Lake containing faint petroglyphs, including at center a design of concentric circles and arcs that has been interpreted as a sun symbol surrounded by a 22° halo with lower tangent arc and parhelia (sundogs). Comparison with halo-model simulations indicates a 40° solar elevation angle at the time of observation (Sassen, 1994).*

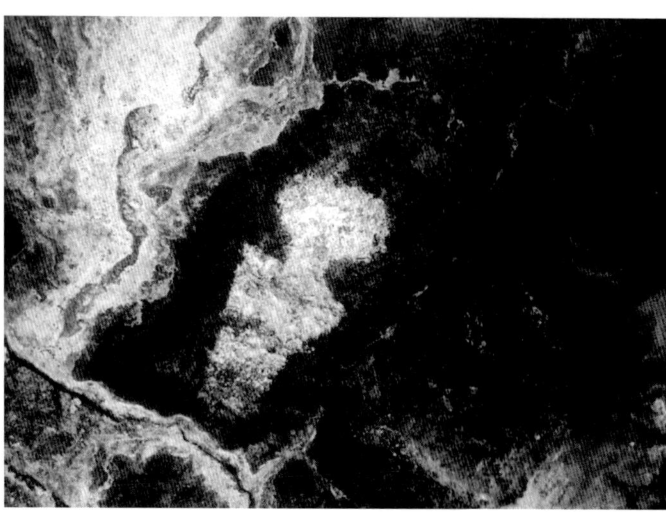

Figure 3. *Fremont-style anthropomorph painted white against the smoke-blackened ceiling of a small cave in the Wasatch Front lying above Willard Bay near the Lake Bonneville bench level.*

Figure 4. *From Connor Springs, early-representational style deer or elk petroglyphs, with odd antlers and a streamline limbless design at right.*

Figure 5. *A fractured petroglyph-covered boulder showing a diversity of vegetative, astronomical, and anthropomorphic representations located above Connor Springs (visible in the distance at top right).*

Figure 6. *Also near Connor Springs, a petroglyph-covered boulder showing mythical images, as well as recent (bright) superimposed Ute nomadic horsemen. The fecund imagery resembles a line of sheep emerging from a womb-like passageway, an unusual ticked spiral, and at lower right, an apparently whimsical birth scene (see text for further interpretation).*

celestial overtones, in which a female figure seems about to cut an umbilical chord attached to an orb half-buried in the Earth. Not too far away, in an overhang of a rocky outcrop, are painted Fremont anthropomorphs, and an atlatl spear-thrower design (figure 7). Finally, at Painted Rock, beside a now-salty spring, are numerous pictographs depicting abstract and Fremont-like figures, and a rare, painted hare (figure 8).

Ranges West of Great Salt Lake

About 40 to 50 kilometers to the northwest of Great Salt Lake lie the Grouse Creek and Raft River Mountains, each range reported to contain at least one pictograph site showing unusual figures (Castleton, 1979). Abstract and vegetative symbols are represented, along with anthropomorphs that are not classic Fremont in design, suggesting influences from outside Utah. Closer to Great Salt Lake are more arid desert mountain ranges, the Hogups and Lakesides, which harbor interesting petroglyph sites. In the southern part of the Hogup Mountains, northeast of Hogup Cave, a single, small site, consisting of scattered boulders, is found not far from a verdant seep. Here, a rainbow and abstract figures are pecked into the boulders, along with exquisitely made mountain sheep, and arrow heads that are made in shapes very similar to those excavated from nearby Hogup Cave (figure 9). These arrow points clearly belong to the Eastgate expanding stem Desert series, through which the petroglyphs can be relatively dated as being about 2,000 years old. The southeastern part of the Lakeside Mountains contains a few boulder

Figure 7. Red paintings of a possible atlatl spear-thrower besides a vertical line of dots, and above a horizontal line (connected to a natural rock crevice) and an abbreviated, horned anthropomorph from a site in the Blue Spring Hills not far from Conner Springs.

Figure 8. A desert hare painted red under a smoke-blackened overhang on a rocky outcrop known as Painted Rock, near Little Mountain northeast of the Promontory Mountains.

Figure 9. Repatinated petroglyphs of finely made arrow heads, and vague stick figures from a boulder outcrop in the Hogup Mountains.

petroglyph sites, and in a shallow cave is a line of dancing Fremont figures similar to that lying due east across Great Salt Lake in Parish Canyon above Centerville (see color photo 1). The petroglyphs include rather bright, recent-appearing figures of mountain sheep, and more ancient figures of apparently mythical content. These figures include a classic "eagle-dancer" or thunderbird figure, and a surrealistic pregnant mountain sheep (figure 10). The repatination of the petroglyphs is so complete that some figures are nearly invisible against the shiny black rocks, unless a polarizing filter is used to view them in oblique sunlight.

Figure 10. Portrayal of an eagle-dancer or thunderbird (?), and surreal combination of long-legged male and pregnant female bighorn mountain sheep engraved onto a boulder on the Lakeside Mountains.

West Desert

Some pictograph sites are found across the West Desert at a greater distance southwest from Great Salt Lake. In Granite Canyon, on the east flank of the Deep Creek Mountains, is a small painted site in a boulder cave at an elevation of about 6,700 feet. This site appears to have been long used, and contains mostly abstract elements along with an "eagle dancer" and small Fremont anthropomorphs. Farther to the west and southwest, in extreme eastern Nevada, are caves and rock shelters harboring some exceptional pictographs, showing both abstract symbols and classic horned Fremont anthropomorphs (figure 11, Color Photo 2). These sites are found along the eastern slopes of the Snake and Antelope Mountains, and more specifically in the Baker Creek, Smith Creek, and Tunnel Canyon drainages. Since these sites occur at relatively high elevations on the barriers, it can be surmised that their ceremonial purposes differed from most Great Salt Lake sites, in that different cultural aspects were involved in the travels of Native Americans to these productive high ranges.

Ranges South of Great Salt Lake

The Cedar, Stansbury, and the Oquirrh Mountains, along the southern margin of Great Salt Lake, also contain sites. The relatively dry Cedar Mountains offers only two small petroglyph sites. These sites show a group of vaguely scratched (historic?) figures on a cliffside, and a few scattered abstract petroglyphs on boulders beside a spring on the

Figure 11. *Red pictograph of a Fremont-style triangular anthropomorph with headdress, holding a vegetative motif and a spiral shield, from the Kachina Rock Shelter in eastern Nevada.*

Figure 12. *Fremont-style rock art panel from a rocky outcrop along the southeastern shore of Stansbury Island, depicting (at right) a line of burden-carrying figures, until recently unknown and hidden behind a rock slab, and fainter abstract and anthropomorphic figures.*

east side of the range. In contrast, an outstanding accumulation of Archaic and transitional Fremont petroglyph sites exists on the southern half of Stansbury Island. The most intriguing is a minor petroglyph site, found on a rocky ridge near the southeastern shore of Great Salt Lake. What makes this site unique, in my experience, is the fact that a large rock slab was placed over a panel of representational figures. To this day, the hidden cliff surface shows much less of the discoloring, patination process than exists on the surrounding cliff face. Hidden under the rock slab are a line of figures (figure 12) that appear to be hump-backed, or more likely burden-carrying anthropomorphs, which are reminiscent of the Kokopeli figures (but without their flutes) of the Anasazi culture to the south. A migration of tadpoles, a symbol of life and metamorphosis, is also shown here (figure 13). Not too far away are several classic Great Basin Desert Archaic boulder sites, and buried, ground-level volcanic boulders with cryptic designs (figure 14). Of the two known sites on the Oquirrh Mountains, the first is a Archaic boulder petroglyph site at Scotts Spring northeast of Tooele. The second is a high-elevation (2.04 km) site involving Archaic petroglyphs beside deep pits dug into a boulder scree slope in an upper branch of Coon Canyon (figure 15). It is tempting to attribute the latter site to animal hunting "magic," although it should be noted that no game figures are preserved here.

Figure 13. *From the same small petroglyph site on Stansbury Island as figure 11, a line of tadpoles that metamorphose into a more frog-like creature at right.*

SIGNIFICANCE OF LOCAL NATIVE AMERICAN ROCK ART

The unusually diverse rock-art styles primarily reflect the geographical setting, that is, the likelihood of intercultural exchanges, and the diversity of habitat that promoted a range of usages by nomadic or semi-nomadic peoples. The long time span of human occupation encompasses the time from the demise of Lake Bonneville to the nomadic horseman of the post-Spanish historic period. During this time, the Native Americans effectively exploited the combination of Basin and Range and wetland resources, which also lead to the varied rock art styles. In general, the known petroglyph sites close to Great Salt Lake and Utah Lake, and on

Figure 14. *A fascinating boulder panel from southern Stansbury Island with artistically rendered cougar and bear (?) footprints, odd abstract elements, and what appears to be a rain curtain and cloud at left. This ground-level boulder was buried during road construction, and briefly excavated for this photograph. Other boulders at this site have been "collected" by local universities.*

Figure 15. A boulder, with a simple snake-like design, abstract, and pit petroglyphs, lying above a hole dug into a steep scree slope in an upper drainage of Coon Canyon in the Oquirrh Mountains.

the desert ranges, are classic Desert Archaic. Some sites show prolonged use with the emergence of representational figures done in the Fremont, or other "mythical" styles, and styles that extend up to the historic period. The "mythical" sites may have served purposes similar to those of the Barrier Canyon adherents. The most ancient figures range from a few thousand, to possibly several thousand years of age, and reflect the concerns of the hunter-gatherer way of life. These sites are not clearly linked to specific endeavors like hunting at strategic game trail locations, such as is indicated in the western Great Basin (Heizer and Baumhoff, 1962), but rather are concentrated where properly patinated rocks occur, especially in the vicinity of springs. Accessibility of sites was also an important factor. It is probably coincidental that most petroglyph sites occur on the southern ends of the ranges that extend into the Great Salt Lake area.

Although a few pictograph sites occur near the northern margin of Great Salt Lake, and show Fremont and uncertain cultural affiliations, the cave pictograph sites in the well-watered Wasatch Range and those in the West Desert are distinctively ceremonial sites emphasizing the Fremont anthropomorph. Many figures, although small in size compared to those in eastern Utah (Schaafsma, 1971), display classic Fremont features like horned headdresses, joined hands, or holding shields or possibly decapitated heads. Such graphic elements, however, do not necessarily imply war-like intentions, but could represent mythical characters, or be symbolic of shamanistic activities. Nonetheless, it is not surprising that the socio-religious focus of these high-elevation, painted caves is different from the sites of the salt flats lying far below.

CONCLUSION

The reasons for the diversity in Native American rock art of the Great Salt Lake region lie in the temporal, cultural, and seasonal concerns of the long-term inhabitants and occasional visitors. This rock art is of great significance because, unlike most cultural debris, the symbols preserve insights into the socio-religious beliefs of the cultures that made them, often created with considerable effort. In many respects, these rock-art sites are unique to Utah, and like Great Salt Lake, are a world-class asset that should be offered the highest degree of protection. Currently, the locations of these sites are known by few people, but increased visitation by a growing local population certainly poses a more imminent threat to their existence than the inexorable weathering that has taken place since the creation of this Native American art. Unfortunately, some sites have already been vandalized, damaged by "collecting," and even obliterated by mining operations.

REFERENCES

Castleton, K.B., 1979, Petroglyphs and Pictographs of Utah, v. 2: Salt Lake City, Utah Museum of Natural History, 341 p.

Eliade, Mircea, 1963, Patterns in Comparative Religion: New York, New American Library, 484 p.

Grant, Campbell, Baird, J.W., and Pringle, J.K., 1968, Rock Drawings of the Coso Range, Publication 4: China Lake, California, Maturango Museum, 145 p.

Heizer, R. F., and Baumhoff, M. A., 1962, Prehistoric Rock Art of Nevada and Eastern California: Los Angles, University of California Press, 412 p.

Jennings, J. D., 1978, Prehistory of Utah and the Eastern Great Basin, *in* University of Utah Anthropological Papers No. 98: Salt Lake City, University of Utah, 263 p.

Sassen, Kenneth, 1991, Rainbows in the Indian rock art of desert western America: Applied Optics, v. 30, p. 3523-3537.

—1994, Possible halo depictions in the prehistoric rock art of Utah: Applied Optics, v. 33, 4756-4760.

Schaafsma, Polly, 1971, The Rock Art of Utah: Papers of the Peabody Museum of Archaeology and Ethnology, v. 65, Harvard, Cambridge, 170 p.

—1980, Indian Rock Art of the Southwest: Albuquerque, University of New Mexico Press, 379 p.

Whitely, D.S., 2000, The art of the Sahaman: Salt Lake City, University of Utah Press, 138 p.